Lecture Notes
in Control and Information Sciences 197

Editor: M. Thoma

J. Henry and J.-P. Yvon (Eds.)

System Modelling and Optimization

Proceedings of the 16th IFIP-TC7 Conference,
Compiègne, France – July 5–9, 1993

Springer-Verlag London Ltd.

ISBN 978-3-540-19893-2   ISBN 978-3-540-39337-5 (eBook)
DOI 10.1007/978-3-540-39337-5

British Library Cataloguing in Publication Data
A catalogue record for this book is available from the British Library

Typesetting: Camera ready by contributors

69/3830-543210 Printed on acid-free paper

FOREWORD

The 16th IFIP Conference on System Modelling and Optimization was held at the University of Technology of Compiègne (UTC), July 5-9, 1993. It was organized jointly by UTC and INRIA (Institut National de Recherche en Informatique et en Automatique).

It was a challenge to organize such a conference in a medium size town after Zürich, Leipzig, Tokyo, ... Nevertheless the attendance exceeded 270 participants; over 200 papers[1] as well as 10 plenary lectures were presented. This success is partly due to the help and support of local authorities: Conseil Régional de Picardie, Conseil Général de l'Oise and Ville de Compiègne.

These proceedings contain a collection of papers resulting from a second selection (reviewing process during the conference). As some authors had commitments to publish their results in other publications, a small amount of good papers is missing.

This conference is the biennial general conference of the IFIP Technical Committee 7 (System Modelling and Optimization) and it reflects the activity of its members and working groups. It was co-sponsored by IFAC, ERCIM, AFCET, SMAI, SMF and benefitted from the financial support of CNRS, Ministère de l'Education Nationale, DRET and EDF.

We express our cordial thanks to members of the International Program Committee and Local Program Committee. Among the many people who helped us in organizing the conference we want to address our special thanks to

- Claude LEMARECHAL and Jacques CARLIER who were the cornerstone of the reviewing process and organization of sessions.
- Corinne FERET who was in charge of registration and financial aspects.
- François TAPISSIER, whose drive and efficiency have been particularly appreciated by most participants.
- and, last but not least, Ghislaine JOLY-BLANCHARD who was the kingpin of the organization.

Jacques HENRY   J.-P. YVON

---

[1](selected among 400 submitted papers)

International Programme Committee

P.	THOFT-CHRISTENSEN	University of Aalborg, DK
A.V.	BALAKRISHNAN	University of California, USA
A.	BENSOUSSAN	INRIA, F
R.E.	BURKARD	Technische Universität Graz, A
J.	DOLEZAL	Czechoslovakia Acad. Sciences, CS
I.V.	EVSTIGNEEV	Academy of Sciences, Russia
E.G.	EVTUSHENKO	Academy of Sciences, Russia
S.D.	FLAM	University of Bergen, N
U.G.	HAUSSMANN	University of British Columbia, CND
J.	HENRY	INRIA-Rocquencourt, F
M.	IRI	University of Tokyo, J
P.	KALL	University of Zurich, CH
A.	KALLIAUER	Oesterr. Elektr. Wirtsch, A
W.	KRABS	Technische Hochschule Darmstadt, D
A.B.	KURZHANSKI	IIASA, A
I.	LASIECKA	University of Virginia, USA
C.	LEMARECHAL	INRIA-Rocquencourt, F
M.	LUCERTINI	Universita di Roma, I
K.	MALANOWSKI	Polish Academy of Sciences, PL
M.	MANSOUR	ETH-Zentrum, CH
M.J.D.	POWELL	University of Cambridge, GB
R.T.	ROCKAFELLAR	University of Washington, USA
W.J.	RUNGGALDIER	Univ. degli studi di Padova, I
H.J.	SEBASTIAN	Tech. Hochschule Leipzig, D
J.	STOER	Universität Würzburg, D
J.P.	VIAL	Université de Genève, CH
J.P.	YVON	UTC et INRIA-Rocquencourt, F
J.	ZOWE	Universität Bayreuth, D

Local Organizing Committee

A.	BAMBERGER	IFP
J.	BLUM	Université de Grenoble
J.D.	BOISSONNAT	INRIA-Sophia Antipolis
J.	CARLIER	UTC
G.	COHEN	Ecole des Mines de Paris
B.	CORNET	Ecole Polytechnique
J.C.	DODU	EDF
N.	EL KAROUI	Université Paris VI
G.	FAYOLLE	INRIA-Rocquencourt
J.	HENRY	INRIA-Rocquencourt
J.B.	HIRIART-URRUTY	Université Paul Sabatier
G.	JOLY-BLANCHARD	UTC
C.	LEMARECHAL	INRIA-Rocquencourt
C.	MALIVERT	Université de Limoges
E.	PARDOUX	Université de Provence
M.C.	PORTMAN	Ecole des Mines de Nancy
E.	ROFMAN	INRIA-Rocquencourt
C.	ROUCAIROL	INRIA-Rocquencourt
J.P.	UHRY	ALMA
J.P.	YVON	UTC & INRIA-Rocquencourt

Organization

C.	GENEST	INRIA-Rocquencourt
F.	TAPISSIER	INRIA-Rocquencourt
C.	DEBLOIS	UTC
C.	FERET	UTC

The conference benefitted from the financial support of :

- Centre National de la Recherche Scientifique (CNRS)
- Conseil Général de l'Oise
- Conseil Régional de Picardie
- Ministère de l'Education Nationale (DAGIC)
- Direction de la Recherche et des Etudes Techniques (DRET)
- Electricité de France (EDF)
- Ville de Compiègne

Some participants were supported by

- Ministère de l'Enseignement Supérieur et de la Recherche (Programme ACCES).
- INRIA-RSI

Sponsorships

- Société Mathématique de France (SMF)
- European Research Consortium for Informatics and Mathematics (ERCIM)
- Association Française des Sciences et Technologies de l'Information et des systèmes (AFCET)
- International Federation of Automatic Control (IFAC)
- Société de Mathématiques Appliquées et Industrielles (SMAI)

Authors index

Contents

I - INVITED PRESENTATIONS

II - CONTRIBUTED PAPERS

1. Non linear programming

1A. Stability

1B. Convex problems

1C. Algorithms

2. Global optimization

3. Game theory

4. Stochastic programming

5. Modelling and parameter estimation

6. Control theory

9C. Algorithm

9D. Application

9E. Shape optimization

10. Discrete optimization

INVITED PRESENTATIONS

SOLUTION METHODS IN STOCHASTIC PROGRAMMING

P. Kall, IOR, University of Zurich [1]

1 Introduction

Many practical decision problems — in particular rather complex ones — use to be modelled according to the following general form

$$
(1.1) \qquad \left\{ \begin{array}{ll} \min & g_0(x) \\ \text{s.t.} & g_i(x) \le 0, \ i = 1, \cdots, m \\ & x \in X \subset \mathbb{R}^n, \end{array} \right.
$$

i.e. as a *mathematical programming* problem. Here it is understood that the set X as well as the functions $g_i : \mathbb{R}^n \to \mathbb{R}, i = 0, \cdots, m$, are given by the modelling process. Typical applications may be found in the areas of industrial production, transportation, agriculture, energy, ecology, engineering, and many others.

However, having the above applications in mind, the inherent assumption for (1.1) that the functions g_i (and the set X) be given, i.e. fixed deterministically, is obviously questionable in many cases. For instance, future productivities in a production problem, inflows into a reservoir connected to a hydropower station, demands at various nodes in a transportation network, etc., will play the role of uncertain parameters which cannot always be wiped out just by replacing them by their mean values or some other (fixed) estimates during the modelling process since, according to simple examples, replacing the parameters by their mean values can yield results, which in practical situations may turn out to be hardly acceptable.

In general, instead of (1.1) we often may rather have to deal with

$$
(1.2) \qquad \left\{ \begin{array}{ll} \text{``}\min\text{''} & g_0(x, \xi) \\ \text{s.t.} & g_i(x, \xi) \le 0, \ i = 1, \cdots, m \\ & x \in X \subset \mathbb{R}^n, \end{array} \right.
$$

where ξ may vary randomly over the set $\Xi \subset \mathbb{R}^k$. More precisely we assume throughout the probability space $(\Xi, \mathcal{F}, P)$ associated with (1.2) to be known, the functions $g_i(x, \cdot) : \Xi \to \mathbb{R}$ to be measurable $\forall x, i$ and the probability distribution P to be independent of x.

However, problem (1.2) is not well defined since the meaning of "min" as well as of the constraints is not clear at all, if we think of taking a decision on x before knowing the realization of ξ. Therefore, socalled *deterministic equivalents* for (1.2) have been introduced which in general (see Wets [36]) may be stated as

---

[1]Institute for Operations Research, University of Zurich, Moussonstr. 15, CH-8044 Zurich

4

$$
(1.3) \quad \begin{cases} \min & E_\xi f_0(x,\xi) \\ \text{s.t.} & E_\xi f_i(x,\xi) \le 0, \ i = 1, \cdots, s \\ & E_\xi f_i(x,\xi) = 0, \ i = s+1, \cdots, \bar{m} \\ & x \in X \subset \mathbb{R}^n, \end{cases}
$$

where the f_i may be constructed in various ways out of the objective and the constraints in (1.2) yielding different types of problems considered in stochastic programming.

To give some examples, how deterministic equivalent problems for (1.2) may be generated, let us choose first with $\alpha \in [0,1]$

$$
(1.4) \quad \begin{cases} f_0(x,\xi) & = \quad g_0(x,\xi) \\ f_1(x,\xi) & = \quad \begin{cases} \alpha - 1 & \text{if } g_i(x,\xi) \le 0, \ i = 1, \cdots, m \\ \alpha & \text{else,} \end{cases} \end{cases}
$$

yielding according to (1.3)

$$
(1.5) \quad \begin{cases} \min_{x \in X} & E_\xi g_0(x,\xi) \\ \text{s.t.} & P(\{\xi \mid g_i(x,\xi) \le 0, \ i = 1, \cdots, m\}) \ge \alpha, \end{cases}
$$

denoted as *chance constrained program* (or problem with joint probabilistic constraints).

If instead of (1.4) we define for $\alpha_i \in [0,1]$, $i = 1, \cdots, m$,

$$
(1.6) \quad \begin{cases} f_0(x,\xi) & = \quad g_0(x,\xi) \\ f_i(x,\xi) & = \quad \begin{cases} \alpha_i - 1 & \text{if } g_i(x,\xi) \le 0 \\ \alpha_i & \text{else} \end{cases} \ , \ i = 1, \cdots, m, \end{cases}
$$

we get the problem with single probabilistic constraints:

$$
(1.7) \quad \begin{cases} \min_{x \in X} & E_\xi g_0(x,\xi) \\ \text{s.t.} & P(\{\xi \mid g_i(x,\xi) \le 0\}) \ge \alpha_i, \ i = 1, \cdots, m. \end{cases}
$$

A different type of problems arises if we introduce a penalty on violating the constraints in (1.2)

$$
(1.8) \quad Q(x,\xi) = k(g_1^+(x,\xi), \cdots, g_m^+(x,\xi))
$$

where usually in applications $k(\cdot) \ge 0$ and $k(0) = 0$, and g_i^+ is defined as

$$
g_i^+(x,\xi) = \begin{cases} 0 & \text{if } g_i(x,\xi) \le 0 \\ g_i(x,\xi) & \text{else.} \end{cases}
$$

With

$$
(1.9) \quad f_0(x,\xi) = g_0(x,\xi) + Q(x,\xi)
$$

we then get the *stochastic program with recourse*

$$
(1.10) \quad \min_{x \in X} E_\xi \{f_0(x,\xi) + Q(x,\xi)\}.
$$

Obviously there are many other possibilities to generate further types of deterministic equivalents for (1.2) by constructing the f_i in different ways out of the objective and the constraints of (1.2).

To be more specific, let the functions $g_i(x, \xi)$ be linear in x, and assume furthermore the sets X and Y to be convex polyhedral. Then we may get for (1.10) problems of the type

$$(1.11) \quad \begin{cases} \min_{x \in X} E_\xi \{c^T x + Q(x, \xi)\} \\ \text{where} \\ Q(x, \xi) = \min\{q^T y \mid Wy \ge h(\xi) - T(\xi)x, \ y \in Y\}, \end{cases}$$

i.e. stochastic *linear* programs with recourse.

Problems (1.5) and (1.7) in the linear case, i.e. for g_i being linear functions in x, become

$$(1.12) \quad \begin{cases} \min_{x \in X} E_\xi c^T(\xi)x \\ \text{s.t.} \ \ P(\{\xi \mid T(\xi)x \ge h(\xi)\}) \ge \alpha \end{cases}$$

and, with $T_i(\cdot)$ and $h_i(\cdot)$ denoting the i-th row and i-th component of $T(\cdot)$ and $h(\cdot)$, respectively,

$$(1.13) \quad \begin{cases} \min_{x \in X} E_\xi c^T(\xi)x \\ \text{s.t.} \ \ P(\{\xi \mid T_i(\xi)x \ge h_i(\xi)\}) \ge \alpha_i, \ i = 1, \cdots, m, \end{cases}$$

the stochastic linear programs with joint and with single chance constraints, respectively.

For these deterministic equivalents for (1.2), formally being mathematical programs, statements on convexity, smoothness etc. are mentioned in the next section which just briefly refers to those properties which are crucial for designing solution procedures for the various problems. Sections 3 and 4 deal with some typical methods for solving rather general classes of chance constrained and recourse problems, repetively.

2 Basic theoretical results

Convexity may be shown easily for the recourse problem (1.10) under rather mild assumptions:

Proposition 1 *If $f_0(\cdot, \xi)$ and $Q(\cdot, \xi)$ are convex in x $\forall \xi \in \Xi$, and if X is a convex set, then (1.10) is a convex program.*

Remark 1 Observe that for $Y = \mathbb{R}^l$ the convexity of $Q(\cdot, \xi)$ can immediately be asserted for the linear case (1.11), and that it holds also for the nonlinear case (1.10) if the functions $g_i(\cdot, \xi)$ are convex and the function $k(\cdot)$ is convex and monotone.

Proposition 2 *(Kall and Stoyan [20]) If $\nabla_x Q(x, \xi)$ exists a.s. (almost surely) and the assumptions of Lebesgue's bounded convergence theorem hold for the difference quotients of Q (with repect to x_i), then $\nabla E_\xi Q(x, \xi)$ exists as well.*

Remark 2 In the linear case the above assumptions are satisfied for $\xi = (h, T)$ having a continuous type distribution, see e.g. Kall [17]. □

For problems with joint chance constraints the situation is much more complicated due to the fac that convexity in this case is by no means selfevident. However under certain assumptions we may also for this type of model get convexity statements. Remind that a probability measure P is *quasi-concave* if, for all convex sets $S_i \in \mathcal{F}$,

$$P(\lambda S_1 + (1 - \lambda)S_2) \geq \min[P(S_1), P(S_2)] \quad \forall \lambda \in [0, 1].$$

Proposition 3 *(Wets [36]) If $g(\cdot, \cdot)$ is jointly convex in (x, ξ) and P is quasi-concave, then the feasible set $\mathcal{B}(\alpha) = \{x \mid P(\{\xi \mid g(x, \xi) \leq 0\}) \geq \alpha\}$ is convex $\forall \alpha \in [0, 1]$.*

Remark 3 The assumption of joint convexity of $g(\cdot, \cdot)$ is so strong that it is even not satisfied in the linear case (1.12), in general. However, if in (1.12) $T(\xi) \equiv T$ (constant) and $h(\xi) \equiv \xi$, then it is satisfied and the constraints of (1.12), F being the distribution function of ξ, read as

$$P(\{\xi \mid Tx \geq \xi\}) = F(Tx) \geq \alpha.$$

□

Hence we stay with the question, when a probability measure — or its distribution function — is quasi-concave. This question was answered first for the subclass of log-concave measures, i.e.

$$P(\lambda S_1 + (1 - \lambda)S_2) \geq P^\lambda(S_1) \cdot P^{1-\lambda}(S_2)$$

for all convex $S_i \in \mathcal{F}$ and $\lambda \in [0, 1]$, in Prékopa [24] and later on for quasi-concave measures in Borell [3] and Rinott [27], yielding

Proposition 4 *Let P on $\Xi = \mathbb{R}^k$ be of continuous type, i.e. have a density f. Then holds:*
P is log-concave iff f is log-concave;
P is quasi-concave iff $f^{-\frac{1}{k}}$ is convex.

In addition to Prop. 3 we get immediately (see e.g. Kall [18])

Proposition 5 *If $g : \mathbb{R}^n \times \Xi \to \mathbb{R}^m$ is continuous, then the feasible set $\mathcal{B}(\alpha)$ is closed.*

For stochastic linear programs with single chance constraints convexity statements have been derived without the joint convexity assumption on $g_i(x, \xi) = h_i(\xi) - T_i(\xi)x$, for special distributions and particular intervals for the values of α_i. More details on this and on stochastic linear programs with recourse may be found in Kall [17]; useful results for the expectation functionals arising in (1.3) are contained in Wets [36].

3   Chance constrained problems

As we have seen in section 2, at least under appropriate assumptions, chance constrained problems as (1.5) or particularly (1.12) appear as ordinary convex smooth mathematical programming Problems. This might suggest that in these cases we may simply apply well known nonlinear programming methods to solve these stochastic programs. However, this viewpoint disregards the fact that applying those methods directly to problems like

$$(3.1) \qquad \begin{cases} \min_{x \in X} E_\xi c^T(\xi)x \\ \text{s.t. } P(\{\xi \mid Tx \geq h(\xi)\}) \geq \alpha \end{cases}$$

we had to deal with evaluating functions like

$$P(\{\xi \mid Tx \geq h(\xi)\})$$

and their gradients repeatedly which means nothing else but performing multivariate numerical integration; up to now this seems to be outside of the set of efficiently solvable problems. Hence, we may try to follow the basic ideas of some of the known nonlinear programming methods, but at the same time we have to find ways to evade the exact evaluation of the integral functions contained in these problems.

Let us concentrate onto the particular stochastic linear program

$$(3.2) \qquad \begin{cases} \min & c^T x \\ \text{s.t.} & P(\{\xi \mid Tx \geq \xi\}) \geq \alpha \\ & Dx = d \\ & x \geq 0. \end{cases}$$

For this problem we know from propositions 3, 4 and 5 in section 2, that it has a convex closed feasible set $\mathcal{B}(\alpha)$ if the distribution function F of P is quasi-concave. Under the assumption that ξ has a (multivariate) normal distribution, we therefore have a convex smooth nonlinear program. For this particular case there have been attempts to adapt penalty and cutting plane methods to solve (3.2). Further on, Prékopa [25] proposed an extension of Zoutendijk's method of feasible directions for the solution of (3.2), which was implemented under the name STABIL by Prékopa et al. [26]. An alternative method following the lines of Veinott's supporting hyperplane algorithm was implemented by Szántai [33]. Finally, J. Mayer developed a special reduced gradient method for (3.2) and implemented it as PROCON [23].

All these approaches have in common that they try to relax the probabilistic constraint

$$P(\{\xi \mid Tx \geq \xi\}) \geq \alpha$$

for the purpose to avoid - whenever possible - the "exact" numerical integration connected with the evaluation of $F(Tx)$ and $\nabla_x F(Tx)$, respectively. To see how this may be realized let us briefly sketch one iteration step of the reduced gradient method's variant implemented in PROCON.

With the notation

$$G(x) := P(\{\xi \mid Tx \geq \xi\})$$

let x be feasible in

(3.3)
$$\begin{cases} \min & c^T x \\ \text{s.t.} & G(x) \geq \alpha \\ & Dx = d \\ & x \geq 0 \end{cases}$$

and - assuming D to have full row rank - let D be partitioned as $D = (B, N)$ into basic and nonbasic parts and accordingly partition $x^T = (y^T, z^T)$, $c^T = (f^T, g^T)$ and a descent direction $w^T = (u^T, v^T)$. Assume further that for some tolerance $\varepsilon > 0$ holds

(3.4)
$$y_j > \varepsilon \quad \forall j \quad \text{(strict nondegeneracy)}.$$

Then the search direction $w^T = (u^T, v^T)$ is determined by the linear program

(3.5)
$$\begin{cases} \max & \tau \\ \text{s.t.} & \begin{bmatrix} f^T u + & g^T v & \leq & -\tau \\ \nabla_y G(x)^T u + & \nabla_z G(x)^T v & \geq & \theta\tau, \text{ if } G(x) \leq \alpha + \varepsilon \\ Bu + & Nv & = & 0 \\ & v_j & \geq & 0, \text{ if } z_j \leq \varepsilon \\ & \|v\|_\infty & \leq & 1, \end{bmatrix} \end{cases}$$

where $\theta > 0$ is a fixed parameter as a weight for the directional derivatives of G and $\|v\|_\infty = \max_j\{|v_j|\}$. From (3.5) we have

$$u = -B^{-1} N v,$$

which renders (3.5) into the linear program

(3.6)
$$\begin{cases} \max & \tau \\ \text{s.t.} & r^T v \leq -\tau \\ & s^T v \geq \theta\tau, \quad \text{if} G(x) \leq \alpha + \varepsilon \\ & v_j \geq 0, \quad \text{if } z_j \leq \varepsilon \\ & \|v\|_\infty \leq 1 \end{cases}$$

where obviously

$$r^T = g^T - f^T B^{-1} N$$
$$s^T = \nabla_z G(x)^T - \nabla_y G(x)^T B^{-1} N$$

are the reduced gradients of the objective and the probabilistic constraint function. Problem (3.6) - and hence (3.5) - is always solvable due to its nonempty and bounded feasible set. Depending on the obtained solution (τ^*, u^{*T}, v^{*T}) to (3.6) the method procedes as follows:

Case 1 $\tau^* = 0$.

Then ε is replaced by 0 and (3.6) is solved again. If then still $\tau^* = 0$, the feasible solution $x^T = (y^T, z^T)$ is obviously optimal.

Otherwise the steps of *case 2* below are carried out, starting with the original $\varepsilon > 0$.

Case 2 $0 < \tau^* \leq \varepsilon$.

Then the following cycle is entered:

Step 1 :   Set $\varepsilon := 0.5\varepsilon$.

Step 2 :   Solve (3.6).

If still $\tau^* \leq \varepsilon$, go to *step 1* ;

otherwise *case 3* applies.

Case 3 $\tau^* > \varepsilon$.

$w^{*T} = (u^{*T}, v^{*T})$ is accepted as search direction.

If a search direction $w^{*T} = (u^{*T}, v^{*T})$ has been found, a line search follows using bisection. Since the line search in this case amounts to determining the intersection of the ray $x + \mu w^*, \mu \geq 0$ with the boundary $bd \, \mathcal{B}(\alpha)$ within the tolerance ε, the evaluation of $G(x)$ becomes important. For this purpose a Monte-Carlo technique proposed by Szántai [32] is used, which allows for computing efficiently upper and lower bounds of $G(x)$ as well as the gradient $\nabla G(x)$.

If the next iterate $\tilde{x}$, resulting from the line search, still satisfies the strict nondegeneracy, the whole step is repeated with the same partition of D into basic and nonbasic parts; otherwise a basis exchange is attempted to reinstall the strict nondegeneracy for a new basis.

4   Recourse problems

Let us consider a recourse problem like

$$
(4.1) \qquad
\begin{cases}
\min & E_\xi \{c^T x + Q(x, \xi)\} \\
\text{s.t.} & Dx = d, \ x \geq 0 \\
\text{where} \\
Q(x, \xi) = \min\{q^T y \mid Wy = h(\xi) - T(\xi)x, \ y \geq 0\}.
\end{cases}
$$

Assume that $h(\cdot)$ and $T(\cdot)$ are linear affine in ξ and that the assumption of *complete fixed recourse* is satisfied, i.e. that

$$
(4.2) \qquad
\begin{cases}
\{t \mid t = Wy, \ y \geq 0\} = \mathbb{R}^m \\
\{u \mid W^T u \leq q\} \neq \emptyset.
\end{cases}
$$

Obviously under these assumptions $Q(x, \cdot) : \Xi \to \mathbb{R}$ is finitely valued and convex in $\xi \ \forall x$. If P were a finite discrete distribution, then (4.1) would be a linear program with dual decomposition structure, as is easily shown.

But even in this favourable situation we may face difficulties since rather often these problems tend to be *very large* in scale. This causes the need for specially taylored methods which really exploit the particular structure.

Nevertheless the fact that for discrete distributions we get at least linear programs to solve, raised the idea to approximate continuous distributions by discrete ones in such a way that the solution of (4.1) would be approximated simultaneously. Later on we sketch just one variant of such an approximation scheme. As mentioned above, the dual decomposition structure is typical for recourse problems (with discrete distribution). Therefore we discuss first how decomposition methods for the solution of these problems have been designed.

4.1 Decomposition

We first remind the basic ideas in Benders' decomposition [1]. For simplicity, and just to present the essential ideas, we restrict ourselves to a support Ξ containing just one realization such that the problem to discuss is reduced to

$$
\begin{aligned}
\min\{c^T x + q^T y\} \\
\text{s.t.} \quad Ax \qquad\quad &= b \\
Tx + Wy &= h \\
x &\geq 0 \\
y &\geq 0.
\end{aligned}
$$

(4.3)

In addition we assume that the problem is solvable and that the set $\{x \mid Ax = b,\ x \geq 0\}$ is bounded. The above problem may be restated as

$$
\begin{aligned}
\min\{c^T x + f(x)\} \\
\text{s.t.} \quad Ax &= b \\
x &\geq 0
\end{aligned}
$$

with

$$
f(x) := \min\{q^T y \mid Wy = h - Tx,\ y \geq 0\}.
$$

Our recourse function $f(x)$ is easily seen to be piecewise linear and convex. It is also immediate that the above problem can be replaced by the equivalent problem

$$
\begin{aligned}
\min\{c^T x + \theta\} \\
\text{s.t.} \quad Ax &= b \\
\theta - f(x) &\geq 0 \\
x &\geq 0;
\end{aligned}
$$

however we do not know the function $f(x)$ explicitly in advance and hence omit this constraint. Then we may try to construct successively new (additional) linear constraints — defining a monotonically decreasing feasible set $\mathcal{B}_1$ of $(n+1)$-vectors $(x_1, \cdots, x_n, \theta)^T$ — such that finally, with $\mathcal{B}_0 := \{(x^T, \theta)^T \mid Ax = b,\ x \geq 0,\ \theta \in \mathbb{R}\}$, the problem $\min_{x \in \mathcal{B}_0 \cap \mathcal{B}_1}\{c^T x + \theta\}$ yields a (first stage) solution of our problem (4.3).

After these preparations we may describe Benders' decomposition as follows:

Step 1 With θ_0 being a lower bound for $\min\{q^Ty \mid Ax = b,\ Tx+Wy = h,\ x \geq 0,\ y \geq 0\}$ solve the program

$$\min\{c^Tx + \theta \mid Ax = b,\ \theta \geq \theta_0,\ x \geq 0\}$$

yielding a solution $(\hat{x}, \hat{\theta})$. Let $\mathcal{B}_1 := \{\mathbb{R}^n \times \{\theta\} \mid \theta \geq \theta_0\}$.

Step 2 Using the last first stage solution $\hat{x}$, evaluate the recourse function

$$\begin{aligned} f(\hat{x}) &= \min\{q^Ty \mid Wy = h - T\hat{x},\ y \geq 0\} \\ &= \max\{(h - T\hat{x})^Tu \mid W^Tu \leq q\}. \end{aligned}$$

Then

a) If $f(\hat{x}) = +\infty$ then there exists a $\tilde{u}$ such that $W^T\tilde{u} \leq 0$ and $(h - T\hat{x})^T\tilde{u} > 0$. However for any x feasible to the recourse constraints $(h - Tx)^T\tilde{u} \leq 0$ has to hold, implying as additional constraint the *feasibility cut*

(4.4) $$- \tilde{u}^TTx + h^T\tilde{u} \leq 0.$$

Redefine $\mathcal{B}_1 := \mathcal{B}_1 \bigcap \{(x^T, \theta) \mid -\tilde{u}^TTx + h^T\tilde{u} \leq 0\}$ and go on to step 3;

b) if $f(\hat{x})$ is finite we have for the recourse problem simultaneously (for $\hat{x}$) a primal optimal basic solution $\hat{y}$ and a dual optimal basic solution $\hat{u}$. By duality it is evident that

$$f(\hat{x}) = (h - T\hat{x})^T\hat{u}$$

whereas for any x holds

$$\begin{aligned} f(x) &= \sup\{(h - Tx)^Tu \mid W^Tu \leq q\} \\ &\geq (h - Tx)^T\hat{u}. \end{aligned}$$

Subtracting the first relation from the second yields

$$\begin{aligned} f(x) &\geq (h - T\hat{x})^T\hat{u} - \hat{u}^TT(x - \hat{x}) \\ &= h^T\hat{u} - \hat{u}^TTx. \end{aligned}$$

The intended constraint $\theta \geq f(x)$ implies the linear constraint — the *optimality cut* —

(4.5) $$\theta \geq h^T\hat{u} - \hat{u}^TTx.$$

This constraint is violated by $(\hat{x}^T, \hat{\theta})^T$ iff $(h - T\hat{x})^T\hat{u} > \hat{\theta}$, in which case it is introduced. Redefine correspondingly $\mathcal{B}_1 := \mathcal{B}_1 \bigcap \{(x^T, \theta) \mid h^T\hat{u} - \hat{u}^TTx \leq \theta\}$ and continue with step 3; otherwise, i.e. if $f(\hat{x}) \leq \hat{\theta}$, stop with $\hat{x}$ being an optimal first stage solution.

Step 3 Solve the updated problem

$$\min\{c^Tx + \theta \mid (x^T, \theta) \in \mathcal{B}_0 \bigcap \mathcal{B}_1\}$$

yielding the optimal solution $(\tilde{x}^T, \tilde{\theta})^T$.
With $(\hat{x}^T, \hat{\theta})^T := (\tilde{x}^T, \tilde{\theta})^T$ return to step 2.

For this method holds

Proposition 6 *Provided that the program (4.3) is solvable and $\{x \mid Ax = b, \ x \geq 0\}$ is bounded, the dual decomposition method yields an optimal solution after finitely many steps.*

We have described this method for the data structure given in the linear program (4.3) by introducing the feasibility and optimality cuts for the recourse function $f(x) := \min\{q^Ty \mid Wy = h - Tx, \ y \geq 0\}$. The modification for a finite discrete distribution with K realizations is immediate. Since then our problem is of the form

$$
\begin{aligned}
\min\{c^Tx &+ \textstyle\sum_{i=1}^{K} p_i q_i^T y_i\} \\
\text{s.t.} \quad Ax &= b \\
T_i x + W y_i &= h_i, \quad i = 1, \cdots, K \\
x &\geq 0 \\
y_i &\geq 0, \quad i = 1, \cdots, K,
\end{aligned}
$$

(4.6)

we simply can introduce feasibility and optimality cuts for all the recourse functions $f_i(x) := \min\{q_i^T y_i \mid W y_i = h_i - T_i x, \ y_i \geq 0\}$, $i = 1, \cdots, K$, as far as necessary; using these cuts as such yields a so-called multicut method, whereas combining them into one cut at a time yields for instance the L-shaped method by Van Slyke and Wets [34].

We may formally simplify the presentations of the feasibility cut as

(4.7) $$g^T x + \alpha \leq 0, \text{ where } g = -T^T \tilde{u}, \ \alpha = h^T \tilde{u}$$

and of the optimality cut as

(4.8) $$g^T x + \alpha \leq \theta, \text{ where } g = -T^T \hat{u}, \ \alpha = f(\hat{x}) - g^T \hat{x}.$$

One cycle of a solution procedure for problem (4.6) then could look as follows:

Given $\mathcal{B}_0$ and the sets $\mathcal{B}_{1i} = \{(x, \theta_1, \cdots, \theta_K) \mid \cdots\}$, $i = 1, \cdots, K$, determined by the cuts generated so far for the particular blocks (and obviously for block i restricting only (x, θ_i)), solve the *master program*

(4.9) $$\min\Big\{ c^T x + \sum_{i=1}^{K} p_i \theta_i \mid (x, \theta_1, \cdots, \theta_K) \in \mathcal{B}_0 \cap \Big(\bigcap_{i=1}^{K} \mathcal{B}_{1i} \Big) \Big\}$$

yielding $(\hat{x}, \hat{\theta}_1, \cdots, \hat{\theta}_K)$ as a solution, with this try to construct further cuts for anyone of the blocks; then

— if there are no further cuts to generate, then stop (optimal solution)

— else repeat the cycle.

The advantage of a method like this lies in the fact that we obviously make use of the particular structure of problem (4.6) in that we have to deal in the master program only with $n + K$ variables instead of $n + \sum_i n_i$, if $y_i \in \mathbb{R}^{n_i}$. The drawback has to be seen as well: It may happen that we have to add very many cuts, and so far we have no reliable criterion to drop cuts which are obsolete for further iterations. Moreover initial iterations often happen to be inefficient. This is not surprising since in the master (4.9) we deal only with

$$\theta_i \geq \max_{j \in J_i}\{(g_j^i)^{\mathrm{T}} x + \alpha_j^i\}$$

for J_i denoting the set of optimality cuts generated for block i so far and the related dual basic solutions $\hat{u}_j^i$ according to (4.8), and not, as we intend to, with

$$\theta_i \geq f_i(x) = \max_{j \in \hat{J}}\{(g_j^i)^{\mathrm{T}} x + \alpha_j^i\}$$

where $\hat{J}$ enumerates *all* dual feasible basic solutions. Hence we are working in the beginning with a peacewise linear convex function $(\max_{j \in \hat{J}_i}\{(g_j^i)^{\mathrm{T}} x + \alpha_j^i\})$ supporting $f_i(x)$ but possibly reflecting rather poorly the shape of f_i. The effect may be — and often is — that even if we start a cycle with an (almost) optimal first stage solution x^* of (4.6), the first stage solution $\hat{x}$ of the master (4.9) may be far away from x^*, and it may take many further cycles to come back towards x^*. The reason for this is now obvious: Our support of $f_i(x)$ associated with the so far available set J_i of optimality cuts is locally, i.e. around x^*, not yet a good approximation for f_i around x^*. Therefore it seems desirable to modify the master program in such a way that, when starting with some first stage iterate z^k, its solution x^k does not move too far away from z^k such that we can expect to improve the approximation to $f_i(x)$ by an optimality cut for block i at x^k. This can be achieved by introducing into the objective of the master the term $\|x - z^k\|^2$ yielding the so-called *regularized* master program

$$(4.10) \qquad \min\Big\{\frac{1}{2\sigma}\|x - z^k\|^2 + c^{\mathrm{T}} x + \sum_{i=1}^{K} p_i \theta_i \mid (x, \theta_1, \cdots, \theta_K) \in \mathcal{B}_0 \cap \Big(\bigcap_{i=1}^{K} \mathcal{B}_{1i}\Big)\Big\},$$

with a control parameter $\sigma > 0$. To avoid too many constraints in (4.10) let us start with some $z^0 \in \mathcal{B}_0$ such that $f_i(z^0) < \infty$ $\forall i$ and $\mathcal{G}^0$ being the feasible set defined by all optimality cuts at z^0. Hence we start (for $k = 0$) with the reduced regularized master program

$$(4.11) \qquad \min\Big\{\frac{1}{2\sigma}\|x - z^k\|^2 + c^{\mathrm{T}} x + \sum_{i=1}^{K} p_i \theta_i \mid (x, \theta_1, \cdots, \theta_K) \in \mathcal{G}^k\Big\}.$$

With

$$F(x) := c^{\mathrm{T}} x + \sum_{i=1}^{K} p_i f_i(x)$$

one cycle of the *regularized decomposition method* (RD) is then for $\gamma \in (0, 1)$ described as

Step 1   Solve (4.11) at z^k getting x^k as first stage solution and $\theta^k = (\theta_1^k, \cdots, \theta_K^k)^{\mathrm{T}}$ as recourse approximates. If for $\hat{F}^k := c^{\mathrm{T}} x^k + p^{\mathrm{T}} \theta^k$ holds $\hat{F}^k = F(z^k)$ then stop (z^k is an optimal solution of (4.6)); otherwise

Step 2   Delete from (4.11) some constraints being inactive at (x^k, θ^k) such that not more than $n + K$ constraints remain.

Step 3 If x^k satisfies the first stage constraints then go to *Step 4*; otherwise add to (4.11) not more than K violated (first stage) constraints yielding the feasible set $\mathcal{G}^{k+1}$, set $z^{k+1} := z^k$, $k := k + 1$, and go to *Step 1*.

Step 4 For $i = 1, \cdots, K$ solve the second stage problems at x^k and

    a)   if $f_i(x^k) = \infty$ then add to (4.11) a feasibility cut,

    b)   else if $f_i(x^k) > \theta_i^k$ then add to (4.11) an optimality cut.

Step 5 If $f_i(x^k) = \infty$ at least for one i then set $z^{k+1} := z^k$ and go to *Step 7*, else

Step 6 If $F(x^k) = \hat{F}^k$ or $F(x^k) \le \gamma F(z^k) + (1 - \gamma)\hat{F}^k$, and if exactly $n + K$ constraints were active at (x^k, θ^k) then set $z^{k+1} := x^k$; otherwise set $z^{k+1} := z^k$.

Step 7 Determine $\mathcal{G}_{k+1}$ as resulting from $\mathcal{G}_k$ after deleting and adding constraints due to *Step 2* and *Step 4*, respectively. With $k := k + 1$ go to *Step 1*.

It can be shown that this algorithm converges in finitely many steps. The parameter σ can be controlled during the procedure such as to increasing it whenever steps (i.e. $\|x^k - z^k\|$) seem too short, and decreasing it when $F(x^k) > F(z^k)$.

It should be mentioned that the RD method has been implemented under the name QDECOM and that, compared to other LP packages, it turned out to be very efficient in solving many test problems of type (4.6).

For further details on the RD method and QDECOM, in particular for a special technique to solve the master (4.11), we refer to the original publication of Ruszczyński [29]; the above presentation is close to the description in his recent paper [30].

4.2 Approximation

Assume that there is an interval

$$(4.12) \qquad \Xi = \times_{i=1}^{k}[a_{i0}, a_{i1}]$$

containing the support of P, that - as in the above linear case -

$$(4.13) \qquad Q(x, \cdot) : \Xi \to \mathbb{R} \quad \text{is convex } \forall x \in X$$

and, finally, that

$$(4.14) \qquad \int_\Xi Q(x, \xi) \, P(d\xi) \quad \text{exists } \forall x \in X.$$

With

$$\hat{\xi} := E\xi$$

by Jensen's inequality [15] follows that

$$(4.15) \qquad Q(x, \hat{\xi}) \le \int_\Xi Q(x, \xi) \, P(d\xi).$$

On the other hand we may use the generalized Edmundson-Madansky (E-M) inequality [5]

$$(4.16) \qquad \int_\Xi Q(x,\xi)\, P(d\xi) \le \sum_\nu Q(x,a_\nu) p^0(a_\nu),$$

where the discrete distribution $p^0(\cdot)$ on the set $\{a_\nu\}$ of vertices of Ξ is uniquely determined by all the joint mixed moments of the components of ξ up to order k, as shown in Frauendorfer [9] and Kall [19]. Hence by (4.15) and (4.16) we have a lower and an upper bound for the expected recourse. If these bounds differ too much, we may consider a partition

$$S = \{\Xi_1, \cdots, \Xi_L\}$$

of Ξ into (halfopen) subintervals Ξ_l and repeat the same considerations with respect to the conditional distribution of ξ with respect to Ξ_l, $l = 1, \cdots, L$.

Hence with

$$p_l = P(\Xi_l) \ (> 0 \ \text{by assumption})$$
$$\hat\xi^l = E(\xi \mid \xi \in \Xi_l)$$
$$\{a_\nu^l\} \ \text{the vertices of } \Xi_l$$
$$p_l^0(a_\nu^l) \ \text{the conditional E-M probabilites on } \Xi_l,$$

we get

$$(4.17) \qquad Q(x,\hat\xi^l) \le \frac{1}{p_l} \int_{\Xi_l} Q(x,\xi)\, P(d\xi)$$

$$(4.18) \qquad \frac{1}{p_l} \int_{\Xi_l} Q(x,\xi)\, P(d\xi) \le \sum_\nu Q(x,a_\nu^l) p_l^0(a_\nu^l),$$

yielding

$$(4.19) \qquad L^S(x) = \sum_{l=1}^L p_l Q(x,\hat\xi^l)$$

as lower bound and

$$(4.20) \qquad U^S = \sum_{l=1}^L p_l \sum_\nu Q(x,a_\nu^l) p_l^0(a_\nu^l)$$

as upper bound for the expected recourse. Obviously $U^S(x)$ and $L^S(x)$ are computationally much more amenable than $\int_\Xi Q(x,\xi)\, P(d\xi)$. Therefore we may solve in a first step with $f(x)$ the first stage objective, e.g. $f(x) = c^T x$,

$$(4.21) \qquad \min_{x \in X} \{f(x) + L^S(x)\},$$

yielding some solution x^S for which

$$(4.22) \quad \begin{cases} f(x^S) + L^S(x^S) & \leq \min_{x \in X} \{ f(x) + \int_\Xi Q(x,\xi)\, P(d\xi) \} \\ & \leq f(x^S) + U^S(x^S). \end{cases}$$

Now either x^S is ε-optimal — with respect to some prescribed tolerance ε — , i.e.

$$(4.23) \qquad\qquad U^S(x^S) - L^S(x^S) \leq \varepsilon,$$

or we may construct a refinement S^2 from the partition $S^1 := S$. Refining the partitions obviously leads to monotonically decreasing upper and increasing lower bounds. Also, from the basic concepts of integration theory, it is evident that successive refinements can be chosen such as to yield two sequences of discrete distributions associated with the lower (Jensen) and upper (E-M) bounds which converge weakly to P (see Billingsley [2], Kall [16]). According to the stability statements proved in Kall [18] and by Robinson and Wets [28] this asserts the desired approximation of the optimal value and the solution set of the recourse problem

$$(4.24) \qquad\qquad \min_{x \in X} \{ f(x) + \int_\Xi Q(x,\xi)\, P(d\xi) \},$$

provided that f be continuous and in addition to (4.13) and (4.14) $Q(x,\cdot)$ satisfies a uniform integrability assumption which in the linear recourse case is trivially given.

Although these stability results are relevant from a theoretical point of view, for computational purposes we are interested to find rules for relatively few but effective improvements. Various partly heuristic refinement strategies have been tested and led to practically satisfactory results as described in Frauendorfer and Kall [11] and in Frauendorfer [9]. In solving the resulting decomposition problems with their special structure it has proven to be advantageous to use specially designed methods as the basis reduction method of Strazicky [31] or the regularized decomposition method of Ruszczyński [29].

However, this only holds for low dimensions of ξ (i.e. $k \leq 5$) in general, since the number of vertices and moment conditions to be handled for the E-M distributions $p^0(\cdot)$ goes with 2^k. If, however, we have the particular case of *simple recourse* , i.e. in (1.11) holds $W = (I, -I)$, $T(\xi) \equiv T$ and $h(\xi) \equiv \xi$, then following essentially the same approximation scheme, the dimension of ξ turns out not to play the same restrictive role as mentioned above, since then instead of computing the upper bound $U^S(x^S)$ in (4.22) it turns out to be as easy to compute the exact value $\int_\Xi Q(x^S, \xi) P(d\xi)$ (see Kall and Stoyan [20]); and the resulting linear programs allow for an efficient pivoting scheme due to their very special data structure, as pointed out by Wets [35].

In general the dimension of ξ may be increased by considering instead of the k-dimensional interval (4.12) a k-dimensional simplex covering the support of P. Then the Jensen inequality holds as well, and there is a discrete distribution on the $k + 1$ vertices for the upper bound (corresponding to the E-M distribution on intervals). Instead of the partitions into intervals we now consider refinements of simplex partitions which may be carried through along similar heuristic refinement strategies as in the former case as proposed by Frauendorfer [10].

In any case the problem arises to determine the probabilities and conditional expectations of the cells of the partitions, which in its strict form means multivariate integration, again. To avoid this, one may use samples of the distribution P to get estimates for these moments. Obviously one has to control the sample size to keep the statistical error small compared to the numerical error of the approximation scheme. This can be achieved as described in Frauendorfer and Kall [11].

If the dimensions of ξ become too high even for the simplicial approximation, so far the only way to deal computationally with recourse problems seems to use stochastic methods, either based on stochastic descent directions or on stochastic cutting planes.

4.3 Stochastic descent methods

We are still dealing with problem (4.24). Just to give the flavour of this type of methods we briefly describe the *stochastic quasi-gradient method* (SQG) of Ermoliev [6]. To simplify the notation we define

$$(4.25) \qquad F(x,\xi) = f(x) + Q(x,\xi)$$

and hence consider the problem

$$(4.26) \qquad \min_{x \in X} E_\xi F(x,\xi),$$

for which we assume that

$$(4.27) \qquad \begin{array}{lll} (i) & E_\xi F(x,\xi) & \text{is finite and convex in } x \\ (ii) & X & \text{is convex and compact.} \end{array}$$

Then, starting from some feasible point $x^0 \in X$, we may define an iterative process by

$$(4.28) \qquad x^{\nu+1} = \Pi_X(x^\nu - \rho_\nu v^\nu),$$

where v^ν is a random vector, $\rho_\nu \geq 0$ is some stepsize and Π_X is the projection onto X, i.e. for $y \in \mathbb{R}^n$ holds with the Euclidean norm

$$(4.29) \qquad \Pi_X(y) = \arg\min_{x \in X} \|y - x\|.$$

Obviously we may not expect any reasonable convergence statement without further assumptions on the search direction v^ν and on the stepsize ρ_ν.

Let therefore v^ν be a socalled *stochastic quasi-gradient*, i.e. assume that

$$(4.30) \qquad E(v^\nu \mid x^0, \cdots, x^\nu) \in \partial_x E_\xi F(x^\nu, \xi) + b^\nu,$$

where ∂_x denotes the subdifferential with respect to x, which coincides with the gradient in the differentiable case. Intuitively we should require that $\|b^\nu\| \xrightarrow{\nu \to \infty} 0$. Since by (4.27) (i) it holds for $g^\nu \in \partial_x E_\xi F(x^\nu, \xi)$ that

$$(4.31) \qquad E_\xi F(x^*, \xi) - E_\xi F(x^\nu, \xi) \geq g^{\nu T}(x^* - x^\nu)$$

for any solution x^* of (4.26), we have from (4.30) that

$$(4.32) \qquad E_\xi F(x^*, \xi) - E_\xi F(x^\nu, \xi) \geq E(v^\nu \mid x^0, \cdots, x^\nu)^T(x^* - x^\nu) + \gamma_\nu,$$

where

$$(4.33) \qquad \gamma_\nu = -b^{\nu \mathrm{T}}(x^* - x^\nu),$$

and hence $\|b^\nu\| \xrightarrow{\nu \to \infty} 0$ would imply $\gamma_\nu \xrightarrow{\nu \to \infty} 0$ as well. Observe that the particular choice of a *stochastic subgradient*

$$(4.34) \qquad v^\nu \in \partial_x F(x^\nu, \xi^\nu)$$

or else more general

$$(4.35) \qquad v^\nu = \frac{1}{N_\nu} \sum_{\mu=1}^{N_\nu} w^\mu, \quad w^\mu \in \partial_x F(x^\nu, \xi^{\nu\mu}),$$

where the ξ^ν or $\xi^{\nu\mu}$ are independent samples of ξ, would yield $b^\nu = 0$, $\gamma_\nu = 0 \ \forall \nu$ provided that the operations of integration and (sub)differentiation may be exchanged as asserted e.g. by Prop. 2 in section 2 for the differentiable case.

Finally, assume that for the stepsize ρ_ν together with v^ν and γ_ν holds

$$(4.36) \qquad \rho_\nu \geq 0, \ \sum_{\nu=0}^{\infty} \rho_\nu = \infty, \ \sum_{\nu=0}^{\infty} E_\xi \{ \rho_\nu |\gamma_\nu| + \rho_\nu^2 \|v^\nu\|^2 \} < \infty.$$

With the choices (4.34) or (4.35), for uniformly bounded v^ν this assumption could obviously be replaced by the stepsize assumption

$$(4.37) \qquad \rho_\nu \geq 0, \ \sum_{\nu=0}^{\infty} \rho_\nu = \infty, \ \sum_{\nu=0}^{\infty} \rho_\nu^2 < \infty,$$

which seems familiar from stochastic approximation.

With these prerequisites it can be shown, that under the assumptions (4.27), (4.30) and (4.36) (or (4.27), (4.34) or (4.35), and (4.37)) the iterative method (4.28) converges almost surely (a.s.) to a solution of (4.26). For details we refer to Ermoliev [6], Gaivoronski [12] and to the related chapters by Ermoliev, Gaivoronski, Marti, Pflug and Uryasev in [8].

Quite a variety of stochastic descent methods has been considered, besides the SQG method. For a detailed presentation on the construction of stochastic descent directions we refer to Marti [21]; the convergence behaviour has been analyzed e.g. by Ermoliev and Norkin [7] and by Marti and Fuchs [22].

4.4 Stochastic cutting planes

For the problem

$$(4.38) \qquad \begin{aligned} \min\{cx &+ \mathcal{Q}(x)\} \\ \text{s.t.} \ Ax &= b \\ x &\geq 0, \end{aligned}$$

where

$$\mathcal{Q}(x) = \int_\Xi Q(x, \xi) dP,$$

we assume for simplicity that

$$(4.39) \qquad \begin{aligned} Q(x,\xi) &= \min\{q^{\mathrm{T}}y \mid Wy = \xi + Tx, \ y \geq 0\} \\ &= \max\{(\xi + Tx)^{\mathrm{T}}u \mid W^{\mathrm{T}}u \leq q\}. \end{aligned}$$

Furthermore we assume relatively complete recourse, i.e. for any x satisfying the first stage constraints the second stage problem is feasible for whatever $\xi \in \Xi$.

In subsection 4.1 on decomposition we described how cuts may be determined. There, with a finite discrete distribution, we had a 'deterministic' method working with a successively increasing lower bound of the true objective function. This was possible because there we could determine the lower bounds of the recourse function $f_i(x) = f(x,\xi_i)$ (see subsection 4.1)

$$\rho_i = \max_{j \in J_i}\{(g_j^i)^{\mathrm{T}}x + \alpha_j^i\}$$

for every of the finitely many realizations ξ_i of ξ and then average them with the given distribution as $\sum_{i=1}^{K} p_i \rho_i$ which then clearly had to be a lower bound for the expected recourse $\sum_{i=1}^{K} p_i f_i(x)$.

Obviously we cannot procede that way if Ξ is not finite; also if e.g. $\dim \Xi = 50$, the components of ξ are independent and have only 3 realizations each, we would have 3^{50} blocks in problem (4.6) which is clearly outside of anything we can handle. One way out of this difficulty is to generate cuts randomly. Therefore we may observe (or generate) at iteration k a realization ξ^k and then construct a cut

$$(g_k^k)^{\mathrm{T}}x + \alpha_k^k \leq \theta.$$

Now we have one cut for the recourse function $Q(x,\xi^k)$ for the particular randomly generated realization ξ^k. If we have collected the $k-1$ cuts constructed before in the analogous way for the random realizations $\xi^1, \cdots, \xi^{k-1}$, we may construct a new 'cut' by taking the statistical mean of all the available cuts, i.e.

$$\frac{1}{k}\sum_{j=1}^{k}\{(g_j^k)^{\mathrm{T}}x + \alpha_j^k\} \leq \theta.$$

Then this is obviously not a cut for the expected recourse (in the sense of an unquestionable lower bound) but just a statistical estimate of it. Nevertheless we may describe a cycle of a stochastic version of Benders' [1] decomposition method as follows:

Step 1 For $\xi^1 := E\xi$ solve problem (4.38) with $Q(x) = Q(x,\xi^1)$ yielding x^1; let k:=1.

Step 2 Generate randomly (and independently from the previous samples) a ξ^k from the distribution of ξ. Compute a dual optimal solution u^k of the recourse problem for realization ξ^k at x^k.

Step 3 With

$$u_j^k \in \arg\max\{(\xi^j - Tx^k)^{\mathrm{T}}u^l \mid 1 \leq l \leq k\}, \ j = 1, \cdots, k$$

determine the k-th cut by

$$(g_k^k)^{\mathrm{T}}x + \alpha_k^k := \frac{1}{k}\sum_{j=1}^{k}(\xi^j - Tx)^{\mathrm{T}}u_j^k$$

Update the previous cuts according to

$$g_j^k := \frac{k-1}{k} g_j^{k-1}, \ \alpha_j^k := \frac{k-1}{k} \alpha_j^{k-1}, \ j = 1, \cdots, k-1.$$

Step 4 With the cuts

$$(g_j^k)^{\mathrm{T}} x + \alpha_j^k \leq \theta, \ j = 1, \cdots, k,$$

solve the master problem (same as (4.9)) to get x^{k+1}. Set $k := k + 1$ and go to *Step 2*.

This method was analyzed by Higle and Sen [13], and a promising statistical convergence behaviour could be proved. The authors modified the method by introducing besides the above cuts — constructed or updated, respectively, in any iteration step — an additional so-called 'incumbent' which is only updated occasionally under certain conditions and has a stabilizing effect to the algorithm. This complemented version is known as *stochastic decomposition* (SD) algorithm and has been implemented by the authors themselves and elsewhere.

An alternate method constructing stochastic cutting planes based on *importance sampling* was proposed by Dantzig and Glynn [4] and further developed by Infanger [14].

For $\dim(\Xi)$ being low, from 'convergence with probability 1' we may not expect an efficiency comparable to an approximation method as discussed in subsection 4.2. For $\dim(\Xi)$ being high (> 10 say) the SD method — similar as the SQG method — still allows to solve problems (at least approximately) which are out of discussion for deterministic methods.

References

[1] J.F. Benders. Partitioning procedures for solving mixed-variables programming problems. *Numer. Math.*, 4:238–252, 1962.

[2] P. Billingsley. *Convergence of Probability Measures*. Wiley, New York, 1968.

[3] C. Borell. Convex set functions in d-space. *Period. Math. Hungar.*, 6:111–136, 1975.

[4] G.B. Dantzig and P.Q. Glynn. *Parallel Processors for Planning Under Uncertainty*. Preprint SOL-88-8R, Dept. Oper. Res., Stanford U., 1989.

[5] H.P. Edmundson. *Bounds on the Expectation of a Convex Function of a Random Variable*. Technical Report Paper 982, The RAND Corporation, 1956.

[6] Y. Ermoliev. Stochastic quasigradient methods and their application to systems optimization. *Stochastics*, 9:1–36, 1983.

[7] Y.M. Ermoliev and V.I. Norkin. *Normalized Convergence in Stochastic Optimization*. Working Paper WP-89-091, IIASA Laxenburg, 1989.

[8] Y. Ermoliev and R.J.-B. Wets, editors. *Numerical Techniques for Stochastic Optimization.* Volume 10 of *Springer Series in Computational Mathematics*, Springer-Verlag, Berlin, 1988.

[9] K. Frauendorfer. Solving SLP recourse problems with arbitrary multivariate distributions - the dependent case. *Math. Oper. Res.*, 13:377–394, 1988.

[10] K. Frauendorfer. *Stochastic Two-Stage Programming.* Volume 392 of *LN Econ. Math. Syst.*, Springer-Verlag, Berlin, 1992.

[11] K. Frauendorfer and P. Kall. A solution method for SLP recourse problems with arbitrary multivariate distributions – the independent case. *Probl. Control & Inform. Th.*, 17:177–205, 1988.

[12] A. Gaivoronski. *Interactive Program SQG-PC for Solving Stochastic Programming Problems on IBM PC/XT/AT Compatibles – User Guide –.* Working Paper WP-88-11, IIASA Laxenburg, 1988.

[13] J.L. Higle and S. Sen. Stochastic decomposition: an algorithm for two stage stochastic linear programs with recourse. *Math. Oper. Res.*, 16:650–669, 1991.

[14] G. Infanger. *Monte Carlo (Importance) Sampling within a Benders' Decomposition Algorithm for Stochastic Linear Programs.* Preprint SOL-89-13, Dept. Oper. Res., Stanford U., 1989.

[15] J.L. Jensen. Sur les fonctions convexes et les inégalités entre les valeurs moyennes. *Acta Mathematica*, 30:173–177, 1906.

[16] P. Kall. Approximations to stochastic programs with complete fixed recourse. *Numer. Math.*, 22:333–339, 1974.

[17] P. Kall. *Stochastic Linear Programming.* Springer-Verlag, Berlin, 1976.

[18] P. Kall. On approximations and stability in stochastic programming. In J. Guddat, H.Th. Jongen, B. Kummer, and F. Nožička, editors, *Parametric Optimization and Related Topics*, pages 387–407, Akademie-Verlag, Berlin, 1987.

[19] P. Kall. Stochastic programs with recourse: an upper bound and the related moment problem. *ZOR*, 31:A119–A141, 1987.

[20] P. Kall and D. Stoyan. Solving stochastic programming problems with recourse including error bounds. *Math. Operationsforsch. Statist., Ser. Opt.*, 13:431–447, 1982.

[21] K. Marti. *Descent Directions and Efficient Solutions in Discretely Distributed Stochastic Programs.* Volume 299 of *LN Econ. Math. Syst.*, Springer-Verlag, Berlin, 1988.

[22] K. Marti and E. Fuchs. Rates of convergence of semi-stochastic approximation procedures for solving stochastic optimization problems. *Optimization*, 17:243–265, 1986.

[23] J. Mayer. *Probabilistic Constrained Programming: A Reduced Gradient Algorithm Implemented on PC.* Working Paper WP-88-39, IIASA Laxenburg, 1988.

[24] A. Prékopa. Logarithmic concave measures with applications to stochastic programming. *Acta Sci. Math. (Szeged)*, 32:301–316, 1971.

[25] A. Prékopa. Eine Erweiterung der sogenannten Methode der zulässigen Richtungen der nichtlinearen Optimierung auf den Fall quasikonkaver Restriktionen. *Math. Operationsforsch. Statist., Ser. Opt.*, 5:281–293, 1974.

[26] A. Prékopa, S. Ganczer, I. Deák, and K. Patyi. The STABIL stochastic programming model and its experimental application to the electricity production in Hungary. In M.A.H. Dempster, editor, *Stochastic Programming*, pages 369–385, Academic Press, London, 1980.

[27] Y. Rinott. On convexity of measures. *Ann. Prob.*, 4:1020–1026, 1976.

[28] S.M. Robinson and R.J.-B. Wets. Stability in two stage programming. *SIAM J. Contr. Opt.*, 25:1409–1416, 1987.

[29] A. Ruszczyński. A regularized decomposition method for minimizing a sum of polyhedral functions. *Math. Prog.*, 35:309–333, 1986.

[30] A. Ruszczyński. *Regularized Decomposition of Stochastic Programs: Algorithmic Techniques and Numerical Results.* Working Paper WP-93-21, IIASA Laxenburg, 1993.

[31] B. Strazicky. On an algorithm for solution of the two-stage stochastic programming problem. *Methods of Oper. Res.*, 19:142–156, 1974.

[32] T. Szántai. *Calculation of the Multivariate Probability Distribution Function Values and their Gradient Vectors.* Working Paper WP-87-82, IIASA Laxenburg, 1987.

[33] T. Szántai. A computer code for solution of probabilistic-constrained stochastic programming problems. In Y.M. Ermoliev and R. J.-B. Wets, editors, *Numerical Techniques for Stochastic Optimization*, pages 229–235, Springer-Verlag, Berlin, 1988. Springer Ser. Computat. Math., vol. 10.

[34] R. Van Slyke and R.J-B. Wets. *L*-shaped linear program with applications to optimal control and stochastic linear programs. *SIAM J. Appl. Math.*, 17:638–663, 1969.

[35] R.J.-B. Wets. Solving stochastic programs with simple recourse. *Stochastics*, 10:219–242, 1983.

[36] R.J.-B. Wets. Stochastic programming. In G.L. Nemhauser et al., editors, *Handbooks in OR & MS, vol. 1*, pages 573–629, Elsevier, 1989. Survey.

SOME USES OF OPTIMIZATION FOR STUDYING THE CONTROL OF ANIMAL MOVEMENT

William S. Levine
Electrical Engineering Department, University of Maryland
College Park, MD 20742

Introduction

There are a number of reasons for believing that optimization and modeling should be useful tools for studying the neural control of movement. If movement is important to reproductive success, as is certainly true for mammals, then the theory of evolution predicts some degree of optimization of movement. But the best reason for using optimization and modeling to study the control of movement is that these tools produce precisely posed questions which can be answered both analytically and experimentally. The two answers can then be compared and the differences used to improve the models and the understanding of the control of movement. It is important to realize that, from the point of view of science, it is more interesting when the analytical and experimental results are different that when they agree. Agreement means that the limit of resolution of the particular experiment has been reached and there is no more to be learned from that experiment. The opposite is true for clinical applications but, for reasons that will be discussed shortly, optimization and modeling have been rarely used clinically.

Thus, the primary objective of most applications of modeling and optimization to motor control is conceptual understanding. The concrete result is usually an unfalsified model. These models can take many different forms. However, only two types of model will be emphasized in this paper. Very detailed attempts to capture as much as possible of the behavior of some portion of the motor control system will be described. In addition, very simplified models whose primary purpose is to provide conceptual understanding will be described. The models that will be discussed in detail are actually optimal control problems that approximate motor control tasks.

After this brief introduction we will begin with a discussion of modeling of the components of the motor control system because this is in many respects the hardest and most important component of the application. We will then describe the optimization aspects. The main point is that these are hard problems. This is demonstrated by means of a simple example. Realistic examples are substantially harder. Finally, we conclude with some comments about the future of optimization and modeling in the study of neural control of movement.

Modeling

There exist dramatically different attitudes toward modeling within the community of researchers on neural control of movement. As a general rule clinicians do not like models. They are taught to be especially pragmatic and to be alert to the differences between individuals. Their experience tends to reinforce this training. For example, anatomical charts can be viewed as simple nonmathematical models of a human. There are generally substantial differences between real people and the anatomical charts. Furthermore, this is not because the charts are poorly prepared. There is so much variability from one human to another that a very accurate chart for one individual would be useless for another. If you doubt this think about the difference in human faces. Evolution, in fact, requires differences between individuals of the same species. These differences are often clinically important.

At the other extreme there are individuals who believe that realistic models of neuromuscular control systems are necessarily very detailed. The basic argument is simple. If one takes only the skeleton of a human being one has a mechanical system composed of several hundred interconnected rigid bodies. Such a system is accurately described by Newtonian mechanics. A Newtonian model of the skeletal system is, at least conceptually, feasible. A high degree of accuracy requires a minimum of approximation. Thus, one logically concludes that very complex mechanical models are necessary merely to account for the basic physics of animal movement.

Before discussing specific models or modeling issues in any depth some brief comments on the following general questions would be useful. Should one use models at all? How complex must a model be to be useful? In reality, everyone working in neuromotor control, even the clinicians, uses models. Clinicians do use anatomical charts, for instance. The real issues are the level of detail, the level of complexity, and the level of belief in the model. Looked at from the point of view of experiment or observation these same issues can be phrased as follows: What details are measurable? How complicated a model can one analyze? How accurate can the model be? These questions must be answered on a case by case basis because the possible uses of modeling are as varied as the situations seen by clinicians.

For example, MacMahon [1] has used an extremely simple model very effectively to determine the optimal stiffness of a running track. The model of the runner was basically a mass and spring. Such a model worked because it captured the essential features of the runner's interaction with the track's stiffness. An even simpler model suffices to determine the proper bank angle of the track. A much more complex model would be needed in order to predict the effect of track stiffness on runner's knees. Of course, one could determine the effect of track stiffness on runner's experimentally without a model at all but you would need to build the track first.

If we temporarily exclude such very simple models from our discussion and concentrate on models that are meant to be reasonably realistic then it is a good idea to discuss separately the issues involved in modeling the skeletal structure, the elastic structures (ligaments and tendons), the actuators (muscles), and the intelligence (neurons).

skeletal structure: Most individual bones have remarkably complicated three dimensional shapes. The joint facets, the places where two or more bones interact, are especially complicated. It is possible to make very accurate measurements of the shape of bones in the laboratory. It is possible to make reasonably accurate measurements of the shape of the bones of a living subject using x-rays or MRI provided there are sufficiently compelling reasons to do so. This data can be used to produce very accurate three dimensional representations of specific bones.

Models of the complete dynamics are virtually nonexistent at this time. Such a model would have to compute the interaction forces between bones that roll and slide one against the other. The mathematical computation of the interaction forces is not always easy when the surfaces are simple planes, cylinders and spheres. When the surfaces are complicated, as bones are, the computations become horrendous. For example, there are normally two points of contact between the tibia and femur. The direction of the contact force vector at each point of contact depends on the directional derivative along the surface and on the other forces acting on the bones. It is all Newtonian mechanics but extremely complex.

Furthermore, it is not possible to accurately measure, in vivo, the movement of bones. The best that can be done is to obtain accurate three dimensional trajectories of various surface markers. Such data is rarely accurate enough to test three dimensional models of the kinematics, much less the dynamics. For these reasons, and one other that is very important and that will be described later, almost all present dynamical models of body movement use pin joints as approximations to the real joints. For planar motion of bones that are long in comparison to the size of their joint facets pin joints are often, but not always, good approximations to the real joint. For example, the hip and ankle of a human or cat moving only in the sagittal plane can be adequately described by a pin joint. The dynamics of the human or cat knee can also be described by a pin joint, again for motion in the sagittal plane. However, the rolling and sliding of the knee must be included in calculations of the muscle moments or large errors result. It is obvious that the hip and knee cannot be modeled by simple pin joints for motion in three dimensions. Surprisingly, recent evidence shows that pin joint models of ankle motion in three dimensions also miss important aspects of the movement [2].

Once you have approximated the real hip, knee, and ankle by pin joints you might as well ignore the relative motion between the flesh of the limbs and the bones. Thus, one approximates the shank by a single rigid segment attached to the thigh, again a single rigid segment, by a pin joint. It is harder to justify lumping the entire foot into a single rigid segment. There are very useful joints between the toes and the foot. There are five lines of bones from the rear of the foot to the toes. These mechanical degrees of freedom are also useful. The main argument for lumping the foot into a single rigid segment is that it is not really possible to measure accurately the non-rigid motion of the foot. Because of the absence of data and the high complexity involved in a more accurate model, I know of no dynamical models of the lower limb that do not lump the foot into a single segment. We have experimented with two segment models of the foot in the hope of capturing some of the uses of the toes but the effort was not successful.

The torso also presents difficult modeling and measurement problems. The vertebrae are good examples of bones whose length is comparable to the size of the joint facets. The relative motion of any one vertebra is highly constrained and not very large. As a result, it is not at all easy to relate measurements made at the surface of the skin with the actual movement of vertebrae. Given the measurement limitations and the difficulty of constructing and analyzing a torso model with many segments the usual practice has been to use one to three rigid segments to approximate the torso. Kane, for example, used a two rigid segment model of the cat's torso to determine the means by which cats right themselves while falling.

Thus, a typical "skeletal" model of a human might contain approximately 15 segments, three for each limb, one or two for the torso, and one for the head. The segment masses and inertias would include the total segment mass, muscles, blood, fat, etc., as well as bones. The motion would usually be restricted to be planar because, as was mentioned earlier, it is very difficult at the present time to do the computations involved in simulating the three dimensional movement of structures as complex as a 10 segment model of a human.

ligaments and tendons: Once you approximate animal joints by simple mechanical structures such as pin joints, there is no need to worry about ligaments which primarily hold animal joints together. Mechanical joints are held together by their mechanical structure. Nonetheless, certain ligaments, notably those of the human knee, have been extensively studied. It is known that they can be approximated by nonlinear springs.

Mechanical models of tendons are very important components of computer models of the dynamics of limb and body movement. Tendons are the connection between muscle and bone. Even in cases where the muscle appears to connect directly to the bone there is some tendon-like connective tissue at the interface. Fortunately, tendon is well-approximated by a nonlinear spring. As a general rule, tendon stiffness is proportional to its cross-sectional area. Moreover, tendons tend to develop so as to match the muscles to which they attach. That is, large strong muscles have thick tendons and small weak muscles have thin tendons. This means that simple scaling rules can be used to provide the parameters of the springs that are used to approximate tendons.

muscles: Any model of the dynamics of human or animal movement must include reasonable dynamic models of muscles to have any possibility of even qualitative agreement with experimental observations. The dynamics of muscle and tendon dominate the dynamics of movement. That is not to say that the masses and inertias of the limbs and torso can be ignored. But it is very rare that limb masses and inertias are omitted from models and quite common for muscle dynamics to be omitted.

Having said that it is essential to model muscle dynamics it must also be emphasized that it is difficult to do. The problem can be divided into two parts: Find the appropriate model structure; find the proper parameters. To understand the problem with the parameters of a model imagine that you have a perfect mathematical model of muscle that depends on only two parameters. Suppose those two parameters are the maximum isometric force that muscle can produce and the mass of the muscle. These are two very important aspects of any muscle and they are more easily measured than many aspects of muscle so they are a good choice of parameters. Of course, neither of these parameters is measurable when we are talking about a muscle that is part of a living creature. It is possible to measure the maximum force that a cooperative subject can exert to rotate a joint. But the interesting joints are activated by several muscles. Thus, the parameters of muscle models are generally educated guesses rather than accurate measurements.

The structure of muscle models is complicated by the fact that muscle is a more or less distributed device. Models of distributed systems are very hard to compute with so most modelers approximate the distributed system by some sort of finite dimensional approximation. Many such models exist. Nowadays the models that attempt to be realistic are all rather similar. We give one example below.

The model structure is shown in Fig. 1. Three of the five mechanical components shown are nonlinear so it is necessary to give the equations describing their behavior. This is done below. This discussion is a summary of the results in [3].

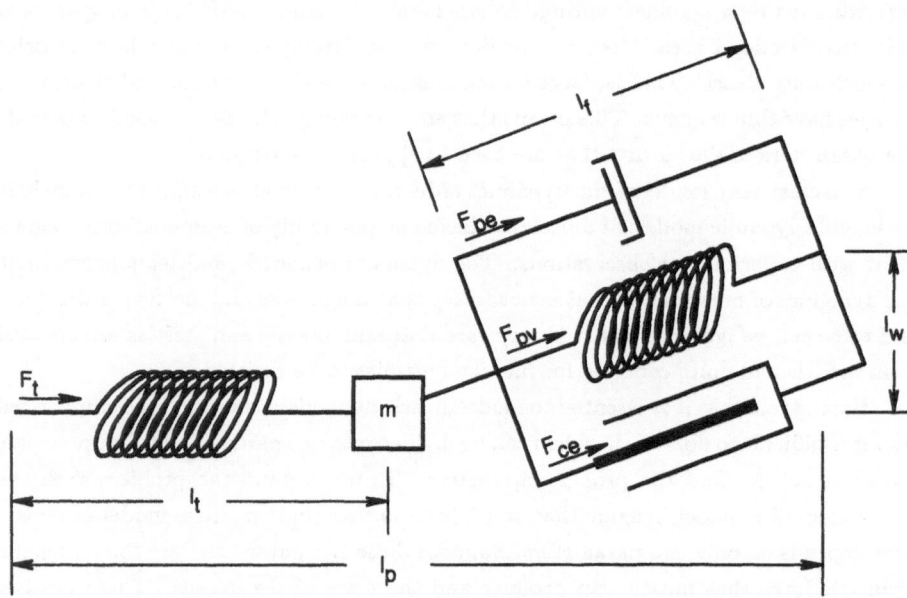

Figure 1. Mechanical model of a muscle and tendon. The model includes the muscle mass, the contractile element, both series and parallel elastic elements and a parallel viscous element.

Parallel elastic element:

A large number of studies (see [4]) indicate that passive muscle fiber force increases exponentially with fiber length (ℓ_f):

$$F_p(\ell_f) = F_{p0}\{\exp[k_{pe}(\ell_f - \ell_{f0})] - 1\} \tag{1}$$

where ℓ_{f0} is the fascicle rest length (i.e., $F_p(\ell_{f0}) = 0$). This relationship is true across a wide variety of species. (See, e.g., [5]).

Series elastic element:

The tendon or aponeurosis is usually combined with the series elastic properties of the muscle fibre and modelled by a series elastic element. This element is described by a nonlinear spring for which the length-tension relationship must be subdivided into two components: 1) an exponential relationship for short lengths, followed by 2) a linear relationship for longer lengths.

$$F_t = \begin{cases} F_{t0}\{\exp[k_{te}(\ell_t - \ell_{t0})] - 1\} & \ell_{t0} \le \ell_t \le \ell_{tc} \\ k_{te}(\ell_t - \ell_{tc} + \frac{F_c}{k_t}) & \ell_t \ge \ell_{tc} \end{cases} \tag{2}$$

where F_t is tendon force, ℓ_t is tendon length, ℓ_{t0} is tendon rest length and ℓ_{tc} and F_c are the tendon length and force at which tendon force shifts from a nonlinear to linear curve. There is substantial evidence for viscous effects in muscle. Hatze [6] citing some of this data, and Otten [7] include a parallel viscous element in their models in addition to explicit models of the velocity dependence of active force generation.

For simplicity, we use a linear model

$$F_{pv} = -B\dot{\ell}_f \tag{3}$$

although there would be no serious difficulty in using a nonlinear one. We are not aware of data regarding reasonable values for B.

Mass:

The mass in the model represents the mass of the muscle itself. Muscle obviously has mass. It is even relatively easy to measure. Unfortunately, this mass is distributed throughout the muscle. Because our model treats this distributed mass as a single point mass it is inappropriate to use the total mass of the muscle. There is a standard result in mechanics [8] that says that a spring having distributed mass m can be approximated by a spring and a point mass equal to $m/3$ when the spring is connected to a heavy mass. Although this is not precisely the situation for muscle we do not need a very accurate estimate. Thus, we use $m/3$, where m is the total mass of the muscle.

The inclusion of muscle mass greatly facilitates the derivation of the dynamical equations describing muscle. The conditions under which the muscle mass can be neglected are rather obvious from the model. The dynamical equations in the case where muscle mass can be neglected can be derived by simply setting $m = 0$ in the equations describing the model.

Contractile element:

The active component of muscle is usually modelled by a contractile element, as shown in Fig. 1. The active force, F_{CE}, produced is dependent upon three main factors: 1) muscle fascicle length, 2) muscle fascicle velocity and 3) amount and timing of activation. As described below, F_{CE} is usually taken to be the product of these three factors, where

$f_\ell(\ell_f)$ = the force-length curve for a muscle at constant maximal activation and
zero shortening velocity (Fig. 2A).

$f_v(\dot{\ell}_f)$ = the force-velocity curve for a muscle at constant maximal activation and
constant length (the length is that which maximized $f_\ell(\ell_f)$) (Fig. 2B)

$a(t)$ = activation, essentially a low-pass version of the input neural signal
produced by the chemical processes involved in the release and reuptake
of calcium by the sarcoplasmic reticulum

F_{CE} = force due to the contractile element.

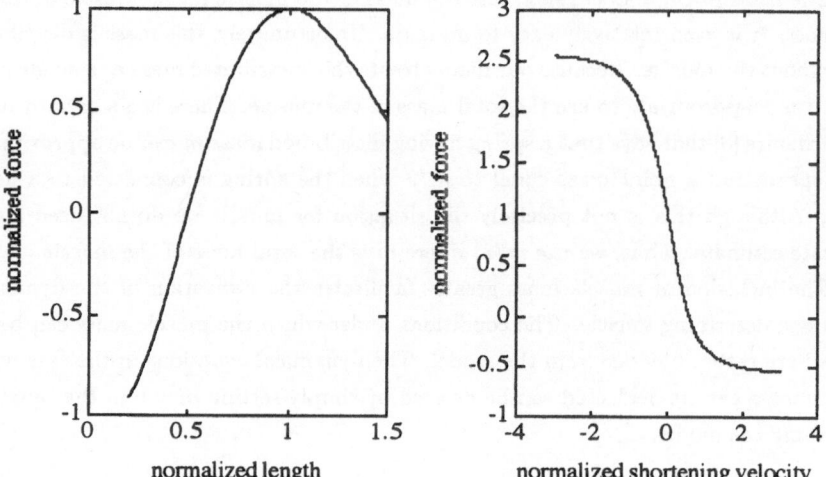

Figure 2A. The normalized force length
curve for the contractile element.

Figure 2B. The normalized force
velocity curve for the contractile
element.

The relation among these factors is

$$F_{CE} = f_\ell(\ell_f)f_v(\dot{\ell}_f)a(t) \tag{4}$$

According to [9], the relationship between sarcomere length and force, $f_\ell(\ell_f)$, at
zero velocity, may be represented by a piece-wise linear function. Because of hetero-
geneities within and among the fibre population, a modified parabola or other smooth
nonlinear function is a more realistic model of the total or average effect of all fibers.

The force produced is assumed to be directly proportional to the number of cross-bridge attachments between actin and myosin. At very short lengths, actin filaments can overlap and interfere with one another, thereby reducing the amount of potential interaction between the myosin cross-bridges and the actin attachment site.

At intermediate lengths, interaction is maximal; it is reduced with increasing length until the force drops to zero when the two filaments no longer overlap.

A convenient choice for this curve is then (Fig. 2A)

$$f_\ell(\ell_f) = F_{iso} \sin(a_1 \bar{\ell}_f^2 + a_2 \bar{\ell}_f + a_3) \tag{5}$$

where

$$F_{iso} = \text{maximum isometric force}$$
$$\bar{\ell}_f = \ell_f/\ell_{iso}$$
$$\ell_{iso} = \text{length at which } F_{iso} \text{ is achieved}$$
$$a_1, a_2, a_3 \quad \text{are curve fitting constants.}$$

Typical values are given in Appendix A.

The next step is to describe the relationship between normalized muscle force and the shortening velocity of muscle fascicles $(f_v(\dot{\ell}_f))$.

A convenient choice for this curve is

$$f_v(\dot{\ell}_f) = 1 + \arctan\left(b_1 \dot{\bar{\ell}}_f^3 + b_2 \dot{\bar{\ell}}_f^2 + b_3 \dot{\bar{\ell}}_f\right) \tag{6}$$

where

$$\dot{\bar{\ell}}_f = \dot{\ell}_f/V_{max}$$
$$b_1, b_2, b_3 \text{ are curve fitting constants.}$$

Typical values of the constants are given in Appendix A.

The final component of the model for the contractile element is the term denoted by $a(t)$ in Eq. (4). The total active force developed within a muscle is dependent upon the total number of activated muscle fibres. Since the development of active tension within a fiber is accompanied by the electrical depolarization of a fiber's membrane, the measured electrical activity of whole muscle (the electromyogram or EMG) would be a plausible indicator of the "active state" of a muscle. Models of this active state may be found in [10], [11] or [12] (see also [13]) for a review of EMG properties and models). However, such measurements are dependent upon the size, shape and type of electrode used, electrode position on the muscle, etc. These measurements are further complicated by the existence of different fiber type populations within the muscle (see [14]) and possible variations in their recruitment order (see [15]). More importantly, the kinetics of calcium release and diffusion complicate the relationship between time course of EMG and time course of developed tension. Furthermore, [16] demonstrated that activation appeared to be length dependent. This might perhaps be due to a length dependent shift in position of the calcium releasing membranous structures with respect to the myofilaments.

Given the complexity of the data it is not surprising that there is considerable debate over how best to model the effect of neuronal input on activation. For example, [6] uses two inputs, one related to the number of muscle fibers stimulated and the other related to the rate of stimulus. More commonly, these two inputs are combined into one. The size principle ([15]) is commonly cited as the basis for combining the two inputs. There does seem to be agreement that there are lags between neuronal input is much smaller than that following a decrease in neuronal input. A simple model for activation is given in Eq. (7).

$$\dot{a}(t) = (u(t) - a(t))(c_1 u(t) + c_2) \tag{7}$$

where

$$0 \le a(t) \le 1$$

$$0 \le u(t) \le 1$$

$\tau_{\text{off}} = 1/c_2$; τ_{off} is the time constant for relaxation

$\tau_{\text{on}} = \dfrac{1}{c_1 + c_2}$; t_{on} is the time constant for excitation

Typical values for c_1 and c_2 are given in Appendix A.

It is easy to verify that Eq. (7) has equilibrium points $a(t) = a_e = u$ whenever $u(t) = u$ is constant. These equilibria are asymptotically stable. The time constant for increasing $a(t)$ is smaller than that for decreasing $a(t)$.

neurons: By now everyone is aware of the renaissance of research on neural networks. It must be said that the "neurons" of neural network research bear only a very general resemblance to the neurons that control animal and human movement. In fact, the majority of the neural network literature deals with networks of simple dynamical systems that can "store" and "retrieve" constants. Real neurons can certainly do that but they can do many other things as well. In particular, they are capable of complicated dynamic behavior.

The mesencephalic cat experiments and their corollary results given some measure of the complexity of the dynamics achievable by real networks of neurons. The basic experiment consists of cutting through the cat's brain at the mesencephalon. The cat is left with only its brain stem. This is enough to support respiration and blood flow but the cat has no sensation and no volition. It is basically a vegetable cat. If this cat is supported over a treadmill and the correct portion of the brain stem is stimulated, then the cat walks. It appears to walk normally. It will even correct for rather large perturbations of its step. There is a great deal of corollary support for the belief that the complex coordination of muscles and limbs required for walking is accomplished by the network of neurons in the spinal cord.

This suggests that one good way to view the spinal cord is as an analog computer. We have proposed, in outline form just such a control scheme for movement. Grossberg and his colleagues [17] have proposed a similar idea in much greater detail. The main reason for mentioning such ideas here is that they reduce the control of movement to finding the right parameters to give to the analog computer, thereby reducing learning to a form of parametric optimization process.

Optimization

We begin this section with a fairly simple example. This example exhibits most of the important difficulties in applying optimization methods to the problems of animal movement. In particular, notice that the right hand side of the vector differential equation of motion is continuous but not continuously differentiable; the value function is discontinuous; and the control has both bang-bang and smooth segments.

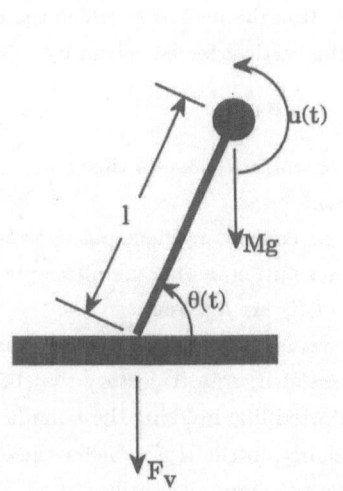

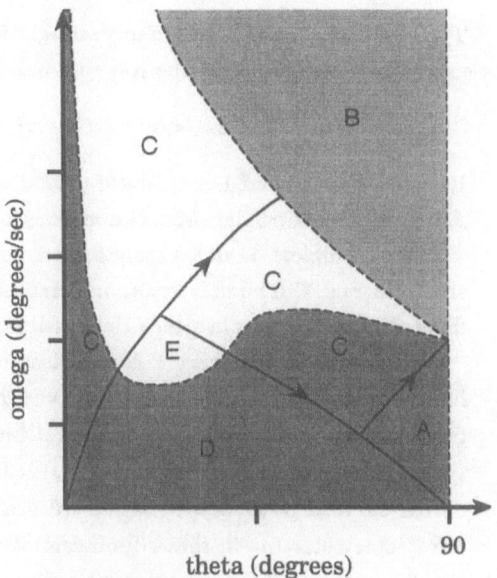

Figure 3. A representation of the physical problem of making the baton "jump".

Figure 4. A representation of the complete feedback solution to the jumping baton problem.

Example ([18]):

One of the simplest possible models of a human or cat performing a maximal height jump is illustrated in Fig. 3. The basic problem is to apply a torque; denoted by $u(t), 0 \leq t < t_f$ with $0 \leq u(t) \leq U_{\max}$ for all t; so as to cause the peak height reached by the center of mass of the rod to be the maximum possible.

Mathematically, the dynamics of the rod must be divided into two epochs. During the first epoch the bottom of the rod is in contact with the ground. The equation of motion is

$$\ddot{\theta} = \begin{cases} \max[\frac{u(t)}{I} - \frac{mgl}{I}, 0] & \theta = 0 \\ \frac{u(t)}{I} - \frac{mgl}{I}\cos\theta & \theta > 0 \\ \text{undefined} & \theta < 0 \end{cases} \tag{8}$$

During the second epoch no portion of the rod touches the ground. The motion of the rod is a rotation about the center of mass which follows a ballistic trajectory that is completely determined by $\theta(t_\ell), \dot{\theta}(t_\ell)$ where t_ℓ is the instant of "lift-off". Thus, we can terminate the optimization at time $t_\ell < t_f$. Ignoring the effect of the rotation of the rod, performance is given by

$$J = \text{ peak height reached by the center of mass } = \ell\sin\theta(t_\ell) + \frac{\ell^2}{2g}\sin^2\theta(t_\ell)\dot{\theta}^2(t_\ell) \tag{9}$$

The calculation of t_ℓ is reasonably straight forward. It is the instant at which the force exerted on the ground by the rod goes negative. The vertical force is given by

$$F_v = m(g + \ell\cos\theta(t)\ddot{\theta}(t) - \ell\sin\theta(t)\dot{\theta}^2(t)) \tag{10}$$

It is easily seen from Eqs. (8) and (10) that the control, $u(t)$, enters the equation for F_v. In fact, making the substitution gives $F_v(\theta, \dot{\theta}, u)$.

The complete feedback solution to the optimal control problem posed above is shown in Fig. 4. The first key to understanding the solution is that the phase plane is divided into 3 regions in which the possible values of F_v are different.

$F_v(\theta, \dot{\theta}, u) < 0$ in region B for all admissible values of u. Thus, the rod cannot have its tip remain on the ground for any $\theta, \dot{\theta}$ in region B. Any trajectory that begins below region B will "jump" as soon as it hits the dotted line marking the boundary of that region (given by $F_v(\theta, \dot{\theta}, u = u_{\max}) = 0$). Similarly, in the region below the lower dotted curve, $F_v(\theta, \dot{\theta}, u = 0) = 0$, $F_v(\theta, \dot{\theta}, u) > 0$ for all admissible values of u. Thus, a trajectory starting in that region cannot end in a jump without exiting that region. Finally, in the region between the two dotted curves the value of u determines the sign of F_v.

The second key is that the height of the jump is monotonically increasing along all trajectories corresponding to $u = u_{\max}$. Furthermore, the height of the jump is a monotone increasing function of $\dot{\theta}$ along the curve $F_v(\theta, \dot{\theta}, u_{\max})$, the lower boundary of region B.

Thus, all trajectories that start in region A are carried to the vertical by their angular momentum regardless of the control, $u(t)$. They fall forward after that so there is no jump and the peak height reached is ℓ, the length of the rod.

All optimal trajectories that commence in region C have $u = u_{\max}$. Any other choice of control will result in a lower jump. Trajectories that begin in region D start with a non-unique optimal control. The optimal strategy is to return to the origin and then apply the unique optimal control $u = u_{max}$ until lift-off. Many controls, can take $\theta, \dot{\theta}$ to 0,0. Trajectories that begin in region E, and especially on the upper right boundary of that region, are the most interesting. The optimal control on the upper right boundary of region E is unique and it is not $u = 0$ or $u = u_{\max}$. It takes the state back to the origin. Elsewhere in region E the optimal control is similar to that in region D except that the choice of controls is smaller.

The value of the Bellman function, the optimal performance as a function of θ and $\dot{\theta}$, is discontinuous along the trajectory shown in Fig. 4 along which $\dot{\theta}$ is decreasing. This demonstrates one difficulty in solving this problem by some conventional optimization algorithm. A second difficulty is the dramatically different dynamic behavior of the system as a function of the achievable values of f_v. Thirdly, the differential equation describing the motion of the baton is discontinuous at $\theta = 0$ and undefined for $\theta < 0$. All of these problems are found in more realistic applications of optimization to the control of movement.

Despite the difficulties many people have tried to use optimization methods to help understand the control of movement. Many people seem to believe that movement is optimally controlled. The problem is that no one knows the performance criterion. Many criteria have been tried. Some of them have been obviously impossible even before the research began. For example, people have used linear programming models to solve the muscle redundancy problem. That is, many combinations of muscle forces will produce the same joint torques. How does the brain choose the optimal combination? The problem cannot be a linear programming problem because even crude experiments show that most muscles are used at levels that are neither maximal nor minimal. Nonetheless, people have tried using linear programming to solve the problem.

A more interesting proposal has been that the brain tries to minimize "jerk", the third derivative of position. Others have proposed, very plausibly, that the brain tries to minimize the energy expended. Unfortunately, it is very difficult to measure or model the rate at which metabolic energy is expended. Even though it seems evident that it is this energy and not the mechanical energy that is minimized, most of the analytical efforts have tried to minimize mechanical energy. The two are not the same and are not simply related.

Both of these proposals expose a facet of the problem of using optimization to understand the control of movement that is both very important and rarely acknowledged. The way to minimize both energy expenditure and jerk is to do nothing. In fact, this is what cats do most of the time. In order to get movement there must be constraints on the solution to the optimal control problem. For example, most of the minimum jerk research involves movements of the hands from one position to another. the end points and sometimes intermediate points are prespecified.

Another approach has been to specify a task that implies a performance measure as well as constraints. For example, Hatze [19] asked his subjects to perform a minimum time kick. We have asked our subjects, both humans and cats, to perform maximal height jumps of various types. We are presently working on problems of pedaling a stationary bicycle at maximum speed. In each of these problems of pedaling a stationary bicycle at maximum speed. In each of these problems the task specification gives both performance criterion and constraints, thereby producing a completely formulated optimal control problem. It is assumed that the subjects actually do perform the task optimally, subject to the various constraints imposed by their physiology and environment. Therefore, the differences between the solution to the mathematical problem that presumably models their task and the actual solution produced by the subjects are due to unmodeled constraints.

More Realistic Example

This is a summary of work reported in [20] and [21]. In order to improve mathematical models of musculoskeletal mechanics and to study inter muscular coordination a model of humans performing maximal height jumps was developed and compared with experimental observations of humans performing the same task.

If simply told to jump as high as possible most humans will start by standing fairly erect. They will then, in one continuous motion, quickly crouch and then smoothly extend their bodies until they leave the ground. The crouching movement will be combined with a downward swing of both arms. The arms will be swung upward during the extension phase of the movement. To simplify the modeling problem, the experimental subjects were asked to hold their arms still at their sides throughout the jump.

The modeling problem is to describe this task as a precisely specified optimal control problem. A mathematical description is given by a performance criterion, equations for the dynamics of the subject, and the constraints. The mathematical model of the dynamics of the subject is based on the skeletal mechanics, as shown in Fig. 5, and the muscle dynamics described earlier. It is assumed that the motion is constrained to the sagittal plane.

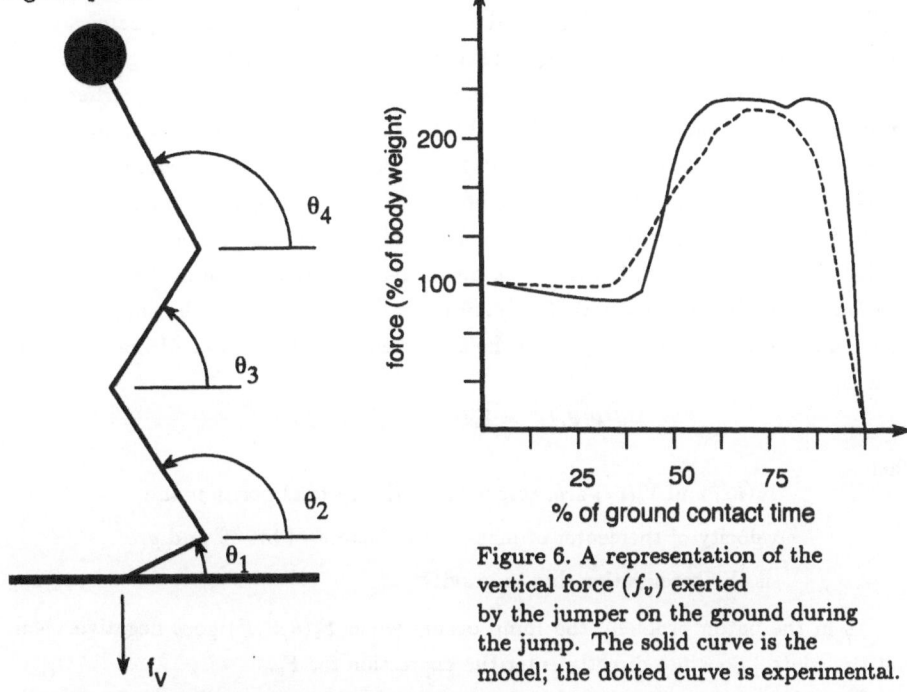

Figure 6. A representation of the vertical force (f_v) exerted by the jumper on the ground during the jump. The solid curve is the model; the dotted curve is experimental.

Figure 5. The idealized model of the jumper.

Thus, the skeletal trajectory is governed by the usual Newtonian equations of motion of a four segment inverted planar pendulum.

$$I(\theta)\dot{\theta}(t) = B(\theta)\dot{\theta}^2 + C(\theta) + D(\theta)P(\theta,\dot{\theta},a(t)) + T(\theta,\dot{\theta})$$

where

$I(\theta)$ is the inertia matrix

$\theta(t)$ is the 4-vector of angles to the horizontal

$B(\theta)\dot{\theta}^2$ is the vector of centripetal forces ($\dot{\theta}^2 \triangleq [\dot{\theta}_1^2 \ \dot{\theta}_2^2 \ \dot{\theta}_3^2 \ \dot{\theta}_1^4]$)

$C(\theta)$ is the vector of gravitational foces

$D(\theta)$ is the 4×8 matrix of muscle moment arms

$P(\theta,\dot{\theta},a)$ is the 8-vector of muscle forces

$a(t)$ is the 8-vector of muscle activations

$T(\theta,\dot{\theta})$ is the 4-vector of external joint torques

Although there are roughly 30 muscles that move a single human leg many of these can be combined when the motion is restricted to be planar. The model lumps the leg muscles into 8 "muscles". For simplicity, each muscle is described by the massless, lossless version of the previously described muscle model. Thus, the dynamics of the jumper are described by 24 states (8 mechanical and 16 muscle). The eight inputs to the system, the controls, are the u's in the muscle model. As explained earlier, these represent an approximation to the Central Nervous System's input to the muscles. The control values are constrained by $0 \leq u(t) \leq 1$. As in the previous example, the foot exerts a vertical force on the ground. A jump occurs when this force changes sign.

It has been shown that, with the exception of trained high jumpers, there is little difference between maximizing the peak height reached by the center of mass and the peak height reached by any part of the subject's body [18]. Thus, the simpler performance measure, maximizing the peak height reached by the center of mass, is used. That is,

$$J(\theta, \dot{\theta}, t_f) = Y_c(t_f) + \dot{Y}_c^2(t_f)/2g$$

where

$Y_c(t_f)$ and $\dot{Y}_c(t_f)$ are, respectively, the vertical position and

velocity of the center of mass at the instant of lift-off and g

is the acceleration due to gravity

As in the baton problem, the jump occurs when $F_v(\theta, \dot{\theta}, P)$ goes negative. Note that the control does not directly enter the expression for F_v.

There is one difficulty that does not appear in the baton problem. The jumper starts with his feet flat on the floor and then, at some time during the foot-on-the-ground phase of the jump, his heels leave the ground while his toes remain on the ground. Fig. 5 shows this heel-off portion of the jump. Conventional optimization algorithms cannot handle non-smooth dynamics so this dynamical "corner" is approximated by a smooth nonlinear stiff spring.

To complete the model it is necessary to specify the parameters of the model, including the initial conditions for the 24 states. The initial joint angles of the model were determined by measurement of the subject's initial joint angles. Their initial angular velocities were zero. The initial muscle forces of the model were chosen to hold the model at the observed equilibrium and minimize a quadratic measure of the force being produced. The parameters of the skeletal model were estimated from direct measurements of the subjects height, weight, and the segment lengths of their limbs. The parameters of the muscle model were estimated from simple measurements of the limbs and from measurements of the maximum joint torque the subjects were able to exert. Because the joint torques are produced by several muscles acting at once, some complicated curve fitting was necessary. It should be obvious that the parameters were not known accurately.

This optimal control problem was then solved by a modified version of the Polak-Mayne algorithm. From prior work [18] it was known that singular optimal controls were likely, especially at the bottom of the counter movement that humans normally perform in trying to jump as high as possible. In an attempt to avoid the computational problems associated with singularity, the subjects were asked to begin their jumps from a relaxed crouch and not perform a counter movement.

The most sensitive test of the agreement between this model and experiment is the vertical force exerted on the ground by the jumper as a function of time. This is an accurately measurable quantity. It is also proportional to acceleration. All of the other mechanical observables are velocities and positions versus time. These are integrals of acceleration and, as a result, some mismodelling disappears with the smoothing.

As can be seen in Fig. 6, the agreement between model and experiment is quite good. In fact, the analytical curve is entirely within the spread of experimental curves [20], [21]. There is one discrepancy that is visible in Fig. 6 but much more obvious elsewhere. Although both model and experiment agree very well with respect to maximum joint extension, the model tolerates larger angular velocities near full extension than do humans. It is conjectured that the joint angle sensors present in humans and other animals trigger protective mechanisms that slow the extension of the joints. The model does not include such protective mechanisms.

At this time, it is believed that the jumping model is at the limit of experimental resolution. That is to say that further elaboration of the model would require more sophisticated experiments.

Conclusions

The ultimate goal of research in the control of movement is to provide the means to improve the life of people who have motor disorders. These include the victims of strokes (250,000 per year in the U.S.), cerebral palsy, and trauma. Optimization could be used to help these people in· a variety of ways, from the fitting of prostheses and orthoses, to the planning and execution of both restorative surgery and rehabilitation. Al of these clinical applications would require models, including parameters, that are accurate enough to predict the clinical outcome for individual patients.

As should be clear from the previous discussion, today's models are not detailed enough nor can their parameters be measured accurately enough for most clinical applications. Recent advances in measurement technology, such as MRI, and in computer technology (faster hardware and automatic equation writing software) are rapidly changing this. Hopefully, within a few years, it will be possible to construct clinically useful models of limb movement.

There are at least two challenging problems in modeling and optimization that must also be solved before these models can really be used in treating patients. The first is the computer modeling of the interaction between complex surfaces. Human joints are much more complicated than man-made joints. Clinically useful models will have to account accurately for the rolling, sliding, and twisting of one joint surface with respect to another. The second is the need for fool-proof algorithms for solving optimization problems that are large and have difficult nonlinearities (discontinuities, definition on subsets of the space). Hopefully, these problems are also near solution.

References

[1] McMahon, T.A. and Greene, P.R. (1979). The influence of track compliance on running. J. Biomech. 12:893-904.

[2] Young, R.P., Scott, S.H., and Loeb, G.E. (1992). An intrinsic mechanism to stabilize posture—joint-angle-dependent moment arms of the feline ankle muscles. Neurosci. Letters.

[3] Levine, W.S., He, J., Loeb, G.E., Rindos, A.J. and Weytjens, J.L.F., (1993) (submitted to J. Biomech.)

[4] Partridge, L.D. and Benton, L.A. (1981). Muscle, the motor, from *Handbook of Physiology—The Nervous System*, Vol. II. Part 1. Williams and Wilkins, MD. 43-106.

[5] Rack, P.M.H. and Westbury, D.R. (1984). Elastic properties of the cat soleus tendon and their functional importance. *J. Physiol.* 347:479-495.

[6] Hatze, H. (1977). A myocybernetic control model of skeletal muscle. Biol. Cybern 25:103-119.

[7] Otten, E. (1988). Concepts and models of functional architecture in skeletal muscle. Exerc. Sport Sci. Rev. 89-137.

[8] Ziegler, F. (1991). Mechanics of Solids and Fluids, Springer-Verlag, 598-600.

[9] Gordon, A.M., Huxley, A.F. and Julian, F.J. (1966). The variation of isometric tension with sarcomere length in vertebral muscle fibres. *J. Physiol.* 184:170-193.

[10] Gottlieb, G.L. and Agarwal, G.C. (1971). Dynamic relationship between isometric muscle tension and the electromyogram in man. J. Appl. Physiol., 30:345.

[11] Meijers, L.M.M., Teulings, J.L.H.M. and Eijkman, E.G.J. (1976). Model of the electromyographic activity during brief isometric contractions. *Biol. Cybern.* 25:7-16.

[12] Hof, A.L. and Van den Berg, J. (1981). EMG to force processing. I: An electrical analogue of the Hill muscle model. *J. Biomech.*, 14 (11): 747-758.

[13] Perry, J. and Bekey, G.A. (1981). EMG-force relationships in skeletal muscle. *CRC Crit. Rev. Biomed. Eng.* (December).

[14] Burke, R.E. (1981). Motor units: Anatomy, physiology and functional organization from *Handbook of Physiology—The Nervous System, Vol. II*, Part 1, Williams and Wilkins, MD, 345-422.

[15] Henneman, E., Somjen, G., and Carpenter, D. (1965). Excitability and inhibitability of motoneurons of different sizes. J. Neurobiol. 28:599-620.

[16] Rack, P.M.H. and Westbury, D.R. (1969). The effects of length and stimulus rate on tension in the isometric cat soleus muscle. *J. Physiol.* 204:443-460.

[17] Gaudiano, P. and Grossberg, S. (1991). Vector associative maps: unsupervised real-time error-based learning and control of movement trajectories. Neural Networks 4, pp. 147-183.

[18] Levine, W.S., Christodoulou, M. and Zajac, F.E. (1983). On propelling a rod to a maximum vertical or horizontal distance, Automatica 19:3, pp. 321-324.

[19] Hatze, H. (1976). The complete optimization of a human motion. Math. Biosci. 28, pp. 99-135.

[20] Pandy, M.G. Zajac, F.E., Sim, E. and Levine, W.S. (1990), An optimal control model for maximum-height human jumping, J. Biomech. 23:12, pp. 1185-1198.

[21] Pandy, M.G. and Zajac, F.E. (1991), Optimal muscular coordination strategies for jumping, J. Biomech. 24:1, pp. 1-10.

Appendix

parameters for f-1 curve:
a_1=-0.906222865
a_2=4.500935629
a_3=-2.023910652

parameters for f-v
b_1=-1.3166
b_2=-0.40268458
b_3=-2.023910652

The tendon and muscle elastic coefficients are derived from Rack and Westbury's measurement from cat soleus muscles and tendons.
$k_{te} = 3.022$/mm
$k_{t\ell} = 31.44$ N/mm

$k_{p0} = 0.8$/mm

Then these values are scaled according to cross sectional area of each muscle to obtain parameters for each muscle.

The parameters for activation dynamics are fiber type dependent. The values for soleus are c_1=19.7, c_2=22, for gastr c_1=24.5, c_2=35.

Deterministic Sampling and Optimization

BERNARD CHAZELLE[*]

Department of Computer Science

Princeton University

Princeton, NJ 08544, USA

Randomization has proven extremely useful in helping to design simple, efficient geometric algorithms; see, e.g., [15, 22, 23, 25, 26, 27, 33, 34, 38, 63] and Clarkson's survey [24]. To understand the computational power of randomness has been a central preoccupation in theoretical computer science. While only few general derandomization techniques have been discovered [2, 43, 62, 67], several clever schemes have been found to reduce the amount of randomness needed by probabilistic algorithms [4, 5, 7, 8, 39, 43, 60]. As was shown by Mulmuley [59], such techniques can be used to prove that most randomized geometric algorithms require few (truly) random bits, but still too many, however, to allow efficient deterministic simulations. We discuss a few specific problems in computational geometry, where randomization (or derandomization) has made a major difference in our understanding of the underlying complexity issues.

*Supported in part by NSF Grant CCR-90-02352 and The Geometry Center, University of Minnesota, an STC funded by NSF, DOE, and Minnesota Technology, Inc.

1. Linear Programming. Randomized geometric algorithms come in two flavors: random sampling-based methods and randomized incremental constructions. The former have been most successful in the design of data structures for geometric searching [24]. The gist of the approach is to sample the input space and compute a solution either recursively or by a naive solution on a small subset: then we use this information to break down the problem into subproblems in divide-and-conquer fashion. Most of the random-sampling methods have been derandomized optimally over the last few years [12, 17, 18, 19, 44, 45, 46, 47, 48, 49], meaning that deterministic algorithms of matching complexity have been found.

A good example of this is linear programming with a fixed number of variables. Consider a linear program with d variables and n constraints. Megiddo [52] was the first to give a deterministic algorithm of complexity $O(C(d)n)$, where $C(d) = 2^{2^d}$. Dyer [30] and Clarkson [20] subsequently improved $C(d)$ to 3^{d^2}. Randomized algorithms with still lower dependency on d were also given in [21, 32, 65]. Among these, Clarkson's algorithm [21] has the best expected complexity, i.e., $O(d^2n + d^{d/2+O(1)} \log n)$. Recently Kalai [40] and independently Matoušek, Sharir and Welzl [50] have developed algorithms with a subexponential dependency on both n and d. See also [36]. In combination with Clarkson's algorithm, one derives a randomized algorithm for linear programming with expected running time $O(d^2n + e^{O(\sqrt{d \ln d})} \log n)$. To match these performance bounds by a deterministic algorithm seems difficult at the present time. Jointly with Matoušek [19], we have shown how to derandomize Clarkson's probabilistic solution [21] to achieve a deterministic running time of $d^{O(d)}n$. Other problems can be solved within the same framework: in particular, in $O(n)$ time we can find the maximum-volume ellipsoid inscribed in an n-facet polyhedron in $\mathbf{R}^d$ or the minimum-volume ellipsoid enclosing n points.

The works of Khachian and Karmarkar established, respectively, that LP is in P and that interior point methods can be very efficient in practice. It remains open whether strongly polynomial methods exist, i.e., methods whose running time does not depend on the size of the elements of the matrix. The sub-exponential algorithms of Kalai [40] and Matoušek et al. [50] are steps in that direction: interestingly, these methods are of the (dual) simplex form with randomized pivoting rules. It might be that a solution to this outstanding open problem is in sight: after the departure from simplex-based methods and the excitement generated by interior-point methods, it would be quite remarkable if, indeed, a polynomial-length pivoting sequence could be exhibited.

2. Derandomizing Incremental Algorithms. Lately, the attention has shifted away from random sampling-based methods to so-called *randomized incremental constructions* [9, 10, 15, 25, 26, 28, 29, 37, 54, 55, 56, 57, 58, 64, 65]. These algorithms examine the input elements in random order and at each step maintain a partial solution based on the portion of input examined so far. This approach has produced surprisingly simple Las Vegas algorithms of low expected complexity. Unfortunately, their derandomization has proven elusive so far. Recently, I developed a derandomization scheme for incremental algorithms and used it to derive an optimal convex hull algorithm in any fixed dimension [13]. One of the major ingredients of my scheme is a method of *deterministic Monte Carlo integration*, which uses recent results in the theory of range spaces of finite VC dimension

[12, 46]. This theory, which originates from machine learning, provides sampling tools of tremendous power and versatility in computational geometry. Jointly with Brönnimann and Matoušek [11] I have extended my derandomization technique to compute the diameter of points in 3-space. We have discovered a deterministic algorithm in $O(n \log^3 n)$, which almost matches the optimal $O(n \log n)$ algorithm of Clarkson and Shor [26].

Almost a decade ago, Megiddo devised a beautiful algorithmic paradigm known as *parametric searching*. The main feature of the paradigm is to turn a *checking* algorithm for an optimization problem into one that actually searches for a solution. Certain conditions must be satisfied, of course, but the setting is general enough to have wide applicability. For example, it can be used to solve the 2-center problem in the plane, compute the diameter of n points in E^3 (as we do in [11]), select distances, compute the kth leftmost vertex in a line arrangement, perform ray-shooting in 3-space, etc. It is an interesting open question how parametric search can be best tailored to geometric settings. It seems that the requirement that the checking algorithm be parallelizable can be weakened in several ways and reduced to a mild form of multidimensional searching.

3. Discrepancy Theory and Sampling.

Discrepancy theory addresses the general issue of approximating one measure by another one. Originally an offshoot of diophantine approximation theory [70], the area has expanded into applied mathematics, and recently, computer science. Besides providing the theoretical foundation for sampling, it holds some of the keys to understanding the computational power of randomization. Here are a few applications of discrepancy theory:

In quasi-Monte Carlo numerical integration one wants to replace a continuous measure (e.g., Lebesgue) by an atomic measure in order to apply finite-element techniques [61]. A classical result of Koksma [41] says that if the function to integrate is of bounded variation then the absolute error in quasi-Monte Carlo integration is proportional to the discrepancy of the sample space. In computer graphics, supersampling is one of the most general lines of attack for antialiasing. The technique draws its efficacy from low-discrepancy 2d point sets [53, 66].

Sampling tools (e.g., ε-approximations and ε-nets) pioneered in the works of Vapnik and Chervonenkis [69] and Haussler and Welzl [38], have had a tremendous impact in the design of probabilistic geometric algorithms and in their derandomization. Recently, Matoušek et al. [51] established a beautiful connection between ε-approximations and low-discrepancy point sets. It appears that discrepancy-type lower bounds impose computational limitations on our ability to perform standard geometric tasks (e.g., range searching), which makes such bounds all the more relevant to algorithm designers. Taking the view that randomization is a computational resource (just like time and space), it is natural to examine how random bits can be recycled, reduced, or simply eliminated. Even et al. [35] recently showed that approximating the uniform distribution for general random variables could be viewed as achieving low rectangle-discrepancy (see also [42]). Indeed, many questions in pseudo-randomness boil down to discrepancy questions in higher dimensions.

We should also mention the classical optimization problem consisting of placing points on a sphere so as to maximize the sum of pairwise distances. A very elegant result of Stolarsky [68] says that the L^2-distance from a given approximation to the optimal solution is precisely the discrepancy between the discrete point measure and the Lebesgue measure of spherical caps.

Of central interest in computational geometry is the set system induced by points and halfspaces. Matoušek, Welzl, and Wernisch [51] have shown that it is possible to two-color any set of n points in E^d, so that within any halfspace no color outnumbers the other by more than $O(n^{1/2-1/2d}\sqrt{\log n})$. It follows from a deep theorem of Beck [6] or, alternatively, one of Alexander [1] that this result is quasi-optimal. Unfortunately, the proofs of these theorems are very complicated and do not provide much intuitive insight into the "large-discrepancy" phenomenon.

We have recently found a surprisingly simple proof of the $\Omega(n^{1/2-1/2d})$ lower bound which has an intuitive probabilistic interpretation [14]. We forsake the Fourier transform and convolution functional approaches of Beck and Alexander to introduce a simple finite-difference operator, from which the discrepancy problem appears to be nothing but the classical *Buffon needle problem* in disguise.

We close this short discussion with an intriguing sampling problem first studied by Alon et al. [3]: If S is a set of n points in E^d, a set W is called a *weak ε-net* for (convex ranges of) S if for any $T \subseteq S$ containing εn points, the convex hull of T intersects W. We recently [16] established the existence of weak ε-nets of size roughly $O(1/\varepsilon^d)$, thus improving a previous bound of Alon et al. [3]. In the case of points uniformly distributed on a circle, we have devised an elegant optimal solution based on tilings of regular polygons in the hyperbolic plane. The fact that in the hyperbolic plane all ideal triangles (i.e., triangles with points at infinity) have the same area brings uniformity conditions which are absent from the Euclidean plane. This suggests new methods for covering hypergraphs using structures in hyperbolic geometry.

References

[1] Alexander, R. *Geometric methods in the study of irregularities of distribution*, Combinatorica 10 (1990), 115–136.

[2] Alon, N., Babai, L., Itai, A. *A fast and simple randomized algorithm for the maximal independent set problem*, J. Alg. 7 (1986), 567–583.

[3] Alon, N., Bárány, I., Füredi, Z., Kleitman, D. *Point selections and weak ε-nets for convex hulls*, manuscript, 1991.

[4] Alon, N., Goldreich, O., Hastad, J., Peralta, R. *Simple constructions of almost k-wise independent random variables*, Proc. 31st Ann. IEEE Symp. Foundat. Comput. Sci. (1990), 544–553.

[5] Beck, J. *An algorithmic approach to the Lovász local lemma. I.* Random Structures & Algorithms 2 (1991), 343–365.

[6] Beck, J., Chen, W.W.L. *Irregularities of distribution*, Cambridge Tracts in Mathematics, 89, Cambridge Univ. Press, Cambridge, 1987.

[7] Berger, B., Rompel, J. *Simulating* $(\log n)^c$-*wise independence in NC,* J. ACM 38 (1991), 1028–1046.

[8] Berger, B., Rompel, J., Shor, P. *Efficient NC algorithms for set cover with applications to learning and geometry*, Proc. 30th Ann. IEEE Symp. Foundat. Comput. Sci., (1989), 54–59.

[9] Boissonnat, J.D., Devillers, O., Schott, R., Teillaud, M., Yvinec, M. *Applications of random sampling to on-line algorithms in computational geometry*, Disc. Comput. Geom., 8 (1992), 51–71.

[10] Boissonnat, J.D., Teillaud, M. *On the randomized construction of the Delaunay tree,* Theoretical Comput. Sci., to appear. Tech Rep. INRIA 1140.

[11] Brönnimann, H., Chazelle, B., Matoušek, J. *Product range spaces, sensitive sampling, and derandomization,* Proc. 34th Annual IEEE Symposium on Foundations of Computer Science, Nov. 1993.

[12] Chazelle, B. *Cutting hyperplanes for divide-and-conquer,* Disc. Comput. Geom., 9 (1993), 145–158.

[13] Chazelle, B. *An optimal convex hull algorithm in any fixed dimension,* Disc. Comput. Geom. (1993), in press. Prelim. version in Proc. 32nd FOCS, 1991.

[14] Chazelle, B. *Geometric discrepancy revisited,* Proc. 34th Annual IEEE Symposium on Foundations of Computer Science, Nov. 1993.

[15] Chazelle, B., Edelsbrunner, H., Guibas, L.J., Sharir, M., Snoeyink, J. *Computing a face in an arrangement of line segments,* Proc. 2nd Ann. ACM-SIAM Symp. Disc. Alg. (1991), 441–448.

[16] Chazelle, B., Edelsbrunner, H., Grigni, M., Guibas, L.J., Sharir, M. *Improved bounds on weak ε-nets for convex sets,* Proc. 25th Ann. ACM Sympos. Theory Comput., May 1993, 495–504.

[17] Chazelle, B., Friedman, J. *A deterministic view of random sampling and its use in geometry,* Combinatorica 10 (1990), 229–249.

[18] Chazelle, B., Matoušek, J. *Derandomizing an output-sensitive convex hull algorithm in three dimensions,* manuscript (1991).

[19] Chazelle, B., Matoušek, J. *On linear-time deterministic algorithms for optimization problems in fixed dimension,* 4th Ann. ACM-SIAM Symp. Disc. Alg. (1993), 281–290.

[20] Clarkson, K.L. *Linear programming in $O(n \times 3^{d^2})$ time*, Inf. Process. Lett. 22 (1986), 21–24.

[21] Clarkson, K.L. *Las Vegas algorithm for linear programming when the dimension is small*, Proc. 29th IEEE Ann. Symp. Foundat. Comput. Sci. (1988), 452–457.

[22] Clarkson, K.L. *A randomized algorithm for closest-point queries*, SIAM J. Comput. 17 (1988), 830–847.

[23] Clarkson, K.L. *New applications of random sampling in computational geometry*, Disc. Comput. Geom. 2 (1987), 195–222.

[24] Clarkson, K.L. *Randomized geometric algorithms*, Euclidean Geometry and Computers, World Scientific Publishing, ed. D. Z. Du and F. K. Hwang, to appear.

[25] Clarkson, K.L., Mehlhorn, K., Seidel, R. *Four results on randomized incremental constructions*, Proc. STACS, 1992.

[26] Clarkson, K.L., Shor, P.W. *Applications of random sampling in computational geometry, II*, Disc. Comput. Geom. 4 (1989), 387–421.

[27] Clarkson, K.K., Tarjan, R.E., Van Wyk, C.J. *A fast Las Vegas algorithm for triangulating a simple polygon*, Disc. and Comput. Geom. 4 (1989), 432–432.

[28] Devillers, O. *Randomization yields simple $O(n \log^* n)$ algorithms for difficult $\omega(n)$ problems*, Int. J. Comput. Geom. Appl., to appear. Tech Rep. INRIA 1412.

[29] Devillers, O., Meiser, S., Teillaud, M. *Fully dynamic Delaunay triangulation in logarithmic expected time per operation*, WADS 91, Vol. LNCS 519, Springer-Verlag 1991. Tech Rep. INRIA 1349.

[30] Dyer, M.E. *On a multidimensional search technique and its application to the Euclidean 1-centre problem,* SIAM J. Comput. 15 (1986), 725–738.

[31] Dyer, M.E. *A class of convex programs with applications to computational geometry,* Proc. 8th ACM Sympos. Comput. Geom., 1992, 9–15.

[32] Dyer, M.E., Frieze, A.M. *A randomized algorithm for fixed-dimensional linear programming,* manuscript, 1987.

[33] Edelsbrunner, H., Guibas, L.J., Hershberger, J., Seidel, R., Sharir, M., Snoeyink, J., Welzl, E. *Implicitly representing arrangements of lines or segments,* Proc. 4th Ann. ACM Sympos. Comput. Geom. (1988), 56–69.

[34] Edelsbrunner, H., Guibas, Sharir, M. *The complexity of many faces in arrangements of lines and of segments,* Proc. 4th Ann. ACM Sympos. Comput. Geom. (1988), 44–55.

[35] Even, G., Goldreich, O., Luby, M., Nisan, N., Velickovic, B. *Approximations of general independent distributions,* Proc. 24th STOC (1992), 10–16.

[36] Gärtner, G. *A subexponential algorithm for abstract optimization problems.* Proc. 33rd IEEE Ann. Symp. Foundat. Comput. Sci., 1992, 464–472.

[37] Guibas, L.J., Knuth, D.,E., Sharir, M. *Randomized incremental construction of Delaunay and Voronoi diagrams,* Algorithmica 7 (1992), 381–413.

[38] Haussler, D., Welzl, E. *Epsilon-nets and simplex range queries,* Disc. Comp. Geom. 2, (1987), 127–151.

[39] Impagliazzo, R., Zuckerman, D. *How to recycle random bits,* Proc. 30th Ann. IEEE Symp. Foundat. Comput. Sci., 1989, 248–253.

[40] Kalai, G., *A subexponential randomized simplex algorithm,* Proc. 24th ACM Symposium on Theory of Computing, 1992, 475–482.

[41] Koksma, J.F. *Een algemeena stelling uit de theorie der gelijkmatige verdeeling modulo 1*, Mathematica B (Zutphen), 11 (1942/43), 7–11. (English treatment in [61]).

[42] Linial, N., Luby, M., Saks, M., Zuckerman, D. *Efficient construction of a small hitting set for combinatorial rectangles in high dimension*, Proc. 25th Ann. ACM Sympos. Theory Comput., May 1993, 258–267.

[43] Luby, M. *A simple parallel algorithm for the maximal independent set problem*, Proc. 17th Ann. ACM Symp. Theory of Comput. (1985), 1–10.

[44] Matoušek, J. *Construction of ε-nets*, Disc. Comput. Geom. 5 (1990), 427–448.

[45] Matoušek, J. *Cutting hyperplane arrangements*, Disc. Comput. Geom. 6 (1991), 385–406.

[46] Matoušek, J. *Approximations and optimal geometric divide-and-conquer*, Proc. 23rd Ann. ACM Symp. Theory of Comput. (1991), 505–511.

[47] Matoušek, J. *Efficient partition trees*, Proc. 7th Ann. ACM Symp. Comput. Geom. (1991), 1–9.

[48] Matoušek, J. *Reporting points in halfspaces*, Proc. 32nd Ann. IEEE Symp. Foundat. Comput. Sci. (Oct.1991), 207–215.

[49] Matoušek, J. *Range searching with efficient hierarchical cuttings*, Proc. 8th Ann. ACM Symp. Comput. Geom. (1992), 276–285.

[50] Matoušek, J., Sharir, M., Welzl, E. *A subexponential bound for linear programming*, Proc. 8th Ann. ACM Symp. Comput. Geom. (1992), 1–8.

[51] Matoušek, J., Welzl, E., Wernisch, L. *Discrepancy and ε-approximations for bounded VC-dimension*, Proc. 32nd Ann. IEEE Symp. Foundat. Comput. Sci. (1991), 424–430.

[52] Megiddo, N. *Linear programming in linear time when the dimension is fixed,* J. ACM 31 (1984), 114–127.

[53] Mitchell, D.P. *Spectrally optimal sampling for distribution ray tracing,* Computer Graphics, 25 (1991), 157–164.

[54] Mulmuley, K. *A fast planar point location algorithm I,* Proc. 29th Ann. IEEE Symp. Foundat. Comput. Sci. (1988), 580–589.

[55] Mulmuley, K. *On obstructions in relation to a fixed viewpoint,* Proc. 30th Ann. IEEE Symp. Foundat. Comput. Sci. (1989), 592–597.

[56] Mulmuley, K. *On levels in arrangements and Voronoi diagrams,* Discrete Comput. Geom. 6 (1990), 307–338.

[57] Mulmuley, K. *A fast planar point location algorithm II,* J. ACM 38 (1991), 74–103.

[58] Mulmuley, K. *An efficient hidden surface removal algorithm,* manuscript.

[59] Mulmuley, K. *Randomized geometric algorithms and pseudo-random generators,* Proc. 33rd Ann. IEEE Symp. Foundat. Comput. Sci. (1992), 90–100.

[60] Naor, J., Naor, M. *Small-bias probability spaces: efficient constructions and applications,* Proc. 22nd Ann. ACM Symp. Theory of Comput. (1990), 213–223.

[61] Niederreiter, H. *Random number generation and quasi-Monte Carlo methods,* CBMS-NSF, 1992.

[62] Raghavan, P. *Probabilistic construction of deterministic algorithms: approximating packing integer programs,* J. Comput. System Sci. 37 (1988), 130–143.

[63] Reif, J.H., Sen, S. *Optimal randomized parallel algorithms for computational geometry,* Proc. 16th Internat. Conf. Parallel Processing, St. Charles, IL, 1987. Full version, Duke Univ. Tech. Rept., CS–88–01, 1988.

[64] Seidel, R. *A simple and fast incremental randomized algorithm for computing trapezoidal decompositions and for triangulating polygons*, Comput. Geom.: Theory and Appl. 1 (1991), 51–64.

[65] Seidel, R. *Small-dimensional linear programming and convex hulls made easy*, Disc. Comput. Geom. 6 (1991), 423–434.

[66] Shirley, P. *Discrepancy as a quality measure for sample distributions*, Proc. Eurographics'91, (1991), 183–193.

[67] Spencer, J. *Ten lectures on the probabilistic method*, CBMS-NSF, SIAM, 1987.

[68] Stolarsky, K.B. *Sums of distances between points on a sphere. II*, Proc. Amer. Math. Soc., 41 (1973), 575–582.

[69] Vapnik, V.N., Chervonenkis, A.Ya. *On the uniform convergence of relative frequencies of events to their probabilities*, Theory Probab. Appl. 16 (1971), 264–280.

[70] Weyl, H. *Über die gleichverteilung von zahlen mod. eins.* Math. Ann., 77 (1916), 111–147.

PARALLEL SEARCH ALGORITHMS FOR DISCRETE OPTIMIZATION PROBLEMS

V. Kumar and Ananth Y. Grama
Department of Computer Science
University of Minnesota
Minneapolis, MN 55455
(612) 625-4002
kumar@cs.umn.edu and *ananth@cs.umn.edu*

Abstract

Discrete optimization problems (DOPs) arise in various applications such as planning, scheduling, computer aided design, robotics, game playing, and constraint directed reasoning. Often, a DOP is formulated in terms of finding a minimum cost solution path in a graph from an initial node to a goal node. It is solved using graph/tree search methods such as backtracking, branch-and-bound, heuristic search, and dynamic programming. Availability of parallel computers has created substantial interest in exploring the use of parallel processing for solving discrete optimization problems. This article provides an overview of our work on parallel search algorithms.

1   Introduction

Discrete optimization problems (DOPs), also referred to as combinatorial optimization problems, form a class of computation-intensive problems of significant theoretical and practical interest. DOPs evaluate objects from a finite or countably infinite set to determine an object that minimizes (or maximizes) a given criterion.

A Discrete Optimization Problem can be expressed as a tuple (S, f). The set S is a finite or countably infinite set of all possible solutions that satisfy certain specified constraints. It is also called the set of **feasible solutions**. The function f is the cost function that maps each element in set S to the set of real numbers R.

$$f : S \rightarrow R$$

The objective is to find a feasible solution x_{opt}, such that $f(x_{opt}) \leq f(x)$ for all $x \in S$.

Problems from various domains can be formulated as DOPs. Some examples are planning and scheduling, optimal layout of VLSI chips, robot motion planning, test pattern generation for digital circuits, and logistics and control.

In most problems of practical interest, the set S is quite large. Consequently, exhaustive enumeration of elements in S to determine x_{opt} is not feasible. Often, elements of S can be viewed as paths in graphs/trees, the cost function can be defined in terms of the cost of the arcs, and the DOP can be formulated in terms of finding a (minimum cost) solution path in the graph from an initial node to a goal node. Branch and bound [32], dynamic programming, and heuristic search [37, 19] methods use the structure of these graphs to solve DOPs without searching the set S exhaustively [27].

Since DOPs fall into the class of NP-hard problems [12], one may argue that there is no point in applying parallel processing to them, as the worst-case run time can never be reduced to a polynomial unless we have an exponential number of processors. However, the average time complexity of heuristic search algorithms for many problems is polynomial [48, 55]. Furthermore, when near or sub-optimal solutions are acceptable, algorithms exist for finding desired solutions in polynomial time (e.g., for certain problems, approximate branch-and-bound algorithms are known to run in polynomial time[54]). In such cases, parallel processing can enable us to solve problems significantly bigger than those solvable using a serial computer. Some applications using search algorithms (e.g., robot motion planning, speech understanding, task scheduling) require real time solutions. For these applications, it is likely that parallel processing is the only way to obtain acceptable performance. Finally, for problems where optimal solutions are highly desirable, parallel search techniques have been effective for solving moderately difficult real world problems (e.g., VLSI floor-plan optimization [2]).

Parallel computers containing thousands of processing elements are now commercially available. The cost of these machines is similar to that of large mainframes, but they offer significantly more raw computing power. Due to advances in VLSI technology and economy of scale, the cost of these machines is expected to go down drastically over the next decade. It may be possible to construct computers comprising of thousands to millions of processing elements at costs ranging from those of high-end workstations to large mainframes. This technology has created substantial interest in exploring the use of parallel processing for search based applications [6, 21, 22, 9, 51, 52, 5, 16, 15].

This paper provides a brief overview of our research [40, 41, 30, 43, 20, 24, 2, 4, 3] on parallel search algorithms for solving DOPs. In particular, we cover parallel formulations of depth-first and best-first search algorithms, and discuss issues of speedup anomalies.

2   Parallel Depth First Search

Depth-first search (DFS), also referred to as Backtracking, is a general technique for solving a variety of discrete optimization problems [23, 4, 2]. Since many of the problems solved by DFS are computationally intensive, there has been a great interest in developing

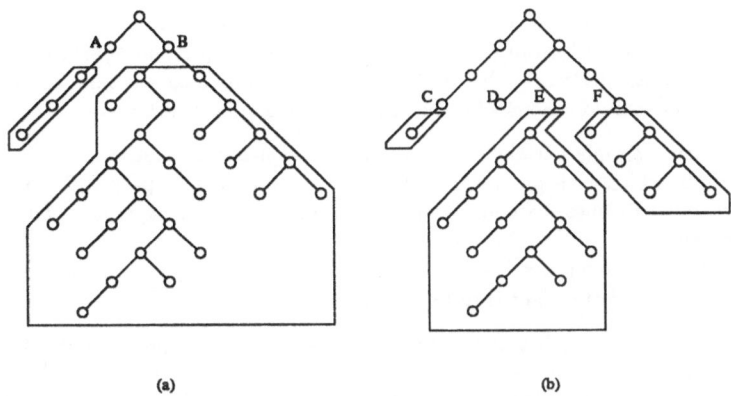

(a)   (b)

Figure 1: Illustration of the unstructured nature of tree search and the imbalance resulting from static partitioning.

parallel versions of DFS. The critical issue in parallel algorithms for DFS is the distribution of the search space among the processors. Consider the tree shown in Figure 1. Note that the left subtree (rooted at node A) is independent of the right subtree (rooted at node B). By statically assigning a node in the tree to a processor, it is possible to expand the whole subtree rooted at that node without communicating with another processor. Thus, it seems that such static allocation should yield a parallel search algorithm.

Let us see what happens if we try to apply this approach to the tree in Figure 1. Assume that we have two processors. The root node is expanded to generate two nodes (A and B), and each of these nodes is assigned to one of the processors. Each processor now searches the subtrees rooted at its assigned node independently. At this point, the problem with static node assignment becomes apparent. The processor exploring the subtree rooted at node A expands considerably fewer nodes than the other processor. Due to this imbalance in the workload, one processor is idle for a significant amount of time, reducing efficiency. Using more processors worsens the imbalance. Consider the partitioning of the tree for four processors. Nodes A and B are expanded to generate nodes C, D, E, and F. Assume that each of these nodes is assigned to one of the four processors. Now, the processor searching the subtree rooted at node E does most of the work, and those searching subtrees rooted at nodes C and D spend most of their time idle. Static partitioning of trees thus yields poor performance because of substantial variation in the size of partitions of the search space rooted at different nodes. Furthermore, since the search space is usually generated dynamically, it is very difficult to estimate the size of the search space beforehand. Therefore, it is necessary to balance search space among processors dynamically.

In **dynamic load balancing**, when a processor runs out of work, it gets more work from another processor **that has work**. Consider **the** two-processor partitioning of the tree in Figure 1(a). Assume that nodes A and B are assigned to the two processors as above. In this case, however, when the processor searching the subtree rooted at node A runs out of work, it requests work from the other processor. Although communication is required to request and transfer work, the work load between processors is now balanced. This section explores several schemes for dynamically balancing load between processors.

We begin with a simple parallel DFS algorithm and use it to motivate others. Consider the scenario in which each processor performs DFS on a disjoint part of the search space. After a processor finishes searching its part of the search space, it requests an unsearched part from other processors. Whenever any processor finds a goal node, all processors terminate. If the search space is finite and has no solutions, then all processors eventually run out of work, and the algorithm terminates.

Since each processor searches the state-space depth-first, unexplored states can be conveniently stored as a stack. Each level of the stack keeps track of untried alternatives. The depth of the stack is the depth of the node currently being explored. Each processor maintains its own local stack on which it executes DFS. When its local stack is empty, a processor requests untried alternatives from another processor's stack. In the beginning, the entire search space is assigned to one processor, and other processors are assigned null spaces (i.e., empty stacks). The search space is distributed among the processors as they request work. We refer to the processor that sends work as the **donor** or **donor processor** and to the processor that requests work as the **recipient** or **recipient processor**.

As illustrated in Figure 2, each processor can be in one of two states: active (i.e., it has work) or idle (i.e., it is trying to get work). An idle processor selects a donor processor and sends it a work request. If the idle processor receives work (space to be searched) from the donor processor, it becomes busy. If it receives a *reject* message (because the donor has no work) it selects another donor and sends it a work request. This process repeats until the processor gets work, or all the processors become idle. When idle, if a processor receives a work request, it returns a *reject* message.

In the active state, a processor does a fixed amount of work (explores a fixed amount of search space) then checks for pending work requests. When a work request is received, the processor partitions its work into two parts and sends one to the requesting processor. If the processor has too little work it is effectively idle, and sends a *reject* message. When a processor exhausts its own search space, it becomes idle.

This process continues until a solution is found or the entire space has been searched. There are several ways of distributing work among processors. A few simple strategies are as follows:

Asynchronous Round Robin In asynchronous round robin (ARR), each processor maintains an independent variable, *target*. Whenever a processor runs out of work, it uses *target* as the label of a donor processor and sends it a work request. The value of the *target* is incremented (modulo P) each time a work request is sent. The initial value of *target* at each processor is set to $((label + 1)$ modulo $P)$ where *label* is the local processor label. Note that work requests are generated independently by each processor. However,

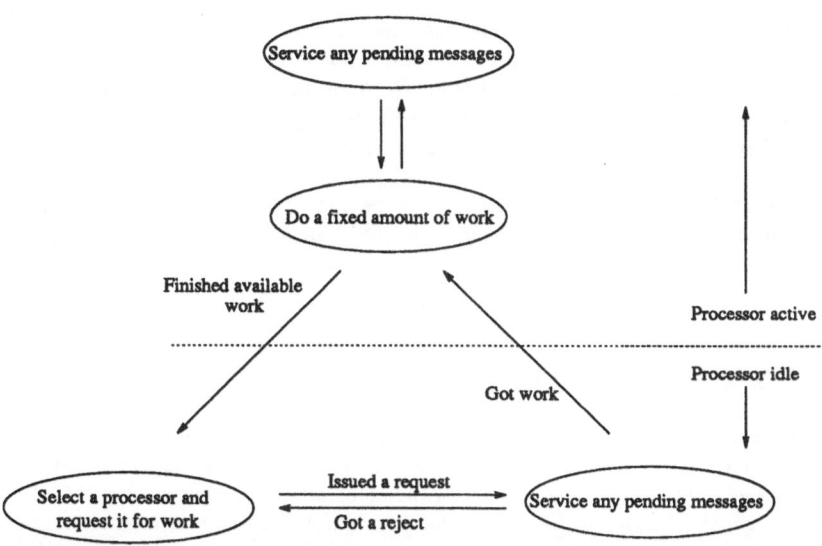

Figure 2: State diagram describing the generic parallel formulation of a subtask distribution scheme using stack splitting.

it is possible for two or more processors to request work from the same donor at nearly the same time. This is undesirable since work requests should be spread uniformly over all processors that have work.

Global Round Robin Global round robin (GRR) uses a single global variable called *target*, stored at processor P_0. Whenever a processor needs work, it requests and receives the value of *target* from P_0; then P_0 increments *target* (modulo P) before responding to another request. The recipient processor now sends a request to the donor whose label is the value of *target* received from processor P_0. GRR ensures that successive work requests are distributed evenly over all processors. A drawback of this scheme is the contention at processor P_0 for access to *target*.

Random Polling Random polling (RP) is the simplest work distribution strategy. When a processor becomes idle, it randomly selects a donor. Each processor is selected as a donor with equal probability, ensuring that work requests are evenly distributed.

In [24, 14, 30, 40], we presented and analyzed parallel formulations of DFS that use these and other schemes for work distribution. To study the effectiveness of these schemes, we have applied these to a number of problems. Parallel formulations of IDA* for solving the 15 puzzle problem yielded near linear speedup on Sequent Balance up to 30 processors and on the Intel Hypercube^{TM} and BBN Butterfly up to 128 processors [29, 30, 40]. Parallel formulation of PODEM, which is the best known sequential algorithm for solving the test pattern generation problem, provided linear speedup on a 128 processor Symult^{TM} [4]. Parallel depth-first branch-and-bound for floorplan optimization for VLSI circuits yielded linear speedups on a 1024 processor Ncube^{TM}, a 128 processor Symult^{TM} and a network of 16 SUN workstations [2]. Linear speedups were also obtained for parallel DFS for the tautology verification problem for up to 1024 processors on the Ncube/10^{TM} and the Ncube/2^{TM} [3, 14, 24]. In [20], we have presented new methods for load balancing of unstructured tree computations on large-scale SIMD machines such as CM-2. The analysis and experiments show that our new load balancing methods provide good speedups for parallel DFS on SIMD architectures.

2.1 Scalability Analysis.

From experimental results for a particular architecture and a range of processors alone, it is difficult to ascertain the relative merits of different parallel algorithm-architecture combinations. This is because the performance of different schemes may be altered in different ways by changes in hardware characteristics (such as interconnection network, CPU speed, speed of communication channels, etc.), number of processors, and the size of the problem instance being solved [26]. Hence any conclusions drawn from experimental results on a specific parallel computer and problem instance are rendered invalid by changes in any one of the above parameters. Scalability analysis of a parallel algorithm and architecture combination has been shown to be useful in extrapolating these conclusions [13, 26, 30].

We have developed a scalability metric, called **isoefficiency**, which relates the problem size to the number of processors necessary for an increase in speedup in proportion to the number of processors used [30]. An important feature of isoefficiency analysis is that it succinctly captures the effects of characteristics of the parallel algorithm as well as the parallel architecture on which it is implemented, in a single expression. The isoefficiency metric has been found to be very useful in characterizing scalability of a number of algorithms [25, 13]. Here, we briefly overview the concept of isoefficiency.

Let W be the total amount of computation done by the optimal or the best known sequential algorithm. We assume that W is proportional to the size of the space searched. The execution time on P processors is defined as T_P. We define speedup S as the ratio $\frac{W}{T_P}$ and efficiency as $\frac{S}{P}$. In general, for a fixed problem size W, increasing the number of processors P causes a decrease in efficiency because parallel processing overhead will increase while the sum of time spent by all processors in meaningful computation will remain the same. Parallel systems that can maintain a fixed efficiency level while increasing both W and P are defined as **scalable** [26].

Note that for a given parallel algorithm, for different parallel architectures, the problem size may have to increase at different rates with respect to P in order to maintain a fixed efficiency. The rate at which W is required to grow with respect to P to keep the efficiency fixed is essentially what determines the degree of scalability of the parallel algorithm for a specific architecture. For example, if W is required to grow exponentially with respect to P, then the algorithm-architecture combination is poorly scalable. This is because in this case it would be difficult to obtain good speedups on the architecture for a large number of processors, unless the problem size being solved is enormously large. On the other hand, if W needs to grow only linearly with respect to P, then the algorithm-architecture combination is highly scalable and can easily deliver linearly increasing speedups with increasing number of processors for reasonable increments in problem sizes. If W needs to grow as $f(P)$ to maintain an efficiency E, then $f(P)$ is defined to be the isoefficiency function for efficiency E and the plot of $f(P)$ with respect to P is defined to be the isoefficiency curve for efficiency E.

A lower bound on any isoefficiency function is that asymptotically, it should be at least linear. This follows from the fact that all problems have a sequential (*i.e.* non decomposable) component. Hence, any algorithm which shows a linear isoefficiency on some architecture is optimally scalable on that architecture. Algorithms with isoefficiencies of $O(P \log^c P)$, for small constant c, are also reasonably optimal for practical purposes. For a more rigorous discussion on the isoefficiency metric and scalability analysis, the reader is referred to [26, 30].

2.1.1 Scalability Analysis of Parallel Depth-First Search.

The primary reason for loss of efficiency in our parallel formulations of DFS is the communication overhead incurred by different processors in finding work and in contention over shared resources. By modeling these, we were able to determine the isoefficiency functions (and thus scalability) of a variety of work-distribution schemes[14, 24, 30] for the ring, cube, mesh, network of workstations, and shared memory architectures. Some of the existing parallel formulations were found to have poor scalability. This motivated the design of substantially improved schemes for various architectures. We established lower bounds on the scalability of any possible parallel formulation for various architectures. For each of these architectures, we determined near optimal load balancing schemes. The performance characteristics of various schemes have been proved and experimentally verified in [14, 24, 30, 40]. Schemes with better isoefficiency functions performed better than those with poorer isoefficiency functions for a wide range of problem sizes and processors. Table 1 presents isoefficiency functions of various parallel formulations on different architectures. For a more detailed discussion see [14, 24].

From our scalability analysis of a number of architecture-algorithm combinations, we have been able to gain valuable insights into the relative performance of parallel formulations for a given architecture. For instance, in [2] an implementation of parallel depth first branch and bound for VLSI floorplan optimization is presented, and speedups obtained on a network of 16 workstations. The essential part of this branch-and-bound algorithm is the GRR load balancing technique. Our scalability analysis can be used to investigate the viability of using a much larger number of workstations for solving this problem. Note from Table 1 that GRR has an overall isoefficiency of $O(P^2 \log P)$ for this platform. Hence, if we had 1024 workstations on the network, we can obtain the same efficiency on

Scheme→ Arch↓	ARR	GRR	GRR-M	Random	Lower Bound
SM	$O(P^2 \log P)$	$O(P^2 \log P)$	$O(P \log P)$	$O(P \log^2 P)$	$O(P)$
Cube	$O(P^2 \log^2 P)$	$O(P^2 \log P)$	$O(P \log^2 P)$	$O(P \log^3 P)$	$O(P \log P)$
Ring	$O(P^3 \log P)$	$O(P^2 \log P)$		$O(P^2 \log^2 P)$	$O(P^2)$
Mesh	$O(P^{2.5} \log P)$	$O(P^2 \log P)$	$O(P^{1.5} \log P)$	$O(P^{1.5} \log^2 P)$	$O(P^{1.5})$
Network of workstations	$O(P^3 \log P)$	$O(P^2 \log P)$		$O(P^2 \log^2 P)$	$O(P^2)$

Table 1: Scalability results of various parallel DFS formulations for different architectures.

a problem instance which is 10240 $(= \frac{1024^2 \log 1024}{16^2 \log 16})$ times bigger compared to a problem instance being run on 16 processors. This result is of significance, since it indicates that it is indeed possible to obtain good efficiencies with large number of workstations. Scalability analysis also sheds light on the degree of scalability of such a system with respect to other parallel architectures such as hypercube and mesh multicomputers. For instance, the best applicable technique implemented on a hypercube has an isoefficiency function of $O(P \log^2 P)$. With this isoefficiency, we would be able to get identical efficiencies as those obtained on 16 processors by increasing the problem size 400 fold (which is $\frac{1024 \log^2 1024}{16 \log^2 16}$). We can thus see that it is possible to obtain good efficiencies even with smaller problems on the hypercube. We can thus conclude from isoefficiency functions that the hypercube offers a much more scalable platform compared to the network of workstations for this problem.

2.2 Other Parallel Formulations of DFS

Wah[53], Finkel[10] and Monien[35] present some of the early work on parallel formulations of DFS. In all of these formulations, the search tree is dynamically partitioned among processors using a load balancing scheme. Scalability of these schemes have been analyzed in [24, 30].

A number of other researchers have also proposed parallel formulations that employ other work distribution schemes. Ichiyoshi el. al. present a scheme [11] in which the search tree is divided into a large number of subtrees. Subtrees are generated hierarchically by the processors organized in the form of a tree and served to processors at the leaf levels for search. In the randomized allocation strategy proposed by Shu and Kale [47], every time a node is expanded, all of the newly generated successor nodes are assigned to randomly chosen processors. The random allocation of subtasks ensures a degree of load balance over the processors. Ranade [39] presents a variant of the above scheme for execution on butterfly networks or hypercubes. This formulation ensures locality of communication by mapping the generated nodes to the two processors which form the sons in a butterfly network. Unlike Shu and Kale's formulation, the communication bandwidth requirements of this formulation are not linear in number of processors. This formulation has an optimum isoefficiency for the hypercube architecture but the maximum efficiency that can be obtained is limited. Scalability analysis of all these schemes is presented in [14, 24].

Ferguson and Korf [8, 38] present a work distribution scheme, called **Distributed Tree Search** (DTS), in which processors are allocated to different parts of the search tree dynamically. This scheme is complementary to our scheme in which parts of the search space are explicitly allocated to the processors. DTS has been shown to be quite effective in the context of parallel α-β search on a 32-processor hypercube. However, no scalability results for this scheme are available.

Saletore and Kale [46] present a parallel formulation in which nodes are assigned priorities and are expanded accordingly. They show that this prioritized DFS formulation yields consistently increasing speedups with increasing number of processors for sufficiently large problems. Jankiram et. al. [1] present a variant of simple backtracking called Randomized Parallel Backtracking. This differs from simple backtracking in that the successor state is randomly chosen here, compared to the left first ordering for the latter. This technique has been observed to perform well for problems for which good heuristics are not available, and the number of the solutions in the space is relatively large.

3 Parallel Best First Search

The A* algorithm is a well known search algorithm that can use problem-specific heuristic information to prune search space. As discussed in [23], A* is essentially a "best-first" branch-and-bound (B&B) algorithm. A number of researchers have investigated parallel formulations of A*/B&B algorithms [22, 28]. An important component of A*/B&B algorithms is the priority queue which is used to maintain the "frontier" (*i.e.*, unexpanded) nodes of the search graph in a heuristic order. In the sequential A*/B&B algorithm, in each cycle a most promising node from the priority queue is removed and expanded, and the newly generated nodes are added to the priority queue.

In most parallel formulations of A*, different processors concurrently expand different frontier nodes. Conceptually, these formulations can be viewed to differ in the data structures used to implement the priority queue of the A* algorithm. Some formulations are suited only for shared-memory architectures, whereas others are suited for distributed-memory architectures as well. The effectiveness of different parallel formulations is also strongly dependent upon the characteristics of the problem being solved. We have investigated a number of parallel formulations of A*[28]. Some of these formulations are new, and the others are very similar to the ones developed by other researchers. We have tested the performance of these formulations on the 15-puzzle, the traveling salesman problem (TSP), and the vertex cover problem (VCP) on the BBN Butterfly multiprocessor. The results for the 15-puzzle and VCP are very similar, but very different from results obtained for the TSP. The reason is that the TSP and VCP generate search spaces that are qualitatively different from each other, even though both problems are NP-hard problems. We have also performed a preliminary analysis of the relationship between the characteristics of the search spaces and their suitability to various parallel formulations [28].

Evett et. al. [7] present the parallel formulation of a variant of A* for SIMD architectures called RA*. In this formulation, each newly generated node is hashed to a unique processor. On running out of memory, nodes with poorer heuristic values are retracted, till such time as they have to be re-expanded because the more promising nodes failed to yield a solution. Manzini [34] independently presents a probabilistic analysis of the performance of a similar scheme. Note that these are among the only parallel formulations of A* (or variants) which maintain the search space in the form of a graph. In contrast, most of the formulations mentioned earlier only work on state-space trees or unfold the search graph into a tree. For problems such as the 15 puzzle, this unfolding does not cause a significant search overhead. However, for some other problems (e.g., finding the shortest path in a reasonably well connected graph), unfolded graphs can be exponentially bigger than the original graphs. Hashing each generated node to a unique processor has some drawbacks. It forces a communication cycle for each node expansion. Hence its efficiency can be poor if node expansion costs are of the same order as node communication costs. It also requires a linear increase in the cross-section communication bandwidth with increase in the number of processors, making it unscalable for large number of processors. In addition, Evett's formulation is forced to retract nodes quite frequently for big search spaces.

4 Speedup Anomalies in Parallel Search

Consider the case in which P processors concurrently perform DFS in disjoint parts of a state-space tree to find a solution in the search space using the work distribution schemes in [40]. The parts of the search spaces searched by different processors are determined dynamically, and are roughly of equal sizes. Since only one solution is needed, the search terminates whenever any of the processors encounters a solution. Depending upon when a solution is first encountered in the space by the processors, the speedup can be superlinear (i.e., $> P$) or sublinear (i.e., $< P$). This phenomenon of speedup greater than P on P processors in isolated executions of parallel DFS has been reported by many researchers [18, 31, 33, 36, 42, 50] for a variety of problems and is referred to by the term speedup anomaly. The speedup can differ greatly from one execution to another, as the actual parts of the search space examined by different processors are determined dynamically, and can be different for different executions. Hence, for some execution sequences the parallel version may find a solution by visiting fewer nodes than the sequential version thereby giving superlinear speedup, whereas for others it may find a solution only after visiting more nodes resulting in sublinear speedup. It may appear that on the average the speedup would be either linear or sublinear.

We have analyzed average speedup in parallel DFS for two different types of models [44, 45, 43]. In the first model, no heuristic information is available. For this model our analysis shows that on the average, the speedup obtained is (i) linear when distribution of solutions is uniform, and (ii) superlinear when distribution of solutions is non-uniform. This model is validated by our experiments on synthetic state-space trees modeling the hackers problem[49], the 15-puzzle problem and the N-Queens problem [17]. (In these experiments, serial and parallel DFS do not use any heuristic ordering, and select successors arbitrarily.) The basic reason for this phenomenon is that parallel search can invest resources into multiple regions of the search frontier concurrently. When the solution density in different regions of the search frontier is nonuniform and these nonuniformities are not known a-priori,then sequential search has equal chance of searching a low density region or a high density region. On the contrary, parallel search can search all regions at the same time, ensuring faster success rate.

In the second model, the search tree contains a small number of solutions and a strong heuristic is available that directs search to regions that contain solutions. There is, however, some probability that the heuristic makes an error and directs search to regions containing no solutions. The work distribution method used for partitioning the tree does not use any heuristic information. However, each processor searches its own space using the heuristic. For this model, our analysis shows that the average speedup is at least linear. This may appear surprising since at any given time most of the processors will be searching spaces that are considered unpromising by the heuristic. An intuitive explanation is that for this model, parallel DFS performs much better than serial DFS when the heuristic makes an error, and thus compensates for the lost performance in the case in which the heuristic is correct. Results from this model have been verified on the parallel formulation of a DFS algorithm called PODEM, which uses very powerful heuristics to order the search tree.

Acknowledgements

This work was supported by IST/SDIO through the Army Research Office grant # 28408-MA-SDI to the University of Minnesota and by the Army Research Office contract number DAAL03-89-C-0038 with the University of Minnesota Army High Performance Computing Research Center.

References

[1] D. P. Agrawal, V. K. Janakiram, and Ram Mehrotra. A randomized parallel branch and bound algorithm. In *Proceedings of International conference on Parallel Processing*, 1988.

[2] S. Arvindam, Vipin Kumar, and V. Nageshwara Rao. Floorplan optimization on multiprocessors. In *Proceedings of the 1989 International Conference on Computer Design (ICCD-89)*, 1989. Also published as MCC Technical Report ACT-OODS-241-89.

[3] S. Arvindam, Vipin Kumar, and V. Nageshwara Rao. Efficient parallel algorithms for search problems: Applications in vlsi cad. In *Proceedings of the Frontiers 90 Conference on Massively Parallel Computation*, October 1990.

[4] S. Arvindam, Vipin Kumar, V. Nageshwara Rao, and Vineet Singh. Automatic test pattern generation on multiprocessors. *Parallel Computing*, 17, number 12:1323–1342, December 1991.

[5] D. J. Challou, M. Gini, and V. Kumar. Parallel search algorithms for robot motion planning. Technical Report CS-TR 92-65, University of Minnesota, Minneapolis, MN, 1992. Also appears in the working notes of the 1993 AAAI Spring Symposium on Innovative Applications of Massive Parallelism.

[6] R. Dehne, A. Ferreira, and A. Rau-Chaplin. A massively parallel knowledge-base server using a hypercube multiprocessor. Technical report, Carleton University, SCS-TR-170, April 1990.

[7] M. Evett, James Hendler, Ambujashka Mahanti, and Dana Nau. Pra*: A memory-limited heuristic search procedure for the connection machine. In *Proceedings of the third symposium on the Frontiers of Massively Parallel Computation*, pages 145–149, 1990.

[8] Chris Ferguson and Richard Korf. Distributed tree search and its application to alpha-beta pruning. In *Proceedings of the 1988 National Conference on Artificial Intelligence*, August 1988.

[9] R. A. Finkel and J. P. Fishburn. Parallelism in alpha-beta search. *Artificial Intelligence*, 19:89–106, 1982.

[10] Raphael A. Finkel and Udi Manber. DIB - a distributed implementation of backtracking. *ACM Transactions of Programming Languages and Systems*, 9 No. 2:235–256, April 1987.

[11] M. Furuichi, K. Taki, and N. Ichiyoshi. A multi-level load balancing scheme for or-parallel exhaustive search programs on the multi-psi. In *Proceedings of the 2nd ACM SIGPLAN Symposium on Principles and Practice of Parallel Programming*, 1990. pp.50–59.

[12] M. Garey and D. S. Johnson. *Computers and Intractability*. Freeman, San Francisco, 1979.

[13] Ananth Grama, Anshul Gupta, and Vipin Kumar. Isoefficiency function: A scalability metric for parallel algor ithms and architectures. *IEEE Parallel and Distributed Technology, Special Issue on Parallel and Distributed Systems: From Theory to Practice*, 1993 (To be Published). Also available as Technical Report TR-93-24, Department of Computer Science, University of Minnesota and from anonymous ftp site ftp.cs.umn.edu (128.101.225.7), file users/kumar/isoeff-tutorial.ps.

[14] Ananth Grama, Vipin Kumar, and V. Nageshwara Rao. Experimental evaluation of load balancing techniques for the hypercube. In *Proceedings of the Parallel Computing 91 Conference*, 1991.

[15] Ananth Y. Grama, V. Kumar, and P. Pardalos. Parallel processing of discrete optimization problems. In *Encyclopaedia of Microcomputers*. Marcel Dekker Inc., New York, 1992.

[16] Ananth Y. Grama and Vipin Kumar. A survey of parallel search algorithms for discrete optimization problems. Technical report, TR-93-11,Department of Computer Science, University of Minnesota, Minneapolis, 1993. Also available from anonymous ftp site ftp.cs.umn.edu (128.101.225.7), file users/kumar/survey_discrete_opt.ps.

[17] Ellis Horowitz and Sartaj Sahni. *Fundamentals of Computer Algorithms*. Computer Science Press, Rockville, Maryland, 1978.

[18] M. Imai, Y. Yoshida, and T. Fukumura. A parallel searching scheme for multiprocessor systems and its application to combinatorial problems. In *IJCAI*, pages 416–418, 1979.

[19] Laveen Kanal and Vipin Kumar. *Search in Artificial Intelligence*. Springer-Verlag, New York, 1988.

[20] George Karypis and Vipin Kumar. Unstructured Tree Search on SIMD Parallel Computers. Technical Report 92–21, Computer Science Department, University of Minnesota, 1992. A short version of this paper appears in the Proceedings of Supercomputing 1992 Conference, November 1992.

[21] V. Kumar, P. S. Gopalkrishnan, and L. Kanal (editors). *Parallel Algorithms for Machine Intelligence and Vision*. Springer Verlag, New York, 1990.

[22] V. Kumar and L. Kanal. Parallel branch-and-bound formulations for and/or tree search. *IEEE Transactions Pattern Analysis and Machine Intelligence*, PAMI-6:768–778, 1984.

[23] Vipin Kumar. Depth-first search. In Stuart C. Shapiro, editor, *Encyclopaedia of Artificial Intelligence: Vol 2*, pages 1004–1005. John Wiley and Sons, Inc., New York, 1987.

[24] Vipin Kumar, Ananth Grama, and V. Nageshwara Rao. Scalable load balancing techniques for parallel computers. Technical report, Technical Report 91-55, Computer Science Department, University of Minnesota, 1991. To appear in Jornal of Distributed and Parallel Computing, 1993.

[25] Vipin Kumar, Ananth Y. Grama, Anshul Gupta, and George Karypis. *Introduction to Parallel Computing: Algorithm Design and Analysis*. Benjamin/Cummings, 1994.

[26] Vipin Kumar and Anshul Gupta. Analyzing the scalability of parallel algorithms and architectures: A survey. In *Proceedings of the 1991 International Conference on Supercomputing*, June 1991. also appear as an invited paper in the Proceedings of 29th Annual Allerton Conference on Communuication, Control and Computing, Urbana,IL, October 1991.

[27] Vipin Kumar and Laveen Kanal. The cdp: A unifying formulation for heuristic search, dynamic programming, and branch-and-bound. In Laveen Kanal and Vipin Kumar, editors, *Search in Artificial Intelligence*. Springer-Verlag, New York, 1988.

[28] Vipin Kumar, K. Ramesh, and V. Nageshwara Rao. Parallel best-first search of state-space graphs: A summary of results. In *Proceedings of the 1988 National Conference on Artificial Intelligence*, pages 122–126, August 1988.

[29] Vipin Kumar and V. N. Rao. Scalable parallel formulations of depth-first search. In Vipin Kumar, P. S. Gopalakrishnan, and Laveen Kanal, editors, *Parallel Algorithms for Machine Intelligence and Vision*. Springer-Verlag, New York, 1990.

[30] Vipin Kumar and V. Nageshwara Rao. Parallel depth-first search, part II: Analysis. *International Journal of Parallel Programming*, 16 (6):501–519, 1987.

[31] T. H. Lai and Sartaj Sahni. Anomalies in parallel branch and bound algorithms. *Communications of the ACM*, pages 594–602, 1984.

[32] E. L. Lawler and D. Woods. Branch-and-bound methods: A survey. *Operations Research*, 14, 1966.

[33] Guo-Jie Li and Benjamin W. Wah. Coping with anomalies in parallel branch-and-bound algorithms. *IEEE Transactions on Computers*, C–35, June 1986.

[34] G. Manzini and M. Somalvico. Probabilistic performance analysis of heuristic search using parallel hash tables. In *Proceedings of the International Symposium on Artificial Intelligence and Mathematics*, Ft. Lauderdale, FL, JAnuary, 1990.

[35] B. Monien and O. Vornberger. Parallel processing of combinatorial search trees. In *Proceedings of International Workshop on Parallel Algorithms and Architectures*, May 1987.

[36] B. Monien, O. Vornberger, and E. Spekenmeyer. Superlinear speedup for parallel backtracking. Technical Report 30, University of Paderborn, FRG, 1986.

[37] Judea Pearl. *Heuristics-Intelligent Search Strategies for Computer Problem Solving*. Addison-Wesley, Reading, MA, 1984.

[38] Curt Powley, Chris Ferguson, and Richard Korf. Parallel heuristic search: Two approaches. In Vipin Kumar, P. S. Gopalakrishnan, and Laveen Kanal, editors, *Parallel Algorithms for Machine Intelligence and Vision*. Springer-Verlag, New York, 1990.

[39] Abhiram Ranade. Optimal speedup for backtrack search on a butterfly network. In *Proceedings of the Third ACM Symposium on Parallel Algorithms and Architectures*, 1991.

[40] V. Nageshwara Rao and V. Kumar. Parallel depth-first search, part I: Implementation. *International Journal of Parallel Programming*, 16 (6):479–499, 1987.

[41] V. Nageshwara Rao and V. Kumar. Concurrent access of priority queues. *IEEE Transactions on Computers*, C-37 (12), 1988.

[42] V. Nageshwara Rao, V. Kumar, and K. Ramesh. A parallel implementation of iterative-deepening-a*. In *Proceedings of the National Conference on Artificial Intelligence (AAAI-87)*, pages 878–882, 1987.

[43] V. Nageshwara Rao and Vipin Kumar. On the efficicency of parallel backtracking. *IEEE Transactions on Parallel and Distributed Systems*, 4(4):427–437, April 1993; available as a technical report TR 90-55, Computer Science Department, University of Minnesota.

[44] V. Nageshwara Rao and Vipin Kumar. Superlinear speedup in state-space search. In *Proceedings of the 1988 Foundation of Software Technology and Theoretcal Computer Science*, December 1988. Lecture Notes in Computer Science number 338, Springer Verlag.

[45] V. Nageshwara Rao and Vipin Kumar. On the efficiency of parallel ordered depth-first search. In *Proceedings of the 1991 Conference on Distributed Memory and Concurrent Computers*, May 1991.

[46] Vikram Saletore and L. V. Kale. Consistent linear speedup to a first solution in parallel state-space search. In *Proceedings of the 1990 National Conference on Artificial Intelligence*, pages 227–233, August 1990.

[47] Wei Shu and L. V. Kale. A dynamic scheduling strategy for the chare-kernel system. In *Proceedings of Supercomputing Conference*, pages 389–398, 1989.

[48] Douglas R. Smith. Random trees and the analysis of branch and bound proceedures. *Journal of the ACM*, 31 No. 1, 1984.

[49] H. Stone and P. Sipala. The average complexity of depth-first search with backtracking and cutoff. *IBM Journal of Research and Development*, May 1986.

[50] Peter Tinker. Performance and pragmatics of an OR-parallel logic programming system. *International Journal of Parallel Programming*, 1988.

[51] Benjamin W. Wah, Guo jie Li, and Chee Fen Yu. Multiprocessing of combinatorial search problems. *IEEE Computer*, pages 93–108, June, 1985.

[52] Benjamin W. Wah, G. J. Li, and C. F. Yu. Multiprocessing of combinatorial search problems. In Vipin Kumar, P. S. Gopalakrishnan, and Laveen Kanal, editors, *Parallel Algorithms for Machine Intelligence and Vision*. Springer-Verlag, New York, 1990.

[53] Benjamin W. Wah and Y. W. Eva Ma. Manip - a multicomputer architecture for solving combinatorial extremum-search problems. *IEEE Transactions on Computers*, c–33, May 1984.

[54] Benjamin W. Wah and C. F. Yu. Stochastic modelling of branch-and-bound algorithms with best-first search. *IEEE Transactions on Software Engineering*, SE-11, September 1985.

[55] Herbert S. Wilf. *Algorithms and Complexity*. Prentice-Hall, 1986.

Min-Max game theory for partial differential equations with boundary/point control and disturbance. An abstract approach.

R. Triggiani
University of Virginia, Charlottesville, VA 22903

1. Introduction

The goal of the present paper is to provide an up-dated review on very recent developments in the area of min-max game theory for "infinite dimensional" systems, with main focus on partial differential equations, where control and disturbance actions are exercised as boundary or point functions. Because of space restrictions, we cannot dwell adequately on the equivalence[B-B.1] of this problem with the so called "H$^\infty$–robust stabilization problem," originally introduced by Zames for finite dimensional systems, within the context of the transfer function approach (as distinct from the state space approach of this paper). Infinite dimensional min-max studies have originated around 1991-2 by various authors, independently, mostly in the case of bounded operators B and G in (2.1) [B-B.2], [C.1], [I.1], [vK.1]. The case of fully unbounded B and G is treated in [M-T.1-4] for 'parabolic' and 'hyperbolic' classes, while [B.1] deals with the hyperbolic class, where either B or G is bounded. Generally the literature gives only statements of *equivalence* of the H$^\infty$–problem and the existence of a suitable solution of a nonstandard Algebraic Riccati Equation, leaving as *open* the non-trivial issue of existence (and uniqueness) [C.1], which cannot be handled by direct methods. Instead, the results below, after [M-T.1-4], give precise statements of existence/uniqueness of the ARE and of the critical value γ_c in terms of the problem data. Progress in [M-T.1-2] [B.1] in the hyperbolic case relies critically on the LQR problem in [FLT.1]; while, progress in [M-T.3-4] in the parabolic case relies critically on the LQR problem in [L-T.1], [F.1], [D-I.1]. This review article on min-max may be viewed as a successor of the review article [La.2] on LQR.

2. Min-max game theory problem (full observation)

Let U (control), Y (state) and V (disturbance) be Hilbert spaces. We consider the following abstract state equation

$$\dot{y}(t) = Ay(t) + Bu(t) + Gw(t) \quad \text{in } [\mathcal{D}(A^*)]' \; ; \quad y(0) = y_0 \in Y \qquad (2.1)$$

Here, the function $u \in L_2(0, \infty; U)$ is the control and $w \in L_2(0, \infty; V)$ is a deterministic disturbance. The dynamics (2.1) is subject to assumptions to be spelled out below for each of the two basic abstract classes (parabolic-like equations; and hyperbolic-like, plate-like, etc. equations), treated in sections 3-4 and sections 5-6, respectively.

For a fixed $\gamma > 0$, we associate with (2.1) the canonical cost functional

$$J(u, w) = J(u, w, y(u, w)) = \int_0^\infty [\| Ry(t) \|_Y^2 + \| u(t) \|_U^2 - \gamma^2 \| w(t) \|_V^2] \, dt \quad (2.2)$$

where $y(t) = y(t; y_0)$ is given by (2.1). The aim of this paper is to study the following game-theory problem:

$$\sup_w \inf_u J(u, w) \quad (2.3)$$

where the infimum is taken over all $u \in L_2(0, \infty; U)$, for w fixed, and the supremum is taken over all $w \in L_2(0, \infty; V)$.

3. Min-max game theory. Case of analytic semigroups

Dynamical assumptions. We return to the dynamics (2.1) complemented now with the following assumptions, (H.1) - (H.3), which will be maintained throughout this section 3 (they model boundary/point action by both the control function u and the disturbance function w, see e.g. [L-T.1, class (H.1)] and section 4 below):

(H.1) A: $Y \supset \mathcal{D}(A) \to Y$ is the infinitesimal generator of a strongly continuous (s.c.) analytic semigroup e^{At} on Y;

(H.2) B is a linear continuous operator $U \to [\mathcal{D}(A^*)]'$, the dual space of $\mathcal{D}(A^*)$, A^* being the adjoint of A in Y, such that

$$A^{-\delta} B \in L(U; Y) \quad \text{for some fixed constant, } \delta < 1 \quad (3.1)$$

(H.3) G is a linear continuous operator $V \to [\mathcal{D}(A^*)]'$ of the same class as B, i.e.

$$A^{-\rho}G \in L(V; Y) \quad \text{for some fixed constant } \rho < 1 \quad (3.2)$$

We shall first consider the general case of A being possibly unstable. Next, we shall indicate the simplifications which occur (in the final results as well as in the proofs) in the case where A is stable (this latter situation arises in canonical parabolic or parabolic-like models, see section 4).

Control-theoretic assumptions. In the general unstable case for A, the theory needs also two additional control theoretic assumptions (they are automatically satisfied in the case where A is stable), which refer to the observation R:

(H.4) (Finite Cost Condition). With $w \equiv 0$, for any $y_0 \in Y$, there exists $\bar{u} \in L_2(0, \infty; U)$ such that for the corresponding solution $\bar{y}$ of (2.1) we have $J(\bar{u}, w = 0, \bar{y}) < \infty$.

(H.5) (Detectability Condition). There exists an operator $K \in L(Y)$ such that the strongly continuous analytic semigroup $e^{(A + KR)t}$ is exponentially stable on Y:

$$\| e^{(A + KR)t} \|_{L(Y)} \leq M_K e^{-\omega_K t}, \, t \geq 0, \, \omega_K > 0 \quad (3.3)$$

Main results. We now state our main result for the present class from [M-T.4].

Theorem 3.1. [M-T.4]. *Assume (H.1) - (H.5). Then there exists a (critical) value $\gamma_c \geq 0$ defined explicitly in terms of the problem data by Eq. (3.23) in Remark 3.1 (see also (3.31) of Remark 3.2) such that:*

(a) *if $0 < \gamma < \gamma_c$, then taking the supremum in w as in (2.3) leads to $+\infty$; i.e. there is no finite solution of the game theory problem (2.3) for all initial conditions $y_0 \in Y$.*

(b) *if $\gamma > \gamma_c$, then:*

(i) *there exists a unique solution $\{u^*(\cdot\,;y_0), w^*(\cdot\,;y_0), y^*(\cdot\,;y_0)\}$ of the game theory problem (2.3);*

(ii) *there exists a unique bounded, nonnegative self-adjoint operator, $P = P^* \in L(Y)$, which satisfies the following Algebraic Riccati Equation ARE_γ, for all x, z $\in \mathcal{D}(\hat{A}^\varepsilon)$, $\forall \varepsilon > 0$ ($\hat{A}$ a translation of -A):*

$$(PAx, z)_Y + (Px, Az)_Y + (Rx, Rz)_Y = (B^*Px, B^*Pz)_U - \gamma^{-2}(G^*Px, G^*Pz)_U \quad (3.4)$$

with the properties that

$$(\hat{A}^*)^\theta P \in L(Y) \quad 0 \leq \theta < 1; \quad B^*P \in L(Y;U); \quad G^*P \in L(Y;V) \quad (3.5)$$

(iii) *the following pointwise feedback relations hold*

$$u^*(t; y_0) = -B^*Py^*(t; y_0) \in L_2(0, \infty; U) \cap C([0, \infty]; U) \quad (3.6)$$

$$\gamma^2 w^*(t; y_0) = G^*Py^*(t; y_0) \in L_2(0, \infty; U) \cap C([0, \infty]; U) \quad (3.7)$$

(iv) *the operator (F stands for "feedback") with maximal domain*

$$A_F = A - BB^*P + \gamma^{-2}GG^*P \quad (3.8)$$

is the generator of a s.c. semigroup e^{A_Ft} on Y which is, moreover, analytic for $t > 0$ and, in fact, for $y_0 \in Y$:

$$y^*(t; y_0) = e^{A_Ft}y_0 = e^{(A - BB^*P + \gamma^{-2}GG^*P)t}y_0 \in L_2(0, \infty; Y) \cap C([0, \infty]; Y) \quad (3.9)$$

Moreover, the semigroup e^{A_Ft} is uniformly stable in Y: there are constants $M_F \geq 1$ and $\omega_F > 0$ such that

$$\|e^{A_Ft}\|_{L(Y)} \leq M_F e^{-\omega_Ft}, \quad t \geq 0 \quad (3.10)$$

(v) *for any $y_0 \in Y$ the value of the game is*

$$(Py_0, y_0)_Y = J^*(y_0) \equiv J(u^*(\cdot\,;y_0), w^*(\cdot\,;y_0), y^*(\cdot\,;y_0)) = \sup_w \inf_u J(u, w, y(\cdot\,;y_0)) \cdot (3.11)$$

(vi) *The operator $A - BB^*P$ with maximal domain generates a s.c. analytic semigroup which is, moreover, stable.*

Conversely, suppose that $P = P^* \geq 0$ is an operator in $L(Y)$ such that:

(a) *the operator $A_F = A - BB^*P + \gamma^{-2}GG^*P$ is the generator of a s.c. uniformly stable semigroup on Y for some $\gamma > 0$; and*

(b) *P is a solution of the corresponding ARE_γ in (3.4), $\forall x, z \in \mathcal{D}(\hat{A}^\varepsilon)$ with the properties that $B^*P, G^*P \in L(Y; \cdot)$.*

Then, the operator $(A - BB^*P)$ *is likewise the generator of s.c. uniformly stable semigroup on Y and, moreover, the game problem (2.3) has a finite value for all* $y_0 \in Y$, *so that then* $\gamma \geq \gamma_c$. $\square$

Remark 3.1. The proof of Theorem 3.1, as given in [M-T.4], is constructive all along, in the sense that all relevant quantities are always expressed explicitly in terms of the problem data: the operators A, B, G, R. However, in the general case where A may be unstable, a pivotal role is played by the algebraic Riccati operator, denoted by $P_{0,\infty}$ (see below), which arises in the corresponding optimal control problem with no disturbance ($w \equiv 0$, this accounts for the subscript o). Such operator $P_{0,\infty}$ is uniquely identified by the problem data: the operators A, B, R, via the corresponding Algebraic Riccati Equation, Eq (3.13) below. As an illustration, we now provide the explicit formula of the critical value γ_c of the parameter γ, in terms of the problem data: A, B, G, R and, hence, $P_{0,\infty}$, through the following sequence of steps [M-T.4]:

(i) First, let $P_{0,\infty}$ be, as mentioned above, the unique Riccati operator (under present assumption (H.5)), corresponding to the optimal control problem: minimize

$$J(u, w \equiv 0, y) = \int_0^\infty [\, \|Ry(t)\|_Y^2 + \|u(t)\|_U^2 \,]\, dt \tag{3.12}$$

over all $u \in L_2(0, \infty; U)$, with y the corresponding solution of (2.1), and no disturbance ($w \equiv 0$). Under present assumptions, this is the unique nonnegative self-adjoint operator in $L(Y)$, which is a solution of the Algebraic Riccati Equation

$$(A^*P_{0,\infty}x, z)_Y + (P_{0,\infty} Ax, z)_Y + (Rx, Rz)_Y = (B^*P_{0,\infty}x, B^*P_{0,\infty}z)_U$$
$$\forall x, z \in \mathcal{D}(\hat{A}^\varepsilon); \text{ or else } \forall x, z \in \mathcal{D}(A_{P_{0,\infty}}), \text{see (3.15)} \tag{3.13}$$

with the regularity properties

$$(\hat{A}^*)^\theta P_{0,\infty} \in L(Y), \quad 0 \leq \theta < 1; \quad B^*P_{0,\infty} \in L(Y; U); \quad G^*P_{0,\infty} \in L(Y; V). \tag{3.14}$$

see [L-T.1-2] for a variational approach, and [F.1] [D-I.1] for a direct approach.

(ii) Next, the operator, with maximal domain,

$$A_{P_{0,\infty}} = A - BB^*P_{0,\infty} \tag{3.15}$$

is the infinitesimal generator of a s.c. analytic semigroup $e^{A_{P_{0,\infty}}t}$ on Y which, moreover, is exponentially stable: there exist constants $M \geq 1$, $k > 0$ such

$$\|e^{A_{P_{0,\infty}}t}\|_{L(Y)} \leq Me^{-kt} \quad t \geq 0. \tag{3.16}$$

(iii) Next, we introduce the operators $\mathcal{L}_{P_{0,\infty}}$ and $\mathcal{W}_{P_{0,\infty}}$ which are defined by

$$(\mathcal{L}_{P_{0,\infty}}g)(t) \equiv \int_0^t e^{A_{P_{0,\infty}}(t-\tau)} Bg(\tau)\, d\tau \, ; \, (\mathcal{W}_{P_{0,\infty}}w)(t) \equiv \int_0^t e^{A_{P_{0,\infty}}(t-\tau)} Gw(\tau)\, d\tau \tag{3.17}$$

as well as their $L_2(0, \infty; \cdot)$ – adjoints given by

$$(\mathcal{W}^*_{P_{0,\infty}} f)(t) \equiv G^* \int_t^\infty e^{A^*_{P_{0,\infty}}(\tau-t)} f(\tau)\, d\tau \; ; \; (\mathcal{L}^*_{P_{0,\infty}} v)(t) \equiv B^* \int_t^\infty e^{A_{P_{0,\infty}}(\tau-t)} v(\tau)\, d\tau \qquad \text{(3.18)}$$

Their regularity properties are (see [M-T.4, Theorem 3.4.1])

$$\mathcal{L}_{P_{0,\infty}}, \mathcal{L}^*_{P_{0,\infty}}: \text{ continuous } L_2(0, \infty; \cdot) \to L_2(0, \infty; \cdot) \qquad \text{(3.19)}$$

$$\mathcal{W}_{P_{0,\infty}}, \mathcal{W}^*_{P_{0,\infty}}: \text{ continuous } L_2(0, \infty; \cdot) \to L_2(0, \infty; \cdot) \qquad \text{(3.20)}$$

(iv) We can now introduce the bounded, self-adjoint operator in $L_2(0, \infty; V)$

$$-S \equiv G^* P_{0,\infty} \mathcal{L}_{P_{0,\infty}} \mathcal{L}^*_{P_{0,\infty}} P_{0,\infty} G - [\mathcal{W}^*_{P_{0,\infty}} P_{0,\infty} G + G^* P_{0,\infty} \mathcal{W}_{P_{0,\infty}}] \in L(L_2(0, \infty; V)) \text{ (3.21)}$$

Moreover, S is non-negative, since [M-T.2, Theorem 5.1] [M-T.4, Proposition 5.3]

$$(Sw, w)_{L_2(0, \infty; V)} = J^0_{w,\infty} (y_0 = 0) + \gamma^2 \|w\|^2_{L_2(0, \infty; V)}$$

$$= \int_0^\infty [\|R y^0_{w,\infty} (t; y_0 = 0)\|^2_Y + \|u^0_{w,\infty} (t; y_0 = 0)\|^2_U]\, dt \qquad \text{(3.22)}$$

where $\{u^0_{w,\infty} (t; y_0), y^0_{w,\infty} (t; y_0)\}$ is the unique optimal pair of the inf-problem in (2.3) for a fixed w.

(v) On the basis of (3.21), we now define the critical value, $\gamma_c \geq 0$, of the parameter, γ, in terms of the problem data by

$$\gamma_c^2 = \sup_{\|w\| = 1} (Sw, w)_{L_2(0, \infty; V)} = \|S\| . \qquad \text{(3.23)}$$

Moreover, P can be written explicitly as the sum of $P_{0,\infty}$ and another operator expressed in terms of S [M-T.4], Eq (13.5)].

Remark 3.2 This remark refers to the case where the generator A is (exponentially) stable: there exist constants $M \geq 1$ and $\omega > 0$ such that

$$\|e^{At}\|_{L(Y)} \leq M e^{-\omega t}, \quad t \geq 0 \qquad \text{(3.24)}$$

[in which case assumptions (H.5) (Finite Cost Condition) and (H.6) (Detectability Condition) are automatically satisfied], which is typical of canonical parabolic-like equations. In this case, then, it is possible to give a simplified, short-cut treatment—which is also more informative and fully explicit—of the min-max game problem (see [M-T.3]). More precisely, one may dispense altogether with the Riccati operator $P_{0,\infty}$ of the optimal control problem (3.12), in expressing all the relevant quantities in terms of the problem data. We illustrate this by providing the explicit definition of γ_c, under the stability assumption (3.24). This proceeds along the following steps [M-T.3]

(i) first, we introduce the operators and their $L_2(0, \infty; \cdot)$ –adjoints

$$(Lu)(t) = \int_0^t e^{A(t-\tau)} Bu(\tau)\, d\tau : \text{ continuous } L_2(0, \infty; U) \to L_2(0, \infty; Y) \qquad \text{(3.25)}$$

$$(Wu)(t) = \int_0^t e^{A(t-\tau)} Gw(\tau) \, d\tau : \text{ continuous } L_2(0, \infty; V) \to L_2(0, \infty; Y) \quad (3.26)$$

$$(L^*f)(t) = B^* \int_t^\infty e^{A^*(\tau-t)} f(\tau) \, d\tau : \text{ continuous } L_2(0, \infty; Y) \to L_2(0, \infty; U) \quad (3.27)$$

$$(W^*v)(t) = G^* \int_t^\infty e^{A^*(\tau-t)} v(\tau) \, d\tau : \text{ continuous } L_2(0, \infty; Y) \to L_2(0, \infty; V) \quad (3.28)$$

(The above regularity results are conservative. Indeed, a *key* feature of the present analytic semigroup case is that L, W, L* and W* are *smoothing* or *regularizing* operators. This is critically used in section 5 of [M-T.3] as in the LQR problem).

(ii) Next, we define the nonnegative self-adjoint operator in L ($L_2(0, \infty; Y)$)

$$S = W^*R^*R \, [I + LL^*R^*R]^{-1} \, W \quad (3.29a)$$

$$= W^*(R^*R)^{\frac{1}{2}} \, [I + (R^*R)^{\frac{1}{2}} LL^*(R^*R)^{\frac{1}{2}}]^{-1}(R^*R)^{\frac{1}{2}} \, W \quad (3.30b)$$

which can be shown to posses the same property (3.22) (see [M-T.3, Eq (2.1.17)]).

$$(Sw, w)_{L_2(0, \infty; V)} = \int_0^\infty [\, \|Ry_{w,\infty}^0 \, (t; y_0 = 0)\|_Y^2 + \|u_{w,\infty}^0 \, (t; y_0 = 0)\|_U^2 \,] \, dt \quad (3.30)$$

Hence, the S in (3.29) coincides with the S in (3.21)

(iii) Then, with reference to (3.29), the critical value γ_c of the parameter γ is defined in terms of the problem data by

$$\gamma_c^2 = \sup_{\|w\| = 1} (Sw, w)_{L_2(0, \infty; V)} = \|S\| \quad \Box \quad (3.31)$$

4. Examples of PDEs which belong to the analytic class

In this section, we illustrate the applicability of Theorem 3.1 (and Remark 3.2) under hypothesis (H.1) for analyticity, to several partial differential equation examples. We shall consider the *worst case scenario where the disturbance acts on the whole boundary, or on the whole interior*. For the case where there are no disturbances, we refer the reader to [L-T.1] and [La.2], where many of the hypotheses verifications can be found and details are hence omitted in this discussion. Further examples are in [M.1].

4.1. Heat Equation with Dirichlet Control and Dirichlet Disturbance

Let $\Omega \subset R^n$ be an open bounded domain with sufficiently smooth boundary Γ. In Ω, we consider the canonical equation with $c \geq 0$

$$y_t = \Delta y + c^2 y \quad \text{in } (0, T] \times \Omega \equiv Q \quad (4.1a)$$

$$y(0, \cdot) = y_0 \quad \text{in } \Omega \quad (4.1b)$$

$$y|_{\Sigma_0} = u + w \quad \text{in } (0, T] \times \Gamma_0 \equiv \Sigma_0 \quad (4.1c)$$

$$y\big|_{\Sigma_1} = w \qquad \text{in } (0, T] \times \Gamma_1 \equiv \Sigma_1 \tag{4.1d}$$

where Γ_0 and Γ_1 are arbitrary open subsets of the boundary Γ, with positive measures, whose union is all of Γ, $u \in L_2(\Sigma_0)$ is the boundary control, $w \in L_2(\Sigma)$ is a boundary disturbance and $y_0 \in L_2(\Omega)$. The cost functional for the min-max game problem is

$$J(u,w,y) = \int_0^\infty \{\|y(t)\|_{L_2(\Omega)}^2 + \|u(t)\|_{L_2(\Gamma_0)}^2 - \gamma^2 \|w(t)\|_{L_2(\Gamma)}^2\}\, dt\,. \tag{4.2}$$

(The theory of Theorem 3.1 actually allows one to penalize $y(t)$ in (4.2) in the norm of $H^{1/2-\rho}(\Omega)$, $\rho > 0$). To show that problem (4.1), (4.2) fits into the abstract setting of Theorem 3.1, we refer to [L-T.1]. Here we have

$$Y = L_2(\Omega) \text{ (or } H^{1/2-\rho}(\Omega)) \,,\ U = L_2(\Gamma_0)\,;\ V = L_2(\Gamma) \tag{4.3}$$

$$Ah \equiv \Delta h + c^2 h,\ \ \mathcal{D}(A) = H^2(\Omega) \cap H_0^1(\Omega) \tag{4.4}$$

$$Bu \equiv -AD_0 u,\ Gw \equiv -AD_1 w,\ R = I \tag{4.5}$$

where, without change of notation, A in (4.5) denotes the isomorphic extension of the selfadjoint operator A in (4.4) from $L_2(\Omega) \to [\mathcal{D}(A)]'$, and D_0 and D_1 (Dirichlet maps) are defined by

$$h = D_0 g \Leftrightarrow (\Delta + c^2)h = 0 \text{ in } \Omega;\ \ h\big|_{\Gamma_0} = g,\ \ h\big|_{\Gamma_1} = 0 \tag{4.6}$$

$$h = D_1 g \Leftrightarrow (\Delta + c^2)h = 0 \text{ in } \Omega;\ \ h\big|_{\Gamma} = g \tag{4.7}$$

If $c = 0$, or if c^2 is less than the first eigenvalue of $(-\Delta)$, then A is stable and (H.5), (H.6) are automatically satisfied. Otherwise, available boundary stabilization theory [L-T.1] guarantees, in the unstable case, that (H.5) and (H.6) are fulfilled, with any Γ_0.

4.2. Heat Equation with Neumann Control and Neumann Disturbance
We now consider problem (4.1) with (4.1c,d) replaced by

$$\frac{\partial y}{\partial \nu}\bigg|_{\Sigma_0} = u + w \text{ on } \Sigma_0\,;\ \ \frac{\partial y}{\partial \nu}\bigg|_{\Sigma_1} = w \text{ on } \Sigma_1 \tag{4.8}$$

where we now take $y_0 \in H^1(\Omega)$. In addition, we replace Eq. (4.2) with the following cost functional which penalizes y in the $H^1(\Omega)$–norm (Theorem 3.1 could also allow y to be penalized in the $H^{3/2-\rho}(\Omega)$–norm)

$$J(u,w,y) = \int_0^\infty \{\|y(t)\|_{L_2(\Omega)}^2 + \||\nabla y(t)|\|_{L_2(\Omega)}^2 + \|u(t)\|_{L_2(\Gamma_0)}^2 - \gamma^2 \|w(t)\|_{L_2(\Gamma)}^2\}\, dt \tag{4.9}$$

The analysis showing that the Neumann problem (4.1a), (4.1b), (4.8), (4.9) with $Y = H^1(\Omega)$ [or $H^{3/2-\rho}(\Omega)$] fits into the setting of Theorem 3.1 is similar to the preceding Dirichlet case, see [L-T.1]. Here, instead, we note that Theorem 3.1 covers also the cost

$$J(u,w,y) = \int_0^\infty \{\|y(t)|_\Gamma\|^2_{L_2(\Gamma)} + \|u(t)\|^2_{L_2(\Gamma_0)} - \gamma^2 \|w(t)\|^2_{L_2(\Gamma)}\} \, dt \qquad (4.10)$$

which penalizes purely boundary values, with $u \in L_2(\Sigma_0)$ and $w \in L_2(\Sigma)$, and R is the Dirichlet trace operator $y \to Ry = y|_\Gamma$: continuous $H^1(\Omega) \to H^{1/2}(\Gamma)$. See the analysis in [L-T.1] for $w \equiv 0$.

4.3 Structurally damped Euler-Bernoulli like equations

Example 4.3.1. A structurally damped Euler-Bernoulli equation with point control and boundary disturbance.

Le Ω be an open bounded domain in R^n, $n \le 3$, with sufficiently smooth boundary Γ. We consider the following model of a structurally damped Euler-Bernoulli equation in the deflection $z(t,x)$ where $\rho > 0$ is any constant

$$z_{tt} + \Delta^2 z - \rho \Delta z_t = \delta(x - x^0)u(t) \quad \text{in } (0, T] \times \Omega = Q \qquad (4.11a)$$

$$z(0, \cdot) = z_0; \ z_t(0, \cdot) = z_1 \quad \text{in } \Omega \qquad (4.11b)$$

$$z|_\Sigma \equiv 0, \ \Delta z|_\Sigma \equiv w \quad \text{in } (0, T] \times \Gamma \equiv \Sigma \qquad (4.11c)$$

with control acting through the Dirac measure δ concentrated at the interior point x^0 of Ω and a disturbance w on the boundary of Ω. Consistently with optimal regularity results [Triggiani JMAA 1991], we consider the following cost functional:

$$J(u,w,z) = \int_0^\infty \{\|z(t)\|^2_{H^2(\Omega)} + \|z_t(t)\|^2_{L_2(\Omega)} + |u(t)|^2_{R^1} - \gamma^2 \|w(t)\|^2_{L_2(\Gamma)}\} \, dt \qquad (4.12)$$

where $\{z_0, z_1\} \in [H^2(\Omega) \cap H_0^1(\Omega)] \times L_2(\Omega)$. Problem (4.11), (4.12) fits into the abstract setting of Theorem 3.1 in the following way (details in [L-T.1]): We let $y = [z(t), z_t(t)]$ and introduce the strictly positive definite self-adjoint operator

$$\mathcal{A}h = \Delta^2 h; \ \mathcal{D}(\mathcal{A}) = \{h \in H^4(\Omega): h|_\Gamma = \Delta h|_\Gamma = 0\} \qquad (4.13)$$

$$Y \equiv \mathcal{D}(\mathcal{A}^{1/2}) \times L_2(\Omega) = \{[H^2(\Omega) \cap H_0^1(\Omega)] \times L_2(\Omega)\}; \ U = R^1, \ V = L_2(\Gamma) \qquad (4.14)$$

$$A \equiv \begin{bmatrix} 0 & I \\ -\mathcal{A} & -\rho \mathcal{A}^{1/2} \end{bmatrix}; \ Bu \equiv \begin{bmatrix} 0 \\ \rho(x - x^0)u \end{bmatrix}; \ R = I \qquad (4.15)$$

$$Gw = \begin{bmatrix} 0 \\ -\mathcal{A}G_2 w \end{bmatrix}, \ z = G_2 v \Leftrightarrow \{\Delta^2 z = 0 \text{ in } \Omega, z|_\Gamma = 0, \Delta z|_\Gamma = v\} \qquad (4.16)$$

A generates a s.c. analytic semigroup of contractions on Y, which is stable.

Example 4.3.2. A structurally damped Euler-Bernoulli like equation with boundary control and boundary disturbance.

$$z_{tt} + \Delta^2 z - \rho \Delta z_t = 0 \qquad \text{in } (0, T] \times \Omega = Q \qquad (4.17a)$$

$$z(0, \cdot) = z_0; z_t(0, \cdot) = z_1 \qquad \text{in } \Omega \qquad (4.17b)$$

$$z|_{\Sigma} \equiv 0 \qquad \text{in } (0, T] \times \Gamma \equiv \Sigma \qquad (4.17c)$$

$$\Delta z|_{\Sigma_0} \equiv u + w \text{ in } \Sigma_0; \ \Delta z|_{\Sigma_1} \equiv w \qquad \text{in } \Sigma_1 \qquad (4.17d)$$

where Γ_0 and Γ_1 are arbitrary open subsets with positive measure whose union is the boundary Γ. We use the cost functional

$$J(u, w, z) = \int_0^\infty \{ \|z(t)\|^2_{H^2(\Omega)} + \|z_t\|^2_{L_2(\Omega)} + \|u(t)\|^2_{L_2(\Gamma_0)} - \gamma^2 \|w(t)\|^2_{L_2(\Gamma)} \} \, dt \quad 4.18$$

where $\{z_0, z_1\} \in [H^2(\Omega) \cap H_0^1(\Omega)] \times L_2(\Omega)$. In order to put problem (4.17) - (4.18) into the abstract setting of Theorem 3.1, we use the same operators $\mathcal{A}$, A and R as defined in (4.13) and (4.15), with state space Y as in (4.14) [L-T.1]. We introduce the operators

$$Bu \equiv \begin{bmatrix} 0 \\ -\mathcal{A}\mathcal{G}_1 u \end{bmatrix} ; \ Gw = \begin{bmatrix} 0 \\ -\mathcal{A}\mathcal{G}_2 w \end{bmatrix} \qquad (4.19)$$

where the Green map $\mathcal{G}_2$ is defined in (4.16), while $\mathcal{G}_1$ is defined by

$$h = \mathcal{G}_1 v \Leftrightarrow \{\Delta^2 h = 0; h|_\Gamma = 0; \Delta h|_{\Gamma_0} = v; \Delta h|_{\Gamma_1} = 0\} \qquad (4.20)$$

5. Min-Max game theory. Abstract trace class

Dynamical assumptions. We return to the dynamics (2.1) which we now complement with appropriate assumptions. These are meant to cover hyperbolic/plate-like/Schroedinger equations, etc., subject to the action of boundary/point control and/or disturbance, see [L-T.1, class (H.2)]. The dynamics (2.1), is subject to the following assumptions, which will be maintained throughout this section.

(H.1) A: $Y \subset \mathcal{D}(A) \to Y$ is the infinitesimal generator of a strongly continuous (s.c.) semigroup e^{At} on Y;

(H.2) B: linear continuous $U \to [\mathcal{D}(A^*)]'$, the dual of $\mathcal{D}(A^*)$ with respect to the Y-topology, and A^* is the Y-adjoint of A;

(H.3) the following abstract trace regularity holds: the (closable) operator $B^* e^{A^* t}$ admits a continuous extension, denoted by the same symbol, from $Y \to L_2(0, T; U)$:

$$\int_0^T \|B^* e^{A^* t} x\|^2_U \, dt \le c_T \|x\|^2_Y \qquad \forall T < \infty; \ x \in Y \qquad (5.1)$$

B^* is the dual of B, $B^* \in L(\mathcal{D}(A^*); U)$ after identifying $[\mathcal{D}(A^*)]''$ with $\mathcal{D}(A^*)$;

(H.4) G is of the same class as B: i.e. G: linear continuous $V \to [\mathcal{D}(A^*)]'$ and

$$\int_0^T \|G^* e^{A^* t} x\|_V^2 \, dt \le c_T \|x\|_Y^2 \quad \forall T < \infty, \ x \in Y \tag{5.2}$$

(H.5) $$R \in L(Y) \tag{5.3}$$

Remark 5.1: Each of the assumptions (H.3) and (H.4) is an abstract trace theory property. Over the past ten years, this property has been proved to hold true for many classes of partial differential equations by purely P.D.E.'s methods (energy methods either in differential or in pseudo-differential form), including: second order hyperbolic equations; Euler-Bernoulli, Kirchoff, and Schroedinger equations; first order hyperbolic systems, etc., all in arbitrary space dimensions and on explicitly identified spaces; see e.g. [L-T.1, class (H.2)]. □

Control theoretic assumptions. We also need the following assumptions

(H.6) (Finite Cost Condition) [F-L-T.1, p316]: With $w \equiv 0$, for any $y_0 \in Y$ there exists $\bar{u} \in L_2(0, \infty; U)$ such that for the corresponding solution $\bar{y}$ of (1.1.1) we have $J(\bar{u}, w = 0, \bar{y}) < \infty$;

(H.7) (Detectability Condition) [F-L-T.1, p319]: There exists a linear, densely defined operator K: $Y \supset \mathcal{D}(K) \to Y$ satisfying the conditions

(i) $\|K^* x\|_Y^2 \le C [\|B^* x\|_U^2 + \|x\|_Y^2] \quad \forall x \in \mathcal{D}(B^*) \subset Y$

(ii) the s.c. semigroup $e^{A_K t}$ on Y, with generator $A_K = A + KR$ or else $A + K(R^* R)^{\frac{1}{2}}$ as guaranteed by [F-L-T.1, Lemma 5.1], is uniformly stable: there exist M_1, $k > 0$ such that $\|e^{A_K t}\|_{L(Y)} \le M_1 e^{-kt}$, $t \ge 0$

Remark 5.2. The Finite Cost Condition holds true in all mixed p.d.e. cases mentioned in Remark 5.1, due to recent corresponding exact controllability/uniform stabilization results, see e.g., [L-T.1, section 7]. □

Results. We now state the main results for this class from [M-T.2].

Theorem 5.1 [M-T.2]. *Assume (H.1) - (H.7). Then there exists an intrinsic (critical) value $\gamma_c \ge 0$ of γ, see Remark 5.3 below, such that:*

(a) if $0 < \gamma < \gamma_c$, then the supremum in w in (2.3) leads to $+\infty$ and the min-max problem has no finite solution for all initial conditions $y_0 \in Y$.

(b) if $\gamma > \gamma_c$, then:

(i) there exists a unique optimal solution $\{u^*(\cdot; y_0), w^*(\cdot; y_0), y^*(\cdot; y_0)\}$ of problem (2.3)

(ii) there exists a unique bounded, positive self-adjoint operator, $P = P^* \in L(Y)$, which satisfies the following Algebraic Riccati Equation, ARE_γ for all $x, z \in \mathcal{D}(A)$; or else for all $x, z \in \mathcal{D}(A_F)$, A_F defined in (5.7) below:

$$(PAx, z)_Y + (Px, Az)_Y + (Rx, Rz)_Y = (B^* Px, B^* Pz)_U - \gamma^{-2} (G^* Px, G^* Pz)_V, \tag{5.4}$$

$$B^*P \in L(\mathcal{D}(A); U) \cap L(\mathcal{D}(A_F); U) \; ; \; G^*P \in L(\mathcal{D}(A); V) \cap L(\mathcal{D}(A_F); V) \quad (5.5)$$

(iv) $\qquad\qquad u^*(t; y_0) = -B^*Py^*(t; y_0) \in L_2(0, \infty; U)$

$$\gamma^2 w^*(t; y_0) = G^*Py^*(t; y_0) \in L_2(0, \infty; V); \qquad\qquad (5.6)$$

(v) the operator (F stands for "feedback")

$$A_F = A - BB^*P + \gamma^{-2}GG^*P, \qquad\qquad (5.7)$$

with maximal domain, is the generator of a s.c. semigroup on Y and, in fact, for $y_0 \in Y$:

$$y^*(t; y_0) = e^{A_F t} y_0 = e^{(A - BB^*P + \gamma^{-2}GG^*P)t} y_0 \in L_2(0, \infty; Y) \cap C([0, \infty]; Y), \quad t \geq 0, \quad (5.8)$$

and, moreover, the semigroup is uniformly (exponentially) stable on Y: there exist constants $C_F \geq 1$ *and* $\omega_F > 0$ *such that*

$$\left\| e^{A_F t} \right\|_{L(Y)} \leq C_F \, e^{-\omega_F t}, \quad t \geq 0; \qquad\qquad (5.9)$$

(vi) for $y_0 \in Y$, *the value of the game is*

$$(Py_0, y_0) = J^*(y_0) = J(u^*(\cdot \, ; y_0), w^*(\cdot \, ; y_0), y^*(\cdot \, ; y_0)) = \sup_w \inf_u J(u, w, y); \qquad (5.10)$$

(vii) the s.c. semigroup $e^{(A - BB^*P)t}$ *is stable, and moreover,*

$$B^*Pe^{(A - BB^*P)t} : \text{ continuous } Y \to L_2(0, \infty; U) \qquad\qquad (5.12)$$

$$G^*Pe^{(A - BB^*P)t} : \text{ continuous } Y \to L_2(0, \infty; V) \qquad\qquad (5.13)$$

Conversely, suppose that $P = P^* \geq 0$ *is an operator in* $L(Y)$ *such that:*
(a) the operator $A_F = A - BB^*P + \gamma^{-2}GG^*P$ *is the generator of a s.c. uniformly stable semigroup on Y for some* $\gamma > 0$; *and*
(b) P is a solution of the corresponding ARE_γ *in (5.4),* $\forall x, z \in \mathcal{D}(A_F)$.
Then, the operator $(A - BB^*P)$ *is likewise the generator of a s.c. uniformly stable semigroup on Y satisfying (5.12), (5.13); moreover, the game problem (2.3) has a finite value for all* $y_0 \in Y$, *so that then* $\gamma \geq \gamma_c$ $\quad\square$

Remark 5.3. The above result, as well as the content of the present remark are contained in [M-T.2], except for the case (described below) where A is also stable, which is treated in [M-T.1]. The main structure of reference [M-T.2] distinguishes a few cases, for the purpose of extracting the best possible results for each case; in particular, a definition of the critical value γ_c, explicitly given in terms of the problem data. We summarize hereafter the discussion from [M-T.1], [M-T.2] regarding γ_c.

a) **Case where both B and G are unbounded:** here, under assumptions (H.1) - (H.7), [M-T.2] shows that there exists a self-adjoint operator $S \in L(L_2(0, \infty; V))$ which satisfies identity (3.22) also in the present case; i.e.

$$(Sw, w)_{L_2(0, \infty; V)} = J^0_{w, \infty} (y_0 = 0) + \gamma^2 \|w\|^2_{L_2(0, \infty; V)} \qquad\qquad (5.14)$$

$$= \int_0^\infty [\, \| R y^0_{w,\infty}\, (t;\, y_0 = 0)\, \|^2_Y + \| u^0_{w,\infty}\, (t;\, y_0 = 0)\, \|^2_U]\, dt$$

where $\{ u^0_{w,\infty}\, (t;\, y_0),\ y^0_{w,\infty}\, (t;\, y_0) \}$ is the unique optimal pair of the inf-problem in (2.3), holding w fixed. Then, as in the case of the analytic class of Remark 3.1, γ^2_c is defined again by (3.23); i.e. by

$$\gamma^2_c = \sup_{\| w \| = 1} \ (Sw,\ w)_{L_2(0,\ \infty;\ V)} \tag{5.15}$$

However, in contrast with the analytic class, in the present general case of B and G satisfying (H.2) - (H.4), there are some technical difficulties in identifying S in (5.14) as given by (3.21); i.e. in asserting again that

$$S = [\, \mathcal{W}^*_{P_{0,\infty}}\, P_{0,\infty}\, G + G^*P_{0,\infty}\, \mathcal{W}_{P_{0,\infty}} \,] - G^*P_{0,\infty}\, \mathcal{L}_{P_{0,\infty}}\, \mathcal{L}^*_{P_{0,\infty}}\, P_{0,\infty}\, G \tag{5.16}$$

Here: $P_{0,\infty}$ is (as in section 3) the unique Riccati operator of the corresponding optimal control problem (3.12), i.e. the unique non-negative self-adjoint operator in $\mathcal{L}(Y)$ which is the solution of the A.R.E.

$$(A^*P_{0,\infty}\, x,\ z)_Y + (P_{0,\infty}\, Ax,\ z)_Y + (Rx,\ Rz)_Y = (B^*P_{0,\infty}\, x,\ B^*P_{0,\infty}z)_U \tag{5.17}$$

$$\forall x,\ z \in \mathcal{D}(A);\ \text{or else}\ \forall x, z \in \mathcal{D}(A_{P_{0,\infty}}),\ A_{P_{0,\infty}} = A - BB^*P_{0,\infty}$$

$$B^*P_{0,\infty} \in \mathcal{L}(\mathcal{D}(A_{P_{0,\infty}});\ U) \cap \mathcal{L}(\mathcal{D}(A);\ U)$$

as guaranteed under present assumptions by [F-L-T.1]. Moreover, the operators $\mathcal{L}_{P_{0,\infty}}$, $\mathcal{W}_{P_{0,\infty}}$ and their L_2 – adjoints are still given by (3.17)-(3.18). For the present class, however, their regularity is a delicate issue, which we now discuss in detail; it is precisely the source of the aforementioned technical difficulties. These consist in justifying a-priori, as bounded operators on the appropriate spaces $L_2(0, \infty; \cdot)$, the basic components of the right hand side of (5.16), i.e. $G^*P_{0,\infty}\, \mathcal{L}_{P_{0,\infty}}$ and $G^*P_{0,\infty}\, \mathcal{W}_{P_{0,\infty}}$. To this end, it would suffice to justify that they are well-defined on dense sets, in which case the identity in (5.16) can then be established on a dense set and hence one can use the optimization problem via (5.14) to obtain (5.16) by continuous extension. A *formal* proof of these facts, say when B = G, or when G^* is dominated by B^*, is offered in Appendix A (which identifies in the use of (A.6) the *formal* step).

In conclusion, when B and G are fully unbounded as in (H.2) - (H.4), the definition of $S \in L (L_2(0, \infty; V))$ rests with identity (5.14), while its explicit representation as in (5.16), which makes explicit the definition of γ_c in terms of the problem data via (5.15), is only formal at present.

b) There are, however, important subcases, where the identification of the operator S in (5.14) in terms of the problem data is fully justified. These are listed below.

1) **Case where B and G are unbounded as in (H.2 - H.4) and, in addition, A is stable:**

$$\| e^{At} \|_{\mathcal{L}(Y)} \leq M e^{-\delta t}\quad \delta > 0,\ t \geq 0 \tag{5.18}$$

In this case, reference [M-T.1] shows that the operator S in (5.16) can be identified in terms of the problem data by the same formula as (3.29), in the corresponding abstract parabolic case; i.e.

$$S = W^* R^* R [I + L L^* R^* R]^{-1} W \qquad (5.19)$$

Here, the operators L, W and their adjoints L^*, W^* are given by (3.25) - (3.28) while their regularity properties, which are non-trivial, continue to hold true for the present class as a result of [L-T.3].

2) **Case where B is unbounded as in** (H.2), (H.3), **while G is bounded.** In this case, we introduce an additional (mild) assumption:

Exact controllability of $\{A^*, (R^*R)^{1/2}\}$, or of $\{A^*, R\}$, from the origin [F-L-T.1 p350]: the dynamical system, say

$$\dot{z}(t) = A^* z(t) + (R^*R)^{1/2} v(t), \; z(0) = 0 \qquad (5.20)$$

is exactly controllable (from the origin) on the space Y over the time interval [0, T], T < ∞, using controls $v \in L_2(0, T; Y)$; i.e. the totality of all solution points z(T) fill all of Y when v runs over $L_2(0, T; Y)$. This verifiable assumption implies that $P_{0,\infty}^{-1} \in L(Y)$ [F-L-T.1], a property needed then to establish the non-trivial result [La.1] that the operator $\mathcal{L}_{P_{0,\infty}}$ satisfies likewise the regularity property (3.19):

$$\mathcal{L}_{P_{0,\infty}}, \; \mathcal{L}_{P_{0,\infty}}^* : \text{continuous } L_2(0, \infty; \cdot) \to L_2(0, \infty; \cdot) \qquad (5.21)$$

while that $\mathcal{W}_{P_{0,\infty}}$, $\mathcal{W}_{P_{0,\infty}}^*$ satisfy the regularity properties (3.20)

$$\mathcal{W}_{P_{0,\infty}}, \; \mathcal{W}_{P_{0,\infty}}^* : \text{continuous } L_2(0, \infty; \cdot) \to L_2(0, \infty; \cdot) \qquad (5.22)$$

is now obvious, since $e^{A_{P_{0,\infty}}t}$ is exponentially stable [F-L-T.1] and G is bounded. Thus, in this case, the right hand side of (5.16) is a-priori well defined as an operator in $L(L_2(0, \infty; V))$ and then, [M-T.2] shows that this right hand side does coincide with the operator S which satisfies (5.14). Then γ_c, as given by (5.15), is now explicitly identified in terms of the problem data: A, B, G, R (hence $P_{0,\infty}$). [B.1] avoids the E. C. assumption on (5.20), with B unbounded and G bounded as it does not identify γ_c in terms of the problem data as in our case a) with B, G unbounded.

3) **The case where B is bounded and G is unbounded as in** (H.4). Here, with no further assumptions, [M-T.2] shows that the regularity (5.22) for $\mathcal{W}_{P_{0,\infty}}$ holds true; and, moreover, that

$$G^* P_{0,\infty} \mathcal{L}_{P_{0,\infty}} : \text{continuous } L_2(0, \infty; U) \to L_2(0, \infty; V) \qquad (5.23)$$

Thus, in this case as well, the right hand side of (5.16) is a-priori well-defined as an operator in $L(L_2(0, \infty; V))$. Moreover, again, [M-T.2] then shows that such right hand side of (5.16) is equal to the operator S which satisfies (5.14). Thus, in this case as well, γ_c is defined explicitly in terms of the problem data via (5.15), (5.16) □

6. Examples of PDE problems which belong to the abstract trace class.

In this section, we illustrate the applicability of Theorem 5.1 (Remark 5.3) to several p.d.e. problems. We consider throughout the *worst case scenario with disturbance w active on the entire boundary*. Further examples are in [M.1]. For the case $w \equiv 0$, we refer to [L-T.1] for verification of the hypotheses.

6.1. Second order hyperbolic equations with Dirichlet boundary control and disturbance.

Let Ω be an open bounded domain in $\mathbb{R}^n$ with sufficiently smooth boundary Γ. We consider the following problem where $c \geq 0$, $u \in L_2(\Sigma_0)$ and $w \in L_2(\Sigma)$.

$$z_{tt} = \Delta z - c^2 z_t \qquad \text{in } (0, T] \times \Omega \equiv Q \tag{6.1a}$$

$$z(0, \cdot) = z_0;\ z_t(0, \cdot) = z_1 \qquad \text{in } \Omega \tag{6.1b}$$

$$z|_{\Sigma_0} = u + w \qquad \text{in } (0, T] \times \Gamma \equiv \Sigma_0 \tag{6.1c}$$

$$z|_{\Sigma_1} = w \qquad \text{in } (0, T] \times \Gamma \equiv \Sigma_1 \tag{6.1d}$$

By optimal regularity theory for this problem (I. Lasiecka-J. L. Lions-R. Triggiani JMPA, 1986), we take $\{z_0, z_1\} \in L_2(\Omega) \times H^{-1}(\Omega)$ with the associated cost functional

$$J(u, w, z) \equiv \int_0^\infty \{\|z(t)\|^2_{L_2(\Omega)} + \|z_t(t)\|^2_{H^{-1}(\Omega)} + \|u(t)\|^2_{L_2(\Gamma_0)} - \gamma^2 \|w(t)\|^2_{L_2(\Gamma)}\} \, dt \tag{6.2}$$

In order to put problem (6.1) - (6.2) into the abstract setting of Theorem 5.1, we introduce the positive self-adjoint operator

$$\mathcal{A}h = -\Delta h, \quad \mathcal{D}(\mathcal{A}) = H^2(\Omega) \cap H^1_0(\Omega) \tag{6.3}$$

and define the operators

$$A = \begin{bmatrix} 0 & I \\ -\mathcal{A} & -c^2 I \end{bmatrix}, \quad Bu = \begin{bmatrix} 0 \\ \mathcal{A}D_0 u \end{bmatrix}, \quad Gw = \begin{bmatrix} 0 \\ \mathcal{A}D_1 w \end{bmatrix}, \quad R = I \tag{6.4}$$

where $\mathcal{A}$ in (6.4) denotes the isomorphic extension of $\mathcal{A}$ in (6.3) from $L_2(\Omega) \to [\mathcal{D}(\mathcal{A}^*)]'$, and D_0 and D_1 are the Dirichlet maps defined in (4.6) and (4.7). Now, let $y(t) = [z(t), z_t(t)]$ and

$$Y = L_2(\Omega) \times H^{-1}(\Omega); \quad U = L_2(\Gamma_0), \quad V = L_2(\Gamma) \tag{6.5}$$

It is shown e.g. in [L.T.1] that the operators A, B, G, satisfy the dynamical hypotheses (H.1) - (H.5) of Theorem 5.1. Also, since R = I, then the Detectability Assumption (H.7) is fulfilled as well. Finally, to test the Finite Cost Condition (H.6), we distinguish this cases

Conservative case: If $c \equiv 0$, in order to satisfy (H.6), then we invoke recent results on exact controllability (or uniform stabilization) of problem (6.1) with $w \equiv 0$, which apply provided that Σ_0 is chosen appropriately. Of the results in the literature concerning exact controllability of (6.1) with $w \equiv 0$, the ones that give "minimal" Σ_0 are given in [C. Bardos-S. Lebeau-J. Rauch *SIAM J. Control,* 1993]. Theorem 5.1 then applies.

Stable case. If $c > 0$, then A in (6.4) is stable as in (5.18), i.e. the s.c. contraction semigroup on $L_2(\Omega) \times [\mathcal{D}(\mathcal{A}^{\frac{1}{2}})]'$ is uniformly stable here, hence on Y.

6.2. Second order hyperbolic equations with Neumann boundary control and disturbance.

The next example refers to the wave equation with boundary damping and both control and disturbance acting in the Neumann boundary conditions and $\bar{\Gamma}_0 \cap \bar{\Gamma}_1 = \phi$

$$z_{tt} = \Delta z \qquad \text{in } (0, \text{T}] \times \Omega \equiv Q \qquad (6.6a)$$

$$z(0, \cdot) = z_0; \, z_t(0, \cdot) = z_1 \qquad \text{in } \Omega \qquad (6.6b)$$

$$z|_{\Sigma_0} = 0 \qquad \text{in } (0, \text{T}] \times \Gamma \equiv \Sigma_0 \qquad (6.6c)$$

$$\frac{\partial z}{\partial \nu}\Big|_{\Sigma_1} = -z_t + u + w \qquad \text{in } (0, \text{T}] \times \Gamma \equiv \Sigma_1 \qquad (6.6d)$$

Here, $Y = H^1_{\Gamma_0}(\Omega) \times L_2(\Omega)$, where $H^1_{\Gamma_0}(\Omega) = \{h \in H^1_0(\Omega) : h_{\Gamma_0} = 0\}$. For $u = w \equiv 0$, the above problem generates a s.c. semigroup on Y, which, moreover, is uniformly stable as in (5.18) under appropriate conditions on $\{\Omega, \Gamma_0, \Gamma_1\}$, the weakest of which are given in (Bardos-Lebeau-Rauch, *SIAM J. Control,* 1993). Hence assumptions (H.5) and (H.6) hold true. Moreover, it can be shown that assumption (H.1) - (H.4) are also fulfilled, and hence Theorem 5.1 and Remark 5.3, case 1), apply.

6.3. Euler-Bernoulli equations with boundary control and disturbance

Let Ω be an open bounded domain in $\mathbb{R}^n$ with sufficiently smooth boundary Γ. We consider the following problem with $c \geq 0$

$$z_{tt} + \Delta^2 z + c^2 z_t = 0 \qquad \text{in } (0, \text{T}] \times \Omega = Q \qquad (6.7a)$$

$$z(0, \cdot) = z_0; \, z_t(0, \cdot) = z_1 \qquad \text{in } \Omega \qquad (6.7b)$$

$$z|_{\Sigma} \equiv 0 \text{ in } \Sigma; \, \frac{\partial z}{\partial \nu}\Big|_{\Sigma} = u + w \qquad \text{in } (0, \text{T}] \times \Gamma \equiv \Sigma \qquad (6.7c)$$

Consistently with optimal regularity results, the cost functional to be minimized is

$$J(u, w, z) = \int_0^\infty \{\|z(t)\|_{L_2(\Omega)}^2 + \|z_t(t)\|_{H^{-2}(\Omega)}^2 + \|u(t)\|_{L_2(\Gamma)}^2 - \gamma^2 \|w(t)\|_{L_2(\Gamma)}^2\} \, dt$$

with initial data $\{z_0, z_1\} \in L_2(\Omega) \times H^{-2}(\Omega)$. It can be shown ([L-T.1]) that problem (6.7) fits into the abstract setting of assumptions (H.1) - (H.5), with R = I, so that assumption (H.7) is also satisfied. Let $y = \{z, z_t\}$, $Y = L_2(\Omega) \times H^{-2}(\Omega)$ and $U = V = L_2(\Gamma)$ and introduce the positive self-adjoint operator

$$\mathcal{A}h \equiv \Delta^2 h, \quad \mathcal{D}(\mathcal{A}) = \{h \in H^4(\Omega) ; \ h|_\Gamma = \frac{\partial h}{\partial \nu}|_\Gamma = 0\} \tag{6.8}$$

and define the operators

$$A = \begin{bmatrix} 0 & I \\ -\mathcal{A} & -c^2 I \end{bmatrix}, \quad Bu = \begin{bmatrix} 0 \\ \mathcal{A} \mathcal{G}_4 u \end{bmatrix}, \quad Gw = \begin{bmatrix} 0 \\ \mathcal{A} \mathcal{G}_4 u \end{bmatrix}, \quad R = I \tag{6.9}$$

$$h = \mathcal{G}_4 v \Leftrightarrow \{\Delta^2 h = 0 \text{ in } \Omega; \ h|_\Gamma = 0, \frac{\partial h}{\partial \nu}|_\Gamma = v\} \tag{6.10}$$

In order to fulfill assumption (H.6) we distinguish two cases

Conservative case. If c = 0, then (H.6) is fulfilled by virtue of the exact controllability/uniform stabilization results of problem (6.7) with $w \equiv 0$ [J. L. Lions, Masson book 1988; Ourada-Triggiani, Diff. & Integr. Eq. 1991].

Stable case. If c > 0, then A in (6.9) satisfies the stability condition (5.18).

6.4. Kirchhoff plate with boundary control and disturbance in the bending moment.

Let Ω be a smooth bounded domain in R^n. We consider

$$z_{tt} + \Delta^2 z - \rho \Delta z_{tt} = 0 \qquad \text{in } (0,T] \times \Omega = Q \tag{6.11a}$$

$$z(0, \cdot) = z_0; \ z_t(0, \cdot) = z_1 \qquad \text{in } \Omega \tag{6.11b}$$

$$z|_\Sigma \equiv 0, \ \Delta z|_\Sigma \equiv u + w \qquad \text{in } (0, T] \times \Gamma \equiv \Sigma \tag{6.11c}$$

$\rho > 0$, with boundary control and disturbance u, w $\in L_2(\Sigma)$, and initial data $\{z_0, z_1\} \in H^2(\Omega) \times H_0^1(\Omega)$. Consistently with optimal regularity theory [L-T.1], we take the following cost functional

$$J(u, w, z) \equiv \int_0^\infty \{\|z(t)\|_{H^2(\Omega)}^2 + \|z_t(t)\|_{H_0^1}^2 + \|u(t)\|_{L_2(\Gamma)}^2 - \gamma^2 \|w(t)\|_{L_2(\Gamma)}^2\} \, dt \tag{6.12}$$

To put problem (6.11) - (6.12) into the abstract setting of Theorem 5.1, we introduce the positive self-adjoint operators

$$\mathcal{A} = \Delta^2 h; \quad \mathcal{D}(\mathcal{A}) = \{h \in H^4(\Omega): h|_\Gamma = \Delta h|_\Gamma = 0\} \tag{6.13}$$

$$\mathcal{A}^{\frac{1}{2}}h = -\Delta h; \quad \mathcal{D}(\mathcal{A}^{\frac{1}{2}}) = H^2(\Omega) \times H_0^1(\Omega) \tag{6.14}$$

and define the operators

$$A \equiv \begin{bmatrix} 0 & I \\ -A & 0 \end{bmatrix}; \quad Bu \equiv \begin{bmatrix} 0 \\ \mathcal{A}G_2 u \end{bmatrix}; \quad Gw = \begin{bmatrix} 0 \\ -\mathcal{A}G_2 w \end{bmatrix}, \tag{6.15a}$$

$$A = (I + \rho \mathcal{A}^{\frac{1}{2}})^{-1}\mathcal{A}, \quad R = I \tag{6.15b}$$

and G_2 is the Green map defined in (4.16) and $G_2 = -\mathcal{A}^{\frac{1}{2}}D_0$ with D_0 defined in (4.6). Let $Y = [z, z_t]$ and define the spaces

$$Y = [H^2(\Omega) \cap H_0^1(\Omega)] \times H_0^1(\Omega) = \mathcal{D}(\mathcal{A}^{\frac{1}{2}}) \times \mathcal{D}(\mathcal{A}^{\frac{1}{4}}), \quad U = V = L_2(\Gamma) \tag{6.16}$$

It is proved in [L-T.1] that problem (6.11) - (6.12) satisfies (H.1) - (H.5) (and plainly (H.7) since R = I). It is shown in [L-T.1] that (6.11) with $w \equiv 0$ is exactly controllable and hence (H.6) is satisfied.

7. The H^∞ – robust stabilization problem with partial observation.

Consider the dynamics (2.1) with cost (2.2), in short, the quadruplet {A, B, G, R} subject to either the assumptions (H.1) through (H.6) of section 3, or else the assumptions (H.1) through (H.7) of section 5. In addition, we associate with it the partial observation

$$z = Cy + \eta \tag{7.1}$$

where $C \in L(Y; Z)$, Z being the Hilbert space of observations, and measurement error η. It is well-known, e.g. [B-B.2], that the solution of the corresponding H^∞–robust stabilization problem with partial observation (7.1) —described by section 4.1 of [B-B.2]—relies on the solution of the following three problems via duality arguments:

(i) Problem 1: this is the min-max problem for the quadruplet {A, B, G, R} which is considered in the present paper, and which culminates with the theory of Theorem 3.1 in the parabolic section 3, and Theorem 5.1 in hyperbolic section 5;

(ii) Problem 2: this is a similar min-max problem, this time, however, for the quadruplet {$\bar{A}$, $\bar{B}$, $\bar{G}$, $\bar{R}$} where:

$$\bar{A} = A^* + \gamma^{-2}PGG^*; \quad \bar{B} = C^*; \quad \bar{G} = PB; \quad \bar{R} = G^* \tag{7.2}$$

Here, P is the algebraic Riccati operator provided by Theorems 3.1 or 5.1 for $\gamma > \gamma_c$ in the theory of Problem 1 for {A, B, G, R}, while C is given by (7.1).

(ii) Problem 3: this is a similar min-max problem. This time however for the quadruple {$\tilde{A}$, $\tilde{B}$, $\tilde{G}$, $\tilde{R}$} where

$$\tilde{A} = A^*; \quad \tilde{B} = C^*; \quad \tilde{G} = R^*; \quad \tilde{R} = G \tag{7.3}$$

As to Problem 3, we note that the control operator $\tilde{B}$ and the disturbance operator $\tilde{G}$ are bounded, while it is the observation $\tilde{R}$ which is now unbounded. It is well-known [L-T.1

p48] that this problem is definitely easier than the one considered in this paper and solved by Theorems 3.1 and 5.1 respectively. As to Problem 2, it is shown in [M-T.2, Section 21] for the abstract trace class of section 5 that the new quadruple $\{\overline{A}, \overline{B}, \overline{G}, \overline{R}\}$ still fits into the setting of section 5; most notably,

$$\overline{A}^{-1}\overline{G} \in L(V; Y) \,; \quad \overline{G}^* e^{\overline{A}^* t}: \text{ continuous } Y \to L_2(0, T; V);$$

finally, that $\{\overline{A}, \overline{B}, \overline{G}, \overline{R}\}$ fits also into the setting of section 3 in the analytic class (where B^*P, G^*P are bounded by (3.5)) is obvious.

Conclusion: Theorems 3.1 and 5.1 permit a solution of the H^∞– robust stabilization problem with partial observation, via Problems 1 through 3 described above.

A. Appendix on Section 5: Formal Proof that $B^*P_{0,\infty} \, \mathcal{L}_{P_{0,\infty}} \in L(L_2(0, \infty; U))$.

With reference to section 5, we consider

$$\dot\zeta = (A - BB^*P_{0,\infty})\zeta + Bg; \quad \zeta(0) = \zeta_0 = 0 \tag{A.1}$$

1. We provide fa ormal proof which yields the following estimate for any $0 < \varepsilon < 1$:

$$\|B^*P_{0,\infty}\zeta\|_{L_2(0, \infty; U)} \le \frac{1}{\sqrt{\varepsilon(1-\varepsilon)}} \|g\|_{L_2(0, \infty; U)} \tag{A.2}$$

i.e., via (3.17), the following equivalent regularity property:

$$B^*P_{0,\infty} \, \mathcal{L}_{P_{0,\infty}}: \text{ continuous } L_2(0, \infty; Y) \to L_2(0, \infty; U) \tag{A.3}$$

We proceed formally. We take the inner product in Y of (A.1) with $P_{0,\infty}\zeta$ and write

$$\frac{1}{2}\frac{d}{dt}(\zeta, P_{0,\infty}\zeta)_Y = (\dot\zeta, P_{0,\infty}\zeta)_Y = ((A - BB^*P_{0,\infty})\zeta, P_{0,\infty}\zeta)_Y + (Bg, P_{0,\infty}\zeta)_Y \tag{A.4}$$

or $\quad \dfrac{1}{2}\dfrac{d}{dt}(\zeta, P_{0,\infty}\zeta)_Y = (A\zeta, P_{0,\infty}\zeta) - \|B^*P_{0,\infty}\zeta\|_U^2 + (g, B^*P_{0,\infty}\zeta)_U .$ $\tag{A.5}$

We now apply (formally) the A.R.E. (5.17) with $x = z = \zeta$ to get

$$2(A\zeta, P_{0,\infty}\zeta)_Y = \|B^*P_{0,\infty}\zeta\|_U^2 - \|R\zeta\|_Y^2 \tag{A.6}$$

Hence, inserting (A.6) into the right hand side of (A.5), yields after a simplification

$$\frac{d}{dt}(\zeta, P_{0,\infty}\zeta)_Y = -\|B^*P_{0,\infty}\zeta\|_U^2 - \|R\zeta\|_Y^2 + 2(g, B^*P_{0,\infty}\zeta)_Y . \tag{A.7}$$

Integrating (A.7) over $0 \le t \le T$, and using (non critically) that $\zeta_0 = 0$, we obtain

$$(\zeta(T), P_{0,\infty}\zeta(T))_Y + \int_0^T \|B^*P_{0,\infty}\zeta\|_U^2 dt + \int_0^T \|R\zeta\|_Y^2 dt = 2\int_0^T (g, B^*P_{0,\infty}\zeta)_U dt \tag{A.8}$$

Dropping on the left of (A.8) the first and third terms, we obtain for any $\varepsilon > 0$:

$$\int_0^T \|B^* P_{0,\infty} \zeta\|_U^2 \, dt \le \varepsilon \int_0^T \|B^* P_{0,\infty} \zeta\|_U^2 \, dt + \frac{1}{\varepsilon} \int_0^T \|g\|_U^2 \, dt \qquad \text{(A.9)}$$

or restricting to $0 < \varepsilon < 1$

$$\int_0^T \|B^* P_{0,\infty} \zeta\|_U^2 \, dt \le \frac{1}{\varepsilon(1-\varepsilon)} \int_0^T \|g\|_U^2 \, dt \qquad \text{(A.10)}$$

Letting $T \; \infty$ as in (A.10), we obtain (A.2) as desired.

2. From (A.1) we write as in (3.17)

$$\zeta(t) = \int_0^t e^{A_{P_{0,\infty}}(t-\tau)} B g(\tau) \, d\tau = (\mathcal{L}_{P_{0,\infty}} g)(t) \qquad \text{(A.11)}$$

$$= -\int_0^t e^{A(t-\tau)} B B^* P_{0,\infty} \zeta(\tau) \, d\tau + \int_0^t e^{A(t-\tau)} B g(\tau) \, d\tau \qquad \text{(A.12)}$$

Then, with $g \in L_2(0, \infty; U)$, we obtain from (A.2) used in the first integral of (A.12), along with assumptions (H.1)-(H.3) used in the second integral of (A.12):

$$\zeta(t) \in C([0, T]; Y) \Leftrightarrow \mathcal{L}_{P_{0,\infty}}: \text{ continuous } L_2(0, T; U) \to C([0, T]; Y) \qquad \text{(A.13)}$$

for any $T < \infty$. Equivalently, via (A.11), by duality see [L-T.1],

$$\int_0^T \|B^* e^{A_{P_{0,\infty}}^* t} x\|_U^2 \, dt \le c_T \|x\|_Y^2, \; x \in Y \qquad \text{(A.14)}$$

Then, since $e^{A_{P_{0,\infty}} t}$ is stable under assumption (H.6) [F-L-T.1], we can invoke [L-T.3] again and conclude that with $g \in L_2(0, \infty; U)$, then:

$$\zeta(t) = (\mathcal{L}_{P_{0,\infty}} g)(t) \in L_2(0, \infty; Y) \cap C([0, \infty]; Y). \qquad \text{(A.15)}$$

References

[B.1]   V. Barbu, H^∞–Boundary Control with State Feedback; the Hyperbolic Case, preprint 1992.

[B-B.1]   T. Basar and P. Bernhard, H^∞– *Optimal Control and Related Minimax Design Problems. A Dynamic Game Approach*, Birkhauser Boston (1991).

[B-B.2]   A Bensoussan and P. Bernhard, Remarks on the theory of robust control, *International Series of Numerical Mathematics*, **107** (1992), pp. 149-166.

[C.1]   R. Curtain, State-space approaches to H-infinity control for infinite dimensional linear systems, *Trans. Inst. Mc* Vol. 13 No. 5 (1991), 253-261.

[Ch.1]   S. Chen, Necessary and sufficient conditions for the existence of positive solutions to Algebraic Riccati Equations with indefinite quadratic term, *Appl. Math. & Opt.*, 26 (1992), 95-110.

[D-I.1]   G. Da Prato and A. Ichikawa, Riccati Equations with Unbounded Coefficients, *Annali di Matem. Pura e Applic.*, **140** (1985), pp. 209-221.

[F.1]   F. Flandoli, Algebraic Riccati Equations Arising in Boundary Control Problems, *SIAM J. Control and Optim.*, 25 (1987), pp. 612-636.

[F-L-F.1]   F. Flandoli, I. Lasiecka, and R. Triggiani, Algebraic Riccati Equations with Non-Smoothing Observation Arising in Hyperbolic and Euler-Bernoulli

Equations, *Ann. di Matem. Pura et Applic.*, **IV** Vol. CLIII (19898), 307-382.

[I.1]   A. Ichikawa, H$^\infty$–Control with State Feedback and Quadratic Games in Hilbert Space, preprint 1991.

[La.1]   I. Lasiecka, Exponential Stabilization of Hyperbolic Systems with Nonlinear, Unbounded Perturbations: A Riccati Operator Approach, *Applicable Analysis*, **42** (1991), pp. 243-261.

[La.2]   Lasiecka, Riccati Equations arising in boundary and point control problems, Springer-Verlag LNICS 185 pp. 23-46. Proceedings of 10th International Conference, Sophia-Antipolis, France, June 9-12, 1992.

[L-T.1]   I. Lasiecka and R. Triggiani, *Differential and Algebraic Riccati Equations with Application to Boundary/Point Control Problems: Continuous Theory and Approximation Theory*, Volume #164 in the Springer-Verlag Lectures Notes LNCIS series (1991), pp. 160.

[L-T.2]   I. Lasiecka and R. Triggiani, The regulatior problem for parabolic equations boundary control, *Appl. Math. & Opt.* 16 (1987), 187-216.

[L-T.3]   I. Lasiecka and R. Triggiani, A lifting theorem for the time regularity of solutions to abstract equations, *Proc. Am. Math. Soc.* 103 (1988), 745-755.

[L-T.4]   I. Lasiecka and R. Triggiani, Riccati Equations Arising from Systems with Unbounded Input-Solution Operator: Applications to Boundary Control Problems for Wave and Plate Problems, *J. of Non-Linear Analysis*, Vol. 20 (1993), 659-695.

[M.2]   C. McMillan, Ph.D. dissertation, Department of Applied Mathematics, University of Virginia, May 1993.

[M-T.1]   C. McMillan and R. Triggiani, Min-Max Game Theory and Algebraic Riccati Equations for Boundary Control Problems with Continuous Input-Solution Map. Part I: the Stable Case, *Marcel Dekker Notes in Pure and Applied Mathematics*, Clement/Lumer Edts. 1993 chapter 32, pp 377-403. *Proceedings of the International Conference on "Evolution Equations in Banach Space,"* Belgium, October 1991.

[M-T.2]   C. McMillan and R. Triggiani, Min-max game theory and algebraic Riccati equations for boundary control problems with continuous input-solution map. Part II: the general case, *Appl. Mathem. & Optimiz.*, to appear. Presented at 10th Internat. Conf., Sophia-Antipolis, France, June 9-12, 1992.

[M-T.3]   C. McMillan and R. Triggiani, Min-Max Game Theory and Algebraic Riccati Equations for Boundary Control Problems with Analytic Semigroups. Part I: the Stable Case, *Marcel Dekker Lecture Notes in Pure and Applied Mathematics*, Elworthy/Everitt/Lee Edits, 1993, Chapter 47, pp. 757-780. Festschrift in honor of L. Markus.

[M-T.4]   C. McMillan and R. Triggiani, Min-Max Game Theory and Algebraic Riccati Equations for Boundary Control Problems with Analytic Semigroups. Part II: the General Case, University of Virginia preprint 1992, *J. of Non Linear Analysis*, to appear. Presented at the SIAM Conference on Control held at Minneapolis, Sept. 1992.

[vK.1]   B. van Keulen, A State-Space Approach to H$^\infty$–Control Problems for Infinite-Dimensional Systems, preprint 1992. Also, Ph.D. Thesis, 1993.

Stochastic Differential Games in Economic Modeling

Alain Haurie*

Université de Genève

Abstract

In this paper we present two continuous-time models of economic competition which are based on a stochastic differential game formalism. We focus our presentation on the modeling possibilities offered by the frameworks of piecewise deterministic and switching diffusion control systems respectively. We develop two duopoly models: a dynamic R&D competition model and a stochastic fishery exploitation model with correlated equilibrium. Some indications on the numerical solution of these games are also given.

1   introduction

The aim of this paper is to to present two stochastic noncooperative differential games which model typical economic competition processes. In both cases the randomness is represented, at least partly, as a controlled jump process. In the first model, which deals with competition through investment in R&D, the jump process describes modal changes corresponding to different competitive situations depending on which firm has made the transition to a new technology. The second model, which deals with a competitive exploitation of a common renewable resource, is defined initially as a controlled diffusion system. The random jumps are then introduced as a *communication device* which permits the players to implement an *equilibrium with memory* which includes some *credible threats* that are activated when a breach of cooperation is suspected.

The first model is based on a *piecewise deterministic* control formalism. The second one uses a *switching diffusion* control formalism.

*Research supported by NSERC-Canada, FCAR-Quebec, and FNRS-Switzerland

2   An R&D competition model

The model presented in this section is a differential game version of the typical *competition through innovation* models studied in industrial organization (see e.g. [34] or [35]).

2.1   Modeling technological innovation

Some firms, e.g. in high-tech industry, spend a large amount as investment in R&D (laboratories, research teams, etc) for the purpose of increasing the probability of getting a competitive hedge due to technological innovation. We assume that this investment leads to the detention of a stock of capital which can be represented by a single variable. Let $x(t) \in \mathbb{R}$ be the level of R&D-capital accumulated by the firm at time t. A state equation describes the investment process

$$\left. \begin{array}{rcl} \dot{x}(t) & = & f(x(t), u(t)) \ = \ u(t) - \gamma x(t) \\[2mm] x(0) & = & x^0, \end{array} \right\} \tag{1}$$

where the control variable $u(t) \in U = [0, \infty)$ is the investment rate in R&D at time t and γ is the depreciation rate of this capital. The equation (1) is basically the capital accumulation representation used in neo-classical economic growth (see e.g. [8]).

An R&D investment policy is represented as a function

$$u(t) : t \mapsto U, \quad t \in [0, \infty).$$

We denote $\mathcal{U}$ the class of all admissible policies $u(\cdot)$. To an initial condition $x(0) = x^0$ and a policy $u(\cdot) \in \mathcal{U}$ corresponds a unique evolution $x(\cdot)$ of the capital stock, defined as the solution of (1).

The R&D capital is an innovation factor. Innovation takes place as a *discrete event* :

> at a random time a qualitative change occurs, either in the production process, or in the design of the product marketed by the firm. This discrete event can be represented as a *stopping time* T with intensity $q(x)$, defined by
>
> $$\mathbb{P}_{u(\cdot)}[T \in (t; t + dt) | T > t] \ = \ q(x(t))dt + o(dt). \tag{2}$$
>
> where $\lim_{dt \to 0} \frac{o(dt)}{dt} = 0$ uniformly in x. The intensity represents the influence of the accumulation of R&D capital on the innovation process (e.g. $q(x)$ is an increasing function of x).

The probability that the innovation takes place between times t and $t + dt$ is given by

$$P_{u(\cdot)}[t < T < t + dt] = q(x(t))e^{-\int_0^t q(x(s),u(s))ds}dt + o(dt) \qquad (3)$$

From (3) we obtain the probability that the innovation occurs before time τ as given by

$$P_{u(\cdot)}[T \leq \tau] = \int_0^\tau q(x(t))e^{-\int_0^t q(x(s),u(s))ds}dt. \qquad (4)$$

Equations (1-4) specify the innovation process of the firm.

2.2 A duopoly model

We now extend the model to the strategic problem of competition through R&D investment. Consider two firms denoted $j = 1, 2$. Firm j's R&D capital is denoted x_j. Its evolution is dictated by the state equation

$$\left.\begin{array}{rcl} \dot{x}_j(t) & = & u_j(t) - \gamma x_j(t), \\ x_j(0) & = & x_j^0, \end{array}\right\} \quad j = 1, 2 \qquad (5)$$

Since there are two firms competing for getting first a market advantage through innovation, we introduce a discrete state variable ξ to describe the technological situation. This discrete state takes 4 possible values $\xi(t) \in S = \{0; 1; 2; 3\}$, where the meaning of each possible state is as follows

0: the two firms are investing in R&D;

1: firm 1 has made the technological change. Firm 2 continues to invest in R&D;

2: firm 2 has made the technological change. Firm 1 continues to invest in R&D;

3: the two firms have made the technological change.

The discrete state will change at random times. The transition probabilities are defined by *transition rates*. The nonzero rates are $q_{01}(x_1)$, $q_{02}(x_2)$, $q_{13}(x_2)$, $q_{23}(x_1)$ which describe the influence of R&D investment on the occurrence of a technological breakthrough in one firm or the other. All the other jump rates are null ($q_{kl} \equiv 0$).

2.2.1 Economics of Competition through R&D

We denote

$$L_j^i(x_j, u_j), \quad i = 0, 1, 2, \tag{6}$$

the R&D cost of firm j, when the discrete state $\xi = i \in S$ prevails. The adjustment cost to firm j, in a transition of the discrete state from $i \in S$ to $k \in S$, is denoted

$$\Phi_j^{ik}(x_j). \tag{7}$$

We may naturally assume that $\Phi_j^{ik}(\cdot) \equiv 0$ when $i = 0, k = 1, j = 2$ as well as when $i = 0, k = 2, j = 1$, and for $i = 1, k = 3, j = 1$ or $i = 2, k = 3, j = 2$, as these cases correspond to situations where the innovation has been made by the competitor.

We now describe the market through the following linear demand law

$$p = A - By^i = A - B(y_1^i + y_2^i), \quad i = 0, 1, 2, 3 \tag{8}$$

where y_j^i is the output level of firm j, when the discrete state $\xi = i \in S$ prevails. Production costs, when $\xi = i \in S$, are given by

$$C_j^i = \alpha_j^i y_j^i + \beta_j^i (y_j^i)^2, \tag{9}$$

where α_j^i and β_j^i are positive constants.

2.3 Cournot Equilibrium

It is easily seen that there is a decoupling, in this model, between the R&D investment decisions and production decisions which will be obtained as in a static duopoly model.

It is possible to compute the profits (before R&D costs) corresponding to a Cournot equilibrium (see e.g. Henderson & Quandt, [29] p. 213)

$$\Pi_j^i = (A - By^i)y_j^i - \alpha_j^i y_j^i - \beta_j^i (y_j^i)^2, \tag{10}$$

where the quantities supplied at equilibrium are given by

$$y_j^i = \frac{2(B + \beta_{j'}^i)(A - \alpha_j^i) - B(A - \alpha_{j'}^i)}{4(B + \beta_j^i)(B + \beta_{j'}^i) - B^2}, \quad j' \neq j. \tag{11}$$

2.4 Dynamic Programming and Stochastic Equilibrium

Each firm chooses an R&D investment policy in order to minimize a total expected discounted cost over an infinite time horizon. At the initial time, $t = 0$, no firm controls a new technology, hence $\xi = 0$. Then, after a random time, there will be a transition toward either state $\xi = 1$ or $\xi = 2$, and, finally, a transition to the terminal (trapping) discrete state $\xi = 3$. When $\xi = 3$, the firms don't invest any more in R&D and the model reduces to a standard static Cournot duopoly. The cost-to-go, for each of the two firms, are then given by

$$v(\cdot, 3, j) = -\int_0^\infty e^{-\rho t} \Pi_j^3 = -\frac{\Pi_j^3}{\rho}. \tag{12}$$

For the other discrete states $\xi = 0, 1$ or 2, the problem can be formulated as follows. Let $v(x, i, j)$ is the expected cost-to-go for firm j when the discrete state is $\xi = i$ and the level of R&D capital is equal to x. Firm 1 must then solve the problem

$$
\left.
\begin{aligned}
v(x_1^0, i, 1) \;=\; & \min_{u_1(\cdot) \in \mathcal{U}_1^i} E_{u_1}[\textstyle\int_0^T e^{-\rho t}(L_1^i(x_1(t), u_1(t)) - \Pi_1^i)dt \\[4pt]
& + E_{\mathbf{u}^{-1}(\cdot) \in \mathcal{U}^i}[e^{-\rho T}(\Phi_1^{i\xi(T)}(x_1(T)) \\[4pt]
& + v(x(T), \xi(T), 1))]|x_1(0) = x_1^0] \\[6pt]
\dot{x}_1(t) \;=\; & u_1(t) - \gamma x_1(t), \quad x_1(0) = x_1^0,
\end{aligned}
\right\} \tag{13}
$$

where the term $\mathbf{u}^{-1}(\cdot) = (u_1(\cdot), u_2{}^*(\cdot))$ represents the R&D investment policy of the two firms. Similarly, firm 2 has to solve

$$
\left.
\begin{aligned}
v(x_2^0, i, 2) \;=\; & \min_{u_2(\cdot) \in \mathcal{U}_2^i} E_{u_2}[\textstyle\int_0^T e^{-\rho t}(L_2^i(x_2(t), u_2(t)) - \Pi_2^i)dt \\[4pt]
& + E_{\mathbf{u}^{-2}(\cdot) \in \mathcal{U}^i}[e^{-\rho T}(\Phi_2^{i\xi(T)}(x_2(T)) \\[4pt]
& + v(x(T), \xi(T), 2))]|x_2(0) = x_2^0] \\[6pt]
\dot{x}_2(t) \;=\; & u_2(t) - \gamma x_2(t), \quad x_2(0) = x_2^0,
\end{aligned}
\right\} \tag{14}
$$

where $\mathbf{u}^{-2}(\cdot) = (u_1{}^*(\cdot), u_2(\cdot))$. These equations (13) and (14) are *dynamic, stochastic equilibrium* conditions for the policy pair $u^*(\cdot) = (u_1^*(\cdot), u_2^*(\cdot))$. Indeed firm j's policy is an optimal reply to the competitor's.

We also notice that this competitive situation only occurs in discrete state $\xi = 0$. In all other cases, there is at most one firm which is continuing to invest in R&D. The set of admissible control $\mathcal{U}^\xi$ is thus depending explicitly on the value taken by the discrete state ξ.

2.4.1 A Deterministic Control and Differential Game Reformulation

When $\xi = 1$ or $\xi = 2$ one can reformulate the stochastic problems (13) and (14) as a deterministic infinite horizon optimal control problem. When $\xi = 0$, the problem can be reformulated as a deterministic differential game over an infinite time horizon. We detail below the different cases.

If $\xi = 1$.

It has been shown in [4] that the stochastic control problem has an equivalent reformulation as the infinite horizon deterministic control problem defined below:
$(u_2^*(\cdot), x_2^*(\cdot))$ is solution of

$$
\left.
\begin{aligned}
v(x_2^0, 1, 2) &= \min_{u_2(\cdot) \in \mathcal{U}_2^1} \int_0^\infty e^{-z_2(t)} \{ L_2^1(x_2(t), u_2(t)) - \Pi_2^1 \\
&\quad + [\Phi_2^{13}(x_2(t)) - \tfrac{\Pi_1^3}{\rho}] q_{13}(x_2(t)) \} dt \\
\dot{x}_2(t) &= u_2(t) - \gamma x_2(t), \quad x_2(0) = x_2^0, \\
\dot{z}_2(t) &= \rho + q_{13}(x_2(t)), \quad z_2(0) = 0.
\end{aligned}
\right\} \tag{15}
$$

Then we compute $v(\cdot, 1, 1)$ through the formula

$$
v(\cdot, 1, 1) = \int_0^\infty e^{-z_2(t)} \{ -\Pi_1^1 - q_{13}(x_2^*(t)) \frac{\Pi_1^3}{\rho} \} dt. \tag{16}
$$

If $\xi = 2$.

Using the same result as above we now compute $(u_1^*(\cdot), x_1^*(\cdot))$ solution of

$$
\left.
\begin{aligned}
v(x_1^0, 2, 1) &= \min_{u_1(\cdot) \in \mathcal{U}_1^2} \int_0^\infty e^{-z_1(t)} \{ L_1^2(x_1(t), u_1(t)) - \Pi_1^2 \\
&\quad + [\Phi_1^{23}(x_1(t)) - \tfrac{\Pi_1^3}{\rho}] q_{23}(x_1(t)) \} dt \\
\dot{x}_1(t) &= u_1(t) - \gamma x_1(t), \quad x_1(0) = x_1^0, \\
\dot{z}_1(t) &= \rho + q_{23}(x_1(t)), \quad z_1(0) = 0.
\end{aligned}
\right\} \tag{17}
$$

Then we compute $v(\cdot, 2, 2)$ through the formula

$$
v(\cdot, 2, 2) = \int_0^\infty e^{-z_1(t)} \{ -\Pi_2^2 - q_{23}(x_1^*(t)) \frac{\Pi_2^3}{\rho} \} dt. \tag{18}
$$

<u>If $\xi = 0$.</u>

This corresponds to the case where both firms have to invest in R&D. We have now to compute a Nash equilibrium for an infinite horizon, deterministic differential game. We thus define $(x_1^*(t), u_1^*(t))$ and $(x_2^*(t), u_2^*(t))$ as solutions to the following system

$$
\left.
\begin{aligned}
v(x_1^0, 0, 1) &= \min_{u_1(\cdot) \in \mathcal{U}_1^0} \int_0^\infty e^{-z_1(t)} \{ L_1^0(x_1^*(t), u_1(t)) - \Pi_1^0 \\
&\quad + [\Phi_1^{01}(x_1^*(t)) + v(x_1^*(t), 1, 1)] q_{01}(x_1^*(t)) + \\
&\quad + v(x_1^*(t), 2, 1) q_{02}(x_2^*(t)) \} dt \\
\dot{x}_1^*(t) &= u_1^*(t) - \gamma x_1^*(t), \quad x_1^*(0) = x_1^0, \\
\dot{z}_1(t) &= \rho + q_{01}(x_1^*(t)) + q_{02}(x_2^*(t)), \\
z_1(0) &= 0,
\end{aligned}
\right\} \tag{19}
$$

and

$$
\left.
\begin{aligned}
v(x_2^0, 0, 2) &= \min_{u_2(\cdot) \in \mathcal{U}_2^0} \int_0^\infty e^{-z_2(t)} \{ L_2^0(x_2^*(t), u_2(t)) - \Pi_2^0 \\
&\quad + [\Phi_2^{02}(x_2^*(t)) + v(x_2^*(t), 2, 2)] q_{02}(x_2^*(t)) + \\
&\quad + v(x_2^*(t), 1, 2) q_{01}(x_1^*(t)) \} dt \\
\dot{x}_2^*(t) &= u_2^*(t) - \gamma x_2^*(t), \quad x_2^*(0) = x_2^0, \\
\dot{z}_2(t) &= \rho + q_{01}(x_1^*(t)) + q_{02}(x_2^*(t)), \\
z_2(0) &= 0.
\end{aligned}
\right\} \tag{20}
$$

If we define the *current-valued Hamiltonians*

$$H_j^i(x^i, u^i, \lambda_j^i, \nu^i) = L_j^i(x_j^*(t), u_j(t)) - \Pi_j^i + \lambda^i(u_j - \gamma x_j) + \nu^i(\rho + \sum_{\ell \in E}(q_{i\ell}(x_1) + q_{i\ell}(x_2)))$$

where $x^i = (x_1^i, x_2^i)$, $u^i = (u_1^i, u_2^i)$, etc.. Then, using again Halkin's minimum principle (see [23]) we obtain the following *pseudo-Hamiltonian systems*:

$$
\left.
\begin{aligned}
H_j^i(x^i(t), u^i(t), \lambda^i(t), \nu^i(t)) &= \min_{u_j \in U^i} H_j^i(x^i(t), (u^i)^{-j}(t), \lambda^i(t), \nu^i(t)) \\
\dot{\lambda}_j^i(t) &= (q^i(x^i(t), u^i(t)) + \rho) \lambda_j^i(t) - \frac{\partial H_j^i}{\partial x_j}(x^i(t), u^i(t), \lambda^i(t), \nu^i(t)) \\
\dot{x}_j^i(t) &= u_j^i(t) - \gamma x_j^i(t),
\end{aligned}
\right\} \tag{21}
$$

where $(u^i)^{-1} = (u_1, u_2^i)$ and $(u^i)^{-2} = (u_1^i, u_2)$.

2.5 Turnpike

The deterministic optimal control or differential game problems considered above are defined over an infinite time horizon. The asymptotic behavior of optimal trajectories in such systems has been studied in ([8]) and, more recently in [7] in the optimal control case. An interesting property of these infinite horizon control systems is that there can exist an attracting *steady state* $\bar{x}^i$ toward which all optimal trajectories will converge whatever be the initial state $x(0)$. This *attractor*, also called the *turnpike* is such that, if the initial condition is precisely $x(^0) = \bar{x}^i$, then the optimal trajectory is stationary, i.e., as long as $\xi(t) = i$,

$$x^{i^*}(t) \equiv \bar{x}^i.$$

Therefore this will be a solution to the hamiltonian systems with $\dot{x} = \dot{\lambda} = \dot{\nu} = 0$.

In [23] and more recently [26] one gives conditions for the asymptotic stability of the equilibrium trajectories, in infinite horizon differential games. In [26] and [25] we show how this asymptotic stability property can be exploited for the numerical computation of the optimal policies.

2.6 Comments on the Form of the Solution

We have formulated a game with only a finite number of possible jumps for the $\xi(\cdot)$-process. It could be easily extended to a situation where the innovation race does not stop. Then the $\xi(\cdot)$-process would evolve over a possibly infinite but countable set of possible states. Again, with each state we would have an associated infinite horizon deterministic differential game, involving the cost functions $v(x, \xi, j)$. A turnpike property is likely to hold for such a system. Then the investment schedules of the two firms will be attracted by a sequence of turnpike values $x_j^{i_n}$, $j = 1, 2$, where $\{i_n : n = 0, 1, ...\}$ represents the sequence of states visited by the $\xi(\cdot)$-process.

3 Communication Equilibrium in a Game of Fisheries Exploitation

In this section we study a stochastic differential game model describing the competitive behavior of two fisheries exploiting the same stock of biomass. This model involves a controlled diffusion process representing the fish species dynamics. Deterministic versions of a fisheries exploitation model have been proposed in e.g. [9] or [31]. In [18] a class of equilibria using *memory strategies* has been studied for deterministic models. In the present model we study a stochastic counterpart to [18].

3.1 State Equations

Let $x(t) \in \mathbb{R}^+$ be the stock of biomass at time t. Two fisheries, $j = 1, 2$, exploit this stock. Let $u_j(t) \geq 0$ be the fishing effort of Fishery j. Let $\pi_j(t)$ be the cumulative profit of Fishery j between time 0 and t. We assume the following dynamics

$$
\begin{aligned}
dx(t) &= f(x(t), \mathbf{u}(t)) \, dt + \sigma d\varepsilon(t) \\
\\
&= (A - Bx(t) - u_1(t) - u_2(t))x(t) \, dt + \sigma d\varepsilon(t) \quad (22)
\end{aligned}
$$

$$
x(0) = x_0 \quad (23)
$$

$$
\begin{aligned}
d\pi_j(t) &= g_j(x(t), \mathbf{u}(t)) \, dt \\
\\
&= \{[a - bx(t)(u_1(t) + u_2(t))]x(t)u_j(t) \\
\\
&\quad - \alpha_j u_j(t)^2 - \beta_j u_j(t)\} \, dt, \quad (24)
\end{aligned}
$$

$$
\pi_j(0) = 0 \quad (25)
$$
$$
j = 1, 2.
$$

The profit rates in Eq. 24 are based on a linear demand function $a - b(q_1 + q_2)$ which determines the price of the fish when the quantity caught by each fishery is $q_j = xu_j$, $j = 1, 2$.

3.2 Equilibria and Cooperative Solutions

3.2.1 Information Structure

We assume that each player (fishery) can observe at any time t the stock level $x(t)$ but is unable to observe the fishing effort of the other player. A strategy for Player j will be a mapping $\tilde{u}_j$ which associates a fishing effort $u_j(t) = \tilde{u}_j(t, \tilde{x}_t) \geq 0$ with the current time t and the past observations

$\tilde{x}_t = \{x(s) : s \leq t\}$ of the biomass stock. A pair of strategies $\tilde{\mathbf{u}} = (\tilde{u}_1, \tilde{u}_2)$ defines an *admissible policy* if it generates a well defined stochastic process $(x(\cdot), u_1(\cdot), u_2(\cdot))$.

This concept of admissible policy allows the players to use *memory strategies*, i.e. to remember some past agreement and to base their current decision on a comparison between what was expected from the agreement and what was achieved. This class of strategies will be made more precise in the forthcoming subsections.

Given the initial stock level x_0 and an admissible policy $\tilde{\mathbf{u}}$ each player expects the following discounted profit

$$V_j^\tau(x_0, \tilde{\mathbf{u}}) = \mathrm{E}_{x_0, \tilde{\mathbf{u}}} \left[\int_\tau^\infty e^{-\rho t} \, d\pi_j(t) \right], \quad j = 1, 2, \tag{26}$$

where ρ is a positive discount rate. For ease of notation we shall write $V_j(x_0, \tilde{\mathbf{u}})$ instead of $V_j^0(x_0, \tilde{\mathbf{u}})$, when $\tau = 0$.

3.2.2   Equilibrium

Definition 2.1   A pair $\tilde{\mathbf{u}}^* = (\tilde{u}_1^*, \tilde{u}_2^*)$ is an *equilibrium* at x_0 if (i) it is admissible and (ii) for any admissible pairs $\tilde{\mathbf{u}}^{*^{-1}} = (\tilde{u}_1, \tilde{u}_2^*)$ or $\tilde{\mathbf{u}}^{*^{-2}} = (\tilde{u}_1^*, \tilde{u}_2)$ the following inequalities hold

$$V_j(x_0, \tilde{\mathbf{u}}^*) \geq V_j(x_0, \tilde{\mathbf{u}}^{*^{-j}}), \quad j = 1, 2. \tag{27}$$

We have defined the equilibrium property as depending on the initial state x^0 since *memory strategies*, are allowed. We also require *subgame perfectness* for an equilibrium in order to make *credible* the implicit threats used in the strategy. The threat is *credible* if its use is a player's optimal reply to the behavior of the other players.

Definition 2.2   A pair $\tilde{\mathbf{u}}^* = (\tilde{u}_1^*, \tilde{u}_2^*)$ is a *subgame perfect equilibrium* on $X \in \mathbb{R}$ if (i) it is admissible at any $x_0 \in X$ and (ii) for any history $\tilde{x}^t$ where $x(t) = x_i$ and for any admissible pairs $\tilde{\mathbf{u}}^{*^{-1}} = (\tilde{u}_1, \tilde{u}_2^*)$ or $\tilde{\mathbf{u}}^{*^{-2}} = (\tilde{u}_1^*, \tilde{u}_2)$ the following inequalities hold

$$V_j^t(x_i, \tilde{\mathbf{u}}^*) \geq V_j^t(x_i, \tilde{\mathbf{u}}^{*^{-j}}), \quad j = 1, 2. \tag{28}$$

In a *subgame perfect* equilibrium, the pair of strategies remains an equilibrium even if the players do not play correctly for a while (e.g. between time 0 and t) and then resume the game driven by the $\tilde{\mathbf{u}}^*$ policy. Subgame perfectness is a concept which has been introduced in the context of games in extensive form [33].

3.2.3 Pure Feedback Policies

Stationary feedback strategies, also called *homogeneous nonrandomized Markov policies*, form a subclass of admissible strategies.

Definition 2.4 A measurable map $\tilde{u}_j : \mathbb{R} \mapsto \mathbb{R}^+$ defines a pure stationary feedback strategy for Player j. The control at time t of Player j is thus defined as $u_j(t) = \tilde{u}_j(x(t))$, $\forall t \geq 0$. Let's introduce the operator $L^{\tilde{u}}$ acting on continuous functions which are a.e. C^2 and which we define by

$$L^{\tilde{u}}(V_j(x)) = \frac{1}{2}\sigma^2 \frac{d^2 V_j(x)}{dx^2} + f(x, \tilde{u}(x,k))\frac{dV_j(x)}{dx}, \quad j = 1, 2$$

Theorem 2.1 If both players use a pair $\tilde{u}$ of feedback strategies then the equations (22,23) admit an a.s. strong solution $x(\cdot)$ which is a Feller process with generator $L^{\tilde{u}}$.

Proof: See Krylov [16].

3.2.4 Memory Strategies Based on Monitoring an Agreement

As in the classical duopoly model of Cournot [10], the conflicting objectives of the two players stem from the declining market price when there is an increase in total supply. Cooperation calls for a mutual restraint in fishing activity. However, if the cooperative solution is not an equilibrium, each fishery will be tempted to increase its own fishing rate when the other one is restraining itself. Due to the randomness in the stock evolution equation (22) each player is unable to detect with certainty a possible deviation by the opponent from an agreed cooperative fishing policy.

We want to design a cooperative policy which retains all the properties of a subgame perfect equilibrium and which generates a pair of outcomes (i.e. expected profits) that dominate the Nash feedback equilibrium defined above. This will be an example of a *communication equilibrium* as discussed in [12] for games in extensive form.

Our approach is inspired from [32] and [15], who studied a repeated Cournot duopoly game with random inverse demand function. A discrete-time dynamic game model, related to the present game of fisheries exploitation, was also studied in [27] and the existence of dominating cooperative equilibria shown using a numerical solution technique. We describe below an extension of the approach used in [27] to the continuous-time diffusion game model under study. A more complete study of that continuous time model is available in [22].

The basic idea for constructing such an equilibrium is to design a new game with an extended state space and to construct a feedback Nash equilibrium for this new game.

Extended State: The extended state will be $s = (x, z, \xi) \in \mathbb{R} \times \mathbb{R} \times \{0, 1\}$. In addition to the stock variable x, the players will also consider the value of a state variable z, which is designed to be a monitoring device, as well as the discrete state variable ξ which defines two possible modes of play for the game. When $\xi = 0$ the mode of play is *retaliatory* or *punitive*; when $\xi = 1$ the mode of play is *conciliatory* or *cooperative*.

Monitoring System: We assume that both players know the fundamental biomass dynamics given in Eq. (22). Let $\gamma_j^1(\cdot, \cdot)$, $j = 1, 2$ be two measurable functions mapping $\mathbb{R}^2$ into $\mathbb{R}^+$. Let a policy pair

$$\bar{u}(t) = (\gamma_1^1(x(t), z(t)), \gamma_2^1(x(t), z(t)))$$

represent a cooperative mode of play that has to be monitored. This policy is a pure feedback in the extended state space (x, z). We define a monitoring system via the auxiliary state equation

$$dz(t) = dx(t) - f(x(t), \bar{u}(t)) \, dt \tag{29}$$
$$z(0) = 0 \tag{30}$$

where $f(.,.)$ is the drift term in the dynamics of the stock, see Eq.(22).

According to this design $z(t)$ will behave as a pure Wiener process as long as the players actually use the policy pair $\bar{u}(t)$. If one player does not behave in accordance with this agreed policy, then the process $z(\cdot)$ will exhibit a drift. A detection of the drift could then trigger a retaliation against the other player who is suspected of cheating.

Switching Between Modes of Play: We introduce a new discrete state variable ξ with value in $\{0, 1\}$. If $\xi(t) = 1$ the current mode of play at time t is cooperative, if $\xi(t) = 0$ the mode of play at time t is retaliatory. The retaliation triggering mechanism changes the mode of play of the game. In the scheme proposed here the mode of play evolves according to a stochastic jump process characterized by jump rates

$$q_{01} = \lim_{dt \to 0} \frac{P[\xi(t + dt) = 1 | \xi(t) = 0]}{dt} \tag{31}$$

$$q_{10}(z) = \lim_{dt \to 0} \frac{P[\xi(t + dt) = 0 | \xi(t) = 1, \ z(t) = z]}{dt}. \tag{32}$$

In Eq. (31) one assumes that the duration of a punitive mode of play is an exponential random variable with mean $\frac{1}{q_{01}}$. In Eq. (32) one assumes that the switch from a cooperative to a punitive mode of play is triggered with a probability which is a function of the observed value of the monitoring variable z. This functional dependence implements the cheating detection scheme discussed above.

Punishment Strategy: During a retaliation episode, the players are no longer monitoring a nominally cooperative policy. Therefore they implement a pair of feedback policies $\gamma_j^0(x)$, $j = 1, 2$ defined over the original x-space.

Resetting the Monitoring Variable: Once a punitive period is terminated, i.e. when there is a jump of $\xi(t)$ to the value 1, then the z variable is reset to a zero value.

Remark 2.1 The extended game, including the switching modes of play, the monitoring system and the jumping rates is entirely built from the data concerning the original game. The policies $\gamma_j^1(x, z)$ and $\gamma_j^0(x)$ are thus in accordance with the information structure of the original game. Indeed, since the monitoring state $z(t)$ is a functional of the past history $\tilde{x}^t$, the proposed policy is in agreement with the general definition of section 2.2.1.

3.2.5 Feedback Policies for the Extended Switching Diffusion Game

The system defined by Eqs.(22)-(23), (29)-(30), (31)-(32) is a particular instance of a switching diffusion control system. For a recent analysis of such systems we refer to [24] and [14]. In these references, only the single player case of optimal control was considered. A general theory of noncooperative switching diffusion games is still to be developed; in particular, the existence of a feedback equilibrium has not yet been proved. However, for the purpose of the present study we can exploit some of the results already available for the case of optimal control of switching diffusion systems.

In a general formulation, such a system is characterized by an hybrid state $(y, \Xi) \in \mathbb{R}^n \times E$, where E is a finite set, and a control $\mathbf{u} \in \mathbb{R}^m$. The continuous state variable has an evolution driven by the Ito equations

$$dy(t) \;=\; f^k(y(t), \mathbf{u}(t)) \, dt + \sigma \, d\varepsilon(t) \tag{33}$$

$$u(t) \;\in\; U^k \subset \mathbb{R}^m \tag{34}$$

$$y(0) \;=\; y_0, \tag{35}$$

where $k \in E$, σ is an $n \times n$ matrix and ϵ a Wiener process. The discrete state variable Ξ evolves according to a jump process with jump rates

$$q_{k\ell} = \lim_{dt \to 0} \frac{P[\Xi(t+dt) = \ell | \Xi(t) = k\,, \, y(t) = y]}{dt}, \quad k, \ell \in E,\, k \neq \ell. \tag{36}$$

A pure feedback policy for such a system is defined as a measurable mapping $\tilde{u} : (y, \Xi) \mapsto U^\Xi$. Introduce the operator $L^{\tilde{u}}$ acting on continuous functions $V(\cdot, y)$ which are a.e. C^2 in y and which we define by

$$L^{\tilde{u}}(V(\cdot,y)) = \frac{1}{2} \sum_{ii'=1}^{n} a_{i,i'} \frac{\partial^2 V(k,y)}{\partial y_i \partial y_{i'}} + \sum_{i=1}^{n} f_i^k(y, \tilde{u}(k,y)) \frac{\partial V(k,y)}{\partial y_i},$$
$$+ \sum_{k\ell \in E} q_{k\ell}(y) V(\ell, y). \tag{37}$$

Theorem 2.2 If one uses a feedback policy $\tilde{u}$ then the equations (33)-(35) and (36) admit an a.s. strong solution $(y(\cdot), \Xi(\cdot))$ which is a Feller process with generator $L^{\tilde{u}}$.

Proof: See the recent study by Ghosh et al. [14].

3.2.6 Cooperative Equilibria with Monitoring

Consider a pair of memory strategies monitoring an agreement as described in section 3.2.4. These strategies are also pure feedback policies for the associated switching diffusion game. Assume now that these policies define a feedback equilibrium for this switching diffusion game. Then we have defined a monitoring scheme and a retaliation scheme which satisfies the conditions for being a subgame perfect equilibrium.

Definition 2.5 A subgame perfect cooperative equilibrium for the stochastic differential game of fisheries exploitation with monitoring and retaliation is defined by

- two modes of play, $\xi = 1$ called cooperative, $\xi = 0$ called punitive

- a monitoring variable $z \in \mathbb{R}$

- a pair of cooperative policies $\gamma_j^1(x, z)$, $j = 1, 2$

- a monitoring scheme as described in section 3.2.4

- a pair of punitive policies $\gamma_j^0(x)$, $j = 1, 2$

- a jump process describing the switches between the two modes of play as described in section 3.2.4

such that these policies constitute a pure feedback equilibrium for the associated switching diffusion stochastic game.

Remark 2.2 There is an implicit fixed-point argument in the definition of a cooperative equilibrium since the cooperative policy which has to satisfy the feedback equilibrium condition is also the policy one wants to monitor. Since an equilibrium is also a concept based on a fixed-point argument (each strategy is the best reply to the other players' strategies), we have a double fixed-point property in the definition of this cooperative equilibrium.

3.3 HJB Equations for Equilibrium Policies

3.3.1 Original Diffusion Stochastic Game

A feedback Nash strategy for each player is subgame perfect and can be computed from the following HBJ equation:

$$\rho \tilde{V}_j(x) = \frac{1}{2}\sigma^2 \frac{\partial^2}{\partial x^2} \tilde{V}_j(0, x)$$

$$+ \max_{u_j} \left[g_j(x, [\tilde{\gamma}^{N(-j)}(x), u_j]) + f(x, [\tilde{\gamma}^{N(-j)}(x), u_j]) \frac{\partial}{\partial x} \tilde{V}_j(x) \right] \quad (38)$$

where $[\tilde{\gamma}^{N(-j)}(x), u_j]$ is the control obtained when all the players use the (non-cooperative) Nash equilibrium policy $\tilde{\gamma}^N(x)$ while Player j uses his control u_j.

3.3.2 Associated Switching Diffusion Stochastic Game

The extended state $((x, z), y)$ is hybrid (continuous and discrete). The discrete state variable evolves according to a random jump Markov process, while x and z obey Ito equations. A new feedback equilibrium can be determined in this extended formulation. Such an equilibrium is characterized by the following coupled HJB dynamic programming equations

$$\rho \tilde{V}_j^*(1, (x, z)) = \frac{1}{2}\sigma^2 \{ \frac{\partial^2}{\partial x^2} \tilde{V}_j^*(1, (x, z)) + \frac{\partial^2}{\partial z^2} \tilde{V}_j^*(1, (x, z))$$

$$+ 2 \frac{\partial^2}{\partial x \partial z} \tilde{V}_j^*(1, (x, z)) \} + q_{10}(z) \left(\tilde{V}_j^*(0, x) - \tilde{V}_j^*(1, (x, z)) \right)$$

$$+ \max_{u_j} \left[g_j(x, [\tilde{\gamma}^{1(-j)}(x, z), u_j]) + f(x, [\tilde{\gamma}^{1(-j)}(x, z), u_j]) \frac{\partial}{\partial x} \tilde{V}_j^*(1, (x, z)) \right.$$

$$\left. + [f(x, [\tilde{\gamma}^{1(-j)}(x, z), u_j]) - f(x, \tilde{\gamma}^1(x, z)] \frac{\partial}{\partial z} \tilde{V}_j^*(1, (x, z)) \right] \tag{39}$$

$$\rho \tilde{V}_j^*(0, x) = \frac{1}{2} \sigma^2 \frac{\partial^2}{\partial x^2} \tilde{V}_j^*(0, x) + q_{01} \left(\tilde{V}_j^*(1, (x, 0)) - \tilde{V}_j^*(0, x) \right)$$

$$+ \max_{u_j} \left[g_j(x, [\tilde{\gamma}^{0(-j)}(x), u_j]) + f(x, [\tilde{\gamma}^{0(-j)}(x), u_j]) \frac{\partial}{\partial x} \tilde{V}_j^*(0, x) \right] \tag{40}$$

where $[\tilde{\gamma}^{1(-j)}(x, z), u_j]$ is the control obtained when all the players use the cooperative policy $\tilde{\gamma}^1(x, z)$, while Player j uses his control u_j. Similarly $[\tilde{\gamma}^{0(-j)}(x), u_j]$ is the control obtained when all other players use the punishment policy $\tilde{\gamma}^0$ and Player j uses his control u_j.

If one can solve these equations and if the monitoring and switching schemes are well designed, we can identify communication equilibria which dominate the feedback Nash equilibrium of the original game (see section 3.3.1).

A numerical approach has been used in [22] to compute *approximate* or *ε-equilibria* which verify this domination property.

3.4 Comments on this Class of Equilibria

The equilibrium which has been defined above and called *cooperative* because its construction involves a preplay agreement is a particular instance of the class of *correlated* and *communication equilibria* introduced by Auman [2] and more recently studied by Forges [12]. The construct of such an equilibrium through the random triggering mechanism seems to be a new approach to the operational use of memory strategies.

4 Conclusion

In this paper we have studied two stochastic differential games describing economic competition processes. The first game falls in the category of *piecewise deterministic differential systems*. The players' strategies are defined as *piecewise open-loop* control laws. An asymptotic stability property, called the *turnpike property* characterizes the behavior of this piecewise open-loop control: it drives the continuous state toward a steady-state (the turnpike); this steady state changes when the modal discrete state changes.

The second game falls in the category of *piecewise diffusion control systems*. The players' strategies are pure feedbacks. Through the use of an observer the players can monitor the observance of an agreement. The discrpancies between the observed trajectory and the expected one influence the probability of entering a retaliatory mode of play. We have shown how to characterize such equilibria. We refer to the more detailed studies cited in the text, for a numerical expiriment where it has been verified that such an equilibrium could be beneficial to both players.

References

[1] AKIAN M., CHANCELLIER J.P. *Dynamic Programming Complexity and Applications*, **Proceedings of the 27th IEEE Conference on Decision and Control, Austin Texas, 1988.**

[2] AUMANN R.J., 1974, *Subjectivity and Correlation in Randomized Strategies*, **J. Economic Theory**, Vol. 1, pp. 67-96.

[3] BERTSEKAS D.P., 1987, **Dynamic Programming: Deterministic and Stochastic Models**, Prentice Hall.

[4] BOUKAS E.K., HAURIE A. & MICHEL P., 1990, *An optimal control problem with a random stopping time*, **J. Optim. Theory Appl. 64,** 471-480.

[5] BOUKAS E.K., HAURIE A. & VAN DELFT CH., 1991, *A Turnpike Improvement Algorithm for Piecewise Deterministic Control*, **Optimal Control Applications and Methods, Vol. 12,** 1-18.

[6] BROCK W.A., 1977, *Differential Games with Active and Passive Variables*, *in* Henn and Moeschlin (Eds.), **Mathematical Economics and Game Theory: Essays in Honor of Oskar Morgenstern**, Springer Verlag, Berlin, 34-52.

[7] CARLSON D.A., HAURIE A. & LEIZAROWITZ A., 1991, **Infinite Horizon Optimal Control: Deterministic and Stochastic Systems**, Springer Verlag.

[8] CASS D. & SHELL K., 1976, **The Hamiltonian Approach to Dynamic Economics**, Academic Press.

[9] CLARK C.W., *Restricted Access to Common-property Resources: a Game Theoretic Analysis*, **Dynamic Optimization and Mathematical Economics**, Edited by P. Liu, Plenum Press, New York, New York, 1980.

[10] COURNOT A., 1838,Recherches sur les principes mathématiques de la théorie des richesses, Hachette, Paris.

[11] FEINSTEIN C.D. & LUENBERGER D.G., 1981, *Analysis of the Asymptotic Behaviour of Optimal Control Trajectories: the Implicit Programming Problem*, **SIAM J. Control Optim. 19**, 561-585.

[12] FORGES F., 1986, *An Approach to Communication Equilibria*, Econometrica, Vol. 54, pp. 1375-1385.

[13] FRIEDMAN J.W., **Oligopoly and the Theory of Games**, North-Holland, Amsterdam, Holland, 1977.

[14] GHOSH M.K.,ARAPOSTATHIS A. AND MARCUS S.I., *Optimal Control of Switching Diffusions with Application to Flexible Manufacturing Systems*, **Proceedings of the 30-th IEEE-Conference on Decision and Control**, Brighton, England, 1991.

[15] GREEN E.J. AND PORTER R.H., *Noncooperative Collusion Under Imperfect Price Information* , Econometrica, Vol. 52, 1984, pp. 87-100.

[16] KRYLOV N.V., **Controlled Diffusion Processes**, Springer-Verlag, Berlin, Germany, 1980.

[17] HALKIN H. *Necessary Conditions for Optimal Control Problems with Infinite Horizon*, **Econometrica, Vol. 42**, 267-273.

[18] HAMALAINEN R., HAURIE A., KAITALA V., *Equilibria and Threat in a Fishery Management Game* , **Optimal Control Applications and Methods**, Vol. 6,pp.315-333, 1985.

[19] HAURIE A., 1989, *Duopole et Percées Technologiques:* un *Modèle de Jeu Différentiel Déterministe par Morceaux*, **L'Actualité Économique, Vol. 65-1**, 105-118.

[20] HAURIE A., *Piecewise Deterministic and Piecewise Diffusion Differential games*, in G. Ricci (ed.), **Decison Processes in Economics**, Springer-Verlag, Berlin, Germany, 1991.

[21] HAURIE A., *From Repeated to Differential Games: How Time and Uncertainty Pervade the Theory of Games*, to appear.

[22] A. HAURIE, J.B. KRAWCZYK, M. ROCHE,1992, *Monitoring Cooperative Equilibria in a Stochastic Differential Game*, **Proceedings 31-st IEEE-CDC**, Tucson Arizona, Dec. To appear in JOTA.

[23] HAURIE A. & LEITMANN G., 1984, *On the Global Stability of Equilibrium Solutions for Open-Loop Differential Games*, **Large Scale Systems 6**, 107-122.

[24] HAURIE A. AND LEIZAROWITZ A., *Overtaking Optimal Regulation and Tracking of Piecewise Diffusion Linear Systems*, **SIAM J. Control and Optimization**, Vol. 30, pp. 816-837, 1992.

[25] HAURIE A. & ROCHE M., 1992, *Un modèle de concurrence par la R&D: Calcul d'un équilibre stochastique*, in **Mélanges en l'honneur des professeurs A. Cottier et G. Mentha**, Université de Genève, 1992.

[26] HAURIE A. & ROCHE M., 1991, *On the Computation of Open-Loop Equilibria in a Class of Differential Games with Random Modal Changes*, **Journal of Economic Dynamics and Control**, to appear, 1993.

[27] HAURIE A. AND TOLWINSKI B., *Cooperative Equilibria in Discounted Stochastic Sequential Games*, **Journal of Optimization Theory and Applications**, Vol. 64, pp. 511-535, 1990.

[28] HAURIE A. & VAN DELFT CH., 1991, *Turnpike Properties for a Class of Piecewise Deterministic Systems Arising in Manufacturing Flow Control*, **Annals of Operations Research 29**, 351-374.

[29] HENDERSON J.M. & QUANDT R.E., 1987, **Microéconomie**, Dunod, Paris.

[30] KUSHNER H.J., **Probability Methods for Approximation in Stochastic Control and for Elliptic Equations**, Academic Press, New York, New York, 1977.

[31] MUNRO G.R., *The Optimal Management of Transboundary Renewable Resources*, **Canadian Journal of Economics**, Vol. 12, pp. 355-376, 1979.

[32] PORTER R.H., *Optimal Cartel Trigger Strategies* , **Journal of Economic Theory**, Vol. 29, pp.313-338, 1983.

[33] SELTEN R., *Reexamination of the Perfectness Concept for Equilibrium Points in Extensive Games*, **International Journal of Game Theory**, Vol. 4, pp. 25-55, 1975.

[34] TIROLE J., 1989, **The Theory of Industrial Organization**, MIT Press.

[35] SCHMALENSEE R., 1990, *Inter-Industry Studies of Structure and Performance*, in **Handbook of Industrial organization**, ed. R. Schmalensee & R. Willig, Amsterdam North Holland.

[36] WHITT W., 1980, *Representation and Approximation of Noncooperative Sequential Games*, **SIAM J. Control**, Vol. 18, pp. 33-48.

STABILITY AND SENSITIVITY ANALYSIS OF SOLUTIONS TO INFINITE-DIMENSIONAL OPTIMIZATION PROBLEMS

K. Malanowski
Systems Research Institute
Polish Academy of Sciences
ul.Newelska 6, 01-447 Warszawa, Poland

Abstract

A survey of stability and sensitivity results for the solutions to parameter depenedent cone constrained optimization problems in abstract
Banach spaces is presented. An application to optimal control problems
for nonlinear ordinary differential equations subject to control and state
constraints is given.

1 Introduction

We are going to present some results of stability and sensitivity analysis of solutions to infinite-dimensional constrained optimization problems with application
to optimal control.

In this analysis local properties of the solutions treated as functions of the
parameter are investigated.

Stability analysis concerns continuity (or Lipschitz continuity) of the solutions in a neighborhood of the reference point, while in sensitivity analysis the
differential properties of the solutions are studied.

In view of the fact that in practical optimization problems the values of
the data usually are not known exactly and/or are subject to disturbances,
stability and sensitivity analysis constitutes a crucial element of the so called
post-optimization analysis, which helps to evaluate the practical usefulness of
the obtained results.

Moreover this analysis finds numerous applications in such areas where we
encounter optimization problems depending on parameters. One can mention
here:

- Parameter estimation for dynamical systems ([4, 5, 15, 19, 20, 41, 45]).

- Game theoretical problems ([18, 21, 28, 59, 64, 65, 81]).

- Hierarchical control ([17, 27, 29, 79]).

- Shape optimization of mechanical structures ([37, 80]).

- Stochastic programming ([44, 69, 70, 71, 72, 73, 74, 76]).

- Convergence analysis of finite dimensional approximations to optimal control problems ([2, 10, 22, 23, 35, 47, 48, 54, 82, 83, 84, 85]).

- Convergence analysis of some optimization algorithms :

    - Sequential quadratic programming ([6, 7, 8, 9, 11, 24, 43, 53, 66]),

    - Augmented Lagrangian ([33, 34, 40, 42, 87]),

    - Continuation methods ([1, 14, 30, 31, 52]),

    - Multiple shooting method ([61]).

The organization of this paper is the following.

In Section 2 a class of parameter dependent cone constrained optimization problems in abstract Banach spaces is introduced. Stability, with respect to the parameter of solutions to these problems is discussed in Section 3.

In Section 4 the notion of the *norm discrepancy*, typical for optimal control of nonlinear systems is introduced, and the stability analysis in presence of the norm discrepancy is discussed.

Section 5 is devoted to sensitivity analysis and in Section 6 the abstract stability and sensitivity results are applied to optimal control problems for nonlinear ordinary differential equations subject to pointwise control and state constraints.

2   Optimization Problems in Banach Spaces

In this section we introduce a class of parameter dependent optimization problems in abstract Banach spaces which we will be interested in.

Let H be a Banach space, called the space of parameters. $G \subset H$ is an open set of admissible parameters.

Moreover there are given two Banach spaces Z and Y (the space of arguments and constraints, respectively) as well as two Hilbert spaces $\widehat{Z}$ and $\widehat{Y}$, which will play the role of the so called pivot spaces, i.e., the following inclusions take place

$$Z \subset \widehat{Z} = \widehat{Z}^* \subset Z^*, \qquad Y \subset \widehat{Y} = \widehat{Y}^* \subset Y^*$$

with all embeddings being dense and continuous.

By $(\cdot, \cdot)_Z$ and $(\cdot, \cdot)_Y$ we will denote the inner products in $\widehat{Z}$ and $\widehat{Y}$, extended by continuity to $Z \times Z^*$ and $Y \times Y^*$, respectively. We put

$$X = Z \times Y, \qquad \widehat{X} = \widehat{Z} \times \widehat{Y}.$$

In Y there is given a closed and convex cone K with the vertex at the origin, which induces a partial order in Y. By $K^+ \subset Y^*$ we denote the cone polar to K and by $\widehat{K}$ the closure of K in $\widehat{Y}$.

On $Z \times G$ there are defined two functions:

$$F(\cdot, \cdot) : Z \times G \to \mathbf{R}^1, \qquad \varphi(\cdot, \cdot) : Z \times G \to Y.$$

For $h \in G$ we consider the following parameter dependent problems of optimization:

$$(P_h) \qquad \min_{z \in Z} F(z, h)$$
$$\text{subject to}$$
$$\varphi(z, h) \in K.$$

Let $\bar{h}$ be a fixed reference value of the parameter. We assume:

(I 1) $F(\cdot)$ and $\varphi(\cdot)$ are two times Fréchet differentiable.

(I 2) There exists a (local) solution $z_{\bar{h}}$ of $(P_{\bar{h}})$.

We are going to investigate existence, local uniqueness as well as local Lipschitz continuity and differentiability proprties of the solutions z_h to (P_h), treated as functions of the parameter h.

The main feature of Problems (P_h) is the presence of *inequality type constraints* represented by the cone K. In contrast to *equality type constraints*, the cone constraints creat serious problems in stability and sensitivity analysis and they do not allow to use in this analysis the classical implicit function theorem.

The simplest class of Problems (P_h) are finite-dimensional mathematical programs. Stability and sensitivity of soltions to mathematical programs has been intensively studied, starting with the pioneering papers by Robinson [66], Levitin [49, 50] and Fiacco [25] and the results are fairly complete. There are several monographs devoted to this subject [12, 26, 31, 51].

In this analysis two types of assumptions play crucial role:

- constraint qualifications (CQ) and

- second order sufficient optimality conditions (SSC).

If appropriate assumptions of those types are satisfied then, for h sufficiently close to the reference value $\bar{h}$ there exist locally unique solutions z_h to (P_h) which are Lipschitz continuous and directionally differentiable functions of the parameter.

The investigation of stability and sensitivity of solutions to general cone constrained optimization problems in Banach spaces were initiated much later and the results are by far less complete. There are some specific features of general problems (P_h) which do not allow to extend to them directly the results known for finite-dimensional mathematical programs:

(1) The unit sphere in a Banach space is not a compact set (also in the weak topology). Hence some compacity arguments, used in finite-dimension can not be repeated. It refers in particular to (SSC) (cf. [62]).

(2) In contrast to finite dimensional situations, (CQ) and (SSC) are not automaticly stable under small perturbations. These stability properties are essential in the whole analysis and must be verified. They depends on the choice of topologies in the respective spaces.

(3) In many cases, in particular in optimal control, Problems (P_h) are well defined and differentiable in a stronger topology, while (SSC) is satisfied in a weaker topology. This phenomenon is called the *norm discrepancy* (cf. e.g., [60]) and it creates serious difficulties in stability analysis.

(4) Sensitivity of solutions to (P_h) is closely related to differentiability properties of the mapping of projection onto the cone K, which in turn depend on the geometry of K. Hence this geometry has to be taken into account.

Let us mention the following approaches that were used in stability analysis of (P_h) in recent years:

1) Application of Robinson's [68] (or related) implicit function theorem for generalized equations [3, 7, 8, 16, 23, 24, 39, 55, 56, 57, 58].
   In this approach stability of the Kuhn-Tucker points is investigated using a technique similar to the classical implicit function theorem. The required assumptions are strong. They include, in particular the uniqueness of the Lagrange multipliers.

2) Application of the open mapping theorem for set-values mapps (the Robinson-Ursescu theorem [67, 86]) [13, 75, 77, 78].
   In principle, the assumptions here do not require the uniqueness of the Lagrange multipliers, but they do not assure the existence of the local solutions for small perturbations of the parameter. So, instead of the solutions, ϵ-solutions are investigated.

3) Composite optimization [38].
   In this approach the original smooth constrained optimization problem is reduced to a non-smooth unconstrained problem, where the objective function is a composition of a smooth and a non-smooth convex function. To this new problem a method based on non-smooth analysis is applied.

The needed assumptions are of the form of the so called growth conditions. They are comletly different and weaker than those used in the other approaches.

We restrict ourselves to presenting the first approach, which is the closest to the classical implicit function theorem. It strongly exploits the concept of linear-quadratic approximations to the original nonlinear optimization problems.

3 Stability Analysis

In addition to (I 1) and (I 2) we assume:

(I 3) The point $z_{\bar{h}}$ is regular in the sense of Robinson, i.e.,

$$D_z \varphi(z_{\bar{h}}, \bar{h}) Z - K + [\varphi(z_{\bar{h}}, \bar{h})] = Y,$$

where $[y] = \{ry \mid r \in \mathbf{R}^1\}$.

Let us introduce the Lagrangian associated with (P_h):

$$\mathcal{L}(\cdot, \cdot, \cdot): \ Z \times Y^* \times G_{\bar{h}} \to \mathbf{R}^1, \qquad \mathcal{L}(z, \lambda, h) = F(z, h) - (\lambda, \varphi(z, h))_Y .$$

By (I 3) there exists a Lagrange multiplier $\lambda_{\bar{h}} \in K^+ \subset Y^*$, associated with $z_{\bar{h}}$, such that the Kuhn-Tucker conditions hold at $x_{\bar{h}} = (z_{\bar{h}}, \lambda_{\bar{h}})$.

We are going to investigate existence, local uniqueness and stability with respect to h of Kuhn-Tucker points $x_h = (z_h, \lambda_h) \in Z \times Y^*$ of (P_h), i.e., of points which satisfy the conditions:

$$
\begin{aligned}
D_z \mathcal{L}(z_h, \lambda_h, h) &= D_z F(z_h, h) + D_z \varphi^*(z_h, h)\lambda_h = 0, \\
(\lambda_h, \varphi(z_h, h))_Y &= 0, \qquad \lambda_h \in K^+ \subset Y^*.
\end{aligned}
$$

If a sufficient optimality condition is satisfied at x_h then z_h becomes the solution and λ_h the associated Lagrange multiplier of (P_h).

In this analysis we will use Robinson's implicit function theorem for generalized equations [68]. To this end we need the following linear-quadratic approximation to $(P_{\bar{h}})$ perturbed by the parameter $\delta = (a, b) \in Z^* \times Y$:

$$(LP_\delta) \qquad \min_{w \in \widehat{Z}} \left\{ \tfrac{1}{2}(w, Qw)_Z - (s + a, w)_Z \right\}$$
$$\text{subject to}$$
$$Dw + e + b \in \widehat{K},$$

where

$$
\begin{aligned}
Q &= D_{zz}^2 \mathcal{L}(z_{\bar{h}}, \lambda_{\bar{h}}, \bar{h}), & s &= -D_{zz}^2 \mathcal{L}(z_{\bar{h}}, \lambda_{\bar{h}}, \bar{h}) z_{\bar{h}} + D_z(z_{\bar{h}}, \bar{h}), \\
D &= D_z \varphi(z_{\bar{h}}, \bar{h}), & e &= \varphi(z_{\bar{h}}, \bar{h}) - D_z \varphi(z_{\bar{h}}, \bar{h}) z_{\bar{h}},
\end{aligned}
$$

Following Robinsin [68], we call $(P_{\bar{h}})$ *strongly regular* at the point $(z_{\bar{h}}, \lambda_{\bar{h}})$ if:

(A) There exists $\rho > 0$ such that for each $\delta \in \mathcal{O}^\rho := \{ \delta \in Z^* \times Y \mid \|\delta\|_{Z^* \times Y} \leq \rho \}$ there is a unique pair $(w_\delta, \mu_\delta) \in X^*$, where w_δ is the solution to (LP_δ) and μ_δ - the associated Lagrange multiplier, and (w_δ, μ_δ) is a Lipschitz continuous function of δ.

Strong regularity plays an analogous role to invertibility of the respective derivative in the classical implicit function theorem. The following fundamental result due to Robinson [68] is an extension of this theorem:

THEOREM 3.1 *Suppose that (A) is satisfied, then there exists a neighborhood $G_{\bar{h}} \subset G$ of $\bar{h}$ and a neighborhood $\mathcal{X}_{\bar{h}} \subset X$ of $x_{\bar{h}}$ such that there is a unique in $\mathcal{X}_{\bar{h}}$ Kuhn-Tucker point $x_h = (z_h, \lambda_h)$ of (P_h), which is a Lipschitz continuous function of h.* ◇

Certainly, if a sufficient optimality condition is satisfied at x_h, z_h becomes the solution and λ_h the associated Lagrane multipier of (P_h).

Theorem 3.1 allows to reduce stability analysis of nonlinear problem (P_h) to such analysis of the linear-quadratic problem (LP_δ), which is much simpler.

It is shown in [55] that the strong regularity (A) holds if the following two conditions are *satisfied and stable* under small perturbations:

(I 4) Linear independence of gradients of active constraints:

$$D_z\varphi(z_{\bar{h}}, \bar{h})Z + (-K + [\varphi(z_{\bar{h}}, \bar{h})]) \cap (K + [\varphi(z_{\bar{h}}, \bar{h})]) = Y.$$

(I 5) Strong second order sufficient optimality condition:

$$(D^2_{zz}\mathcal{L}(z_{\bar{h}}, \lambda_{\bar{h}}, \bar{h})w, w)_Z \geq \gamma\|w\|^2_Z$$

for all $w \in \{w \in \widehat{Z} \mid D_z\varphi(z_{\bar{h}}, \bar{h})w \in K + (-K)\}$.

Hence, by Theorem 3.1 we obtain (cf. Theorem 3.10 in [55]):

THEOREM 3.2 *If assumptions (I 1)-(I 5) are satisfied, then there exist a neighborhood $G_{\bar{h}} \subset G$ of $\bar{h}$ and a neighborhood $\mathcal{Z}_{\bar{h}} \subset Z$ of $z_{\bar{h}}$ such that for each $h \in G_{\bar{h}}$ there is a unique in $\mathcal{Z}_{\bar{h}}$ solution z_h of (P_h) and a unique associated Lagrange multipler λ_h. Moreover, there exists a constant $c > 0$ such that*

$$\|z_{h_1} - z_{h_2}\|_Z, \ \|\lambda_{h_1} - \lambda_{h_2}\|_{Y^*} \leq c\|h_1 - h_2\|_H \qquad \textit{for all } h_1, h_2 \in G_{\bar{h}}.$$

◇

4 Norm Discrepancy

We would like to apply Theorem 3.2 to optimal control problems. However, it turns out that condition (I 5) is too strong to be applicable in optimal control problems for nonlinear systems. It is connected with the phenomenon of the so called *norm discrepancy*. To explain the nature of the norm discrepancy let us consider the following simple example.

EXAMPLE **4.1** Let us choose $Z = L^\infty(0,1)$ and $\widehat{Z} = L^2(0,1)$. Consider the integral functional

$$F(z) = \int_0^1 (z^2(t) - 1)^2 dt.$$

It is easy to see that F is well defined and differentiable on Z, but not on $\widehat{Z}$. However for $z \in Z$

$$(D^2_{zz} F(z)y, y) = 4 \int_0^1 (3z^2(t) - 1)y^2(t)dt$$

is a quadratic form well defined and continuous on $\widehat{Z}$.

Any function z, such that $|z(\cdot)| \equiv 1$ is a global minimizer of F on Z. Let us take as the minimizer $\widetilde{z}(\cdot) \equiv 1$. We have

$$(D^2_{zz} F(\widetilde{z})y, y) = 4 \int_0^1 (3\widetilde{z}^2(t) - 1)y^2(t)dt = 8\|y\|_2^2, \tag{1}$$

i.e., the quadratic form is coercive on $\widehat{Z}$. However, it is not coercive on Z. Namely, *does not exist* $c > 0$ such that

$$(D^2_{zz} F(\widetilde{z})y, y) \geq c\|y\|_\infty^2 \quad \text{for all } y \in Z = L^\infty(0,T).$$

Just this phenomenon is called the *norm discrepancy*.

It can be easily checked (cf.[58]) that coercivity condition (1) is stable under small perturbations of $\widetilde{z}$ in $L^\infty(0,1)$, *but it is not stable under such small perturbations in* $L^2(0,1)$. ◇

The example shows that we can not expect to avoid the norm discrepancy in optimal control problems. Therefore we will reformulate assumption (I 5) in such a way to cover problems with the norm discrepancy. To this end we use the Hilbert spaces $\widehat{Z}$ and $\widehat{Y}$ introduced in Section 2. Instead of (I 5) let us assume:

(I 6) $\lambda_{\bar{h}} \in Y$.

(I 5') There exist neighborhoods $\mathcal{Z}_{\bar{h}} \subset Z$ and $\mathcal{Y}_{\bar{h}} \subset Y$ of $z_{\bar{h}}$ and of $\lambda_{\bar{h}}$, respectively such that for each $(z, \lambda) \in \mathcal{Z}_{\bar{h}} \times \mathcal{Y}_{\bar{h}}$ the following coercivity condition is satisfied:
$$(D^2_{zz} \mathcal{L}(z, \lambda, \bar{h})w, w)_Z \geq \gamma\|w\|_{\widehat{Z}}^2$$
for all $w \in \{w \in Z \mid D_z\varphi(z_{\bar{h}}, \bar{h})w \in K + (-K)\}$.

It turns out that (I 5') is not strong enough to assure strong regularity of $(P_{\bar{h}})$. Instead of (A) we get [56]:

$$\|w_{\delta_1} - w_{\delta_2}\|_{\widehat{Z}}, \|\mu_{\delta_1} - \mu_{\delta_2}\|_{\widehat{Y}} \leq c\|\delta_1 - \delta_2\|_H, \tag{2}$$

provided that

$$(w_{\delta_i}, \mu_{\delta_i}) \in \mathcal{Z}_{\bar{h}} \times \mathcal{Y}_{\bar{h}} \qquad i = 1, 2. \tag{3}$$

In general (2) does not imply (3). In order to get (3) we should assure that convergence in the weaker norm $\widehat{X}$ implies convergence in the stronger norm X. This would hold if (w_δ, μ_δ) belong to a certain set Γ compact in X. We can not expect that it would be satisfied for all variations from any ball either in X or in $\widehat{X}$. Therefore, we must restrict ourselves to a certain set Δ of more regular variations. We assume that such a set exists. Namely:

(I 7) There exists a *sufficiently rich*[1] closed and convex set $\Delta \subset \widehat{X}$ and a convex set Γ compact in X such that for any $\delta \in \Delta$, $(w_\delta, \mu_\delta) \in \Gamma$.

using a modification of Robinson's implicit function theorem we obtain (cf. Theorem 5.4 in [56]):

THEOREM 4.2 *If assumptions (I 1)-(I 4), (I 5'), (I 6) and (I 7) are satisfied then there exist a neighborhood $G_{\bar{h}}$ of $\bar{h}$ and a neighborhood $\mathcal{Z}_{\bar{h}} \subset Z$ of $z_{\bar{h}}$ such that for each $h \in G_{\bar{h}}$ there is a unique in $\mathcal{Z}_{\bar{h}}$ solution z_h of (P_h) and a unique associated Lagrange multiplier λ_h. Moreover, there exists a constant $c > 0$ such that*

$$\|z_{h_1} - z_{h_2}\|_{\widehat{\mathcal{Z}}}, \|\lambda_{h_1} - \lambda_{h_2}\|_{\widehat{Y}} \le c\|h_1 - h_2\|_H \quad \text{for all } h_1, h_2 \in G_{\bar{h}}.$$

$\Diamond$

5 Sensitivity Analysis

In sensitivity analysis we need the Hilbert space structure of the constraints space, for we will use the concept of the orthogonal projection onto a closed convex cone. Therefore we will work in the Hilbert spaces $\widehat{Z}, \widehat{Y}$.

By Theorem 2.3 in [68] the solution z_h to (P_h) is directionally differentiable with respect to h at $\bar{h}$ if and only if the solution y_h to the following linear-quadratic approximation (QP_h) of (P_h) is directionally differentiable at $\bar{h}$:

$$(QP_h) \qquad \min_{w \in \widehat{\mathcal{Z}}} \left\{ \tfrac{1}{2}(w, Qw)_Z - (a_h, w)_Z \right\}$$
$$\text{subject to}$$
$$Dw + b_h \in \widehat{K},$$

where

$$a_h = -D_{zz}^2 \mathcal{L}(z_{\bar{h}}, \lambda_{\bar{h}}, \bar{h})z_{\bar{h}} + F(z_{\bar{h}}, \bar{h}) + D_z \mathcal{L}(z_{\bar{h}}, \lambda_{\bar{h}}, h),$$
$$b_h = -D_z \varphi(z_{\bar{h}}, \bar{h})z_{\bar{h}} + \varphi(z_{\bar{h}}, h).$$

The directional derivatives of z_h and y_h coincide. The same refers to the Lagrange multipliers λ_h and μ_h of (P_h) and (QP_h), respectively. Hence, instead

---

[1]See condition (II 7) in [56]

of investigating differentiability of (z_h, λ_h) we can investigate differentiability of (y_h, μ_h).

Dualizing (QP_h), it can be shown (cf. Section 6 in [56]) that

$$\mu_h = P_{\widehat{K}^+} c_h,$$

where c_h is a continuously differentiable function of h and $P_{\widehat{K}^+}$ denote the orthogonal projection onto $\widehat{K}^+$.

Hence the differentiability properties of μ_h depend on such properties of the orthogonal projection onto a closed convex cone in a Hilbert space. As it was shown by Kruskal [46] such a projection may not be even directionally differentiable. Therefore, we restrict ourselves to a class of cones the projection on which is directionally differentiable.

DEFINITION 5.1 (see [36, 63]) A closed convex cone $\widehat{K} \subset \widehat{Y}$, with the vertex at the origin, is called *polyhedric* if

$$\mathcal{K}(y) := \overline{\left(\widehat{K} + \left[P_{\widehat{K}} y\right]\right)} \cap \left[y - P_{\widehat{K}} y\right]^{\perp} = \overline{\left(\widehat{K} + \left[P_{\widehat{K}} y\right]\right) \cap \left[y - P_{\widehat{K}} y\right]^{\perp}}$$

for all $y \in \widehat{Y}$, where $[y]^{\perp}$ denotes the hyperplane orthogonal to $[y]$. ◇

In addition to (I 1)-(I 7) we assume:

(I 8) The cone $\widehat{K}$ is polyhedric.

Using the chracterization of the directional derivative of the orthogonal projection onto a polyhedric cone in a Hilbert space, given in [36] we obtain:

THEOREM 5.2 *If (I 1) - (I 8) are satisfied, then z_h and λ_h are directionally differntiable at $\bar{h}$ in the sense of spaces $\widehat{Z}, \widehat{Y}$, and the directional derivatives $\delta z_{\bar{h},k}$ and $\delta \lambda_{\bar{h},k}$ at $\bar{h}$ in the direction k, are given, respectively, by the unique solution and the associated Lagrange multiplier of the following linear-quadratic problem of optimization:*

$$\min_{y \in \widehat{X}} \left\{ \tfrac{1}{2}(y, Qy)z + (r, y)z \right\}$$
subject to
$$Dy + d \in \mathcal{K}(y_{\bar{h}}),$$

where $r = D_{zh}^2 \mathcal{L}(z_{\bar{h}}, \lambda_{\bar{h}}, \bar{h})k$, $d = D_h \varphi(z_{\bar{h}}, \bar{h})k$. ◇

In sensitivity analysis of optimization problems an important role is played by the so called optimal value function, which is given by

$$F^0(h) = F(z_h, h).$$

Theorem 5.2 implies the following second order expansion of the optimal value function:

COROLLARY **5.3** *If the assumptions of Theorem 5.2 hold, then for any $k \in H$ and any sufficiently small $\alpha > 0$ we have*

$$
\begin{aligned}
F^0(\bar{h} + \alpha k) \;=\;& F^0(\bar{h}) + \alpha D_h \mathcal{L}[\bar{h}]k + \\
& + \frac{\alpha^2}{2} [\delta z_{\bar{h},k}, k] \begin{bmatrix} D_{zz}\mathcal{L}[\bar{h}] & D_{zh}\mathcal{L}[\bar{h}] \\ D_{hz}\mathcal{L}[\bar{h}] & D_{hh}\mathcal{L}[\bar{h}] \end{bmatrix} \begin{bmatrix} \delta z_{\bar{h},k} \\ k \end{bmatrix} + o(\alpha^2),
\end{aligned}
$$

where $\mathcal{L}[\bar{h}] := \mathcal{L}(z_{\bar{h}}, \lambda_{\bar{h}}, \bar{h})$. $\diamond$

6 Optimal Control Problem

In this section the obtained abstract stability results will be applied to state and control constrained optimal control problems for nonlinear ordinary differential equations.

As the space of parameters we choose the Sobolev space $W^{2,\infty}(0, T; \mathbf{R}^p)$:

$$H = W^{2,\infty}(0, T; \mathbf{R}^p) := \{ h \in L^\infty(0, T; \mathbf{R}^p) \mid \ddot{h} \in L^\infty(0, T; \mathbf{R}^p) \}.$$

As before $G \subset H$ denotes an open set of admissible parameters and for each $h \in G$ we consider the following optimal control problem:

(O_h) find $(u_h, x_h) \in L^\infty(0, T; \mathbf{R}^m) \times W^{1,\infty}(0, T; \mathbf{R}^n)$ such that

$F(u_h, x_h, h) = \min_{u,x} \left\{ F(u, x, h) := \int_0^T f^0(u(t), x(t), h(t)) \, dt \right\}$

subject to

$\dot{x}(t) = f(u(t), x(t), h(t))$ for a.a. $t \in [0, T]$,

$x(0) = x_0(h(0))$,

$\theta(u(t), h(t)) \leq 0$ for a.a. $t \in [0, T]$,

$v(x(t), h(t)) \leq 0$ for all $t \in [0, T]$,

where

$$\theta(\cdot, \cdot) : \mathbf{R}^m \times \mathbf{R}^p \mapsto \mathbf{R}^k, \quad v(\cdot, \cdot) : \mathbf{R}^n \times \mathbf{R}^p \mapsto \mathbf{R}^l.$$

In order to represent (O_h) in the form (P_h) we put:

$Z = L^\infty(0, T; \mathbf{R}^m) \times W^{1,\infty}(0, T; \mathbf{R}^n)$,

$\widehat{Z} = L^2(0, T; \mathbf{R}^m) \times W^{1,2}(0, T; \mathbf{R}^n)$,

$Y = L^\infty(0, T; \mathbf{R}^n) \times \mathbf{R}^n \times L^\infty(0, T; \mathbf{R}^k) \times W^{1,\infty}(0, T; \mathbf{R}^l)$,

$\widehat{Y} = L^2(0, T; \mathbf{R}^n) \times \mathbf{R}^n \times L^2(0, T; \mathbf{R}^k) \times W^{1,2}(0, T; \mathbf{R}^l)$.

$K = K_1 \times K_2 \times K_3 \times K_4$,

$K_1 = \{0\}, \qquad K_2 = \{0\}$,

$K_3 = \left\{ u \in L^\infty(0, T; \mathbf{R}^k) \mid u^i(t) \geq 0, \ i = 1, ..., k \ \text{ for a.a. } t \in [0, T] \right\}$,

$K_4 = \left\{ x \in W^{1,\infty}(0, T; \mathbf{R}^l) \mid x^j(t) \geq 0, \ j = 1, ..., l \ \text{ for all } t \in [0, T] \right\}$

$F(z, h) = F(u, x, h)$,

$\varphi(z, h) = \left(\dot{x} - f(u, x, h), \ x(0) - x_0(h), \ -\theta(u, h), \ -v(x, h) \right)$.

In order to apply Theorem 4.2 we have to verify all assumptions of this theorem. To simplify notation let us put:

$$A(t) := D_x f(u_{\bar{h}}(t), x_{\bar{h}}(t), \bar{h}(t)), \qquad B(t) := D_u f(u_{\bar{h}}(t), x_{\bar{h}}(t), \bar{h}(t)),$$
$$\Theta(t) := D_u \theta(u_{\bar{h}}(t), \bar{h}(t)), \qquad \Upsilon(t) := D_x v(x_{\bar{h}}(t), \bar{h}(t)).$$

We assume:

(II 1) There exists a (local) solution $(u_{\bar{h}}, x_{\bar{h}})$ of $(O_{\bar{h}})$, which satisfies the following regularity condition:

$$(u_{\bar{h}}, x_{\bar{h}}) \in C^0(0, T; \mathbf{R}^m) \times C^1(0, T; \mathbf{R}^n),$$

(II 2) f^0, f, θ, v and $D_x v$ are two times Fréchet differentiable in all arguments, and the respective derivatives are locally Lipschitz continuous in u, x; x_0 is Fréchet differentiable.

Certainly (II 1), (II 2) imply (I 1), (I 2). We introduce constraint qualifications:

(II 3) $v^j(x_0(\bar{h}), \bar{h}) < 0$, for $j = 1, 2, ..., l$.

(II 4) There exists $\eta > 0$ such that

$$|[\Theta_a^*(t), B^*(t)\Upsilon_a^*(t)]\zeta| \geq \eta|\zeta| \tag{4}$$

for all ζ of appropriate dimension and for all $t \in [0, T]$, where $\Theta_a(t)$ and $\Upsilon_a(t)$ denote the submatrices of $\Theta(t)$ and $\Upsilon(t)$ that contain all rows corresponding to the constraints active at $u_{\bar{h}}(t)$ and $x_{\bar{h}}(t)$, respectively.

REMARK 6.1 Constraint qualifications of the form (4) were introduced by Hager (cf. (4.6) in [32]). They have the meaning that all gradients of the active control constraints $\theta^i(u_{\bar{h}}(t), \bar{h}) = 0$ and all gradients of the active state constraints $v^j(x_{\bar{h}}(t), \bar{h}) = 0$, transformed into the space $\mathbf{R}^m$ by means of the mapping $B^*(t) : \mathbf{R}^n \to \mathbf{R}^m$, are jointly linearly independent, uniformly on $[0, T]$. $\diamond$

It can be shown (cf. Lemma 4.3 in [57]) that by (II 3) and (II 4) constraint qualifications (I 4) (i.e., also (I 3)) holds.

Let us define the following Lagrangian associated with (O_h):

$$\mathcal{L} : L^\infty(0, T; \mathbf{R}^m) \times W^{1,\infty}(0, T; \mathbf{R}^n) \times (L^\infty(0, T; \mathbf{R}^n))^* \times$$
$$\times \mathbf{R}^n \times (L^\infty(0, T; \mathbf{R}^k))^* \times (W^{1,\infty}(0, T; \mathbf{R}^l))^* \times G \mapsto \mathbf{R}^1,$$
$$\mathcal{L}(u, x, q, \rho, \kappa, \nu, h) = F(u, x, h) + (q, \dot{x} - f(u, x, h)) +$$
$$+ (\rho, x(0) - x_0(h(0)) + (\kappa, \theta(u, h)) +$$
$$+ (\nu(0), v(x(0), h(0))) + (D_x v^*(x, h)\dot{\nu}, f(u, x, h)).$$

By (I 4) there exist unique Lagrange multipliers $(q_{\bar{h}}, \rho_{\bar{h}}, \kappa_{\bar{h}}, \nu_{\bar{h}})$ associated with $(u_{\bar{h}}, x_{\bar{h}})$. Using (II 1)- (II 4) it can be shown (cf. Corollary 4.6 in [32]) that the multipliers are more regular. Namely:

$$q_{\bar{h}} \in C^1(0, T; \mathbf{R}^n), \quad \rho_{\bar{h}} \in \mathbf{R}^n, \quad \kappa_{\bar{h}} \in C^0(0, T; \mathbf{R}^k), \quad \nu_{\bar{h}} \in C^1(0, T; \mathbf{R}^l),$$

i.e., in particular (I 6) holds.

Let us define the following augmented Hamiltonian:

$$\mathcal{H}(t) = f^0\left(u_{\bar{h}}(t), x_{\bar{h}}(t), \bar{h}\right) - \langle q_{\bar{h}}(t), f\left(u_{\bar{h}}(t), x_{\bar{h}}(t), \bar{h}\right)\rangle +$$
$$+ \langle \kappa_{\bar{h}}(t), \theta(u_{\bar{h}}(t), \bar{h})\rangle + \langle D_x v^*(x_{\bar{h}}(t), \bar{h})\dot{\nu}_{\bar{h}}(t), f(u_{\bar{h}}(t), x_{\bar{h}}(t), \bar{h})\rangle.$$

In addition to (II 1) - (II 4) we assume

(II 5) There exists $\gamma > 0$ such that

$$\langle u, D^2_{uu}\mathcal{H}(t)u\rangle \geq \gamma'|u|^2$$

for all $u \in \mathbf{R}^m$ and all $t \in [0, T]$.

(II 6) The following Riccati equation has a solution bounded on $[0, T]$:

$$\dot{Z}(t) = -Z(t)D^2_{qx}\mathcal{H}(t) - \left(D^2_{xq}\mathcal{H}(t)\right)Z(t) - D^2_{xx}\mathcal{H}(t) +$$
$$+ \left(D^2_{xu}\mathcal{H}(t) + Z(t)D^2_{qu}\mathcal{H}(t)\right)\left(D^2_{uu}\mathcal{H}(t)\right)^{-1}\left(D^2_{ux}\mathcal{H}(t) + D^2_{uq}\mathcal{H}(t)Z(t)\right),$$
$$Z(T) = 0$$

It follows from Theorem 5.2 in [60] that if (I 5) and (I 6) are satisfied then:

$$\left([u, x], \begin{bmatrix} D^2_{uu}\mathcal{L}[\bar{h}] & D^2_{ux}\mathcal{L}[\bar{h}] \\ D^2_{xu}\mathcal{L}[\bar{h}] & D^2_{xx}\mathcal{L}[\bar{h}] \end{bmatrix}\begin{bmatrix} u \\ x \end{bmatrix}\right) =$$
$$= \int_0^T\left([u^T(t), x^T(t)]\begin{bmatrix} D^2_{uu}\mathcal{H}(t) & D^2_{ux}\mathcal{H}(t) \\ D^2_{xu}\mathcal{H}(t) & D^2_{xx}\mathcal{H}(t) \end{bmatrix}\begin{bmatrix} u(t) \\ x(t) \end{bmatrix}\right) dt \geq$$
$$\geq \gamma\left(\|u\|_2^2 + \|x\|_{1,2}^2\right),$$

for all pairs $(u, x) \in L^2(0, T; \mathbf{R}^m) \times W^{1,2}(0, T; \mathbf{R}^n)$ satisfying

$$\dot{x}(t) = A(t)x(t) + B(t)u(t),$$
$$x(0) = 0,$$

where $\mathcal{L}[\bar{h}] := \mathcal{L}(u_{\bar{h}}, x_{\bar{h}}, q_{\bar{h}}, \rho_{\bar{h}}, \kappa_{\bar{h}}, \nu_{\bar{h}}, \bar{h})$ and $\gamma > 0$.

It shows that condition (I 5') is satisfied. Hence, to apply Theorem 4.2 it remains to check condition (I 7). To this end, we have to define the linear-quadratic approximations (LO_δ) of (O_δ) corresponding to problems (LP_δ) in Section 3. As the set Δ of regular variations we choose the set of uniformly bounded and uniformly Lipschitz continuous functions of time δ. It is shown in Proposition 6.7 of [57] that the corresponding solutions to (LO_δ) are also uniformly bounded and uniformly Lipschitz continuous functions. Hence, as the set Γ needed in (I 7) we choose the set of uniformly bounded and uniformly Lipschitz continuous functions, which is compact in $Z \times Y$ by the Arzela-Ascoli theorem. It is shown in Section 7 of [57] that the Lipschitz constants in the definitions of Δ and Γ can be chosen in such a way that condition (I 7) is satisfied. Hence all assumptions of Theorem 4.2 hold. On the other hand it was shown in [36] that the cones K_3 and K_4 are polyhedric. Therefore, by Theorems 4.2 and 5.2 we obtain:

THEOREM 6.2 *If (II 1) - (II 6) are satisfied, then there exist a neighborhood $G_{\bar{h}} \subset G_{\bar{h}}$ of $\bar{h}$ and an open ball $\mathcal{D}_{\bar{h}} \subset Z$ about $(u_{\bar{h}}, x_{\bar{h}})$ such that for any $h \in G'_{\bar{h}}$ there exists a solution (u_h, x_h) of (O_h), unique in $\mathcal{D}_{\bar{h}}$. The solutions are Lipschitz continuous and directionally differentiable functions, in the sense of the space $\widehat{Z}$, of h on $G_{\bar{h}}$.*

The directional derivative is given as a unique solution of an auxiliary linear-quadratic optimal control problem.

The associated Lagrange multipliers are defined uniquely and they are Lipschitz continuous and directionally differentiable functions of h.　　　　　　◇

7 Conclusions

1) The presented approach in stability analysis has been motivated by optimal control problems for ordinary differential equations. However, the general results are formulated in terms of cone-constrained optimization problems. Therefore, at least in principle, there is a hope that the area of applications can be extended to a broader class of dynamics such as functional or partial differential equations. It seems that in such an extention, the main difficulty will be the required regularity of the solutions.

2) Further investigations in the field of stability analysis for infinite dimensional optimization problems should be directed towards weakening the constraints qualifications and second order sufficient optimality conditions. They should somehow follow the results known for finite dimensional mathematical programs. In this analysis a specific structure of the considered problems should be taken into account. It would be desirable if the results have constructive character. In this aspect the approach using the multiple shooting method, proposed by Maurer and Pesch [61] is interesting.

3) An important field of further applications is convergence analysis of finite dimensional approximations to optimal control problems as well as of some numerical algorithms of optimization in Banach spaces.

References

[1] Allgower E. L., Georg K., "Numerical Continuation Methods. An Introduction," *Springer Series in Computational Mathematics Vol 13, Springer Verlag, Berlin, 1990.*

122

[2] Alt W., "On the approximation of infinite optimization problems with an application to optimal control problems," *Appl. Math. Optim.* **12** (1984), 15–27.

[3] Alt W., "Stability of solutions for a class of nonlinear cone constrained optimization problems, Part 1: Basic theory," *Numer. Funct. Anal. and Optimiz.* **10** (1989), 1053–1064.

[4] Alt W., "Stability of solutions for a class of nonlinear cone constrained optimization problems, Part 2: Application to parameter estimation," *Numer. Funct. Anal. and Optimiz.* **10** (1989), 1065–1076.

[5] Alt W., "Stability for parameter estimation in two point boundary value problems," *Optimization* **22** (1991), 99–111.

[6] Alt W., "The Lagrange-Newton method for infinite-dimensional optimization problems," *Numer. Funct. Anal. and Optimiz.* **11** (1990), 201–224.

[7] Alt W., "Parametric programming with applications to optimal control and sequential quadratic programming," *Bayreuther Mathematische Schriften* **35** (1990), 1–37.

[8] Alt W., "Stability of Solutions and the Lagrange-Newton Method for Nonlinear Optimization and Optimal Control Problems," (Habilitationsschrift), *Universität Bayreuth, Bayreuth, 1990.*

[9] Alt W., Malanowski K., "The Lagrange-Newton method for nonlinear optimal control problems", *Comput. Optim. and Applications* **2** (1993), 77-100.

[10] Alt W., Mackenroth U., "Convergence of finite element approximations to state constrained comvex boundary control problems," *SIAM J. Control and Optimization* **27** (1987), 718-736.

[11] Alt W., Sontag R. Tröltzsch F., "An SQP method for optimal control of weakly singular Hammerstein integral equations", *Deutsche Forschungsgemeinschaft, SPP "Anwendungsbezogene Optimierung und Steuerung", Report No. 423, Chemnitz, 1993.*

[12] Bank B., Guddat J., Klatte D., Kummer B., Tammer K., "Non-linear Parametric Optimization" *Birkhäser Verlag, Basel, 1983.*

[13] Barbet L., "Lipschitzian properties of the optimal solutionns for parametric variational inequalities in Banach spaces" (to be published)

[14] Bock H. G.,"Zur numerischen Behandlung zuschtandsbeschränkter Steuerungsprobleme mit Mehrzielmethode und Homotopierverfahren, *ZAMM* **57** (1977), T 266–T 268.

[15] Bock H. G., "Randwertproblemmethoden zur Parameteridentifizierung in Systemen nichtlinearer Differentialgleichungen," *Bonner Mathematische Schriften* **183**, Bonn, 1987.

[16] Bonnans J.-F., Sulem A., "Pseudopower expansion of solutions of generalized equations and constrained optimization problems" (to be published).

[17] Brdyś M., "Theory of Hierarchical Contronl Systems for Complex Slowlyvarying Processes," *Politechnika Warszawska, Prace Naukowe, Elektronika, z 47, Warszawa, 1980* (in Polish).

[18] Chao G. S., Friesz T. L., "Spacial price equilibrium sensitivity analysis," *Transportation Research* **18B** (1984), 423-440.

[19] Colonius F., Kunisch K., "Stability for parameter estimation in two point boundary value problems," *Journal für die Reine und Angewandte Mathematik* **370** (1986), 1–29.

[20] Colonius F., Kunisch K., "Stability of perturbed optimization problems with applications to parameter estimation," *Report Nr. 205, Institut für Dynamische Systeme, Universität Bremen, 1989.*

[21] Defermos S., Nagurney A., "Sensitivity analysis for the asymetric network equilibrium problem," *Math. Programming* **28** (1984), 174-184.

[22] Dontchev A. L., "Perturbations, Approximations and Sensitivity Analysis of Optimal Control Systems," *Lecture Notes in Control and Information Sciences Vol. 52, Springer Verlag, Berlin–Heidelberg–New York, 1983.*

[23] Dontchev A. L., Hager W. W., "Lipschitz stability in nonlinear control and optimization" , *SIAM J. Control and Optimization* **31** (1993), 569–603..

[24] Dontchev A. L., Hager W. W., Poore A B., Yang B., "Optimality, stability and convergence in nonlinear control" (to be published).

[25] Fiacco A. V., "Sensitivity analysis for nonlinear programming using penalty methods," *Mathematical Programming* **10** (1976), 287–311.

[26] Fiacco A. V., "Introduction to sensitivity and stability analysis in nonlinear programming," *Academic Press, New York–London, 1983.*

[27] Findeisen W., Bailey F. N., Brdyś M., Malinowski K., Tatjewski P., Woźniak A., "Control and Coordination in Hierarchical Systems," *J.Wiley & Sons, Chichester–New York–Brisbane–Toronto, 1980.*

[28] Friesz T. L., Tobin R. L., Miller T., "Existence theory for spacially competitive network facility location models," *Annals Oper. Res.* **18** (1989), 267-276.

[29] Gahutu W. H., Looze D. P., "Parametric coordination in hierarchical control," *Large Scale Systems* **8** (1985), 33-45.

[30] Gfrerer H., Guddat J., Wacker Hj., "A globally convergent algorithm based on imbedding and parametric optimization," *Computing* **30** (1983), 225-252.

[31] Guddat J., Guerra-Vazques F., Jongen H. Th., "Singularities, Path Following and Jumps," *J.Wiley & Sons, Chichester, 1990.*

124

[32] Hager W. W., "Lipschitz continuity for constrained processes," *SIAM J. Control and Optimization* **17** (1979), 321–337.

[33] Hager W. W., "Approximations to the multiplier methods," *SIAM J. Numer. Anal.* **22** (1985), 16–46.

[34] Hager W. W., "Multiplier methods for nonlinear optimal control," *SIAM J. Numer. Anal.* **27** (1990), 1061–1080.

[35] Hager W. W., Ianculescu, G. D., "Dual approximations in optimal control," *SIAM J. Control and Optimization* **22** (1984), 423–465.

[36] Haraux A., "How to differentiate the projection on a convex set in Hilbert space. Some applications to variational inequalities," *J.Math.Soc. Japan* **29** (1977), 615–631.

[37] Haslinger J., Neittaanmäki P., "Finite Element Approximation for Optimal Shape Design: Theory and Applications," *J. Wiley $ Sons, Chichester, 1988.*

[38] Ioffe A., "On sensitivity analysis of nonlinear programs in Banach spaces: the approach via composite unconstrained optimization," to appear in *SIAM J. Optimization.*

[39] Ito K., Kunisch K., "Sensitivity analysis of solutions to optimization problems in Hilbert spaces with applications to optimal control and estimation," *J. Diff. Equations* **99**, 1-40.

[40] Ito K., Kunisch K., "The augmented Lagrangian method for equality and inequality constraints in Hilbert spaces," *Mathematical Programming* **46** (1990), 341–360.

[41] Ito K., Kunisch K., "The augmented Lagrangian method for parameter estimation in elliptic systems," *SIAM J. Control and Optimization* **28** (1990), 113–136.

[42] Ito K., Kunisch K., "An augmented Lagrangian technique for variational inequalities," *Appl. Math. Optim.* **21** (1990), 223–241.

[43] de Jong J. L., Machielsen K. C. P., "On the application of sequential quadratic programming to state constrained optimal control," *in: IFAC, Control Applications of Nonlinear Programming, Capri, Italy, 1985.*

[44] Kall P., "On approximations and stability in stochastic programming," *in: J. Guddat, H. Th. Jongen, B. Kummer B., F. Nozicka , (eds), Parametric Optimization and Related Topics, Academie Verlag, Berlin, 1987.*

[45] Kelley C. T., Wright S. J., "Sequential quadratic programming for certain parameter identification problems" (to be published).

[46] Kruskal J. B., "Two convex counterexamples: a discontinuous envelope function and a nondifferentiable nearest-point mapping," *Proc. Amer. Math. Soc.* **23** (1969), 697-703.

[47] Lasiecka I., "Boudary control of parabolic systems: finite element approximation," *Appl. Math. Optim.* **6** (1980), 31-62.

[48] Lasiecka I., "Ritz-Galerkin approximation of the time optimal boudary control problem for parabolic system with Dirichlet boundary conditions," *SIAM J. Control and Optimization* **22** (1984), 477-500.

[49] Levitin E. S., "On the local perturbation theory of a problem of mathematical programming in a Banach space," *Soviet Math. Dokl.* **16** (1975), 1354–1358.

[50] Levitin E. S., "Differentiability with respect to a parameter of the optimal value in parametric problems of mathematical programming," *Cybernetics.* **12** (1976), 46–64.

[51] Levitin E. S., " Perturbation Theory in Mathematical Programming and its Applications," *Nauka, Moscow, 1992*, (in Russian), to be published in English by J. Wiley & Sons.

[52] Lundberg B. N., Poore A. B., "Numerical continuation and singularity detection methods for parametric nonlinear programming," *SIAM J. Optim.* **3** (1993), 134-154.

[53] Machielsen K. C. P., "Numerical solution of optimal control problems with state constraints by sequential quadratic programming in function space," *CWI Tract 53, Amsterdam, 1987.*

[54] Malanowski K., "Convergence of approximations vs. regularity of solutions for convex, control-constrained optimal control problems," *Appl. Math. Opt.* **8** (1981), 69-95.

[55] Malanowski K., "Second order conditions and constraint qualifications in stability and sensitivity analysis of solutions to optimization problems in Hilbert spaces," *Appl. Math. Opt.* **25** (1992), 51-79.

[56] Malanowski K., "Two-norm approach in stability and sensitivity analysis of optimization and optimal control problems," to appear in *Advances in Math. Sc. Appl.*

[57] Malanowski K., "Stability and sensitivity of solutions to nonlinear optimal control problems" , to appear in *Appl. Math. Optim.*.

[58] Malanowski K., "Regularity of solutions in stability analysis of optimization and optimal control problems." , to appear in *Control and Cybernetics.*

[59] Marcotte P., "Network design with congestion effects: a case of bilevel programming," *Math. Programming* **34** (1986), 142-162.

[60] Maurer H., "First and second order sufficient optimality conditions in mathematical programming and optimal control," *Mathematical Programming Study* **14** (1981), 163–177.

[61] Maurer H., Pesch H. J., "Solution differentiability for nonlinear parametric control problems," *Deutsche Forschungsgemeinschaft, SPP "Anwendungsbezogene Optimierung und Steuerung," Report No. 316, München, 1991.*

[62] Maurer H., Zowe J., "First- and second order sufficient optimality conditions for infinite-dimensional programming problems," *Math. Programming* **16** (1979), 98-110.

[63] Mignot F., "Contrôle dans les inéquations variationelles elliptiques," *J. Funct. Anal.* **22** (1976), 130–185.

[64] Outrata J. V., "On the numerical solution of a class of Stackelberg problems," *Zeit.Oper.Res.* **4** (1990), 255–278.

[65] Outrata J. V., "On necessary optimality conditions for Stackelberg problems," *J. Optim. Theory Appl.* **76** (1993), 305–320.

[66] Robinson S. M., "Perturbed Kuhn-Tucker points and rates of convergence for a class of nonlinear-programming algorithms," *Mathematical Programming* **7** (1974), 1–16.

[67] Robinson S. M., "Stability theory for system of inequalities, Part II: Differentiable nonlinear systems, *SIAM J. Numer. Anal.* **13** (1976), 497-513.

[68] Robinson S. M., "Strongly regular generalized equations," *Mathematics of Operatios Research* **5** (1980), 43–62.

[69] Robinson S. M., Wets R. J.-B., "Stability in two-stage stochasic programming," *SIAM J. Control and Optimization* **25** (1987), 1409-1416.

[70] Römisch W., Schultz R., "Stability analysis for stochastic programs," *Annals of Operations Research* **30** (1991), 241-266.

[71] Römisch W., Schultz R., "Lipschitz stability for stochastic programs with complete recourse," *Deutsche Forschungsgemeinschaft, SPP "Anwendungsbezogene Optimierung und Steuerung," Report No. 408, Berlin, 1992d.* (to be published).

[72] Shapiro A., "Asymptotic properties of statistical estimators in stochastic programming," *The Annals of Statistics* **17** (1989), 841-858.

[73] Shapiro A., "On differential stability in stochastic programming," *Math. Programming* **47** (1991), 107-186.

[74] Shapiro A., "Asymptotic analysis of stochastic programs," *Annals of Oper. Res.* **30** (1991), 169-186.

[75] Shapiro A., "Perturbation analysis of optimization problems in Banach spaces," *Numerical Funct. Anal. Optim.* **13** (1992), 97–116.

[76] Shapiro A., "Asymptotic behavior of optimal solutions in stochastic programming," to appear in *Math. Oper. Res.*.

[77] Shapiro A., "Sensitivity analysis of parametrized programs via generalized equations," (to be published).

[78] Shapiro A., Bonnans J. F., "Sensitivity analysis of parametrized programs under cone constraints," *SIAM J. Control and Optimization* **30** (1992), 1409–1421.

[79] Shimizu K., Ishizuka Y., "Optimality conditions and algorithms for parameter design problems with two-level structure," *IEEE Trans. Aut. Contr.* **AC-30** (1985), 986-993.

[80] Sokołowski J., Zolesio J.-P., "Sensitivity Analysis in Shape Optimization," *Springer Series in Computational Mathematics, Vol 16, Springer-Verlag, Berlin, 1992.*

[81] Tobin R. L., Friesz T. L., "Spacial competition facility location models: definition, formulation and solution approach," *Annals Oper. Res.* **6** (1986), 49-74.

[82] Tröltzsch F., "Semidiscrete finite element approximation of parabolic boundary control problems-convergence of switching points," *in: Optimal Control of Partial Differential Equations II, Int. Ser. Num. Math., Vol. 78, Birkhäser, Basel, 1987.*

[83] Tröltzsch F., "Approximation of nonlinear parabolic boundary control problems by the Fourier method — convergence of optimal controls," *Optimization* **22** (1991), 83–98.

[84] Tröltzsch F., "Semidiscrete Ritz-galerkin approximation of nonlinear parabolic boundary control problems-strong convergence of optimal controls," *Deutsche Forschungsgemeinschaft, SPP "Anwendungsbezogene Optimierung und Steuerung," Report No. 325, Augsburg, 1991.*

[85] Tröltzsch F., "On convergence of semidiscrete Ritz-Galerkin schemes applied to boundary control problems of parabolic equations with non-linear boundary conditions," *ZAMM* **72** (1992), 291-301.

[86] Ursescu C., "Multifunctions with closed convex graph," *Czechoslovak Math. J.* **25** (1975), 438-441.

[87] Yang B., "Some Numerical Methods for a Class of Nonlinear Optimal Control Problems," (dissertation) *Colorado State University, Fort Collins, Colorado, 1991.*

APPROXIMATE CONTROLLABILITY FOR SOME NONLINEAR PARABOLIC PROBLEMS

J.I.Díaz

Departamento de Matemática Aplicada
Universidad Complutense de Madrid
28040 Madrid Spain

1 Introduction.

The main goal of this work is to present several results on the controllability of some nonlinear parabolic problems, mainly the semilinear problem

$$\text{(SL)} \quad \begin{cases} y_t - \Delta y + f(y) = v\chi_\omega & \text{in } Q = \Omega \times (0,T) \\ y = 0 & \text{on } \Sigma = \partial\Omega \times (0,T) \\ y(\cdot,0) = y_0(\cdot) & \text{on } \Omega \end{cases}$$

where Ω is a bounded regular set of $\mathbb{R}^n$, ω is an open subset of Ω, χ_ω denotes the characteristic function of ω, $T > 0$ is fixed and the initial datum y_0 is given in a functional space, *e.g.* $y_0 \in L^2(\Omega)$.

The nonlinear term is given by the real function f and the control is represented by the function $v \in L^2(\omega \times (0,T))$. As usual, the study of the semilinear problem is carried out by considering previously some suitable linear problem

$$\text{(L)} \quad \begin{cases} y_t - \Delta y + ay = v\chi_\omega & \text{in } Q \\ y = 0 & \text{on } \Sigma \\ y(\cdot,0) = y_0(\cdot) & \text{on } \Omega, \end{cases}$$

where $a \in L^\infty(Q)$ is given.

Due to the smoothing effect of parabolic equations the notion of controllability (*exact controllability*) must be relaxed: We say that the *approximate controllability property* holds for the problem (SL) (respectively (L)) if given $y_d \in L^2(\Omega)$ and $\varepsilon > 0$ there exist $v \in L^2(\omega \times (0,T))$ and $y(T : v)$ solution of (SL) (respectively (L)) satisfying

$$\|y(T : v) - y_d\|_{L^2(\Omega)} \le \varepsilon.$$

We start, in Section 2, by collecting some abstract and constructive proofs of the approximate controllability for the linear problem (L). The nonlinear case may yield different answers according to the behaviour of the function f near the infinity. This is presented in Section 3 jointly with some remarks about other nonlinear problems.

2 The approximate controllability for linear problems.

The study of the approximate controllability for linear parabolic problems has been developped on different levels of abstraction (a survey containing many references up to 1978 is due to D.Russell [30]). A very elegant proof of this property for the formulation (L) is due to J.L.Lions

Theorem 1 ([22])
The approximate controllability holds for problem (L).

Proof.
By linearity we can assume $y(T : 0) \equiv 0$. Let us show that if $g \in L^2(\Omega)$ satisfies

$$(y(T : v), g) = 0 \quad \forall v \in L^2(\omega \times (0, T)), \tag{1}$$

then necessarily $g \equiv 0$. In that case the conclusion comes from a corollary of the Hahn-Banach Theorem. Here $(\cdot, \cdot)$ denotes the scalar product over $L^2(\Omega)$. Define ψ as the unique solution of the time-reversed problem

$$\begin{cases} -\psi_t - \Delta\psi + a\psi = 0 & \text{in } Q \\ \psi = 0 & \text{on } \Sigma \\ \psi(\cdot, T) = g(\cdot) & \text{on } \Omega. \end{cases}$$

Multiplying by y, integrating by parts and using (1) we get that

$$\int\int_{\Omega \times (0,T)} \psi v \chi_\omega dx dt = 0 \quad \forall v \in L^2(\omega \times (0, T)).$$

In particular, $\psi \equiv 0$ on $\omega \times (0, T)$. Using the Unique Continuation Theorem (due to Mizohata [28] for $a \in C^\infty(Q)$ and Saut-Scheurer [31] for $a \in L^\infty(Q)$) we deduce that $\psi \equiv 0$ on $\Omega \times (0, T)$ and so $g \equiv 0$ on Ω. $\square$

Other variants of the Hahn-Banach Theorem can be used to prove approximate controllability results under some contraints on the controls and/or the state. For instance, in many physical applications only nonnegative controls are admissible. The density of the range set $\{y(\mathrm{T}:v)\}$ when the control acts on Σ was first proved in Díaz [3]. A similar result for the formulation (L) is the following

Theorem 2 ([8])
Let $\mathcal{U}$ be a dense subset of

$$\mathrm{L}_+^2(\omega \times (0,\mathrm{T})) \left(:= \left\{v \in \mathrm{L}^2(\omega \times (0,\mathrm{T})) : v \geq 0 \; a.e.\right\}\right).$$

Then $\{y(\mathrm{T}:v) : y$ solution of (L) , $v \in \mathcal{U}\}$ is a dense subset of $y(\mathrm{T}:0) + \mathrm{L}_+^2(\Omega)$.

Proof.
We start by giving a proof for the case $\omega = \Omega$. Again, without loss of generality we can assume $y(\mathrm{T}:0) \equiv 0$. By linearity $F := \{y(\mathrm{T}:v) : y$ solution of (L), $v \in \mathcal{U}\}$ is such that $\overline{F}$ is a convex set. Then, assumed that there exists $y_1 \in \mathrm{L}_+^2(\Omega) \setminus \overline{F}$, by the Hahn-Banach Theorem (in its geometrical form) we can separate y_1 from $\overline{F}$, i.e. there exists $\alpha \in \mathbb{R}$ and $g \in \mathrm{L}^2(\Omega)$ such that

$$(y(\mathrm{T}:v),g) < \alpha < (y_1,g) \quad \forall v \in \mathcal{U}. \tag{2}$$

Taking $v \equiv 0$ we deduce that $\alpha > 0$. If ψ is given as in the previous proof, multiplying by y, integrating by parts and using (2) we conclude that

$$\int_Q \psi v \mathrm{d}x \mathrm{d}t \leq 0 \quad \forall v \in \mathcal{U}.$$

Then $\psi \leq 0$ in Q which implies $g \leq 0$: A contradiction with (2).
A sketch of the proof for the general case $\omega \subset \Omega$ is as follows: assume that there exists a $g \in \mathrm{L}_+^2(\Omega)$ such that $g \notin \overline{F}$. By the Projection Theorem there exists a unique $u \in \overline{F}$ such that

$$(g-u, p-u) \leq 0 \quad \forall p \in \overline{F}.$$

Moreover, as $\overline{F}$ is a convex closed cone we can take $p = e + u$ and $p = 0$ respectively and obtain

$$(g-u, e) \leq 0 \quad \forall e \in \overline{F} \tag{3}$$

$$(g-u, u) = 0. \tag{4}$$

Now, let $q \in \mathcal{C}([0,\mathrm{T}] : \mathrm{L}^2(\Omega))$ be the solution of the problem

$$\begin{cases} -q_t - \Delta q + aq = 0 & \text{in } Q \\ q = 0 & \text{on } \Sigma \\ q(\cdot, \mathrm{T}) = g(\cdot) - u(\cdot) & \text{on } \Omega. \end{cases} \tag{5}$$

Multiplying (5) by z, with $z \in F$ arbitrary, we obtain

$$0 \geq \int_{\Omega} (g(x) - u(x)) z(x, T) dx = \int_{\omega \times (0,T)} qv dx dt$$

for any $v \in \mathcal{U}$. From the assumption on $\mathcal{U}$ we deduce that $q \leq 0$ on $\omega \times [0, T]$. In particular,

$$0 \leq g(x) \leq u(x) \quad a.e. \ x \in \omega$$

and

$$q \leq 0 \quad \text{on } \partial\omega \times (0, T).$$

Then by the Strong Maximum Principle (on the domain $\omega \times (0, T)$) we deduce that either $q \equiv 0$ on $\Omega \times (0, T)$ or $q < 0$ on $\omega \times (0, T)$. But $q \equiv 0$ implies that $g = u$ which contradicts that $g \notin \bar{F}$. Moreover, we have

$$0 = (g - u, u) = \int_{\omega \times (0,T)} qv_0 dx dt,$$

where we can assume, without loss of generality, that $u = y(T : v_0)$, with $v_0 \in \mathcal{U}$. Now, if $q < 0$ on $\omega \times (0, T)$ we conclude that $v_0 \equiv 0$ on $\omega \times (0, T)$. This implies that $u \equiv 0$ on Ω and from (3) we have that $g \leq 0$ on Ω, *i.e.* $g \equiv 0$ which is a contradiction. $\square$

The rest oh this section will be devoted to the question of the construction of a sequence of control $\{v_k\}_{k \in \mathbb{N}}$ satisfying that $y(T : v_k) \to y_d$ as $k \to \infty$. A first idea introduced in Lions [23] is to consider the auxiliary control problem

$$(\mathcal{P}_k) \quad \begin{cases} \inf\{J_k(v) : v \in L^2(\omega \times (0, T))\}, \\ J_k(v) = \frac{1}{2}\|v\|^2_{L^2(\omega \times (0,T))} + \frac{k}{2}\|y(T : v) - y_d\|^2_{L^2(\Omega)}. \end{cases}$$

Theorem 3 ([23])

Assume (for simplicity) $y_0 \equiv 0$. Then a) Problem $(\mathcal{P}_k)$ has a unique solution v_k and $y(T : v_k) \to y_d$ as $k \to \infty$. b) We have the characterization $v_k = -kp_k\chi_\omega$ where (y_k, p_k) satisfies the optimality system

$$(\mathcal{P}_k^*) \quad \begin{cases} y_t - \Delta y + ay + kp\chi_\omega = 0 & \text{in } Q \\ -p_t - \Delta p + ap = 0 & \text{in } Q \\ y = p = 0 & \text{on } \Sigma \\ y(\cdot, 0) = 0, \ p(\cdot, T) = y(\cdot, T) - y_d(\cdot) & \text{on } \Omega. \end{cases}$$

Idea of the proof.

a) The existence and uniqueness of v_k solution of $(\mathcal{P}_k)$ follow from well-known results ([22]). By Theorem 1 given $\varepsilon > 0$ there exists $v_\varepsilon \in L^2(\omega \times (0,T))$ such that

$$\|y(T:v) - y_d\|_{L^2(\Omega)} \leq \frac{\varepsilon}{2}.$$

Then, as $J_k(v_k) \leq J_k(v_\varepsilon)$ we have

$$k\|y(T:v_k) - y_d\|^2_{L^2(\Omega)} \leq \|v_\varepsilon\|^2_{L^2(\omega \times (0,T))} + \frac{k\varepsilon^2}{4}$$

and so $y(T:v_k) \to y_d$ in $L^2(\Omega)$ as $k \to \infty$.

b) Next, it is enough to remark that the Euler equation associated to $(\mathcal{P}_k)$ is

$$\int_{\omega \times (0,T)} v_k v \, dx dt + k \int_\Omega (y(T:v_k) - y_d) y(T:v) dx = 0 \quad \forall v \in L^2(\omega \times (0,T))$$

and that this is satisfied for the function $-kp_k\chi_\omega$ assuming that (y_k, p_k) satisfies $(\mathcal{P}_k^*)$. □

Remark 1

System (P_k^*) can be treated directly *i.e.* without using the fact that (P_k^*) is the optimality system of the problem (P_k). So, in Lions [23] the existence and uniqueness of a solution (y_k, p_k) of (P_k^*) are shown, as well as that $y_k(T) \to y_d$ in $L^2(\Omega)$ as $k \to \infty$. We also remark that the system (P_k^*) remains still the optimality system of the problem

$$(\mathcal{P}_k^+) \qquad \inf\{J_k(v) : v \in L_+^2(\omega \times (0,T))\}.$$

This is shown in Díaz-Henry-Ramos [8]. The statement of Theorem 3 remains the same. The proof of b) uses the fact that the Euler equation of $(\mathcal{P}_k^+)$ becomes now the variational inequality

$$\int_{\omega \times (0,T)} v_k(v - v_k) dx dt + k \int_\Omega (y(T:v_k) - y_d)(y(T:v) - y(T:v_k)) \, dx \geq 0$$

$\forall v \in L_+^2(\omega \times (0,T))$. □

A second constructive method use some *duality arguments* which are inspired on the HUM (Hilbert Uniqueness Method), introduced by J.L.Lions for the study of the exact controllability. We start by formulating the approximate controllability property in the following terms: *Given $\varepsilon > 0$ and $y_d \in L^2(\Omega)$ find a control $v \in L^2(\omega \times (0,T))$ such that $y(T:v) \in y_d + \varepsilon B$, where B denotes the unit ball in $L^2(\Omega)$.*

As pointed out in Lions [25], it is easy to see that, as a matter of fact, there are *infinitely many controls v* driving the system from the initial datum y_0 to the ball $y_d + \varepsilon B$

at time T. Indeed, let $\delta \in (0, T)$ arbitrary. We take $v = \bar{v}_\delta$ arbitrary in $L^2(\omega \times (0, \delta))$. Let $y_\delta = y(\delta : v)$. Then, according Theorem 1 there exists a control $\hat{v}_\delta \in L^2(\omega \times (\delta, T))$ driving the system from y_δ to $y_d + \varepsilon B$. Then

$$v_\delta(x, t) = \begin{cases} \bar{v}_\delta(x, t) & \text{for } x \in \omega \text{ and } 0 < t < \delta, \\ \hat{v}_\delta(x, t) & \text{for } x \in \omega \text{ and } \delta \leq t < T \end{cases}$$

satisfies the required property (it leads to the system for y_0 to a state $y(T : v_\delta)$ in $y_d + \varepsilon B$). In consequence, it is then natural to ask for the optimal control driving the system from y_0 to the ball $y_d + \varepsilon B$. The problem possed in Lions [25] is the following: *Given $\varepsilon > 0$ and $y_d \in L^2(\Omega)$ find*

$$(\mathcal{P}_\varepsilon) \qquad \inf\{\|v\|_{L^2(\omega \times (0,T))} : y(T : v) \in y_d + \varepsilon B\}.$$

If $\|y_d\|_{L^2(\Omega)} \leq \varepsilon$ this problem has the trivial solution $v = 0$ (since $y(T : 0) = 0 \in y_d + \varepsilon B$). So, in what follows we assume that

$$\|y_d\|_{L^2(\Omega)} > \varepsilon.$$

Theorem 4 ([25])

Problem $(\mathcal{P}_\varepsilon)$ has a unique solution $v_\varepsilon \in L^2(\omega \times (0, T))$. Moreover, $v_\varepsilon = \hat{\varrho}\chi_\omega$, where $\hat{\varrho}$ is the unique solution of the auxilary problem

$$\left. \begin{array}{ll} -\varrho_t - \Delta\varrho + a\varrho = 0 & \text{in } Q \\ \varrho = 0 & \text{on } \Sigma \\ \varrho(\cdot, T) = \varrho_0(\cdot) & \text{on } \Omega \end{array} \right\} \tag{6}$$

and $\varrho_0 = \hat{\varrho}_0$ is given by the minimization problem

$$\begin{cases} \inf\{I(\varrho_0, y_d, \varepsilon) : \varrho_0 \in L^2(\Omega)\}, \\ I(\varrho_0, y_d, \varepsilon) = \frac{1}{2}\int_{\omega \times (0,T)} \varrho^2 dx dt + \varepsilon\|\varrho\|_{L^2(\Omega)} - \int_\Omega y_d\varrho_0 dx. \end{cases} \tag{7}$$

(Here ϱ is the solution of the problem (6)).

Idea of the proof.

Define the functionals

$$F(v) = \frac{1}{2}\int_{\omega \times (0,T)} v^2 dx dt, \quad G(v) = \begin{cases} 0 & \text{if } f \in y^1 + \varepsilon B \\ +\infty & \text{otherwise} \end{cases}$$

$$L \in \mathcal{L}\left(L^2(\omega \times (0, T)) : L^2(\Omega)\right), \qquad Lv = y(T : v).$$

Then problem $(\mathcal{P}_\varepsilon)$ is equivalent to

$$\inf\{F(v) + G(Lv) : v \in L^2(\omega \times (0,T))\}.$$

Using the Fenchel-Rockafellar Duality Theorem (see *e.g.* Ekeland-Temam [18]) we have that

$$\inf\{F(v) + G(Lv) : v \in L^2(\omega \times (0,T))\} = -\inf\{F^*(L^*\varrho_0) + G^*(-\varrho_0) : \varrho_0 \in L^2(\Omega)\},$$

where in general ϕ^* denotes the convex dual of a proper function $\phi : H \to]-\infty, +\infty]$ on a Hilbert space H. It is not dificult to see that

$$L^*\varrho_0 = -\varrho\chi_\omega, \quad G^*\varrho_0 = \int_\Omega y_d\varrho_0 dx + \varepsilon\|\varrho_0\|_{L^2(\Omega)}, \quad F^* = F,$$

and the conclusion holds. $\square$

This second constructive method was sistematically developed in Fabré-Puel-Zuazua [12],[13]. By introducing suitable variants of the functional $I(\varrho_0, y_d, \varepsilon)$ and studying the associated minimization property they obtain the approximate controllability in $L^p(\Omega)$ for $1 \leq p < \infty$ and $\mathcal{C}_0(\Omega)$. Moreover they show that the wanted controls are of the type 'quasi bang-bang' (*i.e.* they take only the values $-k$ and k, for some suitable $k > 0$, except a set of points which at least has empty interior).

Theorem 5 ([12],[13])
Let $a \in L^\infty(Q)$ and denote by $\mathcal{X} = L^p(\Omega)$ with $1 \leq p < \infty$ either $\mathcal{X} = \mathcal{C}_0(\Omega)$ (the space of uniformly continuous functions in Ω that vanish on $\partial\Omega$ endowed with the norm of supremum) and $\mathcal{X}' = L^{p'}(\Omega)$ ($\frac{1}{p} + \frac{1}{p'} = 1$ if $1 < p < \infty$, $p' = \infty$ if $p = 1$) either $\mathcal{X}' = \mathcal{M}(\Omega)$ (the space of bounded measures on Ω) respectively. Let $y_d \in \mathcal{X}$ such that $\|y_d\| > \varepsilon$. Given $\varrho_0 \in \mathcal{X}'$ consider the auxiliar problem

$$\left. \begin{array}{ll} -\varrho_t - \Delta\varrho + a\varrho = 0 & in\ Q \\ \varrho = 0 & on\ \Sigma \\ \varrho(\cdot, T) = \varrho_0(\cdot) & on\ \Omega \end{array} \right\} \tag{8}$$

and define the functional

$$J(\varrho_0, y_d, \varepsilon) = \frac{1}{2}\left(\int_{\omega\times(0,T)} |\varrho| dx dt\right)^2 + \varepsilon\|\varrho_0\|_{\mathcal{X}'} - <y_d, \varrho_0>_{\mathcal{X}\mathcal{X}'}.$$

Then:
1. $J(\cdot, y_d, \varepsilon)$ is a real strictly convex continuous and coercive function. In particular, it achieves its minimum at a unique point $\hat\varrho_0 \in \mathcal{X}'$.

2. If $\hat{\varrho}$ denotes the solution of (8) for $\varrho_0 = \hat{\varrho}_0$ there exists $w \in \text{sign}(\hat{\varrho})\chi_\omega$ such that the solution of

$$\begin{aligned}
y_t - \Delta y + ay &= |\hat{\varphi}|_{L^1(\omega \times (0,T))} w\chi_\omega &&\text{in } Q \\
y &= 0 &&\text{on } \Sigma \\
y(\cdot, 0) &= 0 &&\text{on } \Omega
\end{aligned}\right\}$$

satisfies that $\|y(T) - y_d\|_X \leq \varepsilon.$ □

Remark 2

In Fabré-Puel-Zuazua [14] the optimal control problem $(\mathcal{P}_\varepsilon)$ associated to the L^2-approximate controllability is studied but assuming that $v \in L^r(\omega \times (0,T))$, $2 \leq r \leq +\infty$ and replacing $\|v\|_{L^2(\omega \times (0,T))}$ by $\|v\|_{L^r(\omega \times (0,T))}^2$. □

Remark 3

Many of the above results remain valid for other linear parabolic problems. This is the case of the Stokes problem

$$\begin{aligned}
\vec{y}_t - \Delta\vec{y} &= -\nabla\vec{y} + \vec{v}\chi_\omega &&\text{in } Q \\
\text{div } \vec{y} &= 0 &&\text{in } Q \\
\vec{y} &= \vec{0} &&\text{on } \Sigma \\
\vec{y}(\cdot, 0) &= \vec{y}_0(\cdot) &&\text{in } \Omega
\end{aligned}\right\}$$

where $\vec{v} \in (L^2(\omega \times (0,T)))^n$. The approximate controllability is now formulated as the density of the set $\left\{\vec{y}(T : \vec{v}) : \vec{v} \in (L^2(\omega \times (0,T)))^n\right\}$ in $H = \left\{\vec{w} \in (L^2(\Omega))^n : \text{div } \vec{w} = 0\right\}$. We point out that the property holds even for controls $\vec{v}$ of the type $\vec{v} = (v_1, v_2, 0)$ (see Lions [27] and Fursikov-Imanuvilov [17]). The very special case of the controls $\vec{v} = (v_1, 0, 0)$ leads also to a positive answer for suitable domains Ω of $\mathbb{R}^3$ (Díaz-Fursikov [7]). □

3 The approximate controllability for nonlinear problems.

In order to fix ideas we shall study the approximate controllability for the semilinear problem (SL) of the Introduction. Results for other nonlinear problems will be mentioned at the end of this Section.

As we shall see below, the results are of different nature according to whether the domain of controllability ω satisfies $\omega = \Omega$ or $\omega \subset\subset \Omega$.

3.1 The special case $\omega \equiv \Omega$.

The most favorable situation for which the approximate controllability holds correspond to when we can introduce arbitrary actions (controls) at any point of the domain. In that case it is possible to give a positive answer even in the case for which the existence and uniqueness of the solutions are not assured by the general theory. As pointed out at the Introduction, given y_d and $\varepsilon > 0$ the approximate controllability holds if we find a control v_ε and a function y_ε such that i) y_ε satisfies (SL) and ii) $\|y_\varepsilon(T) - y_d\| \leq \varepsilon$. Thus we merely need to justify the existence of a solution y_ε corresponding to the control v_ε but not the existence and uniqueness for an arbitrary control $v \in L^2(0, T : L^2(\Omega))$.

Theorem 6 ([6])
Let $y_0 \in L^\infty(\Omega) \cap H_0^1(\Omega)$ and assume $f : \mathbb{R} \to \mathbb{R}$ to be continuous. Then the approximate controllability property holds for the problem (SL).

Proof.
Given $y_d \in L^2(\Omega)$ there exist two regular functions u_ε and z_ε such that $u_\varepsilon \in L^2(Q)$ z_ε satisfies (L) with $a \equiv 0$ $v = u_\varepsilon$ and verifies $\|z_\varepsilon(T, \cdot) - y_d\|_{L^2(\Omega)} \leq \varepsilon$. From standard regularity results we know that $z_\varepsilon \in L^2(0, T : H^2(\Omega)) \cap H^1(0, T : L^2(\Omega)) \cap L^\infty(Q)$. Then defining $v_\varepsilon = (z_\varepsilon)_t - \Delta z_\varepsilon + f(z_\varepsilon)$ and $y_\varepsilon = z_\varepsilon$ we have that $v_\varepsilon \in L^2(Q)$ and y_ε satisfies the required condition. □

Remark 4
Theorem 6 admits an arbitrary version ([6]) which is of special interest when the general theory does not assure the global existence of solutions (as, for instance, is the case if $f(s) = -|s|^{p-1}s$ with $p > 1$) or the uniqueness of solutions (case of $f(s) = -|s|^{p-1}s$ with $0 < p < 1$ or the three-dimensional Navier-Stokes problem). A pionneering result (assumming some additional conditions) can be found in Henry [19]. □

A more complicated situation arises when the controls (even actuating in the whole domain) are subject to some constraints. Here we adapt the so-called *cancellation method*, introduced in Henry [19], to the case of nonnegative controls.

Theorem 7 ([8])
Let f be a continuous nondecreasing (or Lipschitz continuous) function such that $f(0) = 0$. Let $\mathcal{U}$ be a dense subset of $L_+^2(\Omega)$. Then the set $\{y(T : v) : y$ solution of (SL) and $v \in \mathcal{U}\}$ is a subset dense of $y(T : 0) + L_+^2(\Omega)$.

Proof.
Without loss of generality we can assume $y_0 \equiv 0$ and thus $y(T : 0) = 0$. By Theorem 2, given $\varepsilon > 0$ there exists $u_\varepsilon \in \mathcal{U}$ such that the solution y of (L), with $a \equiv 0$ and $y_0 \equiv 0$, satisfies $\|y(T : u_\varepsilon) - y_d\|_{L^2(\Omega)} \leq \varepsilon$. Now, let $\hat{u}_\varepsilon \in L_+^\infty(Q)$ with $\|u_\varepsilon - \hat{u}_\varepsilon\|_{L^2(Q)}$ small enough

so that $\|\mathbf{y}(T : u_\varepsilon) - \mathbf{y}(T : \hat{u}_\varepsilon)\|_{L^2(\Omega)} \leq \varepsilon$. From the assumptions on f we know that $f(\mathbf{y}(\cdot : \hat{u}_\varepsilon)) \in L^\infty(Q)$: indeed, if f is not increasing we introduce the change of unknown $\hat{y} = e^{\lambda t} y$ for a suitable $\lambda \in \mathbb{R}$. Now let $v_\varepsilon \in \mathcal{U}$ such that

$$\|v_\varepsilon - f(\mathbf{y}(\cdot : \hat{u}_\varepsilon))\|_{L^2(\Omega)} \leq \varepsilon.$$

Finally, we consider $\tilde{y}$ solution of the auxiliar nonlinear problem

$$\begin{aligned}
\tilde{y}_t - \Delta \tilde{y} + f(\mathbf{y}(\cdot : \tilde{u}_\varepsilon) + \tilde{y}) &= v_\varepsilon + u_\varepsilon - \tilde{u}_\varepsilon && \text{in } Q \\
\tilde{y} &= 0 && \text{on } \Sigma \\
\tilde{y}(\cdot, 0) &= 0 && \text{on } \Omega.
\end{aligned}$$

Then the function $y(\cdot) := \mathbf{y}(\cdot : \tilde{u}_\varepsilon) + \tilde{y}$ satisfies (SL) and

$$\|y(T) - y_d\|$$
$$\leq \|\mathbf{y}(T : \hat{u}_\varepsilon) - \mathbf{y}(T : u_\varepsilon)\| + \|\mathbf{y}(T : u_\varepsilon) - y_d\| + \|\tilde{y}\| \leq 3\varepsilon.$$

The control v_ε is find, through the control of the linear problem, by the *cancellation* (at least approximately) of the nonlinear term. □

3.2 The case $\omega \subset\subset \Omega$ and f sublinear near the infinity.

When $\omega \subset\subset \Omega$ the answers to the approximate controllability question have different nature according to whether the nonlinear term $f(y)$ is sublinear or superlinear at the infinity. The following result concerns the sublinear case. It was obtained in [12], [13] under a global Lipschitz condition on f and later extended in [10] to the present statement.

Theorem 8 ([13],[10])
   Let f be a continous function such that

$$|f(s)| \leq C_1 + C_2 s \quad \text{if} \quad |s| > M, \quad \text{for some } M, C_1, C_2 > 0 \tag{9}$$

and there exists $s_0 \in \mathbb{R}$, $C_3, \delta > 0$ such that

$$|f(s) - f(s_0)| \leq C_3 |s - s_0| \quad \text{for any } s \in (s - \delta, s + \delta). \tag{10}$$

Then the approximate controllability property holds for (SL).

Proof.
   Define the function

$$g(s) = \begin{cases} \dfrac{f(s) - f(s_0)}{s - s_0} & \text{if } s \neq s_0 \\ 0 & \text{if } s = s_0. \end{cases}$$

By (10) there exists $K > 0$ such that $|g(s)| \leq K$ for any $s \in \mathbb{R}$. Given $z \in L^2(Q)$ and $v \in L^2(Q)$ we define the auxiliary functions $e(\cdot : z)$ and $\mathbf{y}(\cdot : z)$ as the solutions of the *linear* problems

$$\left. \begin{array}{ll} e_t - \Delta e + g(z)e = -f(s_0) + g(z)s_0 & \text{in } Q \\ e = 0 & \text{on } \Sigma \\ e(\cdot, 0) = y_0(\cdot) & \text{on } \Omega, \end{array} \right\}$$

and

$$\left. \begin{array}{ll} \mathbf{y}_t - \Delta \mathbf{y} + g(z)\mathbf{y} = v\chi_\omega & \text{in } Q \\ \mathbf{y} = 0 & \text{on } \Sigma \\ \mathbf{y}(0, \cdot) = 0 & \text{on } \Omega. \end{array} \right\}$$

By Theorem 1 we can chose $v = v(z) \in L^2(\omega \times (0, T))$ such that

$$\|\mathbf{y}(T) - y_d + e(T)\|_{L^2(\Omega)} \leq \epsilon.$$

Moreover, the function $y := e + \mathbf{y}$ satisfies

$$\left. \begin{array}{ll} y_t - \Delta y + g(z)y = -f(s_0) + g(z)s_0 + v\chi_\omega & \text{in } Q \\ y = 0 & \text{on } \Sigma \\ y(\cdot, 0) = y_0(\cdot) & \text{on } \Omega, \end{array} \right\} \tag{11}$$

and

$$\|y(T) - y_d\|_{L^2(\Omega)} \leq \epsilon. \tag{12}$$

Consider now the multivalued mapping $\Lambda : L^2(Q) \to \mathcal{P}(L^2(Q))$ given by

$$\Lambda z = \{y : \text{ satisfying (11) and (12)}\}.$$

It can be shown that Λ verifies the assumptions of the Kakutany Fixed Point Theorem and so there exists y solution of (11) with $z = y$ and satisfies (12) which ends the proof by using the definition of g. $\square$

By applying Theorem 2 (instead of Theorem 1) in the above proof we can improve the conclusion of Theorem 8 relative to nonnegative controls.

Corollary 1 ([8])

Let f satisfying (9) and (10). Let $\mathcal{U}$ be a dense subset of $L^2_+(\omega \times (0, T))$. Then $\{y(T : v) : y \text{ solution of (SL)}, v \in \mathcal{U}\}$ is a dense subset of $y(T : 0) + L^2_+(\Omega)$. $\square$

Remark 5

The Kakutany Fixed Point Theorem was also used in Henry [19]. The applicability of other fixed point theorems can be found in Carmichael and Quinn [2]. We also remark that the programme of the above proof can be successfully applied to show the approximate controllability in $L^p(\Omega)$ with $1 \leq p < \infty$ and in $C_0(\Omega)$ (see [13]). Finally, we mention the work [5] where Theorem 8 was extended to a multivalued semilinear equation arising in Climatology. □

Remark 6

Assumptions (9) and (10) holds if, for instance, $f(s) = \lambda|s|^{p-2}s$ with $0 < p < 1$ and $\lambda \in \mathbb{R}$ (notice that this function is not globally Lipschitz). Abstract results on the approximate controllability for some nonlinear parabolic problems, using also some sublinear assumptions on the nonlinear terms, are due to Seidman [32] and Naito and Seidman [29]. □

3.3 The case $\omega \subset\subset \Omega$ and f superlinear near the infinity.

When the nonlinear term $f(y)$ is superlinear near the infinity there appears an *obstruction* over the solutions of the equation and the approximate controllability fails. This fact was first pointed out by A.Bamberger in [19] when

$$f(s) = \lambda|s|^{p-2}s, \qquad p > 1, \quad \lambda > 0 \tag{13}$$

$\Omega = (0,1)$ and the internal control in (SL) is replaced by the homogeneous equation and the boundary control $y_x(0,t) = v(t)$. He uses an energy method to prove that $\|y(T:v)\|_{L^2(\Omega_\varepsilon)} \leq C$ with $\Omega_\varepsilon = (\varepsilon,1)$, $0 < \varepsilon < 1$ and C independent of v. A different technique was used in Díaz [3] for f given by (13), Ω arbitrary and the boundary controls $y(t,x) = v(t,x)$, $(t,x) \in \Sigma$. This thechnique can be easily adapted to the case of internal controls

Theorem 9 ([10])

Let f given by (13) and let $\omega \subset\subset \Omega$. Then for any $v \in L^2(\omega \times (0,T))$ arbitrary we have the estimate

$$|y(x,t:v)| \leq C(p,n) \left(\frac{1}{d(x)^\theta} + \frac{1}{t^{\frac{\theta}{2}}} \right) \qquad a.e.\ (x,t) \in (\Omega \setminus \overline{\omega}) \times (0,T)$$

where $\theta = \frac{2}{p-1}$, $d(x) = dist(x, \partial\omega)$ and $C(p,n)$ is a positive constant independent on v.

Proof.

We introduce the function

$$\mathbf{y}(t,x) = C(p,n) \left(\frac{1}{d(x)^\theta} + \frac{1}{t^{\frac{\theta}{2}}} \right) \qquad \text{for } x \in \Omega \setminus \overline{\omega} \text{ and } t > 0.$$

A careful choice of the constant $C(p, n)$ (see *e.g.* [21]) allows to check that **y** satisfies

$$\mathbf{y}_t - \Delta \mathbf{y} + \lambda |\mathbf{y}|^{p-1} \mathbf{y} \geq 0 \quad \text{in } (\Omega \setminus \overline{\omega}) \times (0, T),$$
$$\mathbf{y} \geq 0 \qquad\qquad\qquad \text{on } \Sigma$$
$$\mathbf{y} \to +\infty \qquad\qquad\quad \text{on } \partial\omega \times (0, T)$$
$$\mathbf{y} \to +\infty \qquad\qquad\quad \text{on } \Omega \times \{0\}.$$

Applying the maximum principle we deduce that $y(x, t : v) \leq \mathbf{y}(x, t)$ for any $t \in (0, T]$ and *a.e.* $x \in \Omega \setminus \overline{\omega}$. In a similar way we prove that $-\mathbf{y}(t, x) \leq y(t, x : v)$ and the conclusion holds. □

Remark 7

As a matter of fact the above estimate can be improved by introducing the function $\mathbf{U}_\infty(x, t)$ solution of the problem

$$\mathbf{U}_t - \Delta \mathbf{U} + \lambda |\mathbf{U}|^{p-1} \mathbf{U} = 0 \quad \text{in } (\Omega \setminus \overline{\omega}) \times (0, T),$$
$$\mathbf{U} = 0 \qquad\qquad\qquad\quad \text{on } \Sigma$$
$$\mathbf{U} \to +\infty \qquad\qquad\quad\; \text{on } \partial\omega \times (0, T)$$
$$\mathbf{U}(\cdot, 0) = y_0(\cdot) \qquad\quad\;\; \text{on } \Omega.$$

Thanks to the assumption $p > 1$ it is possible to show (see [1]) the existence of a minimal solution $\mathbf{U}_\infty$. In fact, we have $\mathbf{U}_\infty > 0$ in $(\Omega \setminus \overline{\omega}) \times (0, T)$, assumed $y_0 \geq 0$. As in Theorem 9 we conclude the estimate

$$y(x, t : v) \leq \mathbf{U}_\infty(x, t) \quad \text{for } a.e.\ x \in \Omega \setminus \overline{\omega} \text{ and } t \in [0, T]$$

where v is again arbitrary in $L^2(\omega \times (0, T))$. We conjecture that (even in this superlinear case) the approximate controllability property holds if we assume the desired state $y_d \in L^2(\Omega)$ such that

$$\mathbf{U}_{-\infty}(x, T) < y_d(x) < \mathbf{U}_\infty(x, T) \quad \text{for } a.e.\ x \in \Omega \setminus \overline{\omega}$$

(here $\mathbf{U}_{-\infty}$ denotes the solution of the above problem replacing $+\infty$ by $-\infty$). □

Remark 8

We also conjecture that if f represents a superlinear source near the infinity (*e.g.* f given by (13) but with $\lambda < 0$ instead $\lambda > 0$) there is not any obstruction and the approximate controllability holds. □

Remark 9

Theorems 8 and 9 show that the approximate controllability property holds or not for (SL) according to whether the function f is sublinear or superlinear near the infinity. This fact contrasts with the occurrence of a free boundary for which the answers are of different nature according to whether f is sublinear (the positive case) or superlinear (the negative case) near the origin (see *e.g.* [20] and [9]). ◻

Remark 10
   The existence of *universal solutions* taking values $+\infty$ or $-\infty$ over $\partial\omega \times (0, T)$ can also be obtained for many other nonlinear equations such as the nonlinear diffusion equation

$$y_t - \Delta \left(|y|^{m-1}y \right) = v\chi_\omega,$$

and the quasilinear equation associated to the $(m + 1)$-Laplacian operator

$$y_t - \Delta_{m+1}y = v\chi_\omega, \qquad \Delta_{m+1}y := \text{div} \left(|\nabla y|^{m-1}\nabla y \right),$$

always under the condition $m > 1$. Other kind of *universal solution* can also be obtained (see [4]) for the Burger equation

$$y_t - y_{xx} + yy_x = v\chi_\omega.$$

The uncontrollability for this equation was also shown in Fursikov-Imanuvilov [16] by using an energy method. We also mention that it is possible to show the exact controllability for the Burger equation over very special functional spaces (see El Badia-Ain Seba [11]). Finally we point out that other nonlinear problems also leads to positive or negative answers to the question of the approximate controllability according the behaviour of the data (see in Díaz [3],[4] a study of the parabolic obstacle problem). ◻

Remark 11
   The approximate controllability for the Navier-Stokes equation is, at the present, an open problem. The interest of this question was already raised in Lions [24] establishing some connections with the study of the turbulence. A partial result is due to Fernández-Cara and Real [15] and shows that the subspace spanned by $\bar{y}(T : v)$ is dense in a suitable Hilbert space. We point out that if the conjecture of the Remark 7 is true then the subspace spanned by $\{y(T : v) : v \in L^2(\omega \times (0, T))\}$ is dense in $L^2(\Omega)$ (even for the superlinear case). ◻

Acknowledgement. *The research of the author is partially supported by the DGICYT (Spain) project PB 90/0620.*

References

[1] C.Bandle, G.Díaz et J.I.Díaz: Solutions d'equations de réaction-diffusion non-linéaires, explosant au bord parabolique. To appear in *C.R.Acad.Sci. de Paris*.

[2] N.Carmichael and M.D.Quinn: Fixed point methods in nonlinear control. In *Distributed Parameter System*. F.Kappel et al. (eds.), Springer-Verlag (1985), 24-51.

[3] J.I.Díaz: Sur la contrôllabilité approchée des inéquations variationelles et d'autre problémes paraboliques non-linéaires. *C.R.Acad.Sci. de Paris*, **312**, serie I, (1991), 519-522.

[4] J.I.Díaz: Sobre la controlabilidad aproximada de problemas no lineales disipativos. In *Jornadas Hispano-Francesas sobre Control de Sistemas Distribuidos*. Univ. Málaga (1991), 41-48.

[5] J.I.Díaz: On the controllability of some simple climate models. In *Environment, Economics and their Mathematical Models*. J.I.Díaz and J.L.Lions (eds.). Masson (1993).

[6] J.I.Díaz and A.V.Fursikov: A simple proof of the controllability from the interior for nonlinear evolution problems. Submitted.

[7] J.I.Díaz and A.V.Fursikov: Approximate controllability of the Stokes system by external local one-dimensional forces. Manuscrit.

[8] J.I.Díaz, J.Henry and A.M.Ramos: Article in preparation.

[9] J.I.Díaz and J.Hernández: Qualitative properties of free boundaries for some nonlinear degenerate parabolic equations. In *Nonlinear Parabolic Equations: Qualitative Properties of Solutions*. L.Boccardo and A.Tesei (eds.). Pitman (1987), 85-93.

[10] J.I.Díaz and A.M.Ramos: Positive and negative approximate controllability results for semilinear problems. In *Actas del XIII CEDYA*. Univ. Politécnica de Madrid (1994).

[11] A.El Badia and B.Ain Seba: Contrôlabilité exacte de l'équation de Burger. C.R.Acad. Sci. de Paris, **314**, serie I, (1992), 373-378.

[12] C.Fabré, J.P.Puel and E.Zuazua: Contrôlabilité approchée de l'équation de la chaleur. C.R.Acad. Sci. de Paris, **315**, serie I, (1992), 807-812.

[13] C.Fabré, J.P. Puel and E.Zuazua: Approximate controllability of the semilinear heat equation. *IMA Preprint Series*, (1992).

[14] C.Fabré, J.P.Puel and E.Zuazua: Contrôlabilité approchée de l'équation de la chaleur linéaire avec des contrôles de norme L^∞ minimale. C.R.Acad. Sci. de Paris, **316**, serie I, (1993), 679-684.

[15] E.Fernández-Cara and J.Real: On a conjeture due to J.L.Lions. To appear in *Nonlinear Analysis. TMA*.

[16] A.V.Fursikov and O.Y.Imanuvilov: On the approximate controllability of the Stokes systems. To appear in *Annales de la Faculté des Sciences de Toulouse*.

[17] A.V.Fursikov and O.Y.Imanuvilov: On the approximate controllability of certain systems simulating a fluid flow. Preprint (1993).

[18] Y.Ekeland and R.Temam: *Analyse Convexe et Problémes Variationelles*. Dunod, Gauthier-Villars, (1974).

[19] J.Henry: *Etude de la contrôlabilité de certains équations paraboliques.* Thèse d'Etat, Université Paris VI (1978).

[20] A.S.Kalsahnikov: Some problems of the qualitative theory of non-linear degenerate second-order parabolic equations. *Russ. Math. Survs.* **42**, (1987), 169-222.

[21] S.Kamin, L.A.Peletier and J.L.Vázquez: Classification of singular solutions of a nonlinear heat equations. *Duke Math.Jour.*, **58**, (1989), 601-615.

[22] J.L.Lions: *Contrôle Optimal des Systems Gouvernés par des Equations aux Derivées Partielles.* Dunod, (1968).

[23] J.L.Lions: Remarques sur la contrôlabilité approchée. In *Jornadas Hispano-Francesas sobre Control de Sistemas Distribuidos.* Univ. de Málaga, (1991), 77-88.

[24] J.L.Lions: Are there connections between turbulence and controllability?. In *Analysis and Optimization des Systems.* Lecture Notes in Control and Information Series **144**, Springer-Verlag, (1990).

[25] J.L.Lions: Exact controllability for distributed systems. Some trends and some problems. In *Applied and Industrial Mathematics.* R.Sigler (ed.), Kluwer (1991), 59-84.

[26] J.L.Lions: Remarks on approximate controllability for parabolic systems. In *Finite Elements in the 90's.*, E.Oñate et al. (eds.), Springer-Verlag, (1991), 612-620.

[27] J.L.Lions: Unpublished manuscrit.

[28] S.Mizohata: Unicité du prologment des solutions pour quelques opérateurs differen-tielles paraboliques. *Mem.Coll. Sci.Univ.Kyoto,* serie **A31**, (1958), 219-239.

[29] K.Naito and T.I.Seidman: Invariance of the approximately reachable set under non-linear perturbations. *SIAM J. Control and Optimization.* **29**, (1991), 731-750.

[30] D.L.Russell: Controllability and stabilizability theory for nonlinear partial differential equations: recents progress and open questions. *SIAM Rev.* **20**, (1978), 639-739.

[31] J.C.Saut and B.Scheurer: Unique continuation for some evolution equations. *J.Differenti Equations,* **66**, (1978), 118-139.

[32] T.I.Seidman: Invariance of the reachable set under nonlinear perturbations. *SIAM J.Control and Optimizations,* **25**, (1987), 1173-1191.

An Approach to Variable Metric Bundle Methods

Claude Lemaréchal and Claudia Sagastizábal

INRIA, BP 105, 78153 Le Chesnay (France)

ABSTRACT

To minimize a convex function f, we state a penalty-type bundle algorithm, where the penalty uses a variable metric. This metric is updated according to quasi-Newton formulae based on Moreau-Yosida approximations of f. In particular, we introduce a "reversal" quasi-Newton formula, specially suited for our purpose. We consider several variants in the algorithm and discuss their respective merits. Furthermore, we accept a degenerate penalty term in the Moreau-Yosida regularization.

Key words. Bundle methods, convex optimization, mathematical programming, proximal point, quasi-Newton algorithms, variable metric.

AMS Subject Classification. Primary: 65K05. Secondary: 90C30, 90C25.

1 Introduction

This paper addresses the numerical minimization of a (finite-valued) convex function $f : \mathbb{R}^N \rightarrow \mathbb{R}$, characterized by a black box which, for any $x \in \mathbb{R}^N$, computes $f(x)$ and some subgradient $g(x) \in \partial f(x)$. Our approach employs bundle methods, so we briefly recall their basic principles here. At the current iteration of the algorithm, the black box has computed the sample values $f(y_i)$ and $g_i \in \partial f(y_i)$ for $i = 1, \ldots, k$; the *cutting-plane* model of f is then

$$\check{f}_k(y) := \max\{f(y_i) + \langle g_i, y - y_i \rangle \ : \ i = 1, \ldots, k\} \,.$$

In the cutting-plane method [CG59], [Kel60], the next iterate y_{k+1} is a minimizer of $\check{f}_k$. However this method is notoriously unstable (see an example of A.S. Nemirovski, described in Section XV.1.1 of [HULL93]). Bundle methods offer a stabilizing device based on the following ingredients:

(i) a sequence $\{x_n\}$ of stabilized iterates;
(ii) a test deciding whether a new stabilized iterate has been found and/or whether the model $\check{f}_k$ should be enriched;
(iii) a sequence $\{M_n\}$ of positive definite matrices defining a scalar product and its associated norm.

A number of different approaches have been developed according to the above principles. We give now a close description of one of them, namely the *proximal* form, which we consider in this paper:

(i) A *candidate* y^c is computed as the minimizer of the penalized model

$$\check{f}_k(y) + \tfrac{1}{2} \langle M_n(y - x_n), y - x_n \rangle \,. \tag{1}$$

(ii) A *nominal decrease*

$$\delta_n := f(x_n) - \check{f}_k(y^c) - \tfrac{1}{2} \langle M_n(y^c - x_n), y^c - x_n \rangle$$

controls the update of x_n and/or the enrichment of $\check{f}_k$. More specifically, a fixed parameter $m \in \,]0, 1[$ being chosen, we perform the descent test

$$f(y^c) \leq f(x_n) - m\delta_n \,. \tag{2}$$

If (2) holds, then we set $x_{n+1} = y_{k+1} = y^c; n$ and k are increased by 1. Otherwise n is kept fixed, we set $y_{k+1} = y^c$ and k is increased by 1; in some improved versions (see the considerations in §6 below), an additional test is made before increasing k.

(iii) The choices of the *norming* $\{M_n\}$ given so far in the literature are:

 – an abstract sequence, as in [Lem78],
 – $M_n \equiv I$, as in [Kiw83],
 – $M_n = \mu_n I$, with heuristic rules for computing μ_n; see [Kiw90], [SZ92].

An essential feature of our present development consists of a quasi-Newton update of M_n [DM77] using the so-called Moreau-Yosida regularization ([Mor65], [Yos64]); we also pay some attention to the updates used by Shor [Sho85].

The next section is devoted to the Moreau-Yosida regularization; we recall a few results, slightly generalized in the sense that we admit semi-positive definite matrices M_n. Then we state the algorithm and give some of its basic properties. In §4 we consider some quasi-Newton formulae, which exploit two possible ideas:

- First, we choose a move in x and we compute the corresponding move in the gradient of a smooth function, namely the regularization $\bar{F}_k$ coming out of (1).
- Second, a "reversal" idea starts from a move in the gradient space; a corresponding move in x is then computed to estimate the curvature of the smoothened objective function.

The last two sections assess the approach, they show that the second idea has several merits related to implementation and convergence issues.

2 Some Basic Results

Given a semi-positive definite matrix M, we denote by

$$H(x) := \inf \{ h(y) + \tfrac{1}{2} \langle M(y-x), y-x \rangle \} \tag{3}$$

the corresponding Moreau-Yosida regularization of a function h. Allowing a degenerate M in the quadratic perturbation departs from the classical framework, but we show that the essential results concerning this regularization can be reproduced.

Notationally, block letters $H, F, G, \ldots$ will be used to designate the regularized versions of objects such as $h, f, g, \ldots$

Theorem 1. *Let h in (3) be a closed convex function.*

(i) If $\operatorname{dom} h^* \cap \operatorname{Im} M = \emptyset$ then $H(x) = -\infty$ for all x.

(ii) If $\operatorname{dom} h^* \cap \operatorname{Im} M \neq \emptyset$ then $H(x) > -\infty$ for all x, and H is a convex function defined on the whole of $\mathbb{R}^N$.

Assume case (ii) and denote by M^- the pseudo-inverse of M. The dual problem of (3)

$$\min_{g \in \operatorname{Im} M} [h^*(g) - \langle g, x \rangle + \tfrac{1}{2} \langle M^- g, g \rangle] = -H(x) \tag{4}$$

has a unique solution $G(x)$. The solution set in (3) is then

$$P(x) = \partial h^*(G(x)) \cap [x - M^- G(x) + \operatorname{Ker} M]; \tag{5}$$

when $\operatorname{ri} \operatorname{dom} h^* \cap \operatorname{Im} M \neq \emptyset$, or when h is polyhedral, this set is nonempty (for all x).

Proof. The function H is the infimum of a function that is jointly convex in (x, y); in view of §IV.2.4 of [HULL93], H is convex (in x); also $H(x)$ is clearly $< +\infty$ (for all x) but we have to cope with the case $H(x) = -\infty$.

For fixed x, denote by $y \mapsto \varphi(y)$ the objective function in (3); it is the sum of two closed convex functions, one of them is finite everywhere. The conjugates of these two functions are respectively (see Example X.1.1.4 in [HULL93], I is the indicator function)

$$h^*(g) \text{ and } \langle g, x \rangle + \tfrac{1}{2} \langle M^- g, g \rangle + I_{\operatorname{Im} M}(g),$$

which have respectively the subdifferentials

$$\partial h^*(g) \text{ and } \{x + M^- g\} + \operatorname{Ker} M. \tag{6}$$

According to Theorem X.2.3.2 of [HULL93] the infimal convolution

$$\varphi^*(s) = \inf_g [h^*(g) + \langle s - g, x \rangle + \tfrac{1}{2} \langle M^-(s - g), s - g \rangle + I_{\operatorname{Im} M}(s - g)]$$

is a closed convex function, which is the conjugate of φ. It follows in particular that $\varphi^*(0) = -H(x)$; to say that this number is not $+\infty$ is to say that there is some $g \in \operatorname{dom} h^*$ such that $0 - g \in \operatorname{Im} M$; (i) and (ii) are proved.

Now (4), precisely, just expresses the relation $\varphi^*(0) = -H(x)$; observing that M^- is an isomorphism on $\operatorname{Im} M$, the infimum is attained at a unique $G(x)$.

Finally, the primal solution set $P(x)$ is $\partial \varphi^*(0)$; by Theorem XI.3.4.1 of [HULL93], this is (5). Under a suitable qualification assumption such as those stated, $\partial \varphi^*(g)$ is the sum of the subdifferentials in (6). Then the optimality condition $0 \in \partial \varphi^*(G(x))$ gives

$$0 \in \partial h^*(G(x)) - x + M^- G(x) + \operatorname{Ker} M;$$

this just expresses the nonemptiness of (5). □

Among other things, this result reveals an important property of our "extended" Moreau-Yosida regularization: finiteness of the value $H(x)$ depends only on the geometry of h and M, but not on the particular value of x. Furthermore, when h is polyhedral (the only case of interest to us), existence of a candidate $y^c \in P(x)$ in (1) also does not depend on the particular x.

Remark 2. Suppose that the dual solution $G(x)$ is available, together with a multiplier $\check{w} \in \operatorname{Ker} M$ of the constraint $g \in \operatorname{Im} M$. This means that there is some $\check{z} \in \partial h^*(G(x))$ such that $0 = \check{z} - x + M^- G(x) + \check{w}$, so that $\check{y} := x - M^- G(x) - \check{w} \in P(x)$; then it is rather clear that the whole $P(x)$ is the closed convex set

$$P(x) = \{y = x - M^- G(x) - w : w \in \operatorname{Ker} M \text{ and } h(y) \le h(\check{y})\}. \tag{7}$$
□

We now check that the well-known regularity properties of H are preserved.

Theorem 3. *Assume that, for all x, the infimum in (3) is attained on some nonempty set $P(x)$. Then the convex function H has at all x a gradient given by*

$$\nabla H(x) = G(x) = M(x - y), \quad \text{for arbitrary } y \in P(x), \tag{8}$$

where $G(x)$ solves (4). Furthermore, ∇H is Lipschitzian. More precisely, for all x_1, x_2:

$$\|\nabla H(x_1) - \nabla H(x_2)\|^2 \leq \frac{1}{\Lambda} \langle \nabla H(x_1) - \nabla H(x_2), x_1 - x_2 \rangle, \tag{9}$$

where Λ is the largest eigenvalue of M.

Proof. Under the stated conditions, Corollary VI.4.5.3 of [HULL93] can be applied. It gives $\nabla H(x) = M(x - y)$, no matter how y is chosen in the optimal set $P(x)$. Then use (7): for such an optimal y and $w \in \text{Ker } M$, we have

$$M(y - x) = M(x - M^- G(x) + w - x) = G(x).$$

As for the Lipschitz property, it is essentially proved in Theorem X.4.3.1 of [HULL93]. □

Remark 4. When h is the piecewise affine function $\check{f}_k$, the regularized value $\check{F}_k(x_n)$ in (1) is easily computed, via the resolution of a quadratic problem. A particular y^c can even be selected: for stability reasons, it is advisable to choose one as close as possible to the current center x_n. For this, it suffices to solve a projection problem onto the polyhedron defined by (7).

When M is positive definite, the solution set $P(x)$ reduces to a singleton, called the *proximal* point of x and denoted by $p(x)$. The update $x_{n+1} = y^c$ therefore appears as the proximal point of x_n, associated with the cutting-plane model $\check{f}_k$. For the general case of a degenerate M, we still use the notation $p(x)$ for an arbitrary point in $P(x)$. Actually, this notation is slightly simplistic; in particular it neglects the matrix M and the function h.

Finally, we recall that, in the classical Moreau-Yosida regularization, minimizing h is equivalent to minimizing H; this property is conserved when M is singular, we omit the proof which is simple. □

3 Description of the Algorithm

In this section we concretize the principles exposed in §1. We state a schematic algorithm and introduce some of its properties. By $M_+ = Up(M, u, v)$ we mean an updated matrix using M and two vectors u and v; for example "Up" may be a formula based on the *quasi-Newton* equation $M_+ u = v$. We will actually consider two variants; in the first, M_+ is imposed to be proportional to the identity matrix, while the second will use a full matrix.

Algorithm 5.

Step 0 (Initialization). Choose $m \in]0,1[$, $x_0 \in \mathbb{R}^N$. Set $y_0 = x_0$, $M_0 = I$, $n = k = 0$.

Step 1 (Computation of the candidate). If the stopping criterion is not satisfied, find $y^c = p(x_n)$ i.e., solve

$$y^c \in \operatorname{Argmin}[\check{f}_k(y) + \tfrac{1}{2}\langle M_n(y - x_n), y - x_n\rangle], \tag{10}$$

and compute

$$\delta_n = \delta_n^k := f(x_n) - \check{f}_k(y^c) - \tfrac{1}{2}\langle M_n(y^c - x_n), y^c - x_n\rangle. \tag{11}$$

Step 2 (Descent-step). If $f(y^c) \le f(x_n) - m\delta_n$, then: update $x_{n+1} = y^c$. Choose two points z and z' and a closed convex function h; compute the corresponding dual solutions $G(z)$ and $G(z')$ from (4); set $u := z' - z$ and $v := G(z') - G(z)$; update $M_{n+1} = Up(M_n, u, v)$ and increase n.

Step 3 (Null-step). Set $y_{k+1} = y^c$, increase k. Loop to 1. □

Before starting the theoretical study of this algorithm, let us mention some implementation issues:

(i) The above description neglects numerical technicalities such as:
  - an explicit stopping test (which can use (14) below),
  - an elaborate choice of the initial matrix,
  - an aggregation mechanism to avoid storing the whole bundle when k becomes large,
  - a safeguard to prevent awkward candidates, since the quadratic problem in Step 1 may have no solution (y^c "at infinity"); the end of this section suggests a possible safeguarding technique.
  - We will see that, for efficiency, some sort of line-search should be inserted before looping to the next iteration; it is in this sense that the above description is only schematic.

(ii) The matrix update in Step 2 will be explained in §4 below; we will consider two possibilities for (z, z', h) and two possibilities for the formula symbolized by "Up".

(iii) Concerning the existence of solutions to (10), we recall here a result of [FW56]: being piecewise quadratic, the objective function has a minimum point if and only if it is bounded from below. A detailed answer to this existence question was given in Theorem 1, which can be particularized now to our present situation:

Proposition 6. *At iteration (n,k), let Γ be the convex hull of the subgradients $g_1, \ldots, g_k$ and assume $\Gamma \cap \operatorname{Ker} M_n \ne \emptyset$. Then (10) has a solution of the form*

$$y^c = x_n - M_n^- G(x_n) + \bar{w}_k \tag{12}$$

with $\bar{w}_k \in \operatorname{Ker} M_n$ and $G(x_n) \in \partial \check{f}_k(y^c)$ is given by (4).
  The following relations hold:

$$\delta_n^k = \tfrac{1}{2}\langle G(x_n), M_n^- G(x_n)\rangle + \varepsilon_n^k \tag{13}$$

with

$$\varepsilon_n^k := f(x_n) - \check{f}_k(y^c) - \langle G(x_n), M_n^- G(x_n) \rangle$$

and, for all $y \in \mathbb{R}^N$,

$$f(y) \geq f(x_n) + \langle G(x_n), y - x_n \rangle - \varepsilon_n^k. \tag{14}$$

Proof. Use Theorem 1 with $h = \check{f}_k$ and $M = M_n$. First of all, the domain of $(\check{f}_k)^*$ is the convex hull Γ of the subgradients g_i making up $\check{f}_k$ (see for example §X.3.4 in [HULL93]). When $\Gamma \cap \mathrm{Ker}\, M \neq \emptyset$, the optimal value in (3) is a finite number and, because $\check{f}_k$ is a polyhedral function, an optimal solution y^c exists; its expression (12) comes from the characterization (5). To obtain the form (13) of δ_n, plug the value (12) of y^c into (11) and use the property $M_n^- \bar{w}_k = 0$.

Finally express that $G(x_n) \in \partial \check{f}_k(y^c)$: for all $z \in \mathbb{R}^N$,

$$f(z) \geq \check{f}_k(z) \geq \check{f}_k(y^c) + \langle G(x_n), z - y^c \rangle$$

and perform some straightforward algebraic manipulations to obtain (14). □

Observe that, when the existence condition $\Gamma \cap \mathrm{Im}\, M \neq \emptyset$ holds, it holds for every subsequent iteration, as long as M is not updated. In the particular case when $0 \in \Gamma$ (which corresponds to $\check{f}_k$ having a minimum point), the existence condition holds at every subsequent iteration.

We recall that the subdifferential of the max-function $\check{f}_k$ is the convex hull of the active subgradients g_i. In other words, for some set of convex multipliers α_i,

$$G(x_n) = \sum_{i \in I_k} \alpha_i g_i \text{ where}$$
$$I_k := \{i = 1, \ldots, k : f(y_i) + \langle g_i, y^c - y_i \rangle = \check{f}_k(y^c)\}. \tag{15}$$

Remark 7. When $G(x_n)$ and ε_n^k are both close to 0, (14) shows that x_n satisfies an approximate optimality property. In view of (13), the aim of the algorithm is thus to force δ_n^k to 0 and to avoid "large" matrices M_n. □

The next result is motivated by the introduction of semi-definite matrices.

Lemma 8. *Let* H *be a closed convex function. If, for some* x_1, x_2, *there are* $g_i \in \partial H(x_i), i = 1, 2$, *such that*

$$\langle g_1 - g_2, x_1 - x_2 \rangle = 0, \tag{16}$$

then H *is affine on the segment* $[x_1, x_2]$ *and* ∂H *is constant on* $]x_1, x_2[$. *If* H *is (finite-valued and) differentiable,* $g_1 = g_2$.

Proof. Take $x := x_1 + \alpha(x_2 - x_1)$ with $\alpha \in [0, 1]$; write the subgradient inequalities

$$H(x) \geq H(x_1) + \alpha \langle g_1, x_2 - x_1 \rangle$$

$$H(x) \geq H(x_2) - (1 - \alpha) \langle g_2, x_2 - x_1 \rangle$$

and obtain by convex combination, using (16),

$$H(x) \geq (1 - \alpha)H(x_1) + \alpha H(x_2).$$

Since the convexity of H gives the converse inequality, H is affine on $[x_1, x_2]$.

Now, restrict the above α to $]0,1[$ and take $g \in \partial H(x)$. Then the affinity of H means that, for any $x' := x_1 + \alpha'(x_2 - x_1)$ with $\alpha' \in]0,1[$,

$$H(x') = H(x) + \langle g, x' - x \rangle.$$

Additionally, for all z,

$$H(z) \geq H(x) + \langle g, z - x \rangle = H(x') + \langle g, z - x' \rangle - \varepsilon,$$

where $\varepsilon := H(x') - H(x) - \langle g, x - x' \rangle = 0$. Thus $\partial H(x) \subset \partial H(x')$; the other inclusion is established likewise, exchanging x and x'.

Finally, if the convex function H is differentiable, it is continuously differentiable and the equality $\nabla H(x) = \nabla H(x')$ extends to the endpoints x_1 and x_2.
□

We conclude this section with a word concerning the computation of y^c. When M_n is singular, the objective function in (10) may be unbounded from below; some safeguarding technique is therefore advisable. Rather than loading the diagonal of M_n, we prefer to perturb the function $\check{f}_k$ temporarily, just to eliminate candidates that are blatantly too far from the current stability center.

Safeguarding Technique Suppose an estimated lower bound for $f(x_{n+1})$ is at hand; we can take for example

$$\ell := f(x_n) - \frac{f(x_{n-1}) - f(x_n)}{m}. \tag{17}$$

Then the constant function of value ℓ can be appended to the affine functions making up $\check{f}_k$; the perturbed model is bounded from below and existence of a (perturbed) proximal point is guaranteed.

If the safeguard is active, $\check{f}_k(y^c) = \ell$, a supposedly very small value; this implies that $\check{f}_k$ does not approximate the actual objective f properly. Thus, our safeguard should not significantly disturb the algorithm.
□

4 Matrix Updates

To compute M_{n+1} in Step 2 of Algorithm 5, we need to specify the pair of vectors u and v, as well as the actual formula for "Up".

4.1 Choice of a Regularizing Scheme

The vectors u and v are uniquely determined from a triple (z, z', h), knowing that

$$u = z' - z \quad \text{and} \quad v = \nabla H(z') - \nabla H(z).$$

We consider two alternatives for (z, z', h).

Model regularization $(x\text{-}\check{f})$ A first natural idea is to take $z := x_n$, $z' := x_{n+1} = y^c$; then we need the two corresponding gradients of some smooth function H. For implementability reasons, H has to be the Moreau-Yosida regularization $\check{F}_k$ of the current model $\check{f}_k$ (the updated model $\check{f}_{k+1}$ could also be taken). Then $G(x_n) = \nabla \check{F}_k(x_n)$ is available and $\nabla \check{F}_k(y^c) = G(y^c)$ is obtained via one more resolution of the quadratic problem:

$$p(y^c) = y^{cc} \in \text{Argmin}\{\check{f}_k(y) + \tfrac{1}{2}\langle M_n(y - y^c), y - y^c\rangle\},$$

or rather of its dual. □

Our second choice manages to regularize f itself, thanks to a backward process. We take $v := g(y^c) - g(x_n)$ and we compute x_-, y_- such that $g(x_n) = \nabla F(x_-)$ and $g(y^c) = \nabla F(y_-)$. This amounts to inverting the proximal mapping, an operation which can be performed explicitly:

Proposition 9. *We use the notation of §2; assume that $z \in \mathbb{R}^N$ is such that $\partial h(z) \cap \text{Im } M$ contains some point G. Then, for any $z_- \in \{z + M^- G\} + \text{Ker } M$,*

$$G = G(z_-) = \nabla H(z_-).\tag{18}$$

In fact, z solves (3) for $x = z_-$, i.e., $p(z_-) = z$.

Proof. With z_- as stated, set $w := z_- - z - M^- G \in \text{Ker } M$ and consider the set

$$\partial h^*(g) - z_- + M^- G + w = \partial h^*(G) - z.$$

This set contains 0 because $G \in \partial h(z)$, i.e., $z \in \partial h^*(G)$. Together with the property $G \in \text{Im } M$, we see that G satisfies the optimality condition of (4) with $x = z_-$.

Furthermore $G = M(z_- - z)$, hence z satisfies the optimality condition for (3). □

This result can be exploited with $h = f$, thus giving our second option:

Objective regularization $(g\text{-}f)$ Suppose M_n is nonsingular. Having on hand the two successive iterates x_n and y^c, we also have the corresponding subgradients $g(x_n)$ and $g(y^c)$ of f. Then we simply compute the points at which these two subgradients are gradients of F. We therefore take

$$u := y^c + M_n^{-1}g(y^c) - [x_n + M_n^{-1}g(x_n)], \quad v := g(y^c) - g(x_n);$$

which can be suitably rewritten as

$$\Delta x := x_{n+1} - x_n , \quad v = g(x_{n+1}) - g(x_n) , \quad u = \Delta x + M_n^{-1} v . \quad (19)$$

□

Knowing that our real problem is to minimize f, i.e., F, this last strategy actually appears as more direct than $(x\text{-}\tilde{f})$: it tries to apply the algorithm

$$x_{n+1} = x_n - \nabla^{-2} F(x_n) \nabla F(x_n) , \quad (20)$$

$\nabla F(x_n)$ and $\nabla^2 F(x_n)$ being replaced by $\nabla \tilde{F}_k(x_n)$ and M_n respectively. Furthermore, we will see in §6.1 that positive definiteness can easily be preserved; we will also see that this additional advantage is desirable.

4.2 Choice of an Explicit Formula

Let us now turn to the possible formulae for "Up". First of all, it is important to remember that the property $v = \nabla H(z') - \nabla H(z)$ ensures $\langle v, u \rangle \geq 0$, and the situation $\langle v, u \rangle = 0$ is described by Lemma 8. An important inequality is

$$\frac{\|v\|^2}{\langle v, u \rangle} \leq \Lambda_n \quad (21)$$

where Λ_n is the largest eigenvalue of M_n. To obtain it, apply (9) with $M = M_n$.
Another useful inequality is

$$\frac{\|M_n u\|^2}{\langle M_n u, u \rangle} \leq \Lambda_n \quad \text{for } u \notin \operatorname{Ker} M .$$

Indeed, set $z := M_n^{1/2} u$ and observe that

$$\frac{\|M_n u\|^2}{\langle M_n u, u \rangle} = \frac{\langle M_n z, z \rangle}{\|z\|^2} .$$

We consider two variants for "Up", based on the quasi-Newton principle.

Diagonal quasi-Newton Variant (dqN) The matrices are restricted to being proportional to the identity: $M_n = \mu_n I$. Given u and v, the updated matrix is $\mu_{n+1} I$, where μ_{n+1} minimizes $1/2 \|v/\mu - u\|^2$; we take

$$\mu_{n+1} := \begin{bmatrix} \dfrac{\|v\|^2}{\langle v, u \rangle} & \text{if } \langle v, u \rangle > 0, \\ 0 & \text{if } \langle v, u \rangle = 0 . \end{bmatrix}$$

If $\langle v, u \rangle = 0$, then $v = 0$ (Lemma 8): the observed curvature of H along u is 0, which explains our choice $\mu_{n+1} = 0$.

□

With relation to Remark 7, we have from (21)

$$\mu_{n+1} \leq \mu_n , \quad (22)$$

an inequality which holds independently of (z, z', h).

Full quasi-Newton Variant (fqN) The updated matrix is computed from the BFGS formula:
$$M_{n+1} := M_n + A - B,$$
where
$$A := \begin{bmatrix} \dfrac{vv^\mathsf{T}}{\langle v, u \rangle} & \text{if } \langle v, u \rangle > 0, \\ 0 & \text{if not} \end{bmatrix}$$
and
$$B := \begin{bmatrix} \dfrac{M_n uu^\mathsf{T} M_n}{\langle M_n u, u \rangle} & \text{if } M_n u \neq 0, \\ 0 & \text{if not.} \end{bmatrix}$$

Note that this variant is robust, since
$$\operatorname{tr} A \leq \Lambda_n \quad \text{and} \quad \operatorname{tr} B \leq \Lambda_n .$$

It is also consistent:

- the choice $A = 0$ was already explained in (dqN),
- the choice $B = 0$ is similar, namely the predicted curvature along u is 0 when $M_n u = 0$.
- As for the quasi-Newton equation, we have
$$M_{n+1}u = M_n u + Au - Bu = Au$$
and this is v in any case. $\quad\square$

Thus, all our formulae introduce smoothness while preserving implementability, two features which are not present in [BGLS93]. Note, however, that we may well have $G(y^c) = G(x_n)$ (think of the very first iteration $n = k = 0$!). Then the variant $(x\text{-}\tilde{f})$ gives $v = 0$ and M_{n+1} degenerates. By contrast, M_{n+1} in $(g\text{-}f)$ can degenerate only when $g(y^c) = g(x_n)$, an unlikely event.

Remark 10. The difference of f-subgradients for v in $(g\text{-}f)$ suggests Shor's r-algorithm [Sho85]. In this variant, the matrix M_{n+1} dilates the space in the direction v. Having a coefficient $\beta_n > 1$, we take (see [Sko73])
$$M_{n+1} := M_n + \begin{bmatrix} 0 & \text{if } v \in \operatorname{Ker} M_n, \\ \dfrac{\beta_n^2 - 1}{\langle v, M_n^{-1} v \rangle} vv^\mathsf{T} & \text{if not.} \end{bmatrix}$$

We mention that this variant accommodates any vector v: we can also take $v = g(y^c)$, as in the ellipsoid-type algorithm [Sho70].

No matter how v is chosen, each matrix M_{n+1} is positive definite if M_n is such; again robustness is preserved:
$$\operatorname{tr} M_{n+1} \leq \operatorname{tr} M_n + (\beta_n^2 - 1)\Lambda_n . \quad\square$$

5 Convergence Issues

As usual with bundle methods, we split our convergence analysis into two parts.

Theorem 11. *Suppose that Algorithm 5 generates a finite sequence $\{x_n, n = 0, 1, \ldots, n_f\}$ and that the last generated matrix M_{n_f} is nonsingular. Then x_{n_f} is optimal.*

Proof. Since M_{n_f} is nonsingular, Algorithm 5 becomes a standard bundle method and the proof of, for example, Theorem XV.3.2.4 of [HULL93] can be reproduced. For the sake of completeness, we give here a simplified version (which cannot be generalized when the bundle is aggregated).

From the definition (11) of δ^k, we have for k large enough and $i = 1, \ldots, k$

$$f(y_i) + \langle g_i, y_{k+1} - y_i \rangle + \tfrac{1}{2} \langle M_{n_f}(y_{k+1} - x_{n_f}), y_{k+1} - x_{n_f} \rangle \le f(x_{n_f}) - \delta^k \quad (23)$$

(δ^k stands for $\delta^k_{n_f}$). On the other hand, non-descent implies

$$f(x_{n_f}) - m\delta_i \le f(y_i) \quad \text{for } i \text{ large enough}$$

and we obtain by addition (neglecting the quadratic term):

$$\langle g_i, y_{k+1} - y_i \rangle \le m\delta^i - \delta^k \quad \text{for large } i \text{ and } k, \quad \text{with } i \le k. \quad (24)$$

Now, there exists by construction an i such that $x_{n_f} = y_i$; taking this i in (23):

$$\langle g_i, y_{k+1} - x_{n_f} \rangle + \tfrac{1}{2} \langle M_{n_f}(y_{k+1} - x_{n_f}), y_{k+1} - x_{n_f} \rangle \le -\delta^k \le 0.$$

Because M_{n_f} is positive definite, this implies the boundedness of $\{y_k\}$, hence, from (15), of $\{g_k\}$: the left-hand side in (24) can be made arbitrarily close to 0. On the other hand, since $\check{f}_k \le \check{f}_{k+1}$, the sequence $\{\delta^k\}$ is decreasing and has a limit, which therefore has to be 0. Then, from (13), $\check{g}_k$ and ε^k tend to 0 and (14) shows that x_{n_f} minimizes f. $\qquad\square$

We now turn to the case of infinitely many descent-steps; our study will be limited to the diagonal variant (dqN). The result below is rather classical ([Kiw90], [SZ92], [CL93]), apart from the possible degeneracy of the quadratic term in the proximal problem (10). The proof suggests that the particular value $1/m$ of the safeguarding parameter in (17) is not totally innocent.

Theorem 12. *Consider Algorithm 5 with the following options:*

– diagonal quasi-Newton update (dqN);
– safeguarded resolution of the quadratic subproblems, as explained at the end of §3.

If an infinite sequence $\{x_n\}$ is generated, it is minimizing: $f(x_n) \to \inf f$.

Proof. The key is (22): $\{\mu_n\}$ is a nonincreasing sequence, hence $t_n := 1/\mu_n$ forms a divergent series. We consider two cases.

Suppose first $\mu_n > 0$ for all n. Then the proof is classical: we can reproduce for example Proposition 2.2 in [CL93] or Theorem XV.3.2.2 in [HULL93].

Now assume $\mu_n = 0$ for some n. In view of (22), $\mu_p = 0$ for all $p \geq n$. As long as the safeguard of §3 does not come into play, y^c minimizes $\check{f}_k$ and the situation is essentially the same as before. From (13), $\delta_n^k = f(x_n) - \check{f}_k(y^c)$. Since $\check{f}_k \leq f$ and the descent test forces $\delta_n^k \to 0$, the conclusion still holds.

The last possibility is when $\mu_n = 0$ with an active safeguard. Then (14) cannot be used because, after perturbation by ℓ, the model $\check{f}_k$ is no longer below f. Rather, combine (17) with the descent test to obtain

$$f(x_{n+1}) \leq f(x_n) - m[f(x_n) - \ell] = f(x_n) - [f(x_{n-1}) - f(x_n)].$$

If this holds infinitely often, $f(x_n) \to -\infty$ and the conclusion still holds. $\square$

We do not know if the above result can be proved for the variant (fqN). The usual technique for BFGS updates is to bound the trace of M_n from above. Here the inequality

$$\operatorname{tr} M_{n+1} \leq \operatorname{tr} M_n + \operatorname{tr} A \leq \operatorname{tr} M_n + \Lambda_n$$

is easily obtained from (21). However, it is not sharp enough to establish the divergence of the series $\{1/\Lambda_n\}$ (a key argument, see for example [BGLS93]).

Our last result is related to speed of convergence. In fact, consider (20), which is the basis for all our development. We are trying to minimize a function F which, despite appearances, depends on the iteration index n, through the matrix M_n. We should therefore check whether the whole idea makes any sense. We do this in an ideal situation: assume that, for given x_n and M_n, the regularized values $F(x_n)$ and $\nabla F(x_n)$ can be exactly computed. Limiting ourselves to the combination (dqN)-$(g$-$f)$, we obtain the following simplification of Algorithm 5:

Algorithm 13.

Step 0 (Initialization). Choose $x_0 \in \mathbb{R}^N$ and compute $g_0 := g(x_0)$. Set $\mu_0 = 1, n = 0$.

Step 1 (Computation of the proximal point). If the stopping criterion is not satisfied, solve

$$x_{n+1} \in \operatorname{Argmin}[f(x) + \tfrac{1}{2}\mu_n \langle x - x_n, x - x_n \rangle].$$

Compute $g_{n+1} := g(x_{n+1})$ and set

$$\Delta x := x_{n+1} - x_n, \quad v := g_{n+1} - g_n, \quad u := \Delta x + \frac{1}{\mu_n} v.$$

Step 2 (Descent-step). Set

$$\mu_{n+1} := \begin{cases} \dfrac{|v|^2}{\langle v, u \rangle} & \text{if } \langle v, u \rangle > 0, \\ 0 & \text{if not.} \end{cases}$$

Increase n and loop to 1. $\square$

The above expression for u comes from (19). We can also write the update as

$$\frac{1}{\mu_{n+1}} = \frac{1}{\mu_n} + \frac{\langle v, \Delta x \rangle}{|v|^2} \qquad (25)$$

whenever $\langle v, u \rangle > 0$.

Theorem 14. *If ∇f is locally Lipschitzian, then $\mu_n \to 0$. Make the following additional assumptions: f has a (unique) minimal point $\bar{x}$ and a quadratic growth condition holds: for some $\alpha > 0$,*

$$f(x) \geq f(\bar{x}) + \alpha |x - \bar{x}|^2 .$$

Then $f(x_n)$ tends to $f(\bar{x})$ q-superlinearly.

Proof. If $\mu_n = 0$ for some n, x_{n+1} is obviously a minimizer of f, so we can assume in Step 2 that (25) holds for all n. If ∇f has the local Lipschitz constant L, we can apply [Pow76] or Theorem X.4.2.2 of [HULL93]: $\langle v, \Delta x \rangle / |v|^2 \geq 1/L$ and $1/\mu_n \to +\infty$.

Now apply the subgradient inequality:

$$f(x_n) \geq f(x_{n+1}) + \langle g_{n+1}, x_n - x_{n+1} \rangle = f(x_{n+1}) + \frac{|g_{n+1}|^2}{\mu_n} .$$

On the other hand, our growth condition implies (Lemma 4.3 of [BGLS93]):

$$\frac{1}{\alpha} |g_{n+1}|^2 \geq f(x_{n+1}) - f(\bar{x})$$

hence

$$f(x_n) - f(\bar{x}) \geq f(x_{n+1}) - f(\bar{x}) + \frac{\alpha}{\mu_n} [f(x_{n+1}) - f(\bar{x})] .$$

The conclusion follows, since $\frac{1}{1 + \alpha/\mu_n} \to 0$. □

Note that we have

$$\alpha |x - \bar{x}|^2 \leq f(x) - f(\bar{x}) \leq L |x - \bar{x}|^2 ,$$

so $\{f(x_n)\}$ and $\{x_n\}$ converge at the same speed. The above result may seem artificial since, in our ideal situation, the ideal value for the penalty is $\mu = 0$. To become really convincing, the proof should be extended to the variant (fqN). Let us say that, at least, (25) gives a constructive (and hopefully reasonable) way of driving μ to 0.

To conclude this section, let us give some comments concerning the assumptions in Theorem 14. The growth condition is assessed by the following result:

Proposition 15. *Let Algorithm 13 be applied to the univariate function $f(x) = 1/3|x|^3$, starting from $x_0 > 0$. Then the convergence of $\{x_n\}$ to the solution 0 cannot be q-superlinear.*

Proof. First draw a picture to see that $x_n > x_{n+1} > 0$ for all n. The next iterate $p(x_n)$ satisfies the relation

$$p^2(x_n) + \mu_n[p(x_n) - x_n] = 0. \qquad (26)$$

Divide succesively by $\mu_n p(x_n)$ and μ_n^2 to obtain

$$\frac{p(x_n)}{\mu_n} = \frac{x_n}{p(x_n)} - 1 \quad \text{and} \quad \frac{x_n}{\mu_n} = \frac{p^2(x_n)}{\mu_n^2} + \frac{p(x_n)}{\mu_n}. \qquad (27)$$

Next, straightforward calculations in (25) and multiplication by $p(x_n)$ give

$$\frac{x_{n+1}}{\mu_{n+1}} = \frac{p(x_n)}{\mu_n} + \frac{p(x_n)}{x_n + p(x_n)}. \qquad (28)$$

Now assume for contradiction that $\{x_n\}$ converges to 0 q-superlinearly, i.e., $p(x_n)/x_n \to 0$. Then $p(x_n)/\mu_n \to +\infty$ and $s_n := x_n/\mu_n \to +\infty$ because of (27); also, from (28), $s_{n+1} = p(x_n)/\mu_n + \varepsilon_n$, where $\varepsilon_n := p(x_n)/[x_n + p(x_n)]$ forms a convergent series. Finally, compute explicitly $p(x_n)$ from (26) and obtain the equalities

$$2s_{n+1} - \varepsilon_n + 1 = \frac{2p(x_n)}{\mu_n} + 1 = \sqrt{1 + 4x_n/\mu_n} = \sqrt{1 + 4s_n}.$$

Using the inequality $\sqrt{1 + 4s_n} \le 1 + 2s_n$ and summing, we see that $\{s_n\}$ is bounded. This is the required contradiction. $\qquad \square$

As for our smoothness assumption on f, its necessity is not obvious: we do not know if the property $\mu_n \to 0$ is really crucial. Such an asymmetry between the two assumptions in Theorem 14 can be explained:

- The growth condition appears natural: if it does not hold, the model F may become a gross approximation of f.
- Likewise, if f does not have a Lipschitzian gradient, its growth prevails over the quadratic perturbation.

6 Implementation Issues

We have already suggested that Algorithm 5 is only schematic, and that some line-search is needed. For this, two strategies may be adopted:

- *Standard line-search.* Once y^c is computed, the next iterate (x_{n+1} or y_{k+1}) is searched along the half-line $\{x_n + t(y^c - x_n) : t > 0\}$.
- *Curved search.* Adjust the norming, replacing the matrix M_n in (10) by M_n/t. Then the candidate y^c depends on $t > 0$, but the mapping $t \mapsto y^c$ is no longer positively homogeneous. The next iterate is searched along a curve, parametrized by the "stepsize" t.

6.1 Extrapolation

When a descent test is accepted, the matrix M_n is going to be updated, but it is advisable to avoid a degenerate M_{n+1}. In fact:

- The proof of Theorem 11 breaks down when M_{n_f} is degenerate; a very first difficulty is that the candidates y^c may become unbounded.
- In Theorem 12, the relevance of a degenerate M_n is questioned: there, such a degeneracy means $M_n = \mu_n I = 0$ forever, unfortunate for an algorithm trying to identify a second order behaviour. Degeneracy appears as clumsiness, at least for the diagonal variant.
- Empirically, a degenerate matrix also presents a serious danger. Suppose an iteration where $M_n = 0$; then the algorithm starts a sequence of pure cutting-plane iterations. We may be in the situation of Nemirovski's counter-example: an enormous number of null-steps becomes necessary until a descent iterate is found.

Inspection of the update formulae in §4 shows that v should be nonzero to yield an invertible M_{n+1}. Guaranteeing this property seems difficult for the (x-$\check{f}$) variant, but it is straightforward with (g-f):

Lemma 16. *Let u and v be given by (19) and take $m' < 1$. Then*

$$\langle g(x_{n+1}), x_{n+1} - x_n \rangle \geq -m'\delta_n \qquad (29)$$

implies $\langle v, u \rangle > 0$.

Proof. The definition of $\check{f}_k$ implies

$$f(x_n) + \langle g(x_n), x_{n+1} - x_n \rangle \leq \check{f}_k(x_{n+1});$$

add $1/2 \langle M_n(x_{n+1} - x_n), x_{n+1} - x_n \rangle$ to both sides, to obtain

$$\langle g(x_n), x_{n+1} - x_n \rangle \leq \check{f}_k(x_{n+1}) + \tfrac{1}{2} \langle M_n(x_{n+1} - x_n), x_{n+1} - x_n \rangle - f(x_n) = -\delta_n .$$

Subtracting from (29), we get

$$\langle v, \Delta x \rangle = \langle g(x_{n+1}) - g(x_n), x_{n+1} - x_n \rangle \geq (1 - m')\delta_n > 0$$

and the conclusion follows due to (19). $\qquad\qquad\square$

To guarantee positive definite matrices, it therefore suffices to find a candidate satisfying (29) as well as the descent condition. This is nothing but a Wolfe type criterion for the stepsize. With $m' \in]m, 1[$, this problem is classical for a standard line-search; as for the curved search, it is solved in (2.17) of [SZ92].

6.2 Interpolation

An awake reader may have already realized that the proof of Theorem 12 is "too easy to be true": its work-horse (22) is a luxury argument. Indeed the same proof would apply if $\{\mu_n\}$ were increasing with a moderate speed.

Remark 17. It can be proved that, if M_n is updated with the symmetric rank-one formula, then the same phenomenon occurs, namely

$$\operatorname{tr} M_{n+1} \le \operatorname{tr} M_n .$$

Moreover, this formula preserves positive definiteness of M_{n+1}. Based on these properties, convergence of Algorithm 5 can be established when SR1 replaces (dqN) for "Up". □

Clearly the algorithm will be in trouble if M_0 is chosen unduly small. To struggle against this misbehaviour, we accept to increase the penalty through a division of the matrix by a short stepsize. Here comes a really delicate point: when the descent test (2) is not satisfied, a decision must be made: either to decrease the stepsize, or to enrich the bundle via a null-step (or even both, why not?). A possible strategy is as follows.

Having y^c, compute the linearization error

$$e^c := f(x_n) - [f(y^c) + \langle g(y^c), x_n - y^c \rangle] .$$

If e^c is large, the new affine piece in $\check{f}_k$ is going to have little influence on the computation of the new candidate. It is therefore reasonable to make a null-step only if

$$e^c \le m'' \delta_n , \tag{30}$$

where m'' is a positive tolerance. When (30) is not satisfied, an interpolation is performed.

Remark 18. An important question is whether this backtracking procedure spoils the bundling mechanism. In other words, will repeated interpolations eventually produce a stepsize $t > 0$ such that (30) holds? It can be proved that the answer is yes; this actually relies on the semismoothness of convex functions [Mif77]. □

6.3 An Improved Algorithm

We end this paper with an example of algorithm including the refinements presented above. It uses the $(g\text{-}f)$ option for the update since the preceeding analysis reveals its definite advantages. Note an important detail: it is M_n/t and not M_n, which we update in the quasi-Newton formula, as in [OS76].

Algorithm 19.
Step 0 (Initialization). Choose $m \in]0, 1[, m' \in]m, 1[, m'' > 0, x_0 \in \mathbb{R}^N$. Set $y_0 = x_0, M_0 = 1, n = k = 0$.

Step 1 (*t*-adjustment). Execute Algorithm 20 to obtain $t > 0$ and y^c.

Step 2 (Descent-step). Update $x_{n+1} = y^c$. Set $\Delta x := x_{n+1} - x_n$, $v := g(x_{n+1}) - g(x_n)$, $u := \Delta x + t M_n^{-1} v$ and update $M_{n+1} = Up(M_n/t, u, v)$. Increase n.

Step 3 (Null-step). Set $y_{k+1} = y^c$, increase k. Loop to 1. $\qquad\qquad$ $\square$

A (dqN) strategy in Step 2 would result in an alternative to the proposals of [Kiw90] and [SZ92]. As for the *t*-adjustement, we have chosen a curved search, which we believe is more natural:

Algorithm 20. The data are $\check{f}_k, x_n, M_n$ and the tolerances m, m', m''.

Step 0. Set $t = 1, t_L = 0, t_R = +\infty$.

Step 1. Compute

$$y^c = y^c(t) := \operatorname{argmin}\left[\check{f}_k(y) + \tfrac{1}{2t}\langle M_n(y - x_n), y - x_n\rangle\right]$$
$$\delta := f(x_n) - \check{f}_k(y^c) - \tfrac{1}{2t}\langle M_n(y^c - x_n), y^c - x_n\rangle$$
$$e^c := f(x_n) - [f(y^c) + \langle g(y^c), x_n - y^c\rangle].$$

Step 2. If $f(y^c) > f(x_n) - m\delta$, go to Step 4.

Step 3. If $\langle g(y^c), y^c - x_n\rangle \geq m'\delta$, stop with a Descent-step.

Otherwise set $t_L = t$, compute a new t in $]t_L, t_R[$ and go to Step 1.

Step 4. If $t_L = 0$ and $e^c \leq m''\delta$, stop with a Null-step.

Otherwise set $t_R = t$, compute a new t in $]t_L, t_R[$ and go to Step 1. $\qquad$ $\square$

References

[BGLS93] J. Bonnans, J.Ch. Gilbert, C. Lemaréchal, and C. Sagastizábal. A family of Variable Metric Proximal methods. Rapport de Recherche 1851, INRIA, 1993.

[CG59] E. Cheney and A. Goldstein. Newton's method for Convex Programming and Tchebycheff approximations. *Numerische Mathematik*, 1:253–268, 1959.

[CL93] R. Correa and C. Lemaréchal. Convergence of some algorithms for convex minimization. Manuscript, INRIA, 78153 Le Chesnay Cedex (France), 1993.

[DM77] J.E. Dennis and J.J. Moré. Quasi-Newton methods, motivation and theory. *SIAM Review*, 19:46–89, 1977.

[FW56] M. Frank and P. Wolfe. An algorithm for quadratic programming. *Naval Research Logistic Quarterly*, 3:95–110, 1956.

[HULL93] J.B. Hiriart-Urruty and C. L-Lemaréchal. *Convex Analysis and Minimization Algorithms*. Springer-Verlag, 1993.

[Kel60] J. E. Kelley. The cutting plane method for solving convex programs. *J. Soc. Indust. Appl. Math.*, 8:703–712, 1960.

[Kiw83] K.C. Kiwiel. An aggregate subgradient method for nonsmooth convex minimization. *Mathematical Programming*, 27:320–341, 1983.

[Kiw90] K.C. Kiwiel. Proximity control in bundle methods for convex nondifferentiable minimization. *Mathematical Programming*, 46:105–122, 1990.

[Lem78] C. Lemaréchal. Bundle methods in nonsmooth optmization. In C. Lemaréchal and R. Mifflin, editors, *Nonsmooth optimization*. Pergamon Press, Oxford, 1978.

162

[Mif77]   R. Mifflin. Semi-smooth and semi-convex functions in constrained optimiza-
          tion. *SIAM Journal on Control and Optimization*, 15:959–972, 1977.

[Mor65]   J.J. Moreau. Proximité et dualité dans un espace hilbertien. *Bulletin de la
          Société Mathématique de France*, 93:273–299, 1965.

[OS76]   S.S. Oren and E. Spedicato. Optimal conditioning of self-scaling variable
          metric algorithms. *Mathematical Programming*, 10:70–90, 1976.

[Pow76]   M.J.D. Powell. Some global convergence properties of a variable metric
          algorithm for minimization without exact line searches. In R.W. Cottle
          and C.E. Lemke, editors, *Nonlinear Programming*, number 9 in SIAM-AMS
          Proceedings. American Mathematical Society, Providence, RI, 1976.

[Sho70]   N. Shor. Utilization for the operation of space dilatation in the minimiza-
          tion of convex function. *Cybernetics*, 6:7–15, 1970.

[Sho85]   N. Shor. *Minimization methods for non-differentiable functions*. Springer-
          Verlag, Berlin, 1985.

[Sko73]   V. Skokov. Note on minimization methods employing space stretching. *Cy-
          bernetics*, 10:689–692, 1973.

[SZ92]   H. Schramm and J. Zowe. A version of the bundle idea for minimizing a
          nonsmooth function: conceptual idea, convergence analysis, numerical re-
          sults. *SIAM Journal on Optimization*, 2(1):121–152, 1992.

[Yos64]   K. Yosida. *Functional Analysis*. Springer Verlag, 1964.

This article was processed using the LaTeX macro package with LMAMULT style

CONTRIBUTED PAPERS

Non linear programming

Stability

PERTURBATION OF STATIONARY SOLUTIONS IN SEMI-INFINITE OPTIMIZATION

Diethard Klatte

Institut für Operations Research, Universität Zürich
Moussonstrasse 15, CH-8044 Zürich, Switzerland

1 Introduction

This paper deals with semi-infinite nonlinear programs under data perturbations. Consider a family of semi-infinite optimization problems

$$\mathrm{P}(t): \qquad f(t,x) \to \min_x \qquad \text{s.t.} \qquad x \in M(t),$$

where t is a parameter varying over T, and M is a multifunction which assigns to each $t \in T$ the solution set $M(t)$ of the system

$$h_i(t,x), i = 1,...,p; \quad g(t,x,y) \geq 0 \; \forall y \in K(t). \tag{1.1}$$

Throughout we shall (at least) suppose that T is an open subset of $\mathbb{R}^m$, $\mathcal{K}$ is a nonempty and compact subset of $\mathbb{R}^s$, and the functions $f : T \times \mathbb{R}^n \to \mathbb{R}$, $h = (h_1,...,h_p) : T \times \mathbb{R}^n \to \mathbb{R}^p$, $g : T \times \mathbb{R}^n \times \mathbb{R}^s \to \mathbb{R}$ as well as the multifunction K defined on T have the following properties:

$$f, h \text{ and } g \text{ are continuous,} \tag{1.2}$$

$$f, h \text{ and } g \text{ are continuously differentiable with respect to } x, \tag{1.3}$$

$$\nabla_x f(.,.), \nabla_x h(.,.) \text{ and } \nabla_x g(.,.,.) \text{ are continuous,} \tag{1.4}$$

$$\emptyset \neq K(t) \subset \mathcal{K} \quad \forall t \in T, \tag{1.5}$$

$$K \text{ is compact-valued and lower semicontinuous on } T. \tag{1.6}$$

For each $(t,x) \in T \times \mathbb{R}^n$, $E(t,x) := \{y \in K(t) | g(t,x,y) = 0\}$ is compact.

In Section 2 it will turn out that the stationary solution set mapping and an appropriate multiplier set mapping are upper semicontinuous with respect to some neighborhood of a given (t^0, x^0), where x^0 is a stationary solution to $\mathrm{P}(t^0)$. This extends results for standard (finite) nonlinear programs (Robinson [19]) and for semi-infinite nonlinear programs in the case

$K(t) \equiv \mathcal{K}$ (Klatte [17]). In our approach, the use of finite-dimensional multipliers and of an extension of the Mangasarian-Fromovitz CQ are essential.

In Section 3 we shall show that this CQ implies for several special cases the pseudo-Lipschitz continuity of the feasible set mapping and, together with some second-order optimality assumption, a "nice" stability behavior for local minimizing sets. We shall avoid the assumptions of the reduction ansatz ([9, 11]) which would allow to reduce the infinite number of constraints to a finite one in some neighborhood of a given (t^0, x^0) satisfying (1.1). The point is that the violation of the assumptions for applying the reduction ansatz might be stable under perturbations of the constraint functions, cf. [12].

2 Upper Semicontinuity of Stationary Points

When $t \in T$ is given, a point x is called a (Karush-Kuhn-Tucker type) *stationary solution* of P(t), if there are a finite, possibly empty, set $\{y^1, ..., y^r\}$ in $E(t, x)$ as well as numbers $u_i \in \mathbb{R}(i = 1, ..., p)$ and $v_j \geq 0(j = 1, ...r)$ such that

$$x \in M(t), \quad \nabla_x f(t, x) + \sum_{i=1}^{p} u_i \nabla_x h_i(t, x) - \sum_{j=1}^{r} v_j \nabla_x g(t, x, y^j) = 0_n. \quad (2.1)$$

The multifunction which assigns to each $t \in T$ the set of all stationary solutions of P(t) is denoted by S.

We shall say that $x \in M(t)$ satisfies the *extended Mangasarian Fromovitz constraint qualification* (EMFCQ), if

$$\nabla_x h(t, x) \text{ has rank } p, \text{ and there is some } \xi \in \mathbb{R}^n \text{ with}$$
$$\xi^T \nabla_x h_i(t, x) = 0 \ (i = 1, ..., p), \ \xi^T \nabla_x g(t, x, y) > 0 \ (\forall y \in E(t, x)). \quad (2.2)$$

The latter definition is due to Jongen, Twilt and Weber [12], it is based on some CQ introduced in Hettich and Zencke [11]. (2.2) is a regularity condition assuring that a local minimizer of (P) is a stationary solution of (P), cf., e.g., [9, 11]. Obviously, EMFCQ implies $p \leq n$ and, if $E(t, x) \neq \emptyset$ then $\xi \neq 0_n$ and $p \leq n - 1$. If $\mathcal{K}$ is finite and the mapping K does not depend on the parameter, Condition (2.2) is the standard Mangasarian-Fromovitz constraint qualification. We note that in the case $K(t) \equiv \mathcal{K}$ there are interesting studies of the topological stability of M under EMFCQ, cf., e.g., [12, 23]. Metric regularity under EMFCQ is discussed in [8].

Following [17], we shall use a concept of so-called reduced multipliers. Set for given (t, x), $G(t, x) := \{\nabla_x g(t, x, y) | y \in E(t, x)\}$, and let cone $G(t, x)$ be its conical hull. Put $m(t, x) := \dim \text{cone}\, G(t, x)$ if $E(t, x) \neq \emptyset$ and $m(t, x) := 0$ if $E(t, x) = \emptyset$. In what follows let Ω be an open subset of $\mathbb{R}^n$. Now we fix any integer d such that $d \geq m(t, x) + 1$ for all $(t, x) \in T \times \Omega$.

For $(t, x) \in T \times \Omega$, the union of all sets $\Lambda^Y(t, x)$ with Y in $[E(t, x)]^r$ and $r \in \{0, 1, ..., d\}$ will be denoted by $\Lambda(t, x)$, where

$$\Lambda^Y(t, x) := \left\{ (u, v) \in \mathbb{R}^p \times \mathbb{R}^d_+ \,\middle|\, \begin{array}{l} (t, x, u, v_{[r]}, Y) \text{ satisfies (2.1),} \\ v_j = 0 \; \forall j : r < j \leq d \end{array} \right\}, \quad (2.3)$$

and $v_{[r]}$ abbreviates the projection of $v \in \mathbb{R}^d$ onto its first r coordinates $(v_1, ..., v_r)$. This defines a multifunction Λ from $T \times \Omega$ to $\mathbb{R}^p \times \mathbb{R}^d_+$. For fixed t and a given stationary solution $x \in S(t) \cap \Omega$, each element of $\Lambda(t, x)$ is called a *reduced multiplier (in $\mathbb{R}^p \times \mathbb{R}^d_+$) associated with (t, x)*.

By construction and by Caratheodory's lemma, we have

$$\Lambda(t, x) \neq \emptyset \iff x \in S(t). \quad (2.4)$$

In [17], the author has shown a generalized Gauvin theorem: $\Lambda(t, x)$ associated with some $(t, x) \in$ graph S is bounded if and only if EMFCQ is satisfied at (t, x). Concerning the implication "$\Lambda(t, x)$ is bounded $\Rightarrow$ EMFCQ holds at (t, x)" the choice of $d \geq m(t, x) + 1$ is essential (cf. [17]), while, by Caratheodory's theorem for cones, $d = m(t, x)$ would be sufficient for the existence of a reduced multiplier. For this reason and for a uniform representation when studying perturbations, we have required $d \geq m(t, x) + 1$ for all (t, x) in $T \times \Omega$.

In the following we extend Theorem 2.3 in Robinson [19] ($K(t) \equiv \mathcal{K}$, $\mathcal{K}$ finite) and Theorem 1 in [17] ($K(t) \equiv \mathcal{K}$). Closedness, upper semicontinuity (in Berge's sense) and lower semicontinuity (in Hausdorff's sense) are used as in [5, Section 2.2]. Recall that for a closed multifunction, upper semicontinuity is equivalent to local boundedness.

Theorem 2.1 *Consider $P(t)$, $t \in T$, under the general assumptions imposed in Section 1. Suppose additionally that K is closed on T. Then the multifunction Λ is closed on $T \times \Omega$.* $\diamond$

Proof. Let $(\bar{t}, \bar{x})$ be any point in $T \times \Omega$. Consider arbitrary sequences $\{(t^k, x^k)\}$, $\{(u^k, v^k)\}$ with

$$(t^k, x^k) \to (\bar{t}, \bar{x}) \text{ and } (u^k, v^k) \in \Lambda(t^k, x^k) \; (\forall k) \quad (2.5)$$

such that (u^k, v^k) converges to some $(\bar{u}, \bar{v})$. Hence, by definition of $\Lambda(t^k, x^k)$, one has $x^k \in M(t^k)$ for each k, and

$$0_n = \nabla_x f(t^k, x^k) + \sum_{i=1}^{p} u_i^k \nabla_x h_i(t^k, x^k) - \sum_{j=1}^{r(k)} v_j^k \nabla_x g(t^k, x^k, y^{jk}) \qquad (2.6)$$

with some $r(k) \in \{0, 1, ..., d\}$ and certain points

$$y^{jk} \in E(t^k, x^k), \ j = 1, ..., r(k). \qquad (2.7)$$

By (1.2), (1.5) and (1.6) we immediately have $\bar{x} \in M(\bar{t})$. Extracting subsequences if necessary, we may assume $r(k) \equiv r$. Assumption (1.2), the compactness of $\mathcal{K}$ and the closedness of K on T imply that

$$(t, x) \mapsto E(t, x) = K(t) \cap \{y \in \mathcal{K} | g(t, x, y) = 0\}$$

is upper semicontinuous (and also closed) on T (cf., e.g., [5, Lemma 2.2.3, Thm. 3.1.2]) . Therefore, each of the r sequences $\{y^{jk}\}$ has a subsequence converging to some element $\bar{y}^j$ of $E(\bar{t}, \bar{x})$, hence we may also assume

$$y^{jk} \to \bar{y}^j \quad (j = 1, ..., r). \qquad (2.8)$$

Thus, by applying (1.4) and by passing to the limit in (2.6), one has

$$0_n = \nabla_x f(\bar{t}, \bar{x}) + \sum_{i=1}^{p} \bar{u}_i \nabla_x h_i(\bar{t}, \bar{x}) - \sum_{j=1}^{r} \bar{v}_j \nabla_x g(\bar{t}, \bar{x}, \bar{y}^j)$$

and $\bar{v}_j = 0$ if $r + 1 \leq j \leq d$. Thus $(\bar{u}, \bar{v}) \in \Lambda(\bar{t}, \bar{x})$. □

Theorem 2.2 *Assume the hypotheses of Theorem 2.1. Let x^0 be a stationary solution of $P(t^0)$. If $x^0 \in M(t^0)$ satisfies EMFCQ, then there exist neighborhoods $\mathcal{U}$ of t^0 and $\mathcal{N}$ of x^0 such that the multifunctions Λ and $t \mapsto S(t) \cap \mathcal{N}$ are upper semicontinuous on $\mathcal{U} \times \mathcal{N}$ and $\mathcal{U}$, respectively.* ◇

Proof. In the case $E(t^0, x^0) \neq \emptyset$, the proof is a copy of that one in [17] and will be hence omitted here. For $E(t^0, x^0) = \emptyset$ we only note that, by the closedness of $E(\cdot, \cdot)$ (cf. the proof of Theorem 2.1), $E(t, x)$ is empty for all (t, x) in some neighborhood $\mathcal{W}$ of (t^0, x^0). Hence for $x \in S(t)$ with (t, x) in $\mathcal{W}$, the condition $g(t, x, y) > 0 (\forall y \in K(t))$ will be automatically satisfied and so the Karush-Kuhn-Tucker system (2.1) reduces to

$$h(t, x) = 0_p, \ \nabla_x f(t, x) + \sum_{i=1}^{p} u_i \nabla_x h_i(t, x) = 0_n.$$

Because of the linear independence of $\{\nabla_x h_i(t^0, x^0)\}_{i=1,\ldots,p}$ then the desired result is classical. This completes the proof. $\square$

The preceding theorem particularly says that the sets of reduced multipliers are closed at all events and compact under regularity. It is a trivial fact from linear programming that under the assumptions of Theorem 2 (b), $S(t)$ may become empty for t near t^0.

We note that the proof of the generalized Gauvin theorem in [17] even provides a simple bound for the set of reduced multipliers $\Lambda(t^0, x^0)$ under EMFCQ at some stationary solution x^0 of $P(t^0)$. Let $E(t^0, x^0)$ be nonempty, let ξ be a EMFCQ vector according to (2.2) and denote by H^- the pseudo-inverse of $\nabla_x h(t^0, x^0)$. Since the set $G(t^0, x^0)$ (cf. the definition above) is compact, $a := \min_{q \in G(t^0, x^0)} \xi^T q$ and $b := \max_{q \in G(t^0, x^0)} \|H^- q\|$ are well defined, and one has for all (u, v) in $\Lambda(t^0, x^0)$,

$$\|u\| \leq \|H^- \nabla_x f(t^0, x^0)\| + b a^{-1} \xi^T \nabla_x f(t^0, x^0), \quad \sum_{j=1}^d v_j \leq a^{-1} \xi^T \nabla_x f(t^0, x^0),$$

by [17, Theorem 1].

3 Stability of Local Minimizers

This section is devoted to the stability behavior of local minimizers of the parametric problem introduced in Section 1. We denote by B_n the closed unit ball in $\mathbb{R}^n$, by $B(z, \varepsilon)$ the closed ε - neighborhood of z both in $\mathbb{R}^n$ and T, and by $\mathrm{cl}\, Q$ the closure of $Q \subset \mathbb{R}^n$. Given $t^0 \in T$, a strict local minimizer x^0 of $\mathrm{SIP}(t^0)$ is called to be *of order* $\kappa \geq 1$ if there are real numbers $\varrho > 0$ and $c > 0$ such that

$$f(x, t^0) \geq f(x^0, t^0) + c\|x - x^0\|^\kappa \qquad \forall\, x \in M(t^0) \cap B(x^0, \varrho).$$

Denote by $\psi_Q(t)$ the set of all global minimizers for $f(., t)$ on $M(t) \cap \mathrm{cl}\, Q$, $t \in T$, $Q \subset \mathbb{R}^n$, and by $\psi_{loc}(t)$ the set of all local minimizers for $f(., t)$ w.r. to $M(t)$.

Following concepts in [1, 4, 15, 20], we shall say that a local minimizer x^0 of $\mathrm{SIP}(t^0)$ is *stable* (w.r. to $\mathrm{SIP}(t)$, $t \in T$), if for some positive real numbers ε', δ' and for each $\varepsilon \in (0, \varepsilon')$ there is some $\delta \in (0, \delta')$ such that with $Q := B(x^0, \varepsilon')$,

$$\psi_Q(t^0) = \{x^0\}, \tag{3.1}$$

172

$$\emptyset \neq \psi_Q(t) \subset \psi_{loc}(t) \qquad \forall t \in B(t^0, \delta'), \tag{3.2}$$

$$\psi_Q(t) \subset B(x^0, \varepsilon) \qquad \forall t \in B(t^0, \delta). \tag{3.3}$$

A stable local minimizer is called to be *stable with rate* r, $r \in (0,1]$, if the relation between ε and δ in (3.3) is quantified (with some $\beta > 0$) by

$$\|x - x^0\| \leq \beta \, d(t, t^0)^r \qquad \forall t \in B(t^0, \delta') \;\; \forall x \in \psi_Q(t). \tag{3.4}$$

Obviously (3.1) – (3.3) include that the multivalued selection ψ_Q of the multifunction ψ_{loc} is continuous at t^0 (in the sense of [5, Chapter 2]).

For both stability properties, we now recall sufficient conditions which specialize more abstract results to SIP(t). Note that by the general assumptions, M is a closed multifunction.

Proposition 3.1 *Let $t^0 \in T$, and let x^0 be a strict local minimizer of* P(t^0). *Suppose that M is lower semicontinuous at (x^0, t^0), i.e., with $x^0 \in M(t^0)$, for each sequence $t^k \to t^0$ there is some sequence $x^k \to x^0$ satisfying $x^k \in M(t^k)$ for k sufficiently large. Then x^0 is stable.* ◇

Proof. The assertion is a specialization of Robinson [20, Thm. 4.3]. □

We note that in the case $K(t) \equiv \mathcal{K}$, EMFCQ at $x^0 \in M(t^0)$ implies the lower semicontinuity of M at (t^0, x^0), cf., e.g., [12, 15]. This is, in fact, also a consequence of well-known (metric) regularity results in the Lipschitz analysis setting (cf., e.g., Auslender [4]) or in the Banach space formulation of (1.1) (Robinson [18]). For a discussion of these connections we refer to [15, 17, 25]. In the general case, the question for "weakest" conditions ensuring lower semicontinuity of M is open. Theorem 3.3 below concerns pseudo-Lipschitz continuity and hence also (however very strong) conditions for lower semicontinuity.

A multifunction Φ from T to $\mathbf{R}^n$ is called *Lipschitzian on* $\tilde{T} \subset T$ if there is some constant $c > 0$ such that $\Phi(t') \subset \Phi(t) + c\|t' - t\|B_n$ for all $t, t' \in \tilde{T}$. Following Aubin [2] and Rockafellar [22], we shall say that Φ is *pseudo-Lipschitzian at* $(t^0, x^0) \in \text{graph}\,\Phi$ *with modulus* $c > 0$, if there exist neighborhoods $\mathcal{W}$ of t^0 and $\mathcal{X}$ of x^0 with $M(t') \cap \mathcal{X} \subset M(t) + c_M\|t' - t\|B_n$ for all $t, t' \in \mathcal{W}$. The next proposition gives the quantified version of Proposition 1. A proof – which even does not utilizes the stucture (1.1) – may be found in the technical report [14], it follows a proposal of Alt [1].

Proposition 3.2 *Let* $t^0 \in T$, *and let* x^0 *be a strict local minimizer to* SIP(t^0) *of order* $\kappa \geq 1$. *Let for some* $\gamma \in (0,1]$ *and* $c_f \in (0,+\infty)$ *as well as for some neighborhoods* $\mathcal{U}$ *of* t^0 *and* $\mathcal{V}$ *of* x^0,

$$|f(x,t^0) - f(x',t)| \leq c_f \left(\|x - x'\| + d(t,t^0)^\gamma \right) \qquad \forall t \in U \quad \forall x, x' \in V.$$

Suppose that M *is pseudo-Lipschitzian at* (x^0, t^0). *Then* x^0 *is stable with rate* $r = \gamma \kappa^{-1}$. $\diamond$

For second-order sufficient conditions ensuring that x^0 is a strict local minimizer of order 2 to P(t^0) (under twice differentiability of the data with respect to x), we refer, e.g., to Shapiro [24], Hettich and Still [10], Kawasaki [13], Klatte [15]. Conditions for strict local minimality of order κ under the reduction ansatz (also for $\kappa = 1$) can be found, e.g., in Hettich and Jongen [9] and Hettich and Zencke [11].

Proposition 3.2 provides (for $\gamma = 1$) Hölder stability of order κ^{-1} for strict local minimizers of order κ. Simple examples show that in the case $\kappa = 2$, stability of order 1 (Lipschitz continuity) may not be expected. Under rather strong constraint qualifications and second-order optimality conditions on P(t^0), Shapiro [25] obtains stability of order 1 for global minimizers of P(t) for $K(t) \equiv \mathcal{K}$. Given a strict local minimizer x^0 of the initial problem P(t^0), in [16] the upper Lipschitz continuity of the Karush-Kuhn-Tucker mapping at (t^0, x^0) is proved under the assumptions of the reduction ansatz (again for $K(t) \equiv \mathcal{K}$). Note that in the case of local minimizers the reduction ansatz (though being very restrictive) is not so bad as announced in Section 1 for a feasible solution in general: it is generically applicable in a neighborhood of (t^0, x^0), cf. again the discussion in [12, Section 1].

We finish the paper by showing that EMFCQ at $x^0 \in M(t^0)$ implies the pseudo-Lipschitz continuity of M at (t^0, x^0). This extends a well-known result for standard (finite) nonlinear programs, cf. Rockafellar [22]. Before stating the theorem, we define

$$G(t,x) := \min_{y \in K(t)} g(t,x,y) \quad \forall (t,x) \in T \times \mathbf{R}^n.$$

In the case that G is Lipschitzian near some given (t^0, x^0), we shall say that G satisfies the *Property A* at (t^0, x^0) iff

$$(\alpha, \beta) \in \partial G(t^0, x^0) \quad \Rightarrow \quad \beta \in \partial_x G(t^0, x^0),$$

where ∂G and $\partial_x G$ denote Clarke's generalized and generalized partial (with respect to x) gradients.

Theorem 3.3 *Consider the feasible set mapping M of the parametric problem $P(t)$, $t \in T$, under the assumptions imposed in Section 1. Let $t^0 \in T$ and $x^0 \in M(t^0)$. Suppose additionally that for some neighborhoods $\mathcal{U}$ of t^0 and $\mathcal{V}$ of x^0, h and g are continuously differentiable on $\mathcal{U} \times \mathcal{V}$ and $\mathcal{U} \times \mathcal{V} \times \mathcal{K}$, respectively. Further suppose that K is Lipschitzian on $\mathcal{U}$ and that G satisfies Property A at (t^0, x^0). If EMFCQ holds at x^0, then M is pseudo-Lipschitzian at (t^0, x^0).* ◇

Proof. G is Lipschitzian on $\mathcal{U} \times \mathcal{V}$, cf., e.g., [3]. For each $t \in T$, Clarke's generalized partial gradient with respect to x is (cf., e.g., [21, Prop. 3H])

$$\partial_x G(t, x) = \text{conv} \{\nabla_x g(t, x, y) | y \in \tilde{E}(t, x)\},$$

where $\tilde{E}(t, x) := \text{argmin}_{y \in K(t)} g(t, x, y)$. (1.1) may be written by

$$h(t, x) = 0_p, \qquad G(t, x) \in \mathbb{R}_+,$$

hence Rockafellar's inverse multifunction theorem [21, Cor. 3.5] applies. Taking Property A into account, one has only to show that for each $\beta \in \partial_x G(t^0, x^0)$ the following holds:

$$\nabla_x h(t^0, x^0)\mu + \beta \cdot \nu = 0_n, \ \mu \in \mathbb{R}^p, \ \nu \in \mathbb{R}_+ \quad \Rightarrow \quad \mu = 0_p, \nu = 0. \quad (3.5)$$

This may be derived from EMFCQ. Indeed, if $E(t^0, x^0) = \emptyset$ then $G(t, x) > 0$ for all (t, x) in some neighborhood of (t^0, x^0), and so the assertion is classical. On the other hand, if $E(t^0, x^0) \neq \emptyset$ then $\tilde{E}(t^0, x^0) = E(t^0, x^0)$, and the dual formulation of EMFCQ at $x^0 \in M(t^0)$ (cf. [17, Lemma 1]) yields that for every $n + 1$ points $y^1, ..., y^{n+1}$ in $E(t^0, x^0)$ the system

$$\sum_{i=1}^{p} \mu_i \nabla_x h_i(t^0, x^0) + \sum_{j=1}^{n+1} \nu_j \nabla_x g(t^0, x^0, y^j) = 0_n, \ \mu \in \mathbb{R}^p, \ \nu \in \mathbb{R}_+^{n+1} \quad (3.6)$$

has only the trivial solution. Since, by Caratheodory's theorem, each β in $\partial_x G(t^0, x^0)$ may be expressed as a convex combination of (at most) $n + 1$ elements of $\{\nabla_x g(t^0, x^0, y) | y \in E(t^0, x^0)\}$, (3.6) implies (3.5). This completes the proof. □

Remark 3.4 Let us discuss some special cases ensuring Property A. Suppose that the assumptions of Theorem 3.3 are fulfilled.

a. It is well-known (cf., e.g., [21, Thm. 5L]) that for $K(t) \equiv \mathcal{K}$, G is Clarke regular at (t^0, x^0), i.e., the ordinary directional derivative of G

exists and is equal to Clarke's directional derivative of G at (t^0, x^0) for any direction in $\mathbb{R}^m \times \mathbb{R}^n$. By [6, 2.3.15], Property A is fulfilled at (t^0, x^0).

b. If G is convex with respect to x (for example, if g is concave with respect to x), then Property A holds, cf. [6, 2.5.3].

c. Let $K(t)$ be given by $K(t) := \{y \in \mathbb{R}^s | \tau_i(t, y) \geq 0, \ i = 1, ..., l\}$, $t \in T$, where $\tau_i(\forall i)$ are C^1 functions from $T \times \mathbb{R}^s$ to $\mathbb{R}$. Suppose that for each $y \in \tilde{E}(t^0, x^0)$, $\{\nabla_y \tau_i(t^0, y) | i : \tau_i(t^0, y) = 0\}$ is linearly independent. Note that, by the assumptions, K is compact-valued and $\emptyset \neq K(t) \subset \mathcal{K}$ ($\forall t \in T$), $\mathcal{K}$ compact. Then G is Clarke regular at (t^0, x^0), cf. [7, Cor. 5.4], and again Property A is fulfilled at (t^0, x^0). $\diamond$

References

[1] W. Alt, Lipschitzian perturbations of infinite optimization problems, in: A.V. Fiacco, Ed., *Mathematical Programming with Data Perturbations* (M. Dekker, New York, 1983) 7-21.

[2] J.P. Aubin, Lipschitz behaviour of solutions to convex minimization problems, *Math. Oper. Res.* **9** (1984) 87-111.

[3] J.P. Aubin and A. Cellina, *Differential Inclusions* (Springer, Berlin, 1984).

[4] A. Auslender, Stability in mathematical programming with nondifferentiable data, *SIAM J. Control Optim.* **22** (1984) 239-254.

[5] B. Bank, J. Guddat, D. Klatte, B. Kummer K. Tammer, *Non-Linear Parametric Optimization* (Akademie-Verlag, Berlin, 1982).

[6] F.H. Clarke, *Optimization and Nonsmooth Analysis* (Wiley, New York, 1983).

[7] J. Gauvin and F. Dubeau, Differential properties of the marginal value function in mathematical programming, *Math. Programming* **19** (1982) 101-119.

[8] R. Henrion, D. Klatte, Metric regularity of the feasible set mapping in semi-infinite optimization, *Preprint Nr. 92-5, Fachbereich Mathematik, Humboldt-Universität zu Berlin, 1992* (To appear in *Appl. Math. Optim.*).

[9] R. Hettich, H.Th. Jongen, Semi-infinite programming: Conditions of optimality and applications, In: J. Stoer (ed.), Optimization Techniques, Part 2, *Lecture Notes in Control and Information Sciences, Vol. 7* (Springer, Berlin, 1978) pp. 1-11.

[10] R. Hettich, G. Still, Second order optimality conditions for generalized semi-infinite programming problems, *Manuscript, Universität Trier, FB 4 and University of Twente, Enschede, Faculty of Applied Mathematics, 1991*.

[11] R. Hettich, P. Zencke, *Numerische Methoden der Approximation und semi-infiniten Optimierung* (B.G. Teubner, Stuttgart, 1982).

[12] H.Th. Jongen, F. Twilt, G.-W. Weber, Semi-infinite optimization: Structure and stability of the feasible set, *J. Optim. Theory Appl* **72** (1992) 529-552.

[13] H. Kawasaki, Second-order necessary and sufficient optimality conditions for minimizing a sup-type function, *Appl. Math. Optim.* **26** (1992) 195-220.

[14] D. Klatte, A note on quantitative stability results in nonlinear optimization, *Seminarbericht Nr. 90 der Sektion Mathematik der Humboldt-Universität zu Berlin* (Berlin, 1987) 77-86.

[15] D. Klatte, Stable local minimizers in semi-infinite optimization: Regularity and second-order conditions, *Manuscript, Institut für Operations Research, Universität Zürich, 1992* (To appear in *J. Comput. Appl. Math.*).

[16] D. Klatte, Stability of stationary solutions in semi-infinite optimization via the reduction approach, in: W. Oettli, D. Pallaschke, Eds., *Advances in Optimization* (Springer, Berlin, 1992) 155-170.

[17] D. Klatte, On regularity and stability in semi-infinite optimization, *Manuscript, Institut für Operations Research, Universität Zürich, 1992* (Submitted to *Set-Valued Analysis*).

[18] S.M. Robinson, Stability theorems for systems of inequalities, Part II: Differentiable nonlinear systems, *SIAM J. Numer. Anal.* **13** (1976) 497-513.

[19] S.M. Robinson, Generalized equations and their solutions, Part II: Applications to nonlinear programming, *Math. Programming Study* **19** (1982) 200-221.

[20] S. M. Robinson, Local epi-continuity and local optimization, *Math. Programming* **37** (1987) 208-223.

[21] R.T. Rockafellar, *The Theory of Subgradients and its Application to Problems of Optimization. Convex and Nonvonvex Functions.* (Heldermann-Verlag, Berlin, 1981).

[22] R.T. Rockafellar, Lipschitzian properties of multifunctions, *Nonlin. Analysis: Theory, Meth. Appl.* **9** (1985) 867-885.

[23] J.-J. Rückmann and G.-W. Weber, Semi-infinite optimization: Excisional stability of the feasible set, *Manuscript, Departments of Mathematics, Humboldt Univ. Berlin and Aachen Univ. of Technology, 1992.*

[24] A. Shapiro, Second-order derivatives of extremal-value functions and optimality conditions for semi-infinite progrema, *Math. Oper. Res.* **10** (1985) 207-219.

[25] A. Shapiro, On Lipschitzian stability of optimal solutions of parametrized semi-infinite programs, *Manuscript, School of Industrial and Systems Engineering, Georgia Institute of Technology, Atlanta, Georgia, 1992* (To appear in *Math. Oper. Res.*).

Convex problems

A Descent Method with Relaxation Type Step

Ulf Brännlund[*]

Department of Mathematics,
Division of Optimization and Systems Theory,
KTH, Stockholm

Abstract

A new bundle method for minimizing a convex nondifferentiable function $f\colon \Re^n \to \Re$ is presented. At each iteration a master problem is solved to get a search direction d. This master problem is a quadratic programming problem of the type

$$\min_d \quad \tfrac{1}{2} d^T d$$
$$\text{s.t.} \quad v_c \geq g_i^T d - \epsilon_i, \, \forall i \in I,$$

where v_c is a parameter, which is an estimate the predicted decrease obtainable from the current iteration point and g_i are ϵ_i-subgradients at the current iteration point.

It is shown that each sequence of $\{x_k\}$ generated by the algorithm minimizes f, i.e. $f(x_k) \downarrow \inf\{f(x) \mid x \in \Re^n\}$, and that $\{x_k\}$ converges to a minimum point whenever f attains its infimum.

Some numerical experiments on some nondifferentiable test problems found in the literature are performed with satisfactory and encouraging results.

1 Introduction

This paper is concerned with a method to minimize a convex nonsmooth function $f\colon \Re^n \to \Re$.

The subdifferential of f at x is defined by

$$\partial f(x) = \{g \in \Re^n \mid f(y) \geq f(x) + g^T(y - x), \, \forall y \in \Re^n\}. \tag{1}$$

An element g of the subdifferential $\partial f(x)$ is called a subgradient of f at x. It is assumed that an "oracle" (a black box), which given an x can deliver $f(x)$ and one element of $\partial f(x)$, is available.

For convex functions the fundamental property of a minimum point x^* is that $0 \in \partial f(x^*)$, i.e.

$$f(x^*) \leq f(y), \, \forall y \in \Re^n \text{ if and only if } 0 \in \partial f(x^*). \tag{2}$$

The first methods for solving convex nonsmooth optimization problems of the above type were the so called relaxation methods, see Polyak [Pol69] and the cutting

---

[*] Research supported by the Swedish Research Council for Engineering Sciences.

plane methods of Kelley [Kel60] and Cheney & Goldstein [ChG59]. The convergence of the former methods depend on that at each iteration the new iteration point is closer to the optimum than the previous. Convergence of the latter methods depend on that at each iteration an approximation of the objective function is improving. Thus, in neither case does convergence depend on that the objective function is monotonically decreasing, and generally it is not.

Bundle methods, whose convergence do depend on a monotonic decrease of the objective function, have received considerable attention ever since the seminal works of Lemaréchal [Lem75] and Wolfe [Wol75]. Bundle methods are related more or less closely to a number of well known methods in differentiable and nondifferentiable optimization among them cutting plane methods, the method of conjugate gradients, to the method of steepest descent, proximal point methods, trust-region methods as well as to relaxation methods. The latter was pointed out by the author in two previous papers contained in Brännlund [Brä93].

In this paper the connection between bundle methods and relaxation methods is made explicit; in section 2 we present a descent algorithm in which the control parameter is an estimate of the optimal function value just as in relaxation methods. In section 3 the convergence of the method is established and in section 5 some numerical experience is reported.

Bundle methods of this so called level type have been suggested previously by Lemaréchal et al. [LNN91]. However, in contrast to theirs, our method is globally convergent without any compactness assumptions and requires bounded storage. As this work was completed we have learned that Kiwiel [Kiw93] also has derived a level method with these properties. We refer to Brännlund [Brä93] for a discussion on different bundle methods and a motivation of our approach.

2 A descent method of level type

We are dealing with an iterative method, in which we at each iteration x_k, have a "bundle" of information consisting of the previous iteration points, $x_1, \ldots, x_k$, together with subgradients generated at these points, $g_i \in \partial f(x_i)$.

We define the linearization error at x_k, if f is linearized in x_i, as

$$\epsilon_i^k = \epsilon(x_k, x_i, g_i) = f(x_k) - (f(x_i) + g_i^T(x_k - x_i)), \tag{3}$$

which is easily shown to be non-negative for a convex function. These linearization errors can be updated recursively.

A vector g is said to be an ϵ-subgradient at x, $g \in \partial_\epsilon f(x)$, if

$$f(y) \geq f(x) + g^T(y - x) - \epsilon, \ \forall y \in \Re^n. \tag{4}$$

A point x is said to be ϵ-optimal if

$$f(y) \geq f(x) - \epsilon\|y - x\| - \epsilon, \ \forall y \in \Re^n. \tag{5}$$

Hence, if $\|g\| \leq \epsilon$ and $g \in \partial_\epsilon f(x)$ then x is ϵ-optimal.

In order to get the search direction, we solve the problem

$$\begin{aligned}
\min_d \quad & \tfrac{1}{2}d^T d \\
\text{s.t.} \quad & v_c \geq g_i^T d - \epsilon_i, \forall i \in I,
\end{aligned} \tag{6}$$

where I is a subset of previously generated subgradients.

Suppose the feasible region of (6) is nonempty, i.e. $|v_c|$ is small enough. For $\hat{d}$ to be the optimal solution of (6) it is necessary and sufficient that there exist multipliers $\hat{\beta}_i \geq 0$, such that

$$\hat{\beta}_i[g_i^T \hat{d} - \epsilon_i - v_c] = 0, \ \forall i \in I, \tag{7}$$

and

$$\hat{d} = -\sum_{i \in I} \hat{\beta}_i g_i. \tag{8}$$

We now give some definitions of some interesting quantities with respect to the solution of the QP problem (6) assuming we have obtained the solution d. We let t be defined by $t = \sum_{i \in I} \beta_i$. Assuming $t > 0$, we define $\alpha_i = \beta_i/t$, and $\sigma = \sum_{i \in I} \alpha_i \epsilon_i$, as well as $p = -d/t = \sum_{i \in I} \alpha_i g_i$. Summing the equations (7) we get the important relations

$$d^T d = -t p^T d = t^2 p^T p = -t(v_c + \sigma). \tag{9}$$

Note that t, and $p^T p$ are not uniquely defined if the Kuhn–Tucker multipliers are not unique, i.e. the mappings $t(v_c)$ and $p(v_c)^T p(v_c)$ are multi-valued. The following proposition states some properties of the solution of (6) and the above definitions.

Proposition 1. *Let $d(v_c)$ be the solution of (6) and $\beta(v_c)$ be a vector of optimal multipliers to the same problem and let $t(v_c) = \sum_{i \in I} \beta_i(v_c)$ and $p(v_c) = -d(v_c)/t(v_c)$. Then,*

(i) $d(v_c)^T d(v_c)$ is a convex non-increasing function of v_c,
(ii) $d(v_c)$ is a continuous function of v_c,
(iii) $t(v_c)$ is non-increasing in v_c,
(iv) $p^T(v_c)p(v_c)$ is non-decreasing in v_c.
(v) $\sigma(v_c)$ is non-increasing in v_c.

We now state the algorithm, whose structure is similar to an ϵ-subgradient algorithm due to Kiwiel [Kiw85].

Algorithm 2.1 *Choose a starting point $x_1 \in \Re^n$ and an upper bound $I_{max} \geq 2$ for $|I_k|$ and parameters $v_1 < 0$, $0 < m_2 < m_1 < 1$, $0 < m_3 < 1 - m_1$, step tolerances $0 < \bar{t} < T$ and a stopping tolerance $\epsilon_s > 0$.*

Step 0 *Compute $f(x_1)$, $g_1 \in \partial f(x_1)$ and put $I_1 = \{1\}$ and $k = 1$.*

Step 1 *Solve the QP (6). If the QP is feasible obtain the search direction d_k and some multipliers β_i, $i \in I_k$. Calculate $t_k = \sum_{i \in I_k} \beta_i$ and let $\alpha_i = \beta_i/t_k$, $p_k = -d_k/t_k$ and $\sigma_k = \sum_{i \in I_k} \alpha_i^k \epsilon_i^k$.*
If $\max\{\sigma_k, \|p_k\|\} \leq \epsilon_s$, then Stop.

Step 2 *If the QP was infeasible or $T\|p_k\|^2 \leq \sigma_k$ or $t_k > T$, then replace v_k by $m_3 v_k$ and return to Step 1.*

Step 3 *Linesearch: Find a step $h_k \leq \max\{1, \bar{t}/t_k\}$ such that for $y_{k+1} = x_k + h_k d_k$ and $g_{k+1} \in \partial f(y_{k+1})$ either of the following two conditions, SS or NS, hold:*

$$\textbf{SS} \ \begin{cases} \text{(i) } f(y_{k+1}) \leq f(x_k) + m_2 h_k v_k \text{ and} \\ \text{(ii) either } h_k t_k \geq \bar{t} \text{ or } \epsilon(x_k, y_{k+1}, g_{k+1}) \geq -m_3 v_k \end{cases}$$

NS $g_{k+1}^T d_k - \epsilon(x_k, y_{k+1}, g_{k+1}) \geq (m_1 + m_3)v_k$

where SS and NS signify that a serious step respectively a null step will be performed at iteration k.

Step 4 *If* $|I_k| = I_{max}$ *then* **Reset:** *Choose* $I \subset I_k$ *with* $|I| \leq I_{max} - 2$ *and* $\max\{i \mid i \in I_k, \ \epsilon_i^k = 0\} \in I$. *Introduce an additional index* $\tilde{k}$ *and define with* p_k *and* σ_k

$$g_{\tilde{k}} = p_k, \ \epsilon_{\tilde{k}}^k = \sigma_k, \ I = I \cup \{\tilde{k}\}$$

Step 5 Update: *If the outcome of the linesearch of Step 3 is a* **Serious Step** *then*

$$\epsilon_i^{k+1} = \epsilon_i^k + f(x_{k+1}) - f(x_k) - h_k d_k^T g_i, \ for \ i \in I, \ \epsilon_{k+1}^{k+1} = 0,$$

and let $v_{k+1} = v_1$.
If the outcome of the linesearch is a **Null Step** *then*

$$\epsilon_i^{k+1} = \epsilon_i^k \ for \ i \in I, \ \epsilon_{k+1}^{k+1} = \epsilon(x, y_{k+1}, g_{k+1}),$$

and let $v_{k+1} = v_k$.
Put $I_{k+1} = I \cup \{k+1\}$. *Set* $k = k + 1$ *and goto Step 1.*

Let us now comment on some of the peculiarities of the algorithm.

Observe that at iteration k, $p_k \in \partial_{\sigma_k} f(x_k)$ which with (4) and (5) motivate the stopping criterion in Step 1.

In Step 2 there are three criteria for deciding when the control $|v|$ is too large, the first for obvious reasons and the others are motivated by Proposition 1. A large $|v_k|$ will yield large t_k, large σ_k and small $\|p_k\|^2$.

The first part of the criterion for a serious step is motivated by that we obtain a sufficient amount of descent from the current iteration point. The second part of **SS** ensures that the step is not too short.

In Step 5 we observe that in case of a nullstep in iteration k we have $v_{k+1} = v_k$, which together with the criterion **NS** in Step 3, ensure that after a nullstep the direction d_{k+1} will be different from d_k.

A linesearch algorithm, which in a finite number of iterations does the job of finding a point y_{k+1} satisfying either condition **SS** or condition **NS** of Step 3, is described in Brännlund [Brä93].

3 Convergence analysis

As was pointed out previously the algorithm is similar to one of Kiwiel [Kiw85], and so is the convergence analysis, for which we have also borrowed some simplifying ideas of Zowe and Schramm [ScZ92]. The convergence result assumes that the stopping tolerance ϵ_s is set to zero. For sake of brevity we leave out some of the proofs of lemmas, which are slight modifications of lemmas in [Kiw85] or [ScZ92]. However, all auxilliary results are stated, leading to the main theorem; that each sequence of $\{x_k\}$ generated by the algorithm minimizes f, i.e. $f(x_k) \downarrow \inf\{f(x) \mid x \in \Re^n\}$, and that $\{x_k\}$ converges to a minimum point whenever f attains its infimum. For a complete account we refer to Brännlund [Brä93].

The relation (5) and the stopping criteria in Step 1 with $\epsilon_s = 0$ yield immediately that if the algorithm stops after a finite number of iterations, say at the kth iteration then x_k is optimal. Thus, from now on we assume that the algorithm does not stop.

The following lemma is the key observation for the convergence analysis.

Lemma 2. *Suppose for some $\bar{x} \in \Re^n$ we have*

$$\liminf_{k \to \infty} \max\{ \|p_k\|, \sigma_k, \|\bar{x} - x_k\| \} = 0, \tag{10}$$

or equivalently there exists a subsequence $\{x_{k_j}\}_{j=1}^{\infty}$ such that $x_{k_j} \to \bar{x}$, and $\|p_{k_j}\| \to 0$, and $\sigma_{k_j} \to 0$. Then $0 \in \partial f(\bar{x})$.

The algorithm could produce a finite sequence $\{x_k\}$, if the algorithm cycled indefinitely between Step 1 and Step 2. The following lemma confirms that if this occurs say at iteration N then x_N is optimal.

Lemma 3. *Suppose that the algorithm performs only a finite number of serious steps, and that the return to Step 1 from Step 2 is performed infinitely often. This can occur in two cases; (a) if at iteration N the algorithm cycles infinitely between Steps 1 and 2, or (b) if only nullsteps are performed after iteration N and the return to Step 1 from Step 2 is performed infinitely often. In both these cases $0 \in \partial f(x_N)$.*

Proof. Note that the notation does not reflect that the QP may be solved several times for each k. The same argument is valid for both cases that we aim to prove. Let in the first case the subscript k denote each time the QP is solved after N and in the second case the regular iteration index.

First, we note that $v_k \uparrow 0$ and that one of the following cases must hold: (i) the QP is infeasible infinitely often, or (ii) $T\|p_k\|^2 \le \sigma_k$ occurs infinitely often, or (iii) $t_k > T$ infinitely often. We treat these three cases separately.

(i) Let $\{k_j\}$ be the indices for which the QP is infeasible. Then, since $f(x_N) \le f(x) - v_{k_j}$ for all $x \in \Re^n$ and $v_{k_j} \uparrow 0$ the conclusion follows from (2).

(ii) Let $\{k_j\}$ be the indices for which $T\|p_{k_j}\|^2 \le \sigma_{k_j}$. Then, since $\sigma_k = -t_k p_k^T p_k - v_k$ from (9), we have $T\|p_{k_j}\|^2 \le \sigma_{k_j} \le -v_{k_j}$, which with $v_{k_j} \uparrow 0$ and Lemma 2 yield the desired conclusion.

(iii) Let $\{k_j\}$ be the indices for which $t_{k_j} \ge T$. Again (9) yields $\sigma_k \le -v_k$ and $t_k p_k^T p_k \le -v_k$, and thus $\sigma_{k_j} \to 0$ and $p_{k_j}^T p_{k_j} \to 0$ since $t_{k_j} > T$ and $v_{k_j} \uparrow 0$. Again, the conclusion follows from Lemma 2.

We may thus assume that the algorithm generates an infinite sequence $\{x_k\}$, such that Step 1 is executed only a finite number of times in each iteration.

For convenience we use the notation

$$h_k' = \begin{cases} h_k & \text{if iteration } k \text{ is a serious step} \\ 0 & \text{otherwise} \end{cases}$$

Thus, at each iteration $x_{k+1} = x_k + h_k' d_k$.

Lemma 4. *Suppose that the sequence $f(x_k)$ is bounded from below. Then,*

$$\sum_{k=1}^{\infty} -h_k' v_k = \sum_{k=1}^{\infty} h_k'(t_k\|p_k\|^2 + \sigma_k) < \infty. \tag{11}$$

If there exists a subsequence such that $x_{k_j} \to \bar{x}$ then (11) is satisfied, $f(x_k) \downarrow f(\bar{x})$ and $h_k' t_k \|p_k\|^2 \to 0$.

Let us now, for purposes of the proof process, introduce a technical assumption.

Assumption 3.1 *There exists a point $\bar{x}$ such that $f(\bar{x}) \leq f(x_k)$ for all k.*

As a consequence of the previous lemma, we have under Assumption 3.1 that the sequence $\{x_k\}$ converges:

Lemma 5. *Suppose Assumption 3.1 holds. Then, the sequence $\{x_k\}_{k=1}^{\infty}$ converges to some $\tilde{x}$, for which $f(\tilde{x}) \leq f(x_k)$ for all k.*

The following lemma establishes that after a nullstep in iteration k is performed the next search direction d_{k+1} is sufficiently different and longer than the previous direction d_k. It is here and in the proof of the following Lemma 7 that main difference from the analysis of Kiwiel's algorithm appears.

Lemma 6. *Suppose the QP (6) is solved at iteration k with $v_k = v$ yielding the quantities d_k, p_k and σ_k and that a nullstep is performed at iteration k. Then,*

$$d_{k+1}^T d_{k+1} \geq d_k^T d_k + \frac{(1 - m_1 - m_3)^2 v^2}{\|g_{k+1}\|^2}.$$

Proof. From the reset rules in Step 4 and the update rule of Step 5 of the algorithm, it is clear that $d_{k+1}^T d_{k+1} \geq \hat{d}^T \hat{d}$, where $\hat{d}$ solves the problem

$$\min \tfrac{1}{2} d^T d$$
$$\text{s.t.} \quad v \geq p_k^T d - \sigma_k$$
$$v \geq g_{k+1}^T d - \epsilon_{k+1}^{k+1}.$$

Clearly, the inequality $v \geq g_{k+1}^T d - \epsilon_{k+1}^{k+1}$ is satisfied with equality at $\hat{d}$. Since $\hat{d}$ is also in same halfspace, $v \geq p_k^T d - \sigma_k$, as d_k, we have from standard projection properties that $\hat{d}^T \hat{d} \geq d_k^T d_k + (\hat{d} - d_k)^T (\hat{d} - d_k)$.

The distance between $\hat{d}$ and d_k, $\|\hat{d} - d_k\|$, is greater than the distance between the two parallel hyperplanes $H_1 = \{ d \mid v = g_{k+1}^T d - \epsilon_{k+1}^{k+1} \}$, and $H_2 = \{ d \mid g_{k+1}^T (d - d_k) = 0 \}$. We have $\hat{d} \in H_1$ and $d_k \in H_2$. Let t_1 and t_2 be the distances from the origin to H_1 and H_2 respectively, i.e.

$$v = g_{k+1}^T (-t_1 g_{k+1} / \|g_{k+1}\|) - \epsilon_{k+1}^{k+1}$$

and

$$g_{k+1}^T (-t_2 g_{k+1} / \|g_{k+1}\| - d_k) = 0.$$

We deduce that the distance between the two hyperplanes is $t_1 - t_2 = -(v - g_{k+1}^T d_k + \epsilon_{k+1}^{k+1}) / \|g_{k+1}\| \geq -(1 - m_1 - m_3) v / \|g_{k+1}\|$.

Combining the above facts, we get

$$\begin{aligned}
d_{k+1}^T d_{k+1} &\geq \hat{d}^T \hat{d} \\
&\geq d_k^T d_k + (\hat{d} - d_k)^T (\hat{d} - d_k) \\
&\geq d_k^T d_k + (t_1 - t_2)^2 \\
&\geq d_k^T d_k + (1 - m_1 - m_3)^2 v^2 / \|g_{k+1}\|^2,
\end{aligned}$$

which is the desired inequality.

Lemma 7. *Suppose that the number of serious steps is finite and thus $x_k = \bar{x}$ for all $k \geq K$ for some large K. Then (10) hold and $0 \in \partial f(x)$.*

Proof. From Lemma 3 we have that if the number of iterates at which the algorithm returns from Step 2 to Step 1 is infinite then the assertion is true.

Suppose on the other hand that the number of returns from Step 2 to Step 1 is finite. Then $v_k = v$, and $T\|p_k\|^2 \geq \sigma_k$ for all $k \geq N$ where N is some finite number. From Lemma 6 and the local boundedness of $\partial f(x)$, we deduce that $d_k^T d_k \to \infty$. Since $t_k p_k^T p_k = -(v_k + \sigma_k) \leq -v_k = -v$ we have that $\{t_k p_k^T p_k\}$ is bounded. Thus, $d_k^T d_k = t_k^2 p_k^T p_k \to \infty$ imply $t_k \to \infty$ and also $p_k^T p_k \to 0$. Thus, the number of returns from Step 2 can not be finite.

Lemma 8. *Suppose that there exist a point $\bar{x} \in \Re^n$ and an infinite subsequence $\{x_{k_j}\}_{j=1}^{\infty}$ such that serious steps are performed at iterations k_j and $x_{k_j} \to \bar{x}$. Then,*

$$\liminf \max\{\|p_{k_j}\|, \sigma_{k_j}\} \to 0 \quad and \quad 0 \in \partial f(\bar{x}).$$

Proof. Let $K_j = \{k \mid k_j \leq k < k_{j+1}\}$ and let $b_j = \min\{\max\{\|p_k\|, \sigma_k\} \mid k \in K_j\}$. By Lemma 2 and the definition of b_j, the lemma is true if there exists a subsequence $\{j_i\}_{i=1}^{\infty}$ such that b_{j_i} converges to zero.

Thus, assume on the contrary that b_j is bounded away from zero. Then, there exists an $\epsilon > 0$ such that

$$\max\{T\|p_k\|^2, \sigma_k\} \geq \epsilon \quad \text{for all } k \in K_j \text{ and large } j.$$

We may assume that $-v_{k_j}$ is also bounded away from zero, say by ϵ_v, i.e. $-v_{k_j} \geq \epsilon_v > 0$ for large j. Otherwise, we may use the same argument as that of Lemma 3 to prove the desired result. Also, since $T\|p_{k_j}\|^2 > \sigma_{k_j}$ at Step 3, we have $T\|p_{k_j}\|^2 \geq \epsilon$.

Since $\bar{x}$ is an accumulation point of x_k, Lemma 4 yields $h'_k t_k \|p_k\|^2 \to 0$. From $h'_k t_k \|p_k\|^2 = \|x_{k+1} - x_k\| \|p_k\|$, we thus obtain $h_{k_j} t_{k_j} \to 0$ and $\|x_{k_j+1} - x_{k_j}\| \to 0$. From the second linesearch condition for a serious step we thus deduce that

$$\epsilon_{k_j+1}^{k_j+1} = \epsilon(x_{k_j}, x_{k_j+1}, g_{k_j+1}) > m_3 \epsilon_v, \quad \text{for large } j. \tag{12}$$

Since $x_{k_j} \to \bar{x}$ and $\|x_{k_j+1} - x_{k_j}\| \to 0$, we have

$$\epsilon(x_{k_j}, x_{k_j+1}, g_{k_j+1}) = f(x_{k_j}) - f(x_{k_j+1}) + g_{k_j+1}^T(x_{k_j+1} - x_{k_j}) \to 0$$

from the continuity of f and the local boundedness of $\partial f(\cdot)$. This contradicts the bound in (12). Therefore b_j can not be bounded away from 0 and we have the desired conclusion.

We are now ready for the main theorem.

Theorem 9. *If the optimal set X^* is nonempty, then $x_k \to x^* \in X^*$. Otherwise $f(x_k) \downarrow \inf\{f(x) \mid x \in \Re^n\} \in [-\infty, \infty)$.*

Proof. Suppose $X^* \neq \emptyset$. Then, Assumption 3.1 holds and thus $\{x_k\}$ converges to some $\bar{x}$ by Lemma 5, which by Lemmas 3, 7 and 8 is optimal.

Suppose $X^* = \emptyset$ and that the assertion is not true, i.e. there exists $\bar{x}$ such that $f(\bar{x}) \leq f(x_k)$ for all k. As above we conclude $x_k \to \bar{x} \in X^*$, which contradicts $X^* = \emptyset$.

4 Other v-control schemes

The updating scheme of v_c in Algorithm 2.1 may not be very efficient, since one of the ideas of level control is to let v be an estimate of $f^* - f(x_k)$ where f^* is the optimal value, and hence the appropriate v ought to increase as $f(x_k)$ gets closer to f^*, which is not the case with Algorithm 2.1. This typically has the effect in Algorithm 2.1 that close to the optimum and after a serious step unnecessarily many QP's has to be solved in order to increase v to an appropriate level.

One obvious remedy is to in Step 5 after a serious step at iteration k set

$$v_{k+1} = \underline{f}_k - f(x_k), \tag{13}$$

where $\underline{f}_k$ is some lower bound of f^* obtained from either the linear programming bound (Cutting-plane) or some other procedure such as a feasible solution to the primal problem in the context of Lagrangian relaxation. With this remedy, the convergence proof remains in all essence valid; either $\underline{f}_k \uparrow f^*$ and the convergence is guaranteed by Lemma 5, since it does not depend on the v updating rule, or the convergence is guaranteed by the analysis in section 3.

Another remedy is to use the same idea as the one suggested by Kiwiel for his corresponding ϵ-subgradient methods. The algorithm remains essentially the same except for Step 2, which is replaced by

Step 2' *If the QP was feasible then set $\tilde{\delta}_k = \max\{T\|p_k\|^2, \sigma_k\}$, otherwise set $\tilde{\delta}_k = -v_k$. If the QP was infeasible or $T\|p_k\|^2 \leq \sigma_k$ or $t_k \geq T$ then set $\delta_k = \tilde{\delta}_k$ and $v_k = -m_3\delta_k$ and return to Step 1,*

and Step 5 in which v and δ are updated by $\delta_{k+1} = \delta_k$ and $v_{k+1} = -m_3\delta_k$.

A monotonically non-increasing sequence of numbers $\{\delta_k\}$ is calculated such that $v_k = -m_3\delta_k$. The modification provides a way to increase v as $f(x_k)$ gets closer to f^*, which is indicated by small values of $\|p_k\|$ and σ_k. The convergence analysis after this modification is straightforward. Either δ_k and v_k will eventually stay constant and from there on the convergence analysis in section 3 hold, or δ_k and v_k will approach zero because of infinitely many returns from Step 2' to Step 1 and one may apply the same argument as that of Lemma 3. We use this alternative updating scheme in the numerical tests in section 5.

5 Numerical experience

In this section we report some numerical experience that we have had with Algorithm 2.1 with some test problems found in the literature.

Algorithm 2.1 has been programmed in the matrix oriented software package MATLAB® [Mat92]. The QPs (the dual of (6)) were solved with the multi-purpose least-squares routine LSSOL [GHM+86] through a so called mex-interface. All floating point calculations were performed in double precision, allowing a machine precision of $2.2 \cdot 10^{-16}$.

We use the following standard parameters $m_1 = 0.2$, $m_2 = 0.1$, $m_3 = 0.4$ and $\bar{t} = 0.00001$. The initial v_1 is chosen to approximately $(f^* - f(x_1))$ and $T = \max\{1, -100v_1/\|g(x_1)\|^2\}$. We choose $I_{max} = n + 4$ in all test examples. We record the number of function and subgradient evaluations, $\#f/g$, the number of QP solutions, $\#QP$, and the number of linesearches, $\#LS$, performed until the algorithm stops for different values of the stopping tolerance ϵ_s.

Shor's test problem: $f(x) = \max\{f_i(x) \mid i = 1, \ldots, 10\}$, $x \in \Re^5$, $x_1 = 0$, where f_i are separable quadratics. For more details see Kiwiel [Kiw85] or Shor [Sho85].

Maxquad: $f(x) = \max\{f_i(x) \mid i = 1, \ldots, 5\}$, $x \in \Re^{10}$, $x_1 = 1$, where f_i convex quadratics. For more details see Kiwiel [Kiw85] or Lemaréchal & Mifflin [LeM78].

Goffin's test problem in $\Re^{50}$ is $f(x) = \max_{1 \le i \le 50} x_i - \sum_{i=1}^{50} x_i$, with starting point $x_1^i = i - 25.5$, with $f(x_1) = 1225$ and optimal value $f^* = 0$.

TR48 is the dual of a linear transportation problem. $x_1 = 0 \in \Re^{48}$ with $f(x_1) = -464816$. For more details, see Lemaréchal & Mifflin [LeM78].

Shor's testproblem					Maxquad				
ϵ_s	$\#f/g$	$\#QP$	$\#LS$	f	ϵ_s	$\#f/g$	$\#QP$	$\#LS$	f
10^{-4}	27	36	26	22.6001783	10^{-4}	66	73	58	-0.84140821
10^{-5}	33	45	32	22.6001640	10^{-5}	94	90	72	-0.84140833
10^{-6}	40	51	36	22.6001627	10^{-6}	117	107	85	-0.84140833
10^{-7}	67	70	51	22.6001621	10^{-7}	140	122	96	-0.84140833
10^{-8}	75	73	53	22.6001621	10^{-8}	140	122	96	-0.84140833

Goffin's testproblem					TR48				
ϵ_s	$\#f/g$	$\#QP$	$\#LS$	f	ϵ_s	$\#f/g$	$\#QP$	$\#LS$	f
10^{-4}	76	92	75	$2 \cdot 10^{-6}$	10^0	188	193	187	-638422.6874
10^{-5}	76	92	75	$2 \cdot 10^{-6}$	10^{-1}	250	260	249	-638564.9611
10^{-6}	81	98	76	$4 \cdot 10^{-7}$	10^{-2}	291	301	287	-638564.9973
10^{-7}	86	101	77	$4 \cdot 10^{-8}$	10^{-3}	308	320	304	-638564.9994
10^{-8}	105	107	80	$3 \cdot 10^{-9}$	10^{-4}	350	365	346	-638565.0000

We have also made some experiments with the algorithm on some problems of finding lower bounds on TSPs, and also on some production planning problems, but for this material we refer the interested reader to Brännlund [Brä93].

6 Discussion and conclusions

We have in this paper presented a new convergent descent method for convex non-differentiable optimization, which is based on level control. The algorithm does not depend on prior knowledge of a lower bound on the optimal function value. However, we believe that such knowledge could improve the performance of the algorithm considerably.

The numerical behavior of the algorithm is satisfactory. The rate of convergence seems to be in the same range as other bundle methods. We are encouraged to develop and attempt other v-control schemes. We are particularly interested in investigating if it is possible to have a pure v-control, i.e. an algorithm without a linesearch, and still maintain the reset strategy. The good performance of the proximal point algorithm of Kiwiel in [Kiw90] and the implicit trust-region algorithm of Schramm and Zowe [ScZ92] indicate that the linesearch is not necessary and that a curved linesearch, in which the QP is solved after every function/subgradient evaluation, may improve performance. Although, in the numerical tests, our algorithm seldom needed more than one function evaluation per linesearch, we feel that it is worth exploring.

Another question which we feel should be investigated is whether or not it is worthwhile to calculate $\underline{f}_k$ from the Cutting-Plane relaxation and use the updating rule (13). Observe that we can still use resets as long as $I_{max} \geq n + 1$.

A case where a good lower bound is known and also can be updated is in the context of Lagrangian relaxation. We believe that many implementations of Lagrangian relaxations might benefit from using an algorithm like ours, instead of standard subgradient techniques.

References

[Brä93]   U. Brännlund, *On relaxation methods for nonsmooth convex optimization*, Ph.D. thesis, Kungliga Tekniska Högskolan, S-100 44 Stockholm, Sweden, 1993.

[ChG59]   E. Cheney and A. Goldstein, *Newton's method for convex programming and Tchebycheff approximation*, Numerische Mathematik 1 (1959) 253–268.

[GHM+86] P. E. Gill, S. J. Hammarling, W. Murray, M. A. Saunders and M. H. Wright. *User's guide for LSSOL (Version 1.0): a Fortran package for constrained linear least-squares and convex quadratic programming*, Report SOL 86-1, Department of Operations Research, Stanford University, 1986.

[Kel60]   J. Kelley, *The cutting plane method for solving convex programs*, Journal of the SIAM 8 (1960) 703–712.

[Kiw85]   K. Kiwiel, *Methods of Descent for Nondifferentiable Optimization*, Springer-Verlag, Berlin, Heidelberg, New York, and Tokyo, 1985.

[Kiw90]   ———, *Proximity control in bundle methods for convex nondifferentiable minimization*, Mathematical Programming 46 (1990) 105–122.

[Kiw93]   K. C. Kiwiel, *A descent proximal level bundle method for convex nondifferentiable optimization*. Working paper, May 1993.

[Lem75]   C. Lemaréchal, *An extension of "Davidon" methods to nondifferentiable problems*, in Mathematical Programming Study, Balinsky and Wolfe, eds., vol. 3. North Holland, 1975, pp. 95–109.

[LeM78]   C. Lemaréchal and R. Mifflin, *A set of nonsmooth optimization test problems*. in Nonsmooth Optimization, C. Lemaréchal and R. Mifflin, eds., Oxford, 1978. IIASA Workshop, Pergamon Press, pp. 151–165.

[LNN91]   C. Lemaréchal, A. Nemirovskii and Y. Nestorov, *New variants of bundle methods*, Tech. Report No 1508, INRIA-Rocquencourt, France, 1991.

[Mat92]   MathWorks Inc., Natick, Mass. USA, MATLAB® *User's guide*, 1992.

[Pol69]   B. Polyak, *Minimization of unsmooth functionals*, USSR Computational Mathematics and Mathematical Physics 9 (1969) 14–29.

[ScZ92]   H. Schramm and J. Zowe, *A version of the bundle idea for minimizing a nonsmooth function: Conceptual idea, convergence analysis, numerical results.*, SIAM Journal on Optimization 2, no. 1 (1992) 121–152.

[Sho85]   N. Shor, *Minimization Methods for Non-Differentiable Functions*, Springer-Verlag, Berlin, 1985.

[Wol75]   P. Wolfe, *A method of conjugate subgradients for minimizing nondifferentiable functions*, Mathematical Programming Study (1975) 145–173.

This article was processed using the LaTeX macro package with LMAMULT style

PROJECTION ONTO AN ACUTE CONE

AND CONVEX FEASIBILITY PROBLEM

Andrzej Cegielski
Institute of Mathematics
Higher College of Engineering
ul. Podgórna 50, PL-65246 Zielona Góra

Consider the following convex feasibility problem:

$$\text{Determine} \quad x^* \in M = \bigcap_{i \in I} M_i,$$

where $M_i \subset \mathbb{R}^n$ are closed, convex subsets, $i \in I = \{1, 2, \ldots, m\}$ and $M \neq \emptyset$. There exist lot of methods which can be applied to solve the above problem. Most of them base on the succesive projections onto sets M_{i_k}, $i_k \in I$, [1], [3], [8], or on the simultaneous projections onto the whole family $\{M_i\}_{i \in I}$ [2], [4], [9], [10]. Papers [2], [7] connect the mentioned above methods in one family. We present below a new method of projection onto an acute cone which solves the convex feasibility problem.

Definition 1. A cone C is said to be *acute* if $\langle x, y \rangle \geq 0$ for all $x, y \in C$. A cone C is said to be *obtuse* if its dual cone $C^* = \{x: \langle x, y \rangle \leq 0, y \in C\}$ is acute.

One can easily prove that C is obtuse in $\text{Lin}C$ if $C^* \cap \text{Lin}C \subset -C$ (see also [6]).

Let P_i denotes the metric projection operator onto the subset M_i. Furthermore, for a given $x \notin M$, denote

$$b_i(x) = P_i(x) - x,$$

$$I(x) = \{i \in I: x \notin M_i\}$$

and

$$B(x) = \{b_i: i \in I(x)\}.$$

Iterative scheme 1.

Choose an arbitrary $x_0 \in \mathbb{R}^n$.

Starting with $x_k \in \mathbb{R}^n$ obtained in the k-th iteration, in the $(k+1)$-st iteration one determines:

1^0 $I_k = I(x_k)$ (if $I_k = \varnothing$, then $x_k \in M$),

2^0 $B_k = B(x_k) = \{b_{ik}: b_{ik} = b_i(x_k), i \in I_k\}$,

3^0 $L_k \subset I_k$ such that

    a) $S_k = \{b_{ik}: i \in L_k\}$ is a linearly independent system,

    b) $\text{cone} S_k$ is obtuse in $\text{Lin} S_k$,

4^0 a vector $t_k \in \text{Lin} S_k$ which is a solution of the system

$$\langle b_{ik}, t \rangle = \langle b_{ik}, b_{ik} \rangle, \quad \text{for } i \in I_k,$$

5^0 $x_{k+1} = x_k + \lambda t_k$, where $\lambda \in [0,2]$.

Remark 1. We show later, how to construct a subset L_k of I_k such that S_k satisfies 3^0a) and 3^0b). To prove the properties of iterative scheme 1 it is sufficient that such subset L_k exists which can be easily seen if one takes $L_k = \{i_k\}$, where $i_k \in \text{Argmax}_{i \in I} \|b_{ik}\|$. It is possible that there exist no others subsets L_k satisfying 3^0. But in this case a good behaviour of classical relaxation algorithms can be observed. Bad behaviour of these algorithms occurs if the projection vectors b_{ik}, $i \in I_k$, generate an obtuse cone in $\text{Lin}\{b_{ik}: i \in I_k\}$ or at least there are few vectors b_{ik} with such property. A subsystem of such projection vectors is selected in iterative scheme 1 and then an appropriate construction is used to obtain a better estimation of $d(x_{k+1}, M)$.

Lemma 1. *For iterative scheme 1 the following inequality is satisfied*

$$\langle z - x_k, t_k \rangle \geq \langle t_k, t_k \rangle,$$

where z is an arbitrary element of the solution set M.

Proof: Let $S_k = \{b_{ik}: i \in L_k\}$ be a linearly independent subsystem of B_k which generates an obtuse cone. Denote

$$N_k = \{y \in \mathbb{R}^n: \langle y - x_k, b_{ik} \rangle \geq \langle b_{ik}, b_{ik} \rangle, i \in L_k\}.$$

Let $z \in M$ and let $x_k \notin M$. Properties of the projection imply that $z \in N_k$. Furthermore, we have

$$N_k = x_k + t_k - (\text{cone} S_k)^*$$

since

$$\langle y - x_k - t_k, b_{ik} \rangle = \langle y - x_k, b_{ik} \rangle - \langle t_k, b_{ik} \rangle = \langle y - x_k, b_{ik} \rangle - \langle b_{ik}, b_{ik} \rangle,$$

for $i \in L_k$. Therefore $z - x_k - t_k \in (\text{cone} S_k)^*$ and

$$\langle z - x_k, t_k \rangle = \langle z - x_k - t_k, t_k \rangle + \langle t_k, t_k \rangle \geq \langle t_k, t_k \rangle$$

since $t_k \in -(\text{cone} S_k)^* \cap \text{Lin} S_k$ and $-(\text{cone} S_k)^* \cap \text{Lin} S_k$ is an acute cone. ∎

Corollary 1. *There holds the following inequality*

$$\| x_{k+1} - z \|^2 \leq \| x_k - z \|^2 - \lambda_k (2 - \lambda_k) \| t_k \|^2.$$

Lemma 2. *Let $S = \{ s_l : l \in L \}$ be a linearly independent system and let $\text{cone} S_k$ be obtuse in $\text{Lin} S_k$. If t is a solution of the system*

$$\langle s_l, t \rangle = \langle s_l, s_l \rangle, \quad l \in L,$$

then the following inequality is satisfied

$$\| t \|^2 \geq \sum_{l \in L} \| s_l \|^2.$$

Proof. First observe that $-t \in (\text{cone} S)^*$. Denote $u_l = s_l / \| s_l \|$, $l \in L$. Let $F = \{ f_l : l \in L \}$ be the dual base to $U = \{ u_l : l \in L \}$ in $\text{Lin} U$, i.e. $f_l \in \text{Lin} U$, $i \in L$, and

$$\langle f_i, u_l \rangle = \delta_{il}, \quad i, l \in L.$$

It is easy to see that

$$\text{cone}(-F) = (\text{cone} U)^* \cap \text{Lin} U,$$

since $\text{cone} U$ is obtuse in $\text{Lin} U$. Let $w \in (\text{cone} U)^* \cap \text{Lin} U$. It follows from the above equality, that

$$w = \sum_{l \in L} \beta_l f_l,$$

where $\beta_l \leq 0$, $l \in L$. Therefore the Schwarz inequality imply

$$\langle f_l, f_l \rangle = \langle f_l, f_l \rangle \langle u_l, u_l \rangle \geq \langle f_l, u_l \rangle^2 = 1$$

and, consequently

$$\langle w, w \rangle - \sum_{l \in L} \langle w, u_l \rangle^2 =$$

$$= \sum_{l \in L} \sum_{j \in L} \beta_l \beta_j \langle f_l, f_j \rangle - \sum_{l \in L} (\beta_l)^2 =$$

$$\sum_{i \in L} (\beta_i)^2 (<f_i, f_i>-1) + \sum_{i,j \in L, i \neq j} \beta_i \beta_j <f_i, f_j> \geq 0,$$

since $\mathrm{cone} F$ is acute.

Now we have $-t=w+v$, where $w \in (\mathrm{cone} U)^* \cap \mathrm{Lin} U$, $v \in (\mathrm{Lin} U)^\perp$, and

$$\| t \|^2 = \| w \|^2 + \| v \|^2 \geq \| w \|^2 \geq \sum_{i \in L} <w, u_i>^2 = \sum_{i \in L} <t, u_i>^2 =$$

$$= \sum_{i \in L} <t, s_i / \| s_i \|>^2 = \sum_{i \in L} \| s_i \|^{-2} <t, s_i>^2 = \sum_{i \in L} \| s_i \|^2. \blacksquare$$

Corollary 2. *For iterative scheme 1 the following inequality is satisfied*

$$\| x_{k+1} - z \|^2 \leq \| x_k - z \|^2 - \lambda_k (2 - \lambda_k) \sum_{i \in J_k} \| b_{ik} \|^2.$$

Let $i_k \in \mathrm{Argmax}_{i \in I} d(x_k, M_i)$.

Lemma 3. *If $i_k \in L_k$ then for iterative scheme 1 the following condition is satisfied*

$$\liminf_k \| t_k \| = 0 \Rightarrow \liminf_k \max_{i \in I} d(x_k, M_i) = 0.$$

Proof. It follows from lemma 2 that $\| t_k \| \geq \max_{i \in I} d(x_k, M_i)$. The lemma is now obvious.

Theorem 1. *Let $\{x_k\}$ be a sequence generated by iterative scheme 1 such that $i_k \in L_k$. If $\lambda_k \in [0,2]$ and $\sum_k \lambda_k (2 - \lambda_k) = \infty$, then x_k converges and $\lim_k x_k \in M$.*

Proof. Lemmas 1 and 3 shows that all conditions of theorem 4 in [12] are satisfied. The mentioned theorem implies the convergence of x_k to an element of M. $\blacksquare$

Now we give a construction of an obtuse cone generated by a subsystem S of a system of vectors $B = B(x)$, for a given $x \notin M$.

Denote $S = \{s_i : l \in L\}$, where $L = \{1, 2, \ldots, r\}$, and $S' = \{s_i : l \in L'\}$, where $L' = \{1, 2, \ldots, r-1\}$, for some $r \geq 1$. We start with the following lemma.

Lemma 4. *If a system $S \subset B$ satisfies the following conditions*
(1') *S' is a linearly independent system,*

(ii') $\text{cone} S'$ *is obtuse in* $\text{Lin} S'$,

(iii) $s_r \in -\text{cone} S' + (\text{Lin} S')^{\perp}$,

then

(i) S *is a linearly independent system,*

(ii) $\text{cone} S$ *is obtuse in* $\text{Lin} S$.

Proof. It follows from the assumptions of the lemma that S' is a base in $\text{Lin} S'$. Let $F' = \{f_l : l \in L'\}$ be a base dual to S' in $\text{Lin} S'$, i.e. f_i satisfies the following system of equations

$$\langle f_i, s_j \rangle = \delta_{ij}, \quad j \in L',$$

and $f_i \in \text{Lin} S'$, $i \in L'$. Furthermore, there holds the equality

$$\text{cone} F' = -(\text{cone} S')^* \cap \text{Lin} S'.$$

From (iii) we have

$$s_r = -s + s^{\perp},$$

where

$$s = \sum_{l \in L'} \beta_l s_l,$$

$\beta_l \geq 0$, $l \in L'$, $s^{\perp} \in (\text{Lin} S')^{\perp}$.

Now observe that $s^{\perp} \neq 0$. Suppose by contradiction that $s^{\perp} = 0$, i.e.

$$s_r = \sum_{l \in L'} \beta_l s_l.$$

By (iii) we have $\beta_l \leq 0$, $l \in L'$. Let $z \in M$. Properties of the projection imply

$$0 < \langle s_r, s_r \rangle \leq \langle z-x, s_r \rangle = \sum_{l \in L'} \beta_l \langle z-x, s_l \rangle \leq 0,$$

a contradiction.

Let

$$\mu_i = \langle f_i, s \rangle / \| s^{\perp} \|^2,$$

$i \in L'$. Then the system of vectors $H = \{h_i : i \in L\}$, where

$$h_i = f_i + \mu_i s^{\perp}, \quad i \in L',$$

$$h_r = s^{\perp} / \| s^{\perp} \|^2$$

is the dual base to S in $\text{Lin} S$, since

$$\langle h_i, s_j \rangle = \langle f_i + \mu_i s^{\perp}, s_j \rangle = \langle f_i, s_j \rangle = \delta_{ij},$$

$i, j \in L'$,

$$\langle h_i, s_r \rangle = \langle f_i + \mu_i s^{\perp}, -s+s^{\perp} \rangle = -\langle f_i, s \rangle + \mu_i \langle s^{\perp}, s^{\perp} \rangle = 0,$$

$i \in L'$,

$$\langle h_r, s_j \rangle = \langle s^{\perp}/\|s^{\perp}\|^2, s_j \rangle = 0,$$

$j \in L'$, and

$$\langle h_r, s_r \rangle = \langle s^{\perp}/\|s^{\perp}\|^2, -s+s^{\perp} \rangle = 1.$$

Observe that $\mu_i \geq 0$, $i \in L'$, since

$$\langle f_i, s \rangle = \langle f_i, \sum_{l \in L} \beta_l s_l \rangle = \sum_{l \in L} \beta_l \langle f_i, s_l \rangle = \beta_i \geq 0,$$

$i \in L'$. Therefore

$$\langle h_i, h_j \rangle = \langle f_i + \mu_i s^{\perp}, f_j + \mu_j s^{\perp} \rangle = \langle f_i, f_j \rangle + \mu_i \mu_j \langle s^{\perp}, s^{\perp} \rangle \geq 0,$$

$i, j \in L'$, since coneF' is acute. Furthermore

$$\langle h_i, h_r \rangle = \langle f_i + \mu_i s^{\perp}, s^{\perp}/\|s^{\perp}\|^2 \rangle = \mu_i \geq 0,$$

$i \in L'$, and

$$\langle h_r, h_r \rangle = \|s^{\perp}\|^{-2} > 0.$$

Hence cone$H$ is acute, i.e. cone$S = -(\text{cone}H)^{*} \cap \text{Lin}H$ is obtuse in Lin$S$. ∎

It follows from lemma 4, that the following algorithm generates a subsystem S of a system of projection vectors B, such that cone$S$ is obtuse in Lin$S$.

Algorithm 1.

1° Set $L:=\emptyset$, $S:=\emptyset$.

2° Set $K:=\emptyset$.

3° If $L \cup K = I$, then the algorithm stops.

4° Choose an arbitrary $i \in I \setminus (L \cup K)$.

5° Check:

$\qquad b_i \in -\text{cone}S + (\text{Lin}S)^{\perp}$;

$\quad$ if yes, then set

$\qquad L := L \cup \{i\}$,

$\qquad S := S \cup \{b_i\}$

$\quad$ and go to do 2°;

if no, then set

$K := K \cup \{i\}$

and go to do 3°.

In the above algorithm we denote: $\mathrm{Lin}\emptyset=0$ and $\mathrm{cone}\emptyset=0$. The inclusion in 5° can be checked by the solution of the system of equations

$$\sum_{i \in L} \gamma_i \langle s_i, s_j \rangle = \langle b, s_j \rangle, \quad j \in L.$$

One can prove that the complexity of algorithm 1 is less than

$$(n + \frac{2}{3}\#S)(\#B - \frac{2}{3}\#S)(\#S)^2.$$

Details see [5].

As a special case of a linearly independent subsystem S of a vector system $B(x)$, such that S generates an obtuse cone, we can take $S = \{s_i : i \in L\}$ such that $\langle s_i, s_j \rangle \leq 0$ for $i, j \in L$, $i \neq j$. In this case the construction of a system S can be described by the following algorithm.

Algorithm 2.

1° Set $L:=\emptyset$, $S:=\emptyset$.

2° Choose an arbitrary $i \in I \backslash L$.

3° Set $L:=L \cup \{i\}$.

4° Check:

$\langle s, b_i \rangle \leq 0$ for all $s \in S$.

If yes, then set

$S:=S \cup \{b_i\}$;

5° Check: $L=I$.

If no, the go to 2°;

if yes, then the algorithm stops.

Iterative scheme 1 for linear inequalities with such choice of S was studied by Todd in [13].

Remark 2. Observe that $x_k + t_k \in N_k$. Therefore lemma 1 and properties of the projection imply

$$x_k + t_k = P_{N_k}(x_k),$$

i.e., for $\lambda_k = 1$, x_{k+1} is the projection of x_k onto a translated acute cone. In the general case, for $\lambda_k \in [0,2]$, x_{k+1} can be expressed in the form

$$x_{k+1} = (1-\lambda_k)x_k + \lambda_k P_{N_k}(x_k).$$

Methods with similar idea (projection onto an acute cone) were also used in some convex minimization problems [5].

References.

[1] S. Agmon, "The relaxation method for linear inequalities", *Canad. J. Math.* 6 (1954) 382-392.

[2] R. Aharoni, and Y. Censor, "Block-iterative projection methods for parallel computation of solutions to convex feasibility problems, *Linear Algebra and Its Applications* 120 (1989) 165-175.

[3] L.M. Bregman, "The relaxation method of finding the common point of conxex sets and its applications to the solution of problems in convex programming" *USSR Comp. Math. & Math. Phys.* 7 (1967) 200-217.

[4] D. Butnariu and Y. Censor, "On the behavior of a block iterative projection method for solving convex feasibility problems", *Int. J. Comp. Math.* 34 (1991).

[5] A. Cegielski, "Relaxation methods in convex optimization problems" (in Polish), WSI, Monografia No. 67, Zielona Góra, 1993.

[6] J.L. Goffin, "The relaxation method for solving systems of linear inequalities", *Mathematics of Operations Research* 5 (1980) 388-414.

[7] S.D. Flåm and J. Zowe, "Relaxed outer projections, weighted averages and convex feasibility, *BIT* 30 (1990) 289-300.

[8] L.B. Gurin, B.T. Polyak, E.V. Raik, *"The method of projections for finding the common point of convex sets"*, *USSR Comp. Math. & Math. Phys.* 7 (1967) 1-24.

[9] A.N. Iusem and A.R. De Pierro, "Convergence results for an accelerated nonlinear Cimmino algorithm", *Numerische Mathematik* 49 (1986) 367-378.

[10] G. Pierra, "Decomposition through formalization in a product space", *Mathematical Programming* 28 (1984) 96-115.

[11] B.T. Polyak, "Minimization of nonsmooth functionals", *USSR Comp Math & Math. Phys.* 9 (1969) 14-29.

[12] D. Schott, "A general iterative scheme with applications to convex optimization and related fields", *Optimization* 22 (1991) 885-902.

[13] M. Todd, "Some remarks on the relaxation method for linear inequalities", Technical Report No. 419, Cornell University, Ithaca, 1979.

A Numerical Approach
to the Design of Masonry Structures *

Michal Kočvara and Jiří V. Outrata **

1 Introduction

Assume that U and Y are Banach spaces, $\Gamma[U \rightsquigarrow Y^*]$ is a proper convex- and closed-valued multifunction and $F[U \times Y \rightarrow Y^*]$ is a continuously differentiable operator. Consider a class of optimum design problems

$$f(u, y) \rightarrow \inf$$

subject to
$$0 \in F(u, y) + N_{\Gamma(u)}(y) \tag{1.1}$$
$$u \in U_{ad}$$
$$y \in V,$$

where u is the *control* variable, y is the *state* variable, $f[U \times Y \rightarrow \mathbb{R}]$ is a continuously differentiable objective function, and $U_{ad} \subset U$, $V \subset Y$ are the sets of admissible controls and states, respectively. $N_{\Gamma(u)}(y)$ is the normal cone to $\Gamma(u)$ at y provided $y \in \Gamma(u)$ and it is the empty set otherwise. In various shape optimization problems in mechanics, Γ is given in the form

$$\Gamma(u) = \{y \in Y \mid A(u)y = b(u), -q(y) \in K\}, \tag{1.2}$$

where the map A assigns to controls u linear continuous operators, K is the cone of nonnegative functions in a *constraint* space Z and the maps A, b and q are continuously differentiable. The equation in (1.2) expresses the so-called equilibrium condition and the "inequality" $-q(y) \in K$ some physical requirement (an obstacle, plastic yielding, ...).

Assume that V consists of the states y for which $q(y) \in \text{int } K$ in a prescribed subdomain of the designed domain. This special requirement can be respected by an exact penalty dependent on the Karush–Kuhn–Tucker (K.K.T.) vector λ assigned to the inequality $-q(y) \in K$. After an appropriate discretization we arrive at a

*Partly supported by the German Scientific Foundation (DFG).
**Institute of Information Theory and Automation, Czech Academy of Sciences, 18208 Praha 8, Pod vodárenskou věží 4, Czech Republic

finite-dimensional optimization problem

$$\phi(u, y, \lambda) \rightarrow \inf$$

subject to

$$0 \in \begin{pmatrix} F(u,y) + A^T(u)\mu + (\nabla q(y))^T \lambda \\ A(u)y - b(u) \\ -q(y) \end{pmatrix} + \begin{pmatrix} 0 \\ 0 \\ N_{\mathbf{R}_+^s}(\lambda) \end{pmatrix} \quad (1.3)$$

$$u \in U_{ad},$$

in which U is replaced by $\mathbb{R}^n$, Y by $\mathbb{R}^m$, Z by $\mathbb{R}^s$ and the range space of $A(u)$ by $\mathbb{R}^p$; for the sake of simplicity we use the same symbols for the original and the discretized sets, maps and variables. $\varphi[\mathbb{R}^n \times \mathbb{R}^m \times \mathbb{R}^s \rightarrow \mathbb{R}]$ is the discretized "augmented" objective, respecting the state constraints of the above type and $\mu \in \mathbb{R}^p$ is the multiplier vector assigned to the (discretized) equality constraint $A(u)y = b(u)$.

Main difficulties in solving (1.3) are due to its first constraint — so called generalized equation ([11]). However, on the basis of the results from [11, 7, 12], necessary optimality conditions for rather similar problems have already been derived in [9] and [10] and in [9] also a numerical approach has been proposed under the restriction that $\nabla_y F(u, y)$ is symmetric over $U \times Y$ and φ does not depend on λ. In fact, this numerical approach has already been applied to shape optimization problems of a simpler structure in [5, 6] and it represents a useful alternative to the classical regularization techniques, cf. [4]. The aim of this contribution is

(i) to extend the numerical approach of [5, 9, 6] to problems of the type (1.3), and

(ii) to apply it to the design of masonry-like structures analyzed in [2].

For the reader's convenience we recall still the definition of the generalized gradient and generalized Jacobian used throughout the paper.

Definition 1.1 Let the operator $F[R^m \rightarrow R^n]$ be Lipschitz near $x_0 \in R^m$ and let σ_F denote the set of points at which F fails to be differentiable. The *generalized Jacobian* of F at x_0, denoted $\partial F(x_0)$, is the set of $[n \times m]$ matrices, given by

$$\partial F(x_0) = \text{conv} \left\{ \lim_{i \to \infty} \nabla F(x_i) \,|\, x_i \to x_0, \ x_i \notin \sigma_F \right\}.$$

For $n = 1$, $\partial F(x_0)$ is termed the *generalized gradient* of F at x_0. Its elements are called *subgradients*.

Remark 1.1 The term subgradient is borrowed from convex analysis, where it is used for the vectors belonging to the subdifferential. Recently, however, it is frequently used also in a nonconvex setting for the elements of various generalized gradients.

The following notation is employed: $\mathcal{N}(C)$ is the null space of a linear operator C, $S'(x_0; d)$ is the directional derivative of an operator S at a point x_0 in the direction d, E is the unit matrix, P_1 denotes the subspace of the polynomials of order one and x^i is the i-th coordinate of a vector $x \in \mathbb{R}^m$. For $x, y \in \mathbb{R}^n$ the inequalities $x \geq y$ ($x > y$) mean $x^i \geq y^i$ ($x^i > y^i$) for all i. H^1, H_0^1 are the usual Sobolev spaces $W^{1,2}, W_0^{1,2}$ and $\overline{A}$ denotes the closure of a set A.

2   Numerical approach

Suppose that φ is continuously differentiable, U_{ad} is a nonempty compact subset of $\mathbb{R}^n$ and for any u from an open set $A \supset U_{ad}$, the following assumptions hold:

(A1) $\Gamma(u) \neq \emptyset$;

(A2) $F(u, \cdot)$ is strongly monotone over $\mathbb{R}^m$, uniformly with respect to $u \in A$, i.e., there exists a constant $\gamma > 0$ such that

$$\langle F(u, y_1) - F(u, y_2), y_1 - y_2 \rangle \geq \gamma \|y_1 - y_2\|^2 \quad \text{for all } y_1, y_2 \in \mathbb{R}^m, u \in A.$$

Under (A1), (A2), the generalized equation in (1.3) defines an operator S which assigns to each $u \in A$ a unique vector $y \in \mathbb{R}^m$ such that with a suitable pair $(\mu, \lambda) \in \mathbb{R}^p \times \mathbb{R}_+^s$, (y, μ, λ) is a solution of this generalized equation. To ensure the unicity of the multipliers μ, λ as well, we impose still the well-known linear independency constraint qualification, namely

(LI) for each $u \in A$, $y = S(u)$, the vectors $A^i(u), i = 1, 2, \ldots, p$, and $\nabla q^j(y)$, $j \in I(u) := \{i \in \{1, 2, \ldots, s\} \mid q^i(y) = 0\}$, are linearly independent.

Then, as shown e.g. in [11], the operator S is locally Lipschitz over A and also the maps Π, Λ assigning to $u \in A$ the respective multipliers μ and λ are single-valued and locally Lipschitz. Therefore, problem (1.3) may be rewritten to the form

$$\Theta(u) := \varphi(u, S(u), \Lambda(u)) \to \inf$$
$$\text{subject to} \qquad\qquad\qquad\qquad\qquad\qquad (2.1)$$
$$u \in U_{ad}.$$

This problem may be numerically solved by available nondifferentiable optimization (NDO) methods, provided Θ is at least directionally differentiable and we are able to compute subgradients from $\partial\Theta(u)$ for arbitrary $u \in U_{ad}$. Fortunately, under (A1), (A2) and (LI), both maps S and Λ are directionally differentiable, whereby these derivatives satisfy a certain system of equations and inequalities, cf. [3, Thm 5.2]. In these relations an important role is played by the index set of *strongly active inequalities*

$$J(u) := \{i \in I(u) \mid \lambda^i > 0\}$$

which will be used also in the further development.

To simplify the notation, for an index set $M \subset \{1, 2, \ldots, s\}$ and a vector $d \in \mathbb{R}^s$, d_M denotes the subvector composed from the components $d^i, i \in M$. Analogously, for a matrix D with s rows, D_K denotes a submatrix composed from the rows $D^i, i \in K$. Further, the arguments at I and J will be omitted whenever it cannot lead to any confusion.

Let us turn our attention to the computation of subgradients. For this purpose we employ the generalized Jacobian Chain Rule of [1], according to which at a fixed $u_0 \in \mathcal{A}$, $y_0 = S(u_0)$ and $\lambda_0 = \Lambda(u_0)$ one has

$$\partial \Theta(u_0) = \nabla_u \varphi(u_0, y_0, \lambda_0) \quad + \quad \{P^T(u_0) \nabla_y \varphi(u_0, y_0, \lambda_0) \,|\, P(u_0) \in \partial S(u_0)\} \\ + \quad \{Q^T(u_0) \nabla_\lambda \varphi(u_0, y_0, \lambda_0) \,|\, Q(u_0) \in \partial \Lambda(u_0)\}. \tag{2.2}$$

The matrices $P(u_0), Q(u_0)$ will be computed exactly according to Definition 1.1. Assume first that at u_0 the strict complementarity holds, i.e. $I(u_0) = J(u_0)$. Then the maps S, Π, Λ are differentiable at u_0 ([7]) and the respective derivatives solve the matrix linear equation

$$D_I(u_0, y_0, \mu_0, \lambda_0) \begin{pmatrix} \nabla S(u_0) \\ \nabla \Pi(u_0) \\ \nabla \Lambda_I(u_0) \end{pmatrix} = \begin{pmatrix} -\nabla_u \mathcal{L}(u_0, y_0, \mu_0, \lambda_0) \\ \nabla_u(A(u_0)y_0) - \nabla b(u_0) \\ 0 \end{pmatrix}, \tag{2.3}$$

where $\mu_0 = \Pi(u_0)$, $\mathcal{L}(u, y, \mu, \lambda) = F(u, y) + (A(u))^T \mu + (\nabla q(y))^T \lambda$ is the *Lagrangian* and

$$D_I(u_0, y_0, \mu_0, \lambda_0) = \begin{pmatrix} \nabla_y \mathcal{L}(u_0, y_0, \mu_0, \lambda_0) & (A(u_0))^T & (\nabla q_I(y_0))^T \\ -A(u_0) & 0 & 0 \\ -\nabla q_I(y_0) & 0 & 0 \end{pmatrix}. \tag{2.4}$$

The derivatives of Λ^i for $i \notin I(u_0)$ are zero. To analyze the general case $I(u_0) \neq J(u_0)$, we fix an index set $\tilde{J}$ satisfying the inclusions

$$J(u_0) \subset \tilde{J} \subset I(u_0) \tag{2.5}$$

and denote $\mathcal{D} = I(u_0) \setminus \tilde{J}, \mathcal{B} = \tilde{J} \setminus J(u_0)$. Let o, o_1, o_2, o_3 be the cardinalities of $I(y_0), J(y_0), \mathcal{D}$ and $\mathcal{B}$, respectively. Matrices $P(u_0) \in \partial S(u_0)$, $Q(u_0) \in \partial \Lambda(u_0)$ can be computed according to the following assertion.

Theorem 2.1 *Let assumptions (A1),(A2) and (LI) hold and consider the point* $(u_0, y_0, \mu_0, \lambda_0)$, *where* $u_0 \in \mathcal{A}$, $y_0 = S(u_0)$, $\mu_0 = \Pi(u_0)$ *and* $\lambda_0 = \Lambda(u_0)$. *Assume that* $\tilde{J}$ *fulfills inclusions (2.5) and the linear system*

$$-(\nabla q_{\mathcal{D}}(y_0))^T y_1^* + (\nabla_y \mathcal{L}(u_0, y_0, \mu_0, \lambda_0))^T y_3^* + (A(u_0))^T y_4^* + (\nabla q_{\tilde{J}}(y_0))^T y_5^* = 0$$
$$y_2^* + \nabla q_{\mathcal{B}}(y_0) y_3^* = 0$$
$$(\nabla_u \mathcal{L}(u_0, y_0, \mu_0, \lambda_0))^T y_3^* + (\nabla_u(A(u_0)y_0) - \nabla b(u_0))^T y_4^* = 0 \tag{2.6}$$

does not possess a solution $(y_1^*, y_2^*, y_3^*, y_4^*, y_5^*) \in \mathbb{R}^{o_2} \times \mathbb{R}^{o_3} \times \mathbb{R}^m \times \mathbb{R}^p \times \mathbb{R}^{o_1+o_3}$ *satisfying the conditions*

$$(y_1^*, y_2^*) \geq 0, \quad (y_1^*, y_2^*) \neq 0, \quad y_3^* \in \mathcal{N}(A(u_0)) \cap \mathcal{N}(\nabla q_J(y_0)). \tag{2.7}$$

Then one has

$$(D_{\tilde{J}}(u_0, y_0, \mu_0, \lambda_0))^{-1} \begin{pmatrix} -\nabla_u \mathcal{L}(u_0, y_0, \mu_0, \lambda_0) \\ \nabla_u(A(u_0)y_0) - \nabla b(u_0) \\ 0 \end{pmatrix} \in \begin{pmatrix} \partial S(u_0) \\ \partial \Pi(u_0) \\ \partial \Lambda_{\tilde{J}}(u_0) \end{pmatrix}, \qquad (2.8)$$

where $D_{\tilde{J}}(u_0, y_0, \mu_0, \lambda_0)$ is given by (2.4) with $I(u_0)$ being replaced by $\tilde{J}$.

Proof. We first show that under the assumptions imposed there exists a direction $h \in \mathbb{R}^n$ such that

$$J(u_0 + \vartheta h) = I(u_0 + \vartheta h) = \tilde{J} \qquad (2.9)$$

for all sufficiently small positive ϑ. By using of directional derivatives of S and Λ, (2.9) may be rewritten to the form

$$\begin{aligned} \nabla q_{\mathcal{D}}(y_0) S'(u_0; h) &< 0 \\ (\Lambda^i)'(u_0; h) &> 0 \quad \text{for all } i \in \mathcal{B}. \end{aligned}$$

Taking into account the characterization of $S'(u_0; h)$, $\Lambda'(u_0; h)$ from [3, Thm. 5.2], we observe that the desired direction exists, whenever the following linear system of equations and inequalities is consistent:

$$\begin{aligned} \nabla_y \mathcal{L}(u_0, y_0, \mu_0, \lambda_0) S'(u_0; h) + (A(u_0))^T \Pi'(u_0; h) \; + & \\ (\nabla q_{\tilde{J}}(y_0))^T \Lambda'_{\tilde{J}}(u_0; h) &= -\nabla_u \mathcal{L}(u_0, y_0, \mu_0, \lambda_0)h \\ A(u_0) S'(u_0; h) &= -\nabla_u(A(u_0)y)h + \nabla b(u_0)h \\ \nabla q_{\tilde{J}}(y_0) S'(u_0; h) &= 0 \\ \nabla q_{\mathcal{D}}(y_0) S'(u_0; h) &< 0 \\ -\Lambda'_{\mathcal{B}}(u_0; h) &< 0. \end{aligned}$$

However, its consistency is according to the well-known Motzkin theorem of the alternative ([8]) equivalent to the inconsistency of (2.6), (2.7), and thus relation (2.9) is fulfilled. Consequently, the maps S, Π, Λ are differentiable at the points $u_0 + \vartheta h$ for all sufficiently small $\vartheta > 0$. With respect to Definition 1.1 it remains to prove that

$$\lim_{\vartheta \searrow 0} \begin{pmatrix} \nabla S(u_0 + \vartheta h) \\ \nabla \Pi(u_0 + \vartheta h) \\ \nabla \Lambda_{\tilde{J}}(u_0 + \vartheta h) \end{pmatrix} = (D_{\tilde{J}}(u_0, y_0, \mu_0, \lambda_0))^{-1} \begin{pmatrix} -\nabla_u \mathcal{L}(u_0, y_0, \mu_0, \lambda_0) \\ \nabla_u(A(u_0)y_0) - \nabla b(u_0) \\ 0 \end{pmatrix},$$

which may be done easily by employing the classical implicit function argument. $\square$

Remark 2.1 An appropriate completion of a matrix from $\partial \Lambda_{\tilde{J}}(u_0)$ by zero elements provides us with a matrix from $\partial \Lambda(u_0)$.

Of course, the satisfaction of the above conditions can hardly be tested in the presented form. It is clear that if $n \geq o_2 + o_3$ (which is almost always satisfied in practical situations), then the inconsistency assumption holds for any choice of $\tilde{J}$ whenever the matrix

$$
\begin{pmatrix}
-(\nabla q_D(y_0))^T & 0 & (\nabla_y \mathcal{L}(u_0, y_0, \mu_0, \lambda_0))^T & (A(u_0))^T & (\nabla q_{\tilde{J}}(y_0))^T \\
0 & E & \nabla q_B(y_0) & 0 & 0 \\
0 & 0 & (\nabla_u \mathcal{L}(u_0, y_0, \mu_0, \lambda_0))^T & (\nabla_u(A(u_0)y_0) - \nabla b(u_0))^T & 0 \\
0 & 0 & A(u_0) & 0 & 0 \\
0 & 0 & \nabla q_J(y_0) & 0 & 0
\end{pmatrix}
$$

has maximal rank. However, this test may also be rather cumbersome. Fortunately, in the solved example the proposed approach worked well with $\tilde{J} = J(y_0)$ and so we were not forced to perform any testing during the iteration process.

On the basis of (2.2) and (2.8), the computation of subgradients of Θ can proceed according to the following statement:

Proposition 2.1 *Let all assumptions of Theorem 2.1 be fulfilled and assume that the triple* $(p_0, q_0, r_0) \in \mathbb{R}^m \times \mathbb{R}^p \times \mathbb{R}^{o_1+o_3}$ *solves the linear system*

$$
(D_{\tilde{J}}(u_0, y_0, \mu_0, \lambda_0))^T \begin{pmatrix} p \\ q \\ r \end{pmatrix} + \begin{pmatrix} \nabla_y \varphi(u_0, y_0, \lambda_0) \\ 0 \\ (\nabla_\lambda \varphi(u_0, y_0, \lambda_0))_{\tilde{J}} \end{pmatrix} = 0.
$$

Then one has

$$
\xi := \nabla_u \varphi(u_0, y_0, \lambda_0) + (\nabla_u \mathcal{L}(u_0, y_0, \mu_0, \lambda_0))^T p_0 - (\nabla_u(A(u_0)y_0) - \nabla b(u_0))^T q_0 \in \partial\Theta(u_0).
$$

The proof is based on the standard "adjoint equation technique", known from optimal control theory. This way of computing subgradients has been applied in the design problem investigated in the next section.

3 Design of masonry structures

We shall investigate masonry-like material that behaves elastically if subject to pressure forces, but is extremely weak if we try to pull it apart. In general, rather small tension suffices to produce a fracture: the material breaks down and shows lines of fracture. However, the appearance of fracture may not be destructive; in most cases it is compatible with global equilibrium of the structure. The mathematical description of the equilibrium of such structures has been done in [2]. Our goal in this paper is to optimize the shape of such structures, i.e., to find such a shape (from a given set of shapes) that some cost functional is minimized.

Let $\Omega \subset \mathbb{R}^2$ be a bounded domain with a Lipschitz boundary $\partial\Omega$. Let $\partial\Omega = \overline{\Gamma}_u \cup \overline{\Gamma}_g$, $\text{meas}(\overline{\Gamma}_u \cap \overline{\Gamma}_g) = \emptyset$, Γ_u of positive measure. Denote by $\mathbb{R}^{2 \times 2}_s$ the set of symmetric 2×2 matrices. We introduce the set of *stress fields*

$$
S(\Omega) = \{\tau : \Omega \to \mathbb{R}^{2 \times 2}_s \mid \tau_{ij} \in L^2(\Omega), \quad i, j = 1, 2\}
$$

and the isomorphism

$$
\beta : S(\Omega) \to S(\Omega)
$$

defined by the generalized inverse Hooke's law and such that

$$\beta_0 \|\sigma\|_\Omega \leq (\beta\sigma, \sigma)_\Omega \leq \beta_1 \|\sigma\|_\Omega \quad \forall \sigma \in S(\Omega).$$

Here, $(\cdot, \cdot)_\Omega$ and $\|\cdot\|_\Omega$ denote $L^2(\Omega)$ inner product and norm, respectively, and β_0, β_1 are some positive constants.

Assume that a body force $F = (F_1, F_2) \in [L^2(\Omega)]^2$ and a surface traction $g = (g_1, g_2) \in [L^2(\Gamma_g)]^2$ are given. We introduce the space of virtual displacements

$$H = \{w \in [H^1(\Omega)]^2 \mid w = 0 \text{ on } \Gamma_u\}$$

and define the set of *statically admissible* stress fields satisfying the *weak equilibrium conditions*

$$E(\Omega) = \{\tau \in S(\Omega) \mid (\tau, e(w))_\Omega = \sum_{i=1}^{2} \int_\Omega F_i w_i \, dx + \sum_{i=1}^{2} \int_{\Gamma_g} g_i w_i \, d\Gamma \quad \forall w \in H\},$$

where

$$(e(w))_{ij} = \frac{1}{2}\left(\frac{\partial w_i}{\partial x_j} + \frac{\partial w_j}{\partial x_i}\right), \quad i, j = 1, 2,$$

is the *strain field*.

The masonry-like materials can be characterized in different ways. The most restrictive one is the requirement that the stress tensor $\tau(x), x \in \Omega$, cannot have negative eigenvalues ([2]). We consider, however, a more simple situation which can be used, e.g., in modeling of supporting pillars in cathedrals: we require that the 'vertical' stress component is nonnegative, i.e., $(\tau)_{22} \geq 0$ a.e. in Ω. In pillars, typically, the external force coming from the weight of the supported roof causes a bending which is compensated by the self-weight of the pillar. Our task is to find the lightest pillar which is still able to support the roof. We introduce the set of admissible stress fields as

$$M(\Omega) = \{\tau \in S(\Omega) \mid (\tau)_{22} \geq 0 \text{ a.e. in } \Omega\}.$$

The problem of masonry structures can be formulated as a variational inequality (VI):

$$\text{Find } \sigma \in E(\Omega) \cap M(\Omega) \text{ such that}$$
$$(\beta\sigma, \tau - \sigma)_\Omega \geq 0 \quad \forall \tau \in E(\Omega) \cap M(\Omega). \tag{3.1}$$

It can be shown that, under some additional conditions on the pressure g, problem (3.1) has a unique solution.

Now, let us introduce the set of *admissible design variables*

$$U_{ad} = \{u \in C^{(0),1}([0, 1]), \underline{u} \leq u \leq \overline{u}, \left|\frac{du}{dx_2}\right| \leq c_1, \left|\frac{d^2u}{dx_2^2}\right| \leq c_2\},$$

where $\underline{u}, \overline{u}, c_1$ and c_2 are given positive numbers. Consider a family of admissible domains $\Omega(u)$ with variable right "vertical" part of the boundary

$$\Omega(u) = \{(x_1, x_2) \mid 0 < \dot{x}_1 < u(x_2), 0 < x_2 < 1\}.$$

The goal of the optimum design problem is to minimize the volume of the structure, denoted by meas $\Omega(u)$. However, to prevent the destruction, we require that the inequality constraint in the definition of $M(\Omega)$ might be strongly active only in a given subset $\Omega_0(u) \subset \Omega(u)$. In other words, denoting by λ the Lagrange multiplier corresponding to this constraint, we require $\lambda = 0$ in $\Omega(u) \setminus \Omega_0(u)$. Of course, in this way one cannot prevent some constraints in $\Omega(u) \setminus \Omega_0(u)$ to be active provided the corresponding multipliers are zero. However, in the examples solved such phenomenon has been observed merely very closely to the boundary of $\Omega_0(u)$. The optimum design problem reads as

$$\varphi(u, \lambda) := \text{meas } \Omega(u) + r \int_{\Omega(u)\setminus\Omega_0(u)} (\lambda(u)) \, dx \to \inf$$

subject to

$\sigma(u)$ solves (3.1) with $\Omega := \Omega(u)$

$\lambda(u)$ is the multiplier corresponding to the inequality in $M(\Omega(u))$

$u \in U_{ad}$,

$$(3.2)$$

where $r > 0$ is the penalty parameter.

Now we discretize problems (3.1) and (3.2) by the finite element method. In particular, we approximate the design function u by a piece-wise linear function u_h, where h is the discretization parameter, introduce the set

$$U_{ad}^h = \{u_h \in U_{ad} \mid u_h \text{ piece-wise linear}\}$$

and construct the polygonal computational domain $\Omega(u_h)$ such that the positions of all "inner" nodes are derived from the positions of the principal moving nodes lying on u_h. Let N be the number of principal moving nodes and n the number of all nodes in the mesh.

To the approximation of the stress field we use piece-wise constant elements, i.e., with each element we associate a vector from $\mathbb{R}^3$ (three entries of a symmetric 2×2 matrix). We denote the global vector of approximated element stresses by $y \in \mathbb{R}^{3p}$, p being the number of elements. Further, we denote by u the vector from $\mathbb{R}^N$ of x_1-coordinates of the principal moving nodes. Again, we use the same symbols for the original and discretized sets, maps and variables. So U_{ad} is the set associated with U_{ad}^h,

$$E(u) = \{v \in \mathbb{R}^{3p} \mid A(u)v = b(u)\}$$

and

$$M(u) = M = \{v \in \mathbb{R}^{3p} \mid \Upsilon^i(v) \leq 0, \quad i = 1, 2, \ldots, p\};$$

here $A(u)$ is the equilibrium matrix, b the right-hand side vector and

$$\Upsilon^i(v) = v^{3i-1}.$$

Let $B(u)$ be the flexibility (symmetric, positive definite) matrix associated with the inverse Hooke's law β. The discretized state problem can be then written as follows:

$$\text{Find } v \in E(u) \cap M \text{ such that}$$
$$\langle B(u)v, w - v \rangle \geq 0 \quad \forall w \in E(u) \cap M. \tag{3.3}$$

Denoting by $\Lambda(u)$ the map assigning to u the Lagrangian multipliers $\lambda^i, i = 1, \ldots, p$, corresponding to inequalities in M, we can write the discretized optimum design problem as

$$\varphi(u, \lambda) \equiv \operatorname{meas} \Omega(u) + \frac{r}{h^2} \sum_{i \in \mathcal{D}_0} \lambda^i \to \inf$$

subject to

$$\lambda = \Lambda(u)$$
$$u \in U_{ad} \tag{3.4}$$

where $\mathcal{D}_0$ contains indices of elements lying in $\Omega(u) \setminus \Omega_0(u)$. This is a problem of type (1.3). Hence, to its numerical solution, we can use the theory from Section 2, provided that assumptions (A1), (A2) and (LI) are fulfilled. Fortunately, (A1) holds for a proper choice of U_{ad}, which has been observed numerically. (A2) follows from the fact that the controls $u \in U_{ad}$ are uniformly Lipschitz continuous, the triangulation cannot degenerate and thus the flexibility matrices $B(u)$ are uniformly positive definite. (LI) is trivially satisfied.

Example 3.1 Let $\Gamma_u = \{(x_1, x_2) \mid 0 \leq x_1 \leq \overline{u}, x_2 = 0\}$ be the part of the boundary $\partial\Omega$ with prescribed zero displacements, and let the constant surface traction $g = (5, 0)$ is prescribed on the left vertical part of $\partial\Omega$, i.e., on the segment $\{(x_1, x_2) \mid x_1 = 0, 0 \leq x_2 \leq 1\}$. The surface traction on the rest of $\partial\Omega$ is prescribed as zero and the body force $F = (0, -10)$ corresponds to the self-weight. The penalty parameter is $r = 1$. The domain is discretized by triangles defined by 7×7 grid of nodes. The set of admissible controls U_{ad} is characterized by numbers

$$\underline{u} = 0.2, \quad \overline{u} = 1.0, \quad c_1 = 5.0, \quad c_2 = 4.0.$$

Thus we have 216 state variables (72 elements $\times$ 3 unknowns per element) and 7 design variables (x_1-coordinates of principal moving nodes). The index set $\mathcal{D}_0$ associated with the set Ω_0 contains indices of elements lying in the left half of Ω. The discretized state (nonlinear programming) problems (3.3) were solved by the SQP (Sequential Quadratic Programming) code NLPQL due to K. Schittkowski [13], while the master optimization problems (3.4) by the NDO code BT (Bundle-Trust) due to H. Schramm and J. Zowe [14]; in particular, by its nonconvex version BTNCLC which can handle linear constraints. Figure 1 shows the optimum shape design obtained after 61 BT iterations together with the values of the 'vertical' components v^{3i-1} of the state vector (left figure) and the corresponding multipliers λ^i (right figure).

Example 3.2 All data are the same as in the previous example, apart from the surface traction g, which applies now only on the upper half of the left vertical part of $\partial\Omega$. The results are shown in Figure 2.

References

[1] F. H. Clarke. *Optimization and nonsmooth analysis.* J. Wiley & Sons, New York, 1983.

[2] M. Giaquinta and E. Giusti. Research on the equilibrium of masonry structures. *Arch. Rational Mech. Anal.*, 88:359–392, 1985.

[3] P. T. Harker and J. S. Pang. Finite-dimensional variational inequality and nonlinear complementarity problems: A survey of theory, algorithms and applications. *Math. Prog.*, 48:161–220, 1990.

[4] J. Haslinger and P. Neittaanmaki. *Finite element approximation for optimal shape design: theory and applications.* J. Wiley & Sons, Chichester, 1988.

[5] M. Kočvara and J. V. Outrata. A nondifferentiable approach to the solution of optimum design problems with variational inequalities. In P. Kall, editor, *Proc. of the 15th IFIP Conf. on System Modelling and Optimization*, Lecture Notes in Control Inf. Sci. 180, pages 364–373, Zurich, Sept. 2–6 1991.

[6] M. Kočvara and J. V. Outrata. Shape optimization of elasto-plastic bodies governed by variational inequalities. In *Boundary Control and Boundary Variation*, Sophia Antipolis, 1992. To appear in Lecture Notes in Control Inf. Sci.

[7] J. Kyparisis. Solution differentiability for variational inequalities. *Math. Prog.*, 48:285–301, 1990.

[8] O. L. Mangasarian. *Nonlinear Programming.* McGraw Hill, New York, 1969.

[9] J. V. Outrata. On necessary optimality conditions for Stackelberg problems. *J. Optim. Theory Appl.*, 76:305–320, 1993.

[10] J. V. Outrata. On optimization problems with variational inequality constraints. To appear in *SIAM J. Optimization*.

[11] S. M. Robinson. Strongly regular generalized equations. *Math. Oper. Res.*, 5:43–62, 1980.

[12] S. M. Robinson. An implicit-function theorem for a class of nonsmooth functions. *Math. Oper. Res.*, 16:282–309, 1991.

[13] K. Schittkowski. NLPQL: a FORTRAN subroutine solving constrained nonlinear programming problems. *Annals Oper. Research*, 5:485–500, 1985/86.

[14] H. Schramm and J. Zowe. A version of the bundle idea for minimizing a nonsmooth function: conceptual idea, convergence analysis, numerical results. *SIAM J. Optimization*, 2:121–152, 1992.

Figure 1: Optimal shape, stresses and multipliers for Example 1.

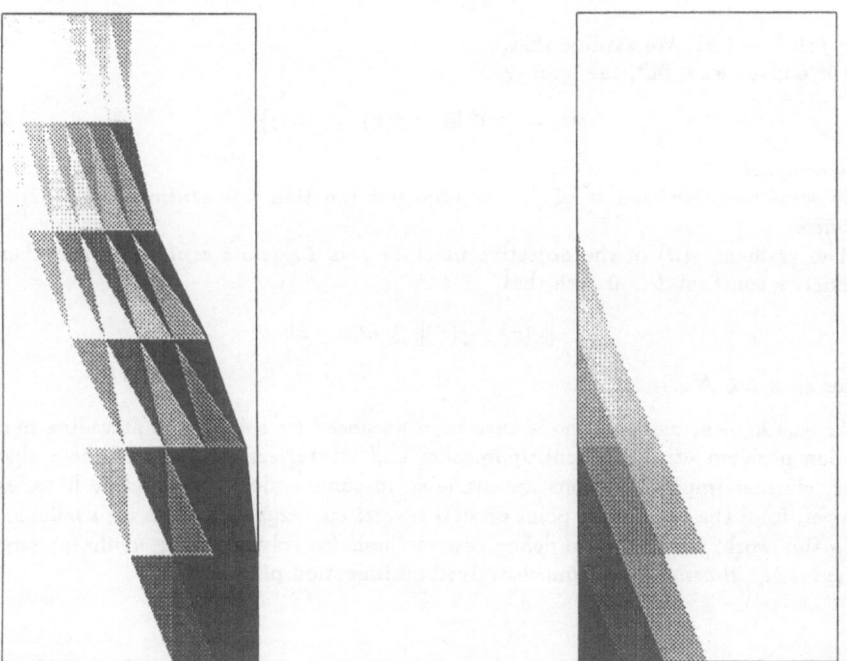

Figure 2: Optimal shape, stresses and multipliers for Example 2.

Algorithms

NONMONOTONE CONJUGATE GRADIENT METHODS FOR OPTIMIZATION

S. Lucidi, M. Roma
Dipartimento di Informatica e Sistemistica
Università di Roma "La Sapienza", via Buonarroti 12, 00185 Roma, Italy.

Abstract

In this paper conjugate gradient methods with nonmonotone line search technique are introduced. This new line search technique is based on a relaxation of the strong Wolfe conditions and it allows to accept larger steps. The proposed conjugate gradient methods are still globally convergent and, at the same time, they should not suffer the propensity for short steps of some classical conjugate gradient methods. Hence, these new methods should be able to tackle efficiently large scale highly nonlinear (possibly ill-conditioned) problems.

1. Introduction

We consider the unconstrained minimization problem

$$\min_{x \in \mathbb{R}} f(x)$$

where $f: \mathbb{R}^n \longrightarrow \mathbb{R}$. We assume that,

i) For a given $x_o \in \mathbb{R}^n$, the level set

$$\Omega_o = \{x \in \mathbb{R}^n : f(x) \leq f(x_o)\}$$

is *compact*.

ii) In some neighborhood $\mathcal{N}$ of Ω_o, the objective function f is *continuously differentiable*.

iii) The gradient $g(x)$ of the objective function f is *Lipschitz continuous* i.e. there exists a constant $L > 0$ such that

$$\|g(x) - g(\bar{x})\| \leq L\|x - \bar{x}\|$$

for all $x, \bar{x} \in \mathcal{N}$.

As well known, many methods have been proposed for solving the preceding minimization problem using different approaches and strategies; for many of these algorithms, efficient implementations are available in some codes of subroutine libraries. Moreover, from the theoretical point of view several convergence results are available.

In this work, our aim is to define new methods for solving *highly nonlinear large scale (possibly ill-conditioned)* unconstrained optimization problems.

Among all the methods proposed, conjugate gradient methods remain one of the most attractive tool for solving nonlinear large problems. As regards highly nonlinear and ill-conditioned problems, recently, nonmonotone strategies have been proposed in the case of Newton-type method (see e.g. [6,7]). They do not require the decrease of the function at each iteration. Numerical experiences have showed that nonmonotonicity is particularly valuable for solving nonlinear large scale unconstrained optimization problems, especially in the case of highly nonlinear and ill-conditioned functions.

Our aim is to join these two approaches and, as first step, we propose a class of globally convergent nonlinear conjugate gradient methods related to Fletcher-Revees method (without restarts) with a new nonmonotone line search.

We consider the Fletcher-Reeves method

$$d_k = \begin{cases} -g_k & \text{for } k = 1 \\ -g_k + \beta_k d_{k-1} & \text{for } k \geq 2, \end{cases} \tag{1.1}$$

$$x_{k+1} = x_k + \alpha_k d_k, \tag{1.2}$$

$$\beta_k = \beta_k^{\text{FR}} := \frac{\|g_k\|^2}{\|g_{k-1}\|^2},$$

where α_k is a steplength obtained by means of a line search and $g_k = \nabla f(x_k)$ denotes the gradient of the function f at point x_k.

From this original version proposed by Fletcher-Reeves many variations have been studied and different implementations have been performed. Moreover there are many theorical discussions about global convergence of this method: Powell [1] has proved the global convergence when an exact line search is performed; Al-Baali [2] has shown that this property along with the descent property still holds for Fletcher-Reeves method when an inexact line search technique is used satisfying the following stepsize rule (*strong Wolf conditions*)

$$f(x_k + \alpha_k d_k) \leq f(x_k) + \gamma \alpha_k g_k' d_k,$$
$$|g_{k+1}' d_k| \leq -\sigma g_k' d_k, \tag{1.3}$$

with $0 < \gamma < \sigma < 1/2$.

Moreover, global convergence of method (1.1) (1.2) with inexact line search satisfying (1.3), has been proved with the following choice of β_k: first Touati-Ahmed and Storey [3] with $0 \leq \beta_k \leq \beta_k^{\text{FR}}$ then Gilbert and Nocedal [4] with $|\beta_k| \leq \beta_k^{\text{FR}}$

Unfortunately, from computational point of view, Fletcher-Reeves method is not very efficient; in particular it is often much slower than Polak-Ribiére and Hestenes-Stiefel methods for which it is not possible to prove global convergence in the general case. In order to explain this inefficiency, Powell [5] showed that under some circumstances, the Fletcher-Reeves method produces very small steps away from the solution and, it is likely, that subsequent steps also are very short causing the algorithm to stall.

In this work we try to overcome this propensity for short steps by using also in this case the nonmonotone approach. In particular we propose a new acceptability criterion

208

Fig.1

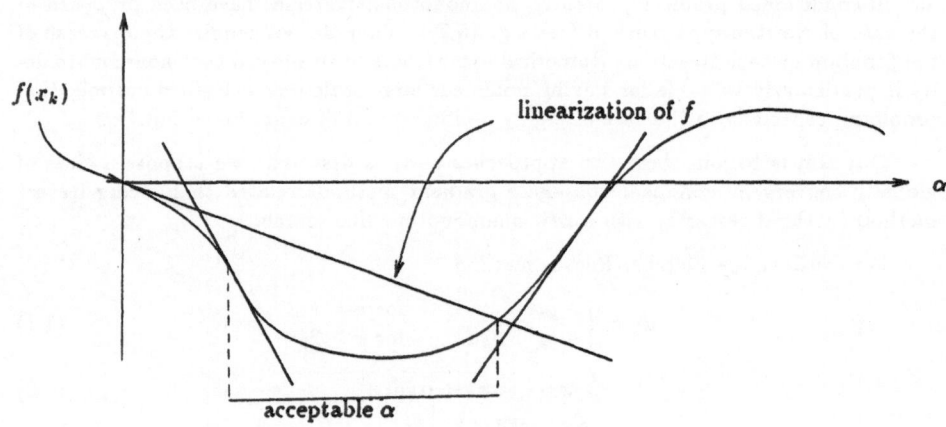

$f(x_k)$

linearization of f

acceptable α

α

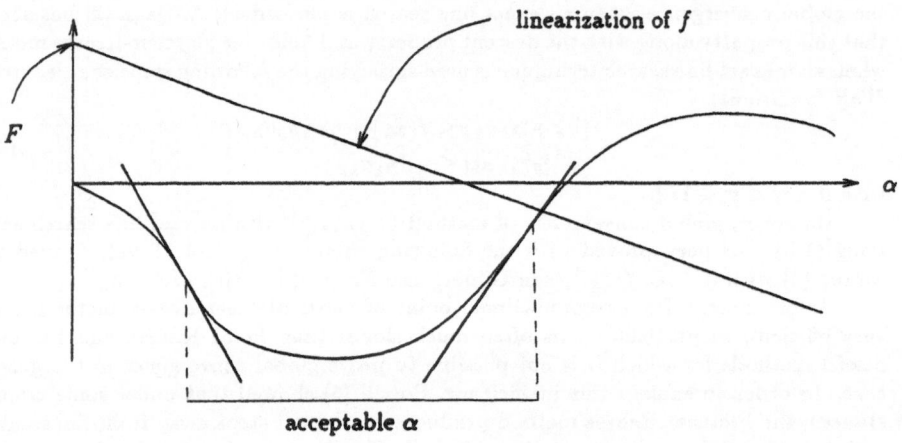

linearization of f

F

acceptable α

α

Effect of nonmonotone line search

of stepsize relaxing the sufficient decrease of the strong Wolfe conditions. Therefore instead of using (1.3) we propose

$$f(x_k + \alpha_k d_k) \leq \max_{0 \leq j \leq M} [f(x_{k-j})] + \gamma \alpha_k g'_k d_k,$$

$$|g'_{k+1} d_k| \leq -\sigma g'_k d_k, \tag{1.4}$$

where M is the number of previous iterates that are taken into account.

The main idea of this approach is to compute the stepsize α_k along the search direction d_k by a nonmonotone line search with respect to the maximum of the function values in a prefixed number of previous iterates. Since it does not enforce monotonicity $f(x_{k+1}) < f(x_k)$, it allows to accept larger values of stepsize α_k and hence the algorithm has the possibility of using *larger steps* (see fig.1 where F indicates the maximum of the previous function values).

2. The new algorithm and its convergence analysis

The class of conjugate gradient method we consider is the following

$$d_k = \begin{cases} -g_k & \text{for } k = 1 \\ -g_k + \beta_k d_{k-1} & \text{for } k \geq 2, \end{cases} \tag{2.1}$$

$$x_{k+1} = x_k + \alpha_k d_k, \tag{2.2}$$

with

$$|\beta_k| \leq \beta_k^{\text{FR}} := \frac{\|g_k\|^2}{\|g_{k-1}\|^2}, \tag{2.3}$$

and where α_k is a steplength obtained by means of a line search satisfying the conditions

$$f(x_k + \alpha_k d_k) \leq \max_{0 \leq j \leq M} [f(x_{k-j})] + \gamma \alpha_k g'_k d_k,$$

$$|g'_{k+1} d_k| \leq -\sigma g'_k d_k. \tag{2.4}$$

M is the number of previous iterates that are taken into account and $0 < \gamma < \sigma < 1/2$.

The global convergence of the algorithm (2.1)-(2.4) has been studied in [8].

First of all, the following theorem concerning the line search technique has been proved.

Theorem 2.1. *Suppose that assumptions i), ii) and iii) hold. Consider any iteration of the form $x_{k+1} = x_k + \alpha_k d_k$ where d_k is a descent direction and α_k is any value satisfying the line search conditions*

$$f(x_k + \alpha_k d_k) \leq \max_{0 \leq j \leq M} [f(x_{k-j})] + \gamma \alpha_k g'_k d_k$$

$$|g'_{k+1} d_k| \leq -\sigma g'_k d_k,$$

with $0 < \gamma < \sigma < 1$ and where M is the number of the previous iterates taken into account. Then there exists a subsequence $\{x_{q(j)}\}$ of $\{x_k\}$ such that

$$\sum_{j=1}^{\infty} \|g_{q(j)-1}\|^2 \cos^2 \theta_{q(j)-1} < \infty \tag{2.5}$$

where

$$\cos \theta_k = -\frac{g'_k d_k}{\|g_k\| \|d_k\|}.$$

$\square$

We remark that this result shows that if in the Wolfe criteria we relax the condition of sufficient decrease of the function values, we obtain a little weaker result; infact, instead of obtaining that the whole sequence $\{x_k\}$ satysfies

$$\sum_{j=1}^{\infty} \|g_k\|^2 \cos^2 \theta_k < \infty$$

we have that there exits a subsequence of $\{x_k\}$ for which (2.5) holds.

However, this weaker property (2.5) is still sufficient to ensure global convergence of the proposed algorithm (2.1)-(2.4) as it is shown by the following theorem proved in [8] .

Theorem 2.2 *Suppose that the assumptions i), ii) and iii) hold; suppose that the sequence $\{x_k\}$ is produced by the algorithm (2.1)-(2.3) where d_k is a descent direction and α_k is any value satisfying the acceptability criterion (2.4) with $0 < \gamma < \sigma < 1/2$, then*

$$\lim_{k \to \infty} \inf \|g_k\| = 0 \qquad (2.6)$$

$\square$

The method of proof is similar to that given in [2] in the case $\beta_k = \beta_k^{\mathrm{FR}}$ and in [4] in the case $|\beta_k| \leq \beta_k^{\mathrm{FR}}$: proceeding by contradiction, if we suppose that

$$\lim_{k \to \infty} \inf \|g_k\| = 0$$

does not hold, i.e. there exists $\epsilon > 0$ such that $\|g_k\| \geq \epsilon$ for all k, then $\|d_k\|^2 \leq ck$ for some constant c which implies (by following similar arguments of [2] and [4]) that the subsequence of points $\{x_{q(j)}\}$ defined in the Theorem 2.1 is such that the series

$$\sum_{j=1}^{\infty} \|g_{q(j)-1}\|^2 \cos^2 \theta_{q(j)-1}$$

diverges. Of course, this contradicts Theorem 2.1.

3. Line search algorithm

The main objective of the classical line searches is to try to locate as much as possible the minimum point along the search direction without particular attention to the lenght of the stepsize and, in some circumstances they may produce short steps. On the contrary, the idea of the nonmonotone approach, especially in the case of the conjugate gradient methods, is to find the stepsize as large as possible which, at the same time, ensures the conditions for the global convergence. Therefore, the very efficient line searches proposed in literature (see e.g. [9]) are not suitable for our nonmonotone approach.

Hence it is necessary to define new line search algorithms which try to accept large steps.

In the sequel we describe a very general algorithm model for locating an α_k which satisfies the conditions

$$f(x_k + \alpha_k d_k) \leq \max_{0 \leq j \leq M} [f(x_{k-j})] + \gamma \alpha_k g_k' d_k \qquad (3.1)$$

$$|g_{k+1}' d_k| \leq -\sigma g_k' d_k. \qquad (3.2)$$

Algorithm model

Data : x_k, $f(x_k)$, $g(x_k)$.

Step 1. Set $\alpha_l = 0$, $\alpha_u = +\infty$, compute $\alpha_0 \in (\alpha_l, \alpha_u)$ and set $\alpha = \alpha_0$.

Step 2. If the conditions

$$f(x_k + \alpha d_k) \leq \max_{0 \leq j \leq M} [f(x_{k-j})] + \gamma \alpha g(x_k)' d_k$$

$$g(x_k + \alpha d_k)' d_k \leq -\sigma g(x_k)' d_k$$

are satisfied go to Step 4; otherwise Step 3.

Step 3. Compute $\hat{\alpha} \in (\alpha_l, \alpha)$ by an *interpolation procedure*, set $\alpha_u = \alpha$, $\alpha = \hat{\alpha}$ and go to Step 2.

Step 4. If the condition

$$g(x_k + \alpha d_k)' d_k \geq \sigma g(x_k)' d_k,$$

is satisfied set $\alpha_k = \alpha$ and *stop*; otherwise compute $\hat{\alpha} \in (\alpha, \alpha_u)$ by an *extrapolation procedure*, set $\alpha_l = \alpha$, $\alpha = \hat{\alpha}$ and go to Step 2.

The following result describes some minimal conditions on the interpolation procedure and the extrapolation procedure which ensure that the preceding algorithm model terminates in a finite number of steps.

Proposition 3.1. *Suppose that assumption i) holds and that d_k is a descent direction. Let $\theta \in (0, 1/2)$ and $\tau > 1$ be two fixed values and assume that the extrapolation procedure gives an $\hat{\alpha}$ such that:*

(i) $\hat{\alpha} \geq \tau \alpha$, *if* $\alpha_u = \infty$

(ii) $\max[(\hat{\alpha} - \alpha), (\alpha_u - \hat{\alpha})] \leq (1 - \theta)(\alpha_u - \alpha)$ *if* $\alpha_u < \infty$,

while, the interpolation procedure gives an $\hat{\alpha}$ such :

(iii) $\max[(\hat{\alpha} - \alpha_l), (\alpha - \hat{\alpha})] \leq (1 - \theta)(\alpha - \alpha_l)$,

where α, α_l, α_u are defined in the Algorithm Model.

Then the Algorithm Model determines, in a finite number of steps, a value α_k which satisfies conditions (3.1) (3.2). □

The easy proof of the preceding result can be found in [10] or [8].

The preceding Algorithm Model and the corresponding Proposition 3.1 could be the basis to define new line search algorithms more suitable for the nonmonotone approach in conjugate gradient methods. In the next section we report some numerical results obtained by an algorithm of this kind.

4. Numerical experiences

We have tested a first tentative algorithm on some of the problems collected in [4] with the same choice of the stop criterion

$$\|g(x_k)\|_\infty < 10^{-5} (1 + |f(x_k)|)$$

and with the same choice of the parameter $\gamma = 10^{-4}$, $\sigma = 0.1$. Concerning nonmonotone line search we have used $M = 3$.

As regards interpolation and extrapolation procedure, we have used a simple quadratic model of the function in the interval of interest to produce a value $\tilde{\alpha}$. Then in order to satisfy the assumptions of Proposition 3.1 and in order to enforce the choice of large steps we have set as trial value $\hat{\alpha}$:

- in the extrapolation procedure

$$\hat{\alpha} = \max[\tilde{\alpha}, \tau\alpha], \qquad \text{if} \qquad \alpha_u = \infty,$$
$$\hat{\alpha} = \max[\alpha + \theta(\alpha_u - \alpha), \min[\tilde{\alpha}, \alpha + (1 - \theta)(\alpha_u - \alpha)]] \quad \text{otherwise}$$

- in the interpolation procedure

$$\hat{\alpha} = \max[\alpha_l + \theta(\alpha - \alpha_l), \min[\tilde{\alpha}, \alpha_l + (1 - \theta)(\alpha - \alpha_l)]]$$

with the following choice of the parameter $\theta = 0.1$ and $\tau = 2$.

The initial trial value α_0 is the following

$$\alpha_0 = \left| \frac{g(x_k)' d_k}{(g(x_k + \delta d_k) - g(x_k))' d_k} \right|$$

where $\delta = 10^{-6}/\|d_k\|$.

The tests were performed on a IBM RISC System/6000, using Fortran in double precision.

In Table 1 we report the numerical results obtained by our nonmonotone conjugate gradient method using the described line search technique. Furthermore we report, also, the results reported in [4] by using the pure Fletcher-Reeves method which is based on the efficient line search algorithm proposed in [9]. Both the methods use the same search direction and differ only on the choice of the stepsize α. Our aim is to compare our approach of trying to perform large step without minimizing the function along the search direction against the classical approach of enforcing a good decrease of the objective function by finding a good approximation of a local minimum point along the search direction.

The structure of the table is the following: in the first three columns our results are reported (number of iterations, number of function evaluations, number gradient evaluations). In the fourth column the results for pure Fletcher-Reeves method are reported according to [4].

Table 1

Functions ($n = 100$)	it	f	g	it/f − g
Calculus of var. 2	405	406	811	405/827
Calculus of var. 3	1270	1271	2541	1313/2627
Penalty 1	22	46	68	10/36
Penalty 2	7	15	22	7/20
Penalty 3	99	114	213	116/236
Ex. Powell Sin.	769	782	1551	1426/2855
Tridiag. 1	70	71	141	70/142
Boundary-value pr.	173	174	347	175/351
Broyden Trid. nonlin.	29	34	63	29/60
Ex. Freud. Roth.	14	19	33	10/41
Wrong ext. Wood	2324	2373	4697	*
Sparse Matrix sqrt	151	159	310	151/306
Extended Powell	769	782	1551	1426/2855
Tridiag. 2	74	75	149	72/146
Trigonometric	186	199	385	202/409

Even if no final conclusion can be drawn, the results seem to be "promising". However, in our opinion, the implemented algorithm does not completely exploit the possibility of the nonmonotone approach.

As final remarks, we note that as future works, there are two main open questions:

- from the computational point of view, new efficient line search algorithms which try to find steps as large as possible could be defined;

- from the theorical point of view, it should be consider new Wolfe conditions as weak as possible which mantain global convergence property.

References

[1] M. J. D. Powell, *Nonconvex minimization calculations and the conjugate gradient method*, in Lecture Notes in Mathematics 1066, Springer-Verlag, Berlin, 1984, pp. 122-141.

[2] M. Al-Baali, *Descent property and global convergence of the Fletcher-Reeves methods with inexact line search*, IMA J. Numer. Anal., 5 (1985), pp. 121-124.

[3] D. Touati-Ahmed, C. Storey, *Efficient Hybrid Conjugate Gradient Techniques*, Journal of Optimization Theory and Applications, 64 (1990), pp. 379-397.

[4] J. C. Gilbert, J. Nocedal, *Global convergence properties of conjugate gradient methods for optimization*, SIAM J. Optimization, 2 (1992), pp. 21-42.

[5] M. J. D. Powell, *Restart procedures for the conjugate gradient method*, Math. Programming, 12 (1977), pp. 241-254.

[6] L. Grippo, F. Lampariello, S. Lucidi, *A nonmonotone linesearch technique for Newton's method*, SIAM J. Numer. Anal. 23, (1986) pp. 707-716.

[7] L. Grippo, F. Lampariello, S. Lucidi, *A class of nonmonotone stabilization methods in unconstrained optimization*, Numer. Math. 59, (1991) pp. 779-805.

[8] S. Lucidi, M. Roma, *Nonmonotone conjugate gradient methods for optimization*, Technical Report IASI-CNR, R. 342, (1992)

[9] J.J. Moré, D. J. Thuente, *Line Search Algorithms with Guaranteed Sufficient Decrease*, to appear in ACM Transaction on Math. Soft.

[10] L. Grippo, *Metodi di ottimizzazione non vincolata*, Report IASI-CNR, RI. 64, (1988) (in Italian)

BARRIER-NEWTON METHODS IN MATHEMATICAL PROGRAMMING

Yuri G. EVTUSHENKO,   Vitali G. ZHADAN
Computing Center, 40 Vavilov Str.
117967 Moscow GSP-1, Russia

Abstract: A space transformation technique is used for the reduction of constrained minimization problems to minimization problems without inequality constraints. The continuous and discrete versions of Newton's method are applied for solving such reduced LP and NLP problems. The space transformation modifies these methods and introduces additional matrices which play the role of a multiplicative barrier, preventing the trajectories from crossing the boundary of the feasible set. The proposed algorithms are based on the numerical integration of the systems of ordinary differential equations. These algorithms do not require feasibility of starting and current points, but they preserve feasibility. The discrete version of the barrier-Newton method has a superlinear convergence rate and a special stepsize regulation gives quadratic convergence.

1. Introduction. Newton's method has been widely used for solving nonlinear, linear programming and optimal control problems. We mention two books [1], [3] and several papers [10], [11], [12], [14], [15]. Due to the extensive activity in this area, our list of references is not complete.

Starting from 1973, we developed a family of numerical methods based on space transformation techniques [2], [4], [5], [6], [7]. Such an approach enabled us to reduce the original NLP problem to a problem without inequality constraints. Initially the space transformation was developed for the gradient-projection method. In [8] we used Newton's method for descent in the space of primal variables, the dual variables were found from the solution of linear algebraic equations. Here we use continuous and discrete versions of the primal-dual Newton's method for solving a reduced problem. The space transformation introduces additional matrices in Newton's method, which play the role of a barrier and prevent the trajectories from crossing the boundary of the feasible set. Therefore, we call such method as a barrier-Newton method. The term "barrier" has often been misunderstood. We do not utilize the barriers which are described, for example, in the book [9]. Numerical methods which are based on penalty functions are inherently unstable, since one has to increase the penalty parameter without bound in order to obtain convergence. In contrast, the multiplicative barriers which we use do not tend to infinity when a current point approaches the boundary of a feasible set. In our algorithms the barrier functions are continuous and equal to zero on a boundary. These barriers imply the feasibility of the trajectories. Therefore we need not introduce any penalty coefficients.

In the first continuous version of the method we use a trajectory-following algorithm. Newton's method is applied to the nonlinear system of equations that is derived from the Kuhn-Tucker stationary condition. The method is stated as initial-value problem involving system of ordinary differential equations. We present a convergence rate analysis for continuous and discrete versions and show how superlinear and quadratic convergence can be attained. In particular case our algorithm bears resemblance to the algorithm of [15], which was developed for LP. However there are significant differences:

1. our algorithms can start the computation from the infeasible points, while the algorithms

preserve feasibility,

2. our algorithms enable us to take different stepsizes in the primal space and in the dual space, which is proved to be important for computation,

3. we do not use any penalty or usual barrier functions in the algorithms.

This paper generalizes our results published in Russian [6], [7]. Proofs will be given for only some results.

2. Local Properties of Barrier-Newton Method.

In this section we consider the following NLP problem:

$$\text{minimize } f(x) \text{ subject to } x \in X = \{x \in R^n : g(x) = 0_m, x \in P\}, \tag{1}$$

where the functions f and g are twice continuously differentiable, $f(x)$ maps R^n onto R^1 and $g(x)$ maps R^n onto R^m , 0_s is the s-dimensional null vector, 0_{sk} is the $s \times k$ rectangular null matrix. The convex set P is assumed to have a nonempty interior.

We consider computational methods for solving Problem (1). Our approach is based on the possibility of solving the system of nonlinear equations which constitute necessary optimality conditions. We introduce a new n-dimensional space with the coordinates $[y^1, \ldots, y^n]$ and make a differentiable transformation from this space to the original one : $x = \xi(y)$. This surjective transformation maps R^n onto P or int$P$, i.e. $P = \text{cl}\xi(R^n)$, where cl$B$ is the closure of B. Consider the transformed minimization problem

$$\text{minimize } \tilde{f}(y) = f(\xi(y)) \text{ subject to } y \in Y, \tag{2}$$

where $Y = \{y \in R^n : \tilde{g}(y) = g(\xi(y)) = 0_m\}$.

Define the Lagrangian functions $L(x, u), \tilde{L}(x, u)$ associated with Problems (1) and (2)

$$L(x, u) = f(x) + u^T g(x), \quad \tilde{L}(y, u) = \tilde{f}(y) + u^T \tilde{g}(y).$$

Let $\tilde{J}(y) = dx/dy$ be the Jacobian matrix of the transformation $\xi(y)$ with respect to y. Then the first-order necessary conditions for local minimum for Problem (2) are

$$\tilde{L}_y(y, u) = \tilde{f}_y(y) + \tilde{g}_y^T(y)u = 0_n, \quad \tilde{g}(y) = 0_m, \tag{3}$$

where $\tilde{f}_y = \tilde{J}^T f_x, \tilde{g}_y = g_x \tilde{J}$.

If $\tilde{J}$ is nonsingular then there exists an inverse transformation $y = \delta(x)$ so it is possible to return from the y-space to the x-space and we obtain in this way a matrix $J(x) = \tilde{J}(\delta(x))$ which is now a function of x. Using this substitution we rewrite expressions (3) in terms of the variable x. They take the form

$$J^T(x)L_x(x, u) = 0_n, \quad g(x) = 0_m, \quad x \in P. \tag{4}$$

As the nullspace of the matrix $J^T(x)$ coincides with the nullspace of the Gram matrix $G(x) = J(x)J^T(x)$, conditions (4) can be written in the form

$$G(x)L_x(x, u) = 0_n, \quad g(x) = 0_m, \quad x \in P. \tag{5}$$

For solving Problem (1) it is possible to solve nonlinear system (4). Some properties of the nonlinear systems, which are obtained after space transformations, were investigated in [6].

For simplicity we consider here only the case where P is a n-dimensional positive orthant, i.e. $P = R^n_+$. In this case we can use the componentwise space transformation $x^i = \xi^i(y^i), 1 \leq i \leq n$. For such transformation the inverse transformation $y = \delta(x)$ is also componentwise type $y^i = \delta^i(x^i), 1 \leq i \leq n$, and the matrix $G(x)$ is diagonal: $G(x) = D(\theta(x))$, where

$$\theta(x) = [\theta^1(x^1), \dots, \theta^n(x^n)], \quad \theta^i(x^i) = [\dot{\xi}^i(\delta^i(x^i))]^2, \quad 1 \leq i \leq n,$$

and $D(z)$ denotes the diagonal matrix containing the components of a vector z. Introduce an additional mapping

$$\phi(z) = [\phi^1(z^1), \dots, \phi^n(z^n)] .$$

From now on we assume that the mappings $\theta(z), \phi(z)$ are continuously differentiable on R^n and for all $1 \leq i \leq n$ satisfy the following conditions:

$$\theta^i(0) = 0, \dot{\theta}^i(0) \neq 0; \text{if } x^i \neq 0, \text{then } \theta^i(x^i) \neq 0, \dot{\theta}^i(x^i) \neq 0;$$

$$(6)$$

$$\phi^i(0) = 0, \dot{\phi}^i(0) \neq 0; \text{if } x^i \neq 0, \text{then } \phi^i(x^i) \neq 0, \dot{\phi}^i(x^i) \neq 0.$$

The necessary conditions (5) can be rewritten in the form

$$D(\theta(x))\phi(L_x(x,u)) = 0_n, \quad g(x) = 0_m, \quad x \in R^n_+. \tag{7}$$

For solving system (7) we use the continuous version of Newton's method. The computation process is described by the system of ordinary differential equations

$$W(x,u)\left(\begin{array}{c} \dot{x} \\ \dot{u} \end{array}\right) = -\left(\begin{array}{c} \alpha D(\theta(x))\phi(L_x(x,u)) \\ \tau g(x) \end{array}\right), \tag{8}$$

where $\alpha > 0, \tau > 0, W$ is a square $(n+m)^2$ matrix,

$$W(x,u) = \left(\begin{array}{cc} M & D(\theta(x))D(\dot{\phi})g_x^T \\ g_x & 0_{nm} \end{array}\right), \quad M = D(\dot{\theta})D(\phi) + D(\theta)D(\dot{\phi})L_{xx}. \tag{9}$$

By following the trajectories, satisfying (8), we can theoretically obtain a solution of the system of nonlinear equations (5). In practice, we build the iterative procedures using a discretization of dynamical systems.

Let e^i denote the n-th order unit vector whose i-th component is equal to one.

Definitions. *The constraint qualification (CQ) for Problem (1) holds at a point x, if all vectors $g^i(x), 1 \leq i \leq m$, and all e^j, such that $x^j = 0$, are linearly independent. We say that x is a regular point for Problem (1), if the CQ holds at x. The strict complementary condition (SCC) holds at a point $[x_*, u_*]$, if $L_{x^i}(x_*, u_*) > 0$ for all i such that $x^i = 0$.*

Define the nullspace of the matrix $g_x(x)D(\theta(x))$ at x:

$$K(x) = \left\{ \bar{x} \in R^n : g_x(x)D(\theta(x))\bar{x} = 0_m \right\}.$$

We will use the following sufficient conditions for a point x to be an isolated local minimum of Problem (1).

Theorem 1 *Sufficient conditions for a point $x_* \in P = R_+^n$ to be an isolated local minimum of Problem (1) are that there exists a Lagrange multiplier vector u_* such that $[x_*, u_*]$ satisfies the Kuhn-Tucker conditions (5), that the SCC holds and that*

$$z^T D(\theta(x_*)) L_{xx}(x_*, u_*) D(\theta(x_*)) z > 0 \tag{10}$$

for all $z \in K(x_*)$, satisfying $D(\theta(x_*))z \neq 0_n$.

Lemma 1 *Let $[x_*, u_*]$ be a Kuhn-Tucker pair, where all the conditions of Theorem 1 are satisfied. Assume that x_* is a regular point for Problem (1). Then $W(x_*, u_*)$ is nonsingular.*

Proof. Without loss of generality, we assume that the first s components of x_* are positive and consequently the remaining $d = n - s$ components are zero. Hence the vector x_* is split in two vectors $x_*^T = [x_B^T, x_N^T], x_B \in R^s, x_N \in R^d, x_B > 0_s, x_N = 0_d$. In a similar way we define the following partitions:

$$L_x = \left(\begin{array}{c} L_x^B \\ L_x^N \end{array} \right), \quad \theta = \left(\begin{array}{c} \theta_B \\ \theta_N \end{array} \right), \quad L_{xx} = \left(\begin{array}{cc} L_{xx}^B & L_{xx}^{BN} \\ L_{xx}^{NB} & L_{xx}^N \end{array} \right), \quad g_x^T = \left(\begin{array}{c} g_{x_B}^T \\ g_{x_N}^T \end{array} \right). \tag{11}$$

To prove this lemma it is enough to show that the homogeneous system

$$W(x_*, u_*)\bar{z} = 0_{n+m}, \quad \bar{z}^T = \left[\bar{x}^T, \bar{u}^T \right] = \left[\bar{x}_B^T, \bar{x}_N^T, \bar{u}^T \right]$$

has only the trivial solution $\bar{z} = 0$. Taking into account conditions (6) and (7), we can rewrite this system in detail

$$D(\theta_B)D(\dot{\phi}_B)\left(L_{xx}^B \bar{x}_B + L_{xx}^{BN} \bar{x}_N + g_{x_B}^T \bar{u} \right) = 0_s, \tag{12}$$

$$D(\dot{\theta}_N)D(\phi_N)\bar{x}_N = 0_d, \tag{13}$$

$$g_{x_B}\bar{x}_B + g_{x_N}\bar{x}_N = 0_m. \tag{14}$$

Here all functions are evaluated at the points $x = x_*, u = u_*$, therefore

$$\theta_N = 0_d, \quad \phi_N \neq 0_s, \quad \theta_B \neq 0_s, \quad \dot{\theta}_N \neq 0_d, \quad \dot{\phi}_B \neq 0_s.$$

Hence (13) yields $\bar{x}_N = 0_d$. Multiplying (12) on the left by $\bar{x}_B^T D^{-1}(\theta_B)D^{-1}(\dot{\phi}_B)$, multiplying (14) on the left by $-u^T$ and adding the results, we obtain $\bar{x}_B^T L_{xx}^B \bar{x}_B = 0$, where $\bar{x}_B$ satisfies the condition $g_{x_B}\bar{x}_B = 0_m$. In view of (10) this takes place only when $\bar{x}_B = 0_s$. So equality (12) is reduced to $g_{x_B}^T \bar{u} = 0_s$.

Let $\bar{u} \neq 0_m$. Then denoting

$$\beta^i = \sum_{j=1}^m \bar{u}^j g_{x^i}^j(x_*), \quad s+1 \leq i \leq n,$$

we obtain

$$\sum_{i=1}^m \bar{u}^j g_{x^i}^j(x_*) - \sum_{i=s+1}^n \beta^i e^i = 0.$$

This equality contradicts CQ at x_* and we conclude that $\bar{u} = 0$. □

We combine vectors x, u and denote

$$z = \begin{pmatrix} x \\ u \end{pmatrix}, \; z_0 = \begin{pmatrix} x_0 \\ u_0 \end{pmatrix}, \; z_* = \begin{pmatrix} x_* \\ u_* \end{pmatrix}, \; R(z) = \begin{pmatrix} D(\theta(x))\phi(L_x) \\ g(x) \end{pmatrix}.$$

Let $x(t, z_0), u(t, z_0)$ denote the solutions of the Cauchy problem (8) with initial conditions $x_0 = x(0, z_0), u_0 = u(0, z_0)$. Using these notations, we rewrite the system of equation (8) as

$$W(z)\frac{dz}{dt} = -D(\gamma)R(z), \quad z(0, z_0) = z_0, \tag{15}$$

where γ has the first n components equal α and all other components equal to τ. For the following, we assume that the initial-value problem under consideration is always uniquely solvable. We denote $\gamma_* = \min[\alpha, \tau]$.

Theorem 2 *Suppose that the conditions of Lemma 1 hold. Then for any $\alpha > 0, \tau > 0$ the pair $z_*^T = [x_*, u_*]$ is asymptotically stable equilibrium point of system (15). If the stepsize h_k is fixed and $0 < h < 2/\gamma_*$, then the discrete version*

$$z_{k+1} = z_k - h_k W^{-1}(z)D(\gamma)R(z) \tag{16}$$

locally converges to the point z_* with at least linear rate. If $W(z)$ satisfies a Lipschitz condition in a neighborhood of z_* and $\alpha = \tau = 1$, then the sequence $\{z_k\}$ converges quadratically to z_*.

3. Nonlocal Convergence Properties. According to theorem 1 the convergence of methods (15), (16) is guaranteed only for a very restricted set of starting points. Now we analyze the global behavior of the method.

Lemma 2 *Let x be a regular point, and let the pair $[x, u]$ be such that $x^i \neq 0, L_x^i(x, u) \neq 0$ for all $1 \leq i \leq n$, and the matrix $M(x, u)$ is nonsingular. Then the matrix $W(x, u)$ is nonsingular.*

Define the nonnegative Lyapunov function

$$F(x, u) = \|D(\theta(x))\phi(L_x(x, u))\| + \|g(x)\|$$

and introduce two sets:

$$\Omega_0 = \left\{ [x, u] : F(x, u) \leq F(x_0, u_0), \quad x \geq 0, \; L_x(x, u) \geq 0 \right\},$$

$$\tilde{\Omega}_0 = \left\{ [x, u] \in \Omega_0 : \quad x > 0_n, \quad L_x(x, u) > 0_n \right\}.$$

Theorem 3 *Suppose that the set Ω_0 is bounded and contains the unique Khun-Tucker pair $[x_*, u_*]$. Suppose also that for any pair $[x, u] \in \tilde{\Omega}_0$ the conditions of lemma 2 are satisfied. Then all trajectories of (15), starting from a pair $[x_0, u_0] \in \tilde{\Omega}_0$, converge to the $[x_*, u_*]$.*

Proof. The system of ordinary differential equations (15) has the first integrals

$$D(\theta(x(t, z_0)))\phi(L_x(x(t, z_0), u(t, z_0))) = D(\theta(x_0))\phi(L_x(x_0, u_0))e^{-\alpha t}, \qquad (17)$$

$$g(x(t, z_0)) = g(x_0)e^{-\tau t}.$$

The solutions of (15) belong to Ω_0 and are therefore bounded. The right-hand sides of (17) are strictly positive and tends to zero only as $t \to \infty$. By moving along the trajectories of (15) we do not violate nonnegativity of x and L_x. Therefore the trajectories do not cross the boundary of the set Ω_0. The transformation functions $\theta(x)$ and $\phi(v)$ thus play the role of the multiplicative barriers preserving nonnegativity. All trajectories that begin inside $\tilde{\Omega}_0$ remain in the interior of $\tilde{\Omega}_0$. According to La Salle's Invariance Principle the solutions $x(t, z_0), u(t, z_0)$ can be prolonged as $t \to \infty$, the positive limit set of the solution is a compact connected set contained in Ω_0 and coincides with the equilibrium pair $[x_*, u_*]$, which is unique on Ω_0. $\square$

4. Linear Programming Problems. Primal and dual linear programming problems are stated in the standard form:

$$\text{minimize } c^T x \text{ subject to } x \in X = \{x \in R^n : g(x) = b - Ax = 0_m, x \geq 0_n\}, \qquad (18)$$

$$\text{maximize } b^T u \text{ subject to } u \in U = \{u \in R^m : v = L_x(x, u) = c - A^T u \geq 0_n\}, \qquad (19)$$

where c and x are n-vectors, b is a m-vector, A is a full rank $m \times n$ matrix, and $v \in R^n$ is the vector of dual slack variables. We assume that the feasible sets X and U are nonempty and that primal and dual nondegeneracy holds. In this case both problems have unique solutions x_* and u_* respectively.

Method (8) can be applied for solving problem (18). Formulas (9) and (15) are simplified in this case:

$$W = \begin{pmatrix} M & -D(\theta)D(\dot{\phi})A^T \\ -A & 0_{mm} \end{pmatrix}, \quad M = D(\dot{\theta}(x))D(\phi(v)), \quad \Gamma = AM^{-1}D(\theta)D(\dot{\phi})A^T.$$

If matrices M and Γ are nonsingular then we can use Frobenius's formula for the inverse matrix:

$$W^{-1} = \begin{pmatrix} M^{-1}[I_n - D(\theta)D(\dot{\phi})A^T\Gamma^{-1}AM^{-1}] & -M^{-1}D(\theta)D(\dot{\phi})A^T\Gamma^{-1} \\ -\Gamma^{-1}AM^{-1} & -\Gamma^{-1} \end{pmatrix}. \qquad (20)$$

Theorem 4 *Let $[x_*, u_*]$ be a nondegenerate optimal pair for the linear programs (18) and (19). Then:*

1. *$W(x_*, u_*)$ is nonsingular;*

2. *$[x_*, u_*]$ is an asymptotically stable equilibrium pair of system (15) and the trajectories $[x(t, z_0), u(t, z_0)]$ of (15) converge to $[x_*, u_*]$ on $\tilde{\Omega}_0$;*

3. *the discrete version (16) locally converges with at least a linear rate to the pair $[x_*, u_*]$;*

4. *if $W(z)$ satisfies the Lipschitz condition in a neighborhood of $z_*, h_k = \alpha = \tau = 1$, then the sequence $\{x_k, u_k\}$ converges quadratically to $[x_*, u_*]$.*

The most interesting property of the proposed method is the following: there exists a set of initial points such that the method solves Problems (18) and (19) in a finite numbers of iterates. It would be extremely interesting to find out the family of the sets of starting points which insures the solution at the prescribed number of iterates. We will show three such sets. For the sake of simplicity we consider here the case where the stepsizes are fixed and transformation functions are linear. We assume also that $\alpha = \tau = 1$, $\theta(x) = x$, $\phi(v) = v$. In this case algorithm (16) is rewritten as follows:

$$D(v_k)x_{k+1} - D(x_k)A^T u_{k+1} = -D(x_k)A^T u_k, \quad Ax_{k+1} = b. \tag{21}$$

Let T_k be a set of pairs $[x, u]$ such that, if $[x_0, u_0] \in T_k$, then algorithm (16) solves Problems (18) and (19) at k iterates.

Introduce the sets of indexes

$$\sigma(x) = \left\{ 1 \le i \le n : x^i = 0 \right\}, \quad \sigma(v) = \left\{ 1 \le i \le n : v^i = 0 \right\}.$$

If all components of x and v are not non-zero, then we write $\sigma(x) = \sigma(v) = \emptyset$. We define the following three sets of pairs

$$\Omega_1 = \left\{ [x, u] : x = x_*, \quad \sigma(x) \cap \sigma(v) = \emptyset \right\}, \Omega_2 = \left\{ [x, u] : u = u_*, \quad \sigma(x) \cap \sigma(v) = \emptyset \right\},$$

$$\Omega_3 = \left\{ [x, u] : \sigma(x) = \sigma(x_*), \quad \sigma(x) \cap \sigma(v) = \emptyset \right\},$$

Theorem 5 *Assume that all conditions of Theorem 4 hold. Let the transformation functions be linear. If $[x_0, u_0] \in \Omega_1$ or $[x_0, u_0] \in \Omega_2$, then method (16) yields an optimal solution in a single step. If $[x_0, u_0] \in \Omega_3$, then method (16) yields an optimal solution at most in two steps.*

5. Interior Point Techniques. In the previous section we did not require feasibility of the starting and current points. Here we consider a particular case of the method in which the trajectories $x(t, z_0)$, $v(t, z_0)$ belong to the positive orthant R_+^n. We define the following sets

$$R_{++}^n = \{ x \in R^n : x > 0_n \}, \quad \text{int} U = \{ u \in R^m : v = c - A^T u > 0_n \},$$

$$\text{ri} X = \{ x \in R_{++}^n : Ax = b \},$$

$$V = \{ v \in R^n : \text{exists } u \in R^m \text{such that } v = c - A^T u \}. \tag{22}$$

We introduce new vectors $q \in R^n$, $p \in R^n$, whose i-th components are $\theta^i/\dot{\theta}^i$, $\phi^i/\dot{\phi}^i$ respectively and denote

$$H = D^{1/2}(q)D^{-1/2}(p), \quad (AH)^+ = HA^T(AH^2A^T)^{-1},$$

$$(AH)^! = (AH)^+AH, \quad \pi(AH) = I_n - (AH)^!.$$

Here the operator $\pi(W)$ projects any n-dimensional vector onto the nullspace of $m \times n$ rectangular matrix W, the matrix $(AH)^!$ projects it onto orthogonal supplement, W^+ is a pseudoinverse matrix.

We say that a point $z \in R^{n+m}, z^T = [x^T, u^T]$ is interior, if $x \in R^n_{++}$ and $u \in \text{int}U$. If moreover $x \in \text{ri}X$ then the point z is strictly interior.

If z is an interior point and conditions (6) are satisfied then the matrices H, AH^2A^T are nonsingular and we can use formula (20) for the inverse matrix. In this case method (15) can be written as follows

$$\frac{dx}{dt} = H\left(\tau(AH)^+(b - Ax) - \alpha\pi(AH)Hp\right), \quad \frac{du}{dt} = (AH^2A^T)^{-1}(\alpha Aq + \tau(b - Ax)). \quad (23)$$

Introduce a new matrix $\Lambda \in R^{d \times n}$. The columns of Λ^T forms a basis for the null space of A, i.e. $A\Lambda^T = 0_{md}$. Now definition (22) can be written equivalently as $V = \{v \in R^n : \Lambda(v - c) = 0_d\}$ and from (23) we obtain

$$\frac{dx}{dt} = H\left[\tau(AH)^+(b - Ax) - \alpha\pi(AH)Hp\right], \quad (24)$$

$$\frac{dv}{dt} = -A^T\frac{du}{dt} = H^{-1}\left[\tau(AH)^+(Ax - b) - \alpha(AH)^{\parallel}Hp\right].$$

Let $x(t, z_0), u(t, z_0), v(t, z_0)$ denote the solutions of systems (23), (24) with initial conditions $x(0, z_0) = x_0 > 0_n, u(0, z_0) = u_0, v(0, z_0) = c - A^Tu_0 = v_0 > 0_n$. In system (24) we used derivatives $\dot{x}$ and $\dot{v}$ instead of $\dot{x}$ and $\dot{u}$ as we did before. We can do this if we are sure that all trajectory $v(t, z_0) \in V$, i.e. the vector $v(t, z_0) - c$ belongs to the set of columns of the matrix A^T. Differentiating d-dimensional vector Λv along the solutions of (24), we obtain $\Lambda\dot{v} = 0_d$. Therefore if $v_0 \in V$, then the trajectory $v(t, z_0) \in V$ for all $t \geq 0$ and we can use system (24) for numerical calculations.

For the sake of simplicity we consider the case where $\theta(x)$ is a homogeneous function of x, i.e. $\theta(\beta x) = \beta^\lambda\theta(x), \lambda = \alpha/\tau$. Then according to Euler's formula we have $x = \lambda q(x), \lambda H^2 = D(x)D^{-1}(p)$. Introduce new vectors $\mu \in R^m, \eta \in R^n$:

$$\mu = \left(AD(x)D^{-1}(p)A^T\right)^{-1}b, \quad \eta = D^{-1}(p)A^T\mu, \quad \psi = D(x)v, \Phi = \sum_{i=1}^n \psi^i. \quad (25)$$

In the case where $v(v) = v$ we have

$$\mu = \left(AD(x)D^{-1}(v)A^T\right)^{-1}b, \quad \eta = D^{-1}(v)A^T\mu. \quad (26)$$

Let e be the vector of ones in R^n. Now systems (23), (24) can be written as

$$\frac{dx}{dt} = \tau D(x)[\eta - e], \quad \frac{du}{dt} = \alpha\mu, \quad \frac{dv}{dt} = -\alpha A^T\mu. \quad (27)$$

The condition $\dot{\theta}(0) > 0$ from (6) is not necessarily fulfilled for homogeneous functions $\theta(x)$, nevertheless the following theorem is valid.

Theorem 6 *Let $[x_*, u_*]$ be a nondegenerate optimal pair for linear programs (18) and (19). Suppose that $D(\theta(x)) = D^\lambda(x), \phi(v) = v, \alpha > 0, \tau > 0$. Assume that Ω_0 is bounded. Let the starting point z_0 be interior, then the trajectories of (27) are such that:*

1. the matrix $AD(x(t, z_0))D^{-1}(v(t, z_0))A^T$ is nondegenerate for all $t \geq 0$,

2. $z(t, z_0) \in \Omega_0$ and $v(t, z_0) \in V$ for all $t > 0$,

3. the objective function $b^T u(t, z_0)$ of the dual problem increases monotonically,

4. the pair $[x(t, z_0), u(t, z_0)]$ is bounded and converges to $[x_*, u_*]$ as $t \to \infty$,

5. all components of vectors $D^\lambda(x(t, z_o))v(t, z_0)$, $Ax(t, z_0)$ change monotonically and

$$D^\lambda(x(t, z_0))v(t, z_0) = e^{-\alpha t} D^\lambda(x_0)v_0,$$

$$Ax(t, z_0) - b = e^{-\tau t}(Ax_0 - b).$$

By applying the Euler numerical integration method to system (27) we obtain the simplest discrete version of the method:

$$x_{k+1} = D(x_k)(e + \tau_k(\eta_k - e)), \quad v_{k+1} = D(v_k)(e - \alpha_k \eta_k), \quad u_{k+1} = u_k + \alpha_k \mu_k, \qquad (28)$$

where $x_0 > 0_n, v_0 > 0_n$, and μ_k and η_k are defined by (25).

As before, if $v_0 \in V$, then $v_k \in V$ for all k and the last formula in (28) can be omitted. The objective function $b^T u_k$ also increases monotonically. From (28) we obtain

$$Ax_{k+1} - b = (1 - \tau_k)(Ax_k - b), \qquad (29)$$

$$\Phi_{k+1} = (1 - \tau_k)\Phi_k + (\tau_k - \alpha_k)\mu_k^T A x_k + \alpha_k \tau_k \mu_k^T (Ax_k - b). \qquad (30)$$

The iterates produced by algorithm (28) are well-defined if vectors x_k, v_k are strictly positive for all k. In order to ensure the positiveness of x_{k+1} and v_{k+1} we have to choose the step lengths α_k and τ_k such that

$$e \geq \alpha_k \eta_k, \quad e \geq \tau_k(e - \eta_k).$$

It is now straightforward to verify that non-negativity conditions hold if α_k and τ_k satisfy

$$0 < \alpha_k \leq \alpha_k^* = 1/[\eta_k^*]_+, \quad 0 < \tau_k \leq \tau_k^* = 1/[1 - \eta_*^k]_+,$$

where $[\alpha]_+ = \max[0, \alpha], \eta_k^*$ and η_*^k are maximal and minimal components of the vector η_k respectively. We will adopt the convention that, if $\eta_*^k \geq 1$, then $\tau_k^* = +\infty$, and, if $\eta_k^* \leq 0$, then $\alpha_k^* = +\infty$.

If we set $\alpha = \tau = 1$ and substitute $b = Ax$ in (28), then formulas (28) coincide with the primal-dual interior point algorithms proposed in [15], if in the latter we ignore the barrier (perturbation) term. In this algorithm the starting point z_0 must be strictly interior. Algorithm (28) does not require feasibility of starting and current points, but according to (29) it preserves feasibility. Another important advantage of algorithm (28) is that it permits us to take different step lengths in the primal space and in the dual space. This property is very useful for computation, especially at the beginning of computation when the starting point is far from the solution and only one maximal stepsize, either α_k^* or τ_k^*, is very large. This advantage disappears as $k \to \infty$ because maximal step lengths tend to one.

We used three classes of procedures for determining the step lengths:

1. step lengths are fixed and small enough, hence the discrete process (28) is close to a continuous one (27);

2. stepsizes are close to one and therefore the discrete process has properties of Newton's method;

3. stepsizes are chosen from steepest descent conditions or from another auxiliary optimization problem.

References

[1] D.Bertsecas, *Constrained Optimization and Lagrange Multiplier Methods*, Academic Press, New York, 1982.

[2] Yu.Evtushenko, *Two numerical methods of solving nonlinear programming problems*, Sov. Math. Dokl., Vol. 15, No. 2 (1974), pp. 420-423.

[3] Yu.Evtushenko, *Numerical Optimization Techniques. Optimization Software*, Inc. Publications Division, New York., 1985.

[4] Yu.Evtushenko, V.Zhadan, *Numerical methods for solving some operations research problems*, U.S.S.R. Comput. Maths. Math. Phys., Vol. 13, No. 3 (1973), pp. 56-77.

[5] Yu.Evtushenko, V.Zhadan, *A relaxation method for solving problems of non-linear programming*, U.S.S.R. Comput. Maths. Math. Phys., Vol. 17, No. 4, (1978), pp. 73-87.

[6] Yu.Evtushenko, V.Zhadan, *Barrier-projective and barrier-Newton numerical methods in optimization (the nonlinear programming case)*, Computing Center of the USSR Academy of Sciences, Reports an Comput. Math., 1991 (in Russian).

[7] Yu.Evtushenko, V.Zhadan, *Barrier-projective and barrier-Newton numerical methods in optimization (the linear programming case)*, Computing Center of the USSR Academy of Sciences, Reports an Comput. Math., 1992 (in Russian).

[8] Yu.Evtushenko, V.Zhadan, *Stable barrier-projection and barrier-Newton methods in nonlinear programming*, To appear in Optimization Methods and Software.

[9] A. Fiacco, G. McCormic, *Nonlinear programming: Sequential unconstrained minimization techniques*, John Wiley and Sons, Inc. N.Y.,1968.

[10] C. Gonzaga, *Path following methods for linear programming*, SIAM Review, Vol.34, No.2, 1992, pp.167-224.

[11] M.Kojima, S.Mizuno, A.Yoshise, *A primal-dual interior point method for linear programming*, in Progress in Mathematical Programming - Interior Point and Related Methods, N.Meggido, ed., Springler-Verlag, Berlin, 1989, Chap. 2.

[12] K.McShane, C.Monma, D.Shanno, An implementation of primal-dual interior point method for linear programming, ORSA J. Comput., 1 (1989), pp.70-89.

[13] S.Mehrotra, On the implementation of a (primal-dual) interior point method, Technical Report 90-03, Department of Industrial Engineering and Management Sciences, Nothwestern University, Evanston, Illinois, 1990.

[14] K.Tanabe, *Differential geometrical methods in nonlinear programming*, Applied Nonlinear Analysis, V.Lakshmikantham ed., Academic Press, New York, 1979.

[15] Y.Zhang, R.Tapia, J.Dennis,*On the Superlinear and quadratic convergence of primal-dual interior point linear programming algorithms*, Technical Report TR90-6, Rice university, Houston, Texas, 1990.

STABLE MULTIPOINT SECANT METHODS WITH RELEASED REQUIREMENTS TO POINTS POSITION

Oleg Burdakov* and Ursula Felgenhauer**

* *Computing Center of the Russian Academy of Sciences*
40, Vavilov street, Moscow 117967,GSP-1, Russia

** *Institute of Numerical Mathematics, Technical University of Dresden,*
Mommsenstrasse 13, Dresden O-8027,Germany

1. Introduction.

The history of solution methods for solving simultaneous nonlinear equations knows several approaches of generalizing the one-dimensional secant method, for example BITTNER 1959 [1], BARNES 1965 , WOLFE 1959 , see also [10], [12].

Consider the problem of solving a system of simultaneous nonlinear equations,

$$f(x) = 0, \qquad (1)$$

where the mapping $f : D \subset \mathrm{R}^n \to \mathrm{R}^n$ is supposed to be sufficiently smooth. As a particular iterative method the sequential $(n+1)$ - point secant method was widely used and investigated. It can be described in the form

$$x_{k+1} = x_k - J_k^{-1} f_k, \qquad (2)$$

where we denote $f_k = f(x_k), f_* = f(x_*)$, and $J_k \in \mathrm{R}^{n \times n}$ satisfy the following secant equations:

$$J_{k+1} s_i = y_i \qquad (3)$$

for all i, $k - n + 1 \le i \le k$ with $s_i = x_{i+1} - x_i$ and $y_i = f_{i+1} - f_i$. It is known, that this method is in general unstable because the search directions s_i may become linearly dependent. A sufficient stability condition is the so-called general position assumption [1],([12]) :

$$\det \left[(\Delta x_i / \|\Delta x_i\|) : k - n + 1 \le i \le k \right] \ge \sigma > 0 \qquad (4)$$

It seems to be obvious that it was the analysis of this relation which suggested a number of stable modifications of the secant methods. They follow the idea to restrict (3) to such a subset of preceeding iterations that the corresponding s_i are certainly (and uniformly) linearly independent. The resulting algorithms then converge locally superlinearly, and they are numerically stable, see [4], [5], [7].

For the construction of the matrices J_k satisfying secant conditions of the type (3) BROYDEN - type update formulas can be used, cf. GAY, SCHNABEL [8], also [3]-[5]. It has to be pointed out, that for nonlinear systems arising from minimization problems there were proposed symmetric variants of matrix updates, [3], [11].

In this paper we will present a modification of the secant method where the information about the nonlinear function f carried by the iteration points x_i and the related function values f_i will be utilized more completely than in earlier versions. For this aim, we leave the secant equation model and instead consider the underlying linear interpolation of f,

$$L_k : D \subset \mathbb{R}^n \to \mathbb{R}^n ; \quad L_k(x_i) = f_i \tag{5}$$

for (some) i, $(k - n) \leq i \leq k$. If the corresponding iteration nodes collaps into a subspace of dimension lower n, we choose an appropriate subset of the last iteration points for the interpolation. It turns out that the analysis of the iterates position is a more general approach than the search direction's analysis; it enables us to release the stability requirements for the points ((4) or modifications, for comparison see [5],[12]).

2. Multiple - point secant methods

2.1 Linear interpolating mappings under released stability conditions

Let $\mathcal{L}(\{x_i\}_0^k)$ denote the set of all linear mappings $L : \mathbb{R}^n \to \mathbb{R}^n$ that interpolate the mapping f at the points $\{x_i\}_0^k$, i.e.

$$L(x_i) = f_i, \ i = 0, \ldots, k. \tag{6}$$

Denote $\mathcal{J}(\{x_i\}_0^k) = \{J \in \mathbb{R}^{n \times n} : J = L', L \in \mathcal{L}(\{x_i\}_0^k)\}$.
It is evident that,

$$\mathcal{L}(X_1) \supset \mathcal{L}(X_2), \quad \forall X_1 \subset X_2,$$

where X_1 and X_2 are nonempty sets of points in $\mathbb{R}^n$. Note also, that if $L \in \mathcal{L}(x_i)$ with $J = L'$, then

$$L(x) = \bar{f} + J(x - \bar{x}), \tag{7}$$

for any $\bar{x}$ and $\bar{f}$ such that

$$\bar{x} = \sum_{i=0}^{k} \lambda_i x_i, \quad \bar{f} = \sum_{i=0}^{k} \lambda_i f_i, \quad \sum_{i=0}^{k} \lambda_i = 1, \quad \lambda_i \in \mathrm{R}^1, \quad i = 0, \ldots, k. \quad (8)$$

The existence of J (and L resp.) mainly depends on the localization of the interpolation nodes. Assume that $\pi = \pi(\{x_i\}_0^k)$ satisfies $\dim \pi = k$ In this case consider a basic system of vectors in the form

$$\Delta x_i = x_{p(i)} - x_{s(i)}; \quad 0 \le s(i), p(i) \le k, \quad 0 \le i \le k-1; \quad (9)$$

such that $\pi = x_{p(k)} + \mathrm{span}\{\Delta x_0, \ldots, \Delta x_{k-1}\}$.
Geometrically the Δx_i correspond to the arcs of a spanning tree in the complete graph with nodes $\{x_i\}_0^k$.

For a given point's enumeration (with the above properties) now define the matrices $(\Delta f_i = f(x_{p(i)}) - f(x_{s(i)}))$:

$$\Delta X = (\Delta x_0, \ldots, \Delta x_{k-1}); \quad \Delta F = (\Delta f_0, \ldots, \Delta f_{k-1})$$

$$\Delta X^\circ = \left(\frac{\Delta x_0}{\|\Delta x_0\|}, \ldots, \frac{\Delta x_{k-1}}{\|\Delta x_{k-1}\|} \right) \quad \text{resp.}$$

In analogy to the "general position" assumption of BITTNER (1959) (see [1]) we formulate the following **released stability condition** on $\{x_i\}_0^k$:

Definition 1: *The points $x_0, \ldots, x_k$ are called to be in σ-stable general position if there exist functions $p, s : \{0, \ldots, k\} \to \{0, \ldots, k\}$ such that for the corresponding secant vectors $\{\Delta x_i\}_0^{k-1}$ from (9) the inequality*

$$\det((\Delta X^\circ)^T \Delta X^\circ) \ge \sigma^2 > 0. \quad (10)$$

is fulfilled.

This definition differs from the classical one in two points: At first, the number of nodes may be less than $n+1$ (where n - the dimension); secondly, the enumeration of the spanning points can be chosen in such a way that the angles between the corresponding secants are maximal.

In the situation that there are given $n+1$ points in (σ-stable) general position, the interpolation function is uniquely determined by $L(x_0) = f_0$ for ex. and

$$L' = J = (\Delta F)(\Delta X)^{-1} \quad . \quad (11)$$

2.2 The case of symmetric Jacobians

For applications in the context of optimization problems with symmetric Jacobian f' there arises the question whether one can construct linear interpolating mappings L with symmetric derivatives $L' = J$. In general, in the case $k = n$, $\{x_i\}_0^n$ – in σ-stable position, it is known that

$$(\Delta F)\,(\Delta X)^{-1} \notin \mathcal{S} = \{A \in \mathbb{R}^{n \times n}: \quad A^T = A\} \tag{12}$$

even for $f' \in \mathcal{S}$; see [11]. Therefore, to enforce the symmetry property, one has to relax the interpolation conditions in an appropriate manner. Instead of the usual "pointwise" interpolation property we will give a subspace formulation which consists in a projected form of (5):

$$\left.\begin{array}{ll} i = 0: & L(x_{p(k)}) = f_{p(k)} \quad ; \\ i = 1, \ldots, k: & \bar{x} \in \pi_i \backslash \pi_{i-1} \Rightarrow \\ & L(\bar{x}) - \bar{f} \perp \Delta x_t \quad \text{for } 0 \le t \le (k-i) \end{array}\right\} \tag{13}$$

More detailed we have

$$e_{k-i}^T \Delta X^T J \Delta X\, e_{k-j} = e_{k-i}^T \Delta X^T \Delta F\, e_{k-j} \tag{14}$$

for $0 \le j \le i \le k - 1$, e_m – the m-th unit vector.

To illustrate the effect of the generalized interpolation conditions assume for the moment that $p(i) \equiv i$ and $\Delta x_i = e_{i+1}\ \forall i$. The relations (14) then reduce to

$$\begin{array}{lll} (J)_{ik} & = & e_i^T J e_k = e_i^T \Delta f_{k-1} \quad \text{for } 1 \le i \le k \le n; \\ (J)_{ik} & = & (J)_{ki} \quad \text{for } k \le i. \end{array}$$

So, the last iteration values f_n, f_{n-1} are used to determine the n-th row and n-th column of J, Δf_{n-2} - for the remaining elements of the $(n-1)$-st row resp. column etc. That means, the choice $p(i) \equiv i$ gives an implementation of the principle that the newer the information the more restrictively it has to be utilized for the construction of the approximative JACOBI matrix.

Now return to $k = n$ with arbitrary ΔX.
The conditions (13) together with $J = J^T$ lead to (see [3], [5])

$$\Delta X^T J \Delta X = \left(\Delta X^T \Delta F\right)^S \tag{15}$$

where

$$(A^S)_{ij} = \begin{cases} a_{ij} & \text{for } i \leq j \\ a_{ji} & \text{for } i > j \end{cases} \tag{16}$$

so that under the stability condition (10) for $\{x_i\}_{i=0}^n$ we get

$$J = \Delta X^{-T} \left(\Delta X^T \Delta F\right)^S \Delta X^{-1} \tag{17}$$

Thus in the symmetric case L not only depends on the points position, but also on its enumeration order.

3. Updating the linear interpolating functions

3.1 BROYDEN-type rank-1 formulas

Suppose there are given a set of points $\Theta_k = \{x_i\}_0^k$ and a linear mapping L_k interpolating f on Θ_k. We consider the situation that one interpolation node x_{k+1} will be added to Θ_k so that

$$x_{k+1} \notin \pi_k = \pi(\Theta_k).$$

Our aim is the construction of a linear mapping L_{k+1} (resp. a matrix J_{k+1} by a simple modification of J_k) interpolating f on $\Theta_{k+1} = \Theta_k \cup \{x_{k+1}\}$. Here and in the following the orthogonal projector on $\pi_k = \pi(\{x_i\}_0^k)$ will be denoted by P_k.

Lemma 1. *Let $L_k \in \mathcal{L}(\{x_i\}_0^k)$ and $x_{k+1} \notin \pi(\{x_i\}_0^k)$. Then the linear mapping*

$$L_{k+1}(x) = f_{k+1} + J_{k+1}(x - x_{k+1}) \tag{18}$$

interpolates f at $\{x_i\}_0^{k+1}$ for any

$$J_{k+1} = J_k + \frac{(f_{k+1} - L_k(x_{k+1}))(u_k + v_k)^T}{\| u_k \|^2}, \tag{19}$$

such that $v_k \in \pi(\{x_i\}_0^{k+1})^\perp$ and $u_k = x_{k+1} - P_k x_{k+1}$.

Formula (19) represents a whole class of matrix updates; and it holds that any matrix $J_{k+1} \in \mathcal{J}(\{x_i\}_0^{k+1})$ with rank $(\Delta J_k) = 1$, can be represented by (19). As a particular vector choice one can use $v_k = 0$, and for

$$s_k = u_k = x_{k+1} - \bar{x}, \, y_k = f_{k+1} - \bar{f} \tag{20}$$

formula (19) can be rewritten in the following form

$$J_{k+1} = J_k + \frac{(y_k - J_k s_k)s_k^T}{s_k^T s_k},$$ (21)

This formula looks like BROYDEN's formula , but the difference is in the choice of s_k and y_k: the vector s_k represents a projected update, and y_k is taken as the corresponding linear combination of function values.

3.2 A symmetrization approach

As we mentioned in ch.2.2, there are applications for which it is useful to dispose of linear approximations to f with symmetric matrices f'. For this purpose we will modify the matrix update formulas, cf. [2] for ex. The interpolation equations (see (6),(13)) then have to be changed by weaker conditions.

We start with the rank-1-formula (21), where $J_k \in S$ (symmetric matrix) is assumed. Projecting J_{k+1} into the matrix subspace

$$S \cap Q[y_k, s_k] \quad \text{where } Q[y, s] = \{A \in R^{n \times n} : As = y\},$$ (22)

we obtain the following symmetric rank-2-version for the update procedure: (for brevity we use $z_k = (y_k - J_k s_k)$)

$$J_{k+1} = J_k + \frac{z_k c_k^T + c_k z_k^T}{s_k^T c_k} - \frac{z_k^T s_k}{(s_k^T c_k)^2} c_k c_k^T .$$ (23)

In analogy to Lemma 1, the following assertion holds:

Lemma 2. *Let L_k be a linear mapping approximating f with respect to $\{x_i\}_0^k, \{f_i\}_0^k$ in the sense (13), and suppose that $x_{k+1} \notin \pi(\{x_i\}_0^k)$, $p_k = f_{k+1} - L_k(x_{k+1})$ – the defect. Then the linear mapping*

$$L_{k+1}(x) = f_{k+1} + J_{k+1}(x - x_{k+1})$$ (24)

approximates f for any

$$J_{k+1} = J_k + \frac{p_k w_k^T + w_k p_k^T}{\|u_k\|^2} - \frac{p_k^T s_k}{\|u_k\|^4} w_k w_k^T$$ (25)

such that $s_k = x_{k+1} - \bar{x}$, $\bar{x} \in \pi(\{x_i\}_0^k)$,

$$w_k = u_k + v_k; v_k \in \pi(\{x_i\}_0^{k+1})^\perp \text{ and } u_k = x_{k+1} - P_k x_{k+1} \quad ,$$

in the following sense:

(i) $\qquad L_{k+1}(x_{k+1}) = f_{k+1} \; ; L_{k+1}(\bar{x}) = L_k(\bar{x})$

(ii) *the subspace-interpolation properties of L_k are conserved, i.e.*

$$(x_i - x_j)^T (L_{k+1}(x_m) - f_m) = (x_i - x_j)^T (L_k(x_m) - f_m)$$

for $i, j, m \leq k$.

3.3 Least - change property

In recent years projective properties of matrix updates such as least change characteristics became an important tool in convergence analysis of quasi-NEWTON methods, cf.for ex. [9], [11]. As we will show, the secant-type matrix updates (19) satisfy the least change qualification for $v_k = 0$.

Lemma 3. *Let $J_k \in \mathcal{J}(\{x_i\}_0^k)$ and $x_{k+1} \notin \pi(\{x_i\}_0^k)$. Let $\| \cdot \|$ and $\| | \cdot \| |$ be arbitrary matrix norms, such that*

$$\| A \cdot B \| \leq \| A \| \cdot \| | B \| | \quad \forall \, A, B \in R^{n \times n} \tag{26}$$

and

$$\| | u\, u^T \| | \leq u^T u \tag{27}$$

for all $u \in R^n$.
Then the solution to

$$\min\{\| J - J_k \| : \; J \in \mathcal{J}(\{x_i\}_0^{k+1})\} \tag{28}$$

is J_{k+1} defined by (19) with $v_k = 0$.
In particular, this matrix solves (28) when $\| \cdot \|$ is the operator norm corresponding to the l_2 (i.e. EUCLIDean) vector norm, and it solves (28) uniquely when $\| \cdot \|$ is the FROBENIUS or "inner product" norm.

Remark, that then (25) satisfies (28) for symmetric matrices with the FROBENIUS-norm by construction.

4. Algorithms

4.1 Choice of the nodes for the current interpolation

Stabilized secant methods were developed in [4],[5] by restricting the seqential multiple secant conditions to a subset of $\{s_i\}$. Analogously, we will describe a choice rule for sets of interpolation nodes Θ_k such that the σ-stable position assumption is fulfilled at every step k, and for the index set T_k related to Θ_k:

$$L_k(x_i) = f_i \qquad \forall i \in T_k .\tag{29}$$

The new iteration point x_{k+1} then will be generated by solving

$$L_k(x_{k+1}) = f_k + J_k(x_{k+1} - x_k) = 0 ;$$

and (after the set T_{k+1} has been determined) the next interpolating function L_{k+1} (resp. its JACOBI-matrix) can be updated in accordance to ch.3.

We introduce a constant $m \leq n$ bounding the "information delay" :

$$\{k\} \subset T_k \subset \{k - m, \ldots, k\} \qquad \forall k \geq 1 .\tag{30}$$

Denote $l_k = |T_k|$ and $T_k = \{t_{k,i} : 1 \leq i \leq l = l_k\}$.

We will suppose that the indices $t \in T_k$ are ordered in dependence of the graph (tree) structure G_k underlied for testing (10).

The graph G_k always can be treated as a directed tree with root (source node) in vertex x_k. Set $t_{k,1} = k$. For a given index set $M \subset T_k$ let $\omega^+(M)$ denote the set of (direct) successors of vertices from $\{x_i : i \in M\}$ in the tree structure. Then for $i = 1, 2, \ldots$ define:

$$t_{k,i+1} = \max\{t \in T_k : x_t \in \omega^+(\{t_{k,1}, \ldots, t_{k,i}\})\} .$$

The subtrees built when $x_t, t = t_{k,i}$ are taken as roots (and all predecessors are deleted) are then denoted by $G_{k,i}$ (with sets of nodes $- \Theta_{k,i}$).

For constructing T_{k+1} we will test the indices from $T_k \backslash \{(k-m)\}$ whether the corresponding nodes can be included in (29) without loss of stability. The test is based on **Def. 1** and makes use of an orthogonalization technique.

Algorithm 1.   Given $x_0 \in R^n$, nonsingular $J_0 \in R^{n \times n}$, a positive constant $\gamma \in (0, 1)$ and $m \geq 1$.
   Set $T_0 = \{0\}$ and $l_0 = 1$. Compute $f_0 = f(x_0)$.

For $k = 1, 2, \ldots$ **do** :

Step 1. If $f_k = 0$ or J_k – singular, then STOP.

Step 2. Compute x_{k+1} that solves $L_k(x) = 0$; $f_{k+1} = f(x_{k+1})$.

Step 3. Set $T_{k+1} = \varnothing$, $\tilde{T} = T_k \backslash \Theta_{(k-m)}$.

Step 4. For $i = 1, 2, \ldots, l_k$: if $t_{k,i} \in \tilde{T}$ do:

Step 4.1 Find $t = j$ solving the problem
$$\|x_t - x_{k+1}\| \longrightarrow \min \quad \text{s.t.} \quad t \in T_{k+1} \cup \{t_{k,i}\} \, .$$

Step 4.2 Compute the orthogonal projection $P(x_{k+1})$ of x_{k+1} onto the linear manifold spanned by $\{x_t : t \in T_{k+1} \cup \{t_{k,i}\}\}$.

Step 4.3 If $\|P(x_{k+1}) - x_{k+1}\| / \|x_j - x_{k+1}\| \geq \gamma$,
then set $T_{k+1} := T_{k+1} \cup \{t_{k,i}\}$, else reduce $\tilde{T} := \tilde{T} \backslash \Theta_{k,i}$.

Step 5. Set $c_k = x_{k+1} - P(x_{k+1})$ (using the last projection from step 4.2) and complete the index set $T_{k+1} := T_{k+1} \cup \{(k+1)\}$, $l_{k+1} = |T_{k+1}|$. Modify the current matrix $J = L'$ by the update formula:

$$J_{k+1} = J_k + \frac{(f_{k+1} - L_k(x_{k+1}))c_k^T}{c_k^T c_k} \tag{31}$$

End.

Remark that the stability test in step 4.3 is based on the geometrical interpretation of the σ-stable general position; the norm quotient used here represents the sinus of the "angle" enclosed between the vector $(x_{k+1} - x_j)$ and the hyperplane $\pi(\{x_t : t \in T\})$ (for the temporarily considered set T from the step before), whereas in (10) we have to do with the appropriated "normed volume" bound.

If the graph G_{k+1} is assembled of the subtree $\tilde{G}$ of G_k with vertices $\{x_t : t \in \tilde{T}\}$ completed by the new node x_{k+1} and the edge (x_{k+1}, x_j) (cf. step 4.1), then the σ-stable position property will hold for all k at least for $\sigma = \gamma^m$.

Consider now the original problem in the form

$$x_* = f^{-1}(0) \quad ,$$

where near the initial iteration point f^{-1} is given approximately by a linear function $\bar{L}$. During the iteration the linear approximation can be actualized by an interpolation rule which interpretes $\{f_i : i \in T'(k)\}$ as the nodes, and for the matrices uses inverse update formulas; [5]. The index set T'_k in analogy to Alg.1 has to be such a subset of $\{(k-m), \ldots, k\}$ that the stability restriction (10) holds for ΔF^o resp.

The above arguments lead to an Algorithm 2 dual to Algorithm 1. For completeness, the given stable secant algorithms can be slightly modified to fit for the case of symmetric Jacobians by replacing the matrix update in **step 5**.

4.2 Convergence result

In this section the locally q-superlinear convergence of the algorithms given in 4.1 will be proved under smoothness assumptions comparable with those used in [10] for the classical secant methods or in [4], [5] for their stabilized versions. We show, that the convergence behavior for the released position assumption is preserved; our attention will be focussed here on **Algorithm 1** as the model case, and modifications for the symmetric variants are shortly described.

Assumptions:

A1 Let x_* be an isolated solution point of (1). There exists an open neighbourhood U of x_* such that $f \in C^1(U)$.

A2 There exists a ball $B \subset U$: $f' \in Lip_\Lambda(B)$, i.e.
$$\|f'(x) - f'(y)\| \le \Lambda \|x - y\| \quad \forall x, y \in B$$

With these smoothness assumptions we are able to show that the stable secant methods converge in the following sense:

Theorem 1. *For* $\bar\delta, \bar\rho$ – *sufficiently small and initial data such that*

$$\|x_0 - x_*\| \le \bar\delta , \quad \|J_0 - f'(x_0)\| \le \bar\rho$$

the iteration converges q-superlinearly, i.e.

$$\lim_{k \to \infty} \|x_{k+1} - x_*\| / \|x_k - x_*\| = 0 .$$

The details of the proof are omitted; let's remark only that it follows the ideas of [3] - [5], [6], [2]. As a first step the linear convergence is proved, cf. [12] or [7]. For the superlinear convergence rate then it is sufficient to prove the DENNIS - MORÉ - characterization for the matrices J_k; i.e. (see [6], [2]):

$$\|(J_k - f'_*)(x_{k+1} - x_k)\| / \|(x_{k+1} - x_k)\| \to 0 \text{ for } k \to \infty . \tag{32}$$

By **Lemma 3**, the matrix update used in **Alg.1** is a least change projection update (with respect to the FROBENIUS-norm) onto the matrix manifolds $Q_k = Q[y_k, s_k]$. Let's interprete $\{Q_k\}$ as a sequence of affine linear subspaces in the related HILBERT space $R^{n \times n}$. It seems to be evident, that in general $\bigcup\{Q_k\}$ will be empty, so that we cannot directly deduce the convergence of $\{J_k\}$ from the least change property. However, the asymptotic behavior of Q_k is of interest in connection with a Theorem of KOSMOL about successive projections in HILBERT spaces, see [9], Theor.3, which generalizes the well known Theorem of JOHN V. NEUMANN:

If there exist $B_k \in Q_k$ $\forall k$ such that $\sum_{k=0}^{\infty} \|B_{k+1} - B_k\| < \infty$,
  then the best-approximation sequence $\{J_k\}$ is bounded for arbitrary J_0,
  and $\|J_{k+1} - J_k\| \to 0$ for $k \to \infty$.

Remark, that the last property is sufficient for (32):

$$(J_k - f'_*)(x_{k+1} - x_k) = (J_k - J_{k+1})(x_{k+1} - x_k) + (f_{k+1} - f_k - f'_*(x_{k+1} - x_k)) ;$$

the second term by **(A2)** and Theor.1 is of order $\|(x_{k+1} - x_k)\|^2$, so that the desired norm behavior would result.

Suggested by the mean value theorem let the matrices be given by

$$B_k = \int_0^1 f'(x_{k+1} - \tau s_k) \, d\tau \tag{33}$$

where s_k is the vector used for the update according to the BROYDEN-type formula (21) in step 5 of **Alg.1**. With the notations (20) then

$$B_k s_k = y_k, \quad \text{i.e.} \quad B_k \in Q[y_k, s_k] .$$

The construction of c_k and T_k together with the q-linear convergence guarantee, that (for sufficiently good inital data) always

$$\|s_k\| = \|c_k\| \le \|x_{k+1} - x_k\| \le (1 + q) q^k \|x_0 - x_*\| .$$

The matrix differences then can be estimated as follows:

$$\|B_k - B_{k-1}\| \le \max_{0 \le \tau \le 1} \|(f'(x_{k+1} - \tau s_k) - f'(x_k - \tau s_{k-1}))\|$$

$$\le 3 \Lambda (1 + q) q^{k-1} \|x_0 - x_*\| .$$

The summability required for ΔB_k follows immediately; the proof of **Theor.1** is completed.

Remark: Analogous results hold for **Algorithm2, 3** and **4**; the methodology for the necessary modifications of the proofs can be adapted from [2] – general quasi-NEWTON methods, or [5] – stable secant methods, where symmetric versions and dual algorithms are analyzed in detail.

References

[1] Bittner L. Verallgemeinerung des Sekantenverfahrens zur näherungsweisen Berechnung der Nullstellen eines nichtlinearen Gleichungssystems, *Wiss. Zeitschr. Techn. Univ. Dresden* ,1959/60, 9, 325-329.

[2] Broyden C.G., Dennis jun. J.E., Moré J.J. On the local and superlinear convergence of quasi-Newton methods, *J.Inst.Math.Appl.*,12, 1973, 223-245.

[3] Burdakov O.P. Methods of the secant type for systems of equations with symmetric Jacobian matrix, *Numerical Funct. Anal. Optim.*,6, 1983, 183-195.

[4] Burdakov O.P. Stable versions of the secant method for solving systems of equations, *U.S.S.R. Comput. Math. and Math. Phys.*, 23, 1983, 1-10.

[5] Burdakov O.P. On superlinear convergence of stable variants of the secant method *Z. angew. Math. und Mech.*, 66, 1986, 615-622.

[6] Dennis jun. J.E., Moré J.J. A characterization of superlinear convergence and its application to quasi-Newton methods,*Math.Comp.*, 28, 1974, 549-560.

[7] Felgenhauer U. On stable convergence of secant methods with projected updates, *Zhurnal vych. matem.i mat. fiziki, (U.S.S.R. Comput. Math. and Math. Phys.)*, 5, 31, 1991, 654-662.

[8] Gay D.M., Schnabel R.B. Solving systems of nonlinear equations by BROYDEN's method with projected updates, in *Nonlinear Programming 3, eds. O.L.Mangasarian, R.R.Meyer, S.M.Robinson, Academic Press, New York*, 1978, 245-281.

[9] Kosmol P. Methoden zur numerischen Behandlung nichtlinearer Gleichungen und Optimierungsaufgaben, *Teubner, Stuttgart* 1989.

[10] Ortega J.M., Rheinboldt W.C. Iterative solution of nonlinear equations in several variables, *Academic Press, New York*, 1970.

[11] Schnabel R.B. Quasi-NEWTON methods using multiple secant equations, *Technical Report CU-CS-247-83, Dept. Computer Science, Univ.of Colorado*, 1983.

[12] Schwetlick H. Numerische Lösung nichtlinearer Gleichungen, *Dtsch. Verlag d. Wissensch., Berlin* 1979.

Dikin's algorithm for matrix linear programming problems

Leonid Faybusovich
Department of Mathematics
University of Notre Dame
Mail distribution Center
Notre Dame,IN, 46556-5683

Abstract

We construct a generalization of affine-scaling vector fields for matrix linear programming problems. We discuss various properties of these vector fields and suggest a generalization of Dikin's algorithm.

Key words: linear programming, interior point methods, matrix problems

1 Introduction

Matrix linear programming problems give a natural example of optimization problems with infinite number of inequality constraints. There are no finite step algorithms for these problems. Thus it is natural to consider interior-point iterative procedures. Various approaches to matrix linear programming problems with semi-definite constraints are discussed in [5]. Path-following interior-point algorithms for this class of problems are considered in [6]. In this paper we introduce and study generalized affine-scaling vector fields. Let $S(n)$ be the set of symmetric real n by n matrices. Consider the following optimization problem:

$$f_\beta(K) = Tr(CK) + \frac{\ln(\det K)}{\beta} \to max, \qquad (1.1)$$

$$Tr(A_iK) = b_i, i = 1, \ldots m, \tag{1.2}$$

$$K \in S(n), K \geq 0. \tag{1.3}$$

Here $C, A_i, i = 1, \ldots m$ are symmetric matrices; $Tr(A)$ stands for the trace of a matrix A and $K \geq 0$ means $< x, Kx > = x^T K x \geq 0$ for any $x \in R^n$. Further, $\beta > 0$ is a scalar parameter and $b_i \in R, i = 1, \ldots m$. If $K(\beta)$ is a solution to the problem 1.1-1.3, it is intuitively clear (and will be proved later on) that $K(\beta)$ should converge when β tends to $+\infty$ to a solution K^* of the linear programming problem

$$Tr(CK) \to max, \tag{1.4}$$

$$Tr(A_iK) = b_i, i = 1, \ldots m, \tag{1.5}$$

$$K \in S(n), K \geq 0. \tag{1.6}$$

It turns out that $K(\beta)$ as a function of β satisfies a system of differential equations which we will call the generalized affine-scaling vector fields. In the present paper we study in detail some properties of these vector fields. We present a version of the Dikin's algorithm [3] for solving the problem 1.4-1.6. Some further properties of these vector fields (e.g. the Hamiltonian structure [4]) will be studied elsewhere.

2 Generalized affine-scaling vector fields

We start with some elementary properties of the function $\phi : K \to \ln \det K$.

Lemma 2.1 Let $K, \xi \in S(n), K$ be positive definite and $\xi \neq 0$. Then

$$Tr(K^{-1}\xi K^{-1}\xi) > 0.$$

Proof: Indeed, $Tr(K^{-1}\xi K^{-1}\xi) = Tr(K^{1/2}(K^{-1}\xi K^{-1}\xi)K^{-1/2}) = Tr([K^{-1/2}\xi K^{-1/2}]^2) \geq 0$. Besides, the equality holds if and only if $K^{-1/2}\xi K^{-1/2} = 0$ or $\xi = 0$. ∎

Proposition 2.2 *The function ϕ is strictly concave on the convex set of positive definite symmetric matrices.*

Proof: Indeed,

$$D\phi(K)(\xi) = Tr(K^{-1}\xi), \tag{2.1}$$

$$D^2\phi(K)(\xi,\eta) = -Tr(K^{-1}\xi K^{-1}\eta), \tag{2.2}$$

$\xi, \eta \in S(n), K > 0$. Here we use notations $D\phi(K), D^2\phi(K)$ for the first and second Frechet derivatives of the function ϕ at a point K. The result follows from Lemma 2.1. ∎

Suppose that the convex set P determined by constraints 1.5, 1.6 is compact. Let $T = span(A_1, \ldots, A_m)$ (the set of all linear combinations of matrices $A_1, \ldots, A_m$) and $T^\perp = \{\xi \in S(n) : Tr(\xi A_i) = 0, i = 1, \ldots, m\}$. It is clear that $S(n) = T \oplus T^\perp$. Let $\pi : S(n) \to T^\perp$ be the projection of $S(n)$ onto $T^\perp$ along T. Consider the map $\psi : int(P) \to T^\perp$

$$\psi(K) = \pi(K^{-1}). \tag{2.3}$$

Here $int(P)$ is the set of positive definite matrices in P. We will denote $P \setminus int(P)$ by ∂P. Observe that $\psi(K) = \pi(\nabla\phi(K))$ (see 2.1.)

Lemma 2.3 *Suppose that K is a positive definite symmetric matrix and $A_1, \ldots, A_m$ are linearly independent symmetric matrices. Then the m by m matrix*

$$\Gamma(K) = \|Tr(A_i K A_j K)\| \tag{2.4}$$

is positive definite.

Proof: It is sufficient to verify that $\xi^T \Gamma(K)\xi > 0$ for any nonzero $\xi \in R^m$. But $\xi^T \Gamma(K)\xi = Tr(\eta K \eta K), \eta = \xi_1 A_1 + \ldots + \xi_m A_m$. The result follows by Lemma 2.1. ∎

Theorem 2.4 *Suppose that the convex set P determined by 1.5, 1.6 is compact, $int(P) \neq \emptyset$ and $A_1, \ldots, A_m$ are linearly independent. Then the map ψ is a diffeomorphism of $int(P)$ onto $T^\perp$.*

Proof: Given $C \in T^\perp$, consider the following extremal problem:

$$f(K) = Tr(CK) + \ln(\det K) \to max, K \in P. \tag{2.5}$$

Since $\ln(\det K) \to -\infty$ when $K \to \partial P, P$ is compact and f is strictly concave, 2.5 has a unique solution $K(C) \in int(P)$ for any $C \in T^\perp$. It is clear that

$\nabla f(K(C)) \in T$. Hence $C + K^{-1}(C) \in T$ or $\pi(C) = -\pi(K^{-1}(C))$. This means that ψ is surjective. If $\psi(K_1) = \psi(K_2) = -C$, then both K_1 and K_2 are solutions to 2.5. We conclude that $K_1 = K_2$ because such a solution is unique. Hence, ψ is injective. It remains to prove that ψ^{-1} is smooth. We have for $\xi \in S(n), K \in int(P)$:

$$D\psi(K)\xi = -\pi(K^{-1}\xi K^{-1}).\eqno(2.6)$$

Suppose that $Tr(A_i\xi) = 0, i = 1, \ldots m$, and $D\psi(K)\xi = 0$. This yields :

$$K^{-1}\xi K^{-1} = \sum_{i=1}^{m} \mu_i A_i$$

for some real μ_i and consequently

$$\sum_{i=1}^{m} \mu_i Tr(A_s K A_i K) = 0, s = 1, \ldots, m.$$

Hence by Lemma 2.3 $\mu_i = 0, s = 1, \ldots m$. In other words, $D\psi(K)$ is injective and hence bijective map from $T^\perp$ to $T^\perp$. Thus ψ^{-1} is smooth by the implicit function theorem. ∎

Remark 2.5 The map ψ is a generalization of a version of the Legendre transform considered in [2].

Let $S^+(n)$ be the set of positive definite symmetric matrices. It is clear that $S^+(n)$ is an open subset in $S(n)$. Consider a Riemannian metric g on $S^+(n)$ defined as follows:

$$g(K; \xi, \eta) = Tr(K^{-1}\xi K^{-1}\eta).\eqno(2.7)$$

Here $K \in S^+(n), \xi, \eta \in S(n)$. It is clear that $int(P)$ is a submanifold of $S^+(n)$ and hence a Riemannian submanifold. Suppose that f is a smooth function on $int(P)$. Our next goal is to describe the gradient $\nabla_g f$ of f relative to the metric g.

Proposition 2.6 Let $A_1 \ldots, A_m$ be linearly independent. Then

$$\nabla_g f(K) = K(\nabla f(K) - \sum_{i=1}^{m} \mu_i(K, f)A_i))K,\eqno(2.8)$$

$K \in int(P)$. *Here* $Df(K)\xi = Tr(\nabla f(K)\xi); \xi, \nabla f(K) \in S(n)$ *and*

$$
\begin{bmatrix} \mu_1(K,f) \\ \vdots \\ \mu_m(K,f) \end{bmatrix} = \Gamma(K)^{-1} \begin{bmatrix} Tr(A_1 K \nabla f(K)K) \\ \vdots \\ Tr(A_m K \nabla f(K)K) \end{bmatrix}, \tag{2.9}
$$

where $\Gamma(K)$ *is defined in 2.4.*

Proof: It is sufficient to verify that $Tr(A_i \nabla_g f(K)) = 0, i = 1, \ldots, m$, and that $g(K; \nabla_g f(K), \xi) = Tr(\nabla f(K)\xi)$ for any $\xi \in T^\perp$. We have

$$
Tr(A_i \nabla_g f(K)) = Tr(A_i K \nabla f(K)K) - \sum_{j=1}^m \mu_j Tr(A_i K A_j K) = 0
$$

by 2.9. Further,

$$
g(K; \nabla_g f(K), \xi) = Tr((\nabla f(K) - \sum_{i=1}^m \mu_i A_i)\xi) = Tr(\nabla f(K)\xi),
$$

since $Tr(A_i \xi) = 0, i = 1 \ldots, m$. ∎

Corollary 2.7 *Let* $f(K) = Tr(KC), C \in S(n)$. *Then*

$$
D(\psi)(K)\nabla_g f(K) = -\pi(C). \tag{2.10}
$$

In other words, all vector fields $\nabla_g F$ *correspond to the constant vector fields under the diffeomorphism* ψ *of* $int(P)$ *onto* $T^\perp$.

Proof: It follows from 2.6, 2.8 and the fact that $\pi(A_i) = 0, i = 1, \ldots, m$. ∎

Remark 2.8 Observe that, if $C, A_i, i = 1, \ldots, m$, are diagonal matrices the vector field 2.8 has an invariant manifold consisting of positive definite diagonal matrices in P. The restriction of 2.8 to this manifold coincides with the standard affine-scaling vector field [2].

Corollary 2.9 *Let* $f(K) = tr(CK), C \in S(n)$. *Then 2.8 has no stationary points in* $int(P)$, *provided* $\pi(C) \neq 0$.

Proof: This immediately follows by 2.10. ∎

Set

$$\nabla_g(K) = V_C(K), \tag{2.11}$$

provided $f(K) = Tr(CK)$.

Corollary 2.10 *For any C, C' vector fields $V_C, V_{C'}$ pairwise commute. In other words, the Lie bracket $[V_C, V_{C'}] = 0$.*

Proof: Since two constant vector fields pairwise commute, this follows from 2.10. ∎

Suppose that $\Gamma(K)^{-1}$ (see 2.4) can be smoothly extended to ∂P. It is possible then using 2.9 to extend smoothly the vector fields $\nabla_g f$ to P. Suppose that this is the case. For any vector subspace $M \in R^n$ denote by $P(M)$ the set $\{K \in P : M \subset KerK\}$.

Proposition 2.11 *The set $P(M)$ is an invariant submanifold for V_C.*

Proof: If $Kx = 0$, then by 2.8 $V_C(K)x = 0$. Hence $K(0) \in P(M)$ implies $K(t) \in P(M)$ for any t. ∎

Example 2.12 Consider the following linear programming problem:

$$Tr(KC) \to max, Tr(K) = 1, K \geq 0.$$

In this case $\Gamma(K) = Tr(K^2)$. Hence

$$V_C(K) = KCK - \frac{Tr(KCK)}{Tr(K^2)}K^2.$$

Proposition 2.13 *Under assumptions of Theorem 2.4 suppose that $K(\beta)$ is a solution to the problem 1.1-1.3. Then*

$$\frac{dK(\beta)}{d\beta} = V_{\pi C}(K(\beta)), \tag{2.12}$$

$\lim K(\beta) = K_0, \beta \to 0$, *where K_0 is a solution to the problem $\ln \det K \to max, K \in P$.*

Remark 2.14 K_0 is a natural analogue of the analytic center of a polyhedron [7].

Proof: Since $K(\beta) \in int(P)$, we should have: $(C + \frac{K(\beta)^{-1}}{\beta}) \in T$ or $\pi(C) = -\frac{\pi(K(\beta)^{-1})}{\beta}$. In other words,

$$D\psi(K(\beta))(\frac{dK(\beta)}{d\beta}) = -\pi C.$$

Comparing this with 2.10 we arrive at 2.11.

Proposition 2.15 *Let K^* be a solution to the problem 1.4-1.6. Then*

$$Tr(CK(\beta)) \leq Tr(CK^*) \leq Tr(CK(\beta)) + n/\beta.$$

In particular, $\lim Tr(CK(\beta)) = Tr(CK^*), \beta \to +\infty.$

Proof: Since $\pi C = -\pi(\frac{K(\beta)^{-1}}{\beta})$, we have

$$C = \sum_{i=1}^{m} w_i A_i - \frac{K(\beta)^{-1}}{\beta} \tag{2.13}$$

for some real w_i. For any $K \in P$ we have

$$Tr(CK) = \sum_{i=1}^{m} w_i Tr(A_i K) - \frac{Tr(CK(\beta)^{-1}K)}{\beta} =$$

$$\sum_{i=1}^{m} w_i b_i - \frac{Tr(K(\beta)^{-1})K)}{\beta} \leq \sum_{i=1}^{m} w_i b_i.$$

Hence $Tr(CK^*) \leq \sum_{i=1}^{m} w_i b_i$. On the other hand, by 2.13

$$Tr(CK(\beta)) = \sum_{i=1}^{m} w_i b_i - \frac{Tr(K(\beta)K(\beta)^{-1})}{\beta}$$

or $\sum_{i=1}^{m} w_i b_i = Tr(CK(\beta)) + n/\beta$. ∎

3 Dikin's algorithm

The next proposition is well-known.

Proposition 3.1 *Let A be n by n matrix and $x_1, \ldots x_n$ be an orthonormal basis in R^n. Then*

$$Tr(A) = \sum_{i=1}^{n} < x_i, Ax_i > .$$

Proof: Let O be an orthogonal matrix such that $x_i = Oe_i, i = 1, \ldots n$, where $e_1, \ldots e_n$ is the standard basis in R^n. We have $< x_i, Ax_i > = < e_i, O^{-1}AOe_i >$. Hence,

$$\sum_{i=1}^{n} < x_i, Ax_i > = Tr(O^{-1}AO) = Tr(A).$$

∎

Corollary 3.2 *Let $x \in R^n, \|x\| = \sqrt{x^T x} = 1$ and K be a nonnegative definite symmetric matrix. Then*

$$< x, Kx > \leq Tr(K). \tag{3.1}$$

Proof: There exist an orthonormal basis $x_1, \ldots x_n$ in R^n such that $x_1 = x$. The result follows by Proposition 3.1. ∎

Proposition 3.3 *Suppose that $K \in S^+(n), X \in S(n)$ are such that*

$$g(K; X - K, X - K) \leq 1, \tag{3.2}$$

(see 2.7). Then $X \geq 0$.

Proof: Suppose that X satisfies 3.2 but is not nonnegative definite. Then there exists $\lambda > 0, x \in R^n, x \neq 0$ such that $Xx = -\lambda x$. One can choose x in such a way that $\|K^{1/2}x\| = 1$. According to 2.7, $g(K; X - K, X - K) = Tr(K^{-1}(X - K)K^{-1}(X - K)) = Tr([K^{-1}(X - K)]^2) = Tr(K^{1/2}(K^{-1}(X - K))^2 K^{-1/2}) = Tr(Y^2)$, where $Y = K^{-1/2}(X-K)K^{-1/2}$. Let $y = K^{1/2}x$. By Proposition 3.1 we have: $< Yy, Yy > = < y, Y^2 y > \leq Tr(Y^2)$. On the other hand, $< Yy, Yy > = < K^{-1/2}(X - K)x, K^{-1/2}(X - K)x > = < K^{-1/2}(-\lambda - K)x, K^{-1/2}(-\lambda - K)x > = \lambda^2 < x, K^{-1}x > + < y, y > + 2\lambda < x, x > = 1 + \lambda^2 < x, K^{-1}x > + 2\lambda\|x\|^2 > 1$. A contradiction. Hence X is nonnegative definite. ∎

Remark 3.4 The same reasoning shows that $g(K; X - K, X - K) < 1$ implies $X \in S^+(n)$.

Let $K \in int(P)$, where P is determined by constraints 1.5, 1.6. Consider a map $\xi_C : int(P) \to int(P)$,

$$\xi_C(K) = K + \gamma \frac{V_C(K)}{\|V_C(K)\|_K}. \tag{3.3}$$

Here $C \in S(n), V_C(K) = K(C - \Lambda(K))K$ was defined in 2.8; $0 < \gamma < 1, \|X\|_K = \sqrt{g(K; X, X)}$. Observe that $g(K; \xi_C(K) - K, \xi_C(K) - K)) = \gamma^2 < 1$. Hence by Proposition 3.3 $\xi_C(K) \in S^+(n)$. Besides, $Tr(A_i \xi_C(K)) = Tr(A_i K) = b_i, i = 1, \ldots m$. In other words, $\xi_C(K) \in int(P)$ if $K \in int(P)$. The Dikin's algorithm for our problem is simply the iteration of the map ξ_C. This is an obvious generalization of the standard Dikin's algorithm [3] for the matrix case. Observe that

$$Tr(C(\xi_C(K) - K)) = \frac{\gamma Tr(CV_C(K))}{\|V_C(K)\|_K}.$$

But

$$Tr(CV_C(K)) = Tr([C - \Lambda(K)]V_C(K)) = \|V_C(K)\|_K^2.$$

Here we used the fact that $\Lambda(K) \in span(A_1, \ldots, A_m)$ (see 2.8) and consequently $Tr(\Lambda(K)V_C(K)) = 0$. We finally obtain:

$$Tr(C\xi_C(K)) - Tr(CK) = \gamma \|V_C(K)\|_K. \tag{3.4}$$

In particular, the cost function of the problem 1.4-1.6 is monotonically increasing along any trajectory $\xi_C^n(K), n = 0, \ldots, K \in int(P)$, of the Dikin's algorithm. This can be used for the analysis of the convergence properties of this procedure. Suppose that

i) $\xi_C^n(K) \to K^*, n \to +\infty$

ii) If $R^n = ImK^* \oplus KerK^*$, the following property holds. For any $x \in ImK^*$ there exists a unique $y \in ImK^*$ such that $x = Ky$. We suppose that $[\xi_C^n(K)]^{-1}x \to y, n \to +\infty$.

Theorem 3.5 *Under assumptions i),ii) we have:*

$$Tr(CK^*) - Tr(C\xi_C^{n+1}(K)) \le (1 - \frac{\gamma}{\sqrt{\dim Ker K^* + \epsilon_n}})(Tr(CK^* - Tr(C\xi_C^n(K)),$$
$$\tag{3.5}$$

where $\epsilon_n \to 0, n \to +\infty$. In other words, the rate of convergence to K^* is at least linear.

Remark 3.6 Compare this with [1].

Proof: Denote $\xi_C^n(K)$ by K_n. We have: $Tr(CK^*) - Tr(CK_i) = Tr((C - \Lambda(K_i))(K^* - K_i)) = Tr(K_i(C - \Lambda(K_i))(K^* - K_i)K_i^{-1})$. Here we used that $Tr(\Lambda(K_i)(K^* - K_i)) = 0$. Let $A = K_i(C - \Lambda(K_i)), B = (K^* - K_i)K_i^{-1}, D = K_i^{-1/2}$. As is easily seen, the matrices DAD^{-1} and DBD^{-1} are symmetric. Now $Tr(AB) = Tr((DAD^{-1})(DBD^{-1})) \leq [Tr(DA^2D^{-1})]^{1/2}[Tr(DB^2D^{-1}]^{1/2} = Tr(A^2)^{1/2}Tr(B^2)^{1/2}$. Observe now that $Tr(A^2) = \|V_C(K_i)\|_{K_i}^2$. We thus obtain:

$$Tr(CK^*) - Tr(CK_i) \leq \|V_C(K_i)\|_{K_i}[Tr(B^2)]^{1/2}. \tag{3.6}$$

Combining this with 3.4, we obtain:

$$Tr(CK^*) - Tr(CK_i) \leq \frac{[Tr(B^2)]^{1/2}}{\gamma}Tr(CK_{i+1} - CK_i),$$

which immediately implies:

$$Tr(CK^*) - Tr(CK_{i+1}) \leq (1 - \frac{\gamma}{[Tr(B^2)]^{1/2}})[Tr(CK^* - Tr(CK_i)]. \tag{3.7}$$

Let $x_1, \ldots, x_n$ be an orthonormal basis of eigenvectors of K^* such that $span(x_1, \ldots, x_l) = KerK^*$. We have

$$Tr(B^2) = \sum_{j=1}^{n} <x_i, B^2 x_i> =$$

$$\sum_{j=1}^{l} <x_j, x_j> + \sum_{j=l+1}^{n} <x_j, (K^* - K_i)K_i^{-1}(K^* - K_i)K_i^{-1}x_j> = l + \epsilon_i.$$

Observe that $(K^* - K_i)K_i^{-1}x_j \to 0$, when $j \to +\infty$, because of the imposed condition ii). Thus 3.7 coincides with 3.5. ∎

References

[1] E.R. Barnes, *A variation on Karmarkar's algorithm for solving linear programming problems* , Math. Programming, 36(1986), 174-182.

[2] D.A. Bayer and J.C. Lagarias, The nonlinear geometry of linear programming.I, Trans. Amer. Math. Soc., 314(1989), 499-526.

[3] I. Dikin, *Iterative solution of problems of linear and quadratic programming,* Soviet. Math. Dokl., 8(1967),674-675

[4] L.Faybusovich, Hamiltonian structure of dynamical systems which solve linear programming problems, Physica D53(1991), 217-232.

[5] R.Fletcher, *Semi-definite constraints in optimization,* SIAM J. Control Optim., 23(1985),493-513.

[6] Yu. E. Nesterov and A.S. Nemirovsky, *Self-concordant functions and polynomial time methods in convex programming.* Technical report, Centr. Econ. and Math. Inst., Moscow,USSR, 1989.

[7] G. Sonnevend, Applications of analytic centers. In: Numerical Linear algebra, Digital Signal Processing and Parallel Algorithms, Springer,1991.

A Variational Principle and a Fixed Point Theorem

Christiane Tammer
Fachbereich Mathematik/Informatik, Martin-Luther-Universität
Halle-Wittenberg, D-06099 Halle, Germany

1   Introduction

In 1976 Brezis and Browder [3] proved a general existence principle for extremal elements on ordered sets. Moreover, they could derive Ekeland's variational principle (Ekeland [4]), the Bishop-Phelps theorem (Bishop and Phelps [1] , [2]) as well as the Kirk-Caristi fixed point theorem (Kirk and Caristi [11]) from their theory and proved the closed relations between the results of Ekeland, Kirk/Caristi and Bishop/Phelps.

The aim of our paper are corresponding results for partially ordered spaces and applications to vector optimization problems. We present a general existence principle for topological spaces which are partially ordered by a convex cone, derive from this principle a corresponding extended Kirk-Caristi fixed point theorem and show the closed relation between this fixed point theorem and a corresponding generalized variational principle. Finitely, we use the obtained results to develope necessary conditions for approximate solutions for a class of vector-valued approximation problems.

2   An existence principle and a Kirk-Caristi-type fixed point theorem

Let us consider a subset D of a linear topological space Y. We denote the topological interior of D by $int\, D$ and the topological closure of D by $cl\, D$.

In order to prove a Kirk - Caristi - type fixed point theorem we will give at first an existence result under the following assumtions:

(A):
$$(X, d) \text{ is a complete metric space,}$$
$$Y \text{ is a linear topological space,}$$
$$K \subset Y \text{ is a convex cone with } k^\circ \in int\, K$$
$$\text{and } K \cap (-K) = \{0\},$$
$$B \subset Y \text{ is a cone with}$$
$$cl\, B + (K \setminus \{0\}) \subset int\, B.$$

(B):
$$f : X \longrightarrow Y$$
is lower semicontinuous with respect to k° and B in the sense that
$$M_r := \{x \in X : f(x) \in rk^\circ - cl\, B\} \text{ is closed for each } r \in R$$
and bounded from below, i. e., $f[X] \subset y + B$ for a certain $y \in Y$.

Under the assumptions (A), (B) and an additional assumption (V) we can show an existence result for solutions of a vector optimization problem :

$$Eff(f[X], K)$$

where

$$Eff(f[X], K) := \{f(\bar{x}) \mid \bar{x} \in X, f[X] \cap (f(\bar{x}) - (K \setminus \{0\})) = \emptyset\}.$$

In order to prove this existence result we need the following assertion given by Takahashi [15] for a real valued function $f : X \longrightarrow R$:

Theorem 1 *[15]: Let (X, d) be a complete metric space and $\varphi : X \longrightarrow R$ lower semicontinuous and bounded from below.*
Under the additional assumption, that for each $u \in X$ with $inf_{x \in X} \varphi(x) < \varphi(u)$ there is an element $v \in X$ with $v \neq u$ and

$$\varphi(v) + d(u, v) \leq \varphi(u),$$

there exists an element $x^{\circ} \in X$ with

$$\varphi(x^{\circ}) = inf_{x \in X} \varphi(x).$$

Applying this result of Takahashi and a suitable functional $z : Y \longrightarrow R$ we will show a existence principle in a linear topological space Y. Here we need the following properties of the functional z:
We say that a functional $z : Y \longrightarrow R$ is K-monotone if for all $y^1, y^2 \in Y$ $y^1 \in y^2 + K$ implies $z(y^1) \geq z(y^2)$. A K-monotone functional z is called even strictly K-monotone if, additionally, $y^1 \in y^2 + (K \setminus \{0\})$ implies $z(y^1) > z(y^2)$.

Further, we will use the following assertion in the proof of our existence result:

Theorem 2 *Assume that Y is a linear topological space, K and B are cones in Y with nonempty interior, $cl\, B + (K \setminus \{0\}) \subset int\, B$ and $k^{\circ} \in int\, K$. Then the functional $\bar{z} : Y \longrightarrow R$ defined by*

$$\bar{z}(y) := inf\{t \in R \mid y \in -cl\, B + tk^{\circ}\}$$

is continuous, subadditive and strictly K-monotone.

The proof follows immediately from Theorem 2.1 and Corollary 2.1 in Gerth/Weidner [7].

Theorem 3 *Assume (A), (B) and*

$$(V): \text{For each } u \in X \text{ with } f(u) \notin Eff(f[X], K)$$
$$\text{there is a } v \in X \text{ with } v \neq u \text{ and}$$
$$f(v) + k^{\circ} d(u, v) \in f(u) - K.$$

Then there exists an element $x^{\circ} \in X$ with

$$f(x^{\circ}) \in Eff(f[X], K).$$

Proof: Consider an arbitrary $\bar{x} \in X$ with $f(\bar{x}) \notin Eff(f[X], K)$. Let us define

$$\bar{f} := f - f(\bar{x})$$

and $z : Y \longrightarrow R$ by

$$z(y) := inf\{t \mid y \in cl \ (\bar{f}(\bar{x}) - B) + tk^o\}. \tag{1}$$

Assumption (V) implies the existence of an element $v \in X$ with $v \neq \bar{x}$ and

$$\bar{f}(v) \in -k^o d(\bar{x}, v) + \bar{f}(\bar{x}) - K. \tag{2}$$

Because of $\bar{f}(\bar{x}) = 0$ Theorem 2 yields that the functional z in (1) is subadditive and strictly K- monotone.

These properties of z together with (2) imply

$$z(\bar{f}(v)) \leq z(-k^o d(\bar{x}, v) + \bar{f}(\bar{x})) \leq z(-k^o d(\bar{x}, v)) + z(\bar{f}(\bar{x})).$$

According to Lemma 4.4 in Tammer [16] we have for $\alpha > 0$

$$z(-\alpha k^o) = -\alpha,$$

and hence

$$z(-k^o d(\bar{x}, v)) = -d(\bar{x}, v).$$

Then we can conclude

$$z(\bar{f}(v)) + d(\bar{x}, v) \leq z(\bar{f}(\bar{x})).$$

Because of assumption (V) we get for each $x \in X$ with $f(x) \notin Eff(f[X], K)$ by using the subadditive, strictly K-monotone functional z according to (1) that there exists an element $v \in X$ with $v \neq x$ and

$$z(\bar{f}(v)) + d(x, v) \leq z(\bar{f}(x)).$$

Further, the lower semicontinuity of f with respect to k^o and B (and hence also of $\bar{f}$) implies the lower semicontinuity of $z \circ \bar{f}$ (compare Lemma 4.1 in Tammer [16]). Now, the assumptions of Theorem 1 are fulfilled and so there exists an element $x^o \in X$ such that

$$z(\bar{f}(x^o)) = min_{x \in X} z(\bar{f}(x)).$$

The strict K-monotonicity of z implies

$$\bar{f}(x^o) \in Eff(\bar{f}[X], K),$$

which means

$$\bar{f}[X] \cap (\bar{f}(x^o) - (K \setminus \{0\})) = \emptyset.$$

Hence

$$f[X] \cap (f(x^o) - (K \setminus \{0\})) = \emptyset$$

and therefore

$$f(x^o) \in Eff(f[X], K).$$

q.e.d.

Now we will use Theorem 3 in order to prove the following fixed point theorem.

Theorem 4 *Assume (A) and (B). Then any mapping $T : X \longrightarrow X$ satisfying*

$$k^\circ d(x, Tx) \in f(x) - f(Tx) - K \qquad \text{for all} \qquad x \in X \tag{3}$$

has a fixed point $x^\circ \in X$, i.e.,

$$Tx^\circ = x^\circ.$$

Proof: Let us suppose that

$$Tx \neq x \qquad \text{for all} \quad x \in X.$$

Then for all $x \in X$ there exists an element $v \in X$ such that

$$x \neq v$$

and

$$f(v) + k^\circ d(x, v) \in f(x) - K$$

because of (3) with $v = Tx$. Now we can conclude from Theorem 3 that there exists an element $x^\circ \in X$ with

$$f(x^\circ) \in Eff(f[X], K).$$

From the definition of the set $Eff(f[X], K)$ we imply that

$$\forall x \in X \qquad \text{it holds} \qquad f(x) \notin f(x^\circ) - (K \setminus \{0\}) \tag{4}$$

and for $x = Tx^\circ \in X$ it follows from (4) that

$$f(Tx^\circ) \notin f(x^\circ) - (K \setminus \{0\}). \tag{5}$$

On the other hand, for $k^\circ \in int\, K$ (especially $k^\circ \neq 0$) and $x^\circ \neq Tx^\circ$ it holds

$$k^\circ d(x^\circ, Tx^\circ) \in K \setminus \{0\}. \tag{6}$$

Relation (3) implies

$$k^\circ d(x^\circ, Tx^\circ) \in f(x^\circ) - f(Tx^\circ) - K$$

and hence

$$f(x^\circ) - f(Tx^\circ) \in k^\circ d(x^\circ, Tx^\circ) + K \subset K \setminus \{0\},$$

since K is a convex cone with $K \cap (-K) = \{0\}$. Thus we get

$$f(Tx^\circ) \in f(x^\circ) - (K \setminus \{0\})$$

in contradiction to (5). q.e.d.

Remarks:

1. For the special case $Y = R$, $K = R_+$ and a functional $f : X \longrightarrow R$ Theorem 4 coincides with the Kirk-Caristi fixed point theorem [11].

2. Other Kirk-Caristi-type fixed point theorems for functions $f : X \longrightarrow Y$ were also proved by Nemeth [13] and Khanh [10] but differ from our result in Theorem 4 in the assumptions on the space Y and on the cone K and in the condition for contraction (3).

3 A variational principle

In this section we will show at first that the fixed point assertion in Theorem 4 implies a variational principle for an objective function which takes it values in a linear topological space. Under the assumptions (A) and (B) we introduce the approximately efficient point set of $f[X]$ with respect to B:

$$Eff(f[X], B_{\epsilon k^\circ}) := \{f(x) \mid x \in X, \quad f[X] \cap (f(x) - \epsilon k^\circ - (B \setminus \{0\})) = \emptyset\}$$

$$\text{where} \qquad B_{\epsilon k^\circ} := \epsilon k^\circ + B$$

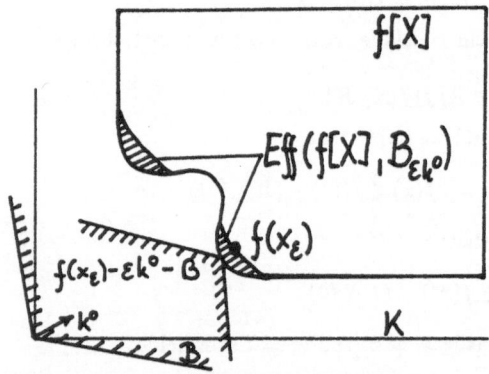

Figure 1: The approximately efficient point set of $f[X]$ with respect to B.

Remarks

1. Beginning with the paper of Loridan [12] several concepts for approximately efficient solutions of a vector optimization problem were published in the last years (compare Nemeth [13], Staib [14], Valyi [18], Gerth [6], Tammer [17] and Helbig/Georgiev/Patewa/Todorow [8]) . Loridan [12] introduces approximately efficient solutions in the case $Y = R^p$ by using the usual ordering cone R_+^p instead of the set B in the definition of the set $Eff(f[X], B_{\epsilon k^\circ})$.

2. In the special case $Y = R^p$ and $K = R_+^p$ condition $cl\ B + (K \setminus \{0\} \subset int\ B$ implies $R_+^p \subseteq cl\ B + (K \setminus \{0\})$ such that our set $Eff(f[X], B_{\epsilon k^\circ})$ is a subset of the set of ϵ-efficient elements defined by Loridan [12].

The next theorem is an assertion about the existence of an efficient solution of a perturbed vector optimization problem and can be considered as an extension of Ekeland's variational principle (Ekeland [4], [5]) to vector optimization. Other extensions of Ekeland's variational principle to vector optimization are given by Loridan [12] for the finite dimensional case and by Nemeth [13], Khanh [10] and Tammer [16] for general spaces. Our following vector-valued variational principle differs essentially from those of Nemeth and Khanh but is rather similar to our former result in [16]. However, we will give here a quite different proof using our Kirk-Caristi-type fixed point theorem from Section 2.

Theorem 5 *Assume (A) and (B). Then for every $\epsilon > 0$ there exists some point $x_\epsilon \in X$ such that*

$$f_{\epsilon k^\circ}(x_\epsilon) \in Eff(f_{\epsilon k^\circ}[X], K), \quad f_{\epsilon k^\circ}(x_\epsilon) \neq f_{\epsilon k^\circ}(x) \quad \forall x \neq x_\epsilon, \tag{7}$$

where

$$f_{\epsilon k^\circ}(x) := f(x) + k^\circ d(x, x_\epsilon)\sqrt{\epsilon}.$$

Proof: Suppose that condition (7) does not hold, i.e., for a certain $\epsilon > 0$ there exists no element x_ϵ with

$$f_{\epsilon k^\circ}(x_\epsilon) \in Eff(f_{\epsilon k^\circ}[X], K)$$

and

$$f_{\epsilon k^\circ}(x_\epsilon) \neq f_{\epsilon k^\circ}(x) \quad \forall x \neq x_\epsilon.$$

This implies that for all $x \in X$ there exists a

$$y \neq x \quad \text{with} \quad f(y) + k^\circ d_1(x, y) \in f(x) - K, \tag{8}$$

where $d_1 := \sqrt{\epsilon} d$.
Let us define $T : X \longrightarrow X$ by $Tx = y$ where y is given according to (8).
Then obviously

$$k^\circ d_1(x, Tx) \in f(x) - f(Tx) - K$$

and hence assumption (3) in Theorem 4 is fulfilled. Now, from Theorem 4 we know that the mapping T has a fixed point which contradicts to our supposition

$$Tx = y \neq x \quad \text{for all} \quad x \in X.$$

$$\text{q.e.d.}$$

In the last proof we have used Theorem 4 to show Theorem 5. In the following we give a second proof of Theorem 4 using Theorem 5 such that we can see that both Theorems 4 and 5 are closely related as it was already mentioned by Brezis and Browder [3] for the real-valued variants of both theorems.

Second proof of Theorem 4: Theorem 5 says that for any $\epsilon > 0$ there exists an element $x_\epsilon \in X$ with

$$f_{\epsilon k^\circ}(x_\epsilon) \in Eff(f_{\epsilon k^\circ}[X], K)$$

and

$$f_{\epsilon k^\circ}(x_\epsilon) \neq f_{\epsilon k^\circ}(x) \quad \forall x \neq x_\epsilon.$$

Hence, there exists no element $x \neq x_\epsilon$ with

$$f_{\epsilon k^\circ}(x) \in f_{\epsilon k^\circ}(x_\epsilon) - K.$$

So for $\epsilon = 1$ we have

$$f(x) + k^\circ d(x, x_\epsilon) \notin f(x_\epsilon) - K \quad \forall x \neq x_\epsilon. \tag{9}$$

Now, if we suppose that

$$Tx_\epsilon \neq x_\epsilon,$$

we get from (9) that

$$f(Tx_\epsilon) + k^\circ d(Tx_\epsilon, x_\epsilon) \notin f(x_\epsilon) - K.$$

This implies

$$k^\circ d(Tx_\epsilon, x_\epsilon) \notin f(x_\epsilon) - f(Tx_\epsilon) - K$$

in contradiction to (3).   q.e.d.

Remarks:

1. The assertions in Theorem 5 hold also for a vector optimization problem

$$Eff(f[U], B_{ek^*})$$

where U is a nonempty closed subset of X (see Tammer [16]).

2. Additionally to the property (7) the point x_ϵ from Theorem 5 can be choosen in such a way that even

   (i) $f(x_\epsilon) \in Eff(f[X], K_{ek^*})$,
   (ii) $d(x_\epsilon, x^\circ) \leq \sqrt{\epsilon}$

where $x^\circ \in Eff(f[X], B_{ek^*})$ (see Tammer [16]).

3. Relations between a vector-valued variational principle and a Kirk-Caristi- type fixed point theorem were also proved by Nemeth [13] and Khanh [10]. The results in the mentioned papers differ from our results in the assumptions, in the assertions and in the proofs.

4 Applications for vector approximation problems

In this section we will show consequences of Theorem 5 for approximately efficient solutions of a vector approximation problem by using the subdifferential of the vector norm given by Jahn [9]:

We assume again (A) with the additional requirement that $(X, \| \cdot \|_X)$ and $(Y, \| \cdot \|_Y)$ are real reflexive Banach spaces and $K \subset Y$ is a closed convex cone which is Daniell (see Jahn [9]) and has a weakly compact base.

Let be $\| \cdot \| \colon X \longrightarrow K$ a vector norm, i.e., $\forall x, y \in X, \forall \lambda \in R$ it holds (compare [9]):

1. $\| x \| = 0 \Longleftrightarrow x = 0$,

2. $\| \lambda x \| = | \lambda | \| x \|$,

3. $\| x + y \| \in \| x \| + \| y \| - K$.

We assume that $\| \cdot \|$ is continuous and consider the vector-valued function

$$F(x) = \| x - a \|$$

with $x, a \in X$.

Let $S \subset X$ be convex and closed. Then the assumptions in (B) are satisfied and $F[S] + K$ is convex.

Now, we will consider the following vector optimization problem:

$$(P): \qquad Eff(F[S], B)$$

In order to derive necessary conditions for approximately efficient elements of the convex vector optimization problem (P) we use the directional derivative of a vector valued function.

The directional derivative of a function $f : S \longrightarrow Y$ at $x^\circ \in S$ in the direction $v \in X$ is defined by:

$$f'(x^\circ)(v) := lim_{t \to +0} 1/t \ (f(x^\circ + tv) - f(x^\circ)).$$

Now we will present a necessary condition for approximately efficient elements of (P) applying Theorem 5 and the additional statements from Remark 1 and 2 in Section 3.

Theorem 6 *Under the assumptions given above for any $\epsilon > 0$ and any approximately efficient element*

$$F(x^\circ) \in Eff(F[S], B_{\epsilon k^\circ})$$

there exists an element $x_\epsilon \in X$ and a linear continuous mapping $M_\epsilon : X \longrightarrow Y$ with

$$F(x_\epsilon) \in Eff(F[S], K_{\epsilon k^\circ}),$$

$$\| x_\epsilon - x^\circ \|_X \leq \sqrt{\epsilon},$$

$$M_\epsilon(x_\epsilon - a) = ||| x_\epsilon - a |||,$$

$$||| x ||| - M_\epsilon(x) \in K \quad \forall x \in X,$$

$$M_\epsilon(v) \in M(v) + K$$

for all linear continuous mappings M from X to Y with $M(x_\epsilon) = ||| x_\epsilon |||$ and $||| x ||| - M(x) \in K \quad \forall x \in X$ and

$$M_\epsilon(v) \notin -\sqrt{\epsilon} k^\circ - int \ K$$

for all feasible directions v at x_ϵ with respect to S having $\| v \|_X = 1$.

Proof: Theorem 5 together with Remark 1 and 2 in Section 3 implies the existence of an element x_ϵ with

$$F(x_\epsilon) \in Eff(F[S], K_{\epsilon k^\circ}),$$

$$\| x_\epsilon - x^\circ \|_X \leq \sqrt{\epsilon}$$

and

$$F_{\epsilon k^\circ}(x_\epsilon) \in Eff(F_{\epsilon k^\circ}[S], K). \tag{10}$$

Now, let be v any feasible direction at x_ϵ with respect to S having $\| v \|_X = 1$. If we put $u = x_\epsilon + tv$, $t > 0$ with $u \in S$ then (10) implies

$$F_{\epsilon k^\circ}(x_\epsilon + tv) \notin F_{\epsilon k^\circ}(x_\epsilon) - (K \setminus \{0\}).$$

This means that

$$F(x_\epsilon + tv) + \sqrt{\epsilon} k^\circ \| tv \|_X \notin F(x_\epsilon) - (K \setminus \{0\}).$$

Using $\| v \|_X = 1$ and the assumption that K is a cone we get

$$1/t \ (F(x_\epsilon + tv) - F(x_\epsilon)) \notin -\sqrt{\epsilon} k^\circ - (K \setminus \{0\}).$$

Now, under the given assumptions the directional derivative of F at x_ϵ exists (see Lemma 2.24 of Jahn [9]). So we can conclude from the definition of the directional derivative that

$$F'(x_\epsilon)(v) \notin -\sqrt{\epsilon} k^\circ - int \ K \tag{11}$$

256

for all feasible directions v at x_ϵ with respect to S having $\| v \|_X = 1$.

Further, under the assumption, that K is a closed convex cone in Y which is Daniell and has a weakly compact base, it follows from Theorem 2.27 of Jahn [9] that there exists a linear continuous mapping M_ϵ with

$$M_\epsilon(x_\epsilon - a) = ||| x_\epsilon - a |||$$

and

$$||| x ||| - M_\epsilon(x) \in K \quad \forall x \in X$$

such that

$$F'(x_\epsilon)(v) = M_\epsilon(v) \tag{12}$$

and

$$F'(x_\epsilon)(v) \in M(v) + K$$

for all linear continuous mappings $M : X \longrightarrow Y$ with $M(x_\epsilon) = ||| x_\epsilon |||$ and $||| x ||| - M(x) \in K$ for all $x \in X$.

From (11) and (12) it follows that

$$M_\epsilon(v) \notin -\sqrt{\epsilon} k^\circ - int\, K$$

for all feasible directions v at x_ϵ with respect to S having $\| v \|_X = 1$.   q.e.d.

Remark:

For the special case $\epsilon = 0$ we get in the last condition

$$M_\epsilon(v) \notin -int\, K$$

for all feasible directions v at x_ϵ with respect to S having $\| v \|_X = 1$. This is a well-known necessary condition for efficient elements of the vector optimization problem $Eff(F[S], K)$ (compare Jahn [9]).

References

[1] Bishop, E.; Phelps, R.R.: A Proof that every Banach Space is subreflexiv. Bull. AMS 67 (1961), 97-98.

[2] Bishop,E.; Phelps, R.R.: The Support Functionals of a Convex Set. In: Proc. Symp. Pure Math. VII, Convexity. AMS (1963),26-36.

[3] Brezis, H.; Browder, F.E.: A General Principle on Ordered Sets in Nonlinear Functional Analysis. Advances in Math. 21 (1976), 355-364.

[4] Ekeland, I: On the Variational Principle. JMAA 47(1974), 324-353.

[5] Ekeland, I.: Nonconvex minimization problems. Bull. Am. Math. Soc. 1(1979), 443-474.

[6] Gerth (Tammer), Chr.: Näherungslösungen in der Vektoroptimierung. Seminarberichte der Sektion Mathematik der Humboldt-Universität zu Berlin 90 (1987), 67-76.

[7] Gerth (Tammer), Chr., Weidner, P.: Nonconvex Separation Theorems and some Applications in Vector Optimization. J. Optim. Theory Appl. 67 (1990), 297-320.

[8] Helbig, S., Georgiev, Patewa, D.; Todorow, M.: ϵ- efficient elements. Lecture on the Conference "Mehrkritcrielle Entscheidung", Fehrenbach, Germany, (1992).

[9] Jahn, J.: Mathematical Vector Optimization in Partially Ordered Linear Spaces. Verl. Peter Lang, 1986.

[10] Khanh, P.Q.: On Caristi-Kirk's Theorem and Ekeland's Variational Principle for Pareto Extrema. Preprint 357 Inst. of Math., Pol. Ac. Sci.(1986).

[11] Kirk, W.; Caristi, J.: Mapping Theorems in Metric and Banach Spaces. Bull. Acad. Poll. Sci. 23 (1975), 891-894.

[12] Loridan, P.: ϵ-solutions in vector minimization problems. J. Optim. Theory Appl. 43 (1984), 265-276.

[13] Nemeth, A.B.: A Nonconvex Vector Minimization Problem. Nonlin. Anal., Theory, Methods and Applications 10(1986), 669-678.

[14] Staib, T.: On two Generalizations of Pareto Minimality. J. Optim. Theory Appl. 59 (1988), 289-306.

[15] Takahashi, W.: Existence Theorems Generalizing Fixed Point Theorems for Multivalued Mappings. In: Thera, M.A. and Baillon, J.B.: Fixed Point Theory and Applications. Longman, Notes in Mathematics Series 252 (1991).

[16] Tammer, Chr.: A Generalization of Ekeland's Variational Principle. Optimization 25 (1992), 129-141.

[17] Tammer, Chr.: Existence results and necessary conditions for ϵ-efficient elements. In: Brosowski, B., Ester, J., Helbig, S., Nehse, R.(Eds.), Multicriteria Decision -Proceedings of the 14th Meeting of the German Working Group "Mehrkriterielle Entscheidung", Lang Verlag, Frankfurt/Main, Bern,(1992),97-110.

[18] Valyi, I. Approximate Saddle-Point Theorems in Vector Optimization. J. Optim. Theory Appl. 55 (1985), 435-448.

Global optimization

On Global Search based on Global Optimality Conditions

Alexander Strekalovsky,
Mathematical Department,
Irkutsk State University,
Gagarin Avenue, 20,
Irkutsk, Russia, 664003.

Abstract

We consider two kinds of nonconvex problems: convex maximization and reverse-convex optimization. Using the new information about the problems in the form of Global Optimality Search Algorithms [1-5], we construct Global Search Algorithms and study their global convergence. Numerical experiments also presented here are rather promising especially for large dimension problems.

1 Introduction

Lets consider the following problems:

$$f(x) \to \max, x \in D; \tag{P}$$

$$g_0(x) \to \min, g_0(x) \geq 0, x \in S; \tag{PR}$$

where f, g are convex functions and g_0 is a continuous function over R^n. D and S may be given by equalities and inequalities.

It is known, for obtaining one global solution of (P) (for example), they use the following information:

a) there exists at least one extremum point of D, which is a solution of (P);

b) each solution of (P) is a stationary point;

c) the duality theory;

In addition to this information we propose to use Global Optimality Conditions (GOS), that use the level surface of convex function.

For example, if z is a solution of (P), then [1-5]

$$\partial f(y) \subset N(y/D) \quad \forall y : f(y) = f(z) \tag{E}$$

It's easy to see, that (E) generalizes Rockafellar's Condition [6]

$$\partial f(y) \subset N(z/D) \tag{1}$$

and is different from Hirriart-Urryty's result [7]

$$\partial_\varepsilon f(y) \subset N_\varepsilon(z/D) \quad \forall \varepsilon > 0$$

It's important to notice, (E) stands for analytic equivalent of obvious inclusion in (P)

$$D \subset \{x \mid f(x) \leq f(z)\}$$

that complete Convex Analysis results, concerning inclusions [6]. Also it's in order to underline, that the approach proposed here for building GO Methods, is quite different from the others, existing in Global Optimization [8], because it's based on GOC of (E)-type, related to known Optimality Conditions [1-8].

Consequently, our approach displays interior harmony expressed particularly by the close connection between the GOC theory and GO-Method Construction.

Numerical experiments presented in [5] and here are rather promising especially for large dimension problems.

2   General Concept of GO Algorithms for (P)

The crucial question is how to use GOC for building GO-Algorithms. In order to answer this question we introduce the following function:

$$\varphi(z) = \sup_{(x,y,y^*)} \{ < y^*, x - y > \mid x \in D, f(y) = f(z), y^* \in \partial f(y)\} \tag{2}$$

As $\varphi(v) \geq 0 \ \forall v \in D$ [5], (E) may be put in the form $\varphi(z) = 0$. Therefore, if $\{z^k\}$ is a maximizing sequence for (P), then [5]:

$$\lim_{k \to \infty} \varphi(z^k) = 0. \tag{E1}$$

Consequently, for building a maximizing sequence $\{z^k\}$, we have to satisfy the equality:

$$< y_k^*, z^{k+1} - y^k > \geq \ \varphi(z^k) - \varepsilon_k, \tag{3}$$

where $z^k, z^{k+1} \in D$, $f(y^k) = f(z^k)$, $y_k^* \in \partial f(y^k)$, $\varepsilon_k \geq 0$, $k = 0, 1, 2, \ldots \ \sum_1^\infty \varepsilon_k < \infty$.

Thus, the essence of GOA consist in maximizing function $\varphi(\cdot)$ on each step with the tolerance ε_k.

We will find the supremum in (2) approximately, dividing this problem into two parts the first being a local search.

For the sake of simplicity we will present our results only for the smooth case.

3   Local Search (L.S.)

We'll present now LS-Algorithms based on (1) in smooth case

$$< f'(z), x - z > \ \le 0 \ \ \forall x \in D. \tag{4}$$

In other words, z is a stationary point and belongs to solution's set of the linearised problem:

$$< f'(z), x > \ \to \max, \ x \in D. \tag{PL}$$

Therefore, for building LS-Algorithm we need the hypothesis:

(HL) : $\forall \delta > 0 \ \ \forall x \in D \ \ \forall y : f(y) = f(x)$, we can find $u = u(y) \in D$:

$$< f'(y), u > \ \ge \sup_x \{< f'(y), x >| \ x \in D\} - \delta.$$

Let $p^0 \in D, \ \delta_s > 0, \ s = 0, 1, 2, \ldots, \ \sum_0^\infty \delta_s < +\infty.$

If we known a point $p^s \in D, s = 0, 1, 2, \ldots$, we define $p^{s+1} \in D$ (according to (HL)), which satisfies the inequality:

$$< f'(p^s), p^{s+1} > \ \ge \sup_x \{< f'(p^s), x >| \ x \in D\} - \delta_s. \tag{5}$$

If ε is a given tolerance and

$$\left\| p^{s+1} - p^s \right\| \le \frac{\varepsilon}{2 \left\| f'(p^s) \right\|} \tag{6}$$

then we do stop.

When $\delta_s \le \varepsilon/2$, due to (5) - (6) we have:

$$\sup_x \{< f'(p^s), x - p^s >| \ x \in D\} \ \le < f'(p^s), p^{s+1} - p^s > +\delta_s \le$$
$$\le \| f'(p^s) \| \cdot \| p^{s+1} - p^s \| + \delta_s \le \varepsilon$$

Thus, the point p^s turns out ε-stationary. All above is correct, whenever

$$\| f'(p^0) \| \ge \xi > 0$$

It's not difficult to prove, that

$$\lim_{s \to \infty} [\sup_x < f'(p^s), x - p^s >] = 0. \tag{7}$$

Obviously, (7) means that $\{p^s\} \in D$ converges to a stationary point.

4 Global Search

Now we will see how to construct a GO-Algorithms using LS and condition (E.1).

In order to do it, we need to introduce the following hypothesis

(HU) : $\forall \delta > 0 \ \forall z \in D \ \forall u \in D$ we can find $w : f(w) = f(z)$, such that

$$< f'(w), u - w > +\delta \geq \sup_y \{< f'(y), u - y >| \ f(y) = f(z)\} \tag{8}$$

For example, for the quadratic $f(\cdot)$ the problem (8) can be solved analytically and exactly [5]. Then, aiming to solve the problem (2) we could unify the hypotheses (HL) and (HU) by the following definition.

Definition. We name the set $(\varepsilon > 0, \delta > 0, 2\delta \leq \varepsilon)$. $\mathcal{R}(z, \varepsilon, \delta) = \{v^1, \ldots, v^N \ | \ f(v^i) = f(z), i = 1, \ldots, N\}$ to be z-resolving, if from inequality

$$f(z) < \sup(f, D) - \varepsilon \tag{9}$$

it follows, that

$$\eta(z) \overset{\triangle}{=} < f'(w^j), u^j - wj > \overset{\triangle}{=} \max_{1 \leq i \leq N} < f'(w^j), u^j - w^j > \ > 0, \tag{10}$$

$$\eta(z) \geq \varphi(z) - \varepsilon, \tag{11}$$

where $(i = 1, \ldots, N)$ (according to (HL) and (HU))

$$< f'(v^i), u^i >\geq \sup_x \{< f'(v^i), x >| \ x \in D\} - \delta, \tag{12}$$

$$< f'(w^i), u^i - w^i > +\delta \geq \sup_y \{< f'(y), u^i - y >| \ f(y) = f(z)\} \tag{13}$$

Now it's natural to suppose, that

(HR) : $\forall \varepsilon > 0 \ \forall z \in D$, s.th. z is a stationary point of (P) (see [4]), there exist $\delta > 0, 2\delta \leq \varepsilon$, and a resolving set $\mathcal{R}(z, \varepsilon, \delta)$.

Then we are able to present *GS*-Algorithm, further called $\mathcal{R}$-algorithm. Let $x^k \in D, \{\varepsilon_k\}, \varepsilon_k > 0, \ \varepsilon_k \downarrow 0 \ (k \to \infty) \ \{\delta_k\}, \ 2\delta_k \leq \varepsilon_k, \ \delta_k > 0, k = 0, 1, 2, \ldots$ We'll describe this algorithm by the following steps.

1. z^k *is a* ε_k - stationary point obtained by LS-Algorithm, beginning from x^k;

2. For z^k (according (HR)) one can construct the resolving set: $\mathcal{R}_k = \mathcal{R}(z^k, \varepsilon_k, \delta_k) = \{v^1, \ldots, v^{N_k} \ | \ f(v^i) = f(z^k), i = 1, \ldots, N_k\}$

3. $\forall i = 1, \ldots, N_k$, according to (HL) one can find a point $u^i \in D$, s.th. (12) takes place with $\delta = \delta_k$

4. $\forall i = 1, \ldots, N_k$, according to (HU) one can find w^i: $f(w^i) = f(z^k)$, s.th. (13) takes place with $\delta = \delta_k$

5. We take the number

$$\eta_k \overset{\triangle}{=} < f'(w^j), u^j - w^j > \overset{\triangle}{=} \max_{1 \le i \le N} < f'(w^j), u^j - wj > \tag{14}$$

6. If $\eta_k > 0$, then we put $x^{k+1} := u^j, k := k + 1$, and go to step 1

7. If $\eta_k \le 0$ and $\varepsilon_k \le \xi$, where ξ is a given tolerance, then stop.

The global convergence of the algorithm has been proved, for example, for the case of quadratic $f(\cdot)$ [5].

5 Numerical experiments

Large number of numerical experiments is presented in [5]. Here we will consider one problem connected with communications network planning, hydraulic or sewage network planning, plant location problems, inventory and production planning (see [8] and the literature therein).

We take the problem in the form:

$$f(x) = \sum_{1}^{n} \alpha_i x_i^{\beta_i} \to \min, \tag{15}$$

$$x \in X = \{x \in R^n \mid x \ge 0, Ax = b\}, \tag{16}$$

where $\alpha_i > 0, 0 < \beta_i < 1, i = 1, \ldots, n$; A is (m×n) matrix, $m < n$, $a_{ij} = \{1\} \vee \{-1\} \vee \{0\}, b \in R_+^n$.

It's easy to see, that $f(\cdot)$ is separable, strictly concave over $int(R_+^n)$ and only concave over $int(R_+^n)$. In addition $f(x)$ is not differentiable, when $x_i = 0$ for at least one $i = 1, \ldots, n$.

For solving the problem of (P) type four following points are the most important:

a) the choice of an algorithm for the problem linearised (PL) (or (12));

b) solution of "the level problem" (8);

c) construction of the resolving set.

Due to the form of the feasible set (16) it's natural to choose the simplex method for a).

One can prove, that for b) in the case $\beta_i = \beta, 0 < \beta < 1, i = 1, \ldots, n$, the level problem (8) has exact solution in the form:

$$w = (f(z)/f(u))^{1/\beta}u, \tag{17}$$

The most difficult was the construction of the resolving set. We will describe it in the simplest case, when a stationary vertex $z^k \in X$ is of the form $z^k = (z_1, \ldots, z_m, 0, \ldots, 0)$, i.e. its basis is $\mathcal{B}(z^k) = \{1, 2, \ldots, m\}, m < n$. The resolving set contains two parts, each of them having own task, $\mathcal{R}_k = \mathcal{R}_1(z^k) \cup \mathcal{R}_2(z^k)$. $\mathcal{R}_1(z^k)$ is responsible for the nearest vertices relatively to z^k. More definitely, the points $v^p \in \mathcal{R}_1(z^k), p = m+1, \ldots, n$, are constructed by using the simplex table $\{\xi_{st}\}$ in the point z^k, $s = 1, \ldots, m, t = 1, \ldots, n$. Taking the separability of $f(\cdot)$ and the equality:

$$f(v^p) = f(z^k), p = m+1, \ldots, n; \tag{18}$$

into account, we put

$$v_s^p = z_s, s = 1, \ldots, m, s \neq r, p > m. \tag{19}$$

the index r is chosen according to the condition $0 < z_r/\xi_{rp} = \min_s\{z_s/\xi_{sp} \mid s = 1, \ldots, m, \xi_{sp} > 0\}$. Furthermore, we put

$$v_j^p = 0, j = m+1, \ldots, n, j \neq p \tag{20}$$

$$v_p^p = (\alpha_r/\alpha_p)^{1/\beta} z_r. \tag{21}$$

Thus, (21) guarantees the equality (18).

The second part $R_2(z^k)$ is responsible for the indexes $j > m$, which are not included in the basis of z^k. $\mathcal{R}_2(z^k)$ aims to construct the set of stationary points for which the basic sets cover the set

$$\{1, \ldots, n\} \backslash \mathcal{B}(z^k).$$

The numerical experiments were done for $\alpha_i = 1.22 \, l_i, \beta_i = 0.52, l_i$ is the length of the branch, $i = 1, \ldots, n$. The result of the problem solution are presented in the following table, where f_0 – initial value of $f(\cdot)$, f_* – global value of $f(\cdot)$, T-EC – the time (in sec.) of solution on EC-1061, T-PC – on PC/AT-286, .loc – the number of obtained local solution, and LP – the number of solved linear programming problems.

n	m	$-f_0$	$-f_*$	T-EC (sec)	T-PC (sec)	loc	LP
8	4	38188.59	31396.99	-	03.02	4	18
-	-	41412.21	31819.34	00.31	-	3	11
-	-	39513.17	31819.34	00.28	-	2	9
-	-	43980.42	31819.34	00.32	-	3	12
12	8	24175.99	24045.15	-	04.06	2	12
-	-	25184.13	24045.15	00.38	-	2	6
-	-	25473.27	24045.16	00.40	-	3	7
17	11	39500.11	38477.61	-	13.12	3	21
-	-	44808.30	38477.59	01.26	-	6	27
-	-	45090.07	38477.59	01.31	-	7	28
-	-	41468.27	38477.59	00.90	-	4	17
32	20	21732.49	20323.54	-	3min 52.13	9	68
-	-	22102.12	20323.66	08.34	-	11	65
-	-	22066.39	20323.66	07.98	-	9	62
-	-	21377.24	20323.66	08.27	-	10	64
64	58	173604.56	173494.13	-	8min 00.81	2	15
-	-	120187.13	118754.43	08.57	-	3	9
-	-	120136.34	118754.43	10.53	-	3	12
-	-	174755.07	173323.22	11.48	-	3	25
-	-	176329.41	173323.00	12.24	-	4	28

References

[1] Strekalovsky A. *On the problem of global extremum* // Soviet Math. Doklady of Science Academy, 35(1987), pp.194-198.

[2] Strekalovsky A. *On the problem of global extremum in nonconvex extremal problems* // Isvestia VUZ, seria mathematika, 1990, No 8, p.74-80.

[3] Strekalovsky A. *On the problem of global extremum condition in Nonconvex Optimization Problems* // Voprosy kibernetiki, Analysis of big systems, Russian Academy of Science, Scientific Soviet on Complex Cybernetics Problem, Moscow, 1992, p.178-197.

[4] Strekalovsky A. *On optimization problems over supplements of convex sets* // Cybernetics and system analysis, 1993, No 1, p. 113-126.

[5] Strekalovsky A. *On global maximum search of convex function over a feasible set* // Journal of numerical math. and math. phizik, 1993, No 3 (to appear)

[6] R.T. Rockafellar. *Convex Analysis*, Princton University Press (1970).

[7] Hiriart-Urruty J.B. *Necessary and sufficient conditions for global optimality* // From convex optimization to nonconvex optimization. "Nonsmooth optimization and related topics" (E.Clarke, V.Demyanov, F.Gianessi editors) Plenum, N.Y., 1989, p.219-239.

[8] Horst R., Tuy H., *Global Optimization. Deterministic Approach*, Springer-Verlag, 1990.

GENETIC SIMULATED ANNEALING
FOR FLOORPLAN DESIGN

Seiichi Koakutsu   Hironori Hirata
Department of Electrical and Electronics Engineering, Chiba University
1-33 Yayoi-cho, Inage-ku, Chiba-shi 263, Japan

Abstract

This paper proposes a genetic simulated annealing incorporating characteristics of genetic algorithms into simulated annealing and applies it to a floorplan
design of VLSI. The proposed method can effectively search wide state space
for an optimal solution using a parallel search starting from many initial points
and genetic operators among those paths. Computational experiments show that
this method is more powerful to obtain a better solution than the conventional
simulated annealing.

1   Introduction

Floorplan design is the first stage in the layout design of VLSI circuits. It is
usually formulated as the problem of arranging a given set of rectangular modules
on the plane to minimize both the area of the enclosing rectangle which should
contain all the modules and the total wire length between modules that should
be connected in the curcuit. At this stage of the layout design, the shapes and
dimensions of the modules are in general not fixed so that designer can manip-
ulate the module's aspect ratio. In this way, floorplan design can be seen as a
combinatorial optimization problem. Simulated annealing (SA) [1-3] and genetic
algorithms (GA) [4-6] are heuristic solution methods for the combinatorial optimization problems, such as floorplan design.

SA is one of the stochastic optimization methods. The annealing means a
process by first heating a solid to its melting point and then slowly cooling the
material at a natural rate. Simulating the "annealing" of a solid into its globally
lowest energy state gives a global minimum. For the combinatorial problems, SA
can obtain an optimum or near optimum solution without being trapped in local

minima by using the stochastic optimization. Theoretically it has been proved that the global minimum of a cost function can always be reached as long as the decrease in temperature is slow enough [7]. For practical problems, however, the global minimum is barely obtained because SA searches the state space of the cost function starting from only one initial point, and its search cannot cover the large region of the state space within a finite computational time.

GA was proposed to apply an evolutionary mechanism to the optimization problems. In GA, a "population" of solutions is first given as the initial points of search. In each generation, genetic opeators: "crossover" and "mutation", are applied to some solutions in order to change the states of the solution, and good solutions survive to the next generation by "survival of the fittest". In GA, each solution of the population evolves to a lower cost valued solution through many generations, and finally GA finds a good solution. In GA, the search can cover a large region of the state space effectively because a number of solutions are maintained and used to search the state space. However, GA cannot search local resions of the state space exhaustively because its crossover-based searching is rather suitable for searching the state space roughly and globally.

In this article, we propose a genetic simulated annealing (GSA) combining SA with GA. GSA aims at seaching wide state space effectively for an optimal solution using a parallel search from many initial points and genetic operators, such as crossover and selection, among those paths. We apply GSA to the floorplan design and compared our method with SA on a set of test examples. We observe the improvement of the average cost by 1.7% – 9.8% over that of SA with the same computation time. The result implies that the GSA's search covers a large region of the state space.

2   Genetic Simulated Annealing

In combinatorial optimization problems, a cost function has many local minima. It is difficlut to obtain a good approximation solution for such problems. In order to obtain an optimum solution, it is important to search wide state space effectively. In this article, we propose GSA incorporating characteristics of GA into SA. GSA aims at searching wide state space effectively using both the local search ability of stochastic optimization based on SA and the global search ability of genetic operators based on GA. GSA has the following four features.

I. Stochastic optimization
GSA uses a stochastic optimization technique. The stochastic optimization used here is the same as that used in SA. General deterministic iterative

improvement methods accept only those solutions that decrease cost value, so that the search fails in certain local optima. To avoid this, SA uses a stochastic optimization technique. In the stochastic optimization, the state transition by which the value of the cost function is increased is also accepted with a certain probability. GSA can aviod local minima using this procedure.

II. Multiple search paths

GSA uses multiple search paths, which corresponds to the population of GA. In GSA, a population of soultions are first given as the initial points of the search and the stochastic optimization is applied to each solution of the population. The search proceeds in parallel along a number of paths and visits many local minima simultaneously. As a result, GSA has a greater chance of obtaining an optimum solution than SA.

III. Selection of search paths

GSA uses a procedure of a selection of search paths, which corresponds to the "survival of the fittest" rule of GA. In this procedure, the solution which has a higher cost value than the average value of the population is selected and is replaced by the solution which has a lower cost value than the average value of the population. This means that the search paths which are not expected to reach good solutions are deleted and are replaced by more successful search paths so that the search paths are gradually concentrated in those regions which are expected to be near good solutions. As a result, GSA can search wide state space effectively for good solutions.

IV. Genetic crossover operators

GSA uses genetic crossover operators in order to generate new solutions. In the above selection of search paths procedure, new solutions are formed by merging selected two solutions: this is done with crossover operators which are the same as those used in GA. The aims of crossover operators are to bring in substructures of solutions from parents into offsprings and to construct a variety of solutions which are hard to be constructed by the operators used in SA. As a result, GSA can search a large region of the state space which cannot be searched by SA.

Fig.1 presents GSA algorithm. First a population $X = \{x_1, \cdots, x_i, \cdots, x_L\}$ of L random points is created as a set of initial solutions for search at 3. The stochastic optimization of each solution is accomplished by procedures 15 to 22. These procedures are the same as those of SA. At 15, one solution x_i is selected at random from the population X and at 16, a candidate solution x_{new} is generated by changing the state of x_i. At 17, cost difference Δf between x_i and x_{new} is

```
 1:  GSA_algorithm(L,N,M,Ts,Te)
 2:  {
 3:      initialize_population(X = { x₁,···,x_L });
 4:      for( T = Ts ; T > Te ; T *= α ) {
 5:        for( loop = 0 ; loop < M ; loop++){
 6:          if( loop % N == 0 ){
 7:            select_two(x_i, x_j ∈ X);
 8:            /* f(x_i) < f_av < f(x_j) */
 9:            x_cr = crossover(x_i, x_j ∈ X);
10:            if( f(x_cr) < f(x_i) )
11:              x_i = x_j = x_cr;
12:            else
13:              x_j = x_i;
14:          }
15:          select_one(x_k ∈ X);
16:          x_new = mutate(x_k);
17:          Δf = f(x_new) − f(x_k);
18:          r = rand( [0,1] );
19:          if(Δf < 0 or r < exp(−Δf/T))
20:            x_k = x_new;
21:          else
22:            ; /* no change */
23:        }
24:      }
25:  }
```

Fig.1 GSA algorithm.

calculated and at 18, x_{new} is accepted with the following probability:

$$\min\{1, \exp(-\Delta f/T)\} \tag{1}$$

where T is a temperature which permits control of the acceptance probability. The genetic selection of search paths is accomplished by procedures 6 to 14. At 7, two solutions x_i and x_j are selected from the population X, where x_j has higher cost value than the average cost f_{av} of the population X and x_i has lower cost value than f_{av}. At 9, x_{cr} is created by the genetic crossover opetators between x_i and x_j. If x_{cr} has the lowest cost value among $\{x_{cr}, x_i, x_j\}$, both x_i and x_j are replaced by x_{cr} at 11. Otherwise, x_j is replaced by x_i at 13.

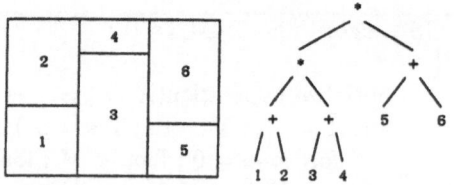

Fig.2. A slicing floorplan and its corresponding slicing tree.

3 Floorplan Design Problem

The floorplan design problem is to place a given set of rectangular modules on the plane and decide the aspect ratios of each modules so as to minimize both the total area of the floorplan and the total wire length between the modules. We use the following cost function:

$$F = A + \lambda W \tag{2}$$

where A is the area of the floorplan, W is the total wire length, and λ is a constant controlling the relative importance of A and W.

In this paper, we restrict our attention on slicing floorplans which can conveniently be represented by slicing trees or Polish expressions [3]. A slicing floorplan is a floorplan which is formed by recursively dividing a rectangle into two smaller rectangles by either a vertical cut or a horizontal cut. The hierarchy of such a slicing floorplan can be represented by a slicing tree. A slicing tree is a binary tree which has internal nodes labeled either * or +, corresponding to either a vertical cut or a horizontal cut, respectively, and leaf nodes labeled with a number corresponding to the module. Here we call $\{*, +\}$ operators, and other numbers operands. Fig.2 shows a slicing floorplan and its corresponding slicing tree. We can also repesent a slicing floorplan by its corresponding Polish expression. In such an expression, $AB+$ means A is placed below B, and $AB*$ means A is placed to the left of B. Here A and B can represent either leaf modules or sub-slicing trees.

Our algorithm requires two class of modification methods which generate new solution states from currently existing ones. One of them is the state transition based on SA, and we call it mutation operator. The other is the genetic crossover operator. Both of them is applied to the Polish expressions and generates resulting new floorplans. The mutation operators take a single individual which is selected at random from the population of the solutions and modify its solution state in a localized manner. So that the mutation operators tend to

	1	2	3	4	5	6	7	8	9	10	11	12	13	14	15
P1	2	6	8	+	*	7	5	*	+	4	1	*	3	*	+
P2	1	4	5	6	+	+	+	8	7	*	3	2	*	+	*
C1:															
P1	2	6	8			7	5			4	1		3		
P2				+	+			+	*			*		+	*
O1	2	6	8	+	+	7	5	+	*	4	1	*	3	+	*
C2:															
P1				+	*			*	+			*		*	+
P2	1	4	5			6	8			7	3		2		
O2	1	4	5	+	*	6	8	*	+	7	3	*	2	*	+
C3:															
P1				+	*			*	+	4	1	*	3	*	+
P2	5	6	8			7	2								
O3	5	6	8	+	*	7	2	*	+	4	1	*	3	*	+

Fig.3. Illustration of the crossover operators.

search the state space rather locally. We use the following three types of moves[3] as our mutation operators:

M1 It swaps two operands.

M2 It switches an operator from * to + (from + to *).

M3 It swaps an operator and a neighboring operand.

In our algorithm, one of those three operators is selected at random and applied. The crossover operators take two individuals as parents and generate an offsping. The purpose of the crossover opetators is to bring in building blocks from parents into offsping and construct new solution state which cannot be accomplished by the simple mutation operators. So that the crossover operators tend to search the state space rather globally because of producing dramatic jumps in the state space. We use the following three types of crossover operators[6]:

C1 First it copies the operands from P1 into the corresponding positions in O1. Then, it copies operators from P2 into the empty positions in O1 without changing their order.

C2 First it copies the operators from P1 into the corresponding positions in O2. Then, it copies operands from P2 into the empty positions in O2 without changing their order.

C3 First it copies the operators from P1 into the corresponding positions in O3. Then, an operator is selected at random from P1. The operands in the subtree rooted that operator are copied from P1 into the corresponding positions in

274

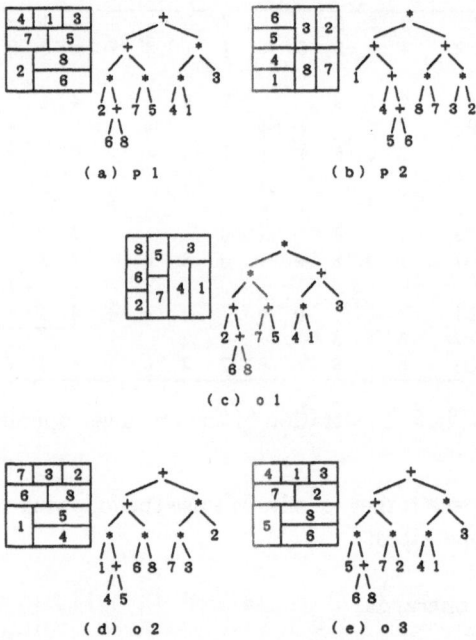

Fig.4. Resulting slicing trees and floorplans by the crossover operators.

O3. The remaining operands are copied from P2 into the empty positions in O3 without changing their order.

Where {P1,P2} represent parents and {O1,O2,O3} represent offsprings. Fig.3 and 4 show the illustration of the crossover operators. In our algorithm, one of those three operators is selected and applied.

4   Computational Experiments

We applied GSA and SA to the following three kinds of floorplan problem instances in our experiments. The first one is simple and its optimal floorplan is known. It has 16 modules which are fixed square of unit area and have wires connecting to its horizontal and vertical neighbors. Therefore, the optimal area is 16, the wire length is 24, and the total cost is $16 + 24 = 40$ with $\lambda = 1$. The second and third one are more realistic ones containing modules whose aspect ratio is not fixed. The second one has 16 modules and 25 wires. The third one has 20 modules and 31 wires. Both of them has a total module area of 100, and has modules whose

Table 1: Comparison of performance.

	Problem 1		Problem 2		Problem 3	
	SA	GSA	SA	GSA	SA	GSA
Chip area	18.300	17.020	105.152	104.894	106.253	106.625
Wire length	31.215	27.655	82.557	79.037	96.470	92.615
Total cost	49.515	44.675	187.709	183.928	202.722	199.240
Best	40.000	40.000	181.250	174.000	187.625	185.375
Worst	59.000	52.500	197.250	191.375	217.812	210.875
Opt.found[%]	15	52	—	—	—	—
Time[sec.]	354.78	363.12	801.66	813.66	1,167.3	1,229.9

aspect ratios can be changed ranging from 0.5 to 2.0. We used the following set of parameters: $\lambda = 1.0, Ts = 500, Te = 0.1, \alpha = 0.9, L = 50$ and $N = 250$, where λ is the weight for wire length in the cost function, Ts is the initial temperature, Te is the final temperature, α is the reduction ratio for the temperature, L is the population size and N is the selection frequency. The inner roop number M for the first, second and third problem instances were $3,000$, $4,000$ and $5,000$, respectively. Our floorplanning programs had been implemented in C in a UNIX environment on a SPARC station 1.

Fig.6, 7 and Table 1 show the results of experiments. GSA and SA were both run 100 times from different initial floorplans on the above promblem instances and the average costs of corresponding solutions are shown in Table 1. The final solutions of GSA are the best ones among the population of solutions. To be fair in our comparisons, the total number of generated new solutions was the same for both GSA and SA. Fig.6 shows the frequency distribution of the results, and Fig.7 shows the best overall solution found by GSA. The results of experiments in Table 1 show that GSA can find the optimal solution with higher probability than SA for the first problem instance. We also see from Fig.6 that GSA can find low cost valued solutions with higher probability than SA for the second and third problem instances. For all problem instances GSA is better than SA in terms of both the average cost of the solutions found and the best-found solutions.

5 Conclusions

This paper proposed GSA combining GA with SA and applied it to the floor-plan design problem of VLSI. GSA aims at searching wide state space effectively using both the local search ability of stochastic optimization based on SA and the global search ability of genetic operators based on GA. The result of the proposed method has been compared with that of SA. We observed the improvement of the average cost by 1.7% – 9.8% over that of SA with same computaion time. The result implies that the GSA's search covers a large region of state space. Theoretial investigations into its convergence feature and searching ability are left for further study.

Acknowledgement. We wish to thank Prof. T. Kurata for his encouragement and helpful discussions.

References

[1] S. Kirkpatrick, C. D. Gelatt, Jr. and M. P. Vecchi, "Optimization by Simulated Annealing," *Science*, 220, pp.671-680, May 1983.

[2] C. Sechen and A. Sangiovanni-Vincentelli, "TimberWolf 3.2 : A new standard cell placement and global routing package," *Proc. 23rd Design Automation Conf.*, pp.432-439, June 1986.

[3] D. F. Wong and C. L. Liu, "A new algorithm for floorplan design," *Proc. 23rd Design Automation Conf.*, pp.101-107, June 1986.

[4] J. H. Holland, *Adaptation in Natural and Artificial Systems*. Ann Arbor, MI:University of Michigan Press (1975).

[5] J. P. Cohoon and W. D. Paris, "Genetic placement," *IEEE trans. Computer-Aided Design*, vol.CAD-6, no.6, pp.956-964, November 1987.

[6] J. P. Cohoon, S. U. Hegde, W. N. Martin and D. S. Richards, "Distributed Genetic Algorithms for the Floorplan Design Problem," *IEEE trans. Computer-Aided Design*, Vol.CAD-10, No.4, pp.483-492, 1991.

[7] N. Metropolis, A. W. Rosenbluth, M. N. Rosenbluth, A. H. Teller and E. Teller, "Equation of State Calculations by Fast Computing Machines," *J. of Chemical Physics*, Vol.21, No.6, pp.1087-1092, 1953.

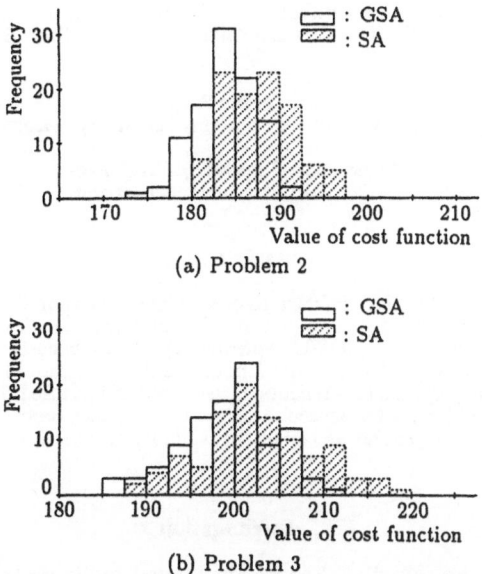

Fig.5. Frequency distribution of results.

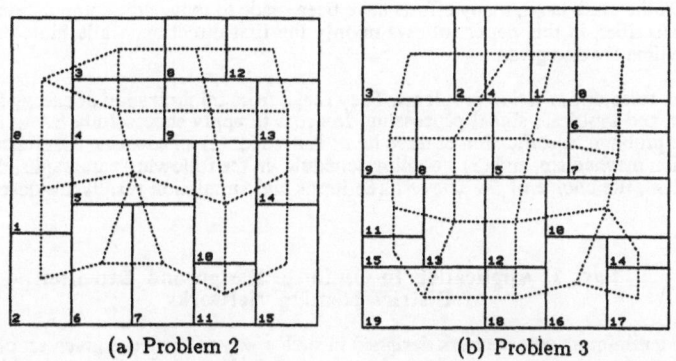

Fig.6. Examples of final floorplan.

Applications of Simulated Annealing to District Heating Network Design and Extension, to CMOS Circuits Sizing and to Filter Bank Design

Wei Li
Signal Processing Laboratory

Yi-bin Peng
Energy Systems Laboratory

Wen-hua Dai
Logic Systems Laboratory

Swiss Federal Institute of Technology
CH-1015 Lausanne, Switzerland

ABSTRACT

This paper describes the new applications of simulated annealing (SA) algorithm to three different domains: (1) design and modification of district heating pipe-networks, (2) sizing of the branch based CMOS(Complementary Metal-Oxide-Semiconductor) digital circuits, and (3) design of finite wordlength, parallel-structured filter banks. In each application, the appropriate cost function is described and justified. The modifications of the basic SA algorithm are introduced to adapt to each studied problem. Results show that the SA algorithm is general and gives satisfactory results in difficult optimization environment.

Introduction

The concept of simulated annealing (SA) was first introduced in combinatorial optimization by Kirkpatrick [1]. The SA is a stochastic optimization algorithm. It can avoids local minima by allowing up-hill moves. Asymptotic convergence of the algorithm was shown by Hajek [2]. The SA algorithm was generalized to minimize functions of continuous variables in [3]. Insights in the choices and significance of important parameters such as initial temperature, final temperature and scale were given in [4]. In recent years, the researches concentrate on two major directions. On one hand, the applications of SA methods have been spreading to new engineering optimization problems. On the other hand, many efforts have been made to reduce the computational complexity. The work described in this paper follows mainly the first direction, while hints on reducing the computation time are also given.

We deal with three optimization problems. They range from minimization of energy systems cost to VLSI design and multirate signal processing. In order to apply successfully SA to an engineering optimization problem, specification of three items is needed: (1) an adequate cost function definition, (2) a transition mechanism, and (3) a cooling schedule. In the following paragraphs, the formulation of the problem, the choice of the above three items and simulation results are described for each application.

Part 1. Application to Optimize Design and Extension of District Heating Networks

Determining a minimum cost network designed in such a way as to meet a given set of specifications (prescribed flow requirements, technological limits of each element of the system, achieving a quality of service, etc. ...), is a fundamental class of problems which arise in a wide variety of contexts of applications. Those applications may relate to transportation science, engineering distribution systems, energy networks, water distribution networks etc. A numerical method based on simulated annealing is developed and applied to resolve problems concerning the optimization of District Heating (DH) networks. This is in order to numerically determine the optimum values of a series of discrete variables, in case of initial design and/or on-line modification and extension of DH networks.

A DH system delivers the thermal energy produced in one or several centralized heat generation plant(s) to a relatively large number of geographically scattered heat consumers. This system is essentially composed of three principal elements: heat generation plant(s), distribution network composed of pairs of supply and return pipes, heat consumers. The hot water or vapor is circulated in the pipes by electric pump(s) so as to transfer the required thermal energy. At conditions of normal operation, each element is characterized by a certain number of parameters which should be maintained within their fixed technological limits. The relevant compatibility nature of these limits will give the maximum supply capacity of the studied DH network.

From an initial configuration, the distribution network develops generally step by step, this way to satisfy the growing number of heat consumers. The present study is limited to the heat-pipe networks; this means the problem of the production capacity is not treated. The optimization problem of interest is pointing to the total sum of Present Worth of Actual Cost (PW of AC) of the network expansion over a given time period, as resulting from the needed replacement of some existing heat-pipes or the new pipes for extension purpose.

The figure below gives a simplified illustration of the DH network development context. Heat-pipes of a DH network can be classified into two different categories : primary heat-pipe (also called transport heat-pipe) and secondary heat-pipe (or derivation heat-pipe).

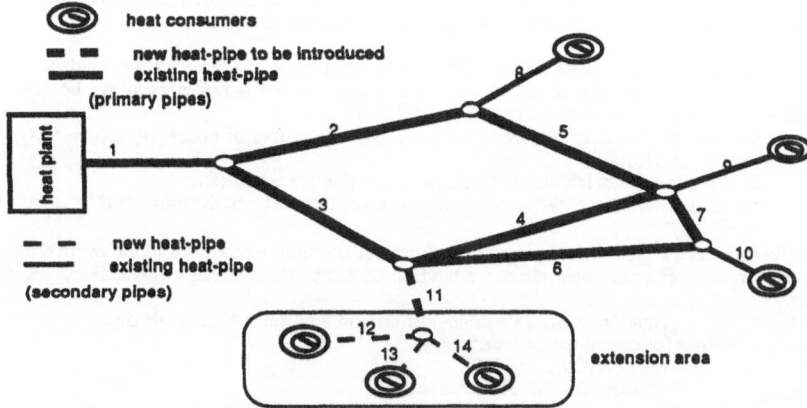

Figure 1.   Schematic representation of a simple DH network in extension

The calculation of the total PW of AC can be divided into the following terms [10] :
-    PW of AC of investment costs K_{in}
-    PW of AC of the residual values of investment K_r
-    PW of AC of operating costs (pumping costs K_p and thermal loss costs K_h)

The best network expansion strategy obeys the criterion of minimum global sum of all the PW of AC, that is to say :
$$\text{Minimize } C\ (x, d, l) \quad = \quad K_{in}(n)\ -\ K_r(n)\ +\ K_p(n)\ +\ K_h(n)$$
where the C (x, d, l) represents the objective function; and x, d, l are the free variables to be optimized :

{x ∈ X}   :   indices of different development scenarios;

{l ∈ L}   :   vector of possible lengths for all the concerned pipelines;

{d ∈ D}   :   vector of possible diameters for all the concerned pipelines;

n   :   given time period of analysis (in years), here a period of 10 years is assumed.

In the application example presented in this paper, a given extension scenario and the pre-determined lengths for those concerned heat-pipes are introduced and are considered as known values; consequently, only the diameters of the heat-pipes are to be optimized. Our study consists in finding out a series of optimum diameter values for the heat-pipes, with the aim to minimize the global economical expenditure in context of the DH network design and extension.

The proposed optimization method is focused on the research of diameter values of a certain number of pipes. These involved pipes may represent either some existent pipes to be modified or the ones to be newly introduced. The optimization algorithm consists in forcing the evolution of the highly combinatorial system, by accepting or rejecting a large number of randomly generated elementary transformations due to diameter values; along the procedure a significantly defined cost value of the system (the objective function value) is continuously modified. The elementary transformation is guided in the way to simulate the evolution of a physical system towards its thermodynamical equilibrium in relation to temperature. The successive diminution of the process control parameter (i.e. the simulated temperature) leads the system to approach the global minimum of the objective function.

For DH heat-pipe networks, the diameters for heat-pipes can only take nominal discrete values corresponding to available pipes on the market. The range of eligible diameters is generally limited by practical engineering considerations, particularly by technological limits of the pumping system, topographical characteristics and amounts of the heat demands. For the purpose of verification, two types of optimization process are simulated [3] : one is to consider the diameter values as discrete variables, the other is to consider them as continuous variables; a marginal comparison between the results of those two processes is carried out, this way to validate the results.

Theoretically, the convergence of the algorithm is assured when the temperature tends infinitively towards zero. Apart from the initial temperature, three other parameters must be taken into consideration to control the numerical simulation process:

- the criterion for temperature decrement, i.e. the number of necessary moves (N_t) at each temperature level;
- the decrement rate leading to the diminution of temperature (R_t);
- the stop criterion, i.e. the number of necessary temperature decrements (N_{stop}).

These values do have a great influence on the numerical simulation process and the needed computer time to convergence. For our application, we have fixed these values through calculational experience as follows :

N_t	$\geq$	(number of free variables)*(maximum number of possible discrete values for one of the free variables)
R_t		between 0.85 and 0.90
N_{stop}		between 20 and 50 decrements

• *Adaptive margins for random moves in the simulation process*

The margins of random moves for each discrete free variable can be dynamically adapted to the numerical simulation process, i.e. their limits are steadily reduced to maintain the acceptance rate of moves greater than a prefixed small percentage. This is an optional possibility to accelerate the convergence.

• *Test of convergence*

Beginning with the initial temperature, we carry out several SA processes, introducing each time a different initial configuration and different "cooling" parameters as mentioned above. If the solutions obtained remain identical or quasi-identical (no significant differences), the simulation process and the results are assumed to be acceptable; in the opposite case, the simulation process and the results are certainly false, if that, the values of some or all the SA parameters have to be re-adjusted in order to correct the simulation process.

• *Results*

This method has been used to determine the optimum modifications to be introduced to the DH network of Lausanne (Switzerland). A solution for the modifications to the DH network of interest was originally projected by the operating utility based on operating experience. A so-called optimum solution was found out by the proposed method; this solution is, of course, based on some hypotheses and the assumed economical conditions which should be considered as factors of uncertainty. A comparison of the above mentioned projected solution with the one found using the proposed optimization algorithm (for a given economical context) :

Table 1. Comparison between the projected solution and the calculated optimum solution

PW of AC (SFr. 1990)	projected solution	"optimum" solution
of investment	1,065,000	970,000
of residual investment	-720,000	-685,400
of pumping energy	25,000	20,000
of thermal losses	210,000	187,000
total sum	580,000	491,600
difference (≈15%)		(-) 88,400

The computer time (on IBM-PC 386) required for the above application case, as an example, is about 6 hours, which can be considered as acceptable.

Part 2. Application to Sizing the CMOS Digital Circuits

Optimal floor plan and minimal channel routine of the VLSI (Very Large Scale Integration) circuit are two typical problems successfully solved by SA methods [6]. Other complex problems existed in VLSI design domain are rarely attacked by the methods since the cost function definition of the problems is difficult. In this section, SA algorithm is used for CMOS digital circuits optimal sizing. It is a classical but still interesting problem in the field.

The CMOS digital circuit sizing can be described as a problem of geometrical widths distribution on all of transistors inside the circuit. Signal propagation delays of the sized circuit will be greatly reduced at the price of some area increase. A commonly used criterion for measuring circuit sizing quality is $A*T$. A is the chip area of the circuit and T is the worst case delay of the circuit. Minimizing the $A*T$ of a sized circuit is the objective of above described optimization problem. Generally, the sizing tasks can be carried out at three different level, i.e., gate-level, transistor-level and branch-level [7] (see Fig. 2).

During the last decade, most sizing efforts were concentrated on gate-level in which the number of variables for sizing are effectively reduced in comparison with transistor-level sizing. The sizing problem was formulated as a critical-path analysis problem [8] and solved by analytical nonlinear optimization techniques. Two benefits of critical-path based sizing are obvious: the first is the low computational complexity and the second is the possibility to add delay constraints to the sizing process. However, there still exist several drawbacks: firstly, the number of the paths will increase sharply with increased circuit size, thus the sizing time can not anymore be kept linear for large scale circuits. Secondly, the false critical path, which can never occur under real operating conditions, tend to hide the real critical paths in the circuits; Thirdly, the circuit area can not be effectively minimized since the sizing is done at gate-level, finer consideration of saving area at transistor-level can not be done.

In this paper, a novel method is proposed in which the circuit sizing is done at branch-level, and formulated as a NP-complete problem. The branches' widths serve as a group of independent variables and their geometric changes are continuous between the user defined minimum and maximum size. The simulated annealing algorithm proposed by Corana [3] is applied to find the global minimum $A*T$ product for a sizing circuit. If we define respectively $w0(i)$ and $w(i)$ as the initial and current size of branch i, and denote by k a user-defined area control factor and by K the total number of branches inside the circuit, then $A = \sum_K w0(i) + k*[\sum_K w(i) - \sum_K w0(i)]$. The cost function is defined as below:

$$Cost = [\sum_{i \in nodes} \sum_{j \in Brs(i)} weight(j)*T_d[Br(j)]]*A.$$

where $Brs(i)$ is the set of branches with output node i, $weight(j)$ is the sizing weight of branch j and $T_d[Br(j)]$ is the delay time of branch j.

Two special features are reflected in the cost function definition: firstly, it is a path-independent definition. Therefore, problems associated with path-dependent sizing are avoided; secondly, the time complexity of the cost function is $O(K)$. This is a very favorable feature for large circuits.

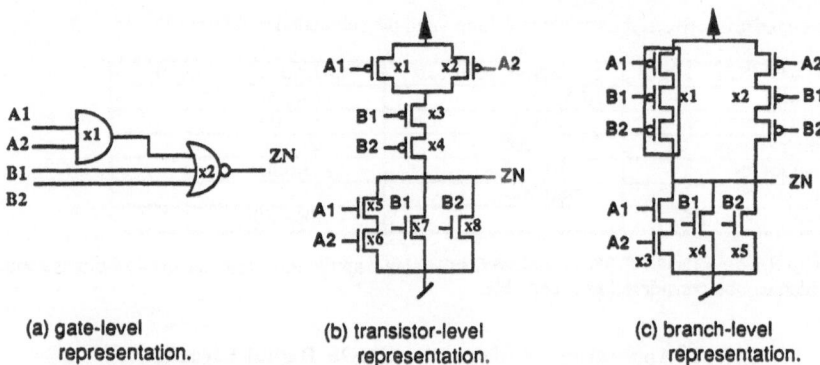

| (a) gate-level representation. | (b) transistor-level representation. | (c) branch-level representation. |

Figure 2. Three sizing levels of a CMOS digital circuit, the number of variables for sizing is 2, 8, 5 respectively.

The employed algorithm is a homogeneous SA algorithm [6] with the following main features:
  (1) The control parameter, the analog of temperature in a physical annealing process, is decreased at each step during the course of the algorithm;
  (2) Finite length Markov chain is performed at each fixed control parameter;
  (3) The generation step is given by the uniform distribution on the neighborhoods of current configuration, and the acceptance step is given by the Metropolis criterion.

The algorithm has been implemented in SASIMOS (Simulated Annealing Based Sizing of CMOS) tool. Experiments of sizing some benchmark circuits have been done. In all cases, high speed versions of the circuits were obtained with minimal area consumption. Fig. 3 shows a typical result for sizing a 8-bit adder, in which the three different sizing results are compared. The OPTIMOS is another analytical sizing tool which was previously developed in Logic Systems Laboratory [9].

In conclusion, SA algorithm applied to a path-independent cost function for sizing CMOS digital circuit is proposed. Furthermore, the sizing is done at branch-level representation of the circuit. Although the transistor-level representation of the circuit can be transferred to branch-level representation, how to size the circuit at transistor-level without changing its topology is still an open issue.

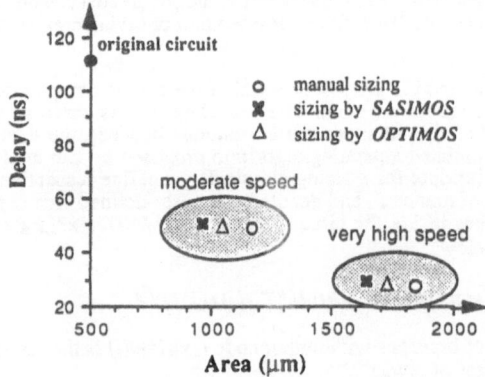

Figure 3. Optimization of a 8-bit adder

Part 3. Application to Design of Parallel Multiresolution Filter Banks with Finite Wordlength Coefficients

Filter banks have been widely used in speech and image compression. As they were first introduced for speech processing applications, their design does not take into account the statistics of images. Design of filter banks for image coding applications is investigated here. Images are sources which exhibit high non-stationarity. To employ the spatial and spectral non-uniformity of images, the filters should be localized in spatial domain as well as in spectral domain. Natural images typically contain a large DC component. When subband samples are coarsely quantized, significant aliasing will occur if the high frequency bands do not have zero response to DC component. Furthermore, the exact reconstruction is required at the output of the filter bank in absence of quantization.

A design method is proposed based on spatial domain analysis with cost function defined as a weighted sum of the following three items: (a) reconstruction error, (b) joint localization in spatial and spectral domains, and (c) strict confinement of DC to the lowest band. As the floating point arithmetic is costly both in computation and memory, filter coefficients with finite wordlength or even powers-of-two values are desired. Therefore, each filter coefficient is constrained to be finite wordlength. For a detailed description of the cost function, readers can refer to [11].

The proposed SA algorithm is an improved version of the algorithm reported in [3]. Let N_S be the number of cycles between step adjustments, N_T the number of step adjustments between temperature reduction, and r_T the temperature reduction factor. A temperature reduction occurs every $N_S * N_T$ cycles of moves along every direction, and after N_T step adjustments. Values of these parameters are set as suggested in [3]:
$N_S = 20$, $N_T = \max(100, 5*\text{Vector\_dimension})$, $r_T = 0.85$.

New points are generated by applying random moves to a certain coefficient. Let ΔE be the energy change caused by a certain move; this move is accepted or rejected based on the Metropolis criterion:

if $\Delta E <= 0$, accept the move;

else if $\Delta E > 0$, accept the move with probability $P = e^{-\Delta E/T}$.

The new trial component z of a filter coefficient vector in the L dimensional space is generated by

$$z_{new} = Q[z_{old} + v*r]$$

where $Q[z]$ is the rounded value of z to the nearest discrete value, r is a random number and v is the current scale for the component.

Traditionally, random numbers uniformly distributed between -1 and 1 are used for perturbation. Bilbro [12] proposed to use Gaussian distribution instead of uniform distribution for random moves in image restoration applications, leading to increased design speed. Inspired by this idea, we propose a two-modes SA algorithm. A typical annealing curve is shown in Fig 5. The whole annealing process can be divided into three stages. In the first one, at very high temperatures, the average energy is maintained to a maximum average value. The transition to the second stage occurs when the annealing curve reaches a given high temperature T_1. In this stage, the average energy decreases dramatically with the decrease of the temperature. The annealing curve makes another transition at a very low temperature T_2, which initiates the third stage. In this third stage, practically no more up-hill move will be accepted, and the average energy of the system will decrease slowly and tend to the global minimum. In more details, during the first stage, the system can move randomly through all of its states, and the probability to be in a state is equal for each state. So, random numbers uniformly distributed in the interval [-1, 1] are used for random moves. In the second and third stages, the local topology of the system becomes much more important. Good points are more likely to lie in the neighborhood of current points. This consideration leads to adoption of Gaussian distributed random numbers with unit variance for random moves in the second and third stages. This scheme is extremely efficient in the third stage, because the new trial points

284

will be more concentrated around the current point. The difference between both random number generators at the same scale is shown in Fig. 6. Note that the change from one mode to the other is automatic, dictated by the acceptance ratio at each temperature.

As a design example, Figure 4 shows the amplitude frequency response of a 3-band filter bank with 15 taps for each analysis filter. All the coefficients are represented in 8 bits. Figure 4 gives the annealing curve for this design example.

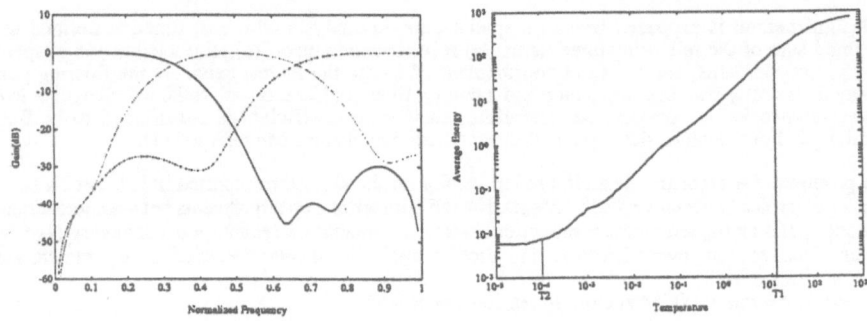

Fig. 4. Frequency-magnitude responses
of the analysis bank.

Fig. 5. Annealing curve.

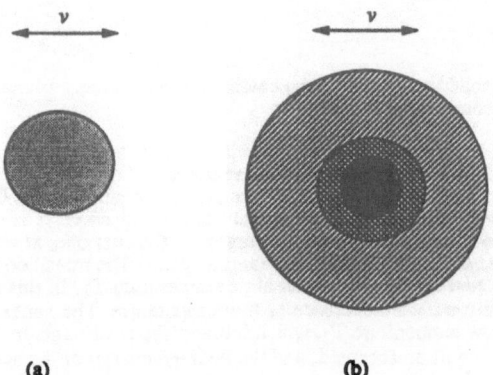

Fig. 6: (a) Perturbation is uniform; (b) Perturbation is Gaussian. ν represents the current scale.

Conclusion

Three different optimization problems are described and solved by using simulated annealing algorithm. The general formalism of simulated annealing makes it easy to adapt to different problems. The experiences show that SA performs well in difficult optimization environment, e.g., high dimension, discrete variables, conflicting constraints. The high computational complexity is still the major obstacle to fit with on-line applications.

Acknowledgment

This paper is the result of fruitful collaboration among three different laboratories at Swiss Federal Institute of Technology (Lausanne). The authors are grateful to Prof. M. Kunt (Signal Processing Laboratory), Prof. G. Sarlos (Energy Systems Laboratory) and Prof. J. Zahnd (Logic Systems Laboratory) for their permission and support of the reported work. Special thanks go to Dr. P.-A. Haldi (Energy Systems Laboratory), C. Piguet, J.-M. Masgonty, Ph. Mosch and C. Arm (CSEM) for the valuable discussions and their helpful suggestions.

References

[1] S. Kirkpatrick, C.D. Gelatt, Jr., M.P. Vecchi, "Optimization by simulated annealing," *Science* Vol. 220, 1983, pp.671-680.
[2] B. Hajek, "Cooling schedules for optimal annealing," *Mathematics of Operations Research*, Vol. 13, 1988, pp. 311-329.
[3] A. Corana, M. marchesi, C. Martini, and S. Ridella, "Minimizing multimodal functions of continuous variables with the 'simulated annealing' algorithm," *ACM Trans. on Mathematical Software*, Vol. 13, No. 3, 1987, pp. 263-280.
[4] S.R. White, "Concepts of scale in simulated annealing," *Proc. of the IEEE International Conference on Computer Design, ICCD84*, New York 1984, pp. 646-651.
[5] Lance A. Glasser *et al.*, "Delay and power optimization in VLSI circuits," *Proc. of 21st IEEE DAC*, 1984, paper 32.1.
[6] P.J.M. Van Laavhoven *et al.*, "Simulated Annealing: Theory and Applications," *D. Reidel Publishing Company*, pp. 7-82 , 1987.
[7] C. Piguet *et al.*, "Conception des circuits ASIC neumériques CMOS," Dunod, Paris,1990.
[8] Lucas S. Heusler, "Transistor Sizing for Timing Optimization of Combinational Digital CMOS Circuits," Σ Hartung-Gorre Verlag, 1990.
[9] S. Zaker and J.Zahnd, "OPTIMOS: A Branch-Level Digital Circuit Optimizer," *EDAC-EUROASIC 1993*, Paris.
[10] Y.-B. Peng *et al.*, "A numerical method to optimize the modifications to district heating networks in extension," *Proc. of Unichal'93*, Paris, 1993.
[11] W. Li *et al.*, "Design of perfect reconstruction, finite wordlength, and parallel-structured filter banks," in *proc. Eusipco'92*, pp. 229-232, Brussels, 1992.
[12] G.L. Bilbro, "General method for accelerating simulated annealing algorithmes for bayesian image restoration.

Game theory

Game Theory

GALE'S FEASIBILITY THEOREM ON NETWORK FLOWS
AND A BARGAINING SET FOR COOPERATIVE TU GAMES

Irinel Dragan
University of Texas at Arlington
Department of Mathematics
Arlington, Texas 76019-0408, U.S.A.

The bargaining set theory of cooperative TU games has been initiated by R. J. Aumann and M. Maschler in [2]; a good account of the state-of-the-art in the field has been given by M. Maschler in [8]. A new concept of bargaining set for TU games will be introduced in the present paper and a combinatorial characterization of the elements belonging to this set will be given, based upon a result due to D. Gale ([6]), applied to a network associated with the bargaining procedure. The case of three person games is fully discussed, to allow the comparison with Aumann/Maschler results given in [2].

1. The bargaining model.

Consider a cooperative TU game in coalitional form, that is a pair $G = (N, v)$, where N is a finite set of players $|N| = n$ and $v : 2^N \to R$ is the so-called characteristic function, subject to $v(\phi) = 0$. For any coalition S, $S \subseteq N$, $S \neq \phi$, the number $v(S)$ is the worth of S. A coalition structure is any partition of N; let

$$F_\rho = \{x : x \in R^n : x(S) = v(S), \forall S \in \rho\} \tag{1.1}$$

be the set of admissible payoffs for the coalition structure ρ, where $x(S)$ is the sum of x_i, $\forall i \in S$. Denote by F the set of all admissible payoffs which is the union of all F_ρ for all coalition structures ρ. The core of G is defined by means of the excess function $e(S, x) = v(S) - x(S)$, $\forall S \subseteq N$, $x \in F$, as

$$C(G) = \{x \in F : e(S, x) \leq 0, \forall S \subseteq N\}, \tag{1.2}$$

(see Aumann/Dreze, [1]). For some coalition structure ρ, an $x \in F_\rho$ which belongs to the core is considered stable, because for any coalition S we have $x(S) \geq v(S)$, i.e., in x, the total win of players in S is at least equal to the worth of S. However, if $x \notin C(G)$ then it may happen that x is still stable depending on the meaning given to the word stable. We shall introduce a stability principle somehow similar to the one due to Aumann/Maschler, but having a strong combinatorial character.

A partial coalition structure of N is a set of pairwise disjoint and non-empty coalitions that may cover N or not.

A bargaining proposal (b.p.) against $x \in F_\rho$, $x \notin C(G)$, is a partial coalition structure $\mathcal{S} = (T_1, ..., T_q)$ subject to

$$e(T_h, x) > 0, \quad h = 1, ..., q. \tag{1.3}$$

Let $K_h \subset T_h$, $K_h \neq \phi$, $h = 1, ..., q$, be a family of coalitions having members in all blocks of $\mathcal{S}$ and K be the union of all K_h. Obviously, we have $T_0 \cap K = \phi$, where $T_0 = N - (T_1 \cup ... \cup T_q)$.

A <u>bargaining distribution for $\mathfrak{S}$ initiated by K</u> is any $y \in R^{n-t_0}$, $t_0 = |T_0|$, such that

$$y(T_h) = v(T_h), \ h=1,...,q, \qquad y_i > x_i \forall i \in K, \qquad y_i \geq x_i, \ \forall i \in N - T_0. \tag{1.4}$$

Now, the meaning of the concept of bargaining proposal is shown by:

<u>Lemma 1.1</u>: For a partial coalition structure $\mathfrak{S} = (T_1,...,T_q)$, there exists a bargaining distribution initiated by some coalition K with $T_h \cap K \neq \phi$ $h=1,...,q$, if and only if $\mathfrak{S}$ is a bargaining proposal.

<u>Proof</u>: Obviously, (1.3) follows from (1.4), using the definition of the excess function. Conversely, if (1.3) hold, then we can define $y_i = x_i + e(T_h, x)/|K_h|, \forall i \in K_h, h=1,...,q$, and $y_i = x_i$ otherwise, and (1.4) will hold.

In words, a coalition K could find a better payoff than x if and only if there is a partial coalition structure where K has members in all blocks and the excesses of these blocks are positive. In this case, K may initiate a bargaining proposal, namely that partial coalition structure.

Note that the pair $(y;\mathfrak{S})$ is similar to an Aumann/Maschler objection; the differences are: no rationality conditions are imposed to x in $F_\wp$, then K may not be a subset of a block in $\wp$, and $(y;\mathfrak{S})$ is not an objection against another group of players but it is an objection against $(x;\wp)$.

Note also that for a bargaining proposal $\mathfrak{S}$, there are a lot of bargaining distributions initiated by a given K, namely they are solutions of the linear system (1.4) or equivalently

$$\alpha(T_h) = e(T_h, x), \ h=1,...,q, \qquad \alpha_i > 0, \ \forall i \in K, \qquad \alpha_i \geq 0, \forall i \in N - T_0 \tag{1.5}$$

connected to (1.4) by $\alpha_i = y_i - x_i, \forall i \in N - T_0$.

Bargaining proposals and bargaining distributions have been considered in earlier papers of the author ([4], [5]); these terms have been used for avoiding any confusion with the Aumann/Maschler objections.

Now, if $x \in F_\wp$ does not belong to the core, then obviously there are bargaining proposals because at least one excess is positive. We introduce a new countering procedure which will allow us to define a new stability principle, even though there are coalitions able to improve their payoffs relative to x. The new assumption of our bargaining model is that before announcing their bargaining proposal the players in $N - T_0$ agree upon the following commitment: a group of players $\check{K} \subseteq N - T_0$ will be able to leave $\mathfrak{S}$ and join coalitions of another partial coalition structure only if $\check{K}$ contains players of all blocks of $\mathfrak{S}$, and for each T_h the players in $\check{K} \cap T_h$ will compensate T_h with $e(T_h, x)$ from their extra wins in their new coalitions. If a partial coalition structure $\mathcal{U}$ would offer to $\check{K}$ such extra wins, perhaps more, then we shall consider that $\mathcal{U}$ can counter $\mathfrak{S}$.

Consider a partial coalition structure $\mathcal{U} = (U_1,...,U_p)$, different of $\mathfrak{S}$, such that

$$(N - U_0) \cap T_h \neq \phi, \ h=1,...,q, \qquad U_0 = N - (U_1 \cup ... \cup U_p) \tag{1.6}$$

and

$$e(U_j, x) > 0, \ j=1,...,p. \tag{1.7}$$

Note that " different of $\mathfrak{S}$" means that no block in $\mathcal{U}$ is a block in $\mathfrak{S}$. Obviously, a family of coalitions $\tilde{K}_h \subseteq (N-U_0) \cap T_h$, $\tilde{K}_h \neq \phi$, $h=1,...,q$, whose union is $\tilde{K}$, may be willing to move to $\mathcal{U}$, if each group $\tilde{K}_h$ is able to satisfy the commitment of compensating T_h with $e(T_h, x)$. The bargaining model becomes clear by the following definitions:

A <u>compensatory bargaining counter distribution</u> (c.b.c.d.) for a partial coalition structure $\mathcal{U}$, subject to (1.6) and (1.7), against $\mathfrak{S}$, is any $z \in R^{n-u_0}$, $u_0 = |U_0|$, satisfying

$$\beta(U_j) = e(U_j, x), \ j=1,...,p, \qquad \beta_i \geq 0, \forall i \in N-U_0, \tag{1.8}$$

where $\beta_i = z_i - x_i$, $\forall i \in N-U_0$, and

$$\sum \beta_i \geq e(T_h, x), \ h=1,...,q, \tag{1.9}$$

where the sum is extended to all $i \in (N-U_0) \cap T_h$.

Clearly, (1.8) are conditions imposed to a payoff and (1.9) are saying that the former members of T_h could compensate all members of T_h with $e(T_h, x)$ from their extra wins in $\mathcal{U}$. Therefore, if there is such a $z \in R^{n-u_0}$, then $\mathcal{U}$ can counter $\mathfrak{S}$.

A <u>compensatory bargaining counter proposal</u> (c.b.c.p.) is any partial coalition structure $\mathcal{U}$ subject to (1.6) and (1.7) for which there exists a compensatory bargaining counter distribution against $\mathfrak{S}$.

We shall be able in the second section to characterize the compensatory bargaining counter proposals, i.e., those $\mathcal{U}$ subject to (1.6), (1.7) for which (1.8), (1.9) is consistent. This will be possible by using a feasibility theorem for network flows due to D. Gale. Now, we can introduce the stability principle:

An admissible payoff x for a coalition structure $\mathfrak{S}$ is <u>compensatory stable</u> (c-stable) if either $x \in C(G)$, or $x \notin C(G)$ and for each b.p. there is a c.b.c.p. against the b.p.

The <u>compensatory bargaining set</u> M_c of G is the set of c-stable payoffs.

2. Gale's feasibility theorem and the compensatory bargaining counter proposals.

Consider a bipartite graph $\Gamma = (U, T, E)$, where $U \cup T$ is the set of vertices, $U \cap T = \phi$, and E is the set of edges; if $U = (u_1,...,u_p)$ and $T = (t_1,...,t_q)$, then E contains only edges (u_j, t_h), $u_j \in U$, $t_h \in T$. Now, build a network $\mathfrak{R}(U,T)$ as follows: to Γ add all edges (u_j, t_h) which were not in E, define a capacity function by $c(u_j, t_h) = \infty$ if $(u_j, t_h) \in E$ and $c(u_j, t_h) = 0$ otherwise; a supply function is defined on the set of vertices, namely $\sigma(u_j) > 0$ is the supply available at u_j, $j=1,...,p$, and $\sigma(t_h) = -\delta(t_h) < 0$ is showing the demand $\delta(t_h)$ at t_h, $h=1,...,q$.

A flow $f = (f_{jh})$ in $\mathfrak{R}(U,T)$ is a function defined on the set of edges in the network and satisfying

$$\sum_h f_{jh} \leq \sigma(u_j), \ j=1,...,p, \qquad \sum_j f_{jh} \geq \delta(t_h), \ h=1,...,q, \tag{2.1}$$

$$0 \le f_{jh} \le c(u_j,\ t_h),\ j=1,...,p,\ h=1,...,q,$$

where f_{jh} is the flow on edge $(u_j,\ t_h)$.

It is used to say that a network with demands and supplies is feasible if there is a flow, i.e., the linear system (2.1) is consistent (see [6]). A feasibility theorem for such networks has been given by D. Gale in [7], for more general networks than our above network. If Gale's theorem is applied to our network $\Re(U,T)$, we get:

Theorem 2.1: The network $\Re(U,T)$ is feasible if and only if for each subset $X \subset U$ (including the empty set), we have

$$\sum_{u_j \in X} \sigma(u_j) \ge \sum_{t_h \in I(X)} \delta(t_h) \tag{2.2}$$

where $I(X)$ is the set of vertices in T adjacent to vertices in X in the original graph Γ.

Note that the network considered by Gale has no edges with infinite capacities, so that the above statement may not be considered as a valid application of Gale's theorem (see [6], Th. 1, pg. 38). In this case, a proof similar to Gale's proof could be given, namely: we should impose the condition that in the extended network the capacity of every cut which contains no edge of infinite capacity is at least equal to the capacity of the cut comprising all exit edges; by examining all cuts and eliminating the redundant conditions we end up with (2.2). Note also that algorithmically we should not check (2.2) to determine the feasibility, but we should rather find a maximum flow in the extended network and see whether all exit edges are saturated.

Let $\mathfrak{S}=(T_1,...,T_q)$ be a bargaining proposal against a payoff $x \in F_\rho$ which does not belong to the core; let $\mathcal{U}=(U_1,...,U_p)$ be a partial coalition structure subject to

$$(N-U_0) \cap T_h \ne \phi,\ h=1,...,q, \qquad (N-T_0) \cap U_j \ne \phi,\ j=1,...,p, \tag{2.3}$$

and

$$e(U_j,x)>0,\ j=1,...,p. \tag{2.4}$$

Note that beside (1.6) and (1.7) we have imposed to $\mathcal{U}$ the second group of conditions (2.3); we shall show later that we can confine ourselves to such partial coalition structures without any loss of generality.

For our pair of partial coalition structures ($\mathcal{U}$, $\mathfrak{S}$) we can associate a graph $\Gamma=(U,T,E)$ as follows: each block $U_j \in \mathcal{U}$ will be represented by a vertex $u_j \in U$ and each block $T_h \in \mathfrak{S}$ will be represented by a vertex $t_h \in T$; we take an edge $(u_j,\ t_h)$ in E if $U_j \cap T_h \ne \phi$. Note that the first conditions (2.3) are saying that each vertex t_h is connected to at least one vertex in $\mathcal{U}$ and the second group of conditions (2.3) are saying that each vertex u_j is connected to at least one vertex in T. Now, we build a network $\Re(U,T)$, by adding first edges as shown above and defining similarly their capacities and by taking: $\sigma(u_j)=e(U_j,x)$, $j=1,...,p$, and $\delta(t_h)=e(T_h,x)$, $h=1,...,q$. Now, in this network we have the result:

Theorem 2.2: Let $\mathfrak{S}$ be a b.p. and $\mathcal{U}$ be a partial coalition structure subject to (2.3) and (2.4); then $\mathcal{U}$ is a c.b.c.p. against $\mathfrak{S}$ if and only if the network $\Re(U,T)$ associated with ($\mathcal{U}$, $\mathfrak{S}$) is feasible.

Proof: Suppose that $\mathcal{U}$ is a c.b.c.p. against $\mathfrak{S}$, that is (1.8), (1.9) is consistent; let β_i, $\forall i \in N - U_0$, be a solution of this linear system. Define $f_{jh} = 0$, if $U_j \cap T_h = \phi$; define $f_{jh} = \sum \beta_i$, where the sum extends to all $i \in U_j \cap T_h$, if $U_j \cap T_h \neq \phi$. From this definition and (1.8), (1.9) we get

$$\sum_h f_{jh} = \sum_{i \in (N-T_0) \cap U_j} \beta_i \leq \sum_{i \in U_j} \beta_i = e(U_j, x), \, j = 1, \ldots, p,$$

$$\sum_j f_{jh} = \sum_{i \in (N-U_0) \cap T_h} \beta_i \geq e(T_h, x), \, h = 1, \ldots, q,$$

and if $c_{jh} = 0$, i.e., $U_j \cap T_h = \phi$, we have $f_{jh} = c_{jh} = 0$, where $c_{jh} = c(u_j, t_h)$. Hence f defined above satisfies (2.1), i.e., the network is feasible. Conversely, suppose that the network is feasible, i.e., (2.1) is consistent; let f_{jh}, $j = 1, \ldots, p$ and $h = 1, \ldots, q$, be a solution of this linear system. Define β_i, $i \in N - U_0$ as follows:

$$\beta_i = 0, \quad if \quad i \in N - U_0, \, i \notin T_0, \tag{2.5}$$

$$\beta_i = f_{jh} / |U_j \cap T_h| + d_j / |(N - T_0) \cap U_j| \quad if \quad i \in U_j \cap T_h \neq \phi,$$

where

$$d_j = e(U_j, x) - \sum_h f_{jh}, \, j = 1, \ldots, p. \tag{2.6}$$

As all flows and all differences d_j, $j = 1, \ldots, p$, are nonnegative by (2.1) with $\sigma(u_j) = e(U_j, x)$, $j = 1, \ldots, p$, and $\delta(t_h) = e(T_h, x)$, $h = 1, \ldots, q$, we have $\beta_i \geq 0$, $\forall i \in N - U_0$. For $U_j \cap T_h \neq \phi$ we get from (2.5) and (2.6) that for every U_j we have:

$$\sum_{i \in U_j} \beta_i = \sum_{h/U_j \cap T_h \neq \phi} \sum_{i \in U_j \cap T_h} \beta_i = \sum_h f_{jh} + d_j = e(U_j, x), \, j = 1, \ldots, p. \tag{2.7}$$

On the other hand, we get from (2.5):

$$\sum_{i \in U_j \cap T_h} \beta_i \geq f_{jh}; \tag{2.8}$$

as $\beta_i = 0$ if $i \in N - U_0$, $i \notin T_0$, from (2.1) and (2.8) we get for each T_h:

$$\sum_{i \in (N-U_0) \cap T_h} \beta_i \geq \sum_{j/U_j \cap T_h \neq \phi} f_{jh} = \sum_j f_{jh} \geq e(T_h, x), \, h = 1, \ldots, q, \tag{2.9}$$

because $f_{jh} = 0$ when $U_j \cap T_h = \phi$. Now, (2.7) and (2.9) show that the numbers β_i, $\forall i \in N - U_0$, defined by (2.5), satisfy (1.8) and (1.9), hence $\mathcal{U}$ is a c.b.c.p. against $\mathfrak{S}$.

From Theorems 2.1 and 2.2 follows:

Theorem 2.3: Let $\mathfrak{S}$ be a b.p. against $x \in F_e$, $x \in C(G)$,, and $\mathcal{U}$ be a partial coalition structure subject to (2.3) and (2.4); then $\mathcal{U}$ is a c.b.c.p. against $\mathfrak{S}$, if and only if for each proper subset X of blocks in $\mathcal{U}$ (including $X = \phi$), we have

$$\sum_{U_j \notin X} e(U_j, x) \geq \sum_{T_h \in I(X)} e(T_h, x) \tag{2.10}$$

where $I(X)$ is the set of blocks in $\mathfrak{S}$ having common players with the blocks in X.

Note that Theorem 2.3 is a combinatorial characterization of those c.b.c.p. which are subject also to the second group of conditions (2.3). The following Lemma shows that we can confine ourselves to such partial coalition structures when we try to discover whether $\mathfrak{S}$ could be countered.

Lemma 2.4: For a b.p. $\mathfrak{S}$, there is a c.b.c.p. if and only if there is a c.b.c.p. satisfying $(N-T_0) \cap U_j \neq 0$, $j=1,...,p$.

Proof: For a partial coalition structure $\mathcal{U}^*$, which does not satisfy this condition for all $j=1,...,p$, even though $\mathcal{U}^*$ satisfies the other conditions (2.3) and (2.4), the linear system (1.8) and (1.9), where U_j are replaced by U_j^*, has the property: if $(N-T_0) \cap U_j^* = \phi$ for some j, then the subsystem (1.9) does not contain unknowns β_i with $i \in U_j^*$, while that part of (1.8) which contains these unknowns is consistent and does not contain other variables. Therefore, those blocks of $\mathcal{U}^*$ which have nonempty intersection with $N-T_0$ form a c.b.c.p. whenever $\mathcal{U}^*$ forms a c.b.c.p.

Note that in checking whether a partial coalition structure $\mathcal{U}$ subject to (2.3) and (2.4) is a c.b.c.p. against a given b.p. $\mathfrak{S}$, we may not check whether (2.10) holds; instead we may prefer to find a maximum flow in the extended network $\mathfrak{N}(U,T)$ and check whether this flow saturates the exit arcs. The proof of Theorem 2.1 shows that these are two alternative methods. Anyway, in general, we shall meet huge problems; however, better alternatives are available for important particular cases, as we shall show in the next section, based upon the above results.

3. A characterization for the compensatory bargaining set.

Recall from the first section that for an admissible payoff $x \in F_\varphi$, where φ is a coalition structure, we may have a c-stable x if $x \in C(G)$, or if $x \notin C(G)$ but for every b.p. against x there is a c.b.c.p. against the b.p.; all c-stable payoffs form the compensatory bargaining set M_c. While definition (2.1) is characterizing nicely the elements of M_c belonging to the core, a characterization of the c-stable payoffs which do not belong to the core is now possible due to Theorem 2.3 of the second section. We get as a corollary:

Theorem 3.1: An admissible payoff vector $x \in F_\varphi$, $x \notin C(G)$, is an element of the bargaining set M_c, if and only if for each partial coalition structure $\mathfrak{S}=(T_1,...,T_q)$ subject to $e(T_h, x)>0$, $h=1,...,q$, there is a partial coalition structure $\mathcal{U}=(U_1,...,U_p)$ subject to $T_h \cap (\cup U_j) \neq \phi$, $h=1,...,q$, and $U_j \cap (\cup T_h) \neq \phi$, $j=1,...,p$ and $e(U_j, x)>0$, $j=1,...,p$, such that for each proper subset X of blocks in $\mathcal{U}$, (including $X=\phi$), we have

$$\sum_{U_j \notin X} e(U_j, x) \geq \sum_{T_h \in I(X)} e(T_h, x),$$

where $I(X)$ is the subset of blocks in $\mathfrak{S}$ having common players with the blocks in X.

Note that from this result we can derive some other results helpful in checking whether a given payoff belongs to M_c. The case of superadditive games will be considered. Recall that $G=(N,v)$ is superadditive, if for any pair of disjoint coalitions S_1 and S_2 we have $v(S_1)+v(S_2)\leq v(S_1 \cup S_2)$. An important subclass of the set of superadditive games is that of convex games; recall that $G=(N,v)$ is convex, if for any pair of coalitions T_1 and T_2 we have $v(T_1)+v(T_2)\leq v(T_1 \cup T_2)+v(T_1 \cap T_2)$. We need the simple result:

Lemma 3.2: If G is a superadditive game, then for any pair of disjoint coalitions S_1 and S_2, we have $e(S_1,x)+e(S_2,x)\leq e(S_1 \cup S_2,x)$, for any $x \in F$; the inequality holds for a finite set of pairwise disjoint coalitions, too.

Proof: The inequality follows from the definition of the excess function and the superadditivity of the game; an induction is needed for more than two coalitions.

From Lemma 3.2 we shall be able to derive a result saying that for superadditive games threats by means of multicoalitional b.p.'s and countering by multicoalitional c.b.c.p.'s are meaningless.

Lemma 3.3: In a superadditive game G, any b.p. $\mathfrak{S}=(T_1,..,T_q)$ with $q>1$ against a payoff vector $x \in F$, $x \in C(G)$, is countered by $\mathcal{U}=(U_1)$, where U_1 is the union of all T_h, $h=1,..q$.

Proof: Obviously, U_1 satisfies (2.3); from Lemma 3.2 we get the inequality $e(U_1,x)\geq \sum_h e(T_h,x)$, in which the right-hand side is positive by (1.3), so that (2.4) holds either. The same inequality shows that (2.10) holds ($X=\phi$ is the only X to consider), hence by Theorem 2.3 our $\mathcal{U}=(U_1)$ is a c.b.c.p. against $\mathfrak{S}$.

Lemma 3.4: In a superadditive game G, any b.p. $\mathfrak{S}=(T_1)$ which is countered by a c.b.c.p. $=(U_1,..,U_p)$ with $p>1$, is also countered by $\mathcal{U}^*=(U^*)$, where U^* is the union of all U_j, $j=1,..p$.

Proof: Obviously, $\mathcal{U}^*$ satisfies (2.3); from Lemma 3.2 we get the inequality $e(U^*,x)\geq \sum_j e(U_j,x)$, in which the right-hand side is positive from (2.4), because U^* satisfies also (2.4). The same inequality together with (2.10) for $X=\phi$, where the right-hand side is $e(T_1,x)$, is giving $e(U^*,x)\geq e(T_1,x)$; therefore, by Theorem 2.3 our $\mathcal{U}^*$ is a c.b.c.p. against $\mathfrak{S}$.

Now, from Lemmas 3.3 and 3.4 we prove easily the following:

Theorem 3.5: In a superadditive game G, an admissible payoff $x \in F$, $x \in C(G)$, belongs to the compensatory bargaining set $M_c(G)$, if and only if for each coalition T^* with $e(T^*,x)>0$ there is a coalition U^* with $U^* \cap T^* \neq \phi$ and $e(U^*,x)>0$ such that $e(U^*,x)\geq e(T^*,x)$.

Proof: If $x \in M_c(G)$ and T^* is a coalition with $e(T^*,x)>0$, then to the b.p. $\mathfrak{S}^* = (T^*)$ corresponds a c.b.c.p. $\mathcal{X}$ countering $\mathfrak{S}^*$; then, by Lemma 3.4, $\mathfrak{S}^*$ is also countered by $\mathcal{X}^* = (U^*)$, where U^* is the union of coalitions in $\mathcal{X}$. Conversely, suppose that for some $x \in F$, $x \notin C(G)$, any b.p. made of one coalition is countered by a c.b.c.p. made also of one coalition; consider a b.p. $\mathfrak{S} = (T_1, ..., T_q)$. If $q=1$, then $\mathfrak{S}$ can be countered by a c.b.c.p. made of one coalition; if $q>1$, then by Lemma 3.3, $\mathfrak{S}$ can also be countered by $\mathcal{X} = (U_1)$, where U_1 is the union of all T_h, $h=1,...,q$. Hence any b.p. can be countered and x belongs to $M_c(G)$.

Consider now the 3-person case, because in [2] this is a fully discussed case and the reader can compare easier the results given by the two theories; for this reason, we take the game $(G): v(1)=v(2)=v(3)=0$, $v(12)=a$ $v(23)=b$, $v(13)=c$, and $v(123)=d$, where a, b, c, and d are nonnegative numbers, the same game as in [2]. We intend to determine the coalitionally rational noncore elements of the compensatory bargaining set, because the coalitional rationality was a condition imposed in [2]. Recall that for a coalition structure $\wp$, an admissible payoff vector x is coalitionally rational, if we have $x(S) \geq v(S)$ for every $S \subseteq S_k \in \wp$. We shall be based upon the auxiliary result:

Theorem 3.6: For the 3-person game (G), an admissible payoff vector x, $x \notin C(G)$, which is coalitionally rational, belongs to the compensatory bargaining set $M_c(G)$, if and only if $H = \max\limits_{S} e(S,x)>0$ is reached for at least two coalitions.

Proof: If $x \in F$ is coalitionally rational we should have $x_i \geq 0$, $i=1,2,3$, so that $e(i,x)=-x_i \leq 0$, $i=1,2,3$, hence no singleton can be a coalition in a b.p. or a c.b.c.p. It follows that each b.p. and each c.b.c.p. is made of one coalition only, each one of cardinality at least two. Two coalitions of this size will always have at least one player in common so that to check whether x is c-stable, for each coalition T with $e(T,x)>0$, $|T| \geq 2$, (three such coalitions may exist at most), we should check whether there is a coalition U with $e(U,x)>0, |U| \geq 2$, (two such coalitions may exist at most), such that $e(U,x) \geq e(T,x)$. If yes, U forms a c.b.c.p. against the b.p. formed by T. From these remarks, the result trivially follows.

If Theorem 3.6 is used for all possible coalition structures $\wp$, we get:

Theorem 3.7: For the 3-person game $v(1)=v(2)=v(3)=0$, $v(12)=a$, $v(23)=b$, $v(13)=c$, and $v(123)=d$, with a, b, c and d nonnegative, we have:

(A) if $\wp=(1,2,3)$, $x=(0,0,0)$, then $x \in M_c-C(G)$ if and only if $H=\max(a,b,c,d)>0$ and at least two of the four numbers equal H;

(B) if $\wp$ consists of one coalition of cardinality two and one singleton and (G) satisfies $a+b+c \geq 2d$ and the triangle inequalities $a \leq b+c$, $b \leq a+c$, $c \leq a+b$, then in $M_c-C(G)$ we have:

$(0,\ 1/2(a+b-c),\ 1/2(-a+b+c))$ for $\wp=(23,1)$, if $b<a+c$,
$(1/2(a-b+c),\ 0,\ 1/2(-a+b+c))$ for $\wp=(13,2)$, if $c<a+b$,
$(1/2(a-b+c),\ 1/2(a+b-c),\ 0)$ for $\wp=(12,3)$, if $a<b+c$;

(C)   if $\wp$ consists of one coalition of cardinality two and one singleton and (G) satisfies $a+b+c<2d$ and the inequalities $a\leq d$, $b\leq d$, $c\leq d$, then in $M_c-C(G)$ we have:

$(o,d-c,b+c-d)$ and/or $(o,a+b-d,d-a)$ for $\wp=(23,1)$ if $b<d$ and $d\leq b+c$, or $d\leq a+b$

$(a+c-d,o,d-a)$ and/or $(d-b,o,b+c-d)$ for $\wp=(13,2)$ if $c<d$ and $d\leq a+c$, or $d\leq b+c$

$(d-b,a+b-d,0)$ and/or $(a+c-d,d-c,0)$ for $\wp=(12,3)$ if $a<d$ and $d\leq a+b$,, or $d\leq a+c$

(D)   if $\wp=(123)$, then $M_c-C(G)=\phi$.

Note that the 3-person game (G) has c-stable admissible payoffs outside the core in some cases, hence the concept of compensatory bargaining set is meaningful. On the other hand, as Theorem 3.7 shows, there are coalition structures, like $\wp=(1,2,3)$ for the 3-person game, for which no coalitionally rational payoff belongs to M_c; indeed, this is the case if one of the numbers a, b, c, d is greater than the other three. The existence problem seems to be very difficult.

For details concerning the 3-person case, see also [3].

REFERENCES

1.   Aumann, R. J. and Dreze, J., (1974), Cooperative games with coalition structures, Int. Journal of Game Theory, Vol. 3, 4, 217-237.

2.   Aumann, R. J. and Maschler, M., (1964), The bargaining set for cooperative games, in Advances in Game Theory, M. Drescher, L. S. Shapley, A. W. Tucker, eds., Annals of Math. Studies, No. 52, 443-476.

3.   Dragan, I., (1988), The compensatory bargaining set of a cooperative n-person game with side payments, Univ. Texas at Arlington, TR #256.

4.   Dragan, I., (1988), An existence theorem for the modified bargaining set of a cooperative n-person convex game, Libertas Math., 8, 55-64.

5.   Dragan, I., (1985), A combinatorial approach to the theory of the bargaining sets, Libertas Math., 5, 133-150.

6.   Ford, L. R. and Fulkerson, D. R., (1962), Flows in networks, Princeton Univ. Press.

7.   Gale, D., (1957), A theorem on flows in networks, Pacific J. Math., Vol. 7, 1073-1082.

8.   Maschler, M., (1992), The bargaining set, kernel and nucleolus, in Handbook of Game Theory, Vol. I, R. J. Aumann, S. Hart, eds., Chapter 18, 591-667.

One approach to allocating the damage to environment

Victor V. Zakharov

Faculty of Applied Mathematics
S.-Petersburg University
198904, St.Peterhoff, S.-Petersburg, Russia

Abstract

A game-theoretic approach is presented for the problem of allocating the damage to environment from pollution by several enterpises. This approach is bases on the cost allocation method proposed in the paper as "a fair distribution" for allocation problem. The present method is shown to be applicable to all cooperative games with monotonic characteristic functions. Properties of the fair distribution are discussed. It is shown that fair imputation belongs to the cores of cooperative games being constructed.

1. Introduction.

The cost allocation problems were first treated by engineers and economists of the Tennessee Valley Authority (TVA) in the 1930 (see Ransmeier 1942, Straffin and Heaney, 1981). Later different methods of allocating the joint costs were introduced and considered in the papers Legrous (1986), Moulin (1988), Sharkey (1982), Shubik (1962), Spinetto (1975), Syed (1987). Tijs and Drissen (1986), Young (1985).

Considering cost allocation problem one defines a cost game as a pair $\langle I, v \rangle$ where $I = \{1, \ldots, n\}$ denotes the set of players. The players can be united in coalitions $S \subset I$. The cost or characteristic function $v : 2^I \to R$ assigns to any nonempty coalition S the minimum costs $v(S)$ which are charged to coalition S if the players in S unite in their activity. If S is empty, then $v(\emptyset) = 0$. It is naturally to suppose that function v is subadditive, that is $v(S \cup T) \leq v(S) + v(T)$ for all disjoint coalitions $S, T \subset I$.

Any cost allocation method is a mapping $M : V \to R^n$, where V is the set of all subadditive cost functions. The i-th component of the vector $M(v) = (M_1(v), \ldots, M_n(v))$ is the cost for i-th player.

Assume for each $i \in S$

$$v^i(S) = v(S) - v(S \setminus i).$$

This quantity defines the marginal cost contribution of the player i to coalition S.

One of the well-known cost allocation methods is Shapley method. The Shapley value is the average marginal contribution to all coalition of the game.

It is given by

$$\Phi_i^v = \sum_{S:i \in S} \frac{|S - i|\,|I - S|}{|I|} v^i(S).$$

Schmeidler (1969) introduced the game theoretic concept of the nucleolus

$$N(v) = \left\{ x : x \in X(v), \ \max_{S \subset I} \left[\sum_{i \in S} x_i - v(S) \right] \leq \right.$$

$$\left. \leq \max_{S \subset I} \left[\sum_{i \in S} x_i' - v(S) \right] \ \forall x' \in X(v) \right\}$$

where

$$X(v) = \left\{ x \in R^n : \sum_{i \in I} x_i = v(I), \ 0 \leq x_i \leq v(i) \right\}.$$

Methods based on separable and nonseparable costs assign to a subadditive cost function v such that $\sum_{i \in I} v^i(I) < v(I)$, give the cost allocation according to the formula

$$A_i(v, w) = v^i(I) + \frac{w_i}{\sum_{i \in I} w_i} \left[v(I) - \sum_{i \in I} v^i(I) \right]$$

where $w = (w_1, \ldots, w_n)$ is a vector of weight coefficients. There are several known separable cost methods with different weight vectors.

Syed (1987) introduced General Cost Allocation Method which can be applied also in the case $\sum_{i \in I} v^i(I) > v(I)$.

In the next section we consider method which is named *fair cost allocation method*. Difference between this method and others consists in the fact that it can be applied in the case when cost functions are not subadditive or superadditive.

2. Fair Cost Allocation Method

Following Chistyakov (1993) consider dynamic process of searching compromise in a cooperative game. Let $\langle I, w \rangle$ be cooperative game and $A : W \to W$ be an operator where W is a set of all monotonic characteristic functions, i.e. $w(S) \geq w(T) \geq 0, T \subset S \subset I$.

Let $w(S)$ be the maximal damage that the coalition S is capable of causing under the assumption that the players (enterprises) not belonging to S do not cause damage. Suppose that the enterprises are faced to pay a penalty equal to economic measure of the damage.

For any cost function $w \in W$ we define operator $\bar{A}$ as follows

$$\bar{A}(w(S)) = \bar{v}(S) = \frac{\sum\limits_{R \subset I : R \cap S \neq \emptyset} w(R)}{\sum\limits_{R \subset I} w(R)}.$$

In the case when $w(S)$ is interpreted as the minimal damage which must be covered by coalition S we determine the following operator

$$\underline{A}(w(S)) = \underline{v}(S) = \frac{\sum\limits_{R \subset S} w(R)}{\sum\limits_{R \in I} w(R)}.$$

The quantity $\bar{v}(S)$ is interpreted in this problem as the maximal penalty to the members of the coalition $S, \underline{v}(S)$ is interpreted as the minimal penalty to the members of one, if the damage caused is equal to 1.

Now we can state the following assumptions.

Assumption 2.1. The characteristic function $\bar{v}$ is concave. That is for any coalitions S and T contained in I, $\bar{v}(S \cup T) \leq \bar{v}(S) + \bar{v}(T) - \bar{v}(S \cap T)$.

Proof.

$$\sum_{R \subset I : R \cap (S \cup T) \neq \emptyset} w(R) \leq \sum_{R \subset I : R \cap S \neq \emptyset} w(R) + \sum_{R \subset I : R \cap T \neq \emptyset} w(R) -$$

$$- \sum_{R \subset I : R \cap (S \cap T) \neq \emptyset} w(R) - w((S \cup T) \setminus (S \cap T)) \leq$$

$$\leq \sum_{R \subset I : R \cap S \neq \emptyset} w(R) + \sum_{R \subset I : R \cap T \neq \emptyset} w(R) - \sum_{R \subset I : R \cap (S \cap T) \neq \emptyset} w(R).$$

Deviding both parts of the inequality by $\sum_{R \subset I} w(R)$ get inequality we need.□

Assumption 2.2. The characteristic function $\underline{v}$ is convex, i.e. $v(S \cup T) \geq \underline{v}(S) + \underline{v}(T) - \underline{v}(S \cap T)$.

Proof.

$$\sum_{R \subset (S \cup T)} w(R) \geq \sum_{R \subset S} w(R) + \sum_{R \subset T} w(R) - \sum_{R \subset S \cap T} w(R) + \sum_{R \subset (S \cup P)} w(R),$$

for any $P \subset (T \setminus S)$.

Hence

$$\sum_{R \subset (S \cup T)} w(R) \geq \sum_{R \subset S} w(R) + \sum_{R \subset T} w(R) - \sum_{R \subset (S \cap T)} w(R).□$$

Assumption 2.3. For any monotonic function $v(S)$ and for any $S \subset I$ the following inequality holds

$$\underline{v}(S) \leq \bar{v}(S).$$

Now describe *the Fair Cost Allocation Method.* For each of the coalition $S \subset I$ we determine the distribution of unit of the penalty as follows

$$a_i^S = \frac{w(S) - w(S \setminus i)}{\sum\limits_{j \in S}(w(S) - w(S \setminus j))}, \quad i \in S.$$

If $S = \{i\}$, we set $a_i^S = 1$. The quantity a_i^S can be interpreted as maximal share of the penalty for the i-th player in the coalition S.

We regard the distribution of the penalty if the i-th player recieves a penalty being proportional to the quantity

$$\xi_i = \frac{\sum\limits_{S \subset I : i \in S} a_i^S \cdot w(S)}{\sum\limits_{S \subset I} w(S)}. \tag{1}$$

Show that $\sum_{i \in I} \xi_i = 1$. Really,

$$\sum_{i \in I} \xi_i = \sum_{i \in I} \frac{\sum\limits_{S \subset I : i \in S} \frac{w(S) - w(S \setminus i)}{\sum\limits_{j \in S}(w(S) - w(S \setminus j))} w(S)}{\sum\limits_{S \subset I} w(S)} =$$

$$= \frac{\sum\limits_{S \subset I} w(S) \cdot \sum\limits_{i \in I} \frac{w(S) - w(S \setminus i)}{\sum\limits_{j \in S}(w(S) - w(S \setminus j))}}{\sum\limits_{S \subset I} w(S)} = \frac{\sum\limits_{S \subset I} w(S)}{\sum\limits_{S \subset I} w(S)} = 1.$$

Definition 2.1. The allocation of any joint costs θ is said to be *fair distribution* if the i-th player receives a penalty x_i^0 equels to

$$x_i^0 = \xi_i \theta = \frac{\sum\limits_{S \subset I : i \in S} a_i^S w(S)}{\sum\limits_{S \subset I} w(S)} \cdot \theta.$$

3. Properties of The Fair Distribution.

As we see in the second section *The Fair Distribution* allocate each unit of the damage between the players according to (1). Define a mapping M^θ as follows

$$M^\theta(w) = \xi\theta = (\xi_1\theta, \ldots, \xi_n\theta).$$

Consider the properties of cost allocation method M_θ.

Definition 3.1. The method M^θ is said to be efficient if

$$\sum_{i\in I} M_i^\theta(w) = \theta.$$

Theorem 3.1. The fair cost allocation method is efficient.

Proof.

$$\sum_{i\in I} M_i^\theta(w) = \theta \sum_{i\in I} \xi_i = \theta.$$

Definition 3.2. The method M^θ is said to be individually rational cost allocation method if for each monotonic function w the following inequalities hold

$$\underline{v}(i)\theta \le M_i^\theta(w) \le \bar{v}(i)\theta.$$

Theorem 3.2. The method M^θ is individually rational.

Proof.

$$\underline{v}(i) = \frac{w(i)}{\sum_{R\subset I} w(R)} \le \frac{w(i) + \sum_{S\subset I : i\in S, S\neq\{i\}} a_i^S \cdot w(S)}{\sum_{R\subset I} w(R)} = M_i^1(w) =$$

$$= \frac{\sum_{S:i\in S} a_i^S w(S)}{\sum_{R\subset I} w(R)} \le \frac{\sum_{S:i\in S} w(S)}{\sum w(R)} = \bar{v}(i).$$

Multiply all parts of inequality by θ and get inequalities we need.

Theorem 3.3. For any monotonic characteristic function w the *fair distribution* ξ belongs to the cores of the games $\langle I, \bar{v}\rangle$ $\langle I, \underline{v}\rangle$, i.e.

$$\underline{v}(S) \le \sum_{i\in S} \xi_i \le \bar{v}(S), \quad \forall S \subset I.$$

Proof.

$$\underline{v}(S) = \frac{\sum\limits_{R \subset S} w(R)}{\sum\limits_{R \subset I} w(R)} \leq \frac{\sum\limits_{R \subset S} w(R) + \sum\limits_{i \in S} \sum\limits_{R: i \in R \not\subset S} a_i^R w(R)}{\sum\limits_{R \subset I} w(R)} =$$

$$= \sum\limits_{i \in S} \xi_i \leq \frac{\sum\limits_{R \subset S} w(R) + \sum\limits_{R: R \not\subset S, R \cap S \neq \emptyset} w(R)}{\sum\limits_{R \subset I} w(R)} = \frac{\sum\limits_{R \subset I: R \cap S \neq \emptyset} w(R)}{\sum\limits_{R \subset I} w(R)} = \bar{v}(S).$$

Hence

$$\underline{v}(S) \leq \sum\limits_{i \in S} \xi_i \leq \bar{v}(S).$$

Corollary 3.1. The fair distribution ξ includes in the cores of the games $\langle I, \underline{v} \rangle$ and $\langle I, \bar{v} \rangle$.

4. Conclusion.

In this paper we described cost allocation method which can be applied in the case when cost functions are monotonic but not necessary subadditive. When we introduced operator $\bar{A}$ or $\underline{A}$ we in reality had introduced principle of optimality. This principle of optimality consist in the way of defining the operator. Operators considered above transformed the cooperative game to the games with nonempty cores.

As we see the imputation $x^0 = (x_1^0, \ldots, x_n^0)$ corresponding fair cost allocation method belongs to the cores of the games $\langle I, \bar{v}\theta \rangle$ and $\langle I, \underline{v}\theta \rangle$. It means that there is no coalition who can say that her penalty is great, or one of any other coalition is small.

References.

S.Chistyakov, *Dynamic aspects in solution of classic cooperative games.* Report of Russian Academy of Sciences. 330 (1993), N6.

P.Legros, *Allocating joint costs by means of the nucleous,* Intern. J. Game Theory 15 (1986), 109-119.

H.Moulin, *Axioms of Cooperative decision making,* Cambrige University Press, 1988.

J.S. Rausmeier, *The Tennessee Valley Authority: A case study in the economics of multiple purpose stream planning,* Vanderbilt University Press, Tenn., 1942.

D. Schmeidler, *The nucleous of characteristic function game*, SIAM Journal of Applied Math. 17 (1969), 1163-1170.

L.S.Shapley, *A value of n-person games*, Ann. of Mathem. Studies 28 (1953), 307-317.

W.W.Sharkey, *The theory of natural monopoly*, Cambridge University Press, 1982.

M.Shubik, *Incentive, dencentralized control, the assigment of joint cost and internal pricing*, Management Sci. 8 (1962), 325-345.

P.R.Spinetto, *Fainess in cost allocation and cooperative games*, Desision Sci. 4 (1975), 482-491.

P.D.Straffin and J.P.Heaney, *Game theory and the Tennesse Vally Autority*, Intern. J. Game Theory 10 (1981), 35-43.

S.A. Syed, *General separable cost allocation method*, Mathematica Japonica. 32 (1987), 1081-1086.

S.H.Tijs and T.S.H.Driessen, *Game theory and cost allocation problems*, Management Sci. 32 (1986), 1015-1028.

H.P.Young, *Cost allocation*, Proceedings of Symposia in Applied Mathematic 33 (1985) 69-94.

V.V.Zakharov and L.A.Petrosjan, *A game theoretic approach to the problem of environmental protection*, Vestnic Leningr. Univ. Math. 14 (1982).

Local Vector Optimization within a Configuration Process

Birger Funke
ELITE Foundation
Korneliuscenter, Promenade 9
52076 Aachen, Germany

This paper is a result of investigations made in the research project PROKON, which is supported by the German Ministry for Research and Technology.
One aim of this joint research project is to eleborate a tool-kit "KONWERK" to build Expert Systems related to application classes.
In the research group in Aachen we do investigations in order to incorporate optimization into the reasoning process.

The configuration process:

The configuration (construction) task is the complete determination of a system consisting of several parts (subsystems) and taking into account different restrictions.
Let us consider a standard configuration process.
The configuration process is performed step by step. Each step is one of the following actions:
1) determine a parameter
2) specialize the system/a part of a system (subsystem)
3) devide a system/subsystem into parts
4) aggregate parts.
There is an agenda of all actions which have to be carried out during the configuration process. If this agenda is empty the problem is solved and the process will finish.

A set of constraints ensures that all choosen parameters and objects are consistent (fit together). However, if an inconsistant state is reached normally a backtracking to the last decision occurs (chronological backtracking) [Cunis et al.].
During such a process the configurer again and again has to choose certain objects out of a set of local consistant objects taking into account several objectives.

Depending on the kind of system the user of the expert system wants to configure, there are often many possible decisions (choose an object out of the given set of objects), leading to a consistent configuration.

In order to find not only a consistent but an optimal system, the decisionmaker has to take into consideration the objectives to the entire system or to the subsystems. This means in general to solve a Multi-Criteria Optimization Problem.

Configuration of a Computer consisting of a CPU, a HardDiscDrive and a FloppyDiscDrive only:
The System becomes divided into its parts:

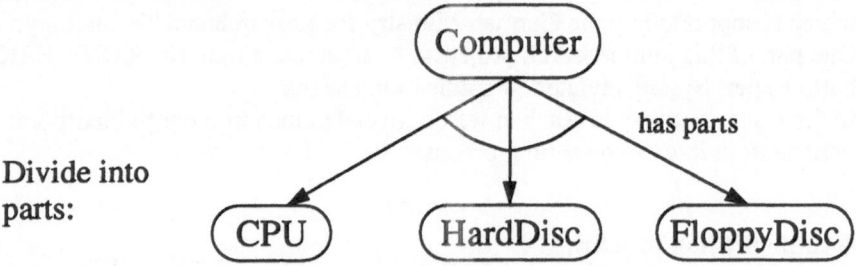

There are different objects to each part of the system.

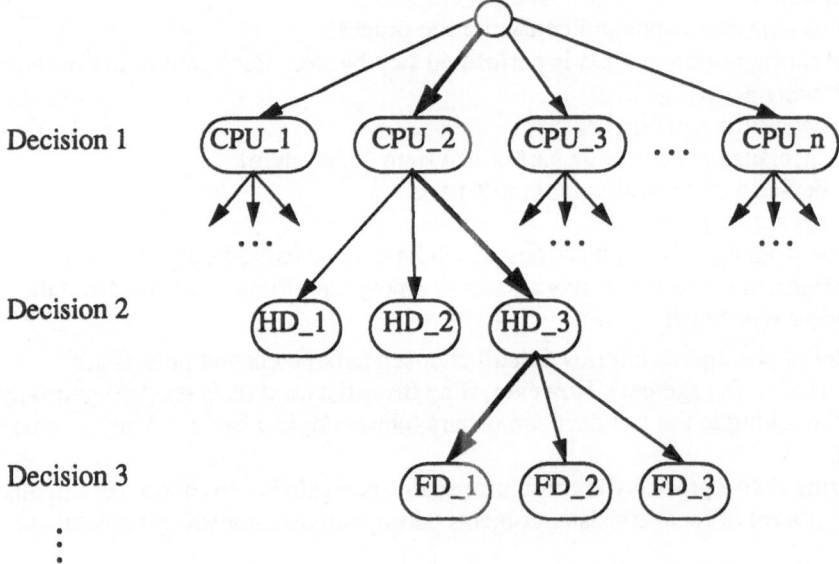

If a backtracking occurs during the configuration process the last decision is removed and a new decision becomes necessary.

Because configuration problems are nonpolynomial we will introduce a heuristical approach to the problem.

Optimization within the Configuration Process:

In general there are two different approaches in order to find an optimal configured system:
1) optimization based control of the configuration process in order to provide a global optimal solution (e.g. Branch and Bound method -> np-complete!)
2) optimization based selection of objects or parameters during the configuration process (local optimization)

Of course it is possible to combine both.

However, in this paper we consider the last approach only.

Local Vector Optimization Problem:

If there is to choose one object out of a set of objects there can be some objectives given. If it is not possible to compare these objectives and to calculate an aggregated objective function the selection could base on the results of a vector optimization [Sebastian 90].

The optimization problem is characterized by a vector of objectives and a finite set of alternatives.

The task:
$$z_1(a) \to \max$$
$$z_2(a) \to \max$$
$$...$$
$$z_K(a) \to \max \quad \text{and} \quad (a \in A).$$

$$Z: = \{z_1, z_2, ..., z_K\} \quad \text{set of objectives,}$$
$$A: = \{a_1, a_2, ..., a_N\} \quad \text{set of alternatives.}$$

In general there is no »perfect solution« $a^* \in A$ that:
$$z_k(a^*) \geq z_k(a_i) ; (\forall k \in \{1, 2, ..., K\}, \forall i \in \{1, 2, ..., N\}).$$

In this paper a method is introduced dealing with the computation of the solution-set (Pareto-set) of this finite discrete vector optimization problem.

This method essentially consists of direct comparisons between all the alternatives referring to the different objective functions.

Therefore, the method allows the calculation of the Pareto-set even if there are no general informations available concerning connections of the known properties of the alternatives to the objective functions.

Domination between alternatives:

We define a function R_k <u>referring to one objective</u> k:

R_k: A×A → {"dominating", "equal", "is dominated", "not comparable" }

$R_k(a_i , a_j) =$ "dominating" , if $z_k(a_i)$ »better« then $z_k(a_j)$
$R_k(a_i, a_j) =$ "equal" , if $z_k(a_i)$ is »equal« $z_k(a_j)$
$R_k(a_i, a_j) =$ "is dominated by" , if $z_k(a_i)$ is »worse« then $z_k(a_j)$
$R_k(a_i, a_j) =$ "not comparable" , otherwise

<u>Example:</u>
Let be $W(z_k) \subseteq R$, $W(z_k) \subseteq R$, $(k \in \{1, 2, ..., K\})$. Now we could set each z_k as follows:

$R_k(a_i, a_j) =$ "dominating" , if $z_k(a_i) > z_k(a_j)$
$R_k(a_i, a_j) =$ "equal" , if $z_k(a_i) = z_k(a_j)$
$R_k(a_i, a_j) =$ "is dominated by" , if $z_k(a_i) < z_k(a_j)$
$R_k(a_i, a_j) =$ "not comparable" , never

Now we can define another function R <u>referring to</u> the set of <u>all objectives</u>:

$R(a_i , a_j) =$ "dominating" , if
      (1) $R_k(a_i, a_j) \in \{$ "dominating", "equal" $\}$
          $\forall (k \in \{1, 2, ..., K\})$ and
      (2) $R_k(a_i, a_j) =$ "dominating"
          for at least one $(k \in \{1, 2, ..., K\})$

$R(a_i, a_j) =$ "equal" , if
          $R_k(a_i , a_j) =$ "equal", $\forall (k \in \{1, 2, ..., K\})$
$R(a_i, a_j) =$ "is dominated by", if
          $R_k(a_i, a_j) =$ "dominated"
$R(a_i, a_j) =$ "not comparable" , otherwise $(a_i, a_j \in A)$

The General Algorithm:

As we have a finite set of alternatives it is possible to compare the alternatives between one another:

(1) let i=1, A'=A
(2) compare a_i with all alternativs a_j of A' with j>i
    and delete all alternatives of A' dominated by a_i A',
    discontinue and delete a_i of A', if a_i is dominated by any a_j
(3) set i=i+1
(4) if i ≤ card(A'), go to (2)
(5) set P=A'

The set P is the solution set.

The number of necessary comparisons grows by the square of the number of alternatives (worst case: (card(A)-1)*card(A)/2).
A priori all alternatives have to be compared with the others, but in general it is possible to reduce the computational effort by using the transitivity of the relation of dominance --> alternatives dominated by others can be disregarded.
A further reduction of the computational effort is possible considering special tasks.

Local Vector Optimization within a Backtracking Algorithm:

As described above during the configuration process a backtracking to previous decisions is possible and can occur several times.
If the number of alternatives is large (» 1000, when the number of objective functions K is about 4 or greater) it would be critical to do the optimization within the backtracking-algorithm.
However, just after a backtracking the difference between the new situation and the situation before is often very small. Because of this similarity it is possible to use the results of the first optimization to find out the new set of optimal alternatives (in the sense of the Pareto optimality).
Therefore we can store the results of the first computation and consider special algorithms for four similar situations:
1) the computation of the Pareto-set after removing one alternative
2) the computation of the Pareto-set after adding one new alternative
3) the computation of the Pareto-set after adding one objective
4) the computation of the Pareto-set after removing one objective.

Let be P' the earlier computed Pareto-set.
For the described special algorithms we use the following relations:

$P = Pa(A, Z) = Pa(P, Z)$.
Using the set P' it could be possible to find a subset $T \subseteq A$ that holds $P \subseteq T \subseteq A$.
Then follows $P = Pa(A, Z) = Pa(T, Z) = Pa(P, Z)$.

1) Removing an alternative a_i:

The Pareto-set is influenced by removing a_i from the set of alternatives only if a_i was an element of P', too. In this case the Pareto-set P has to be built from all alternatives of the set P' except a_i and additionally from all other alternatives which were dominated by a_i only.
If A^i is the set of all alternatives of A dominated by a_i only, then we consider the subset

$T = A^i \cup (P \setminus \{a_i\})$, then $Pa(A \setminus \{a_i\}, Z) = Pa(T, Z)$.

All elements of P' except a_i $Pa(A, Z) \setminus \{a_i\} \subseteq Pa(A \setminus \{a_i\}, Z)$ are elements of P, too.
The computational effort is (worst case) not greater than:

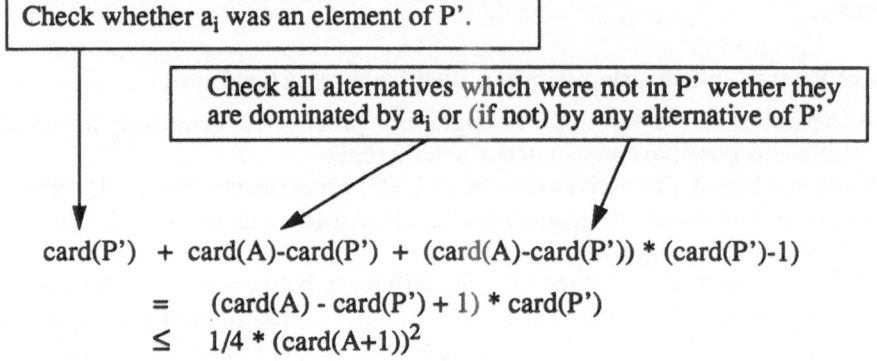

card(P') + card(A)-card(P') + (card(A)-card(P')) * (card(P')-1)

$=$ (card(A) - card(P') + 1) * card(P')

$\leq$ 1/4 * (card(A+1))^2

The same consideration we can use to the other cases:

2) Adding an alternative a_{N+1}:

Either a_{N+1} is dominated by an alternative of the set P' and therefore it cannot be an element of the Pareto-set P or
the new alternative a_{N+1} becomes a new element of the Pareto-set:

$\rightarrow a_{N+1}$ must be added to the Pareto-set P' and

$\rightarrow$ all alternatives of the "old" Pareto-set P', which are dominated by a_{N+1} have to be removed.

It follows that $Pa(A \cup \{a_{N+1}\}, Z) = Pa(P' \cup \{a_{N+1}\}, Z)$.

3) Adding an objective:

All alternatives included in P' are also elements of the new Pareto-set P.
It is possible that other alternatives will become Pareto-elements, too.
Therefore we have to compute

$Pa(P' \cup Pa(A \setminus P', Z \cup \{z_{K+1}\}), Z \cup \{z_{K+1}\})$ or simply
$Pa(A, Z \cup \{z_{K+1}\})$.

4) Removing an objective:

The Pareto-set we have to compute is completly included in the set P'.
Therefore it is: $Pa(A, Z \setminus \{z_i\}) = Pa(P', Z \setminus \{z_i\})$.

Implementation:

We implemented two main functions for our optimization module:
The first one is an implementation of the general algorithm explained above.
If this function finds that one alternative is dominated by any other it marks the dominated alternative by the name of the dominating one.
The second function deletes one specified alternative a_i from the given set.
Furthermore it marks all alternatives which were marked as dominated by a_i for use in further computations, because they could be only dominated by a_i.
Then it gives a special mark to all alternatives of the set P' except the alternative a_i itself. So it is possible to oppress the repeated comparison of elements of the Pareto-set P' one another. It is possible to call this second function again and again to remove more then only one alternative.
At last the first function is called again which will use the new given marks and compute the new Pareto-set P.
We can compute the Pareto-set for a full set of alternatives. If backtracking occurs we delete one alternative (or if necessary more alternatives by repeating the function call). Then we compute the vector optimization again using the method described above with in general much less computational effort.

Results:

The following figure compares the computational effort of the special algorithm to delete an alternative with the general algorithm.

The used data are random numbers.

The results are very different but as a rule the number of necessary comparisons is much less then in the case of the general algorithm.

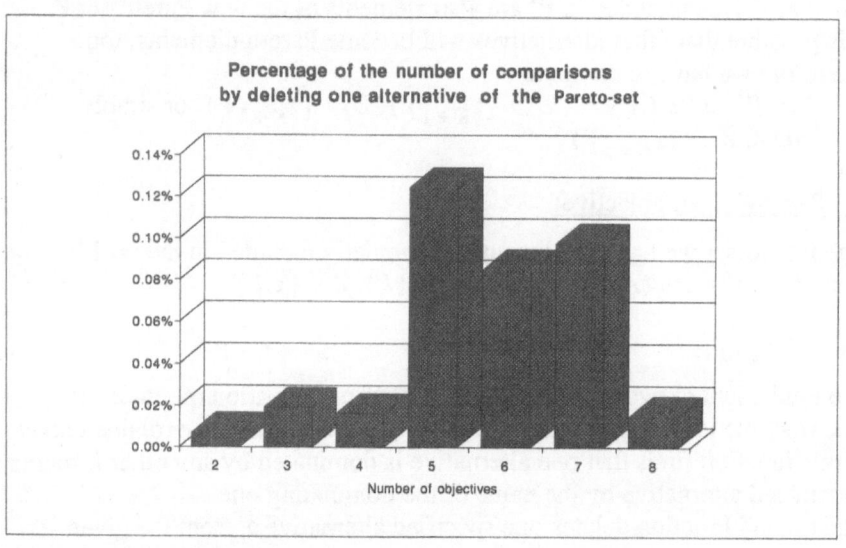

References:

[Cunis et al.]   Cunis,R.;Günter A.; Strecker, H.(eds.):
                 "Das Plakon-Buch", Springer, Informatik Fachberichte, 1991.

[Geber 85]   Geber, Wolfgang:"Diskrete Vektoroptimierung" ,
                 Lang, Frankfurt a.M.; Bern; New York, 1985.

[Göpfert 90]   Göpfert, Alfred: "Vektoroptimierung: Theorie, Verfahren und
                 Anwendungen", BSB Teubner, Leipzig, 1990.

[Sebastian 90]   Sebastian, Hans-Jürgen: "Modelling an Expert System Shell for
                 the Configuration of Technical Systems including Optimization -
                 based Reasoning." in: Wissenschaftliche Zeitschrift der TH
                 Leipzig(14), S.241-255,1990.

Stochastic programming

STOCHASTIC EXTREMA, SPLITTING RANDOM ELEMENTS
AND MODELS OF CRACK FORMATION

Igor V. Evstigneev
University of Bonn, Adenauerallee 24-26,
Bonn 1, D5300, Germany

Priscilla E. Greenwood
University of British Columbia, Mathematics Road 121-1984
Vancouver, B.C. V6T 1Z2, Canada

1. We present a construction of splitting random elements for random fields over discrete partially ordered sets. The idea of the construction is prompted by certain considerations in the dynamic theory of fracture [Sih, 1973]. The mathematical model we deal with describes (in a very simplified form) the process of crack propagation in a material containing random irregularities.

2. We start with general definitions. Let $(\Omega, \mathcal{F}, P)$ be a complete probability space, $(T, \leq)$ a finite or countable partially ordered set and $\xi = (\xi_t(\omega))$ $(t \in T, \omega \in \Omega)$ a random field over T with values in a measurable space. For every $t \in T$, we consider the sets $M_1(t) = \{a \in T : a \leq t\}$, $M_2(t) = \{b \in T : b \geq t\}$ and the σ-algebras $\mathcal{A}_i^\xi(t) = \sigma\{\xi_c, c \in M_i(t)\}$, $i = 1, 2$. Here and in what follows, the symbol $\sigma\{\mathcal{H}\}$ represents the smallest σ-algebra with respect to which all random elements (or events) belonging to the class $\mathcal{H}$ are measurable and which contains the class $\mathcal{N} = \{\Gamma \in \mathcal{F} : P(\Gamma) = 0\}$ of all negligible events.

If $\mathcal{C}_1$ and $\mathcal{C}_2$ are two families of subsets of Ω, then $\mathcal{C}_1\mathcal{C}_2$ stands for the class of sets $\Gamma \subseteq \Omega$ representable in the form $\Gamma = \Gamma_1 \cap \Gamma_2$, where $\Gamma_1 \in \mathcal{C}_1$ and $\Gamma_2 \in \mathcal{C}_2$. A random element $\tau(\omega) \in T(\omega \in \Omega)$ is called splitting with respect to the field ξ if, for each $t \in T$, we have

$$\{\omega : \tau(\omega) = t\} \in \mathcal{A}_1^\xi(t)\mathcal{A}_2^\xi(t) \tag{1}$$

[cf. (Evstigneev, 1988)]. Splitting random elements possess the following important property. Suppose for the moment that the field ξ is Markov, i.e., for any $t \in T$, the σ-algebras $\mathcal{A}_1^\xi(t)$ and $\mathcal{A}_2^\xi(t)$ are conditionally independent given the σ-algebra $\mathcal{A}_3^\xi(t) = \sigma\{\xi_t\}$:

$$\mathcal{A}_1^\xi(t) \perp \mathcal{A}_2^\xi(t) | \mathcal{A}_3^\xi(t), \quad t \in T.$$

Then, for every splitting random element τ, we have

$$\mathcal{A}_1^\xi(\tau) \perp \mathcal{A}_2^\xi(\tau) | \mathcal{A}_3^\xi(\tau), \tag{2}$$

where

$$\mathcal{A}_i^\xi(\tau) = \sigma\left\{\{\tau = t\} \cap \Gamma : t \in T, \Gamma \in \mathcal{A}_i^\xi(t)\right\}.$$

If, in addition, the σ-algebra $\mathcal{F}$ is generated by $\mathcal{A}_1^\xi(t)$ and $\mathcal{A}_2^\xi(t)$ for each $t \in T$, then the converse is true: (2) implies (1). Thus, the Markov property of the field holds at those, and only those, random elements which are splitting.

3. We describe a method for constructing splitting random elements. The method is based on solving extremal problems related to the field. These are problems of minimization of certain random functionals on random subsets of T. Let a set $Z(\omega) \subseteq T$ be associated with each $\omega \in \Omega$. We say that $Z(\omega)$ is a random set if $\{\omega : t \in Z(\omega)\} \in \mathcal{F}$ for all $t \in T$. We call $Z(\omega)$ splitting (with respect to the field ξ) if $\{\omega : t \in Z(\omega)\} \in \mathcal{A}_1^\xi(t)$ $\mathcal{A}_2^\xi(t)$, $t \in T$. With each $u \in T$ and $i = 1, 2$, we associate the event $\Omega_i^Z(u) = \{\omega : Z(\omega) \cap M_i(u) \neq \emptyset\}$ and the σ-algebra $\mathcal{A}_i^\xi(u)|_{\Omega_i^Z(u)}$ consisting of events $\Gamma \in \mathcal{F}$ such that $\Gamma \cap \Omega_i^Z(u) = \Gamma' \cap \Omega_i^Z(u)$ for some $\Gamma' \in \mathcal{A}_i^\xi(u)$.

A random set $Z(\omega)$ is called honest if, for all $u \in T$, $i = 1, 2$ and $t \in M_i(u)$, we have

$$\{\omega : t \in Z(\omega)\} \in \mathcal{A}_i^\xi(u))|_{\Omega_i^Z(u)}. \tag{3}$$

The meaning of requirement (3) is: if we know that $Z(\omega)$ intersects $M_i(u)$, we can describe $Z(\omega) \cap M_i(u)$ using only the information contained in the σ-algebra $\mathcal{A}_i^\xi(u)$. A random element $\tau(\omega) \in T$ is called honest if (3) holds for the singleton $Z(\omega) = \{\tau(\omega)\}$ [compare with the notion of an honest time (Pittenger and Shih, 1973)].

Next, suppose that, for each $\omega \in \Omega$, a real-valued functional $F(\omega, t)$ is defined on the set $Z(\omega)$. This functional is called honest if

$$\{\omega : F(\omega, a) - F(\omega, b) \geq 0, \quad a, b \in Z(\omega)\} \in \mathcal{A}_i^\xi(u)|_{\Omega_i^Z(u)}$$

for all $u \in T$, $i = 1, 2$ and $a, b \in M_i(u)$. Assume that, for each $\omega \in \Omega$, the fulfillment of the inequalities

$$F(\omega, a) \geq F(\omega, t_0), \quad a \in Z(\omega), \quad a \leq t_0,$$
$$F(\omega, b) \geq F(\omega, t_0), \quad b \in Z(\omega), \quad b \geq t_0,$$

is sufficient for the point $t_0 \in Z(\omega)$ to be a minimum point of $F(\omega, \cdot)$ on $Z(\omega)$. Then we say that the functional $F(\omega, \cdot)$ is sufficient on the set $Z(\omega)$.

The following theorem is a tool for constructing splitting random elements [cf. (Evstigneev, 1988), Theorem 3.1].

Theorem 1. *Let* $Z(\omega) \subseteq T$ *be a splitting random set and* $F(\omega, t), t \in Z(\omega)$, *an honest functional. Suppose that, for each* $\omega \in \Omega$, *the functional* $F(\omega, \cdot)$ *is sufficient on* $Z(\omega)$. *Then*

$$\overline{Z}(\omega) = \{t \in Z(\omega) : F(\omega, t) = \min_{s \in Z(\omega)} F(\omega, s)\}$$

is an honest splitting random set. If for every $\omega \in \Omega$ *the set* $\overline{Z}(\omega)$ *consists of one element* $\tau(\omega)$, *then* $\tau(\omega)$ *is an honest splitting random element.*

We note that Theorem 1 can be applied in the special cases when $Z(\omega)$ does not depend on ω or satisfies the condition $\{\omega : t \in Z(\omega)\} \in \sigma\{\xi_t\}$; any of these requirements guarantees the splitting property of $Z(\omega)$.

4. Fix two natural numbers M and N. Denote by T the set of all mappings t of $\mathbf{M} = \{0, 1, \ldots, M\}$ into $\mathbf{N} = \{0, 1, \ldots, N\}$ with the natural partial ordering ($t \leq s \iff t(m) \leq s(m)$ for all $m \in \mathbf{M}$). Let $\mathbf{Z} = \{\ldots, -1, 0, 1, \ldots\}$ and let $\mathcal{J}$ be the collection of

all intervals $J = [p,q] = \{m \in \mathbf{Z} : p \le m \le q\}$, where $p, q \in \mathbf{M}$. Denote by W_J the class of all functions w with the domain dom $w = J$, taking values in $\mathbf{N}$. For $w \in W_{[p,q]}$, we define $G_w = \{(p, w(p)), \ldots, (q, w(q))\}$. We denote by L_w the polygonal line in the plane $\mathbf{R}^2$ obtained as the union of segments with the ends $(j, w(j)), (j+1, w(j+1))$ $(j = p, \ldots, q-1)$.

Consider a thin plate having the form of a rectangle $R = \{(a,b) \in \mathbf{R}^2 : 0 \le a \le M, 0 \le b \le N\}$. Suppose that the material of the plate is non-homogeneous and contains random irregularities. Under the influence of exogenous forces, a stress field appears in the plate, which leads to its fracture. A crack starts at some point $x \in X \equiv \{(m,n) \in \mathbf{Z}^2 : 0 \le m \le M, 0 \le n \le N\}$ and propagates to the left-hand and the right-hand sides of the rectangle R. Possible configurations of a fully developed crack are represented by polygonal lines L_t, $t \in T$. Configurations of the crack at intermediate stages of its development are described by polygonal lines L_w, $w \in W_J$ $(J \in \mathcal{J})$. In what follows, using the expression "the crack w", we mean the configuration of the crack described by the polygonal line L_w.

For $J = [p,q] \in \mathcal{J}$, we define

$$J^{(1)} = [p^-, q], J^{(2)} = [p, q^+], \tilde{J} = [p^-, q^+], \tag{4}$$

where $p^- = \max(p-1, 0)$, $q^+ = \min(q+1, M)$. Let $J \subseteq J'$ be two intervals and $w \in W_J$, $w' \in W_{J'}$ two functions that the restriction $W'|_J$ of the function w' to the interval J coincides with w. If $J' = \tilde{J}$, we call w' an elementary extension of w. If $J' = J^{(1)}$ (resp. $J' = J^{(2)}$), we say that w' is a left (resp. right) elementary extension of w.

5. We are given:

1) a random field $\xi_x(\omega)$, $x \in X$ (with values in a measurable space), describing the local structure of the material at points $x \in X$;

2) a real-valued function $H_x(\omega) \ge 0$ of $\omega \in \Omega$ and $x \in X$ interpreted as the amount of energy needed for a crack to start at the point x;

3) a real-valued functional $\psi(\omega, J, w)$ of $\omega \in \Omega$, $J \in \mathcal{J}$ and $w \in W_{J^{(1)}} \cup W_{J^{(2)}}$ representing the amount of energy needed for the left or right elementary extension of $w|_J$ to $w \in W_{J^{(1)}}$ or to $w \in W_{J^{(2)}}$.

We suppose that, for each $x \in X$, the random variable $H_x(\omega)$ is measurable with respect to the σ-algebra $\sigma\{\xi_x\}$. The following locality property of the functional ψ is postulated: for all $J \in \mathcal{J}$, $i = 1, 2$ and $a, b \in W_{J^{(i)}}$, we have,

$$\{\omega : \psi(\omega, J, a) - \psi(\omega, J, b) \ge 0\} \in \sigma\{\xi_{(m,n)} : $$
$$\min(a(m), b(m)) \le n \le \max(a(m), b(m)), m \in J^{(i)}\}.$$

It is also assumed that $\psi(\omega, J, w) = 0$ if $w \in W_{J^{(i)}}$ and $J^{(i)} = J$.

For $J \in \mathcal{J}$ and $w \in W_J$, we define

$$\Psi(\omega, J, w) = \psi(\omega, J, w|_{J^{(1)}}) + \psi(\omega, J, w|_{J^{(2)}}).$$

The functional Ψ expresses the value of energy needed for the elementary extension of the crack $w|_J$ to the crack w. This energy is equal to the sum of energies required for the left and the right elementary extensions of $w|_J$ to $w|_{J^{(1)}}$ and $w|_{J^{(2)}}$.

The process of crack formation goes as follows. The crack starts at the point $\lambda_0(\omega) = (\mu_0(\omega), \nu_0(\omega)) \in X$ which minimizes the initiation energy $H_x(\omega)$, $x \in X$; this minimum is assumed to be unique. (An important special case: $\lambda_0(\omega) = (0, m_0)$, "an incision"). We model the process of crack propagation as a sequence of stages. At stage $k(k = 0, 1, \ldots, M)$ the crack is represented by a polygonal line $L_{\tau_k(\omega)}$, where $\tau_k(\omega)$ is a random function, $\tau_k(\omega) = \tau_k(\omega, m), m \in \Delta_k(\omega)$, defined on the interval $\Delta_k(\omega) = [(\mu_0(\omega) - k) \vee 0, (\mu_0(\omega) + k) \wedge M]$. We have $\tau_k(\omega)|_{\Delta_{k-1}(\omega)} = \tau_{k-1}(\omega)$. The initial function $\tau_0(\omega)$ is given by $\lambda_0(\omega) = (\mu_0(\omega), \nu_0(\omega))$: the domain of $\tau_0(\omega)$ is the degenerate interval $\Delta_0(\omega) = [\mu_0(\omega), \mu_0(\omega)]$ and its value equals $\nu_0(\omega)$. By virtue of (4), $\Delta_k(\omega) = \widetilde{\Delta}_{k-1}(\omega)$ $(k = 1, \ldots, M)$ and, consequently, $\tau_k(\omega)$ is an elementary extension of $\tau_{k-1}(\omega)$. It is assumed that this extension is energetically optimal, i.e., for all $\omega \in \Omega$ and $k = 1, \ldots, M$, the function $\tau_k(\omega)$ minimizes the functional $\Psi(\omega, \Delta_{k-1}(\omega), w)$ on the set of $w \in W_{\Delta_k(\omega)}$, satisfying $w|_{\Delta_{k-1}(\omega)} = \tau_{k-1}(\omega)$. This minimum is assumed to be unique. Thus, at every stage, the crack grows to the left and to the right, choosing energetically optimal directions. It reaches the left-hand and the right-hand sides of the plate R not later than the Mth stage, because $\Delta_M(\omega) = [0, M]$. The function $\tau(\omega) = \tau_M(\omega)$ describes the final configuration of the crack. [Compare the above construction with (Sih, 1973), p. 318.]

6. We define recursively a sequence of sets $Z_k(\omega) \subseteq T$, $k = 0, \ldots, M + 1$, and functionals $F_k(\omega, t)$, $t \in Z_k(\omega)$, $k = 0, \ldots M$. We put

$$Z_0(\omega) = T, \quad F_0(\omega, t) = \min_{x \in G_t} H_x(\omega).$$

Suppose that we have defined the set $Z_k(\omega)$ and the functional $F_k(\omega, t), t \in Z_k(\omega)$. Then we set

$$Z_{k+1}(\omega) = \{t \in Z_k(\omega) : F_k(\omega, t) = \min_{t' \in Z_k(\omega)} F_k(\omega, t')\} (k \leq M),$$

$$F_{k+1}(\omega, t) = \Psi(\omega, \Delta_k(\omega), t|_{\Delta_{k+1}(\omega)}), t \in Z_k(\omega) \ (k < M).$$

We consider the random field $\xi = (\xi_t)_{t \in T}$ where $\xi_t = (\xi_x)_{x \in G_t}$. In the theorem formulated below the honest and splitting properties are with respect to ξ.

Theorem 2. *The random sets $Z_k(\omega)$ $(k = 0, \ldots, M + 1)$ and the random element $\tau(\omega)$ are splitting. The functionals $F_k(\omega, t), t \in Z_k(\omega)$ $(k = 0, \ldots, M)$, the sets $Z_k(\omega)(k = 0, \ldots, M + 1)$ and the random element $\tau(\omega)$ are honest. For all $k = 0, \ldots, M$ and $\omega \in \Omega$, the functional $F_k(\omega, \cdot)$ is sufficient on $Z_k(\omega)$.*

Remark 1. The set $Z_k(\omega)$ consists of those $t \in T$ for which $t|_{\Delta_{k-1}(\omega)} = \tau_{k-1}(\omega)$ $(k = 1, \ldots, M + 1)$. We have $Z_{M+1}(\omega) = \{\tau(\omega)\}$.

Remark 2. Suppose that the field $(\xi_t)_{t \in T}$ is Markov. (This is so, e.g., if $(\xi_{(m,n)})_{n \in \mathbb{N}}$, $m = 0, \ldots, M$, are $M + 1$ independent Markov chains.) Then, by virtue of the splitting property of τ established in Theorem 2, $\mathcal{A}_1^\xi(\tau) \perp \mathcal{A}_2^\xi(\tau) | (\tau, \xi_\tau)$. This means that L_τ splits R into two conditionally independent parts.

Theorems 1 and 2 follow from results obtained in our monograph, Evstigneev and Greenwood (1992), Theorems 3.1, 3.2 and 10.1.

References

Evstigneev, I.V. (1988) Stochastic extremal problems and the strong Markov property of random fields. Uspekhi Mat. Nauk, v. 43, p. 3 - 41. (In Russian.) English transl. in Russian Math. Surveys, 1988, v. 43, p. 1-49.

Evstigneev, I. V. and P. E. Greenwood (1992) Markov fields over countable partially ordered sets: Extrema and splitting. Discussion Paper No. A-371 (University of Bonn). To appear as a Memoir of the Amer. Math. Soc.

Pittenger, A. O. and C. T. Shih (1973) Coterminal families and the strong Markov property. Trans. Amer. Math. Soc., v. 182, p. 1–42.

Sih, G. C., Ed. (1973) Dynamic crack propagation. Noordhoff Publ., Leyden.

STOCHASTIC OPTIMIZATION ALGORITHMS FOR REGENERATIVE DEDS

A.A. Gaivoronski[1], E. Messina[2]

[1] ITALTEL
Castelletto di Settimo Milanese
20019 Milano, Italy
[2] Information Sciences Dept., Universitá degli studi di Milano
via Comelico 39-41, 20135 Milano, Italy

Abstract

In this paper we are concerned with the application of stochastic programming techniques for the parametric optmization and analysis of Discrete Events Dynamic Systems (DEDSs). In particular, we consider optimization of steady state behavior of DEDSs which sample path depends discontinuosly on control parameters.

This is a difficult problem, since in order to evaluate the steady state performance measure it is necessary to make an observation on the infinite time horizon, which is impossible. On the other hand, in the general case any observation on the finite time interval gives a biased estimates of the performance measure.

Here we consider an important subclass of DEDSs for which we developed a new algorithm for optimizing their steady state performance, by adapting the methods of stochastic optimization. Such subclass is the set of regenerative DEDSs.

1 Introduction and Problem Definition

A large class of dynamic systems, such as material flow in production or assembly lines, message flow in communication networks, jobs in multiprogrammed computer systems and generally queueing networks can be characterized as Discrete Event Dynamic Systems (DEDSs), driven by a sequence of discrete events in time.

The interesting as well as difficult part of analysis of the performance characteristics of such systems arises from the complex interaction of such discrete events
over time. Furthermore, we assume that the systems are affected by the presence
of uncertainty which can be modelled through uncontrollable stochastic parameters ω which take values from a properly defined probability space Ω, making
the evaluation of the steady state performance measure of such systems certainly
not trivial.

Many systematic ways of analysis have been developed in the last years; among
them probabilistic models, such as queueing models, have been the most successful ones. However, in studying very complex stochastic systems which cannot be
treated in an analytic manner, Discrete Event Simulation is the only possibility.
Anyway, some difficulties arise while studying the stationary behaviour of DEDSs
due to the complexity of their models and the amount of computer time required.

Let us consider DEDS with performance measure $f(t,\omega)$, which evolves on semi-
finite time horizon $t \in [0,\infty)$. The value of $f(t,\omega)$ can be observed through
a simulation of an appropriate DEDS model, in particular a Petri Net model.
Suppose that the problem consists of estimating the steady state performance of
the system, which can be represented as follows:

$$F = \lim_{t \to \infty} \frac{f(t,\omega)}{t} \tag{1}$$

From here onwards we assume that the properties of underlying stochastic process is such that this limit exists and does not depend on ω.

This is a difficult problem since in order to make one observation of F it is
necessary to perform a simulation run of infinite duration, which is obviously
impossible to afford. Nevertheless, it is possible to identify a subclass of DEDSs,
namely the regenerative DEDSs, for which it is easier to perform the steady state
performance analysis, by exploiting their particular properties and by adapting
existing algorithms used for studying the transient behaviour of DEDSs. Such a
class consists of the set of systems having the regenerative property [CRAIG75].
One of the possible formulations of this property is the following.

Regenerative Property. It exists a sequence of times $t_k(\omega)$, $k = 1, \ldots$ such that
$t_k(\omega) \to \infty$ as $k \to \infty$ and

$$\Delta f_k(\omega) = f(t_{k+1},\omega) - f(t_k,\omega), \quad k = 1, \ldots \tag{2}$$

are i.i.d. random variables and

$$\Delta t_k(\omega) = t_{k+1} - t_k, \quad k = 1, \ldots \tag{3}$$

are i.i.d. random variables.

We assume here that $t_0 = 0$ and $f(t_0, \omega) = 0$.

Normally the times t_k are associated with arrivals of the process at some fixed state s_r, called *regenerative state*. Many important DEDSs fall in this class, for example those described by a Markov process with finite state space. In this case any state which the process visits infinite number of times can be selected as the regenerative state.

In this paper we develop an algorithm for the optimization of steady state performance of regenerative DEDS, based upon approach which has shown its effectiveness in the case of transient analysis of DEDS. Such algorithm combines the advantages of both analytical and simulation approaches while minimizing their disadvantages.

Let us consider first the problem of estimation of steady state performance measure for regenerative DEDS. In order to estimate F it is enough to observe the time t_1 of the first and the subsequent returns to the regenerative state s_r and the value of the performance measure at this time, obtaining:

$$F = \lim_{t \to \infty} \frac{f(t, \omega)}{t} = \frac{E(f(t_1, \omega))}{E(t_1)} \tag{4}$$

That is, the estimation on the infinite time horizon is reduced to the estimation on a finite time interval of random length. One of the possible estimation schemes is the following:

Algorithm 1: (Regenerative Estimation)

1. Initialization. Choose a regenerative state s_r and a number N of regenerative periods (the number of returns to s_r).

2. Observation. Start the process at time $t_0 = 0$ from s_r and perform a simulation run until the N-th return to s_r. Let t_N be the time of the N-th return and f_N be the value of the performance measure at this time.

3. Estimation. Compute the average values $\hat{f}_N = \frac{f_N}{N}$ and $\hat{t}_N = \frac{t_N}{N}$. The unbiased estimate $\hat{F}_N$ of the steady-state performance measure F is given by the relation $\hat{F}_N = \frac{f_N}{t_N}$.

Let us turn now to the problem of optimization. Suppose that f depends also on a vector x of parameters which belongs to some set X of $\Re^n$. We use sensitivity estimates of f in order to choose controlled parameters $x \in X$ in such a way as to obtain the optimal values of the steady-state performance measure F.

In order to do that we use statistical estimates of the gradient (sensitivities) of

F; more formally, we are interested in the estimation of the gradient $F_x(x)$:

$$F(x) = \lim_{t \to \infty} \frac{f(x, t, \omega)}{t}, \tag{5}$$

$$F_x(x) = \frac{d}{dx} \lim_{t \to \infty} \frac{f(x, t, \omega)}{t}. \tag{6}$$

The regeneration property allows to reduce the problem of sensitivity analysis of the unobservable steady-state performance measure to the sensitivity analysis of the observable transient performance measure. Indeed combining (4) and (6) and taking into account that the times t_k associated with arrivals of the process at s_r now depend not only on ω but also on x, we obtain:

$$
\begin{aligned}
F_x(x) &= \frac{d}{dx} \frac{E f(t_1(x, \omega), \omega)}{E t_1(x, \omega)} \\
&= \frac{E t_1(x, \omega) \frac{d}{dx}(E f(t_1(x, \omega), \omega)) - E f(t_1(x, \omega), \omega) \frac{d}{dx} E t_1(x, \omega)}{(E t_1(x, \omega))^2}
\end{aligned} \tag{7}
$$

Thus, in order to estimate the gradient $F_x(x)$ we need to estimate $E f(t_1(x, \omega), \omega)$, $E t_1(x, \omega)$, $\frac{d}{dx} E f(t_1(x, \omega), \omega)$ and $\frac{d}{dx} E t_1(x, \omega)$. The estimation of the last two quantities is the subject of the sensitivity analysis of the transient behaviour of $DEDS$ ([GAIV92, GLYN86, HO87, HOLI88, PFLU88, SURI87, RUBI86]). In the case of Petri Net models we can utilize here the score function (likelihood ratio) type estimates developed in [ARCH92]. It is important to note that, in order to obtain unbiased estimates of all these quantities it is enough to observe the system during finite time intervals.

The sensitivity estimates obtained can be used in order to choose controlled parameters $x \in X$ in such a way as to obtain the optimal values of the steady-state performance measure $F(x)$ defined in (5).

2 Algorithm for Solution of Optimization Problem

Let us consider the following optimization problem:

$$\min_{x \in X} F(x) \tag{8}$$

where $F(x)$ is defined in (5).

Due to the presence of random parameters this problem belongs to the class of stochastic programming problems. In order to obtain the solution of problem (8) we can apply the method of projection of stochastic quasi-gradients ([ERMO92]) which starts from some initial point x^0 and generates the sequence of points x^s according to the following recursive rule

$$x^{s+1} = \prod_X(x^s - \rho_s\xi^s) \tag{9}$$

where ρ_s is the stepsize, $\prod_X(.)$ is the projection operator on the set X and ξ^s is a statistical estimate of the gradient of the function $F(x)$, defined as:

$$E(\xi^s|x^0,\ldots,x^s) = F_x(x^s) + g_s \tag{10}$$

where g_s vanishes as s tends to infinity. Under mild assumptions on the problem properties and on the stepsize, the sequence x^s tends to the solution of the problem (8) provided that the property (10) of the step direction ξ^s is satisfied. More precisely, the following result holds:

Theorem 1 *Suppose that the following conditions are satisfied:*

1. $X \subset \Re^n$ *is a convex and compact set;*
2. $F(x)$ *is a convex function on some open set which contains X;*
3. $E(\|\xi^s - F_x(x^s) - g_s\|^2|x^0,\ldots,x^s) = C < \infty$, $g_s \mapsto 0$ *a.s.;*
4. $\rho_s \geq 0$; $\sum_{s=0}^{\infty}\rho_s = \infty$, $\sum_{s=0}^{\infty}\rho_s^2 < \infty$

then the sequence x^s generated by (9)-(10) has accumulation points and all such points belong to the set X^* of solutions of problem (8):

$$X^* := \{x^* \in X \text{ such that } F(x^*) = F^*\} \tag{11}$$

where

$$F^* = \min_{x \in X} F(x). \tag{12}$$

The proof of this theorem can be found in [ERMO88].

The statistical estimate ξ^s can be obtained using the results presented above. More precisely, suppose that ζ^s, χ^s, γ^s and κ^s are such that

$$E(\zeta^s|x^0,\ldots,x^s) = E(f(t_1(x^s,\omega),\omega) \tag{13}$$

$$E(\chi^s|x^0,\ldots,x^s) = E(t_1(x^s,\omega)) \tag{14}$$

$$E(\gamma^s|x^0,\ldots,x^s) = \frac{d}{dx}(Ef(t_1(x^s,\omega),\omega)) \tag{15}$$

$$E(\kappa^s|x^0,\ldots,x^s) = \frac{d}{dx}(Et_1(x^s,\omega)) \tag{16}$$

The value of ζ^s and χ^s can be obtained using Algorithm 1, while γ^s and κ^s can be obtained through the algorithms of the sensitivity analysis mentioned above. Now according to (7) the candidate for ξ^s could be the following:

$$\xi^s = \frac{\chi^s \gamma^s - \zeta^s \kappa^s}{(\chi^s)^2} \tag{17}$$

However, there are fundamental problems with such choice of step directions. They have their source in the fact that, generally speaking, with such choice of ξ^s the bias g_s from (10) will not vanish as $s \mapsto \infty$.

Hence, if we want to apply the method (9)-(10) we are left with the following alternatives:

- Choose a very large number N of regeneration cycles used for obtaining estimates by Algorithm 1. This will lead to small bias in (17) but could result in very long, even unaffordable requirements of simulation time.

- Choose the number of regeneration cycles affordable but this may lead to a large bias in estimates ξ^s from (17) and, as a consequence, the solution obtained will be very imprecise.

This argument is made in order to illustrate the necessity to develop a new algorithm which takes explicitly into account the properties of regenerative DEDS. In what follows we propose one such algorithm which performs optimization in parallel with more and more precise regenerative estimations, and obtains the optimal solution of problem (8) during a single simulation run.

Algorithm 2: (Optimization of regenerative DEDS)

1. Initialization. Choose the initial point $x^0 \in X$, the regenerative state s_r, the number of regeneration cycles $N \geq 1$, the initial values of the auxiliary variables $a^0 = 0$, $b^0 = 0$, $c^0 = 0$, $d^0 = 0$, where $a^0, b^0 \in \Re^1$ and $c^0, d^0 \in \Re^n$, and the normalization constant $M > 0$, whose selection will be explained in the next section.
   Start the simulation from s_r at $t = 0$.

2. Simulation. Continue the simulation until N returns to s_r. Suspend the simulation at the N-th return and compute the estimates ζ^s, χ^s, γ^s and κ^s, defined in (13)- (16), as described in Algorithm 1 and in sensitivity algorithms from ([ARCH92, GAIV92, HO87, RUBI86, SURI87]).

3. Update the values of the auxiliary variables:

$$a^{s+1} = (1 - \delta_{1s})a^s + \delta_{1s}\zeta^s \tag{18}$$

$$b^{s+1} = (1 - \delta_{2s})b^s + \delta_{2s}\chi^s \tag{19}$$

$$c^{s+1} = (1 - \delta_{3s})c^s + \delta_{3s}\gamma^s \tag{20}$$

$$d^{s+1} = (1 - \delta_{4s})d^s + \delta_{4s}\kappa^s \tag{21}$$

where $\delta_{is}, i = 1,\ldots,4$ are auxiliary step sizes.

4. Compute an estimate of the gradient of $F(x)$ at the current point:

$$\xi^s = \begin{cases} \overline{\xi^s} & \text{if } \|\xi^s\| < M \\ 0 & \text{otherwise} \end{cases} \tag{22}$$

where

$$\overline{\xi^s} = \frac{b^s c^s - a^s d^s}{(b^s)^2} \tag{23}$$

5. Obtain the new approximation to the optimal solution:

$$x^{s+1} = \prod_X (x^s - \rho^s \xi^s) \tag{24}$$

6. Check the stopping criterion. If it is satisfied then terminate the iterations, otherwise go to step 2.

3 Convergence of Algorithm 2

The main focus will be devoted to step 3 and 4 of Algorithm 2 which are the main novelty of the algorithm compared with (9)-(17). In fact, in order to prove the convergence of Algorithm 2 it is necessary to show that (22) is the asymptotically unbiased statistical estimate of the gradient of the performance measure $F(x)$, formally:

$$\lim_{s \to \infty} (E(\xi^s | x^0, \ldots, x^s) - F_x(x^s)) = 0 \tag{25}$$

In fact, we are going to show that even the following stronger condition is satisfied:

$$\lim_{s \to \infty} (F_x(x^s) - \xi^s) = 0 \quad a.s.$$

Note that equation (25) is verified if the auxiliary variables defined in (18)-(21) satisfy the following conditions:

$$\|a^s - Ef(t_1(x^s,\omega),\omega)\|^2 \mapsto 0 \quad a.s. \tag{26}$$

$$\|b^s - Et_1(x^s,\omega)\|^2 \mapsto 0 \quad a.s. \tag{27}$$

$$\|c^s - \frac{d}{dx}Ef(t_1(x^s,\omega),\omega)\|^2 \mapsto 0 \quad a.s. \tag{28}$$

$$\|d^s - \frac{d}{dx}Et_1(x^s,\omega)\|^2 \mapsto 0 \quad a.s. \tag{29}$$

To this end it is necessary to impose appropriate conditions on the step sizes in (18)-(21). These conditions are summarized in the following theorem:

Theorem 2 *Suppose that assumptions 1, 2, 4 of Theorem 1 are satisfied together with the following conditions:*

1. *Function $\frac{d}{dx}Ef(t,\omega)$ satisfies Lipschitz condition with respect to $t \in [0,\infty)$, i.e.*

$$\|\frac{d}{dx}Ef(y,\omega) - \frac{d}{dx}Ef(z,\omega)\| \le L_1\|y - z\|, \quad y,z \in [0,\infty) \tag{30}$$

Function $\frac{d}{dx}Et_1(x,\omega)$ satisfies Lipschitz condition on X, i.e.

$$\|\frac{d}{dx}Et_1(y,\omega) - \frac{d}{dx}Et_1(z,\omega)\| \le L_2\|y - z\|, \quad y,z \in X \tag{31}$$

2. *The normalization constant M from (22) is selected in such a way that*

$$M > \sup_{x\in X}\|F_x(x)\| \tag{32}$$

3. *The estimates ζ^s, χ^s, γ^s and κ^s defined in (13)-(16) are such that:*

$$E((\zeta^s - Ef(t_1(x^s,\omega),\omega))^2|x^0,\ldots,x^s) < C < \infty \tag{33}$$

$$E((\chi^s - Et_1(x^s,\omega))^2|x^0,\ldots,x^s) < C < \infty \tag{34}$$

$$E(\|\gamma^s - \frac{d}{dx}Ef(t_1(x^s,\omega),\omega)\|^2|x^0,\ldots,x^s) < C < \infty \tag{35}$$

$$E(\|\kappa^s - \frac{d}{dx}Et_1(x^s,\omega)\|^2|x^0,\ldots,x^s) < C < \infty \tag{36}$$

4. *The step sizes δ_{is}, $i = 1,\ldots,4$, are such that:*

$$\frac{\rho_s}{\delta_{is}} \mapsto 0 \quad as \ s \mapsto \infty, \quad 1 \ge \delta_{is} \ge 0, \quad \sum_{s=1}^\infty \delta_{is} = \infty, \quad \sum_{s=1}^\infty \delta_{is}^2 < \infty. \tag{37}$$

Then all accumulation points of the sequence x^s generated by algorithm (22)-(24) belong to the set of optimal solutions of the problem (8).

In order to prove this theorem we need the following Lemma from ([ERMO92]).

Lemma 1 *Let w^s be a sequence such that*

$$w^{s+1} \leq (1 - \beta_{1s})w^s + \beta_{2s}\nu^s, \quad E(\nu^s | w^1, \ldots, w^s) = \epsilon_s, \quad w^s \geq 0, \quad \frac{\epsilon_s \beta_{2s}}{\beta_{1s}} \mapsto 0 \ a.s.$$

$$\beta_{1s} \leq 1, \quad \sum_{i=1}^{\infty} \beta_{1i} = \infty, \quad \sum_{i=1}^{\infty} \beta_{2i}^2 < \infty, \quad E(\|\nu^s - \epsilon_s\|^2 | w^1, \ldots, w^s) < C < \infty$$

Then: $\qquad \lim_{s \to \infty} w^s = 0 \qquad\qquad a.s.$

<u>Proof of Theorem 2</u> We present the proof for the case when condition 3 is substituted by somewhat more stringent condition:
the sequences

$$\zeta^s - Ef(t_1(x^s, \omega), \omega) \tag{38}$$

$$\chi^s - Et_1(x^s, \omega) \tag{39}$$

$$\gamma^s - \frac{d}{dx}Ef(t_1(x^s, \omega), \omega) \tag{40}$$

$$\kappa^s - \frac{d}{dx}Et_1(x^s, \omega) \tag{41}$$

are bounded a.s.. The more general case described in condition 3 is treated similarly.
If all conditions of Theorem 1 are satisfied, then convergence of Algorithm 2 will follow from that theorem. It is left to show that condition 3 of Theorem 1 is satisfied. In order to do this we are going to prove (26)-(29). Let us prove, for example (27), i.e.

$$\|b^s - Et_1(x^s, \omega)\|^2 \mapsto 0 \qquad a.s..$$

Although $b^s \in \Re^1$ we are going to use norms and scalar products in order to made the proof extendable to all four relations (26)-(29).
Let us define

$$Et_1(x, \omega) = \varphi(x), \quad \|b^s - \varphi(x^s)\|^2 = w^s \tag{42}$$

Since

$$b^{s+1} - \varphi(x^{s+1}) =$$
$$(1 - \delta_{2s})(b^s - \varphi(x^s)) + \delta_{2s}(\chi^s - \varphi(x^s)) + \varphi(x^s) - \varphi(x^{s+1}),$$

we have the following expression for w^{s+1}

$$
\begin{aligned}
w^{s+1} \;=\; & (1-\delta_{2s})^2 w^s + 2\delta_{2s}(1-\delta_{2s})(b^s - \varphi(x^s), \chi^s - \varphi(x^s)) + \\
& + \delta_{2s}^2 \|(\chi^s - \varphi(x^s))\|^2 + \|\varphi(x^s) - \varphi(x^{s+1})\|^2 + \\
& + 2(1-\delta_{2s})(b^s - \varphi(x^s), \varphi(x^s) - \varphi(x^{s+1})) + \\
& + 2\delta_{2s}(\chi^s - \varphi(x^s), \varphi(x^s) - \varphi(x^{s+1}))
\end{aligned}
\tag{43}
$$

Due to condition 1 of this theorem and condition 1 of Theorem 1 there exists a constant K such that

$$
\|\varphi(x^s) - \varphi(x^{s+1})\| \le K\|x^s - x^{s+1}\| \le K\rho_s\|\xi^s\|
$$

Applying condition 2 we obtain the following estimate:

$$
\|\varphi(x^s) - \varphi(x^{s+1})\| \le KM\rho_s
\tag{44}
$$

Due to (39) there exists a constant R_1 such that $\|\chi^s\| \le R_1$. Since b^s is a convex combination of χ^i,(for $i = 1,\ldots,s-1$) and b^0, $\varphi(x)$ is bounded on X. Hence, due to condition 1, we may assume without loss of generality that:

$$
\|b^s - \varphi(x^s)\| < R
$$

for some $R < \infty$. This together with (43)-(44) yield:

$$
\begin{aligned}
w^{s+1} \;\le\; & (1-\delta_{2s})w^s + 2\delta_{2s}(1-\delta_{2s})(b^s - \varphi(x^s), \chi^s - \varphi(x^s)) + \\
& + \delta_{2s}^2 \|(\chi^s - \varphi(x^s))\|^2 + K^2 M^2 \rho_s^2 + \\
& + 2KM(1-\delta_{2s})R\rho_s + 2KM\delta_{2s}\rho_s\|(\chi^s - \varphi(x^s))\|
\end{aligned}
$$

Observe now that all conditions of Lemma 1 are satisfied with: $\beta_{1s} = \beta_{2s} = \delta_{2s}$ and

$$
\begin{aligned}
\nu^s \;=\; & 2(1-\delta_{2s})(b^s - \varphi(x^s), \chi^s - \varphi(x^s)) + \delta_{2s}\|\chi^s - \varphi(x^s)\|^2 + \\
& + K^2 M^2 \frac{\rho_s^2}{\delta_{2s}} + 2KM(1-\delta_{2s})R\frac{\rho_s}{\delta_{2s}} + 2KM\rho_s\|\chi^s - \varphi(x^s)\|.
\end{aligned}
$$

Therefore,

$$
\|b^s - \varphi(x^s)\| \mapsto 0 \quad a.s..
$$

Similarly, all relations (26)-(29) are satisfied which yields :

$$
\|\xi^s - F_x(x^s)\| \mapsto 0 \quad a.s..
$$

Thus, condition 3 of Theorem 1 is satisfied together with all other conditions. Applying now Theorem 1 we obtain required convergence of Algorithm 2.

4 Conclusions

In this paper we have proposed an algorithm for the steady state optimization of a particular class of systems, namely regenerative DEDS. Since the regenerative cycle can be of considerable length, in order to reduce the required simulation time it could be possible to choose a set of m approximately regenerative states $S = \{s_r,\ r = 1, \ldots, m\}$ instead of a single one. However, in this case it has to be taken into account that the estimate of the gradient of the performance measure $F(x)$, as defined in our algorithm, can be biased, i.e.

$$\lim_{s \to \infty} |F_x(x^s) - \xi^s| \le C, \quad C > 0$$

It follows that the resulting solution x^* may be an approximation of the optimal one. Future investigations are under way with the aim of obtaining the upper bound C.

References

[ARCH92] F. Archetti, A. Gaivoronski, A Sciomachen, *Sensitivity Analysis and Optimization of Stochastic Petri Nets* **Journal of Discrete Event Dynamic Systems: Theory and Applications**, vol. 3, pp. 5-37, 1993.

[CRAIG75] M. A. Crane, D. L. Iglehart, *Simulating Stable Stochastic Systems. III Regenerative Processes and Discret Event Simulations*, **Oper. Research**, vol. 23, pp. 33-45, 1975.

[ERMO92] Yu Ermoliev, A. Gaivoronski, *Stochastic Programming techniques for Optimisation for Discrete Event Systems* **Annals of Oper. Research 39**, 1992.

[ERMO88] Yu Ermoliev, R. J-B Wets, eds. *Numerical Techniques for stochastic optimisation* Springer Verlag, Berlin 1988.

[GAIV92] A. Gaivoronski, L. Shi and R. Sreenivas *Augmented Infinitesimal Perturbation Analysis: an Alternative Explanation* **Journal of DEDS** 1992.

[GLYN86] P.W. Glynn, *Optimisation of Stochastic Systems* in **Proceedings of 1986 Winter Simulation Conference**, 1986.

[HO87] Y.C. Ho, *Performance Evaluation and Perturbation Analysis of Discrete Event Dynamic Systems* **IEEE Transaction on Automatic Control**, vol A.C. 32, No. 7, 1987, p 563-572.

[HOLI88] Y. C. Ho, S. Li, *Extentions of Infinitesimal Perturbation Analysis* **IEEE Transactions on Automatic Control**, vol. AC-33, 1988, p. 427-438.

[KALL76] P. Kall *Stochastic Linear Programming* Springer Verlag, Berlin, 1976.

[PREK73] A. Prekopa *Contribution to the Theory of Stochastic Programming* **Mathematical Programming** 4, 202-221, 1973.

[PFLU88] G. Ch. Pflug, *Derivatives of Probability Measures - Concepts and Applications to the Optimisation of Stochastic Systems* in **Discrete Event Systems: Models and Applications, IIASA Conference, Sopron, Hungary, August 3-7, 1987; P. Varaja and A.B. Kurzhanski (eds.), Lecture Notes in Control and Information Sciences**, Springer Verlag, 1988, p. 162-178.

[RUBI86] R. Y. Rubistein. *The Score Function Approach of Sensitivity Analysis of Computer Simulation Models.* **Math. and Computation in Simulations**, vol 28, 1986, p. 351-379.

[SURI87] R. Suri, *Infinitesimal Perturbation Analysis of General Discrete Event Systems*, **J. Assoc. Comput. Mach**, 34, 1987, p 686-717.

[WETS82] R. J-B Wets, *Stochastic Programming: Solution Techniques and Approximation Schemes* in **Mathematical Programming; The State of the Art 1982**, A. Bachem, M. Groetschel and B. Korte (eds.), Springer Verlag, Berlin, 566-603.

STOCHASTIC DYNAMIC OPTIMIZATION: MODELLING AND METHODOLOGICAL ASPECTS

K. Frauendorfer

Institute for Operations Research
University of St. Gallen
Bodanstr. 6, CH-9000 St. Gallen

Abstract. We consider two important problem classes within stochastic dynamic optimization: the stochastic multistage recourse problem with stochastic dependent random vectors and the discrete control problem with Markovian structure and with full state information. We apply barycentric approximation schemes and discuss the corresponding approximate problems.

Keywords: Approximation, Sequential Stochastic Decision Process, Discretization.

1. Stochastic models of discrete time

In many practical situations (e.g. in finance, economics) decisions have to be made periodically under uncertainties about future outcomes and events. Usually, the decisions have to be chosen in such a way that total cost or profit have to be optimized with respect to certain constraints over the underlying planning horizon. These situations are preferably modelled as stochastic dynamic optimization problems which represent a very general class of models within the field of mathematical optimization. Methodologies for these models are often based on the dynamic programming technique, which on the one hand requires the Markovian property of the (state) equation and stochastic independence of the random parameters or disturbances; on the other hand dynamic programming has a wider scope of applicability since it can handle difficult constraint sets such as discrete sets. Furthermore, dynamic programming leads to a globally optimal solution as opposed to variational mathematical programming techniques, for which this cannot be guaranteed in general. From a computational viewpoint the curse of dimensionality amounts to computational complexities which in many cases are insurmountable. Apart from special cases this requires to discretize the spaces so that the minimization is carried out for an affordable number of discretization points.

We shall concentrate on two classes of stochastic dynamic problems: the stochastic multistage (recourse) model with stochastic dependent random vectors and the discrete control problem with Markovian structure and with full state information. We focus on the barycentric approximation technique (due to [8]) and the structural properties needed for minorization and majorization of the underlying dynamic stochastic problem. The presented investigations are based on recent results of [9] and of [11] and may be seen as extensions. (We assume measurability and integrability of the involved functions.)

The stochastic multistage recourse problem is given as

$$\min\ \rho_0(u_0) + \int_{\Theta \times \Xi} [\sum_{t=1}^{T} \rho_t(u_0, u_1, \cdots, u_t, \eta_1, \cdots, \eta_t)] dP(\eta, \xi) \qquad (1.1)$$

$$s.t.:\ f_0(u_0) \le 0$$
$$f_t(u_0, u_1, \cdots, u_t, \xi_1, \cdots, \xi_t) \le 0 \quad t = 1, \cdots, T$$
$$u(\cdot)\ \text{nonanticipative}\ .$$

and the stochastic control problem over a discrete finite time horizon is

$$\min\ E_\pi [\sum_{t=0}^{T-1} \rho_t(\eta_t, x_t, \pi(x_t)) + b(x_T)]$$
$$s.t.\ x_{t+1} = f_t(x_t, \pi(x_t), \xi_t) \quad t = 0, 1, \cdots, T-1 \qquad (1.2)$$

In the multistage case (see e.g. [3],[4],[5],[6],[7],[9],[12-27]) $u_t \in \mathbb{R}^{n_t} (t = 1, \cdots, T)$ denote the decisions selected at time t after $(\eta_t, \xi_t) \in \mathbb{R}^{K_t} \times \mathbb{R}^{L_t}$ is observed but prior to the 'future' outcomes $(\eta_{t+1}, \xi_{t+1}), \cdots, (\eta_T, \xi_T)$. According to this rule, we obtain a *policy* (or *recourse function*) $u(\cdot)$ which is called *nonanticipative*. The sets of feasible decisions u_t at time $t = 0, 1, \cdots, T$ may depend on previous decisions and on observations up to time t and are given through the system of inequalities in (1.1). The sequence of observations $(\eta_1, \xi_1), \cdots, (\eta_t, \xi_t)$ and decisions $(u_0, u_1, \cdots, u_t)$ up to time t determines a cost $\rho_t(u_0, u_1, \cdots, u_t, \eta_1, \cdots, \eta_t)$. The objective is to find a nonanticipative recourse function $u(\cdot)$ that minimizes the expected value of the overall cost and satisfies the constraints in (1.1). The (induced) probability space associated with stochastic dependent random elements $(\eta, \xi) = (\eta_t, \xi_t;\ t = 1, \cdots, T)$ is denoted $(\Theta \times \Xi, \mathcal{B}^{K+L}, P)$. $\mathcal{B}^{K+L}$ is the Borel field on $\Theta \times \Xi$ and P denotes a regular probability measure on $(\Theta \times \Xi, \mathcal{B}^{K+L})$. Let further $(\eta^t, \xi^t) \in \Theta^t \times \Xi^t := (\Theta_1 \times \Xi_1) \times \cdots \times (\Theta_t \times \Xi_t)$, and in particular $\Theta \times \Xi := \Theta^T \times \Xi^T$. The induced marginal probability spaces are denoted $(\Theta_t \times \Xi_t, \mathcal{B}^{K_t+L_t}, P_t)$ and $(\Theta^t \times \Xi^t, \mathcal{B}^{K^t+L^t}, P^t)$.

For the control case (see e.g. [1],[2],[5],[6],[7],[11],25]) let X_t and U_t denote the state space and the control space both equipped with the Borel field. The objective is to minimize the total expected cost over all admissible *control policies* $\pi = \{\pi_t; t = 0, \cdots, T-1\}$ where $\pi_t : X_t \to U_t$. The dynamical system is characterized through the state equations in (1.2). Again $(\eta_t, \xi_t) \in \mathbb{R}^{K_t} \times \mathbb{R}^{L_t}$ denote the random disturbances of the dynamical system. Here a control action $u_t = \pi_t(x_t)$ is chosen prior to the outcome (η_t, ξ_t). Further, $(\eta_t, \xi_t)\ t = 0, \cdots, T-1$ are assumed to be stochastically independent, but may be stochastically dependent on the current state and on the current control action (x_t, u_t). The corresponding probability measure is therefore a *controlled transition probability* and will be denoted $P(\eta_t, \xi_t | x_t, u_t)$. In (1.2) also the state x_t is random as it depends on the uncertain outcome ξ_t; additionally, $\{x_t\}$ depends on the control policy π_{t-1} so that the expectational operator E in (1.2) has to be taken with respect to the underlying policy π.

Remark 1.1: For the ease of exposition, let $u^t := (u_0, u_1, u_2, \cdots, u_t)$ be the sequence of decisions and $\eta^t := (\eta_1, \eta_2, \cdots, \eta_t)$, $\xi^t := (\xi_1, \xi_2, \cdots, \xi_t)$ be the sequence of random outcomes (disturbances) up to time t; note that $u^t := (u^{t-1}, u_t)$, $\eta^t := (\eta^{t-1}, \eta_t)$, $\xi^t := (\xi^{t-1}, \xi_t)$. Further we set $n := n_0 + n_1 + \cdots + n_T$, $K := K_1 + \cdots + K_T$, $L := L_1 + \cdots + L_T$, $n^t := n_0 + n_1 + \cdots + n_t$, $K^t := K_1 + \cdots + K_t$, $L^t := L_1 + \cdots + L_t$.

2. Barycentric approximation in the recourse case

Considering the dynamic version of the stochastic multistage problem, we obtain:

$$\phi_T(u^{T-1}, \eta^T, \xi^T) := \min \{\rho_T(u^{T-1}, u_T, \eta^T) \mid f_T(u^{T-1}, u_T, \xi^T) \leq 0\} \qquad (2.1.1)$$

and for $t = T, \cdots, 1$

$$\phi_{t-1}(u^{t-2}, \eta^{t-1}, \xi^{t-1}) :=$$

$$\min \rho_{t-1}(u^{t-2}, u_{t-1}, \eta^{t-1}) +$$
$$+ \int \phi_t(u^{t-2}, u_{t-1}, \eta^{t-1}, \xi^{t-1}, \eta_t, \xi_t) \, dP_t(\eta_t, \xi_t \mid \eta^{t-1}, \xi^{t-1})$$

$$(2.1.2)$$

$$s.t. \quad f_{t-1}(u^{t-2}, u_{t-1}, \xi^{t-1}) \leq 0$$

(let (η^0, ξ^0) be a fictitious degenerate random vector).

The integration area in (2.1.2) depends on (η^{t-1}, ξ^{t-1}). We choose $\Theta_t(\eta^{t-1}, \xi^{t-1})$, $\Xi_t(\eta^{t-1}, \xi^{t-1})$ uniquely for a given (η^{t-1}, ξ^{t-1}), in such a way that $\Theta_t(\eta^{t-1}, \xi^{t-1})$ and $\Xi_t(\eta^{t-1}, \xi^{t-1})$ are regular simplices $(t = 1, \cdots, T)$ covered by $\Theta_t \times \Xi_t$. To simplify the writing of integral terms, we neglect $\Theta_t(\eta^{t-1}, \xi^{t-1}) \times \Xi_t(\eta^{t-1}, \xi^{t-1})$ below the integral, and emphasize that the dependency on (η^{t-1}, ξ^{t-1}) is expressed through the regular conditional probability measure $P_t(\eta_t, \xi_t \mid \eta^{t-1}, \xi^{t-1})$. Let $Q_t^l(\eta_t, \xi_t \mid \eta^{t-1}, \xi^{t-1})$ and $Q_t^u(\eta_t, \xi_t \mid \eta^{t-1}, \xi^{t-1})$ be the barycentric approximations of $P_t(\eta_t, \xi_t \mid \eta^{t-1}, \xi^{t-1})$ (due to [9]).

Further in [9], auxiliary functions $\psi_{t-1}(u^{t-2}, \eta^{t-1}, \xi^{t-1})$ and $\Psi_{t-1}(u^{t-2}, \eta^{t-1}, \xi^{t-1})$ have been introduced given through the recursions

$$\psi_T(u^{T-1}, \eta^T, \xi^T) := \phi_T(u^{T-1}, \eta^T, \xi^T) =: \Psi_T(u^{T-1}, \eta^T, \xi^T) \qquad (2.2.1)$$

$$\psi_{t-1}(u^{t-2}, \eta^{t-1}, \xi^{t-1}) :=$$
$$:= \min \rho_{t-1}(u^{t-2}, u_{t-1}, \eta^{t-1}) + \qquad (2.2.2)$$
$$+ \int \psi_t(u^{t-1}, u_{t-1}, \eta^{t-1}, \xi^{t-1}, \eta_t, \xi_t) \, dQ_t^l(\eta_t, \xi_t \mid \eta^{t-1}, \xi^{t-1})$$

$$s.t. \quad f_{t-1}(u^{t-2}, u_{t-1}, \xi^{t-1}) \leq 0$$

$$\Psi_{t-1}(u^{t-2}, \eta^{t-1}, \xi^{t-1}) :=$$
$$:= \min\ \rho_{t-1}(u^{t-2}, u_{t-1}, \eta^{t-1}) + \tag{2.2.3}$$
$$+ \int \Psi_t(u^{t-1}, u_{t-1}, \eta^{t-1}, \xi^{t-1}, \eta_t, \xi_t)\ dQ_t^u(\eta_t, \xi_t \mid \eta^{t-1}, \xi^{t-1})$$
$$s.t.\quad f_{t-1}(u^{t-2}, u_{t-1}, \xi^{t-1}) \le 0$$

In [9] we have investigated convex stochastic multistage problems and required the following assumptions:

i) $\Theta_t \times \Xi_t$ is compact, convex and covers the support of the random vector (η_t, ξ_t) $(t = 1, \cdots, T)$;

ii) $(u^t, \eta^t) \to \rho_t(u^t, \eta^t)$ are continuous saddle functions on $\mathbb{R}^{n^t} \times \Theta^t$ (convex in u^t, concave in η^t) for $t = 1, \cdots, T$;

iii) $\{u_0 \mid f_0(u_0) \le 0\}$ and $\{(u^t, \xi^t) \mid \xi^T \in \Xi^T,\ f_t(u^t, \xi^t) \le 0\}$ $(t = 1, \cdots, T)$ are compact, convex subsets of $\mathbb{R}^{n^t} \times \Xi^t$.

iv) *Strict nonanticipativity:* For any $(\eta^t, \xi^t) \in \Theta^t \times \Xi^t$ and for any decision u^{t-1} (feasible with respect to ξ^t), there exists some v_t – dependent on (u^{t-1}, η^t, ξ^t) – for which $f_t(u^{t-1}, v_t, \xi^t) < 0$ holds;

v) *Inheritance of the saddle property:* The functionals $(E_t \phi_t)(u^{t-1}, \eta^{t-1}, \xi^{t-1})$ are continuous saddle functions - convex in (u^{t-1}, ξ^{t-1}), concave in η^{t-1} - with respect to the domains.

Assumptions i)-v) ensure that the optimization problems in (2.2) are solvable and that the corresponding value functions are continuous saddle functions. If one is interested in problems which are *nonconvex* or *discrete* with respect to the decisions u^t, assumptions ii),iii),iv) and v) may be weakened to ii'),iii'),iv') and v') and the results of [9] are still valid.

ii') $\eta^t \to \rho_t(u^t, \eta^t)$ are continuous concave functions on Θ^t for any feasible u^t $(t = 1, \cdots, T)$;

iii') $\{u_0 \mid f_0(u_0) \le 0\}$ and $\{(u^t, \xi^t) \mid \xi^T \in \Xi^T,\ f_t(u^t, \xi^t) \le 0\}$ $(t = 1, \cdots, T)$ are compact subsets of $\mathbb{R}^{n^t} \times \Xi^t$;

iv') *Solvability:* The optimization problems (2.2) are solvable for any admissible parameter (u^{t-1}, η^t, ξ^t);

v') *Inheritance of the saddle property:* The functionals

$$(\eta^{t-1}, \xi^{t-1}) \to (E_t \phi_t)(u^{t-1}, \eta^{t-1}, \xi^{t-1})$$

are continuous saddle functions - convex in (ξ^{t-1}), concave in η^{t-1} - with respect to their domains for any feasible u^{t-1}.

Remark 2.1: Obviously, ii)-v) imply ii')-v'); in general, assumption iv') will be harder to verify than iv).

Theorem 2.1 : Under the assumptions i),ii'),iii'),iv') and v') it holds for any $(\eta^t, \xi^t) \in \Theta^t \times \Xi^t$ and any (feasible) decision u^{t-1} $(t = 1, \cdots, T)$:

$$\psi_t(u^{t-1}, \eta^t, \xi^t) \le \phi_t(u^{t-1}, \eta^t, \xi^t) \le \Psi_t(u^{t-1}, \eta^t, \xi^t) . \tag{2.3}$$

Proof: Follows immediately from Lemma 4.1 and Lemma 4.2 in [9]. //

Based on the barycentric approximations $Q_t^l(\eta_t, \xi_t \mid \eta^{t-1}, \xi^{t-1})$, $Q_t^u(\eta_t, \xi_t \mid \eta^{t-1}, \xi^{t-1})$ at time t, we have constructed discrete probability measures Q^l and Q^u for the entire planning horizon so that $(\Theta \times \Xi, B^{K+L}, Q^l)$, $(\Theta \times \Xi, B^{K+L}, Q^u)$ may be accepted as *barycentric approximations* of $(\Theta \times \Xi, B^{K+L}, P)$. In particular, the supports of Q^l and Q^u were interpreted as barycentric scenario trees and their paths as barycentric scenarios. These probability measures entail stochastic multistage problems whose solutions and values represent approximates for problem (1.1):

$$\min\ \rho_0(u_0) + \int_{\Theta \times \Xi} [\sum_{t=1}^{T} \rho_t(u_0, u_1, \cdots, u_t, \eta_1, \cdots, \eta_t)] dQ^l(\eta, \xi) \qquad (2.4)$$

$$s.t. : f_0(u_0) \leq 0$$
$$f_t(u_0, u_1, \cdots, u_t, \xi_1, \cdots, \xi_t) \leq 0 \quad t = 1, \cdots, T$$
$$u(\cdot)\ \text{nonanticipative}\ .$$

$$\min\ \rho_0(u_0) + \int_{\Theta \times \Xi} [\sum_{t=1}^{T} \rho_t(u_0, u_1, \cdots, u_t, \eta_1, \cdots, \eta_t)] dQ^u(\eta, \xi) \qquad (2.5)$$

$$s.t. : f_0(u_0) \leq 0$$
$$f_t(u_0, u_1, \cdots, u_t, \xi_1, \cdots, \xi_t) \leq 0 \quad t = 1, \cdots, T$$
$$u(\cdot)\ \text{nonanticipative}\ .$$

Under the assumptions i)-v), i),ii')-v') respectively, the optimal values of (2.4) and (2.5) are lower and upper bounds for the optimal value of (1.1). Problems (2.4) and (2.5) may be treated as deterministic mathematical programs (convex, nonconvex or discrete) due to the discreteness of the probability measure $Q^l(\eta, \xi)$ and $Q^u(\eta, \xi)$. Therefore, besides the dynamic programming method also mathematical programming methodologies can be applied.

3. Barycentric approximation in the discrete control case

The dynamic version of the discrete control problem (1.2) reads as

$$\phi_T(x_T) = b(x_T) \qquad (3.1.1)$$

and backwards for $t = T - 1, \cdots, 0$

$$\phi_t(x_t) := \min_{u_t \in U_t} \int \rho_t(x_t, u_t, \eta_t) + \phi_{t+1}(f_t(x_t, u_t, \xi_t))\ dP_t(\eta_t, \xi_t | x_t, u_t) \qquad (3.1.2)$$

ending with the optimal value $\phi_0(x_0)$ of problem (1.2).

For any t, for any state x_t and for any control action u_t we denote with $\Theta_t(x_t, u_t)$, $\Xi_t(x_t, u_t)$ regular simplices in $\mathbb{R}^{K_t}, \mathbb{R}^{L_t}$, which cover the support of (η_t, ξ_t). Let $P_t(\eta_t, \xi_t \mid x_t, u_t)$ be regular conditional probability measures for any $(x_t, u_t) \in X_t \times U_t$ and let $Q_t^l(\eta_t, \xi_t \mid x_t, u_t)$, and $Q_t^u(\eta_t, \xi_t \mid x_t, u_t)$ be the associated barycentric approximations on $(\Theta_t(x_t, u_t) \times \Xi_t(x_t, u_t), B^{K_t + L_t}) - t = 0, \cdots, T - 1$. Applying [9] both barycentric approximations are uniquely determined through the moment functionals (for $k = 1, \cdots, K_t$, $l = 1, \cdots, L_t$)

$$m_{t,k,l}(x_t, u_t) := \int \eta_{t,k} \cdot \xi_{t,l} \, dP_t(\eta_t, \xi_t \mid x_t, u_t) \qquad (3.2.1)$$

$$m_{t,k,0}(x_t, u_t) := \int \eta_{t,k} \, dP_t(\eta_t, \xi_t \mid x_t, u_t) \qquad (3.2.2)$$

$$m_{t,0,l}(x_t, u_t) := \int \xi_{t,l} \, dP_t(\eta_t, \xi_t \mid x_t, u_t) \qquad (3.2.3)$$

which are finite as the support of (η_t, ξ_t) is bounded. Further, in the case that the integrand $\rho_t(x_t, u_t, \eta_t) + \phi_{t+1}(f_t(x_t, u_t, \xi_t))$ is a continuous saddle function in (η_t, ξ_t) we have

$$\int [\rho_t(x_t, u_t, \eta_t) + \phi_{t+1}(f_t(x_t, u_t, \xi_t))] dQ_t^l(\eta_t, \xi_t \mid x_t, u_t) \le$$

$$\le \int [\rho_t(x_t, u_t, \eta_t) + \phi_{t+1}(f_t(x_t, u_t, \xi_t))] dP_t(\eta_t, \xi_t \mid x_t, u_t) \le \qquad (3.3)$$

$$\le \int [\rho_t(x_t, u_t, \eta_t) + \phi_{t+1}(f_t(x_t, u_t, \xi_t))] dQ_t^u(\eta_t, \xi_t \mid x_t, u_t),$$

We define auxiliary functions $\psi_{t-1}(x_{t-1})$, $\Psi_{t-1}(x_{t-1})$ given through the recursions :

$$\psi_T(x_T) = b(x_T) \qquad (3.4.1)$$

and backwards for $t = T - 1, \cdots, 0$

$$\psi_t(x_t) := \min_{u_t \in U_t} \int \rho_t(x_t, u_t, \eta_t) + \psi_{t+1}(f_t(x_t, u_t, \xi_t)) \, dQ_t^l(\eta_t, \xi_t \mid x_t, u_t) \qquad (3.4.2)$$

Similarly,

$$\Psi_T(x_T) = b(x_T) \qquad (3.5.1)$$

and backwards for $t = T - 1, \cdots, 0$

$$\Psi_t(x_t) := \min_{u_t \in U_t} \int \rho_t(x_t, u_t, \eta_t) + \Psi_{t+1}(f_t(x_t, u_t, \xi_t)) \, dQ_t^u(\eta_t, \xi_t \mid x_t, u_t) \qquad (3.5.2)$$

Let be

$$(E_t^l \rho_t + \psi_{t+1} \circ f_t)(x_t, u_t) := \qquad (3.6.1)$$

$$:= \int \rho_t(x_t, u_t, \eta_t) + \psi_{t+1}(f_t(x_t, u_t, \xi_t)) \, dQ_t^l(\eta_t, \xi_t \mid x_t, u_t)$$

$$(E_t^u \rho_t + \Psi_{t+1} \circ f_t)(x_t, u_t) := \tag{3.6.2}$$

$$:= \int \rho_t(x_t, u_t, \eta_t) + \Psi_{t+1}(f_t(x_t, u_t, \xi_t)) \ dQ_t^u(\eta_t, \xi_t | x_t, u_t)$$

Theorem 3.1 : Assume that for $t = 0, \cdots, T - 1$ the optimization problems in (3.4) and (3.5) are solvable $\forall x_t \in X_t$ and that

$$(\eta_t, \xi_t) \to \rho_t(x_t, u_t, \eta_t) + \Psi_{t+1}(f_t(x_t, u_t, \xi_t))$$

are continuous saddle functions $\forall (x_t, u_t) \in X_t \times U_t$. Then it holds that the auxiliary functions are minorants and majorants for the value functions of the given problem; i.e.

$$\psi_t(x_t) \le \phi_t(x_t) \le \Psi_t(x_t) . \tag{3.7}$$

Proof: Trivially, for $t = T$ we have $\psi_T \circ f_{T-1} = \phi_T \circ f_{T-1} = \Psi_T \circ f_{T-1}$ yielding

$$(E_{T-1}^l \rho_{T-1} + \psi_T \circ f_{T-1})(x_{T-1}, u_{T-1}) = (E_{T-1}^l \rho_{T-1} + \phi_T \circ f_{T-1})(x_{T-1}, u_{T-1}) \le$$
$$\le (E_{T-1} \rho_{T-1} + \phi_T \circ f_{T-1})(x_{T-1}, u_{T-1}) \le \tag{3.8}$$
$$\le (E_{T-1}^u \rho_{T-1} + \phi_T \circ f_{T-1})(x_{T-1}, u_{T-1}) = (E_{T-1}^u \rho_{T-1} + \Psi_T \circ f_{T-1})(x_{T-1}, u_{T-1})$$

This implies

$$) \quad \psi_{T-1}(x_{T-1}) \le \phi_{T-1}(x_{T-1}) \le \Psi_{T-1}(x_{T-1} \tag{3.9}$$

Now, for $t = 1, \cdots, T - 1$, it follows easily through induction (backwards):

$$(E_{t-1}^l \rho_{t-1} + \psi_t \circ f_{t-1})(x_{t-1}, u_{t-1}) \le (E_{t-1}^l \rho_{t-1} + \phi_t \circ f_{t-1})(x_{t-1}, u_{t-1}) \le$$
$$\le (E_{t-1} \rho_{t-1} + \phi_t \circ f_{t-1})(x_{t-1}, u_{t-1}) \le \tag{3.10}$$
$$\le (E_{t-1}^u \rho_{t-1} + \phi_t \circ f_{t-1})(x_{t-1}, u_{t-1}) \le (E_{t-1}^u \rho_{t-1} + \Psi_t \circ f_{t-1})(x_{t-1}, u_{t-1})$$

The first and last inequalities hold as $\psi_t(x_t)$ is a lower approximate for $\phi_t(x_t)$ and $\Psi_t(x_t)$ is an upper approximate for $\phi_t(x_t)$. The second and the third inequalities follow from barycentric approximations (3.3) of the saddle functions

$$(\eta_{t-1}, \xi_{t-1}) \to \rho_{t-1}(x_{t-1}, u_{t-1}, \eta_{t-1}) + \phi_t(f_{t-1}(x_{t-1}, u_{t-1}, \xi_{t-1})) \tag{3.11}$$

(3.4), (3.5) and (3.10) imply the assertion (3.7). //

Remark 3.1: We note that

$$(\eta_{t-1}, \xi_{t-1}) \to \rho_{t-1}(x_{t-1}, u_{t-1}, \eta_{t-1}) + \psi_t(f_{t-1}(x_{t-1}, u_{t-1}, \xi_{t-1})) \tag{3.12.1}$$
$$(\eta_{t-1}, \xi_{t-1}) \to \rho_{t-1}(x_{t-1}, u_{t-1}, \eta_{t-1}) + \Psi_t(f_{t-1}(x_{t-1}, u_{t-1}, \xi_{t-1})) \tag{3.12.2}$$

are not necessarily saddle functions. Therefore, the expectation functionals $(E_{t-1}^l \rho_{t-1} + \psi_t \circ f_{t-1})(x_{t-1}, u_{t-1})$ and $(E_{t-1}^u \rho_{t-1} + \Psi_t \circ f_{t-1})(x_{t-1}, u_{t-1})$ are not necessarily bounding $(E_{t-1} \rho_{t-1} + \psi_t \circ f_{t-1})(x_{t-1}, u_{t-1})$ and $(E_{t-1} \rho_{t-1} + \Psi_t \circ f_{t-1})(x_{t-1}, u_{t-1})$.

4. Conclusions

Based on barycentric approximations of the conditional probability measures and of the transition probability measures we have derived approximate problems for the multistage recourse problem and for the discrete stochastic control problem. Under the assumption that the integrands represent saddle functions (with respect to the random vectors) we obtain lower and upper bounds for the optimal value which release the goodness of the underlying approximation.

In the *multistage case* the approximate recourse functions of (2.2) and (2.3) correspond to distinguished paths which may be interpreted as scenario tree. The corresponding recourse decisions depend on the history (i.e. on outcomes of the random parameter in the past). This enables to treat the corresponding optimization problem (2.2),(2.3) as mathematical program with the possibility to use its methodologies as solution procedures. In [9] we have presented an error analysis over the planning horizon, which gives information on the goodness of the approximation at each time t.

In the *discrete control case* the approximate policies of (3.4) and (3.5) are given as optimal solutions with respect to the given state. Again, at each time we obtain lower and upper values for the given state, anticipating optimal policies in the future, which provide information on the goodness of the current approximation. The policies depend only on the state as the random disturbances have supposed to be independent of the past. This necessitates that for solving the approximate problems (3.4) and (3.5) one has to rely on the dynamic programming technique.

In [11] the authors have considered the case where the random disturbances are independent of state, control and history. Therein convex stochastic control problems are considered with discrete finite and infinite planning horizon. The proposed approximation method is based on applying the classical inequalities of Jensen and Edmundson-Madansky to the expectation functionals. The relations of these bounds to barycentric approximations have been outlined in [8]. In the case that the random disturbances are independent of state, control and history the discrete control model takes also the features of the multistage model; in particular, the approximate problems (3.4) and (3.5) may be formulated as deterministic mathematical programs. The associated optimal solutions represent policies, which are defined only on some distinguished states (which are induced by the support of the discrete probability measures and by the optimal control action of the previous period). Proper interpolations with respect to state yield then approximate policies which gives decision rules on the entire state space.

The goodness of the approximate problems (2.2), (2.3) and (3.4), (3.5) may be improved through simplicial partitions of the underlying probability spaces (in a similar way as in [8]). If the sequences of partitions represent refinements then monotonicity of the auxiliary functions follows immediately from Theorem 16.2 in [8], implying that the corresponding approximate values behave also monotonously. Convergence can be proven via epi-convergence of the involved value functions if the refinements become 'arbitrarily' small. This seems to be more of theoretical importance as one is heavily faced with the curse of dimensionality. Practically, the challenge is seeing more in the selection of computational activities, which provides us with valuable information on how to reduce the error significantly.

References

1. M. AOKI: *Optimization of Stochastic Systems - Topics in Discrete-Time Dynamics*; Academic Press 1989

2. D. P. BERTSEKAS: *Dynamic Proramming - Deterministic and Stochastic Models*; Prentice-Hall 1987

3. J. BIRGE and R. J.-B. WETS: *Stochastic Programming, Part I, II*; Proceedings of the 5th International Conference on Stochastic Programming, Ann Arbor, Michigan, August 13-18, 1989; Annals of Operations Research 30-31 (1991)

4. G.B. DANTZIG and G. INFANGER: Multistage Stochastic Linear Programs for Portfolio Optimization; Technical Report SOL 91-11, Stanford University (1991)

5. M.A.H. DEMPSTER (ed.): *Stochastic Programming*; Proceedings of the 1st Internatinal Conference on Stochastic Programming, Oxford 1974; Academic Press (1980)

6. M.A.H. DEMPSTER: On Stochastic Programming II: Dynamic Problems Under Risk; Research Report DAL TR 86-5, Dalhousie University (1986)

7. Y. ERMOLIEV and R. J.-B. WETS: *Numerical Techniques for Stochastic Optimization*; Springer-Verlag (1988)

8. K. FRAUENDORFER: *Stochastic Two-Stage Programming*; Lecture Notes in Economics and Mathematical Systems 392, Springer-Verlag (1992)

9. K. FRAUENDORFER: Multistage Stochastic Programming: Error Analysis for the Convex Case; (to appear in *Zeitschrift for Operations Research*)

10. H.I. GASSMANN: MSLiP: A Computer Code for the Multistage Stochastic Linear Programming Problem; *Math. Prog.* 47 (1990) 407-423

11. O. HERNANDEZ-LERMA and W. J. RUNGGALDIER : Monotone Approximations for Convex Stochastic Control Problems; (to appear in *Journal of Estimation, Control and Systems* 1993)

12. F.V. LOUVEAUX: A Solution Method for Multistage Stochastic Programs with Recourse, with Application to an Energy Investment Problem; *Operations Research* 28/4 (1980) 889-902

13. J.M. MULVEY and H. VLADIMIROU: Solving Multistage Stochastic Networks: An Application of Scenario Aggregation; *Networks* 21/6 (1991) 619-643

14. P. OLSEN: Multistage Stochastic Programming with Recourse: The Equivalent Deterministic Problem; *SIAM J. Control and Optimization* 14/3 (1976) 495-517

15. P. OLSEN: When is a Multistage Stochastic Programming Well-Defined?; *SIAM J. Control and Optimization* 14/3 (1976) 518-527

16. P. OLSEN: Multistage Stochastic Programming with Recourse as Mathematical Programming in an L_p-Space; *SIAM J. Control and Optimization* 14/3 (1976) 528-537

17. P. OLSEN: Discretization of Multistage Stochastic Programming Problems; *Math. Prog. Study* 6 (1976) 111-124

18. S.M. ROBINSON: Extended Scenario Analysis; in: J. Birge and R. J.-B. Wets (eds.): *Stochastic Programing, Part II*; Annals of Operations Research 31 (1991) 385-398

19. R.T. ROCKAFELLAR: Integral Functionals, Normal Integrand and Measurable Selections; in: *Nonlinear Operators and the Calculus of Variations*; Lecture Notes in Mathematics 543, Springer-Verlag (1976) 157-207

20. R.T. ROCKAFELLAR and R. J.-B. WETS: Nonanticipativity and L^1-Martingales in Stochastic Optimization Problems; *Math. Prog. Study* 6 (1976) 170-187

21. R.T. ROCKAFELLAR and R. J.-B. WETS: The Optimal Recourse Problem in Discrete Time: L^1-Multipliers for Inequality Constraints; *SIAM J. Control and Optimization* 16/1 (1978) 16-36

22. R.T. ROCKAFELLAR and R. J.-B.- WETS: Generalized Linear-Quadratic Problems of Deterministic and Stochastic Optimal Control in Discrete Time; *SIAM J. Control and Optimization* 28/4 (1990) 810-822

23. R.T. ROCKAFELLAR and R. J.-B. WETS: Scenarios and Policy Aggregation in Optimization under Uncertainty; *Math. of OR* 16 (1991) 119-147

24. A. RUSZCZYNSKI: Parallel Decomposition of Multistage Stochastic Programming Problems; *Mathematical Programming* 58/2 (1993) 201-228

25. P. VARAIYA and R. J.-B. WETS: Stochastic Dynamic Optimization: Approaches and Computation; IIASA Working Paper (1988)

26. S.W. WALLACE and T. YAN: Bounding Multistage Stochastic Linear Programs from above; Working Paper (1989)

27. R. J.-B- WETS: The Aggregation Principle in Scenario Analysis and Stochastic Optimization; in: S. Wallace (ed.): *Algorithms and Model Formulations in Mathematical Programming*; Springer-Verlag, NATO ASI 51 (1989) 91-113

Modelling
and
parameter estimation

Modelling, Control and Optimization of Mechanical Systems with Open Chains in Robotics

P. Buráň, J. Doležal

Institute of Information Theory and Automation
Academy of Sciences of the Czech Republic
182 08 Prague, Czech Republic

Introduction

The contribution follows two main aims. First, to present the developed and implemented graphical system for modelling and analysis of mechanical systems with open chains which play an important role in robotics. Only having an efficient simulation environment available it is possible to consider problems of control and subsequent optimization of such mechanical systems. The second aim is just to describe incorporation of control schemes and optimization strategies possibly within such environment.

The graphic environment called SiMeS, provides a tool for analysis and visualization of any robotic system with an *open chain* mechanical structure representation of motion. It includes all basic mechanical variables which occur during the motion. Environment can be used both as educational tool for complicated motions presentation and for concrete control results representation in different fields of industry and science.

Modelling-Oriented Graphical Environment

The environment contains a reasonably general model description, which enables to create open chain mechanical models with arbitrary number of degrees of freedom. User writes the necessary data into the interactive table and can easily configure also complicated mechanical structures. SiMeS offers several possible ways of depicting and visualization mechanical movements. This include both a simple trajectory and complicated spatial pictures. User can control the complexity of the created picture by switching the generated points. Multiple diagrams can be arbitrarily combined using available pre computed quantities.

For an arbitrary given structure it is then possible to represent time courses of a real machine position, velocities, accelerations, forces and energies. Position courses can be represented either in spatial view or in partial views from the system axis directions. The example of some views are in Figs. 1 and 2. Another available and convenient option

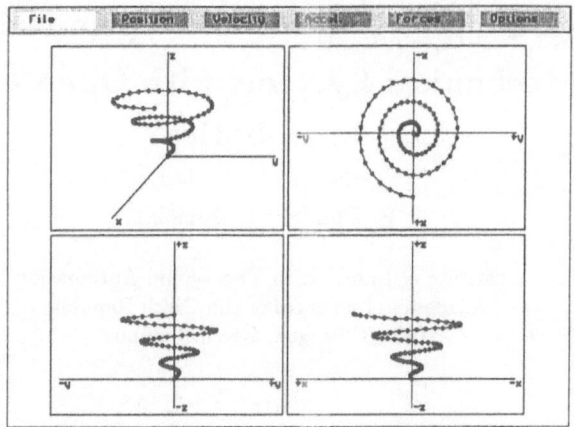

FIGURE 1 Position vectors in all views

is a motion animation and creation of *stroboscopic* prints of the movement in question. Alternatively the environment is able to visualize all other mechanical variables in partial links of a mechanical structure.

The environment was developed as a tool for the authors to deal with more complicated structure of mechanical systems. As no affordable commercial product was available some effort has to be invested to this software. It worth to note that existing commercial software is usually considerably expensive and thus not readily available especially in academic environment.

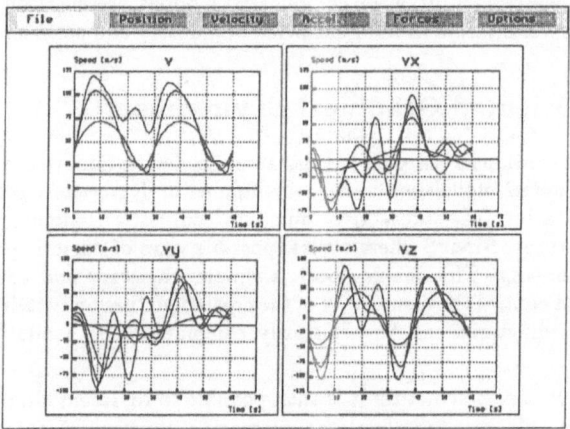

FIGURE 2 Multiple velocity diagrams

Simulation Methodology

The solution of these problems consists of a simplification of a real machine (industrial robot) description and its replacement by a mechanical model, which has to be simple enough to allow satisfactory and efficient treatment. However, it should contain every important parameter and variable, followed by a solution of the resulting model. The solution yields kinematic and dynamic description of the original system. For model building different methods can be used [1], [2].

The internal model includes the so-called A-matrix method [1]. A transformation of the space H is a 4×4 matrix that can represent translation, rotation, stretching and perspective transformations. All standard transformations in robotics may be interpreted as a product of rotation and translation transformations. If they are interpreted from left to right, then the rotations and translations are in terms of the currently defined coordinate frame. The description of object A in terms of object B by means of a homogeneous transformation may be inverted to obtain the description of object B in terms of object A.

Any manipulator can be considered to consist of a series of links connected together by joints. Using homogeneous transformations we can describe the relative position and orientation between coordinate frames embedded in each link of the manipulator. For a given six link manipulator we can write

$$T_6 = A_1 A_2 A_3 A_4 A_5 A_6 \tag{1}$$

where A_1 describes the position and orientation of the first link, A_2 describes the position and orientation of the second link with respect to the first, etc. For instance the cylindrical system with three degrees of freedom is described as

$$Cyl(z, \alpha, r) = Trans(0, 0, z) Rot(z, \alpha) Trans(r, 0, 0) \tag{2}$$

Using the described transformation matrices, we can write basic kinematic equations. Given a point ^ir described with respect to link i, its position in base coordinates is

$$r = T_i \, {}^i r \tag{3}$$

The kinetic energy of a particle of mass dm located on link i at ^ir is

$$dK_i = \frac{1}{2} \text{Trace} \left[\sum_{j=1}^{i} \sum_{k=1}^{i} \frac{\partial T_i}{\partial q_j} ({}^i r \, dm \, {}^i r^T) \frac{\partial T_i}{\partial q_k}^T \dot{q}_j \dot{q}_k \right] \tag{4}$$

and the kinetic energy of link i is then

$$K = \sum_{i=1}^{6} K_i = \frac{1}{2} \sum_{i=1}^{6} \text{Trace} \left[\sum_{j=1}^{i} \sum_{k=1}^{i} \frac{\partial T_i}{\partial q_j} J_i \frac{\partial T_i}{\partial q_k}^T \dot{q}_j \dot{q}_k \right] \tag{5}$$

After creating and solving the model the developed system enables simulation of practically any possible situation that can arise in the respective case. It is thus possible to simulate any real machine behaviour for the given initial conditions. Related preliminary results in this respect are described in [3].

Control and Optimization of Mechanical Systems

When the problem of control of the above systems is to be investigated it is necessary to replace the simple mechanical variables by different variables which are typical for the action parts of the systems, e, g. electric, hydraulic and pneumatic drives, and the simulation tasks can be augmented to encompass the associated control problems. Such problems can be solved for different *controller configurations* and usage of the simulation environment SiMeS will contribute to computer-aided design of the respective control loops.

Last but not least is the optimization option for such systems. Again, after adapting the appropriate numerical algorithms and incorporating them to the existing environment it is possible to solve and analyze with extensive graphical support a number of important optimization problems connected with the considered class of mechanical systems. Most frequent *optimization* problems deals minimal time and/or minimum energy issues.

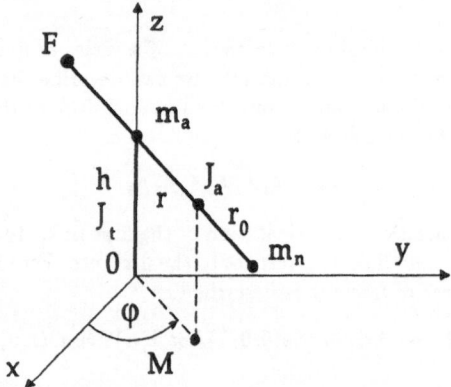

FIGURE 3 Cylindrical robot

For optimization problems solution the following type of dynamic system is considered. Given kinematic structure with n degrees of freedom. Then state variables are defined as

$$
\begin{aligned}
x_1(t) &= q_1(t) \\
x_3(t) &= q_2(t) \\
&\;\;\vdots \\
x_{2n-1}(t) &= q_n(t)
\end{aligned}
\tag{6}
$$

where $x_i(t)$ are state variables and $q_i(t)$ are generalized coordinates of the main partial kinematic motions and

$$
\begin{aligned}
x_2(t) &= \dot{q}_1(t) \\
x_4(t) &= \dot{q}_2(t) \\
&\;\;\vdots \\
x_{2n}(t) &= \dot{q}_n(t)
\end{aligned}
\tag{7}
$$

Control variables are defined as

$$u_1 = F_1(t)$$
$$u_2 = F_2(t)$$
$$\vdots$$
$$u_n = F_n(t)$$

(8)

where $u_i(t)$ are control variables and $F_i(t)$ are generalized forces (corresponding to forces or torques) with control constraints

$$|u_i(t)| \leq F_{i_{max}}$$

(9)

for $i = 1 \ldots n$.

In general, both these optimal control problems (minimal time and/or minimal energy) are in their nature problems of the calculus of variation. They may be considered as limiting cases of optimal programming problems for multistage systems in which time increment between steps becomes small compared to time horizons of interest. To comply with the standard notation we can rewrite above system equations in the form

$$\dot{x} = f[x(t), u(t), t]$$

(10)

where $x(t_0)$ is given, $t_0 \leq t \leq t_f$, an n-vector function $x(t)$ is determined by an m-vector function $u(t)$ [4]. The problem is to find a function u(t) that minimize a performance index of the form

$$J = \varphi[x(t_f), t_f] + \int_{t_0}^{t_f} L[x(t), u(t), t]dt$$

(11)

Let us define multiplier (co state) functions as

$$\dot{\lambda}^T = -\frac{\partial H}{\partial x} = -\frac{\partial L}{\partial x} - \lambda^T \frac{\partial f}{\partial x}$$

(12)

with boundary conditions

$$\lambda^T(t_f) = \frac{\partial \varphi}{\partial x}$$

(13)

and where Hamiltonian function is defined by

$$H[x(t), u(t), \lambda(t), t] = L[x(t), u(t), t] + \lambda^T(t)f[x(t), u(t), t]$$

(14)

Then along the optimal trajectory the following necessary optimality condition holds (assuming for simplicity that no constraint of type (15) is active)

$$\frac{\partial H}{\partial u} = 0$$

(15)

for $t_0 \leq t \leq t_f$.

Optimal control problems arising in robot control usually assume the prescribed terminal state. Then in the case of fixed time minimal energy problem additional constraining functions for terminal state determination are specified

$$\psi[x(t_f), t_f] = 0$$

(16)

where ψ is a q-vector. However, in minimal time optimization problems the terminal time t_f is not specified and t_f is regarded as a control parameter. It satisfies the following condition

$$\left(\frac{\partial \phi}{\partial t} + \lambda^T f + L \right)_{t=t_f} = 0$$

(17)

350

Implementation of Gradient Algorithm

In [4] a general approach to the solution of optimization problems is described based on *influence* functions. Without any doubts such approach is possible, however the resulting computational burden is practically prohibited in real-world situations. The approach was used in the past in chemical engineering [5] and robotics [6, 7]. This methodology has shown also feasible for more general problems involving constraints [8, 9].

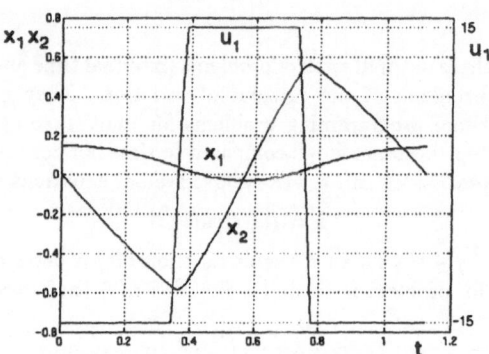

FIGURE 4 Translation history of the cylindrical robot

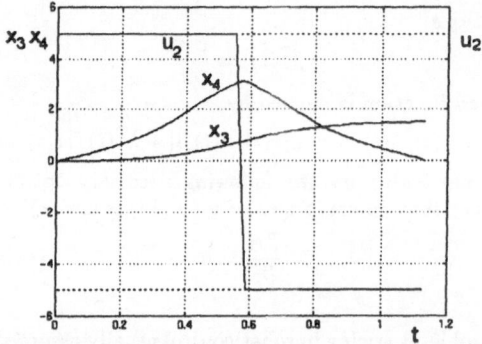

FIGURE 5 Rotation history of the cylindrical robot

When dealing with mechanical systems one has to realize the common features of many on the first glance distinct areas. Namely the joint proportionality of effort and result in biology, chemical or mechanical engineering results in bi linear or more generally in control linear system which have vast practical applications. However this maybe simplifying feature causes numerous other problems. Especially when the issue of optimization has to be investigated.

As a rule *bang-bang* and not so seldom *singular* control problems (necessary optimality conditions do not give the answer any more) were then studied from various points of view. Recall just the monographs [10, 11]. To have an idea of what degree problems are to be expected in singular case recall the quiet old paper [12] giving conditions of equivalence for singular and state-constrained control problems.

No wonder that alternative approaches appeared mostly taking advantage of the known character of optimal control strategy. Mostly parametric optimization methods were applied [13] to singular problems. Some typical examples in this respect are discussed in [14]. It is interesting that further investigations in [15] has shown that not all solutions presented in [13] are optimal. So there is still enough room in this topis also for theoretical work.

From computational point of view it can be concluded that enormously increasing computer power calls for reevaluation of some methods previously being denoted as inefficient. Usually such methods are simple and straightforward enough that many-iteration-process is on today's computer feasible. This observation authors practically verified on some method for parametric optimization. Therefore the previously mentioned simple gradient method has been tested on several case known from literature.

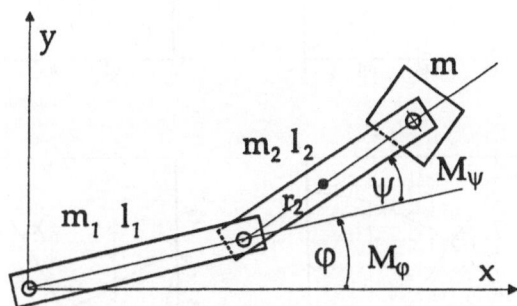

FIGURE 6 Articulated arm robot

Example 1: Cylindrical robot

In Example 1 from [14] one has to find both energy-optimal and time-optimal solutions for a cylindrical robot – see Fig. 3 for details. The system was considered with state variables $x_1 = r$, $x_2 = \dot{r}$, $x_3 = \varphi$, $x_4 = \dot{\varphi}$, initial state $(r_a; 0; \varphi_a; 0)$, terminal state $(r_b; 0; \varphi_b; 0)$ (for presented case $r_a = r_b$). For time-optimal control was $L(x, u, t) = 1$, $|u_1| \leq F_{max}$, $|u_2| \leq M_{max}$. For energy-optimal control $L(x, u, t) = 0.5(c_1 u_1^2 + c_2 u_2^2)$.

Gradient algorithm was implemented in MATLAB 4.0 on HP Apollo 720 workstation. Then also hundreds of iterations do not cause any severe problem. As a rule for energy-optimal solution considerably fewer iterations are needed as compared with time-optimal

352

case. For percentage accuracy this means tenth, resp. hundreds of iterations. The time-optimal solution obtained is in Figs. 4 and 5. One can conclude fairly successful outcome of our attempts.

Example 2: Articulated arm

Case with two rotations (articulated arm) according to [12] is denoted as Example 2 in Fig. 6. State variables are $x_1 = \varphi$, $x_2 = \dot{\varphi}$, $x_3 = \psi$, $x_4 = \dot{\psi}$, initial state $(0; 0; 0; 0)$, terminal state $(\varphi_b; 0; 0; 0)$. For time-optimal control was again $L(x, u, t) = 1$, $|u_1| \leq M_{\varphi_{max}}$, $|u_2| \leq M_{\psi_{max}}$. For energy-optimal control $L(x, u, t) = 0.5(c_1 u_1^2 + c_2 u_2^2)$.

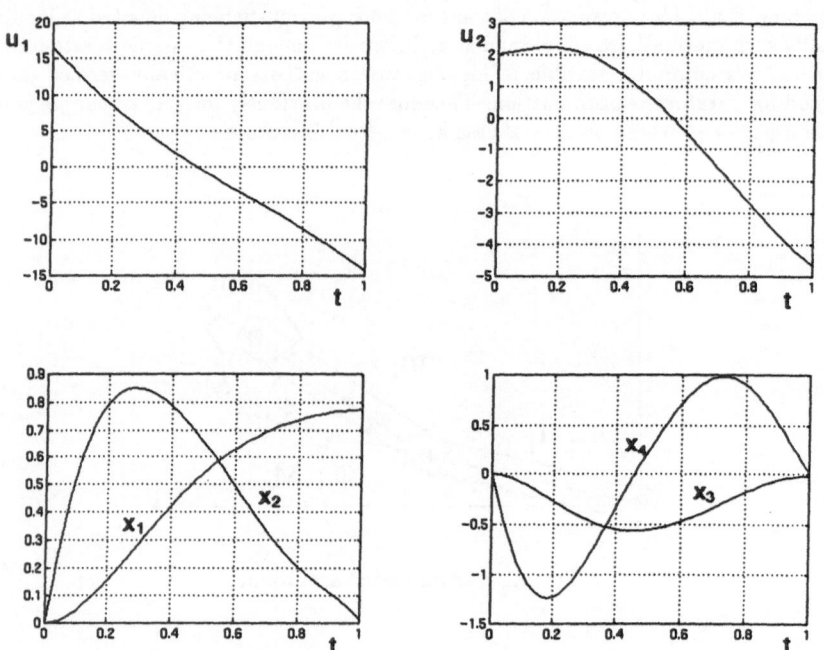

FIGURE 7 Energy-optimal solution for articulated arm

Obtained results for energy-optimal formulation are summarized in Fig. 7. Analogously Fig. 8 depicts time-optimal case considered also in [14] and [15].

Conclusions

Besides the presentation of simulation environment for mechanical systems considerable effort has been devoted to contribute to the dilemma of using classical, as here, or alternative, more problem oriented parametric optimization methods. Our experience shows,

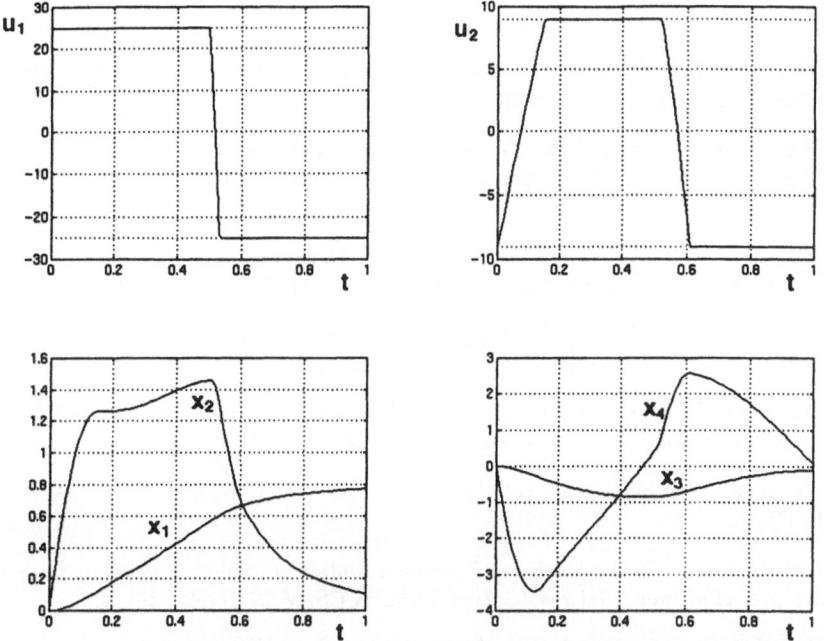

FIGURE 8 Time-optimal solution for articulated arm

that if one does not strictly adheres to rather *small* number of iterations, one at the end is successful when solving singular and thus very difficult for convergence optimal control problems. To give more distinctive conclusion need further extensive experimentation in this area.

References

[1] PAUL, R. P.: *Robot Manipulators: Mathematics, Programming, and Control.* MIT Press, Cambridge, Mass., 1981.

[2] RÁNKY, P. G., HO, C. Y.: *Robot modelling: Control and applications with software.* IFS (Publications) Ltd, England and Springer-Verlag, Berlin Heidelberg New York Tokyo, 1985.

[3] BURÁŇ, P.: *Modelling and simulation of mechanical systems with open chains.* Symposium on Modelling and Simulation of Systems, Olomouc, ČSFR, May 1993.

[4] BRYSON, A. E., HO, Y. C.: *Applied optimal control: Optimization, estimation and control.* Ginn and Company, Waltham, Massachusetts, 1969.

[5] DOLEŽAL, J.: *Optimization approach to the determination of multifunctional catalysts in chemical engineering.* Automatizace **21** (1978), 3-8. In Czech.

[6] NEUMAN, P.: *Optimal control of industrial robots.* Automatizace **23** (1980), 152-156. In Czech.

[7] NEUMAN, P.: *The application of optimum control of nonlinear dynamic systems for the control of industrial robots.* Acta Technica ČSAV **27** (1982), 291-305.

[8] DOLEŽAL J.: *On the solution of optimal control problems involving parameters and general boundary conditions.* Kybernetika **17** (1981), 71–81.

[9] DOLEŽAL J.: *On optimal control problems with general boundary conditions.* J. Optimiz. Theory Applics. **35** (1981), 453–458.

[10] BELL, D.L., JACOBSON, D.H.: *Singular Problems in Optimal Control.* Academic Press, New York 1975.

[11] GABASOV, R., KIRILLOVA, F.M.: Singular Optimal Control. Nauka, Moscow 1973. In Russian.

[12] JACOBSON, D.H., LELE, M.M., SPEYER J.L.: *A transformation technique for optimal control problems with state variable inequality constraint.* IEEE Trans. Autom. Contr. **AC-14** 1969, 457-464.

[13] GOH, C.J., TEO, K.L.: *Control parameterization: a unified approach to optimal control problems with general constraints.* Automatica **24** 1988, 3-18.

[14] GEERING, H.P., GUZZELLA, L., HEPNER, S.A.R., ONDER CH.H.: *Time-optimal motions of robots in assembly tasks.* IEEE Trans. Automat. Contr. **AC-31** (1986), 512-518.

[15] VAN WILLIGENBURG, L.G., LOOP, R.P.H.: *Computation of time-optimal controls applied to rigid manipulators with friction.* Int. J. Control **54** (1991), 1097-1117.

A NEW APPROACH TO SOLVING ALGEBRAIC SYSTEMS BY MEANS OF SUB-DEFINITE MODELS

A.S. Narin'yani   A.L. Semenov   A.B. Babichev
T.P. Kashevarova   A.S. Leshchenko
Novosibirsk Division of the Russian
Research Institute of Artificial Intelligence
pr. Lavrent'eva 6, Novosibirsk, Russia, 630090
e-mail: semenov@isi.itfs.nsk.su

Abstract

This paper describes a novel approach to solving systems of algebraic equations and inequalities that is based on *subdefinite calculations*. The use of these methods makes it possible to solve overdetermined and underdetermined systems, as well as systems with imprecise and incomplete data. *The UniCalc solver*, also described in this paper, was developed on the basis of this approach. To illustrate the capabilities of UniCalc, we give examples of problems solved with its help.

1 Introduction

Solving systems of nonlinear algebraic equations attracts considerable interest, and numerous methods exist for solving such systems. As a rule, these algorithms are based on approximate methods of numerical mathematics; in the following they are referred to as the classical methods. (Examples of such methods are the Newton's method, bisection methods, relaxion methods, gradient methods and etc.) Most of these methods are iterative and require a good initial approximation. In addition, the classical methods are good only for systems that are fully defined. However, the systems that describe phenomena of real life may be either underdetermined or overdetermined; they may include inequalities, uncertain and imprecise values. To solve such classes of problems, several approah exist.

One is *interval analysis* [1, 2] allowing classical methods to be adapted to systems with imprecise data and making it possible to tackle problems unsuitable for classical methods.

Another approach to problems with uncertain data is *fuzzy set theory* [3] which uses some function called *the membership function* describing the degree of uncertainty.

To solve real systems containing additional conditions, an approach called *constraint propagation* is used [4]. This method makes it possible to consider certain constraints (e.g. inequalities) but it fails with systems with imprecise data. One way to obviate this drawback is to combine constraint propagation method with interval mathematics [5].

This paper considers an approach for solving algebraic equations created in the framework of artificial intelligence research [6, 7]. This approach is called *the method of subdefinite calculations* and is free of almost all drawbacks of the abovementioned approaches.

It is applicable to a wide class of systems and does not require any initial approximations. Based on the method of subdefinite calculations, the UniCalc solver was developed in the Novosibirsk Division of the Russian Research Institute of Artificial Intelligence.

The rest of the paper is organized as follows. Sections 2 and 3 present the algorithm of subdefinite calculations and some implementation questions. Sections 4, 5, and 6 consider both the UniCalc solver as a whole and some of its components. Numerical experiments with UniCalc are described in section 7. Section 8 provides a summary, UniCalc's technical characteristics and some conclusions.

2 The Algorithm

2.1 Concepts and Designations

Let a calculation model M be determined by the set of variables X and the set of relations over these variables R. We shall denote the model by $M = (X, R)$. Let A be the value domain of the model's variables, and let $^*A$ denote the set of all nonempty subsets of the set A. The values $^*a \in {}^*A$ containing one member only are called *precise*, while other values are called *subdefinite*. The value $^*a$ corresponding to the entire set A is *the fully indefinite value*. Let us map uniquely each variable x in the set X into a variable $^*x$ whose value domain is the set $^*A$. This maps the set of variables X onto a set $^*X$ of variables $^*x$. The variables $^*x$ will be called *subdefinite variables*.

The *subdefinite description of the model* M is a set (M, h), where $h = (^*a_1, \ldots, ^*a_n)$ is a vector of subdefinite values. Narin'yani showed in [6, 7] that for subdefinite descriptions it is possible to construct a finite automaton that generates finite sequences of states. The sequences are terminated either with the 'end' state, or the 'conflict' state. In the first case, the automaton produces a vector h which is an n-dimensional parallelepiped in the space A^n containing the set of the values of the variables X that satisfy the model M.

The process of inference/calculations over models usually involves interpreting each relation with a help of a set of functions allowing us to find the values of other variables from known values of some variables.

Since the variables $^*X$ rather than X are used while working with subdefinite descriptions, it is necessary to define operations over subdefinite values and subdefinite functions. Narin'yani showed [6] that if we map each m-ary operation s over values from A into an operation $^*s$ over undefined values from $^*A$ according to the formula

$$^*s(^*a_1, \ldots, ^*a_m) = \{a = s(b) \mid b \in {}^*a_1 \times {}^*a_2 \times \ldots \times {}^*a_m\}, \qquad (1)$$

and if each m-ary function f of variables in X is mapped into a function $^*f$ of variables in $^*X$ by the formula

$$^*f(^*x_1, \ldots, ^*x_m) = \{x = f(b) \mid b \in {}^*x_1 \times {}^*x_2 \times \ldots \times {}^*x_m \cap R\}, \qquad (2)$$

where R is an additional relation connecting the variables $x_1, \ldots, x_m$ (for uncorrelated variables $R = {}^*A^m$), then it is possible to use the inference/calculation apparatus for the initial models for inference/calculations on subdefinite models.

2.2 The Calculation Algorithm for Subdefinite Models

Suppose we have a model with a subdefinite description (in [7], such models are called *GCM, generalized calculation models*). According to section 2.1, the inference/calculation procedure for such models may be described as a procedure of calculating the interpretation functions of subdefinite variables. If a certain relation of the initial model relating variables $x_1, \ldots, x_m$ is interpreted by a set of the following functions:

$$x_i := f_i(x_1, \ldots, x_{i-1}, x_{i+1}, \ldots, x_m), \ i = 1, \ldots, m \tag{3}$$

then, using formula 2, the corresponding interpretation functions for the subdefinite model may be defined as follows:

$$^*x_i := {}^*f_i(^*x_1, \ldots, {}^*x_{i-1}, {}^*x_{i+1}, \ldots, {}^*x_m) \cap {}^*a_i, \ i = 1, \ldots, m \tag{4}$$

where $^*a_i$ is the current value of $^*x_i$.

The inference/calculation process formulated in [7] for GCM may be presented in the form of the following algorithm (let $^*X_t$ denote the set of the variables in $^*X$ whose values have become more precise at the step t, and let $F \mid {}^*X_t$ denote the set of interpretation functions having at least one argument which belongs to $^*X_t$):

Step 1. $t = 1$; $^*X_t = \{ \ ^*x_1, \ldots, {}^*x_n \ \}$. (Note that all model variables have values: they are either initialized or their value is fully indefinite, i.e., the set A.)

All the model interpretation functions are placed in set $F \mid {}^*X_1$ and all the members of this set form the set of active functions.

Step 2. Combine the function set $F \mid {}^*X_t$ with the active interpretation functions.

Step 3. Choose an arbitrary subset in the produced set. This subset is joined to the working function set and removed from the active function set.

Step 4. Remove an arbitrary subset of functions from the working function set. The functions of the subset are calculated at this step. The results of these calculations are compared with the values of subdefinite variables in the common memory, and variables that have changed their values form the set $^*X_{t+1}$ defining set the $F \mid {}^*X_{t+1}$.

If a result of calculating some interpretation function yields the value of a variable that is equal to the empty set the model is incompatible. The algorithm terminates.

Step 5. If the set of active functions, the working function set and the set $^*X_{t+1}$ are simultaneously empty, go to Step 7.

Step 6. $t = t + 1$ and go to step 2.

Step 7. Output the values in the set $^*X = \{ \ ^*x_1, \ldots, {}^*x_n \ \}$.

This algorithm has the following features:

- The model can be underdetermined or overdetermined and the parameters of the model may be imprecise or unknown;

- No distinction is made between the model's arguments and results: all variables have values that are subdefinite in various degrees; the calculations are performed for all variables of the model;

- This algorithm determines a parallel, asynchronous, undetermined process with flow data-driven control;

- According to formula 4, subdefiniteness of the variables never increases and the calculation is converging. For all models having only finite subdefinite values, this procedure terminates in a finite number of steps.

Consider an example. Let the model be specified by two equations:

$$8a - b = 15, \quad 6b + 4c = 230$$

and suppose that it is known that a, b, and c are integers in the range from 1 to 100. Hence, the subdefinite value of each of these variables can be represented by a set of integers with the initial state $A = \{1, ..., 100\}$. This relations are interpreted by the following functions:

$$
\begin{aligned}
f_1 &: \quad a := (b + 15)/8; \\
f_2 &: \quad b := 8a - 15; \\
f_3 &: \quad b := (230 - 4c)/6; \\
f_4 &: \quad c := (230 - 6b)/4.
\end{aligned}
$$

Introducing the subdefinite description, we have:

$$
\begin{aligned}
{}^*\!f_1 &: \quad {}^*\!a := \{a = (b + 15)/8 \mid b \in {}^*\!b\} \cap {}^*\!a; \\
{}^*\!f_2 &: \quad {}^*\!b := \{b = 8a - 15 \mid a \in {}^*\!a\} \cap {}^*\!b; \\
{}^*\!f_3 &: \quad {}^*\!b := \{b = (230 - 4c)/6 \mid c \in {}^*\!c\} \cap {}^*\!b; \\
{}^*\!f_4 &: \quad {}^*\!c := \{c = (230 - 6b)/4 \mid b \in {}^*\!b\} \cap {}^*\!c.
\end{aligned}
$$

In compliance with the above algorithm, we obtain

$$
\begin{aligned}
{}^*\!X_1 &= \{{}^*\!a = \{1, \ldots, 100\}, {}^*\!b = \{1, \ldots, 100\}, {}^*\!c = \{1, \ldots, 100\}\}; \\
F \mid {}^*\!X_1 &= \{{}^*\!f_1, {}^*\!f_2, {}^*\!f_3, {}^*\!f_4\}.
\end{aligned}
$$

The table bellow illustrates the algorithm's execution:

Subdefinite values	Working function	Active functions
${}^*\!a = \{2, 3, \ldots, 14\}$	${}^*\!f_1$	${}^*\!f_2, {}^*\!f_3, {}^*\!f_4$
${}^*\!b = \{1, 9, \ldots, 97\}$	${}^*\!f_2$	${}^*\!f_3, {}^*\!f_4$
${}^*\!b = \{1, 9, 17, 25, 33\}$	${}^*\!f_3$	${}^*\!f_1, {}^*\!f_4$
${}^*\!c = \{8, 20, 32, 44, 56\}$	${}^*\!f_4$	${}^*\!f_1, {}^*\!f_3$
${}^*\!a = \{2, 3, 4, 5, 6\}$	${}^*\!f_1$	${}^*\!f_2, {}^*\!f_3$
${}^*\!b = \{1, 9, 17, 25, 33\}$	${}^*\!f_3$	${}^*\!f_2$
${}^*\!b = \{1, 9, 17, 25, 33\}$	${}^*\!f_2$	

Therefore, the algorithm yields the following result:

$$\{{}^*\!a = \{2, 3, 4, 5, 6\}, {}^*\!b = \{1, 9, 17, 25, 33\}, {}^*\!c = \{8, 20, 32, 44, 56\}\}.$$

To find an every separate solution one should combine the values from different sets and verify the system. In our case such solutions are:

$$\{2, 1, 56\}, \{3, 9, 44\}, \{4, 17, 32\}, \{5, 25, 20\}, \{6, 33, 8\}$$

3 Implementation of the Algorithm

3.1 Implementation of Subdefinite Numbers

The implementation of the algorithm just described assumes the choice of a set A corresponding to the class of models; in addition acceptable operations and functions over values in A and variables in X are to be extended for objects in $^{*}A$ and $^{*}X$. The active data type apparatus [8] has been proposed and used to implement various types of subdefinite data (integers, reals and booleans, sets, objects of planimetry, etc.) [9].

We consider only models over real numbers; therefore the set A is the field of real numbers R or the ring of integer numbers Z. We assume that all the arithmetic operations as well as exponentiation and root extraction are defined over values in the set A. The set of acceptable functions includes exponential, logarithmic and all trigonometric functions. Since computer number representations are finite, the set A will be represented by finite intervals in R or Z which are bounded by some numbers $MinA$ and $MaxA$ playing the role of $-\infty$ and $+\infty$, respectively (these numbers differ for real and integer sets A). The operations on elements of A are defined as follows: if $a, b \in A$ and $\odot$ is a operator, then

$$a \odot b = \max(a \odot b, MinA) \text{ if } a \odot b < 0$$
$$a \odot b = \min(a \odot b, MaxA) \text{ if } a \odot b \geq 0$$

Assuming these conventions and considering $^{*}A$ as the set of all intervals in the set A, we can extend all operations on elements of A to operations on $^{*}A$ defined by formula 1, using the appropriate operations of interval mathematics. Note that any nonempty subset of set A may be embedded into some interval. Therefore, assuming that $^{*}A$ is the set of all intervals of A, we can only increase subdefiniteness of solutions without losing any of them. Alefeld [2] showed that interval continuous analytic functions are monotonic by inclusion, i.e., if f is a continuous analytic function of interval variables $X_1, \ldots, X_n$ and $X_1 \subseteq Y_1, \ldots, X_n \subseteq Y_n$, then $f(X_1, \ldots, X_n) \subseteq f(Y_1, \ldots, Y_n)$. In view of this property, to extend functions over variables from A, it is possible to use their interval extensions in the corresponding continuity domains, since the interval thus obtained will always include the set determined by formula 2. Considering the possibility of using intervals with 'infinite' limits, we can continue interval expansion to discontinuous functions, assuming, for example, that $1/0 = (0, MaxA)$ and $1/(-1, 1) = (MinA, MaxA)$. Such an approach makes it possible to avoid the divide-by-zero situation that is fatal for other methods.

3.2 Representation of Models

The most appropriate method for the algorithm described above is the network representation of models. The virtual flow data-driven processor [10, 11] is based on such a representation using the apparatus of active data types. It is the kernel of the UniCalc system. In this processor, a network is represented as a bipartite oriented graph with two types of vertices: objects and operators. Objects represent model variables, and operators represent functional links between objects (interpretation functions). The outgoing edges point to the operators whose arguments are the corresponding objects, and incoming edges specify objects which provide values to these operators. A network is associated with a discipline of its execution, which substantiates the inference algorithm. This substantiation depends on the computer's architecture, its computational capabilities, as well as a number of other factors. For instance, sets may be selected at step 3 according to certain priorites, from a queue or randomly. Interpretation functions are computed at step 5 sequentially or in parallel, etc. In the current implementation of the processor, the interpretation functions are selected depending on their ordinal number and are computed sequentially. If one uses computers designed for parallel processing or transputers, the strategy of network execution should be changed.

4 The UniCalc Solver

To use the algorithm described above for practical purposes, we have developed the Uni-Calc solver whose nucleus is the flow processor considered in section 3. The UniCalc solver was designed to solve arbitrary systems of algebraic and algebraic-differential rela-tions. Here, a relation is considered to be an equation, inequality or a logical expression. According to the algorithm used, the system to be solved can be either overdetermined or underdetermined, and the system's parameters (coefficients, variables, initial conditions for the Cauchy problem) can be imprecise and expressed as intervals. Such a system may contain only integer and real variables or combine both integer and real variables.

As a result of solving algebraic systems, we either find a parallelepiped that contains all roots of the system, or a message about the system's incompatibility is issued. If the system has a single root, then the parallelepiped will be reduced to a point (with a given accuracy). If the system has several roots, to locate each of them it is necessary to add the appropriate relations, or use the built-in tool for automatic root locating.

The solver is an integrated environment supporting input of the system to be solved, its modification, calculations, viewing results, specifying accuracy, etc. To input and modify the system UniCalc has the built-in text editor. To write the problems, a source language close to the common used mathematical notation is provided. All problems are processed with the above algorithm. To translate the source input into a network, the solver includes a translator and pre-processors, in particular, for symbolic transformations and solution of systems of algebraic-differential relations.

The UniCalc's user interface offers a number of services to support problem solving. If solving problem requires large computation times, it is possible to suspend calculations to see the intermediate results, and depending on convergence rate, either to continue computation or to stop it. A feature for locating roots is used when exact solutions are in large intervals. This process involves dichotomic division of the obtained interval for the selected variable. This tool is also useful to find global function extreme values.

5 Symbolic Pre-processor

The main objective of this pre-processor is optimization of the network and performing some types of symbolic manipulations (currently these calculations cover only differen-tiation). The system to be solved by UniCalc is represented as a functional network whose size depends upon the number of variables and expressions in the system. The pre-processor translates the source expressions into its internal representation simplifying the source expressions and storing only one copy of each subexpression. Note that reduc-ing similar terms optimizes networks and sometimes narrows down intervals. In addition, storing only one copy of each term reduces the number of intermediate calculations and reduces subdefiniteness of results, as well as computation time. Prior to constructing a network it is possible to view, print out, or store the expressions obtained as a result of reductions and symbolic transformation.

The symbolic pre-processor also incorporates some features eliminating the drawbacks exposed in using UniCalc, one of which is low speed on systems of linear algebraic equations. To eliminate this, symbolic variable substitution is performed to obtain a triangular (trapezoid in general case) system of equations where some variables are expressed via others. If the system is subdefinite, an interval solution will be obtained.

6   Differential Equations

UniCalc can solve systems of algebraic-differential equations supported by the differential equation pre-processor. The system of equations applied to the input of this pre-processor is a mathematical model containing ordinary differential equations of the first order with initial data, algebraic equations, inequalities (linear and nonlinear), logical expressions. Differential equations must be solved for the derivative. Parameters of differential and algebraic equations as well as initial data may be specified imprecisely.

Presently, UniCalc uses the conventional numerical approach to solving systems of ODEs. In this approach, the derivatives are approximated by finite differences, following which the procedure is iterated over the integration domain with a certain step. To approximate the derivatives, we use explicit and implicit Euler schemes, as well as explicit Runge-Kutta schemes of various orders. Thus, the system of algebraic-differential equations is reduced to a system of algebraic relations (possibly with interval parameters), and the algebraic system thus obtained is solved with the help of the basic method of the solver. Note that in the case of implicit methods we do not need the initial approximation.

The user can choose the finite difference scheme from the options offered by the system, the accuracy for an integration step, and the result output points. The results are produced as arrays of intervals for each unknown function of the system. The calculation process can be accompanied by a plot in which approximate solution of each function is represented as a band-curve bounded by the upper and lower boundaries of intervals.

7   Numerical Experiments

To estimate the efficiency of the solver and determine the range of possible applications, many problems have been tested, including linear and nonlinear systems of equations and inequalities, mixed systems, various integer problems, optimization problems, interval problems, systems of differential equations, etc.

NONLINEAR SYSTEMS. Among the numerous nonlinear systems solved with Uni-Calc (including almost all tests considered in papers [12, 13]), we want to point out the problem offered in [14] as a test problem. This problem concerns combustion of propane in air to form ten products. The system of equations is as follows:

$$f_1 = n_1 + n_4 - 3 = 0$$
$$f_2 = 2n_1 + n_2 + n_4 + n_7 + n_8 + n_9 + 2n_{10} - R = 0$$
$$f_3 = 2n_2 + 2n_5 + n_6 + n_7 - 8 = 0$$

$$f_4 = 2n_3 + n_9 - 4R = 0$$

$$f_5 = K_5 n_2 n_4 - n_1 n_5 = 0$$

$$f_6 = K_6 n_2^{1/2} n_4^{1/2} - n_1^{1/2} n_6 \left(\frac{p}{n_T}\right)^{1/2} = 0$$

$$f_7 = K_7 n_1^{1/2} n_2^{1/2} - n_4^{1/2} n_7 \left(\frac{p}{n_T}\right)^{1/2} = 0$$

$$f_8 = K_8 n_1 - n_4 n_8 \left(\frac{p}{n_T}\right) = 0$$

$$f_9 = K_9 n_1 n_3^{1/2} - n_4 n_9 \left(\frac{p}{n_T}\right)^{1/2} = 0$$

$$f_{10} = K_{10} n_1^2 - n_4^2 n_{10} \left(\frac{p}{n_T}\right) = 0$$

$$f_{11} = n_T = \sum_{i=1}^{10} n_i$$

where K_i, p and R are constants. We need to find a solution where all the n_i are positive.

It is noted in the paper that in the original formulation this is a hard problem and it was reduced to another equivalent system which was solved. In contrast to this, UniCalc easily solved the original problem. Determining the initial intervals of the variables took 2.5 secs, and finding the final solution, about 3 mins on an IBM PC AT–286.

OPTIMIZATION PROBLEMS. Currently, UniCalc does not have any special facility to solve optimization problems. However, its algorithm is general enough to solve some problems of this type. Both problems of unconstrained and constrained optimization are considered. The first 15 problems offered in [15] as test problems were successfully solved and one of them was to find a minimum of the Rosenbrock's function of 15-th order.

One way of stating an optimization problem for UniCalc is its mathematical statement. For example, to find a maximum of a polynomial in the UniCalc language we write

```
(* Find a maximum of the function *)
f(x) := x^6 - 26 * x^3 + 45 * x^2 - 10 * x + 1;
fmax = f(x) ;
    (* The first derivative is zero *)
dif(f(), x) = 0;
    (* At the point of maximum, the second derivative is negative *)
dif(f(), x:2) < 0;
```

The results are $x = 1.2176$, $fmax = 11.86$.

INTEGER PROBLEMS. UniCalc can be used to solve various integer problems which are difficult for other solvers, for example, integer optimization problems (linear and nonlinear), Diophantine equations, integer factoring, etc.

PROBLEMS WITH SUBDEFINITE DATA. UniCalc is good to solve research problems and problems with imprecise data. For example, let we have the following problem.

A ball falling freely from height $h = 2$ meters collides elastically with a stationary plane inclined at angle α to the horizon. It is known that the angle α is in the interval from 25 to 35 degrees, and the restitution coefficient Rc is in the range from 0.75 to 0.85. Determine the possible directions of the ball's velocity vector at the end of the collision such that the maximal height $h1$ after the collision is greater then or equal to 0.5 meter.

```
V   =   sqrt(2 * @g * h);
Vt  =   V * sin(alpha);
Un  = -Rc * V * cos(alpha);
Ut  =   Vt;
ctg(beta) = abs(Un) / abs(Vt);
(Ut^2  + Un^2 ) * cos(alpha + beta)^2 = 2 * @g * h1;
h1  >=   0.5;
        (* Initialization *)
alphadeg := [25, 35];
Rc       := [0.75, 0.85];
h        := 2;
betadeg  := [0, 90];
        (* radian to degree conversion *)
convert(angle) := angle * 180 / @pi;
alphadeg        = convert(alpha);
betadeg         = convert(beta);
```

The solutions are:

```
alphadeg := [25, 27.39],    betadeg :=  [28.75, 32.14],
h1 = [0.5, 0.56],           Rc := [0.78, 0.85],
Un = [-4.82, -4.44],        Ut = [2.65, 2.88],
V =  [6.26, 6.26],          Vt = [2.65, 2.88].
```

Note that the values obtained for Rc and alphadeg differ from the initial ones due to the fact that UniCalc does not distinguish between input and output data. The initial values were updated according to the problem statement.

8 Conclusions and Future Research

This paper describes the method of subdefinite calculations and the UniCalc solver based on this apparatus. The practical applicability and usefulness of this approach have been proved by successfully solving many different problems. We consider that our solver can be useful to a wide circle of users: students, engineers, economists, research workers, etc.

The UniCalc solver runs under MS-DOS version 3.0 or higher on the IBM PC family of computers. It requires 512 Kb of RAM and 600 Kb of free disk space. Such a configuration is suitable for up to 300 relations and variables. UniCalc does not require a math-coprocessor, but will use one if it is available to speed up the calculation.

Research on the apparatus of subdefinite models continues. Our next goals include

1. Improving the solver's capabilities (introducing the object-oriented representation of models, improving efficiency, porting UniCalc to Unix and MS Windows).

2. Theoretic trends. These works include in particular

   - Other representations of subdefinite data (for example, by multiintervals) with an appropriate modification of the calculation procedures;

   - Studying methods for implementing the considered algorithm on parallel computers and transputers;

• Extending the notion of subdefiniteness to other objects (for example, subdefinite relations and subdefinite functions).

References

[1] Moore R.E. Interval Analysis. Englewood Cliffs, New Jersey, Prentice-Hall, 1966. — 145 p.

[2] Alefeld G., Herzberger Ju. Introduction in Interval Computations. Academic Press, New York, 1983.

[3] Zahde L.A. Fuzzy Sets. Information and Control. v.8, N3, 1965. pp. 338 – 353.

[4] Leler W. Constraint Programming Languages. Their Specification and Generation. Reading, Massachussets, Addison-Wesley, 1988. — 202 p.

[5] Hyvonen E. Constraint Reasoning Based on Interval Arithmetic. In: Proceedings of IJCAI – 91. 1991. pp. 1193 – 1198.

[6] Narin'yani A.S. Subdefinite Models and Operations with Subdefinite Values. Preprint, USSR Acad. of Sciences, Siberian Division, Computer Center; N400. Novosibirsk, 1982. — 33 p. (In Russian).

[7] Narin'yani A.S. Subdefiniteness in Knowledge Representation and Processing Systems. Transactions of USSR Acad. of Sciences, Technical Cybernetics, N5, 1986. pp. 3 – 28. (In Russian).

[8] Narin'yani A.S. Active Data Types for Representing and Processing of Subdefinite Information. In: Actual Problems of the Computer Architecture Development and Computer System Software. Novosibirsk, 1983. pp. 128 – 141. (In Russian).

[9] Telerman V.V. Active Data Types. Preprint, USSR Acad. of Sciences, Siberian Division, Computer Center; N792. Novosibirsk, 1988. — 30 p. (In Russian).

[10] Narin'yani A.S., Telerman V.V., Dmitriev V.E. Virtual Data-Flow Machine as Vehicle of Inference/Computations in Knowledge Bases. In: Artificial Intelligence II. Methodology, Systems, Application. Ed. by Ph. Jorrand, V. Sgurev. North-Holland, 1987. pp. 149 – 155.

[11] Dmitriev V.E. Technological Complex for Producing Problem-oriented S-processors. In: Designing Software Tools for Intelligent Problems. Novosibirsk: USSR Acad. of Sciences, Siberian Division, Computer Center. 1988. pp. 103 – 111. (In Russian).

[12] Kearfott R.B. Some Tests of Generalized Bisection. ACM Trans. Math. Soft. N3, 1987. pp. 197 – 220.

[13] Vrahatis M.N. Solving Systems of Nonlinear Equations Using the Nonzero Value of the Topological Degree. ACM Trans. Math. Softw. N4, 1988. pp. 312 – 329.

[14] Meintjes K. and Morgan A.P. Chemical Equilibrium Systems as Numerical Test Problems. ACM Trans. Math. Softw. N2, 1990. pp. 143 – 151.

[15] More J.J., Garbow B.S., Hillstrom K.E. Testing Unconstrained Optimization Software. ACM Trans. Math. Softw. N1, 1981. pp. 17 – 41.

USE OF CONVEX ANALYSIS FOR THE MODELLING OF BIOCHEMICAL REACTION SYSTEMS

Stefan Schuster[1], Thomas Höfer[1],
Claus Hilgetag[1] and Ronny Schuster[2]

[1]Fachbereich Biologie, Institut für Biophysik,
Humboldt-Universität zu Berlin, Invalidenstr. 42, D-10099 Berlin, Germany
[2]Institut für Biochemie,
Humboldt-Universität zu Berlin, Hessische Str. 3/4 , D-10099 Berlin, Germany

1. Introduction

In the mathematical description of biochemical reaction systems, investigation of steady-state fluxes and conservation relations among the concentrations of reacting species plays an important role. This investigation mainly starts from the analysis of the null spaces of the stoichiometry matrix, N, and its transpose, N^T, respectively [1-3]. Very frequently, only non-negative null-space vectors are of interest [4,5]. Thus, solutions to systems of equations and inequalities have to be detected. Solution procedures are provided in convex analysis [6-8].

As for conservation relations, non-negativity is of importance for the interpretation of these relations in terms of conservation of chemical units (e.g. atom groups). In a previous paper [9], semi-positive conservation relations have been studied in detail and an algorithm for determining the set of all of these relations to a given reaction system was given. The results are briefly reviewed in Section 2.

In a recent paper [10], we investigated the problem of how to determine the set of all flux values of a biochemical reaction system that are admissible in steady state when all fluxes are non-negative. Sign constraints for fluxes are of importance when the reaction rates are defined as unidirectional rates [4], in the modelling of tracer dynamics, in the case of irreversible reactions and if the orientation of flux is implied by the biological function the system has to fulfil.

In many biochemical systems, however, the fluxes of some reactions are not restricted with respect to their sign, for example the fluxes through the reactions shared by glycolysis and gluconeogenesis. It is therefore an intriguing goal to generalize the formalism presented in [10] so as to cope with the situation that some reactions proceed in a fixed direction whereas others may proceed in either direction. This generalization will be given in Section 3.

In several situations (for example in optimization studies), it is meaningful to consider the equilibrium constants fixed while the kinetic parameters are variable. It can be shown that under these side conditions, the region accessible for logarithmic steady-state concentrations is a polyhedron. Section 4 is devoted to the study of this concentration polyhedron.

2. Extreme semi-positive conservation relations

If the rank, v, of the stoichiometry matrix, N, of the reaction system under study, is smaller than the number of concentration variables, n (which equals the number of lines of N), there is an $(n-v) \times n$ matrix D which has rank $n-v$ and fulfils the equation

$$D N = 0 . \tag{2.1}$$

Matrix D expresses the conservation relations among concentrations [1-4]. These relations reduce the number of independent variables. In fact, on simulating the dynamic behaviour of the system, one can eliminate $n-v$ variables.

Choice of the conservation matrix is not unique because each matrix $\hat{D} = P \, D$ with P being any non-singular $(n-v) \times (n-v)$ matrix is a conservation matrix as well. Conservation relations are frequently brought in relation to conservation of atoms or atom groups [11,12]. Consider, for example, the esterification of formic acid and methanol,

$$HCO_2H + CH_3OH \longrightarrow HCO_2CH_3 + H_2O . \tag{2.2}$$

A feasible conservation matrix is

$$D = \begin{pmatrix} 1 & 0 & 1 & 0 \\ 0 & 1 & 1 & 0 \\ 1 & 0 & 0 & 1 \end{pmatrix} . \tag{2.3}$$

Its rows correspond to conservation of the formyl group, methyl group and the hydroxy group originating from the acid.

A necessary condition for a conservation relation to represent conservation of real molecular moieties is that all coefficients in this relation be non-negative,

$$D \geq 0 . \tag{2.4}$$

In order to exclude the trivial case that all coefficients are zero, we include tha additional condition that at least one coefficient be positive. We call relations having these two properties "semi-positive conservation relations". Since, accordingly, the column vectors of D corresponding to such a type of relation fulfil a homogeneous equation/inequality system, they constitute a pointed convex polyhedral cone [6]. Convex analysis shows that such cones can be represented as non-negative linear combination of generating vectors, which are unique up to scalar multiples [6]. That is, the cone, K, of all semi-positive conservation relations for a given reaction system can be written as

$$K = \{ y \in \mathbb{R}^n \, \big| \, y = \sum_{k=1}^{p} \eta_k e^{(k)} , \, \eta_k \geq 0 , \, k=1,...,p \} , \tag{2.5}$$

where $e^{(k)}$ are the generating vectors and p is their number. By definition, a generating vector of a cone is a vector belonging to this cone which cannot be represented as convex linear combination of two different vectors belonging to this cone as well. The conservation relations corresponding to the generating vectors are to be called extreme semi-positive conservation relations.

A complete set of generating vectors can be found by an algorithm developed in convex analysis [7,8]. A variant of this algorithm specified so as to be applicable to the problem dealt with in this section was given in [9]. It is important to note that the number of generating vectors can be greater than the dimension of the cone. This dimension, in turn, is less than, or equal to, n-v. For reaction (2.2), for example, we have $n=4$, $v=1$, $\dim(K)=3$ and $p=4$. In many systems, all generating vectors correspond to maximal conserved moieties, but there are exceptions where only some of them do so [9]. For the above system, the algorithm yields, as generating vectors, the row vectors of matrix (2.3) and, in addition, the vector $(0 \quad 1 \quad 0 \quad 1)^T$, which corresponds to the hydrogen originating from the hydroxy group of methanol.

3. Decomposition of steady-state fluxes into elemental modes

3.1. All reactions are "irreversible"

Let V denote the flux vector of the reaction system under study. In what follows, we only consider steady states of the system, which implies

$$N V = 0 . \tag{3.1a}$$

The problem of finding all semi-positive steady-state flux vectors is isomorphic to the problem of finding all semi-positive conservation relations because these vectors are determined by the equation/inequality system (3.1a) together with

$$V \geq 0 , \quad V \neq 0 . \tag{3.1b,c}$$

Transposition of the stoichiometry matrix and replacement of V by any column vector of D leads to the system (2.1), (2.4) and the conditon that no column vector of D be the null vector. Accordingly, the algorithm mentioned in the previous section can be used for the present problem, just by transposing N.

The generating vectors of the flux cone can be interpreted as elemental modes of the system, which have the property that a maximum number of reactions have zero fluxes and which are compatible with the steady-state and sign conditions. Every functional mode of the system can then be obtained by superposition of the elemental modes .

In a recent paper [10], we generalized the problem by allowing for the case that some flux values are fixed, i.e., they are known from measurement or calculation. In this situation, the system (3.1) has to be replaced by

$$N^1 V^1 = - N^2 V^2 , \tag{3.2}$$

$$V^2 = \text{const.} \tag{3.3}$$

$$V \geq 0 , \tag{3.4}$$

where the vector V^2 and the stoichiometry matrix N^2 correspond to those fluxes that are known. The set of solutions to system (3.2, 3.3, 3.4), which we denote by P, is the intersection of cone K and a hyperplane defined by V^2. This intersection is a convex polyhedral set which may or may not be bounded [cf. 6]. Every point of such a set is the sum of a convex combination of the vertices and a non-negative linear combination of its extreme rays. Fig. 1 shows an example.

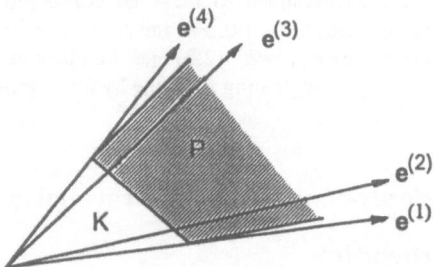

Figure 1. Example of a convex cone, K, which may correspond, in appropriate spaces, to the set of all semi-positive conservation relations, or to the set of all non-negative fluxes. The polyhedral region P contains the flux vectors admissible if two flux values are fixed.

The admissible set of vectors V^1 can be found by using a relatively simple extension of the algorithm for detecting the extreme semi-positive fluxes. Since N^2 and V^2 are constant, we can write eq. (3.2) as

$$N^1 \, V^1 = - \, W \tag{3.5}$$

with $W = N^2 \, V^2$. By introducing an auxiliary flux variable $\bar{v}$ we can write the equation/inequality system (3.2), (3.3), (3.4) as

$$(N^1 \quad W) \begin{pmatrix} V^1 \\ \bar{v} \end{pmatrix} = 0, \quad V^2 \geq 0, \quad \bar{v} \geq 0. \tag{3.6a,b,c}$$

Since the system (3.6) coincides with the system (3.2), (3.3), (3.4) if $\bar{v}=1$, it suffices to determine all solution vectors for the homogeneous system (3.6) and to transform these so as to obtain $\bar{v}=1$. Let K be the cone of solutions to system (3.6). The solution set, P, of the system (3.2, 3.3, 3.4) can be represented by the intersection of cone K with the hyperplane defined by $\bar{v}=1$. Thus, the vertices of P are determined by the intersection of the generating vectors of K and this hyperplane.

Since the generating vectors can be multiplied by any positive scalar, one can calculate the interesection points by dividing all generating vectors by their bottom-most elements (representing $\bar{v}$), provided they are non-zero. All remaining vectors, the last elements of which are zero, do not intersect the hyperplane $\bar{v}=1$. They are extreme rays of polyhedron P when shifted from the origin to appropriate intersection points.

It is worth mentioning that the dimension of cone K can have any dimension from zero to $r-v$, depending on how the null-space is situated relative to the positive orthant.

3.2. Some reactions are "reversible"

Now, we wish to consider the case that some of the reactions are not restricted to have a fixed orientation of net flux. Lacking a better notion, we call these reactions reversible, although those reactions that have a fixed sign of flux for some of the reasons mentioned in the Introduction may also be reversible in a physico-chemical sense. By "reversible reactions", we here mean reactions that can proceed in either direction under physiological conditions, such as the reactions shared by glycolysis and gluconeogenesis.

The flux vector can now be decomposed into two subvectors V^i and V^r with

$$N V = 0 , \quad V^i \geq 0 . \tag{3.7a,b}$$

This equation/inequality system determines a convex polyhedral cone, $\mathcal{C}$, (which is not necessarily pointed) in the V-space. Every point of this cone can be represented as the sum of a convex combination of fundamental vectors, f_k, and a linear combination of basic vectors, b_m, [cf. 6],

$$\mathcal{C} = \{V: V = \sum_{k=1}^{p} \eta_k f_k + \sum_{m=1}^{s} \lambda_m b_m , \quad \eta_k \geq 0 \ \forall k \} . \tag{3.8}$$

The basic vectors, b_m, are those extreme rays of cone $\mathcal{C}$ for which also the negative vector, $-b_m$, is contained in $\mathcal{C}$.

A special case of the equation/inequality system is the situation when the subvector V^i coincides with the total flux vector V. This case has been dealt with in an earlier paper [10] (cf. previous section). Cone $\mathcal{C}$ is then a pointed cone. In the more general situation when some reactions can proceed in either direction, we can distinguish the following cases:

a) There are only fundamental vectors, but no basic vectors (such as for an unbranched reaction chain in which one reaction is irreversible). The fundamental vectors are then unique up to scalar multiples, and their number may be greater than the dimension of the cone.

b) Cone $\mathcal{C}$ has fundamental vectors as well as basic vectors. The basic vectors are then not unique (unless there is only one of them), because they span a subspace of kernel(N), which means that any two basic vectors can be replaced by two different linear combinations of these basic vectors. Importantly, the fundamental vectors then are not unique either (unless there is only one of them), since any of them can be replaced by a convex combination of this vector and any of the basic vectors. Whereas all the basic vectors are linearly independent of each other, the fundamental vectors may not.

c) There are only basic vectors. These are then not unique. Their number equals the dimension of cone $\mathcal{C}$, which coincides with kernel(N).

A complete set of fundamental and basic vectors can be detected by the following algorithm, which we derived by adaptation of an algorithm given in [7], to the problem here considered.

We start with a matrix $T^{(0)}$ which contains an identity matrix and the transpose of N,

$$T^{(0)} = \begin{pmatrix} I & \begin{array}{c} N_r^T \\ N_i^T \end{array} \end{pmatrix} . \tag{3.9}$$

The decomposition of N into N_i and N_r is done according to the decomposition of V into V^i and V^r. That is, N_i corresponds to the fluxes that are subject to sign restriction.

Now one successively calculates new matrices, $T^{(1)}$, $T^{(2)}$,..., called tableaux. All of these tableaux consist of two submatrices each, $B^{(j)}$ and $F^{(j)}$. $B^{(0)}$ and $F^{(0)}$ are those submatrices of $T^{(0)}$ that contain N_r and N_i, respectively. In each step, one has to distinguish two cases:

(i) $t^{(j)}_{i,r+j+1}=0$ for all rows $t^{(j)}_{i\cdot}$ that belong to $B^{(j)}$, or $B^{(j)}$ is void.

(ii) There is an index i_0 with $t^{(j)}_{i_0,r+j+1}\neq 0$ and $t^{(j)}_{i_0\cdot} \in B^{(j)}$.

In the first case, we define, for each row i belonging to $F^{(j)}$, the index set $S(\alpha)$ as the set of the column indices, h, of all elements $t^{(j)}_{\alpha,h}$ of the left-hand side part of $T^{(j)}$ that are zero:

$$S(\alpha) = \{h: h \leq r, t^{(j)}_{\alpha,h} = 0\}. \tag{3.10}$$

$T^{(j+1)}$ is then obtained as the union of $B^{(j+1)} = B^{(j)}$ and

$$F^{(j+1)} = \{t^{(j)}_{i\cdot} \in F^{(j)}: t^{(j)}_{i,r+j+1}=0\} \cup \{t^*: t^* = \left|t^{(j)}_{i,r+j+1}\right|\cdot t^{(j)}_{k\cdot} + \left|t^{(j)}_{k,r+j+1}\right|\cdot t^{(j)}_{i\cdot}\ \forall\ i,k:$$
$$t^{(j)}_{i\cdot}, t^{(j)}_{k\cdot} \in F^{(j)},\ t^{(j)}_{i,r+j+1}\cdot t^{(j)}_{k,r+j+1}<0,\ S(i)\cap S(k)\not\subseteq S(h)\ \forall\ h\neq i,k,\ t^{(j)}_{h\cdot}\in F^{(j)}\}. \tag{3.11}$$

In case (ii), $T^{(j+1)}$ is obtained as

$$B^{(j+1)} = \{t^*: t^* = t^{(j)}_{i_0,r+j+1}\cdot t^{(j)}_{i\cdot} - t^{(j)}_{i,r+j+1}\cdot t^{(j)}_{i_0\cdot},\ t^{(j)}_{i\cdot}\in B^{(j)},\ i\neq i_0\ \}, \tag{3.12}$$

$$F^{(j+1)} = \{t^*: t^* = \operatorname{sgn} t^{(j)}_{i_0,r+j+1}(t^{(j)}_{i_0,r+j+1}\cdot t^{(j)}_{i\cdot} - t^{(j)}_{i,r+j+1}\cdot t^{(j)}_{i_0\cdot}),\ t^{(j)}_{i\cdot}\in F^{(j)}\ \}. \tag{3.13}$$

The algorithm stops when $j=r$, or even before, when all columns with a column index $>r$ are zero vectors. Those rows of $B^{(r)}$ and $F^{(r)}$ that originate from the identity matrix, constitute the sought basic vectors and fundamental vectors, respectively.

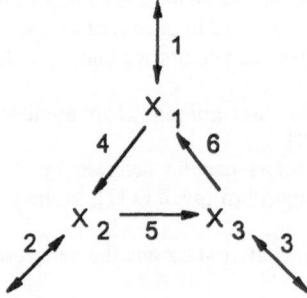

Figure 2. Reaction system containing a cycle of irreversible reactions.

We demonstrate this algorithm by way of the example shown in Fig. 2, which may serve as a model of the PEP/PYR/OAA cycle [13]. The starting tableau reads (zeroes are omitted for clarity's sake)

$$
\mathbf{T}^{(0)} = \left(
\begin{array}{cccccccccc}
1 & & & & & & \vdots & 1 & & \\
& 1 & & & & & \vdots & & 1 & \\
& & 1 & & & & \vdots & & & 1 \\
\cdots & \cdots & \cdots & \cdots & \cdots & \cdots & \cdots & \cdots & \cdots & \cdots \\
& & & 1 & & & \vdots & -1 & 1 & \\
& & & & 1 & & \vdots & & -1 & 1 \\
& & & & & 1 & \vdots & 1 & & -1
\end{array}
\right)
\begin{array}{c}
\mathbf{B}^0 \\ \\ \\ \\ \mathbf{F}^0
\end{array}
\qquad (3.14)
$$

$$
\mathbf{I} \qquad\qquad\qquad \mathbf{N}^T
$$

We first examine the left-hand side column of $\mathbf{N}^T$. Since $n_{11}=1$, case (ii) applies. $\mathbf{B}^{(1)}$ is obtained by combining the first row with the second and third rows according to formula (3.11). $\mathbf{F}^{(1)}$ results from combination of the first row with all rows of $\mathbf{F}^{(0)}$ according to formula (3.12). This gives

$$
\mathbf{T}^{(1)} = \left(
\begin{array}{cccccccc}
1 & & & & & \vdots & 1 & \\
& 1 & & & & \vdots & & 1 \\
\cdots & \cdots & \cdots & \cdots & \cdots & \cdots & \cdots & \cdots \\
1 & & 1 & & & \vdots & 1 & \\
& & & 1 & & \vdots & -1 & 1 \\
-1 & & & & 1 & \vdots & & -1
\end{array}
\right).
\qquad (3.15)
$$

Finally, we obtain

$$
\mathbf{T}^{(3)} = \left(
\begin{array}{cccccccc}
1 & -1 & & 1 & & \vdots & 0 & 0 & 0 \\
& 1 & -1 & & 1 & \vdots & 0 & 0 & 0 \\
-1 & & 1 & & & 1 & \vdots & 0 & 0 & 0
\end{array}
\right).
\qquad (3.16)
$$

$\mathbf{T}^{(3)}$ only consists of $\mathbf{F}^{(3)}$. The fundamental vectors are

$$
\mathbf{f}^{(1)T} = (\; 1 \;\; -1 \;\; 0 \;\; 1 \;\; 0 \;\; 0 \;), \qquad (3.17a)
$$

$$
\mathbf{f}^{(2)T} = (\; 0 \;\; 1 \;\; -1 \;\; 0 \;\; 1 \;\; 0 \;), \qquad (3.17b)
$$

$$
\mathbf{f}^{(3)T} = (\; -1 \;\; 0 \;\; 1 \;\; 0 \;\; 0 \;\; 1 \;). \qquad (3.17c)
$$

$\mathbf{f}^{(1)}$, for example, represents the elemental mode $v_1=-v_2=v_4$, $v_3=v_5=v_6=0$. The fact that no basic vector is obtained corresponds to the feature that no elemental mode of the system shown in Fig. 2 can be inverted to obtain another elemental mode. It is somewhat surprising that we do not come up with the cyclic mode $v_4=v_5=v_6$. It can, however, be obtained by

372

addition of $f^{(1)}$, $f^{(2)}$ and $f^{(3)}$. This example shows that the number of elemental modes in the sense that a maximum number of rates is zero, may be greater than the number of generating vectors. If reactions 1, 2, and 3 are cancelled from the network, one obtains only one fundamental vector, which represents the abovementioned cyclic flux. Consequently, enlarging a network by additional reactions does not simply lead to additional elemental modes, but it may entail replacement of some elemental modes by others.

4. The admissible set of steady-state concentrations

In the previous sections, we have derived conclusions by only using knowledge of the stoichiometric structure of the system and the sign of fluxes. In many situations, one has more knowledge than this structural information, in that also the equilibrium constants of reactions are known. If all rate functions are of generalized mass-action type (which applies to all reversible enzyme kinetic functions), the concentrations X_i fulfil the inequalities [3]

$$\sum_i n_{ij} \log X_i - \log q_j \begin{cases} \leq 0 & \text{if } v_j > 0 \\ = 0 & \text{if } v_j = 0 \\ \geq 0 & \text{if } v_j < 0 \end{cases}, j = 1,\dots,r \qquad \begin{matrix} \text{(4.1a)} \\ \text{(4.1b)} \\ \text{(4.1c)} \end{matrix}$$

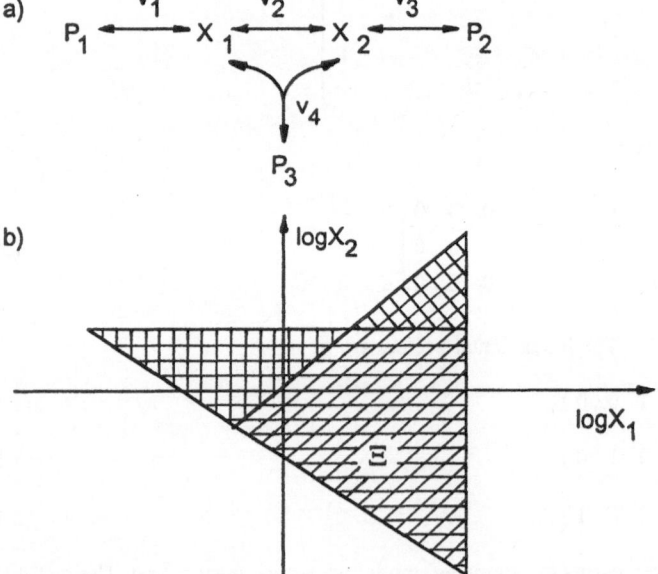

Figure 3. Exemplifying reaction system (a) and the corresponding region admissible for logarithmic concentrations (b). The subregions corresponding to the different sign patterns are marked by different hatching.

In (4.1a,c), we admit equality in order to allow for the case of infinitely fast reactions (quasi-equilibrium reactions).

Together with the steady-state equation (3.1a), system (4.1) determines the admissible region, Ξ, for the logarithmic steady-state concentrations. Fig. 3 shows an example. If the sign of all ν_j were prescribed, (4.1) would be a linear inequality system, so that Ξ would be a convex polyhedron. As we do not, *a priori*, know the signs of the ν_j, relations (4.1a,b,c) define r^3 different convex polyhedra. The admissible region is the union of all of these polyhedra and is, hence, a polyhedron as well, which may not be convex. In general, many of the single polyhedra are empty because condition (3.1a) excludes many sign patterns. In fact, polyhedron Ξ can be constructed by this composition procedure, which is rather tedious, though. It therefore deserves further study to develop a straightforward algorithm for determing the vertices of polyhedron Ξ.

We here go beyond the scope of convex analysis, but many assertions derived in that theory can nevertheless be used. This is true for the following interesting relationship between existence of conservation relations (cf. Section 2) and the shape of polyhedron Ξ. The three following statements are equivalent:

(i) There exists a linear dependence

$$\sum_i d_{ki} n_{ij} = 0 , \ j=1,...,r^{\cdot}. \tag{4.2}$$

(ii) If $y^0 \in \Xi$, then all points $y^0 + \rho \, d_k$ with $\rho > 0$ lie in Ξ as well.

(iii) The same property as (ii), but with $\rho > 0$ replaced by $\rho < 0$.

This theorem is an extended version of Theorem 1 given in a previous paper [14], where it was phrased for the polyhedron defined by the condition that all fluxes be positive. The proof to the theorem given above is a straightforward extension of the proof to Theorem 1 in [14].

The theorem implies that polyhedron Ξ is unbounded if, and only if, the system involves conservation relations. If none of the conservation relations is semi-positive, the concentrations are bounded below by non-zero limits.

5. Discussion

While linear algebra is used in the mathematical description of chemical reaction systems since the beginning of the century, convex analysis entered the scene only very recently. The reason for including inequalities into the analysis is that

(i) concentrations are always non-negative;

(ii) conservation relations have to be semi-positive in order to express conservation of chemical units;

(iii) for many biochemical reactions, the orientation of flux is fixed due to irreversibility of the reaction or to the biological functioning.

The algorithm for determining the extreme semi-positive conservation relations can be applied to quasi-equilibrium approximation, which is often used for simplifying the modelling of (bio)chemical reaction systems [15]. Mathematically, this approximation is based on calculation of linear combinations of concentrations (for example, the sum of free enzyme and enzyme-substrate complex), which are usually interpreted as slowly varying pools. In order that the newly calculated variables are suscseptible to chemical interpretation, they have to be related to semi-positive conservation relations of the fast subsystem.

The generating vectors of the flux cone may be interpreted as elemental modes of the system. The idea to define fundamental modes has been put forward by Leiser and Blum [13], but these authors have not given a mathematical definition. In the case that the flux cone is pointed, the elemental modes are unique. In the case that cone C is not pointed because it contains basic vectors, this favourable property is lost. It therefore deserves further studies on how the set of elemental modes can be made unique, for example by imposing the side condition that all basic vectors be orthogonal to the fundamental vectors.

The algorithms for determining the complete set of extreme semi-positive conservation relations, the polyhedral set of steady-state fluxes with or without some flux values fixed and the cone of fluxes with or without some fluxes being reversible have been translated into computer programs (written in Turbo-Pascal, C, and Turbo-Pascal, respectively) by the authors and are available on request.

Investigation of the concentration polyhedron Ξ is of valuable help in optimization studies [14]. For example, minimization of intermediate concentrations has frequently been considered to be an advantage of metabolic channelling (direct transfer of intermediates between enzymes) [16,17]. This hypothesis has, however, sometimes been cast into doubt [18]. By analysing the admissible region for steady-state concentrations, it can be shown that metabolic channelling has no further effect on mimimization of concentrations if an optimal state has been attained in the free-diffusion system [19]. It can further be shown that in states of minimum intermediate concentrations, time hierarchy obtains in that some reactions attain quasi-equilibrium [14].

References

1. F. Horn and R. Jackson, Arch. Rational Mech. Anal. 47:81-116 (1972).
2. C. Reder, J. theor. Biol. 135:175-201 (1988).
3. S. Schuster and R. Schuster, J. Math. Chem. 3:25-42 (1989).
4. B.L. Clarke, in: *Advances in Chemical Physics*, ed. I. Prigogine and S.A. Rice, Wiley, New York, vol. 43, p. 1 (1980).
5. B.L. Clarke, Cell Biophys. 12:237-253 (1988).
6. R. Rockafellar, *Convex Analysis*, Princeton University Press, Princeton (1970).
7. F. Nožička, J. Guddat, H. Hollatz and B. Bank, *Theorie der linearen parametrischen Optimierung*, Akademie-Verlag, Berlin (1974).
8. S.N. Chernikov, *Linear Inequalities*, (in Russian), Nauka, Moscow (1968). German translation: *Lineare Ungleichungen*, Deutscher Verlag der Wissenschaften, Berlin (1971).
9. S. Schuster and T. Höfer, J. Chem. Soc. Faraday Trans. 87:2561-2566 (1991).
10. R. Schuster and S. Schuster, Comp. Appl. Biosci. 9:79-85 (1993).
11. D.J.M. Park Jr., Comput. Chem. 12:175-188 (1988).
12. P. Cavallotti, G. Celeri and B. Leonardis, Chem. Engng. Sci. 35:-2297-2304 (1980).
13. J. Leiser and J.J. Blum, Cell Biophys. 11:123-138 (1987).
14. S. Schuster and R. Heinrich, J. Math. Biol. 29:425-442 (1991).
15. M. Schauer and R. Heinrich, Math. Biosci. 65:155-171 (1983).
16. J. Ovádi, J. theor. Biol. 152:1-22 (1991).
17. P. Mendes, D.B. Kell and H.V. Westerhoff, Eur. J. Biochem. 204:257-266 (1992) .
18. A. Cornish-Bowden and M.L. Cárdenas, Eur. J. Biochem. 213:87-92 (1993).
19. R. Heinrich and S. Schuster, J. theor. Biol. 152:57-61 (1991).

ROBUST SURVIVAL MODEL AS AN OPTIMIZATION PROBLEM

P. Kovanic and R. A. Barack[1]:

Summary

A new approach to the identification of a nonlinear multidimensional analytical survival model has been developed based on the gnostical theory of uncertain data. No a–priori statistical model of random data components is assumed; on the contrary: the estimate of the most important characteristic of data uncertainty, its distribution function, is a result of the analysis. This estimation process represents a multidimensional constrained optimization problem. The form of the distribution function is thus determined by the data and may differ from all standard statistical distributions. The estimation of deterministic model parameters can be performed simultanously with estimation of the distribution of indeterministic data components. All results are robust not only with respect to a–priori statistical assumptions (there are none applied) but also to outliers and peripheral data clusters. Both uncensored and censored data are taken in account for estimation. The method has been successfully implemented and applied to practical problems connected with the evaluation of the reliability of truck components.

Problem

We wish to model (and to estimate) the distribution of time–to–failure (survival or life time) for some particular device using measured life times and other parameters (eg. load to which the object is subjected). We suppose that the task possesses some standard features of nonlinear regression, namely:

1. Given the structure of a positively valued regression function $t(z, \theta)$, ($z > 0, \theta \in R_M$) of a covariate z (load) and of an unknown (vector) parameter θ that has to be estimated.

2. A k–th observed value of the time to failure for the load Z_k is assumed to have the multiplicative model $T_k = t(Z_k, \theta) \cdot \rho_k$, where the ρ_k (the rezidual) states for the k–th realization of the (scalar, positive) random variable modelling the data uncertainty. The model may also be considered in its logarithmized form, i.e. $Y_k = y(Z_k, \theta) + \varepsilon_k$.

[1]ÚTIA ČSAV, P. O. Box 18, 182 08 Prague, Czechoslovakia
Telex: 122018 atom c, Fax: (42)(2)847452, Phone: (42)(2)8152230
E-mail: kovanic@cspgas11.bitnet

3. Some data may be censored (some experiments interrupted before the failure occurs). The result then yields the information 'time to failure is greater than T' instead of 'time to failure equals T'.

Simultaneously, our further assumptions try to reflect some nonstandard, but realistic conditions:

4. The distribution law of variable ρ is not specified, only existence of its continuous density function is assumed. We cannot expect any 'popular' form of distribution function (d.f.), neither symmetry nor any such similar simplifying property.

5. The unknown distribution function is not necessarily defined on an infinite domain. Unknown bounds of their support may also be object of the estimation.

6. There may be constraints on the values of some or all estimated parameters dictated by the nature of the process under consideration.

7. The survival experiments are often time-consuming and expensive. Thus, the number of data may be rather small.

8. Robustness with respect to a priori assumptions of a statistical nature are desirable, as is robustness with respect to outlying data, because the data on life times are usually widely spread.

Under such conditions, the commonly used and standard statistical approaches can hardly be applied.

It was therefore decided to make use of a new non–statistical approach called *the gnostical theory of uncertain data.* (Kovanic 1986). The new method has been previously applied to with excellent results (Kovanic, Volf 1992).

Main principles of gnostics

The theory is general, and was developed without relation to a particular field of application. It is a mathematical theory which originated at the intersection of several abstract scientific disciplines and, therefore, it is not very easily explainable (particularly with a limited reference to mathematics). The first gnostical axiom states that the structure of real data is isomorphic with the Cartesian product of a couple of commutative groups. This is a quite natural statement resulting from the assumption that real data are not merely some arbitrary numbers but products of a well-known technology called *quantification.* This technology dealing with measuring and counting has its own axiomatics ensuring that the quantification (as a mapping of empirical structures of real quantities into the structure of real numbers) is consistent. However, the theory of measuring assumns that the process is quite precise and leaves the case of uncertain disturbance to postmeasuring treatment, mainly performed by statistical methods. In the Gnostical model of uncertain data, a real datum is considered to be a one-dimensional projection of a two-dimensional vector. One component depicts the "ideal" (unknown) quantity to be estimated, the second one representing the unknown disturbance - uncertainty. The group nature of both components of this vector can be shown to result from the axiomatics of measuring theory. The first axiom can be thus reformulated as the requirement that both ideal and disturbing data component are *measurable* in the sense of measurment theory. Students of statistics should note that this measureability substantially differs from the sense this notion has within the frame-work of probability theory. The most important of these differences is that the mentioned "practical" measurability can be more easily verified than some assumption of the "i.i.d." character of a random sample of random functions. There is no randomness in gnostics. In gnostics, data error bears no relationship to a random process. Each change in the value of a datum has its cause, and it could be rationally explained, if only there were sufficient information. *Data uncertainty is thus a lack of information* and it is therefore highly *subjective*.

Since the pair (*ideal value, disturbance*) can be considered a vector, it follows from the first gnostical axiom that the effect of the contamination of the data can be represented by an operator "rotating" the this vector. The operator effecting this rotation is called the *quantification operator*. Gnostical theory also represents the estimation process by an operator. The *second gnostical axiom* chooses among possible operators the simplest ones satisfying a symmetry requirement. It is proved that this axiom is equivalent to the assumption that the path of a datum is one of two analytical functions, the first one - connected with the disturbed quantification process - being the Minkowskian circle while the second one, corresponding to the optimum estimation process, is an ordinary Eucleidian circle. This approach represents a complete theory of **individual** data including following important results:

- New formulae for evaluation of data error, entropy increase, and for information loss caused by contamination. Instead of linear and quadratic functions resulting from the Eucleidian geometry accepted by statistics and other theories of uncertainty, the new functions correspond to some special cases of Riemannian geometries.

- A new formula for computing data weight decreased by uncertainty.

- An entropy $\Longleftrightarrow$ information conversion law according to which the compensation of changes in both quantities takes place on the level of their second derivatives,

- Two special forms of the distribution functions of individual uncertainty.

- Variation theorems for geodesic lines (circular paths constituting the ideal gnostical cycle) proving its optimality.

- A (Lorentz invariant) mapping between structures of gnostical models of quantified uncertain data and uncertainty operators and structures of uncharged relativistic particles and their energy-momentum tensors.

The last result makes it natural to choose the third gnostical axiom - the data composition law - as a generalized analog of the relativistic law of the energy- momentum conservation. The gnostical theory of small data samples based on this axiom yields procedures which are naturally robust due to the (nonlinear, Riemannian) data errors and to the data composition law that is not additive neither to data nor to their squares.

The implication given by gnostics that there exist commonalities between uncertain data and relativistic particles may confuse and frustrate some practitioners. However, they do apply without hesitation or confusion the operations of the arithmetical mean and standard deviation and use least squares regression models inherently based on covariance matrices. These fundamental notions of classical statistics were borrowed from Newtonian mechanics. The arithmetical mean of data is the analogue of the coordinate of the center of gravity; the sum of data square errors are calculated in the same way as are diagonal components of the energy-momentum tensor of a system of mass points. Covariances are the images of nondiagonal components of this tensor. The additive composition law adopted in statistics both for data and square errors is thus motivated by the Newtonian version of the energy-momentum conservation law. As can be shown, this is also valid from the gnostical point of view; but only under the condition of very small relative data errors. Newtonian mechanics viewed by relativistic eyes is a special case of relativistic mechanics under the conditions of small (with respect to light speed) relative velocities. This velocity corresponds to relative data errors in gnostics. Gnostics has developed formulae valid for relative errors of any arbitrary size. These formulae are akin to recent developments in physics as classical statistical formulae are to the medieval mechanics. Real data (e.g. from economy or reliabilty) are often highly contaminated. The message is therefore clear: *do not use medieval statistical formulae when treating highly contaminated data!* These cannot provide reliable results. Instead, apply modern gnostical data treatment technology!

Gnostical data distribution functions

There are two kinds of distribution functions (d.f.) supported by the gnostical theory: 'local' and 'global'.

Consider the i-th real-valued observation A_i (an 'additive' datum) together with its 'multiplicative' equivalent

$$Z_i = exp(A_i) \tag{1}$$

having a strictly positive value. For a positive real scale parameter s, a real variable $z \rangle 0$ and for a sample of N data, define the auxiliary quantities

$$q_i(z, s) = (Z_i/z)^{2/s} \tag{2}$$

for use in the calculation of N 'fidelities'

$$f_i(z, s) = 2/(1/q_i(z, s) + q_i(z, s)) \tag{3}$$

and 'irrelevancies'

$$h_i(z, s) = (1/q_i(z, s) - q_i(z, s))/(1/q_i(z, s) + q_i(z, s)). \tag{4}$$

Within the frame-work of the gnostical theory, irrelevance plays the role of the distance between z and Z, the fidelity being the weight of the datum Z_i.

Introduce the arithmetical mean

$$\bar{f}(z,s) = \sum_{i=1}^{N} f_i(z,s)/N \tag{5}$$

of the fidelities and define the symbol $\bar{h}(z,s)$ for the irrelevances analogously. Let $w(z,s)$ be the weighting function defined by the relation

$$w^2(z,s) = (\bar{f}(z,s))^2 + (\bar{h}(z,s))^2 \tag{6}$$

The distribution function generated by the individual datum Z_i is then

$$L_i(z,s) = (1 + h_i(z,s))/2 \tag{7}$$

having the density

$$l_i(z,s) = \frac{d(L_i(z,s))}{dz} = f_i^2(z,s)/(zs). \tag{8}$$

For a real p $(0 < p < 1)$ define the following real functions

$$H(p) = -p\ln(p) - (1-p)\ln(1-p) \tag{9}$$

$$I(p) = H(1/2) - H(p). \tag{10}$$

The quantity $I(L_i(z,s))$ is interpreted in the gnostical theory as the information loss due to the uncertainty which causes the datum value Z_i be observed instead of the value z.

The local d.f. $L(z,s)$ is simply the arithmetical mean of the d.f.'s (7) of individual data:

$$L(z,s) = \sum_{i=1}^{N} L_i(z,s)/N. \tag{11}$$

The global d.f. $G(z,s)$ is obtained using the weighting function w (6),

$$G(z,s) = \sum_{i=1}^{N} L_i(z,s)/w(z,s). \tag{12}$$

Under the condition of a weak influence of uncertainty (small errors of data, a small value of the scale parameter s), the two functions differ only slightly. However, their behaviour is quite different under gross errors.

In the sense of being a monotonous function for an arbitrary data sample, the local d.f. always exists. In a special case of availability of the statistical model of data, the d.f. $L(z,s)$ (11) can be interpreted as a nonparametric estimate of the probability d.f. and the function $l_i(z,s)$ (8) as a proper kernel of Parzen's type (Parzen 1962). In this case, the gnostical theory is used only as a background motivating the choice of the special kernel (8). For this purpose, gnostical theory generates remarkably nice and smooth density curves even in the case of small data samples. If it is of interest, the asymptotic features of the estimate can then be studied by established statistical methods. But in the more general case of data not having a statistical model, formula (11) still has significance as a continuous model of the data sample's distribution and as an estimate of the expectation of an another datum of the same origin. The steep descent of the gnostical kernel (8) has an important consequence: the individual subclusters of data influence each other only weakly. This enables the characterization of local details of the data sample and opens an efficient method of cluster analysis.

Unlike the local d.f., the global function $G(z,s)$ has a theoretical support only for homogenous data samples, i.e. for data with a unimodal density function. When applied to the multimodal case, this function may lose the fundamental feature of a d.f., its monotony. This allows the hypothesis on the homogenity of the data sample to be tested. The global d.f. has no known statistical analogy. Its practical importance is connected with its remarkable robustness with respect both to outlying data and outlying subclusters of data: in estimation of probability for extremal quantiles, the 'central', 'main' part of the data sample plays the dominating role. The global d.f. thus characterizes the overall distribution law of the data. Practical consequences of this include the d.f.'s good performance in application to small samples and its applicability to samples generated by different distribution laws (documented in Baran (1988)).

The local and global d.f.s differ substantially in their dependence on the scale parameter (s). Let $F(N)$ be the 'empirical' distribution function of the data sample. The function $F(N)$ has the known form of an irregular staircase. The local d.f. $L(z,s)$ of the same sample can be made to approach $F(N)$ as close as required choosing a sufficiently small positive value for the scale parameter. By contrast, the maximum distance of the global d.f. $G(z,s)$ has a minimum for a 'best' $\tilde{s}$, which can be taken as a robust estimate of the scale parameter. In this case, the d.f. $G(z,\tilde{s})$ is as close as possible to the empirical distribution function $F(N)$.

Main ideas of the new identification approach

As a nonstatistical alternative, the mentioned *gnostical* estimates of distribution functions have been applied. These d.f. are parametrized only by data (residuals ρ_k are data in our case) and by a positive scale parameter s that is not necessarily constant. In the considered case, it is supposed that $s_k = S_0 * exp(S_1 * Z_k)$ with some unknown constants S_0 and S_1. The real domain of the d.f. is supposed to have a general form $\mathcal{D} := (R_L, R_U)$ with some unknown – finite or infinite – bounds R_L and R_U. The parameter vector to be estimated is therefore $\Theta = \langle S_0, S_1, R_L, R_U, A, B, C, D, E, \rangle$. There are two important reasons for application of the gnostical d.f. of the *global* type:

1. No apriori assumptions on the form of the d.f. are necessary. This d.f. is completely defined by the parameters R_L, R_U, S_0 and S_1 and by the data set $\{\langle \rho_k, I_k \rangle, k = 1, .., N\}$ where N is the number of data and I_k the indicator of the

data censorship.

2. As already mentioned, this d.f. is robust with respect to outlying data and even to peripheral data subclusters.

To optimize all parameters, the criterion function to be used is the evaluation of the goodness–of–fit of the empirical distribution function (EDF) (of model residuals ρ_k) by the gnostical d.f. The standard type of the EDF is the step function but its application (connected with Kolmogorov–Smirnov goodness–of–fit test and evaluation) would involve non–smoothness of the function to be minimized. This is why a new type of the empirical d.f., the point empirical d.f. (PEDF) was developed based on a continuous approximation of the binomial d.f. The criterion function can be then simply the sum of squares of differences between residuals and corresponding values of the PEDF.

Main steps of the procedure

The process includes following steps:

1. Setting or calculating initial conditions as a first approximation to Θ.

2. Evaluation of model residuals using uncensored data.

3. Calculating of point estimates of the empirical d.f. (PEDF) of residuals of these data.

4. Correcting the PEDF with respect to censored data.

5. Evaluation of next approximation to Θ using an optimization procedure and the condition of improving fit of the EPDF by the global gnostical d.f. This step thus modifies both surviving model and the d.f. of residuals.

6. Repetition from 2) till stabilization of results.

7. Verification of the homogeneity of the residual distribution represented by the local gnostical d.f.

8. Substitution of resulting Θ into procedures for estimation of quantiles and of probability of life–times.

The well–known procedure of Schittkowski (1985/1986) yielding a local minimum has been proved to satisfy all requirements of the difficult task under consideration. Repeating runs with different initial conditions were used to verify the global optimality.

The final result analyses is then verified using the gnostical analyzer - a program enabling to get a deep inside into small and strongly dispersed data samples.

The new method of identification of reliability model has been successfully implemented for evaluation of life-time tests of truck components in a big Czech car factory.

References:

Baran, R.H. (1988). Comments on "A New Theoretical and Algorithmical Basis for Estimation, Identification and Control" by P. Kovanic. Automatica, 24 283-287.

Kovanic, P. (1986). "A New Theoretical and Algorithmical Basis for Estimation, Identification and Control", Automatica, 22, 657-674.

Kovanic P., Volf P. (1992), Robust Identification of Survival Models, The Journal of General Systems (submitted)

Parzen, E. On estimation of a probability density function and mode. Ann. Math. Stat., 35, 1065-1076.

Schittkowski, K. (1985/6). "NLPQL: A Fortran Subroutine Solving Constrained Nonlinear Programming Problems", Annals of Operation Research,Vol.5, 485-500.

Maximum-volume ellipsoids contained in bounded convex sets: application to batch and on-line parameter bounding

L. Pronzato
Laboratoire I3S, CNRS URA-1376, Sophia Antipolis,
06560 Valbonne, France
and
E. Walter
Laboratoire des Signaux et Systèmes,
CNRS-ESE, Plateau de Moulon,
91192 Gif-sur-Yvette Cedex, France

Abstract—Two algorithms are proposed for computing the (unique) ellipsoid with maximum volume contained in a bounded convex set. The first one can be used on-line and only applies when the set is a polyhedron. The second one is off-line and applies to any bounded convex set. Both are based on a procedure for computing a minimum-volume outer ellipsoid, developed in the field of experimental design, which is recalled.

Keywords—Convex sets, ellipsoids, parameter bounding, polyhedra.

1 Introduction and problem statement

Bracketing a bounded convex subset S of $\mathcal{R}^p$ between an inner ellipsoid and an outer ellipsoid provides a simple characterization of its properties. For instance, guaranteed upper and lower bounds on the volume of S are then immediately obtained.

As another example of application of these techniques, assume that a control vector u has to be chosen so as to satisfy a target property for any x in a given prior feasible set

$$p(u,x) \in T, \forall x \in \mathcal{F}, \tag{1}$$

where x and $p(.,.)$ are vector-valued and where T and $\mathcal{F}$ are bounded convex sets. Solving (1) for u may prove exceedingly difficult, and it is often easier to consider either of the two following sufficient conditions for (1) to be satisfied.

$$
\begin{array}{lll}
\text{(i)} & p(u,x) \in T, \forall x \in \mathcal{O}, & \text{with } \mathcal{F} \subset \mathcal{O}, \\
\text{(ii)} & p(u,x) \in \mathcal{I} \subset T, \forall x \in \mathcal{F}. &
\end{array}
$$

The sets $\mathcal{I}$ and $\mathcal{O}$ may respectively be an inner ellipsoid for T and an outer ellipsoid for $\mathcal{F}$.

In [12], an algorithm from the field of experimental design has been used to construct the minimum-volume ellipsoid containing S. In [11], an algorithm is proposed to compute the maximum-volume ellipsoid contained in a bounded convex polyhedron. The later algorithm is extended in this paper to the cases where (i) the polyhedron is determined on-line or (ii) S is any bounded convex set.

Some basic results and algorithms concerning inner and outer ellipsoids are recalled in Section 2. On-line problems (i.e. when the constraints defining S are taken into account one at a time) are considered in Section 3. General bounded convex sets are treated off-line in Section 4, where an illustrative example is presented.

2 Basic results

From the Loewner-Behrend theorem [1], there exists a unique ellipsoid $\mathcal{E}_o^*(S)$ of minimal volume containing S, and, from the Danzer-Zaguskin theorem [2] there also exists a unique ellipsoid $\mathcal{E}_i^*(S)$ of maximal volume contained in S.

In what follows, $\mathcal{E}(c, A)$ denotes the ellipsoid defined by

$$\mathcal{E}(c, A) := \{x \in \mathcal{R}^p \mid (x - c)^T A(x - c) \leq p\},$$

Ξ is the set of all normalized distributions ξ of weights on S,

$$\int_S \xi(dx) = 1, \xi(dx) \geq 0,$$

and $M(\xi)$ and $c(\xi)$ are defined as

$$M(\xi) := \int_S xx^T \xi(dx), c(\xi) := \int_S x\xi(dx).$$

An algorithm [12, 14] for the construction of $\mathcal{E}_o^*$, which is part of the algorithms to presented in the next sections, is now recalled.

MIVOE: (MInimum Volume Outer Ellipsoid algorithm)

Step i Choose $\epsilon \, (0 < \epsilon << 1)$ and a discrete distribution $\xi^0 \in \Xi$ such that $M(\xi^0)$ is invertible. Set $k = 0$.

Step ii Compute

$$x^+ := \arg\max_{x \in S} d(x, \xi^k), \tag{2}$$

with

$$d(x, \xi) := x^T M^{-1}(\xi)x + \frac{(x^T M^{-1}(\xi)c(\xi) - 1)^2}{1 - c^T(\xi)M^{-1}(\xi)c(\xi)} - (p + 1).$$

If $d(x^+, \xi^k) < \epsilon$, take $\mathcal{E}(c^*, A^*)$ as an approximation of $\mathcal{E}_o^*(\mathcal{S})$, with

$$c^* := c(\xi^k), A^* := (M(\xi^k) - c(\xi^k)c^T(\xi^k))^{-1},$$

stop.

Step iii Find the best weights λ_i in the sense of the criterion

$$\Phi(\xi) := \ln \det[M(\xi) - c(\xi)c^T(\xi)]$$

of a normalized distribution whose support points x_i are x^+ and those of ξ^k. Denote the resulting distribution by ξ^{k+1}. Remove the support points with zero weight from ξ^{k+1}, $k \leftarrow k+1$, go to Step ii. □

Step ii corresponds to the maximization of a convex quadratic function over $\mathcal{S}$. Local methods may thus not converge to the global optimum. When $\mathcal{S}$ is a bounded polyhedron, the only candidates for x^+ (2) correspond to its vertices, so that the global solution is obtained at a low computational cost. Step iii corresponds to the maximization of a concave function of the weights $\lambda_i, i = 1, \ldots, n$, with the constraints $\lambda_i \geq 0, \sum_{i=1}^n \lambda_i = 1$. A constrained Newton method (which amounts to solving a sequence of convex quadratic programming problems) is thus especially indicated. The following expressions for the gradient and Hessian of $\Phi(\xi)$ allow an easy implementation of the algorithm.

$$\frac{\partial \Phi(\xi)}{\partial \lambda_i} = x_i^T M^{-1}(\xi)x_i + \frac{(x_i^T M^{-1}(\xi)c(\xi) - 1)^2 - 1}{1 - c^T(\xi)M^{-1}(\xi)c(\xi)},$$

$$\frac{\partial^2 \Phi(\xi)}{\partial \lambda_i \partial \lambda_j} = -(x_j^T M^{-1}(\xi)x_i)^2 + \frac{2x_j^T M^{-1}(\xi)x_i}{1 - c^T(\xi)M^{-1}(\xi)c(\xi)} \times$$
$$[c^T(\xi)M^{-1}(\xi)(x_i + x_j) - (x_i^T M^{-1}(\xi)c(\xi))(x_j^T M^{-1}(\xi)c(\xi)) - 1] +$$
$$\frac{(x_i^T M^{-1}(\xi)c(\xi))(x_j^T M^{-1}(\xi)c(\xi))}{(1 - c^T(\xi)M^{-1}(\xi)c(\xi))^2} \times$$
$$[2c^T(\xi)M^{-1}(\xi)(x_i + x_j) - (x_i^T M^{-1}(\xi)c(\xi))(x_j^T M^{-1}(\xi)c(\xi)) - 4].$$

The Newton method can be initialized with the following distribution,

$$\xi = (1 - \alpha)\xi^k + \alpha\xi_{x^+},$$

where ξ_{x^+} gives a unit mass to the single support point x^+ (2) and α is given by

$$\alpha = \frac{d(x^+, \xi^k)}{(p+1)(d(x^+, \xi^k) + p)}.$$

This value of ξ corresponds to the distribution ξ^{k+1} that would be chosen by a Fedorov-like algorithm [4]. Note that optimizing the weights is then not necessary to insure global monotone convergence. It is, however, highly recommended to obtain a satisfactory speed of convergence. Applications of this algorithm to the parameter-bounding context can be found in [12].

Assume now that S is a bounded convex polyhedron, defined by

$$S = \{x \in \mathcal{R}^p \mid a_i^T x \le b_i, i = 1, \ldots, m\}. \tag{3}$$

The following algorithm has been suggested in [11] to obtain $\mathcal{E}_i^*(S)$. It will also be part of the algorithms presented in the next sections.

MAVIE: (MAximum Volume Inner Ellipsoid algorithm)

Step i Choose $c^0 \in \text{int}(S)$, ϵ' ($0 < \epsilon' << 1$), set $k = 0$.

Step ii Compute the m vectors

$$v_i^k = \frac{pa_i}{b_i - a_i^T c^k}, i = 1, \ldots, m, \tag{4}$$

and determine the minimum volume ellipsoid $\mathcal{E}(c^{*k}, A^{*k})$ containing them (using MIVOE).

Step iii Compute

$$e^k = -\frac{pA^{*k}c^{*k}}{p - c^{*k^T}A^{*k}c^{*k}},$$
$$c^{k+1} = c^k + e^k.$$

— If $\|e^k\| < \epsilon'$, compute

$$B^{k+1} = (A^{*k^{-1}} - \frac{c^{*k}c^{*k^T}}{p})\frac{(p - c^{*k^T}A^{*k}c^{*k})}{p},$$

take $\mathcal{E}(c^{k+1}, B^{k+1})$ as an approximation of $\mathcal{E}_i^*(S)$ and stop;

— else $k \leftarrow k + 1$, go to Step ii.   □

In [10] the sequence of ellipsoids $\mathcal{E}(c^k, B^k)$ generated by MAVIE is proved to converge to $\mathcal{E}_i^*(S)$. The volumes of these ellipsoids increase monotonously. This algorithm was independently suggested in [7], but its convergence was left as an open question.

The choice of c^0 at Step i may be non-trivial when the vertices of S are not known. However, c^0 can be obtained through the construction of a series of outer ellipsoids. For instance, the following procedure, based on the shallow-cut ellipsoid method [5], yields an ellipsoid contained in S.

Step 0-i Choose c^0 and B^0 such that $S \subset \mathcal{E}(c^0, B^{0-1})$, set $k = 0$, compute

$$\rho = \frac{1}{(p+1)^2}, \sigma = \frac{p^3(p+2)}{(p+1)^3(p-1)}, \tau = \frac{2}{p(p+1)}, \zeta = 1 + \frac{1}{2p^2(p+1)^2}.$$

Step 0-ii Compute

$$r = \min_i \left[b_i - a_i^T c^k - \frac{\sqrt{p}}{p+1} \sqrt{a_i^T B^{k-1} a_i} \right],$$

and let j be the argument of the minimum. If $r \geq 0$, stop: we have

$$\mathcal{E}(c^k, \frac{1}{(p+1)^2} B^{k-1}) \subset \mathcal{S}$$

(and also $\mathcal{S} \subset \mathcal{E}(c^k, B^{k-1})$).

Step 0-iii Compute

$$c^{k+1} = c^k - \rho \frac{\sqrt{p} B^k a_j}{\sqrt{a_j^T B^k a_j}},$$

$$B^{k+1} = \zeta \sigma (B^k - \tau \frac{B^k a_j a_j^T B^k}{a_j^T B^k a_j}),$$

$k \leftarrow k + 1$, go to Step 0-ii. □

This procedure terminates in a finite number of steps (note that Step 0-ii is a corrected version of the one in [5]). The condition $\mathcal{S} \subset \mathcal{E}(c^0, B^{0-1})$ of Step 0-i is easily fulfilled by choosing $B^0 = \beta I_p$, with I_p the p-dimensional identity matrix and β large enough.

Remark 1 *(i) We assumed at Step ii that the minimum-volume outer ellipsoid can be obtained without any approximation using MIVOE. In practice, MIVOE contains an ϵ-stopping rule so that this ellipsoid is not obtained exactly. Practical rules for choosing an ϵ' for MAVIE can be derived from the general ideas presented in [9].*

(ii) The distribution ξ^0 used to initialize MIVOE can be taken equal to the optimal distribution obtained from the previous call to MAVIE.

(iii) The determination of $\mathcal{E}_i^*(\mathcal{S})$ does not require the calculation of the vertices of $\mathcal{S}$. The complexity is related to the dimension of the vector λ of weights in MIVOE. This dimension increases at most linearly with the number of constraints that define $\mathcal{S}$.

MAVIE computes $\mathcal{E}_i^*(\mathcal{S})$ by taking all the constraints defining $\mathcal{S}$ into account as a whole. If one wishes to reduce $\mathcal{S}$ progressively by the introduction of new constraints, an on-line determination of $\mathcal{E}_i^*(\mathcal{S})$ is of interest.

3 On-line algorithm

In this section, we restrict our attention to the case where $\mathcal{S}$ is a bounded polyhedron. A recursive algorithm for the determination of an inner ellipsoid has already been suggested in [8], in the context of parameter bounding. However, it does not yield the maximum-volume inner ellipsoid. Moreover, the ellipsoid obtained tends to vanish after a modest number of iterations (i.e. a few additional constraints defining S).

Assume that the constraints $a_i^T x \leq b_i$ defining S satisfy $b_i \neq 0$, and can thus be transformed into $a_i^T x \leq p$. This can always be obtained by a suitable translation in $\mathcal{R}^p$. Let S^m be the polyhedron defined by the first m constraints,

$$S^m = \{x \in \mathcal{R}^p \mid a_i^T x \leq p, i = 1, \ldots, m\},$$

and let $\mathcal{L}^m$ be the set of extreme points of the convex closure of the a_i's, $i = 1, \ldots, m$. The algorithm is as follows.

OMAVIE: (On-line MAVIE)

Step i Consider the first k a_i's that span $\mathcal{R}^p(k \geq p)$. Compute $\mathcal{E}_i^*(S^k) = \mathcal{E}(c^k, B^k)$ using MAVIE and determine $\mathcal{L}^k$.

Step ii Let $a_{k+1}^T x \leq p$ be the new constraint.

— If $a_{k+1} \in \overline{\mathcal{L}^k}^c$ (the convex closure of $\mathcal{L}^k$), set $\mathcal{E}_i^*(S^{k+1}) = \mathcal{E}_i^*(S^k)$, $\mathcal{L}^{k+1} = \mathcal{L}^k$,

— else take $\mathcal{L}^{k+1}$ as the set of extreme points of $\mathcal{L}^k \cup \{a_{k+1}\}$.

— If

$$a_{k+1}^T c^k + \sqrt{p}\sqrt{a_{k+1}^T B^{k-1} a_{k+1}} \leq p, \tag{5}$$

set $\mathcal{E}_i^*(S^{k+1}) = \mathcal{E}_i^*(S^k)$,

— else, compute $\mathcal{E}_i^*(S^{k+1})$ by applying MAVIE to the polyhedron defined by the extreme points in $\mathcal{L}^{k+1}$.

Step iii $k \leftarrow k + 1$, go to Step ii. $\qquad\square$

Consider the set $\mathcal{C}_c(\mathcal{R}^p)$ of all convex compact subsets S of $\mathcal{R}^p$ that contain the vector c in their interior ($c \in \text{int}(S)$), and the transformation $T_c(.)$ given by

$$\begin{aligned} \mathcal{C}_c(\mathcal{R}^p) &\longrightarrow \mathcal{C}_c(\mathcal{R}^p) \\ S &\longmapsto T_c(S) := \{\phi \in \mathcal{R}^p \mid (\phi - c)^T(x - c) \leq p, \forall x \in S\}. \end{aligned}$$

Then $T_c(.)$ defines a relation of duality in the following sense: $\forall S \in \mathcal{C}_c(\mathcal{R}^p), T_c(T_c(S)) = S$, and for any given polyhedron $\mathcal{K}$ in $\mathcal{C}_c(\mathcal{R}^p)$, $T_c(.)$ defines a one-to-one relation between the q-faces of $\mathcal{K}$ and the $(p - q - 1)$-faces of $T_c(\mathcal{K})$. (For a polyhedron in $\mathcal{R}^p$, a vertex is a 0-face, an edge is a 1-face, a hyperplane is a $(p-1)$-face, a q-face is a face of a $(q+1)$-face.) MAVIE (and thus OMAVIE) is based on the construction of $\mathcal{E}_o^*(T_c(S^k))$, with $c \in S^k$. The potential vertices of $T_c(S^k)$ are given by

$$v_i = \frac{pa_i}{p - a_i^T c}, i = 1, \ldots, k,$$

where the a_i's define S^k through the constraints $a_i^T x \leq p$. From the duality property mentioned above, the active constraints correspond to the a_i's associated to the vertices of $T_c(S^k)$. $\mathcal{E}_o^*(T_c(S^k))$ only depends on these vertices, which are the extreme points points among the v_i's (i.e. the points lying on the boundary of the convex closure of the v_i's). One can easily check that for any c in $\text{int}(S^k)$, these extreme v_i's in the dual space are associated to the extreme a_i's in the primal space. Only the vectors in $\mathcal{L}^k$ have thus to be stored. Now, if $a_{k+1} \in \overline{\mathcal{L}^k}^c$ (Step ii), then $S^{k+1} = S^k$, otherwise, $\mathcal{L}^k$ has to be updated. However, it remains to be tested whether $\mathcal{E}_i^*(S^k)$ is cut by the new constraint. This corresponds to (5).

Remark 2 *(i) The information to be stored grows with the number of constraints on S. However, simple tests indicate at Step ii whether new constraints can be rejected or not.*

(ii) This algorithm constructs $\mathcal{E}_i^*(S)$ without requiring a characterization of S through its vertices. However, the set $\mathcal{L}^k$ corresponds to the vectors a_i's associated with active constraints. The determination of $\mathcal{L}^k$ (see e.g. [3]) could thus be replaced by a recursive determination of S^k, based for instance on the algorithm described in [15].

We consider now the general case where S is any convex subset of $\mathcal{R}^p$.

4 Bounded convex sets: general case

When some of the constraints defining S are nonlinear, we are unable to generate a sequence of ellipsoids contained in S. However, using a relaxation procedure, an algorithm converging globally to the maximum-volume inner ellipsoid can still be derived.

S corresponds to the convex closure of its extreme points, the number of which may be infinite. Let $\mathcal{X}(S)$ denote the set of these points. To any $x \in \mathcal{X}(S)$, one can associate (at least) one supporting hyperplane $\mathcal{H}$ tangent to S. Let $a^T x = b$ be the equation of one of these hyperplanes, with a and b such that $a^T x \leq b, \forall x \in S$. S is thus included in any polyhedron $\mathcal{P}(\mathcal{H}_1, \ldots, \mathcal{H}_m)$ defined by m such hyperplanes, i.e. given by (3). The relaxation procedure consists in taking only a finite number of hyperplanes into account at each iteration, thereby constructing a sequence of polyhedra containing S and a sequence of inner ellipsoids for these polyhedra.

RMAVIE: (Relaxed MAVIE)

Step i Choose m extreme points of S, $m \geq p$, such that the corresponding supporting hyperplanes $\mathcal{H}_i$ yield a bounded polyhedron $\mathcal{P}^m = \mathcal{P}(\mathcal{H}_1, \ldots, \mathcal{H}_m)$ (i.e. the vectors $a_i, i = 1, \ldots, m$ must span $\mathcal{R}^p$). Choose ϵ'' $(0 < \epsilon'' << 1)$, set $k = m$.

Step ii Determine $\mathcal{E}_i^*(\mathcal{P}^k) = \mathcal{E}(c^k, B^k)$ using MAVIE.

Step iii Compute

$$x^- = \arg \min_{x \in \mathcal{X}(S)} (x - c^k)^T B^k (x - c^k). \tag{6}$$

— If $(x^- - c^k)^T B^k (x^- - c^k) > p - \epsilon''$, take $\mathcal{E}(c^k, B^k)$ as an approximation of $\mathcal{E}_i^*(\mathcal{S})$, stop.

— else determine a supporting hyperplane $\mathcal{H}_{k+1}$ passing through x^-, take $\mathcal{P}^{k+1} = \mathcal{P}(\mathcal{H}_1, \ldots, \mathcal{H}_k, \mathcal{H}_{k+1})$, $k \leftarrow k + 1$, go to Step ii. □

The global convergence of this algorithm follows from the convexity of $\mathcal{S}$ and the convergence of MAVIE [10]. The main difficulty lies in the computation of x^- (6), which corresponds to a non-convex minimization problem. Global optimization methods by deterministic methods [6] (e.g. based upon interval analysis [13]) are advisable. It is sometimes enough to consider a combination of local minimizations only (one minimization per constraint defining $\mathcal{S}$). Note that a precise determination of the global minimum is not crucial (it is enough to determine x^- such that $(x^- - c^k)^T B^k (x^- - c^k) \leq p - \epsilon''$).

Remark 3 *(i) During the successive calls to MAVIE, the initial vector c^0 can be taken as the center c^k of the ellipsoid previously determined. The number of vectors v_i^k (4) is increased by one at each iteration. When MAVIE calls MIVOE for the first time (see Remark 1 (ii)), the initial distribution ξ^0 can be chosen so as to give the same weights to the new vector v_i^k and the vectors that were support points for the previous ellipsoid.*

(ii) Remark 1 (i) applies for this algorithm too.

(iii) The sequence of volumes $(\text{vol } \mathcal{E}(c^k, B^k))_k$ is monotonously decreasing (if one assumes that all ellipsoids are determined exactly).

Example 1: Consider the set

$$\mathcal{S} = \{x \in \mathcal{R}^2 \mid x_1 x_2 \leq 1, \, x_2 + (x_1 - 2)^2 \leq 2, \, x_1 \leq 3\}.$$

This set is presented on Figure 1 (solid line), together with the sequence of ellipsoids $(\mathcal{E}_i^*(\mathcal{P}^k))_k$ generated by RMAVIE when the initial extreme points of $\mathcal{S}$ are $(\frac{3}{2}, \frac{2}{3})$, $(1.5, 1.75)$, $(2.5, 0.4)$, $(2.5, 1.75)$. □

5 Conclusions

When the set to be characterized by an inner ellipsoid is a bounded convex polyhedron, the maximum volume inner ellipsoid can be determined without requiring the knowledge of the vertices of the polyhedron. This situation is met for instance in parameter bounding when the model structure is linear in the parameters. The only approach suggested so far [8] to the best of our knowledge, does not yield an ellipsoid with maximum volume. Moreover, the inner ellipsoid obtained tends to vanish quickly when the number of constraints increases. Of course, this is not the case with the maximum-volume inner ellipsoid obtained with OMAVIE. The information to be stored is minimized by detecting whether a new linear constraint may become active.

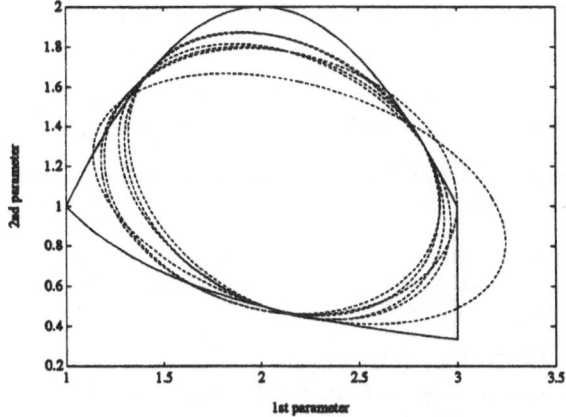

Figure 1: Convex set (solid line) and sequence of ellipsoids converging to the maximum-volume inner ellipsoid.

The case of non-polyhedric bounded convex sets has also been considered. The determination of the inner ellipsoid with maximum volume then requires the sequential determination of violated constraints (i.e. constraints crossed by the ellipsoid).

References

[1] M. Berger. *Géométrie*. CEDIC/Nathan, Paris, 1979.

[2] L. Danzer, B. Grünbaum, and V. Klee. Helly's theorem and its relatives. In V. Klee, editor, *Proceedings of Symposia in Pure Mathematics, Vol. VII: Convexity*, pages 101–180. Am. Math. Soc., Providence, 1963.

[3] L. Devroye. A note on finding convex hulls via maximal vectors. *Information Processing Letters*, 11(1):53–56, 1980.

[4] V. Fedorov. *Theory of Optimal Experiments*. Academic Press, New York, 1972.

[5] M. Grötschel, L. Lovász, and A. Schrijver. *Geometric Algorithms and Combinatorial Optimization*. Springer, Berlin, 1980.

[6] R. Horst and H. Tuy. *Global Optimization*. Springer, Berlin, 1990.

[7] L. Khachiyan and M. Todd. On the complexity of approximating the maximal inscribed ellipsoid for a polytope. Technical Report 893, School of Operations Research

and Industrial Engineering, College of Engineering, Cornell University, Ithaca, New York, 1990.

[8] J. Norton. Recursive computation of inner bounds for the parameters of linear models. *International Journal of Control*, 50(6):2423–2430, 1989.

[9] E. Polak. *Computational Methods in Optimization, a Unified Approach*. Academic Press, New York, 1971.

[10] L. Pronzato and E. Walter. Maximum-volume ellipsoids contained in convex sets. Technical report, Laboratoire des Signaux et Systèmes, CNRS/ESE, F-91192 Gif-sur-Yvette Cedex, France, 1992.

[11] L. Pronzato and E. Walter. Volume-optimal inner and outer ellipsoids, with application to parameter bounding. In *Proc. 2nd European Control Conf.*, Groningen, June 1993. to appear.

[12] L. Pronzato and E. Walter. Minimum-volume ellipsoids containing compact sets. application to parameter bounding. *Automatica*, to appear, 1994.

[13] H. Ratschek and J. Rokne. *New Computer Methods for Global Optimization*. Ellis Horwood limited, Chichester, 1988.

[14] D. Titterington. Optimal design: some geometrical espects of *D*-optimality. *Biometrika*, 62(2):313–320, 1975.

[15] E. Walter and H. Piet-Lahanier. Exact recursive polyhedral description of the feasible parameter set for bounded error models. *IEEE Transactions on Automatic Control*, 34:911–915, 1989.

OPTIMAL IDENTIFICATION OF THE FLOTATION PROCESS

Janusz GAJDA
Electrical Department, University of Mining and Metallurgy
al. Mickiewicza 30, PL-30059 Kraków

1. INTRODUCTION

The velocity of the mixture flow is one of the essential parameters of the flotation process, which affects its effectiveness. The accuracy analysis of the experimentally determined model describing the flotation mixture transportation with consideration given to the reasons limiting this accuracy is presented in the paper. The velocity of the flow as a one of the model coefficients is sought first. A model of the process is determined on the base of measuring experiment results conducted on the object. Then the influence of the following factors on the model accuracy is investigated: properties of identified object, signal exciting the object, applied measuring instruments, methods and data processing algorithms. Optimal object identification means the determination of its model with the best accuracy and conditioned by above mentioned properties [1], [2].

Minimization of the model error in the technical parameter set of applied measuring methods and instruments, under the assumption, that the object properties are partially known and under technical limitations in the choice of measuring instruments, is the aim of this paper.

2. MEASURING METHOD

Flotation process is based on the ore enrichment of metal by physical separation of suspended impurities. Separation is made in the flotation machine, in which flow the mixture of ore, impurities, water, air and chemical components improving the effectiveness of the process. The applied measuring method of the velocity of flotation mixture flow depends on both the object excitation by heat impulse and the determination of the coefficients of the model describing the heat transport in such environment by output observation.

Fig. 1 shows the scheme of the applied measuring system. Because of the space-distributed parameters of a model describing the heat transportation phenomena, the location z of the temperature sensor is the additional factor which influences the sought model accuracy [5], [6]. The measured signal from the temperature sensor is converted in to the digital form and loaded in the microcomputer realizing the identification algorithms. The coefficients of the sought model are the result of such a measuring data processing.

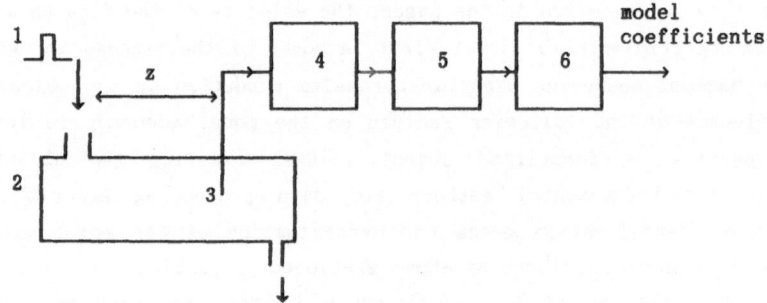

Fig.1. Measuring system scheme.
1 - exciting signal (heat impulse),2- flotation machine,
3- temperature sensor, 4- supply system, 5- analog/digital
converter, 6- microcomputer, z- sensor location.

Model (1) describes the phenomena of the heat transport.

$$T_m(z,t) = T_p + e^{\frac{Vz}{2D}} \int_0^t K_v(t-\tau)\, f(\tau)\, d\tau \qquad (1)$$

where the following denotations are made:

$T_m(z,t)$ - temperature of the floated mixture as a function of time t
and temperature sensor location z,

$$K_v(t)= \frac{z}{2(\pi t^3 D)^{0.5}} \exp\left[- \frac{z^2}{4Dt} - \frac{v^2}{4D} t \right] - \text{impulse response of the}$$

model,

$t \in (0, t_r)$ - time,

t_r - process observation time,

T_p - initial temperature,

$f(t)= k\delta(t)$ - exciting signal in the Dirac impulse form with
amplitude k.

Velocity of the heat transport corresponding to the sought velocity of the floated mixture flow V, heat diffusion coefficient D, initial temperature T_p and exciting signal amplitude k, are the coefficients of model (1). The values of these coefficients have been determined on the base of measurement experiments conducted on the investigated object (Table 1).

Table 1

coefficient	value
V [ms^{-1}]	0.08
D [$m^2 s^{-1}$]	0.067
T_p [$^\circ C$]	0.032
k [$^\circ C$]	14.3

A good fitness between object and proposed model has been achieved (Fig.2).

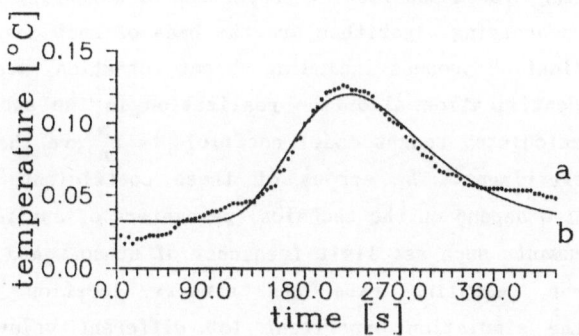

Fig.2. Temperature of the floated mixture vs. time.
a- experimental result,
b- calculated response of the model (1).

3. MODEL QUALITY EVALUATION

Because of the importance of the model coefficients accuracy, from the point of view of identification aim, the model quality criteria are determined in the model coefficients domain. The relative error of n-th model coefficient p_n^m is defined by the expression (2).

$$\delta p_n = \frac{p_n^m - p_n^o}{p_n^o} \tag{2}$$

where:

δp_n - relative error of n-th sought model coefficient,

p_n^m, p_n^o - n-th coefficient of sought model (superscript "m") and reference model (superscript "o"), respectively.

Usefulness of the expression (2) is limited by the indeterminacy of reference model coefficients p_n^o. In practice only their approximated values are known (for instance intervals including reference model coefficients). In the discussed case concerning the flotation machine, the reference model identical with the sought model form (1) has been assumed with coefficients equal to values in Table 1.

The evaluation of the sought model accuracy has been performed on the base of simulation tests by use of the method described in [2], [3]. The assumed reference model representing identified object as well as the model of applied measuring system considering properties of measuring instruments, methods and data processing algorithms are the base of such investigations. The whole identification process including object actuation, measuring data conversion and identification algorithm realization is the subject of the simulation. The calculated sought model coefficients p_n^m are the results of the simulation experiment. The errors of these coefficients defined by expression (2), also depend on the technical parameters of applied measuring methods and instruments such as: limit frequency of sensors and converters, quantization error, sampling rate and sensors location. Many time repetitions of the simulation experiment for different values of these technical parameters allow to determine their influence on the errors of sought model coefficients. Such investigation is called the model measurability analysis.

Joint criteria of the sought model quality considering the errors of all its coefficients are defined as functionals determined on a measurability matrix M.

$$M = E \, [\delta \bar{p} \, \delta \bar{p}^T]$$ (3)

where:

E[ ] - expected value, $\delta \bar{p}^T = [\delta p_1 \; \delta p_2 \; \ldots \; \delta p_N]$,

N - number of sought model coefficients.

The necessity of expected value calculation is caused by the stochastic disturbances appearing in measuring signals and by the measuring errors.

Diagonal elements of matrix (3) are the relative mean-square errors of the respective coefficients of the sought model. They include information about both the bias errors and the variance of these coefficients. The both kinds of errors are treated in the same way.

The geometrical interpretation and properties of the proposed quality criteria have been presented in [3]. The constructed on the base of matrix (3) N-dimensional ellipsoid (4)

$$\delta \bar{p}^T M^{-1} \delta \bar{p} = 1$$ (4)

describes the indeterminacy area of sought model coefficients.

In the two-dimensional case (N=2) the indeterminacy area is forming an ellipse. The volume of (4) depends on the variance of model coefficients. Minimization of the variance causes its reduction (surface reduction in two-dimensional case). Simultaneously the determinant of matrix (3) is reduced. The length of the area (4) projection on the coordinate system axis corresponding to the individual model coefficient, depends on its root-mean-square estimation error. The length of the maximum axis of area (4) corresponds to the maximum eigenvalue of matrix (3). The mentioned above properties of the matrix (3) and the area (4) have been used to the evaluation of the model (1) quality.

4. IDENTIFICATION PROCESS ANALYSIS AND OPTIMIZATION

The evaluation of the influence of measuring system parameters (Fig.1)

on the coefficients errors of sought model (1) has been a main goal of the
conducted analysis. The exemplary results of the analysis concerning some
chosen technical parameters are presented in Fig.3. The model error has
been evaluated by the use of matrix M trace.

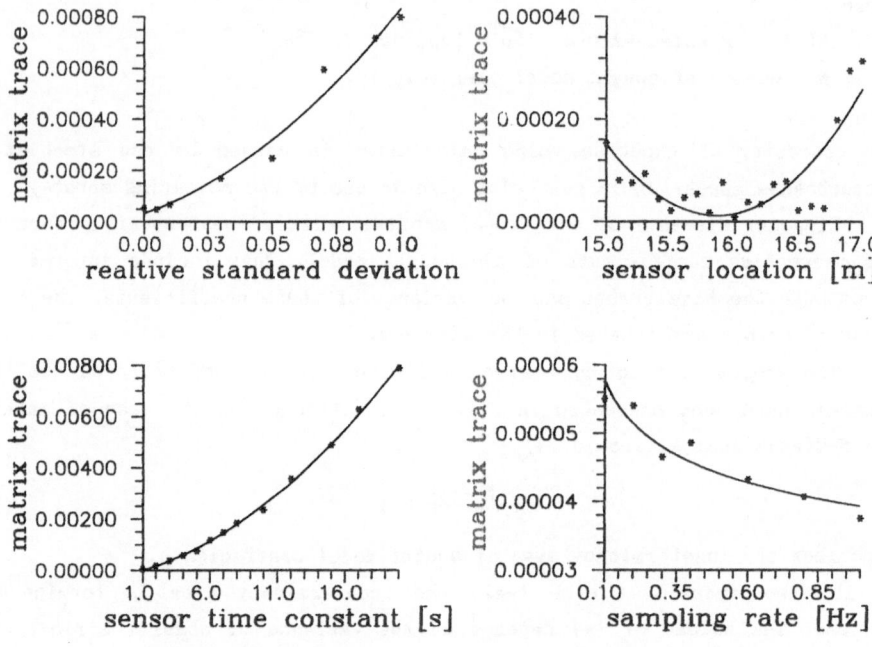

Fig.3. Sought model (1) errors vs. the technical parameters of the
applied measuring system presented in Fig.1.

It is obvious, that characteristics presented in Fig.3 depend also on
the assumed values of reference model coefficients. Because the limited

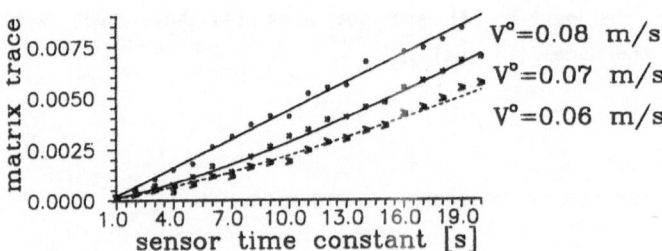

Fig.4 Sought model error vs. time constant of temperature sensor for
different values of reference model coefficient V^o (flow velocity).

knowledge concerning the identified object, only the intervals including these coefficients may be estimated. The conducted simulation tests allow to determine the presented in Fig.3 characteristics for different values of reference model coefficients belonging to the corresponding intervals (Fig.4). Such characteristics define the maximum and minimum value of the sought model error for different value of the chosen technical parameter (time constant of temperature sensor) and for determined changeability interval of reference model coefficient.

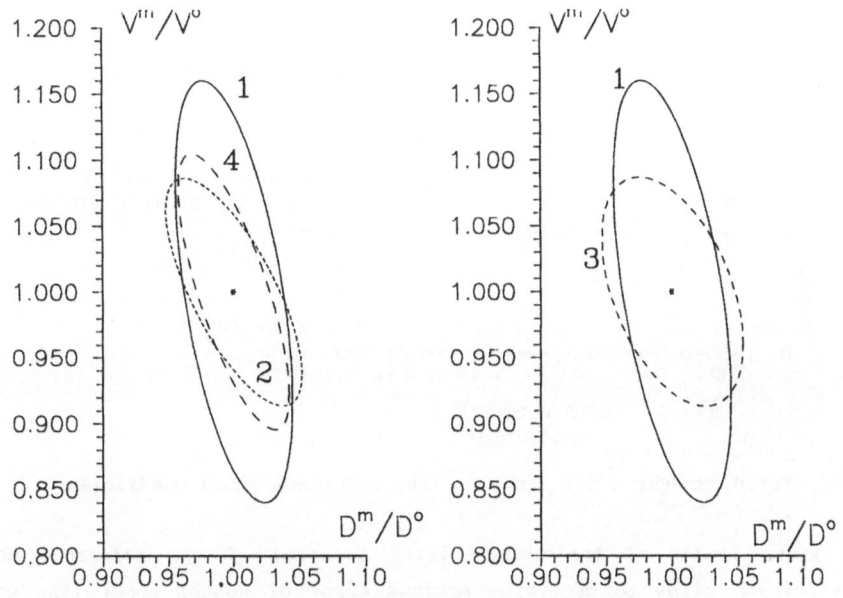

Fig. 5. Indeterminacy area of the model (1) coefficients.
1 - before the optimization, after the optimization according to the: 2 - matrix (3) trace, 3 - maximum eigenvalue, 4 - matrix (3) determinant.

The conducted analysis allow to indicate the technical parameters of applied measuring system, which essentially influence the model quality. The proper choice of these parameters with consideration given to existing technical limitations is the subject of performed identification process optimization. The optimization has been performed in the presence of additive stochastic disturbances, with 0.2% relative standard deviation, included in the temperature signal. The trace, determinant and maximum eigenvalue of matrix (3) have been used as the optimum criterion in three

performed optimization experiments . The obtained results may be compared by making use of indeterminacy area of sought model coefficients (Fig.5):
- velocity of the floated mixture flow V^m is determined with the greater error than diffusion coefficient D^m,
- in all optimization experiments the error of V^m coefficient has been reduced nearly twice,
- specific properties of applied optimum criteria result in different indeterminacy areas of optimal model coefficients.

The optimization results are also influenced with the assumed values of reference model coefficients. Fig. 6 presents the sought model error vs. reference model coefficient V^o for the optimal technical parameters of measuring system.

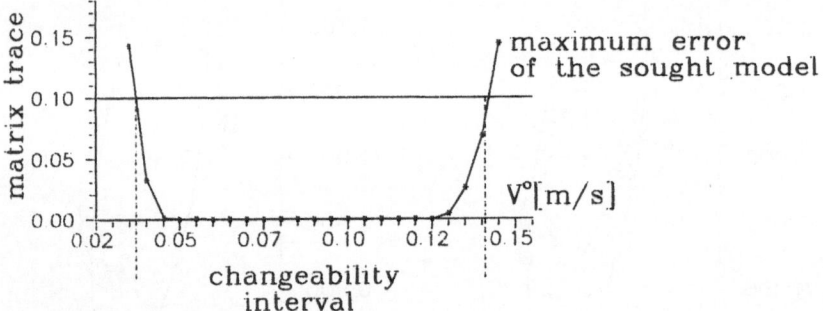

Fig.6. Sought model error vs. the reference model coefficient V^o.

The known limits of the changeability interval of the reference model coefficients allow to determine maximum error of sought model (the worst case analysis).

5. CONCLUSIONS

The time constant of applied temperature sensor and sensor location influence the sought model error most significantly. The proper choice of these technical parameters, in a way minimizing the model error is possible only as a result of optimization process.
The simulation methods significantly raise the effectiveness of such the

experiments. The results presented in paper have been achieved in Department of Applied Physics of Technical University Delft.

6. REFERENCES

[1] J.Gajda, "An Optimization of the Identification Process", Journal of Systems Analysis, Modelling, Simulation, vol.8, no 10, Berlin 1991.
[2] J.Gajda, "Modelling and Optimization of the Identification Process", 15-th IFIP Conference on System Modelling and Optimization, Zurich 1991.
[3] J.Gajda "An Influence of the Measuring Equipment Properties upon the Optimal Excitation Signal", IEEE Instrumentation / Measurement Technology Conference, New York 12-14 May, 1992.
[4] P.C.Kaptur, A.Dey, S.P.Mehrota "Identification of feed and simulation of industrial flotation circuits", Int. Journal of Mineral Processing, vol. 31, Elsevier Science Publishers B.V. Amsterdam, 1991.
[5] R.K.Mehra, "Optimization of Measurement Schedules and Sensor Design for Linear Dynamic Systems", IEEE Trans. on Automatic Control, vol. AC-21, no.1, 1976.
[6] S.Omatu, S.Koide, T.Soeda, "Optimal Sensor Location Problem for Linear Distributed Parameter System", IEEE Trans. on Automatic Control, vol. AC-23, no. 4, 1978.

Computer model for simulating control of the water and electrolyte state in the human body

H.Scharfetter, H. Hutten
Institute of Biomedical Engineering, Graz University of Technology,
Inffeldgasse 18, A-8010 Graz

ABSTRACT:

An existing multicompartmental model for simulating the exchange processes during hemodialysis has been modified and extended. Especially new aspects of the control of volumes and electrolytes have been integrated. The model contains compartments for the following subatances: Na^+, K^+, Cl^-, HCO_3^-, CO_2, H^+, urea, creatinine and vitamine B12. Volumes are calculated for blood-plasma, erythrocytes, interstitial fluid and intracellular space. Since few parameters of the system are not exactly known, they can be fitted for achieving good agreement between simulated and measured data. A first evaluation of the model was performed by using the data measured from 8 hemodialysis patients.

INTRODUCTION:

Dialysis is a life-supporting therapy for patients with end-stage renal failure. The aim is the elimination of toxic substances (e.g. urea, creatinine), electrolytes (e.g. sodium, potassium), of water and non-specified medium sized molecules. The elimination process is accomplished by an exchange between the blood and the dialysate across special membranes. Until today dialysis is accompanied with a multitude of different acute complications. Typical examples are vertigo, vomitting, cramps, and blood pressure crises. Preliminary results have shown that at least the frequency of acute complications can be diminished, if the exchange process is adjusted to individual requirements by the application of profile dialysis. It is performed by a sequence of short intervals with different and adjusted process parameters, e.g. composition of the dialysate. The aim is to avoid any situation that can be related with acute complications. Such situations can occur if deviations of internal quantities such as the water distribution or electrolyte distribution exceed certain limits.

As a basis for the individual adjustment, a published computer-model of the exchange processes during hemodialysis [8] has been extended and evaluated. This model considers the kinetics of the exchange processes between different distribution spaces within the patient as well as between blood and the dialysate. The model can be used to simulate a patient's state during the dialysis-session if some data (e.g. body-weight, age, blood-electrolytes) are known at the beginning of the treatment. Individual parameters can be adjusted individually after a single extended measurement-series.

METHODS:

The model has been realised as a multicompartmental model. A compartment means a state variable of the physiological system, most commonly the concentration of a certain substance if it follows approximately a uniform kinetics in a distribution space. Thus the entire body, which is built very inhomogeneously, can be subdivided into a small number of rather homogeneous compartments.

In the model, changes of the state variables occur by substance exchange between different spaces. Such exchange processes are described by passive mechanisms like diffusion, osmosis and convection, but active transport of ions and chemical reactions must also be taken into account. The model contains compartments for Na^+, K^+ and Cl^-, for urea, creatinine and vitamine B12, furthermore for HCO_3^-, CO_2 and H^+. Volumes are calculated for blood-plasma, erythrocytes, interstitial space and intracellular fluid. Additionally some variables

404

that are of interest for clinical application, like standard bicarbonate and the resting potential of cell membranes are included. Urea and creatinine kinetics can be considered by two body compartments [4] as it is shown in fig. 1:

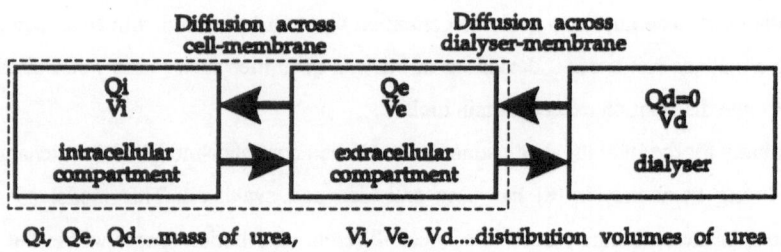

Qi, Qe, Qd....mass of urea,   Vi, Ve, Vd....distribution volumes of urea

Fig. 1: Model for urea exchange. The diffusional exchange is   bidirectional depending on the concentration gradient. The dashed rectangle marks the two body compartments.

The mass balance is expressed by the following equations:

$$\frac{d(V_{ex}C_{i,ex})}{dt} = -K_i C_{i,ex} - \frac{d(V_{in}C_{i,in})}{dt} \tag{1}$$

$$\frac{d(V_{in}C_{i,in})}{dt} = T_i(rC_{i,ex} - C_{i,in}) \tag{2}$$

V_{in}, V_{ex} *intracellular and extracellular volume [l]*

$C_{i,in}, C_{i,ex}$ *intracellular and extracellular concentration of substance i [mmol/l]*

T_i *diffusion transfer coefficient across cell membrane [l/min]*

r *coefficient for the conversion to free plasma water concentration*

K_i *Clearance for substance i, related to the blood flow through the dialyser.*

The volume-regulation of human body-cells, interstitial space and blood is closely related with the electrolyte and acid-base-state. Different mechanisms of exchange are depicted in fig. 2 and 3. Fig. 2 shows the active (pump) and passive (diffusion, filtration) fluxes of the small ions K^+, Na^+ and Cl^-, as well as the water shifts across the cell membrane.

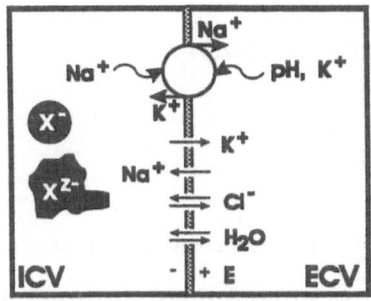

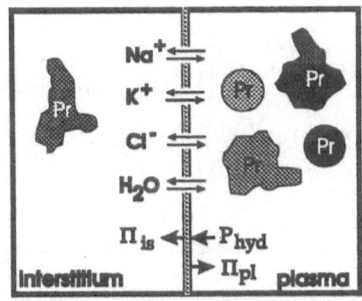

Fig.2 Exchange mechanisms across the cell membrane. The dark particles marked with X are nondiffusible ions. ICV = intracellular volume, ECV = extracellular volume. E = resting membrane potential

*Fig. 3 Exchange mechanisms across the capillary wall. The dark particles marked with **Pr** are nondiffusible substances (proteins), which establish the oncotic pressure gradient.*

Some charged proteins cannot penetrate the cell membrane, thus a Donnan equilibrium is established. Fig. 3 illustrates the exchange processes across the capillary wall, for which only passive mechanisms are considered. Different concentrations of nondiffusible proteins on both sides of the membrane maintain the oncotic pressure gradient (Π_{is},Π_{pl}) which compensates the hydrostatic pressure P_{hyd} in the vessel.

The exchange of ions across the cell-membrane is maintained by electrodiffusion and active pump mechanisms [5,7], and can be described for any ion k by the total mass flux $\dot{Q}_k$:

$$\dot{Q}_k = \varphi_{k,act} + \varphi_{k,ediff} + \varphi_{k,filt} \tag{3}$$

For the electrodiffusive ionic fluxes $\varphi_{k,ediff}$ the Goldman constant field equations are used, which are extended by terms that describe filtration processes [7]. The filtration flux $\varphi_{k,filt}$ is approximated by the Stavermann formula [1]. The active transport of Na^+ and K^+ is performed by the so called Na-K-pump which is assumed to show a 2:3 stoichiometry. The pump flux can be approximated by a nonlinear equation considering the dependence on extra- and intracellular ion concentrations [7]. The relation between maximal pump flux and pH is nearly linear in the physiological pH-range [2,5], thus a linear actvity-factor is provided. The active pump flux for Na^+, $\varphi_{Na,act}$, is calculated by the equation:

$$\varphi_{Na,act} = \dot{\varphi}_{Na,act}^{pH0} \cdot \left[\frac{c_{Na,in}}{c_{Na,in} + K_{Na}}\right]^3 \left[\frac{c_{K,ex}}{c_{K,ex} + K_K}\right]^2 (0.517\,\text{pH} - 2.927) \qquad (4)$$

$$K_{Na} = 0.2\left(1 + \frac{c_{K,in}}{8.33}\right) \qquad K_K = 0.1\left(1 + \frac{c_{Na,ex}}{18.5}\right) \qquad (5)$$

$\dot{\varphi}_{Na,act}^{pH0}$ is obtained from steady-state-considerations, while the intracellular ion concentrations are estimated using lookup tables [8]. Water exchange is controlled by osmosis, which is described by equation (6). For this osmotic relation Na^+, K^+, Cl^-, urea and unspecified indiffusible ions in the intracellular volume are considered. L_p is the filtration coefficient of the cell membrane [l/(min mmHg)], σ_i is the osmotic reflexion coefficient of substance i:

$$\dot{V}_{in} = L_p RT \left[\sum_i \sigma_i (c_{i,in} - c_{i,ex})\right] \qquad (6)$$

For the substance-flows across the capillary membrane only passive mechanisms have to be considered, thus osmotic and diffusion-dependencies are sufficient [5,8].

Control of the acid-base-state (pH, CO_2, HCO_3^-) requires 5 compartments for each of the three components. Convective, diffusive and chemical processes must be considered adequately. The structure of the acid-base model is depicted in fig. 4 [6].

In the intracellular subsystem the endogenous production of CO_2 takes place and is described by a constant (empirically found mean value). CO_2 is removed from the intracellular space by subsequent diffusion into the interstitium and the capillaries where different chemical reactions take place. Blood is described by two compartments for each of the three components, one for the body capillaries and one for the lung. The latter subsystem provides the gas exchange. Alveolar ventilation, is controlled by the concentration of CO_2, O_2 and by pH. That mechanism can be described by simple equations of first order. The mass balance equations are different for every distribution space, one example is shown for the volume "body capillaries and veins" [6].

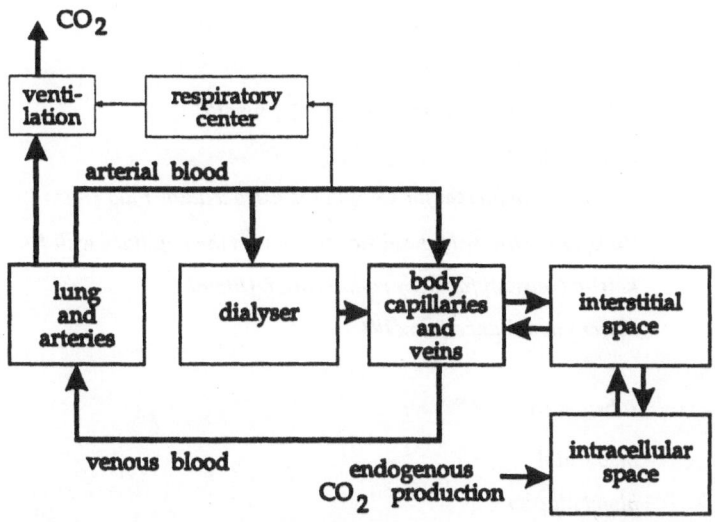

Fig. 4: Acid-base model. The body is subdivided into four distribution spaces for the substances CO_2, HCO_3^- and H^+. The fifth distribution space is represented by the dialyser.

For describing the chemical reactions the mass action law (Henderson-Hasselbalch-equation) and the kinetics of the buffer systems in the human body must be taken into account:

$$\frac{d}{dt}c_{CO2,b} = \frac{1}{V_b}\dot{Q}_{CO2,convect} + \frac{1}{k}\dot{Q}_{chem} + \frac{T_{CO2}}{k\beta V_b}(c_{CO2,is} - c_{CO2,b}) \tag{7}$$

$$\frac{d}{dt}c_{HCO3,b} = \frac{1}{V_b}\dot{Q}_{HCO3,convect} - \frac{1}{l}\dot{Q}_{chem} + \frac{T_{HCO3}}{l V_b}(c_{HCO3,is} - hrc_{HCO3,b}) \tag{8}$$

$$\frac{d}{dt}c_{H,b} = \frac{1}{V_b}\dot{Q}_{H,convect} - \frac{1}{BV_b}\dot{Q}_{chem} + \frac{T_H}{BV_b m}(c_{H,is} - hrc_{H,b}) \tag{9}$$

$$\dot{Q}_{i,convect} = (HMV - \dot{Q}_{bl})c_{i,la} + (\dot{Q}_{bl} - \dot{Q}_f)c_{i,dial} - HMV \cdot c_{i,b} \tag{10}$$

$$\dot{Q}_{chem} = T_{chem}(c_{HCO3,b} - c_{CO2,b}\frac{K}{c_{H,b}}) \tag{11}$$

c	*molar plasma concentration [mmol/l]*
V_b	*estimated entire volume of body capillaries [l]*
$\dot{Q}$	*mass flux [mmol/min]*

k, l	*correction factors for virtual distribution volume*
h	*Donnan factor*
m	*reference H^+ concentration, $10^{-7.4}$ [mmol/l]*
r	*conversion factor*
β	*solubility coefficient for CO_2 in the extracellular fluid [mmol/(l mmHg)]*
T_i	*transfer coefficient for substance i across the capillary wall [l/min]*
T_{chem}	*kinetic constant for CO_2 equilibration [l/min]*
B	*buffer capacitance [mmol/l]*

Indices:

i	*substance i*
is	*interstitial*
b	*blood plasma*
dial	*dialyser*
la	*lung and arteries*
chem	*chemical reaction*
diff, f, convect	*diffusion, ultrafiltration, convection*

The third space of gas exchange is the dialyser, requiring three additional compartments. It represents the distortion term in the system.

Because the model is mimicking the structure of the physiological system, the identification problem for the system-parameters is reduced to the estimation of only three individual coefficients. The other parameters are well known from literature. Those free parameters are the effective global cell-surface, the intracellular Na^+-concentration and the conversion-factor r for the plasma-water-concentration of electrolytes. Sensitivity-analysis of the model showed that those parameters are the most critical ones in the whole system. Fitting of the model to individual patients is achieved by a least-squares-minimization of the differences between predicted and measured blood-analysis data. For this purpose the Levenberg-Marquardt algorithm from the "Optimization-Toolbox" of MATLAB was used. The clinical measurements were performed by analyzing blood from 8 stable patients undergoing chronic hemodialysis. 6 to 7 arterial samples and one venous sample were collected from the dialyser at non equidistant times. For the ion concentrations, three subsequent measurements from every sample were performed and the mean value was calculated. Concentrations of Na^+ and K^+ were determined with the electrolyte analyzers AVL 988 and AVL 984, the acid-base state was measured with the blood gas analyzer AVL Compact 1. The dialyser performance could be calculated from one pair of arterial and venous samples.

RESULTS:

After fitting of the aforementioned individual parameters, it was possible to simulate the state variables of several patients with good acccuracy. The diagrams in fig. 5 and 6 show the simulated and measured courses of Na^+- and K^+-concentrations during a conventional bicarbonate-dialysis. The patient, from whom those data were obtained, showed no complications during the treatment. As can be seen, the concentrations of Na^+ and K^+ decrease steadily. The deviations of the simulated Na^+-curve from the measurement data might be caused by the fact, that the sodium concentration of the dialysate had not been kept

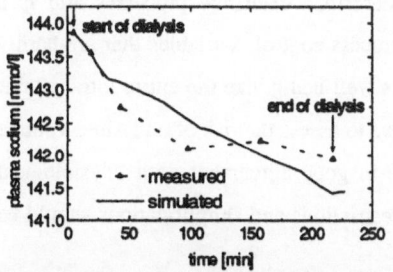
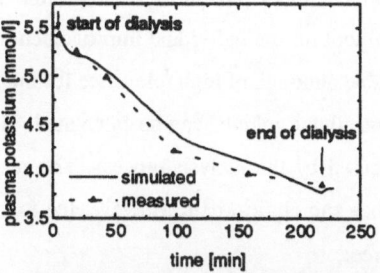

Fig. 5: Time course of the plasma sodium concentration during dialysis.

Fig. 6: Time course of the plasma potassium concentration during dialysis.

exactly constant by the machine. However, the deviations are lower than 0.3 % and it is not possible to decide, whether they are caused by simplifications in the model or by measurement errors. Fig. 7 illustrates the time course of volume-shifts in the patient's body. Both, the intravasal and the interstitial volume decrease as expected, whereas more than 0.5 l of water are shifted to the intracellular volume. This effect is mainly caused by the sodium exchange and should be prevented as far as possible.

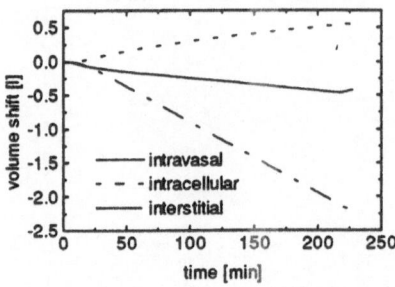

Fig. 7: Time course of the water volume shifts during dialysis

DISCUSSION:

The model renders possible the simulation of the described exchange processes and is a powerful tool for the individual improvement of the process control. Variables that are hardly measurable although of high relevance for the patient's well-being, like the entire intracellular and extracellular volume, can be calculated. This allows to assess the risk of undesired volume shifts caused by the dialysis-process. The results are in good agreement with physiological experience and clinical data. However the following restrictions and simplifications should be considered:

Exchange processes that follow different dynamics in different tissues are described in a form that is based on homogeneous tissue with uniform properties. This simplification is justified if only global state variables like the whole-body intracellular fluid are of interest.

The slopes of the simulated electrolyte curves deviate slightly from that of the measured ones. Further investigations must reveal whether the deterioration of the dialyser's performance during dialysis may explain this difference.

No hemodynamical effects are implemented yet, and the regulation of the blood pressure and its influence on filtration processes in the body are not considered. It can be assumed that the influence of those effects is negligible for the discussed applications.

Acknowledgements :The authors appreciate the support given by Prof. Dr. H. Holzer and Dr. G. Wirnsberger of the Medical Clinic LKH Graz.

REFERENCES:

1] Adam G., Läuger P., Stark G., "Physikalische Chemie und Biophysik", Springer-Verlag Berlin Heidelberg New York, 1988, pp 307-310.

[2] Forbush B., "Rapid Release of ^{42}K and ^{86}Rb from Na,K-ATPase from Occluded State of the Na-K-Pump in the Presence of ATP or ADP", *J Biol Chem*, vol. 262 (23), pp 11104-11115 (1987).

[3] Gordon L.G.M., Macknight D.C., "Contributions of Secondary Active Transport Processes to Membrane Potentials", *J Membrane Biol*, vol. 120, pp 141-154 (1991).

[4] Maher J.F. (ed), "Replacement of Renal Function by Dialysis", (Third Edition), Kluwer Academic Publishers, Dordrecht-Boston-Lancaster 1989, pp 106.

[5] Mann H., Stiller S., "Einsatz des Modells Künstliche Niere zur Planung und Kontrolle der Dialysetherapie", *Abschlußbericht zum BMFT-Vorhaben MSO 516*, Aachen 1983, pp 92-105.

[6] Reindl C., "Mathematisches Modell zur Beschreibung des Säure-Basen-Status während der Dialyse", Diplom thesis, Graz 1993.

[7] Strieter J., Stephenson J.L., Palmer G.L., Weinstein A.M., "Volume-activated Cloride Permeability Can Mediate Cell Volume Regulation in a Mathematical Model of a Tight Epithelium", *J Gen Physiol, vol.* 96, pp 319-344 (1990).

[8] Thews O., Hutten H., "A comprehensive model of the dynamic exchange processes during hemodialysis", *Med Prog Technol*, vol. 16, pp 145-161 (1990).

Subcutaneous Insulin Absorption Model for Parameter Estimation from Time-Course of Plasma Insulin

Z. Trajanoski, P. Wach, P. Kotanko*, F. Skrabal*

Institute of Biomedical Engineering, Graz University of Technology, Inffeldgasse 18,
A-8010 Graz
*Department of Internal Medicine, Krankenhaus der Barmherzigen Brüder,
Marschallgasse 12, A-8020 Graz

Abstract - **A mathematical model for the absorption of subcutaneously administered soluble insulin is presented. A dissociation-diffusion model with distributed parameter was transformed into a lumped parameter model by spatial discretization. Simulations of various insulin injections with this model demonstrate volume and concentration dependant absorption as experimentally observed. The model can be used for simulating absorption following multiple injections or continuous insulin infusion within a wide range of insulin concentrations and volumes. Parameter estimation from experimental data confirm that this model can be employed for individual therapy adjustment.**

INTRODUCTION

To avoid or delay late complications of diabetes mellitus due to microvascular manifestation the glucose control should be maintained as close to normal as possible. However, despite considerable research a closed loop system for glucose control in patients with diabetes mellitus is clinically not yet available. Due to the life time of glucose sensors the closed loop system can only be used for short term blood glucose control. The insulin therapies nowadays apply the subcutaneous (s.c.) route for insulin administration, i.e. continuous s.c. insulin infusion (CSII) or multiple s.c. insulin injections therapy (MIT) without feed-back from a glucose sensor. Hence, to achieve normoglycaemia, it is important to predict plasma insulin concentration after s.c. administration of insulin. Unfortunately, the absorption of insulin from the s.c. tissue is extremely variable [3], and the knowledge of factors influencing the insulin pharmacokinetics is of great importance. Among these factors, the size of dose, the concentration and different insulin preparations are the major determinants. A model for the absorption of s.c. applied insulin with special emphasis on this factors can therefore be a useful tool in routine applications. However, the models used in the past [1, 6, 7, 10] are oversimplified and they reproduce the process of insulin absorption insufficiently. More elaborate modelling techniques are required to predict time-course of plasma insulin following various insulin treatments.

In this study a comprehensive model formulated in terms of physical and pharmacokinetic principles [13] was adapted in order to enable simulation of s.c insulin absorption in the range of therapeutic concentrations and volumes. The presented model is adequate for parameter estimation from time-course of plasma insulin, thus enabling its use for routine clinical investigations and individual therapy adjustment.

Glossary

C_h, C_d Concentration of hexameric and dimeric insulin in the subcutaneous tissue, U ml^{-1}

I Concentration of plasma insulin, mU l^{-1}

D Diffusion constant, cm^2 min^{-1}

Q Equilibrium constant between hexameric and dimeric insulin, ml^2 U^{-2}

P Equilibration rate constant, min^{-1}

B Absorption rate constant, min^{-1}

K_e Plasma insulin elimination rate constant, min^{-1}

V_{sc} Volume of the subcutaneous insulin depot, ml

V_p Volume of the plasma insulin compartment, litre

∇^2 Laplacion operator

A testable hypothesis of the absorption process in terms of a mathematical model was introduced by Mosekilde *et al.* [13]. In this model three processes i.e. equilibration between hexameric and dimeric insulin according to the mass balance law:

$$C_h = QC_d^3 \tag{1}$$

insulin diffusion in the s.c. tissue,

$$\frac{\partial C_h}{\partial t} = D\nabla^2 C_h \tag{2}$$

and insulin binding in the tissue characterize insulin kinetics. Recent investigations with soluble human insulin and insulin analogues with reduced self-association support this hypothesis [5, 9]. The mathematical representation of these processes is given by a set of three non-linear coupled partial differential equations [13]. The equations can be solved numerically using appropriate boundary and initial conditions from measured disappearance of radiolabeled injected insulin. However, monitoring disappearance of radiolabeled insulin is an expensive technique, whereas monitoring plasma insulin appearance is a common clinical investigation. Therefore this model had to be modified to enable parameter estimation from the simpler experimental protocol.

It has been presumed, that only dimeric insulin molecules are absorbed and that insulin binding in the s.c. tissue is neglegible in the range of therapeutic concentrations (40 U/ml and 100 U/ml). The equations corresponding to the modified model are:

$$\frac{\partial C_h}{\partial t} = P(QC_d^3 - C_h) + D\nabla^2 C_h \tag{3}$$

$$\frac{\partial C_d}{\partial t} = -P(QC_d^3 - C_h) + D\nabla^2 C_d - BC_d \tag{4}$$

Furthermore it has been assumed that the injected volume builds up a spherical depot, and that this depot is symetrically widened through diffusion of insulin in the tissue.

In analogy to an electric network the model with distributed parameters (3, 4) was transformed into a lumped parameter model for the absorption of s.c. injected soluble insulin. The s.c. depot was devided into 15 shells and each shell represents two compartments (hexameric and dimeric insulin compartment) with homogenous concentration and constant volume.

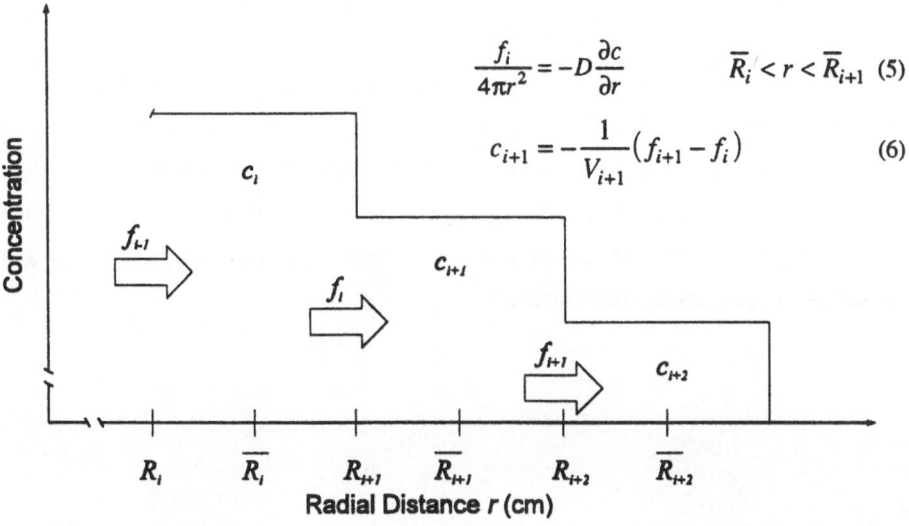

$$\frac{f_i}{4\pi r^2} = -D\frac{\partial c}{\partial r} \qquad \overline{R}_i{}' < r < \overline{R}_{i+1} \quad (5)$$

$$c_{i+1} = -\frac{1}{V_{i+1}}(f_{i+1} - f_i) \qquad (6)$$

Fig. 1: Spatial discretization of the diffusion equation (2). The s.c. depot was devided into 15 shells with constant volume V_i, concentration c_i and insulin flux f_i from the shell with radius $\overline{R}_i$ to the shell with radius $\overline{R}_{i+1}$. The radius $\overline{R}_i$ equals to a radius which devides the shell i into two equal parts.

The diffusion equation (2) was spatially discretized as shown in Fig. 1. It has been assumed that the concentration of hexameric and dimeric insulin in the 15th shell is zero and that there is no insulin flux from the shell $i = -1$. In each compartment dissociation of insulin and absorption of dimeric insulin molecules were considered according to the equations (3) and (4). Injections of different volumes were simulated by assigning appropriate initial values to each compartment. At the beginning the concentrations of the innermost shells equals the injected concentration of hexameric and dimeric insulin calculated by equation (1). CSII was simulated by supplying a series of infusion pulses with different frequencies to the s.c. tissue.

Combined with a single compartment model for distribution and elimination of insulin,

$$\frac{dI}{dt} = \frac{1}{V_p} \int_{V_x} BC_d dV - K_e I \tag{7}$$

plasma insulin concentrations after injection and infusion of various concentrations and volumes were simulated. Parameter values reported in the literature were used for the simulations [13]. For the distribution-elimination model 12 litre distribution volume and elimination rate of 0.09 min^{-1} [10] were assumed. Additionally parameter estimation from experimental data was performed using the Levenberg-Marquardt method for non-linear least-square optimization [8]. Data from plasma insulin measurements from 8 normal weight healthy volunteers after a glucose-clamp experiment were used [4]. In this case, the physicochemical parameters (P, D, Q) were held constant and the absorption rate B and elimination rate constant K_e were estimated.

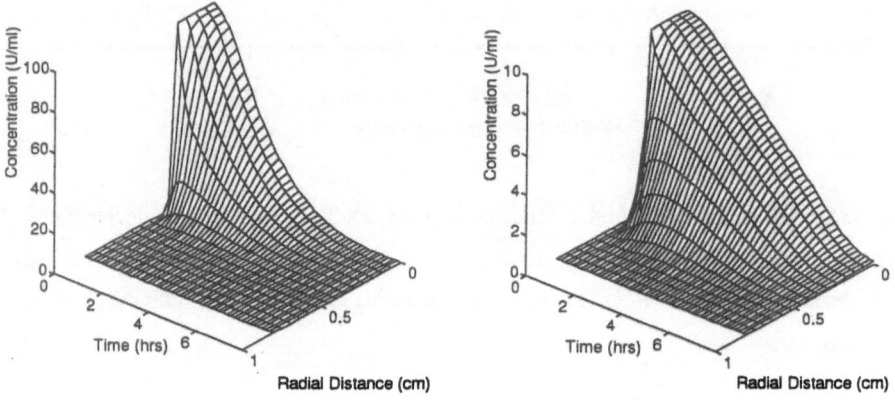

Fig. 2: Spatio-temporal distribution of hexameric (left panel) and dimeric (right panel) insulin concentration in the s.c. depot after injection of 0.1 ml, 100 U/ml

Since complete sensitivity analysis for this complex model is difficult to perform, only a partial sensitivity analysis was performed by varying one parameter and keeping all other parameters constant. Variation of the parameters over a range of reported variability [3, 13] did not lead to unphysiological changes of the absorption curves. However, to ensure convergence and numerical stability a rather small time step (0.2 min) has to be chosen.

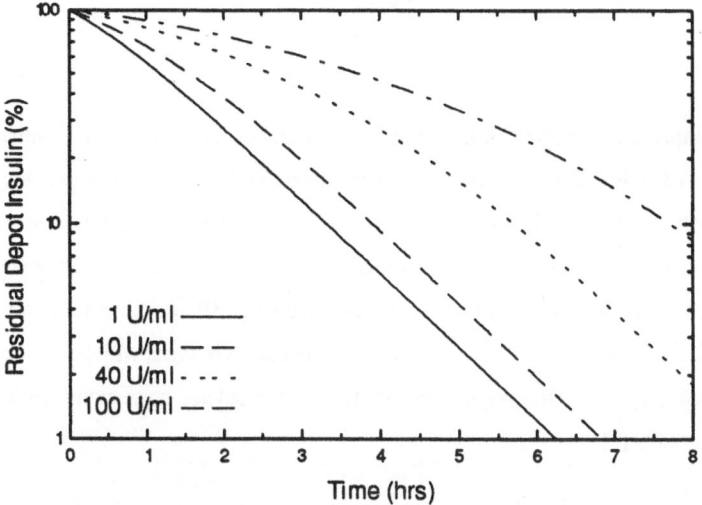

Fig. 3: Absorption of soluble insulin after injection of 0.2 ml and various concentrations.

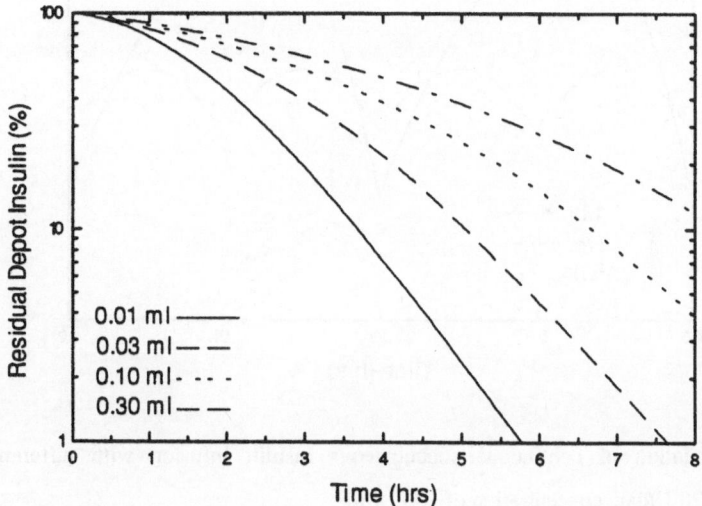

Fig. 4: Absorption of soluble insulin following s.c. injection of 0.01, 0.03, 0.1 and 0.3 ml and concentration of 100 U/ml.

RESULTS

The spatio-temporal distribution of the concentration in the s.c depot after a typical insulin bolus can be seen in Fig. 2. Due to diffusion in the s.c tissue, the concentration is lowered and the balance between hexameric and dimeric insulin is shifted towards dimeric insulin. With decreasing concentration of the injected insulin, the initial concentration of dimeric insulin is greater (Equation (1)) and faster absorption takes place (Fig. 3). On the other hand diffusion reduces more rapidly the concentration for smaller injection volumes, which are considerably faster absorbed (Fig. 4). The simulations of these phenomena correspond with reported experimental results [3].

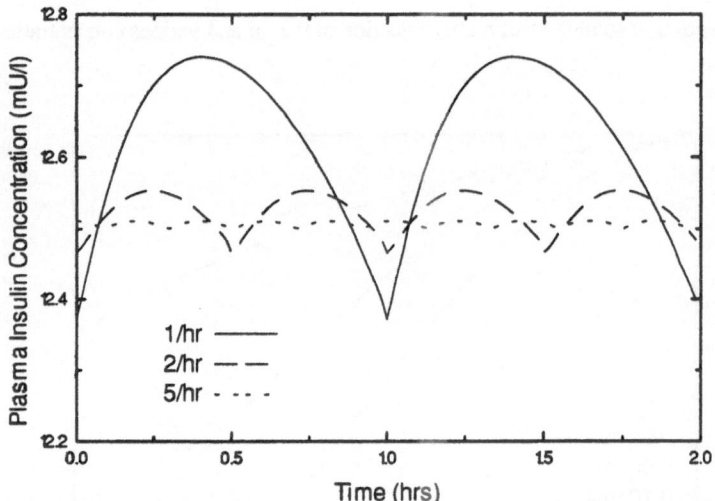

Fig. 5: Simulation of continuous subcutaneous insulin infusion with different puls frequencies - 20 U/day, concentration of 100 U/ml.

Simulations of CSII with different puls frequencies (Fig. 5) demonstrate that even lower repetition frequencies are sufficient to achieve almost constant plasma insulin concentration. The buffer function of the s.c. depot smooths out the variation of the delivery rate.

Fig. 6 shows experimental data and model fit after estimation of the absorption and elimination rate constant. The estimated parameters were within the range of inter- and intrapersonal variability (B= 0.017 min^{-1}, K_e=0.076 min^{-1}).

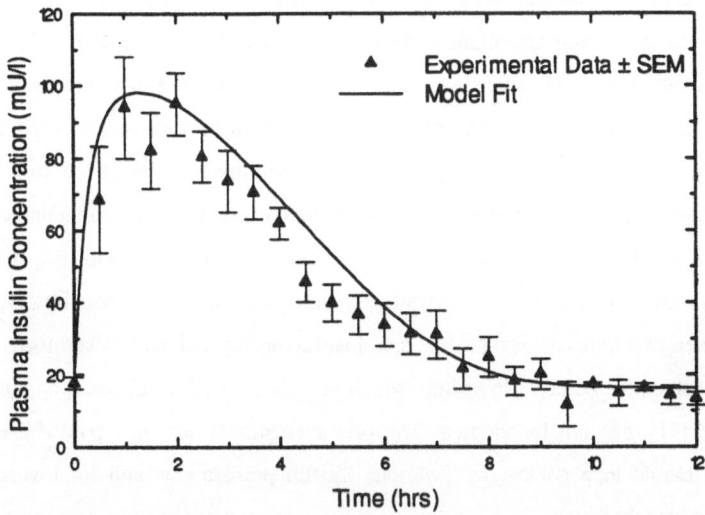

Fig. 6: Calculated plasma insulin concentration after parameter estimation from experimental data [4]. Shown are mean values ± SEM, n=8.

DISCUSSION

The simulation results have shown that the presented model is adequate to quantitate insulin absorption from the s.c. tissue and plasma insulin concentration over a wide range of administered volumes and concentrations. At present the use of the model for therapy adjustment of individual subject is of greater clinical interest. Metabolic parameters like insulin sensitivity and glucose effectiveness [2] can be calculated using data from a intravenous glucose tolerance test. For patients with diabetes mellitus this routine test is

420

modified by simultaneous intravenous insulin infusion. The data from this test can be used for determination of parameters from the plasma insulin compartment (V_p, K_e). Additional s.c. administration of insulin and measurement of plasma insulin concentration enables estimation of the absorption rate constant in a simple, less invasive manner. The model can be employed to simulate plasma insulin concentratioin following different insulin delivery schedules (CSII, MIT), concentrations, as well as volumes. Furthermore it can be easily modified in order to study the pharmacokinetics of novel insulin preparations (dimeric and monomeric insulin analogues [5]) and lessen the number of experimental studies with radiolabeled insulin. Finally the model can be used for designing new algorithms for glucose control. Recent developments have shown that extracorporal measurement of s.c. tissue glucose concentration can be performed with microdialysis [12]. Hence, a closed loop system with s.c. measured glucose and s.c. infused insulin with conventional insulin pumps seems feasible. However, due to the inherent nonlinearities of the s.c. insulin absorption and delay time in the microdialysis system, the control algorithms which have been used for closed loop control of plasma glucose [11], can not be adopted. The presented model can be a powerful tool for simulating s.c. closed loop control with various insulin preparations and for investigating different control algorithms.

ACKNOWLEDGMENT - This work was supported by the Austrian Science Foundation, Grant no. S49/06.

REFERENCES

1. BERGER M, AND RODBARD D: Computer simulation of plasma insulin and glucose dynamics after subcutaneous insulin injection. *Diabetes Care*. 12: 725-736, 1989

2. BERGMAN RN, IDER YZ, BOWDEN CR, AND COBELLI C: Quantitative estimation of insulin sensitivity. *Am. J. Physio.* 5(6), E667-E677, 1979

3. BINDER C, LAURITZEN T, FABER O, AND PRAMMING S: Insulin pharmacokinetics. *Diabetes Care*. 7, 188-199, 1984

4. BOTTERMANN P, WAHL K, ERMLER R, LEBENDER A, AND GYARAM H: Action profiles and plasma concentrations of insulin after s.c. application of different insulin preparations. In *Computer Systems for Insulin Adjustment in Diabetes Mellitus. Proc Int Symp Computer Syst for Insulin Adjustment in Diabetes Mellitus*. Beyer J, Albisser M, Schrezenmeir J, Lehmann L, Eds. Switzerland, Pancienta, 85-109, 1985

5. BRANGE J, OWENS DR, KANG S, AND VØLUND A: Monomeric insulins and their experimental and clinical implications. *Diabetes Care*. 13: 923-954, 1990

6. DE MEIJER PH, RUSSEL FG, VAN LIER HJ, AND VAN GINNEKEN CA: A comparison of three mathematical models to describe the disappearance curves of subcutaneously injected 125I-labeled insulin. *Br. J. Clin. Pharmacol.* 27(4): 461-467, 1989

7. FISCHER U, FREYSE E-J, JUTZI E, BESCH W, RASCHKE M, HOFER S, AND ALBRECHT G: Absorption rates of subcutaneously injected insulin in dog as calculated from the plasma insulin levels by means of a simple mathematical model. *Diabetologia*. 24: 196-201, 1983

8. FLETCHER R: Practical methods of optimization. Chichester, John Wiley & Sons, 1980

9. KANG S, BRANGE J, BURCH A, VØLUND A, AND OWENS DR: Subcutaneous insulin absorption explained by insulin's physicochemical properties. *Diabetes Care*. 14, 942-47, 1991

10. KRAEGEN EW, AND CHISHOLM DJ: Insulin responses to varying profiles of subcutaneous insulin infusion: kinetic modelling studies. *Diabetologia*. 26: 208-213, 1984

11. MEHDI D, LISSANE S, HUMBERT C, AND MUSS JP: Adaptive control of glucose concentration in diabetic subject's blood. In *Modelling and Control in Biomedical systems, Proc First IFAC Symposium on Modelling and Control in Biomedical Systems, Venice, Italy*, 1988, Cobelli C, Mariani E, Eds. Pergamon Press, Oxford, 191-195, 1989

12. MEYERHOFF C, BISCHOF F, STERNBERG F, ZIER H, AND PFEIFFER EF: On line continuous monitoring of subcutaneous tissue glucose in men by combining portable glucosesensor with microdialysis. *Diabetologia*. 35: 1087-1092

13. MOSEKILDE E, JENSEN KS, BINDER C, PRAMMING S, AND THORSTEINSSON B: Modeling absorption kinetics of subcutaneous injected soluble insulin. *J. Pharmacokinet. Biopharm.* 17, 67-87, 1989

Control theory

DUALITY AND OPTIMALITY CONDITIONS FOR INFINITE DIMENSIONAL OPTIMIZATION PROBLEMS

Armin Hoffmann

Institute of Mathematics, TU Ilmenau, Postbox 327, D-98684 Ilmenau

Abstract

Using a nonsymmetric duality for abstract continuous convex control problems optimality conditions are derived for calculating the primal and dual solutions in the case of linear on state depending dual operators. Functional and pointwise conditions are considered. *Subject*: 49K22, 49K27, 49N15, 90C42. *Keywords*: abstract optimal control, nonsymmetric duality, sufficient conditions of optimality, optimal control involving integral equations.

1. Introduction and problems

In this article we consider a general abstract optimal control problem and some specializations. It is known that sufficient conditions of optimality directly follow from duality results. We developed in [7] a general theory of duality for such problems similar to that one given by KLÖTZLER in [12]. However, the general duality results (existence properties) can not be used to determine the solutions of the primal and dual problems. We want to derive some optimality conditions for the general and some special problem to calculate the dual and primal solution.

Let X, Y and V be BANACH spaces, Y^* be the dual BANACH space of Y, $M \subset X$, $U(x) \subset V$ be non empty, K be a closed convex cone in Y with its vertex at 0 and K^+ be the nonnegative polar cone of K. We consider the general abstract optimal control problem

$$
\begin{aligned}
f(x, u) &\longrightarrow \quad \inf \\
g(x, u) &\in \quad K \\
x &\in M, \ u \in U(x)
\end{aligned}
\tag{1}
$$

where the point to set image mapping $\Phi : M \longrightarrow \mathcal{P}(Y \times \mathbb{R})$ according to $\Phi(x) := \bigcup_{u \in U(x)} \{(g(x, u), f(x, u))\}$ and the extended image mapping $\Phi^+(x) := \Phi(x) - K \times (-\mathbb{R}_+)$ satisfy the following

Assumption 1.

    (S) Φ *is* HAUSDORFF-*continuous on* M,

    (B) $\Phi(x)$ *is bounded for any* $x \in M$,

(K) $\overline{\Phi^+(x)}$ is convex for any $x \in M$,

(A) $g(u, U(x)) - K$ is closed for any $x \in M$.

The assumption (K) is for classical optimal control problems a priori satisfied [22, p.141] , [10, §5.3.2]. The same holds for the following type of optimal control problems [7, ch. 1.2]

$$\int_0^1 F(x(t), u(t), t) \, dt \qquad \longrightarrow \quad \inf$$

$$x(t) - \int_0^1 G(x(s), u(s), s, t) \, ds \quad = \quad 0 \qquad , \qquad (2)$$

$$x(\cdot) \in \mathcal{C}([0,1], \mathbb{R}^n)$$

$$u(\cdot) \in \{u(\cdot) \in \mathcal{L}_\infty([0,1], \mathbb{R}^m) \, | u(t) \in U(t), t \in [0,1] \; a.e.\}$$

where F, G are continuous operators on sufficiently large subsets of $\mathbb{R}^n \times \mathbb{R}^m \times [0,1]$ or $\mathbb{R}^n \times \mathbb{R}^m \times [0,1]^2$, $U(t) \subset U$ for all $t \in [0,1]$ $a.e.$, $U \subset \mathbb{R}^m$ is bounded. For simplicity we assume $K = \{0\}$ and $M = X$. It is known that the property (K) is responsible for the validity of the global minimum principle with respect to the control u for the problem (2), cf.[7, ch. 1] and (1), [10]. The assumption (A) ensures the stability of (1) in the sense of [6], i.e. the optimal value p_0 of (1) is equal to the optimal value of some asymptotic extension of (1) with respect to the control u [7, ch. 1]. The global PONTRJAGIN maximum (minimum) principle holds for a lot of classical optimal control problems. In the above frame work this means

$$\forall u \in U(\overline{x}) : f(\overline{x}, u) - \langle y^*, g(\overline{x}, u) \rangle \geq f(\overline{x}, \overline{u}) - \langle y^*, g(\overline{x}, \overline{u}) \rangle \qquad (3)$$

where $(\overline{x}, \overline{u})$ is a solution of (1). For certain classical optimal control problems a generalization could be shown according to

$$\forall x \in M, \; u \in U(\overline{x}) :$$
$$f(x, u) - \langle L(x), g(x, u) \rangle \geq f(\overline{x}, \overline{u}) - \langle L(\overline{x}), g(\overline{x}, \overline{u}) \rangle$$
$$slack \; condition : \qquad \langle L(\overline{x}), g(\overline{x}, \overline{u}) \rangle = 0 \qquad (4)$$
$$condition \; of \; positivity : \; \forall x \in M : L(x) \in K^+$$

by [14], [23], [24] and [25]. There, $L : M \longrightarrow K^+$ is the gradient (or subgradient) $L(x)_{(t)} := \nabla_\xi S(t, \xi)_{|\xi = x(t)}$ of some marginal functional S with respect to the initial value ξ of the state x at the starting time t of the associated classical optimal control problem. In the case K={0} we can omit the slack condition and the condition of positivity. KLÖTZLER [12] could embed this principle in a generalized frame work of a non symmetric duality. The dual problem is obtained by the construction (cf. [19], [20], [6])

$$\psi(L) := \inf_{\substack{x \in M \\ u \in U(x)}} (f(x, u) - \langle L(x), g(x, u) \rangle) \longrightarrow \sup \qquad (5)$$
$$L \in G_{\mathcal{F}}$$

By using the HAMILTON-JACOBI inequality and the partial integration (or GAUSS's Theorem for $t \in \mathbb{R}^r$) the dual functional ψ can explicitly represented with respect to S and additional constraints (HAMILTON - JACOBI inequality). $G_{\mathcal{F}}$ should design some topological properties of L like continuity, locally lipschitzian property etc.and the condition of positivity of L. **Strong duality** is defined by

$$d_0 := \sup_{L \in G_{\mathcal{F}}} \psi(L) = \inf_{\substack{g(x,u) \in K \\ x \in M, u \in U(x)}} f(x,u) =: p_0. \tag{6}$$

In the case of the existence of a solution L of the dual problem (5) and a solution $(\overline{x}, \overline{u})$ of the primal problem (1) strong duality is equivalent to the validity of (4). Some applications to geometrical problems can be found in [13].

In the general setting of (1) the reduction of (5) doesn't work. The gradient structure of the operator L could not be proved. Under the Assumption 1 in [7, ch. 2, 3] duality results are proved for some topological properties of L.

2. Strong duality for the general problem - existence

The following strong duality results formulated in the setting of (4) can be found in a more general formulation in [7, ch. 3].

Theorem 1. (cf.[7, Th. 1.12]) *If the* Assumption 1 *is satisfied and there is some* $C(x) \geq 0$ *such that*

$$\forall x \in M, \ u \in U(x) :$$
$$f(x,u) + C(x) \inf_{z \in K} \|g(x,u) - z\| \geq p_0 \tag{7}$$

then there is some operator $L : M \to K^+$ *such that*

$$\forall x \in M, \ u \in U(x) : f(x,u) - \langle L(x), g(x,u) \rangle \geq p_0$$
$$\langle L(\overline{x}), g(\overline{x}, \overline{u}) \rangle = 0. \tag{8}$$

Proof. (guide) We separate the point $(0, p_0)$ from the set $\overline{\Phi^+(x)}$ for any $x \in M$ using usual separation theorems [10]. The y^*—part of the normal of such separating hyperplane defines our operator $L : M \to K^+. \diamondsuit$

Strong duality in the sense "sup" $=$ "inf" can be proved without regularity / stability conditions like (7).

Theorem 2. (cf.[7, Th. 3.4]) *If the* Assumption1 *is satisfied then there is some sequence of locally* LIPSCHITZ*ian operators* $L_n : M \to K^+$, $n = 1, 2, \ldots$ *such that (cf. [4])*

$$\forall x \in M, \ u \in U(x) : f(x,u) - \langle L_n(x), g(x,u) \rangle \geq p_0 - \frac{1}{n}$$
$$0 \leq \langle L_n(\overline{x}), g(\overline{x}, \overline{u}) \rangle \leq \frac{1}{n}. \tag{9}$$

Proof. (Guide) We show that the inverse of the multifunction of x to the set of all y^* satisfying $\forall u \in U(x) : f(x,u) - \langle y^*, g(x,u) \rangle \geq p_0 - \frac{1}{n}$ is open. Using selection results [1], [5] we obtain some L_n with (9).$\diamondsuit$

If we use the stronger regularity conditions like the following assumption then the strong duality with "max" $=$ " min " can be shown.

Assumption 2. $(\overline{x}, \overline{u})$ *is a strict solution of (1) and there is some* $y^* \in K^+$ *such that*

$$\forall \epsilon > 0 \exists \delta(\epsilon) > 0 \forall x \in M, z \in K, u \in U(x) : \|x - \overline{x}\| \le \delta(\epsilon) \Rightarrow$$
$$f(x, u) - \langle y^*, g(x, u) - z \rangle + \epsilon \|g(x, u) - z\| \ge f(\overline{x}, \overline{u}).$$

Remark 1. TICHOMIROV [22] call the last property " *the problem (1) admits the local* LIPSCHITZ*ian extension of the* LAGRANGE*an".*

Theorem 3. (cf.[7, Th. 3.2]) *If the Assumptions 1 and 2 are satisfied then there is some continuous operator* $L : M \to K^+$ *such that (8) holds.*

Proof. (guide) We construct with the help of the multifunction of x to the set of all y^* satisfying the assumption2 a multifunction Λ from M to K^+ being convex valued. Using the ROBINSON - URSESCU Theorem [18] and some special selection theorem [7, Th. 2.8] we prove that Λ is semicontinuous on M. The statement follows from MICHAEL's Selection Theorem [15]◊

Remark 2. If some stability condition like (7) with constant being independent on x then in all above theorems the operators $L, L_n : M \to K^+$ are bounded respectively uniformly bounded on M [7, Cor. 3.1.1, Th. 3.3, Cor. 3.4.1].

The main disadvantage of the above theorems is that no further structural properties could be proved. This is due to the used tools like selection theorems. If we use instead of Assumption 2 more complicated conditions containing some further structural informations then it seems (?) to be possible to prove duality results with additional structural statements by taking similar tools. Now we want to go a different way to get more structure in duality statements being useful for calculating the primal solution $(\overline{x}, \overline{u})$ and the dual solution L. For simplicity we assume that L has *affine structure* and the inequalities in (4) are *locally satisfied with respect to* x in some neighbourhood W of $\overline{x}$ w.r.t. M.

3. Functional conditions

3.1. Necessary conditions for strong duality

Let $(\overline{x}, \overline{u})$ be a solution of (1). Assume f and g to be twice FRÉCHÉT-differentiable with respect to x on an open M comprehending subset of X and assume $U(x) = U$ for all such x. We define the following notions and abbreviations

$$
\begin{aligned}
T(M, x) \quad &\cdots \quad \text{the tangent cone of } M \text{ at } x \\
Z(M, x) \quad &\cdots \quad \text{the cone of admissible directions of } M \text{ at } x \\
T^+(M, x) \quad &:= \quad \{y^* \,|\, \langle y^*, y \rangle \ge 0 \,\forall y \in T(M, x)\} \\
\Theta_0(x, u) \quad &:= \quad f(x, u) - \langle y^*, g(x, u) \rangle \\
\Theta_1(x, u) \quad &:= \quad \nabla_x f(x, u) - \sigma^* g(x, u) - \nabla_x g(x, u)^* y^* \\
\hat{\Theta}_1(x, u) \quad &:= \quad \nabla_x f(x, u) - \nabla_x g(x, u)^* y^* \\
\Theta_2(x, u)(h, h) \quad &:= \quad \nabla_{xx} f(x, u)(h, h) - \langle y^*, \nabla_{xx} g(x, u)(h, h) \rangle \\
& \qquad -2 \langle \sigma h, \nabla_x g(x, u) h \rangle \\
L(x) \quad &:= \quad y^* + \sigma^*(x - \overline{x}) \text{ with } x \in W \cap M
\end{aligned}
$$

where $h \in X$, σ^*, $\nabla_x g(x, u)^*$ are the dual linear operators of the linear and continuous operators $\begin{array}{c} \sigma : X \to Y^* \\ h \to \sigma h \end{array}$ and $\begin{array}{c} \nabla_x g(x, u) : X \to Y \\ h \to \nabla_x g(x, u)h \end{array}$. We get from strong duality the

Theorem 4. *If (4) is satisfied for some L with the affine structure $L(x) := y^* + \sigma^*(x - \overline{x})$ local with respect to x in some neighbourhood W of $\overline{x}$ then the following conditions are true*

1. **Minimum condition:**

$$\forall u \in U : \Theta_0(\overline{x}, u) \geq \Theta_0(\overline{x}, \overline{u}) \tag{10}$$

2. **Adjoint condition:**

$$\forall h \in T(M, \overline{x}) : \hat{\Theta}_1(\overline{x}, \overline{u})h \geq 0 \tag{11}$$

3. **Mixed adjoint condition:**

$$\begin{array}{c} \forall u \in U, \forall h \in T(M, \overline{x}) : \\ \Theta_0(\overline{x}, u) = \Theta_0(\overline{x}, \overline{u}) \Longrightarrow \Theta_1(\overline{x}, u)h \geq 0 \end{array} \tag{12}$$

4. **Second order condition:**

$$\forall h \in Z(M, \overline{x}) : \hat{\Theta}_1(\overline{x}, \overline{u})h = 0 \Longrightarrow \Theta_2(\overline{x}, \overline{u})(h, h) \geq 0 \tag{13}$$

5. **Mixed second order condition:**

$$\begin{array}{c} \forall u \in U, \forall h \in Z(M, \overline{x}) : \\ [\Theta_0(\overline{x}, u) = \Theta_0(\overline{x}, \overline{u}) \wedge \Theta_1(\overline{x}, u)h = 0] \Longrightarrow \\ \Theta_2(\overline{x}, u)(h, h) \geq 0 \end{array} \tag{14}$$

Proof. [7, ch.5.1] The proof is straight forward. $\Diamond$

Remark 3. It suffices [7] to demand some kind of uniform GÂTEAUX differentiability.

Example 1. *Let*

$X := l_2, ; \ U := [-1, 1] , \ Y = l_2 \cong Y^* , \ M := \left\{ x \in l^2 \, \middle| \, 0 \leq x_j \leq \tfrac{1}{j}, j = 1, 2, \ldots \right\}$,

$f(x, u) := u \|x\|^2 , \ g(x, u) := (u + x_2, \tfrac{u}{2} + x_3, \cdots, \tfrac{u}{n-1} + x_n, \cdots)$. *We consider the problem (1) with $K = \{0\}$ and $U(x) \equiv U$. Simple considerations yield the solution $\overline{x} = (1, \tfrac{1}{2}, \tfrac{1}{4}, \cdots, \tfrac{1}{2(n-1)}, \cdots)$, $\overline{u} = -\tfrac{1}{2}$ with the optimal value $f(\overline{x}, \overline{u}) = p_0 = -\tfrac{1}{2} + \tfrac{\pi^2}{48}$. The adjoint condition (11) implies $y_1^* \geq -\tfrac{1}{2}$, $y_{j-1}^* = -\overline{x}_j$, $j = 2, 3, \ldots$ and together with the minimum condition (10) we find $y_1^* = -\tfrac{1}{2} + \tfrac{\pi^2}{8}$. Using the condition (13) we find for σ*

$$\sigma = \begin{pmatrix} a_1 & b_1 & d_3 & d_4 & \cdots & d_j & \cdots \\ a_2 & b_2 & c_3 & 0 & \cdots & 0 & \cdots \\ a_3 & b_3 & 0 & c_4 & \cdots & 0 & \cdots \\ \vdots & \vdots & \vdots & \vdots & \ddots & \vdots & \vdots \\ a_j & b_j & 0 & 0 & \cdots & c_j & \cdots \\ \vdots & \vdots & \vdots & \vdots & \vdots & \vdots & \ddots \end{pmatrix} \tag{15}$$

with $a_i \in \mathbb{R}$, $c_j \in \left(-\infty, -\frac{1}{2}\right]$, $d_j \in \mathbb{R}_+$ for $i = 1, 2, \ldots$ and $j = 3, 4, \ldots$. The mixed adjoint equation (12) has to be satisfied for all $u \in U$ in this example because $f(\overline{x}, u) - \langle y^*, g(\overline{x}, u) \rangle \equiv f(\overline{x}, \overline{u})$ on U. A special solution for (12) with (15) using only one parameter c_j is given by $a_1 = 1$, $a_2 = 2$, $a_i = 0$, $b_1 = 1$, $b_2 = b_i = 0$ for $i = 3, 4, \ldots$ and $d_j = \frac{1-c_j}{j-1}$ for $j = 3, 4, \ldots$. It is elementary to show that for $c_j \leq -1$ the condition (14) can be satisfied.

In the following chapter we give sufficient conditions of optimality. We apply these conditions at the above Example 1.

3.2. Sufficient conditions for strong duality

We assume that the BANACH space can continuously embedded in some normed spaces X_i, $i = 1, 2$, i.e. $\exists C_i > 0 \; \forall h \in X : \|h\|_i \leq C_i \|h\|$. Let $M_\delta := \{x \in M \mid \|x - \overline{x}\| < \delta\}$, $U_\delta := \bigcup\limits_{x \in M_\delta} U(x)$ and f, g be twice continuously differentiable w.r.t. x for any $u \in U$.
In the contrary to the previous chapter we again assume that the control set depends on x. Using the remainder $R_i^f(\overline{x}, u, x - \overline{x})$ and $R_i^g(\overline{x}, u, x - \overline{x})$ of the i-th order, $i = 1, 2$, we define remainders being independent on u by

$$\left.\begin{aligned} R_i^f(\overline{x}, x - \overline{x}) &:= \sup_{u \in U_\delta} \left| R_i^f(\overline{x}, u, x - \overline{x}) \right| \\ R_i^g(\overline{x}, x - \overline{x}) &:= \sup_{u \in U_\delta} \| R_i^g(\overline{x}, u, x - \overline{x}) \| \end{aligned}\right\} \quad i = 1, 2.$$

Assumption 3.

1. $\exists D_1, D_2, \delta > 0 \; \forall y^* \in Y^*, h \in X : \|h\| < \delta \Longrightarrow$

$$\sup_{u \in U_\delta} |\langle y^*, \nabla_x g(\overline{x}, u) h \rangle| \leq D_1 \| y^* \| \, \| h \|_1$$

$$\sup_{u \in U_\delta} |\langle y^*, \nabla_{xx} g(\overline{x}, u)(h, h) \rangle| \leq D_2 \| y^* \| \, \| h \|_2^2$$

2.

$$\lim_{\|h\| \to 0} \frac{R_i^f(\overline{x}, h)}{\| h \|_i^i} = 0 \, , \; i = 1, 2 \quad \text{and} \quad \lim_{\|h\| \to 0} \frac{R_i^g(\overline{x}, h)}{\| h \|_i^i} = 0 \, , \; i = 1, 2.$$

Theorem 5. (In x local sufficient condition of optimality and strong local duality) Let $\delta > 0, \overline{x} \in M$, $\overline{u} \in U(\overline{x})$, $f(\cdot, u)$, $g(\cdot, u)$ be twice differentiable for any $u \in U$, (S) and Assumption 3 be satisfied. If there are some $y^* \in Y^*$, some continuous linear operator $\sigma^* : X \to Y^*$, positive constants α, β and a mapping $r : M_\delta - \overline{x} \to \mathbb{R}$ such that
$\forall x \in M_\delta \; \forall u \in U(x) : h := x - \overline{x}$

$$\underbrace{\Theta_0(\overline{x}, u) + \Theta_1(\overline{x}, u) h - \Theta_0(\overline{x}, \overline{u})}_{=: \varphi_1(x, u)} \leq \alpha \|h\|_1 \Longrightarrow$$

$$\Theta_0(\overline{x}, u) + \Theta_1(\overline{x}, u) h - \Theta_0(\overline{x}, \overline{u}) \geq -r(h)$$
$$\varphi_2(x, u) := \Theta_2(\overline{x}, u)(h, h) \geq \beta \|h\|_2^2 + r(h)$$

(16)

then there is some $\epsilon > 0$ such that

$$f(x,u) - \langle y^* + \sigma^*(x - \overline{x}), g(x,u)\rangle - f(\overline{x},\overline{u}) \geq \min\left\{\frac{\alpha}{2}\|x - \overline{x}\|_1, \frac{\beta}{2}\|x - \overline{x}\|_2^2\right\} \quad (17)$$

for any $x \in M_\epsilon$ and any $u \in U(x)$.

Proof. (Guide, cf.[7, ch.5.1]) The proof is divided in two parts. In the first part it has to show that for all u and x such that $\Theta_0(\overline{x},u) + \Theta_1(\overline{x},u)h - \Theta_0(\overline{x},\overline{u}) \geq \alpha\|h\|_1$ the condition (17) is fulfilled w.r.t. the first norm. We use TAYLOR expansion up to the first order. In the second part we show the above implication by considering the TAYLOR expansion up to the second order. The detailed proofs are straight forward. The Assumption 3 is essential.$\Diamond$

The advances of the "different norm strategy" are explained in [16]. The condition (17) ensures together with the admissibility of $(\overline{x},\overline{u})$ that $(\overline{x},\overline{u})$ is a solution (local w.r.t. x) of the problem (1) and that strong duality with $L(x) := y^* + \sigma^*h$ holds if we again take local considerations with respect to x in some neighbourhood of $\overline{x}$ relatively to M. The following continuation of the Example 1 demonstrates the working of the above conditions. Because of $X = Y = l_2 \cong Y^*$ we take $\|\ \|_i = \|\ \|$ for $i = 1, 2$.

Example 2. (continuation of Example 1) *We obtain* $\varphi_1(x,u) = -h_1 - h_2\left(\frac{1}{2} + \frac{\pi^2}{8}\right)$ *and*

$\varphi_2(x,u) = u\left(h_1^2 + h_2^2\right) - h_1 h_2 - h_2^2 + u\sum_{j=3}^{\infty} h_j^2 - \sum_{j=3}^{\infty} c_j h_j^2$. *Since $h_i \leq 0, i = 1, 2$ we find for any*

$\beta > 0$ *some $\delta(\beta) > 0$ such that for any $\|h\| < \delta(\beta)$ the inequality $\varphi_1(x,u) \geq -\left(h_1^2 + h_2^2\right) +$ $h_1 h_2 + h_2^2 + \beta\left(h_1^2 + h_2^2\right) =: r(h)$ is satisfied. If we put $c_j := -1 - \beta, \beta > 0, j = 3, 4, \ldots$ then*

we get (17) according to $\varphi_2(x,u) - r(h) = \beta\left(h_1^2 + h_2^2\right) + \sum_{j=3}^{\infty} h_j^2(u + 1 + \beta) \geq \sum_{j=1}^{\infty} h_j^2 \beta =$

$\beta\|h\|^2$. *Therefore, $(\overline{x},\overline{u})$ is a solution for our Example 1 local w.r.t. x , global w.r.t. u and $L(x) = y^* + \sigma^*h$ is the associated (local) dual solution.*

4. Pointwise conditions

4.1. Necessary condition for strong duality

For simplicity we consider only the problem (2) without state constraints. The general case is dealt with in [8] and [9]. In the classical theory of optimal control the adjoint variable lies in the space of bounded variations $B([0,1], \mathbb{R}^n)$. Without state constraints the following assumptions for the affine structure of L yields useful results. Let

$$L(x)_{(t)} := \int_0^t \left(\chi(s) + \sigma(s) \circ (x(s) - \overline{x}(s))\right) ds \quad (18)$$

where χ and σ are $\mathcal{L}_p$-vector and matrix functions, resp. for some $p > 1$. $t \to U(t)$ is a measurable multifunction (cf.[10]).

Theorem 6. [9, Th.4](Generalized PONTRJAGIN principle) *If* $(\overline{x}, \overline{u})$ *is a solution of (2) then for the validity of (4) under the above structural assumption (18) the following pointwise optimality condition is satisfied:*

$$\forall t \in [0,1] \, a.e. \forall \xi \in \mathbb{R}^n \forall \mu \in U(t):$$

$$F(\xi, \mu, t) - F(\overline{x}(t), \overline{u}(t), t) + \int_0^1 \chi(s)^T \circ (G(\xi, \mu, t, s) - G(\overline{x}(t), \overline{u}(t), t, s)) \, ds + \tag{19}$$

$$(\xi - \overline{x}(t))^T \circ (\chi(t) + \sigma(t) \circ (\xi - \overline{x}(t))) \geq 0$$

Proof. (Guide) We consider the following arrangement of (4) for any $\epsilon > 0$:

$$
\begin{aligned}
0 \;\leq\; & \tfrac{1}{\epsilon} \left(f(x_\epsilon, u_\epsilon) - \langle y^* + \sigma(x_\epsilon - \overline{x}), g(x_\epsilon, u_\epsilon) \rangle - f(\overline{x}, \overline{u}) \right) \\
=\; & \tfrac{1}{\epsilon} \left(f(x_\epsilon, u_\epsilon) - f(\overline{x}, \overline{u}) - \langle y^*, g(x_\epsilon, u_\epsilon) - g(\overline{x}, \overline{u}) \rangle \right. \\
& \left. - \langle \sigma(x_\epsilon - \overline{x}), g(x_\epsilon, u_\epsilon) - g(\overline{x}, \overline{u}) \rangle \right)
\end{aligned}
\tag{20}
$$

where

$$
u_\epsilon(t) := \begin{cases} \tilde{u}(t) & \text{for } t \in [\tau, \tau + \epsilon) \\ \overline{u}(t) & \text{elsewhere,} \end{cases}
$$

τ is LEBESGUE's point of $\tilde{u}$ in $(0,1)$, $\mu := \tilde{u}(\tau)$ and $\overline{u}(t) \in U(t)$,

$$
x_\epsilon(t) := \begin{cases} \xi & \text{for } t \in [\tau, \tau + \epsilon) \\ \hat{x}(t) & \text{for } t \in [\tau - \delta(\epsilon), \tau) \cup [\tau + \epsilon, \tau + \epsilon + \delta(\epsilon)) \\ \overline{x}(t) & \text{elsewhere,} \end{cases}
$$

$x_\epsilon \in \mathcal{C}([0,1], \mathbb{R}^n)$. The limit for $\epsilon \longrightarrow +0$ yields the above statement. We need FUBINI's Theorem and the equi-integrability [17]. Therefore the $\mathcal{L}_p$–assumptions for χ and σ. Further we have to choose $\delta(\epsilon) \leq \epsilon^2$. The term in the third row of (20) is of order two w.r.t. ϵ and yields 0 in the limit. $\Diamond$

Corollary 1. *If the functions F and G are once (twice) differentiable with respect to ξ then holds the adjoint equation (adjoint inequality of second order for all $w \in \mathbb{R}^n$)*

$$\nabla_\xi F(\xi, \mu, t) + \int_0^1 \chi(s)^T \circ \nabla_\xi G(\xi, \mu, t, s) \, ds - \chi(t) = 0$$

$$\left(w^T \circ \left[\nabla_{\xi\xi} F(\xi, \mu, t) + \int_0^1 \chi(s)^T \circ \nabla_{\xi\xi} G(\xi, \mu, t, s) \, ds - 2\sigma(t) \right] \circ w \geq 0 \right) \tag{21}$$

at $\xi = \overline{x}(t)$ and $\mu = \overline{u}(t)$ for $t \in [0,1]$ a.e..

Corollary 2. *It holds the global pointwise minimum principle*

$$\forall t \in [0,1] \, a.e. \forall \mu \in U(t): F(\overline{x}(t), \mu, t) - F(\overline{x}(t), \overline{u}(t), t) +$$
$$\int_0^1 \chi(s)^T \circ (G(\overline{x}(t), \mu, t, s) - G(\overline{x}(t), \overline{u}(t), t, s)) \, ds \geq 0. \tag{22}$$

Remark 4. If we have state constraints we have to assume a richer structure of L e.g. we have to add some jumping parts like ("$\circ$" - matrix multiplication)

$$\sum_{t_i < t} \left[(l_i + s_i \circ (x(t_i) - \overline{x}(t_i))) + \int_0^1 \rho_i(s) \circ (x(s) - \overline{x}(s)) \, ds \right]$$

where $l_i \in IR^n$, $s_i \in IR^{n \times n}$, $i = 1, 2, \ldots$, the series fulfills some convergence properties and the matrix functions ρ_i are of $\mathcal{L}_p$. In this case pointwise jumping conditions can arise at some points t_i (cf. [9, Th.4]) where the trajectory does not belong to the interior of $M(t)$. If we have $K \neq \{0\}$ then additional pointwise conditions of positivity occur. For necessary optimality conditions of (2) cf. e.g. [11], [21].

Now we want to demonstrate the above condition at an easy example.

Example 3. $f(x, u) := \int\limits_0^1 \left[0.896 u^2(t) - x^2(t) \right] dt \longrightarrow \inf$, subject to $g(x, u)_{(t)} := x(t) -$ $\int\limits_0^1 t \left[x(s) u(s) + x^2(s) \right] ds = 0$ for all $t \in [0, 1]$ a.e., $x \in C([0, 1], \mathbb{R})$, $u \in \mathcal{L}_\infty((0, 1), \mathbb{R})$, $u(t) \in [-1, 1]$ for all $t \in [0, 1]$ a.e.. The solution of the state equation has the form $x(t) = \alpha t$. It follows therefore $x(t) = 3 \left[1 - \int\limits_0^1 su(s) ds \right] t := x_u(t)$. Thus the solution of the problem is given by the solution of $f(x_u, u) \to \inf$, $u \in U$. A conjecture can elementary be calculated straight forward. We get $\overline{u}(t) = \begin{cases} -5t & \text{for } t \in [0, 0.2) \\ -1 & \text{for } t \in [0.2, 1) \end{cases}$ and $\overline{x}(t) = 4.48t$ on $[0, 1]$ with the optimal value $p_0 = -5.9136$. From the Corollary 1 we easy find with the adjoint equation (inequality) (21) that $\chi(t) = -2\overline{x}(t) + \lambda \left[\overline{u}(t) + 2\overline{x}(t) \right] = 2 \left[\overline{u}(t) + \overline{x}(t) \right] = \begin{cases} -1.04t & \text{for } t \in [0, 0.2) \\ -2 + 8.96t & \text{for } t \in [0.2, 1) \end{cases}$ where $\lambda = \int\limits_0^1 s\chi(s) ds = 2$ and that $\sigma(t) \leq 1$ on $[0, 1]$. The minimum principle (22) is satisfied. Using the transformations $z = \xi - \overline{x}(t)$ and $v = \mu - \overline{u}(t)$ then we find from (19) that for $t \in [0, 1]$ a.e., $|z| < \delta$ and $v \in [-1, 1] - \overline{u}(t)$ holds

$$\left[\sqrt{1 - \sigma(t)} z + \frac{v}{\sqrt{1 - \sigma(t)}} \right]^2 + \left[0.896 - \frac{1}{1 - \sigma(t)} \right] v^2 + 2 \left[0.896\overline{u}(t) + \overline{x}(t) \right] v \geq 0. \quad (23)$$

We obtain generally that $\sigma(t) \leq 1 - \frac{1}{0.896} \leq -.1161$ is possible on $[0, 1]$. However we can find a better approach if we divide the interval $[0, 1]$ in two parts. For $t \in [0.2, 1]$ holds $v \in [0, 2]$ i.e. $v \geq 0.5v^2$. Thus we obtain on $[0.2, 1]$ from (19)

$$\left[\sqrt{1 - \sigma(t)} z + \frac{v}{\sqrt{1 - \sigma(t)}} \right]^2 + \left[\overline{x}(t) - \frac{1}{1 - \sigma(t)} \right] v^2 \geq 0.$$

This implies $\sigma(t) \leq 1 - \frac{1}{\overline{x}(t)} \leq 1 - \frac{1}{4.48t}$ i.e. on $[0.2, 1]$ the formula $\sigma(t) \leq 0$ is possible. Finally, we have a conjecture for the dual operator

$$L(x)_{(t)} = \int\limits_0^t \left(\chi(s) + \sigma(s) \circ (x(s) - \overline{x}(s)) \right) ds$$

$$\chi(t) = \begin{cases} -1.04t & \text{for } t \in [0, 0.2) \\ -2 + 8.96t & \text{for } t \in [0.2, 1) \end{cases} \quad \text{and} \quad (24)$$

$$\sigma(t) = \begin{cases} -\Delta & \text{for } t \in [0, 0.2 + \rho) \\ 0 & \text{for } t \in [\rho + 0.2, 1) \end{cases} \quad \text{with } \Delta \geq 0.1161, \ \rho \geq 0$$

which contains the two parameters Δ and ρ.

4.2. Sufficient conditions of optimality

The following simple sufficient condition can be stated whenever the dual operator is identically constant i.e. $L(x) \equiv y^*$ (cf. [3])

Theorem 7. *If $L(x) = y^*$ for all x in some neighbourhood W of $\overline{x}$ and the inequality (19) is valid and if $(\overline{x}, \overline{u})$ is admissible for (2) then $(\overline{x}, \overline{u})$ are solutions of (2) local w.r.t. x, global w.r.t. u and y^* is the associated local dual operator.*

Proof. The proof is given straight forward by the integrating of (19) w.r.t. t over $[0, 1]$ and using FUBINI's Theorem.$\Diamond$

If we have the case that $L(x) \not\equiv$ constant then the Theorem 7 cannot applied. The integrating of (19) w.r.t. t over $[0, 1]$ and using FUBINI's Theorem yields

$$A(x, u) := \quad f(x, u) - f(\overline{x}, \overline{u}) - \langle y^*, g(x, u) - g(\overline{x}, \overline{u}) \rangle -$$
$$\int_0^1 (x(t) - \overline{x}(t))^T \circ \sigma(t) \circ (x(t) - \overline{x}(t))\, dt \geq 0.$$

The term

$$B(x, u) := \int_0^1 (\sigma(t) \circ (x(t) - \overline{x}(t)))^T \circ \int_0^1 (G(x(s), u(s), s, t) - G(\overline{x}(s), \overline{u}(s), s, t))\, ds\, dt$$

can not be reproduced. For the sufficiency it has to be proved that $A(x, u) + B(x, u) \geq 0$ for all admissible (x, u) of (2). We can apply the Theorem 5 or we prove directly that $A(x, u) + B(x, u) \geq 0$. Sufficient pointwise conditions so far (except the above trivial case) are unknown. Since the character of the state equation is not pointwise in the contrary to the classical optimal control problem (differential state equation) we think that pointwise sufficient conditions for the general case of (2) are impossible. Additional conditions to G and F seem to be necessary. Simple examples shows that the condition "$A(x, u) \geq 0$ for all admissible (x, u) of (2)" is not sufficient. We continue our Example 3.

Example 4. *We want to show the statement $A(x, u) + B(x, u) \geq 0$ for all admissible (x, u) of (2) directly locally w.r.t. x and globally w.r.t. u. We choose some $\delta > 0$ and ξ with $\|\xi\| < \delta$. The amount of δ has to be calculated! Now we use*

$$\sigma(t) \; = \left\{ \begin{array}{ll} -\Delta & for\ t \in [0, 0.2 + \rho) \\ 0 & for\ t \in (\rho + 0.2, 1] \end{array} \right.$$

with $\Delta \geq 0.12$ and $\rho = 0.2$ and divide the interval $[0.1]$ in the two parts $[0, 0.4]$ and $(0.4, 1]$. With respect to these intervals we consider all functions f, g, ξ, μ, σ and χ as elements of the associated $\mathcal{L}_2([0, 0.4], \mathbb{R})$ and $\mathcal{L}_p([0.4, 1], \mathbb{R})$. Using the HILBERT space norms for the functions ξ, μ separately on both intervals we get after elementary estimations that $A(x, u) + B(x, u)$ is larger or equal as a quadratic function in $\mathbb{R}^4$ of these fore norms. We find that for $\delta = 1$ the quadratic function is positive definite whenever the parameter $\Delta \in (0.26, 2)$.

Example 5. *If we change in the Example 4 the set $U := \{-1, 0, 1\}$ and the parameter of u^2 in the objective to 0.897 then we analogously find out that $\overline{u}(t) =$*
$$\begin{cases} 0 & \text{for } t \in [0, 0.1) \\ -1 & \text{for } t \in (0.1, 1] \end{cases}, \ \overline{x}(t) = 4,485t, \ p_0 = -5,89775\overline{075}, \ \chi(t) = 2\left[\overline{x}(t) + \overline{u}(t)\right].$$ *The second order adjoint inequality is satisfied for $\sigma \leq 1$, but the generalized principle (19) can not be satisfied. We formally find*

$$\sigma(t) \leq \ 1 - \tfrac{1}{8.97(0.1-t)} < 0 \text{ for } t \in [0, 0.1)$$
$$\sigma(t) \leq \ 1 - \tfrac{1}{8.97(t-0.1)} < 0 \text{ for } t \in (0.1, 0.2]$$
$$\sigma(t) = \ 0 \text{ for } t \in (0.2, 1],$$

but σ doesn't belong to the $\mathcal{L}_p$−space for any $p \geq 1$. Local strong duality with the described affine structure can atmost hold in the sense of Theorem 2.

Example 6. *If we consider the case $U \subset \mathbb{R}$ open and bounded and take the parameter of u^2 in the objective smaller than one then the problem (2) has not any solution. In the existing stationary point the condition (19) can be satisfied for some $\sigma(t) < 0$. Therefore, (19) cannot be sufficient!*

References

[1] AUBIN, J. P. / CELLINA, A.: Differential inclusions. Grundlehren des mathematischen Wissens 264. Springer-Verlag, Berlin-Heidelberg- New York-Tokyo 1984.

[2] ANDREJEWA, J. A. / KLÖTZLER, R.: Zur analytischen Lösung geometrischer Optimierungsaufgaben mittels Dualität bei Steuerproblemen I, II . Z. Angew. Math. Mech. 64(1984),35-44(part I), 147-153(part II).

[3] CARLSON, D. A.:Sufficient conditions for optimality and supported trajectories for optimal control problems governed by VOLTERRA integral equations . Proc.Opt. Days 1986, LN Econ. Math. Syst. 302(1988),274-282.

[4] CLARKE, F. H.: Optimization and nonsmooth analysis. John Wiley and Sons, Inc., New York 1983.

[5] DOMMISCH, G.: Zur Existenz LIPSCHITZ-stetiger Auswahlfunktionen für mehrdeutige Abbildungen. Diss. A, Humboldtuniversität zu Berlin 1986.

[6] GOL'STEJN, E. G.: Dualitätstheorie in der nichtlinearen Optimierung und ihre Anwendung. Akademie-Verlag, Berlin 1975.

[7] HOFFMANN, A.: Beiträge zur Dualitätstheorie für abstrakte Optimalsteuerprobleme in BANACHräumen. Diss. B, TH Ilmenau 1987.

[8] HOFFMANN, A.: Pointwise optimality criteria for optimal control problems governed by integral state relations I. Wiss. Z. TH Ilmenau, 33(1987)6,75-85.

[9] HOFFMANN, A.: Pointwise optimality criteria for optimal control problems governed by integral state relations II. Wiss. Z. TH Ilmenau, 37(1991)3,129-138.

[10] IOFFE, A. D. / TICHOMIROV, V. M.: Teorija ekstremal'nych zadač. Monografija, Izd. Nauka, Moskva 1974.

[11] KEMPTER, D., LORENTZ, R.: Zur Theorie der abstrakten Steuerungsaufgben in BANACHräumen und Steuerungsaufgaben in konkreten Funktionenräumen. Diss. A, TH Ilmenau 1978.

[12] KLÖTZLER, R.: On a general conception of duality in optimal control. Proceedings Equadiff 4, Prague (ČSSR) 1977, 189 - 196.

[13] KLÖTZLER, R.: Globale Optimierung in der Steuerungstheorie. Z. Angew. Math. Mech. 63(1983)5, T305-T312.

[14] KROTOV, V. F.: Metody rešenija variacionnych zadač na osnove dostatočnych uslovij absoljutnogo minimuma I, II. Avtomat. i. Telemeh. (1962)12, 1571 -1589 (part I), (1963)5, 581 - 589 (part II).

[15] MICHAEL, E.: Continuous selections 1. Ann. of Math. 63(1956), 361 - 382.

[16] MAURER, H.: First- and second-order sufficient optimality conditions in mathematical programming and optimal control. Math. Programming Stud. 14(1981),163-177.

[17] NATANSON, I. P.: Theorie der Funktion einer reellen Veränderlichen. Akademie-Verlag, Berlin 1975, 4. Auflage.

[18] ROBINSON, S. M.: Regularity and stability for convex multivalued functions. Math. Oper. Res. 1(1976)2,130 - 143.

[19] ROCKAFELLAR, R. T.:Augmented LAGRANGE multiplier functions and duality in nonconvex programming. SIAM J. Control Optim. 16(1978), 571-583.

[20] ROCKAFELLAR, R. T.: Conjugate convex functions in optimal control and the calculus of variations, J. Math. Anal. and Appl. 32(1970), 174 - 222.

[21] SCHMIDT, W.: Durch Integralgleichungen beschriebene optimale Prozesse in BA-NACHräumen - notwendige Optimalitätsbedingungen. Z. Ang. Math. Mech. 62(1982), 65 - 75.

[22] TICHOMIROV, V. M.: Grundprinzipien der Theorie der Extremalwertaufgaben. Teubner-Texte zur Mathematik, BSB B. G. Teubner Verlagsgesellschaft, Leipzig 1982.

[23] VINTER, R. B.: A necessary and sufficient condition for optimality of dynamic programming type, making no a priori assumption on the controls. J. Control Opt. 16(1978), 571 - 583.

[24] ZEIDAN, V.: A modified HAMILTON - JACOBI approach in the generalized problem of BOLZA. Appl. Math. Optim. 11(1984), 97 - 109.

[25] ZEIDAN, V.: First- and second-order sufficient conditions for optimal control and the calculus of variations. Appl. Math. Optim. 11(1984), 209 - 226.

E-Mail: *armin.hoffmann@mathematik.tu-ilmenau.de*

SOLUTION DIFFERENTIABILITY FOR PARAMETRIC NONLINEAR CONTROL PROBLEMS WITH INEQUALITY CONSTRAINTS

Helmut Maurer and Hans Josef Pesch

Westfälische Wilhelms-Universität Münster, Institut für Numerische und instrumentelle Mathematik, Einsteinstrasse 62, 48149 Münster, Germany.

Technische Universität München, Mathematisches Institut, Arcisstrasse 21, 80333 München, Germany.

Abstract: *This paper considers parametric nonlinear control problems subject to mixed control-state constraints. The data perturbations are modeled by a parameter p of a Banach space. Using recent second-order sufficient conditions (SSC) it is shown that the optimal solution and the adjoint multipliers are differentiable functions of the parameter. The proof blends numerical shooting techniques for solving the associated boundary value problem with theoretical methods for obtaining SSC. In a first step, a differentiable family of extremals for the underlying parametric boundary value problem is constructed by assuming the regularity of the shooting matrix. Optimality of this family of extremals can be established in a second step when SSC are imposed.*

Key Words: Parametric control problems, mixed control-state constraints, second-order sufficient conditions, multipoint boundary value problems, shooting techniques, Riccati equation.

1 Introduction

This paper is concerned with parametric nonlinear control problems subject to mixed control-state constraints. The following parametric control problem will be referred to as $OC(p)$ where p is a parameter belonging to a Banach space P: minimize the functional

$$J(x, u, p) = \int_a^b L(x(t), u(t), p)\, dt \qquad (1)$$

subject to

$$\dot{x}(t) = f(x(t), u(t), p) \quad \text{for a.e.} \quad t \in [a, b] \, , \qquad (2)$$

$$x(a) = \varphi(p), \quad x(b) = \psi(p) \qquad , \qquad (3)$$

$$C(x(t), u(t), p) \leq 0 \quad \text{a.e.} \quad t \in [a, b] \, . \qquad (4)$$

We shall not treat the most general case and assume that the control variable u and the inequality constraint (4) are *scalar*. Extensions to the vector-valued case are possible. The functions $L : I\!R^{n+1} \times P \to I\!R$, $f : I\!R^{n+1} \times P \to I\!R^n$, $\varphi, \psi : P \to I\!R^n$ and $C : I\!R^{n+1} \times P \to I\!R$ are assumed to be C^2-functions on appropriate open sets. The admissible class is that of piecewise continuous control functions. Later on conditions will be imposed such that the optimal control is continuous and piecewise of class C^1.

The problem $OC(p_0)$ corresponding to a fixed parameter $p_0 \in P$ is considered as the *unperturbed* or *nominal* problem. It will be ensured by second-order sufficient conditions (SSC) that $OC(p_0)$ has a local minimum $x_0(t)$, $u_0(t)$, $\lambda_0(t)$ where $\lambda_0(t)$ denotes the adjoint function which will be defined below. Our aim is to embed the unperturbed solution into a piecewise C^1-family of optimal solutions $x(t, p)$, $u(t, p)$, $\lambda(t, p)$ for the perturbed problem $OC(p)$ with p in a neighborhood of p_0.

This solution differentiability problem has been considered for *pure* control constraints in [2] - [5] using optimization techniques in Hilbert spaces or Banach spaces. The approach developed in this paper is different and melts finite-dimensional numerical solution techniques with recent theoretical results for weak second-order sufficient conditions.

Solution differentiability is obtained in *two steps*. In a *first step*, a C^1-family of extremals $x(t, p)$, $\lambda(t, p)$ is constructed which satisfies the first order necessary conditions for $OC(p)$. This is achieved by setting up a suitable parametric boundary value problem and by imposing the regularity of the Jacobian for the shooting method. The *second step* consists in showing that this C^1-family of extremals is indeed optimal by requiring second-order sufficient conditions. Both steps are connected by the fact that a direct line can be traced from the variational system of the unperturbed boundary value problem to SSC by introducing a Riccati ODE. Some aspects of the first step have been considered already in [1], [10], [11].

Only the main results are given in this paper. Detailed proofs and more advanced numerical examples will appear elsewhere.

2 The parametric boundary value problem for $OC(p)$

Necessary optimality conditions for control problems with mixed constraints can be found in [8]. The Hamiltonian for the unconstrained problem (1) - (3) is

$$H(x, \lambda, u, p) = L(x, u, p) + \lambda^* f(x, u, p), \quad \lambda \in I\!R^n , \tag{5}$$

whereas the *augmented* Hamiltonian for the constrained problem $OC(p)$ is defined by

$$\tilde{H}(x, \lambda, \mu, u, p) = H(x, \lambda, u, p) + \mu C(x, u, p), \quad \mu \in I\!R . \tag{6}$$

The adjoint function $\lambda : [a, b] \to I\!R^n$ and the multiplier function $\mu : [a, b] \to I\!R$ with $\mu \geq 0$ and $\mu C = 0$ are determined by a suitable boundary value problem (BVP) which we shall set up now. First, a careful list of assumptions is given which is needed for a suitable numerical analysis of the problem in conjunction with SSC.

The structure of the unperturbed solution (x_0, u_0):

The *active set* or *boundary* of the inequality constraint $C(x, u, p_0) \leq 0$ is supposed to consist of one *boundary arc*, i.e. we have

$$\{t \in [a, b] \mid C(x_0(t), u_0(t), p_0) = 0\} = [t_1^0, t_2^0], \quad a < t_1^0 < t_2^0 < b . \tag{7}$$

The generalization to several boundary arcs is immediate. The points t_1^0, t_2^0 are called *junction points* with the boundary arc. The case that the active set may also contain isolated points τ_i (*contact points*) will not be considered here. Contact points are spurious under the assumptions introduced below and hence are not stable with respect to perturbations. From (7) we can expect that the perturbed solution $x(t, p)$, $u(t, p)$ has one boundary arc for $t_1(p) \leq t \leq t_2(p)$ with $t_i(p_0) = t_i^0$, $i = 1, 2$. It will be shown that the junction points $t_1(p)$, $t_2(p)$ are C^1-functions of the parameter p.

C^1-regularity of the Hamiltonian on interior arcs:

Let $\lambda_0 : [a, b] \to I\!R^n$ be the adjoint function associated with (x_0, u_0) which will be a solution of the unperturbed BVP (13) - (16) to be defined below. The following assumption guarantees that the control is a C^1-function on interior arcs.

(A1) (a) **(Strict Legendre-Clebsch condition)**

$$H_{uu}(x_0(t), \lambda_0(t), u_0(t), p_0) \geq c > 0 \quad \text{for} \quad a \leq t \leq t_1^0 \text{ and } t_2^0 \leq t \leq b .$$

(b) **(C^1-regularity of the Hamiltonian)**

There exists a uniquely defined C^1-function $u(x, \lambda, p)$ such that

$$u(x, \lambda, p) = arg \min_{u \in I\!R} H(x, \lambda, u, p)$$

holds for all (x, λ, p) in a neighborhood of the trajectory

$$x_0(t), \lambda_0(t), p_0 \quad \text{for} \quad a \leq t \leq t_1^0 \quad \text{and} \quad t_2^0 \leq t \leq b .$$

The strict Legendre-Clebsch condition **(A1)(a)** excludes all control problems with control appearing linearly, i.e. bang-bang or singular controls. It should be noted that the C^1-regularity of the Hamiltonian does not follow from part (a). Consider e.g. $L(x, u, p) = (u^2 - p)^2$, $f(x, u, p) = u$ and $p_0 = 1$. Here (a) holds but (b) is violated since any control $u(x, p) = \pm\sqrt{p}$ is optimal.

It follows by definition that $u_0(t) = u(x_0(t), \lambda_0(t), p_0)$. The function $u(x, \lambda, p)$ in part (b) can be determined locally by the identity $H_u(x, \lambda, u(x, \lambda, p), p) = 0$. By differentiation we obtain the following partial derivatives in view of assumption (a):

$$u_x = -(H_{uu})^{-1} H_{ux} , \quad u_\lambda = -(H_{uu})^{-1} f_u^* . \quad u_p = -(H_{uu})^{-1} H_{up} . \tag{8}$$

Regularity conditions on boundary arcs:

The following assumption is the counterpart to assumption **(A1)**:

(A2) (a) $C_u(x_0(t), u_0(t), p_0) \neq 0$ for $t_1^0 \leq t \leq t_2^0$.

(b) The equation $C(x, u, p) = 0$ can be solved for a uniquely defined C^1-function $u = u_b(x, p)$ in a neighborhood of $x_0(t), p_0$ for $t_1^0 \leq t \leq t_2^0$.

Simple examples show that condition (b) is stronger than (a). The function $u_b(x, p)$ is called the *boundary control*. By definition we have $u_0(t) = u_b(x_0(t), p_0)$. Since $C(x, u_b(x, p), p) \equiv 0$ differentiation yields in view of (a)

$$\frac{\partial u_b}{\partial x} = -C_u^{-1} C_x \,, \quad \frac{\partial u_b}{\partial p} = -C_u^{-1} C_p \,. \tag{9}$$

Assumption **(A2)** enables us to compute the multiplier function μ for the augmented Hamiltonian $\tilde{H} = H + \mu C$. On the boundary the optimal control satisfies the condition $\tilde{H}_u = H_u + \mu C_u = 0$. In terms of the variables x, λ, p the multiplier μ can then be expressed as

$$\mu(x, \lambda, p) = -H_u(x, \lambda, u_b(x, p), p) / C_u(x, u_b(x, p), p) \,. \tag{10}$$

Joining interior and boundary arcs:

It is easy to see that assumptions **(A1)**, **(A2)** imply the continuity of the control at junction points t_1, t_2. This leads to the condition

$$C(x(t_i), u(x(t_i), \lambda(t_i), p), p) = 0 \,, \quad i = 1, 2 \,, \tag{11}$$

where $u(x, \lambda, p)$ is the minimizing function in **(A1)(b)**. Furthermore it will be required that the unperturbed solution $x_0(t), u_0(t)$ has a *non-tangential* junction with the boundary:

(A3) $\frac{d}{dt} C(x_0(t), u_0(t), p_0)|_{t_i^0} \neq 0 \,, \quad i = 1, 2 \,.$

Here the derivative is understood as derivative from the left at t_1^0 and as derivative from the right at t_2^0. This condition is essential for constructing perturbed extremal solutions. Conditions (11) and **(A3)** imply that the multiplier μ in (10) satisfies

$$\mu(t_i) = 0 \quad (i = 1, 2) \,, \quad \dot{\mu}_0(t_1^0) > 0 \,, \quad \dot{\mu}_0(t_2^0) < 0 \,, \tag{12}$$

where μ_0 denotes the multiplier corresponding to (x_0, u_0).

Under assumptions **(A1)** and **(A2)** the following *parametric* boundary value problem $BVP(p)$ arises for determining the trajectory $x(t)$ and the adjoint function $\lambda(t)$ with one boundary arc in $[t_1, t_2]$:

ODE

$$\dot{x} = \begin{cases} f(x, u(x, \lambda, p), p) & \text{for } t \notin [t_1, t_2] , \\ f(x, u_b(x, p), p) & \text{for } t \in [t_1, t_2] , \end{cases} \tag{13}$$

$$\dot{\lambda} = \begin{cases} -H_x(x, \lambda, u(x, \lambda, p), p) & \text{for } t \notin [t_1, t_2] , \\ -\tilde{H}_x(x, \lambda, \mu(x, \lambda, p), u_b(x, p), p) & \text{for } t \in [t_1, t_2] , \\ \mu(x, \lambda, p) \quad \text{from} \quad (10) . \end{cases} \tag{14}$$

Boundary and junction conditions:

$$x(a) = \varphi(p) , \quad x(b) = \psi(p) , \tag{15}$$

$$\tilde{C}(x(t_i), \lambda(t_i), p) = 0 \quad (i = 1, 2) , \quad \tilde{C}(x, \lambda, p) := C(x, u(x, \lambda, p), p) . \tag{16}$$

The differentiability properties of $u(x, \lambda, p)$ and $u_b(x, p)$ imply that any solution $x(t)$ and $\lambda(t)$ of $BVP(p)$ is a C^1-function on $[a, b]$. It should be noted that, in addition, the sign condition $\mu(t) = \mu(x(t), \lambda(t), p) \geq 0$ for $t_1 \leq t \leq t_2$ must be checked for optimal candidates $x(t)$ and $\lambda(t)$.

The shooting procedure treats the initial value $\lambda(a)$ and the junction points t_1 and t_2 as unknown parameters

$$s = (s_\lambda, t_1, t_2) \in \mathbb{R}^{n+2} , \quad s_\lambda \in \mathbb{R}^n , \quad t_1, t_2 \in \mathbb{R} . \tag{17}$$

Let $x(t, s, p)$ and $\lambda(t, s, p)$ denote the solution of ODEs (13) and (14) with initial conditions

$$x(a, s, p) = \varphi(p) , \quad \lambda(a, s, p) = s_\lambda . \tag{18}$$

Then the solution of $BVP(p)$ is equivalent to solving the n+2 nonlinear equations

$$F(s, p) := \begin{pmatrix} x(b, s, p) - \psi(p) \\ \left(\tilde{C}(x, \lambda, p)|_{(t_i, s, p)} \right)_{i=1,2} \end{pmatrix} = 0 \tag{19}$$

for the shooting parameter s as a function of p near p_0. The function $\tilde{C}$ has been introduced in (16).

Conditions for the regularity of the Jacobian of F at (s_0, p_0) with $s_0 = (\lambda_0(a), t_1^0, t_2^0)$ will be given in the next section.

3 Second-order sufficient solution conditions and solution differentiability

Second order sufficient conditions (SSC) in a weak form have recently been derived in [7], [9], [12]. The notation "weak" refers to the fact that these conditions take into account the boundary of the inequality constraint in contrast to *strong* SSC developed in [6]. We shall follow the presentation in [7] and indicate the connection between SSC and the variational system associated with ODEs (13) and (14).

In the following, all terms with an upper or lower index zero are evaluated at the unperturbed trajectory x_0, u_0, λ_0, p_0. The notation y and η is used for n-vectors or $n \times n$-matrices which can be interpreted as variational quantities associated with x and λ. The variational system for (13) and (14) at $p = p_0$ is composed by $2n$ linear ODEs

$$\dot{y} = A^0(t)y + B^0(t)\eta , \quad \dot{\eta} = W^0(t)y - A^0(t)^*\eta . \tag{20}$$

On *interior arcs* $t \notin [t_1^0, t_2^0]$ the $n \times n$-matrices herein are given by

$$\left. \begin{array}{rcl} A^0(t) & = & \frac{d}{dx}f(x, u(x, \lambda, p), p) = f_x - f_u H_{uu}^{-1} H_{ux} \\ B^0(t) & = & \frac{d}{d\lambda}f(x, u(x, \lambda, p), p) = -f_u H_{uu}^{-1} f_u^* \\ W^0(t) & = & -\frac{d}{dx}H_x(x, \lambda, u(x, \lambda, p), p) = -H_{xx} + H_{xu}H_{uu}^{-1}H_{ux} \end{array} \right\} \tag{21}$$

These expressions make use of the derivatives u_x and u_λ from (8).

On the *boundary arc* $t \in [t_1^0, t_2^0]$ the matrices are

$$\left. \begin{array}{rcl} A^0(t) & = & \frac{d}{dx}f(x, u_b(x, p), p) = f_x - f_u C_u^{-1} C_x \\ B^0(t) & = & \frac{d}{d\lambda}f(x, u_b(x, p), p) = 0 \\ W^0(t) & = & -\frac{d}{dx}\tilde{H}_x(x, \lambda, \mu(x, \lambda, p), u_b(x, p), p) \\ & = & -\tilde{H}_{xx} + \tilde{H}_{xu}C_u^{-1}C_x + C_x^* C_u^{-1}(\tilde{H}_{ux} - \tilde{H}_{uu}C_u^{-1}C_x) \end{array} \right\} \tag{22}$$

These formulae can be found in [10], (58).

For (20) the corresponding Riccati matrix ODE in a symmetric $n \times n$-matrix function $Q(t)$ is

$$\dot{Q} = -QA^0(t) - A^0(t)^*Q - QB^0(t)Q + W^0(t) . \tag{23}$$

Along interior arcs $t \notin [t_1^0, t_2^0]$ this can be rewritten using (21) as

$$\dot{Q} = -Qf_x^0 - (f_x^0)^*Q - H_{xx}^0 + (H_{xu}^0 + Qf_u^0)(H_{uu}^0)^{-1}(H_{ux}^0 + (f_u^0)^*Q) . \tag{24}$$

On the boundary arc $t \in [t_1^0, t_2^0]$ the Riccati ODE (23) reduces to a *linear* ODE

$$\dot{Q} = -QA^0(t) - A^0(t)^*Q + W^0(t) . \tag{25}$$

Hence solutions $Q(t)$ of (23) are understood as piecewise C^1-matrices.

A direct computation reveals that SSC in Theorem 2 of [12] holds if the Riccati ODE (23) has a finite solution $Q(t)$ on $[a, b]$. A direct proof of SSC under more restrictive assumptions has been given in [7].

Theorem 1: (Second-order sufficient conditions)

Let (x_0, u_0) be feasible for $OC(p_0)$. Assume that there exists an absolutely continuous function $\lambda_0 : [a, b] \to I\!\!R^n$ such that

(a) **(A1)** and **(A2)** hold for $p = p_0$,

(b) (x_0, λ_0) is a solution of $BVP(p_0)$ with $\mu_0(t) \geq 0$ for $t_1^0 \leq t \leq t_2^0$ where the multiplier μ_0 is defined by (10),

(c) the Riccati ODE (23) has a finite symmetric solution $Q(t)$ in $[a, b]$.

Then (x_0, u_0) provides a local minimum for $OC(p_0)$. Moreover, $u_0(t)$ is continuous and is a C^1-function for $t \neq t_i^0$ $(i = 1, 2)$ while x_0 and λ_0 are C^1-functions in $[a, b]$.

Now we can state the main result of this paper.

Theorem 2: (Solution differentiability)

Let (x_0, u_0) be feasible for $OC(p_0)$ with the boundary structure (7). Let (x_0, λ_0) be a solution of $BVP(p_0)$ such that the following assumptions hold:

(a) **(A1)** - **(A3)** are satisfied,

(b) the multiplier μ_0 in (10) satisfies the strict complementarity condition $\mu_0(t) > 0$ for $t_1^0 < t < t_2^0$,

(c) the Riccati ODE (23) has a finite symmetric solution $Q(t)$ in $[a, b]$.

(d) the $n \times n$ matrix $y(b)$ is regular when $y(t)$ and $\eta(t)$ are solutions of (20) with initial conditions $y(a) = O_n$, $\eta(a) = I_n$.

Then there exist a neighborhood $V \subset P$ of $p = p_0$ and C^1-functions

$$x, \lambda : [a, b] \times V \to I\!\!R^n, \quad t_i : V \to I\!\!R \quad (i = 1, 2)$$

and a function

$$u : [a, b] \times V \to I\!\!R$$

which is of class C^1 for $t \neq t_i(p)$ $(i = 1, 2)$, such that the following statements hold:

(1) $x(t, p_0) = x_0(t)$, $\lambda(t, p_0) = \lambda_0(t)$ and $u(t, p_0) = u_0(t)$ for $t \in [a, b]$,

(2) the Jacobian of F in (19) at (s_0, p_0) is regular and $x(t, p)$ and $\lambda(t, p)$ solve $BVP(p)$,

(3) the triple $x(\,\cdot\,,p)$, $\lambda(\,\cdot\,,p)$, $u(\,\cdot\,,p)$ satisfies the second-order sufficient conditions in Theorem 1 for every $p \in V$ and hence the pair $x(\,\cdot\,,p), u(\,\cdot\,,p)$ provides a local minimum for $OC(p)$.

The solution differentiability provides a theoretical basis for performing a *sensitivity analysis* where the perturbed solution is approximated by a first order Taylor expansion according to

$$x(t,p) \approx x_0(t) + \frac{\partial x}{\partial p}(t,p_0)(p-p_0)\,, \quad \lambda(t,p) \approx \lambda_0(t) + \frac{\partial \lambda}{\partial p}(t,p_0)(p-p_0)\,.$$

The "variations"

$$y(t) := \frac{\partial x}{\partial p}(t,p_0)\,, \quad \eta(t) := \frac{\partial \lambda}{\partial p}(t,p_0) \tag{26}$$

are $n \times n$-matrices of class C^1. By differentiating the boundary value problem (13) - (16) one obtains a *linear* inhomogeneous BVP for $y(t)$ and $\eta(t)$.

4   An example with a perturbed control constraint

Consider the following *non-convex* control problem with perturbation $p \in I\!R$: minimize

$$\int\limits_0^1 (u^2 - 10x^2)dt \tag{27}$$

subject to

$$\dot{x} = x^2 - u\,, \quad x(0) = 1\,, \quad x(1) = 1\,, \tag{28}$$
$$x + u \le p. \tag{29}$$

By studying the unconstrained problem (27), (28) first, it can be seen that the constraint (29) becomes active for the nominal parameter $p = p_0 = 5.9$. The unperturbed solution (x_0, u_0) has one boundary arc $[t_1^0, t_2^0]$ with $0 < t_1^0 < t_2^0 < 1$. The BVP (13) - (16) is evaluated as

$$(\dot{x}, \dot{\lambda}) = \left\{ \begin{array}{ll} (x^2 - 0.5\lambda, 2x(10 - \lambda)) & ,\ t \notin [t_1, t_2] \\ (x^2 + x - p_0, 2x(10 - \lambda) - \mu) & ,\ t \in [t_1, t_2] \\ \mu = \lambda + 2(x - p_0) & . \end{array} \right\} \tag{30}$$

$$x(0) = 1\,, \quad x(1) = 1\,, \quad x(t_i) + 0.5\lambda(t_i) = p_0 \quad (i = 1,2)\,.$$

The shooting procedure yields

$$\lambda_0(0) = -5.324898490\,, \quad t_1^0 = 0.6735245190\,, \quad t_2^0 = 0.8988553586\,.$$

Assumptions **(A1)** and **(A2)** are trivially satisfied whereas assumption **(A3)** holds with $C(x, u, p) = x + u - p$ and

$$\dot{C}_0(t_1^0) = 1.848039743 , \quad \dot{C}(t_2^0) = -1.673196320 .$$

It can easily be verified that the Riccati equation (23)

$$\dot{Q} = \begin{cases} -4x_0(t)Q + 20 - 2\lambda_0(t) + 0.5Q^2 & , \quad t \notin [t_1^0, t_2^0] \\ (-2 - 4x_0(t))Q + 18 - 2\lambda_0(t) & , \quad t \in [t_1^0, t_2^0] \end{cases}$$

has a bounded solution in $[0, 1]$. Hence all assumptions for Theorem 2 are met and we can embed (x_0, u_0) into a C^1-family $x(t, p), u(t, p)$ of solutions to the perturbed problem (27) - (29). The "variations" $y(t) = \partial x(t, p_0)/\partial p$ and $\eta(t) = \partial \lambda(t, p_0)/\partial p$ in (26) satisfy a linear inhomogeneous BVP which is obtained by differentiating (30). The solution is $y(0) = 0$, $\eta(0) = -0.2717521464$ which can be used to compute

$$\frac{dt_1}{dp}(p_0) = 0.4330244699 , \quad \frac{dt_2}{dp}(p_0) = -0.4869079015 .$$

References

[1] BOCK, H.G., *Zur numerischen Behandlung zustandsbeschränkter Steuerungsprobleme mit Mehrzielmethode und Homotopieverfahren*, ZAMM, Vol. **57**, pp. T 266 - T 268, 1977.

[2] DONTCHEV, A.L., HAGER, W.W., POORE, A.B., and YANG, B., *Optimality, Stability and Convergence in Nonlinear Control*, preprint, June 1992.

[3] ITO, K., and KUNISCH, K., *Sensitivity Analysis of Solutions to Optimization Problems in Hilbert Spaces with Applications to Optimal Control and Estimation*, Journal of Differential Equations, Vol. **99**, pp. 1-40, 1992.

[4] MALANOWSKI, K., *Second-Order Conditions and Constraint Qualifications in Stability and Sensitivity Analysis of Solutions to Optimization Problems in Hilbert Spaces*, Applied Mathematics and Optimization, Vol. **25**, pp. 51-79, 1992.

[5] MALANOWSKI, K., *Regularity of Solutions in Stability Analysis of Optimization and Optimal Control Problems*, Working Papers No. 1/93, Systems Research Institute, Polish Academy of Sciences, 1993.

[6] MAURER, H., *First and Second Order Sufficient Optimality Conditions in Mathematical Programming and Optimal Control*, Mathematical Programming Study, Vol. **14**, pp. 163-177, 1981.

[7] MAURER, H., *The Two-Norm Approach for Second-Order Sufficient Conditions in Mathematical Programming and Optimal Control*, Preprints "Angewandte Mathematik und Informatik" der Universität Münster, Report No. 6/92-N, 1992.

[8] NEUSTADT, L.W., *Optimization: A Theory of Necessary Conditions*, Princeton University Press, Princeton, 1976.

[9] ORRELL, D., and ZEIDAN, V., *Another Jacobi Sufficiency Criterion for Optimal Control with Smooth Constraints*, Journal of Optimization Theory and Applications, Vol. **58**, pp. 283-300, 1988.

[10] PESCH, H.J., *Real-Time Computation of Feedback Controls for Constrained Optimal Control Problems, Part 1: Neighbouring Extremals*, Optimal Control Applications & Methods, Vol. **10**, pp. 129-145, 1989.

[11] PESCH, H.J., *Real-Time Computation of Feedback Controls for Constrained Optimal Control Problems, Part 2: A Correction Method Based on Multiple Shooting*, Optimal Control Applications & Methods, Vol. **10**, pp. 147-171, 1989.

[12] PICKENHAIN, S., *Sufficiency Conditions for Weak Local Minima in Multidimensional Optimal Control Problems with Mixed Control - State Restrictions*, Zeitschrift für Analysis und ihre Anwendungen, 1992.

[13] ZEIDAN, V., *Sufficiency Criteria via Focal Points and via Coupled Points*, SIAM J. Control and Optimization, Vol. **30**, pp. 82-98, 1992.

SPECTRAL IDEMPOTENT ANALYSIS AND
ESTIMATES OF THE BELLMAN FUNCTION

Sergei N. Samborski

Université de Caen, Mathématiques, 14032 Caen Cedex, FRANCE

The idempotent analysis is an analysis of functional spaces and mappings acting on them in the case when the values of functions belong to an idempotent semiring, i.e. to a commutative semiring with operations of "addition" $\oplus$ and "multiplication" $\odot$, where the operation $\oplus$ is assumed to be idempotent : $a \oplus a = a$. The last requirement which replaces the assumption of t invertibility of the addition in the classical functional analysis, permits to obtain unexpectedly deep analogues of constructions and theorems of classical analysis [1]. For instance, it is well known that the spectral analysis of linear operators plays a fundamental role in the study of an asymptotic behavior of their iterations. The same role in the study of iterations of operators, which are "linear" in the sense of the new operations $\oplus$ and $\odot$, is played by the spectral idempotent analysis. Because these "linear" operators arise in optimization problems, results of this type are applicable to estimations of the corresponding Bellman functions or Pareto sets.

1. Endomorphisms of semimodules

Suppose we are given two sets X (the state space) and U (the control space) and a mapping $f : X \times U \longrightarrow X$. The choice of the initial state $x \in X$ and the collection of controls $\{u_1, \ldots, u_n\}$ determines an admissible trajectory $(x_0, x_1, \ldots, x_n)$ in X where $x_i = f(x_{i-1}, u_i)$. Let also a criterion Φ on the set of admissible trajectories be given. Consider two simple examples.

1.1. The criterion Φ with values belonging to $\mathbb{R} \cup (-\infty)$ has the form.

$$\Phi(x_0, \ldots, x_{k-1} ; u_1, \ldots, u_k) = \sum_{i=0}^{k-1} \varphi(x_i, u_{i+1}) \qquad (1)$$

For every $k \in \mathbb{N}$ let $\omega_k(x)$ is the maximal value of the criterion Φ on the set of all admissible k-step trajectories which start from the point x. This is the Bellman function. The following recursive relation is well known

$$\omega_k(x) = (A\, \omega_{k-1})(x) = (A^k\, \omega_0)(x) \qquad (2)$$

where the mapping A given by

$$(A\omega)(x) = \sup_{u \in \mathcal{U}} (\varphi(x,u) + \omega(f(x,u))) \tag{3}$$

Introduce some notations. We shall denote by $\mathcal{R}$ the set $\mathbb{R} \cup (-\infty)$ endowed with a structure of idempotent semiring by means of operations $a \oplus b = \max(a,b)$, $a \odot b = a+b$. The neutral elements with respect to these operations will be denoted by (this is obviously $-\infty$) and by 1 (this is 0). It is convenient to introduce a metric in $\mathcal{R}$, for example the following one : $(a,b) = |\exp a - \exp b|$. The set of continuous functions on X with values in the metric space $\mathcal{R}$ has a natural structure of a semimodule over $\mathcal{R}$ (i.e., we can add (in the sense of $\oplus$) two functions pontwise and multiply (in the sense of $\odot$) functions on scalars (elements of $\mathcal{R}$)). The mapping A of semimodule V into itself is called endomorphism if $A(\lambda \odot f_1 \oplus f_2) = \lambda \odot A(f_1) \oplus A(f_2)$ for any $f_1, f_2 \in V$, $\lambda \in \mathcal{R}$. An endomorphism A is called an "integral endomorphism with the kernel a" if

$$(Af)(x) = \underset{y \in X}{\text{Sup }} (a(x,y) \odot f(y)) \overset{\text{def}}{=} \oint_{y \in X} a(x,y) \odot f(y)$$

where $a : X \times X \longrightarrow \mathcal{R}$.

Return to our control problem. A "change of variable" shows that the mapping A from (3) is an integral endomorphism. Under some assumptions on f and φ (see Theorem 1 below) it is possible to affirm that A is an integral endomorphism of the semimodule $C(X,\mathcal{R})$ of $\mathcal{R}$-valued continuous functions on X.

1.2. Let $\mathcal{R}_*^m = \mathbb{R}^m \cup (-\infty)$. Suppose that there is a partial order $\leq$ in $\mathcal{R}_*^m$ compatible with the componentwise operations $\odot$ and $\odot$ in $\mathcal{R}_*^m$: 1) $\alpha \leq \beta \Rightarrow \alpha \oplus \gamma$ $\forall \alpha,\beta,\gamma \in \mathcal{R}_*^m$; 2) $\alpha \leq \beta \Rightarrow \lambda \odot \alpha \leq \lambda \odot \beta$ $\forall \alpha,\beta \in \mathcal{R}_*^m$, $\lambda \in \mathcal{R}$; 3) $\lambda \leq \mu \Rightarrow \lambda \odot \alpha \leq \mu \odot \alpha$ $\forall \lambda,\mu \in \mathcal{R}$, $\alpha \in \mathcal{R}_*^m$. If α is a subset of $\mathcal{R}_*^m$ then $\max \alpha$ denotes the set of all maximal elements of the closure of α in the topology in $\mathcal{R}_*^m$ induced from $\mathbb{R}^m$.

Suppose now that the criterion Φ in (1) takes its values in $\mathcal{R}_*^m$, this means that the mapping φ acts from $X \times U$ into $\mathcal{R}_*^m$. As usually we the subset $\max(\cup_{\pi_k} \Phi(\pi_k))$ call the Pareto set $\omega_k(x)$ for the criterion Φ on the set of admissible trajectories where π_k runs over all admissible k-step trajectories starting at the point x. It is easy to receive the recursion relation (2) but this time with the mapping A defined by the rule

$$(A\omega)(x) = \max(\bigcup_{u \in U} (\varphi(x,u) + \omega(f(x,u)))) \tag{4}$$

Définition 1. The semiring $P(\mathcal{R}^m, \leq)$ is defined as the set of bounded subsets α in $\mathcal{R}_*^m$ such that $\max \alpha = \alpha$ with the operations $\alpha \oplus \beta = \max(\alpha \cup \beta)$ and $\alpha \odot \beta = \max(\alpha+\beta) = \max\{a+b | a \in \alpha, b \in \beta\}$.

Let $C(X,P(\mathfrak{R}^m,\leq))$ be the semimodule of continuous mappings from a compact metric set X into $P(\mathfrak{R}^m,\leq)$ (with the Hausdorff metric for the closures of elements of $P(\mathfrak{R}^m,\leq)$). Then the mapping A from (4) determines the endomorphism of $C(X,P(\mathfrak{R}^m,\leq))$.

Hence the study of iterations of endomorphisms A from (3) or (4) gives estimates for the Bellman function or for the Pareto sets of the corresponding control problems.

2. Spectral functions.

Theorem 1. *Suppose the kernel* $\quad a: X\times X \longrightarrow \mathfrak{R} \quad$ *of an integral endomorphism* A *is a bounded function uniformly continuous in the first argument and equicontinuous in the second one. Then :* 1) A *is a continuous endomorphism of the semimodule* $C(X,\mathfrak{R})$ *;* 2) *There exist a subsemimodule* $\mathfrak{I}$ *of the semimodule* $C(X,\mathfrak{R})$ $(\mathfrak{I}\neq\mathbb{O})$ *and an element* $\lambda\in\mathfrak{R}$ *such that*

$$(Af)(x) = \lambda\odot f(x)$$

for all $f\in\mathfrak{I}$, *and the maximal element among such* λ *can be obtained by the formula*

$$\lambda = \mathrm{Sup}_{i\in\mathbb{N}}\,\lambda(i)$$

where $\lambda(i) = i^{-1}.\mathrm{Tr}(A^i)$ *and trace* (Tr) *of an integral endomorphism* B *with kernel* b *is determined as*

$$\mathrm{Tr}\,B = \oint_{x\in X} b(x,x).$$

It is known [2] that if an endomorphism A satisfies the condition of Theorem 1 and in addition the kernel a is upper semicontinuous on $X\times X$ then the endomorphism A is compact in $C(X,\mathfrak{R})$.

Suppose that A is a compact endomorphism in $C(X,\mathfrak{R})$. For $\lambda\in\mathfrak{R}\backslash\{\mathbb{O}\}$ (the case $\lambda=\mathbb{O}$ will be considered further on) we shall define the *resolvent* R_λ in the following way :

$$R_\lambda = \bigoplus_{n\geq 0} \lambda^{-n}\odot A^n \;(= \sup_{n\geq 0}\,(-n\lambda+A^n))$$

(there and further $\lambda^m = \lambda\odot...\odot\lambda$ (m times), $\lambda^{-1} = -\lambda$, $\lambda^0 = \mathbb{1}$). The resolvent R_λ is an integral endomorphism with the kernel

$$r_\lambda(x,y) = \bigoplus_{n\geq 0} \lambda^{-n}\odot a^n(x,y),$$

where a^n is the kernel of the endomorphism A^n.

Consider the following equations

$$\lambda\odot y = Ay \oplus f \qquad\qquad (5)$$

$$\lambda \odot y = Ay \qquad (6)$$

$$(y, f \in C(X, \mathfrak{R}_*)).$$

Proposition 1. *The function* y *from* $C(X, \mathfrak{R}_*)$ *is a solution of equation* (5) *that is not identically zero if and only if it can be represented in the following form*

$$y = g \oplus R_\lambda f,$$

where $R_\lambda f$ *is bourded and the function* g *is a solution of equation (5) belonging to* $C(X, \mathfrak{R}_*)$.

It is easy to see that boundedness of the function $R_\lambda f$ depends only on the boundedness of the kernel $r_\lambda(.,.)$ of the resolvent on $X \times Y$, where $Y = \text{Supp } f$. And so the equation (5) is (or is not) simultaneously solvable for all functions f having a common support Y. Thus, for any function from $C(X, \mathfrak{R}_*)$, one of the following possibilities occurs :

a) both equations (5) and (6) are not solvable ;
b) the equation (5) is solvable, and the equation (6) is not solvable ;
c) both equations (5) and (6) are solvable.

The set of elements $\lambda \neq \mathbb{O}$ for which the possibility a) or the possibility b) occurs is an open set. Hence, these possibilities continue to be satisfied under any sufficiently small change of λ. The corresponding values of $\lambda \neq \mathbb{O}$ will be called *regular* elements and the others (which are also nonzero) will be called *spectral elements with respect to* Y. The set of spectral elements is closed in $\mathbb{R} = \mathfrak{R}_* \backslash \{\mathbb{O}\}$.

Definition 2. The union Λ of all spectral elements with respect to all possible supports Y will be called the *spectral set* of A.

Theorem 2. *The spectral set* Λ *of a compact endomorphism* A *consists of eigenvalues of* A *and it is either finite or countable with the unique limit point* $\mathbb{O}$.

We suppose that X is a compact set, A is a compact endomorphism and Ω_λ is the projector (endomorphism of $C(X, \mathfrak{R}_*)$) onto its eigensubsemimodule corresponding to an eigenvalue $\lambda > \mathbb{O}$ from the spectral set of the endomorphism A ($\Omega_\lambda^2 = \Omega_\lambda$, $\Omega_\lambda A = A\Omega_\lambda = \lambda \odot \Omega_\lambda$). The projector Ω_λ is an integral endomorphism with the kernel

$$\omega_\lambda(x,y) = \lim_{n \to \infty} \lambda^{-n} \odot a^n(x;y).$$

Let $x \in X$ and Y be a subset of the set X. Let $\Lambda_Y(x)$ denote the set of eigenvalues of the endomorphism A for which the inequality

$$\mathbb{O} < \oint_{y \in Y} \omega_\lambda(x,y) < \infty$$

is valid. In the case when the equality $\mathbb{O} = \oint_{y \in Y} \omega_\lambda(x,y)$ holds for all $\lambda > \mathbb{O}$, the set $\wedge_Y(x)$ will be assumed to coincide with the one point set $\{\mathbb{O}\}$.

Proposition 2. *For every fixed* $x \in X$, $Y \subseteq X$, *the set* $\wedge_Y(x)$ *is exactly a one-point set.*

Hence the function $\wedge_Y(.) : X \longrightarrow \wedge \cup \{\mathbb{O}\} \subset \mathfrak{R}$ is well defined. It will be called *the spectral function* of the endomorphism A corresponding to the subset Y.

Theorem 3. *Suppose* X *is compactum,* A *is a compact endomorphism of the semimodule* $C(X, \mathfrak{R})$ *and* $\wedge_Y(.)$ *is the spectral function of* A *corresponding to* $Y \subseteq X$. *Then :*

i) *for any continuous function* $f : X \longrightarrow \mathfrak{R}$ *with the support* Y *there exists a constant* $\bar{c} \neq \mathbb{O}$

that for all positive integers n *and for all* $x \in \mathrm{Supp} \wedge_Y(.)$ *the following estimate is satisfied ;*

$$(A^n f)(x) \leq \bar{c} \odot \wedge_Y^n(x)$$

is satisfied ;

ii) *if for a bounded function* $f : X \longrightarrow \mathfrak{R}$ *with the support* Y *there exists a positive integer* S *such that the condition is valid*

$$\mathrm{Supp}\, A^s f = \mathrm{Supp}\, A^{s+1} f$$

holds, then for every $x \in X$ *there exist a number* $\underline{c} > \mathbb{O}$ *such that for all positive integer* n *the following estimate*

$$(A^{s+n} f)(x) \geq \underline{c} \odot \wedge_Y^{s+n}(x)$$

is valid.

3. Estimates of Pareto sets in multicriteria problems

Let G be a closed subgroup of the group $\mathbb{R}^n$ with respect to addition.

Définition 3. The semiring $P(G, \mathfrak{R})$ is defined as the set of bounded functions from G to $\mathfrak{R}$ with the operations

$$(\varphi \oplus \psi)(g) = \varphi(g) \oplus \psi(g), \quad g \in G$$

$$(\varphi \odot \psi)(g) = \oint_{h \in G} \psi(g-h) \odot \psi(h) \stackrel{\mathrm{def}}{=} \sup_{h \in G}(\psi(g-h) \odot \varphi(h)).$$

Let e be a mapping from G to $\mathfrak{R}$ such that

$$\oint_{h \in G} e\,(g\text{-}h) \odot e(h) = e(g).$$

Définition 4. The semiring $P_e\,(G,\mathfrak{R})$ is defined as the set of functions from G to $\mathfrak{R}$ of the form

$$\oint_{h \in G} e\,(g\text{-}h) \odot \psi(h)$$

for all possible $\psi \in P(G,\mathfrak{R})$.

Proposition 3. *The semiring* $P(\mathfrak{R}^m, \leq)$ *(see* Definition 1) *is isomorphic to the semiring* $P_e\,(\mathbb{R}^{m\text{-}1}, \mathfrak{R})$ *for some* $e : \mathbb{R}^{m\text{-}1} \longrightarrow \mathfrak{R}$ *depending on the partial order* $\leq$.

Example. Suppose $n = 2$ and the partial order $\leq$ coincides with the usual componentwise partial order in $\mathbb{R}^2$. We shall take e to be the mapping $x \longrightarrow -|x|$. Then the semiring $P(\mathfrak{R}^2, \leq)$ is isomorphic to the semiring of $\mathfrak{R}$-values functions ψ on $\mathbb{R}^1$ of Lipschitz class with constant 1 satisfying $\operatorname{Sup}_g (\psi(g) + |g|) < \infty$. The isomorphism indicated in proposition 3 can be defined as follows : for a set α we construct the set $\overline{\alpha} = \{a \in \mathbb{R}^2 / \exists b \in \alpha : a \leq b\}$, then rotate it in the plane $\mathbb{R}^2$ by the angle $\pi/4$ and take the rotated set. We get a Lipschitz function.

Let char G be the set of characters of the group G, i.e. the set of such maps χ from G to $\mathfrak{R}$ that $\chi(a+b) = \chi(a) \odot \chi(b)$. We define Ω_e to be subset of char G consisting of elements χ such that $\oint_{g \in G} \chi\,(g) \odot e(g) < \infty$. The set char G is a subset of linear bounded mappings from $\mathbb{R}^n$ into $\mathbb{R}$; thus, char G is a topological vector space with the topology induced from the adjoint space $(\mathbb{R}^n)^*$. It is easy to see that Ω_e is a closed subset containing the zero character in the topological vector space char G over $\mathbb{R}$.

Définition 5. *The Fourier transform* Fa *for each function* $a \in P_e(G, \mathfrak{R})$ *is defined to be the* $\mathfrak{R}$-*valued fonction*

$$(Fa)\,(\chi) = \oint_{g \in G} \chi\,(g) \odot a(g) = \operatorname*{Sup}_{g \in G}\,(\chi(g) + a(g))$$

on Ω_e.

In the case of Example the Fourier transform coincides with the Legendre transform.

Denote by Conc (Ω_e) the semiring of concave $\mathfrak{R}$-valued function on Ω_e with the operations of pointwise maximum ($\oplus$) and pointwise addition ($\odot$).

Proposition 4. *The Fourier transform is a homomorphism from the semiring* P_e $(G, \mathfrak{R})$ *into the semiring* Conc (Ω_e) *with the following properties :*

a) F *is an epimorphism ;*

b) *the mapping given by the formula* $(\mathcal{L} f)$ $(g) = \text{Inf}_{\chi \in \Omega_e} (-\chi(g) + f(\chi))$ *is a semiring monomorphism from* Conc (Ω_e) *into* $P_e(G, \mathfrak{R})$ *and the one-sided inverse of* F ;

c) *in the inverse image* F^{-1} (a) *of each element* a∈ Conc (Ω_e) *as a subset of the partially ordered set* $P_e(G, \mathfrak{R})$, *there exist a maximal element* $\mathcal{L}$ (a) *and a minimal element* $\mathcal{H}$ (a).

Let X be compact and A be an endomorphism of the semimodule $C(X, P_e(G, \mathfrak{R}))$. To this endomorphism we assign a collection

$$\{A_\chi : C(X, \mathfrak{R}) \longrightarrow C(X, \mathfrak{R}) \mid \chi \in \Omega_e\} \tag{7}$$

of endomorphisms defined in the following way. Let a∈ $C(X, P_e(G, \mathfrak{R}))$. For a fixed x the element a(x) is an element of $P_e(G, \mathfrak{R})$ and F(a(x)) (.) is a function of $\chi \in \Omega_e$, we get an element F(a(x)) (χ) that belongs to $\mathfrak{R}$. Thereby for every function a∈ C $(X, P_e(G, \mathfrak{R}))$ a function asssigning to the pair (x, χ) the element F(a(x)) (χ) of $\mathfrak{R}$ is defined and this function will be denoted in the sequel by Ψa. It determines the map ψ.

For an endomorphism A on the semimodule $C(X, P_e(G, \mathfrak{R}))$ there exist a unique endomorphism F(A) such that the following commutation relation $\psi A = F(A)\psi$ holds. Now suppose that the operators belonging to the collection (7) act as follows :

$$A_\chi v (x, \chi) = (F(A)v) (x, \chi).$$

Assumption. We suppose that the endomorphism A is such that the endomorphisms A_χ are compact.

This assumption is not restrictive for endomorphisms connected with multicriteria problems of p.1. According to p.2. for each endomorphism A_χ (for a fixed $\chi \in \Omega_e$) there corresponds the spectral function $\lambda_Y (. \chi)$ connected with a subset $Y \subset X$.

Proposition 5. *The functions* $\chi \longrightarrow \lambda_Y (x, \chi)$ *are concave on* Ω_e *for each fixed* x.

Définition 6. The spectral function of an endomorphism A of the semimodule $C(X, P_e(G, \mathfrak{R}))$ corresponding to a subset $Y \subset X$ is defined as the mapping $\Lambda_Y : X \longrightarrow P_e(G, \mathfrak{R})$ given by the following rule :

$$\Lambda_Y (x) = \mathcal{L} (\lambda_Y (x, .))$$

where $\lambda_Y(., \chi)$ is the spectral function of the endomorphism A_χ acting in $C(X, \mathfrak{R})$ and $\mathcal{L}$ is the inverse Fourier transform that exist due to property b) from Proposition 4.

For each element $f \in P_e(G, \mathfrak{R})$ denote by $\mathcal{M}(f)$ the minimal element $f' \in P_e(G, \mathfrak{R})$ such that $F(f) = F(f')$. From the proposition 4 it follows that $\mathcal{M}(f) = \mathcal{H}(Ff)$ and mapping $\mathcal{M}$ is well defined.

Theorem 4. *Let* A *be an endomorphism of the semimodule* $C(X, P_e(G, \mathfrak{R}))$ *over the semiring* $P_e(G, \mathfrak{R})$, *satisfying assumption,* X *be compactum and the function* $v \in C(X, P_e(G, \mathfrak{R}))$ *satisfy the relation*

$$\text{Supp}(A^s v) = \text{Supp}(A^{s+1} v)$$

for some positive integer s. *Then for every* $x \in X$ *there exist* $\bar{a}, \underline{a} \in P_e(G, \mathcal{R})$ *with* $\underline{a} > \mathbb{O}$ *such that for all positive integer* k *the estimate*

$$\mathcal{M}(\underline{a} \odot (\wedge_Y^K(x)) \le (A^{s+k} v)(x) \le \bar{a} \odot \wedge_Y^K(x)$$

holds, where $\wedge_Y(.)$ *is the spectral function of the endomorphism* A *corresponding to the subset* $Y = \text{Supp } v$, $\le$ *is the partial order in the semiring* $P_e(G, \mathfrak{R})$ *and the mapping* $\mathcal{M}$ *is as defined above.*

Concluding remarks

The existence of "eigenfunctions" for finite X was proved in 1967 by I. Romanovski [3]. For compactum X first results of this type were published in 1987, [4]. In the simplest case when the spectrum of correponding endomorphism consist of one eigenvalue λ and Supp $f = X$ (in this case $\wedge_Y$ (x) $\equiv \lambda$) two-sided estimates for the Bellman function, like them in Theorem 3, were obtained earlier in [5]. Connexion of the analogue of the Fourier transform for semirings with the Legendre transform apperently has been part of the folklore, (see, however, [6]). The theorems 2 and 3 was obtained in collaboration with S. Lesin, the theorem 4 generalizes the corresponding result for finite X which was obtained earlier with A. Tarashchan [7].

REFERENCES

1 . V.P. Maslov and S.N. Samborski (Ed.), Idempotent Analysis, Advances in Soviet Mathematics, Vol. 13, AMS, 1992.

2 . V.N. Kolokoltsov, On Linear, Additive and Homogeneous Operators in Idempotent Analysis, in [1], p. 87-101.

3 . I.V. Romanovskii, Optimization of stationary control of discrete deterministic processes in dynamic programming, Kibernetika (Kiev) 2, 1967, p. 66-78 (English translation in Cybernetics, 3, 1967).

4 . P.I. Dudnikov and S.N. Samborskii, Endormorphisms of semimodules over semirings with an idempotent operation, Preprint N° 87-48, Inst. Mat. Akad. Nauk Ukrain., Kiev, 1987 (Russian).

5 . I.V. Romanovskii, Asymptotic behavior of a discrete deterministic process with continuous state space. Optimal Planirovaniye, 8. "Nauka", Novosibirsk, 1967, p. 171-193 (Russian).

6 . V.P. Maslov, On a new superposition principle for optimization problems, Russian Math. Surveyes, 42, 1987.

7 . S.N. Samborskii and A. Tarashchan, On semirings arising in multicriteria optimization problems and problems of analysis of Computational media. Soviet Math. Doklady, 40, N° 2, 1990.

BOUNDARY VALUE PROBLEMS FOR STATIONARY
HAMILTON-JACOBI AND BELLMAN EQUATIONS

Victor P. Maslov

Institute of Problems in Mechanics, Russian Academy of Sciences,
pr. Vernadskogo 101, 117526, Moscow, RUSSIA

Sergei N. Samborski

Université de Caen, Mathématiques, 14032 Caen Cedex, FRANCE

There is a well known connection between the classical variational calculus and problems of optimal control synthesis : the Hamilton-Jacobi equation (HJE) corresponds to the Bellman equations (BE). Meanwhile in problems of Control Theory there are peculiarities arising from the essence of initial problems : the solutions of HJE or BE are not usual differentiable functions, they may take infinite values and they may be discontinuous.

This produces the following question : *in what sense the solutions of* HJE *and* BE *schould be understood ?*

Under certain conditions the classical theory expresses an extremely fruitful connection between HJE and the corresponding Hamilton system. But in control problems the Hamiltonians are usually nondifferentiable. The natural approach is to approximate such Hamiltonians by the differentiable ones with the help of convergence which ensures a convergence of solutions of the corresponding Hamilton systems to the optimal trajectories. In this way we come to the following question : *when is the approximation of Hamiltonians admissible ?*

Let us consider the following simplest example, which is a standard problem of the dynamic programming :

$$\begin{cases} \Phi(q(.), u(.)) = \int_0^t L(q(\tau), u(\tau)) \, d\tau + g(q(t)) \to \mathrm{Inf} \\[2ex] \dot{q} = f(q(\tau), u(\tau)), q(0) = x, q(t) \in \Gamma \subseteq \partial\Omega, \\[2ex] q(\tau) \in \Omega \subseteq \mathbb{R}^n, u(\tau) \in \mathcal{U} \subseteq \mathbb{R}^m \quad \forall \tau \in [0, t] \end{cases} \qquad (1)$$

where $u(.)$ is a control, $q(.)$ is a desired trajectory, t is not fixed, $\partial\Omega$ is the boundary of Ω.

The Hamiltonian of this problem is of the form

$$H(q, p) = \mathrm{Inf}_{u \in \mathcal{U}} \, \mathcal{H}(q, p, u)$$

where

$$\mathcal{H}(q, p, u) = p.f(q, u) + L(q, u)$$

are the Hamiltonians corresponding to a fixed value of the control u. The corresponding BE has the form

$$H(q, DS(q)) = 0 \qquad (2)$$

with the boundary value condition g on Γ (under certain conditions for g) the goal of this equation is to determine the function S ~~if assigns~~ such that ~~to~~ for every x ∈ Ω a value S (x) equal either to the infimum of values of the functional Φ on admissible trajectories starting from x or to ∞ in the case when such trajectories do not exist (the Bellman function, an analogue of the "action" function in Mechanics). Thus the desired function always exists, but its connection with the equation (2) is not obvious in the points of nondifferentiability.

Assuming that we know answers to the posed questions we can write the following scheme for solving the control problem :

Initial Control Problem = Family of Hamiltonians $\mathcal{H}(. , . , u)$

Inf ↓

Hamiltonian H (. , .)

↓

(Approximating by smooth Hamiltonians)

↓

Hamilton system

solving ↓

Bellman function (or its approximations) = solution of BE

However there exists another possibility to find the optimal trajectories and Bellman function without answering to the above mentioned questions. This is the Pontryagin maximum principle which is widely used. Namely, instead of the Hamilton system corresponding to H we can consider the family of Hamilton systems that corresponds to the family {$\mathcal{H}(. , . , u$} (u∈ 𝒰). Only after that we take Inf, i.e. the family is supplemented with a "selection rule" at every moment. This rule states that if a solution from the family corresponds infinitisimally to the optimal trajectory then the corresponding value of the parameter u* satisfies the relation

$$H(q, p) = Inf_{u∈𝒰} \mathcal{H}(q, p, u) = \mathcal{H}(q, p, u^*) \qquad (3)$$

This scheme can be expressed as follows :

Initial Control Problem = Family of Hamiltonians $\mathcal{H}(. , . , u)$

↓

Family of Hamilton systems

Inf ↓

Family of Hamilton systems and Relation (3)

solving ↓

Bellman function, optimal trajectories.

We have given the two schemes to emphasize that both suggest the same actions but in a different order. In the first scheme we pass to Inf and then to the Hamilton system ; in the second scheme this is done in reverse order.

The obvious "advantage" of the Pontryagin scheme is using classical concepts of solution and differentiability, whereas the first scheme requires, in addition, a definition or a generalization of these concepts.

The obvious disadvantage of this scheme lies in the fact that the basic object (i.e., the family of ordinary differential equations together with relations containing the operations min or max) can hardly be considered a natural mathematical object because it combines in itself depressingly different mathematical structures. This produces inconveniences and artificial complications both on the theoretical level and when organizing calculations.

For this reason the computing practise often uses the algorithms which essentially implement the more advantageous first scheme, refering, however to the maximum principle as to a "theoretical basis". Why the transposition of two basic actions (taking Inf and passing to the Hamilton system) does not change the result ? The answer is connected with the special role that the operation Inf (or Min) plays for HJE and BE. To explain this role we shall fix the Hamiltonian H (while not fixing the set Γ and the terminal g) and consider the set of "solutions" of the equation (2), i.e., the set of Bellman functions S of the corresponding control problems. Then for any two solutions S_1, S_2 and a constant λ the functions (min (S_1, S_2) and $\lambda + S_1$ are again solutions. Moreover if we fix Γ then the map $\xi : g \to S$ has the following property : ξ (min (g_1, g_2)) = min (ξ (g_1), ξ (g_2)).

Hence the new "linear" operations $\oplus$ = min and $\odot$ = + play the same role for HJE as the usual + and . for linear equations. More precisely : the set of solutions of HJE forms a functional semimodule over the semiring $\mathbb{R} \cup \infty$ with the operations ($\oplus$, $\odot$).

This observation suggests also a method of construction of the spaces such that the convergence hese spaces (together with the corresponding notion of differentiability) permits us to implement the above first scheme of constructing optimal trajectories and the Bellman function and to give a basis to often used calculation schemes without any reference to the maximum principle. This also allows us to include the problems of optimal control into the classical range of questions of Hamilton and Jacobi Theory but in the new spaces adapted to the algebraic structure of the investigated equations. The construction of these new spaces follows the well known S.L. Sobolev scheme (which is so fruitful in linear problems), but on the basic of the new "linear" structure ($\oplus$, $\odot$). The elements of obtained spaces are on the same level of abstraction as the elements of the spaces L^2 or W^1. The use of the construction which completely corresponds to the integral in the space L^2 permits us to introduce a distinguished system of generators (we call them "basic solutions") in the set of solutions of a fixed HJE and to write arbitrary solutions in terms of boundary value conditions by means of an "integral" over these "basic solutions" (an analogue of the Green operator).

Surely, the questions formulated above have a long history and there is a vast bibliography suggesting different approches. We will not discuss the advantages and disadvantages of these approaches. Note however that our main purpose is a construction allowing to include the considered range of questions into a common context of fundamental ideas of Hamilton and Jacobi Theory. We feel that the idea of transfering the structural load to the spaces, where the corresponding differential operators act, is the most promising for this purpose. At the same time this idea is developed least of all (see, however, [1 - 2].

1. The spaces $\mathcal{P}(\Omega)$.

Let the $\mathcal{R} = \mathbb{R} \cup (\infty)$ be endowed with a structure of an idempotent semiring, i.e. with the operations $\oplus$, $\odot : \mathcal{R} \times \mathcal{R} \to \mathcal{R}$ defined by the rules $a \oplus b = \min (a\ b)$, $a \odot b = a + b$. The neutral elements with respect to the operations $\oplus$ and $\odot$ will be denoted by $\mathbb{O}$ (this is obviously, ∞) and by $\mathbf{1}$ (this is 0). In the set $\mathcal{R}$ it is convenient to introduce a metric compatible with the algebraic structure, for example, the following one :

$$\rho (a, b) = |\ \exp\text{-}a - \exp\text{-}b\ |.$$

Suppose that Ω is an open domain in $\mathbb{R}^n$, $\overline{\Omega}$ is its closure and $\partial\Omega$ is its boundary. Introduce the set $\Phi (\Omega)$ (of "test functions") consisting of continuous $\mathcal{R}$-valued functions determined on $\overline{\Omega}$ that are differentiable at the points of Ω, where their values are finite (i.e., differ from $\mathbb{O}$).

Definition 1. In the set $\mathcal{B}(\Omega, \mathcal{R})$ of lower semi bounded mappings from $\overline{\Omega}$ in $\mathcal{R}$ determine the following "scalar product"

$$(f, g) = \mathrm{Inf}_{x \in \Omega} (f(x) + g(x)) \overset{\text{def}}{=} \oint_{\Omega} f \odot g$$

and the following equivalence relation :

$$f \sim g \quad \text{if and only if} \quad (f, \varphi) = (g, \varphi) \text{ for any } \varphi \in \Phi(\Omega).$$

Definition 2. Introduce the following weak convergence :

$$f_i \to f \text{ if } \rho ((f_i, \varphi), (f, \varphi)) \to 0 \text{ for any } \varphi \in \Phi(\Omega)$$

in the set of equivalence classes. Let $\mathcal{P}(\Omega)$ denote the obtained space.

$\mathcal{P}(\Omega)$ has a natural structure of partially ordered semimodule over the semiring $\mathcal{R}$: for any $f, g \in \mathcal{P}(\Omega), \lambda \in \mathcal{R}$: we have $\lambda \odot f \oplus g \in \mathcal{P}(\Omega)$ and $f \geq g$ if $(f, \varphi) \geq (g, \varphi) \ \forall \ \varphi \in \Phi(\Omega)$.

Let $\Gamma_1, \ldots, \Gamma_m$ ($m \in \mathbb{N}$) be pair wise non-intersecting differentiable submanifolds in $\mathbb{R}^n$ and the set Γ be the union of their closures : for example, let Γ be the boundary of a domain with corners in $\mathbb{R}^n$. We choose the set of $\mathcal{R}$-valued continuous functions on Γ that are differentiable at the points where their values are finite to be the space $\Phi(\Gamma)$ of "test functions". Then one can determine the space $\mathcal{P}(\Gamma)$ in the same manner as above.

460

2. $\mathcal{P}(\Omega)$-solutions of differential equations.

Suppose that $\mathcal{H}$ is a differential expression of the form

$$(\mathcal{H} y)(x) = - H(x, Dy(x)),\qquad (4)$$

where y is a differentiable function on $\overline{\Omega}$ and Dy is its gradient, H (Hamiltonian) acts from $\overline{\Omega} \times \mathbb{R}^n$ into $\mathbb{R}$ continuously.

Definition 3. An element $y \in \mathcal{P}(\Omega)$ is called a weak solution of the equation $\mathcal{H}(y) = F$, where $F \in \mathcal{P}(\Omega)$, equal to $g \in \mathcal{P}(\Gamma)$ on $\Gamma \subset \partial \Omega$, if there exists a sequence of functions $\{y\}_{i=1}^{\infty}$ differentiable on Ω such that the restrictions of y_i to Γ weakly converge to g, y_i weakly converge to y, and $\mathcal{H}(y_i)$ weakly converge to F. An element $y \in \mathcal{P}(\Omega)$ will be called a $\mathcal{P}(\Omega)$-solution (or, simply, a solution) of the equation $\mathcal{H}(y) = F$ equal to g on $\Gamma \subseteq \partial \Omega$ if y is a weak solution equal to g on Γ and if for any other weak solution $\overline{y}$ equal to g on Γ the inequality $y \geq \overline{y}$ holds.

Examples 1. Let $\overline{\Omega} = [-1, 1] \subset \mathbb{R}$ and $\mathcal{H}(y) = -dy/dx = 0$. The solution of this equation equal to a in 1 is a function f equal to a in 1 and $\mathbb{O}(= \infty)$ in $[-1, 1[$. The solution equal to b in -1 is a function $x \to b$ ($x \in [-1, 1]$) and a solution equal to a in 1 and b in -1 exists if only $a \leq b$ and it is the function equal to a in 1 and to b for $x \in [-1, 1[$.

2. Let $\overline{\Omega} = [0, 1] \times [-1, 1] \subset \mathbb{R}^2$, $\mathcal{H}(y) = -\partial y/\partial x_1 = 0$. The solution equal to 0 in (0, 0) is the function equal to 0 for $x_2 = 0$ and $\mathbb{O}$ for $x_2 \neq 0$.

3. Let $\overline{\Omega} = [-1, 1]$. The solution of the equation $-(dy/dx)^2 + 1 = 0$ equal to 0 in the points -1 and 1 is the function $x \to - |x| + 1$.

Condition A. The differential operator $\mathcal{H}$ of type (4) (or Hamiltonian H) satisfies this condition if for any $x \in \Omega$ the set

$$\Lambda_x = \{ p \in \mathbb{R}^n \mid H(x, p) \leq 0 \}$$

is convex.

Theorem 1. *Suppose that the differential operator $\mathcal{H}$ of type (4) satisfies condition A. Then the set of all solutions of the equation $\mathcal{H}(y) = 0$ is a closed subsemimodule of the semimodule $\mathcal{P}(\Omega)$.*

3. Basic solutions

Let A be a set and $\{f_\alpha \mid \alpha \in A\}$ be a family of elements (equivalence classes) from $\mathcal{P}(\Omega)$. This family is said to be integrable, if there exists an element from $\mathcal{P}(\Omega)$, denoted by $\oint_{\alpha \in A} f_\alpha$ in the sequel, such that for any function $\varphi \in \Phi(\Omega)$ the equality

$$(\oint f_\alpha, \varphi) = \operatorname{Inf}_{\alpha \in A}(f_\alpha, \varphi)$$

holds $((.\,,.)$ is the scalar product in the sense of definition 1).

Definition 4. The $\mathcal{P}(\Omega)$-solution of the equation $\mathcal{H}(y) = 0$ equal to $\mathbf{1}$ $(= 0)$ in $x \in \partial\Omega$ is called a basic solution corresponding to x.

Condition B. For any $x \in \partial\Omega$ there exists a weak solution of the inequality $\mathcal{H}(y) \geq 0$ equal to $\mathbf{1}$ in x.

Theorem 2. *Suppose that the differential operator $\mathcal{H}$ of type* (4) *satisfies the conditions A and B. Then :*

i)$\forall\ x \in \partial\Omega$ *there exists a basic solution* W $(.\,,z)$ *corresponding to* z ;

ii) *if* $\Gamma \subseteq \partial\Omega$ *and* y *is a solution of the equation* $\mathcal{H}y = 0$ *equal to* g *on* Γ, *then*

$$y(x) = \oint_{z \in \Gamma} W(x, z) \odot g(z) \tag{5}$$

in particular , the "compatibility condition"

$$g(x) - g(z) \leq W(x, z)$$

holds ;

iii) *if* $\Gamma \subseteq \partial\Omega$ *and* g *is a function on* Γ *satisfying the "compatibility condition" then the formula* (5) *gives the solution of the equation* $\mathcal{H}y = 0$ *equal to* g *on* Γ.

Example. Let $\Omega = [-1, 1]$, $\mathcal{H}(y) = -(dy/dx)^2 + 1 = 0$. The basic solutions are the following : W $(x, -1) = x + 1$; W $(x, 1) = -x + 1$. The solution of $\mathcal{H}(y) = 0$ with boundary conditions y $(-1) = y(1) = 0$ is the function $x \to -\,|x|+1 = \min(W(x, -1), W(x, 1))$.

4. Approximations

Theorem 3. Let $\{H_i : \overline{\Omega} \times \mathbb{R}^n \to \mathbb{R}\ |\ i = 1, 2,...\}$ *be a sequence of continuous Hamiltonians converging to a Hamiltonian* H $: \overline{\Omega} \times \mathbb{R}^n \to \mathbb{R}$ *uniformly on* $\overline{\Omega} \times \mathbb{R}^n$, *for any* $i \in \mathbb{N}$ *the Hamiltonian* H_i *satisfy the Condition A and* $H_i \leq H$. *Let* $z \in \partial\Omega$, H *satisfies the Condition B and let* $W_i(.\,,z)$, W $(.\,,z)$ *be the basic solutions of the equations* $\mathcal{H}_i(y) = 0$, $\mathcal{H}(y) = 0$ *corresponding to* z *(they exist according to theorem* 2). *Then the solution* W $(.\,,z)$ *is limit in* $\mathcal{P}(\Omega)$ *of the solutions* $W_i(.\,,z)$ *when* $i \to \infty$.

5. Formula for Basic Solutions.

Because of theorems 2 and 3 it is sufficient to consider only basic solutions for only smooth Hamiltonians. Let H be a twice differentiable Hamiltonian in the domain Ω with a smooth boundary. Consider the Hamilton system of 2n ordinary differential equation in $\Omega \times \mathbb{R}^n$ associate to the Hamiltonian H :

$$\dot{p} = - H_q (q , p) , \dot{q} = H_p(q , p) \tag{6}$$

Theorem 4. *Let* H *be a twice differentiable Hamiltonian in a bounded domain* $\Omega \subset \mathbb{R}^n$ *with a differentiable boundary, let the differential operator* $\mathcal{H}$ *of the type* (4) *satisfy the Condition* A *and* B *and the function* $(q , p) \to p$. *Let* $\partial H/\partial p$ - H (q , p) *be lower semibounded. Consider*

$$W (x , z) = \mathrm{Inf} \left(\int_0^t p (\tau, p_0, z) \frac{\partial H}{\partial p} (p (\tau, p_0, z), q (\tau, p_0, z)) \, d\tau \right) \tag{7}$$

where Inf *is taken over all* $t \geq 0$ *and over all trajectories* $(q (. , p_0, z), p (. , p_0, z))$ *of the system* (6) *with the initial data* $p (0) = p_0, q (0) = z$ *satisfying the condition* H $(z, p_0) = 0$ *and* $q (t, p_0, Z) = x$; W $(x, z) = \Phi$ *if such trajectories do not exist.*

Then the equivalence class in $\mathcal{P} (\Omega)$ *determined by* W $(. , z)$ *is a basic solution of the equation* $\mathcal{H}$ $(y) = 0$ *corresponding to* $z \in \partial \Omega$.

6. Bellman function.

The theorems 2 - 4 are applicable to the finding of the Bellman function for the problem (1), where $\overline{\Omega}$ is a compactum with a smooth boundary and $\mathcal{U}$ is a convex set. Suppose that admissible control functions u (.) are piece-wise continuous and that the function L is lower semibounded.

Proposition 1. *The Bellman function of the problem* (1) *coincides* (*in* $\mathcal{P} (\Omega)$) *with the element given by the formula*

$$\oint_{z \in \Gamma} W (. , z) \odot g (z),$$

where W $(. , z)$ *is the basic solution corresponding to* $z \in \Gamma$. *This basic solution* W $(. , z)$ *is a limit in* $\mathcal{P} (\Omega)$ *of the basic solutions* $W_i (. , z)$ *corresponding to smooth Hamiltonians* H_i *which approximate the Hamiltonian* H $(q , p) = \mathrm{Sup}_{u \in \mathcal{U}} (p . f (q , u) - L (q , u))$ *in the sense of the theorem 3) . The basic solution* $W_i (. , z)$ *may be obtained by the formula* (7).

7. Spaces $\mathcal{P}^1 (\Omega)$

In any equivalence class $f \in \mathcal{P} (\Omega)$ let $\overline{f}$ denote the representative given by the formula

$$\bar{f}(x) = \underset{\varphi \in f}{\text{Inf}} \ \varphi(x)$$

(it is the unique lower semicontinuous function in f). If $\bar{f}$ is bounded from above then the equivalence class $-f$ be defined uniquely for f as the class which contains the function $-\bar{f}$.

Definition 5. The space $\mathcal{P}_*(\Omega)$ of scalar functions is defined as a subsemimodule of the semimodule $\mathcal{P}(\Omega)$ consisting of such $f \in \mathcal{P}(\Omega)$ that $-f$ exists and the equality $f = -(-f)$ holds ; the convergence is weak.

Let ξ, η be vector fields on Ω. They will be called equivalent if for any smooth local coordinates the components ξ_i and η_i are equivalent for each $i = 1,..., n$. If the components belong to $\mathcal{P}_*(\Omega)$ we say that the equivalence class belongs to the space $\mathcal{P}_*(T\Omega)$ of vector fields on Ω. The convergence in $\mathcal{P}_*(T\Omega)$ is defined as the componentwise weak convergence.

Example. The sequence $f_n = (1/n) \sin nx$ weakly converge to 0. The sequence of vector fields Df_n (gradients) does not converge although in any coordinates the components converge (but the limit object is not a vector field from $\mathcal{P}_*(T\mathbb{R})$).

Definition 6. A sequence of function f_i (differentiable on Ω) $\mathcal{P}^1$- converges to the equivalence class $f \in \mathcal{P}_*(\Omega)$ if f is a weak limit of f_i and the vectors fields Df_i weakly converge to an element of $\mathcal{P}_*(T\Omega)$.

Proposition 2. *If for two $\mathcal{P}^1$- convergent sequences $\{f_i\}$ and $\{g_i\}$ we have* $\lim f_i = \lim g_i = f \in \mathcal{P}$ (Ω) *then* $\lim Df_i = \lim Dg_i$, *i.e. the weak limit of the gradients depends only on f but not on the sequence which approximates f.*

Definition 7. A function f (equivalence class) belongs to the space $\mathcal{P}^1(\Omega)$ if it is a $\mathcal{P}^1$ limit of a sequence of differentiable functions. The convergence is determined by $\mathcal{P}^1$ convergence in the sense of definition 6.

By virtue of proposition 2 for functions (equivalence classes) from the space $\mathcal{P}^1(\Omega)$ the vector field, called the gradient Df of the function f, is uniquely defined (as a class of equivalent vector fields).

It is easy to see that the defined gradient generalizes the usual one, that many nondifferentiable functions ($x \to |x|$, for example) are differentiable in the new sense and that $\mathcal{P}^1$ is a semimodule over the semiring $\mathcal{R}$.

The definition of the weak convergence in the space $\mathcal{P}(\Omega)$ and consequently, in the spaces $\mathcal{P}_*(\Omega)$, $\mathcal{P}^1(\Omega)$ depends on the choice of the set Φ of test-functions (Definition 1-2). Change now this set as follows. Let $\Phi_0(\Omega)$ be the subset of the set $\Phi(\Omega)$ consisting of functions tending to $\mathbb{O}$ (i.e., ∞) as the argument tends to the boundary $\partial\Omega$ of the domain Ω.

Let $\mathcal{P}_0(\Omega)$, $\mathcal{P}^1{}_0(\Omega)$ denote the spaces introduced by the definitions 2, 7 but with the weak convergence defined by the set of test-functions $\Phi_0(\Omega)$.

Proposition 3. *There exists a subsemimodule in $\mathcal{P}^1{}_0(\Omega)$ (denote it $\mathcal{A}$) maximal with respect to* *inclusion, containing all differentiable functions, in which the operation* $\oplus$ *is continuous in the topology induced from* $\mathcal{P}^1{}_0$ (Ω) *(it means* : $f_i \to f \in \mathcal{A}$, $g_i \to g \in \mathcal{A} \Rightarrow f_i \oplus g_i \to f \oplus g$).

Example. The function $x \to |x|$ does not belong to $\mathcal{A}$, but the function $x \to -|x|$ belongs to $\mathcal{A}$.

8. $\mathcal{P}^1$ solutions.

Return to the differential operator $\mathcal{H}$ of type (4). The construction of the generalized gradient D and of the spaces $\mathcal{P}^0$, $\mathcal{P}^1$ gives us a possibility to determine solutions of the equation $\mathcal{H}$ $(y) = 0$ in a very natural way : they are functions $y \in \mathcal{P}^1$ (Ω) such that the equality - H $(x, Dy(x)) = 0$ holds in the sense of elements of $\mathcal{P}^0$.

Definition 8. The solution $y \in \mathcal{A}$ of the equation $\mathcal{H}$ $(y) = 0$ with the boundary value condition g on $\partial\Omega$ is called stable if there exists a $\lambda_0 < 0$ such that for every $\lambda \in [\lambda_0, 0]$ there exists a solution $y_\lambda \in \mathcal{A}$ of the equation $\mathcal{H}$ $(y) = \lambda$ with the boundary value condition g on $\partial\Omega$ such that we have $y_0 = \lim\limits_{\lambda \to 0, \lambda < 0} y_\lambda$ in

the metric of C (Ω)

Theorem 5. *Suppose that* Ω *is bounded and* $H : \overline{\Omega} \times \mathbb{R}^n \to \mathbb{R}$ *is a continuous mapping. Let the following conditions be satisfied :*

i) *the mapping* $\mathcal{H} : \mathcal{P}^1$ $(\Omega) \to \mathcal{P}^0$ (Ω) *is continuous* ;

ii) *the set* $\Lambda_{x,\lambda} = \{p \in \mathbb{R}^n \,|\, H(x, p) \leq \lambda\}$ *is compact and strictly convex for all* $x \in \Omega$ *and all* λ *from some interval* $[0, \lambda_0]$ *where* $\lambda_0 > 0$.

iii) *there exists a solution* $\overline{y} \in \mathcal{P}^1$ *of the inequality* $\mathcal{H}$ $(y) \geq 0$ *with the boundary value condition*

g *on* $\partial\Omega$.

Then the equation $\mathcal{H}$ $(y) = 0$ *has a stable solution* $y \in \mathcal{A}$ *with the boundary value condition* g *on* $\partial\Omega$ *which is the unique stable solution from* $\mathcal{A}$ *equal to* g *on* $\partial\Omega$.

Remark. The conditions of Theorem 5 imply that the condition A and B hold. The obtained $\mathcal{P}^1$ solution is a $\mathcal{P}^0$ solution in the sense of Definition 3. Note that the convexity of the sets $\Lambda_{x,\lambda}$ follows from the condition i). The strict convexity of these sets in ii) implies the existence of $\mathcal{A}$-solutions in Ω for very general boundary value conditions on $\partial\Omega$. The condition ii) may be weakened for $g \in \mathcal{A}$ (Γ).

REFERENCES

[1]   V. P. **Maslov** and S.N. **Samborskii** (Ed.). *Idempotent Analysis*, Advances in Soviet

Mathematics, vol. 13, AMS, 1992.

[2]   V.N. Kolokoltsov and V. P. Maslov, *Idempotent calculus as the apparatus of optimization theory*, Functional Anal. Appl. 23, 1989.

OPTIMAL, PIECEWISE CONSTANT CONTROL
WITH BOUNDED NUMBER OF DISCONTINUITIES

Jacek Kabziński

Institute of Automatic Control, Technical University

PL 90-924 Łódź, ul. Stefanowskiego 18/22, POLAND

1. Introduction

In many practical control problems the optimal control has a bang-bang structure. The number of discontinuities associated with the optimal policy often determines the possibility of control's technical application. In many practical control mechanisms the control variable has only a finite number of settings and there exists an upper bound on a number of switchings.

One of the problems from practical point of view is a 'chattering control' phenomenon: the optimal policy consists in switching the control variable indefinitely rapidly, so as to simulate an intermediate setting.

If we apply a computer to calculate optimal control 'on line', then the resulting control is piecewise constant and the number of discontinuities is upper bounded because of necessary computation time. If the plant is a fast one this limitation is important.

Piecewise constant control over N sampling intervals of equal length appears if we apply parameterization schemes to convert optimal control problems into finite-dimensional optimization problems. Under this approach a nonlinear programming problem is solved and the number of unknowns depends on the number of control discontinuities.

Because of the above mentioned situation this paper is devoted to the control problems with a bound imposed on a number of optimal control discontinuities. The purpose of the paper is to investigate these problems as optimal control problems - not as nonlinear programming problems.

First we formulate the problem precisely and we address the question of existence of the optimal control.

Then we discuss necessary conditions for optimality. Pontryagin maximum principle is not applicable to our problem. In spite of this we are able to formulate the necessary conditions in style of the maximum principle.

2. Problem formulation

We consider the control system governed by the equation

$$\dot{x}(t) = f(t, x(t), u(t)),$$ (1)

where $x(t) \in \mathbf{R}^n$ denotes the state vector and $u(t) \in \mathbf{R}^m$ - the control vector. We are to minimize the performance index

$$J(u(\cdot), x(\cdot)) = \int_{T_{0u}}^{T_{1u}} f_0(t, x(t), u(t)) dt + g(x(T_{0u}, T_{0u}, x(T_{1u}), T_{1u}),$$ (2)

where $f_0(\cdot, \cdot, \cdot)$ and $g(\cdot, \cdot, \cdot, \cdot)$ are given scalar functions, subject to the following constraints:

i)   the equation (1) is satisfied a.e. in $[T_{0u}, T_{1u}]$,

ii)   boundary conditions are given by

$$h_0(T_{0u}, x(T_{0u})) = 0, \qquad h_1(T_{1u}, x(T_{1u})) = 0,$$ (3)

where

$$h_0: \mathbf{R} \times \mathbf{R}^n \rightarrow \mathbf{R}^{s_0}, \qquad h_1: \mathbf{R} \times \mathbf{R}^n \rightarrow \mathbf{R}^{s_1}$$ (4)

are given functions,

iii)   the control function belongs to the set $\mathbf{U}$ of admissible controls satisfying following conditions:

- every $u(\cdot) \in \mathbf{U}$ is defined, piecewise constant, left hand continuous on a bounded interval $[T_{0u}, T_{1u}]$,

- every $u(\cdot) \in \mathbf{U}$ is allowed to be discontinuous at any point from a given set $TD = \{t_{d1}, \ldots, t_{d\rho}\}$ and to posses no more then r other points of discontinuity in $(T_{0u}, T_{1u}]$, (it means that every $u(\cdot) \in \mathbf{U}$ has no more then $\underline{r} = r + \rho + 1$ continuity intervals in $(T_{0u}, T_{1u}]$),

- for every $u(\cdot) \in \mathbf{U}$, $u(t) \in U_i \subset \mathbf{R}_m$ for every t from the i-th continuity interval of $u(\cdot)$.

Let us assume that the functions $f(\cdot, \cdot, \cdot)$ and $f_0(\cdot, \cdot, \cdot)$ are continuous with respect to all variables and continuously differentiable in x and t and $h_0(\cdot, \cdot)$, $h_1(\cdot, \cdot)$, $g(\cdot, \cdot, \cdot, \cdot)$ are continuously differentiable.

The above problem formulation allows to consider as special cases minimization over switching points with given control values and minimization over control values with given switching points.

The set of all piecewise constant, bounded functions i.e. the minimal set of admissible controls according to Pontryagin theory is not contained in $\mathbf{U}$. Therefore the maximum principle cannot be applied to the above problem.

The other way to obtain solution with bounded number of discontinuities is to penalize control switchings in the performance index. This approach was investigated by J.M. Blatt [1] where an elementary version of necessary conditions called 'indifference principle' was derived.

3. Existence of optimal control

Every standard proof [2] of optimal control existence theorem consists of following steps:

1. Show, that the infimum of the performance index is finite and select minimizing sequence of controls and trajectories.

2. Demonstrate that Arzeli-Ascoli theorem implies that trajectories from the minimizing sequence are uniformly convergent and the limit trajectory satisfies boundary conditions.

3. Using convexity conditions and some kind of lemma on measurable selections prove that the limit trajectory is generated by an admissible control.

In the problem considered in this paper every control $u(\cdot) \in U$ is determined by a finite set of points of discontinuity and constant control values. This fact implies a general existence theorem without any convexity conditions. We are able to give the set U of admissible controls a topology of a compact metric space. So the sequence of minimizing controls is convergent to an admissible control. Under standard assumptions corresponding trajectories are uniformly convergent and the performance index is continuous functional over compact domain. So the existence of optimal control follows by the Weierstrass theorem.

4. Minimization over switching times

Let us consider a special case of the problem defined in section 2, where the set T_D is empty, $\rho = 0$ and the sets U_i are reduced to single points in $\mathbf{R}^m$:

$$U_i = u_i \in \mathbf{R}^m. \tag{5}$$

It means that the optimal policy consists in choosing switchings from one control value to the other.

Let us introduce the function

$$H: \mathbf{R} \times \mathbf{R}^n \times \mathbf{R}^n \times \mathbf{R}^m \times \mathbf{R} \to \mathbf{R}, \tag{6}$$

$$H(p_0, p, x, u, t) = \langle p, f(t, x, u) \rangle - p_0 f_0(t, x, u).$$

Necessary conditions are given by the following theorem.

__Theorem 1.__ Let the control $\hat{u}(\cdot)$ and the trajectory $\hat{x}(\cdot)$ defined on $[\hat{T}_0, \hat{T}_1]$ be the optimal solution for the problem formulated above. Then there exist:

- a real number $p_0 \geq 0$,
- an absolutely continuous function $p: [\hat{T}_0, \hat{T}_1] \to \mathbf{R}^n$,
- vectors $l_0 \in \mathbf{R}^{s0}$, $l_1 \in \mathbf{R}^{s1}$,

such that

a) p_0, $p(\cdot)$, l_0, l_1 do not vanish simultaneously,

b) the function $p(\cdot)$ satisfies:

$$\dot{p}(t) = -H_x(p_0, p(t), \hat{x}(t), \hat{u}(t), t)$$

$$p(\hat{T}_0) = h_{0_x}^T(\hat{T}_0, \hat{x}(\hat{T}_0))l_0 + p_0 g_{x_0}(\hat{x}(\hat{T}_0), \hat{T}_0, \hat{x}(\hat{T}_1), \hat{T}_1) \tag{7}$$

$$p(\hat{T}_1) = -h_{1_x}^T(\hat{T}_1, \hat{x}(\hat{T}_1))l_1 - p_0 g_{x_1}(\hat{x}(\hat{T}_0), \hat{T}_0, \hat{x}(\hat{T}_1), \hat{T}_1)$$

c) $\hat{H}(t) = H(p_0, p(t), \hat{x}(t), \hat{u}(t), t)$ is continuous on $[\hat{T}_0, \hat{T}_1]$ and

$$\hat{H}(\hat{T}_0) = -\langle h_{0_t}(\hat{T}_0, \hat{x}(\hat{T}_0)), l_0 \rangle - p_0 g_{T_0}(\hat{x}(\hat{T}_0), \hat{T}_0, \hat{x}(\hat{T}_1), \hat{T}_1)$$

$$\hat{H}(\hat{T}_1) = \langle h_{1_t}(\hat{T}_1, \hat{x}(\hat{T}_1)), l_1 \rangle + p_0 g_{T_1}(\hat{x}(\hat{T}_0), \hat{T}_0, \hat{x}(\hat{T}_1), \hat{T}_1) \tag{8}$$

$$\hat{H}(t) = \hat{H}(\hat{T}_1) - \int_t^{\hat{T}_1} H_t(p_0, p(s), \hat{x}(s), \hat{u}(s), s) ds.$$

□

    The most important condition for calculation of the optimal control is the continuity of Hamiltonian on the optimal path. This implies that for every switching time t_i $i = 1, \ldots, r$

$$H(p_0, p(t_i), \hat{x}(t_i), \hat{u}_i, t_i) = H(p_0, p(t_i), \hat{x}(t_i), \hat{u}_{i-1}, t_i) \tag{9}$$

where $\hat{u}_i$ is the optimal control value after and $\hat{u}_{i-1}$ before the i-th switching. One can say that the hamiltonian is indifferent to the change in the optimal control value, and so the above theorem is called the indifference principle.

    The proof of the indifference principle is based on application of control 'needle variation' demonstrated on figure 1. One can also use time transf6ormations originated from Dubovitsky and Miliutin [3] and apply Ioffe-Tichomirov extremum principle [4].

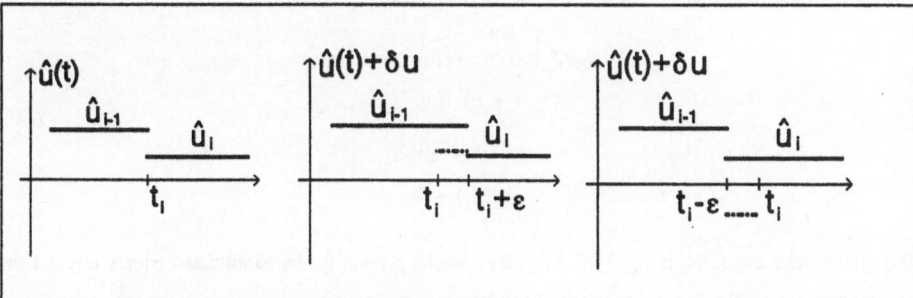

Figure 1. Optimal control and possible control variations.

A careful look at the switching time variation gives the well known [7] formula

$$\frac{\partial J(\hat{x}(\cdot),\hat{u}(\cdot))}{\partial t_i} = H(p_0,p(t_i),\hat{x}(t_i),\hat{u}_i,t_i) - H(p_0,p(t_i),\hat{x}(t_i),\hat{u}_{i-1},t_i). \tag{10}$$

One can incorporate this formula into any gradient minimization algorithm, but one has to remember that minimization variables t_i , $i=1,...,r$ are bounded by $T_0 \leq t_1 \leq ... \leq t_r \leq T_1$ and by the boundary conditions(3).

An application of the indifference principle gives a set of algebraic nonlinear equations to be solved.

Example 1.

Consider a control system governed by the equations

$$\dot{x}_1(t) = -x_2^2(t) + u^2(t) \tag{11}$$
$$\dot{x}_2(t) = u(t).$$

We are to fined the time optimal control subject to:

$$x_1(0) = x_2(0) = 0, \quad x_1(T_1) = 1, \quad |u(t)| \leq 1. \tag{12}$$

For $t<1$ we get $-x_2^2(t)+u^2(t) \leq 1$, hence $x_1(T_1)=1$ holds only for $T_1 > 1$. Let us consider a sequence of controls $\{u_k(\cdot)\}$, such that the control $u_k(\cdot)$ takes (in turn) values 1 and -1 on successive intervals $1/k$ long. From $x_1(T_{1,k})=1$ we get

$$1 < T_{1,k} \leq \frac{k^2}{k^2-1} \tag{13}$$

therefore $T_{1,k} \to 1$ as $k \to \infty$. Optimal control with a finite number of switchings does not exist. We will find an optimal control with no more then r switchings and with admissible values 1 and -1. We notice that for admissible controls u(t) and -u(t) we get the same value of the final time T_1. From the indifference principle we get (cases $p_0=0$ and $p_1=0$ could be eliminated)

$$H(p_0,p,x,u,t) = -1 + p_1(-x_2^2+u^2) + p_2 u,$$
$$\dot{p}_1(t) = 0,$$
$$\dot{p}_2(t) = 2x_2(t)p_1, \tag{14}$$
$$p_2(T_1) = 0.$$

The indifference condition is $p_2(t_i)=0$ for every switching point t_i. As hamiltonian equals zero on the optimal path we obtain $r+1$ equations ($t_{r+1}=T_1$):

$$-1+p-1(-x_2(t_i)+1) = 0. \qquad (15)$$

The last equation follows from $x_1(t_{r+1})=1$. These $r+2$ equations are easily solved for the unknowns p_1, $t_1, \dots, t_{r+1}$. For example we get:

for $r=1$: $t_1=0.3473$, $T_1=1.0419$,

for $r=2$: $t_1=0.2028$, $t_2=0.6083$, $T_1=1.0139$,

for $r=3$: $t_1=0.1438$, $t_2=0.4315$, $t_3=0.7192$, $T_1=1.0069$.

5. Minimization over control values

Let us now assume that $r=0$ and all admissible switchings are from a given set $T_D=\{t_{d1},\dots,t_{d\rho}\}$. The constant control values u_i are now the minimization variables. Control variation consists in changing the control value u_i over all, not necessarily small interval $(t_{di-1},t_{di}]$. If we assume that the set of admissible control values U_i is open and that functions f, f_0, are differentiable with respect to u, then by changing the control value to $u_i+\varepsilon$ we obtain that [7]

$$\frac{\partial J(\hat{x}(\cdot),\hat{u}(\cdot))}{\partial u_i} = \int_{t_{d\,i-1}}^{t_{di}} \frac{\partial}{\partial u} H(p_0,p(t),\hat{x}(t),\hat{u}_i,t)dt. \qquad (16)$$

If the differentiability assumptions are impossible we have to compare the optimal control $\hat{u}(\cdot)$ with the disturbed control $\hat{u}(\cdot)+\delta u(\cdot)$, such that a small parameter ε will arrive in the right side of disturbed equation (1). If we assume that for every $t \in (t_{di-1},t_{di}]$ the set $f(t,\hat{x}(t),U_i)$ is convex then for any $v \in U_i$ and $\varepsilon > 0$ there exists $u_\varepsilon \in U_i$ such that:

$$f(t,\hat{x}(t),u_\varepsilon) = \varepsilon f(t,\hat{x}(t),v) +(1-\varepsilon)f(t,\hat{x}(t),\hat{u}_i), \qquad (17)$$

so

$$f(t,\hat{x}(t),u_\varepsilon) - f(t,\hat{x}(t),\hat{u}_i) = \varepsilon[f(t,\hat{x}(t),v) - f(t,\hat{x}(t),\hat{u}_i)]. \qquad (18)$$

It follows that we are able to introduce a small parameter into right side of the equation (1), in spite that v is no close to $\hat{u}_i$ at all.

Motivated by the above deduction we formulate following theorem:

Theorem 2. Let the control $\hat{u}(\cdot)$ taking the values

$\hat{u}_0$ for $t \in [\hat{T}_0,t_{d1}]$,

$\hat{u}_1$ for $t \in (t_{d1},t_{d2}]$,

. . . .

$\hat{u}_\rho$ for $t \in (t_{d\rho},\hat{T}_1]$

and the trajectory $\hat{x}(\cdot)$ defined on $[\hat{T}_0,\hat{T}_1]$ be the optimal solution for the problem formulated above. Then there exist:

- a real number $p_0 \geq 0$,
- an absolutely continuous function $p:[\hat{T}_0, \hat{T}_1] \to \mathbf{R}^n$,
- vectors $l_0 \in \mathbf{R}^{s0}$, $l_1 \in \mathbf{R}^{s1}$,

such that

a) p_0, $p(\cdot)$, l_0, l_1 do not vanish simultaneously,

b) the function $p(\cdot)$ satisfies:

$$\dot{p}(t) = -H_x(p_0, p(t), \hat{x}(t), \hat{u}(t), t)$$

$$p(\hat{T}_0) = h_{0_x}^T(\hat{T}_0, \hat{x}(\hat{T}_0))l_0 + p_0 g_{x_0}(\hat{x}(\hat{T}_0), \hat{T}_0, \hat{x}(\hat{T}_1), \hat{T}_1) \tag{19}$$

$$p(\hat{T}_1) = -h_{1_x}^T(\hat{T}_1, \hat{x}(\hat{T}_1))l_1 - p_0 g_{x_1}(\hat{x}(\hat{T}_0), \hat{T}_0, \hat{x}(\hat{T}_1), \hat{T}_1)$$

c) $\hat{H}(t) = H(p_0, p(t), \hat{x}(t), \hat{u}(t), t)$ satisfies

$$\hat{H}(\hat{T}_0) = -\langle h_{0_t}(\hat{T}_0, \hat{x}(\hat{T}_0)), l_0\rangle - p_0 g_{T_0}(\hat{x}(\hat{T}_0), \hat{T}_0, \hat{x}(\hat{T}_1), \hat{T}_1)$$

$$\hat{H}(\hat{T}_1) = \langle h_{1_t}(\hat{T}_1, \hat{x}(\hat{T}_1)), l_1\rangle + p_0 g_{T_1}(\hat{x}(\hat{T}_0), \hat{T}_0, \hat{x}(\hat{T}_1), \hat{T}_1) \tag{20}$$

d) if for $i = 0, \ldots, \rho$, for every (t, x) from a certain neighborhood of the optimal path, for any $u_1, u_2 \in U_i$, for every $a \in [0,1]$ there exists $u \in U_i$, such that

$$f(t, x, u) = a f(t, x, u_1) + (1-a) f(t, x(t), u_2)$$

$$\int_{t_i}^{t_{i+1}} f_0(t, x, u)dt \leq a \int_{t_i}^{t_{i+1}} f_0(t, x, u_1)dt + (1-a) \int_{t_i}^{t_{i+1}} f_0(t, x, u_2)dt, \tag{21}$$

then the following maximum condition is satisfied:

$$\int_{t_i}^{t_{i+1}} H(p_0, p(t), \hat{x}(t), \hat{u}_i, t)dt = \max_{u \in U_i} \int_{t_i}^{t_{i+1}} H(p_0, p(t), \hat{x}(t), u, t)dt. \tag{22}$$

□

The proof of the above theorem is also based on Ioffe-Tichomirov extremum principle.

While applying theorem 2 we meet the same difficulties as with Pontryagin maximum principle, so we rather recommend to use a clever numerical procedure then to look for analytical formulae for the optimal control from theorem 2.

Example 2.

Calculation of the performance index gradients on the optimal path from example 1 gives:

$i = 0$, $\quad \hat{u}_i = 1$, $\quad \partial J/\partial u_i = 0.8216$,

$i = 1$, $\quad \hat{u}_i = -1$, $\quad \partial J/\partial u_i = -1.6433$,

$$i = 2, \qquad \hat{u}_i = 1, \qquad \partial J/\partial u_i = 1.6433,$$
$$i = 3, \qquad \hat{u}_i = -1, \qquad \partial J/\partial u_i = -3.5470.$$

So it is impossible to improve the final time by changing control values without violating the constraint $|u(t)| \leq 1$.

Example 3.

Let us consider a simple problem described by:

$$\dot{x}(t) = x(t) + u(t), \quad x(0) = 1,$$

$$J = \int_0^1 u^2(t)\,dt + x^2(1). \tag{23}$$

The optimal control equals

$$u_o(t) = -\frac{2e^{2-t}}{1+e^2}. \tag{24}$$

It changes from -1.7616 for $t=0$ to -0.6481 for $t=1$ and produces the performance index value $J_o = 1.7616$.

We are to find optimal piecewise constant control with the only points of discontinuity $t_{d1},\ldots,t_{d\rho}$ distributed uniformly in [0,1]. From theorem 2 we get

$$H = -u^2 + px + pu, \tag{25}$$
$$\dot{p} = -p, \qquad p(1) = -2x(1).$$

Then it follows from maximum condition that for $i=0,\ldots,\rho$ the optimal control value equals

$$\hat{u}_i = \frac{\rho+1}{2}(e^{-t_{di}} - e^{-t_{d\,i+1}})p_0. \tag{26}$$

The last equation to solve is $p_0 = -2ex(1)$. The value $x(1)$ follows from

$$x(t_{i+1}) = e^{\frac{1}{\rho+1}}x(t_i) + (e^{\frac{1}{\rho+1}}-1)\hat{u}_i. \tag{27}$$

For $\rho=1$ we get:

$$\hat{u}_0 = -1.4081, \qquad \hat{u}_1 = -0.8540$$

and the performance index value $J_1 = 1.7893 = 1.016\,J_0$.

For $\rho=3$:

$$\hat{u}_0 = -1.5648, \qquad \hat{u}_1 = -1.2187, \qquad \hat{u}_2 = -0.9491, \qquad \hat{u}_3 = -0.7392$$

and the performance index value $J_2 = 1.7686 = 1.004 \, J_o$.

If we consider the above problem under the condition $|u(t)| \leq 1.2$, then it follows after similar calculation, for $\rho = 1$, that:

$$\hat{u}_0 = -1.2, \qquad \hat{u}_1 = -1.0108$$

and the performance index value $J_3 = 1.8378 = 1.027 \, J_1 = 1.043 \, J_0$.

6. Conclusions

Theorems 1 and 2 give general necessary conditions for optimality of the control policy. For optimality of a switching time it is necessary that hamiltonian is indifferent for the fluctuation in the optimal control value. We note that the maximum principle implies the indifference principle, but not vice versa. The indifference principle requires exploration of many more candidate optimal paths, but this does not make it unworkable. For optimality of a constant control value it is necessary to maximize an integral of hamiltonian over the interval between switchings.

The presented results could be generalized in many directions. For example concurrent minimization over switching points and the constant control values or a problem with the upper bounds the number of discontinuities imposed only on some control entries [5].

Acknowledgements

The presentation of this paper is partially sponsored by Stefan Batory Foundation in Warsaw.

References

1. Blatt J.M., *Optimal control with a cost of switching control*, J. Austral. Math. Soc. vol. 20 ser. B, 1976, 316-332.
2. Cesari L., *Optimization Theory and Applications*, Springer Verlag, New York 1981.
3. Girsanov V.I., *Lectures on Mathematical Theory of Extremum Problems*, Springer Verlag, Berlin, 1972.
4. Ioffe A.D., Tichomirov M.W., *Theory of Extremum Problems*, Nauka, Moskwa, 1974
5. Kabziński J., *Optimal control with a bounded number of switchings, Part I, II*, Bull. Pol. Acad. Sci., Technical Sci. , vol 32, 1984, 341-353.
6. Pontryagin L.s., Boltyanski R.V., Mischenko E.F., *The Mathematical Theory of Optimal Processes*, Interscience Publ. Inc., New York, 1962
7. Sargent R.W.H., Sullivan G.R., *The development of an efficient optimal control package*, in Optimization Techniques, edited by J. Stoer, Springer Verlag, Berlin, 1978, 158-168.

OBSERVERS FOR POLYNOMIAL VECTOR FIELDS OF ODD DEGREE

Konstantin E. Starkov

Institute of Control Sciences, Russian Academy of Sciences

ul. Profsoyuznaya, 65, Moscow, SU- 117342, Russia,

Fax: (7-095) 420-20-16

In this paper we have dealt with constructing of observers for polynomial vector fields of odd degree. This problem dates back to D. Luenberger who suggested the first construction of the observer for linear autonomous systems. Then the polynomial version of this problem was studied in [1-3].

Consider a polynomial vector field f on R^m of odd degree d and a linear observation law $y = h(x) = Cx$; C is a $(p \times m)$- matrix; $y(t) = Cx(t)$ is an output.

At first, we assume that f is homogeneous of odd degree $d \geqslant 3$.

Suppose that there exists a positive definite symmetric matrix P such that

1) $PC^T C = C^T CP$;

2) the cone

$$Con = \{(x, e) \in R^m \times R^m \mid e^T P(f(x + e) - f(x)) \geqslant 0\}$$

is satisfied to the following conditions

$$Con \cap \{KerC \times R^m\} \subset \{0 \times R^m\}; \quad Con \cap \{R^m \times KerCP\} \subset \{R^m \times 0\}.$$

The main result of this paper is the following assertion.

Theorem1. Assume that the vector field f is homogeneous and f and C are satisfied to conditions 1) and 2). Then there exists a positive number $\ae$ such that the system

$$(1) \quad \dot{z} = F(z) = f(z) - \ae\{\|y\|^{d-1} + \|PC^T(Cz - y)\|^{d-1}\}PC^T(Cz - y)$$

serves as a global observer for the vector field f.

Sketch of the proof. Denote by the e the vector of error:
$e = z - x$. Then we have:

(2) $\dot{e} = f(x + e) - f(x) - æ(\|Cx\|^{d-1} + \|PC^T Ce\|^{d-1})PC^T Ce$

Consider the following quadratic form: $V(x,e) = 0.5e^T Pe$.

By differentiation of V along the vector field of (2) we can obtain

$$\dot{V} = e^T P\{f(x + e) - f(x)\} - æ\|Cx\|^{d-1}\|CPe\|^2 - æ\|C^T CPe\|^{d-1}\|CPe\|^2$$

Now consider a number of objects:

1) projections $\pi_i: R^{2m} \to R^m$; $i = 1,2$; $\pi_1(x,e) = x$; $\pi_2(x,e) = e$

2) cones $Q_i = \pi_i(Con)$; $i = 1,2$;

3) domains $D_1 = \{\|x\| \geqslant 1; \|e\| \leqslant 1\}$; $D_2 = \{\|x\| < 1; \|e\| > 1\}$;

4) sets $D_i^r = \{(x,e)| \|e\| = r\} \cap D_i$; $i = 1,2$.

It is obvious that inclusions from the assertion of Theorem 1 are satisfied if and only if $Q_1 \cap KerC = \{0\}$; $Q_2 \cap KerCP = \{0\}$. This remark is used in the proof.

Further, let $(x,e) \in D_1 \cup D_2$. Now if (x,e) does not belong to $(D_1 \cup D_2) \cap Con$ then $\dot{V}(x,e) < 0$.

Suppose that $(x,e) \in D_1 \cap Con$. Then we use the following unequality in D_1

$$|e^T P(f(x + e) - f(x))| \leqslant \sum_{|J|=2}^{d-1} |A_J(e)| \|x\|^{d-1}; \quad J = (J_1,\ldots,J_m)$$

for some polynomials A_J.

Now we can deduce that there exists $æ_1 > 0$ such that
$$\dot{V}|D_1^r \cap Con < 0;$$
here and below by $\dot{V}|$ we denote the restriction of the V on the set.

Assume that $(x,e) \in D_2 \cap Con$. If we make analogous manipulations with $\dot{V}$ we can obtain that there is $æ_2 > 0$ such that
$$\dot{V}|D_2^r \cap Con < 0.$$

Further, by use of homogeneity we have that if $\mathscr{x} = \max(\mathscr{x}_1, \mathscr{x}_2)$ then

(3) $\dot{V}|\{(x,e)\mid \|e\| > 0\} < 0.$

At last, it is easy to see that from (3) we can deduce the desired assertion.

This theorem can be generalized for unhomogeneous case by means of homogenezation. For each polynomial f_1 we take into consideration its homogenezation

$$\omega_{f_1}(x_0,x) = x_0^d f_1(x/x_0)$$

and a new pair " a vector field- an observation law":

(4) $X = \Omega_f(X) = (0,\omega_f(X))^T; \ X = (x_0,x)^T, \ , Y = H(x) = (x_0,Cx)^T,$

Now putting $x_0 = 1$ for outputs of (4) , we get necessary estimations of the state vector of the initial unhomogeneous system.

For example, consider the system

(5) $\dot{x} = F(x) = f(x) + g(y), \quad , y = h(x) = Cx;$

where f is a homogeneous polynomial vector field of odd degree d; $g(y)$ is a vector polynomial of degree less than d. Suppose, that the pair (f,h) in (4) is satisfied to the conditions of Theorem1. We construct the pair (Ω_F, H) which is also satisfied to the conditions of the theorem.

Note here that the observer (2) is noninitialized and robust, but has a property of bad convergence of the error $e = z - x$ in a small neighbourhood of the point $e = 0$.

At last, we connect our results with results of [4].

Theorem 2. Suppose that conditions of Theorem1 are satisfied.

Then there exists the global stabilizing feedback for the system

478

$$\dot{x} = f(x) + C^T u.$$

This theorem is a consequence of the theorem from [4].

Example, (the system for which the observer of (1)- type exists)

$$\dot{x}_1 = x_1 x_2 x_3 + x_2^2 x_3 - x_2 x_3^2$$

$$\dot{x}_2 = - x_1 x_2 x_3 + x_1 x_3^2 - x_1^2 x_3$$

$$\dot{x}_3 = - x_3^3 .$$

References

[1] Kawaji S. Umetsu Y. State observer for quadratic systems//Int.J. Contr.1983. v. 38. N3. p.577–588.

[2] Starkov K.E. Designing of observers for polynomial systems//Avtom. i Telem. 1991. N2. p. 64–73.

[3] Starkov K.E. Observability, state estimation and Carleman linearization//Avtom. i Telem. 1991. N11. p.80–86.

[4] Andreini A. Bacciotti A. Stefani G. Global stabilizability of homogeneous vector field of odd degree//Syst.&Contr. Lett. 1988. v.10. p. 251–256.

ALGORITHM FOR SOLUTION OF OPTIMIZATION PROBLEMS UNDER MULTIPLE SCENARIOS OF UNCONTROLLED INPUTS

Michał Warchoł, Krzysztof Malinowski
Instytut Automatyki Politechniki Warszawskiej
ul Nowowiejska 15/19,   00-661 Warszawa,   POLAND

1. Introduction.

In this paper we propose an algorithm to compute control decisions in presence of multiple scenarios of uncontrolled future inputs. Such algorithm may be useful in case when we want to apply model-based predictive control. In the simplest and the most popular case [Balchen et.al. 1992], [Keyser et.al. 1988], model-based predictive control consists of repeating open-loop optimization at each instant of intervention. However, this may not be the wisest thing to do in the case, when there is a high level of uncertainty i.e. in the case when the single forecast of future uncontrolled input may very much differ from the actual input [Malinowski 1993].

Then it can be useful to consider model-based predictive repetitive control design; where computations of control for the next interval are performed by considering a multiple scenario model of uncertainty. Assuming such model, considering of a finite number of scenarios, the required control computation requires solution of multi-stage optimization problem. The use of clasical solution technique may require considerable time. In this paper we want to propose an algorithm that could be capable of exploiting the fact, that there is a limited number of future scenarios of uncontrolled inputs, while the dimension of the state space can be significant.

480

The proposed approach proved to be successful when applied to repetitive model-based predictive control of a reservoir during flood [Warchol and Malinowski, 1993].

2. Problem formulation.

Let us consider the following discrete time dynamic system

$$x_{k+1} = f_k(x_k, u_k, v_k) \quad k = 0, ..., N-1 \tag{1}$$

x_k is state vector; u_k is controlled input vector; v_k is uncontrolled external input (disturbance). The uncontrolled input
is represented by a finite number of scenarios $\{v\}_i$ (fig 1).
Each scenario can be assumed as a sequence of values of input
in consecutive time instants (finite sequence, since we consider finite time horizon). To each scenario the probability
p_i is attached.

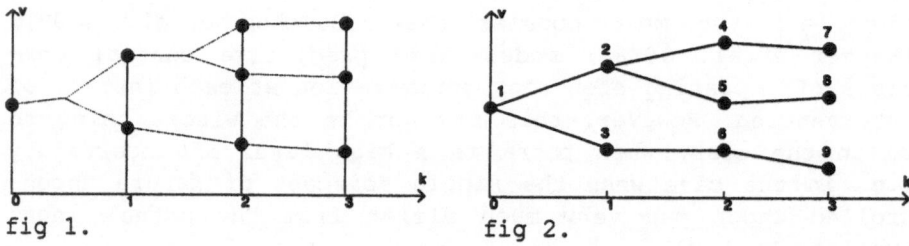

fig 1.   fig 2.

The objective is to minimize the performance index

$$F(\{x\}, \{u\}) \tag{2}$$

Where $\{u\}$ and $\{x\}$ are, respectively, controlled input trajectory and state trajectory, i.e. e.g. $\{x\} = \{x_0, ..., x_{N-1}\}$. As the
performance index value depends on unknown disturbance scenario, the actual objective will be to minimize a priori
evaluation of F, depending on all possible scenarios $\{u\}_i$, and
$\{x\}_i$ that can result from operation of the control scheme in
the future. This a priori evaluation of F can be an expected
value of F, but other a priori evaluations are possible.

3. The graph of possible decision situations.

In what follows a graph of possible decision situations will be helpful. The nodes of this graph correspond to situations, in which the decisions will be taken, and the arcs correspond to possible evolution of the situation due to uncontrolled external input. The example graph of decision considered at time 0 is presented in fig 2. It is assumed that information, which will be available to the controller in the future will be sufficient to identify external inputs, leading to that node. So, for example, at time instant 1 the controller will know whether the current node is node 2 or node 3.

Each node is related to the respective control decision u^i (superscript indices denote the numbers of nodes, while subscript indices will refer to time instances or components of vectors). Then the minimization may be performed with respect to real vectors u^i instead of control rules. It is so, since the values of control rules are important only for those states, which can occur and one should consider only finite number of interesting values (input, state and control) in the future. To each node of the graph state x^i and control u^i can be attached and the following relationship can be made:

$$x^i = f_{k(j)}(x^j, u^j, v^{j,i})$$ (3)

Where j is a number of parent to node i; $k(j)$ represents time instant corresponding to node j; $v^{j,i}$ is value of disturbance, under which situation j at time $k(j)$ evolves into situation i at time $k(i)$.

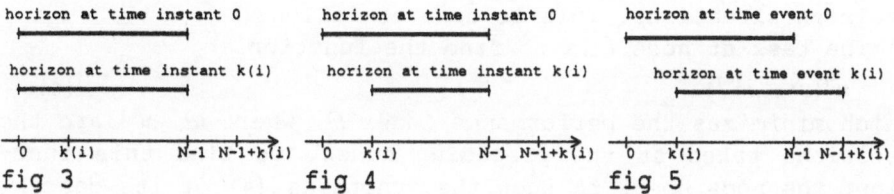

fig 3   fig 4   fig 5

To take into account the future decisions let us consider, for example, node i at time instant $k(i)$. In this case the decision taken at this node will minimize certain performance index. There are two ways to create this performance index

1. In the performance index attached to a parent node one can eliminate all scenarios that now cannot occur. In this case each node takes into account the full time horizon (fig 3).

2. After reduction of the set of scenarios, like in point 1, the parent node of the considered node is disregarded, and only horizon from $k(i)$ to N-1 is taken into account at node i (fig 4). New performance index can be attached to this node.

It is not useful to consider horizon from time event $k(i)$ to time event N-1+$k(i)$ (fig 5), because in this case one should consider also decisions taken later then during N-th time interval and the control horizon would expand more and more.

In this way one can form performance indices for each node (each situation). These will be functions of states, control values and uncontrolled input values. Using equations (3) and taking into account future scenarios of uncontrolled inputs as well as future control decisions we can express performance indices denoted as $F^i()$, in terms of controls only. Thus, in each node minimization of function will be performed to calculate optimal control u^i for this node.

4. Problem solution.

The proposed algorithm will transform the control decision problem into a set of non-linear equations in control decisions u^i, asserted with the nodes of the graph of possible decision situations. This is done as follows.

The task at node i is to find the function

$$u^{*i}(u^0,...,u^{i-1}) \tag{4}$$

which minimizes the performance index F^i, where $u^0,...,u^{i-1}$ are the decisions taken at the preceding nodes. To find this function, the node needs to know the functions (4) of its descendants, because the value of F^i depends upon decisions of the descendant nodes.

4.1 Case of smooth performance index.

Node i, in order to calculate its decision function (4), has to perform minimization of the function

$$F^i(u^0,...,u^{d^i},u^i,u^{*m^{i,1}}(u^0,...,u^i),...,u^{*m^{i,k^i}}(u^0,...,u^i),$$

$$u^{*m^{i,1}+1,j}(u^0,...,u^i),...,u^{*l^{i,1},j}(u^0,...,u^i),... \tag{5}$$

$$...,u^{*m^{i,k^i}+1,j}(u^0,...,u^i),...,u^{*l^{i,k^i},j}(u^0,...,u^i))$$

related to u^i, where d^i represents index of the parent node of node i; $u^0, ...,u^{d^i}$ are ancestor decisions; k^i is number of child (direct descendant) nodes of node i; $m^{i,j}$ is an index of j-th child of node i; $l^{i,j}$ is a maximum index of nodes, being descendants of the j-th child of node i; $u^{*m^{i,j}}(u^0,...,u^i)$ are j-th child decisions, $u^{*j,i}(u^0,...,u^i)$ are descendant decisions (j-th node), transformed by respective child to depending only on $u^0,...,u^i$. Consider the following algorithm for indexing nodes of the graph (fig 6).

1. Indexing starts from the root node.
2. To index any branch (node with its descendants) we first assign index to a node at the root of the branch, then we assign indices sub-branch, starting from first child, then to sub-branch, starting from second child and so on.

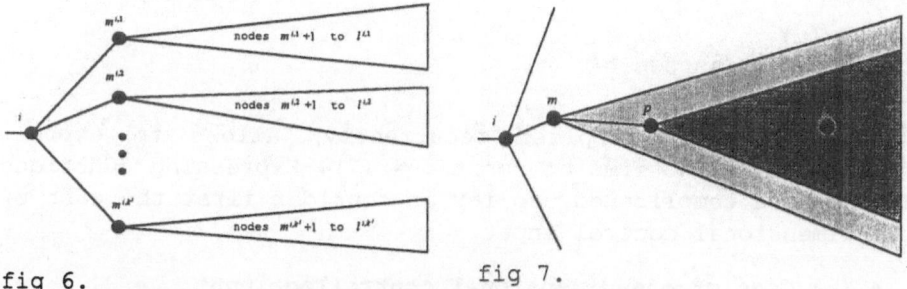

fig 6. fig 7.

If function (5) is smooth then its derivative w.r.t. u^i will be equal to zero for $u^i=u^{*i}(u^0,...,u^{d^i})$

$$F^i_{u^i}(...)+F^i_{u^{m^{i,1}}}(...)\cdot u^{*m^{i,1}}{}'_{u^i}(...)+,...,+F^i_{u^{m^{i,k^i}}}(...)\cdot u^{*m^{i,k^i}}{}'_{u^i}(...)+$$

$$+F^i\,_{u^{m^{i,1}+1}}{}'(...) \cdot u^{*m^{i,1}+1,j}{}_{u^i}{}'(...)+,...,+F^i\,_{u^{l_{i,1}}}{}'(...) \cdot u^{*l_{i,1},j}{}_{u^i}{}'(...)+ \tag{6}$$

$$+F^i\,_{u^{m^{i,k^i}+1}}{}'(...) \cdot u^{*m^{i,k^i}+1,j}{}_{u^i}{}'(...)+,...,+F^i\,_{u^{j^{i,k^i}}}{}'(...) \cdot u^{*j^{i,k^i}}{}_{u^i}{}'(...)=0$$

where arguments of F^i are similar, like in expression (5) with substitution of u^i for $u^{*i}(u^0,...,u^{d})$. Apostrophes denote derivatives, and subscript indices under the apostrophes denote argument in which differentiation is performed. In equation (6) only the first component of the left side is known. The most difficult problem is calculation of the derivatives

$$u^{*m^{i,j}}{}_{u^i}{}'(...),\;\; 0<j\le k_i \tag{7}$$

and

$$u^{*j,j}{}_{u^i}{}'(...) \tag{8}$$

One should express functions (7) and (8) in terms of

$$u^{*j}(...)$$

to transform the problem into a set of equations. First, let us notice that

$$u^{*j,j}(u^0,...,u^j)) = u^{*j,m}(u^0,...,u^i,u^{*m}(u^0,...,u^i)) \tag{9}$$

where m is direct descendant (child) of node i on the path from a node i to a node j (fig 7). If this child has a index j, then

$$u^{*j}(...)$$

can be used instead of

$$u^{*j,m}(...).$$

Relationship (9), applied recurrently, allows to express functions (8) in terms of functions (7). Expressing functions (7) is more complicated, so let as consider first the case of one-dimensional control input.

4.1.1 Case of one-dimensional controlled input.

Let as denote

Def 1.

Vector derivative order is an n-dimensional vector $k=[k_1,...,k_n]$, where value k_i of a given component is equal to the order of derivation with respect to the i-th argument.

Def 2.

Let us define ordering relation $\leq$ between n-dimensional vectors k and r

$$k \leq r \Leftrightarrow \mathop{\forall}_{i=1,\dots,n} k_i \leq r_i$$

Let as denote the operator of differentiation of a function $f()$ with respect to arguments denoted by vector differential order p as $\text{deri}(f(),p)$. Denote left-hand side of equation (6) as $w^i(u^0,\dots,u^d)$ and $\text{deri}(w^i(u^0,\dots,u^d),p)$ as $w^i_p(u^0,\dots,u^d)$. The following theorem allow us to express functions (7) in useful form

Theorem 1.

1. Function $w^i_p(u^0,\dots,u^d)$ can be expressed as the sum of products of the following elements

   a) $\text{deri}(F^i{}_j'(),r)$, $j \geq i$

   b) $\text{deri}(u^{*i}(),r)$, $r \leq p$

   c) $\text{deri}(u^{*j,j}(),r)$, $j > i$

2. $w^i_p()$ depends on $\text{deri}(u^{*i}(),p)$ linearly.

Thus, to get (7), one should to take derivatives of equation (6) for node j, and then solve the resulting linear equation. In the final expression derivatives of decisions of the descendants can occur and can be obtained recurrently in the same way.

Then, the following algorithm can be used to form the set of equations in optimal decisions.

Algorithm I

1. Each node first stores equation (6).
2. If this equation does not involve neither functions (7) nor functions (8) (this holds for last-time-instant nodes), stored equation is the final equation for the node, else the node asks its children about functions (7) and functions (8).

3. The node asked about function $\text{deri}(u^{*i}(),p)$ differentiates its final equation with respect to arguments indicated by p and solves the resulting linear equation. If the equation involves derivatives of decision of descendant nodes functions then the considered node asks recurrently his children about these functions.

4. The node asked about function $\text{deri}(u^{*j,j}(),r)$, $j>i$, first applies equation (9), and then, if the resulting equation consist derivatives of decision function of the descendant nodes, it recurrently asks children about these functions.

Points 1 and 2 consist the main task of each node. Points 3 and (or) 4 are performed only if a descendant asks about the derivatives of $u^i()$.

4.1.2 Case of multi-dimensional controlled input.

Let us now consider equation (6) when u^i is multidimensional, and let the size of u^i be denoted by N^i. This equation expresses a necessary condition of minimum of the goal function. Now one should to store a set of equation, because we are looking for minimum with respect to N^i variables. Equation (6) now evolves to a set of N^i equation, where the r-th equation has the form

$$F^i{}_{u^i,r}'(...) + \sum_{m=1}^{N^{m^{i,1}}} F^i{}_{u^{m^{i,1}},m}'(...) \cdot u^{*m^{i,1}}[m]_{u^i,r}'(...) +,...,+ \sum_{m=1}^{N^{m^{i,k^i}}} F^i{}_{u^{m^{i,k^i}},m}'(...) \cdot u^{*m^{i,k^i}}[m]_{u^i,r}'(...) +$$

$$+ \sum_{m=1}^{N^{m^{i,1}+1}} F^i{}_{u^{m^{i,1}+1},m}'(...) \cdot u^{*m^{i,1}+1,j}[m]_{u^i,r}'(...) +,...,+ \sum_{m=1}^{N^{i^{i,1}}} F^i{}_{u^{i^{i,1}},m}'(...) \cdot u^{*i^{i,1},j}[m]_{u^i,r}'(...) + \qquad (10)$$

$$+ \sum_{m=1}^{N^{m^{i,k^i}+1}} F^i{}_{u^{m^{i,k^i}+1},m}'(...) \cdot u^{*m^{i,k^i}+1,j}[m]_{u^i,r}'(...) +,...,+ \sum_{m=1}^{N^{i^{i,k^i}}} F^i{}_{u^{i^{i,k^i}},m}'(...) \cdot u^{*i^{i,k^i},j}[m]_{u^i,r}'(...) = 0$$

Now, while differentiating this equation, it is requested to specify index of the node and index of co-ordinate of the node input. So under each apostrophe two symbols, separated by a colon are given. Besides, after each symbol, representing control, its co-ordinate in square brackets is written. It causes also derivatives of control.

To get an algorithm, similar to Algorithm I, described in 3.1.1, let us denote the left-hand side of equation (10) as $w^i[r](u^0,...,u^d)$ and $\text{deri}(w^i[r](u^0,...,u^d),p)$ as $w^i{}_p[r](u^0,...,u^d)$. The following theorem, corresponding to theorem 1, holds

Theorem 2.

1. $w^i{}_p[r](u^0,...,u^d)$ can be expressed as the sum of products of the following elements

a) $\mathrm{deri}(F^i{}^l_{j,q}(),\mathbf{r})$, $j \geq i$

b) $\mathrm{deri}(u^{*i}[q](),\mathbf{r})$, $\mathbf{r} \leq \mathbf{p}$

c) $\mathrm{deri}(u^{*j,i}[q](),\mathbf{r})$, $j > i$

2. $w^i{}_p[r]()$ depends on symbols $\mathrm{deri}(u^{*i}[q](),\mathbf{p})$ linearly.

Relation $\leq$ is now defined as follows

$$\mathbf{k} \leq \mathbf{r} \Leftrightarrow \underset{i=1,\dots,n}{\forall} \underset{j=1,\dots,N^i}{\forall} k_{i,j} \leq r_{i,j}$$

Theorem 2 allows to apply the algorithm, same as Algorithm I for getting set of equation, but taking into account, that

1. Set of equations is expected for each node,
2. It is required to solve a set of linear equations instead of to solving a single linear equation to get derivative of control function.

4.2 Case of not smooth performance index.

To make the above calculations, the functions F^i should be differentiable many times. If the expression of the type

$$\begin{cases} f_1(\mathbf{u}) & \text{for } g(\mathbf{u}) < 0 \\ f_2(\mathbf{u}) & \text{for } g(\mathbf{u}) > 0 \end{cases} \tag{11}$$

occurs in the performance index then the differentiation can be impossible in points in which $g(\mathbf{u})=0$. Meanwhile, the solution often occurs just in such points. To solve this problem one should

1. Modify problem by replacing expression (11) with

$$\frac{f_2(\mathbf{u})e^{\lambda g(\mathbf{u})} + f_1(\mathbf{u})e^{-\lambda g(\mathbf{u})}}{e^{\lambda g(\mathbf{u})} + e^{-\lambda g(\mathbf{u})}} \tag{12}$$

2. Increase λ and examine behaviour of the function

$$\frac{\log(\|g(\mathbf{u})\|\sqrt{\lambda})}{\log \lambda} \tag{13}$$

It can be shown that if $g(\mathbf{u})=0$ at the solution of the original optimisation problem then the value of (13) approaches $-1/2$ as λ increases, otherwise the value of (13) approaches $1/2$.

3. Solve the problem again, now using $f_1(\mathbf{u})$ or $f_2(\mathbf{u})$ instead of (11), in the case, where $g(\mathbf{u}) \neq 0$ and use equation

$$g(u^0,...,u^{d^i},u^{*i}(u^0,...,u^{d^i}),$$

$$u^{*m^{i,1}}(u^0,...,u^{d^i},u^{*i}(u^0,...,u^{d^i})),...,u^{*m^{i,k^i}}(u^0,...,u^{d^i},u^{*i}(u^0,...,u^{d^i})),$$

$$u^{*m^{i,1}+1,j}(u^0,...,u^{d^i},u^{*i}(u^0,...,u^{d^i})),...,u^{*l^{i,1},j}(u^0,...,u^{d^i},u^{*i}(u^0,...,u^{d^i})),... \qquad (14)$$

$$...,u^{*l^{i,k^i}+1,j}(u^0,...,u^{d^i},u^{*i}(u^0,...,u^{d^i})),...,u^{*l^{i,k^i},j}(u^0,...,u^{d^i},u^{*i}(u^0,...,u^{d^i})))) = 0$$

instead of equation (6) in the case, when $g(\mathbf{u})=0$.

5. Example results.

The simple problem can be considered in order to illustrate the presented approach. Let us consider the following system

$$x[1] = x[0]^2 - 2u[0] + v[0]$$

$$x[2] = 2x[1]^2 - 3u[1] + v[1] \qquad x[0] = 1$$

with two possible disturbance scenarios

$$\{v[0],v[1]\}=\{1,1\}$$

$$\{v[0],v[1]\}=\{0,0\}$$

and horizon $N=2$. The goal at time instant $k=0$ let be the minimization of function

$$Q[0] = E(x[2]-2)^2$$

and minimization of function

$$Q[1] = Ex[2]^2$$

at time instant $k=1$.

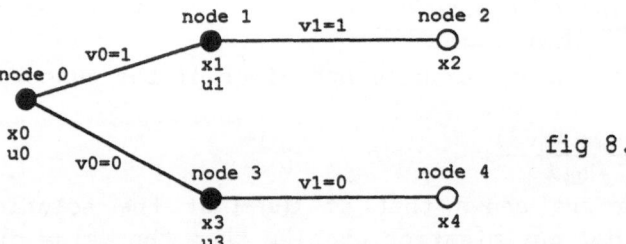

fig 8.

Let us number the nodes of the graph of possible decision situation as shown on figure 8. Below all indices will cause number of nodes, with which considered symbol is attached. A *Mathematica* program will be used for symbolic operations therefore indices will be simple printed after symbols (e.q. $u1$ instead of u_1).

Let us first define a state values

```
x0=1
1
x1:=x0^2-2u0+1
x2:=2x1^2-3u1+1
x3:=x0^2-2u0
x4:=2x3^2-3u3
```

and a performance meters

```
Q0:=(x2-2)^2+(x4-2)^2
Q0
           2        2              2         2
(-1 + 2 (2 - 2 u0)  - 3 u1)  + (-2 + 2 (1 - 2 u0)  - 3 u3)
Q1:=x2^2
Q1
          2        2
(1 + 2 (2 - 2 u0)  - 3 u1)
Q3:=x4^2
Q3
          2       2
(2 (1 - 2 u0)  - 3 u3)
```

The nodes 1 and 3 will minimize $Q1$ and $Q3$ with respect its controls. This results in the following expressions are equal to zeros

```
e1=D[Q1,u1]
                  2
-6 (1 + 2 (2 - 2 u0)  - 3 u1)
e3=D[Q3,u3]
                 2
-6 (2 (1 - 2 u0)  - 3 u3)
```

The node 0 will consider the influence of its decision with children node decisions

```
Q0p=Q0/.{u1->u1o[u0],u3->u3o[u0]}
            2            2              2            2
(-1 + 2 (2 - 2 u0)  - 3 u1o[u0])  + (-2 + 2 (1 - 2 u0)  - 3 u3o[u0])
e0=D[Q0p,u0]
                   2
2 (-1 + 2 (2 - 2 u0)  - 3 u1o[u0]) (-8 (2 - 2 u0) - 3 u1o'[u0]) +
                   2
2 (-2 + 2 (1 - 2 u0)  - 3 u3o[u0]) (-8 (1 - 2 u0) - 3 u3o'[u0])
```

To get equation attached to node 0 the node must know derivatives $u1o'[u0]$ and $u3o'[u0]$.

```
rule1=Solve[D[eq1,u0]==0,Derivative[1][u1o][u0]]
                  -(16 - 16 u0)
{{u1o'[u0] -> -------------+}}
                      3
rule3=Solve[D[eq3,u0]==0,Derivative[1][u3o][u0]]
                  -(8 - 16 u0)
{{u3o'[u0] -> ------------}}
                     3
```

Applying rules $rule1$ and $rule3$ to last expression the node 0 can get expression

```
e0/.rule1/.rule3
                                                      2
{{2 (16 - 8 (2 - 2 u0) - 16 u0) (-1 + 2 (2 - 2 u0)  - 3 u1o[u0]) +
                                                      2
    2 (8 - 8 (1 - 2 u0) - 16 u0) (-2 + 2 (1 - 2 u0)  - 3 u3o[u0])}}
```

which results in the following expression equal to zero

```
(e0/.rule1/.rule3)/.{u1o[u0]->u1,u3o[u0]->u3}
                                                      2
{{2 (16 - 8 (2 - 2 u0) - 16 u0) (-1 + 2 (2 - 2 u0)  - 3 u1) +
                                                      2
    2 (8 - 8 (1 - 2 u0) - 16 u0) (-2 + 2 (1 - 2 u0)  - 3 u3)}}
```

The research reported in this paper has been supported by the Polish Committee for Scientific Research (KBN) under KBN Project 3 0219 91 01.

References.

Balchen J.G., D. Lingquist and S. Strand 1992, State-Space Predictive Control, *J. Chemical Engineering Science*, Vol.47, No.4, pp.787-807

Keyser de R.M., G.A. Van der Velde, F.A. Dumortier 1988, A Comparative Study of Self-Adaptive Long-Rang Predictive Control Methods, *Automatica*, 24, pp.149-163

Malinowski K., Warchol M. 1993, Computation of Releases from the Storage Reservoir During Flood by Repetitive Stochastic Design, *Submitted for publication*

Malinowski K. 1993, Repetitive Optimization for Predictive Control of Dynamic Systems under Uncertainty, *Control Applications of Optimization*, Birkhäuser Publishers

OPTIMIZATION OF THE STAGE SEPARATION AND THE FLIGHT PATH OF A FUTURE LAUNCH VEHICLE

Kurt Chudej

Lehrstuhl für Höhere und Numerische Mathematik
Sonderforschungsbereich 255: Transatmosphärische Flugsysteme
Technische Universität München
Barer Straße 23, D-80290 München, Germany

1   Introduction

This presentation combines the simultaneous optimization of the staging and the whole ascent trajectory of both stages of a future launch vehicle by an indirect method.

Usually ascent optimization in flight mechanics is done for a given structure mass of the considered space vehicle. In the development phase of a future two-stage space vehicle one therefore solves a huge amount of trajectory optimization problems for different mass designs. Often certain functional relationships between the mass of important components are available. Then it seems desirable to be able to include in the optimization process of the flight path also a variation of the mass ratio of both stages. The appearance of only piecewise defined model functions, which are often derived from given table data, is today a common fact in realistic models. In order to apply an indirect optimization method it was therefore necessary to extend the necessary conditions of optimal control theory.

This general approach is applied to a model of a future hypersonic two-stage space transportation system (e.g. [5, 8, 7, 15, 17]). The whole space system will be launched horizontally. The lower stage is powered by a hydrogen fed turbo-ramjet propulsion system. Thrust depends on velocity and altitude and is modeled according to given performance maps. In an altitude of 35 km both stages separate. The lower stage returns to the launch site and the upper stage approaches the prescribed target orbit. The upper stage is in a conventional way rocket propelled. Typical state and control constraints – i.e. the important dynamic pressure constraint, bounds on the lift coefficent and the mass flow rate, etc. – are considered. The motion of the vehicle is described by a 3-dimensional mass point model over a spherical, rotating Earth with no wind in the atmosphere. Lift and drag are described by a Mach number dependent quadratic polar.

The optimization combines a trajectory optimization of the ascent of both stages and a staging optimization. The optimization of the stage separation includes the variation of the mass ratio of the lower and upper stage with respect to equality constraints describing a functional relationship of the structure and fuel mass of both stages.

2   Mathematical Model

Performance Index and Staging Condition

The objective of the optimization is to maximize the payload of the two-stage hypersonic space vehicle by varying the mass ratio of the lower stage versus the upper stage as well as the flight path. The following mass model is used:

The prescribed total mass $m(0)$ at take-off is divided into the structure and propellant portion of both stages and the payload.

$$m(0) = m_{\text{structure,I}} + m_{\text{fuel,I}} + m_{\text{structure,II}} + m_{\text{fuel,II}} + m_{\text{payload}} \tag{1}$$

The variable mass ratio of the two stages is constrained by equality constraints.

$$m_{\text{structure,I}} = \Psi_{\text{I}}(m_{\text{fuel,I}}) \tag{2}$$
$$m_{\text{structure,II}} = \Psi_{\text{II}}(m_{\text{fuel,II}}) \tag{3}$$

The mass laws (2, 3) for the lower and upper stage are modeled according to Shau [18] and Schöttle [16], [17]. It is assumed that the fuel of both stages is totally consumed[1]: for stage I when reaching the staging point t_s and for stage II when going into orbit at the final time t_f. This yields the formulas for the payload and the staging conditions.

$$m_{\text{payload}} = m(t_f) - \Psi_{\text{II}}(m(t_s^+) - m(t_f)) \overset{!}{=} \max \tag{4}$$
$$m(t_s^+) = m(t_s^-) - \Psi_{\text{I}}(m(0) - m(t_s^-)) \tag{5}$$

Formula (4) serves as a Mayer-type performance index and formula (5) as a prescribed state-discontinuity at the a priori unknown optimal staging time t_s.

Note that at the beginning of the optimization only $m(0)$ and the functions Ψ_{I} and Ψ_{II} are known, but that the optimal values of $m(t_s^-), m(t_s^+)$ and $m(t_f)$ are unknown.

Equations of Motion

The equations of motion of a point mass over a spherical and rotating Earth with no wind in the atmosphere (Miele [12]) are used to describe the position and velocity of the space craft.

$$\dot{v} = \frac{[T(v,h)\,\delta\,\cos\varepsilon - D(v,h;C_{\text{L}})]}{m} - g(h)\sin\gamma + \tag{6}$$

$$+ \omega^2 R \cos\Lambda\,(\sin\gamma\cos\Lambda - \cos\gamma\sin\chi\sin\Lambda)$$

$$\dot{\gamma} = \frac{[T(v,h)\,\delta\,\sin\varepsilon + L(v,h;C_{\text{L}})]\cos\mu}{mv} - \left[\frac{g(h)}{v} - \frac{v}{R}\right]\cos\gamma + \tag{7}$$

$$+ 2\,\omega\cos\chi\cos\Lambda + \omega^2\cos\Lambda\,(\sin\gamma\sin\chi\sin\Lambda + \cos\gamma\cos\Lambda)\frac{R}{v}$$

$$\dot{\chi} = \frac{[T(v,h)\,\delta\,\sin\varepsilon + L(v,h;C_{\text{L}})]\sin\mu}{mv\cos\gamma} - \cos\gamma\cos\chi\tan\Lambda\frac{v}{R} + \tag{8}$$

$$+ 2\,\omega\,(\sin\chi\cos\Lambda\tan\gamma - \sin\Lambda) - \omega^2\cos\Lambda\sin\Lambda\cos\chi\frac{R}{v\cos\gamma}$$

$$\dot{h} = v\sin\gamma \tag{9}$$

---

[1] A certain amount of fuel for the return of the lower stage is part of $m_{\text{structure,I}}$.

$$\dot{\Lambda} = \cos\gamma \sin\chi \frac{v}{R} \tag{10}$$

$$\dot{\theta} = \cos\gamma \cos\chi \frac{v}{R\cos\Lambda} \tag{11}$$

$$\dot{m} = -\frac{T(v,h)}{g_0\, I_{sp}(v,h)}\,\delta \tag{12}$$

The thrust model is based on tabular data of the maximum thrust $T(v,h)$ and the specific

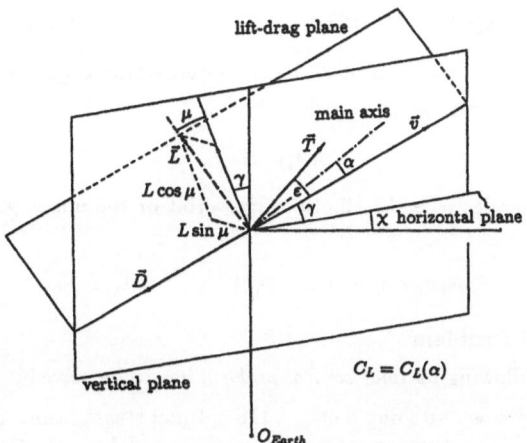

Figure 1: State and Control Variables

impuls $I_{sp}(v,h)$ of a turbo and ramjet engine, which are approximated by C^∞-functions through a nonlinear least squares approach. The used drag and lift model consists of a Mach number dependent quadratic polar.

$$
\begin{aligned}
L(v,h;C_L) &= q(v,h)\, F\, C_L \\
D(v,h;C_L) &= q(v,h)\, F\, C_D(M;C_L) \\
C_D(M;C_L) &= C_{D_0}(M) + k(M)\, C_L^2
\end{aligned}
$$

The following abbreviations are used: $q(v,h) = \varrho(h)\, v^2/2; \ M = v/a(h)$.

Control and State Constraints

The linear control throttle setting δ and the non-linear control lift coefficient C_L are subject to constraints.

$$0 \le \delta \le 1 \tag{13} \qquad\qquad |C_L| \le C_{L,max} \tag{14}$$

Due to its importance on the flight path the dynamic pressure q is limited by q_{max}.

$$q(v,h) \le q_{max} \tag{15}$$

Initial and Final Conditions

The initial conditions describe a horizontal launch at Istres (France). The launch direction $\chi(0)$ is optimized.

$$
\begin{array}{llrll}
v(0) &=& v_0 & \quad(16) \\
\gamma(0) &=& 0 & \quad(17) \\
m(0) &=& m_0 & \quad(18)
\end{array}
\qquad\qquad
\begin{array}{llrll}
h(0) &=& 0 & \quad(19) \\
\Lambda(0) &=& \Lambda_0 & \quad(20) \\
\theta(0) &=& 0 & \quad(21)
\end{array}
$$

The engine switching point t_e is constrained by

$$\underline{M} \leq M(v(t_e), h(t_e)) \leq \overline{M} \quad(22) \qquad\qquad \underline{h} \leq h(t_e) \leq \overline{h} \quad(23)$$

due to the limited operational range of the two airbreathing engines of the lower stage. The stage separation altitude is prescribed.

$$h(t_s) = h_s \tag{24}$$

The final conditions describe the elliptic target orbit of the upper stage (70 × 450 km, 28.5° inclination).

$$\psi_i(v, \gamma, \chi, h, \Lambda)|_{t_f} = 0 \quad , \quad i = 1, \dots, 5 \tag{25}$$

Optimal Control Problem

In summary, the following *optimal control problem* has to be solved:

Find the optimal engine switching point t_e, the optimal staging time t_s, the optimal final time t_f and piecewise continuous control functions $u : [0, t_f] \to U \subset \mathbb{R}^k$, such that the performance index

$$J[u; t_e, t_s, t_f] = \varphi(x(0), x(t_s^-), x(t_s^+), x(t_f)) \tag{26}$$

is minimized s.t.

equations of motion	$\dot{x} = f(x, u)$,	$f : \mathbb{R}^{n+k} \to \mathbb{R}^n$ (27)
initial conditions	$\hat{\psi}_1(x(0)) = 0$,	$\hat{\psi}_1 : \mathbb{R}^n \to \mathbb{R}^{p_1}$ (28)
interior point condition	$\tilde{\psi}(x(t_e)) = 0$,	$\tilde{\psi} : \mathbb{R}^n \to \mathbb{R}^{\tilde{p}}$ (29)
staging condition	$\bar{\psi}(x(t_s^-), x(t_s^+)) = 0$,	$\bar{\psi} : \mathbb{R}^{2n} \to \mathbb{R}^{\bar{p}}$ (30)
final condition	$\hat{\psi}_2(x(t_f)) = 0$,	$\hat{\psi}_2 : \mathbb{R}^n \to \mathbb{R}^{p_2}$ (31)
state and control constraints	$S(x, u) \leq 0$,	$S : \mathbb{R}^{n+k} \to \mathbb{R}^s$ (32)

3 Numerical Solutions

The necessary conditions of optimal control theory (Bryson & Ho [1]; Jacobson, Lele & Speyer [6]; Norris [13]; Maurer [9, 10]) are applied for a solution by an indirect method. Due to the combined appearance of piecewise defined model functions and state constraints new generalized jump conditions of the adjoint variables have to be considered (Maurer [11]; Bulirsch & Chudej [3]).

In summary, the set of all necessary conditions leads to a multipoint boundary-value problem with jump conditions of the following type:

Find the n-dimensional vector function $z(t)$ and the parameters $\tau_1, \ldots, \tau_s, t_f$ satisfying

$$\dot{z}(t) = F(t, z(t)) = \begin{cases} F_0(t, z(t)) & \text{if } 0 < t < \tau_1 \\ \vdots \\ F_s(t, z(t)) & \text{if } \tau_s < t < t_f \end{cases}$$

$$r_i(t_f, z(0), z(t_f)) = 0 \qquad 1 \le i \le \bar{n}$$
$$r_i(\tau_{j_i}, z(\tau_{j_i}^-), z(\tau_{j_i}^+)) = 0 \qquad \bar{n} < i \le (n+1)(s+1)$$
$$j_i \in \{1, \ldots, s\}$$

where F is a combination of the right hand sides of the state and adjoint equations and some so-called trivial equations of type $\dot{\sigma} = 0$ for each jump parameter σ.

This multipoint boundary-value problem with jump conditions is solved by the multiple shooting method (e.g. Bulirsch [2], [19], Oberle [14], Hiltmann [4]). GBS-Extrapolation methods or high order Runge-Kutta-Fehlberg methods with step size control are used as initial value problem solvers.

The following table and the diagrams show the results of the optimization for the considered model of the hypersonic space vehicle:

$m_{\text{structure,I}}$	162 787.40 [kg]
$m_{\text{fuel,I}}$	52 624.66 [kg]
$m_{\text{structure,II}}$	20 520.59 [kg]
$m_{\text{fuel,II}}$	69 088.93 [kg]
m_{payload}	4 978.42 [kg]
m_0	310 000.00 [kg]

Table 1: Optimized structure and fuel mass

4 Conclusion

The computed solutions for a realistic model of a hypersonic space vehicle demonstrate the feasibility of simultaneously optimizing the stage separation and the flight path by an indirect method. This is enabled by using state of the art optimization codes such as multiple shooting and a generalization of the necessary conditions of optimal control theory. The later is necessary in order to include in the problem formulation simultaneously such important features as piecewise defined model functions and state constraints.

Acknowledgement

This work was supported by the Deutsche Forschungsgemeinschaft (DFG) through the Sonderforschungsbereich 255: Transatmosphärische Flugsysteme.

The author wishes to thank Prof. Dr. Dr. h.c. R. Bulirsch who stimulated and supported this work.

496

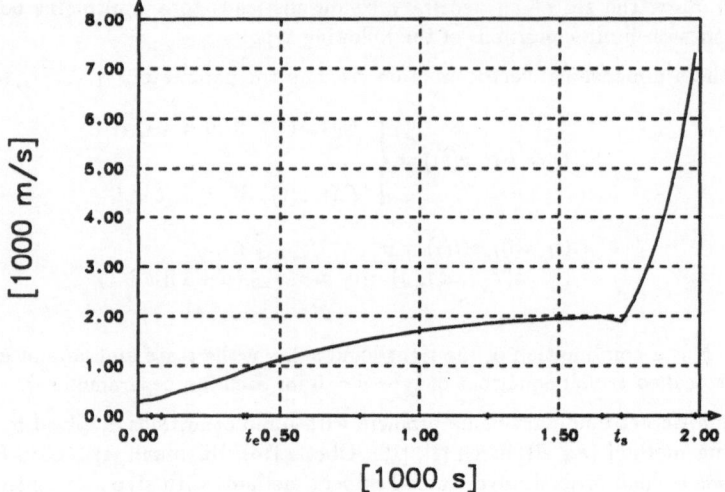

Figure 2: velocity v

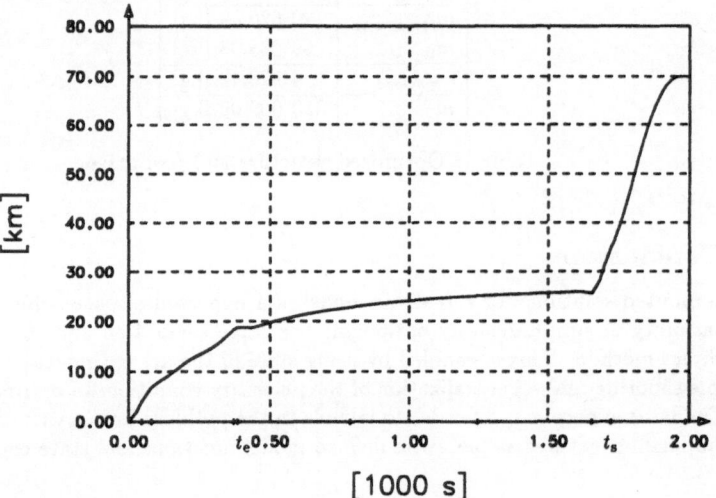

Figure 3: altitude h

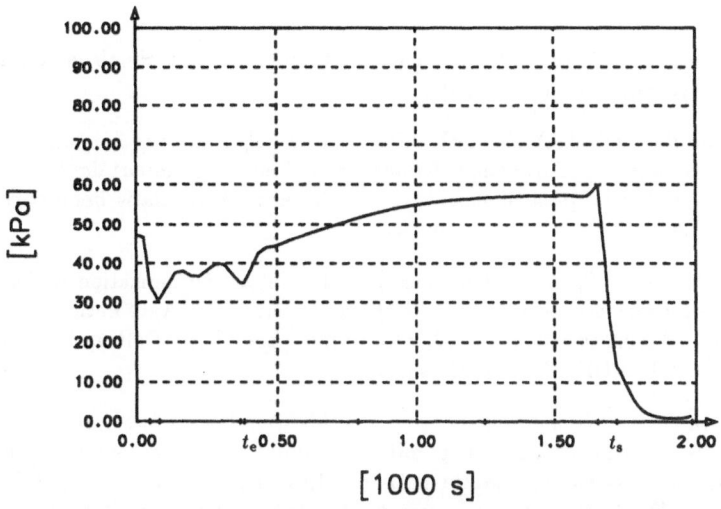

Figure 4: dynamic pressure q

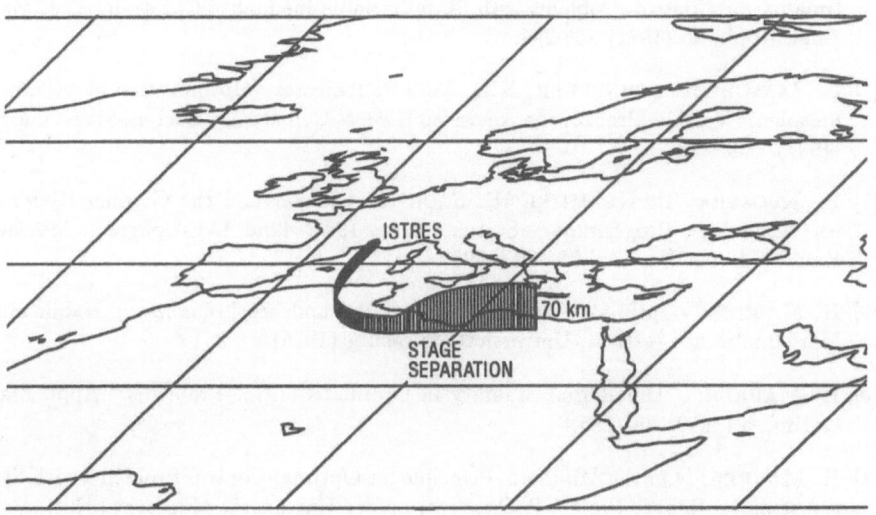

Figure 5: Launch and stage separation of the hypersonic space vehicle

References

[1] A.E. BRYSON; Y.-C. HO: Applied Optimal Control, 2nd ed. Hemisphere Publishing Corp., Washington, D.C. (1975).

[2] R. BULIRSCH: Die Mehrzielmethode zur numerischen Lösung von nichtlinearen Randwertproblemen und Aufgaben der optimalen Steuerung. Report der Carl-Cranz-Gesellschaft e.V., Oberpfaffenhofen (1971); Nachdruck: Mathematisches Institut, TU München (1985).

[3] R. BULIRSCH; K. CHUDEJ: Guidance and Trajectory Optimization under State Constraints. Preprints of the 12th IFAC Symposium on Automatic Control in Aerospace - Aerospace Control '92. Ottobrunn, Eds. D.B. DeBra, E. Gottzein. Düsseldorf: VDI/VDE-GMA (1992) 533-538.

[4] P. HILTMANN: Numerische Lösung von Mehrpunkt-Randwertproblemen und Aufgaben der optimalen Steuerung mit Steuerfunktionen über endlichdimensionalen Räumen. Dissertation, Mathematisches Institut, TU München (1990); Report No. 448, DFG-Schwerpunktprogramm: Anwendungsbezogene Optimierung und Steuerung, Mathematisches Institut, TU München (1993).

[5] E. HÖGENAUER: Raumtransporter. Z. Flugwiss., 11 (1987) 309-316.

[6] D.H. JACOBSON; M.M. LELE; J.L. SPEYER: New Necessary Conditions of Optimality for Control Problems with State-Variable Inequality Constraints. J. Math. Anal. Appl., 35 (1971) 255-284.

[7] C. JÄNSCH, K. SCHNEPPER, K.H. WELL: Trajectory Optimization of a Transatmospheric Vehicle. Proc. of the American Control Conference, Boston, Massachusetts (1991) 2232-2237.

[8] H. KUCZERA; P. KRAMMER; P. SACHER: Sänger and the German Hypersonics Technology Programme - Status Report 1991. 42nd IAF-Congress, Montreal, Kanada, Paper-No. IAF-91-198 (1991).

[9] H. MAURER: Optimale Steuerprozesse mit Zustandsbeschränkungen. Habilitation, Mathematisches Institut, Universität Würzburg (1976).

[10] H. MAURER: Differential Stability in Optimal Control Problems. Appl. Math. Optim., 5 (1979) 283-295.

[11] H. MAURER: On the Minimum Principle for Optimal Control Problems with State Constraints. Report No. 41, Rechenzentrum der Universität Münster (1979).

[12] A. MIELE: Flight Mechanics I, Theory of Flight Paths. Addison-Wesley, Reading, Massachusetts (1962).

[13] D.O. NORRIS: Nonlinear Programming Applied to State-Constrained Optimization Problems. J. Math. Anal. Appl., 43 (1973) 261-272.

[14] H.J. OBERLE: Numerische Berechnung optimaler Steuerungen von Heizung und Kühlung für ein realistisches Sonnenhausmodell. Habilitation, Institut für Mathematik, TU München (1982); Report TUM-Math-8310, Institut für Mathematik, TU München (1983).

[15] G. SACHS; W. SCHODER: Optimal Separation of Lifting Vehicles in Hypersonic Flight. Proc. of the AIAA Guidance, Navigation and Control Conference. New Orleans, Louisiana, Paper-No. AIAA-91-2657 (1991) 529-536.

[16] U.M. SCHÖTTLE: Flug- und Antriebsoptimierung luftatmender aerodynamischer Raumfahrtträger. Dissertation, Institut für Raumfahrtsysteme, Universität Stuttgart (1988).

[17] U.M. SCHÖTTLE; H. GRALLERT; F.A. HEWITT: Advanced air-breathing propulsion concepts for winged launch vehicles. Acta Astronautica, 20 (1989) 117-129.

[18] G.-C. SHAU: Der Einfluß flugmechanischer Parameter auf die Aufstiegsbahn von horizontalstartenden Raumtransportern bei gleichzeitiger Bahn- und Stufungsoptimierung. Dissertation, Fakultät für Maschinenbau und Elektrotechnik, TU Braunschweig (1973).

[19] J. STOER; R. BULIRSCH: Numerische Mathematik 2, 3. Auflage, Springer, Berlin (1990).

A Notation

State Variables

v velocity
γ path inclination
χ azimuth inclination
h altitude
Λ geographical latitude
θ geographical longitude
m mass

$x = (v, \gamma, \cdots, m)^T$

Control Functions

C_L lift coefficient
μ bank angle
δ mass flow
ε thrust angle

$u = (C_L, \mu, \delta, \varepsilon)^T$

Other Important Quantities

a speed of sound
D, L drag and lift force
C_D, C_L drag and lift coefficient
f right hand side of o.d.e.
F reference area
g gravitational acceleration

I_{sp}	specific impuls
J	performance index
m_0	prescribed initial total mass
$m_{fuel,I}, m_{fuel,II}$	fuel mass of stage I / stage II
$m_{structure,I}, m_{structure,II}$	structure mass of stage I / stage II
$m_{payload}$	payload
M	mach number
Ψ_I, Ψ_{II}	structure mass model
$\hat{\psi}_1, \hat{\psi}_2, \tilde{\psi}$	boundary and interior point conditions
$\bar{\psi}$	staging conditions
q	dynamic pressure
r_0	Earth's radius, $R = r_0 + h$
S	state and control constraints
T	thrust force
t	time
t_e	switching time of engines (stage I)
t_s	separation time of stage I and stage II
t_f	final time
ϱ	atmospheric density
ω	angular velocity

A real-time optimal control algorithm for water treatment plants

R.Kora (*)(**), P.Lesueur(*)(**), P.Villon(*)(**)

(*) : Lyonnaise des Eaux Dumez, Laboratoire d'Informatique Avancée de Compiègne, Technopolis Bât 3, Z.A.C. de Mercières, F-60200 Compiègne, France.
(**) : Université de Technologie de Compiègne, B.P. 649, F-60206 Compiègne Cedex, France.

Abstract : The goal of this paper is to present an adaptive optimal control algorithm for water treatment plants based on tree-searching algorithms, especially dynamic programming (D.P.) and A* algorithm. A modelisaton in discrete terms leads to a discrete time optimal control problem with boolean control variables. The Monte-Carlo method is added to D.P. in order to avoid combinatorial explosion. The results of the behavior and the robustness of the final algorithm are discussed. Since this algorithm requires a prediction of water consumption, a method that forecasts the water demand is also presented.

Keywords : Dynamic programming, Discrete optimization, A*, Least Squares approximation.

1. INTRODUCTION

A water treatment plant is composed of two steps of pumping, besides the water treatment step. The cost of pumping represents the largest part of the distribution network management cost. Therefore, an optimal control of the energetic cost of pumping is necessary. Softwares that optimize the pumping cost of the distribution network already exist [3][7]. However, there are strong relationships between this cost and the plant's pumping cost. That's why an optimal control algorithm for water treatment plants is required.
Expert systems have been previously developped, but the high degree of maintenance needed by this kind of application is a serious drawback to its use on site. An optimization method needing less maintenance has been prefered.
In this paper, a method responding to this need is presented. First, we present the physical problem and its mathematical modelisation. In second time, a water demand forecast method, required as input of the optimization algorithm will be exposed. Third, the optimal control algorithm is explained. Finally, the numerical results are discussed.

2. DESCRIPTION OF THE PHYSICAL PROBLEM

A water treatment plant can be represented like this :

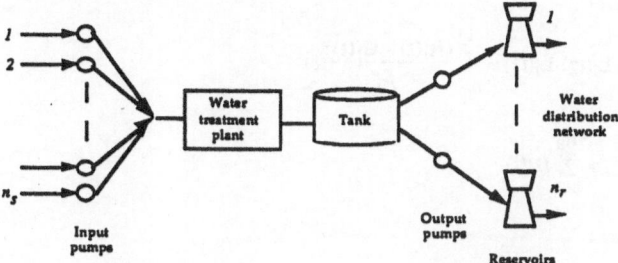

Figure 1 : Diagram of a water treatment plant.

There are two pumping steps : the input step and the output step. For each part, constraints are given by the maximum and minimum levels of the reservoirs and the buffer tank. The water treatment plant contains several on-off pumps connected to the water sources and output pumps corresponding to each reservoir. The water sources have different concentrations of pollutants. The problem is to command the pumps with a minimal energetic cost under the following constraints:

1) Keep the concentration of pollutants at the input of the water treatment plant quasi constant in time.
2) The reservoirs and the tank must never be empty.
3) Satisfy the demand of the customers.

2.1.Modelisation

We use the following notations :

n_s : number of water sources,
n_r : number of reservoirs,
f_{in} : plant input flow,
f_{out}: plant output flow,
$Lmin_{tank}, Lmax_{tank}$: minimum and maximum level for the tank,
$Lmin_i, Lmax_i$: minimum and maximum level for reservoir i,
C_i : water consumption for reservoir i,
$c_{s,i}$: cost for source i,
$c_{r,i}$: cost for reservoir i.

The control variables at time t are :

$fs_i(t)$: flow between source i and the plant (discrete value),
$ft_i(t)$: flow between the tank and reservoir i (discrete value).

The state variables at time t are :

$L_{tank}(t)$: level of water in the tank,
$c_{Nh4,i}(t)$, $c_{Fe,i}(t)$: concentrations of pollutants (Nh4 and Fe) in the water
coming in the buffer tank,
$L_i(t)$: level of water in reservoir i.

The state equations are :
- for the buffer tank :

$$L_{tank}(t_2)-L_{tank}(t_1) = \int_{t_1}^{t_2} \frac{(f_i(t) - \sum_{i=1}^{n_r} ft_i(t))}{S_{tank}} dt$$

- for reservoir i :

$$L_i(t_2)-L_i(t_1) = \int_{t_1}^{t_2} \frac{(ft_i(t) - C_i(t))}{S_i} dt$$

- for the plant

$$f_{in} = f_{out} = \sum_{i=1}^{n_s} fs_i(t)$$

The constraints are :

$Lmin_{tank} \leq L_{tank}(t) \leq Lmax_{tank}$.
$Lmin_i \quad \leq \quad L_i(t) \quad \leq Lmax_i$, for $i=1,...,n_r$

The cost function is :

$$J(fs(t),ft(t)) = \sum_{i=1}^{n_s} (\int_0^T c_{s,i}\, dt) + \sum_{i=1}^{n_r} (\int_0^T c_{r,i}\, dt)$$

Source costs $c_{s,i}$ and reservoir costs $c_{r,i}$ are developped subsequently.

In fact, the controls are not continuous variables, but discrete ones. Therefore, this modelisation will be transformed into discrete variables and equations.

The system can be divided in two sub-systems : the input and output parts. They are similar though they do not have the same costs nor the same constraints. On one hand, for the input part, the energetic cost is insignificant. The prime necessity is the continuity of the input flow and the continuity of the input pollutants concentration. On the other hand, for the output part, the energetic cost is very important, as well as the variation of the pumps control.

A sub-system is composed of three parts : sources, pumps and reservoirs. Dividing the system in two sub-systems allows a great reduction of the problem complexity. With this representation, the buffer tank stands twice : once in the input part, as the only reservoir, and once in the output part, as the only source. It represents the relationship between the two sub-systems since the output pumps total flow is also the water consumption of the buffer tank considered as a reservoir.

2.2. Time Discretisation

The time interval $[T_b, T_e]$ is subdivided on N subintervals with $t_0 = T_b$ and $t_N = T_e$.

For one sub-system, the model becomes :
- control variables :
  - $f_{i,j}$: flow of reservoir i during the time interval $[t_{j-1}, t_j]$, $j \in \{1,...,N\}$
- state variables :
  - $L_{i,j}$: level of water in reservoir i at time t_j, $j \in \{0,N\}$
  - $c_{Nh4,i,j}$, $c_{Fe,i,j}$: concentration of pollutants (NH4 and Fe) in the water
    received by reservoir i between time t_{j-1} and time t_j (used
    for the input part only).
- state equation :

$$L_{i,j} = L_{i,j-1} + \frac{(f_{i,j} - C_{i,j}) * (t_j - t_{j-1})}{S_i}$$

with $C_{i,j}$: water consumption for reservoir i between t_{j-1} and t_j, and S_i : surface of reservoir i.
- constraints :
  - $Lmin_i \le L_{i,j} \le Lmax_i$, $j \in [0,N]$

The cost function at time t_j is :

$$c_j = c_{j-1} + \sum_{k=1}^{n_r} c_{r,k,j}$$

where $c_{r,k,j}$ is the cost function for reservoir k at time t_j.

The cost function for reservoir k at time t_j is decomposed as below :
$c_{r,k,j} =$

$\quad W_{1,k} * E(f_{k,j}, j) +$
$\quad W_{2,k} * (f_{k,j} - f_{k,j-1})^2 +$
$\quad W_{3,k} * ((c_{Nh4,k,j} - c_{Nh4,k,j-1})^2 + (c_{Fe,k,j} - c_{Fe,k,j-1})^2)) +$
$\quad W_{4,k} * |L_{k,j} - Lobj_{k,j}| +$
$\quad W_{5,k} * S(f_{k,j}) * (1 - \delta(f_{k,j}, f_{k,j-1}))$ with $\delta(a,b) = 1$ if $a=b$ and $\delta(a,b) = 0$ otherwise.

The first term is the energetic cost; the second one is the water flow variation cost; the third one is the pollutants concentration variation cost; the fourth one is the goal cost and the last one is a stop or start cost applied to few pumps which are very powerful. Coefficients $W_{1,k}$, $W_{2,k}$, $W_{3,k}$, $W_{4,k}$ and $W_{5,k}$ are heuristic weights attached to each term.

We obtain discrete time optimal control problem with discrete control variables. We use tree-search algorithms to solve it. The graph contains N levels corresponding to N time intervals. To each combination of the state variables corresponds a graph node. The complexity a priori of the problem is $O(n^N)$, with n number of sons of a particular node and N number of time steps. Therefore, this is not a polynomial problem.

A tree-search algorithm will be used, but before to present it, we can remark that a precise forecast of the water consumption is required to get acceptable results. So first, a consumption forecast method will be explained.

3. WATER CONSUMPTION FORECAST

Since the optimal control algorithm is restarted every hour, a reconstruction method of water demand forcast is necessary. For this, we dispose of flow mesures at every quarter of hour.

The principe of the method is to predict the daily consumption curve as a linear combination of independant shape functions defined on the interval $[T_b, T_e]$ (i.e. 0 to 24 hours), using a least squares method.

3.1. Modelisation and resolution

The prediction method builts a reference curve $C_{ref}(s)$ whith a predictor-corrector method using a least squares criterion. If t is the reconstruction instant, T_b and T_e the beginning and the end of the current day, then :

$$C_{ref}(s) = \begin{cases} C_{mes}(s) & \text{for } T_b \le s < t \\ C_{for}(s) & \text{for } t \le s \le T_e \end{cases}$$

where C_{mes} represents the real water consumption between T_b and t and C_{for} represents the demand forecast for the interval $[t, T_e]$. At the begining, when $t = T_b$, we compute C_{for} using historic data [6].

The method uses a family of continous functions $C(s) = [c_1(s), c_2(s) ...., c_n(s)]^t$
$(s \in [T_b, T_e])$.

The criterion to minimize is :

$$(1) \quad J(\alpha(t)) = \int_{T_b}^{T_e} (C_{ref}(s) - \alpha^t C(s))^2 ds$$

$$= \int_{T_b}^{t} (C_{mes}(s) - \alpha^t C(s))^2 ds + \int_{t}^{T_e} (C_{for}(s) - \alpha^t C(s))^2 ds$$

If α is the solution of problem (1), the prediction will be :

$$C_{ref}(s) = \alpha^t C(s), \quad s \in [T_b, T_e]$$

Developping $J(\alpha)$, a quadratic form is obtained :

$$J(\alpha) = \alpha^t A(t) \alpha - 2 b^t(t) \alpha + r(t)$$

where :

$$A = \int_{T_b}^{T_e} C(s) C(s)^t ds, \quad b = \int_{T_b}^{T_e} C(s) C_{ref}(s) ds, \quad A \in M_{nn}(R), b \in R^n$$

If functions c_i are independants, problem (1) has a unic solution given by :

$$\alpha^* = A^{-1} b$$

and the demand forecast is :

$$C_{ref}(s) = (\alpha^*)^t C(s)$$

3.2. The choice of the shape functions family

A statistical study of water consumption curves shows that each day of the week pursue a particular form. Consumption curves for Sundays are different from curves for Mondays. For a fixed type of day (i.e. Monday), consumption peaks heappen at the same instants but they have different amplitudes. At figure 2, we represent the mean shape of a number of curves for each day of a week. The curves are previously normalised.The Saturday and Sunday shapes are easily distinguished from week days.

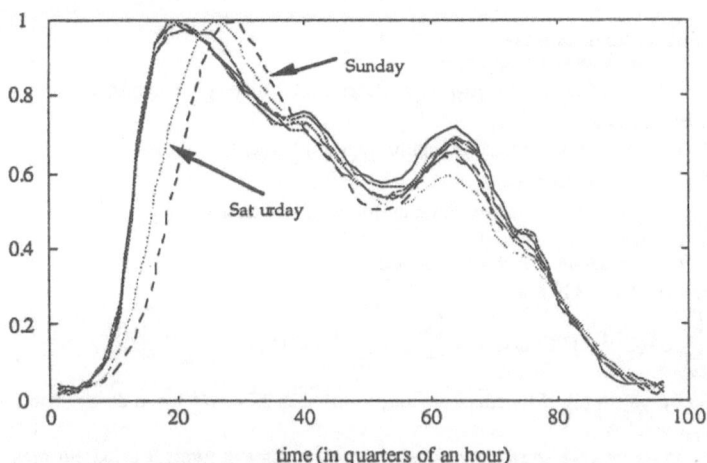

time (in quarters of an hour)

figure 2 : Shapes of consumption curves

The profile of the day whose water demand is predicted is included in the functions family. A constant function is used to regulate the amplitude of the final demand forecast.

Using this family of two functions, the predicted curve will have the same shape as the profile curve (witch is a normalised curve). In order to obtain a more supple shape family and to obtain a better prediction during the reconstruction, the profile of the day is divided in several sections corresponding to the consumption peaks. Each section will be a function of the family. These functions are non null at the corresponding section and null elsewhere. At figure 3, the function family is reperesented for Mondays.

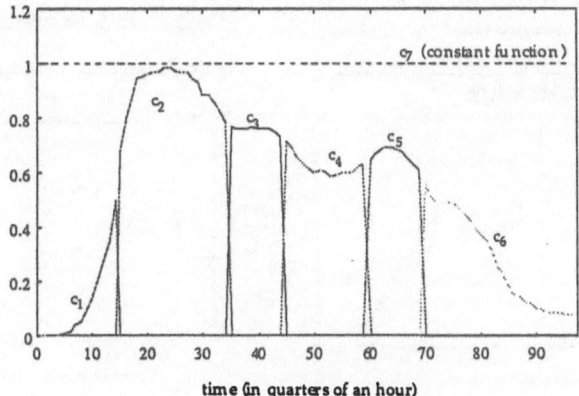

time (in quarters of an hour)

figure 3 : Shape family for Mondays

3.3. The discrete case application

This method must be transformed for the discrete case, because consumption curves are represented by vectors of discrete values. The problem to be solved is :

$$(2) \qquad J(\alpha) = \sum_{i=1}^{m}(C_{ref,i} - \alpha^t \, C^i)^2$$

where :

- m : number of mesures for a day,
- t : reconstruction time (also mesure order),
- $C \in M_{nm}(R)$ functions family, the rows C_i represents a function c_i, C^i represents column i of C
- $\alpha \in R^n$: coefficients applied to C,
- $C_{mes} \in R^{t-1}$: vector of mesured consumption values till order t,
- $C_{for} \in R^{n-t+1}$: demand forecast,
- $C_{ref} = (C_{mes}{}^t, C_{hist}{}^t)^t \in R^n$: reference vector for the least square method.
- n : number of functions,

Developping $J(\alpha)$, a quadratic function is obtained:

$$J(\alpha) = \alpha^t A \, \alpha - 2 \, b^t \, \alpha + r$$

where $A = \sum_{i=1}^{m} C^i \, (C^i)^t = C \, C^t$ and $b = \sum_{i=1}^{m} C_{ref,i} \, C^i = C \, C_{ref}$.

The solution of problem (2), if functions are independants, is $\alpha^* = A^{-1} \, b$ and the demand forecast is :

$$C_{prd} = C^t \, \alpha^*.$$

In the following figures, we show the reconstruction of a demand forecast using our method during a period of 24 hours :

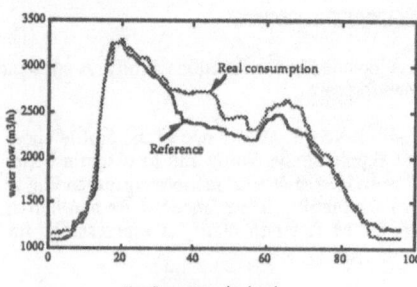

figure 4 : Reference and real consumption curves at 0 h:

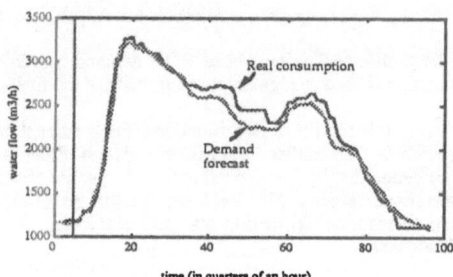

figure 5 : Demand forecast at 1 hour

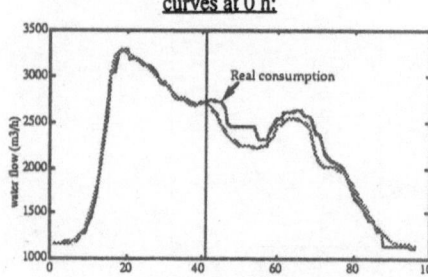

figure 6 : Demand forecast at 10 hour

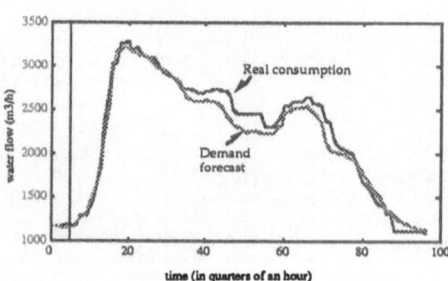

figure 7 : Demand forecast at 13 hour

4. THE CONTROL OPTIMIZATION ALGORITHM

The final goal is to find a 24h control strategy to apply to the pumping stations in order to :
- *satisfy the water consumption*
- *minimize the energetic cost*
- *limit the water flow variations and water quality variations*
- *limit frequent start and stop of some pumps*

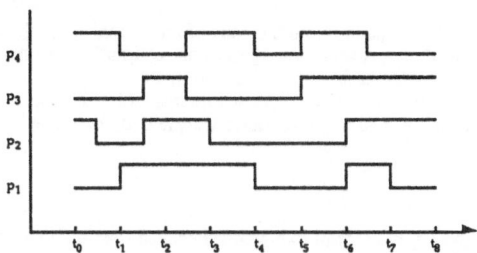

figure 8 : A example pump control chronogramme

A particular state of the system is caracterized by the levels of reservoirs and pollutants concentrations at time step i, $1 \le i \le N$. With n pumping possible combinations, the state S_i at step i leads to n other new states. The state graph containing all possible control sequences is represented on figure 9. To obtain the optimal pump control, we have to compute a minimal cost path from $t_0 = T_b$ to $t = T_e$.

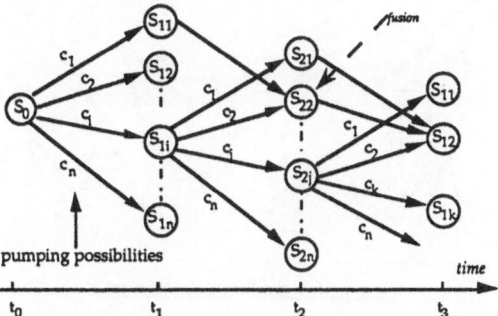

figure 9 : An example of the state tree

The different kinds of methods of tree-searching [5] that might be applied can be roughly classified in two groups.

The first group, the branch and bound method or A* like methods, need an upper and lower bounds for evaluation procedure. This method gives an optimal solution in several cases. The quality of the sub-optimal solution obtained depends on the accuracy of bounds [8]. These methods correspond to a depth first search in the state tree.

The second group is composed of dynamic programming (D.P.) methods. The main idea is the Bellman principle [2] : "Any optimal path is composed of sub-paths which are themselves optimal". These methods correspond to a breath first search in the state tree. Traditionnally they are used when the the state space cardinal is not too large [4].

508

<u>For our problem, a A* method was first implemented.</u>
For each node n are calculated :

       • c(0,n) : the cost beetwen the initial state and node n (exact partial cost)

       • $\hat{c}$(n,T) : the estimation of the cost beetwen node n and the final state (lower bound residual cost)

       • f(n) = c(0,n) + $\hat{c}$(n,T) : evaluation function

The selection of the best nodes is made through means of an estimation of the residual cost. By a short calculus, it is easy to show that a residual energetic cost estimation is obtained by subtracting the energetic evaluation of the volumes in the reservoirs from the cost. By this, nodes whose reservoirs are full are not disadvantaged by comparison with nodes whose reservoirs are empty.
To reduce the complexity, heuristical cutting procedures were added, for example forbidden transitions :

p1	p2	p3
1	0	0

→✗►

p1	p2	p3
0	1	1

This method can't avoid the combinatorial explosion, because the function cost contains other costs besides the energetic one, which are not easy to estimate in $\hat{c}$(n,T).

<u>Dynamic Programming was then implemented.</u>

The Monte-Carlo (random exploration) method was then added to the algorithm (D.P.M.C.). A fixed number of sons was created randomly among all possible ones :

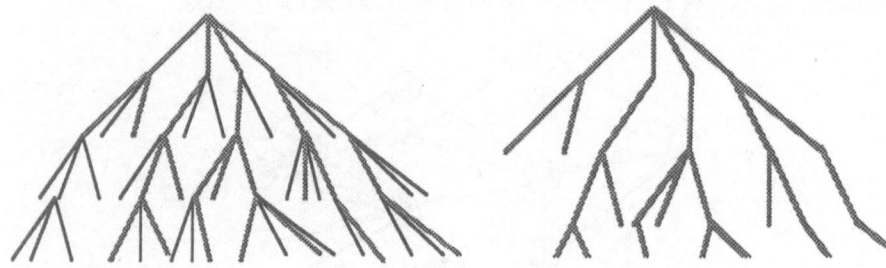

<u>figure 10</u> : random exploration

Figure 10 represents the complete state tree at left and the effectively radomly explored tree et right. The advantages of this approach are :
-First, it allowed to simulate all the graph and not only a fixed part.
-Second, it assumed that doing several replicates would not give a sub-optimal solution that is too far from the optimal one.
-Finally, coefficients of the random path generator that were corresponding to the heuristic constraints mentionned above, enabled us to replace "strong" heuristic constraints by "weak" ones, resulting in a smaller loss of optimality.
The evaluation function f, seen before, was used to limit the complexity. At each step of time, a large (fixed) number of acceptable nodes are built. Instead of keeping all of them, a constant number of them with better f are conserved (we use a heap sort). Figure 11 shows the limitation of the breadth of the state tree :

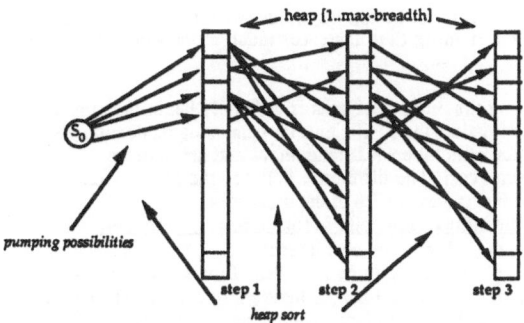

figure 11 : state tree exploration

The control was calculated every hour and the algorithm was working in closed-loop. It was working on a finite horizon, with a fixed end, because of the periodical nature of the problem (24 hours). Heuristical cutting procedures were added, but in some cases they were too restrictive and led to dead ends, while in other cases, they were not sufficient to avoid a combinatorial explosion.

5. RESULTS OF THE OPTIMIZATION ALGORITHM (DPMC)

The behavior of the algorithm as a function of its parameters was studied. Five different examples were tested with ten replicates using different maximum-breadth number and different number of sons, and the cost function defined for this application. Plots of the results that were performed for summer days on the output part of the plant (maximum number of sons =12) using the real water consumption follow:

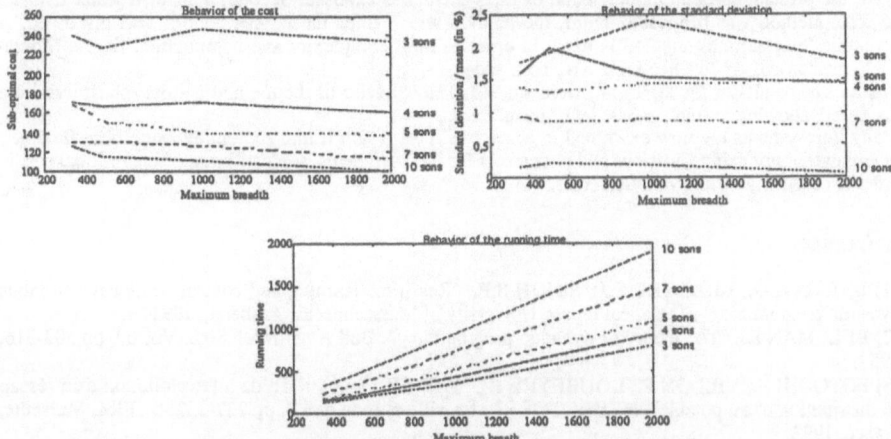

Figure 12 : Behavior of the algorithm versus its parameters

The results show first that the method is working correctly. Since the mean cost shows no great reduction when the maximum breadth increases, it means that the residual cost has a good evaluation. When the number of sons increases, the cost decreases, indicating that the optimal solution has more chance to be selected when the number of sons increases. It is interesting to note that when the number of sons is high, the cost decreases when the maximum breadth increases. It comes from the fact that the algorithm must choose one son out of the number of sons. When this number is high, the choice is difficult to make, explaining why the cost decreases.

The standard deviation shows no variation with the maximum breadth, but it decreases with the number of sons increase. A high number of sons is therefore preferable since it has a lower cost and a lower standard deviation, but it can not be too high, otherwise the selection of the best nodes will not work correctly. Moreover, the running time increases much more with the maximum breadth than with the number of sons.

In conclusion, increasing the number of sons is definitely better than increasing the maximum breadth. To lower the standard deviation, the coefficients for generating random numbers were increased. Therefore, they made the generation of numbers less randomous and hence decreased the standard deviation with no great change in the mean cost. The difference between the lower bound and the suboptimal computed solution is 10% for the bad cases and 3% in the best cases.
We have also studied the influence of noise in forecasting consumption:
As the noise increases, the sub-optimal cost increases also, but, heuristically, we can say that there is no big change in the control. The error in consumption forecast does not effect the control in significant terms. However, the number of failed runs (no node acceptable at a stage of the tree) increased drastically with an error between 10 and 20%. In fact what happens is that the solution found at the last pass is very close to the limits of the acceptability domain. An error in the consumption drives it out of of this domain, and at the next pass, a no acceptable node can be found.
To prevent this, it is possible to have two sets of constraints, one for building the tree and one for the real situations. It is easier to modify the duration of the time step when the solution is very close to the constraints, so that the impact of the error is weakened by a shorter period of application.
Using the forecast of water consumption given by the method exposed in 3, results showed no change and seemed acceptable for the users.
The results obtained with the optimal control method (DPMC) were also compared with the results of a neural network method ([6]).

6. CONCLUSION

In this article, an algorithm for the control of pumps in water treatment plants was introduced. This algorithm has been implemented in C for two water treatment plants of the group Lyonnaise des Eaux-Dumez. The users say that it is an efficient tool.
First, the presentation and modelisation of the system was exposed. Second, a adapted water demand forecast method was proposed. Third, the method which finds the pumps control was discussed. A Dynamic Programming method is used. In order to limit complexity and computation time a Monte-Carlo method and A*-like heuristics have been added.
Finally, the results of the algorithm were studied. The behavior of the method and its robustness were discussed, allowing to find the best set of parameters.
Many imrovements are now examined to accelerate it or to turn it into something even more flexible : random generator using gaussians and not crenels for its coefficients, sophistication of the residual cost in order to consider goal and variation costs.

References

[1] BARTO A.G., BRADTKE S.J., SINGH S.P., "Real-time learning and control using asynchronous dynamic programming", Technical report, University of Massachusetts, Amherst,, 1991.
[2] BELLMAN R., "The theory of dynamic programming", Bull.Amer.Math.Soc., Vol 60, pp 503-516, 1957.
[3] FOTOOHI F., VILLON P., LOUBEYRE R., "SAPHIR : un outil d'aide à l'exploitation d'un réseau de distribution d'eau potable", HYDROTOP 92 : La ville et l'eau, vol 2, pp.227-232, SAFIM, Marseille, France, 1992.
[4] HAREL D., Algorithmics : the spirit of computing, Addison-Wesley, Reading,Massachussets,1987.
[5] KNUTH D.E., The art of computer programming : fundamental algorithms, Addison-Wesley, Reading, Massachussets, 1973.
[6] LANGLOIS T., CANU S., "Reinforcement learning and dynamic programming : a comparison.", Neuro-Nîmes 92 , EC2, Nîmes, France, 1992.
[7] LOUBEYRE R., JARRIGE P.A., DEMBELE N., "Small water distribution network optimization", Proceedings AWWA Houston, April 1991.
[8] NILSSON N.J., Principles of artificial intelligence, Springer-Verlag, Berlin, Germany, 1987.
[9] PRESS W.H., FLANNERY B.P., TEUKOLSKY S.A., VETTERLING W.T., Numerical recipes in C : the art of scientific computing, pp 244-255, Cambridge University Press, Cambridge, U.K.,1989.

Differential games

STRONGLY TIME CONSISTENT OPTIMALITY PRINCIPLES FOR THE GAME WITH DISCOUNT PAYOFFS

Leon A.Petrosjan

Faculty of Applied Mathematics, St.Petersburg University

198904, Bibliotechnaya pl.,2, Stary Petergof, St.Petersburg, Russia

The problem of time consistency (dynamic stability) for the n-person differential games with discount payoffs was first mentioned in [1], where it was proved that even the Pareto optimal solutions may be time inconsistent in this case. The reason is that in a discount payoff case the payoffs of the players in subgames acquiring along an optimal path essentially change their structure implying the time inconsistency of the chosen optimality principle (OP). Till the last time no attempts have been made for the regularization of the OP's in discount payoff case. We refer to the report [2] on V International Symposium on Dynamic Games where this question was once more stated. Here we try to use the approach of S. Chistjakov and our's (see [3], [4], [5]) to construct a family of strongly time-consistent (dynamic stable) optimality principles in the case under discussion.

Consider a nonzero-sum differential game with n players, $N = \{1, 2, \ldots, n\}$, described by the system of differential equations

$$\dot{x}_i = f^i(x, u_1, \ldots, u_n) \tag{1.1}$$

where $x_i \in R^n$,

$$u_i \in U_i \subset CompR^\ell, x_i^0 = x_i(0)$$

on the time interval $[0, \infty)$. The payoff of the player i is defined as follows.

$$H_i(x_0; x_{(t)}) = \int_0^\infty \exp^{-\lambda_i t} h_i(x(t))dt, h_i > 0 \tag{1.2}$$

where $x(t)$ is trajectory realized in the game.

The resulting game we denote by $\Gamma(x_0)$.

In order to pose the problem correctly, it is necessary to introduce various types of players strategies. There are different approaches. In this work we confine ourselves to the open-loop strategies. By the open-loop strategy of the i-th player, we mean any measurable open-loop control $u_i(t)$, defined on the time interval $[0, \infty)$ with values in the set U_i. We impose restrictions (see, for example [5]) on the system (1.1) to guarantee (for all measurable open-loop controls: $u_1, \ldots, u_i(t), \ldots, u_n(t)$) the existence, uniqueness and continuation (over the time interval $[0, \infty)$) of the solution of the system (1.1). The n-tuple $u(t) = \{u_i(t), \ldots, u_n(t)\}$ is called the "situation of the game".

For the analysis of the differential game the choice of the appropriate optimality principle and the proof of the existence of this principle is of the highest importance.

1. One of the optimality principles is the Pareto-optimality.

Pareto-optimal or ϵ-Pareto-optimal situations always exist in differential games. Recall the definition of Pareto-optimality. We shall say that the situation $u^*(t) = \{u_1^*(t), u_2^*(t), \ldots, $ is Pareto-optimal, if from the existence of the situation $\bar{u}(t) = \{\bar{u}_1(t), \bar{u}_2(t), \ldots, \bar{u}_n(t)\}$, such that $H_i(x_0, x^*(t)) \leq H_i(x_0, \bar{x}(t)), i = 1, \ldots, n$, where $x^*(t)$ is the trajectory in the situation $u^*(t)$, and $\bar{x}(t)$ is the trajectory in the situation $\bar{u}(t)$, it follows that

$$H_i(x_0, x^*(t)) = H_i(x_0, \bar{x}(t)), i = 1, \ldots, n.$$

There may be many Pareto-optimal situations and many corresponding Pareto-optimal trajectories.

The payoffs of the players in various Pareto-optimal situations may differ. Therefore, we designate the following: $P_1(x_0)$ is the set of Pareto-optimal situations in the game $\Gamma(x_0)$, $P_2(x)$ is the set of Pareto-optimal trajectories in the game $\Gamma(x_0)$, $P_3(x)$ is the set of the all possible vector-payoffs of the players in the various Pareto-optimal situations from the set $P_1(x_0)$, so the set $P_3(x_0)$ consists of all possible vectors having the form $\{H_i(x_0, u^*(t)), \ldots, H_i(x_0, u^*(t)), \ldots, H_n(x_0, u^*(t))\}$ where u^* is the Pareto-optimal situation. The main questions which arise during investigation of Pareto-optimal situations and trajectories are questions of dynamic stability and the strong dynamic stability or questions of the time consistency and the strong time consistency.

For differential games with a prescribed duration we can show (see [5]) that Pareto-optimality is the dynamic stable principle of optimality. However, strong dynamic stability does not occur. We will be interested in the question of the dynamic stability of Pareto-optimal solutions in the case of the game $\Gamma(x_0)$ in which the duration is not prescribed in advance.

The definition of dynamic stability (time consistency) for the game $\Gamma(x_0)$ will differ slightly from such definition for a game with a prescribed duration. This is because there is a discount factor, $e^{-\lambda_i t}$. This factor has special economic meaning, showing that the significance of the income to a player diminishes as the moment when the player is to receive this income recedes.

Before the introduction of the dynamic stability (time consistency) concept, we will define a subgame of the game $\Gamma(x_0)$. Let $\bar{u}(t)$ be some fixed situation in the game $\Gamma(x_0)$, and $\tilde{x}(\tau)$ the corresponding trajectory, $\tau \in [0, \infty)$

We define the subgame $\Gamma(\tilde{x}(\tau))$ as follows: the equations of motion of the game $\Gamma(x_0)$ coincide with those for the game $\Gamma(x_0)$ and have the form (1.1). However, the development of the game begins from point $\tilde{x}(\tau)$ and occurs on the time interval $[\tau, \infty)$. Thus, we consider the equations (1.1) in the game $\Gamma(\tilde{x}(\tau))$ with the starting condition $\tilde{x}(\tau)$. The open-loop strategy $u_i^\tau(t)$ of the i-th player in subgame $\Gamma(\tilde{x}(\tau))$ represents the truncation of an open-loop strategy $u_i(t)$ of the i-th player on the time interval $[\tau, \infty)$. Thus, naturally, any trajectory $x^\tau(t)$ in the subgame $\Gamma(\tilde{x}(\tau))$ is the truncation of the trajectory $x(t)$ in the game $\Gamma(\tilde{x}(\tau))$ on the time interval $[\tau, \infty)$.

The payoff of the i-th player in the subgame $\Gamma(\tilde{x}(\tau))$ is defined as follows: let $u^\tau(t)$ be the situation in the subgame $\Gamma(\tilde{x}(\tau))$ and $x^\tau(t)$ be the trajectory in this situation. Then the i-th players payoff equals:

$$K_i(\tilde{x}(\tau); x^\tau(t)) = \int_\tau^\infty e^{-\lambda_i(t-\tau)} h_i(x^\tau(t)) dt, i = 1, \ldots, n$$

The definition of the subgame of the game $\Gamma(x_0)$ differs from the definition accepted in the literature in that the payoff function of i-th player in the subgame $\Gamma(\tilde{x}(\tau))$ does not coinside with the truncation of the payoff function of the i-th player in the game $\Gamma(x_0)$ on the time interval $[0, \infty)$, but rather differs from this truncation by the discount multiplier $e^{\lambda_i \tau}$.

This seems a natural formulation given that player i, beginning the subgame at the moment τ, must value the payoffs which he will be able to receive in the future just as in the game $\Gamma(x_0)$, i.e. the player must appraise the payoffs with the same discount factor.

At the same time, such a nonstandard definition of the subgame forces us to address the complications of the dynamic stability (time consistancy) concept.

Let $u^*(t) \in P_1(x_0)$; $x^*(t) \in P_2(x_0)$ is realised in the situation $u^*(t)$; $K^*(x_0, x^*(t)) \in P_3(x_0)$ represents the vector-payoff of players on the trajectory $x^*(t)$; and

$K_i^*(x_0, x^*(t)) = \int_0^\infty e^{-\lambda_i t} h_i(x^*(t)) dt, i = 1, \ldots, n$

Definition: The Pareto-optimal situation $u^*(t)$ is called dynamic stable (time consistent) if the truncation of this situation $u^{\tau*}(t) \in P_1(x^*(\tau))$ for any τ from $[0, \infty)$, where $P_1(x^*(\tau))$ is the Pareto-optimal set of situation in subgame $\Gamma(x^*(\tau))$. The payoff of the i-th player in the game with infinite time horizon we interprete in the following way: $h_i(x(t)) dt$ is the income received by the player i on the time-interval $[t, t + dt]$ along the trajectory $x(t)$; $e^{-\lambda_i t}$ is the factor which characterizes the degree of utility of income, received in the moment t by the i-th player in the beginning of the process (at moment $t = 0$). That is, in the beginning of the process, the utility for the i-th player of receiving income on the interval $[t, t + dt]$ is not equal to the real income $h_i(x(t)) dt$, but is rather equal to the smaller quantity $e^{-\lambda_i t} h_i(x(t)) dt$.

Such a formulation of income utility expresses the natural condition that utility of income received in the future significantly drops when the time increases.

At the same time, with the transition from the main game to the concrete subgame $\Gamma(x(\tau))$, the players' utilities (payoff functions) begin running from the moment τ which is the beginning of the given subgame.

Therefore, the payoff of the i-th player in the subgame $\Gamma(x(\tau))$ we interpret as follows: $h_i(x(t))dt$, when $t > \tau$ is the income player i receives on the interval $[t, t + dt]$ in the subgame $\Gamma(x(\tau))$, $e^{-\lambda_i(t-\tau)}$ is the factor characterizing the degree of utility of the income to be received of moment $t > \tau$ to the i-th player in the beginning of the subgame at $t = \tau$. that is, at the beginning of the subgame $\Gamma(x(\tau))$, at $t = \tau$, the utility to the i-th player of receiving the income is not equal to the real income $h_i(x(t))dt$, but it is equal to the quantity $e^{-\lambda_i(t-\tau)}h_i(x(t))dt$.

Such a formula for defining the utility for the players in the subgame expresses the loss of utility of income received at moment t, relative to the starting condition of the given subgame.

Maintaining the dynamic stability (time-consistency) of the Pareto-optimal trajectory demands the conservation of the Pareto-optimality on the interval of this trajectory, beginning from the moment τ in the subgame $\Gamma(x^*(\tau))$.

2. Regularization of the Pareto optimality principle.

Definition. Let $0 = t_0 < t_1 < \ldots < t_k < t_{k+1} < \ldots, t_{k+1} - t_k = \delta$ be a partition of the time interval $[0, \infty)$. The Pareto-optimal situation $u^*(t)$ is called δ-time consistent (δ-dynamic stable) if the truncation of this situation $u^{\tau*}(t) \in P_1(x^*(t_k))$ for any $t_k, k = 1, 2, \ldots$, and $P_1(x^*(t_k))$ is the set of all Pareto-optimal situations in the subgame $\Gamma(x^*(t_k))$.

The Pareto-optimal situation $u^*(t)$ is called δ-strongly time consistent (δ-strongly dynamic stable) if any situation of the form

$$\bar{u}(t) = \begin{cases} \bar{u}(t), & t \in [0, t_k); \\ \hat{u}(t), & t \in [t_k, \infty), \hat{u}(t) \in P_1(x^*(t_k)) \end{cases}$$

$k = 1, \ldots$, belongs to the set $P_1(x_0)$, where $P_1(x^*(t))$ is the set of all Pareto-optimal situations in the subgame $\Gamma(x^*(t_k))$.

Consider the set $\tilde{P}_1(x_0)$ of all possible $\bar{u}(t)$ having the following property:

Let $\bar{x}(t)$ be a corresponding trajectory, consider for each $k = 0, 1, \ldots$ the subgames $\Gamma(\bar{x}(t_k))$, and sets $P_1(\bar{x}^*(t_k))$, then

$$\bar{u}(t) = u_k^*(t),$$

for $t \in [t_k, t_{k+1})$, where $u_k^*(t) \in P_1(\bar{x}(t_k))$ is a Pareto optimal situation in the subgame $\Gamma(\bar{x}(t_k))$.

We call the set $\tilde{P}_1(x_0)$ the regularization of the Pareto-optimal set $P_1(x_0)$. Clearly $P_1(x_0) \subset \tilde{P}_1(x_0)$

Consider the corresponding set of trajectories $\bar{x}(t)$, $\tilde{P}_2(x_0)$, clearly $P_2(x_0) \subset \tilde{P}_2(x_0)$

Let $\bar{x}(t) \in \bar{P}_2(x_0)$.

Consider the sets $\bar{P}_1(\bar{x}(t_k))$ for $k = 1, 2, \ldots$. We can see that for any $\bar{x}(t)$ the situation $\tilde{u}(t)$ of the form

$$\tilde{u}(t) = \begin{cases} \bar{u}(t), & t \in [0, t_k); \\ \hat{u}(t), & t \in [t_k, \infty), \hat{u}(t) \in \bar{P}_1(\bar{x}(t_k)) \end{cases}$$

$k = 1, \ldots$, belongs to the set $\bar{P}_1(x_0)$. This is clear from the definition of the set $\bar{P}_1(x_0)$. For $\bar{u}(t)$ $(\bar{u}(t) \in \bar{P}_1(x_0))$ we have

$$\bar{u}(t) = u_k^*(t), t \in [t_\ell, t_{\ell+1}), \ell = 1, \ldots, k-1,$$

where $u_k^*(t) \in \bar{P}_1(\bar{x}(t_k))$ and for $\hat{u}(t) \in \bar{P}_1(\bar{x}(t_k))$ we have that

$$\hat{u}(t) = u_\ell^*(t), t \in [t_\ell, t_{\ell+1}), \ell = k, \ldots,$$

where $u_\ell^*(t) \in P(\hat{x}(t_\ell))$, $\hat{x}(t_k)$ is the trajectory in the subgame $\Gamma(\bar{x}(t_k))$ corresponding to $\hat{u}(t)$. If now we consider the trajectory

$$\bar{\bar{x}}(t) = \begin{cases} \bar{x}(t), & t \in [t, t_k); \\ \hat{x}(t), & t \in [t_k, \infty) \end{cases},$$

then this trajectory corresponds to the situation $\tilde{u}(t)$, and thus we shall have

$$\tilde{u}(t) = u_\ell^*(t),$$

for $t \in [t_\ell, t_{\ell+1}), \ell = 0, 1, \ldots$ where $u_\ell^*(t) \in \bar{P}_1(\bar{x}(t_\ell))$.

The last condition means that the regularized Pareto-optimality principle $\bar{P}_1(x_0)$ is δ-strongly time-consistent.

Theorem 1. *The $OP\bar{P}_1(x_0)$ is δ-strongly time-consistent.*

3. Cooperative optimality principles.

Here we shall consider core as OP in the game, but all the results remain valid for any other subset of imputations, considered as optimality principle.

Consider n–person differential game $\Gamma(x_0)$

$$\dot{x} = F(x, u_1, \ldots, u_n), \ u_i \in U \subset CompR^\ell, x(t_0) = x_0$$

with payoffs

$$K_i(x_0, u_1, \ldots, u_n) = \int_{t_0}^\infty e^{\lambda_i(t-t_0)} h_i(x(t)) dt, \ h_i > 0, \ \lambda_i > 0$$

Cooperative form of $\Gamma(x_0)$.

Consider the n–tuple of open loop controls $u_1^*, \ldots, u_n^*$ such that the corresponding trajectory maximizes the sum of the payoffs (utilities)

$$\max_u \sum_{i=1}^{n} K_i(x_0; u_1, \ldots, u_n) = \sum_{i=1}^{n} K_i(x_0; u_1^*, \ldots, u_n^*) =$$

$$\sum_{i=1}^{n} \int_{t_0}^{\infty} e^{-\lambda_i(t-t_0)} h_i(x^{1*}(t)) dt = V(N; x_0),$$

where N is the set of all players in $\Gamma(x_0)$. The trajectory $x^{1*}(t)$ is called conditionally optimal. Let $V^1(S; x_0)$ be the characteristic function ($S \subset N$) and $C^1(x_0)$ the core. Consider the family of subgames $\Gamma(x^{1*}(t))$, along $x^{1*}(t)$, $t \in [t_0, \infty)$, corresponding cores and c.f. $V^1(S; x^{1*}(t))$. The payoff functions in the subgames have the form

$$K_i^t(x^*(t), u_i, \ldots, u_n) =$$

$$\int_{t}^{\infty} e^{-\lambda_i(\tau-t)} h_i(x^{1*}(\tau)) d\tau,$$

and differs by multiplier $e^{\lambda_i t}$ from the payoff functions in the subgames defined for the games without discount factor. This essentially changes the relative weights of payoff functions of different players, when the game develops, and thus the whole game itself.

Consider the partition of the time interval $[t_0, \infty)$ by the points $t_0 = \Theta_0 < \Theta_1 < \ldots < \Theta_k < \Theta_{k+1} < \ldots$, where $\Theta_{k+1} - \Theta_k = \delta > 0$ does not depend upon k, the subgame $\Gamma(x^{1*}(\Theta_1))$,c.f. $V^2(S; x^{1*}(\Theta_1))$ and core $C^2(x^{1*}(\Theta_1))$. Let $x^{2*}(t)$, $t \geq \Theta_1$ be the conditionally optimal trajectory in the subgame $\Gamma(x^{1*}(\Theta_1))$, i.e. such that

$$\max_u \sum_{i=1}^{n} K_i^{\Theta}(x^{1*}(\Theta_1), u_1, \ldots, u_n) =$$

$$= \sum_{i=1}^{n} K_i^{\Theta}(x^{1*}(\Theta_1), u_i^*, \ldots, u_n^*) =$$

$$= \sum_{i=1}^{n} \int_{\Theta_1}^{\infty} e^{-\lambda_i(\tau-\Theta_1)} h_i(x^{2*}(\tau)) d\tau.$$

Then consider the subgame $\Gamma(x^{2*}(\Theta_2))$, c.f. $V^3(S; x^{2*}(\Theta_2))$, core $C^3(x^{2*}(\Theta_2))$.

Continuing in the same manner we get the sequence of subgames $\Gamma(x^{k*}(\Theta_k))$, c.f. $V^{k+1}(S; x^{k*}(\Theta_n))$, cores $C^{k+1}(x^{k*}(\Theta_k))$ and conditionally optimal trajectories $x^{(k+1)*}(t), t \geq \Theta_k$.

Definition. The trajectory $x^*(t) = x^{k*}(t)$, $t \in [\Theta_{k-1}, \Theta_k)$, $k = 1, 2, \ldots$ is called optimal in the game $\Gamma(x_0)(x^{k*}(\Theta_k) = x^{(k+1)*}(\Theta_k))$.

In this formalization of the cooperative game we suppose that the players starting the game agree to use the optimal trajectory $x^*(t)$, $t \geq \Theta_0$, $t_0 = \Theta_0$. Our formalization depends upon $\delta > 0$, and we denote the cooperative form of $\Gamma(x_0)$ by $\Gamma^\delta(x_0)$.

The vector function $(\beta_1(\tau), \ldots, \beta_i(\tau), \ldots, \beta_n(\tau))$ is called the utility distribution procedure (UDP) in $\Gamma^\delta(x_0)$ if

$$\sum_{i=1}^n \int_{\Theta_k}^{\Theta_{k+1}} \beta_i(t)dt = \sum_{i=1}^n \int_{\Theta_k}^{\Theta_{k+1}} e^{-\lambda_i(t-\Theta_k)}h_i(x^{(k+1)*}(t))dt =$$

$$= V^{k+1}(N; x^{(k+1)*}(\Theta_k)) - V^{k+1}(N; x^{(k+1)*}(\Theta_{k+1})), \beta_i(t) \geq 0,$$

$$i = 1, \ldots, n, \ k = 0, 1, \ldots \tag{2.1}$$

Let $\xi^k \in C^{k+1}(x^{(k+1)*}(\Theta_k))$, $k = 0, 1, \ldots$ be any imputation in the subgame $\Gamma(x^{k*}(\Theta_k))$ belonging to the core of this subgame. Define the function $\beta_i(t)$, $t \in [\Theta_k, \Theta_{k+1})$ by the formula

$$\beta_i(t) = \frac{\xi_i^k \sum_{i=1}^n \int_{\Theta_k}^{\Theta_{k+1}} e^{-\lambda_i(t-\Theta_k)}h_i(x^{(k+1)*}(t))dt}{V^{k+1}(N; x^{k*}(\Theta_k))(\Theta_{k+1} - \Theta_k)} =$$

$$= \frac{\xi_i^k[V^{k+1}(N; x^{(k+1)*}(\Theta_k)) - V^{k+1}(N; x^{(k+1)*}(\Theta_{k+1}))]}{V^{k+1}(N; x^{k*}(\Theta_k))\delta} \geq 0 \tag{2.2}$$

$i = 1, \ldots, n, \ k = 0, 1, \ldots$. The functions $\{\beta_i(t)\}$, from (2.2) $t \geq t_0$ constitute for each $\xi^k \in C^{k+1}(x^{(k+1)*}(\Theta_k))$, $k = 0, 1, \ldots$ a UDP in $\Gamma^\delta(x_0)$. Let $\bar\xi = \int_{t_0}^\infty \beta(t)dt$. Denote by $\bar C^{k+1}(x^{(k+1)*}(\Theta_k)$ the set of all such vectors $\bar\xi$, for all possible UDP's $\beta(t)$, for different $\xi^k \in C^{k+1}(x^{(k+1)*}(\Theta_k))$, $k = 0, 1, \ldots$. Consider $\bar C(x_0)$ as optimality principal (OP) in $\Gamma^\delta(x_0)$), and call it the regularized core (RC). Define $\bar C(x^*(\Theta_k))$ for subgame $\Gamma^\delta(x^*(\Theta_k))$, in the similar way.

Definition. The $\bar C(x_0)$ is called δ-STC if for all $\beta(t)$ we have

$$\left\{ \int_{\Theta_0}^{\Theta_k} \beta_i(t)dt \right\} + \bar C(x^*(\Theta_k)) \subset \bar C(x_0) = \bar C(x^*(\Theta_0)) \tag{2.3}$$

In (2.3) under the summation $a + A$, where a is a vector from R^n and A a set from R^n we understand the set of all vectors of the form $a + b$, where $b \in A$

Theorem 1. The (RC) $\bar C(x_0)$ is a δ-STC optimality principle in $\Gamma^\delta(x_0)$.

Proof. Suppose

$$\alpha \in \left\{ \int_{\Theta_0}^{\Theta_k} \beta_i(t)dt + \bar C(x^*(\Theta_k)) \right\},$$

then $\alpha = \{\alpha_i\}$ can be represented in the form

$$\cdot \alpha_i = \int_{\Theta_0}^{\Theta_k} \beta_i(t)dt + \alpha_i^{\Theta_k},$$

where $\alpha_i^{\Theta_k} \in \bar{C}(x^*\Theta_k))$. By the definition of $\bar{C}(x^*(\Theta_k))$ (as the OP in the subgame $\Gamma(x^*(\Theta_k)))$

$$\alpha_i^{\Theta_k} = \int_{\Theta_k}^{\infty} \tilde{\beta}_i(t)dt,$$

where $\tilde{\beta}_i(t)$ is defined by (2.2) on the time interval $t \geq \Theta_k$ for some collection of

$$\bar{\xi}^\ell = \{\bar{\xi}_i^\ell\}, \bar{\xi}^{(\ell+1)*} \in C^{\ell+1}(x^*(\Theta_\ell)), \ell = k, k+1, \ldots.$$

If we define now

$$\tilde{\beta}_i(t) = \begin{cases} \beta_i(t)dt, & \Theta_0 \leq t < \Theta_k; \\ \tilde{\beta}_i(t), & \Theta_k \leq t < \infty, \end{cases}$$

then $\alpha_i = \int_{\Theta_q}^{\infty} \tilde{\beta}_i(t)dt$, where $\tilde{\beta}(t) = \{\tilde{\beta}_i(t)\}$ is the UDP in $\Gamma^\delta(x_0)$ defined by (2.1),(2.2) for some special collection of

$$\bar{\xi}^\ell \in C^{\ell+1*}(x^{(\ell+1)*}(\Theta_\ell)),$$

for $0 \leq \ell < k$, and $\bar{\xi}^\ell \in C^{\ell+1}(x^{(\ell+1)*}(\Theta_\ell))$ for $\ell \geq k$. Thus $\alpha \in \bar{C}(x^*(\Theta_0))$ by the definition.

References

[1] R. H. Strotz, Myopia and Inconsistency in Dynamic Utility Maximization, Review of Economic Studies, Vol XXIII, 1955–56.

[2] V. Kaitala, M. Pohjola, Sustainable International Agreements on Green House Warning – a Game Theory Study, In Proceedings of the V International Symposium on Dynamic Games Applications, Geneva, 1992.

[3] L.A. Petrosjan, Solutions of n–person differential games. Dynamic control, Sverdlovsk, 1979, p. 208–210. (Russian)

[4] L.A. Petrosjan, The time–consistency of the optimality principles in non–zero sum differential games. Dynamic Games in Economic Analysis. Springer–Verlag, 1991, p.299–311.

[5] L. A. Petrosjan, N. N. Danilov, Cooperative differential games and applications, Tomsk, 1986, p.274.

THE DETECTION PROBABILITIES IN SIMULTANEOUS PURSUIT GAMES

T.V. Slobodinskaia
St.Petersburg Technological Institute,
Moskowsky 26, 198031
St.Petersburg,Russia

Solving applied problems, formulated as zero-sum games, the choice of payoff function is a free choice of a researcher. The question arises how the optimal result changes when the payoff function is replaced by one having similar interpretation. Therefore it is interesting to compare the detection probabilities in simultaneous pursuit games and simultaneous search games. In our paper the detection probabilities are calculated provided the players use the strategies, optimal in pursuit game. The comparative analysis of obtained results under the optimal strategies in search games is done. The problems considered are close to [1]-[4].

A. The simultaneous game on an closed interval.

The player P (pursuer or searcher) chooses a point $z = (z_1, \ldots, z_m), z_i \in [-1; 1], i = 1, \ldots, m$ and the player E (evader or hider) chooses a point $y \in [-1; 1]$. If we consider the game as a pursuit game, the payoff of the player E is equal to $K(z_1, \ldots, z_m; y) = \min_{i=1,\ldots,m} \rho(z_i, y)$. If we consider the game as a search game, the payoff of P is equal to 1, when $\min_{i=1,\ldots,m} \rho(z_i, y) \leq l$ and equal to 0, when $\min_{i=1,\ldots,m} \rho(z_i, y) > l$, where $\rho(z_i, y)$ is euqlidean distance, $l > 0$ a given number (detection radius).

It is known (see [2],[3]) that both a pursuit game and search game have saddle points in mixed strategies. Let $\mu(\nu)$ and $X(Y)$ be mixed strategies of the players $P(E)$ in pursuit and in the search games correspondingly. Denote optimal strategies by (∗) and the detection probability by P_g.

The optimal strategy μ^* of the player P in the pursuit game prescribes equal probabilities $\frac{1}{2}$ to two collections of points

$$\left(-1, -\frac{2m-5}{2m-1}, \ldots, -\frac{1}{2m-1}, \frac{3}{2m-1}, \ldots, \frac{2m-3}{2m-1}\right)$$

and

$$\left(-\frac{2m-3}{2m-1}, -\frac{2m-7}{2m-1}, \ldots, -\frac{3}{2m-1}, \frac{1}{2m-1}, \ldots, \frac{2m-5}{2m-1}, 1\right).$$

For the player E is optimal the choice of the points

$$\left(-1, -\frac{2m-3}{2m-1}, \ldots, -\frac{2m-2i-1}{2m-1}, \ldots, -\frac{1}{2m-1}, \frac{1}{2m-1}, \ldots, \frac{2m-3}{2m-1}, 1\right)$$

with equal probabilities $\frac{1}{2m}$ (strategy ν^*).

Describe optimal strategies of the players in the search game.

Let

$$n = \begin{cases} \frac{1}{\ell}, & \text{if } \frac{1}{\ell} \text{ is an integer;} \\[2mm] [\frac{1}{\ell}] + 1, & \text{if } \frac{1}{\ell} \text{ is not an integer.} \end{cases}$$

Consider points

$$\bar{x}_i = -1 + \ell + \frac{2(1-\ell)}{n-1} \cdot (i-1), i = 1, \ldots, n; \bar{y}_j = -1 + \frac{2(j-1)}{n-1}, j = 1, \ldots, n.$$

Then we have

$$\rho(\bar{x}_{i-1}, \bar{x}_i) \leq 2\ell, i = 2, 3, \ldots, n; \rho(\bar{y}_{j-1}, \bar{y}_j) > 2\ell, j = 2, 3, \ldots, n.$$

For the player P any choice with equal probabilities of any subset of m different points x_i will be optimal (strategy X^*). The optimal strategy Y^* of the player E consists in the choice of the points $\bar{y}_j$ with equal probabilities. The value of the search game is equal to $\frac{m}{n}$ if $m < n$ and to 1 if $m \geq n$.

Calculate P_S for the strategies μ^*, ν^*.

The distance between two neighbour points from the different spectral collections of points in the strategy μ^* is equal to $\frac{2}{2m-1}$. Hence

1.If $\frac{2}{2m-1} \leq \ell$, then $P_S(\bar{\mu}^*, y) = 1$ for any pure strategy y of the player E.

2.If $\frac{1}{2m-1} \leq \ell < \frac{2}{2m-1}$, then $P_S(\mu^*, y) \geq \frac{1}{2}$.

3.If $\frac{1}{2m-1} > \ell$, then the player E has a pure strategy $\bar{y}$, such that $P_S(\mu^*, \bar{y}) = 0$.

The distance between two neighbour spectral points of the strategy ν^* is equal to $\frac{2}{2m-1}$. Hence

1.If $\ell \geq \frac{1}{2m-1}$ then the player P has a pure strategy $\bar{x} = (\bar{x}_1, \ldots, \bar{x}_m)$ such that $P_S(\bar{x}, \nu^*) = 1$ This takes place if, for instance,

$$\bar{x}_i = -1 + \ell + \frac{2(1-\ell)}{m-1} \cdot (i-1), i = 1, \ldots, m.$$

2.If $0 < \ell < \frac{1}{2m-1}$, then $P_S(x, \nu^*) \leq \frac{1}{2}$ for any pure strategy x of the player P. Indeed, any point x_i of his m-collection of the points can "catch" in it's ℓ-neighbourhood no more then one spectral point of the strategy ν^* and, hence, $P_S(x, \nu^*) \leq \frac{m}{2m} = \frac{1}{2}$.

3.

$$P_S(\mu_o, \nu_o) = \begin{cases} 1, & \text{when } \ell \geq \frac{2}{2m-1}, \\[2mm] \frac{1}{2}, & \text{when } 0 < \ell < \frac{2}{2m-1}. \end{cases}$$

Let do comparative analysis of guaranted payoffs of the players P and E using strategies μ^*, X^* and ν_o, Y^* for the different values of the parameters ℓ and m.

1.The value of a search game is equal to 1 if $m \geq n \geq \frac{1}{\ell}$. But if the player P uses strategy μ^* in a search game, he can be sure in the payoff equal to 1 only if $\ell \geq \frac{2}{2m-1}$. The difference $\frac{2}{2m-1} - \frac{1}{m} = \frac{1}{m(2m-1)} > 0$ is positive, but small for a large m.

2.The value of a search game is equal to $\frac{m}{n}$, if $m < n$ and

$$\frac{1}{\ell} \leq n < \frac{1}{\ell} + \ell \leftrightarrow \frac{m\ell}{\ell+1} < \frac{m}{n} \leq m\ell.$$

Hence, $\frac{m}{n} > \frac{1}{2}$ if $\ell > \frac{1}{2m-1}$. But if player P uses strategy μ^*, he can be sure in detection of the player E with the probability equal to $\frac{1}{2}$ if $\frac{1}{2m-1} \leq \ell < \frac{2}{2m-1}$. Thus, the detective probability of the strategy X^* is larger, then the detective probability of the strategy μ^*. Let compare them for some values of the parameters ℓ and m. Let $\ell = \frac{1}{2m-1}$. Then $\frac{1}{2} < \frac{m}{n} \leq \frac{1}{2} + \frac{1}{2 \cdot (2m-1)}$. Hence

$\frac{1}{2} < \frac{m}{n} < 1$ if $m = 1$; $\frac{1}{2} < \frac{m}{n} < \frac{1}{2} + \frac{1}{6}$ if $m = 2$; $\frac{1}{2} < \frac{m}{n} < \frac{1}{2} + \frac{1}{10}$ if $m = 3$; $\frac{1}{2} < \frac{m}{n} < \frac{1}{2} + \frac{1}{16}$. if $m = 4$ and so on.

Thus, using strategies X^* or μ^* there is a small difference for the detection probabilities if $\ell = \frac{1}{2m-1}$ and this difference becomes smaller with the increasing of m.

Let $\ell = \frac{3}{2(2m-1)} \in [\frac{1}{2m-1}; \frac{2}{2m-1}]$; Then $\frac{3m}{4m+1} < \frac{m}{n} \leq \frac{3m}{4m-2}$. Hence $\frac{6}{9} < \frac{m}{n} \leq 1$ if $m = 2$; $\frac{9}{13} < \frac{m}{n} \leq \frac{6}{10}$ if $m = 3$, $\frac{12}{17} < \frac{m}{n} \leq \frac{6}{7}$ if $m = 4$ and so on. $\lim_{n \to \infty} \frac{m}{n} = \frac{3}{4}$. And in any case we have $P_S(X^*, y) > \frac{2}{3}$. But using strategy μ^* in a search game, P can guarantee the detection probability not smaller then $\frac{1}{2}$ only. And with the increasing of ℓ the difference between $\frac{m}{n}$ and $\frac{1}{2}$ will increase.

Compare strategies ν^* and Y^* of the player E.

1. If $n = 2m$, then the strategies ν^* and Y^* coincide and $P_S(z, \nu^*) = P_S(z, Y^*) \leq \frac{1}{2}$ for any pure strategy $z = (z_1, \ldots, z_m)$ of the player P.

2. Let $n \neq 2m$. Using strategy Y^* E will be detected with the probability equal to 1 if $m \geq n \geq \frac{1}{\ell}$, i.e. if $\ell = \ell_1 \geq \frac{1}{m}$. But if E uses strategy ν^*, then P has a pure strategy, which guarantees him the detection probability equal to 1 if $\ell = \ell_2 \geq \frac{1}{2m-1}$. $\ell_1 > \ell_2$, hence the detection radius increases. But the difference $\ell_1 - \ell_2 = \frac{m-1}{m(2m-1)}$ is small for large m.

3.Let $n \neq 2m$. $P_S(z, \nu^*) \leq \frac{1}{2}$ for any pure strategy $z = (z_1, \ldots, z_m)$ of the player P, if $0 < \ell < \frac{1}{2m-1}$. If P chooses any collection of m points from the $2m$ spectral points of the strategy ν^*, then $P_S(z, \nu^*) = \frac{1}{2}$. At the same time, $P_S(z, Y^*) \leq \frac{m}{n}$ and $\frac{m\ell}{\ell+1} < \frac{m}{n} \leq m\ell$, hence, if m is fixed, $\lim_{\ell \to 0} \frac{m}{n} = 0$.

B.The simultaneous game on a circle , the case of one pursuer.

The player P chooses a point $z \in S(0; R)$ where $S(0; R)$ is a circle of the radius R with the center O. The player E chooses a point $t \in S(O; R)$. The payoff of the player E in the pursuit game is equal to $K(z, t) = \rho(z, t)$. $P_S(z, t) = 1$ if $\rho(z, t) \leq \ell$ and $P_S(z, t) = 0$, if $\rho(z, t) > \ell$ in the search game.

The pursuit game has a saddle point. For P the choice of the point O (the center of the circle) with the probability equal to 1 is an optimal (strategy μ^*). For the player E the choice of any two points being on the opposite ends of the diameter of the circle

$S(0;R)$ with equal probabilities $\frac{1}{2}$ is optimal (strategy ν^*). The value of the game is equal R (see [1]).

Optimal strategies in the search game are unknown.

It is obvious that

1.
$$P_S(\mu^*,t) = \begin{cases} 1, & \ell \geq R, \\ 0, & \ell < R. \end{cases}$$

2. $P_S(z,\nu^*) \leq \frac{1}{2}$, when $\ell < R$ for any pure strategy z of the player P.

C. The simultaneous game on a circle, the case of two pursuers.

The player P chooses two points (z_1, z_2), $z_i \in S(O; R)$, $i = 1, 2$ and the player E chooses one point $t \in S(O; R)$. The payoff of the player E in the pursuit game is equal to $K(z_1, z_2; t) = \min_{i=1,2} \rho^2(z_i, t)$. $P_S(z,t) = 1$, if $\min_{i=1,2} \rho(z_i,t) < \ell$ and $P_S(z,t) = 0$, if $\min_{i=1,2} \rho(z_i,t) < \ell$ in the search game.

The pursuit game has a saddle point in mixed strategies. The following strategies are optimal. For the player P, the uniform distributions of the points z_1, z_2 on the circle of radius $r_0 = \frac{2R}{\pi}$ with its center at the point O, the points z_1 and z_2 being the opposite ends of the diameter of this circle is optimal (strategy μ^*). For the player E, the uniform distribution on the circle of radius R with its center at the point O (strategy ν^*). The value of the game is equal to

$$\Phi(r_0, R) = r_0^2 + R_2 - \frac{4}{\pi}r_0 R = R^2(1 - \frac{4}{\pi^2})$$

(see [1]).

The optimal strategies in the search game are unknown, but the estimations of the value of the search game may be derived (see [2]).

Calculate $P_S(\mu^*, Y)$. Denote by $s(\Gamma)$-the length of a part of a circle with the radius r_0 and center O situated within the circle $K(t, \ell)$.

1. If $\ell \leq r_0$, then E has such a pure strategy t (the choice of the point O with the probability equal to 1), that $P_S(\mu^*, t) = 0$.

2. If $r_0 < \ell < R$ and $s(\Gamma) \geq \pi r_0$, then $P_S(\mu^*, t) = 1$ for any pure strategy t of the player E.

3. If $r_0 < \ell < R$ and $s(\Gamma) < \pi r_0$, then

$$P_S(\mu^*, t) = \frac{2s(\Gamma)}{2\pi r_0} = \frac{s(\Gamma)}{2R}$$

It is obvious, that $\min_t P_S(\mu^*, t)$ will be obtained in the point $\bar{t}$, laying on the boundary of the circle $S(O, R)$. Let $\bar{t}$ be such a point . Choose co-ordinate system as is shown on Figure.1. Then $s(\Gamma) = 2\varphi r_0$, where φ is a Polar angle of intersection of the circles of radius r_0 and ℓ with its centers at the points O and $\bar{t}$ correspondingly. The equations of this circles are $x^2 + y^2 = r_0^2$ and $(x - R)^2 + y^2 = \ell^2$. In Polar coordinates, this equations have the form $\rho = r_0$ and $\rho^2 - 2R\rho \cos\varphi + R^2 - \ell^2 = 0$.

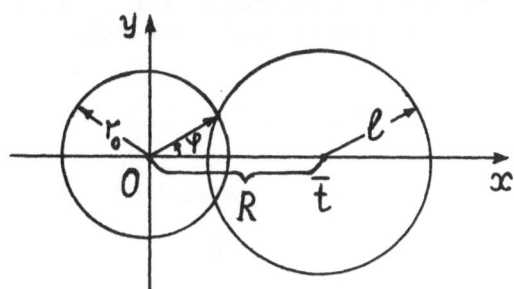

Figure 1

Then

$$r_0^2 - 2Rr_0 \cos\varphi + R^2 - \ell^2 = 0,$$

i.e.

$$\cos\varphi = \frac{R^2 - \ell^2 + r_0^2}{2Rr_0}.$$

Thus $s(\Gamma) = 2r_0 \arccos \frac{R^2 - \ell^2 + r_0^2}{2Rr_0}$ and hence

$$P_S(\mu^*, t) \geq \frac{r_0 \arccos\left(\frac{R^2 - \ell^2 + r_0^2}{2Rr_0}\right)}{R} = \frac{2}{\pi} \arccos \frac{(\pi^2 + 4)R^2 - \pi\ell^2}{4\pi R^2}$$

If, for example, $\ell = \frac{R\sqrt{3}}{2} + \epsilon$, then

$$P_S(\mu^*, t) \geq \frac{2}{\pi} \arccos \frac{(\pi^2 + 4)R^2 - \frac{3\pi^2 R^2}{4}}{4\pi R^2} \approx 0,656$$

If $\ell = \frac{R}{\sqrt{2}} + \epsilon$, then

$$P_S(\mu^*, t) \geq \frac{2}{\pi} \arccos \frac{\pi^2 + 8}{8\pi} \approx 0,496.$$

4. If $\ell \geq R$, then the strategy μ^* will guarantee the detection probability equal to 1 only when $\ell \geq \sqrt{R^2 - r_0^2}$. In this case the player P has a pure strategy which guarantees the detection probability equal to 1, when $\ell = R + \epsilon$ for any $\epsilon > 0$. This is the choice of the center O (point z_1) and any other point $z_2 \in S(O, R)$.

Calculate $P_S(z_1, z_2; \nu^*)$. Remember that ν^* is the uniform distribution on the circle of radius R with its center at the point O.

1. If $\ell > R$, then the player P has a pure strategy $z_1 = z_2 = 0$, such that $P_s(\bar{z}_1, \bar{z}_2, \nu^*) = 1$;

2. If $\ell \leq R$, then $P_S(z_1, z_2; \nu^*) \leq \frac{s(\Gamma)}{\pi R}$, where $s(\Gamma)$ is the length of a part of a circle with the radius R and center O laying within the circle with the radius ℓ and center $\bar{z}$. The point $\bar{z}$ is laying on the boundary of the circle $S(O; R)$ (see Figure 2).

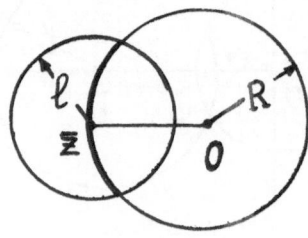

Figure 2.

And we have

$$s(\Gamma) = 2R \arccos(1 - \frac{\ell^2}{2R^2}).$$

Hence,

$$P_S(z_1, z_2; \nu^*) \leq \frac{2 \arccos(1 - \frac{\ell^2}{2R^2})}{\pi}.$$

Summary
One could expect that the difference in detection probabilities for optimal pursuit and search strategies will be small for small ℓ (detection radius), but the above examples show that this is not always case. This suggests that the choosing of the payoff function must be done with special care by the applied mathematitian, because the intuitive interpretational background is not enough and may lead to the very different optimal behaviour.

References

[1] Petrosjan L.A. Differential pursuit games. Leningrad: Isdat. Leningrad Univ., 1977(Russian).

[2] Petrosjan L.A., Zenkevich N.A. Optimal game-theoretic search. Leningrad: Isdat. Leningrad Univ., 1984(Russian).

[3] Slobodinskaia T.V. In the book Petrosjan L.A., Tomsky G.V. "Differential pursuit games with incomplete information", Irkutsk, 1984. p.178-179 (Russian).

[4] Gal S. Search games. N.Y.: Acad. Press, 1980.

The α-core of the Positional Differential Cooperative Game

M.S. RADJEF
LAMOS
Laboratory of Modelisation and Optimisation of Systems
University of Bejaia
06 000 Bejaia ALGERIA

Abstract.

In this paper we study the properties of the α-core solution of a differential cooperative game. In order to guarantee the stability of optimal principles, the quasi-movements [8] are used as solutions to the differential system describing the game. The sufficient conditions for existence of a dynamically stable α-core are given. For a linear quadratic differential game, we derive the explicit form of the strategies of α-core.

1. Introduction.

Let us consider an N-person differential game ($N \geq 2$) whose the dynamic is described by the following differential vector equation

$$(\Sigma) \qquad \frac{dx}{dt} = f(t,x,v_1,\ldots,v_N), \qquad x[t_0]=x_0 \qquad (1.1)$$

Here $x \in \mathbb{R}^n$ is the phase vector, v_i is the command action of the i-th player. The final time $T>t_0 \geq 0$ of the game is fixed. The strategies V_i are identified as the functions $v_i(t,x)$ satisfying only the inclusion $v_i(t,x) \subseteq Q_i$ in every possible position (t,x) of the game. This is denoted $V_i - : -v_i(t,x)$. $\mathcal{V}_i$ is the set of i-th player's strategies V_i, Q_i a non empty closed and bounded subset of $\mathbb{R}^{m_i}$; $Q=Q_1 x \ldots x Q_N$, $V=(V_1,\ldots,V_N)$ and $\mathbb{N}=\{1,\ldots,N\}$ the set of players.

In order to guarantee the existence of the solutions (quasi-movements [8]) to the system (1.1), we assume the following hypothesis:

Conditions 1.1.
   a) the function $f(.):[0,T] x \mathbb{R}^n x Q \longrightarrow \mathbb{R}^n$ is continuous;
   b) for any bounded domain G of the position space (t,x), there exists a constant $\lambda = \lambda(G)>0$ such that

$$\| f(t^{(1)},x^{(1)},v) - f(t^{(2)},x^{(2)},v) \| \leq \lambda(G) \left(\| x^{(1)}-x^{(2)} \| + | t^{(1)}-t^{(2)} | \right)$$

for any position $(t^{(r)},x^{(r)}) \in G$, $(r=1,2)$ and any $v \in Q$;

c) there exists a constant $\gamma > 0$ such that $\| f(t,x,v) \| \leq \gamma (1 + \| x \|)$ for any $t \in [0,T]$, $v \in Q$ and $x \in \mathbb{R}^n$.

Let $K \subseteq \mathbb{N}$ a certain coalition of players, the strategy of the coalition K is denoted by $V_k = \{V_i, i \in K\}$ with $V_i \in \mathcal{V}_i$, $i \in K$.
The system (1.1) can be rewritten as follows:

$$\frac{dx}{dt} = f(t,x,v_K,v_{\mathbb{N} \backslash K}), \qquad x[t_o] = x_o; \qquad (1.2)$$

v_K $(v_{\mathbb{N} \backslash K})$ is the command action of the coalition K (respectively of the counter-coalition $\mathbb{N} \backslash K$).

Under the conditions 1.1, for any pair of strategies $(V_K, V_{\mathbb{N} \backslash K})$, there correspond [8] sets of quasi-movements $\mathcal{X}[t_o,x_o,V_K]$, $\mathcal{X}[t_o,x_o,V_{\mathbb{N} \backslash K}]$ and $\mathcal{X}[t_o,x_o,V_K,V_{\mathbb{N} \backslash K}]$ which are non-empty compact subsets of the space $C_n[t_o,T]$ of continuous functions in $[t_o,T]$. The intersections of thus subsets with the hyperplan $\{t=T\}$ are denoted respectively by $X[T,t_o,x_o,V_K]$, $X[T,t_o,x_o,V_{\mathbb{N} \backslash K}]$ and $X[T,t_o,x_o,V_K,V_{\mathbb{N} \backslash K}]$ and are non-empty and compact subsets of $\mathbb{R}^n$. These subsets are said attainability domains of the system (1.1) which are generated from the position (t_o,x_o) by, respectively, the strategies V_K, $V_{\mathbb{N} \backslash K}$ and $(V_K,V_{\mathbb{N} \backslash K})$.

The aim of each i-th player is to choose the strategy $V_i - : - v_i(t,x) \subseteq Q_i$ which maximizes the payoff function

$$J_i(V_1, \ldots, V_N) = F_i(x[T]), \quad i \in \mathbb{N}, \qquad (1.3)$$

where the function $F_i(.) : \mathbb{R}^n \longrightarrow \mathbb{R}$ are definite and continuous on the attainability domains of the system (1.1).
Hereafter, we denote the game by

$$J = \left\langle \Sigma, \mathbb{N}, \left\{ \mathcal{V}_i \right\}_{i \in \mathbb{N}}, \{J_i\}_{i \in \mathbb{N}} \right\rangle. \qquad (1.4)$$

2. Definitions and properties of the α-coeur.

In the classical theory of cooperative games we distinguish different optimality principles (optimum of Pareto, Slater, core, vector of Shapley, Neumann-Morgenstern, ...). The developpement of the differential game theory has made necessary the generalization of the notions of the "static" game theory.
The important point here is that the approach proposed by Krassovski [1] in the study of the zero-sum differential positional game, and,

applied to non-antagonistic games is that for each collection of strategies there corresponds a whole set of values of the players payoff functions. This yields naturally another generalization of the solution concepts of the games theory.
In this paper the α-core solution [3] is considered.

Definition 1. An N-tuple of strategies $V^{\alpha}=(V_1^{\alpha},\ldots,V_N^{\alpha})\in \mathcal{V}$, $\mathcal{V}=\prod_{i\in N}\mathcal{V}_i$ belongs to the α-core $C_{\alpha}(t_0,x_0)$ of the differential game (1.4) with the initial position $(t_0,x_0)\in [0,T[\times\mathbb{R}^n$, if for any coalition $K\subseteq N$ and its strategy $V_k\in\mathcal{V}_k=\prod_{i\in K}\mathcal{V}_i$, there exists a strategy $V_{N\setminus K}\in\mathcal{V}_{N\setminus K}$ of the counter-coalition $N\setminus K\subseteq N$ such that the system of inequalities

$$\min_{x[.]} F_i(x[T,t_0,x_0,V_k,V_{N\setminus K}]) \geq \min_{x[.]} F_i(x[T,t_0,x_0,V^{\alpha}]) \qquad i\in K$$

whith at least one of these stringent, is not satisfied.

In other words, $V^{\alpha}\in C_{\alpha}(t_0,x_0)$ if any coalition of players deviate from their strategies inclued in $\mathcal{V}^{\alpha}$, using strategies $V_k\in\mathcal{V}_k=\prod_{i\in K}\mathcal{V}_i$, $V_k\neq V_K^{\alpha}$, the counter-coalition $N\setminus K$ can play strategies $V_{N\setminus K}\in\mathcal{V}_{N\setminus K}$ such that: either all players of the coalition K obtain the same gain value as under V^{α}, ie

$$\min_{x[.]} F_i(x[T,t_0,x_0,V_k,V_{N\setminus K}]) = \min_{x[.]} F_i(x[T,t_0,x_0,V^{\alpha}]) \qquad i\in K;$$

or there exists, at least one player (i_0) of the coalition K, who obtains a gain value with $(V_k,V_{N\setminus K})$ less than the gain value guaranteed by the strategies V^{α}, ie

$$\min_{x[.]} F_{i_0}(x[T,t_0,x_0,V_k,V_{N\setminus K}]) < \min_{x[.]} F_{i_0}(x[T,t_0,x_0,V^{\alpha}])$$

Proposition 1. For each strategy V^{α} of the α-core, there corresponds a unique gain value for each player, ie
$$\max_{x[.]} F_i(x[T,t_0,x_0,V^{\alpha}])= \min_{x[.]} F_i(x[T,t_0,x_0,V^{\alpha}])= F_i(x[T,t_0,x_0,V^{\alpha}]), \quad i\in N$$
for each quasi-movement $x[.,t_0,x_0,V^{\alpha}] \in \mathcal{X}[t_0,x_0,V^{\alpha}]$.

Proof. Assume the counter statement: there exists an index $j\in N$ such that

$$\max_{x[.]} F_j(x[T,t_0,x_0,V^{\alpha}]) > \min_{x[.]} F_j(x[T,t_0,x_0,V^{\alpha}]) \qquad (2.2)$$

The domain of accessibility

$$X[T,t_0,x_0,V^\alpha]=X=\{x[T,t_0,x_0,V^\alpha]\}, \; x[.]\in X[t_0,x_0,V^\alpha]\}$$

is closed and bounded [8];the function $F_j(.)$ is continuous on $X[T,t_0,x_0,V^\alpha]$, then there exists a point $x^*\in X[T,t_0,x_0,V^\alpha]$ such that

$$\max_{x[.]} F_j(x[T,t_0,x_0,V^\alpha]) = \max_{x\in X} F_j(x) = F_j(x^*) \qquad (2.3)$$

On the other hand there exists [9] a strategy $V^\circ \in \mathcal{V}$ such that

$$x^*=x[T,t_0,x_0,V^\circ]$$

for all quasi-movements $x[.,t_0,x_0,V^\circ] \in \mathcal{X}[t_0,x_0,V^\circ]$. Then

$$\min_{x[.]} F_i(x[T,t_0,x_0,V^\circ]) = F_i(x[T,t_0,x_0,V^\circ])= F_i(x^*), \quad i\in N \qquad (2.4)$$

for all quasi-movements $x[.,t_0,x_0,V^\circ] \in \mathcal{X}[t_0,x_0,V^\circ]$.

From (2.3) and (2.4), the relation (2.2) can be rewritten as follows

$$\min_{x[.]} F_j(x[T,t_0,x_0,V^\circ]) > \min_{x[.]} F_j(x[T,t_0,x_0,V^\alpha])$$

As $V^\alpha \in C_\alpha(t_0,x_0)$, then, by definition, there exists an index $q\in N, q\neq j$ such that

$$F_q(x^*)= \min_{x[.]} F_q(x[T,t_0,x_0,V^\circ]) < \min_{x[.]} F_q(x[T,t_0,x_0,V^\alpha]) \qquad (2.5)$$

From (2.4) and (2.5), we deduce

$$F_q(x^*)= \min_{x[.]} F_q(x[T,t_0,x_0,V^\circ]) < \min_{x[.]} F_q(x[T,t_0,x_0,V^\alpha]) = \min_{x\in X} F_q(x).$$

That means that we found $x^*\in X$ such that

$$F_q(x^*) < \min_{x\in X} F_q(x),$$

which is absurd.

Definition 2. An N-tuple of strategies $V^p \in \mathcal{V}$ is said to constitute an optimum of Pareto for the game (1.4),if for each N-tuple of strategies $V\in \mathcal{V}$, $V\neq V^p$, the system of inequalities

$$\min_{x[.]} F_j(x[T,t_0,x_0,V]) \geq \min_{x[.]} F_j(x[T,t_0,x_0,V^p]) \; ,j\in N$$

whith at least one of these stringent,is not statisfied.

Proposition 2. If an N-tuple of strategies V^α belongs to the α-core of the game (1.4), then V^α is an optimum of Pareto.

Proof. Let $V^\alpha \in C_\alpha(t_0,x_0)$ and $K=N$ the coalition comprised of all players. By definition of the α-core,for each N-tuple of strategies $V\in \mathcal{V}$ the system of inequalities

$$\min_{x[.]} F_j(x[T,t_o,x_o,V]) \geq \min_{x[.]} F_j(x[T,t_o,x_o,V^\alpha]) , \qquad j\in\mathbb{N}$$

whith at least one of these stringent,is not satisfied.

That means that V^α is an optimum of Pareto.

Remark 1. As Pareto-optimal strategy is also Slater-optimal, then the α-core ($C_\alpha(t_o,x_o)$) is includ in the set γ^P of Pareto-optimal strategies and includ in the set γ^S of Slater-optimal strategies, ie

$$C_\alpha(t_o,x_o) \subseteq \gamma^P \subseteq \gamma^S.$$

Proposition 3. Let $V^\alpha \in C_\alpha(t_o,x_o)$ and $V^g=(V_1^g,\ldots,V_N^g)$ the N-tuple of strategies, where V_i^g is maximin strategy of the i-th player, $i\in\mathbb{N}$:

$$F_i^g(t_o,x_o)=\max_{V_i} \min_{x[.]} F_i(x[T,t_o,x_o,V_i]) = \min_{x[.]} F_i(x[T,t_o,x_o,V_i^g])$$

$$F_i^\alpha(t_o,x_o)= F_i(x[T,t_o,x_o,V^\alpha])$$

Then

$$F_i^\alpha(t_o,x_o) \geq F_i^g(t_o,x_o), \quad i\in\mathbb{N}$$

Proof. Let V^α an N-tuple of strategies of the α-core $C_\alpha(t_o,x_o)$ and consider the coalition $K=\{i\}$ comprised only of the player $\{i\}$,its maximin strategy V_i^g. By definition of $V^\alpha \in C_\alpha(t_o,x_o)$, there exists a strategy $V_{\mathbb{N}\setminus i}$ of the counter-coalition $\mathbb{N}\setminus\{i\}$ such that

$$\min_{x[.]} F_i(x[T,t_o,x_o,V_i^g,V_{\mathbb{N}\setminus i}]) \leq \min_{x[.]} F_i(x[T,t_o,x_o,V^\alpha]) \qquad (2.6)$$

As [8] $\qquad X[T,t_o,x_o,V_i^g,V_{\mathbb{N}\setminus i}] \leq X[T,t_o,x_o,V_i^g]$

Then, and from (2.6)

$$F_i^g(t_o,x_o)= \min_{x[.]} F_i(x[T,t_o,x_o,V_i^g]) \leq \min_{x[.]} F_i(x[T,t_o,x_o,V_i^g,V_{\mathbb{N}\setminus i}]) \leq$$

$$\leq \min_{x[.]} F_i(x[T,t_o,x_o,V^\alpha]) = F_i^\alpha(t_o,x_o).$$

Thus $\qquad F_i^g(t_o,x_o) \leq F_i^\alpha(t_o,x_o), \qquad i\in\mathbb{N}.$

Remarque 2. The proposition 3 suggests that using the strategy of α-core is more beneficial that pursueing the player's own maximin strategies.

Let us consider the particular case of the game (1.4)

$$J_a= \left\langle \{1,2\}, \Sigma\text{-:-} \frac{dx}{dt}=f(t,x,v_1,v_2), \{\gamma_1,\gamma_2\}, F(x[T]) \right\rangle \qquad (2.7)$$

where $J_1=F_1(x[T])=F(x[T])$, $J_2=F_2(x[T])=-F(x[T])$. The game J_a is said the two-person zero sum positional differential game.

Definition 3. A pair of strategies (V_1°,V_2°) is said to constitute a saddle point for the game (2.7) if the inequalities:

$$F(x[T,t_o,x_o,V_1^\circ]) \geq F(x[T,t_o,x_o,V_1^\circ,V_2^\circ]) \geq F(x[T,t_o,x_o,V_2^\circ]) \qquad (2.8)$$

are satisfied for all quasi-movements $x[.,t_o,x_o,V_1^\circ]$, $x[.,t_o,x_o,V_1^\circ,V_2^\circ]$, $x[.,t_o,x_o,V_2^\circ]$.

Remark 3. From inequalities (2.8), we can deduct easily that for each saddle point there corresponds a unique value of the game J_a:

$$\min_{x[.]} F(x[T,t_o,x_o,V_1^\circ,V_2^\circ]) = \max_{x[.]} F(x[T,t_o,x_o,V_1^\circ,V_2^\circ]) = F(x[T,t_o,x_o,V_1^\circ,V_2^\circ])$$

for all quasi-movements $x[.,t_o,x_o,V_1^\circ,V_2^\circ]$.

Proposition 4. The saddle point of the game J_a belongs to the α-core.

Proof. Assume that (V_1°,V_2°) is a saddle point of the game J_a. Let us show that (V_1°,V_2°) belongs to the α-core $C_\alpha(t_o,x_o)$.

a) Let $K=\{1\}$, V_1- a strategy of the player $\{1\}$ and assume that the player $\{2\}$ chooses as strategy $V_2=V_2^\circ$. As [8]

$$X[T,t_o,x_o,V_1,V_2^\circ] \subseteq X[T,t_o,x_o,V_2^\circ], \qquad \text{then}$$

$$\min_{x[.]} F_1(x[T,t_o,x_o,V_1,V_2^\circ]) \leq \max_{x[.]} F_1(x[T,t_o,x_o,V_1,V_2^\circ]) \leq$$

$$\leq \max_{x[.]} F_1(x[T,t_o,x_o,V_2^\circ]) \leq F_1(x[T,t_o,x_o,V_1^\circ,V_2^\circ]) = \min_{x[.]} F_1(x[T,t_o,x_o,V_1^\circ,V_2^\circ])$$

Thus, if the player $\{1\}$ decides to form a coalition alone and to play a strategy $V_1 \neq V_1^\circ$, then it is sufficient for the second player to keep its strategy V_2° of the pair $V^\circ=(V_1^\circ,V_2^\circ)$ in order to discourage the player $\{1\}$ since he obtains a gain value smaller than under V°.

b) Assume that the player $\{2\}$ decides to form a coalition alone $k=\{2\}$ and to play a strategy V_2. If the player $\{1\}$ keeps his strategy $V_1=V_1^\circ$. Then

$$\min_{x[.]} F_2(x[T,t_o,x_o,V_1^\circ,V_2]) = \min_{x[.]} \{-F_1(x[T,t_o,x_o,V_1^\circ,V_2])\} =$$

$$= -\max_{x[.]} F_1(x[T,t_o,x_o,V_1^\circ,V_2]). \qquad (2.8)$$

On the other hand

$$\max_{x[.]} F_1(x[T,t_o,x_o,V_1^\circ,V_2]) \geq \min_{x[.]} F_1(x[T,t_o,x_o,V_1^\circ,V_2]) \geq$$

$$\geq \min_{x[.]} F_1(x[T,t_o,x_o,V_1^\circ]) \geq \min_{x[.]} F_1(x[T,t_o,x_o,V_1^\circ,V_2^\circ]). \qquad (2.9)$$

Multiplying by (-1) all members of the relation (2.9), and taking into account (2.8), we obtain

$$- \max_{x[.]} F_1(x[T,t_o,x_o,V_1^\circ,V_2]) \leq - \min_{x[.]} F_1(x[T,t_o,x_o,V_1^\circ,V_2^\circ]) =$$

$$= - \max_{x[.]} F_1(x[T,t_o,x_o,V_1^\circ,V_2^\circ]);$$

and

$$\min_{x[.]} F_2(x[T,t_o,x_o,V_1^\circ,V_2]) \leq \min_{x[.]} F_2(x[T,t_o,x_o,V_1^\circ,V_2^\circ]).$$

Then, the player {2} obtains a gain value less than that obtained under (V_1°,V_2°).

c) Let $K=\{1,2\}$. Assume that the coalition K has a pair of strategies $V=(V_1,V_2)$ which verify the two inequalities

$$\min_{x[.]} F_1(x[T,t_o,x_o,V_1,V_2]) \geq \min_{x[.]} F_1(x[T,t_o,x_o,V_1^\circ,V_2^\circ]) \qquad (2.10)$$

$$\min_{x[.]} F_2(x[T,t_o,x_o,V_1,V_2]) \geq \min_{x[.]} F_2(x[T,t_o,x_o,V_1^\circ,V_2^\circ]) , \qquad (2.11)$$

whith at least one of those stringent.
The inequality (2.11) can be rewritten as follows

$$\min_{x[.]} - F_1(x[T,t_o,x_o,V_1,V_2]) \geq \min_{x[.]} - F_1(x[T,t_o,x_o,V_1^\circ,V_2^\circ]).$$

Adding (2.10) and (2.11), we obtain

$$0 = \min_{x[.]} \{F_1(x[T,t_o,x_o,V_1,V_2]) - F_1(x[T,t_o,x_o,V_1,V_2])\} \geq$$

$$\geq \min_{x[.]} F_1(x[T,t_o,x_o,V_1,V_2]) + \min_{x[.]} (- F_1(x[T,t_o,x_o,V_1,V_2])) >$$

$$> \min_{x[.]} F_1(x[T,t_o,x_o,V_1^\circ,V_2^\circ]) + \min_{x[.]} (- F_1(x[T,t_o,x_o,V_1^\circ,V_2^\circ])) =$$

$$= F_1(x[T,t_o,x_o,V_1^\circ,V_2^\circ]) - F_1(x[T,t_o,x_o,V_1^\circ,V_2^\circ]) = 0.$$

This is absurd.
Then there is not any pair of strategies $V=(V_1,V_2)$ which verify the system of inequalities (2.10) and (2.11) whith at least one of those stringent. This means that $V^\circ=(V_1^\circ,V_2^\circ)$ belongs to α-core.

3. SUFFICIENT CONDITIONS.

In order an optimal solution to be efficient and consistent within a differential game, it is important that this solution satisfies the same optimality principle over the whole progression period of the game. This property was introduced in the differential games [4] as the dynamically stable solution (time consistent). This property is even more important as far as the α-core is concerned since this latter relies on the principle of cooperation under threat. At any one position of the game, one or a group of players may break the agreed order choosing a different strategy from that satisfying the optimality principle defined when the game started.

In the following, the conditions of existence of a dynamically stable α-core are discussed.

Definition 1. We say that the N-tuple of strategies $V^\alpha = (V_1^\alpha, \ldots, V_N^\alpha)$ of the α-core $C_\alpha(t_0, x_0)$ of the game (1.4) with the initial position (t_0, x_0) is dynamically stable, if for every position of the game $(t_*, \hat{x}[t_*]) \in [t_0, T] \times X[t_*, t_0, x_0, V^\alpha]$, we have $V^\alpha \in C_\alpha(t_*, \hat{x}[t_*])$.

The α-core $C_\alpha(t_0, x_0)$ is said dynamically stable if each strategy of the α-core is dynamically stable.

Theorem 1. Assume that there exist an N-tuple of strategies $V^\alpha \in V$ and "N" continuous functions $\mathcal{S}_i(.): [0, T] \times \mathbb{R}^n \longrightarrow \mathbb{R}$ such that

1°/ for every $x \in \mathbb{R}^n$ and $i \in N$

$$\mathcal{S}_i(T, x) = F_i(x), \qquad (3.1)$$

2°/ for every $t_* \in [0, T]$ and $i \in N$

$$\underset{x[.] \in X[t_0, x_0, V^\alpha]}{\text{Sup}} \quad \overline{\lim_{t \longrightarrow t_*+o}} \quad [\mathcal{S}_i(t, x[t]) - \mathcal{S}_i(t_*, x[t_*])](t - t_*)^{-1} =$$

$$= \underset{x[.] \in X[t_0, x_0, V^\alpha]}{\text{Inf}} \quad \underline{\lim_{t \longrightarrow t_*+o}} \quad [\mathcal{S}_i(t, x[t]) - \mathcal{S}_i(t_*, x[t_*])](t - t_*)^{-1} = 0 \qquad (3.2)$$

3°/ in each position $\{t_*, x_*\} \in [0, T] \times \mathbb{R}^n$ and for every $i \in N$

$$\underset{x[.] \in X[t_*, x_*, V_i]}{\text{Sup}} \quad \overline{\lim_{t \longrightarrow t_*+o}} \quad [\mathcal{S}_i(t, x[t]) - \mathcal{S}_i(t_*, x_*)](t - t_*)^{-1} \leq 0 \qquad (3.3)$$

for all strategies $V_i \in V_i$,

then the N-tuple of strategies V^α belongs to the α-core of the game (1.4) with the initial position (t_0, x_0). In addition, V^α is dynamically stable.

Proof. Let $(t_*, \hat{x}[t_*]) \in [t_0, T) \times X[t_*, t_0, x_0, V^\alpha]$ an initial position of the game (1.4). The relation (3.2) implies [5]

$$\mathcal{Z}_i(t,x[t]) = \mathcal{Z}_i(t_*,\hat{x}[t_*]), \qquad i \in \mathbb{N} \qquad (3.4)$$

for each $t \in [t_*,T]$ and $x[.] \in \mathcal{X}[t_*,\hat{x}[t_*],V^\alpha]$. Putting t=T in (3.4) and using (3.1), we obtain

$$\mathcal{Z}_i(t_*,\hat{x}[t_*]) = \mathcal{Z}_i(T,x[T,t_*,\hat{x}[t_*],V^\alpha]) = F_i(x[T,t_*,\hat{x}[t_*],V^\alpha])$$

for all quasi-movements $x[.] \in \mathcal{X}[t_*,\hat{x}[t_*],V^\alpha]$ and $i \in \mathbb{N}$. Then

$$\mathcal{Z}_i(t,\hat{x}[t]) = F_i(x[T,t_*,\hat{x}[t_*],V^\alpha]) = \min_{x[.]} F_i(x[T,t_*,\hat{x}[t_*],V^\alpha]), \quad i \in \mathbb{N}. \qquad (3.5)$$

The relation (3.3) gives [9]

$$\max_{x[.]} \max_{t_* \leq t \leq T} \mathcal{Z}_i(t,x[t,t_*,x_*,V_i]) \leq \mathcal{Z}_i(t_*,x_*) \qquad \qquad i \in \mathbb{N}$$

for all strategies $V_i \in \mathcal{V}_i$, $i \in \mathbb{N}$.

In particular

$$\max_{x[.]} \max_{t_* \leq t \leq T} \mathcal{Z}_i(t,x[t,t_*,\hat{x}[t_*],V_i]) \leq \mathcal{Z}_i(t_*,\hat{x}[t_*])$$

for all strategies $V_i \in \mathcal{V}_i$ and $i \in \mathbb{N}$. Then

$$\max_{x[.]} \mathcal{Z}_i(T,x[T,t_*,\hat{x}[t_*],V_i]) \leq \mathcal{Z}_i(t_*,\hat{x}[t_*]) \qquad (3.6)$$

From (3.1), (3.5)-(3.6), we obtain

$$\max_{x[.]} F_i(T,x[T,t_*,\hat{x}[t_*],V_i]) \leq \min_{x[.]} F_i(x[T,t_*,\hat{x}[t_*],V^\alpha]) \qquad (3.7)$$

for all strategies $V_i \in \mathcal{V}_i$ and $i \in \mathbb{N}$.

Let us consider the strategy $\bar{V}=(V_{K\backslash i}, V_i, V_{\mathbb{N}\backslash K}^\alpha)$.

As $\quad X[T,t_*,\hat{x}[t_*],\bar{V}] \subseteq X[T,t_*,\hat{x}[t_*],V_i]$, then we have

$$\min_{x[.]} F_i(x[T,t_*,\hat{x}[t_*],\bar{V}]) \geq \min_{x[.]} F_i(x[T,t_*,\hat{x}[t_*],V_i]) \qquad (3.8)$$

The relations (3.7) and (3.8) give

$$\min_{x[.]} F_i(x[T,t_*,\hat{x}[t_*],V^\alpha]) \geq \min_{x[.]} F_i(x[T,t_*,\hat{x}[t_*],\bar{V}]), \qquad \qquad i \in K.$$

Then the system of inequalities

$$\min_{x[.]} F_i(x[T,t_*,\hat{x}[t_*],\bar{V}]) \geq \min_{x[.]} F_i(x[T,t_*,\hat{x}[t_*],V^\alpha]), \qquad i \in K$$

whith at least one of these stringent, is not satisfied.

4. Example.

In the particular case where the game (1.4) is linear quadratic [6,7], we give the explicit form of strategies of the α-coeur.

Assume that Σ is described by the linear differential equation

$$\frac{dx}{dt} = A(t)x + \sum_{i \in N} B_i(t)v_i, \qquad x(t_0) = x_0 \qquad (4.1)$$

where $x \in \mathbb{R}^n$, $v_i \in \mathbb{R}^n$, $T > 0$, the elements of matrices $A(t)$ and $B_i(t)$ are the continuous functions in $[0,T]$. The set of the i-th player's strategies is

$$\mathcal{V}_i = \{V_i - : -v_i(t,x) \ / \ v_i(t,x) = P_i(t)x\},$$

where $P_i(t)$ a matrix whose elements are the continuous functions in $[0,T]$.

The i-th player payoff function is defined by the relation

$$J_i(V) = x'(T)C_i x(T) + \int_{t_0}^{T} \{x'(t)M_i x(t) + \sum_{k=1}^{N} v_k'(t)D_{ki}v_k(t)\}dt,$$

where C_i, M_i and D_{ki} are symmetric and constant matrices.

For each given sequence of positive numbers $\{\lambda_1, \ldots, \lambda_N\} = \lambda$, we define the matrices

$$C(\lambda) = \sum_{i=1}^{N} \lambda_i C_i \ ; \ M(\lambda) = \sum_{i=1}^{N} \lambda_i M_i \ ; \ D_j(\lambda) = \sum_{i=1}^{N} \lambda_i D_{ij} \ .$$

By the symbol $C < 0$, we designate the negative definite matrix.

Assertion. If $D_{ij} < 0$ $(i, j \in N, \ i \neq j)$ and there exist $\lambda_i > 0$, $i \in N$ such that

$$C(\lambda) < 0, \ M(\lambda) < 0, \ D_j(\lambda) < 0, \qquad j \in N, \qquad (4.2)$$

then there exists a strategy V^λ in the α-core of the game (4.1) with any initial position.

Proof. Under the hypothesis (4.2), there exists [6,7] a situation $V^\lambda = (V_1^\lambda, \ldots, V_N^\lambda)$ Pareto-optimal of the form

$$V_i^\lambda - : - -D_i^{-1}(\lambda)B_i'(t)\theta(t)x, \qquad i \in N,$$

where $\theta(t)$, $t_0 \leq t \leq T$ is matrix solution of the system

$$\frac{d\theta}{dt} + \theta A(t) + A'(t)\theta + M(\lambda) - \theta \sum_{i=1}^{N} B_i(t)D_i^{-1}(\lambda)B_i'(t)\theta = 0$$

$$\theta(T) = C(\lambda)$$

This solution can be extendable to the interval $[0,T]$ [2].

Let us establish that V^λ belongs to the α-core. Let $K \subseteq N$ a certain coalition, which does not coincide with the set of players N, and pursues an arbitrary strategy $V_1 = \{V_1, \ 1 \in K\}$, $V_1 - : -v_1(t,x) = P_1(t)x$, $1 \in K$. Let us consider a given player j of the counter-coalition $N \setminus K$ and represent the payoff functions of members of K as

$$J_i(V_k,V_j,V^{\lambda}_{\mathbb{N}\setminus K\setminus j})=x'(T)C_i x(T)+\int_{t_o}^{T}\{x'(t)\hat{M}_i x(t)+v'_j(t)D_{ij}v_j(t)\}dt, \quad i\in K.$$

where $V^{\lambda}_{\mathbb{N}\setminus K\setminus j}=\{V^{\lambda}_q, q\in\mathbb{N}\setminus K\setminus j\}$ and

$$\hat{M}_i(t)=M_i(t)+\sum_{r\in K}P'_r(t)D_{ir}P_r(t)+\theta(t)\sum_{q\in\mathbb{N}\setminus K\setminus j}B_q(t)D_q^{-1}(\lambda)D_{iq}D_q^{-1}(\lambda)B'_q(t)\theta(t)$$

$x(t)$ is solution of the system (4.1) with

$v_q=-D_q^{-1}(\lambda)B'_q(t)\theta(t)x, \quad q\in\mathbb{N}\setminus K\setminus j, \quad v_r=P_r(t)x \ (r\in K)$ and $v_j=P_j(t)x.$

There exists [9] a number $\beta_i>0$ such that for every $\beta>\beta_i$ and $V_j^*-:-\beta x,$ the inequality

$$J_i(V_j^*,V_k,V^{\lambda}_{\mathbb{N}\setminus K\setminus j}) \leq J_i(v^{\lambda}) \tag{4.3}$$

is met.

With $\qquad \beta^* = \max_{i\in K} \beta_i,$ and $V_j^*-:-\beta_j^* x$

the inequality (4.3) holds for every $i\in K$. Consequently, for the strategy $V_k\in\mathcal{V}_k$ choosen arbitrarily,we could exhibit a strategy of the counter-coalition $\hat{V}_{\mathbb{N}\setminus K}=\{V_j^*,V^{\lambda}_{\mathbb{N}\setminus K\setminus j}\}$ such that

$$J_i(V_k,\hat{V}_{\mathbb{N}\setminus K}) \leq J_i(v^{\lambda}), \quad i\in K.$$

Then $v^{\lambda}\in C_{\alpha}(t_o,x_o).$

REFERENCES

1. KRASSOVSKI N.,SOUBBOTINE A. Jeux différentiels. Ed. Mir,Moscou,1977
2. LEE E.B., MARKUS L. Foundations of optimal control theory.- Ed. John Wiley and Sons Inc. N.Y.,London, 1967.
3. MOULIN H. Théorie des jeux pour l'économie et la politique. Ed. Hermann, Paris, 1981.
4. PETROSJAN L.A. Stability of solutions in non-antagonistic differential multi-person games. Vestnik Leningr. Universiteta, N°19, 1977. (In russian).
5. RADJEF M.S. Sur les conditions d'existence d'une imputation stable dans un jeu différentiel coopératif de position. to appair in RAIRO APII (Automatique Productique Informatique Industrielle).
6. STARR A.W.,HO Y.C. Nonzero-sum differential game. J. Optimiz. Theory and Applic., 1969, 3, N°3, p.184-204.
7. ZHUKOVSKII V.I. On the many players differential games. Izv. Akad. Nauk SSSR. Tekhn. Kibernetika, 1971, N°3, p.3-13.(In russian).
8. ZHUKOVSKII V.I.,RADJEF M.S. Les quasi-mouvements et leurs propriétés. J. Mathematica Balkanica, V.5, 1991.
9. ZHUKOVSKII V.I., TENIANSKI N.Y. Equilibrium Situations in Multi-criteria Dynamical Systems. Eds Univ. of Moscow, 1984.

Stochastic control
and
application to economy

Generalized Bellman-Hamilton-Jacobi Equations for Piecewise Deterministic Markov Processes[*]

J.J. Ye

Department of Mathematics and Statistics,
University of Victoria, Victoria, B.C., Canada V8W 3P4

July 20, 1993

1 Introduction

Piecewise deterministic Markov processes (PDPs), first introduced by M.H.A. Davis [4], are continuous time homogeneous Markov processes consisting of deterministic motion between random jumps. PDPs, with stochastic jump processes and deterministic dynamical systems as special cases, is a class of stochastic models useful for formulating dynamic optimization problems for stochastic systems not involving diffusion. A complete theory of PDPs can be found in Davis [5].

Suppose that the state space for our PDP is a subset in $\mathbb{R}^n$ with interior $E^0 = \{x \in \mathbb{R}^n : \psi(x) > 0\}$, where $\psi : \mathbb{R}^n \to \mathbb{R}$ is some C^1 function. Motion of a PDP x_t in E is determined by three so-called local characteristics: a draft $f(x)$, a jump rate $\lambda(x)$ and a transition measure $Q(dy; x)$. let $\phi(t, x)$ be the flow of f, i.e. the solution of the ODE $d\phi(t, x)/dt = f(\phi(t, x)), \phi(0, x) = x$. Then $x_t := \phi(t, x)$ for $t < \mathbf{T}_1$ where $\mathbf{T}_1$ is a positive random variable with survivor function

$$F(t) = P_x[\mathbf{T}_1 > t] = \begin{cases} \exp[-\int_0^t \lambda(\phi(s, x))ds] & t < t_*(x) \\ 0 & t \geq t_*(x), \end{cases}$$

where $t_*(x) := \inf\{t > 0 : \phi(t, x) \in \partial E\}$. Having realized $\mathbf{T}_1 = T_1$, we select a random variable Z with distribution $Q(\cdot; \phi(T_1, x))$ and set $x_{T_1} := Z$. The process now restarts from x_{T_1} according to the same recipe. Thus x_t moves along integral curves of f between jumps which occurs at times $T_1, T_2, \ldots$ with rate $\lambda(x_t)$ in E^0; a jump must occur should x_t hit ∂E.

Now suppose that f, λ, Q of $\{x_t\}$ depend on a *control action* v from a compact set U. The set of admissible controls may be different for interior and boundary states. We assume that $v \in U_0 \subset \mathbb{R}^m$ if $x \in E^0$ and $v \in U_\partial \subset \mathbb{R}^l$ if $x \in \partial E$. Therefore, we shall distinguish transition measure $Q_0(dy; x, v)$, for $x \in E^0$, $v \in U_0$, describing jumps from interior points, from $Q_\partial(dy; x, v)$, for $x \in \partial E$, $v \in U_\partial$, describing jumps from boundary points. The usual class of controls for Markovian problems, namely state feedback function of the form $u_t = \tilde{u}(x_t)$, is not appropriate choice here since the equation $\frac{d}{dt}x_t = f(x_t, \tilde{u}(x_t))$ may fail to have a solution if $\tilde{u}(\cdot)$ is insufficiently smooth. Following Vermes [10], we use instead "piecewise feedback open loop" controls: at T_k we choose a function $u_k : \mathbb{R}_+ \to U_0$

---

[*]Research supported in part by NSERC under grant WFA0123160

which will be the control until the next jump time T_{k+1}; Thus for $t \in [T_k, T_{k+1})$, $x(t)$ satisfies $\frac{d}{dt}x(t) = f(x(t), u_k(t - T_k))$, and the solution is well-defined for just measurable $u_k(\cdot)$. When x_t hit the boundary ∂E, we use feedback control $u_\partial : \partial E \to U_\partial$.

The PDP *optimal control problem* is to minimize the *infinite-horizon discounted cost*

$$J_x(u) := E_x^u[\int_0^\infty e^{-\delta t} l_0(\mathbf{x}_t, u_t) dt + \sum_i e^{-\delta T_i} l_\partial(\mathbf{x}_{T_i^-}, u_\partial(\mathbf{x}_{T_i^-})) 1_{\{\mathbf{x}_{T_i^-} \in \partial E\}}].$$

the last term represents a penalty incurred only when the boundary ∂E is hit and control u_∂ is used; this arises naturally in many problems.

The value function $V : E \longrightarrow \mathbb{R}_+$ is defined by

$$V(x) := \inf_{u \in \mathcal{C}} J_x(u) \qquad \text{for all } x \in E,$$

where $\mathcal{C}$ is the class of piecewise open loop controls in E^0 and feedback control in ∂E.

The Bellman-Hamilton-Jacobi (BHJ) equation associated with the problem defined above is

$$\max_{v \in U_0}\{-\nabla V(z) \cdot f(z, v) - \lambda(z, v) \int_{E^0} (V(y) - V(z)) Q_0(dy; z, v) + \delta V(z) - l_0(z, v)\} = 0. \qquad (1)$$

It is an intego-differential equation. Since value functions for most of optimal control problems are not smooth, several versions of generalized BHJ equations have been studied in the literature. The *generalized* Bellman-Hamilton-Jacobi (BHJ) equations related to this control problem were studied by Vermes, Dempster and Ye. Vermes [10] showed that the value function is a solution to a limiting form of the BHJ equation. Dempster and Ye [8] showed that under the assumption of this paper, the value function is a Lipschitz continuous solution of the generalized BHJ equation involving the Clarke generalized gradient. And if in addition the state space E is bounded, it is also a unique solution of the generalized BHJ equation involving the Clarke generalized gradient among the class of "regular functions". In this paper, we characterized the value function as the unique lower Dini solution and the unique viscosity solution of the BHJ equation. One of the consequences is that when the value function is regular in the sense of Dempster and Ye [8], it is the unique solution of the generalized BHJ equation involving Clarke generalized gradient without the boundedness assumption on the state space E. The advantage of working with the generalized BHJ equation involving Dini derivatives is that it provides an optimal piecewise feedback controls and a procedue for approximating optimal controls. Results on optimal synthesis for the optimal piecewise feedback controls are also given.

We make the following assumptions throughout:-

(A1) The *control sets* U_0, U_∂ are compact.

(A2) The vector field $f : E \times U_0 \longrightarrow \mathbb{R}^n$ is bounded, continuous and Lipschitz continuous in $x \in E^0$ uniformly in $v \in U_0$.

(A3) The jump rate $\lambda : E^0 \times U_0 \longrightarrow \mathbb{R}_+$ is bounded, continuous and Lipschitz continuous in $x \in E^0$ uniformly in $v \in U_0$.

(A4) As mentioned above, the transition measure Q may be expressed in terms of $Q_0 := Q|_{E^0} : E^0 \times U_0 \longrightarrow \mathbb{P}(E^0)$ and $Q_\partial := Q|_{\partial E} : \partial E \times U_\partial \longrightarrow \mathbb{P}(E^0)$. $Q_0 : E^0 \times U_0 \longrightarrow \mathbb{P}(E^0)$ is bounded, continuous relative to the weak* topology on $\mathbb{P}(E^0)$ and Lipschitz continuous in

$x \in E^0$ (i.e. for all $\theta \in C(E^0)$ the map $x \longmapsto \int_{E^0} \theta(y) Q_0(dy; x, v)$ is continuous and Lipschitz) uniformly in $v \in U_0$. Q_θ is defined on $\partial E \times U_\theta$ and has an extension to $E \times U_\theta$ such that the extension $Q_\theta : E \times U_\theta \longrightarrow I\!\!P(E^0)$ is bounded, continuous and Lipschitz continuous in $x \in E$ uniformly in $v \in U_\theta$.

(A5) The set of *admissible controls* $u := (u_0, u_\theta) \in C \subset C_0 \times C_\theta$ is defined in terms of the set of interjump *open loop* measurable (deterministic) *control functions*

$$C_0 := \{u_0 \in \mathcal{L} : u_0(\tau, z) : I\!\!R_+ \times E^0 \longrightarrow U_0\},$$

where τ represents the *time elapsed* since the last jump and z represents the *post jump state*, and the set of measurable *feedback boundary controls*

$$C_\theta : \{u_\theta \in \mathcal{L} : u_\theta : \partial E \longrightarrow U_\theta\}$$

for which $P_x^u[\lim_n \mathbf{T}_n = \infty] = 1$ for all $x \in E$, where for initial state x, $P_x^u(\cdot)$ is the probability measure (on path space) induced by u and $\mathcal{L}$ denotes the set of all measurable functions between a given domain and range.

(A6) The *running cost* $l_0 : E^0 \times U_0 \longrightarrow R_+$ is bounded, continuous and Lipschitz continuous in $x \in E^0$ uniformly in $v \in U_\theta$. The *boundary (jump) cost* $l_\theta : \partial E \times U_\theta \longrightarrow I\!\!R_+$ is continuous and has an extension to $E \times U_\theta$ such that the extension $l_\theta : E \times U_\theta \longrightarrow I\!\!R_+$ is bounded, continuous and Lipschitz continuous in $x \in E$ uniformly in $v \in U_\theta$.

(A7) There exists $\alpha > 0$ such that for all $x \in \partial E$ and all $v \in U_0$

$$f(x, v) \cdot n(x) \geq \alpha > 0,$$

where $n(x) := -\nabla \psi(x) / \|\nabla \psi(x)\|$ is the unit outward normal to $\partial E \in I\!\!R^n$ at the point $x \in \partial E$ and $\cdot$ denotes inner product.

(A8) The set

$$N_\theta(x) := \{(f(x, v), \lambda(x, v), l_0(x, v) + \lambda(x, v) \int_{E^0} \theta(y) Q_0(dy; x, v)); v \in U_0\}$$

is convex for all $x \in E^0$ and $\theta \in C(E^0)$.

(A9) $\inf_{x \in E^0, v \in U_0} \lambda(x, v) + \delta > \lambda_+^0$, where $\xi_+ := \max\{\xi, 0\}$ and

$$\lambda^0 := \sup_{\substack{x, y \in E^0 \\ v \in U_0}} (x - y) \cdot (f(x, v) - f(y, v)) / \|x - y\|^2.$$

2 The generalized BHJ equations

In this section, we characterize the value function as the unique lower Dini solution, the unique viscosity solution of the BHJ equation and the unique solution of the BHJ equation involving Clarke generalized gradient among the class of regular functions.

Theorem 1 *The value function $V(z)$ is the unique Lipschitz continuous solution to the generalized BHJ equation involving the lower Dini derivatives on E^0*

$$\min_{v\in U_0}\{V_-(z; f(z,v)) + \lambda(z,v)\int_{E^0}(V(y) - V(z))Q_0(dy; z, v) - \delta V(z) + l_0(z,v)\} = 0 \qquad (2)$$

with boundary condition

$$V(z) = \min_{v\in U_\vartheta}\{l_\vartheta(z,v) + \int_{E^0} V(y)Q_\vartheta(dy; z, v)\} \qquad (3)$$

on ∂E, where $V_-(z; d)$ is the lower Dini derivative at z in direction d, i.e.

$$V_-(z; d) := \liminf_{t\downarrow 0} \frac{V(z + td) - V(z)}{t}.$$

The following definition of viscosity solution of (1) was first defined by Soner in [9]. In [9], Soner considered a different PDP optimal control problem where controlled trajectories were forced to remain in the bounded domain.

Definition 1 A function $\phi: E^0 \longrightarrow R$ is said to be a *viscosity solution of the HJB equation* (1) on E^0 if condition (i) and (ii) below are satisfied.

(i) Given any $z \in E^0$ $\forall \psi \in C^\infty(R^n)$ and point $k \in R$ such that $\psi(z) > 0$ and the function $z \longrightarrow \psi(z)[\phi(z) - k]$ achieves its maximum over E^0 at z, then

$$\max_{v\in U}\{-p \cdot f(z,v) - \lambda(z,v)\int_{E^0}(V(y) - V(z))Q_0(dy; z, v) + \delta V(z) - l_0(z,v)\} \leq 0,$$

where $p := \frac{\phi(z)-k}{\psi(z)}\nabla\psi(z)$.

(ii) Given any point $z \in E^0$, any $\psi \in C^\infty(R^n)$ and point $k \in R$ such that $\psi(z) > 0$ and the function $z \longrightarrow \psi(z)[\phi(z) - k]$ achieves its minimum over E^0 at z, then

$$\max_{v\in U}\{-p \cdot f(z,v) - \lambda(z,v)\int_{E^0}(V(y) - V(z))Q_0(dy; z, v) + \delta V(z) - l_0(z,v)\} \geq 0.$$

Theorem 2 *The value function is the unique Lipschitz continuous viscosity solution of the generalized BHJ equation* (1) *with boundary condition* (3).

The following regularity condition is introduced in Dempster [6].

Definition 2 Let function $\phi: E \longrightarrow R$ be Lipschitz continuous. We say ϕ is *regular* if, and only if, for all $x \in E$ and all $d \in R^n$ $\phi_-(x; d) = \phi_0(x; d)$, where $\phi_0(x; d)$ is the *lower generalized directional derivative* of ϕ at x in direction d, i.e.,

$$\phi_0(x; d) := \liminf_{\substack{x'\to x \\ t\downarrow 0}} \frac{\phi(x' + th) - \phi(x')}{t}.$$

Remark 1 If a function ϕ is regular in the sense of Clarke [2], i.e., the directional derivative $\phi'(x; d)$ exists for any direction d and $\phi'(x; d) = \phi^0(x; d)$, $-\phi$ is regular in the above sense.

In view of the fact that $\phi_0(x; d) = \min_{\xi\in\partial\phi(x)} \xi \cdot d$ (cf. [1], proposition 2.12]), the following theorem is a direct consequence of Theorem 1.

Theorem 3 *If V is regular, then V is a unique solution of the BHJ equation involving the Clarke generalized gradient on E^0:*

$$\min_{\substack{\xi \in \partial V(z) \\ v \in U_0}} \left\{ \xi \cdot f(z,v) + \lambda(z,v) \int_{E^0} (V(y) - V(z)) Q_0(dy; z, v) - \delta V(z) + l_0(z,v) \right\} = 0$$

with boundary condition (3) on ∂E.

The proof of Theorem 1 and 2 is based on the reduction of a PDP optimal control problem to the following deterministic optimal control problem with a boundary condition.

(P_z) minimize $J(z, u(\cdot)) := \int_0^{t_*^u(z)} e^{-\Lambda_t^u(z)} f_0(x(t), u(t)) dt + e^{-\Lambda_{t_*}^u(z)} F(x(t_*^u(z)))$

over the class Ω of all *admissible pairs* $(x(\cdot), u(\cdot))$

such that $u : [0, t_*^u(z)] \to I\!\!R^m$ is measurable,

$\quad\quad u(t) \in U_0 \subset I\!\!R^m \quad \forall t \in [0, t_*^u(z)],$

$\quad\quad \dot{x}(t) = f(x(t), u(t)) \quad$ a.e. $t \in [0, t_*^u(z)],$

$\quad\quad x(0) := z \in E^0,$

where $\Lambda_t^u(z) := \int_0^t \bar{\lambda}(x(s), u(s)) ds$ and $t_*^u(z)$ is called the *boundary hitting time* of the trajectory $x(t)$ corresponding to control u for initial state z defined by

$$t_*^u(z) := \inf\{t > 0 : x(t) \in \partial E\}.$$

In the case where the trajectory for initial state z never reaches the boundary of E, $t_*^u(z) = \inf \emptyset = \infty$ by convention and $[0, t_*^u(z)]$ should be viewed as $[0, \infty)$. Where there is no confusion, we will simply use $t_*(z)$ even t_* instead of $t_*^u(z)$.

Precisely, we have the following reduction result due to Dempster and Ye [8].

Proposition 1 *The value function of a PDP optimal control problem coincides with the value function for problem (P_z) with problem data defined by*

$$f_0(x, v) := l_0(x, v) + \int_{E^0} V(y) Q_0(dy; x, v) \lambda(x, v) \tag{4}$$

$$F(x) := \min_{v \in U_\partial} \{ l_\partial(x, v) + \int_{E^0} V(y) Q_\partial(dy; x, v) \} \tag{5}$$

$$\bar{\lambda}(z, v) := \lambda(z, v) + \delta. \tag{6}$$

The PDP optimal control $u = (u_0, u_\partial)$ is equivalent to choosing for each possible postjump state $z \in E^0$, an optimal interjump control function $u_{0(\cdot)}(z)$ in the deterministic control problem with a boundary condition (P_z) with problem data defined by (4), (5) and (6) and for each $z \in \partial E$, a boundary control action $u_\partial(z)$ which solves the following optimization problem:

$$\min_{v \in U_\partial} \{ l_\partial(z, v) + \int_{E^0} V(y) Q_\partial(dy; z, v) \}. \tag{7}$$

Proof of Theorem 1 and 2 The Lipschtiz continuity of the value function under the assumptions of the paper is known (*cf.* Dempster and Ye [8]).

By Proposition 1, the value function $V(z)$ for the PDP optimal control problem coincides with one for the deterministic optimal control problem (P_z) with (4) (5) (6). As for the standard optimal control problems (See e.g. Theorem 4.1 of Berkovitz), one can show that $V(z)$ is a solution to the generalized HBJ equation involving the lower Dini derivatives

$$\min_{v \in U_0}\{V_-(z; f(z,v)) - \bar{\lambda}(z,v)V(z) + f_0(z,v)\} = 0 \quad \forall z \in E^0 \tag{8}$$

with boundary condition

$$V(z) = F(z) \quad \forall z \in \partial E. \tag{9}$$

Substituting (4) (5) (6) into (8) (9) , we obtain (2) and (3).

Under the assumption of the paper, by [III of Crandall and Lions [3]], the viscosity soltion to the BHJ equation for (P_z) is unique. Therefore the uniqueness of the lower Dini Solution and Theorem 2 follows from the fact that a lower Dini solution is also a viscosity solution.

∎

3   Optimal synthesis

A *piecewise feedback control* for the PDP optimal control problem is a pair $\mu = (\mu_0, u_\partial)$ such that μ_0 is a *piecewise feedback interior control* in the following sense: for a fixed $z \in E^0$, $\mu_0(\cdot, z) : E^0 \to U_0$ is a measurable function and $u_\partial : \partial E \longrightarrow U_\partial$ is the feedback boundary control.

In this section, we show that the generalized BHJ equation involving the lower Dini derivatives provides an optimal piecewise feedback controls and a procedue for approximating optimal controls. The results are based on the following verification theorem and the reduction result Proposition 1.

Theorem 4 (A Verification Theorem) *Let W be a Lipschitz solution to the generalized HJB equation involving the lower Dini derivatives*

$$\min_{v \in U_0}\{W_-(z; f(z,v)) - \bar{\lambda}(z,v)W(z) + f_0(z,v)\} = 0 \quad \forall z \in E^0 \tag{10}$$

with boundary condition

$$W(z) = F(z) \quad \forall z \in \partial E. \tag{11}$$

Let (x^*, u^*) be an admissible pair for (P_{z_0}). If the following condition is satisfied

$$W'(x^*(t); f(x^*(t), u^*(t))) - \bar{\lambda}(x^*(t), u^*(t))W(x^*(t)) + f_0(x^*(t), u^*(t)) = 0 \quad a.e. \ t \in [0, t_*^{u^*}(z_0)] \tag{12}$$

and $|W(x(t))| < \infty$ for all admissible trajectory $x(t)$. Then (x^*, u^*) solves (P_{z_0}).

Proof Let $(x(\cdot), u(\cdot))$ be any admissible pair for (P_{z_0}). Then $\theta(t) := W(x(t))$ is a composition of two Lipschitz maps and hence is Lipschitz. Therefore, at t where $\dot{x}(t)$ and $\dot{\theta}(t)$ exist, we have by virtue of Lipschitz continuity of W and the HJB equation (10) that

$$\begin{aligned}
\dot{\theta}(t) &= W'(x(t); f(x(t), u(t))) \\
&\geq W_-(x(t); f(x(t), u(t))) \\
&\geq \bar{\lambda}(x(t), u(t))W(x(t)) - f_0(x(t), u(t)).
\end{aligned}$$

Consequently,

$$\int_0^{t_*^u(z_0)} e^{-\Lambda_t^u(z_0)} f_0(x(t), u(t)) dt + e^{-\Lambda_{t_*}^u(z_0)} F(x(t_*^u(z_0)))$$

$$\geq \int_0^{t_*^u(z_0)} e^{-\Lambda_t^u(z_0)} [\bar{\lambda}(x(t), u(t)) W(x(t)) - \frac{dW(x(t))}{dt}] dt + e^{-\Lambda_{t_*}^u(z_0)} F(x(t_*^u(z_0)))$$

$$= -\int_0^{t_*^u(z_0)} d[e^{-\Lambda_t^u(z_0)} W(x(t))] + e^{-\Lambda_{t_*}^u(z_0)} F(x(t_*^u(z_0)))$$

$$= W(z_0).$$

The last equation in the case $t_* = +\infty$ was justified by virtue of the boundedness assumption $|W(x(t))| < \infty$.

On the other hand, by virtue of equation (12), we have

$$-\frac{dW(x^*(t))}{dt} + \bar{\lambda}(x^*(t), u^*(t)) W(x^*(t)) = f_0(x^*(t), u^*(t)) \quad a.e. \ t \in [0, t_*^{u^*}(z_0)].$$

Multiply both sides of the above equation by $e^{-\Lambda_{t_*}^{u^*}(z_0)}$ and integrating from 0 to $t_*^{u^*}(z_0)$, we have by virtue of (11) that

$$W(z_0) = \int_0^{t_*^{u^*}(z_0)} e^{-\Lambda_t^{u^*}(z_0)} f_0(x^*(t), u^*(t)) dt + e^{-\Lambda_{t_*}^{u^*}(z_0)} F(x(t_*^{u^*}(z_0))).$$

Substituting the above equation into the right hand side of (13), we deduce that (x^*, u^*) is an optimal solution of (P_{z_0}) and the proof of the theorem is complete. ∎

For any $x \in E$, define

$$Q(x) := \{v : \min_{u \in U_0}\{V_-(x; f(x, u)) + \lambda(x, u) \int_{E^0} (V(y) - V(x)) Q_0(dy; x, u) + l_0(x, u)\}$$

$$= V_-(x; f(x, v)) + \lambda(x, v) \int_{E^0} (V(y) - V(x)) Q_0(dy; x, v) + l_0(x, v)\},$$

where $V(x)$ is the value function. We also define $G(x) := f(x, Q(x))$. Suppose that the following differential inclusion has an absolutely continuous solution for any initial state $z \in E^0$:

$$\dot{x}(t) \in G(x(t)) \quad a.e. \ t \in [0, t_*(z)] \tag{13}$$

$$x(0) = z. \tag{14}$$

An optimal piecewise feedback control $\mu^* = (\mu_0^*, u_\partial^*)$ can be constructed as follows: For any $z \in E^0$, let $\phi_{(\cdot)}(z)$ be an absolutely continuous solution of the differential inclusion (13) with initial condition (14). Take $u_{0t}^*(z) := \mu_0^*(\phi_t(z), z)$ to be the measurable selection of the set valued map $t \longrightarrow Q(\phi_t(z))$ such that

$$\frac{\partial}{\partial t} \phi_t(z) = f(\phi_t(z), u_{0t}^*(z)) \quad a.e. \ t \in [0, t_*(z)]$$

$$\phi_0(z) = z;$$

For any $z \in \partial E$, take $u_\partial^*(z)$ to be the minimizer of the following function:

$$u \longrightarrow l_\partial(z, u) + \int_{E^0} V(y) Q_\partial(dy; z, u).$$

By virtue of Theorem 4, $u_{0t}(z)$ is an optimal control of problem (P_z) with data defined by (4) (5) and (6). Therefore by Proposition 1, the piecewise feedback control $\mu^* = (\mu_0^*, u_\partial^*)$ is optimal.

We now show that under our assumptions, the solution $\phi_t(z) := \psi(t)$ of the differential inclusion (13) with initial condition (14) exists and can be obtained as a uniform limit of a sequence of piecewise linear functions $\{E_n(t)\}$. First, we construct the sequence $\{E_n(t)\}$ by the following algorithm due to Berkovitz [1]:

Algorithem 1 Let $n > 0$ be any integer. Let $v_0 \in Q(\xi_0)$, where $\xi_0 := z$. For $0 \leq t \leq \frac{1}{n}$, define

$$E_n(t) := \xi_0 + f(\xi_0, v_0)(t - 0).$$

Suppose $E_n(t)$ has been defined for $0 \leq t \leq \frac{i}{n}$. Let $\xi_i = E(\frac{i}{n})$ and $v_i \in Q(\xi_i)$. Then for $\frac{i}{n} \leq t \leq \frac{i+1}{n}$, define

$$E_n(t) := \xi_i + f(\xi_i, v_i)(t - \frac{i}{n}).$$

Note that in the above procedure, if $t_*^n := \inf\{t > 0 : E_n(t) \in \partial E\} < \infty$, then we take

$$E_n(t) := E_n(t_*^n) \qquad \forall t \geq t_*^n.$$

Therefore $E_n(t)$ is defined for $0 \leq t < \infty$.

Definition 3 A set-valued map G from E to $\mathbb{R}^n$ is said to satisfy property (Q) at point $x \in E$ if

$$G(x) = \bigcap_{\delta > 0} \text{clco}\{G(N_\delta(x) \cap E)\},$$

where $N_\delta(x)$ is the $\delta-$ neighborhood of x defined by $N_\delta(x) := \{y : \|y - x\| < \delta\}$.

The following result shows that Berkovitz's algorithm works for our problem which may have unbounded time horizon.

Theorem 5 *In addition to assumptions* (A1)–(A9), *suppose that* $f(x, u)$ *satisfies the following growth condition:* $\|f(x, u)\| \leq c + d\|x\|$, *where* $c > 0$ $d > 0$ *are constants and the set-valued map* G *satisfies property* (Q). *Then* $\{E_n(t)\}$ *defined by Algorithm 1 converges to an optimal interjump trajectory* $\psi(t)$ *uniformly on any compact subinterval of* $[0, \infty)$.

Before the proof of Theorem 5, let us state the following lemma (cf. Berkovitz [1]).

Lemma 1 *Given* $z \in E^0$ *and* $v_0 \in Q(z)$. *let* N *be any positive integer. Then there exist constants* $R_N > 0$ *and* $M_N > 0$ *such that for all* n *and all* $0 \leq t \leq N$, *we have* $\|E_n(t)\| \leq R_N$ *and* $\|\dot{E}_n(t)\| \leq M_N$.

Proof of Theorem 5. For each integer $N > 0$, by Lemma 1, the functions $\{E_n\}$ are uniformly bounded and equicontinuous on $[0, N]$. So by Ascoli's theorem, there exists a subsequence that we label as $\{E_n^N\}$, that converges uniformly on $[0, N]$ to a continuous function ψ^N. Since $E_n^N(0) = z$ for all $n = 1, 2, 3, \ldots$, we have $\psi^N(0) = z$. Thus, we may choose successive subsequences, each a subsequence of the previous sequence, with the property that

(1) $E_n^N \longrightarrow \psi^N$ uniformly on $[0, N]$,

(2) $E_n^N(0) = \psi^N(0) = z \ \forall a = 1, 2, 3, \ldots,$

(3) $\psi^M = \psi^N$ on $[0, N]$ for all $M \geq N$.

The diagonal process gives a subsequence $\{E_n\}$ of the original sequence such that $E_n \to \psi$ uniformly on any compact subinterval of $[0, \infty)$, where ψ is defined by

$$\psi(t) = \psi^N(t) \qquad \forall \, t \in [0, N].$$

Since $\dot{E}_n(t) = f(\xi_i, v_i) \ \forall t \in [\frac{i}{n}, \frac{i+1}{n}]$ and f is bounded by assumption $(A2)$, the integrals $\int_0^t \dot{E}_n(s)ds, 0 \leq t \leq N$ are uniformly bounded. Therefore by using Dunford Pettis criterion we can show that $\psi(t)$ is absolutely continuous and the derivatives $\dot{E}_n^N(t) \to \dot{\psi}^N(t)$ weakly in $L^1[0, N]$. Hence by Mazur's Theorem, there exists a sequence of funtions $\psi_{n_j}^N$ defined by the formulars:

$$\psi_{n_j}^N = \sum_{i=1}^{k(n_j)} \alpha_{n_j i} E_{n_j+i}^N, \qquad \alpha_{n_j i} \geq 0 \ \sum_{i=1}^{k(n_j)} \alpha_{n_j i} = 1,$$

where $n_{j+1} > n_j + k(n_j)$ and such that $\dot{\psi}_{n_j}^N \longrightarrow \dot{\psi}^N$ in $L_1[0, N]$. An elementary argument shows that the uniform converges of E_n^N to ψ^N on $[0, N]$ implies that $\psi_{n_j}^N \longrightarrow \psi^N$ uniformly on $[0, N]$. Since $\dot{\psi}_{n_j}^N \longrightarrow \dot{\psi}^N \in L_1[0, N]$, there exists a subsequence that we label as $\psi_{n_j}^N$ such that $\dot{\psi}_{n_j}^N \longrightarrow \dot{\psi}^N$ almost everywhere on $[0, N]$. By repeating this process for sucessive N, choosing each subsequences $\{\psi_{n_j}^N\}$ from the previous sequence and diagonalizing, we have a sequence $\dot{\psi}_{n_j} \longrightarrow \dot{\psi}$ a.e. $\in [0, \infty)$.

Let t_0 be a point such that $\dot{\psi}_{n_j}(t_0) \longrightarrow \dot{\psi}(t_0)$ and N be any positive integer such that $t_0 \in [0, N]$.

Let $\delta > 0$ be given. Since $\|\dot{E}_n(t)\| \leq M_N$ for all $t \in [0, N]$ and all n, there exists $\eta > 0$ such that if $|t' - t''| < \eta$ and $t', t'' \in [0, N]$, $\|E_n(t') - E_n(t'')\| < \delta/2$ for all n. Also there exists a positive integer N_1 such that for $n > N_1$, $\|E_n(t) - \psi(t)\| < \delta/2$ for all $t \in [0, N]$.

There exist a positive integer N_2 such that for $n \geq N_2$, $\frac{1}{n} < \eta$ and $\frac{i}{n} \leq t_0 < \frac{i+1}{n}$ for some i. Hence for $n > M := \max\{N_1, N_2\}$,

$$\left\| E_n\left(\frac{i}{n}\right) - \psi(t_0) \right\| \leq \left\| E_n\left(\frac{i}{n}\right) - E_n(t_0) \right\| + \| E_n(t_0) - \psi(t_0) \| < \delta.$$

That is

$$E\left(\frac{i}{n}\right) \in N_\delta(\psi(t_0)).$$

Since by our definition of E_n, $\dot{E}_n(\frac{i}{n} + 0) \in G(E_n(\frac{i}{n}))$, we have $\dot{E}_n(\frac{i}{n} + 0) \in G(N_\delta(\psi(t_0)))$. By our choice of n and definition of E_n, we have $\dot{E}_n(\frac{i}{n} + 0) = \dot{E}(t_0)$. Therefore $\dot{E}_n(t_0) \in G(N_\delta(\psi(t_0))$ for $n > M$. This in turn implies that there exists a positive integer $J = J(\delta)$ such that for $j > J$,

$$\dot{\psi}_{n_j}(t_0) \in \text{co}\{G(N_\delta(\psi(t_0)))\}.$$

Since $\dot{\psi}_{n_j}(t_0) \longrightarrow \dot{\psi}(t_0)$, we have $\dot{\psi}(t_0) \in \text{clco}\{G(N_\delta(\psi(t_0)))\}$. Since $\delta > 0$ are arbitrary, we have

$$\dot{\psi}(t_0) \in \bigcap_{\delta > 0} \text{clco}\{G(N_\delta(\psi(t_0)))\}.$$

Finally since property (Q) holds for G, we get that

$$\dot{\psi}(t_0) \in G(\psi(t_0)).$$

In view of the discussion preceeding Algorithm 1, $\psi(t)$ is an optimal interjump trajectory. The proof of the theorem is therefore complete. ∎

References

[1] L.D. Berkovitz, Optimal feedback controls, *SIAM Journal of Control and Optimization* **27** (1989), 991-1006.

[2] F.H. Clarke, *Optimization and Nonsmooth Analysis*, John Wiley & Sons, New York, 1983.

[3] M.C. Crandall and P.-L. Lions, Viscosity solutions of Hamilton-Jacobi equations. *Trans. Amer. Math. Soc.* **227**(1983), 1-42.

[4] M.H.A. Davis, Piecewise-deterministic Markov processes: A general class of non-diffusion stochastic models. *J. Roy. Statist. Soc.* **B46**(1984), 353-388.

[5] M.H.A. Davis, *Markov Models and Optimization*, Chapman and Hall, London. To appear.

[6] M.A.H. Dempster, Optimal control of piecewise deterministic Markov processes, *Proceedings of the Imperial College Workshop on Applied Stochastic Analysis* London, April 5-7, 1989 (M.H.A. Davis and R.J.Elliott, eds.), Gordon & Breach, London, 1991.

[7] M.A.H. Dempster and J.J. Ye, Generalized Bellman-Hamilton-Jacobi optimality conditions for a control problem with a boundary condition, Submitted for publication.

[8] M.A.H. Dempster and J.J. Ye, Necessary and sufficient optimality condition for control of piecewise deterministic Markov processes, *Stochastics and Stochastics reports* **40**(1992), 125-145.

[9] H.M. Soner, Optimal control with state space constraint II. *SIAM J. Control and Optimization* **24**(1986), 1110-1122.

[10] D. Vermes, Optimal control of piecewise deterministic Markov processes, *Stochastics* **14** (1985), 165-208.

[11] J.J. Ye. Optimal control of piecewise deterministic Makov processes. Ph.D. Dissertation, Dept. of Math., Stats. & C.S., Dalhousie University, Halifax, Canada, 1990.

AN APPLICATION OF IMPULSE CONTROL METHOD TO TARGET ZONE PROBLEM

Monique JEANBLANC-PICQUÉ

Equipe d'analyse et probabilités

Université d'Evry-Val d'Essonne

1   Introduction

In this paper, we study the behavior of exchange rate when allowed to fluctuate within prespecified margins. This kind of regime is known as Target Zone. For example, the European Monetary System (EMS) allows a variation of 2.25% around a central parity. The exchange rate is allowed to float more or less freedly within the Target Zone, but is prevented from moving outside the band by foreign exchange interventions. The current value of the Exchange rate is supposed (from economic arguments) to depend of a term called "the fundamental" which corresponds to the exchange of foreign money of the countries and of the anticipations on the variation of the exchange rate. Under some economic hypotheses, it is proved that the relation between the fundamental (which is modeled as a generalized Brownian motion) and the exchange rate is represented by a S-shaped curve, with a tangency property of the fluctuation band. The existence of an exchange rate band tends to stabilize the exchange rate behavior within the band (the so-called "honeymoon" stabilizing effect). When the exchange rate is near the boundary, people expect the central bank will intervene. An important paper of Krugman [5] presents a model of the Target Zone, under the following assumptions :

- the interventions are only marginal interventions, which means that there is no intervention inside the band,

- the size of interventions remains infinitesimal.

This model is based on regulated Brownian Motion.

Many empirical studies on the EMS suggest that most interventions of central banks occur when the exchange rate is inside its fluctuation band. We propose in this paper a target zone allowing intra-marginal interventions. In a first part, we restate the exchange rate model. Then we give a mathematical approach to the problem: how to force a generalized Brownian motion to stay in a given band when the objective is to minimize a cost fonction associated with each intervention. In the Krugman model, the cost is not defined, but it corresponds to an intervention cost proportional to the size of the intervention (See Avezani [1] for details). We study here the case of a cost made of two parts : a fixed cost plus a cost proportional to the size of the intervention. In the last part, we apply these results to the exchange rate problem with a particular attention to the fact that the exchange rate must keep the same value before and after intervention, in order to carry out speculation opportunities. All details and proofs can be found in [4] .

2   The exchange rate

The basic model of the exchange rate [2, 5, 7] states that the logarithm of the exchange rate at time t, X_t is equal to the fundamental K_t plus a term proportional to the expected percentage change in the exchange rate

$$X_t = K_t + \theta E[dX_t/dt|\mathcal{F}_t] \, , \, \theta > 0. \tag{1}$$

The fundamental is the sum of two components

$$K_t = M_t + V_t$$

where M_t is the difference between domestic and foreign money supply and V_t is a composit money demand shock called the volatility. The money supply is under control of the central banks, whereas the velocity is supposed to be an exogeneous process given by a generalized Brownian motion

$$K_t = k + \eta t + \sigma W_t. \tag{2}$$

When there are no interventions, $dM_t = 0$. When there are interventions, the central banks buy (or sell) a quantity of foreign money and, in the case of no-infinitesimal intervention $M_{t+} - M_{t-} \neq 0$. The target zone is an interval $[L, U]$.

We suppose, as [2, 5, 7] that X is a C^2-function of K:

$$X_t = g(K_t).$$

The drift term of X is $\frac{1}{\theta}(X_t - K_t)$ (see 1). It is then possible, using Ito's formula for $g(K_t)$ to determine the function g :

$$g(k) = k + \theta\eta + Ae^{\mu_1 k} + Be^{\mu_2 k} \tag{3}$$

where μ_i are the roots of $\frac{1}{2}\sigma^2 x^2 + \eta x - \frac{1}{\theta}$. We obtain the S-shaped curve called the saddlepath solution. In particular, the saddlepath exchange rate for a free float is given by

$$X_t = K_t + \theta\eta.$$

We suppose that in the target zone situation, the saddlepath solution is tangent to the boundary of the target zone (See figure 1) (this condition is called the "smooth-pasting" condition in the literature). If not, it follows from [5] that the points $(k_l, g(k_l))$ and $(k_u, g(k_u))$ such that $g'(k_u) = g'(k_l) = 0$ are inside the target zone, i.e. $L \leq g(k_l) \leq g(k_u) \leq U$. In this case, it suffices to reduce the target zone to recover the first case. In order to realize this smooth-pasting condition, we have to solve two equations to determine the value of the two constants A and B.

The previous works on the target zone assume that interventions in the foreign exchange market occur only when the exchange rate reaches the bounds of the target zone. Let us recall that the authorities can act only on the fundamental. If the fundamental is prevented from moving outside the range $[k_l, k_u]$, the exchange rate remains in the target zone. Since the exchange rate must remain the same after an intervention (otherwise, economic agents can use this opportunuity), the size of the interventions must be infinitesimal. These conditions lead to a reflection strategy which involves local times. The controlled process is

$$dM_t = dL_t - dU_t$$

where L_t (resp U_t) are increasing processes which increases only when X_t reaches L (resp U), i.e., when K_t raeches k_l (resp. k_u). If we associate a cost function to each intervention, this cost can not involve a fixed part, otherwise the total cost will be infinite.

Our aim is to present a different control strategy associated with a cost function which includes a fixed cost.

3   Impulse control for Brownian motion

3.1   The model

A probability space $(\Omega, \mathcal{F}, \mathcal{F}_t, W_t, P)$ and a one-dimensional Brownian motion W_t are given.

Let η and σ be two constants and define a generalized Brownian K_t

$$K_t = k + \eta t + \sigma W_t. \tag{4}$$

The control must be such that the controlled process

$$Y_t = K_t + \sum_{T_i < t} \zeta_i$$

remains in a given band $[a, b]$, where

$T = (T_i, i \geq 1)$ is an increasing sequence of stopping times

$\zeta = (\zeta_i, i \geq 1)$ is a sequence of $\mathcal{F}_{T_i}$-measurable random variables (the impulses).

The cost function is

$$J(k; T, \zeta) = E_k\left[\sum_{i \geq 1} e^{-\lambda T_i}(\gamma + c|\zeta_i|)\right] \tag{5}$$

where $(\lambda, \lambda > 0, \gamma, \gamma > 0, c, c > 0$ are given parameters.

Definition 1 *A pair* (T, ζ) *is admissible for the initial point* k *if*

- $a \leq Y_t \leq b, \quad \forall t \in \mathbb{R}^+$;

- $J(k; T, \zeta) < \infty$.

We denote by $\mathcal{A}(k)$ the set of admissible pairs with initial point k. It is easy to prove

Lemma 1 *The set of admissible pairs is not empty.*

The objective is to minimize the cost function over all the admissible strategies.

3.2   The value function

Let us denote by V the value function defined on $[a, b]$

$$V(x) = \inf\{J(x; T, \zeta); (T, \zeta) \in \mathcal{A}(x)\}. \tag{6}$$

and by L the operator on C^2-functions

$$L(\psi) := \eta\psi' + \frac{1}{2}\sigma^2\psi'' - \lambda\psi. \tag{7}$$

Suppose that there exists an optimal control for each initial point $k \in [a, b]$. If the initial point is x, we can choose between two different strategies :
 - follow the optimal strategy. The cost fonvtion is equal to $V(x)$
 - jump at y and follow the optimal strategy from y. The cost is the cost due to the jump and the cost of the y-optimal strategy, i.e., $\gamma + c|x - y| + V(y)$. Since the x-optimal strategy is the first one, we must have

$$V(x) \leq V(y) + \gamma + c|x - y|.$$

We are lead to establish the following lemma :

Lemma 2 *Let ψ be a $C^2([a, b])$-function such that*

$$\begin{cases} L\psi(x) & \geq 0, \quad \forall x \in [a, b] \\ \psi(x) & \leq \psi(y) + \gamma + c|x - y|, \\ & \forall(x, y) \in [a, b]^2. \end{cases}$$

Then $\psi(k) \leq J(k; T, \zeta)$ for each admissible strategy in $\mathcal{A}(k)$.

Proof:
The proof follows from Ito's formula applied to $e^{-\lambda t}\psi(Y_t)$.

We can conjecture ([3, 6]) that an optimal strategy is on the form

$$\begin{cases} T_i^* & = \inf\{t > T_{i-1}^* | Y_t = a \text{ or } b\} \\ Y_{T_i^*+} & = Y_{T_i^*} + \zeta_i^* = \beta \, \mathbb{1}_{Y_{T_i^*}=b} + \alpha \, \mathbb{1}_{Y_{T_i^*}=a}. \end{cases} \tag{8}$$

The controller does not act when the process is inside the band. When the process reaches the boundary, the controller throws the process inside the band. The size of the jumps is time independant.

Let us remark that if this policy is optimal, arguing as before lemma 2, we obtain that

$$\begin{cases} V(b) = V(\beta) + \gamma + c(b - \beta) \\ V(a) = V(\alpha) + \gamma + c(\alpha - a) \end{cases}$$

Furthermore, since

$$\forall \epsilon > 0 \ V(b) = V(\beta) + \gamma + c(b - \beta) \leq V(\beta \mp \epsilon) + \gamma + c(b - \beta \mp \epsilon)$$

(and a similar formula for (a, α)), we obtain that

$$V'(\beta) = c, \qquad V'(\alpha) = -c.$$

Lemma 3 *Let ψ be a C^2-function on $[a, b]$ such that there exists a pair (α, β) with $a < \alpha \leq \beta < b$ and*

$$\begin{align} \psi(b) &= \psi(\beta) + \gamma + c(b - \beta), & (9) \\ \psi(a) &= \psi(\alpha) + \gamma + c(\alpha - a), & (10) \\ \psi'(\beta) &= c, & (11) \\ \psi'(\alpha) &= -c, & (12) \\ L\psi(x) &= 0, \quad \forall x \in [a, b]. & (13) \end{align}$$

Then ψ is the value function and the optimal strategy is given by (8).

4   Existence of an optimal control

We prove the existence of a function V and two parameters (α, β) which satisfy (9 - 13).
We proceed in three steps:

- We solve (13). We obtain a set of functions which depends on two parameters (μ, ν) on the form $\mu e^{\lambda_1 x} + \nu e^{\lambda_2 x}$, where λ_i are the roots of $\frac{1}{2}\sigma^2 x^2 + \eta x - \lambda$.

- We prove that to each choice of (α, β) we can associate a pair $(\mu(\alpha, \beta), \nu(\alpha, \beta))$ such that the function $\mu(\alpha, \beta)e^{\lambda_1 x} + \nu(\alpha, \beta)e^{\lambda_2 x}$ satisfies (11) and (12).

- We prove that we can choose (α, β) such that this function satisfies (9) and (10).

We obtain the following result:

Theorem 1 *There exists a pair (α, β) which defines an optimal control of the form*

$$T_i^* = \inf\{t > T_{i-1}^* | Y_t = a \text{ or } b\}$$

$$Y_{T_i^*+} = Y_{T_i^*} + \zeta_i^* = \beta \, \mathbb{1}_{Y_{T_i^*}=b} + \alpha \, \mathbb{1}_{Y_{T_i^*}=a}.$$

The set (a, α, β, b) is called an optimal strategy. In order to apply our results to exchange rate, we need to prove a result which states that if a is given as well as the size of the a-jump (i.e. $\alpha - a$), there exists b (and β) such that the band $[a, b]$ admits an optimal strategy on the form (a, α, β, b)

5   Target zone

5.1   The model

We suppose here that the central bank intervenes only at some fixed level. We define a cost for each policy of the central bank in order to reduce the problem to an optimal strategy problem. We are working with the same cost function (5) as in section 1.

Let us denote by $\mathcal{G}$ the curve defined by $x = g(k)$.
The exchange rate must be continuous, even if the fundamental has jumps, thus we have to solve the following problem: find (a, b) such that $g(a) = g(\alpha_{a,b})$ and $g(b) = g(\beta_{a,b})$. The central banks fixe the band $[a, b]$, such that if the fundamental remains in the band $[a, b]$ the exchange rate remains in the target zone $[L, U]$. If we denote by K_U (resp. K_L) the point where $g(K_U) = U$, $g'(K_U) \neq 0$ (resp. $g(K_L) = L$, $g'(K_L) \neq 0$) the exchange rate remains in $[L, U]$ as soon as $K_U \leq a \leq b \leq K_L$.

5.2   The fundamental's band

Taking a point $(a, g(a))$ on $\mathcal{G}$, $K_U \leq a \leq k_l$, we can define α, $k_l \leq \alpha \leq k_u$ such that $g(\alpha) = g(a)$ (do not forget that the exchange rate is not affect under an intervention). To each (a, α), we associate a pair (b, β) such that the strategy (a, α, β, b) is optimal. If $b(b) = g(\beta)$, we are done. Otherwise, we prove that there exists a such that this condition is satisfied with $k_l \leq \beta \leq k_u \leq b \leq K_L$.

Theorem 2 *There exists a band (a, b) such that the interventions take place when the fundamental reaches the boundary of this band and follows the optimal policy described in (8). Under this strategy, the exchange rate remains in the target zone.*

6 Bibliography

References

[1] Avesani R. (1991) Optimal interventions in exchange rate bands. *Preprint. University of Trento.*

[2] Flood R., Garber P. (1991) The linkage between speculative attack and target zone: models of exchange rates. *Quarterly J. of Econ.* to appear

[3] Harrison M.J., Selke T., and Taylor A. (1983) Impulse Control of a Brownian Motion. *Math. Oper. Res.* 8, pp. 454-466.

[4] Jeanblanc-Picqué M. (1993) Impulse control Method and Exchange rate. *Mathematical finance* 3, pp. 161-177.

[5] Krugman P. (1991) Target zones and exchange Dynamics. *Quarterly J. of Econ.* 106, pp. 669-682

[6] Sulem A. (1986) A Solvable one-dimensional Model of a diffusion inventory system. *Math. Oper. Res.* 11, pp. 25-133.

[7] Svensson L. (1991) Target zones and interest rate variability. *Journal of International Economics* 31, pp. 27-54.

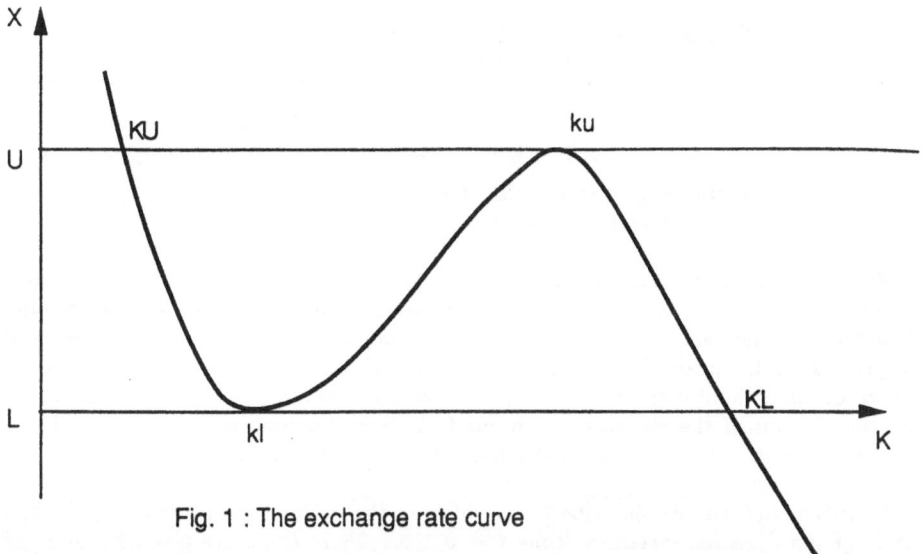

Fig. 1 : The exchange rate curve

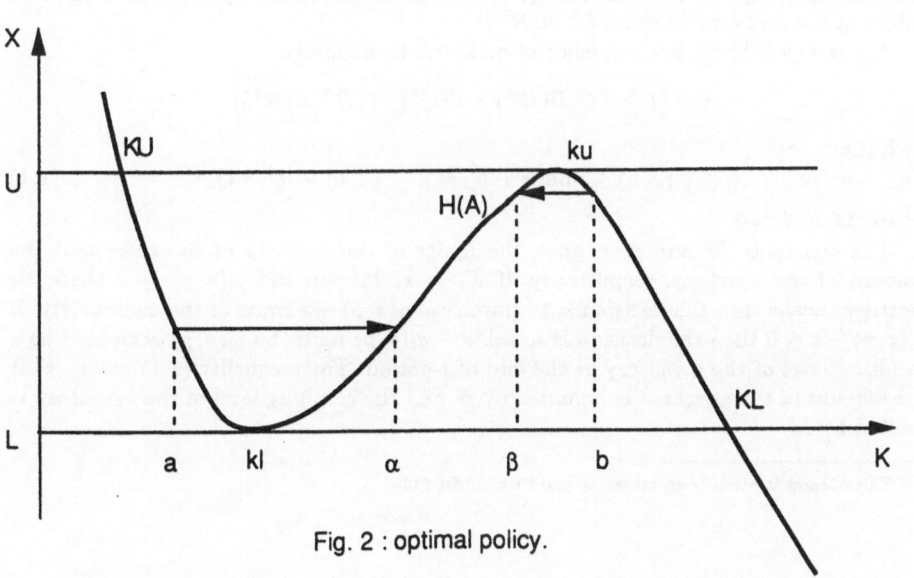

Fig. 2 : optimal policy.

OPTIMAL STRATEGY IN A TRADING PROBLEM
WITH STOCHASTIC PRICES

BY

R. REMPAŁA

Institute of Mathematics, Polish Academy of Sciences
00-950 Warszawa, ul. Śniadeckich 8, Poland

Abstract. A stochastic, dynamic version of a one-commodity stock-exchange problem is considered. The market prices are described by a Markov chain. At the beginning of each period the decision maker knows the level of his inventory of a commodity and the present market price. Then he decides to buy or sell some amount of the commodity. The resulting amount is treated as the next period inventory. The capacity of the store and the amount of the purchase are limited. A linear holding cost is introduced. An optimal purchase-sell policy which minimizes expected cost is given.

1. Introduction. We describe the problem in the optimal control framework. The states of the dynamic system at time $t = 0, 1, \ldots, T$, $T < \infty$, are given by pairs of real-valued random variables (P_t, X_t). The process $\{P_t\}$ represents the market prices. It is a nonnegative Markov process with a measurable state space $(Q, \mathcal{B}(Q))$, where Q is a bounded Borel subset of R^+. The process is defined in terms of transition functions $\{G_t\}$. For $t = 0, 1, \ldots$, $G_t(p, \Gamma)$ is defined for $p \in Q$ and $\Gamma \in \mathcal{B}(Q)$. For each p, $G_t(p, \cdot)$ is a probability measure on $\mathcal{B}(Q)$ and for each $\Gamma \in \mathcal{B}(Q)$, $G_t(\cdot, \Gamma)$ is a $\mathcal{B}(Q)$ measurable function on Q. The second process $\{X_t\}$ represents the inventory levels and it takes values in the measurable space $(R, \mathcal{B}(R^+))$.

A strategy (policy) is a sequence of measurable mappings

$$y_t : (R^+ \times Q, \mathcal{B}(R^+) \times \mathcal{B}(Q)) \to (R^+, \mathcal{B}(R^+))$$

such that

$$0 \leq y_t(x, p) \leq \min(x + N, M), \quad (x, p) \in R^+ \times Q, \tag{1.1}$$

where $M \geq N > 0$.

The constants M and N express the limits of the capacity of the store and the amount of the purchase, respectively. If $X_t = x$, $P_t = p$ and $y_t(x, p) \geq x$ then the strategy means that the decision is to purchase $y_t(x, p) - x$ units of the commodity. If $y_t(x, p) - x \leq 0$ then the decision is to sell $x - y_t(x, p)$ units. So $y_t(x, p)$ is treated as a resulting level of the inventory at the end of t-period. The inequality (1.1) means that the amount of the purchase is bounded by N and the resulting level of the inventory is limited by M.

This research is partially supported by grant No. 2P301 01004.

If a policy $y = (y_0, y_1, \ldots)$ is given then the corresponding inventory process X_0, $X_1, \ldots$ is determined by the equations

$$X_0 = x_0, \quad P_0 = p_0, \quad M \geq x_0 \geq 0, \ p_0 \geq 0,$$
$$X_{t+1} = y_t(X_t, P_t), \quad t = 0, 1, \ldots \tag{1.2}$$

For a policy y and the corresponding processes $\{X_t\}$, $\{P_t\}$ we define the finite horizon cost functional. Let n be a positive integer and let

$$J(n, y) = E \sum_{t=0}^{n} \{P_t(y_t(X_t, P_t) - X_t) + h_t X_{t+1}\} - P_{n+1} X_{n+1} \tag{1.3}$$

where $h_t \geq 0$ denotes the unit holding cost.

The control problem for the system (1.1)–(1.2) and the cost (1.3) is to determine the strategy y^*, called optimal, such that for all y satisfying (1.1)

$$J(n; y) \geq J(n, y^*). \tag{1.4}$$

Observe that in the problem the strategies truncated to the subscripts $t = 0, 1, \ldots, n$ are relevant only.

The purpose of this article is to provide a solution for the problem (1.1)–(1.4). We show that the optimal order policy is similar to the (s, S) polices found in the inventory control literature (Ref. [4, 2]) except that the reorder level s is a function of the price $(s(p), \ p \in Q)$ and order-up-to level S is a function of the price and of the entering inventory x. More precisely, we prove that for every t there exists a function $s_t(p)$ such that

$$S_t(x, p) = \min(x + N, s(p)) \quad \text{if } x \leq s_t(p)$$

and

$$S_t(x, p) = s_t(p) \quad \text{if } x > s_t(p).$$

Dynamic programming is used to derive this optimal policy.

2. Optimal Strategy Theorem.

The main objective of this section is to specify the structure of the optimal policy for the problem (1.1)–(1.4).

We recall the data of the problem. The constants $M, N, h_0, h_1, \ldots, h_n$ are such that $M \geq N > 0$, $h_t \geq 0$, $t = 0, 1, \ldots, n$; G_t, $t = 0, 1, \ldots, n$, denote the transition functions describing the process $\{P_t\}$; Q is a state space of the process $\{P_t\}$. It is assumed that Q is a bounded Borel subset of $\mathbf{R}^+$; $x_0 \geq 0$, $p_0 \geq 0$ denote the initial values of the corresponding processes.

Theorem 2.1. *There exists a sequence of functions $s_0(p), s_1(p), \ldots, s_n(p)$ such that an optimal policy of the problem (1.1)–(1.4) is of the form*

$$y_t^*(x, p) = \begin{cases} \min(x + N, s_t(p)) & \text{if } x \leq s_t(p), \\ s_t(p) & \text{if } x > s_t(p). \end{cases}$$

In the following sections we apply the dynamic programming algorithm (cf. Zabczyk, Ref. [5]) to prove Theorem 2.1 and the next Theorem 2.2 which specifies the structure of $s_t(p)$. To this end we first define by induction auxiliary sequences of functions and sets. (Similar auxiliary sequences are discussed in the author's paper Ref. [3] and in Magirou, Ref. [1].)

Let

$$s_0^n(p) = - \int_Q q G_n(p, dq), \quad p \in Q,$$

$$A_0^n = \{p \in Q : p + h_n + s_0^n(p) \le 0\}.$$

Suppose that for $k = t - 1$ with $t - 1 \ge 0$ we have defined fnctions $s_0^{n-(t-1)}, \ldots, s_{t-1}^{n-(t-1)}$ and sets $A_0^{n-(t-1)}, \ldots, A_{t-1}^{n-(t-1)}$. Then for $k = t$ put

$$s_0^{n-t}(p) = - \int_Q g G_{n-t}(p, dq),$$

$$s_i^{n-t}(p) = s_0^{n-t}(p) + \int_{A_{i-1}^{n-(t-1)}} (q + h_{n-(t-1)} + s_{i-1}^{n-(t-1)}(q)) G_{n-t}(p, dq) \qquad (2.1)$$

$$i = 1, \ldots, t,$$

and define

$$A_i^{n-t} = \{p \in Q : p + h_{n-t} + s_i^{n-t}(p) \le 0\}, \quad i = 0, 1, \ldots, t. \qquad (2.2)$$

Remark 2.1. It follows from the inductive definitions that for $t = 0, 1, \ldots, n$ and $i = 0, 1, \ldots, t$.

(i) the function s_i^{n-t} is uniquely determined by G_{n-t+j}, $j = 0, \ldots, i$, and h_{n-t+j}, $j = 1, \ldots, i$.

(ii) the set A_i^{n-t} is uniquely determined by G_{n-t+j}, h_{n-t+j}, $j = 0, \ldots, i$.

The lemma below is provided in order to specify a property of $\{s_i^{n-t}\}$ and $\{A_i^{n-t}\}$.

Lemma 2.1. *For every superscript* $n - t$ *with* $t = 0, 1, \ldots, n$ *we have*

(i) $s_t^{n-t} \le s_{t-1}^{n-t} \le \ldots \le s_0^{n-t} \le 0$ *and so*
(ii) $A_0^{n-t} \subset A_1^{n-t} \subset \ldots \subset A_t^{n-t} \subset Q$.

Proof. Observe that $s_0^{n-t}(p) \le 0$, for $p \in Q$ and $t = 0, 1, \ldots, n$. Thus for $k = t = 0$ the assertions (i) and (ii) are true.

Assume that (i) is true for $k = t - 1 \ge 1$. This implies that

$$p + h_{n-(t-1)} + s_i^{n-(t-1)}(p) \le p + h_{n-(t-1)} + s_{i-1}^{n-(t-1)} \qquad (*)$$

which gives

$$A_{i-1}^{n-(t-1)} \subset A_i^{n-(t-1)}. \qquad (**)$$

Now let us consider $k = t$. By the definition of s_i^{n-t}, $(*)$ and $(**)$ we have

$$s_i^{n-t}(p) = s_0^{n-t}(p) + \int_{A_{i-1}^{n-(t-1)}} (g + h_{n-(t-1)} + s_{i-1}^{n-(t-1)}(g)) G_{n-t}(p, dg)$$

$$\ge s_0^{n-t}(t) + \int_{A_i^{n-(t-1)}} (g + h_{n-(t-1)} + s_i^{n-(t-1)}(g)) G_{n-t}(p, dg) = s_{i+1}^{n-t}(p).$$

So the assertion (i) of the lemma is clear. By arguments similar to (*) and (**) applied to the superscript $n - t$ the assertion (i) implies (ii).

Now we are in a position to formulate Theorem 2.2 which completes the assertion of Theorem 2.1.

Theorem 2.2. *Let t^* be the nonnegative integer such that $M - (t^* + 1)N \leq 0$ and $M - t^* N > 0$. Let A^{n-t} be given by (2.2). Then*

$$
\mathbf{s}_{n-t}(p) = \begin{cases} M & \text{for } p \in A_0^{n-t}, \\ M - iN & \text{for } p \in A_i^{n-t} \setminus A_{i-1}^{n-t} \text{ and } 0 < i \leq j = \min(t^*, t), \\ 0 & \text{for } p \in Q \setminus A_j^{n-t}. \end{cases}
$$

As was mentioned, the methods of dynamic programming will be used in order to prove Theorems 2.1 and 2.2. We write the dynamic programming equations (Lemma 2.2) in a form which is appropriate for our problem. The proof is by standard backward induction (cf. [5, Thm. 3.4]).

Lemma 2.2. *Let $V_0, \ldots, V_{n+1}$ be measurable functions defined on $[0, M] \times Q$ by*

$$
V_{n+1}(x, p) = -px
$$

$$
V_t(x, p) = \inf_{0 \leq z \leq \min(N+x, M)} \left[p(z - x) + hz + \int_Q V_{t+1}(z, q) G_t(p, dq) \right], \quad (2.3)
$$

$$
t = n, \ldots, 0.
$$

A policy y^* such that

$$
V_t(x, p) = p(y_t^*(x, p) - x) + h_t y_t^*(x, p) + \int_Q V_{t+1}(y_t^*(x, p), q) G(p, dq) \quad (2.4)
$$

is the optimal for the problem (1.1)–(1.4).

In the next section we prove Theorems 2.1 and 2.2 using Lemma 2.2.

3. Solutions of dynamic programming equations. The following lemma will be proved by induction with respect to $t = n, n - 1, \ldots, 0$. For $x \in [0, M]$, $p \in Q$ let

$$
e_{n-t}(x, p) = \int_Q V_{n-(t-1)}(x, q) G_{n-t}(p, dq)
$$

and put $j = j(t) = \min(t^*, t)$. We recall that t^* was defined in Theorem 2.2, $V_{n-(t-1)}$ denotes the value function given by (2.3) and the functions s_i^{n-1}, $i = 0, \ldots, j$, are described by (2.1).

Lemma 3.1. (a) *For every $x \in [0, M]$, $e_{n-t}(x, \cdot)$ is a bounded function on Q. For $p \in Q$, $e_{n-t}(\cdot, p)$ is a continuous, convex, piecewise linear on $[0, M]$ with the following slopes:*

$$
s_j^{n-t}(p) \quad \text{for } 0 \leq x \leq M - jN \quad \text{and}
$$
$$
s_i^{n-t}(p) \quad \text{for } M - (i+1)N < x \leq M - iN \text{ and } i = j-1, \ldots, 0.
$$

564

(b) *The following policy satisfies the equations* (2.4):

$$y_{n-t}^*(x,p) = \begin{cases} \min(x + N, M) & \text{for } p \in A_0^{n-t}, \\ \min(x + N, M - iN) & \text{for } p \in A_i^{n-t} \setminus A_{i-1}^{n-t} \text{ and } i = 1,2,\ldots,j, \\ 0 & \text{for } p \in Q \setminus A_j. \end{cases}$$

Proof. The proof is divided into four steps. The method of backward induction will be used, and (2.1), Lemma 2.1 and the sets (2.2) will be needed.

Step 1. Let $t = n$. Then by definition

$$e_n(x,p) = \int_Q V_{n+1}(x,g)G_n(p,dq) = -\int_Q gxG_n(p,dq) = x \cdot s_0^n(p).$$

Hence it is clear that part (a) of the lemma is satisfied.

Observe that in this case

$$V_n(x,p) = \inf_{0 \le z \le \min(x+N,M)} [-px + (p + h_n + s_0^n(p))z].$$

This gives

$$y_n^*(x,p) = \begin{cases} \min(x + N, M) & \text{for } p \in A_0^n, \\ 0 & \text{for } p \in Q \setminus A_0^n. \end{cases}$$

which proves the lemma in the case $t = n$.

Step 2: Induction hypothesis. Assume that the lemma is true for some $t \ge 0$. So

$$V_{n-t}(x,p) = -px + py_{n-t}^*(x,p) + h_i y_{n-t}^*(x,p) + e_{n-t}(y_{n-t}^*(x,p),p). \qquad (3.1)$$

By induction assumption, for every $p \in Q$, $V_{n-t}(\cdot,p)$ is continuous piecewise linear. Moreover, there exist bounded functions $a_i^{n-t} : Q \to \mathbf{R}$ and $s_i^{n-t} : Q \to \mathbf{R}$, $i = 0,1,\ldots,j$, such that $e_{n-t}(x,p) = s_i^{n-t}(p)x + a_i^{n-t}(p)$ on corresponding intervals. So we can specify the function V_{n-t} for different p. Observe that for $p \in A_0^{n-t}$, $y_{n-t}^*(x,p) = \min(x+N,M)$ and so (3.1) takes the form

$$-px + (p + h_{n-t} + s_0^{n-t}(p))M + a_0^{n-t}(p) \quad \text{for } M - N < x \le M$$

and

$$-px + (p + h_{n-t} + s_i^{n-t})(x + N) + a_i^{n-t}(p)$$
$$\text{for } \max(0, M - (i+2)N) < x \le \max(0, M - (i+1)N) \text{ and } i = 0,1,\ldots,j.$$

For $p \in A_i^{n-t} \setminus A_{i-1}^{n-t}$, $i = 1,2,\ldots,j$, we have $y_{n-t}^*(x,p) = \min(x + N, M - iN)$ and so V_{n-t} equals

$$-px + (p + h_{n-t} + s_i^{n-t}(p))(M - iN) + a_i^{n-t}(p) \quad \text{for } M - (i+1)N < x \le M$$

and

$$-px + (p + h_{n-t} + s_k^{n-t}(p))(x + N) + a_k^{n-t}(p)$$
$$\text{for } \max(0, M - (k+2)N) < x \le \max(0, M - (k+1)N) \text{ and } k = i, i+1,\ldots,j.$$

Finally, for $p \in Q \setminus A_j^{n-t}$, $y_{n-t}^*(x,p) = 0$ and so $V_{n-t}(x,p) = -px + e_{n-t}(0,p)$.

Step 3. Now we can specify the function e_{n-t-1} given by

$$e_{n-t-1}(x,p) = \int_Q V_{n-t}(x,q) G_{n-t-1}(p,dq).$$

to do this, we use Lemma 2.1(ii) and the results from Step 2.

Consider $x \in (M - N, M]$. It is clear that there exists a bounded function a_0^{n-t-1} such that

$$e_{n-t-1}(x,p) = \int_Q -gx G_{n-t-1}(p,dq) + a_0^{n-t-1}(p) = x s_0^{n-t-1}(p) + a_0^{n-t-1}(p).$$

Similarly, there exists a_{i+1}^{n-t-1} such that for $x \in (\max(0, M - (i+2)N), M - (i+1)N]$ we have

$$
\begin{aligned}
e_{n-t-1}(x,p) = & \int_Q -gx G_{n-t-1}(p,dg) + \int_{A_0^{n-t}} x(g + h_{n-t} + s_i^{n-t}(q)) G_{n-t-1}(p,dq) \\
& + \int_{A_1^{n-t} \setminus A_0^{n-t}} x(g + h_{n-t} + s_i^{n-t}(q)) G_{n-t-1}(p,dq) \\
& + \ldots + \int_{A_i^{n-t} \setminus A_{i-1}^{n-t}} x(q + h_{n-t} + s_i^{n-t}(q)) G_{n-t-1}(p,dq) + a_{i+1}^{n-t-1}(p) \\
= & \, x s_0(p) + x \int_{A_i^{n-t}} (q + h_{n-t} + s_i^{n-t}(g)) G_{n-t-1}(p,dq) + a_{i+1}^{n-t-1}(p) \\
= & \, x s_{i+1}^{n-t-1}(p) + a_{i+1}^{n-t-1}(p), \quad \text{for } i = 0,1,\ldots,k
\end{aligned}
$$

with $k = j$ if $j + 1 \leq t^*$ and $k = j - 1$ otherwise. This proves the assertion (a) of the lemma. In fact, the function e_{n-t-1} is of the desired form. The properties of V_{n-t} imply that the function is continuous for every p and bounded on $[0, M] \times Q$. The convexity follows from Lemma 2.1(i).

Step 4. Now we are in a position to prove part (b) of the lemma for $t - 1$. Let now $j = j(t+1) = \max(t^*, t+1)$. Using part (a) of the lemma and Lemma 2.2 it is sufficient to solve the following minimization problem:

$$\inf_{0 \leq z \leq \min(x+N,M)} \{-px + L(z,p)\}$$

where

$$L(z,p) = pz + h_{n-t-1} z + e_{n-t-1}(z,p).$$

It is clear that $L(\cdot, p)$ is convex piecewise linear with the following slopes:

$$s_i^{n-t-1}(p) + p + h_{n-t-1}$$
$$\text{for } \max(0, M - (i+1)N) < z \leq M - iN, \ i = 0,1,\ldots,j(t+1)$$

Observe that by the definition of A_i^{n-t-1}, $i = 0,1,\ldots,j$, and Lemma 2.1 the function $L(x,p)$ is nonincreasing with respect to x for $p \in A_0^{n-t-1}$ and so $y_{n-t-1}^*(x,p) = \min(x + N, M)$. For $p \in A_i^{n-t-1} \setminus A_{i-1}^{n-t-1}$, $i = 1,\ldots,j$, the function is nonincreasing for $x \leq M - iN$

and nondecreasing for $x \geq M - iN$. So $y^*_{n-t-1}(x, p) = \min(x + N, M - iN)$. Finally, for $p \in Q \setminus A^{n-t-1}_j$ the function $L(\cdot, p)$ is nondecreasing and hence $y^*_{n-t-1}(x, p) = 0$ which complete the proof of the lemma.

Proofs of Theorems 2.1 *and* 2.2. The proofs are simple consequences of Lemma 3.1(b) and Lemma 2.2.

4. Final Remarks. The structure of y^*_{n-t} given by Lemma 3.1(b) is described by the sets $A^{n-t}_0, \ldots, A^{n-t}_j$ with $j = \min(t^*, t)$. We recall that t^* is the nonnegative integer such that $M - (t^* + 1)N \leq 0$ and $M - t^*N > 0$. Remark 2.1(ii) states that the set A^{n-t}_i is uniquely determined by G_{n-t+k}, h_{n-t+k} with $k = 0, 1, \ldots, i$. Thus, y^*_{n-t} is not affected by future data beyond the horizon t^*.

Observe that if $M = N$ then $t^* = 0$ and

$$y^*_{n-t} = \begin{cases} M & \text{for } p \in A^{n-t}_0, \\ 0 & \text{for } p \in Q \setminus A^{n-t}_0. \end{cases}$$

The relation $p \in A^{n-t}_0$ means that

$$p + h_{n-t} \leq -s^{n-t}_0(p) = \int_Q g G_{n-t}(p, dg).$$

In this case if the expected price in next period is greater than the present price and the unit holding cost then it is optimal to purchase up to level M. Otherwise it is optimal to sell everything. That is, the store should be full or empty.

REFERENCES

[1] Magirou, V. F. (1982), *Stockpiling under price uncertainty and storage capacity constraints*, European J. Oper. Res. 11, 233–246.

[2] Parlar, M. and Rempała, R. (1992), *Stochastic inventory problem with picewise quadratic holding cost function containig a cost-free interval*, J. Optim. Theory Appl. 75 (1), 133–153.

[3] Rempała, R. (1992), *Forecast horizon in a dynamic family of one-dimensional control problems*, Dissertationes Math. 315.

[4] Scarf, H. (1960), *The optimal of (s, S) policies for the dynamic inventory problem*, in: Mathematical Methods in the Social Sciences, Arrow, K. J., Karlin, S. and Suppes, S. (eds.), Stanford University Press, 196–202.

[5] Zabczyk, J. (1984), *Lecture in stochastic control*, University of Warwick, Control Theory Center, Report No. 125.

EVALUATING WELFARE LOSSES DUE TO THE USE OF APPROXIMATED STOCHASTIC OPTIMAL CONTROL ALGORITHMS: AN APPLICATION TO THE BANCA D'ITALIA QUARTERLY MODEL.

Andrea Cividini and Stefano Siviero
Servizio Studi
Banca d'Italia
Roma, Italy

Stochastic optimal control of large-size nonlinear econometric models can require enormous computing resources. As an alternative to the full stochastic approach, the standard deterministic control (neglecting the stochastic nature of estimated models) is often used. In recent articles, Hall and Stephenson have proposed an algorithm that takes into account, at least partially, the stochastic nature of optimal control with econometric models, without calling for enormous increases in computing time. Their algorithm, however, neglects the role of the variance. In this paper, the full stochastic optimal control algorithm described in detail in Cividini-Siviero (1992) is used to solve realistic problems of stabilizing the Italian economy towards a steady-state growth path. All problems use the version of the Banca d'Italia Quarterly Model that has been modified so as to ensure the existence of a long-run equilibrium. The welfare losses due to the approximated algorithms are then compared for two different optimal control exercises.

1. INTRODUCTION

Optimal control techniques are increasingly used to search for suitable and effective economic policy measures. The popularity of these techniques is due to the fact that, in response to an explicit specification of the policy-maker's preferences, they help to select the best policy and to rank all the alternatives.
However, a number of practical difficulties are directly associated with specific features of the application of optimal control techniques to economic problems: in particular, a major problem arises from the size of macroeconometric models, which can contain several hundred equations. The degree of interdependence, the complexity of the dynamic structure and the degree of nonlinearity can pose serious difficulties in solving optimal control problems[1]. Moreover, econometric models are, by their very nature, stochastic; this feature requires that uncertainty be explicitly modelled and accounted for. The implementing of optimal control in its full

---

1. For the computational aspects, see for example Petersen and Cividini (1989) and Cividini (1990).

stochastic form thus calls for a large number of simulations and for this reason the much simpler deterministic optimal control is very often used, even though it is known to be incorrect.

A number of efforts have been made to find approximate solutions that are both sufficiently close to the full stochastic solution and computationally convenient. One such an example is the algorithm proposed by Hall and Stephenson, where a deterministic optimal control problem is solved after modifying the linear term for the effect of nonlinear deterministic bias. This algorithm implicitly redefines the objective function by neglecting the variance term.

In this paper, we use the Banca d'Italia Quarterly Model of the Italian Economy to investigate the welfare losses that can be attributed to the use of approximated (i.e., sub-optimal) algorithms for policy design with a large-size econometric model. The version of the model used in this paper has been modified to allow for a steady-state growth path equilibrium to exist.

While several algorithms were compared in Cividini-Siviero (1992) by using of a fictitious two-equation mildly nonlinear model having a closed form, the aim of the present paper was to produce a realistic policy design. The experiments in this paper give realistic information about the actual losses that can be incurred in policy design when algorithms other than the full stochastic one are used.

2. ALTERNATIVE APPROACHES TO THE SOLUTION OF STOCHASTIC OPTIMAL CONTROL PROBLEMS

In this paper we are concerned with the standard finite-horizon optimal control problem with a quadratic objective function and a set of constraints specified by a stochastic nonlinear model.

To simplify the exposition, this paragraph describes the simple one target-one instrument case; it is also assumed that the instrument can be moved at no cost[2].

The problem can thus be written as follows:

$$\min_{x_t} J^F = \min_{x_t} E\left\{ \sum_{t=1}^{T} \alpha_t (y_t - \bar{y}_t)^2 \right\} \tag{1}$$

subject to the nonlinear model $f_t(Y_t, X_t, x_t, \theta, \varepsilon_t) = 0$ [2]

where y_t = objective variable, $y_t \in Y_t$

Y_t = set of endogenous variables

$\bar{y}_t$ = target value for y_t

x_t = instrument variable

---

2. The more general case of n targets, m instruments and linear constraints specified explicitly is fully described in Cividini and Siviero (1992).

x_t = set of predetermined variables, other than x_t

ε_t = set of structural disturbances

θ = set of parameters

α_t = non-negative weights, exogenously given

[1, T] = time horizon of the optimization problem

Using the well-known relationship

$$E(y_t^2) = (E(y_t))^2 + Var(y_t) \tag{3}$$

the objective function can be written as:

$$J^F = \sum_{t=1}^{T} \alpha_t (E(y_t) - \bar{y}_t)^2 + \sum_{t=1}^{T} \alpha_t \, Var(y_t) \tag{4}$$

The objective function can thus be thought of as being made up of two parts:

- a quadratic component penalizing the deviations of the expected values of the objective variable from its target;
- a linear component penalizing the variability of the objective, i.e., the deviation of the variance from zero.

In view of the complexity of full stochastic solutions[3], optimal control problems are usually solved by neglecting the stochastic nature of the problem, i.e., by means of standard deterministic control. This approximation has the attractive feature that it requires only deterministic model solutions and is therefore computationally tractable even for large-size econometric models. However, the deterministic control introduces two implicit modifications to the original objective function:

- it neglects the deterministic simulation bias $d_t = E(y_t) - \hat{y}_t$ where $\hat{y}_t$ is the deterministic model solution implicity defined by $f_t(\hat{Y}_t, x_t, x_t, \theta, 0) = 0$;

- it neglects the role of the variance.

Thus the objective function reduces to:

$$J^d = \sum_{t=1}^{T} \alpha_t [\hat{y}_t - \bar{y}_t]^2 \tag{5}$$

---

3. Algorithms of this kind usually work by iterating over linearizations of the entire model, using standard dynamic control theory to optimize the stochastic linearized model at each iteration. See for instance Chow (1976).

The objective function [5] is equivalent to the original function [4] only in linear models with additive uncertainty, where $d_t = 0$ and the degree of uncertainty cannot be affected by policy actions, so that certainty equivalence results would apply.
In the general case of nonlinear models the implicit modifications introduced in the objective function are a reasonable approximation only when nonlinearity is mild. We therefore implemented an operational iterative procedure to solve the full stochastic optimal control problem [4]. Without going into the details, it is worth noting that our full stochastic algorithm works by linearizing the relationships between the instruments and the expected value and the variance of the objectives. We achieve this by using stochastic simulations to compute numerically the multiplier matrices, mapping changes of x_t into changes of $E(y_t)$ and changes of $Var(y_t)$[4].
There is clearly a tradeoff between saving computer resources and keeping the problem in its full form. Hall and Stephenson have recently (1990) proposed a simplified algorithm that transforms the original stochastic problem into a sequence of deterministic ones. An iterative procedure solves, at each iteration, a deterministic optimal control problem, in which the objective function is modified to include the bias of the deterministic simulation computed around the solution to the previous deterministic problem by means of a single stochastic simulation[5]. In this approach the objective function is equivalent to the first term of the original one:

$$J^{HS} = \sum_{t=1}^{T} \alpha_t (E(y_t) - \bar{y}_t)^2 \qquad [6]$$

Clearly, the variance plays no role in such an iterative procedure, so it is still not possible to solve the optimal control problem by reducing the degree of uncertainty.
Since economic theory is frequently concerned with the effects of risk aversion in the policy-maker's objective function, it seems interesting to keep the stochastic optimal control problem in its full form. Moreover, as suggested by Mitchell (1979), different weights can be given to the first and second components of the loss function in eq. [4]:

$$J^M = \lambda \sum_{t=1}^{T} \alpha_t (E(y_t) - \bar{y}_t)^2 + \sum_{t=1}^{T} \alpha_t \, Var(y_t) \qquad [7]$$

---

4. In the general case, with n instruments and T periods, a full stochastic approach requires $nT+1$ stochastic simulations at each iteration. This is so because it is necessary to compute the partial derivatives of all the objective variables and their variances with respect to all the instruments in each time period.

5. It is necessary to iterate because the bias is a function of the control variables. For more details see Cividini and Siviero (1992).

where λ is the risk aversion parameter: "in general, the higher is λ the less the policy-maker is concerned with the predictability (variance) of the target" (Mitchell, 1979, p. 914)[6].

3. A BRIEF DESCRIPTION OF THE BANCA D'ITALIA QUARTERLY MODEL

The Quarterly Model of the Italian Economy (henceforth, BIQM) is a large size econometric model consisting, in its present version, of about 800 equations, 100 of which are stochastic. It has been estimated using quarterly data over the period 1970-1990 and is routinely used for forecasting and policy analysis.
The model displays Keynesian features in the short-run, with output, unemployment and excess capacity determined by aggregate demand. The basic Keynesian scheme is, however, enriched by dynamic mechanisms derived from the neo-classical growth theory and the evolution of macro theory in recent decades. These factors exert most of their influence in the long-run[7].
While a full description of the BIQM is well beyond the scope of this paper, we feel it useful to discuss, albeit briefly, the modifications made to the original version in order to obtain a model having a long-run equilibrium. As the economic literature suggests, it is assumed that the most suitable equilibrium concept in the context of models having (exogenous) technical progress and capital accumulation is that of a steady-state growth path (SSGP). Along this path, all variables (exogenous as well as endogenous) grow at a constant rate, which is a function of three basic parameters: the rate of growth of population (n), that of foreign prices (π), and that of productivity (γ)[8].
The original BIQM is not compatible with a steady-state growth path, as several equations violate the requirements for first-degree homogeneity: this would prevent the endogenous variables from settling down on a path characterized by constant rates of growth. The model has thus been modified by eliminating all causes of non-homogeneity, including time-trends, constants and seasonal effects.
The issue arises of whether any analysis performed with the modified model is consistent with the structure of the Italian economy as described by the original BIQM: extensive comparison of

6. Mitchell gives examples, taken from the economic literature, for both extreme cases $\lambda \to \infty$ and $\lambda = 0$. Notice that, with $\lambda \to \infty$, we are back to Hall and Stephenson's problem.

7. Given that the basic features of the present version of the model do not differ greatly from those of the previous version, readers can consult Banca d'Italia (1986) and Galli et al. (1989) for further details.

8. Further details on the requirements for a SSGP to exist can be found in a number of papers by P. Malgrange and other authors.

the differences are limited and it is quite safe to use the
modified version for long-run analysis[9].
The modified model has two principal advantages over the original
BIQM used for forecasting and policy analysis: it can be used for
control exercises spanning extremely long time horizons, and it
suggests natural target values for most of the endogenous
variables, given that, along a SSGP, the rate of growth of all
variables is a known constant.

4. DESIGN OF OPTIMAL CONTROL EXPERIMENTS

Two different kinds of optimal control experiments have been
designed, both using the modified BIQM described above. In all
cases, the instruments are the average tax rate on personal income
(TMEDQ)[10], the exchange rate (EXCHUS) and the T-Bill interest rate
(TABOT).
There are five objectives: the rate of growth of output (XDOT),
that of prices (PDOT), the rate of unemployment (URED), the ratio
of goverment debt to GDP (DSSPIL) and the ratio of foreign debt to
GDP (DEBBAL).
In all experiments the target values for output and prices are
their steady-state growth rates (about .8 per cent per quarter for
output, 1 per cent for prices); for unemployment, the target value
is the so-called NAIRU (or Natural Rate), implicitly defined by the
Phillips Curve; target values of 60 per cent and 5 per cent have
been chosen for the ratios of government debt and foreign debt to
GDP, respectively.
Costs are associated not only with deviations of objectives from
their target values, but also with excessive changes of the
instruments from one period to the next.
The two experiments are designed as follows:

---

9. In Siviero (1993), the two models are compared in terms of
   multiplier analysis and degree of nonlinearity; in most cases,
   differences are qualitatively negligible and even
   quantitatively very small. The paper discusses in great detail
   all the changes to the original BIQM and presents the
   techniques used to find a SSGP.

10. While the choice of treating TABOT and TMEDQ as separate
    instruments is somewhat questionable in the light of economic
    theory, it is often argued that these two policy instruments
    can, to some extent, be independently manoeuvred, at least in
    the short-run. As the main objective of this paper, is to
    compare the performances of optimal control algorithms, rather
    than to design feasible policies for the stabilization of the
    Italian Economy, these issues will not be discussed here. For an
    example of plausible stabilization design, where the
    interdependence of TABOT and EXCHUS is explicitly modelled, see
    Cagliesi (1993), where an initial attempt has been made to apply
    the deterministic optimal control algorithm and the modified
    BIQM described above to an actual policy planning problem.

1) All exogenous variables grow at rates compatible with a steady-state growth path starting from the first quarter of 1990. The objective of this experiment is to control the economy so as to reach the steady-state, starting from historical values up to 1989.4. The time horizon is 12 quarters.

2) Starting from a steady-state growth path equilibrium, a (sustained) 1 per cent shock is given to the level of all foreign prices, keeping the growth rate unaffected thereafter. Optimal control techniques are then aimed at bringing the model back to its steady-state growth path. The time horizon is 12 quarters.

The values of the weights in the cost function are not given here: it can be said, however, that variable weights are used, giving a higher cost to deviations from the objectives towards the end of the optimization period. Costs on moving the instruments are also imposed.

With regard to the stochastic simulations needed for both the Hall and Stephenson and full stochastic algorithms, previous experience (reported in Cividini-Siviero (1992)) and several experiments with the BIQM suggest that the method of antithetic variates can be very effective in reducing the number of simulations, while still giving a good estimate of the expected value of the endogenous variables[11].

5. EMPIRICAL RESULTS

In order to evaluate the results associated with the three alternative optimization algorithms described above (deterministic, Hall and Stephenson and full stochastic), the two optimal control problems listed in section 4 were first solved; the value of the cost function associated with the different instrument paths was then computed using a high number of replications (250+250 antithetic), drawn with a different seed from that used for the optimization exercises.

In the paper by Cividini and Siviero (1992), it was found that the differences between the three algorithms were limited in the simple model used. The approximation by Hall and Stephenson appeared to be closer to the true full stochastic solution than to the deterministic one, even more so with low simulation bias. The issue then was whether an actual econometric model displays more or less deterministic simulation bias than the simple one used in our previous paper.

Even though all the empirical evidence seems to suggest that the degree of nonlinearity (and hence of deterministic bias) is low for estimated econometric models, the results of the exercises described in the previous section seem to suggest that the Hall and Stephenson algorithm is, in the case of the BIQM, much closer to the deterministic than to the stochastic solution.

Let us examine the results of the two optimal control exercises in turn.

---

11. For both the experiments a number of 50+50 antithetic replications was used.

574

In the first experiment, starting from historical values, where output is below the target value and foreign and public debt are well above, all algorithms tend to move EXCHUS up (so as to boost output and reduce foreign debt), to decrease TABOT (to reach the targets for XDOT, URED and DSSPIL), and to increase TMEDQ (aimed at reducing DSSPIL) (see figure 1).

Figure 1

First experiment - Instruments values

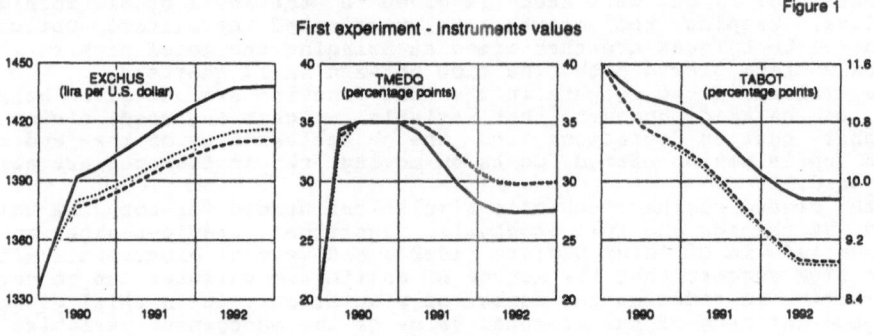

——— full stochastic .... Hall and Stephenson ----- deterministic

However, the extent of movements in the variables differs considerably according to which algorithm is used.
With the deterministic algorithm, TABOT goes from 11.6 (before optimization starts) to 8.9 (in 1992.4). The same happens with Hall and Stephenson's algorithm. The full stochastic solution reduces TABOT to only 9.8 by 1992.4, with the difference between this solution and the other two widening over time.
Similarly, the devaluation of the exchange rate is only 5.6 per cent with the deterministic solution, marginally higher with the Hall and Stephenson solution (6.1 per cent), and almost 8 per cent with the full stochastic solution.
The tax rate moves in similar fashion in the all cases only so long as it remains within the boundary of less than 35 per cent; thereafter, the three cases tend to differentiate.
As far as the objectives are concerned, the differences in the values of instruments do not affect all the objectives in the same way. In any case, switching from one algorithm to another seems implicitly to redefine the priorities among the objectives.
Reductions in the value of the cost function are not dramatic when one switches from the deterministic to the full stochastic solution, so that the issue is whether the features of the BIQM justify the additional computational cost that the latter solution implies. However, the Hall and Stephenson solution is no better than the deterministic one; in fact it is marginally worse when performances are evaluated with the objective function inclusive of the variance term.
As expected, in experimenting with values of λ greater than one (i.e, following Mitchell's suggestion), the results are somewhere between those obtained with the full stochastic and the Hall-Stephenson algorithms.
Also, in evaluating the cost functions, as one would expect a

priori, each procedure minimizes its own: in all cases, the cost function associated with Hall-Stephenson is very close to the deterministic.
The aim of the second experiment was to compute the optimal policy mix which allows the model to reach a new steady-state growth path equilibrium after a permanent exogenous shock. A few remarks are needed here. In the absence of shocks the deterministic solution would be achieved, since the target values coincide with the steady-state values. Similarly, the Hall-Stephenson results do not differ greatly from the deterministic results. On the other hand with the full stochastic algorithm the instruments would have to be moved even in the absence of any shock, since the steady-state values would not minimize the loss function, on account of the presence of the variance term. Hence the movement of the instruments in the latter case is only marginally influenced by the need to contrast the shock, since this is overshadowed by the preexistent need to minimize the variance. Given these factors, interpretation of the results must be cautious; they are nevertheless useful to compare relative performances with a different set of variables.
With the Hall and Stephenson and the deterministic algorithms, there is an appreciation of the currency (clearly to counterbalance inflationary impulses), whereas TABOT and TMEDQ hardly move (see figure 2).

Figure 2

Second experiment - Instruments values

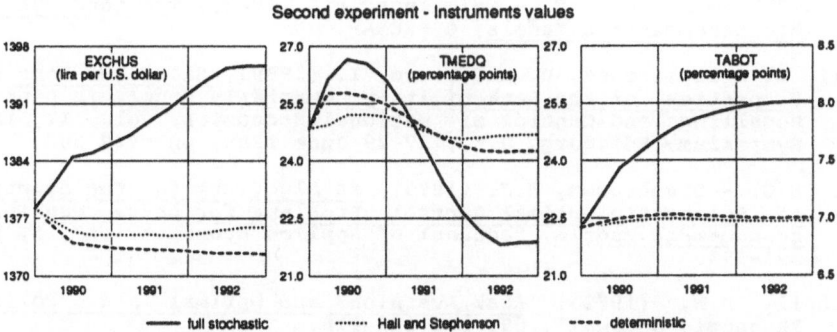

—— full stochastic ········ Hall and Stephenson ---- deterministic

With the full stochastic algorithm, the currency depreciates by 1.3 per cent and TABOT increases by more than 1 percentage point. Once again, the differences are far from negligible.
The above results show how the Hall-Stephenson algorithm is, in these experiments, only marginally different from the deterministic. On the other hand, the full stochastic solution is radically different, but the question arises whether this treatment of uncertainty is indeed plausible. One further experiment we will perform in the future will consider an additional source of uncertainty: in particular, exogenous variables will not follow a deterministic trend but will be generated by stochastic processes allowing for random variability around the deterministic trend.

REFERENCES

Banca D'Italia, 1986, Modello Trimestrale dell'Economia Italiana, Vol 1 and II, Banca d'Italia, Temi di discussione, 80.

Cagliesi, G., 1993, Target Zones: A Theoretical and Empirical Approach to Exchange Rate Determination and the Use of Optimal Monetary Policies, University of Pennsylvania, Unpublished Ph.D. Dissertation.

Chow, G.C. (1976), An Approach to Feedback Control of Nonlinear Econometric Systems, "Annals of Economic and Social Measurement", 5, 3, pp. 297-309.

Cividini, A. (1990), An Optimal Control Algorithm for Analyzing Econometric Models, in "Proceedings of the IFIP TC7 Conference on Modelling the Innovation: Communications, Automation and Information Systems", edited by M. Carnevale - M. Lucertini - S. Nicosia, Amsterdam, North-Holland.

Cividini, A. - Siviero, S. (1992), Implementing Stochastic Optimal Control of Nonlinear Models: A Comparison with Alternative Solution Methods, Banca d'Italia, "Temi di discussione" n.179.

Deleau, M. - Le Van, C. - Malgrange, P. (1988), The Long Run of Macroeconometric Models, CEPREMAP.

Galli, G. - Terlizzese, D. and Visco, I. (1989), Short and Long Run Properties of the Bank of Italy Quarterly Model, in Dinamic Modelling and Control of National Economies, vol. II, IFAC Symposium, Edimburg, U.K., 27-29 June 1989, pp. 449-506.

Hall, S.G. - Stephenson, M.J. (1990), An Algorithm for the Solution of Stochastic Optimal Control Problems for Large Nonlinear Econometric Models, "Journal of Applied Econometrics", 5, pp. 393-399.

Mitchell, D.W. (1979), Risk Aversion and Optimal Macro Policy, "Economic Journal", 89, pp. 913-918.

Petersen, C.E. - Cividini, A. (1989), Vectorization and Econometric Model Simulation, "Computer Science in Economics and Management", 2, pp. 103-117.

Siviero, S. (1993), Analysis of Long-Run Properties and Cycle with the Bank of Italy Quarterly Model, mimeo.

Distributed parameter systems

Identification

Unique continuation Mizohata theorem and inverse problems for heat equations.

Abdellatif El Badia

Université de Technologie de Compiègne
Dépt. Génie Info. Division Maths Appl.
B.P. 649. 60206 Compiègne cédex, France.

Abstract. This paper deals with the unique determination of certain unknown coefficients in the heat equations from some observations, using the unique continuation Mizohata theorem and Goursat's problem.

Key words. Inverse problem, Unique Mizohata's continuation theorem.

1. Introduction. The purpose of this paper is to prove that one can uniquely determine certain unknown coefficients in the heat equations from some observations.

Problem 1.

Let $p \in C^1(0,1)$, $u_0 \in L^2(0,1)$, $g \in L^2(0,T)$ and $\mathcal{P}(p, u_0, g)$ be the heat equation

$$\partial_t u - (\partial_{xx} - p(x))u = 0 \quad \text{in } (0,1)\text{x}(0,T) \tag{1}$$

with the boundary conditions

$$\partial_x u(0,t) = 0 \qquad \text{on } (0,T) \tag{2}$$

$$\partial_x u(1,t) = g(t) \qquad \text{on } (0,T) \tag{3}$$

and the initial condition

$$u(x,0) = u_0(x) \quad \text{in } (0,1). \tag{4}$$

As is known, there exists a unique solution $u \in L^2(0,T;H^{3/2}(0,1))$ (see [6]) for given initial coefficients and boundary values (p, u_0, g). However, in our case the coefficients (p, u_0) are unknown.

The problem is the following:

We consider the system $\mathcal{P}(p, u_0, g)$ where the function g is known. Given the observation $\mathcal{C}u = (u(0,t); u(x,T))$, the question is: can p and u_0 be uniquely determined?

Problem 2

Let Ω be the square $(0,1)\text{x}(0,1)$, we denote by $\Sigma_0 = \{0\}\text{x}(0,1)$, $\Sigma_1 = (0,1)\text{x}\{0\}$, $\Sigma_2 = (0,1)\text{x}\{1\}$, $\Sigma_3 = \{1\}\text{x}(0,1)$ and let $T > 0$,

For $p \in C^1(0,1)$, $f \in L^2(\Omega)$, $g_1 \in L^2(\Sigma_1)$, $g_2 \in L^2(\Sigma_2)$ and $g_3 \in L^2(\Sigma_3)$, we consider the heat equation $\mathcal{P}(p,f,g_1,g_2,g_3)$

$$\partial_t u - (\Delta - p(x))u = 0 \text{ in } \Omega x(0,T) \tag{5}$$

with boundary conditions

$$\partial_x u = 0 \qquad \text{on } \Sigma_0 \tag{6}$$

$$\partial_y u = g_1 \qquad \text{on } \Sigma_1 \tag{7}$$

$$\partial_y u = g_2 \qquad \text{on } \Sigma_2 \tag{8}$$

$$\partial_x u = g_3 \qquad \text{on } \Sigma_3 \tag{9}$$

and the initial condition

$$u(x,y,0) = f(x,y) \quad \text{in } \Omega. \tag{10}$$

As is known, the system $\mathcal{P}(p,f,g_1,g_2,g_3)$ has a unique solution $u \in L^2(0,T;H^{3/2}(\Omega))$ (see [6]) for given coefficients (f , g_i , p) ; $i = 1,2,3$.

The function g_1 is known but the functions p, f, g_2 and g_3 are unknown. Then the problem is the following:

Given the observation $\mathcal{C}u = (u(\xi,t), u(x,y,T))$ where $(\xi,t) \in \Sigma_1$, can f, p, g_2 and g_3 be uniquely determined?

The key of the problems 1-2 is Goursat's problem and the unique continuation Mizohata theorem (see [7], [11]).

2.Main Results

THEOREM 1. *Let the operator*

$$\mathcal{C} : L^2(0,T;H^{3/2}(0,1)) \cap C^0(0,T;L^2(0,1)) \rightarrow L^2(0,T)xL^2(0,1)$$

$$u \rightarrow (u(0,t), u(x ,T)).$$

Let u and v be the solutions of the problems $\mathcal{P}(p, u_0 g)$ and $\mathcal{P}(q, v_0 g)$ respectively, then

$(p, u_0)=(q, v_0)$ *if and only if* $\mathcal{C}u = \mathcal{C}v \quad \forall g \in L^2(0,T) \cdot$

THEOREM 2. *Let the operator*

$$\mathcal{C} : L^2(0,T;H^{3/2}(\Omega)) \cap C^0(0,T;L^2(\Omega)) \rightarrow L^2(\Sigma_1)xL^2(\Omega)$$

$$u \quad\longrightarrow\quad (u(\xi,t),\, u(x,y,T)).$$

Let u and v be the solutions of the problems $\mathcal{P}(p,f,g_1,g_2,g_3))$ and $P(q,h,g_1,k_2,k_3))$ respectively, then $(p,\, f,g_2,g_3)=(q,\, h,k_2,k_3)$ if and only if $\mathcal{B}u=\mathcal{B}v \quad \forall\, g_1 \in L^2(\Sigma_1)$ •

There are some related parers. S Kitamura and S. Nakagiri considered in [4] the heat equation $\partial_t u - \partial_x(\sigma(x)\partial_x u) - p(x)u$ $(0<t<\infty,\, 0<x<1$) and gave a sufficient condition for (σ,p) to be determined from the observation $\mathcal{B}u = u(x,t)$. They also studied the problem to determine (σ,p) from the observation $\mathcal{B}u = u(a,t)$ for some $a \in [0,1]$, assuming σ and p to be constant functions. A. Pierce considered in [9] the heat equation $\partial_t u - \partial_{xx}u - p(x)u$ $(0<t<\infty,$ $0<x<1$) with a homogeneous boundary condition of the third kind on $x = 0$: $(\partial_x - h)u(0,t) = f$, and with the homogeneous boundary condition of the same kind on $x = 1$: $(\partial - H)u(1,t) = 0$. He showed that under such a situation the values $u(0,t)$ and $f \neq 0$ determine (p,h,H), by virtue of the inverse spectral theory of Gelfand-Levitan [3], Levitan-Gasymov [5] T. Suzuki, considered in [12] the heat equation $\partial_t u - \partial_{xx}u - p(x)u$ $(0<t<\infty,\, 0<x<1$), with the initial condition $u(x,0) = a(x)$, and with homogeneous boundary conditions: $(\partial_x - h)u(0,t) = (\partial_x - H)u(1,t) = 0$. He considered the observation $\mathcal{B}u = u(x_p,t)$ and the number $N = \#\{\varphi_n\ /\ <a,\varphi_n> = 0\}$. He studied the injectivity of the mapping $(a,p)\to u(x_p,t)$ and he showed the following results

The mapping : $(a,p)\to u(1,t)$ is injective if and only if $N = 0$

The mapping : $(a,p)\to u(x_p,t);\, x_p \in [1/2,1]$ is injective if and only if $N < \infty$

The mapping : $(a,p)\to u(1/2,t)$ is injective if and only if $N \leqslant 1$

The mapping : $(a,p)\to u(x_p,t)$ is not injective if $xp \in [0,1/2[$.

We remark about this results that it is difficult to calculate the number N, since the eigenfunctions φ_n are unknown.

W. Rundell studied in [10] the injectivity of the mapping : $p\to u(0,t)$ for the heat equation $(\partial_t - \partial_{xx} + p(x))u = 0$, with boundary conditions $\partial_x u(0,t) = \partial_x u(1,t) = 0$ and with the initial condition $u(x,0) = f(x)$, assuming that $f \in C^2(0,1)$ and using the integral equation

$$\int_0^x k(x,s)f(s)ds = 0,$$ where k is given by lemma 1.

In this paper we give a new identification method of the coefficients p and u_0 in the heat equation $(\partial_t - \partial_{xx} + p(x))u = 0$, with boundary conditions $\partial_x u(0,t) = 0$, $\partial_x u(1,t) = g$ and with the initial condition $u(x,0) = u_0(x)$, using the Goursat's problem and an integral equation , from the observation $\mathcal{B}u = (u(0,t);\, u(x,T))$. We also showed the injectivity of the mapping

$(p,u_0) \rightarrow \mathcal{C}u$, using a continuation Mizohata theorem and we give some extensions for the heat equation in the square $(0,1)x(0,1)$.

To be able to prove theorems 1 and 2, we need the following premilinaries:

3. Preliminaries.

LEMMA 1 Let $p, q \in C^1(0,1)$, the Goursat's problem

$$\partial_{xx}k(x,s)-\partial_{ss}k(x,s) = (p(s)-q(x))k(x,s) \ in \ 0 < s < x < 1$$

$$\partial_s k(x,0) = 0 \qquad\qquad in \ (0,1)$$

$$k(x,x) = \frac{1}{2}\int_0^x (q-p)(t)dt \qquad\qquad in \ (0,1),$$

has a unique solution u.

Proof. (see [8, p 136])

PROPOSITION 1. Let $p \in C^1(0,1)$ and let u be the solution of the problem $\mathcal{Q}(p)$

$$\partial_t u - (\partial_{xx} - p(x))u = 0 \qquad\qquad in \ (0,1)x(0,T)$$

$$\partial_x u(0,t) = 0 \qquad\qquad in \ (0,T)$$

then the fonction

$$T(u)(x,t) = u(x,t) + \int_0^x k(x,s)u(s,t)ds$$

is the unique solution of $\mathcal{Q}(q)$ such that $T(u)(0,t) = u(0,t)$.

Proof. It is easy to check that $T(u)$ is solution of $\mathcal{Q}(q)$. One just has to set $T(u)=v$ and we obtain

$$\partial_t v = \partial_t u + \int_0^x k(x,s)\partial_t u(s,t)ds \ = \partial_t u + \int_0^x k(x,s)(\partial_{ss} - p(s))u(s,t) \ ds$$

$$= \partial_t u + \int_0^x \{ \partial_{ss}k(x,s) -k(x,s)p(s) \} \ u(s,t)ds$$

$$+ \partial_s u(x,t)k(x,x) - \partial_s u(0,t)k(x,0) - \partial_s k(x,x)u(x,t) + \partial_s k(x,0)u(0,t)$$

$$\partial_{xx}v = \partial_{xx}u + \int_0^x \partial_{xx}k(x,s)u(s,t)ds \ + \frac{d}{dx}\{ k(x,x)u(x,t) \} + \partial_x k(x,x)u(x,t).$$

Thus we have

$$\partial_t v - \partial_{xx} v + q(x)v = \partial_t u - \partial_{xx} u + p(x)u + (q-p)(x)u$$

$$+ \int_0^x \{ \partial_{ss}k(x,s) - \partial_{xx}k(x,s) + (q(x) - p(s))k(x,s) \} u(s,t) \, ds$$

$$- 2\frac{d}{dx}(k(x,x) + \partial_s k(x,0)u(0,t).$$

Then by lemma 1, one gets

$$\partial_t v - \partial_{xx} v + q(x)v = 0 \text{ and } \partial_x v = \partial_x u + k(0,0)u(\xi,t) = 0.$$

The uniqueness is given by the unique continuation Mizohata theorem [7],[11].

PROPOSITION 2. If the function ψ satisfies the equations

$$\partial_t \psi - (\partial_{xx} - p(x))\psi = 0 \qquad \text{in } (0,1)\text{x}(0,T) \qquad (11)$$

$$\partial_x \psi(0,t) = \partial_x \psi(1,t) = 0 \qquad \text{in } (0,T) \qquad (12)$$

$$\psi(1,t) = 0 \qquad \text{in } (0,T) \qquad (13)$$

then $\qquad \psi = 0 \qquad$ in $(0,1)\text{x}(0,T)$. $\qquad (14)$

Proof. The proof is very simple. Consider

$$\varphi'' + (\lambda - p)\varphi = 0 , \ \varphi'(0) = \varphi'(1) = 0,$$

and denote by λ_n the eigenvalues of the above Sturm-Liouville problem and by φ_n its normalized eigenfunctions, forming a basis in $L^2(0,1)$.

we set

$$\psi(x,0) = \psi^0, \qquad (15)$$

then the eigenfunction expansion of the solution of (11),(12) and (15) is given by

$$\psi(x,t) = \sum_{n \geqslant 0} (< \psi^0, \varphi_n > e^{-\lambda_n t}) \, \varphi_n(x).$$

Since $\psi(1,t) = 0$ in $(0,T)$. From the uniqueness theorem for Dirichlet series, it follows that

$$< \psi^0, \varphi_n > \varphi_n(1) = 0 \ \forall n \in \mathbf{N}.$$

But $\varphi_n(1) \neq 0$, then $\psi^0 = 0$ and consequently $\psi = 0$.

PROPOSITION 3. *Let $p \in C^1(0,1)$ and let u be the solution of the problem $\mathcal{H}(p)$*

$$\partial_t u - (\Delta - p(x))u = 0 \qquad \text{in } \Omega x(0,T)$$

$$\partial_x u = 0 \qquad \text{on } \Sigma_0$$

then the function

$$T(u)\,(x,y,t) = u(x,y,t) + \int_0^x k(x,s)u(s,y,t)ds$$

is the unique solution of $\mathcal{H}(q)$ such that $T(u) = u$ on Σ_0.

Proof. The proof is very easy. We argue as in proposition 2.

4. Proof of theorem1. Let $p, q \in C^1(0,1)$, $u_0, v_0 \in L^2(0,1)$, $g \in L^2(0,T)$, let u and v be the solutions of $\mathcal{P}(p,u_0, g)$ and $\mathcal{P}(q, v_0,g)$ respectively such that $(u(0,t);u(x,T)) = (v(0,t); v(x,T))$. The functions u and v are also solutions of the problems $\mathcal{Q}(p)$ and $\mathcal{Q}(q)$ respectively.

Obviously, $(p,u_0) = (q,v_0) \Rightarrow \mathcal{B}u = \mathcal{B}v \ \forall \ g \in L^2(0,T)$.

Reciprocal. Using the condition $u(0,t) = v(0,t)$, then by lemma 1 and by proposition 1, the function v is given by

$$v(x,t) = T(u)\,(x,t) = u(x,t) + \int_0^x k(x,s)u(s,t)ds$$

where $T(u)$ is the unique solution of $\mathcal{Q}(q)$ such that $u(0,t) = v(0,t)$.

Using $u(x,T) = v(x,T) \ \forall \ g \in L^2(0,T)$, then by proposition 1, we obtain

$$\int_0^x k(x,s)u(s,T)ds = 0 \ \forall \ g \in L^2(0,T).$$

Let us assume for the moment that

$$\int_0^x k(x,s)\xi(s)ds = 0 \ \forall \xi \in L^2(0,1),$$

then $k(x,s) = 0$ and consequently $p = q$, $u_0 = v_0$, which proves the theorem1.

To prove $\int_0^x k(x,s)\xi(s)ds = 0 \ \forall \xi \in L^2(0,1)$, is sufficient to show that the set

$\mathscr{A} = \{ u(x,T), u$ is solution of $\mathscr{P}(p,g,f) / g \in L^2(0,T) \}$

is dense in $L^2(0,1)$. This result makes the object of the following lemma:

LEMMA 2. Let u be the solution of the system $\mathscr{P}(p,f,g)$, then the set

$A = \{ u(x,T) / g \in L^2(0,T) \}$ is dense in $L^2(0,1)$.

Proof. By translation, it is sufficient to prove that the set

$\mathscr{N} = \{ \bar{u}(x,T) / g \in L^2(0,T) ;$ where $\bar{u}$ is the solution of the problem $\mathscr{P}(p,0,g) \}$

is dense in $L^2(0,1)$.

Now: let $\rho \in \mathscr{N}^\perp$, and let us define ψ as the solution of

$$-\partial_t \psi - (\partial_{xx} - p)\psi = 0 \quad \text{in } (0,1) \times (0,T) \tag{16}$$

$$\partial_x \psi(0,t) = \partial_x \psi(1,t) = 0 \quad \text{in } (0,T) \tag{17}$$

$$\psi(x,T) = \rho \quad \text{in } (0,1). \tag{18}$$

Then, if we multiply (16) by $\bar{u}$ and if we integrate by parts, we obtain

$$\int_0^T g(t)\psi(1,t)dt = 0, \forall g \in L^2(0,T).$$

and consequently $\psi(1,t) = 0$ on $(0,T)$.

Then we obtain the following system

$$-\partial_t \psi - (\partial_{xx} - p)\psi = 0 \quad \text{in } (0,1) \times (0,T)$$

$$\partial_x \psi(0,t) = \partial_x \psi(1,t) = 0 \quad \text{in } (0,T)$$

$$\psi(1,t) = 0 \quad \text{in } (0,T).$$

Thus, by proposition 2, $\psi = 0$ in $(0,1) \times (0,T)$. Hence $\rho = 0$ and then the set $\mathscr{N}$ is dense in $L^2(0,1)$.

We shall now prove that $\int_0^T k(x,s)\xi(s)ds = 0 \ \forall \xi \in L^2(0,1)$.

As a matter of fact, by lemma 2, for any function ξ in $L^2(0,1)$, there exists a sequence $g_n \in L^2(0,T)$ such that $u_n = u(x,T;g_n)$ converges towards ξ in $L^2(0,1)$.

Then

$$\int_0^x k(x,s)\xi(s)ds = \int_0^x k(x,s)(\xi - u_n + u_n)ds ,$$

hence

$$\left| \int_0^x k(x,s)\xi(s)ds \right| \leqslant \left| \int_0^x k(x,s)(\xi - u_n) \ ds \right| + \left| \int_0^x k(x,s)u_n(s)ds \right|$$

$$\leqslant \left| \int_0^x k(x,s)(\xi - u_n) \ ds \right|, \quad \text{since} \int_0^x k(x,s)u_n(s)ds = 0$$

$$\leqslant c \left(\int_0^1 |\xi - u_n|^2 \ ds \right)^{1/2} \quad \text{by Cauchy-Schwarz}$$

Then $\qquad \int_0^x k(x,s)\xi(s)ds = 0.$

5. Proof of theorem 2. Let p and q $\in C^1(0,1)$, f and h $\in L^2(\Omega)$, g_1 and $k_1 \in L^2(\Sigma_1)$, g_2 and $k_2 \in L^2(\Sigma_2)$, g_3 and $k_3 \in L^2(\Sigma_3)$. Let u and v be the solutions of the problems $\mathscr{P}(p,f, g_1,g_2,g_3)$ and $\mathscr{P}(q,h,g_1,k_2,k_3)$ respectively such that u = v on Σ_0 and u(x,y,T) = v(x,y,T) . The functions u and v are also solutions of the problems $\mathscr{H}(p)$ and $\mathscr{H}(q)$ respectively.

Obviously, $(p, f,g_2,g_3)=(q, h,k_2 kg_3) \Rightarrow \mathscr{C}u = \mathscr{C}v \ \forall \ g_1 \in L^2(\Sigma_1),$

Reciprocal. Using u = v on Σ_0, then by proposition 3, we obtain

$$v(x,y,t) = T(u) (x,y,t) = u(x,y,t) + \int_0^x k(x,s)u(s,y,t)ds$$

where T(u) is the solution of $\mathscr{H}(q)$ such that u = v on Σ_0.

Using u(x,y,T) = v(x,y,T), $\forall \ g_1 \in L^2(\Sigma_1)$, then by proposition 3

$$\int_0^x k(x,s)u(s,y,T)ds = 0 \ \forall \ g_1 \in L^2(\Sigma_1).$$

By integration with respect to y, we have

$$\int_0^1 \int_0^x k(x,s)u(s,y,T)dsdy = 0 \ \forall \ g_1 \in L^2(\Sigma_1).$$

Let us assume for the moment that

$$\int_0^1 \int_0^x k(x,s)\xi(s,y)\,ds\,dy = 0 \quad \forall \xi \in L^2(\Omega).$$

Then $k(x,s) = 0$ and consequently

$$p = q, \quad u = v, \quad g_2 = k_2, \quad g_3 = k_3 \quad \text{and } f = h.$$

To prove $\int_0^1 \int_0^x k(x,s)\xi(s,y)\,ds\,dy = 0 \quad \forall \xi \in L^2(\Omega)$, it is sufficient to show that the set

$$\mathcal{B} = \{ u(x,y,T), \text{ u solution of } \mathcal{P}(p,f,g_1,g_2,g_3)) \ / \ g_1 \in L^2(\Sigma_1) \}$$

is dense in $L^2(\Omega)$. This result is shown in the following lemma .

LEMMA 3. *The set*

$$\mathcal{B} = \{ u(x,y,T) ; \text{ where u is the solution of the system } \mathcal{P}(p,f,,g_1,g_2,g_3)/ \ g_1 \in L^2(\Sigma_1) \} \text{ is dense in } L^2(\Omega).$$

Proof. By translation, it is sufficient to show that the space

$$\mathcal{M} = \{ \bar{u}(x,y,T) \ / \ g_1 \in L^2(\Sigma_1) ; \text{ where } \bar{u} \text{ is solution of the problem } \mathcal{P}(p,0,g_1,0,0) \}$$

is dense in $L^2(\Omega)$.

As in theorem 1, using Hahn - Banach theorem and the Mizohata's continuation theorem, we prove that the space $\mathcal{M}$ is dense in $L^2(\Omega)$.

Now, we shall prove that

$$\int_0^1 \int_0^x k(x,s)\xi(s,y)\,ds\,dy = 0 \quad \forall \xi \in L^2(\Omega).$$

As for theorem 1, by lemma 3, for any function $\xi \in L^2(\Omega)$, there exists a sequence $g_n \in L^2(\Sigma_1)$ such that $u_n = u(x,y,T;g_n)$ which converges towards ξ in $L^2(\Omega)$. Then

$$\int_0^1 \int_0^x k(x,s)\xi(s,y)\,ds\,dy = \int_0^1 \int_0^x k(x,s)(\xi(s,y) - u_n + u_n)\,ds\,dy = 0, \text{ hence}$$

$$\left| \int_0^1 \int_0^x k(x,s)\xi(s,y)\,ds\,dy \right| \leqslant \left| \int_0^1 \int_0^x k(x,s)(\xi(s,y) - u_n)\,ds\,dy \right| + \left| \int_0^1 \int_0^x k(x,s)(\xi(s,y)u_n)\,ds\,dy \right|$$

But $\int_0^1 \int_0^x k(x,s)(\xi(s,y)u_n)dsdy = 0$, hence

$$\left| \int_0^1 \int_0^x k(x,s)\xi(s,y)dsdy \right| \leqslant \left| \int_0^1 \int_0^x k(x,s)(\xi(s,y) - u_n)dsdy \right|$$

$$\leqslant c \left(\int_0^1 \int_0^x (\xi(s,y) - u_n)^2 dsdy \right)^{1/2} \quad \text{by Cauchy - Schwarz}$$

Then $\qquad\qquad \int_0^1 \int_0^x k(x,s)\xi(s,y)dsdy = 0.$

6. Identification method.

Let us now go back to the problem $\mathcal{P}(p,u_0,g)$

We are given a function $r \in C^1(0,1)$, and the observation $\mathcal{C}u = (u(0,t); u(x,T))$, one solves the Cauchy problem

$$\partial_t w - (\partial_{xx} - r(x))w = 0 \qquad \text{in } (0,1)x(0,T) \tag{19}$$

$$\partial_x w(0,t) = 0 \qquad \text{in } (0,T) \tag{20}$$

$$w(0,t) = u(0,t) \qquad \text{in } (0,T). \tag{21}$$

By proposition 1, we have

$$u(x,t) = w(x,t) + \int_0^x k(x,s)w(s,t)ds, \text{ where } k \text{ is given by lemma 1.}$$

Then the identification problem is reduced to determine k and p from the following system:

$$u(x,T) = w(x,T) + \int_0^x k(x,s)w(s,T)ds$$

$$\partial_{xx}k(x,s) - \partial_{ss}k(x,s) = (p(s)-r(x))k(x,s) \qquad \text{in } 0 < s < x < 1$$
$$\partial_s k(x,0) = 0 \qquad \text{in } (0,1)$$

$$k(x,x) = \frac{1}{2}\int_0^x (r-p)(t)dt \qquad \text{in } (0,1)$$

where r is given, $u(x,T)$ is observed and $w(x,T)$ is calculated by (19)-(21);

7. Remark.

For the wave equation we have obtained in [2] the following result:

THEOREM 3. *Let* $p \in C^l(0,1)$, $T \geqslant 2$ *and let be the mapping*

$$B_1 : C^l(0,T) \rightarrow \left\{ L^2(0,T) \rightarrow L^2(0,T)xL^2(0,1) \right\}$$

$$p \rightarrow \quad \left\{ g \rightarrow (\partial_x u(0,t), u(x,T)) \right\}$$

where u is the solution of the problem,

$$\partial_{tt}u - (\partial_{xx} - p(x))u = 0 \text{ in } (0,1)x(0,T), \ u(0,t) = 0, \partial_x u(1,t) = g \text{ in } (0,T) \text{ and}$$

$$u(x,0) = \partial_t u(0,t) = 0 \text{ in } (0,1).$$

Then B_1 is injective ▬

REFERENCES

[1] A. El Badia, Application d'un théorème d'unicité du type Mizohata à un problème inverse pour l'équation de la chaleur. C.R.A.S. Paris, Serie I, t 316, pp 363-368, 1993

[2] A. El Badia, Application d'un théorème d'unicité du type Mizohata à un problème inverse pour l'équation des ondes C.R.A.S. Paris, Serie I, t 316, pp 895-900, 1993

[3] I.M. Ggelfand and B.M. Levitan On the determinatiopn of a differential equation from its spectral fonction American.Math.Soc.Transl.Series 2,1, (1955), pp. 235-304.

[4] S.Kitamura and S.Nakajiri. Identifiability of spatially-varying and constant parameters in distributed systems of parabolic type. SIAM. J. Control and Optimisation. n° 15, (1977), pp 785-802.

[5] B.M. Levitan and M.G.Gasymov. Determination of a differential equation by two of its spectra. Russian. Survey, 19-2, (1964), pp. 1-63.

[6] J.L.Lions et E.Magenes. Problèmes aux limites non homogènes et applications. t 2; Paris, Dunod, 1968.

[7] S. Mizohata. Unicité du prolongement des solutions pour quelques opérateurs différentiels . paraboliques. Mem. Coll. Sci. Uni. Kyoto. Sér. A 31 (3), (1958), pp. 216-239.

[8] E. Picard. Leçons sur quelques types simples d'équations aux dérivées partielles avec des apllications à la physique mathématique. Gauthier-Villars. 1950.

[9] A. Pierce. Unique identification of eigenvalues and coefficients in a parabolic problem. SIAM.J.Cont. and Opt. Vol 17, N° 4, july (1979).

[10] W. Rundell, The use of integral operators in undetemined coefficient problems for partial differential equations, Appl. Ann. Vol. 18, (1984), pp. 309-324,

[11] J.C. Saut et B. Scheurer. Unique continuation for some evolution equations. J.Diff.Equ. 66, (1987), pp. 118-149.

[12] T. Suzuki. Inverse problems for heat equations on compact intervals and on circles, I, J.Math. Soc. Japan. Vol 38, n° 1.)1986). pp. 39-65.

IDENTIFICATION OF PARAMETERS IN NON LINEAR PROBLEMS WITH MISSING DATA. APPLICATION TO THE IDENTIFICATION OF POLLUTIONS IN A RIVER.

AINSEBA B.E[*]., KERNEVEZ J.P.[*], LUCE R.[**]
*UNIVERSITÉ DE TECHNOLOGIE DE COMPIÈGNE.
LABORATOIRE DE MATHÉMATIQUES APPLIQUÉES,
RUE PERSONNE DE ROBERVAL, 60200 COMPIÈGNE, FRANCE
**UNIVERSITÉ DE PAU & U.R.A.~CNRS~1204, I.P.R.A.,
LABORATOIRE DE MATHÉMATIQUES APPLIQUÉES,
AVENUE DE L'UNIVERSITÉ, 64000 PAU, FRANCE

Abstract. The aim of this paper is to look at the identification of some parameters which appear in the expression of the right-hand term of non linear P.D.E.'s. The studied models are with missing data because the initial conditions are not well known. Two approaches are possible. One is classical and uses the least square method, this approach needs to compute the parameters and the missing terms. The other is the Sentinel Method more convenient for our situation. We compare both methods on an example of identification of pollution terms in a river.

1. Presentation of the problem.

The aim of this study is to estimate the pollution amplitudes λ_i which are flowed into a part Ω of a river. We dispose of measurements on some concentration substrates (Deficit in Biological Oxygen, Nitrogen,...) on a part ω of Ω, ω should be the smallest as possible because of the cost measurements. The general form of the pollution model is given by (A. OKUBO[10], L.C. WROBEL[11]):

$$(1) \quad \begin{cases} \dfrac{\partial y}{\partial t} - \mathcal{A}y + F(y) = \sum\limits_{i=1}^{n} \lambda_i f_i(t)\delta(x - x_i) \\ y(x,0) = \xi \\ y = 0 \text{ on } \Gamma \times]0,T[\end{cases}$$

where $\mathcal{A} = -V.\nabla + div(\nabla_D)$, V represents the smooth velocity field of the river, D the dispersion coefficients and ∇_D the dispersion directional gradient. The non linear reaction rate F, is supposed to be bounded and C^1. The f_i's represent the manner on which the pollutions are flowed at the points x_i's in the river. The initial concentration of pollutant $\xi = \xi_0 + \tau \hat{y}_0$ is partially or completely unknown; ξ_0 represents the known part of this initial data and $\tau \hat{y}_0$ the unknown part which does not interest us. The parameters τ and λ_i's are supposed small. We observe y on an open subset ω of the river Ω, during a time-interval $[0,T]$.

2. Identification with the Least Square Method.

The classical least square method consists in estimating the pollution amplitudes together with the unknown initial concentration of pollutant.

The cost function is given in this case by:

(2)
$$J(\xi, \lambda) = \int_{\omega \times]0,T[} (y - y_{obs})^2 \, dx dt$$

and we look for

$$\inf_{\xi, \lambda} J(\xi, \lambda)$$

THEOREM 1. *Suppose that Ω is an open set of $\mathbf{R}^N$, $N \geq 2$. Then there exist $(\xi^*, \lambda^*) \in G \times \mathbf{R}^n$ satisfying*

$$J(\xi^*, \lambda^*) \leq J(\xi, \lambda) \; \forall (\xi, \lambda) \in G \times \mathbf{R}^n$$

where G is a space defined by completion of $L^2(\Omega)$ (J.L. LIONS[8])

To establish this result; we need the following lemma:

LEMMA 1. *Consider the following linear problem*

$$\begin{cases} \dfrac{\partial y}{\partial t} - Ay = h(x,t) + \sum_{i=1}^{n} \lambda_i f_i(t) \delta(x - x_i) \\ y(x,0) = \xi \\ y = 0 \text{ on } \Gamma \times]0,T[\\ \|\xi\|_{H^\beta} \leq R, \; |\lambda|_{\mathbf{R}^n} \leq R \end{cases}$$

where h is a bounded function of $L^2(\Omega \times]0,T[)$ and $\beta \succ 0$. Let $(\overline{\xi}, \overline{\lambda})$ the optimal pair.

$$J(\overline{\xi}, \overline{\lambda}) \leq J(\xi, \lambda) \; \forall \|\xi\|_{H^\beta} \leq R \text{ and } \forall |\lambda|_{\mathbf{R}^n} \leq R$$

Then $\forall \varepsilon \succ 0$, $\exists R \succ 0$ independent of h such that

$$\left| J(\overline{\xi}, \overline{\lambda}) - \inf_{\xi, \lambda} J(\xi, \lambda) \right| \leq \varepsilon$$

Remarks:
- The optimal pair exists because the pair (ξ, λ) belongs to a ball of radius R.
- $\inf_{\xi, \lambda} J(\xi, \lambda)$ exists (J.L. LIONS [9], B.E. AINSEBA [1]).

Proof of the lemma. Let h be a function of $H^{-2\gamma, -\gamma}(\Omega \times]0,T[)$ with $\gamma \in \left]0, \frac{1}{4}\right[$ and G_R the application:

$$G_R : H^{-2\gamma, -\gamma}(\Omega \times]0,T[) \longrightarrow L^2(\omega \times]0,T[)$$
$$h \longrightarrow y_{\omega \times]0,T[}(h, \overline{\xi}, \overline{\lambda})$$

The optimality condition for $(\overline{\xi}, \overline{\lambda})$ is:

$$\int_{\omega \times]0,T[} (y(h, \overline{\xi}, \overline{\lambda}) - y_{obs})(y(0, \xi, \lambda) - y(0, \overline{\xi}, \overline{\lambda})) dx dt \geq 0 \ \forall \|\xi\|_{H^\beta} \leq R$$
$$\text{and } \forall |\lambda|_{\mathbf{R}^n} \leq R$$

Writing the optimality conditions for two values h_1, h_2 of h and taking respectively (ξ, λ) equal to $(\overline{\xi}^2, \overline{\lambda}^2)$ and $(\overline{\xi}^1, \overline{\lambda}^1)$, we obtain by adding:

$$\int_{\omega \times]0,T[} (y(h_1, \overline{\xi}^1, \overline{\lambda}^1) - y(h_2, \overline{\xi}^2, \overline{\lambda}^2)) \, y(0, \overline{\xi}^1 - \overline{\xi}^2, \overline{\lambda}^1 - \overline{\lambda}^2) \, dx dt \leq 0.$$

Thus we can prove that the G_R are equicontinuous in R:

$$\int_{\omega \times]0,T[} (y(h_1, \overline{\xi}^1, \overline{\lambda}^1) - y(h_2, \overline{\xi}^2, \overline{\lambda}^2))^2 dx dt =$$
$$\int_{\omega \times]0,T[} (y(h_1, \overline{\xi}^1, \overline{\lambda}^1) - y(h_2, \overline{\xi}^2, \overline{\lambda}^2))(y(0, \overline{\xi}^1 - \overline{\xi}^2, \overline{\lambda}^1 - \overline{\lambda}^2) + y(h_1 - h_2, 0, 0)) dx dt$$
$$\leq \int_{\omega \times]0,T[} (y(h_1, \overline{\xi}^1, \overline{\lambda}^1) - y(h_2, \overline{\xi}^2, \overline{\lambda}^2)) y(h_1 - h_2, 0, 0) dx dt$$
$$\leq \left(\int_{\omega \times]0,T[} (y(h_1, \overline{\xi}^1, \overline{\lambda}^1) - y(h_2, \overline{\xi}^2, \overline{\lambda}^2))^2 dx dt \right)^{\frac{1}{2}} \left(\int_{\omega \times]0,T[} y(h_1 - h_2, 0, 0)^2 dx dt \right)^{\frac{1}{2}}$$

So

$$\int_{\omega \times]0,T[} (y(h_1, \overline{\xi}^1, \overline{\lambda}^1) - y(h_2, \overline{\xi}^2, \overline{\lambda}^2))^2 dx dt \leq \int_{\omega \times]0,T[} y(h_1 - h_2, 0, 0)^2 dx dt$$
$$\leq C \left(\|h_1 - h_2\|_{L^2(\Omega \times]0,T[)} \right)^2$$

Where C is an constant independent of R.

In the linear case, the minimum is reached for $(\xi, \lambda) \in G \times \mathbf{R}^n$; but H^β is dense in L^2, which is dense in G, so:

$$\lim_{R \to \infty} J(\overline{\xi}, \overline{\lambda}) = \inf_{\xi \in H^\beta, \lambda \in \mathbf{R}} J(\xi, \lambda) = \inf_{\xi \in G, \lambda \in \mathbf{R}} J(\xi, \lambda)$$

It results that $G_R(h)$ converges in $L^2(\Omega)$, for each h, when $R \longrightarrow \infty$. We conclude with the Ascoli theorem that the G_R's converge uniformly on every compact of $H^{-2\gamma, -\gamma}(\Omega \times]0,T[)$. The embedding of $L^2(\Omega \times]0,T[)$ in $H^{-2\gamma,-\gamma}(\Omega \times]0,T[)$ being compact, the lemma is proved for every bounded function F of $L^2(\Omega \times]0,T[)$.

Proof of theorem 1. Consider the following linear problem:

(3)
$$\begin{cases} \dfrac{\partial y}{\partial t} - Ay + F(z) = \sum_{i=1}^{n} \lambda_i f_i(t) \delta(x - x_i) \\ y(x, 0) = \xi \\ y = 0 \text{ on } \Gamma \times]0, T[\end{cases}$$

where z is a given function in $L^2(\Omega \times]0, T[)$.

Take $R \geq 0$ and a small non negative real β. Consider the multi-application $\mathcal{V}_\epsilon$ of $L^2(\Omega \times]0, T[)$ in itself, defined by:

$$\mathcal{V}_\epsilon = \{y \text{ solution of (3)}/ \exists \xi_\epsilon, \lambda_\epsilon \text{ s.t. } \|\xi_\epsilon\|_{H^\beta} \leq R, \; |\lambda_\epsilon|_{\mathbf{R}^n} \leq R$$
$$\text{and } |J(\xi_\epsilon, \lambda_\epsilon) - \inf_{\xi \in G, \lambda \in \mathbf{R}^n} J(\xi, \lambda)| \prec \epsilon\}$$

To establish the theorem 1 we have to prove that $\mathcal{V}_\epsilon$ admits a fixed point. This can be done by using the Kakutani fixed point theorem[4]. It will be enough to prove that $\mathcal{V}_\epsilon$ is a correspondence of a compact convex set into itself, hemi-continuous superiorly, with convex, closed and none empty values.

It's obvious that $\mathcal{V}_\epsilon(z)$ is convex. Compact because for every sequence y_n of $\mathcal{V}_\epsilon(z)$, there exists a sequence (ξ_n, λ_n) bounded in $H^\beta \times \mathbf{R}^n$, from which we can extract a convergent sub sequence in $L^2 \times \mathbf{R}^n$. Thus we obtain a convergent sub sequence y_{n_k} of y_n.

$\mathcal{V}_\epsilon(z)$ is none empty for every z in $L^2(\Omega \times]0, T[)$; since $F(z)$ is bounded, we can use lemma 1 and choose R enough large to approach the minimum up to ϵ.

In order to establish that $\mathcal{V}_\epsilon$ is upper semi-continuous, we should reason by absurd: for each neighbourhood $\mathcal{H}$ of $\mathcal{V}_\epsilon(z)$ in $L^2(\Omega \times]0, T[)$, we could find a sequence z_n converging to z, from which we can extract a sub sequence written z_n converging to z almost everywhere and a sequence $y_n \in \mathcal{V}_\epsilon(z_n)$, but $y_n \notin H$.

F being continuous and bounded, $F(z_n)$ converge to $F(z)$ in $L^2(\Omega \times]0, T[)$. Denote by $\xi_n = y_n(0, x)$, we have $\|\xi_n\|_{H^\beta} \leq R$ and $|\lambda_n|_{\mathbf{R}^n} \leq R$. So we can extract a sub sequence ξ_n converging to ξ weakly in $H^\beta(\Omega)$ and strongly in $L^2(\Omega)$. Also we can extract a sub sequence of λ_n converging to λ. Since the closed ball, of radius R, in H^β (respectively in $\mathbf{R}^n$) is convex, and weakly closed, then $\|\xi\|_{H^\beta} \leq R$ (respectively $|\lambda|_{\mathbf{R}^n} \leq R$).

Let $y(z, \xi, \lambda)$ be the solution of (3), corresponding to z, ξ, λ. Then y_n converge to y in $L^2(\Omega \times]0, T[)$ and therefore in $L^2(\omega \times]0, T[)$. It results that $y(z, \xi, \lambda) \in \mathcal{V}_\epsilon(z)$ but $y_n \notin H$!

Finally when z spans $L^2(\Omega \times]0, T[)$, $F(z)$ is bounded in $L^2(\Omega \times]0, T[)$ and so $y(z, \xi, \lambda)$ is bounded in $L^2(\Omega \times]0, T[)$. The range of $\mathcal{V}_\epsilon$ is a compact of $L^2(\Omega \times]0, T[)$ included in a convex ball of $L^2(\Omega \times]0, T[)$. So the convex envelope of the $\mathcal{V}_\epsilon$ range is compact. If we restrict $\mathcal{V}_\epsilon$ to $conv(Im(\mathcal{V}_\epsilon))$, we can apply the Kakutani fixed point theorem. So there exists a fixed point of the multi-application $\mathcal{V}_\epsilon$. The minimum is attained.

3. Identification with the Sentinel Method.

3.1. Definition. The Sentinel method (J.L. LIONS [6][7])consists in the construction of a functional S:

$$S(\tau, \lambda) = \int_{(0,T) \times \omega} w \, y \, dx dt$$

where w is so determined that the functional S be sensible to the parameters to identify and insensible to the others.

If we want to identify the parameter λ_1, we look for w solution of the following problem:

$$(P) \left| \begin{array}{l} \left| \dfrac{\partial S(0,0)}{\partial \tau} \right| \leq \epsilon, \left| \dfrac{\partial S(0,0)}{\partial \lambda_i} \right| \leq \epsilon \text{ for } i \neq 1, \left| \dfrac{\partial S(0,0)}{\partial \lambda_1} - \alpha_1 \right| \leq \epsilon \\ w \text{ of minimal norm in } L^2(\omega \times {]0,T[}) \end{array} \right.$$

When w is found, we consider the Taylor expansion of S in the neighborhood of $(0,0)$

$$S(\tau,\lambda) = S(0,0) + \tau \frac{\partial S}{\partial \tau} + \lambda_1 \frac{\partial S}{\partial \lambda_1} + \sum_{i=2}^{n} \lambda_i \frac{\partial S}{\partial \lambda_i} + O(\tau,\lambda)$$

So we have an estimation of λ_1 when $\epsilon \to 0$, since $S(0,0)$ can be calculated and $S(\tau,\lambda) = \int_{(0,T) \times \omega} w \, y_{obs} \, dx dt$ is known from the measurement. To identify the other parameters λ_i, just substitute the index 1 by the index i in the previous explanation.

3.2. Calculation of w.

PROPOSITION 1. *Solving problem (P) is equivalent to find w of minimal norm govervening the system (4) to a state satisfying (5)*

$$(4) \qquad \left\{ \begin{array}{l} -\dfrac{\partial p}{\partial t} - A^*p + F'(\bar{y})p = w \, \chi_{\omega \times (0,T)} \\ p(x,T) = 0 \\ p(x,t) = 0 \text{ on } \Gamma \times {]0,T[} \end{array} \right.$$

$$(5) \qquad \left\{ \begin{array}{l} | \, p(x,0) \, | \leq \epsilon, \\ | \int_0^T f_1(t) \, p(x_1,t) \, dt - \alpha_1 \, | \leq \epsilon \\ | \int_0^T f_i(t) \, p(x_i,t) \, dt \, | \leq \epsilon \quad i = 2, n \end{array} \right.$$

where $\bar{y}$ is the solution of (1) for λ and τ equal 0.

Proof. We note y_μ the derivative of y, solution of (1), with respect to μ (where $\mu \in \{\tau, \lambda_1, .., \lambda_n\}$). By multiplying (4) by y_μ and integrating by parts we obtain (5).

THEOREM 2. *For every open subset ω of $\Omega \subset \mathbf{R}^N$ ($N \geq 2$), the space :*

$$\mathcal{A} = \left\{ p(x,0), \int_0^T f_i(t) \, p(x_i,t) \, dt \ (\ i = 1,n) \ \Big| w \in L^2(\omega \times \,]0,T[) \right\}$$

is dense in $L^2(\Omega) \times \mathbf{R}^n$

Proof. To demonstrate the density, it is sufficient to use the Hahn-Banach theorem and remark that $\mathcal{A}^\perp = 0$. Let $(\rho_0, \varsigma_1, .., \varsigma_n) \in \mathcal{A}^\perp$ and ρ be solution of

(6)
$$\begin{cases} \dfrac{\partial \rho}{\partial t} - A\rho + F'(\bar{y})\rho = \sum_{i=1}^n \varsigma_i f_i(t)\delta(x - x_i) \\ \rho(x,0) = \rho_0 \\ \rho(0,t) = 0 \end{cases}$$

by multiplying the first equation of (6) by p solution of (4) and after integration by parts, we obtain:

$$\int_\Omega p(x,0) \, \rho_0(x) \, dx + \sum_{i=1}^n \varsigma_i \int_0^T f_i(t) \, p(x_i,t) \, dt = \int_{\omega \times]0,T[} \rho \, w \, dxdt$$

and thus

$$\int_{\omega \times]0,T[} \rho \, w \, dxdt = 0 \ \forall w \in L^2(\omega \times \,]0,T[)$$

then $\rho = 0$ on $\omega \times \,]0,T[$. Mizohata theorem applied on $\Omega - \{x_i\}_{i=1,n}$ gives $\rho = 0$ on $\Omega - \{x_i\}_{i=1,n}$. It's clear now, using the continuity of ρ that ρ vanishes.

Remark: We can establish this theorem in the one dimensional case provided we make some assumptions about the relative positions of the pollutions and the observatory ω (B.E. AINSEBA[2]). The above proof is no longer valid in this case because of the domain $\Omega - \{x_i\}_{i=1,n}$ is not connex.

4. Linear case.

Consider (1) where we suppose that $F(y) = Ky$ and $\xi_0 = 0$ in order to simplify the following.

Denote by C the operator:

$$C : (\xi, \lambda) \longrightarrow y_{|\omega \times]0,T[}$$

The least square method consists in minimizing the cost function:

$$J(\xi, \lambda) = \int_{\omega \times]0,T[} (y - y_{obs})^2 \, dxdt$$

The solution is given by:

$$C^*C(\xi^*, \lambda^*) = C^* y_{obs}$$

where C^* is the adjoint operator of C:

$$C^*: \quad w \longrightarrow (p(x,0), \int_0^T f_1(t)\, p(x_1,t)dt, \int_0^T f_i(t)\, p(x_i,t)dt \ i = 2,n)$$

The sentinel method consists in computing w satisfying:

$$\|C^*w - (0,\alpha_1,0,..,0)\| \leq \varepsilon$$

the w characterized by the Hilbert Uniqueness Method (J.L. LIONS [8]) is given by $w = Cg$, where g satisfies

$$(C^*C + \varepsilon I)g = (0,\alpha_1,0,..,0)$$

Next, we compute λ_1 by

$$\lambda_1 = \frac{\int_{\omega \times]0,T[} w\, y_{obs}\, dxdt}{\alpha_1}$$

It is easily seen that the λ^* computed by the least square method is the same that the one computed by the sentinel method up to ε:

$$\lambda_1 = \frac{\langle g, C^* y_{obs}\rangle}{\alpha_1} = \frac{\langle g, C^*C(\xi^*,\lambda^*)\rangle}{\alpha_1} = \frac{\langle ((0,\alpha_1,0,..,0) - \varepsilon g),(\xi^*,\lambda^*)\rangle}{\alpha_1}$$
$$= \frac{\alpha_1 \lambda_1^* - \varepsilon \langle g, (\xi^*,\lambda^*)\rangle}{\alpha_1} = \lambda_1^* + O(\varepsilon)$$

When $\varepsilon = 0$, to compute w, we need to solve an exact controlabillity parabolic problem!

5. Numerical results.

The numerical computations of the initial concentration of pollutant ξ and of the amplitudes λ_i using the least square method is done by the minimization of the discretized cost function (2) using a conjugate gradient algorithm. The discretized non linear equation (1) is resolved by a Newton method.

The calculation of w to compute the sentinel is done by minimizing the discretized of the following cost function:

$$J(w) = \frac{1}{2}\left(\int_\Omega p(x,0)^2 dx + (\int_0^T f_1(t)p(x_1,t)dt - \alpha_1)^2 + \sum_{i=2}^n (\int_0^T f_i(t)p(x_i,t)dt)^2 \right)$$

The equation (1) and (4) are discretized with a finite-difference approximation of level 1 in time and with a P_1 finite-element method in space.

As an example, we took a section of a river of 4650 meters length and 60 meters width (fig 1). We suppose that the fluid flow in the river is ideal and irrotational, so $V = \nabla\Phi$, where Φ is the velocity potential:

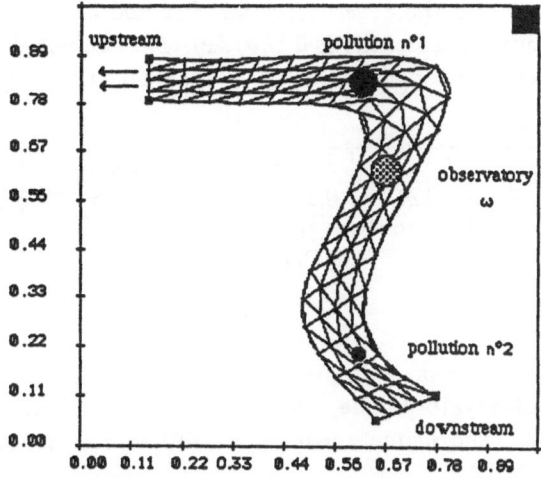

FIG. 1. *disposition of the observatory and of the pollution in the river*

$$\begin{cases} \Delta_D \Phi = 0 \text{ in } \Omega \\ \Phi \text{ is given on } \Gamma \end{cases}$$

The maximal value of the velocity is $7km/day$. The longitudinal dispersion coefficient D_1 is $8m^2/s$. The transversal dispersion coefficient is given by the OKUBO rule ([10]):

$$\frac{D_2}{D_1} = \left(\frac{L_2}{L_1}\right)^{4/3}$$

where L_2, L_1 are respectively the width and the length of the river.

The non linear function F is

$$F(y) = 0.19y + \frac{y}{1+y}$$

In the numerical results given below, we compare the least square method and the sentinel method when the measurements are supposed corrupted by some noise.

Test 1 :

noise in %	0	1	5	10	20	50
λ_1^S	0.124	0.117	0.12	0.128	0.13	$-7.4\,10^{-2}$
λ_1^{lsm}	0.10027	0.1005	0.1108	0.2099*	$7.21\,10^{-2*}$	$-5.9\,10^{-2*}$

TABLE 1

---

[0] *The optimization fails

The pollutions and the observatory are disposed as in figure 1. The exact value of the parameter to identify is $0.1 kg/day$. The computed w minimizing the sentinel cost function has a norm $\|w\|_{L^2(\omega \times]0,T[)} = 65.63$.

Test 2 :

noise in %	0	1	2	3	4	5	10
λ_1^S	0.11	0.124	0.155	0.146	$7.3\,10^{-2}$	$5.5\,10^{-2}$	0.151
λ_1^{lsm}	0.1001	$9.8\,10^{-2}$	0.1008	$9.6\,10^{-2}$	$4.6\,10^{-2}$	0.16	0.22

TABLE 2

The situation is the same as previously but the pollutions are situated upstream from the observatory. The values of the both parameters is $0.1 kg/day$.

To identify λ_1, the w minimizing the sentinel cost function has a norm $\|w\|_{L^2(\omega \times]0,T[)} = 134.12$.

noise in %	0	1	5	10
λ_2^S	0.11	$1.58\,10^{-2}$	$-6.01\,10^{-2}$	-2.31
λ_2^{lsm}	0.101	0.14	0.69	2.69

TABLE 3

We can also identify λ_2, the corresponding w has a norm $\|w\|_{L^2(\omega \times]0,T[)} = 882.73$.

Remarks:

- In the previously results given, the chosen noise is uniformly centered around the exact observation. But the nature of the noise influences the results. For instance, the identification of the parameters is worse with decentered noise.

- The identification of the parameters is better when the norm of w is small.

- In order to identify a parameter of pollution, it is recommended to observe downstream from the source.

- By the least square method, the initial concentration of pollutant is not well identified.

6. Conclusion.

In real situations the location of pollution sources is given and the observatory is chosen downstream from these sources. The monitoring has to be repeated many times. Then the computational cost is better when using the sentinel method; we have to make once anoptimization, then one scalar product at each new measurement. On the other hand the least square method needs to do an optimization for each new measurement.

The ideas given in this paper still hold for other situations of differential or partial differential equations with incomplete data [2], [3].

REFERENCES

[1] B.E. AINSEBA *Contrôlabilité exacte, Identifiabilité, Sentinelles*, Thèse Université de Technologie de Compiègne 1992.

[2] B.E. AINSEBA, J.P. KERNEVEZ, R. LUCE *Application des sentinelles à l'identification des pollutions dans une rivière*, submitted to RAIRO.

[3] B.E. AINSEBA, J.P. KERNEVEZ, R. LUCE *Identification de paramètres dans des problèmes non linéaires à données imcomplètes*, submitted to RAIRO.

[4] J.P. AUBIN *L'analyse non linéaire et ses motivations économiques*, Ed. Masson, 1984.

[5] J.P. KERNEVEZ *Enzyme mathematics*, North-Holland 1980.

[6] J.L. LIONS *Sentinelles pour les systèmes distribués*, Ed. Masson 1992.

[7] J.L. LIONS Pages:819–823,865–870 *Sur les sentinelles des systèmes distribués: Cas des conditions initiales incomplètes* (P 819-823). *Conditions frontières, termes sources, coefficients incomplétement connus* (P 865-870) C.R.A.S., Vol 307, 1988.

[8] J.L. LIONS *Contrôlabilité exacte, perturbations et stabilisation des systèmes distribués*, Vol 1 et 2, Ed. Masson 1988.

[9] J.L. LIONS *communication personnelle*.

[10] A. OKUBO *Diffusion and ecological problems: Mathematical models*, Springer Verlag, 1980.

[11] L.C. WROBEL, C.A. BREBBIA *Water pollution: Modelling, measuring and prediction*, Elsevier, 1991.

BOUNDARY CONDITION IDENTIFICATION FROM NOISY POINTWISE MEASUREMENTS FOR ELLIPTIC DISTRIBUTED PARAMETER SYSTEMS

L. Carotenuto[1], D Famularo[1], G. Raiconi[2]

[1] Università della Calabria, D.E.I.S., Arcavacata di Rende (CS), Italy.
[2] Università di Salerno, Dip. di Informatica e Applicazioni, Salerno, Italy.

ABSTRACT.
The identification of a boundary condition for a system governed by the Laplace equation in a bounded domain is considered .The unknown variable is the normal derivative of the system state (flux) along part of the boundary. The data are noisy measurements of the system state taken at discrete interior points, together with the state or the flux along the remainder of the boundary. The estimate is obtained according to a maximum-a-posteriori criterion: it comes from regarding the unknown function as a gaussian process in space, for which the prior mean and the covariance kernel are known. The Boundary Element Method is used to approximate the functional relation between the boundary conditions and the model output. The results of the numerical experiments show that the identification method works well also when the unknown function is deterministic: in this case the estimate actually minimizes a weighed sum of the output error and of regularization terms, the parameters of the prior probability distribution playing the role of weighing coefficients.

INTRODUCTION

The identification of boundary conditions in distributed parameter systems (DPS) governed by elliptic or parabolic partial differential equations (PDE) has received less attention than the parameter estimation problem, although in several applications the characteristics of the medium, which are reflected into the parameters of the PDE, may be known with sufficient accuracy, whereas the external input is unknown because, for example, it acts on an inaccessible part of the boundary. The problem is considered by Lions [1] who points out, for the Laplace equation in a regular domain, that the knowledge of the potential and of the flux along a part of the boundary uniquely defines the potential along all the boundary, but in a highly unstable fashion. In order to overcome such difficulty, Lions regards the identification problem as an optimal control problem, the unknown boundary value being the "control", constrained in a suitable admissible set, with cost function consisting of a weighed sum of the norm of the reconstruction error and of a regularization functional. In this way a unique estimate is obtained, continuously dependent on the data. Most of the subsequent works [2,3,4] follow this appoach, although different types of available data, and different discretization and optimization procedures are considered. The choice of the admissible set and of the regularization functional is somehow arbitrary, and should summarize the a-priori information the researcher has about the problem. A different way to keep into account prior information is proposed by Tarantola [5] in solving ill-posed inverse problems: such information is used to define the prior statistics of a stochastic model of the unknown parameters, and the estimate is obtaned by maximizing the a-posteriori probability, given the data. In this paper we consider a system governed by the Laplace equation in a bounded domain of the plane: on some part of the boundary both potential and flux are unknown. The data consist of the values of the potential, corrupted by noise, at discrete points in the interior of the domain, as well as the potential or the flux on the rest of the boundary. The problem of estimating the unknown boundary condition is stated according to the approach of Tarantola [5]; the boundary element method is chosen as the most appropriate way to obtain a finite dimensional, and actually computable, approximate solution. The characteristic features of the estimation procedure are finally discussed, in light of the results of extensive numerical experiments.

MODELLING AND PROBLEM STATEMENT.

Let the system under investigation be modelled by the Laplace equation:

$$\frac{\partial^2 u}{\partial x_1^2} + \frac{\partial^2 u}{\partial x_2^2} = 0$$

$$x = [x_1 \quad x_2] \in \Omega \tag{1}$$

Ω compact subset of R^2, with piecewise smooth boundary Γ.

Let the boundary be subdivided into three disjoint parts, say Γ_1, Γ_2, Γ_3: on Γ_1 we know the potential u , on Γ_2 we know the outward normal derivative of u: $q = \partial u/\partial n$, whereas on Γ_3 neither u nor q is known. Noisy measurements of u are taken at interior points x_j, $j = 1, 2,.., N$. Then the data can be summarized as

$$u(x) = \bar{u}_1(x), \quad x \in \Gamma_1 \tag{2a}$$

$$q(x) = \bar{q}_2(x), \quad x \in \Gamma_2 \tag{2b}$$

$$d_j^{ob} = d_j^m + e_j, j = 1, 2, ..., N$$

where d_j^m are the "model outputs": $d_j^m = u(x_j)$, and e_j are the measurement errors.

The identification problem consists in obtaining an estimate $\hat{q}_3$ of the flux q_3 along Γ_3.

The relation between the model outputs d_j^m and $q_3(x)$ can be formally obtained considering that the boundary value problem

$$\frac{\partial^2 u}{\partial x_1^2} + \frac{\partial^2 u}{\partial x_2^2} = 0, \quad x \in \Omega \ .$$

$$u(x) = \bar{u}_1(x), \quad x \in \Gamma_1$$

$$q(x) = \bar{q}_2(x), \quad x \in \Gamma_2$$

$$q(x) = q_3(x), \quad x \in \Gamma_3$$

is well posed: then d_j^m is a continuous linear functional on $L^2(\Gamma_1) \times L^2(\Gamma_2) \times L^2(\Gamma_3)$ that can be written as:

$$d_j^m = F_j^1(\bar{u}_1) + F_j^2(\bar{q}_2) + F_j^3(q_3) \tag{3}$$

where the functional F_j^k depends on the location of the measurement point x_j'. Clearly $F_j^1(\bar{u}_1)$ is the model output corresponding to $u_1 = \bar{u}_1$, $q_2 = q_3 = 0$, and likewise for the other terms. Following [5], the unknown $q_3(x)$, $x \in \Gamma_3$, is modelled as a Gaussian random function; the prior information is described by a Gaussian probability with mean $\bar{q}_3(x)$ and covariance kernel $C(x,y)$, $x \in \Gamma_3$, $y \in \Gamma_3$, that defines a positive definite invertible covariance operator C. The measurement errors e_j are assumed to be independent Gaussian random variables with zero mean and variance $E[e_j^2] = \sigma_d^2$. The mean of the Gaussian probability of q_3, conditioned on the measurements, gives the least square a posteriori estimate $\hat{q}_3(x)$, $x \in \Gamma_3$ as the (unique) minimizer of the misfit functional [5]

$$J(q_3) = \left(\sigma_d^2\right)^{-1} \Sigma_{1,N} \left[d_j^{ob} - F_j^1(\bar{u}_1) - F_j^2(\bar{q}_2) - F_j^3(q_3) \right]^2 +$$

$$\int_{\Gamma_3} (q_3(x)) - \bar{q}_3(x) \ C^{-1}(q_3(\cdot) - \bar{q}_3(\cdot)) \ (x) \ d\Gamma_3(x).$$

which is a weighed sum of the output error and of a "covariance related" norm.

BOUNDARY ELEMENT APPROXIMATION.

In order to obtain in real instances an estimate of the unknown flux, some approximation must be introduced. The boundary element method (BEM) [6] seems the most appropriate tool to solve numerically the estimation problem, that is the minimization of $J(q_3)$ over a finite dimensional space, mainly because it enables to represent explicitly the functionals F_j^3. It is based on the expression of the solution of the Laplace equation at an interior point y as a weighted boundary integral of u and q:

$$u(y) = - \int_\Gamma q^*(y, x) \, u(x) \, d\Gamma(x) + \int_\Gamma u^*(y, x) \, q(x) d\Gamma(x) \tag{4}$$

with $u^*(y,x)$ fundamental solution of (1), given by

$$u^*(y, x) = - (4\pi)^{-1} \, \log[(x_1 - y_1)^2 + (x_2 - y_2)^2]$$

and $q^*(y,x)$ derivative of $u^*(y,x)$ in the direction of the outward normal to Γ in x.
Then the components of the model output are given by:

$$d_j^m = - \int_\Gamma q^*(x_j, x) \, u(x) d\Gamma(x) + \int_\Gamma u^*(x_j, x) \, q(x) d\Gamma(x) \tag{5}$$

for $j = 1, 2,.., N$.

If the point y belongs to the boundary, an integral relation of the type of equation (4) holds [6]:

$$c(y) \, u(y) = - \int_\Gamma q^*(y, x) \, u(x) d\Gamma(x) + \int_\Gamma u^*(y, x) \, q(x) d\Gamma(x) \tag{6}$$

Let an approximation scheme for u and q be chosen, based on the discretization of Γ into elements and on the selection of appropriate basis functions: then, denoting by U_k, Q_k the nodal value vectors of the approximations of $u(\cdot)$ and $q(\cdot)$ on Γ_k, the approximate version of equations (5) is

$$D^m = - (H_1^m U_1 + H_2^m U_2 + H_3^m U_3) + G_1^m Q_1 + G_2^m Q_2 + G_3^m Q_3 \tag{7}$$

where $D^m = [d_1^m ... \, d_N^m]'$, and H_k^m, G_k^m are suitable matrices whose entries are obtained through element integrals involving the fundamental solution, its derivatives and the basis functions. Clearly they depend on the location of measurements.

Likewise equation (6) is replaced by the linear algebraic equation

$$H_1 U_1 + H_2 U_2 + H_3 U_3 = G_1 Q_1 + G_2 Q_2 + G_3 Q_3 \tag{8}$$

where the matrices H_k, G_k depend only on the geometry of the domain and on the discretization scheme.
Keeping into account the known boundary conditions (2a,b), we solve (8) with respect to Q_1, U_2, U_3:

$$[Q_1' \, U_2' \, U_3']' = [-G_1 : H_2 : H_3]^{-1} \, (-H_1 \, \bar{U}_1 + G_2 \, \bar{Q}_2 + G_3 \, Q_3)$$
$$= - M_1 \, \bar{U}_1 + M_2 \, \bar{Q}_2 + M_3 \, Q_3 \tag{8a}$$

with obvious definition of the matrices M_k. By replacing into (7) we obtain

$$D^m = (-H_1^m - B^m M_1)\bar{U}_1 + (G_2^m + B^m M_2)\bar{Q}_2 + (G_3^m + B^m M_3)Q_3 \qquad (9)$$

$$B^m = [G_1^m : -H_2^m : -H_3^m]$$

Equation (9) is the explicit discrete version of equations (3). It clarifies some advantages of the BEM approach: once the matrices M_k have been computed, the model output can be computed for any input flux, without solving linear equations; changing the location of some measurement point requires only the recomputation (by quadrature) of the relevant rows of matrices G_k^m, H_k^m.

Where the approximation of the "covariance" norm in $J(q_3)$ is concerned, assume that the B.E. approximation of q_3 is given by

$$q_3(x) = [\phi_1(x) \dots \phi_r(x)] Q_3, \ x \in \Gamma_3$$

and likewise for $\bar{q}_3(x)$.

Let $\theta_i(\cdot) = C^{-1}(\phi_i(\cdot))$, that is θ_i is the solution of the integral equation

$$\phi_i(x) = \int_\Gamma C(x, z) \theta_i(z) d\Gamma_3(z).$$

Then we have

$$\int_{\Gamma_3} (q_3(x) - q_3(x)) C^{-1} (q_3(\cdot) - q_3(\cdot)) (x) d\Gamma_3(x) = (Q_3 - \bar{Q}_3)'S(Q_3 - \bar{Q}_3)$$

with $[S]_{ij} = \int_{\Gamma_3} \theta_i(x)\phi_j(x) d\Gamma_3(x)$

APPLICATIONS

The numerical experiments are designed to analyse the dependence of the estimate on the choice of the prior statistics, for several spatial distributions of the unknown flux.

Geometry and Boundary Element approximation. The domain is taken to be the unit square: $\Omega = [0,1] \times [0,1]$. The boundary is partitioned into:

$$\Gamma_1 = \{x: x_2 = 0\}, \Gamma_2 = \{x: x_1 = 0\} \cup \{x: x_1 = 1\}, \Gamma_3 = \{x: x_2 = 1\}$$

and we assume

$$u = 0 \text{ on } \Gamma_1, q = 0 \text{ on } \Gamma_2, q \text{ and } u \text{ unknown on } \Gamma_3.$$

Each side is subdivided into twenty equal intervals, with nodes taken at the midpoints of the intervals. Both u and q are approximated by piecewise linear continuous functions, with discontinuous derivative at the nodes. The integrals in equations (5), (6) then become sums of element integrals which are evaluated by exact integration using the fundamental solution and its normal derivative. Since we assume $u_1 = 0$ and $q_2 = 0$, only the matrix M_3 in (8a) is relevant for the problem, and is evaluated once for all. Likewise, when the location of the measurement points is assigned, the matrices G_3^m, B^m can be computed. Clearly the evaluation of the model output D^m requires, for every Q_3, only a matrix-vector multiplication.

Prior statistics. The probabilistic approach requires that the unknown function is regarded as a gaussian random process in space, for which an a-priori characterization is given, on the basis of the available information: then we assign the mean $E[q_3(y)] = \bar{q}_3$, and the covariance kernel $E[(q_3(y) - \bar{q}_3)(q_3(z) - \bar{q}_3)] = \sigma^2 \exp(- |y-z| / L)$, where y (z) is a local coordinate on Γ_3 : $y = x_1$ when $x_2 = 1.0$. Such choice of the covariance kernel defines the covariance operator C as:

$$(Ch)(y) = \sigma^2 \int_0^1 \exp(- |y - z| / L) \, h(z) \, dz = g(y)$$

The norm weighed with C^{-1} is given by [5]

$$\| g \|_{C^{-1}}^2 = 1/(2\sigma^2) \left[\frac{1}{L} \int_0^1 g^2(y) dy + L \int_0^1 g_y^2(y) \, dy + g^2(0) + g^2(1) \right]$$

The discrete version of the norm is obtained by replacing g(y) with its piecewise linear approximation, that gives

$$\| g \|_{C^{-1}} = 1/(2\sigma^2) \, \tilde{g} \, '[(1/L)K_1 + LK_2 + K_3] \, \tilde{g}$$

where $\tilde{g}$ is the vector of nodal values and K_1, K_2, K_3 are appropriate matrices, depending of the mesh size. In the finite dimensional approximation the prior mean is represented by the vector Q^{pr}: $[Q^{pr}]_i = \bar{q}_3$.

The data. The vector of observations D^{obs} is built up by simulation, using equation (9), with $U_1 = 0$, $Q_2 = 0$, and Q_3 varying from one experiment to another: the simulation is performed with a mesh (fifty intervals per side) finer than that used to build up the model to be identified. The effect of measurement errors is simulated by adding independent, zero mean gaussian random numbers to the above data . Two sets of measurement points are employed for the identification :

a) $x_1 = (0.025 + 0.05i , 0.7)$, i = 0, 1,..., 19;

b) $x_1 = (0.025 + 0.10i , 0.7)$, i = 0, 1,..., 9.

Two types of input flux are used in the simulations:

i) $q_3(y) = 1 + \sin(3\pi y)$;

ii) $q_3(y)$: realization of a stochastic process in space with correlation function $C(y,z) = \exp(-|y - z|/0.5)$.

Identification procedure and results. The numerical value of the estimate is obtained by minimizing the finite dimensional approximation of the functional $J(q_3)$: with the above choice of the prior statistics it takes the form:

$$J_a(Q_3) = 1/(\sigma_d^2) \| D^{obs} - (G_3^m + B^m M_3)Q_3 \|^2 +$$
$$1/(2\sigma^2)(Q_3 - Q^{pr})'[(1/L)K_1 + L K_2 + K_3](Q_3 - Q^{pr}).$$

Then the solution of the identification problem is taken to be the minimizer, $\hat{Q}_3$, of J_a, which is unique because K_1 is positive definite. The explicit expression of $\hat{Q}_3$ is

$$\hat{Q}_3 = Q^{pr} + \{(G_3^m + B^m M_3)'(G_3^m + B^m M_3)' + \sigma_d^2(2\sigma^2)$$
$$[(1/L)K_1 + L K_2 + K_3]\}^{-1}(G_3^m + B^m M_3)' [D^{obs} - (G_3^m + B^m M_3)' Q^{pr}].$$

The form of J_a suggests that the proposed solution of the identification problem can be regarded from two different points of view: if it is plausible to assume that q_3 is a random function, that its properties are well fitted by the assumed model, and that $\bar{q}_3$, σ^2 and L summarize the a-priori information, then $\hat{q}_3 = [\phi_1(x) ... \phi_r(x)]\hat{Q}_3$ is an approximation of the maximum a-posteriori estimate. If such assumptions are not plausible, then $\hat{q}_3$ is the function that, within the chosen approximating space, achieves the best compromise between the fitting of the measurements, the "closeness" to the given constant $\bar{q}_3$, and the "smoothness". In this case the parameter σ_d^2/σ^2 determines the relative weight of the data fitting term and of the regularization term, whereas L determines the relative weight of the norm of the function and the norm of the derivative.

All the numerical experiments have been performed using both 20 and 10 measurement points: since the improvement obtained by doubling the number of data is not substantial, the results that will be shown are obtained with 10 measurements.

Experiment 1. The input flux is the deterministic function $q_3(y) = 1 + \sin(3\pi y)$; figure 1 shows the estimates $\hat{q}_3$, obtained with $\bar{q}_3 = 0$, $L = 0.1$, σ^2 ranging from 10^{-7} to 10, when the data are practically error free ($\sigma_d^2 = 10^{-10}$). The effect of changing the ratio σ_d^2/σ^2 is clear: the highest value the ratio makes the regularization too effective, whereas the lowest one reveals the intrinsic instability of the inverse problem; in a wide intermediate range the reconstruction is practically exact. When the measurement error is appreciable ($\sigma_d^2 = 10^{-3}$, maximum relative error of the specific sample, 12%) the sensitivity of the reconstruction to both L and σ^2 is more complex (figures 2 and 3): low values of σ^2 or high values of L produce very smooth estimates that do not capture the variability of the input; at the other extreme, large oscillations occur. With $\sigma_d^2 = 10^{-2}$ (figures 4 and 5) this behavior is amplified: anyhow the estimate reproduces the salient features of the true input in a range of values of the parameters σ^2 and L. Note that the realization of the measurement errors used in the experiments and shown in the figures is particularly ill-behaved, mainly because its average is nonzero. The dependence of the estimate on σ^2 and L is synthetized in table 1: for each pair (σ^2, L) one hundred sets of data are built up, with different realizations of the measurement error. The corresponding estimates and mean square errors ($e = \| \hat{q}_3 - q_3 \|$) are computed and the values of e are compared with a threshold chosen in such a way that it is certainly exceeded if the estimate has a "phase inversion" and/or a wrong mean value: the entries of the tables give the percentage of times the threshold is exceeded.

Table 1

L/σ^2	0.1	0.5	1.0	5.0	10.
0.1	87	30	20	20	42
0.5	100	46	31	20	23
1.0	100	50	35	19	20
5.0	100	62	46	23	19
10.	100	75	51	27	21

$$\sigma_d^2 = 10^{-3}$$

L/σ^2	0.1	0.5	1.0	5.0	10.
0.1	100	90	71	52	54
0.5	100	100	90	57	54
1.0	100	100	93	61	53
5.0	100	100	100	68	61
10.	100	100	100	78	64

$$\sigma_d^2 = 10^{-2}$$

Experiment 2. The input flux is a realization of a random process with $\bar{q}_3 = 0$, $\sigma^2 = 1$, $L = 0.5$. Figure 6 shows the estimate $\hat{q}_3$ obtained using the same parameters as in the simulation that generated q_3, together with the estimates corresponding to $\sigma^2 = 10^{-3}$ and to $\sigma^2 = 10^3$, when the data are practically error free ($\sigma_d^2 = 10^{-10}$). In a wide range of σ^2 the estimation procedure actually performs a smooth interpolation of the "true" input; with the highest value of σ^2 we obtain an oscillating estimate, that anyhow follows the pattern of the "true" input. When appreciable measurement errors are introduced (figure 7), relatively low values of σ^2 give "flat" estimates, whereas the highest one produces a clear phase inversion.

CONCLUDING REMARKS.

The method proposed for the identification of an unknown boundary condition is based on the minimization of a weighed sum of the euclidean norm of the output error and of a positive definite quadratic functional. The estimate thus obtained can be given two interpretations, depending on the prior information available. If it is plausible to assume that the unknown function is a gaussian stationary process in space with known prior mean and covariance kernel, and that the measurement errors are zero mean gaussian random variables of known variance, then the estimate maximizes the a-posteriori gaussian probability. In the specific examples the covariance kernel is taken to be exponential, thus depenting on two parameters: the variance and the "correlation length". If the above assumptions are not plausible, the estimate is merely the minimizer of the sum of two terms: the norm of the output error and a regularization functional which penalizes both the displacement of the estimate from a constant value and its derivative. The parameters that in the first approach had the meaning of error variance, input variance and correlation length now become coefficients that weigh the various terms and enable to obtaina a family of candidate solutions of the identification problem. The reduction to finite dimension of the originally infinite dimensional problem is accomplished via boundary element approximation: once some relevant matrices are computed, the model outputs generated by different boundary inputs are obatained pratically without computational effort. The examples here shown are purely numerical ad refer to extremely simple model and geometry, but the results are encouraging for applications to more complex problems, such as the reconstruction of boundary tractions from measurements of the stress in the interior of an elastic body, a problem that is important in the recognition of the fine form by tactile sensors.

REFERENCES

[1] Lions J.L.. *Some aspects of modelling problems in distributed parameter systems*. In "Distributed Parameter Systems Modelling and Identification", A.Ruberti Ed., Springer Verlag, Berlin, 1978

[2] Sunahara Y.. *Some new aspects in parameter estimation for infinite dimensional systems*. Proceedings of the IMACS/IFAC Int.Symp. on Distributed Parameter Systems, Hiroshima, pp 15-18,1987

[3] Ohnaka K.,Uosaki K..*Boundary element approach for identification of boundary conditions in distributed parameter systems*. Int. J. of Control, V.41, pp 981-990, 1985.

[4] Hu S:, Wang K..*On identification of unknown boundary conditions for a class of 2nd order elliptic systems*. Proceedings of the IMACS/IFAC Int.Symp. on Distributed Parameter Systems, Hiroshima, pp 337-344, 1987.

[5] Tarantola A. Inverse Problem Theory. Elsevier, Amsterdam, 1987.

[6] Brebbia, C.A., Telles J.C.F., Wrobel L.C..Boundary element techniques. Springer Verlag, Berlin, 1984.

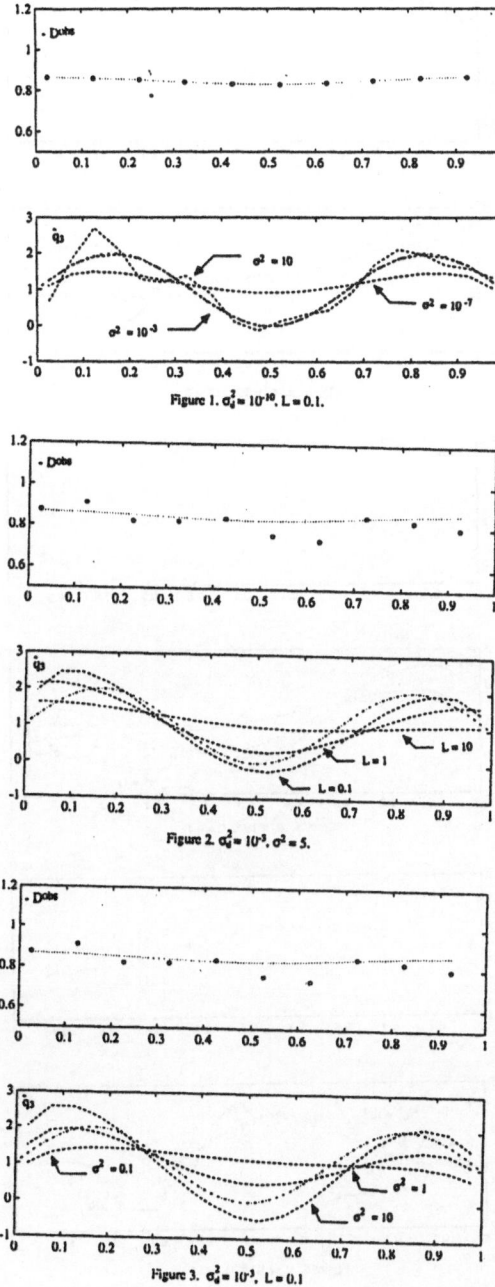

Figure 1. $\sigma_d^2 = 10^{-10}$, L = 0.1.

Figure 2. $\sigma_d^2 = 10^{-5}$, $\sigma^2 = 5$.

Figure 3. $\sigma_d^2 = 10^{-5}$, L = 0.1.

610

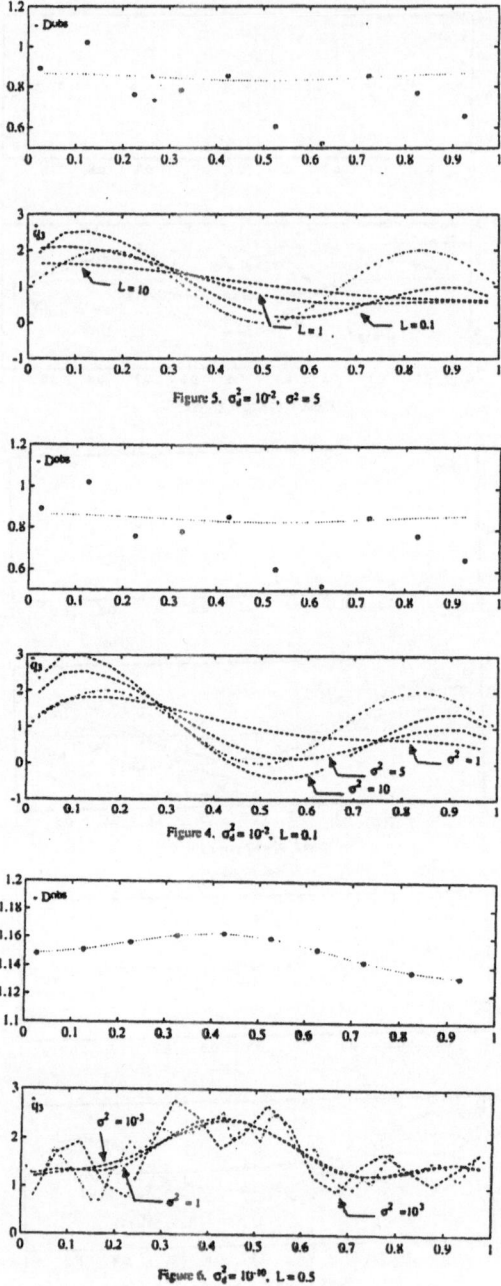

Figure 5. $\sigma_d^2 = 10^{-2}$, $\sigma^2 = 5$

Figure 4. $\sigma_d^2 = 10^{-2}$, $L = 0.1$

Figure 6. $\sigma_d^2 = 10^{-10}$, $L = 0.5$

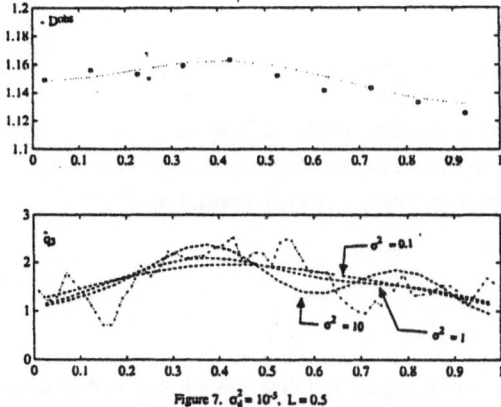

Figure 7. $\sigma_d^2 = 10^{-5}$, $L = 0.5$

INVERSE 2D PHASE CHANGE PROBLEM

C.Bénard, B.Guerrier, H.G.Liu & X.Wang

FAST - URA 871 (CNRS, Univ. Pierre et Marie Curie, Univ. Paris Sud)

Bat. 502, Campus Universitaire, 91405, Orsay, France

1. INTRODUCTION

This study deals with the identification of the space and time dependent solid/fluid interface, in a phase change process. In many industrial applications (such as casting, crystal growth, welding), the position of the phase transition interface is unknown. Indeed, the coupled physical phenomena occuring in the fluid (convection, surface effects, action of external forces on the liquid bulk) are often partially unknown and cannot be easily modelled. On the other hand, the direct measurement of the interface is impracticable. The purpose of the tracking inverse problem is to identify the shape and evolution of the interface using modelling and measurements in the solid phase alone.

Many studies of inverse problems in phase change process have been devoted to the control problem, which consists in searching for the boundary conditions in order to generate a prescribed solid-liquid interface (Colton and Reemtsen 1984, Jochum 1980, 1982, Knaber 1985, Reemtsen and Kirsch 1984, Zabaras et al., 1992). Fewer results are available for the tracking problem considered here (Banks and Kojima 1989, Katz and Rubinsky 1984, Hsu and al., 1986). The results presented here extend those given in Afshari et al. (1989), Bénard et al.(1991), Wang et al. (1992).

In this paper a numerical study, based on a regularisation method, is analysed. The unknown moving boundary is obtained from measurements collected in a finite number of points of the fixed boundary. The inverse problem is solved by minimization of a penalized least squares criterion defined on a sliding time horizon. The impact of the different parameters, such as regularisation coefficient, length of the observation horizon, and number of measurements available are thoroughly analyzed.

The problem studied is defined in section 2. The method used for solving the inverse problem is described in section 3, and results are analysed in section 4. In section 5, we compare the previous algorithm with the method used in Blum (1989) and Mannikko et al. (1992). In this method, the initially non-linear two phase problem is transformed into a linear one, by resolution of a linear diffusion problem in an extended fictitious domain.

2. DEFINITION OF THE PROBLEM

2.1 Problem geometry

A rectangular enclosure of aspect ratio A (heigth H along Oy, width L=H/A along Ox) is filled with a material at uniform initial temperature Ti, with Ti<Tf, where Tf is the melting temperature. At t=0, the temperature T(x=0,y) becomes higher than Tf, and the solid begins to melt. Temperature and flux measurements are performed on the back face Γ_0, as shown in Figure 1. The dimensionless time, distance and temperature are given respectively by (where a is the solid diffusivity):

$$t = t_{dim}.a/H^2, \qquad x = x_{dim}/H, \qquad y = y_{dim}/H, \qquad T = (T_{dim} - Tf)/(Tf - Ti)$$

The interface position is denoted by s(y,t). The dimensionless state equation in the time dependent solid domain is given by:

$$\frac{\partial^2 T}{\partial x^2} + \frac{\partial^2 T}{\partial y^2} = \frac{\partial T}{\partial t} \qquad \text{with } 0 < y < 1 \text{ and } s < x < 1/A \qquad (1)$$

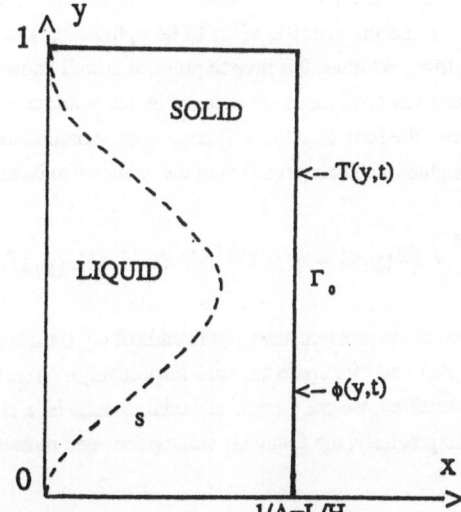

Figure 1

2.2 Definition of the reference

This study is limited to the identification of "smooth" interfaces: the time and space evolution of the interface to be identified, s(t,y), is:

$$s(y,t) = 0.16 \, (1-\exp(-t)) \, (1-\cos 2\pi y)/2 \qquad (2)$$

A preliminary resolution of the direct problem (Equation 1) is made on the time horizon Γ ($\Gamma = 1$), with the following boundary conditions:

. Face Γ_0 : $T(x=1/A, y, t) = -1$

. Horizontal boundaries (y=0 and y=1): adiabatic conditions

. Moving interface defined by Equation 2: Temperature = 0

By solving this direct problem, we get the flux on Γ_0 which will be considered as the "measured output" in the identification procedure. Ny measurements equally spaced will be used in the inverse problem, with a sampling time interval $\Delta t=0.01$.

3. RESOLUTION METHOD

3.1 Criterion

The identification of s(y,t) is obtained by minimisation of a criterion based on the distance between the output measurement ($\widetilde{\Phi}(x=1/A,y,t)$, flux measured on the back face Γ_0), and the model output (Φ (x=1/A,y,t) obtained by direct simulation with the following input: T(x=1/A,y,t) (Temperature measured on the back face Γ_0) and the estimate s(y,t) to be optimized. Due to the fact that a diffusive system behaves as a low pass filter, this inverse problem is well known to be an ill-posed problem (Hadamard, 1923), and the continuous dependence of the solution on the data must be restored by suitable assumptions. We have used the well-known regularisation method (Tikhonov et al, 1977): in the criterion, a weighted stabilizing function of the unknown to be identified is added to the output error function.

$$J(s) = \int \int \left(\phi(y,t) - \bar{\phi}(y,t) \right)^2 \, dy \, dt \; + \; \Omega(s(y,t)) \tag{3}$$

By space and time discretisation of the problem, the identification of s(y,t) leads to the identification of the vector $s_{i,j}$ (with subscripts i and j for space and time respectively). To reduce the dimension of the unknown vector to be identified, we use a sequential minimisation on a sliding time horizon, $\tau=(r-1) \Delta t$, with $\tau \ll \Gamma$. More precisely, the following assumptions and discretisations have been chosen:

-Space discretisation of the interface: because of the "smoothness" of the interface, a cubic spline interpolation is used to develop s in space. Ten splines have been used, as it can be shown that this approximation leads to a relative error less than 2.10^{-4} between the real interface and its approximation (This relative error has been estimated on the whole horizon).

-Time representation of the interface: On the minimisation horizon $\tau=(r-1) \Delta t$, the interface s is constant (Beck and Murio,1986).

With these assumptions, only ten parameters have to be identified at each step of the minimisation procedure: at time t=(m+1)Δt, the interface s$_{m+1,i}$ (1≤i≤10) is obtained by minimisation of the discretized version of J, defined on the interval [t=(m+1)Δt , t=(m+r)Δt]. The optimization process is then resumed on the next time interval [t=(m+2)Δt , t=(m+r+1)Δt].

$$Jd(s_{m+1}) = \frac{1}{r\ ny} \sum_{i=1}^{ny} \sum_{j=1}^{r} (\phi_{i,m+j} - \tilde{\phi}_{i,m+j})^2 + \alpha\Omega d(s) \tag{4}$$

The regularisation is obtained by penalisation of the time or/and space derivative of s(y,t). The discretized Ω d is given by:

$$\Omega da(s) = \sum_{k=1}^{ns} (s_{k,m+1} - s_{k,m})^2 \qquad \text{or} \qquad \Omega db(s) = \sum_{k=1}^{ns-1} (s_{k+1,m+1} - s_{k,m+1})^2$$

or

$$\Omega dc(s) = \sum_{k=1}^{ns-1} ((s_{k+1,m+1} - s_{k+1,m}) - (s_{k,m+1} - s_{k,m}))^2$$

The minimization of Jd is performed by the Gauss-Newton method.

3.2 Direct model

The identification algorithm requires, at each iteration, the resolution of the direct problem: the model output (Flux on the boundary Γ_ϕ) has to be evaluated from the temperature on the boundary Γ_0 and the estimation of the interface position to be optimized. The numerical scheme is the one proposed by Bénard et al. (1984). It is based on Patankar (1980) finite volume method. Since the solid domain is nonrectangular, a nonorthogonal coordinate transformation is used to map the irregular physical cavity into a fixed rectangular computational space (with H=1 and L=1/A). This transformation gives a transformed heat conduction equation with additional non diagonal terms, that can be solved, at each discretization time step by a usual ADI method. The direct model is characterized by the discretization parameters δt, δx and δy. Based on previous results (Afshari et al. 1989, Bénard et al. 1991, Wang et al.,1992), we have used the following values: δt=0.025, δx≈0.16/A, δy≈0.1.

4. ROLE OF THE PARAMETERS OF THE PROBLEM

In order to test the method accuracy, we first consider a case with no noisy data. The impact of ny, the number of measurements, is studied in section 4.1. In section 4.2, the role of the length of the identification horizon (τ= (r-1)Δt) is analysed.

616

4.1 Exact data - Role of ny

For practical reasons, the number of measurements cannot be very large. The aim of this first test is to evaluate the minimum number of captors required, in order to get enough information to solve the identification problem. This test was performed with A=6.25 and Ω d=Ω da. Figure 2 gives the results obtained for ny=4, 6 and 10 (t=0.8). These results show the great sensitivity of the identification to ny: Although the value of the criterion Jd is not very different in the three cases, the method converges to a wrong solution, as soon as ny becomes smaller than 6. On the contrary, for ny=10, the results show a good agreement between the exact and identified interface.

Because of the smoothness of the reference studied, this difficulty can be overcome by use of "fictitious" measurements: the number of outputs is extended to ny*=10, by spline interpolation from the ny real data. As shown on Figure 3, the results are greatly improved and the interface shape can be reconstructed with ny≫6.

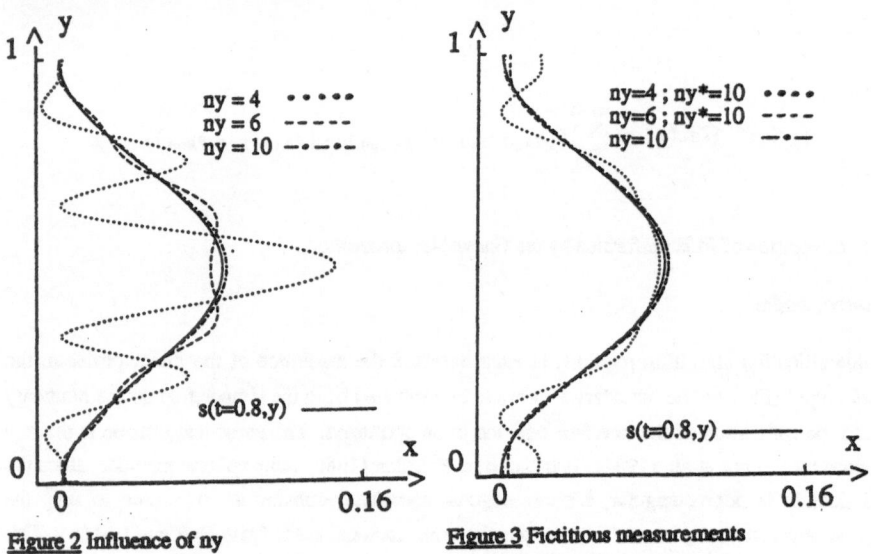

Figure 2 Influence of ny   Figure 3 Fictitious measurements

Reference interface: full line - Identified interface: dotted lines

4.2 Exact data - Role of the length of the horizon and regularisation parameter.

Because of the assumption "s is constant on τ=(r-1) Δt", which can introduce a bias on the identified solution, the impact of the combined influence of r and α must be carefully analysed. To figure out the quality of the identification, a relative distance is defined on the whole horizon Γ = 1:

$$ER = \sqrt{\frac{\sum_i \sum_j (s_{i,j} - s_{i,j}^{identified})^2}{\sum_i \sum_j s_{i,j}^2}}$$

The behavior of ER, as a function of α, is typical of regularisation method: when α increases starting from a small value, ER decreases thanks to the filtering effect of Ω (α,s). It increases again when the filtering becomes too important.

Some results are given in Table 1, for various r and α, with α varying around its optimal value. This test was performed for A=2.5, and ny=10. The real and identified interfaces are compared in Figures 4 and 5, for r=5 and α= 0.01 and 0.1 (Ω d = Ω dc). The observation of these results leads to the following conclusions:

- Because of the decrease in the aspect ratio (A=2.5 instead of 6.25), the measurements give more "filtered" information on the evolution of the moving boundary. Comparison of Figure 2 and 4 shows that the quality of the identification is not affected.

- The assumption "s is constant on τ" is valid in our test case.

- The solution is not very sensitive to the choice of r and α, and r can be reduced to 5 without any significant loss of precision, but with a significant reduction in the CPU time. Figure 5 illustrates the impact of the filtering due to a too strong regularisation, which leads to a small delay at the beginning of the melting process.

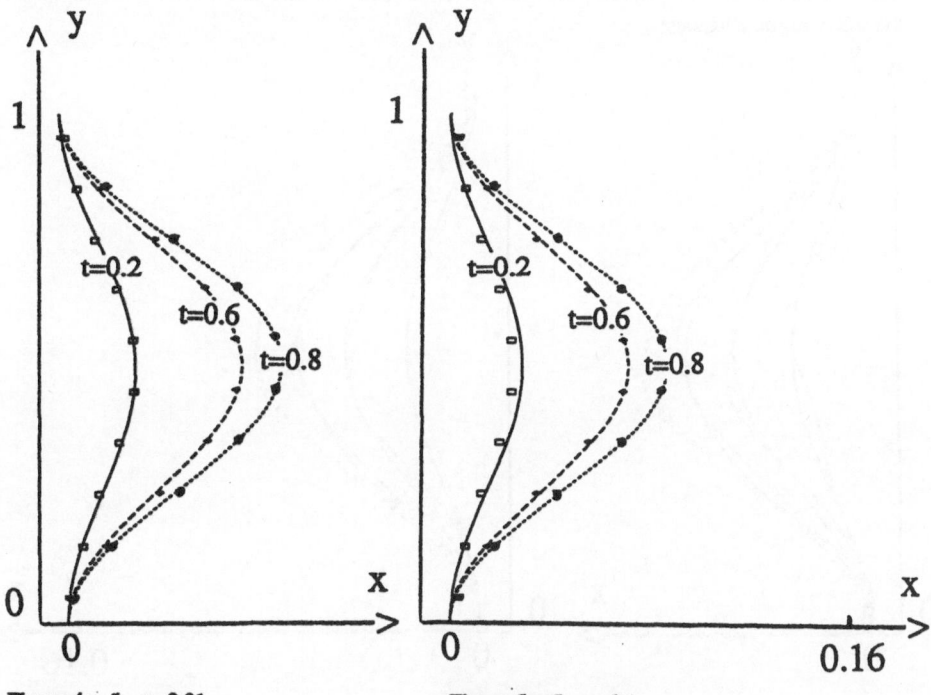

Figure 4 r=5 - α=0.01   Figure 5 r=5 - α=0.1

Reference interface: full line - Identified interface: discret poin ts

618

r / α	10	5	3
10^{-1}	$4\ 10^{-2}$	$6\ 10^{-2}$	10^{-1}
10^{-2}	$5\ 10^{-2}$	$5\ 10^{-2}$	$8\ 10^{-2}$
10^{-3}	$6\ 10^{-2}$	$7\ 10^{-2}$	10^{-1}
10^{-4}	10^{-1}	10^{-1}	10^{-1}

Table 1 Influence of r and α

4.3 Noisy data

Sections 4.1 and 4.2 have shown the accuracy of the identification method in the case of noiseless data. In this section, the method robustness in the case of noisy data will be checked. To simulate noisy measurements, a white gaussian noise B(t) is added to the ny measurements of flux and temperature on the face Γ_0. The noisy measurements, $\widetilde{\Phi}_b$ and $\widetilde{T}_b$, are defined as follows:

$$\widetilde{\Phi}_b(i{=}1 \text{ à ny},t) = \widetilde{\Phi}(i{=}1 \text{ à ny},t) + \sigma|\Phi(i,t)|_{max}\ B(t)$$

$$\widetilde{T}_b(i{=}1 \text{ à ny},t) = \widetilde{T}(i{=}1 \text{ à ny},t) + \sigma B(t)$$

Three noise levels have been considered: σ= 0.02, 0.05 and 0.1. The analysis of the results leads to the following conclusions:

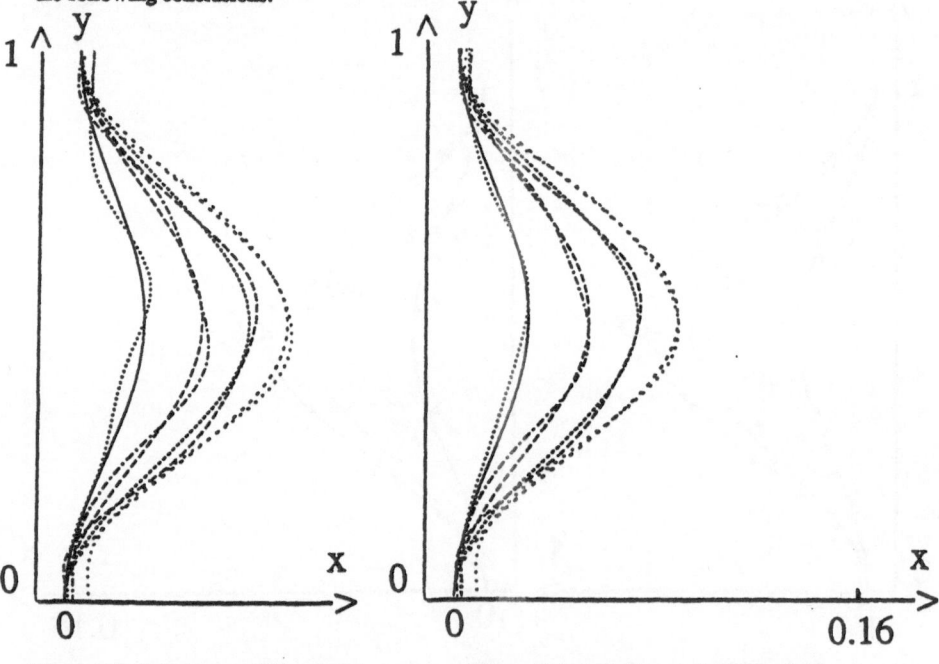

Figure 6 Ω d=Ω da - σ= 0.02 - α=0.2   Figure 7 Ω d=Ω db - σ= 0.02 - α=0.03

Comparison of reference and identified interface (t=0.2, 0.4, 0.6 and 0.8)

- Regularisation on the space or time derivative of s:

The regularisation was made either on the time derivative of s (Figure 6 for σ=0.02, Ω d=Ω da), or on the space derivative of s (Figure 7 for σ =0.02, Ω d=Ω db): as can be shown, the two regularization terms give quite similar results, as soon as α is set to its optimal value.

- Regularisation parameter

In the case of very noisy measurements, a larger value of α has of course to be used to stabilize the solution. Figures 8 and 9 give the results of the identification, for σ=0.05 and 0.1 and for Ω d = Ω dc. As long as σ is small enough, the regularity of the space profile of the unknown interface can be identified with a satisfactory precision by the appropriate choice of α. For σ =0.1, the amplification of the measurement noise is too sensitive and the interface cannot be correctly identified anymore.

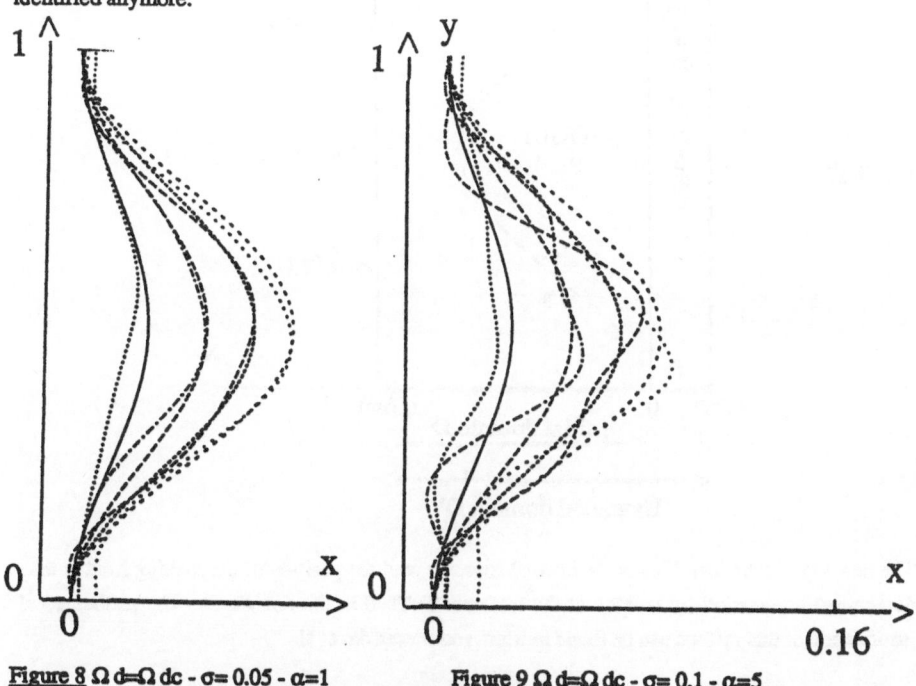

Figure 8 Ω d=Ω dc - σ= 0.05 - α=1   Figure 9 Ω d=Ω dc - σ= 0.1 - α=5

Comparison of reference and identified interface (t=0.2, 0.4, 0.6 and 0.8)

5. USE OF A LINEAR QUADRATIC CRITERION

To decrease the number of parameters si,j to be identified at each step of the identification procedure described above, we have chosen to use a sliding time horizon τ, with the assumption that s is constant on τ. In this section, we have tested the method presented in Blum (1989) and Mannikko and al. (1993). The initial non linear phase change problem is changed into a linear one in the following way: the initial domain D (rectangular enclosure with height 1 and width 1/A in our example) is extended to a new domain D', as shown in Figure 10.

620

The boundary condition on the extended boundary (Γ' in our example) is set to the melting temperature, Tf=0. The diffusion equation is considered in this new fixed domain, with introduction of an internal heat source in the extended part of the domain, $u(x_0 < x < 0, y, t)$.

$$\frac{\partial^2 T}{\partial x^2} + \frac{\partial^2 T}{\partial y^2} + u = \frac{\partial T}{\partial t} \tag{5}$$

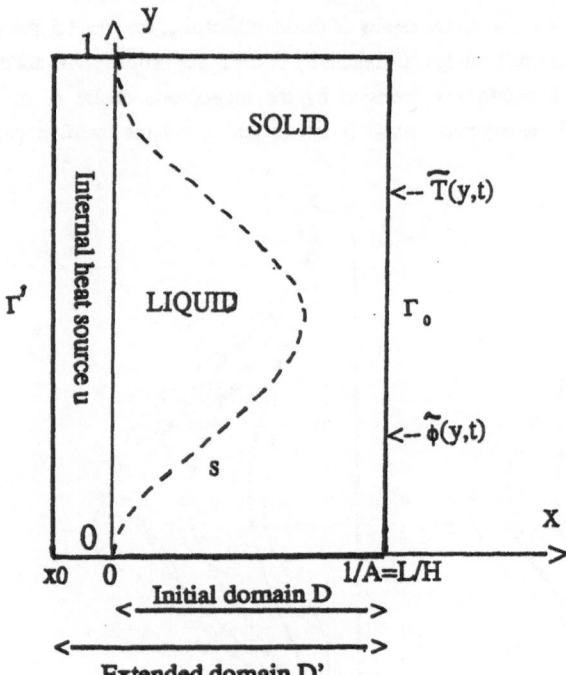

Figure 10

The new input to be identified is the internal source u, and the position of the melting front is then determined a posteriori by looking at the isotherm T=Tf=0 in the domain D. The mathematical justification of this method can be found in Blum and Mannikko et al.

As before, the identification of u is made by minimization of a regularised criterion:

$$J = \int \int (\phi(y,t) - \bar{\phi}(y,t))^2 dy dt + \Omega(u) \tag{6}$$

Because of the linearity of this new optimisation problem, the classical method of linear quadratic optimal control can be used. The minimization of J(u) was made on the whole horizon, by resolution of a non stationary Ricatti equation. A finite difference scheme is used to discretize Equation 5, with 11 nodes in the x direction (10 nodes in the domain D, and one node in the extended part of the domain), and 10 nodes in the y direction. Once the temperature is calculated on the nodes of domain D, the isotherm T=Tf=0 is obtained by linear interpolation.

Figure 11 gives the time evolution of the melting front for y=0.45, and Figure 12 shows the comparison between the real and identified interface profile, for t=0.2, 0.3 and 0.4, in the case of noiseless data.

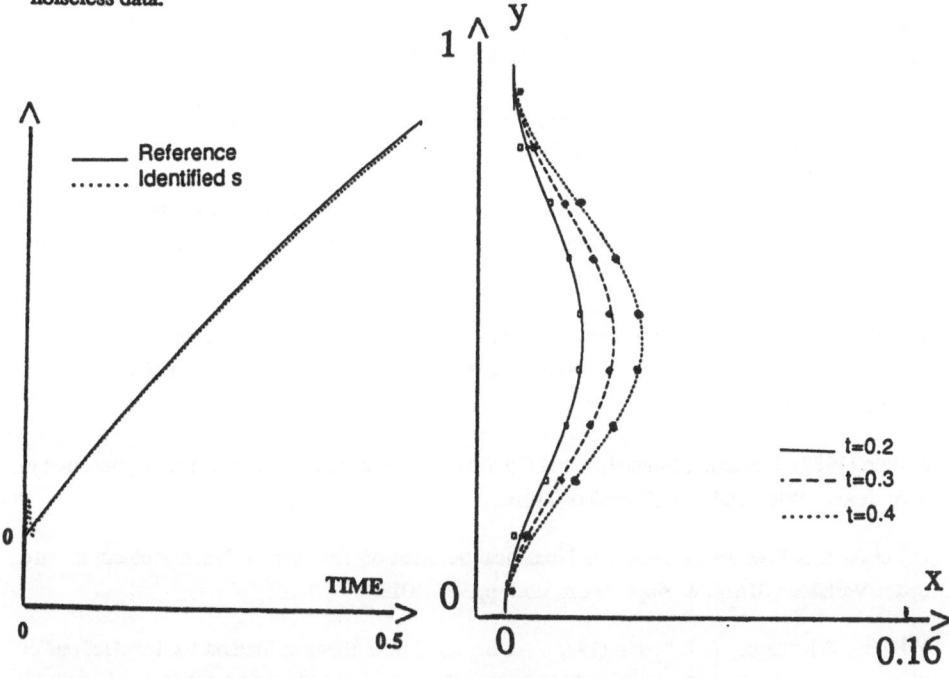

Figure 11 Time evolution of s(y=0.45)   **Figure 12** Interface profile

The good precision of the identified time evolution of s(y=0.45) can be seen on Figure 11, except at the very beginning of the identification (t«0.015). The spatial profile of the melting front is also well predicted, except at the vicinity of the two horizontal boundaries: this phenomena can be explained by the test reference chosen. Indeed, the interface is very close to zero in these areas, and a more refined space discretisation should be needed, in order to get a better approximation of the isotherm T=0.

6. CONCLUSION

Numerical resolution of a 2D inverse phase change problem has been proposed. Two different methods have been tested. The first one use a sequential minimisation procedure, on a sliding time horizon. The second one is based on the transformation of the initial non linear phase change problem in a linear one, which is solved by linear quadratic optimal control method. They both give satisfactory results in the case of a regular interface.

Comparison of these two methods for identification of various shapes of interfaces, and detection of local irregularity will be the next step of this work.

REFERENCES

A.Afshari, C.Bénard, C.Duhamel & B.Guerrier (1989), On-line Identification of the State of the Surface of a Material Undergoing Thermal Processing, Proc. 5th IFAC Symposium on Control of Distributed Parameter Systems, Perpignan (France), pp.459-463.

T.Banks & F.Kojima (1989), Boundary Shape Identification Problems in Two Dimensional Domains Related to Thermal Testing of Materials, Quart. Appl. Math., vol.47, pp.273-293.

C.Bénard, D.Gobin & A.Zanoli (1986), Moving Boundary Problem: Heat Conduction in the Solid Phase of a Phase Change Material during Melting Driven by Natural Convection in the Liquid, I.J.H.M.T., vol.29, 11, pp.1669-1681.

C.Bénard & A.Afshari (1991), Front Tracking for the Control of Solid-Liquid Phase Change Process, Proc. 7th Int. Conf. on Num. Methods in Thermal Problem, Standford, vol.7, 1, pp.186-198.

J.Blum (1989), Numerical Simulation and Optimal Control in Plasma Physics with Application to Tokamaks, J.Wiley and Sons, Gauthiers-Villars.

D.Colton & R.Reemtsen (1984), The Numerical Solution of The Inverse Stefan Problem in Two Space Variables, SIAM. J. Appl. Math., no.5, pp.996-1013.

Y.F.Hsu, B.Rubinsky & K.Mahin (1986), An Inverse Finite Element Method for the Analysis of Stationnary Arc Welding Processes, ASME J., Heat Transfer, vol.108, pp.734-741.

P.Jochum (1980), The Inverse Stefan Problem as a Problem of Nonlinear Approximation Theory, J. Approximation Theory, Vol.30, pp.81-98.

P.Jochum (1982), To The Numerical Solution of an Inverse Stefan Problem in Two Space Variables, Numerical Treatment of Free Boundary Value Problem, Ed Albrecht J. et al., ISNM 58, Birkhauser-Verlag, Basel.

M.A.Katz & B.Rubinsky (1984), An Inverse Finite Element Technique to Determine the Change of Phase Interface Location in One Dimensional Melting Problem, Num. Heat Transfer, vol.7, pp.269-283.

P.Knaber (1985), Contol of Stefan Problem by Means ol Linear-Quadratic Defect Minimization, Num. Math., vol.46, pp.429-442.

T.Mannikko, P.Neittaanmaki & D.Tiba, A Rapid Method for the Identification of the Free Boundary in Two-Phase Stefan Problems, To be published.

S.V.Patankar (1980), Numerical Heat Transfer and Fluid Flow, Hemisphere Publishing Corp., Mc Graw Hill.

R.Reemsten & A.Kirsch (1984), A Method for the Numerical Solution of the One Dimensional Inverse Stefan problem, Num. Math., vol.45, pp.253-273.

A.N.Tikhonov & V.Y.Arsenine (1977), Solutions of Ill-Posed Problems, V.H.Winston and Son.

X.Wang, M.M.Rosset-Louerat & C.Bénard (1992), Inverse Problem: Identification of a Melting Front in the 2D Case, Int. Series of Num. Math., vol.107, Birkhauser Verlag, Basel

N.Zabaras, Y.Ruan & O.Richmond (1992), Design of Two Dimensional Stefan Processes with Desired Freezing Front Motion, Num. Heat Transfer, vol.21, 3, pp.307-326.

Control theory – Controllability

A LAGRANGIAN ALGORITHM FOR STATE CONSTRAINED BOUNDARY CONTROL PROBLEMS

Maïtine Bergounioux
Département de Mathématiques et d'Informatique
U.F.R. Sciences - Université d'Orléans
B.P. 6779 - 45067 Orléans Cedex 2

Key Words : Optimal Control , Lagrange multipliers , Augmented Lagrangian , Qualification Condition

1 Introduction

In this work we investigate boundary control problems governed by linear elliptic (or parabolic) equations and involving constraints on both the control and the state . The case where the control is distributed has been studied in [1, 2] for the elliptic case and in [3, 4] for the parabolic case .

The aim of this paper is to present a Lagrangian Algorithm to solve such boundary problems . We limit ourselves to the stationnary case because the evolution problems may be treated in the same way .

Let Ω be an open , bounded subset of $\mathbb{R}^n$ ($n \leq 3$) with a C^∞ , n-1 dimensional boundary Γ , such that Ω is locally on the same side of Γ .
Let us consider the problem

$$\min \ J(y,v) \ = \{\frac{1}{2}\int_\Omega (y - z_d)^2 \ dx \ + \ \frac{M}{2}\int_\Gamma \ v^2 \ d\Gamma\} \qquad (\mathcal{P})$$

$$-\Delta y \ = \ f \ \text{in} \ \Omega \qquad (1.1)$$

$$y \ = \ v \ \text{on} \ \Gamma \qquad (1.2)$$

$$v \in U_{ad} \qquad (1.3)$$

$$y \in K \qquad (1.4)$$

where $f \in \mathbb{L}^2(\Omega)$, $v \in \mathbb{L}^2(\Gamma)$ and $M > 0$. U_{ad} and K are non empty, closed, convex subsets of respectively , $\mathbb{L}^2(\Gamma)$ and $\mathbb{L}^2(\Omega)$.

Remark 1.1 *We could also consider a Neumann boundary condition (see [1]) . The functional frame is a little different but the algorithm we get is not very far from the one we get for Dirichlet boundary conditions .*

The system (1.1) and (1.2) has a unique solution $y = T(v)$ which belongs to $\mathbf{W} = \{ y \in \mathbb{L}^2(\Omega) \mid \Delta y \in \mathbb{L}^2(\Omega) \}$. Problem $\mathcal{P}$ has a unique solution $(\bar{y}, \bar{v})$, because the cost functional J is coercive , strictly convex and the feasible domain is non empty, closed and convex .

We may , then , assuming a qualification condition , obtain decoupled optimality conditions. More precisely we have the following result :

Theorem 1.1 *Assume :*

$$\exists v_o \in U_{ad} , \; \exists \rho > 0 , \; \exists R > 0 \; such \; that$$
$$\forall \eta = (\chi, \xi) \in \mathcal{B}(\mathcal{E}_\Omega) \times \mathcal{B}(\mathcal{E}_\Gamma), \; \exists v_\eta \in \mathcal{B}^2(v_o, R) \cap U_{ad} \; such \; that \; y_\eta \in K$$
(1.5)

where

y_η *is the solution of* $\begin{cases} -\Delta y_\eta &= f - \rho\chi \quad in \; \Omega \\ y_\eta &= v_\eta - \rho\xi \quad on \; \Gamma \end{cases}$,

$\mathcal{E}_\Omega$ *and* $\mathcal{E}_\Gamma$ *are dense subsets of respectively* $\mathbb{L}^2(\Gamma)$ *and* $\mathbb{L}^2(\Omega)$,
$\mathcal{B}(\mathcal{E}_\Omega)$ *and* $\mathcal{B}(\mathcal{E}_\Gamma)$ *are the unit-balls of respectively* $\mathcal{E}_\Omega$ *and* $\mathcal{E}_\Gamma$,
and $\mathcal{B}^2(v_o, R)$ *is the* $\mathbb{L}^2(\Gamma)$- *ball centered in* v_o , *of radius R.*
Then , there exists $\bar{s}$ *in* $\mathcal{E}'_\Omega$ *and* $\bar{q}$ *in* $\mathcal{E}'_\Gamma$ *such that*

$$\forall y \in K^* \quad \int_\Gamma (-\frac{\partial \bar{p}}{\partial n} + \bar{q})(y - \bar{y}) \, d\Gamma \int_\Omega [\bar{s} + \bar{p}][-\Delta(y - \bar{y})] \, dx \; \geq 0 \quad (1.6)$$

$$\forall v \in U_{ad}^* \quad \int_\Gamma (M\bar{v} - \bar{q})(v - \bar{v}) \, dx \; \geq 0 \quad (1.7)$$

where
$\bar{p}$ *is the adjoint state defined by*

$$\begin{cases} -\Delta \bar{p} &= \bar{y} - z_d \quad in \; \Omega \\ \bar{p} &= 0 \quad\quad on \; \Gamma \end{cases} , \quad (1.8)$$

$K^* = \{y \in K \mid \Delta(y - \bar{y}) \in \mathcal{E}_\Omega \; and \; (y - \bar{y})_{|\Gamma} \in \mathcal{E}_\Gamma \}$, *and*
$U_{ad}^* = \{ v \in U_{ad} \mid v - \bar{v} \in \mathcal{E}_\Gamma \}$.

Proof .- See [1] . □

Remark 1.2 *We may consider two choices of the subspaces $\mathcal{E}_\Omega$ and $\mathcal{E}_\Gamma$ that lead to more or less strong optimality systems . More precisely :*
Case (i) : if $\mathcal{E}_\Omega = \mathbb{L}^2(\Omega)$ and $\mathcal{E}_\Gamma = \mathbb{L}^2(\Gamma)$ then $\bar{s} \in \mathbb{L}^2(\Omega)$, $\bar{q} \in \mathbb{L}^2(\Gamma)$, $K^* = K$ and $U_{ad}^* = U_{ad}$.
Case (ii) : if $\mathcal{E}_\Omega = \mathcal{C}^o(\Omega)$ and $\mathcal{E}_\Gamma = \mathcal{C}^o(\Gamma)$ then $\bar{s} \in \mathcal{M}(\Omega)$, $\bar{q} \in \mathcal{M}(\Gamma)$, where $\mathcal{M}(\Omega)$ and $\mathcal{M}(\Gamma)$ are Radon-measures spaces .

The relations (1.6) and (1.7) mean that $(\bar{y}, \bar{v})$ is a saddle point of the Lagrangian $\mathcal{L}$ of the problem, on the space $\mathcal{E}'_\Omega \times \mathcal{E}'_\Gamma$. So we may apply the Uzawa method to compute this saddle point as we have already done in the distributed case (see [3]).

2   The Lagrangian Algorithm

To simplify the presentation of the algorithm , we suppose from now that assumption (1.5) is ensured with $\mathcal{E}_\Omega = \mathbb{L}^2(\Omega)$ and $\mathcal{E}_\Gamma = \mathbb{L}^2(\Gamma)$ (case (i)).

Then we know that $(\bar{y}, \bar{v}) \in \mathcal{D}$ is the solution of $(\mathcal{P})$ if and only if there exists $\bar{s}$ in $\mathbb{L}^2(\Omega)$ and $\bar{q}$ in $\mathbb{L}^2(\Gamma)$ such that :

$$\forall y \in K \qquad \int_\Gamma (-\frac{\partial \bar{p}}{\partial n} + \bar{q})(y - \bar{y}) \, d\gamma + \int_\Omega (\bar{s} + \bar{p}) \, [-\Delta(y - \bar{y}) \, dx \geq 0 \, ,$$

$$\forall v \in U_{ad} \qquad \int_\Gamma (N\bar{v} - \bar{q})(v - \bar{v}) \, d\gamma \geq 0 \, .$$

These equations mean that $(\bar{y}, \bar{v}, \bar{s}, \bar{q})$ is the saddle point on $K \times U_{ad} \times \mathbb{L}^2(\Omega) \times \mathbb{L}^2(\Gamma)$ of the lagrangian defined as following :

$$\forall (y, v, s, q) \in \mathbf{W} \times \mathbb{L}^2(\Gamma) \times \mathbb{L}^2(\Omega) \times \mathbb{L}^2(\Gamma)$$

$$\mathcal{L}(y, v, s, q) = J(y, v) - \int_\Omega s \, (\Delta y + f) \, dx + \int_\Gamma q \, (y - v) \, d\gamma \, .$$

So we want apply the Uzawa method to compute this saddle point . Nevertheless it is better, for numerical reasons, to compute the saddle point of the augmented lagrangian defined by :

$$\forall (y, v, s, q) \in \mathbf{W} \times \mathbb{L}^2(\Gamma) \times \mathbb{L}^2(\Omega) \times \mathbb{L}^2(\Gamma)$$

$$\mathcal{L}_r(y, v, s, q) = \mathcal{L}(y, v, s, q) + \frac{r_1}{2} \int_\Omega (\Delta y + f)^2 \, dx + \frac{r_2}{2} \int_\Gamma (y - v)^2 d\gamma \ .$$

where $r = (r_1, r_2), r_1 > 0$ and $r_2 > 0$.

The problem is that the control and the state are coupled again via the boundary condition which appears in the augmentation term. So, as we have done it for the distributed case, we propose a block relaxation method coupled with the Uzawa method (see [5]). We get then, the following algorithm :

Algorithm (A)

Step 1. Initialization.

Let us give s_o in $\mathbb{L}^2(\Omega)$, q_o in $\mathbb{L}^2(\Gamma)$ and v_{-1} in $\mathbb{L}^2(\Gamma)$
(for example $s_o = 0$, $q_o = 0$ and $v_{-1} = 0$) .

Step 2. s_n , q_n and v_{n-1} being given , find $y_n \in K$ solution of

$$\min \ \{ \frac{1}{2} \int_\Omega (y - z_d)^2 dx + \int_\Omega s_n(-\Delta y) \, dx + \int_\Gamma q_n y d\Gamma$$
$$+ \frac{r_1}{2} \int_\Omega (\Delta y + f)^2 dx + \frac{r_2}{2} \int_\Gamma (y - v_{n-1})^2 \, d\Gamma \ , \ y \in K \}$$

also find $v_n \in U_{ad}$ solution of

$$\min \ \{ \frac{M}{2} \int_\Gamma v^2 \, d\Gamma - \int_\Gamma q_n v \, d\Gamma + \frac{r_2}{2} \int_\Gamma (y_n - v)^2 d\Gamma \ , \ v \in U_{ad} \} \ .$$

Step 3. If $-\Delta y_n - f = 0$ and $(y_n - v_n)_{|\Gamma} = 0$, then STOP , else go to Step 4.

Step 4. Set

$s_{n+1} = s_n + \rho_1(-\Delta y_n - f)$, where $\rho_1 \geq a_o > 0$
$q_{n+1} = q_n + \rho_2(y_n - v_n)$, where $\rho_2 \geq a_o > 0$
and go to Step 2.

$\square$

In this very case , where the multipliers are regular , we may ensure the convergence of algorithm (A).

Theorem 2.1 *Assume there exists a saddle point of $\mathcal{L}_r$ on $K \times U_{ad} \times \mathbb{L}^2(\Omega) \times \mathbb{L}^2(\Gamma) : (\bar{y}, \bar{v}, \bar{s}, \bar{q})$ and let be ρ_1 in $]0, r_1]$ and ρ_2 in $]0, r_2]$.*
Then :

1. *v_n converges to $\bar{v}$ strongly in $\mathbb{L}^2(\Gamma)$.*
2. *y_n converges to $\bar{y}$ strongly in $\mathbf{W}$.*
3. *s_{n+1} - s_n converges to 0 strongly in $\mathbb{L}^2(\Omega)$.*
4. *q_{n+1} - q_n converges to 0 strongly in $\mathbb{L}^2(\Gamma)$.*

$\square$

3 Numerical Results

Algorithm (A) has been tested for the boundary problem ($\mathcal{P}$) for the following choices of K and U_{ad} ;

$$K = \{\, y \in \mathbb{L}^2(\Omega) \mid \varphi(x) \le y(x) \le \psi(x) \ \text{ a.e. on } \Omega \,\} \cap \mathbf{W} \ ,$$

$$U_{ad} = \{\, v \in \mathbb{L}^2(\Gamma) \mid \alpha(x) \le y(x) \le \beta(x) \ \text{ a.e. on } \Gamma \,\} \ ,$$

where φ , ψ $\in \mathbb{L}^2(\Omega)$ and α , β $\in \mathbb{L}^2(\Gamma)$.

We present some results for the 2D-case :

- $\Omega = (]0, 1[\times]0, 1[) - ([\frac{1}{3}, \frac{2}{3}] \times [\frac{1}{3}, \frac{2}{3}])$, (Ω is a square with a hole in the middle).

- $f \equiv 1$, $z_d \equiv 1$, $U_{ad} = [\text{-10,10}]$, $\varphi \equiv -10$, and $\psi \equiv 10^3$.

3.1 Results

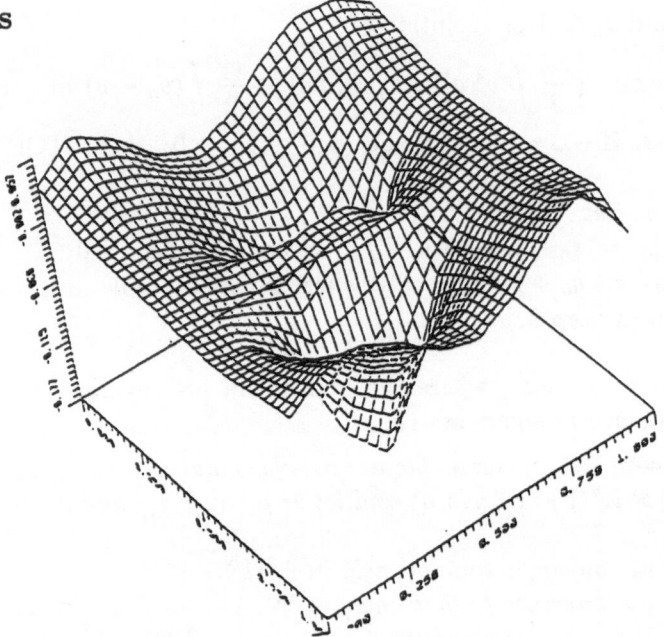

Fig 1. State

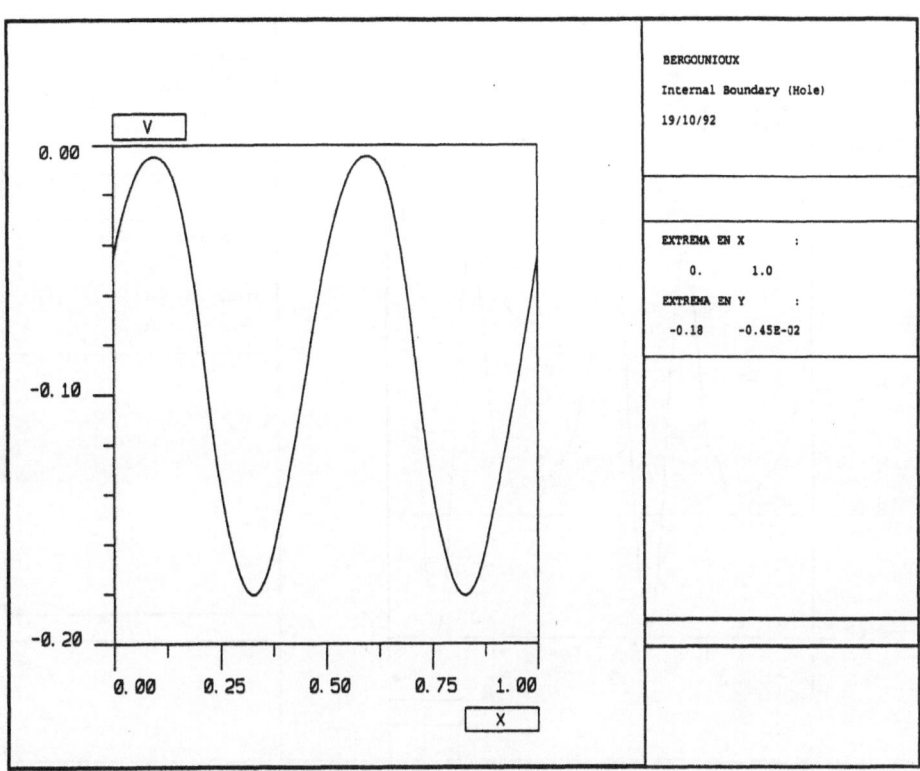

Fig 2. Control on internal boundary (hole)

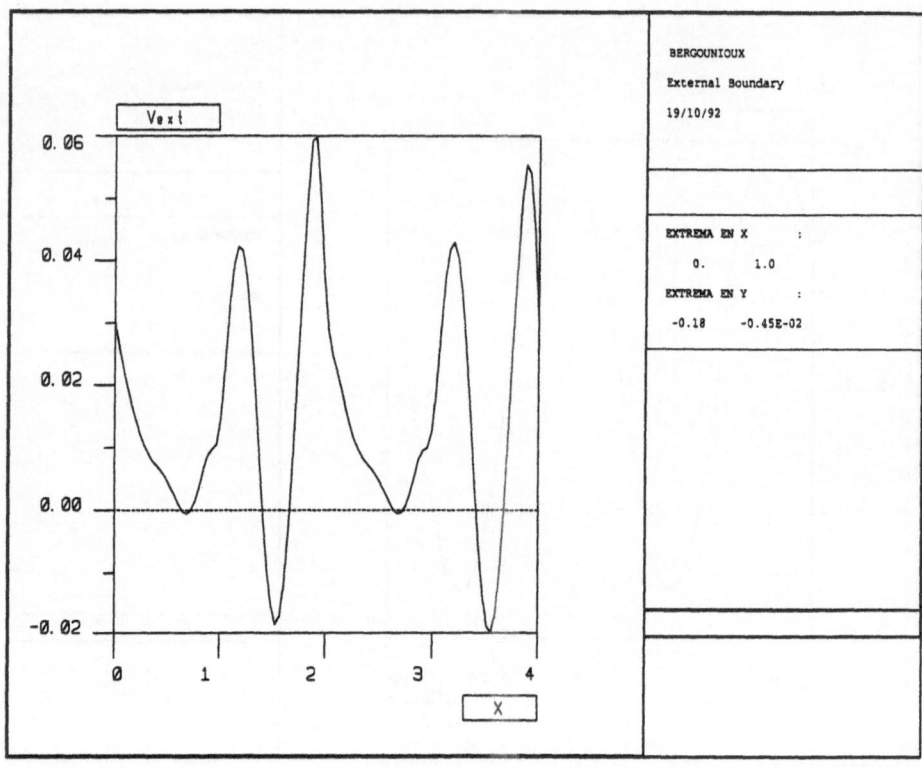

Fig 3. Control on external boundary

3.2 Convergence Rates

The best choice of parameters seems to be $r_1 = r_2 = 0.8$. We have got the result in 26 iterations for a precision of 1E-06. The number of iterations in the subroutine solving the two minimization problems of Step 2. of algorithm (A) has been limited to 5.

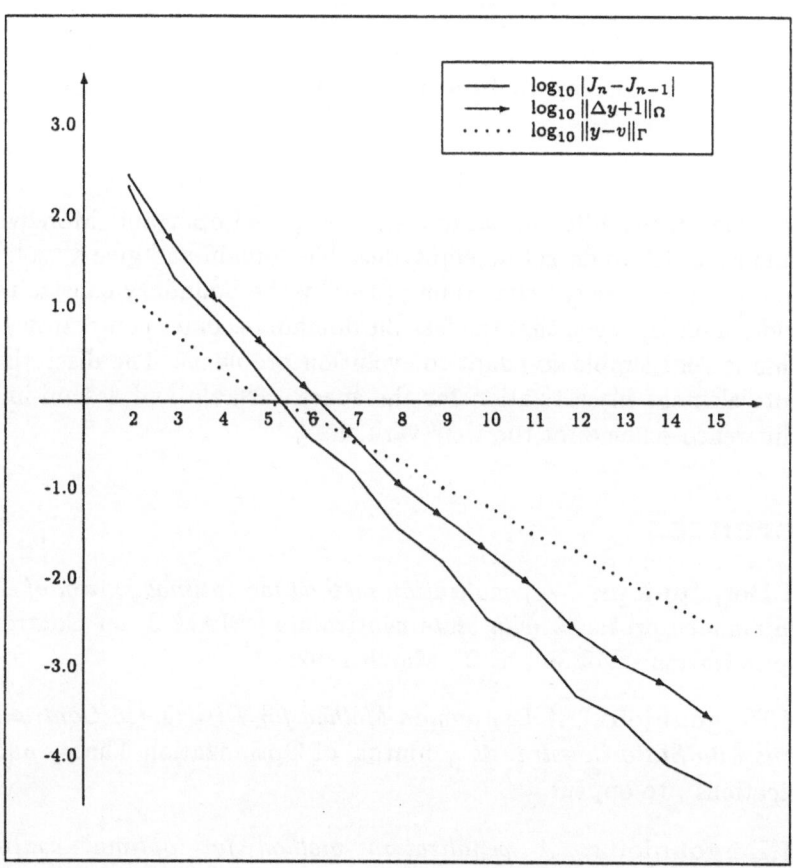

Fig 4. Convergence of the different errors

Number of iterations	Value of J	$\|\Delta y + 1\|_\Omega$	$\|y - v\|_\Gamma$
1	237.311	1761.65	22.4528
4	2.937	16.245	2.507
7	0.530634	0.45838	0.37011
10	0.461421	0.2033 E-01	0.6620 E-01
15	0.458106	0.3809 E-03	0.3622 E-02
20	0.458050	0.1974 E-04	0.2025 E-03

Table 1. Errors in function of the iteration number

The choice of the different parameters r_i is quite important. Moreover the algorithm is "able" to detect an empty feasible domain and give a "solution" satisfying the less "heavy" constraints (usually the boundary constraint and so provides a set U_{ad} such that the feasible domain becomes non-empty). The algorithm is very simple to adapt to evolution problems. The discretization is a finite element discretization for the space variable and a semi-implicit finite difference scheme for the time variable.

References

[1] **M.Bergounioux** , *A penalization method for optimal control of elliptic stationnary problems with state constraints* , SIAM J. on Control and Optimization - Vol 30 , n° 2 , March 1992

[2] **M.Bergounioux** , *A Lagrangian Method for Distributed Control Problems with State Constraints* , Journal of Optimization Theory and Applications , to appear

[3] **M.Bergounioux**, *A penalization method for optimal control of parabolic problems with state constraints* , Applied Mathematics and Optimization , to appear

[4] **M. Bergounioux - T. Männikkö - D. Tiba** , *On a non qualified optimal control problem* , Report n° 148 , University of Jyväskylä - June 1992

[5] **M.Fortin - R.Glowinski** , *Méthodes de lagrangien augmenté - Applications à la résolution de problèmes aux limites* , MMI , Dunod 1982

[6] **P. Ciarlet** *Basic Error Estimates for Elliptic Problems* Handbook of Numerical Analysis. Vol II. Finite Element Methods - Part 1. North-Holland 1991.

[7] **J.L. Lions** , *Contrôle optimal des systèmes gouvernés par des Equations aux Dérivées Partielles* , Dunod- Gauthier-Villars , Paris 1968 .

[8] **J.L. Lions** , *Contrôle des systèmes distribués singuliers* , Gauthier-Villars , Paris 1983

Duality for Nonconvex Optimal Control Problems Governed by Parabolic, Periodical Systems in Complex Hilbert Spaces. Necessary and Sufficient Optimality Conditions.

Andrzej Rogowski

Institute of Mathematics, Łódź University

ul. Stefana Banacha 22

PL 90-238 Łódź, Poland

1 Introduction.

In the paper we consider the following optimal control problem:
minimize

$$\tilde{J}(x, u) = \int_0^T \tilde{L}(t, x(t), u(t)) \, dt, \tag{1}$$

in $x : [0, T] \to Y$, $u : [0, T] \to U$ subject to

$$x'(t) + A(t)x(t) = B(t)u(t) \quad \text{for } t \in [0, T] \text{ a.e.,} \tag{2}$$
$$x(0) = x(T), \tag{3}$$
$$u(t) \in M \quad \text{for } t \in [0, T] \text{ a.e.,} \tag{4}$$

where $T > 0$, Y and U are separable Hilbert spaces, not necessarily real, $M \subset U$ is convex and closed, $A(t)$ is a family of linear, closed and densely defined operators on Y and $B(t) : M \to Y$ are continuous and compact for $t \in [0, T]$. The function $\tilde{L}(t, \cdot, \cdot)$ is concave-convex.

Problems of such type are widely investigated by Barbu and Precupanu in chap.4 of [B-P] (see also references therein), but only for $\tilde{L}$ convex in (x, u) and Y, U real. The main tool used in [B-P] is an approximation method.

In the paper presented here we explore the duality method for complex spaces, which gives information about solutions of problem (1)–(4) by investigation the dual control problem (the easier one in the considered case). We also state sufficient and necessary optimality conditions (the variational principle) and give the metod of approximation to solution of problem (1)–(4). Another attempt to that problem can be found in e.g. [M], [T].

To be precise, assume that we are given complex Hilbert spaces X, Y, U such that X is continuously and densely imbedded in Y. We identify Y with its own antidual Y^* and hence $X \subset Y \subset X^*$, where X^* is the antidual space to X and both inclusions are continuous and densely defined. We denote by $< v, x >_X$ the scalar product (anty-linear in x) for $x \in X$ and $v \in X^*$; if $v \in Y$ this is equal to the ordinary scalar product $< v, x >_Y$ in Y. The norm in X is denoted by $|\cdot|_X$. In the same way we denote norms and scalar products in other spaces. Let $W(0, T) = \{x \in L^2(0, T; X) : x' \in L^2(0, T; X^*)\}$, where x'

denotes the X^*-distributional derivative of x (see [L-M],[B-P] for the properties of such spaces).

If the integrand $\tilde{L}$ appearing in (1) need not be continuous in u then problem (1)–(4) can always be considered as a special case of the following one defined on the whole space U:

minimize

$$J(x,u) = \int_0^T L(t, x(t), u(t)) \, dt, \tag{5}$$

in $x : [0,T] \to Y$, $u : [0,T] \to U$ subject to

$$x'(t) + A(t)x(t) = B(t)u(t) \quad \text{for } t \in [0,T] \text{ a.e.,} \tag{6}$$
$$x(0) = x(T), \tag{7}$$

Here L can be taken as follows:

$$L(t,x,u) = \begin{cases} \tilde{L}(t,x,u) & : u \in M \\ +\infty & : \text{otherwise} \end{cases} \tag{8}$$

Throughout the paper we need the following assumptions:

(H1) $\tilde{L} : [0,T] \times X \times U \to R$ is an $\mathcal{L} \otimes \mathcal{B}_{X \times U}$ measurable function, concave and upper semicontinuous (u.s.c.) in the second and convex and lower semicontinuous (l.s.c.) in the last variable;

(H2) $\tilde{L}$ satisfies the growth conditions of the form

(a) $\tilde{L}(t,x,u) \geq -a|x|_X^2 + b|u|_U^r - d(t)$ for a.e. $t \in [0,T]$

(b) $\tilde{L}(t,x,v) \leq \varphi(t,u) - a_1|x|_X^2 + d_1(t)$ for $x \in X$, $t \in [0,T]$ a.e., where $\varphi(t,u)$ is $\mathcal{L} \otimes \mathcal{B}_U$ measurable, l.s.c. in u and such that $\varphi^*(t,u^*) \geq Re < u^*, \gamma(t) >_U + \eta(t)$ for some $\eta \in L^1(0,T)$, $\gamma \in L^{r_1}(0,T;U)$ with $r_1 \geq r \geq 2$, d, d_1–integrable functions,

(H3) $\{A(t) : 0 \leq t \leq T\}$ is a family of linear continuous operators from X into X^* satisfying

(a) for every $x \in X$, the functions $t \to A(t)x$ and $t \to A^*(t)x$ are X^*-measurable over $[0,T]$ and for some $c > 0$

$$|A(t)x|_{X^*} \leq c|x|_X \quad \text{for all } x \in X, \quad \text{a.e. } t \in [0,T], \tag{9}$$

(b) there exists $\alpha > 0$ such that

$$Re < A(t)x, x >_X \geq \alpha|x|_X^2 \quad \text{for all } x \in X, \quad \text{a.e. } t \in [0,T]; \tag{10}$$

(H4) $\{B(t) : 0 \leq t \leq T\}$ is a family of linear, continuous operators from U into Y such that

(a) the functions $t \to B(t)u$ and $t \to B^*(t)y$ are measurable for $u \in U$, $y \in Y$,

(b) $B^*(t) : Y \to U^*$ are compact and satisfy

$$\left|B^*(t)y\right|_{U^*} \leq \beta|y|_Y \quad \text{for all} \quad y \in Y, \quad \text{a.e.} \quad t \in [0,T]. \tag{11}$$

Let $W_{per}(0,T)$ denote the closed subspace of $W(0,T)$ consisting of all elements satisfying (3). We shall reformulate problem (5)–(7) into the equivalent form, more covenient for investigation:

(P) Minimize $J(x,u)$ over $(x,u) \in K(0,T)$ where

$$K(0,T) = \left\{(x,u) \in W_{per}(0,T) \times L^r(0,T;U) : (x,u) \text{ satisfies } (2)\right\}. \tag{12}$$

It is obvious, that under above assumptions $K(0,T)$ is a closed, linear subspace of $W_{per}(0,T) \times L^r(0,T;U)$.

Our first result concerns the duality principle. Define the dual problem

(P_D) Minimize

$$J_D(p) = \int_0^T L_D\left(t, B^*(t)p(t), -p'(t) + A^*(t)p(t)\right)dt$$

subject to $p \in W_{per}(0,T)$.

Here $L_D : [0,T] \times U^* \times X^* \to R$ is defined as the modified, partial Fenchel conjugate to some $G : [0,T] \times X^* \times U \to \bar{R}$ in the following way

$$-L_D(t, u^*, q) = \sup_{u \in U}\left(Re < u^*, u >_U - G(t, q, u)\right) \tag{13}$$

where

$$G(t, q, u) = \sup_{x \in X}\left(Re < q, x >_X + L(t, x, u)\right). \tag{14}$$

2 Duality.

Here we establish the duality principle for problems **(P)** and **(P_D)**.

After calculation, applying hypotheses (H1)–(H4), integration by parts in $W(0,T)$, properties of space $W_{per}(0,T)$ and of modified Fenchel conjugate we obtain

Theorem 1 (Duality principle) *The following relation holds*

$$\inf_{p \in W_{per}} J_D(p) = \inf_{(x,u) \in K} J(x,u).$$

Proof. First notice that the operators $W_{per} \ni x \mapsto x' + A(\cdot)x$ and $W_{per} \ni p \mapsto p' - A^*(\cdot)p$ are "onto" $L^2(0,T;X^*)$ [L-M]. Then using (13), (14) observe that the following chain of equalities takes place

$$\inf_{p \in W_{per}} J_D(p) = \inf_{p \in W_{per}} \int_0^T L_D\left(t, B^*(t)p(t), -p'(t) + A^*(t)p(t)\right)dt =$$

$$\inf_{p \in W_{per}} \inf_{u \in L^r(U)}\left(\int_0^T \left(-Re < B^*(t)p(t), u(t) >_U + G\left(t, -p'(t) + A^*(t)p(t), u(t)\right)\right)dt\right) =$$

$$\inf_{p \in W_{per}} \inf_{u \in L^r(U)} \left(\int_0^T \left(-Re < p(t), B(t)u(t) >_Y + G\left(t, -p'(t) + A^*(t)p(t), u(t)\right)\right) dt \right) =$$

$$\inf_{p \in W_{per}} \inf_{(x,u) \in K} \left(\int_0^T \left(-Re < x'(t) + A(t)x(t), p(t) >_X + \right.\right.$$

$$\left.\left. G\left(t, -p'(t) + A^*(t)p(t), u(t)\right)\right) dt \right) =$$

$$\inf_{(x,u) \in K} \inf_{p \in W_{per}} \left(\int_0^T \left(Re < p'(t) - A^*(t)p(t), x(t) >_X + \right.\right.$$

$$\left. G\left(t, -p'(t) + A^*(t)p(t), u(t)\right)\right) dt +$$

$$\left. Re < p(0), x(0) >_Y - Re < p(T), x(T) >_Y \right) =$$

$$\inf_{(x,u) \in K} \left(- \sup_{q \in L^2(X^*)} \int_0^T \left(-Re < q(t), x(t) >_X - G\left(t, q(t), u(t)\right)\right) dt \right) =$$

$$\inf_{(x,u) \in K} \int_0^T L\left(t, x(t), u(t)\right) dt = \inf_{(x,u) \in K} J(x,u).$$

$\square$

It is worth to notice that we apply perturbation with respect to the second variable of L (comp.(13), (14)).

3 Existence of the minimum and variational principle.

Let $a' = \frac{1}{4a}$, $b' = \frac{1}{r'(rb)^{\frac{1}{r-1}}}$, $a'' = a'\alpha$, $b'' = b'\beta$, $\tilde{a} = a''$, $\tilde{b} = b''\theta^{-1}$, where $|x|_Y \le \theta |x|_X$ for $x \in X$.

The growth conditions (H2) together with (H3) and (H4) imply that J_D is bounded below in $W_{per}(0,T)$. Precisely, the following theorem takes place:

Theorem 2 *If $r > 2$ or $\tilde{a} > \tilde{b}$ in the case of $r = 2$, the there exists $\bar{p} \in W_{per}(0,T)$ solving problem* (P_D).

Proof. Elementary calculus using (H2a) and above notations shows that

$$L_D(t, u^*, q) \ge -b'|u^*|_{U^*}^{r'} + a'|q|_{X^*}^2 - d(t),$$

By the uniform coercivity of $A(t)$ (see (10)), (H4b) and continuouity of the imbedding X into Y we see that

$$\begin{aligned} J_D(p) &\ge \int_0^T \left(-b''|p(t)|_Y^{r'} + a''|p(t)|_X^2 - d(t) \right) dt \ge \\ &\quad -\tilde{b}\|p\|_{L^2(0,T;X)}^{r'} + \tilde{a}\|p\|_{L^2(0,T;X)}^2 - d_0 \ge \\ &\quad Q\|p\|_{L^2(0,T;X)}^2 - d_0. \end{aligned}$$

for some $Q > 0$; here $1 < r' \leq 2$, $\frac{1}{r'} + \frac{1}{r} = 1$. Therefore J_D is bounded below. Moreover, by (H3a), continuity of the operator $p \to p' - A^*(\cdot)p$ (see [L-M]) from $W_{per}(0, T)$ (with the topology induced from $L^2(0, T; X)$) into $L^2(0, T; X^*)$ we see that

$$\|p'\|^2_{L^2(0,T;X^*)} \leq \|A^*(\cdot)p\|^2_{L^2(0,T;X^*)} + \|p' - A^*(\cdot)p\|^2_{L^2(0,T;X^*)} \leq c_1 \|p\|^2_{L^2(0,T;X)}$$

Hence the minimizing sequence for J_D exists and is bounded in $W_{per}(0, T)$ and therefore we can extract a subsequence $\{p_n\}$ weakly convergent to some $\bar{p} \in W_{per}(0, T)$. By the properties of The space $W(0, T)$ (see [L–M]) we check that

$$|p_n(0) - \bar{p}(0)|_Y \leq D \left(\|p_n - \bar{p}\|^2_{L^2(0,T;X)} + \|p'_n - \bar{p}'\|^2_{L^2(0,T;X^*)} \right) \leq E_1.$$

This implies existence of a subsequence denoted again $\{p_n\}$ such that $p_n(0) \to \bar{p}(0)$ weakly in Y. Moreover

$$\begin{aligned}
|p_n(t) - \bar{p}(t)|^2_Y &= |p_n(0) - \bar{p}(0)|^2_Y + \int_0^T \frac{d}{dt} < p_n(\tau) - \bar{p}(\tau), p_n(\tau) - \bar{p}(\tau) >_Y d\tau \\
&= |p_n(0) - \bar{p}(0)|^2_Y + 2 \int_0^T Re< \frac{d}{dt}(p_n(\tau) - \bar{p}(\tau)), p_n(\tau) - \bar{p}(\tau) >_X d\tau \\
&\leq E_1^2 + 2 \left(\|p_n - \bar{p}\|_{L^2(0,T;X)} \|p'_n - \bar{p}'\|_{L^2(0,T;X^*)} \right) \leq E.
\end{aligned}$$

By the absolute continuity of p_n and $\bar{p}$ in the space of absolutely continuous functions in X^* we see that

$$\begin{aligned}
< p_n(t) - \bar{p}(t), x >_X &= < p_n(0) - \bar{p}(0), x >_X + \int_0^T < p'_n(\tau) - \bar{p}'(\tau), x >X d\tau \\
&= < p_n(0) - \bar{p}(0), x >_Y + \int_0^T < p'_n(\tau) - \bar{p}'(\tau), x >X d\tau \to 0.
\end{aligned}$$

Therefore

$$\begin{aligned}
|< p_n(t) - \bar{p}(t), y >_Y| &\leq |< p_n(t) - \bar{p}(t), x >_X + < p_n(t) - \bar{p}(t), y - x >_Y \\
&\leq |< p_n(t) - \bar{p}(t), x >_X| + E|y - x|_Y \to E\varepsilon
\end{aligned}$$

for some $x \in X$. By the arbitrariness of y and $\varepsilon > 0$ this implies that $p_n(t) \to \bar{p}(t)$ weakly in Y for $t \in [0, T]$. By (H4b) $B^*(t)p_n(t) \to B^*(t)\bar{p}(t)$ strongly in U^* and therefore, by growth conditions and Fatou Lemma, $J_D(\bar{p}) = \inf_{p \in W_{per}} J_D(p)$. $\square$

The next theorem gives us the variational principle.

Theorem 3 *If $\bar{p} \in W_{per}(0, T)$ solves (P$_D$) and $\partial G = \partial qG \times \partial_u G$ then there exists $(\bar{x}, \bar{u}) \in K$ solving problem (P) and satisfying inclusions:*

$$B^*(t)p(t) \in \partial_u L(t, x(t), u(t)) \text{ for a.e. } t \in [0, T] \tag{15}$$

$$-p'(t) + A^*(t)p(t) \in \partial_x(-L)(t, x(t), u(t)) \text{ for a.e. } t \in [0, T] \tag{16}$$

where ∂_u, ∂_x denote partial subdifferentials in the sense of convex analysis.

Proof. Denote by J_{Dp} the convex functional

$$J_{Dp}(u^*) = \int_0^T -L_D(t, B^*p(t) + u^*(t), -p'(t) + A^*(t)p(t))dt \tag{17}$$

defined on $L^{r'}(0,T;U^*)$. It is easily seen, by the growth conditions (H2) that J_{D_p} is continuous on $L^{r'}(0,T;U^*)$. Therefore $\partial J_{D_p}(0)$ is nonempty and there exists $\bar{u} \in L^r(0,T;U)$ such that $\bar{u} \in \partial J_{D_p}(0)$ This means that

$$\int_0^T G(t, -\bar{p}'(t) + A^*(t)\bar{p}(t), \bar{u})dt - \int_0^T L_D(t, B^*\bar{p}(t), -p'(t) + A^*(t)p(t))dt$$

$$= \int_0^T Re < B^*(t)\bar{p}(t), \bar{u}(t) >_U.$$

Since for every $u \in L^r(0,T;U)$ there exists $x \in W_{per}(0,T)$ solving (2) hence, by (14), (13) and integration by parts in $W(0,T)$

$$\int_0^T L_D(t, B^*\bar{p}(t), -\bar{p}'(t) + A^*(t)\bar{p}(t))dt =$$

$$= \int_0^T G(t, -\bar{p}'(t) + A^*(t)\bar{p}(t), \bar{u}(t))dt - \int_0^T Re < \bar{p}(t), \bar{x}'(t) + A(t)\bar{x}(t) >_X dt$$

$$= \int_0^T G(t, -\bar{p}'(t) + A^*(t)\bar{p}(t), \bar{u}(t))dt - \int_0^T Re < -\bar{p}'(t) + A^*(t)\bar{p}(t), \bar{x}(t) >_X dt$$

$$\geq J(\bar{x}, \bar{u}) \geq \inf_{(x,u)\in K} J(x, u)$$

Taking into account the Duality principle we see that all above inequalities are actually equalities. Hence $(\bar{x}, \bar{u}) \in K$ is a minimizing pair for J. Moreover, this equalities together with (13), (14), imply that $\bar{p}$ and $(\bar{x}, \bar{u})$ satisfy the following inclusions

$$B^*(t)\bar{p}(t) \in \partial_u G(t, -\bar{p}'(t) + A^*(t)\bar{p}(t), \bar{u}(t)) \tag{18}$$

$$-p'(t) + A^*(t)p(t) \in \partial_q G(t, -\bar{p}'(t) + A^*(t)\bar{p}(t), \bar{u}(t)) \text{ for a.e. } t \in [0,T] \tag{19}$$

If $\partial G = \partial q G \times \partial_u G$ then the above inclusions are equivalent to (15) and (16) (comp. [A–E]). This ends the proof. □

Remark. Condition that $\partial G = \partial_q G \times \partial_u G$ is satisfied e.g. in the following cases:

(i) L is srtictly concave in x and strictly convex in u;

or

(ii) $G = \tilde{G} + \chi_D$, where χ_D is the characteristic function of some set $D = D_1 \times D_2$, with D_1, D_2 being closed and convex , $D_1 \subset X^*$, $D_2 \subset U$. □

Now, it is easy to see, that combining Theorems 1–3 we can state

Theorem 4 *There exist $(\bar{x}, \bar{u}) \in K$ and $\bar{p} \in W_{per}(0,T)$ solving problems (P) and (P_D) respectively, and satisfying (2), (3), (15), (16).* □

The last theorem estimates the duality gap for a minimizing sequence.

Theorem 5 *Let* $\{p_n\}$, $n \in N$ *be a minimizing sequence for* J_D *in* $W_{per}(0,T)$. *Then, there exists a sequence* $(x_n, u_n) \in K$ *such that*

$$u_n(t) \in \partial_{u^*}(-L_D)(t, B^*(t)p_n(t), -p_n' + A^*(t)p_n(t)) \tag{20}$$
$$x_n'(t) + A(t)x_n(t) = B(t)u_n(t) \tag{21}$$

minimizing J *over* K. *If* $\alpha = \inf_{(x,u) \in K} J(x,u)$ *then*

$$\alpha \le J(x_n, u_n) \le J_D(p_n) \quad \text{for } n \in N. \tag{22}$$

Moreover, any sequence satisfying (20), (21) satisfies (22).

Proof. Let $\{p_n\}$ be a minimizing sequence for J_D in $W_{per}(0,T)$. Arguing as in the proof of Theorem 3 there exists $u_n \in L^r(0,T;U)$ satisfying inclusion $u_n \in \partial J_{Dp_n}(0)$, $x_n \in W_{per}(0,T)$ solving the equation $x_n'(t) + A(t)x_n(t) = B(t)u_n(t)$ and such that $J(x_n, u_n) \le J_{Dp_n}(t)$. $\qquad\square$

4 Example.

Consider the following nonconvex control problem:
    minimize

$$\int_0^T \int_\Omega g(t, x(t,z), \nabla x(t,z), u(t)) dz dt \tag{23}$$

in $x \in L^2(0,T; H_0^1(\Omega))$, $u \in L^2(0,T; R^N)$, $\Omega \subset R^n$
subject to the constraints

$$x_t - \Delta x = \sum_{j=1}^N u_j(t)\chi_j(t,z) \text{ on } Q = [0,T] \times \Omega, \tag{24}$$

$$x(t,z) = 0 \text{ on } \Sigma = \partial\Omega, \ t \in [0,T] \text{ a.e.,} \tag{25}$$

$$x(0,z) = x(T,z) \text{ on } \Omega, \tag{26}$$

$$|u_j(t)| \le 1 \text{ for } j = 1, \ldots, N, \ t \in [0,T] \text{ a.e.} \tag{27}$$

Here $g(t, x, y, u)$ is a Caratheodory function on $R^{1+1+n+N}$ concave in (x,y), convex in u, $\{\chi_j(t)\}_{j=1}^N$ are the characteristic functions of a family of disjoint measurable subsets $\Omega_j(t)$ which cover the domain Ω for $t \in [0,T]$ a.e.

This problem can be written as a problem of type (P), where $X = H_0^1(\Omega)$, $Y = L^2(\Omega)$, $U = R^N$ and consequently $X^* = H^{-1}(\Omega)$, $U^* = R^N$. $A(t) : H_0^1(\Omega) \to H^{-1}(\Omega)$, $B(t) : R^N \to L^2(\Omega)$, $L : [0,T] \times H_0^1(\Omega) \times R^N \to R$ are defined as follows: $A(t) = -\Delta$, $(B(t)u)(z) = \sum_{j=1}^N u_j\chi_j(t,z)$, $z \in \Omega$, $t \in [0,T]$ a.e.,

$$L(t,x,u) = \begin{cases} \int_\Omega g(t, x(z), \nabla x(z), u) dz & : |u| \le 1 \\ +\infty & : \text{otherwise,} \end{cases}$$

is concave in x and convex in u. If additionally there exist $a, a_1, b \in R$, $r \ge 2$, and integrable functions d, e such that $g(t, x, y, u) \ge -a\left(|x|^2 + |y|^2\right) + b|u|^r - d(t)$ and for every $|u| \leq 1$ $g(t,x,y,u) \leq -a_1 \left(|x|^2 + |y|^2\right) + e(t)$ then all assumptions (H1)–(H4) are satisfied and therefore

there exist $(\bar{x}, \bar{u}) \in K(0,T)$ (comp.(12)) and $\bar{p} \in W_{per}(0,T)$ being solutions to problems (23)–(27) and (P$_D$) satisfying

$$p_t + \Delta p \in -\partial_x(-L)(t, x(t), u(t)),$$
$$p(t,z)\chi_j(t,z) - \lambda u(t) \in \partial_{u_j} g(t, x(t,z), \nabla x(t,z), u(t)) \text{ a.e. on } Q,$$
$$\lambda(1 - |u|) = 0,$$
$$p(t,z) = 0 \text{ on} \Sigma,$$
$$p(0,z) = p(T,z) \text{ for } z \in \Omega.$$

$\square$

REFERENCES

[A-E] J.P.Aubin and I.Ekeland, Applied Nonlinear Analysis, John Wiley & Sons, New York (1984).

[B-P] V.Barbu and T.Precupanu, Covexity and Optimization in Banach Spaces, D.Reidel Publ., Dordrecht (1986).

[L-M] J.-L.Lions and E.Magenes, Problèmes aux limites non homogènes et applications, Dunod, Paris (1968).

[M] K.Malanowski, *Second-Order Conditions and Constraint Qualifications in stability and Sensivity Analysis of Solutions to Optimization Problems in Hilbert Spaces,* Appl. Math. Optim. **25** (1992), 51–79.

[T] F.Tröltzsch, *Approximation of Non-Linear Parabolic Boundary Control Problems by the Fourier Method—Convergence of Optimal Controls,* Optimization **22** (1991), 83–98.

EXACT CONTROLLABILITY OF THE WAVE EQUATION IN A POLYGONAL DOMAIN WITH CRACKS BY ACTING ON A NEIGHBOURHOOD OF THE BOUNDARY

Mary Teuw NIANE

Université de Saint-Louis, BP 234, Saint-Louis, SENEGAL

Ousmane SECK

Université de Dakar, ENSUT, BP 5085, Dakar, SENEGAL

Abstract: We prove the exact controllability of the wave equation in a polygonal domain with cracks by acting on a neighbourhood of the boundary.

I. NOTATIONS

Let Ω be a bounded polygonal domain in $\mathbf{R}^2$ whose boundary $\partial\Omega$ is the union of the edges Γ_j, $j = 0,..,N$. We denote by s_j the vertex between Γ_{j-1} and Γ_j for $1 \leq j \leq N$ and by s_0 the vertex between Γ_N and Γ_0. We also denote by ω_j (resp ω_0) the measure of the internal angle between Γ_{j-1} and Γ_j for $1 \leq j \leq N$ (resp between Γ_N and Γ_0).

Let $S = \left\{ j \in \left\{ 0,..,N \right\} / \omega_j = 2\pi \right\}$. S is the set of cracks vertices and we assume that S is non empty.

We denote by ν_j the outward normal vector field on Γ_j, by τ_j the tangent unit vector field on Γ_j pointing toward s_j and by (r_j, θ_j) the locar polar coordinates at s_j.

Let ε, δ, γ be three real numbers such that $0 < \varepsilon < \delta < \gamma$, let x_0 be any point of $\mathbf{R}^2$ and $T > 0$. We set

$$\Gamma_\delta = \partial\Omega \setminus \bigcup_{j \in S} \left(\Gamma \cap B(s_j, \delta) \right), \qquad \Gamma_{\delta\varepsilon}^j = \partial\Omega \cap \left(B(s_j, \delta) \setminus B(s_j, \varepsilon) \right)$$

$$\Gamma(x_0) = \bigcup_{0 \leq j \leq N} \left\{ x \in \Gamma_j, (x - x_0) \cdot \nu_j > 0 \right\}, \quad \Gamma_\delta(x_0) = \Gamma_\delta \cap \Gamma(x_0)$$

$$\Gamma^* = \Gamma_\delta(x_0) \cup \left\{ s_j, \ j \in S \right\}, \qquad O_1 = \bigcup_{x \in \Gamma^*} B(x, \gamma)$$

$$Q = \left]0, T\right[\times \Omega, \quad \Sigma = \left]0, T\right[\times \partial\Omega, \quad R(x_0) = \max \left\{ \|x - x_0\|, \ x \in \overline{\Omega} \right\}$$

II. STATEMENT OF THE PROBLEM

In this paper, we study the exact controllability of the wave equation by acting on a neighbourhood of the boundary. The problem consists of searching for a function v with its support in a neighbourhood of the boundary such that if y is the solution of the problem

$$\begin{cases} y'' - \Delta y = v \text{ in } Q \\ \gamma y = 0 \text{ on } \Sigma \\ y(0) = y_0 \ , \ y'(0) = y_1 \text{ in } \Omega \end{cases} \tag{2.1}$$

then $y(T) = y'(T) = 0$ at time T.

Such a result has been established by E.Zuazua [1] from the H.U.M method of J.L.Lions [2]. The main step of his proof is that if Γ_0 is a part of the boundary and if there exists a positive real C such that for all $(\phi_0, \phi_1) \in H_0^1(\Omega) \times L^2(\Omega)$, if ϕ is the unique solution of the homogeneous wave equation

$$\begin{cases} \phi'' - \Delta\phi = 0 \text{ in } Q \\ \gamma\phi = 0 \text{ on } \Sigma \\ \phi(0) = \phi_0 \ , \ \phi'(0) = \phi_1 \text{ in } \Omega \end{cases} \tag{2.2}$$

such that the inequality

$$\int_0^T \int_{\Gamma_0} \left(\frac{\partial\phi}{\partial\nu} \right)^2 d\sigma dt \geq C \, E_{0,1}(\phi_0, \phi_1) \tag{2.3}$$

where $E_{0,1}(\phi_0, \phi_1)$ is the energy defined by

$$E_{0,1}(\phi_0, \phi_1) = \frac{1}{2} \int_\Omega \left(|\nabla\phi_0|^2 + |\phi_1|^2 \right) dx \tag{2.4}$$

is verified, then for all neighbourhood 0 of Γ_0 in Ω, there exists $c_1 > 0$ such that

$$\int_0^T \int_0 \left(|\phi|^2 + |\phi'|^2 \right) dxdt \geq c_1 E_{0,1}(\phi_0, \phi_1) \tag{2.5}$$

When the domain Ω has cracks, $\frac{\partial\phi}{\partial\nu}$ may not be square integrable near a crack tip. Then the calculations that had allowed to obtain (2.3) are no longer valid. We avoid those difficulties by establishing an integration by parts formula which permits us to establish an inequality like (2.5). By combining this inequality and a compactness argument, we obtain the equivalence of the norm of $\chi_0\phi$ in $L^2(Q)$ and the norm of $L^2(\Omega) \times H^{-1}(\Omega)$. Therefore, the standard programme of the H.U.M method allows us to get the exact controllability on a neighbourhood of the boundary with controls in L^2.

III. INTEGRATION BY PARTS FORMULA

Let $f \in L^2(\Omega)$ and $u \in H_0^1(\Omega)$ be the unique solution of the Dirichlet problem

$$\begin{cases} -\Delta u = f \\ \gamma u = 0 \end{cases} \tag{3.1}$$

According to P.Grisvard [3], [4], u is given by

$$u = u_R + \sum_{j \notin S} c_j \sqrt{r_j} \sin\left(\frac{\theta_j}{2}\right) \eta_j \tag{3.2}$$

where $u_R \in H^{3/2+\alpha}(\Omega) \cap H_0^1(\Omega)$ is the regular part of u (α may be small) , c_j is a real number and η_j is a cut-off function with its support in a neighbourhood of s_j.

We have the following integration by parts formula [5]

Theorem 1

For all $f \in L^2(\Omega)$ and $u \in H_0^1(\Omega)$ the unique solution of (3.1) given by (3.2), we have

$$\int_\Omega -\Delta u \ (x-x_0).\nabla u \ dx \ = \ \frac{\pi}{4} \sum_{j \in S} (s_j - x_0).\tau_j \ c_j^2$$

(3.3)

$$- \frac{1}{2} \int_{\Gamma_\delta} (x-x_0).\nu \left(\frac{\partial u}{\partial \nu}\right)^2 d\sigma - \frac{1}{2} \sum_{j \in S} \lim_{\varepsilon \to 0} \int_{\Gamma_{\delta\varepsilon}^j} (x-x_0).\nu_j \left(\frac{\partial u}{\partial \nu_j}\right)^2 d\sigma$$

Proof

Let $0 < \varepsilon < \delta$. We perform the integrations by parts in the domains Ω_ε which do not include the opened disks $B(s_j,\varepsilon)$ and where u is regular.

Let $\Omega_\varepsilon = \Omega \setminus \bigcup_{j \in S} B(s_j,\varepsilon)$ and $\gamma_\varepsilon^j = \Omega \cap \partial B(s_j,\varepsilon)$.

Then we have $\partial\Omega = \Gamma_\delta \cup \bigcup_{j \in S} \Gamma_{\delta\varepsilon}^j \cup \bigcup_{j \in S} \gamma_\varepsilon^j$

By applying the Green formula, we obtain

$$\int_{\Omega_\varepsilon} -\Delta u \ (x-x_0).\nabla u \ dx = - \frac{1}{2} \int_{\Gamma_\delta} (x-x_0).\nu \left(\frac{\partial u}{\partial \nu}\right)^2 d\sigma + I_1 + I_2 \quad \text{where}$$

$$I_1 = - \frac{1}{2} \sum_{j \in S} \int_{\Gamma_{\delta\varepsilon}^j} (x-x_0).\nu_j \left(\frac{\partial u}{\partial \nu_j}\right)^2 d\sigma \quad \text{and}$$

$$I_2 = - \frac{1}{2} \sum_{j \in S} \int_{\gamma_\varepsilon^j} \left[(x-x_0).\nu_j \left(\frac{\partial u}{\partial \nu_j}\right)^2 + 2(x-x_0).\tau_j \frac{\partial u}{\partial \nu_j} \frac{\partial u}{\partial \tau_j} - (x-x_0).\nu_j \left(\frac{\partial u}{\partial \tau_j}\right)^2 \right]$$

A direct calculation shows that $I_2 = \frac{\pi}{4} \sum_{j \in S} (s_j - x_0).\tau_j \ c_j^2 + O(\varepsilon)$.

As $-\Delta u \ (x-x_0).\nabla u \in L^1(\Omega)$, we have

$$\lim_{\varepsilon \to 0} \int_{\Omega_\varepsilon} -\Delta u \ (x-x_0).\nabla u \ dx = \int_\Omega -\Delta u \ (x-x_0).\nabla u \ dx.$$

We deduce from this equality that the limit of I_1 exists and therfore the theorem is proved.

IV. EQUIVALENCE OF NORMS

In this paragraph, we establish, from the integration by parts formula (3.3), the equivalence of norms which is the basis of the H.U.M method.

We consider a neighbourhood V of Γ^* and let $O = V \cap \Omega$.

Remark: O is not only a neighbourhood of $\Gamma_\delta(x_0)$, but it is also a neighbourhood of the crack tips.

The main result of this paragraph is the following one:

Theorem 2

There exists $T_0 > 0$ such that for all $T > T_0$, the function defined on $L^2(\Omega) \times H^{-1}(\Omega)$ by

$$N_1(\phi_0, \phi_1) = \left\{ \int_0^T \int_O |\phi|^2 \, dx \, dt \right\}^{1/2} \tag{4.1}$$

where ϕ is the solution of (2.2), is a norm equivalent to the norm defined by

$$E_{0,-1}(\phi_0, \phi_1) = \frac{1}{2} \left(\| \phi_0 \|^2_{L^2(\Omega)} + \| \phi_1 \|^2_{H^{-1}(\Omega)} \right) \tag{4.2}$$

The proof of this result is performed on many steps and use the two following lemmas.

Lemma 1

There exists $T_0 > 0$ such that foe all $T > T_0$, the function defined on $H_0^1(\Omega) \times L^2(\Omega)$ by

$$N_2(\phi_0, \phi_1) = \left\{ \int_0^T \int_O \left(|\phi|^2 + |\phi'|^2 \right) \, dx \, dt \right\}^{1/2} \tag{4.3}$$

is a norm equivalent to the norm defined by the enrgy.

Proof:

We denote by (,) the inner product in $L^2(\Omega)$.

First of all, we developp $\int_Q (\phi'' - \Delta\phi)(x - x_0) . \nabla\phi \, dx \, dt = 0$. Using the integration by parts formula (3.3), we obtain

$$(\phi', (x-x_0) . \nabla\phi) \big|_0^T + \frac{1}{2} \int_Q |\phi'|^2 \, dx \, dt + \frac{\pi}{4} \sum_{j \in S} (s_j - x_0) . \tau_j \int_0^T c_j^2 \, dt$$

$$- \frac{1}{2} \int_0^T \int_{\Gamma_\delta} (x-x_0) . \nu \left(\frac{\partial u}{\partial \nu} \right)^2 d\sigma dt$$

$$- \frac{1}{2} \sum_{j \in S} \lim_{\varepsilon \to 0} \int_0^T \int_{\Gamma_{\delta\varepsilon}^j} (x-x_0) . \nu_j \left(\frac{\partial u}{\partial \nu_j} \right)^2 d\sigma dt = 0$$

then

$$- \frac{\pi}{4} \sum_{j \in S} (s_j - x_0) \cdot \tau_j \int_0^T c_j^2 \, dt + \frac{1}{2} \int_0^T \int_{\Gamma_\delta(x_0)} (x-x_0) \cdot \nu \left(\frac{\partial u}{\partial \nu} \right)^2 d\sigma dt$$

$$\tag{4.4}$$

$$+ \frac{1}{2} \sum_{j \in S} \lim_{\varepsilon \to 0} \int_0^T \int_{\Gamma_{\delta\varepsilon}^j} (x-x_0) \cdot \nu_j \left(\frac{\partial u}{\partial \nu_j} \right)^2 d\sigma dt \geq (T - T_{01}) \, E_{0,1}(\phi_0, \phi_1)$$

where $T_{01} > 0$.

After this, we let

$$O_2 = \bigcup_{x \in \Gamma^*} B\left(x, \frac{\delta + \gamma}{2} \right) , \quad O_3 = \bigcup_{x \in \Gamma^*} B\left(x, \frac{3\gamma + \delta}{4} \right)$$

and we consider a cut-off function η such that supp $(\eta) \subset O_3$

and $\eta_{|O_2} = 1$.

Then we developp the expression $\int_Q (\phi'' - \Delta\phi) \, \eta \, (x-x_0) \cdot \nabla\phi \, dxdt = 0$.

Noting that $\eta = 1$ on $\Gamma_{\varepsilon\delta}^j$ and $\eta = 1$ on $\Gamma_\delta(x_0)$, we obtain

$$- \frac{\pi}{4} \sum_{j \in S} (s_j - x_0) \cdot \tau_j \int_0^T c_j^2 \, dt + \frac{1}{2} \int_0^T \int_{\Gamma_\delta(x_0)} (x-x_0) \cdot \nu \left(\frac{\partial u}{\partial \nu} \right)^2 d\sigma dt$$

$$\frac{1}{2} \sum_{j \in S} \lim_{\varepsilon \to 0} \int_0^T \int_{\Gamma_{\delta\varepsilon}^j} (x-x_0) \cdot \nu_j \left(\frac{\partial u}{\partial \nu_j} \right)^2 d\sigma dt =$$

$$\tag{4.5}$$

$$\int_Q ((x-x_0) \cdot \nabla\phi)(\nabla\eta \cdot \nabla\phi) \, dxdt - \frac{1}{2} \int_Q \text{div}(\eta(x-x_0)) |\nabla\phi|^2 dxdt$$

$$+ \int_Q \eta |\nabla\phi|^2 dxdt + (\phi', (x-x_0) \cdot \nabla\phi) |_0^T + \frac{1}{2} \int_Q \text{div}(\eta(x-x_0)) \, |\phi'|^2 dxdt$$

We deduce from (4.4) and (4.5) the existence of $T_{02} > 0$ such that

$$\int_0^T \int_{O_3} \left(|\nabla\phi|^2 + |\phi'|^2 \right) dxdt \geq (T - T_{02}) \, E_0(\phi_0, \phi_1). \tag{4.6}$$

To complete the proof, we developp $\int_Q (\phi'' - \Delta\phi) \, \eta \, \phi \, dx \, dt = 0$ and obtain :

$$\int_Q \eta |\nabla\phi|^2 dxdt = (\phi', \eta\phi) |_0^T + \int_Q \eta |\phi'|^2 dxdt + \frac{1}{2} \int_Q \Delta\eta |\phi|^2 \, dxdt \tag{4.7}$$

then

$$\int_Q \eta |\nabla \phi|^2 dxdt \le 2R(x_0) E_0(\phi_0, \phi_1) + K \int_0^T \int_0 \left(|\phi|^2 + |\phi'|^2 \right) dx \, dt \qquad (4.8)$$

Therefore the lemma can be derived from (4.8) and (4.6).

Lemma 2

There exists a constant C > 0 such that

$$\int_0^T \int_0 |\phi|^2 \, dxdt \le C \int_0^T \int_0 |\phi'|^2 \, dxdt \qquad (4.9)$$

Proof.

Let us assume that this relation is not true. Then there exists a sequence $(\phi_{0n}, \phi_{1n}) \in H_0^1(\Omega) \times L^2(\Omega)$ such that

$$\int_0^T \int_0 |\phi_n|^2 \, dxdt = 1 \qquad (4.10)$$

and

$$n \int_0^T \int_0 |\phi_n'|^2 \, dxdt \le 1. \qquad (4.11)$$

According to lemma 1 and (4.10) , the sequence (ϕ_{0n}, ϕ_{1n}) is bounded in $H_0^1(\Omega) \times L^2(\Omega)$. Then we can assume that it converges weakly to an element (ϕ_0, ϕ_1) in this espace.

As the imbedding of $H_0^1(\Omega) \times L^2(\Omega)$ into $L^2(\Omega) \times H^{-1}(\Omega)$ is compact, $(\phi_{0n}, \phi_{1n})_{n \in \mathbb{N}}$ converges to (ϕ_0, ϕ_1) strongly in $L^2(\Omega) \times H^{-1}(\Omega)$.

We consider the solution $\phi \in C(0,T;H_0^1(\Omega)) \cap C^1(0,T;L^2(\Omega))$ of the problem

$$\begin{cases} \phi'' - \Delta\phi = 0 \text{ dans } Q \\ \gamma\phi = 0 \text{ sur } \Sigma \\ \phi(0) = \phi_0 , \phi'(0) = \phi_1 \text{ dans } \Omega \end{cases} \qquad (4.12)$$

We have

$$\int_0^T \int_0 |\phi_n - \phi|^2 dxdt \le K \left[||\phi_{0n} - \phi_0||^2_{L^2(\Omega)} + ||\phi_{1n} - \phi_1||^2_{H^{-1}(\Omega)} \right] \qquad (4.13)$$

that shows that ϕ_n converges strongly to ϕ in $L^2(0,T;L^2(\Omega))$ and furthermore, we have

$$\int_0^T \int_0 |\phi|^2 \, dxdt = 1. \tag{4.14}$$

On the other hand, according to (4.11), the sequence $(\phi'_n)_{n \in \mathbb{N}}$ converges to 0 in $L^2(0,T;L^2(0))$ and according to lemma 1, for all $m,n \in \mathbb{N}$, we have

$$\int_0^T \int_0 \left(|\phi_n - \phi_m|^2 + |\phi'_n - \phi'_m|^2 \right) dxdt \geq k \, E_{0,1}(\phi_{0n} - \phi_{0m}, \, \phi_{1n} - \phi_{1m})$$

where $k \in \mathbb{R}_+^*$.

So the sequence (ϕ_{0n}, ϕ_{1n}) is a Cauchy sequence in $H_0^1(\Omega) \times L^2(\Omega)$, therefore it converges to (ϕ_0, ϕ_1) in this space and ϕ'_n converges to ϕ' in $L^2(0,T;L^2(0))$ so that $\phi' = 0$ in $]0,T[\times 0$. Therefore ϕ is constant on $]0,T[\times 0$ and as $\gamma\phi = 0$ on $0 \cap \partial\Omega$, we have $\phi = 0$ in $]0,T] \times 0$, which is false according to (4.14).

Proof of theorem 2

From lemmas 1 and 2, we derive that the application

$$(\phi_0, \phi_1) \longrightarrow \left\{ \int_0^T \int_0 |\phi'|^2 \, dxdt \right\}^{1/2} \tag{4.15}$$

defines on $H_0^1(\Omega) \times L^2(\Omega)$ a norm which is equivalent to the norm defined by the energy.

Now, if $(\phi_0, \phi_1) \in L^2(\Omega) \times H^{-1}(\Omega)$, we consider the solution $\psi_0 \in H_0^1(\Omega)$ of the problem

$$\begin{cases} -\Delta\psi_0 = \phi_1 \text{ dans } \Omega \\ \gamma\psi_0 = 0 \text{ sur } \partial\Omega \end{cases} \tag{4.16}$$

and we let

$$\psi(t,x) = \int_0^t \phi(s,x) \, ds + \psi_0(x) \tag{4.17}$$

The function ψ is the solution of the homogeneous wave equation (4.12) with initial data (ψ_0, ϕ_0) and verifies $\psi' = \phi$. According to what preceeds, one can deduce that there exists a constant $C_1 > 0$ such that

$$\int_0^T \!\! \int_0 \, |\phi|^2 dxdt = \int_0^T \!\! \int_0 \, |\psi'|^2 dxdt \geq C_1 \left\{ ||\psi_0||^2_{H^1_0(\Omega)} + ||\phi_0||^2_{L^2(\Omega)} \right\} \qquad (4.18)$$

As $-\Delta : H^1_0(\Omega) \longrightarrow H^{-1}(\Omega)$ is an isomorphism, we deduce from (4.18) that there exists a constant $C > 0$ such that

$$\int_0^T \!\! \int_0 \, |\phi|^2 dxdt \geq C \left\{ ||\phi_1||^2_{H^{-1}(\Omega)} + ||\phi_0||^2_{L^2(\Omega)} \right\} \qquad (4.19)$$

Remark: The inequality

$$\int_0^T \!\! \int_0 \, |\phi|^2 dxdt \leq C' \quad E_{0,-1}(\phi_0, \phi_1)$$

is obvious and can be obtain easily by developping the solution ϕ according to the eigenfunctions of $-\Delta$.

V. EXACT CONTROLLABILITY ON A NEIGHBOURHOOD OF THE BOUNDARY

Now we are able to apply the H.U.M method and give the main result of this study:

Theoreme 3

There exists $T_0 > 0$ such that for all $T > T_0$ and for all $(y_0, y_1) \in H^1_0(\Omega) \times L^2(\Omega)$, there exists $v \in L^2(Q)$ with its support in $]0,T[\times 0$ such that if y is the unique solution of

$$\begin{cases} y'' - \Delta y = \chi_0 \, v \quad \text{in } Q \\ \gamma y = 0 \text{ in } \Sigma \\ y(0) = y_0 \, , \, y'(0) = y_1 \text{ in } \Omega \end{cases} \qquad (5.1)$$

then $y(T) = y'(T) = 0$.

Proof :

Let $(\phi_0, \phi_1) \in L^2(\Omega) \times H^{-1}(\Omega)$ and $\phi \in C(0,T;L^2(\Omega) \cap C^1(0,T;H^{-1}(\Omega))$ the solution of the problem

$$\begin{cases} \phi'' - \Delta\phi = 0 \text{ in } Q \\ \gamma\phi = 0 \text{ on } \Sigma \\ \phi(0) = \phi_0 \ , \ \phi'(0) = \phi_1 \text{ in } \Omega \end{cases} \tag{5.2}$$

We define $y \in C(0,T;H_0^1(\Omega)) \cap C^1(0,T;L^2(\Omega))$ as the unique solution of the problem

$$\begin{cases} y'' - \Delta y = \chi_0 \ \phi \quad \text{in } Q \\ \gamma y = 0 \text{ in } \Sigma \\ y(T) = \ y'(T) = 0 \text{ in } \Omega \end{cases} \tag{5.3}$$

and we let

$$\Lambda \ (\phi_0,\phi_1) = (-y'(0), \ y(0)) \tag{5.4}$$

Then we have

$$< \Lambda(\phi_0,\phi_1), \ (\phi_0,\phi_1) >_{L^2(\Omega)\times H_0^1(\Omega), L^2(\Omega)\times H^{-1}(\Omega)} = \int_0^T \int_\Omega \mid \phi \mid^2 \text{ dxdt} \tag{5.6}$$

Therefore theorem 2 shows that $\Lambda : L^2(\Omega) \times H^{-1}(\Omega) \longrightarrow L^2(\Omega)\times H_0^1(\Omega)$ is an isomorphism. That completes the proof.

REFERENCES

[1] E. Zuazua : in J.L.Lions Contrôlabilité exacte, Perturbations et Stabilisation des systèmes distribués, Tome 1, Masson, 1986.

[2] J.L.Lions :Contrôlabilité exacte, Perturbations et Stabilisation des systèmes distribués, Tome 1, Masson, 1986.

[3] P.Grisvard: Elliptic problems in non smooth domains,Pitman, 1985

[4] P.Grisvard : Contrôlabilité exacte des solutions de l'équation des ondes en présence de singularités, J. Math Pures et Appl, 68, p. 215-259

[5] M.T.Niane, O.Seck: Contrôlabilité exacte frontière de l'équation des ondes en présence de fissures par adjonction de contrôles internes au voisinage des sommets de fissures, C.R.Acad.Sci, Paris , t316, p 695-700, 1993.

Optimal Control Problems of Quasilinear Parabolic Equations

Eduardo Casas
Dpto. de Matemática Aplicada y Ciencias de la Computación
E.T.S.I. de Caminos, Universidad de Cantabria, 39071 Santander, Spain,

Luis Alberto Fernández
Dpto. de Matemáticas, Estadística y Computación
Facultad de Ciencias, Universidad de Cantabria, 39071 Santander, Spain,

and

Jiongmin Yong
Departament of Mathematics
Fudan University, Shanghai 200433, China

1 Introduction

In this work we are concerned with optimal control problems governed by quasilinear parabolic equations in divergence form of the type

$$y_t(x,t) - div_x \left((1 + |\nabla_x y(x,t)|)^{\alpha-2} \nabla_x y(x,t) \right). \tag{1}$$

The aim is to prove existence of solutions and derive some optimality conditions in a rigorous way.

Apparently, these questions have been treated in some previous studies such as [1] and [8]. However, equations modelled by (1) do not enter into the framework of these works, because neither satisfy the condition ii) of Lemma 3.3 nor (C2) in [1] (and consequently, the condition $H(A)_4 - (6)$ imposed in [8] which coincides with the first one).

Our procedure for deriving the optimality conditions is the same as in the elliptic case, see [2], [3] and the bibliography cited there (in [4] state-constrained problems were considered by using a different approach). Nevertheless, in the parabolic case new difficulties arise by the (possible) non-uniqueness of solution for the linearized problem that is (in general) of degenerate type.

Only an outline of the proofs is provided. The detailed version will appear in a forthcoming paper.

2 Setting of the problem

Let Ω be a bounded open subset of R^n with Lipschitz continuous boundary Γ and $T \in (0, +\infty)$. In $Q_T = \Omega \times (0, T)$ we consider the differential operator

$$Ay = -div_x\{a(x, t, \nabla_x y)\} + a_0(x, t, y)$$

where the coefficients $a : Q_T \times R^n \longrightarrow R^n, a = (a_1, \ldots, a_n)$ and $a_0 : Q_T \times R \longrightarrow R$ satisfy:

$$\begin{cases} a_j(\cdot, \cdot, \eta) \text{ is a measurable function in } Q_T \\ a_j(x, t, \cdot) \text{ belongs to } C^1(R^n) \quad 1 \leq j \leq n \end{cases} \tag{2}$$

$$\begin{cases} a_0(\cdot, \cdot, y) \text{ is a measurable function in } Q_T \\ a_0(x, t, \cdot) \text{ belongs to } C^1(R) \end{cases} \tag{3}$$

$$\sum_{i,j=1}^{n} \frac{\partial a_j}{\partial \eta_i}(x, t, \eta) \xi_i \xi_j \geq \Lambda_1(k + |\eta|)^{\alpha-2} |\xi|^2 \tag{4}$$

$$\sum_{i,j=1}^{n} \left| \frac{\partial a_j}{\partial \eta_i}(x, t, \eta) \right| \leq \Lambda_2(k + |\eta|)^{\alpha-2} \tag{5}$$

$$0 \leq \frac{\partial a_0}{\partial y}(x, t, y) \leq h_0(|y|) \tag{6}$$

$$a_0(x, t, 0) = a_j(x, t, 0) = 0, \quad j = 1, \ldots, n \tag{7}$$

for some $\alpha > \max\{2n/(n+2), 1\}$, $k \geq 0$, $\Lambda_1 > 0$, $\Lambda_2 > 0$, some positive increasing function h_0, all $(x, t) \in Q_T$, $y \in R$ and $\eta, \xi \in R^n$.

Now, the following quasilinear parabolic initial-boundary value problem can be considered

$$\begin{cases} y_t + Ay = u & \text{in } Q_T, \\ y(x, t) = 0 & \text{on } \Sigma = \Gamma \times (0, T), \\ y(x, 0) = y_0(x) & \text{in } \Omega. \end{cases} \tag{8}$$

In the study of this type of parabolic problems is useful to introduce the Banach space

$$W^\alpha(0, T) = \{ y \in L^\alpha(0, T; W_0^{1,\alpha}(\Omega)) : y_t \in L^{\alpha'}(0, T; W^{-1,\alpha'}(\Omega)) \}$$

equipped with the norm

$$\|y\|_{W^\alpha(0,T)} = \|y\|_{L^\alpha(0,T;W_0^{1,\alpha}(\Omega))} + \|y_t\|_{L^{\alpha'}(0,T;W^{-1,\alpha'}(\Omega))}$$

where $\alpha' = \alpha/(\alpha - 1)$.

Using a classical cut off procedure for the term a_0 we derive the following result about the existence and uniqueness of solution for problem (8) as well as the continuous dependence of the solution with respect to the datum u:

Theorem 1 *Given $y_0 \in L^\infty(\Omega)$ and $u \in L^\infty(Q_T)$, there exists a unique solution y_u of (8) in the space $W^\alpha(0, T) \cap L^\infty(Q_T)$. Moreover, if we designate by y_m the solution of (8) corresponding to (y_0, u_m) and suppose that $u_m \to u$ weakly* in $L^\infty(Q_T)$ as $m \to +\infty$, then we have*

$$y_m \to y_u \quad \text{in } L^\alpha(0, T; W_0^{1,\alpha}(\Omega)) \cap C([0, T]; L^2(\Omega)) \cap L^q(Q_T)$$

for all $q \in [1, +\infty)$.

We are concerned with the following control problem:

$$(P) \quad \min_{u \in K} J(u) = \int_{Q_T} L(x, t, y(x, t), u(x, t)) dx dt,$$

where K is a nonempty, convex, bounded and weak*-closed subset of $L^\infty(Q_T)$ and $L : Q_T \times R \times R \longrightarrow R$ is a measurable function that satisfies the following assumptions

$$L(x, t, \cdot, \cdot) \in C^1(R^2) \quad \text{for each} \quad (x, t) \in Q_T, \quad L(\cdot, \cdot, 0, 0) \in L^1(Q_T) \quad \text{and} \tag{9}$$

$$\left| \frac{\partial L}{\partial u}(x, t, y, u) \right| + \left| \frac{\partial L}{\partial y}(x, t, y, u) \right| \le h_1(|y| + |u|) + h_2(x, t) \tag{10}$$

for all $(x, t, y, u) \in Q_T \times R \times R$, where h_1 is a positive increasing function and $h_2 \in L^2(Q_T)$.

As usual, the existence of (P)-solutions can be deduced by taking minimizing sequences

Theorem 2 *Assuming that L is convex with respect to u, then (P) has at least one solution.*

A key point of the proof is the convexity assumption on L, which implies the lower semicontinuity of functional J; see for instance [5]. Needless to say that the differentiability hypotheses assumed on the coefficients of the operator and function L can be removed at this level and replaced by some other weaker conditions. However, they are essential to derive the optimality conditions.

3   Optimality conditions

In order to derive some optimality conditions satisfied by a solution $\bar{u}$ of (P), we begin studying the differentiability of the relation between the control and the state: $u \longrightarrow y_u$. In this study some difficulties arise by the (possible) non-uniqueness of solution for the linearized problem that is (in general) of degenerate type. Nevertheless, there are some special situations in which this difficulty does not appear. These can be summarized as follows:

Theorem 3 *Let us consider $F : L^\infty(Q_T) \longrightarrow L^2(0, T; H_0^1(\Omega))$ (resp. $L^\alpha(0, T; W_0^{1,\alpha}(\Omega))$, if $\alpha < 2$) the mapping defined by $F(u) = y_u$.*

a) *If $\alpha = 2$, then F is Gâteaux differentiable.*

b) *If $\alpha \ne 2$ and $k \ne 0$, then F is Gâteaux differentiable at the points $u \in L^\infty(Q_T)$ such that $|\nabla_x y_u| \in L^\infty(Q_T)$.*

Moreover if $z = DF(u)v$, then $z \in W^2(0, T)$ and it is the unique solution in this space of the problem

$$\begin{cases} z_t - div_x \left\{ \frac{\partial a}{\partial \eta}(x, t, \nabla_x y_u) \nabla_x z \right\} + \frac{\partial a_0}{\partial y}(x, t, y_u) z = v & \text{in } Q_T, \\ \\ z(x, t) = 0 & \text{on } \Sigma, \\ \\ z(x, 0) = 0 & \text{in } \Omega. \end{cases} \tag{11}$$

Let us mention just the main points of the proof. Under the hypotheses of the theorem it is immediate to verify the boundedness of the coefficients of the linearized operator and deduce the existence and uniqueness of solution for (11) in $W^2(0,T)$, which is a crucial point in the proof.

Given $u, v \in L^\infty(Q_T)$ and $0 < \lambda < 1$, we consider the problem

$$
\begin{cases}
y_t + Ay = u + \lambda v & \text{in } Q_T, \\
y(x,t) = 0 & \text{on } \Sigma, \\
y(x,0) = y_0(x) & \text{in } \Omega
\end{cases}
\tag{12}
$$

By Theorem 1, for each λ there exists a unique solution $y^\lambda \in W^\alpha(0,T) \cap L^\infty(Q_T)$. If we introduce the function

$$
z^\lambda = \frac{y^\lambda - y_u}{\lambda},
$$

it is clear that for each $\lambda \in (0,1)$, z^λ verifies

$$
\begin{cases}
z_t^\lambda - div_x \left\{ \dfrac{a(x,t,\nabla_x y^\lambda) - a(x,t,\nabla_x y_u)}{\lambda} \right\} + \dfrac{a_0(x,t,y^\lambda) - a_0(x,t,y_u)}{\lambda} = v & \text{in } Q_T, \\
z^\lambda(x,t) = 0 & \text{on } \Sigma, \\
z^\lambda(x,0) = 0 & \text{in } \Omega.
\end{cases}
\tag{13}
$$

Using the hypotheses (2)–(7) together with the mean value theorem it can be proved that

$$
z^\lambda \longrightarrow z \quad \text{weakly in } L^2(0,T;H_0^1(\Omega)) \ (\text{resp. } L^\alpha(0,T;W_0^{1,\alpha}(\Omega)) \ \text{ if } \alpha < 2)
$$

for some element z, as $\lambda \to 0$. Passing to the limit in (13), we deduce that the limit point z is the solution of problem (11).

Finally, the strong convergence of $\{z^\lambda\}_{\lambda>0}$ towards z can be also derived.

With the aid of Theorem 3 we deduce the Gâteaux differentiability of the cost functional J in some cases and therefore, we can easily obtain the following optimality conditions

Theorem 4 *Let $\bar{u}$ be an optimal control for (P) and $\bar{y} = y_{\bar{u}} \in W^\alpha(0,T) \cap L^\infty(Q_T)$. Let us suppose that it is verified one of the following conditions*

a) $\alpha = 2$, *or*

b) $\alpha \neq 2$, $k \neq 0$ *and* $|\nabla_x \bar{y}| \in L^\infty(Q_T)$.

Then, there exists a unique element $\bar{p} \in W^2(0,T)$ such that

$$
\begin{cases}
-\bar{p}_t - div_x \left(\left[\dfrac{\partial a}{\partial \eta}(x,t,\nabla_x \bar{y}) \right]^T \nabla_x \bar{p} \right) + \dfrac{\partial a_0}{\partial y}(x,t,\bar{y})\bar{p} = \dfrac{\partial L}{\partial y}(x,t,\bar{y},\bar{u}) & \text{in } Q_T, \\
\bar{p}(x,t) = 0 & \text{on } \Sigma, \\
\bar{p}(x,T) = 0 & \text{in } \Omega,
\end{cases}
\tag{14}
$$

and

$$
\int_{Q_T} \left(\bar{p} + \frac{\partial L}{\partial u}(x,t,\bar{y},\bar{u}) \right)(u - \bar{u})dxdt \geq 0 \quad \forall u \in K
\tag{15}
$$

Let us remark that $\alpha = 2$ includes the linear and semilinear cases.

When $\alpha \neq 2$ and $k \neq 0$, in order to apply Theorem 4 it is necessary to have that $|\nabla_x \bar{y}| \in L^{\infty}(Q_T)$. Among the regularity results in this direction, we do not know any that can be used directly in our framework. Nevertheless, for some operators it is possible to combine the classical results of Ladyzhenskaya, Solonnikov and Ural'tseva [7] with a regularization of the state equation, the Ekeland's variational principle and the above differentiability result to establish the optimality conditions for some particular operators by means of a passage to the limit process. As an example, let us state the following result:

Theorem 5 *Let be $\alpha \neq 2$, $k \neq 0$ and assume that*

a) *the operator A is given by*

$$Ay = -div_x\{(k + |\nabla_x y|)^{\alpha-2}\nabla_x y\},$$

b) $\Gamma \in C^{2,\sigma}$, *with $\sigma \in (0,1)$.*

Now, let $\bar{u}$ be a solution of (P), with $\bar{y}$ its associated state.
Then, there exists an element $\bar{p} \in L^{\infty}(0,T;L^2(\Omega))$ with the following properties:

$$
\begin{cases}
\bar{p} \in L^{\alpha\wedge2}(0,T;W_0^{1,\alpha\wedge2}(\Omega)), & (1 + |\nabla_x\bar{y}|)^{\alpha-2}\nabla_x\bar{p} \in L^{\alpha'\wedge2}(Q_T)^n \\[2mm]
\bar{p}_t \in L^{2\wedge\alpha'}(0,T;W^{-1,2\wedge\alpha'}(\Omega)), & (1 + |\nabla_x\bar{y}|)^{\frac{\alpha-2}{2}}\nabla_x\bar{p} \in L^2(Q_T)^n
\end{cases}
\tag{16}
$$

where $\alpha \wedge 2 = \min\{\alpha, 2\}$, such that

$$
\begin{cases}
-\bar{p}_t - div_x\left\{M(\nabla_x\bar{y})(x,t)\nabla_x\bar{p}\right\} = \dfrac{\partial L}{\partial y}(x,t,\bar{y},\bar{u}) & \text{in } Q_T, \\[3mm]
\bar{p}(x,t) = 0 & \text{on } \Sigma, \\[3mm]
\bar{p}(x,T) = 0 & \text{in } \Omega,
\end{cases}
\tag{17}
$$

and

$$
\int_{Q_T}\left(\bar{p} + \frac{\partial L}{\partial u}(x,t,\bar{y},\bar{u})\right)(u-\bar{u})dxdt \geq 0 \quad \forall u \in K,
\tag{18}
$$

where

$$
M(\nabla_x\bar{y}) = \frac{(\alpha-2)}{|\nabla_x\bar{y}|}[k + |\nabla_x\bar{y}|]^{\alpha-3}\nabla_x\bar{y}\nabla_x\bar{y}^T + [k + |\nabla_x\bar{y}|]^{\alpha-2}I
$$

and I denotes the identity matrix $n \times n$. Moreover, we have

$$
\|\bar{p}\|^2_{L^{\infty}(0,T;L^2(\Omega))} + \int_{Q_T}(k + |\nabla_x\bar{y}|)^{\alpha-2}|\nabla_x\bar{p}|^2dxdt \leq
$$

$$
\leq (2 + \frac{1}{\Lambda_1})\left(\int_{Q_T}\frac{\partial L}{\partial y}(x,t,\bar{y},\bar{u})\bar{p}\,dxdt\right)
\tag{19}
$$

Let us point out that such element $\bar{p}$ may be not unique.

4 Sketch of the proof of Theorem 5

For each $\epsilon > 0$, let us introduce the following regularization of the state equation:

$$\begin{cases} y_t - div_x(a^\epsilon(\nabla_x y)) = G^\epsilon[u] & \text{in } Q_T, \\ y(x,t) = 0 & \text{on } \Sigma, \\ y(x,0) = G_0^\epsilon[y_0](x) & \text{in } \Omega. \end{cases} \tag{20}$$

where

$$a^\epsilon(\eta) = (k + \sqrt{|\eta|^2 + \epsilon^2})^{\alpha-2}\eta, \forall \eta \in R^n$$

$$G^\epsilon[u](x,t) = (\rho_\epsilon^{n+1} \star \tilde{u})(x,t) \cdot \phi_\epsilon(t) \quad \text{and} \quad G_0^\epsilon[y_0](x) = (\rho_\epsilon^n \star \tilde{y}_0)(x) \cdot \zeta_\epsilon(x),$$

for all $(x,t) \in Q_T$ with

$$\tilde{u}(x,t) = \begin{cases} u(x,t) & \text{if } (x,t) \in Q_T \\ 0 & \text{if } (x,t) \notin Q_T \end{cases}, \quad \tilde{y}_0(x) = \begin{cases} y_0(x) & \text{if } x \in \Omega \\ 0 & \text{if } x \notin \Omega \end{cases},$$

$\{\rho_\epsilon^m\} \subset C_0^\infty(R^m)$ designates an approximation of the identity, $\star$ denotes the corresponding convolution product in each case and

$$\phi_\epsilon \in C_0^\infty(0,T), \quad 0 \le \phi_\epsilon(t) \le 1 \quad \text{and} \quad \phi_\epsilon(t) \to 1 \quad \text{as } \epsilon \to 0$$

$$\zeta_\epsilon \in C_0^\infty(\Omega), \quad 0 \le \zeta_\epsilon(x) \le 1 \quad \text{and} \quad \zeta_\epsilon(x) \to 1 \quad \text{as } \epsilon \to 0. \tag{21}$$

Due to the smoothness of the functions a^ϵ it is possible to write out the equation of (20) in a developed form by expanding the divergence term. Moreover, thanks to the regularity of $G^\epsilon[u]$ and the compatibility condition between the boundary and the initial datum $G_0^\epsilon[y_0]$, we can apply [7, Theorem 4.1, pp. 558-559] and deduce that problem (20) has a unique solution $y_u^\epsilon \in C^{2+\sigma,1+\sigma/2}(\overline{Q_T})$. The relevant fact here is that $|\nabla_x y_u^\epsilon| \in L^\infty(Q_T)$.

It is not difficult to prove that the previous established initial-boundary value problems approximate to the initial one in the following sense:

Proposition 1 *Let us suppose that $\{u_\epsilon\}$ is a sequence in $L^\infty(Q_T)$ such that*

$$u_\epsilon \longrightarrow u \quad weakly^* \text{ in } L^\infty(Q_T) \quad as \ \epsilon \to 0 \tag{22}$$

Let us denote by y_ϵ the solution of (20) corresponding to $G^\epsilon[u_\epsilon]$. Then,

$$y_\epsilon \longrightarrow y_u \quad in \ L^\alpha(0,T;W_0^{1,\alpha}(\Omega)) \cap C([0,T];L^2(\Omega)) \cap L^q(Q_T) \quad as \ \epsilon \to 0 \tag{23}$$

for all $q \in [1,+\infty)$, where y_u is the solution of (8) with $Ay = -div_x\{(k + |\nabla_x y|)^{\alpha-2}\nabla_x y\}$.

By other hand, let us define the set

$$K_1 = \{u \in K : \|u - \bar{u}\|_{L^\infty(Q_T)} \le 1\} \tag{24}$$

where $\bar{u}$ is a fixed optimal control for (P), and introduce the following family of problems:

$$(P_\epsilon') \quad \min_{u \in K_1} J_\epsilon(u)$$

where the cost functional is given by

$$J_\epsilon(u) = \int_{Q_T} L(x,t,y_u^\epsilon(x,t),u(x,t))dxdt.$$

Now, we can prove in a straightforward way the following approximation result for the control problems:

$$\lim_{\epsilon \to 0} \inf_{u \in K_1} J_\epsilon(u) = \min_{u \in K} J(u) = J(\bar{u}) \tag{25}$$

Using this convergence we know that

$$0 \leq \delta_\epsilon^2 \equiv J_\epsilon(\bar{u}) - \inf_{u \in K_1} J_\epsilon(u) \longrightarrow 0 \quad \text{as} \quad \epsilon \to 0 \tag{26}$$

Therefore, by Ekeland's variational principle [6], for each $\epsilon > 0$ we can find $u_\epsilon \in K_1$ verifying the following properties

$$\|u_\epsilon - \bar{u}\|_{L^\infty(Q_T)} \leq \delta_\epsilon \tag{27}$$

$$J_\epsilon(u_\epsilon) \leq J_\epsilon(\bar{u}) \tag{28}$$

$$J_\epsilon(v) - J_\epsilon(u_\epsilon) \geq -\delta_\epsilon \|v - u_\epsilon\|_{L^\infty(Q_T)} \quad \forall v \in K_1 \tag{29}$$

Last relation means that u_ϵ is an optimal control for the following problem

$$(P_\epsilon') \quad \min_{v \in K_1} \left(J_\epsilon(v) + \delta_\epsilon \|v - u_\epsilon\|_{L^\infty(Q_T)} \right).$$

By virtue of the boundedness of the spatial gradient of $y_{u_\epsilon}^\epsilon$ in Q_T and the convexity of the norm, it is possible to obtain the first order optimality system corresponding to (P_ϵ') for each $\epsilon \in (0,1)$ as in Theorem 4, taking now into account that the necessary condition is

$$J_\epsilon'(u_\epsilon)(v - u_\epsilon) + \delta_\epsilon \|v - u_\epsilon\|_{L^\infty(Q_T)} \geq 0 \quad \forall v \in K_1.$$

Consequently, we obtain the following result for each $\epsilon \in (0,1)$:

Theorem 6 *Let y_ϵ be the state corresponding to u_ϵ, i.e. $y_\epsilon = y_{u_\epsilon}^\epsilon$.*
Then, there exists a unique element $p_\epsilon \in W^2(0,T)$ such that

$$\begin{cases} -\dfrac{\partial p_\epsilon}{\partial t} - div_x \left(\left[\dfrac{\partial a^\epsilon}{\partial \eta}(\nabla_x y_\epsilon) \right]^T \nabla_x p_\epsilon \right) = \dfrac{\partial L}{\partial y}(x,t,y_\epsilon,u_\epsilon) & \text{in } Q_T, \\ p_\epsilon(x,t) = 0 & \text{on } \Sigma, \\ p_\epsilon(x,T) = 0 & \text{in } \Omega, \end{cases} \tag{30}$$

and

$$\int_{Q_T} \left(p_\epsilon G^\epsilon[v - u_\epsilon] + \frac{\partial L}{\partial u}(x,t,y_\epsilon,u_\epsilon)(v - u_\epsilon) \right) dx\,dt + \delta_\epsilon \|v - u_\epsilon\|_{L^\infty(Q_T)} \geq 0 \quad \forall v \in K_1 \tag{31}$$

To complete the proof of Theorem 5 it is enough to combine Proposition 1 with an estimation of the norms of the sequence $\{p_\epsilon\}_{\epsilon \in (0,1)}$ in $L^2(0,T;H_0^1(\Omega)) \cap L^\infty(0,T;L^2(\Omega))$ (resp. $L^\alpha(0,T;W_0^{1,\alpha}(\Omega)) \cap L^\infty(0,T;L^2(\Omega))$ if $\alpha < 2$) independent of ϵ and to pass to the limit in the relations (30) and (31) as $\epsilon \to 0$. Finally, to recover the set K instead of K_1 in the condition (18) it is sufficient for each $u \in K$ to pick $\tau > 0$ such that $v = \bar{u} + \tau(u - \bar{u}) \in K_1$.

Imposing some additional assumptions on the coefficients, it can be showed that Theorem 5 continues to hold for operators

$$y_t - div_x(a(x,t,\nabla_x y)) + a_0(x,t,y),$$

with the adjoint state equation given as in (14).

The proof uses essentially the same argumentation, the main difference being the definition of the regularized coefficients a_0^ϵ and a_j^ϵ that must be done via the classical convolution products with an approximation of the identity.

References

[1] N. U. Ahmed. Optimal control of a class of strongly nonlinear parabolic systems. *J. of Math. Anal. & Appl.*, 61:188–207, 1977.

[2] E. Casas and L.A. Fernández. Optimal control of quasilinear elliptic equations with non differentiable coefficients at the origin. *Revista Matemática de la Universidad Complutense de Madrid*, 4(2-3):227–250, 1991.

[3] E. Casas and L.A. Fernández. Distributed control of systems governed by a general class of quasilinear elliptic equations. *J. of Differential Equations*, 1993.

[4] E. Casas and J. Yong. Maximum principle for state-constrained optimal control problems governed by quasilinear elliptic equations. *Technical Report 1063, IMA Preprint Series*, November 1992.

[5] B. Dacorogna. *Direct Methods in the Calculus of Variations*. Springer–Verlag, Berlin–Heidelberg, 1989.

[6] I. Ekeland. Nonconvex minimization problems. *Bull. Amer. Math. Soc.*, 1(3):76–91, 1979.

[7] O.A. Ladyzhenskaya, V.A. Solonnikov, and N.N. Ural'tseva. *Linear and Quasilinear Equations of Parabolic Type*. Am. Math. Soc, Providence, R. I., 1968.

[8] N. S. Papageorgiou. On the optimal control of strongly nonlinear evolution equations. *J. of Math. Anal. & Appl.*, 164:83–103, 1992.

CHARACTERIZATION OF CONTROLLABILITY VIA A REGULAR FUNCTION : EXAMPLE OF THE VIBRATING STRING

G. MONTSENY, P. BENCHIMOL, L. PLANTIE

Laboratoire d'Automatique et d'Analyse des Systèmes du CNRS

7, Avenue du Colonel Roche

31077 Toulouse Cédex, France

Abstract

We study, in the case of a point-controlled vibrating string, two *real* valued functions of a structural parameter which reflect, in some way, the degree of controllability of the system. These functions may be seen as realistic connections between *non-robust* binary notions of controllability and the continuous solutions of well-posed control problems. The classical controllability property may be characterized via a regularity argument : the system is controllable if and only if the considered functions are differentiable and non-controllability corresponds to cusp points with negative concavity.

Introduction

Control theory for dynamic systems is closely associated to basic controllability notions. In the infinite dimensional case, there exist various definitions [4], more or less similar to the finite dimensional one, and corresponding to different concepts for the reachable set. Unfortunately, they are often *non-robust* in the sense that they may no longer be verified for some arbitrarily small variation of system parameters. Consider, for example, the string equation in the domain]0,1[, with an *internal* point actuator : it is classical that such a system is controllable [9] if and only if the support of the actuator is an irrational number. From a practical viewpoint, such an assertion is obviously not sufficient. Indeed, the dependance of any fondamental property of a system should preferably be continuous with respect to structural parameters [1], at least almost everywhere. Here, the lack of continuity is clearly due to the fact that controllability is a *binary* concept (togother with the infinite-dimensionality of the state space). And although rational position of the actuator does lead to non-controllability, such a distinction is not realistic in practice because small variations of the position parameter *cannot be significant*.

On the other hand, the *quantitative* concept of *degree of controllability* has been early introduced by Kalman, Ho and Nerendra [6], associated to the question : how controllable is the system ? In the finite-dimensional case, many papers [5], [7], [12], [14], [15] developped various definitions based on the determinant of the controllability matrix. They possesse the following common properties :

- they are continuous with respect to the system parameters,
- they vanish when the system is not controllable,
- they are independant of the initial state of the system.

However, if one try to generalize these definitions to the infinite dimensional case (via convenient adaptations), the aforementioned particularities generate major drawbacks :

- the continuity is generally lost due to the non robustness of the controllability notions.
- the area of the unit sphere (of initial states) is infinite and it becomes difficult to realize integration on such domain (unless considering a convenient weight measure).

In the case of the point controlled vibrating string (generalization to many other systems are of course possible), we define and study two degrees of controllability, derived from particular real valued functions of a structural parameter. The two difficulties previously mentioned are solved as follow :

- the final state control problem considered in the classical definitions of controllability and degree of controllability, which is generally ill-posed in the distributed case [10], is replaced by two well-posed control problems. The corresponding degrees of controllability depend on an additional parameter which reflects in practice the price one should pay for controlling the system (the more one accepts to pay, the more the system will be controllable).
- In an infinite dimensional state space, balls are not compact ; they physically constitue "too large" sets. So, we must consider that the supposed random initial state is constrained by the *finite variance* additional hypothesis. Physically, such an hypothesis means that the high order modes of the system are of decreasing (and summable) mean energy ; this seems us to be a very realistic and little restrictive condition.

The first definition (degree of *active* controllability) is associated to the classical control problem (P_ϵ) : $\min_u \{\|X(T)\|^2 + \epsilon \int_0^T u^2(t)dt\}$, seen as a regularization of the exact control problem (P_0) : $\min_u \{\int_0^T u^2(t)dt, \ X(T) = 0\}$. We show that the considered function is in fact continous, almost everywhere differentiable and exhibits cusp points when the system is not controllable. So, it may be interpreated like a characterization of *binary* controllability, via *smoothness* arguments. Furthermore, it also gives quantitative informations in terms of *efficiency* of the control. Its computation is essentially based on the knowledge of the eigen-modes of the system, which is of course rather restrictive in practice. The second definition (degree of passive controllability) is associated to a *passive* control problem corresponding to a static dissipative feedback (analogous to a viscous damper). The computation is only based on simple numerical simulations, which permits very easy adaptations to various systems. These two approaches, at first sight rather different, lead to very close results. In the case of a boundary control, one could easily show that they are intimately correlated. Indeed, the two considered control problems become equivalent : the *transparent* boundary condition (which in the physical field corresponds to the well-known notion of *matched impedance*) is obtained by a static boundary feedback ; the induced control u is in fact the solution of the exact optimal control problem (P_0) (with $T = 2$). This perfect analogy then suggests the question, which was one of the origins of this paper : is it possible, *in some sense*, to adapt such a property to internal controls ? this question, not yet completely solved, has a partial (positive) response in the present analysis.

1 Preliminaries and problem statement

Consider the point-controlled string equation on $]0, 1[\times]0, T[$:

$$\partial_t^2 \theta - \partial_x^2 \theta = u\, \delta_a \ , \quad \theta(0,t) = \theta(1,t) = 0 \ ,$$
$$\theta(x,0) = \theta_0(x) \ , \quad \partial_t \theta(x,0) = \theta_1(x) \ . \tag{1.1}$$

The associated energy is classically defined by : $\mathcal{E}(t) = \frac{1}{2}(\|\theta(.,t)\|^2_{H_0^1(0,1)} + \|\partial_t\theta(.,t)\|^2_{L^2(0,1)})$.
If we set for $n = 1$ to $+\infty$ and $\varphi_n(x) = \sqrt{2}\,\sin(n\pi x)$:

$$\xi_n(t) = (\theta(t) \mid \varphi_n)_{L^2(0,1)} \ , \quad \xi = (\xi_n)_{n\geq 1} \ , \quad \rho_n = (\theta_0 \mid \varphi_n)_{L^2(0,1)} \ , \quad \rho = (\rho_n)_{n\geq 1} \ ,$$
$$\eta_n = (\theta_1 \mid \varphi_n)_{L^2(0,1)} \ , \qquad \eta = (\eta_n)_{n\geq 1} \ , \quad \gamma_n = \varphi_n(a) \ , \qquad \gamma = (\gamma_n)_{n\geq 1} \ , \tag{1.2}$$

$\Omega = diag(-n^2\pi^2)$ (the diagonal matrix with terms $n^2\pi^2$) , then, it is well-known that
system (1.1) is equivalent to :

$$\ddot{\xi} = \Omega\,\xi + \gamma\,u \ , \quad \xi(0) = \rho \ , \quad \dot{\xi}(0) = \eta \ , \tag{1.3}$$

with $(\xi_n, \dot{\xi}_n) \in h^1 \times l^2$ (the new state space). So, we have : $\mathcal{E}(t) = \frac{1}{2}\sum_{n\geq 1}(n^2\pi^2\xi_n^2(t) + \dot{\xi}_n^2(t))$.
In the particular case $u = 0$ and with the above notations, the solution of (1.1) is given
by : $\theta(x,t) = \sum_{n\geq 1}\xi_n(t)\sqrt{2}\sin(n\pi x)$, with $\xi_n(t) = \rho_n\cos(n\pi t) + \frac{\eta_n}{n\pi}\sin(n\pi t)$.

"Active" control : The aim of the control function u is to bring the string energy
around zero. For this reason, we consider the two following classical control problems :

$$(P_0) \qquad \min_u \ \{\|u\|^2_{L^2(0,T)} \ , \ \mathcal{E}(T) = 0\} \qquad (exact\ control\ on\ [0,T]) \ ,$$
$$(P_\epsilon) \qquad \min_u \ \{\mathcal{E}(T) + \epsilon\|u\|^2_{L^2(0,T)}\} \ , \ \epsilon > 0 \ (weighted\ cost\ functional) \ . \tag{1.4}$$

Problem (P_ϵ) may be viewed as a regularisation [10] of problem (P_0) : if (P_0) admits
a unique solution u_0, then the unique solution u_ϵ of (P_ϵ) verifies : $u_\epsilon \longrightarrow u_0$ strongly
in $L^2(0,T)$. But if (P_0) have no solution, u_ϵ does not converge when $\epsilon \longrightarrow 0$. It has
been developped the so-called *Hilbert Uniqueness Method* [9], suitable for such problems.
This method permits to construct by completion of a normed space of regular func-
tions, the Hilbert space H_a of *exactly controllable initial states* : $\forall(\theta_0, \theta_1) \in H_a$, $\exists u \in$
$L^2(0,T)$, $\mathcal{E}(T) = 0$. The space H_a is well-defined and *dense* in $H_0^1 \times L^2$ if and only if a is
a *strategic point* [9], i.e. $a \notin \mathbb{Q}$. Furthermore, if $a \neq a'$, then $H_a \neq H_{a'}$ [8]. Remark that
we may similarly conclude that system (1.1) is *approximately controllable* if and only if
a is irrational. So, if we want to exhibit the controllability property with respect to the
parameter a, a convenient *controllability criterion* could be the characteristic function of
the set $\mathbb{Q}\cap[0,1]$ (when it is equal to 1, the system is not controllable). Note that such
a function is equal to zero *almost everywhere*, which constitues a major drawback from
both mathematical and physical viewpoints (in fact, such a function should be considered
as null).

"Passive" control : This control is realized by a static feedback (corresponding to viscous point damping) :
$$u(t) = \alpha\,\partial_t\theta(a,t), \ with \ \alpha \leq 0 \ . \tag{1.5}$$
It easy to show that the energy of the system is decreasing : $\dot{\mathcal{E}}(t) \leq 0$, and

$$\frac{1}{|\alpha|}\int_0^t u^2(\tau)d\tau = \mathcal{E}(0) - \mathcal{E}(t) \ . \tag{1.6}$$

The parameter α may be optimized by solving the problem :
$$\min_\alpha \ \{\mathcal{E}(T)\}. \tag{1.7}$$

One may show that the solution of (1.7) is independant of a and T, and is given by $\alpha = 2$ (which, from a physical viewpoint, corresponds to the *impendance matching*).

Initial state : Convenient definitions of degrees of controllability should preferably be independant of the choice of the initial state. In the finite dimensional case, tools like the determinant of the controllability matrix were used [5]. This is impossible in the distributed case and we must consider other approaches.

From practical considerations, the initial state of the system is generally unknown *a priori*. Thus, we may consider it as a random variable. Furthermore, the initial energy of the system must be finite. So, suitable definitions of the degree of controllability will be stated from mean performance criteria. Remark that a detailed description of the random initial state may eventually be obtained directly from the modelling step and non necessarily through the knowledge of the eigen-modes of the system (which are generally very difficult to determine in practical situations).

Let $(\Omega, \mathcal{T}, p)$ be a probability space. We make the assumption that the initial state of (1.1) is a random variable : $(\theta_0, \theta_1) \in L^2(\Omega \; ; \; H_0^1 \times L^2)$ or equivalently : $(\rho, \eta) \in L^2(\Omega \; ; \; h^1 \times l^2)$. In the aim of simplification, we suppose that $E(\rho_n) = E(\eta_n) = 0$ (E denoting the mathematical expectation). Setting : $\sqrt{E(\rho_n^2)} \equiv r_n$ and $\sqrt{E(\eta_n^2)} \equiv s_n$, the finite energy hypothesis leads us to : $\mathcal{E}(0) = E(\mathcal{E}(0)) = \frac{1}{2} \sum_{n \geq 1} (n^2 \pi^2 r_n^2 + s_n^2) < +\infty \iff (r_n) \in h^1$, $(s_n) \in l^2$ (we consider in fact that $\mathcal{E}(0)$ is deterministic, but this assumption could easily be dropped). The sequences (r_n) and (s_n) discribe the mean harmonic containt of the initial state. Practically, it would be judicious to choose for instance : $r_n = 0$, $s_n \sim \frac{1}{n}$. Indeed, such a decreasing behavior forces the state $\theta(x, t)$ to be almost everywhere differentiable, with only lipschitz cusp points.

Definition of the degrees of controllability :

Definition 1 *the degree of <u>active</u> controllability relative to system (1.1) and problem (P_ϵ) is defined by the function :*
$$F_\epsilon = (f_\epsilon + \epsilon \, g_\epsilon)/\mathcal{E}(0) \, , \quad f_\epsilon \; : \; a \in]0,1[\longmapsto E[\, \mathcal{E}(T) \,] \quad g_\epsilon \; : \; a \in]0,1[\longmapsto E[\, \|u_\epsilon\|_{L^2(0,T)}^2 \,] , \tag{1.8}$$
with u_ϵ solution of (P_ϵ).

Definition 2 *the degree of <u>passive</u> controllability relative to system (1.1) is defined by the function :*
$$G_T \; : \; a \in]0,1[\longmapsto E[\, \mathcal{E}(T) \,]/\mathcal{E}(0) , \tag{1.9}$$
with u defined by (1.5).

Remark 1 *functions f_ϵ, g_ϵ correspond respectively to the mean residual energy of the string and the mean energy of the control u_ϵ. The function G_T corresponds to the mean residual energy of the string at time T.*

From (1.5), and the properties of the mathematical expectation, it is obvious that $F_\epsilon(a)$, $G_T(a) \in [0,1]$. On the other hand, it is natural to consider that system (1.1) is *well-controllable* when $F_\epsilon(a)$ or $G_T(a)$ are small. Note that, contrarily to the classical definitions [7], [12], [14], $F_\epsilon(a) = 1$ and $G_T(a) = 1$ only if the system is *not at all* controllable, *i.e.* the input function u does not permit any effect on the state of the system.

2 Degree of active controllability

In order to give more accurate expressions, we consider in the sequel :
$$T = 2 \; . \tag{2.1}$$
Indeed, in this case, the set of functions $\{\sin(n\pi t)\}_{n\geq 1} \cup \{\cos(n\pi t)\}_{n\geq 1} \cup \{\frac{1}{\sqrt{2}}\}$ constitues an orthonormal basis of $L^2(0, T)$ which leads to simplifications. All the calculations necessary in this section are standard and have been omitted. For a more detailed exposition, see [11].

Proposition 1 [11] *the functions f_ϵ, g_ϵ defined by (1.8) take the form :*

$$f_\epsilon(a) = \frac{1}{2}\Big(\sum_{n\geq 1} \frac{\epsilon^2 n^6 \pi^6}{(\sin^2(n\pi a) + \epsilon \, n^2\pi^2)^2} r_n^2 + \sum_{n\geq 1} \frac{\epsilon^2}{(\sin^2(n\pi a) + \epsilon)^2} s_n^2 \Big) \; , \tag{2.2}$$

$$g_\epsilon(a) = \frac{1}{2}\Big(\sum_{n\geq 1} \frac{n^2\pi^2 \sin^2(n\pi a)}{(\sin^2(n\pi a) + \epsilon \, n^2\pi^2)^2} r_n^2 + \sum_{n\geq 1} \frac{\sin^2(n\pi a)}{(\sin^2(n\pi a) + \epsilon)^2} s_n^2 \Big) \; . \tag{2.3}$$

In the aim of simplification, we now consider the case : $r_n = 0$, $n = 1, +\infty$. Then, $F_\epsilon(a)$ takes the form :

$$F_\epsilon(a) = \; \epsilon \, \frac{\displaystyle\sum_{n\geq 1} \frac{s_n^2}{\sin^2(n\pi a) + \epsilon}}{\displaystyle\sum_{n\geq 1} s_n^2} \; . \tag{2.4}$$

<u>Behaviors when $\epsilon \to 0$</u> : Given $a \in]0, 1[$, let us define the set :

$$I = \{n \in \mathbb{N}^*, \sin(n\pi a) = 0\} \; . \tag{2.5}$$

Proposition 2 *with the previous notation,*

$$f_0(a) \equiv \lim_{\epsilon\to 0} f_\epsilon(a) = \frac{1}{2}\sum_{n\in I} s_n^2 \; , \quad g_0(a) \equiv \lim_{\epsilon\to 0} g_\epsilon(a) = \frac{1}{2}\sum_{n\notin I} \frac{s_n^2}{\sin^2(n\pi a)} \; , \tag{2.6}$$

$$F_0(a) \equiv \lim_{\epsilon\to 0} F_\epsilon(a) = \frac{f_0(a)}{\mathcal{E}(0)} \; \; .$$

The proof is straightforward and essentially based on the permutation of $\lim$ and $\sum$. □

The mean control cost for problem (P_ϵ) (given by (2.6)) has a finite limit when $\epsilon \to 0$ only if $(\frac{s_n}{\sin(n\pi a)})_{n\notin I} \in l^2$. When $a \notin \mathbb{Q}$, we obtain the characterization of the exactly controllable initial state space previously established in [8] (restricted to $\theta_0 = 0$). The state constraint of the problem (P_0) is then satisfied if $a \notin \mathbb{Q}$ and $g_0(a) < +\infty$. We can add that, in the case $a \in \mathbb{Q}$, such a space could be characterized by : $s_n = 0$ *if* $n \in I$, and $(\frac{s_n}{\sin(n\pi a)})_{n\notin I} \in l^2$. This relation means that all the non controllable modes of the system must initially be at rest. On the other hand, result (2.6) is not surprising : $f_0(a)$ is equal to the mean energy of the uncontrolled modes. Indeed, this function is equal to zero if $a \notin \mathbb{Q}$. If $a \in \mathbb{Q}$, it is generally different from zero and represents the minimum possible value of the residual energy (independently of the control cost). Then, the function $F_0(a)$ could be considered as a controllability criterion : when it is equal to zero, the system is controllable. Note that such a criterion is more precise than the simple characteristic function of $\mathbb{Q} \cap [0, 1]$: indeed, it gives *quantitative* information at the non-controllability points. However, its lack of continuity remains unconvenient. We give in figure 2 and 1 the graphs of F_0 and F_ϵ when $s_n = \frac{1}{n}$. Finally, we have the following property :

Proposition 3 *for $a \in]0,1[$ and g_0 defined by (2.6),*

$$\forall a \in \mathbb{Q} , \ \forall (s_n) \in l^2 , \quad g_0(a) < +\infty ,$$
$$\forall a \notin \mathbb{Q} , \ \exists (s_n) \in l^2 , \quad g_0(a) = +\infty . \tag{2.7}$$

Proof :

1. $a \in \mathbb{Q}$. Then, the set : $E = \{\sin^2(n\pi a)\}_{n \notin I}$ is finite and $(s_n^2) \in l^2$.

2. $a \notin \mathbb{Q}$. We may use a lemma established in [8] : for any point $a \in]0,1[$, there exists a strictly increasing *integer* sequence (q_m) such that : $0 < |\sin(q_m \pi a)| \le \frac{\pi}{q_m}$.

   Let $F = \{q_m\}_{m \in \mathbb{N}}$ and $s = (s_n)$ defined by : $s_n = 0$ if $n \notin F$ and $s_n = \frac{1}{n}$ if $n \in F$; clearly, $s \in l^2$; furthermore $g_0(a) = \frac{1}{2} \sum_{n \in F} \frac{s_n^2}{\sin^2(n\pi a)} \ge \sum_{n \in F} \frac{1}{\pi^2} = +\infty$, since F is infinite. $\square$

This result is equivalent to the ill-posedness of problem (P_0). If we consider the Hilbert space H_a of exactly controllable initial states as the new state space associated to (1.1), then problem (P_0) becomes well-posed. But it should be noted that $H_a \ne H_{a'}$ if $a \ne a'$; and it could be shown from appropriate examples, that the solution u_0 of (P_0) is not continuous with respect to a. On the other hand, result (2.7) could be strengthened. Indeed, from (2.6), it is easy to show that if $a \in \mathbb{Q}$, the solution u_ϵ of (P_ϵ) is strongly convergent in $L^2(0,T)$ when ϵ goes to zero, for any initial state $(\theta_0, \theta_1) \in H_0^1 \times L^2$. This is not the case when $a \notin \mathbb{Q}$. So, in some sense and if the initial energy of high order modes is small, it would be preferable to consider (P_0) with a *non strategic point a*, the constraint being reduced to the controllable modes only. Remark that these degenerate behaviors confirm that only the problem (P_ϵ), $\epsilon > 0$, is to be considered in practice.

Regularity of F_ϵ, $\epsilon > 0$: The previous analysis has shown how the function F_ϵ degenerates when $\epsilon \to 0$. We now analyse its regularity properties when $\epsilon > 0$. We first need the following result :

Lemma 1 [11] *let φ be a continous 1-periodic function, with a symmetric graph with respect to point $(\frac{1}{2}, 0)$: $\varphi(\frac{1}{2} - x) = -\varphi(\frac{1}{2} + x)$, and $\alpha = (\alpha_n)$ a sequence belonging to l^2. Then, the function defined by : $g(x) = \sum_{n \ge 1} \alpha_n \varphi(nx)$, $x \in [0,1]$, belongs to $L^2(0,1)$.*

Let us now state the main result of this section :

Theorem 1 *for any $\epsilon > 0$, the function F_ϵ is continuous. Furthermore, if $|s_n| \le \dfrac{k}{n^{\frac{3}{4}+\nu}}$, $\nu > 0$, then F_ϵ is almost everywhere differentiable.*

Proof :

1. The convergence of the series is uniform with respect to a.

2. We show in fact that $F'_\epsilon \in L^2(0,1)$. Differentiation in $L^2(0,1)$ defines a closed operator ; so, it is sufficient to prove that the series : $\displaystyle\sum_{n\geq 1} \frac{d}{da}\left[\frac{\epsilon s_n^2}{(\sin^2(n\pi a)+\epsilon)}\right] \equiv S(a)$ belongs to $L^2(0,1)$. Via elementary calculation, it is easy to verify that $S(a)$ takes the form : $S(a) = \displaystyle\sum_{n\geq 1} n\pi s_n^2 \varphi(na)$, with φ possessing all the properties of lemma 1. Furthermore, if $\mid s_n \mid \leq \dfrac{k}{n^{\frac{3}{4}+\nu}}$, $\nu > 0$, then it is obvious that : $(n\pi s_n^2)_{n\geq 1} \in l^2$. $\quad\square$

This last result is very convenient. First, the continuity of F_ϵ means that the mean energy criterion $f_\epsilon + \epsilon g_\epsilon$ does not exhibit any essential differences between all the possible values of parameter a. Furthermore, small variation of this parameter will not *substantially* affect the control u_ϵ (recall that (P_ϵ) is well-posed) neither the mean value of the cost functionnal. This is in conformity with the intuitive understanding of the phenomenon. For any fixed ϵ, we may interpreat F_ϵ like a continuous quantitative measure of controllability, relative to problem (P_ϵ) : "when $F_\epsilon(a)$ is small, system(1.1) is *well-controllable* and this remains true for small variations of a". Secondly, with an additional hypothesis, F_ϵ is almost everywhere differentiable (we cannot expect more). This property strengthens the legitimacy of F_ϵ.

Particular case $s_n \sim 1/n$: This particular case corresponds to "regular" initial states, with only *lipschitz* cusp points (which do not constitues a major restriction from a practical viewpoint). Under this hypothesis, we may discribe in a simple way all the cusp points of the function F_ϵ and connect them to the controllability property. if we set $s_n = \frac{1}{n}$, the function F_ϵ becomes :

$$F_\epsilon(a) = \frac{\epsilon}{2} \sum_{n\geq 1} \frac{\frac{1}{n^2}}{\sin^2(n\pi a)+\epsilon} . \tag{2.8}$$

Lemma 2 *The function F_ϵ takes the form :* $F_\epsilon(a) = 4\pi^2\epsilon^2 \displaystyle\sum_{k\geq 0} \frac{\gamma^k(\epsilon)}{1-\gamma^2(\epsilon)}\varphi(ka)$, *where* $\gamma(\epsilon) = (1+4\epsilon) - \sqrt{4\epsilon(4\epsilon+2)}$, $0 < \gamma(\epsilon) < 1$, φ *is 1-periodic*, $\varphi(x) = \epsilon^2(\frac{1}{6} - 2x(1-x))$ *if $x \in [0,1]$.*

Proof : using the Fourier's series [13] and interchange the order of sommation [11].

Proposition 4 *The function F_ϵ has a cusp point (with negative concavity) at point a if and only if $a \in \mathbb{Q}\cap]0,1[$.*

Proof: Let $a = \dfrac{p}{q} \in]0,1[$ $p, q \in \mathbb{N}$; then it easy to verify : $\varphi'_-(ka) = -\varphi'_+(ka) = 2k\pi^2$, φ'_- , φ'_+ denoting respectively the right and left half derivatives. Furthemore, if $a \notin \mathbb{Q}$, $\varphi(kx)$ is differentiable at point a. Hence, F_ϵ is differentiable at point a ; on the other hand, $\dfrac{\gamma^k(\epsilon)}{1-\gamma^2(\epsilon)} > 0$, so if $a \in \mathbb{Q}\cap]0,1[$, it is obvious that : $F'_{\epsilon-}(ka) > F'_{\epsilon+}(ka)$ which completes the proof. $\quad\square$

We may easily conclude :

Theorem 2 *In the case $s_n \sim \frac{1}{n}$, the system (1.1) is controllable if and only if its degree of active controllability is differentiable at point a.*

3 Degree of passive controllability

The mathematical study of the function G_T is yet to be completed. So, we only give partial results. Nevertheless, the numerical computation of G_T exhibits a very close analogy with F_ϵ. Concerning the regularity of G_T, continuity with repect to a may be obtained from Fourier transform technics ; differentiability is actually under study. In the aim of simplification, we make the same hypothesis as in **2**, that is : $r_n \equiv 0$. we have first the folowing result.

Proposition 5 *let I defined by (2.5). We have for $a \in]0,1[$:*

1. $\forall a \notin \mathbb{Q}, \quad \lim\limits_{T \to +\infty} \mathcal{E}(T) = 0 \qquad p - a.e. , \quad \lim\limits_{T \to +\infty} G_T(a) = \lim\limits_{T \to +\infty} E(\mathcal{E}(T)) = 0 ,$

2. $\forall a \in \mathbb{Q}, \quad \lim\limits_{T \to +\infty} \mathcal{E}(T) = \sum\limits_{n \in I} \eta_n^2 \quad p - a.e. , \quad \lim\limits_{T \to +\infty} G_T(a) = \lim\limits_{T \to +\infty} E(\mathcal{E}(T)) = \sum\limits_{n \in I} s_n^2 .$

Proof :

1. using Lasalle's principle and compactness of trajectories [2], [3], and the dominated convergence theorem (permutation of lim and E).

2. we may split the system (1.3), (1.5) into two decoupled sub-systems corresponding respectively to the sets of eigen-functions : $\{\varphi_n , n \in I\}$ and $\{\varphi_n , n \notin I\}$. The first system is conservative ; on the second, we may apply the same arguments as in 1. $\square$

Corollary 1 *So, we have :* $\lim\limits_{T \to +\infty} G_T(a) = \lim\limits_{\epsilon \to 0} F_\epsilon(a) = F_0(a) , \forall a \in]0,1[$.

Let us now decribe the principal steps of the computation of G_T. Denoting $X = (\theta, \partial_t \theta)$, the system (1.1), (1.5) becomes : $\dot{X} = A\,X$, $X(0) = X_0$, whose solution is written : $X(t) = \Phi(t)\,X_0$. Consider the operator $C : l^2 \to H_0^1 \times L^2$, defined by $C\eta = (0, \theta_1)$ in (1.1), (1.2), (1.3) . We may write :

$$G_T = E(\|\Phi(T)C\,\eta\|^2) = \sum_{i,j} K_{ij} E(\eta_i\,\eta_j) , \qquad (3.1)$$

with $K = C^* \,\Phi^*(T)\,\Phi(T)\,C$.

The computation of G_T is then based on the knowledge of $\Phi(T)$ and the probabilistic correlations : $E(\eta_i\,\eta_j)$. A numerical finite dimensional approximation of $\Phi(T)$ has been done using Riemann Invariant technics.

In figure 3 we can see the function G_T, $T = 4$, $T = 25$. The comparison with figure 1, reveals a very close analogy between G_T and F_ϵ. A better comparison is obtained if we consider the function : $H_{T,\lambda} = G_T + \lambda\,E(\int_0^T u^2(t)\,dt)$, $\lambda > 0$. From (1.6) it is easy to see that $H_{T,\lambda} = (1 - \lambda)\,G_T + \lambda\,\mathcal{E}(0)$. Figure 4 shows that graphs of F_ϵ, $H_{T,\lambda}$, for $\epsilon = 0.04$, $T = 4$, $\lambda = 0.05$.

Conclusion

Binary notions of controllability for distributed systems are unsufficient from a practical viewpoint : they are often non robust, they do not give any quantitative information on the efficiency of the control and they lead to heavy difficulties when the eigen-modes of the system are unknown. On the contrary, the quantitative degrees of controllability defined in the present paper lead to robustness (continuity with respect to the parameter) and are clearly linked to optimal design problems. According to these degrees of controllability, a sentence like "the system is or is not controllable" could be replaced by "the system is more or less controllable". So, "points of non controllability", which constitue a dense subset in our example, may advantageously be replaced by "zones of bad controllability". The degree of active controllability is very interesting because it directly measures the efficiency of the optimal control. The degree of passive controllability is only based on numerical simulation of the system, which presents a real advantage. Some problems are not yet completely solved, namely the question of the equivalence (in some sense) between the two degrees of controllability and, of course, the generalization of such concepts to abstract dynamic systems. Nevertheless, the particular case considered here (as an extension of the matching impedance properties) illustrates the legitimacy of this kind of approach.

References

[1]   A.A ANDRONOV, L.S. PONTRYAGIN (1956). *Systèmes grossiers.* Œuvres complètes d'Andronov. Editions de l'académie des sciences d'URSS.

[2]   F. CONRAD (1992). *Stabilization of second order evolution equations by unbounded non linear feedback.* Research report, INRIA, may 1992.

[3]   F. CONRAD (1993). *Private correspondence.*

[4]   R.F. CURTAINS, A.J. PRITCHARD (1978). *Infinite dimensional linear systems theory.* Lecture notes in control and information sciences. vol. 8. Springer-Verlag.

[5]   X. GUANGQIAN, P.M. BAINUM (1992). *Actuator placement using degree of controllability for discrete-time systems.* Transaction of ASME, vol. 114, pp. 508-516.

[6]   R.E. KALMAN, Y.C. HO, K.S. NARENDRA (1961). *Controllability of linear dynamical systems.* Contributions to Differential Equations, vol. 1, No. 2, pp. 182-213.

[7]   C.D. JOHNSON (1969). *Optimization of a certain quality of complete controllability and observation for linear dynamical systems.* ASME journal of basic engineering, vol.191, pp. 228-238.

[8]   S. JAFFARD. *Sur le contrôle ponctuel des cordes vibrantes et des poutres.* Preprint.

[9]   J.L. LIONS (1989). *Contrôlabilité exacte. Perturbations et stabilisation de systèmes distribués,* tome 1. Masson

[10]   G. MONTSENY (1990). *Commande de systèmes à paramètres répartis : un exemple de problème mal posé.* RAIRO APII, 24, p323-336.

[11]   G. MONTSENY, P. BENCHIMOL (1992). *Degree of controllability for point controlled vibrating string.* LAAS Report N° 92361. Submitted to Automatica.

[12]   P.C. MULLER, H.I. WEBER (1972). *Analysis and optimization of certain qualities of controllability and observability for linear dynamical systems.* Automatica, vol. 8, pp. 237-246.

[13]   M.R. SPIEGEL (1980). *Laplace transforms*. Mc Graw-Hill

[14]   R. TOMOVIC (1965). *Controllability, Invariancy, and sensitivity*. Proceedings third Allerton conference, pp. 17-26.

[15]   C.N. VISWANATHAN, R.W. LONGMAN (1979). *A definition of the degree of controllability-A criterion for actuator placement*. Proceedings of second VPI & SU/AIAA Symposium on dynamics and control of large flexible spacecraft, Blacksburg, VA, pp. 369-381.

Figures

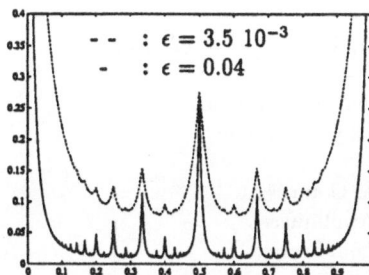

Figure 1: *function F_ϵ*

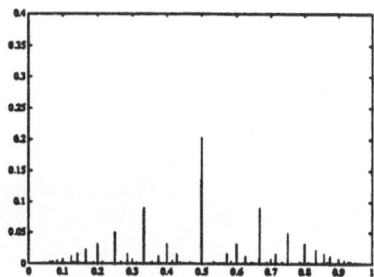

Figure 2: *function F_0*

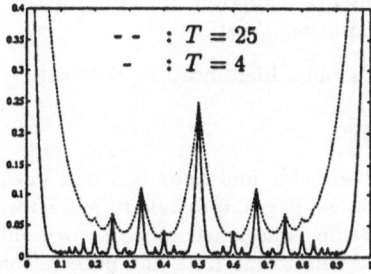

Figure 3: *function G_T*

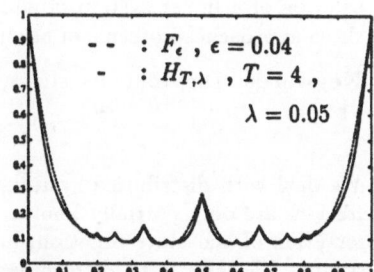

Figure 4: *comparison F_ϵ and $H_{T,\lambda}$*

Algorithm

NUMERICAL METHODS
TO COMPUTE SENTINELS FOR PARABOLIC SYSTEMS
WITH AN APPLICATION
TO SOURCE TERMS IDENTIFICATION

Timo Männikkö[1], Olivier Bodart[2], Jean-Pierre Kernévez[2]

[1]University of Jyväskylä, Department of Mathematics,
P.O. Box 35, FIN-40351 Jyväskylä, Finland

[2]Université de Technologie de Compiègne,
Division Mathématiques Appliquées,
BP 649, F-60206 Compiègne, France

Abstract. We apply the method of sentinels to the identification of source terms in parabolic systems. We present two numerical approaches; the first one is based on the solution of an optimal control problem. and the second one is based on the solution of a linear system of equations. In numerical experiments, we compare these approaches in terms of accuracy and computational cost.

Keywords. Distributed systems, Exact controllability, Identification, Sentinels.

1. INTRODUCTION

We deal with distributed systems whose right-hand side and boundary and initial conditions are only partially known. Since, in practice, it is impossible to get a total observation of the state function, also the observation is assumed to be known only partially. The aim is to determine the right-hand side terms from the observations despite the uncertainties on the boundary and initial data.

The method of sentinels has been designed to achieve this goal. Sentinels, introduced by J.L. Lions in [8]–[9], are weighted integrals whose values are selectively sensitive to some specific unknown terms and insensitive to the others.

Let Ω be an open bounded domain in $\mathbb{R}^2$ and let $T > 0$. Let us denote $Q = \Omega \times]0, T[$ and $\Sigma = \partial\Omega \times]0, T[$. Moreover, let δ_x denote the Dirac distribution at point $x \in \mathbb{R}^2$. The state function y is supposed to solve the parabolic system

$$(1.1) \quad \begin{cases} \dfrac{\partial y}{\partial t} - \Delta y + \sigma y = \xi + \displaystyle\sum_{i=1}^{n} \delta_{a_i} s_i & \text{in } Q, \\ \qquad\qquad y = g + \tau_1 \hat{g} & \text{on } \Sigma, \\ \quad y(\cdot, 0) = y^0 + \tau_0 \hat{y}^0 & \text{in } \Omega, \end{cases}$$

where Δ is the Laplace operator and $\sigma \geq 0$. The distribution ξ is given in a suitable space, $(a_i)_{i=1}^n$ is a given set of points of Ω and functions $s_i \in L^2(]0,T[)$ are unknown. Furthermore, $g \in L^2(\Sigma)$ and $y^0 \in L^2(\Omega)$ are given, but $\hat{g} \in L^2(\Sigma)$ and $\hat{y}^0 \in L^2(\Omega)$ are unknown, as are also $\tau_0, \tau_1 \in \mathbb{R}$.

Using the transposition method, it can be easily shown that, for given $(s_i)_{i=1}^n$, $\tau_1 \hat{g}$ and $\tau_0 \hat{y}^0$, the system (1.1) admits a unique weak solution $y \in L^2(Q)$. We assume that this state y is observed in $\mathcal{O} \times]0,T[$, where $\mathcal{O}$ is an open subset of Ω. The aim is then to determine the functions s_i using that observation.

The paper is organized as follows. In Section 2 we present briefly the theory of sentinels and their use in parameter identification. In Section 3 we present two numerical approaches for computing sentinels. Finally, in Section 4 we give some numerical test results.

2. SENTINELS AND SOURCE TERMS IDENTIFICATION

When identifying the functions s_i, we use the notion of sentinels, as introduced in Lions [8]–[9]. Let $(e_j)_{j=0}^\infty$ be an orthonormal basis of $L^2(]0,T[)$. Then we can write

$$(2.1) \qquad s_i(t) = \sum_{j=0}^\infty s_{ij} e_j(t),$$

where s_{ij} are real numbers. Let us denote $\tau = (\tau_0, \tau_1)$. We then consider the functional

$$(2.2) \qquad \mathcal{S}((s_{ij}), \tau) = \int_{\mathcal{O} \times]0,T[} w(x,t) y(x,t) \, dx \, dt,$$

where $w \in L^2(\mathcal{O} \times]0,T[)$ is to be determined.

Definition. *Let c be a (nonzero) real number. Functional $\mathcal{S}$ is said to be a sentinel for s_{kl} (where k and l are fixed) with the sensitivity c, if*

$$(2.3\text{i}) \qquad \frac{\partial \mathcal{S}(0,0)}{\partial \tau_0} = 0,$$

$$(2.3\text{ii}) \qquad \frac{\partial \mathcal{S}(0,0)}{\partial \tau_1} = 0,$$

$$(2.3\text{iii}) \qquad \frac{\partial \mathcal{S}(0,0)}{\partial s_{ij}} = c'_{ij} \quad \forall \, (i,j),$$

where $c'_{kl} = c$ and $c'_{ij} = 0$ for $(i,j) \neq (k,l)$.

The conditions (2.3) obviously have to be reformulated in such a way that they yield a characterization of the function w. That will be done by introducing the solution of an adjoint state equation.

Proposition 2.1. *Let $q \in H^{2,1}(Q)$ be the solution of the parabolic system*

$$(2.4) \qquad \begin{cases} -\dfrac{\partial q}{\partial t} - \Delta q + \sigma q = w\chi_{\mathcal{O}} & \text{in } Q, \\ q = 0 & \text{on } \Sigma, \\ q(\cdot, T) = 0 & \text{in } \Omega, \end{cases}$$

where $\chi_{\mathcal{O}}$ denotes the characteristic function of $\mathcal{O}$. The conditions (2.3) are equivalent to

$$(2.5\text{i}) \qquad\qquad\qquad q(\cdot, 0) = 0 \quad \text{a.e. in } \Omega,$$

$$(2.5\text{ii}) \qquad\qquad\qquad \frac{\partial q}{\partial \nu} = 0 \quad \text{a.e. on } \Sigma,$$

$$(2.5\text{iii}) \qquad\qquad\qquad \int_0^T q(a_i, t) e_j(t)\, dt = c'_{ij} \quad \forall\, (i,j).$$

Proof. The result is easily proved by substituting w in (2.2) by the left-hand side of (2.4) and integrating by parts. (We note that, due to the Sobolev imbedding theorem, q is continuous with respect to space variable, so that the integral in (2.5iii) is well defined.) $\qquad\square$

Conditions (2.5) express an exact controllability type problem where the desired state is $\{0, 0, (c'_{ij})\}$. For such problems the Hilbert Uniqueness Method (HUM) of J.L. Lions, introduced in [6]–[7], can be used. However, the reachability of $\{0, 0, (c'_{ij})\}$ is not ensured due to the regularizing effects of the heat equation. Thus, the approximate controllability has to be proved for this problem, so that (2.3) can be satisfied with arbitrary accuracy.

Theorem 2.2. *The set $\{\{q(\cdot, 0), \partial q/\partial \nu, (\int_0^T q(a_i, t) e_j(t)\, dt)\} \mid w \text{ spans } L^2(\mathcal{O}\times]0, T[)\}$ is dense in $L^2(\Omega) \times L^2(\Sigma) \times \{(\alpha_{ij}) \mid \sum_{j=0}^{\infty} \alpha_{ij}^2 < \infty\}$.*

Sketch of the proof. We use the theorem of Hahn-Banach. Let us consider a triplet $\{\rho^0, \rho^1, (\mu_{ij})\} \in L^2(\Omega) \times L^2(\Sigma) \times \{(\alpha_{ij}) \mid \sum_{j=0}^{\infty} \alpha_{ij}^2 < \infty\}$ such that

$$(2.6) \qquad \left(\{\rho^0, \rho^1, (\mu_{ij})\},\ \{q(\cdot, 0), \frac{\partial q}{\partial \nu}, (\int_0^T q(a_i, t) e_j(t)\, dt)\} \right) = 0$$

for any $w \in L^2(\mathcal{O}\times]0, T[)$. (Here $(\cdot, \cdot)$ denotes the inner product in $L^2(\Omega) \times L^2(\Sigma) \times \{(\alpha_{ij}) \mid \sum_{j=0}^{\infty} \alpha_{ij}^2 < \infty\}$.) Let ρ solve the system

$$(2.7) \qquad \begin{cases} \dfrac{\partial \rho}{\partial t} - \Delta \rho + \sigma \rho = \displaystyle\sum_{i=1}^{n} \sum_{j=0}^{\infty} \mu_{ij} \delta_{a_i} e_j & \text{in } Q, \\ \rho = -\rho^1 & \text{on } \Sigma, \\ \rho(\cdot, 0) = \rho^0 & \text{in } \Omega. \end{cases}$$

Multiplying (2.7) by q and integrating by parts, the condition (2.6) is shown to be equivalent to

$$(2.8) \qquad \rho = 0 \text{ a.e. in } \mathcal{O} \times]0, T[.$$

Then, applying the uniqueness theorem of J.C. Saut and B. Scheurer (see [12]), it can be shown that (2.8) implies $\{\rho^0, \rho^1, (\mu_{ij})\} = 0$. (To be in the framework of that theorem it must be shown that $\rho \in L^2(\delta, T; H^2_{loc}(\Omega \setminus \{\cup a_i\}))$ for any $\delta > 0$, which is a non-obvious regularity result (see Bodart and Fabre [2]).) $\qquad \square$

When applying the sentinels to the identification of the source terms, we first define a set of sentinels, each of them being sensitive to one basis function for one source (i.e., for one s_{ij}) and insensitive to everything else. Then, for fixed k and l, we have

$$(2.9) \qquad \begin{aligned} \mathcal{S}((s_{ij}), \tau) &= \mathcal{S}(0,0) + \tau_0 \frac{\partial \mathcal{S}(0,0)}{\partial \tau_0} + \tau_1 \frac{\partial \mathcal{S}(0,0)}{\partial \tau_1} + \sum_{i=1}^{n} \sum_{j=0}^{\infty} s_{ij} \frac{\partial \mathcal{S}(0,0)}{\partial s_{ij}} \\ &\approx \mathcal{S}(0,0) + s_{kl} c. \end{aligned}$$

Hence, the coefficient s_{kl} is estimated by

$$(2.10) \qquad s_{kl} \approx \frac{1}{c} \Big[\mathcal{S}((s_{ij}), \tau) - \mathcal{S}(0,0) \Big],$$

where the first term on the right-hand side is obtained from the observation of y, and the second term is calculated by using the solution of (1.1) with $((s_{ij}), \tau) = (0,0)$.

Since it is not reasonable to compute an infinite set of sentinels, we use instead of (2.1) the approximation

$$(2.11) \qquad s_i(t) \approx \sum_{j=0}^{N} s_{ij} e_j(t)$$

with some suitable N, so that only a finite number of sentinels is needed for the reconstruction of source terms. Natural choice for basis functions e_j is to use the standard "hat functions" centered on points $t_j = j \Delta t$, where $\Delta t = T/N$.

3. Numerical Methods

In our first approach we apply an optimal control algorithm: We want to find w such that conditions (2.5) are satisfied at machine precision. The algorithm consists then in solving the problem

$$(3.1) \qquad \begin{aligned} \min_{w \in L^2(\mathcal{O} \times]0, T[)} \Big\{ &\frac{\varepsilon}{2} \int_{\mathcal{O} \times]0,T[} w^2 \, dx \, dt + \frac{1}{2} \int_{\Omega} q(x,0)^2 \, dx + \frac{1}{2} \int_{\Sigma} \left(\frac{\partial q}{\partial \nu} \right)^2 d\Sigma \\ &+ \frac{1}{2} \sum_{i=1}^{n} \sum_{j=0}^{N} \left(\int_0^T q(a_i, t) e_j(t) \, dt - c_{ij}^l \right)^2 \Big\}, \end{aligned}$$

where q is the solution of (2.4) and $\varepsilon > 0$ is some real number. This approach does not give control of minimum norm, but the precision on the desired state is satisfactory. The implementation of such an algorithm is quite easy, as the gradient of the cost function is obtained by the standard state–adjoint state formulation: In this case the adjoint equation is

$$
(3.2) \quad
\begin{cases}
\dfrac{\partial p}{\partial t} - \Delta p + \sigma p = \displaystyle\sum_{i=1}^{n} \sum_{j=0}^{N} \left(\int_0^T q(a_i, t) e_j(t)\, dt - c'_{ij} \right) \delta_{a_i} e_j & \text{in } Q, \\[4mm]
p = -\dfrac{\partial q}{\partial \nu} & \text{on } \Sigma, \\[3mm]
p(\cdot, 0) = q(\cdot, 0) & \text{in } \Omega,
\end{cases}
$$

and the gradient is then given by

$$
(3.3) \qquad \left(J'_\varepsilon(w), \hat{w} \right) = \int_{\mathcal{O} \times]0, T[} (\varepsilon w + p)\hat{w}\, dx\, dt,
$$

where J_ε denotes the cost function in (3.1).

The second approach consists in discretizing the insensitivity conditions (2.3). We consider the decomposition of $\tau_0 \hat{y}^0$ on an basis of $L^2(\Omega)$, and we ask the sentinel to be insensitive to the variations of the finite number of coordinates on that basis. Similar decomposition is done to $\tau_1 \hat{g}$. More precisely, we use the approximations

$$
(3.4) \qquad \tau_0 \hat{y}^0 \approx \sum_{j=1}^{m_0} \tau_{0j} \hat{y}^0_j, \qquad \tau_1 \hat{g} \approx \sum_{j=1}^{m_1} \tau_{1j} \hat{g}_j,
$$

where $(\hat{y}^0_j)_{j=1}^{\infty}$ and $(\hat{g}_j)_{j=1}^{\infty}$ are the two bases in question (deduced from the finite element approximation), and τ_{0j}, τ_{1j} are the unknown coordinates. The conditions (2.3i) and (2.3ii) are then replaced by

$$
(3.5\mathrm{i}) \qquad \frac{\partial S(0,0)}{\partial \tau_{0j}} = 0 \quad \forall\, j \in \{1, \dots, m_0\},
$$

$$
(3.5\mathrm{ii}) \qquad \frac{\partial S(0,0)}{\partial \tau_{1j}} = 0 \quad \forall\, j \in \{1, \dots, m_1\},
$$

and the condition (2.3iii) remains unchanged.

Let the functions $\varphi_{0j}, \varphi_{1j}$ and ψ_{ij} solve the systems

$$
(3.6) \quad
\begin{cases}
\dfrac{\partial \varphi_{0j}}{\partial t} - \Delta \varphi_{0j} + \sigma \varphi_{0j} = 0 & \text{in } Q, \\[3mm]
\varphi_{0j} = 0 & \text{on } \Sigma, \\[2mm]
\varphi_{0j}(\cdot, 0) = \hat{y}^0_j & \text{in } \Omega,
\end{cases}
$$

$$(3.7) \quad \begin{cases} \dfrac{\partial \varphi_{1j}}{\partial t} - \Delta \varphi_{1j} + \sigma \varphi_{1j} = 0 & \text{in } Q, \\[2mm] \varphi_{1j} = \hat{g}_j & \text{on } \Sigma, \\[2mm] \varphi_{1j}(\cdot, 0) = 0 & \text{in } \Omega, \end{cases}$$

$$(3.8) \quad \begin{cases} \dfrac{\partial \psi_{ij}}{\partial t} - \Delta \psi_{ij} + \sigma \psi_{ij} = \delta_{a_i} e_j & \text{in } Q, \\[2mm] \psi_{ij} = 0 & \text{on } \Sigma, \\[2mm] \psi_{ij}(\cdot, 0) = 0 & \text{in } \Omega, \end{cases}$$

respectively. Let us denote $(\phi_i)_{i=1}^{M} = \{(\varphi_{0j}|_{\mathcal{O} \times]0,T[}), (\varphi_{1j}|_{\mathcal{O} \times]0,T[}), (\psi_{ij}|_{\mathcal{O} \times]0,T[})\}$, where $M = m_0 + m_1 + n(N+1)$. Using the same uniqueness result as before, it is easy to show (see Kernévez [5]; cf. also Männikkö [10]) that the functions (ϕ_i) are linearly independent, and thus, they form a basis in an M-dimensional subspace of $L^2(\mathcal{O} \times]0,T[)$. Our second approach is then based on the following result:

Theorem 3.1. *There exists a unique function $w \in L^2(\mathcal{O} \times]0,T[)$ such that conditions (3.5i), (3.5ii), (2.3iii) hold, and it is of minimal norm. This w belongs to the finite dimensional subspace of $L^2(\mathcal{O} \times]0,T[)$ spanned by the basis (ϕ_i). The coordinates (λ_i) of w on that basis are obtained by solving a linear system of equations $A\lambda = b$, where the elements of the matrix $A = (a_{ij})$ are given by $a_{ij} = \int_{\mathcal{O} \times]0,T[} \phi_i \phi_j \, dx \, dt$, and the right-hand side vector is $b = \{0, 0, (c'_{ij})\}^T$.*

Sketch of the proof. This result can be proved by using the method of HUM. The conditions (3.5) are equivalent to

$$(3.9\text{i}) \qquad \int_{\Omega} q(x,0) \hat{y}_j^0(x) \, dx = 0 \quad \forall\, j \in \{1, \ldots, m_0\},$$

$$(3.9\text{ii}) \qquad \int_{\Sigma} \frac{\partial q}{\partial \nu} \hat{g}_j \, d\Sigma = 0 \quad \forall\, j \in \{1, \ldots, m_1\},$$

where q solves the adjoint equation (2.4). The discrete HUM optimality system is as follows: Let φ be the solution of

$$(3.10) \quad \begin{cases} \dfrac{\partial \varphi}{\partial t} - \Delta \varphi + \sigma \varphi = \displaystyle\sum_{i=1}^{n} \sum_{j=0}^{N} \alpha_{ij} \delta_{a_i} e_j & \text{in } Q, \\[4mm] \varphi = \displaystyle\sum_{j=1}^{m_1} \beta_{1j} \hat{g}_j & \text{on } \Sigma, \\[4mm] \varphi(\cdot, 0) = \displaystyle\sum_{j=1}^{m_0} \beta_{0j} \hat{y}_j^0 & \text{in } \Omega, \end{cases}$$

where α_{ij}, β_{1j} and β_{0j} are real numbers. Let us denote $(\lambda_i)_{i=1}^{M} = \{(\beta_{0j}), (\beta_{1j}), (\alpha_{ij})\}$. Let A be the $M \times M$ symmetric positive definite matrix whose elements are defined by

$$(3.11) \qquad a_{ij} = \int_{\mathcal{O} \times]0,T[} \varphi_i \varphi_j \, dx \, dt.$$

Here φ_i is the solution of (3.10) when λ is the ith vector of the canonical basis of $\mathbb{R}^M$ (i.e., $\varphi_i = \phi_i$). This matrix corresponds to the discretization of the continuous HUM operator. By first solving the equation

$$(3.12) \qquad A\lambda = \{0, 0, (c'_{ij})\}^T,$$

and then taking $w = \varphi|_{\mathcal{O} \times]0,T[}$, we obtain a function w with the desired properties. Moreover, according to (3.6)–(3.8) and (3.10), it holds $w = \sum_{i=1}^M \lambda_i \phi_i$. Finally, it can be shown by several methods that this algorithm gives the control w of minimal norm for the discrete problem. $\qquad \square$

The matrix A is nonsingular but, since it corresponds to the discretization of the HUM operator, it is often ill-conditioned, and therefore, the problem is regularized by using the matrix $A + \varepsilon I$ (with some small $\varepsilon > 0$) instead.

4. NUMERICAL RESULTS

In our numerical example we consider the following problem: The domain $\Omega \subset \mathbb{R}^2$ is presented in Figure 1; the coordinates of the corners are $(0.054, 0.107)$, $(-0.035, 0.752)$, $(0.912, 0.484)$ and $(0.933, 0.293)$ (clockwise starting from lower left corner). The domain is discretized by using the finite element method (FEM). The mesh is given also in Figure 1; there are 36 rectangular elements and 135 nodal points (marked with small circles). The time interval is $]0, 1[$, which is discretized by using the finite difference method (implicit Euler scheme). The number of time levels is $N = 10$.

There are two point sources at $(0.599, 0.528)$ and $(0.488, 0.814)$ (marked with asterisks in Figure 1), and the observatory $\mathcal{O}$ consists of the four elements in the leftmost part of the domain (shaded in Figure 1). Furthermore, we have $\sigma = 1$, $\xi = 0$, $g = 0$ and $y^0 = 0$ in system (1.1). In this example we have also $\tau_1 \hat{g} = 0$, so that there is no boundary uncertainty. (Numerical experiments with nonzero $\tau_1 \hat{g}$ are currently in progress).

For test purposes we choose some suitable functions s_i (see below) and then construct the observation of y such that it corresponds to these functions. In this way we may compare the computed solution with the exact solution. The problem has been solved with both approaches presented in Section 3. The regularization parameter ε is 10^{-9} in both cases. In the second approach we have used the FEM basis functions as the approximation basis $(\hat{y}_j^0)$.

The results obtained with the second approach (linear system of equations) are presented in Figures 2 and 3, where s_1 and s_2 are given as functions of time. The solid curves represent the exact expressions $s_1(t) = 116(1 + \sin(2\pi t))$ and $s_2(t) = -41(\sin t + \sin(3t)/3 + \sin(5t)/5 + \sin(7t)/7 + \sin(9t)/9)$, and the small circles show the calculated approximations. As can be seen, the calculated solutions are very near the exact ones, especially in the case of s_1.

The first approach (optimal control problem) gave similar results; there is very little difference in the given accuracy of the methods. However, the computational cost was not the same. During the first approach, partial differential equations of the type (1.1) were solved 6100 times, and this required about $3.5 \cdot 10^9$ flops. For the second approach the corresponding values were 103 equations and $6.5 \cdot 10^7$ flops. Also in other test

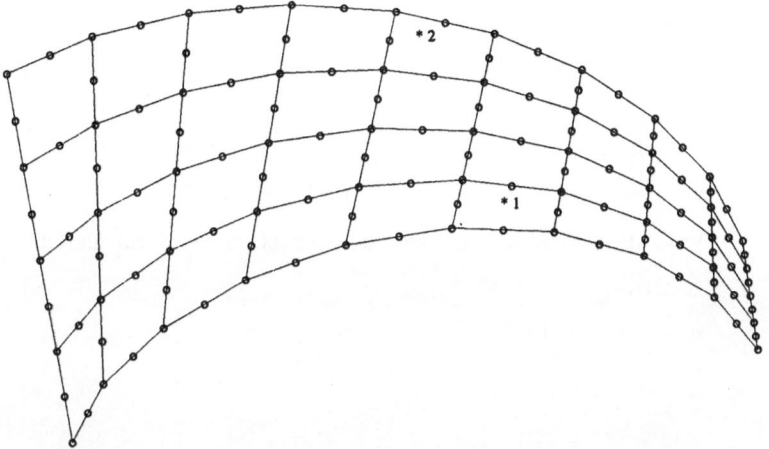

FIGURE 1. Domain Ω, sources a_1 and a_2, and observatory $\mathcal{O}$.

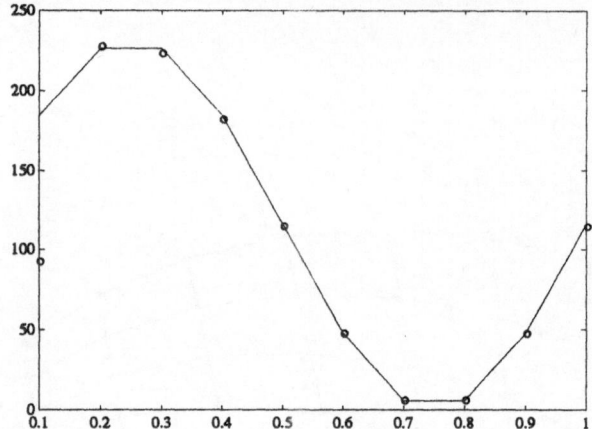

FIGURE 2. Function s_1; solid line = exact, circles = calculated.

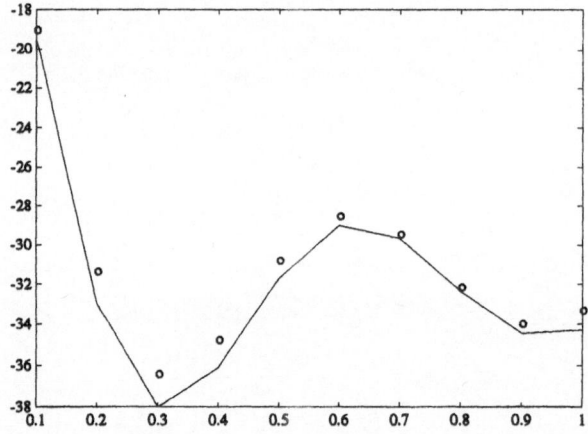

FIGURE 3. Function s_2; solid line = exact, circles = calculated.

examples the methods behaved very similarly, i.e., the accuracy is almost the same but the computational cost is different.

REFERENCES

[1] Bodart, O., "Ph.D. Thesis", Université de Technologie de Compiégne (to appear).

[2] Bodart, O. and Fabre, C., *Contrôle insensibilisant la norme de la solution d'une équation de la chaleur semi-linéaire*, C. R. Acad. Sci. Paris, Sér. I Math. **316** (1993), 789–794.

[3] Bodart, O. and Kernévez, J.P., *Sentinels in Rivers*, in "Jornadas Hispano-Francesas sobre Control de Sistemas Distribuidos", (Proceedings), Málaga, Spain, October 25–26, 1990, pp. 69–76.

[4] Bodart, O., Kernévez, J.P. and Männikkö, T., *Sentinels for Distributed Environmental Systems*, in "Modelling, Identification and Control", (ed. Hamza, M.H.), Proceedings of the 11th IASTED International Conference, Innsbruck, Austria, February 10–12, 1992, pp. 219–222.

[5] Kernévez, J.P., "Méthodes numériques de calcul de sentinelles. Application à l'identification de sources de pollution", Collection RMA, Masson, Paris (to appear).

[6] Lions, J.L., *Exact Controllability, Stabilization and Perturbations for Distributed Systems*, SIAM Rev. **30** (1988), 1–68.

[7] Lions, J.L., "Contrôlabilité exacte, perturbations et stabilisation des systèmes distribués", Tomes 1 & 2, Collection RMA, Masson, Paris, 1988.

[8] Lions, J.L., *Sur les sentinelles des systèmes distribués, Le cas des conditions initiales incomplètes*, C. R. Acad. Sci. Paris, Sér. I Math. **307** (1988), 819–823.

[9] Lions, J.L., "Sentinelles pour les systèmes distribués à données incomplètes", Collection RMA, Masson, Paris, 1992.

[10] Männikkö, T., *Method of Sentinels for Eliminating Uncertainties and Noise in the Observation*, Advances in Math. Sci. and Appl. (to appear).

[11] Männikkö, T., Kernévez, J.P. and Lions, J.L., *Sentinels for Monitoring the Environment*, in "Intelligent Process Control and Scheduling, Modelling and Control of Water Resources Systems and Global Changes", (eds. Kerckhoffs, E.J.H. et al.), Proceedings of the 1991 European Simulation Symposium, Ghent, Belgium, November 6–8, 1991, pp. 229–234.

[12] Saut, J.C. and Scheurer, B., *Unique Continuation for Some Evolution Equations*, J. Diff. Eq. **66** (1987), 118–139.

OPTIMAL PERIODIC CONTROL
FOR DISTRIBUTED-PARAMETER SYSTEMS
VIA REDUCED GRADIENT METHODS

Krystyn Styczeń and Krystyna Nitka-Styczeń

Institute of Engineering Cybernetics, Technical University of Wrocław

Janiszewskiego 11/17, 50-372 Wrocław, Poland

Abstract. This paper deals with optimal periodic control problems for distributed-parameter systems with many kinds of controls (lumped and distributed). Descent optimization methods reduced to the control space are proposed. The formulas of partial gradients with respect to lumped and distributed controls are derived and their applicability to technological and biological problems is pointed-out.

1. INTRODUCTION

Time-periodic control is useful for the optimization of many distributed-parameter processes occurring in

• physico-chemical industrial systems, where this control may optimize processes performed in tubular apparatus such as tubular chemical reactors, separation columns, heat exchangers etc., [5], [9],

• biochemical systems, where this control may optimize the biomass and metabolites production [8],

• ecological and social service systems, where this control may optimize the exploitation of age-structured populations [2], [6].

Potential applications of such a control are yet more wide, because periodic phenomena are immanent for time-space processes in the cell biology, the regional economic development, the growth of microbiological populations with diversified physiological characteristics, etc. [2], [4], [7], [8]. The time-periodic control may, for example, enhance the process performance or its selectivity, or diminish the consumption of raw materials and energy.

As distinct from the case of lumped-parameter systems [5], [9], methods for the determination of optimal time-periodic control (OTPC) for distributed-parameter processes are less investigated [3], [5], [9]. In view of the complexity of OTPC problems, there is a special need of the design of constructive computational algorithms determining the optimal control.

We propose a family of descent methods based on the performance index gradient reduced to the control space. The most difficult constraints, i.e. the process equations with boundary conditions, are here resolved with respect to the state trajectory, while additional unsolvable constraints are approximately taken into account with the help of the shifted penalty function [13]. Such methods are applicable even for complicated OTPC problems containing complex non-linearities and controls of different kinds (lumped and distributed).

We always omit the argument of time-periodic functions treated as elements of appropriate function spaces, while we point out the argument of mappings of the spatial variable z. We call F(G)-differentiable mappings differentiable in the Fréchet (Gateaux) sense. We exploit the following basic spaces and functions:

- $C_\tau^{r,n}$ the space of r-times continuously differentiable n-dimensional τ-periodic functions $\phi^T = (\phi_1, ..., \phi_n)$ endowed with the norm $\sum_{i=1}^n \| \phi_i \|_\infty$,

- $H_\tau^{r,n}$ the Sobolev space of r-times differentiable n-dimensional τ-periodic functions,

- $C^r(I; \mathcal{X})$ the space of r-times continuously differentiable mappings defined on the interval $I = [0, 1]$ with values in a Banach space $\mathcal{X}$, end endowed with the norm $\sum_{p=0}^r \max_{z \in I} \| x^{(p)}(z) \|_\mathcal{X}$,

- $C_K(I; \mathcal{X})$ the discrete space of vector-valued functions

$$x(z) = \sum_{k=0}^{K-1} x_k \sqrt{2} \sin k\pi z, \quad z \in I, \quad x_k \in \mathcal{X},$$

endowed with the norm $\max_{z \in I} \| x(z) \|_\mathcal{X}$,

- α_τ the time-averaging functional defined as follows:

$$\alpha_\tau : C_\tau^{0,p} \to R^p, \quad \alpha_\tau \circ \phi = \frac{1}{\tau} \int_0^\tau \phi(t) dt.$$

We note that the above spaces of continuous functions may be regarded as Banach algebras with multiplication of elements defined with the help of the product of continuous functions, which justifies the use of the scalar product in such spaces.

2. OPTIMAL PERIODIC CONTROL OF PISTON FLOW PROCESSES

Optimal time-periodic control (OTPC) problem for piston flow processes may be formulated as the following initial-value control problem for systems governed by first-order abstract differential equations: minimize the objective functional

$$J(\tau, x(\cdot), u, v, w(\cdot)) = F_0(y) \tag{1}$$

being a scalar function of the vector of averaged process characteristics

$$y = \alpha_\tau \circ \left[\int_0^1 g(x(z), w(z)) dz + h_0(u, v) + h_1(x(1), v) \right] \tag{2}$$

subject to the constraints

$$vx'(z) + D_t x(z) = f(x(z), w(z)), \quad z \in I, \tag{3}$$

$$x(0) = Bu + b, \tag{4}$$

$$F_1(y) = 0, \quad F_2(y) \leq 0, \tag{5}$$

$$\big(x(z)(t), u(t), v(t), w(z)(t) \big) \in X \times U \times V \times W, \quad z \in I, \ t \in T_\tau, \tag{6}$$

where $I = [0,1]$, $T_\tau = [0,\tau]$, τ is the operation period, and

$$x(\cdot) \in \mathcal{X}_\tau = C^1(I; C_\tau^{0,n})$$

is the abstract state trajectory of the process,

$$u \in \mathcal{U}_\tau = C_\tau^{1,m}, \quad v \in \mathcal{V}_\tau = C_\tau^{0,1}, \quad w(\cdot) \in \mathcal{W}_\tau = C_K(I; C_\tau^{0,m_1})$$

are lumped and distributed process controls, which may be conventionally called the inlet, flow, and spatial control, respectively,

$$F_r : R^q \to R^{q_r}, \quad r = 0,1,2, \quad g : C_\tau^{0,n} \times C_\tau^{0,m_1} \to C_\tau^{0,q},$$

$$h_r : C_\tau^{0,n_r} \times C_\tau^{0,1} \to C_\tau^{0,q}, \quad f : C_\tau^{0,n} \times C_\tau^{0,m_1} \to C_\tau^{0,n},$$

$$q_0 = 1, \quad n_0 = m, \quad n_1 = n,$$

and

$$D_t : C_\tau^{0,n} \to C_\tau^{0,n}, \quad \mathbf{D}(D_t) = C_\tau^{1,n},$$

is the differentiation operator for τ–periodic functions of time, B is an $n \times m$ real matrix, $b \in R^n$, and X, U, V, W are parallelepipeds in R^n, R^m, R_+, R^{m_1}, respectively.

The variable z denotes in the formulation (1)–(6) the space coordinate, while the equation (3) is equivalent to a system of first-order partial differential equations depicting piston flow processes, and may be regarded as the abstract description of the space-time dynamics of such processes. The vector y contains usually the average yield of the desired product, the average consumption of raw materials and energy etc.

We consider the distributed controls $w(\cdot)$ as elements of the discrete control space of abstract sinusoidal functions, which is useful for the construction of computational optimization algorithms. Moreover, this type of distributed controls will guarantee the existence of strong solutions of the process equation understood as the functions $x(\cdot) \in \mathcal{X}_\tau$ such that $x(z) \in \mathbf{D}(D_t)$ for all $z \in I$, and (3) is satisfied in the sence of the space $C(I; C_\tau^{0,n})$.

We always assume that the functions F_r, g, h_r, f are continuously differentiable on the sets of admissible values of their arguments.

<u>Lemma 1</u>: The linear unbounded operator

$$A(v) = -v^{-1}D_t : \ C_\tau^{0,n} \to C_\tau^{0,n}, \quad \mathbf{D}(A(v)) = C_\tau^{1,n},$$

is for every admissible control v the infinitesimal generator of a strongly continuous semigroup $T(z,v)$ of bounded operators on the space $C_\tau^{0,n}$.

<u>Proof:</u> Denote $|\phi_i(t_i)| = \max_{t \in T_\tau} |\phi_i(t)|$. Then $\dot{\phi}_i(t_i) = 0$, because the function $|\phi_i(t)|$ attains its maksimum at t_i. This implies the following equality for all $\lambda > 0$:

$$|\lambda \phi_i(t_i) - v^{-1}(t_i)\dot{\phi}_i(t_i)| = \lambda |\phi_i(t_i)|.$$

Thus we obtain

$$\| \lambda\phi_i - v^{-1}\dot\phi_i \|_\infty \geq | \lambda\phi_i(t_i) - v^{-1}(t_i)\dot\phi_i(t_i) | = \lambda \| \phi_i \|_\infty,$$

and

$$\| \lambda\phi - v^{-1}\dot\phi \|_{C_\tau^{0,n}} \geq \lambda \| \phi \|_{C_\tau^{0,n}},$$

which means that the operator $A(v)$ is dissipative in the space $C_\tau^{0,n}$. Since the function v is positive, the differential equation $\lambda\phi - v^{-1}\dot\phi = \psi$ is solvable for all $\psi \in C_\tau^{0,n}$ and $\lambda > 0$. The lemma follows now from the Lumer-Phillips theorem [10]. $\quad\square$

<u>Lemma 2</u>: If the quadruple $s = (x(\cdot), u, v, w(\cdot)) \in \mathcal{X}_\tau \times \mathcal{U}_\tau \times \mathcal{V}_\tau \times \mathcal{W}_\tau$ solves the equations (3) and (4), then these equations are uniquely resolvable with respect to the state trajectory as a function of controls in a certain vicinity of the solution s, i.e. there exists the quadruple $(x(\cdot) + \Delta x(\cdot), u + \delta u, v + \delta v, w(\cdot) + \delta w(\cdot))$ solving the equations (3) and (4), and satisfying the estimation

$$\Delta x(\cdot) = \delta x(\cdot) + o(\| \delta u \|_{\mathcal{U}_\tau} + \| \delta v \|_{\mathcal{V}_\tau} + \| \delta w(\cdot) \|_{\mathcal{W}_\tau}).$$

<u>Proof</u>: Using the semigroup $T(z, v)$ we write the equations (3) and (4) in the integral form $P(s) = 0$, where the mapping $P : \mathcal{X}_\tau^0 \times \mathcal{U}_\tau \times \mathcal{V}_\tau \times \mathcal{W}_\tau \to \mathcal{X}_\tau^0$, $\mathcal{X}_\tau^0 = C(I; C_\tau^{0,n})$, is defined as follows:

$$P(s)(z) = x(z) - T(z, v)(Bu + b) - \int_0^z T(z - \zeta, v) f(x(\zeta), w(\zeta)) d\zeta, \quad z \in I.$$

The mapping P is continuously F-differentiable with respect to all its arguments, and the partial F-derivative $P_x(s)$ is continuously invertible on the space $\mathcal{X}_\tau^0$ by virtue of the Picard-Banach version of the contraction principle. The lemma is now a consequence of continuous differentiability of the distributed control $w(\cdot) + \delta w(\cdot)$ [10]. $\quad\square$

<u>Theorem 1</u>: The partial F-differentials of the objective functional of the OTPC problem (1)–(6) reduced to the spaces of lumped and distributed controls can be expressed in the form

$$J_u(\tau, s)\delta u = \alpha_\tau \circ \left[(\mu^T h_{ou}(u, v) + v\lambda^T(0)B)\delta u \right], \tag{7}$$

$$J_v(\tau, s)\delta v = \alpha_\tau \circ \left[(\mu^T(h_{0v}(u, v) + h_{1v}(x(1), v)) - \lambda^T(1)x(1) + \lambda^T(0)Bu + \int_0^1 \lambda'^T(z)x(z)dz)\delta v \right], \tag{8}$$

$$J_{w_k}(\tau, s)\delta w_k = \alpha_\tau \circ \left[(\int_0^1 H_w(z)\sqrt{2}\sin k\pi z dz)\delta w_k \right], \quad k = 0, 1, ..., K - 1, \tag{9}$$

where

$$H(x(z), w(z), \mu, \lambda(z)) = \mu^T g(x(z), w(z)) + \lambda^T(z) f(x(z), w(z))$$

is the abstract Hamiltonian function, and the multipliers $\mu \in R^q$ and $\lambda(\cdot) \in \mathcal{X}_\tau^0$ satisfy the relationships

$$\mu^T = F_{0y}(y), \quad v\lambda'(z) + D_t\lambda(z) = -H_x^T(z), \quad z \in I, \quad v\lambda(1) = h_{1x}^T(x(1), v), \tag{10}$$

i.e. the function $\lambda(\cdot)$ is the solution of the final-value abstract differential equation in the space $C(I; C_r^{0,n})$.

Proof: We connect with the OTPC problem (1)–(6) the following Lagrange functional:

$$L(\tau, y, x(\cdot), u, v, w(\cdot), \mu, \lambda(\cdot))$$

$$= F_0(y) + \mu^T \Big[\alpha_\tau \circ \big(\int_0^1 g(x(z), w(z)) dz + h_0(u, v) + h_1(x(1), v) \big) - y \Big]$$

$$+ \alpha_\tau \circ \int_0^1 \lambda^T(z) [f(x(z), w(z)) - v x'(z) - D_t x(z)] dz + \alpha_\tau \circ [v \lambda^T(0)(Bu + b - x(0))],$$

which can be rewritten in the equivalent form

$$L = F_0(y) - \mu^T y + \alpha_\tau \circ \int_0^1 H(x(z), w(z), \mu, \lambda(z)) dz$$

$$+ \alpha_\tau \circ \Big[\mu^T \big(h_0(u, v) + h_1(x(1), v) \big) - \lambda^T(1) v x(1) + \int_0^1 \lambda'^T(z) v x(z) dz + \lambda^T(0) v(Bu + b) \Big].$$

Using the identity $J = L$ holding for all admissible solutions of the OTPC problem, and denoting by L^- the perturbed Lagrange functional, we obtain from the first-order Taylor expansion of this functional

$$L^- - L = F_{0y}(y) - \mu^T \delta y + \alpha_\tau \circ \int_0^1 \big(H_x(z) + v \lambda'^T(z) \delta x(z) \big) dz$$

$$+ \alpha_\tau \circ \Big[\big(\mu^T h_{1x}(x(1), v) + v \lambda^T(1) \big) \delta x(1) \Big] + L_u \delta u + L_v \delta v$$

$$+ L_w \delta w(\cdot) + o(\| \delta u \|_{\mathcal{U}_\tau} + \| \delta v \|_{\mathcal{V}_\tau} + \| \delta w(\cdot) \|_{\mathcal{W}_\tau}).$$

The theorem is now a consequence of the definition of the multipliers μ and $\lambda(\cdot)$, and the particular form of the distributed control $w(\cdot)$. ☐

The transformation of the time $t \to \tau t$ allows us to obtain the derivative of the objective functional with respect to τ from the equality $J_\tau = L_\tau$, which yields

$$J_\tau = \frac{1}{\tau^2} \alpha_1 \circ \int_0^1 x^T(z) D_t \lambda(z) dz. \tag{11}$$

Since the optimization problem (1)–(6) is defined on the product of Banach spaces of continuous time-periodic functions, the direction of steepest descent $-\delta \nu$ may be found from the relationship [12]

$$\langle J_\nu^*, \delta \nu \rangle_\Gamma = \| J_\nu^* \|_{\Gamma^*}^2, \tag{12}$$

where $\nu = (u, v, w(\cdot))$, and $\Gamma = \mathcal{U}_\tau \times \mathcal{V}_\tau \times \mathcal{W}_\tau$. Taking into account that the partial gradients (7)–(9) are defined with the help of integrals of continuous time-periodic functions, we can set $\delta \nu = \beta J_\nu$, where a constant β is chosen to satisfy the equality (12), and J_ν is understood as the vector time-continuous function determining the dual functional J_ν^*. The choice of the direction $\delta \nu$ may be precisely justified with the help of the general form of linear bounded functionals on the space $C_\tau^{0,p}$ [1].

3. REDUCED GRADIENT METHODS

Using quadratic penalty functionals we incorporate the constraints (5) and (6) into the objective functional setting

$$\tilde{F}_0(y) = F_0(y) + \rho\Big(|\,F_1(y) - \vartheta_1\,|^2 + \varsigma(F_2(y) - \vartheta_2) + y_{q+1}\Big),$$

where

$$y_{q+1} = \alpha_\tau \circ \Big[\int_0^1 \big(\gamma(x(z), X) + \gamma(w(z), W)\big)dz + \gamma(u, U) + \gamma(v, V)\Big]$$

is the additional process characteristics (the measure of the deviation of the instantaneous constraints for the state and control), ρ is the penalty coefficient, ϑ_τ are shifts for the constraints (5), and ς, γ are penalty functions defined as follows:

$$\varsigma(x) = x^T \max(0, x) \quad \text{for } x \in R^n, \quad \gamma = \sum_{j=1}^p [\varsigma(\phi_j(t) - \phi_j^{max}) + \varsigma(-\phi_j(t) + \phi_j^{min})] \quad \text{for } \phi \in C_\tau^{0,p}.$$

Then the whole problem (1)–(6) is resolvable with respect to the state trajectory as a function of process controls, and the reduced gradients (7)–(9) may be used for the process optimization. To this end, however, the multiplier $\lambda(\cdot)$ must be found as the solution of the final-value abstract differential equation (10), which requires the computation and storage of the state trajectory $x(\cdot)$.

Since discrete representations of the distributed functions $x(\cdot)$ and $\lambda(\cdot)$, and their processing may be complicated, we propose the following practical approach for the determination of reduced gradients: solve the initial-value state equation (3) step by step exploiting the difference approximation, store the state $x(1)$ only, and solve the final-value equation (10) step by step with the aid of the difference approximation in the inverse direction. We operate in such a method solely on instantaneous values $x(z), \lambda(z)$ of the functions $x(\cdot)$, $\lambda(\cdot)$, or on few such values if we apply generalized difference approximations to abstract differential equations (3) and (10).

We summarize the above approach in the following form:

Algorithm 1: (approximate-descent algorithm for optimal periodic control of piston flow processes)

Initialization: choose an admissible point (τ_0, ν_0), and auxiliary parameters $\rho, \vartheta_{\tau 0}, \epsilon_0$.

1^0 Set $h = 0$, and $\tau_{h0} = \tau_0$, $\nu_{h0} = \nu_0$, $\epsilon_h = \epsilon_0$, $\rho_h = \rho_0$, $\vartheta_{h\tau 0} = \vartheta_0$.

2^0 Set $j = 0$.

3^0 The state trajectory loop: using difference schemes

$$x(z_l) - x(z_{l-1}) = \Delta z \sum_{k=l-1}^l \xi_k v^{-1}\big[-D_t x(z_k) + f(x(z_k), w(z_k))\big],$$

$$x(0) = Bu_{hj} + b, \quad l = 1, 2, ..., l_0, \quad \xi_l = 1 - \xi_{l-1},$$

$$\lambda(z_{l-1}) - \lambda(z_l) = -\Delta z \sum_{k=l-1}^{l} \xi_k v^{-1} \left[-D_t \lambda(z_k) - H_x^T(z_k) \right],$$

$$\lambda(1) = v^{-1} h_{1x}^T(x(1), v), \quad l = l_0, l_0 - 1, ..., 1, \quad \xi_{l-1} = 1 - \xi_l,$$

compute $x(1)_{hj}$ (choose $\xi_l = 1$ ($\xi_l = 0$) for the explicit (implicit) scheme), and next find the descent direction $(J_\tau(\tau_{hj}, \nu_{hj}), J_\nu(\tau_{hj}, \nu_{hj}))$ according to the formulas (11), and (7)–(9), respectively.

4^0 The period and control loop: perform the line search

$$\sigma_{hj} = \arg \min_{\sigma > 0} J(\tau_{hj} + \sigma J_\tau, \nu_{hj} + \sigma J_\nu),$$

determine the improved solution

$$\tau_{h,j+1} = \tau_{hj} + \sigma_{hj} J_\tau, \quad \nu_{h,j+1} = \nu_{hj} + \sigma_{hj} J_\nu,$$

and compute $\delta_{hj} = \| (J_\tau^*, J_\nu^*) \|$. If $\delta_{hj} > \epsilon_h$ then set $j = j + 1$, and go to 3^0. Otherwise set $\epsilon_h = \epsilon_h/2$.

5^0 The penalty loop: modify ρ_h and v_{hrj} according to general rules of the shifted penalty method [13], set $h = h + 1$, and go to 2^0.

Though the resolvability of the process equation (3) is guaranteed in the Banach space of appropriate continuous functions (B-space), it is desirable to imbed this problem into the following Hilbert space of square integrable functions (H-space):

$$\mathcal{X}_\tau \times \mathcal{U}_\tau \times \mathcal{V}_\tau \times \mathcal{W}_\tau \hookrightarrow W_1^2(I; H_\tau^{0,n}) \times H_\tau^{1,n} \times H_\tau^{0,1} \times L_K^2(I; H_\tau^{0,m_1}).$$

We admit this way more general types of solutions including discontinuous controls, and we are able to construct more efficient descent directions emploing the scalar product (conjugated or quasi-Newton directions). The analysis of properties of such methods may be facilitated by the fact that the F-gradient in the B-space constitutes also the G-gradient in the H-space under discussion due to the continuity of solution and the general form of G-differentials in the space of square-integrable functions. Thus the theory of convergence of gradient-type methods for weakly differentiable functionals on H-spaces may be used [12], [11].

4. OPTIMAL PERIODIC CONTROL OF DISPERSIVE FLOW PROCESSES

OTPC problem for a class of dispersive flow processes may be formulated as the following boundary-value control problem for systems governed by second-order abstract differential equations: minimize the objective functional

$$J_\tau(\tau, x(\cdot), u, w(\cdot)) = F_0(y) \tag{13}$$

being a scalar function of the vector (2) satisfying (5), and subject to the constraints

$$- Dx''(z) + \bar{v}x'(z) + D_t x(z) = f(x(z), w(z)), \quad z \in I, \tag{14}$$

$$\bar{v}x(0) - Dx'(0) = \bar{v}(Bu + b), \quad Dx'(1) = 0, \tag{15}$$

$$(x(z)(t), u(t), w(z)(t)) \in X \times U \times W, \quad z \in I, \ t \in T_\tau, \tag{16}$$

where

$$x(\cdot) \in \mathcal{X}_\tau = C^2(I; H_\tau^{1,n}), \ u \in \mathcal{U}_\tau = H_\tau^{1,n}, \ w(\cdot) \in \mathcal{W}_\tau = C_K(I; H_\tau^{1,m_1}),$$

$$D_t : H_\tau^{1,n} \to H_\tau^{1,n}, \ \mathbf{D}(D_t) = H_\tau^{2,n},$$

and D is the dispersion coefficient.

It is possible to symmetrize the equations (14) and (15) with the help of the regularized operator of the fractional differentiation for periodic functions [11]. Such an approach proves the unique resolvability of the process equations (14) and (15) with respect to the state trajectory as a function of controls u and $w(\cdot)$. The performance index gradient reduced to the control space can be expressed in the form

$$J_u(\tau, s)\delta u = \alpha_\tau \circ \left[(\mu^T h_{ou}(u, \bar{v}) + \bar{v}\lambda^T(0)B)\delta u \right],$$

$$J_{w_k}(\tau, s)\delta w_k = \alpha_\tau \circ \left[(\int_0^1 H_w(z)\sqrt{2}\sin k\pi z dz)\delta w_k \right], \quad k = 0, 1, ..., K - 1,$$

where $\mu^T = F_{oy}(y)$ and the multiplier $\lambda(\cdot) \in \mathcal{X}_\tau^0$ is the solution of the following boundary-value abstract second-order differential equations in the space $C(I; H_\tau^{1,n})$:

$$D\lambda''(z) + \bar{v}\lambda'(z) + D_t\lambda(z) = -H_x^T(z), \quad z \in I, \tag{17}$$

$$\bar{v}\lambda(1) + D\lambda'(1) = h_{1x}^T(x(1), \bar{v}), \quad D\lambda'(0) = 0. \tag{18}$$

Computational algorithms for solving of the process equations (14), (15), and the adjoint equations (17), (18) may be here based on difference approximation schemes, or on the symmetrization method transforming the equations under discussion to a contractive mapping in a suitably chosen space.

The considered above OTPC problems with one spatial coordinate have been formulated as non-periodic initial-value or two-points boundary-value abstract control problems. This formulation has allowed us to show the existence of the performance index gradient reduced to the control space under simple assumptions. If we deal with processes described by more general operators constituting, for example, systems of Agmon type, then other formulations of OTPC problems must be used, for instance, abstract control problems with periodic boundary conditions. The reduction of such problems to the space of lumped and distributed controls requires additional assumptions guaranteeing the so-called non-resonant case.

5. APPLICATIONS

An important problem connected with the optimization of many industrial production systems consists in the minimization of raw materials and energy consumption for a given yield of the output product. The OTPC problem of such a type for a single chemical reaction performed in a plug flow tubular reactors can be formulated as the following particular case of the problem (1)–(6):

$$n = m = m_1 = 1, \quad q_0 = 1, \quad B = 1, \quad b = 0,$$

$$F_0(y) = y_1, \quad F_1(y) = y_2, \quad F_2(y) = 0,$$

$$y = \alpha_\tau \circ \Big(\int_0^1 c_1 w(z) dz + c_2 v u, \; v(u - x(1)) - \bar{y}_2 \Big)^T,$$

$$f = -\kappa \exp\left(-\beta/w(z)\right) x^\sigma(z),$$

where the control u is the input concentration, the control v is the flow rate, the control $w(\cdot)$ is the temperature distribution, c_1 and c_2 are cost coefficients for the input substance and the heat energy, $\bar{y}_2$ is the given mean production, and κ, β, σ are process parameters. The partial gradients reduced to the control space may be here written as follows:

$$J_u = (c_2 + 2\rho(y_2 - \vartheta) + \lambda(0))v,$$

$$J_v = (c_2 + \lambda(0))u + 2\rho(y_2 - \vartheta)(u - x(1)) - \lambda(1)x(1) + \int_0^1 \lambda'(z)x(z)dz,$$

$$J_{w_k} = \int_0^1 \big[c_1 - \kappa\beta\lambda(z)\exp(-\beta/w(z))x^\sigma(z)/w^2(z)\big]\sqrt{2}\sin k\pi z\, dz,$$

while the adjoint equation is of the form

$$v\lambda'(z) + D_t\lambda(z) = \lambda(z)\kappa\exp(-\beta/w(z))\sigma x^{\sigma-1}(z), \quad \lambda(1) = -1.$$

Another class of OTPC problems arises from the optimization of exploitation of age-structured populations. A simple problem of this kind may be stated as the following particular case of the problem (1)–(6):

$$n = m = m_1 = q_0 = 1, \quad B = 1, \quad b = 0,$$

$$F_0(y) = y, \quad F_1(y) = F_2(y) = 0,$$

$$y = \alpha_\tau \circ \Big[\int_0^1 \big(c_1(z)w^\sigma(z) + c_2(z)x(z) + c_3(w(z) - \bar{w}(z))^2\big)dz - c_4 v x(1) \Big],$$

$$f = \beta(z)w(z),$$

where the control u means the breeding rate, the control v is the feeding rate, the control $w(\cdot)$ is the external exchange of the population, $\beta(\cdot)$ is the mortality function, and other parameters have similar meaning as in [4]. The partial gradients reduced to the control space take here the form

$$J_u = v\lambda(0), \quad J_v = -(c_4 + \lambda(1))x(1) + \lambda(0)u + \int_0^1 \lambda'(z)x(z)dz,$$

$$J_{w_k} = \int_0^1 \big[c_1(z)\sigma w^{\sigma-1}(z) + 2c_3(w(z) - \bar{w}(z)) + \lambda(z)\beta(z)\big]\sqrt{2}\sin k\pi z\, dz,$$

while the adjoint variable satisfies the equation

$$v\lambda'(z) + D_t\lambda(z) = -c_2(z), \quad \lambda(1) = -c_4.$$

References

[1] P. L. Butzer and H. Berens. *Semi-Groups of Operators and Approximation*. Berlin, Springer-Verlag, 1967.

[2] D. A. Carlson and A. Haurie. *Infinite Horizon Optimal Control, Theory and Applications*. Berlin, Springer-Verlag, 1987.

[3] F. Colonius. *Optimal periodic control of quasilinear systems in Hilbert spaces, in Optimal Control of Partial Differential Equations, II*, pages 57–66. International Series on Numerical Mathematics. Basel, Birkhauser-Verlag, 1987.

[4] N. Derzko and S.P. Sethi. Distributed parameter systems approach to the optimal cattle ranching problem. *Optimal Control Applications and Methods*, vol. 1, pp. 3–10, 1980.

[5] J. M. Douglas. *Process Dynamics and Control*. Englewood Cliffs, New Jersey, Prentice Hall, 1972.

[6] A. Haurie, S. Sethi, and R. Hartl. Optimal control of an age-structured population model with applications to social services planning. *Large Scale Systems*, vol. 6, pp. 133–158, 1984.

[7] M.A. Hjortso. Periodic forcing of microbial cultures: A model for induction synchrony. *Biotechnology and Bioengineering*, vol. 30, pp. 825–835, 1987.

[8] V. V. Kafarov, A. Ju. Vinarov, and L. S. Gordeev. *Modelling of Biochemical Reactors*. Moscow, Lesprom, 1979 (in Russian).

[9] Yu. Sh. Matros. *Catalytic Processes under Unsteady-State Conditions*. Amsterdam, Elsevier, 1989.

[10] A. Pazy. *Semigroups of Linear Operators and Applications. Partial Differential Equations*. New York, Springer-Verlag, 1983.

[11] K. Styczeń. *Optimal Cyclic Control for Nonlinear Dynamic Systems*. Scientific Papers of the Inst. of Eng. Cybern. of Technical University of Wrocław no. 91, Monographs no. 23. Wydawnictwo Politechniki Wrocławskiej, Wrocław, 1993 (in Polish).

[12] M. M. Vainberg. *The Variational Method and the Method of Monotone Operators in the Theory of Nonlinear Equations*. Moscow, Nauka, 1972 (in Russian).

[13] A. Wierzbicki. *Models and Sensitivity of Control Systems*. Amsterdam, Elsevier, 1984.

ADOMIAN'S METHOD APPLIED TO
IDENTIFICATION AND OPTIMAL CONTROL PROBLEMS

Yves CHERRUAULT

Université PARIS VI - MEDIMAT- 15 rue de l'Ecole de Médecine - 75270 PARIS.

ABSTRACT : _In this paper we study applications of a decomposition method (due to Adomian) to problems related to identification of models and optimal control of systems._

1 - INTRODUCTION

In the eighties George ADOMIAN [3] ,[4] has proposed a "curious" method making use of special kinds of polynomials for solving nonlinear functional equations. This technique gives the exact solution as a series of functions. But at the beginning of the method nothing was done about convergence. From this time many works has been done and they allow to find analogies with the classical iterative methods and more generally with the decompositional techniques for which the fixed point theorem can be used [5] , [6] . Results of convergence are given in [10], [6] , [5] and are now available for different kinds of equations : algebraic, differential [1] , integral [6] , partial differential [9] , [10], [11] .

Let us first recall the basic principles of the Adomian' method on the general functional equation :

(1) $y - N(y) = f$

where f is given function belonging to an Hilbert H and where N is a non linear operator from H into H. We are looking for a function y satisfying (1) which is the canonical form of Adomian. If an equation is not written under the form (1) we have to do transformations (integrations, ...) for obtaining this canonical form.

For solving (1) we express y as a series of functions

(2) $\quad y = \sum_{i=0}^{\infty} y_i$

and we decompose N as follows :

(3) $\quad N(y) = \sum_{i=0}^{\infty} A_n$

The $A_n's$ are called Adomian's polynomials and they are obtained from the relationships :

(4) $\quad z = \sum_{i=0}^{\infty} \lambda^i y_i \quad , \quad N(\Sigma \lambda^i y_i) = \sum_{i=0}^{\infty} \lambda^i A_i$

where λ is a parameter introduced for convenience. We deduce the A_n owing to the formula :

(5) $\quad n! \, A_n = \frac{d^n}{d\lambda^n} \left[N(\Sigma \lambda^i y_i) \right]_{\lambda=0} \quad ; \quad n = 0,1,2, \ldots$

Recent results [1] ,[2] ,[12] prove that we can find better formulae than (5). We shall give one later . One can even propose formulae allowing to calculate A_{n+k} from A_n [12] .

Without considering convergence problems we can put (2) and (3) into (1) and we have :

(6) $\quad \sum_{i=0}^{\infty} y_i - \sum_{i=0}^{\infty} A_i = f$

Relationship (6) may be satisfied if we set :

(7) $\quad \begin{cases} y_0 = f \\ y_1 = A_0 \\ \vdots \\ y_n = A_{n-1} \\ \vdots \end{cases}$

These formulae give the basic expressions of the Adomian's technique.

With another formalism one can prove convergence of Adomian's method when N is a contraction (sufficient condition). Furthermore it can be proved that A_n depends only on $y_0, y_1, \ldots, y_n$. It is also possible to obtain the Adomian's polynomials in function of only y_0. Of course we have the same result for u_n. Indeed we have [12] :

$$(8) \quad u_{n+1} = F_n(u_0) = \sum_{\ell=1}^{n} N^{(n+1-\ell)}(u_0) \sum_{\substack{p_1+\ldots+p_n=\ell \\ p_1+\ldots+np_n=n}} a^{(\ell)}_{p_1,\ldots,p_n} [N^{(1)}(u_0)]^{p_1} \ldots [N^{(n)}(u_0)]^{p_n}$$

where the $a^{(\ell)}_{p_1,\ldots,p_n}$ can be explicitly calculated. With this formula convergence results can be proved without using fixed point theorem. With reasonable hypothesis on $N(y_0), \ldots, N^{(P)}(y_0)$ we may prove convergence of $\sum_n A_n$.

Properties of substituted series [6] also allow to prove convergence of Adomian's method (i.e. the series $\sum y_i$ converges). For the following we must have in mind that the Adomian's method gives the exact solution of the functional equation as a series $y = \sum_{i=0}^{\infty} y_i = f + \sum_{i=0}^{\infty} A_i$

The solution will be obtained as far as the A_n will be calculated. In practice we shall only use a truncated series because in most cases it is impossible to obtain all the A_n and to calculate the sum of the series.

REMARK : In [1] a result is given proving that Adomian's method is equivalent to the iterative Taylor's method for the resolution of non linear differential equations with $f = 0$.

Let us insist on the fact that the solution obtained by "Adomian" is not a discretized approximation but take explicitly into account all the variables of the problem and also the parameters included in the equations. This property will be very useful when applied to identification and optimal control problems.

2 - MODELS IDENTIFICATION :

Modelling (in biology for instance) gives rise to identification problems in nonlinear (or linear) differential systems.

Consider the differential system :

(9) $\begin{cases} \dot{x}_i = f_i(x_1, \ldots, x_n, \beta_1, \ldots, \beta_p, t) \\ x_i(0) = \alpha_i \quad ; \quad i = 1, 2, \ldots, n \end{cases}$

in which the $\beta_i's$ are unknown parameters to be identified from experimental date (observation). Integrating (9) between 0 and t leads to the canonical formulation

(10) $\quad \vec{x} = \vec{\alpha} + \int_0^t \vec{f}(x_1, \ldots, x_n, \beta_1, \ldots, \beta_p, t)dt$

with $\vec{\alpha} = (\alpha_1, \ldots, \alpha_n)$, $\vec{f} = (f_1, \ldots, f_n)$ et $\vec{x} = (x_1, \ldots, x_n)$.

Using Adomian's method for solving (9) leads to $\vec{x} = (x_i)$ where the x_i are given by :

(11) $\quad x_i = \sum_{j=0}^{\infty} y_j^i(\beta_1, \ldots, \beta_p) \quad , \quad i = 1, \ldots, n$

where the y_j^i are defined as functions depending _explicitly_ on $\beta_1, \beta_2, \ldots, \beta_p$ and obtained from the Adomian's polynomials associated to the nonlinearity of the problem.

Let :

(12) $\quad \vec{z} = B \cdot \vec{x}$ be an observation where B is a constant, known, matrix (n x n).

For identifying the β_i we use the functional :

(13) $\quad J = \sum_{i=1}^{n} \sum_{j=1}^{m} [(Bx)_i(t_j) - (Bx)_i^c(t_j)]^2$

where the t_j are known times of measurements and where $(Bx)_i^c$ is the i-th component of $\vec{Bx}$ calculated from the system (10). Using classical methods leads to $(\vec{Bx})^c$ depending <u>implicitly</u> on $\beta_1,\ldots, \beta_p$. But with the decomposion technique and owing to (11) we obtain a $(\vec{Bx})^c$ depending <u>explicitly</u> on the $\beta_i's$.

The unknown parameters are then obtained by minimizing J : Min J
$\{\beta_i\}$

This optimization problem may be solved by using Alienor method [7] , [8] . This technique allows to find a global optimum. It is based on a reducing transformation [7] allowing to express the β_i in function of a single variable $\beta_i = h_i(\theta)$, $i = 1,\ldots,p$.

The initial problem :

$$(14) \quad \text{Min J} \atop \{\beta_i\}$$

is then <u>approached</u> by a minimization problem depending on a single variable θ :

$$(15) \quad \underset{\theta}{\text{Min}} \, J^*(\theta) \quad \text{where} \quad J^*(\theta) = J(h_1(\theta),\ldots, h_p(\theta)) \, .$$

<u>REMARK</u> : Our identification method is based on Adomian decomposition technique and avoids numerical resolution of the differential system for a sequence of $(\beta_1,\ldots, \beta_p)_i$, $i = 1,\ldots$.

3 - <u>APPLICATIONS TO OPTIMAL CONTROL OF SYSTEMS</u> :

Consider a general optimal control problem :

$$(16) \quad \underset{\vec{u}(t)}{\text{Min}} \int_0^T g(\vec{x}, \vec{u}) \, dt$$

where T is fixed and with the following constraints on $\vec{x}$ and $\vec{u}$:

$$(17) \begin{cases} \dot{x}_i = f_i(x_1,\ldots, x_n, u_1,\ldots, u_m, t) \\ x_i(0) = \alpha_i \, , \, i = 1,\ldots, n \end{cases}$$

We are looking for a control $\vec{u}(t)$ minimizing the functional (16) and satisfying the differential system (17). Using the decomposition method for solving (17) (expressed in the canonical form) leads to :

$$(18) \qquad x_i = \sum_j v_j^i (u_1, \ldots, u_m) \quad , \ i = 1, \ldots, n$$

where the sum in (18) will be truncated to a finite number of terms. Putting (18) into (16) gives a new problem of minimization.

$$(19) \qquad \underset{u_1, \ldots, u_m}{\text{Min}} \int_0^T g(\Sigma \, v_j^1, \ldots, \Sigma v_j^n \, , \, u_1, \ldots, u_m, \, t) dt$$

which is a classical minimization problem because the functional (19) depends explicitly on $u_1, \ldots, u_m$. Then it becomes possible to use a global optimization method such as Alienor [8] for solving (19). Like previously we avoid the numerical resolution of (17). Furthermore we have transformed an optimal control problem into a classical optimisation problem.

REMARK 3.1 : It may be difficult to obtain the Adomian's polynomials directly in function of $u_1, \ldots, u_m$. For simplifying the obtention of A_n we can use approximations for u_i :

$$(20) \qquad u_i = \sum_{p=1}^q c_p^i \, \theta_p(t)$$

where the $\theta_p(t)$ are known functions (polynomials, exponentials, spline functions, ...). In that case the problem consists in determining the parameters c_p^i and the difficulties of calculus vanish. The functional in (20) has to be minimized according to the c_p^i's which explicitly appears in the function to minimize.

REMARK 3.2 : The problem of existence and uniqueness of the solution of (16) (17) becomes very simple and it depends only on the properties of g and of x_i given by the formulae (18).

For instance if g is continuous according to its variables and if $x_i(u_1, \ldots, u_n)$ obtained from Adomian's method are also continuous then J admits at least a minimum on a bounded set $|u_j| \leq M$, $j = 1,\ldots, m$ (theorem of Weierstrass). In this case we have existence of an optimal control. The uniqueness will depend on the properties of g and $x_i(u_1,\ldots, u_m)$. It will be necessary to examine this property in each particular case (g = polynomial,...) and it is not possible to give general results.

4 - CONCLUSIONS :

Decomposition methods such as Adomian's method are new and original methods for solving nonlinear functional equations. Contrarily to classical numerical methods they do not approximate or linearize the nonlinear operators. Furthermore they do not discretize space or time. The power of such methods has greatly increases since we are able to propose simple and efficient formulae for calculating the Adomian's polynomials. Applications of these decomposition methods to identification and optimal control allow to obtain functionals depending explicitly on the variables or parameters to determine. Thus we obtain classical optimization problems for which the numerical methods of the literature can be used without difficulties.

REFERENCES

[1] K. ABBAOUI, Y. CHERRUAULT :"Convergence of Adomian's method applied to differential equations" - *To appear in* Computers and Mathematics with applications.

[2] K. ABBAOUI, Y. CHERRUAULT : "Adomian's method applied to biological problems (identification, control)"- *To appear.*

[3] G. ADOMIAN : "Nonlinear stochastic systems theory and applications to physics"- Kluwer 1989 .

[4] G. ADOMIAN : "A review of the decomposition method and some recent results for nonlinear equations"- Computers Math. Applic., vol. 21, N°5 , 101-127 , 1991 .

[5] Y. CHERRUAULT : "Convergence of Adomian's method" - Kybernetes 18(2) , 31-38 (1989) .

[6] Y. CHERRUAULT, G. SACCOMANDI, B. SOME : "New results for convergence of Adomian's method applied to integral equations" - Mathl. Comput. Modelling 16(2) , 85-93 (1992).

[7]   Y. CHERRUAULT : "A new method for global optimization" - Kybernetes 19(3), 19-32 , (1990) .

[8]   Y. CHERRUAULT : "New deterministic methods for global optimization and applications to biomedicine"-Int. J. Biomed. Comput. 27(3,4), 215-229,1991.

[9]   Y. CHERRUAULT, Th. MAVOUNGOU : "Numerical study of Fisher's equation by Adomian's method" - *To appear in* Mathl. and Comput. Modelling.

[10] L. GABET : "Modélisation de la diffusion de médicaments à travers les capillaires et dans les tissus à la suite d'une injection et Esquisse d'une théorie décompositionnelle et application aux équations aux dérivées partielles"-Thèse de l'Ecole Centrale, ler juillet 1992.

[11] L. GABET : "The decomposition method and linear partial differential equations" - Mathl. Comput. Modelling 17(6), 11-22 , 1993 .

[12] S. GUELLAL, Y. CHERRUAULT : "New formulaes for calculing Adomian's polynomials and applications to convergence" - *To appear*.

Application

A Remark on Stabilization of the SCOLE Model
with an a priori Bounded Boundary Control

Bopeng RAO

Département de Mathématiques, Université de Nancy I, U. R. A. CNRS 750

Projet Numath, INRIA Lorraine, B.P.239, 54506 Vandœuvre-lès-Nancy, France

1. Introduction

Let H be a real Hilbert space with inner product $(\cdot, \cdot)_H$ and norm $\| \cdot \|_H$ and let E be a second real Hilbert space with inner product $(\cdot, \cdot)_E$ and norm $\| \cdot \|_E$. Also let A be a linear maximal monotone operator in H. Finally let B denote a linear bounded operator from E to H. We consider the abstract control problem:

$$\frac{du}{dt} + Au = Bf(t), \qquad u(0) = u_0 \in H. \tag{1}$$

Unlike the standard control problem, we restrict ourselves to the case where the control $f(t)$ satisfies the *a priori* constraint: $\|f(t)\|_E \le r$, $r > 0$. As suggested for finite-dimensional systems in Gutman–Hagander [1], the saturating control $f(t)$ can be chosen as follows:

$$f(t) = \begin{cases} -\dfrac{rB^*u}{\|B^*u\|_E}, & \text{if} \quad \|B^*u\|_E \ge r; \\[3mm] -B^*u, & \text{if} \quad \|B^*u\|_E < r, \end{cases} \tag{2}$$

so that the closed-loop control system (1)–(2) is dissipative. Furthermore, the following abstract result has been established in Slemrod [5]:

Theorem 1. For each $u_0 \in H$, there exists a unique weak solution of the system (1)–(2) for all $t \in \mathbb{R}^+$ with 0 a stable equilibrium. If in addition $E = \mathbb{R}$, $(\lambda I + A)^{-1}$ is compact for all real $\lambda > 0$, and the only solution of the equation: $B^*e^{-tA}\psi_0 = 0$ is $\psi_0 = 0$, then $\|u(t)\|_H \to 0$ as $t \to +\infty$, for all $u_0 \in H$.

Notice that the system (1)–(2) defines a nonlinear semigroup of contractions on H in the case $E = \mathbb{R}$. This result can be thus applied in the study of the following boundary control system:

$$\begin{cases} y_{tt}(x,t) + y_{xxxx}(x,t) = 0, & 0 < x < L, \\ y(0,t) = y_x(0,t) = 0, \\ y_{tt}(L,t) - y_{xxx}(L,t) = f_2(t), \\ y_{xtt}(L,t) + y_{xx}(L,t) = f_1(t), \end{cases} \qquad (3)$$

where $f_1(t), f_2(t)$ denote respectively the moment control and the force control. This system (called the SCOLE model) characterizes the vibrations of a clamped elastic beam together with the oscillations of the rigid body linked to the beam. For further descriptions of the physical structure, we refer to Littman–Markus [2].

The goal of this work is to investigate the strong stabilization of the system (3) only by an *a priori* bounded moment control, or only by an *a priori* bounded force control. Recall that we can't apply simultaneously an *a priori* bounded moment control and an *a priori* bounded force control at the end $x = L$. Since this will imply that $E = \mathbb{R}^2$ and the system (3) does not generate one semigroup of contractions (*cf.* Slemrod [5]).

2. Formulation of the Problem

Let us introduce the energy space:

$$H = \{u = (y, z, \xi, \eta) \in H^2(0, L) \times L^2(0, L) \times \mathbb{R}^2 \quad \text{such that} \quad y(0) = y_x(0) = 0\},$$

endowed with the inner product:

$$\langle u, \tilde{u} \rangle = \int_0^L (y_{xx}\tilde{y}_{xx} + z\tilde{z})dx + \xi\tilde{\xi} + \eta\tilde{\eta}, \qquad \forall u, \tilde{u} \in H.$$

Next we define the unbounded operator A in H and the bounded operators B_1, B_2 from $\mathbb{R}$ to H by letting:

$$D(A) = \begin{pmatrix} u = (y, z, \xi, \eta) \in H^4(0, L) \times H^2(0, L) \times \mathbb{R}^2 \quad \text{such that} \\ y(0) = y_x(0) = z(0) = z_x(0) = 0, \; \xi = z(L), \; \eta = z_x(L) \end{pmatrix},$$

$$Au = (-z, y_{xxxx}, -y_{xxx}(L), y_{xx}(L)), \qquad \forall u = (y, z, \xi, \eta) \in D(A),$$

$$B_1 f = (0, 0, 0, f), \quad \forall f \in \mathbb{R}; \quad B_1^* u = \eta, \quad \forall u = (y, z, \xi, \eta) \in H,$$

$$B_2 f = (0, 0, f, 0), \quad \forall f \in \mathbb{R}; \quad B_2^* u = \xi, \quad \forall u = (y, z, \xi, \eta) \in H.$$

Now following a procedure of Slemrod [5], we introduce the auxiliary functions:

$$z(t) = y_t(t), \quad \xi(t) = y_t(L, t), \quad \eta(t) = y_{xt}(L, t), \quad u(t) = \big(y(t), z(t), \xi(t), \eta(t)\big).$$

Then we formulate the system (3) into the form:

$$\frac{d}{dt} u(t) + Au(t) = B_1 f_1(t) + B_2 f_2, \qquad u(0) = u_0 \in H. \tag{4}$$

3. Stabilization with an a priori Bounded Moment Control

In this case the *a priori* bounded controls are chosen:

$$f_1(t) = \begin{cases} -r \operatorname{sgn} y_{xt}(L, t), & \text{if} \quad |y_{xt}(L, t)| \geq r; \\ -y_{xt}(L, t), & \text{if} \quad |y_{xt}(L, t)| < r, \end{cases} \qquad f_2(t) = 0. \tag{5}$$

We obtain thus the following closed-loop control system:

$$\frac{d}{dt} u(t) + Au(t) = B_1 f_1(t), \qquad u(0) = u_0 \in H. \tag{6}$$

Theorem 2. Assume that $L < 3$. Then for all $u_0 \in H$, the system (6) admits a unique weak solution satisfying:

$$\|u(t)\|_H \to 0 \quad \text{as} \quad t \to +\infty.$$

Proof. The operator A is maximal monotone in the space H with compact resolvent $(\lambda I + A)^{-1}$ for all $\lambda > 0$ (*cf.* Slemrod [5]). Applying theorem 1, it is sufficient to prove that the only solution of the equation $B_1^* e^{-tA} \psi_0 = 0$ is $\psi_0 = 0$. To this end, we first establish for smooth initial data $\psi_0 \in D(A)$ the energy estimate:

$$\int_0^T \|\psi(t)\|_H^2 dt \leq C \|\psi_0\|_{D(A)}^2, \tag{7}$$

where $\psi(t) = e^{-tA} \psi_0$ satisfying $B_1^* \psi(t) = 0$.

Let $\psi(t) = \big(w(t), v(t), a(t), b(t)\big)$, then one has:

$$\begin{cases} w_t(x, t) - v(x, t) = 0, & 0 < x < L, \\ v_t(x, t) + w_{xxxx}(x, t) = 0, \\ a_t(t) - w_{xxx}(L, t) = 0, \\ b_t(t) + w_{xx}(L, t) = 0. \end{cases}$$

Since $\psi(t) \in D(A)$ and $B^*\psi(t) = 0$, we have:

$$a(t) = w_t(L,t), \quad b(t) = w_{xt}(L,t) = 0,$$

which imply that

$$\begin{cases} w_{tt} + w_{xxxx} = 0, & 0 < x < L, \\ w(0,t) = x(0,t) = 0, \\ w_{tt}(L,t) - w_{xxx}(L,t) = 0, \\ w_{xt}(L,t) = w_{xx}(L,t) = 0. \end{cases} \tag{8}$$

We first multiply the equation (8) by x^2 and integrate by parts so that we get

$$-\left[\int_0^L x^2 w_t dx \right]_0^T = \int_0^T w_{xxx}(L,t)dt + 2\int_0^T w_x(L,t)dt \tag{9}$$

$$= [w_t(L,t)]_0^T + 2T w_x(L,t).$$

Since $\|\psi(t)\|_H^2 = \|\psi_0\|_H^2$ for all $t \geq 0$, it follows from (9) that

$$|w_x(L,t)| \leq \frac{C}{T}\|\psi_0\|_H,$$

which implies that $w_x(L,t) = 0$ for all $t \geq 0$.

Next multiplying the equation (8) by xw_x and integrating by parts, we obtain:

$$\frac{1}{2}\int_0^T \int_0^L w_t^2 dx dt + \frac{3}{2}\int_0^T \int_0^L w_{xx}^2 dx dt = \frac{L}{2}\int_0^T w_t^2(L,t)dt - \left[\int_0^L w_t x w_x dx \right]_0^T \tag{10}$$

Multiplying the equation (8) by w and integrating by parts, we have:

$$\int_0^T \int_0^L w_t^2 dx dt - \int_0^T \int_0^L w_{xx}^2 dx dt = \int_0^T w_{xxx}(L,t)w(L,t)dt + \left[\int_0^L w_t w dx \right]_0^T \tag{11}$$

Using the boundary condition $w_{tt}(L,t) = w_{xxx}(L,t)$, it follows that

$$\int_0^T w_{xxx}(L,t)w(L,t)dt = [w_t(L,t)w(L,t)]_0^T - \int_0^T w_t^2(L,t)dt. \tag{12}$$

Now combining (10)–(12), we obtain:

$$\int_0^T \int_0^L w_t^2 dx dt + (3-L)\int_0^T \int_0^L w_{xx}^2 dx dt \leq C\|\psi_0\|_H^2. \tag{13}$$

Replacing w by w_t in (13), we obtain that

$$\int_0^T \int_0^L (w_t^2 + w_{tt}^2) dx\, dt + (3 - L) \int_0^T \int_0^L (w_{xx}^2 + w_{xxt}^2) dx\, dt \leq C\|\psi_0\|^2_{D(A)}. \qquad (14)$$

Finally thanks to the conditions $w_t(0, t) = w_{xt}(0, t) = 0$, it follows from (14) that

$$\int_0^T \|\psi(t)\|_H^2 dt = \int_0^T \left(\int_0^L (w_t^2 + w_{xx}^2) dx + w_t^2(L, t) \right) dt \leq C\|\psi_0\|^2_{D(A)} \qquad (7)$$

for all $L < 3$.

Using the fact that $\|\psi(t)\|_H^2 = \|\psi_0\|_H^2$ for all $t \geq 0$, and passing to the limit in (7) as $T \to +\infty$, we conclude that for any smooth initial data $\psi_0 \in D(A)$, the only solution of the equation $B_1^* e^{-tA} \psi_0 = 0$ is $\psi_0 = 0$.

Now let $\psi_0 \in H$, then for any $h > 0$ we define:

$$\psi_h = \int_0^h e^{-\tau A} \psi_0 d\tau.$$

Then we have:

$$B_1^* e^{-tA} \psi_h = \int_0^h B_1^* e^{-(t+\tau)A} \psi_0 d\tau = 0.$$

Since $\psi_h \in D(A)$ (*cf.* Pazy [3]), thus we deduce that $\psi_h = 0$ for any $h > 0$, which implies that $\psi_0 = 0$. The proof of theorem 2 is complete.

Remark 1. For $L = 1$, theorem 2 was proved in Slemrod [5] by means of a careful study of the eigenvalues of the operator A, completed with the numerical calculations. Here we ameliorate this result. Our method is very simple.

Remark 2. Of course, the estimate (7) does not allow to obtain any uniform decay of energy, but it is sufficient for obtaining the strong stabilization of the system (6). On the other hand, we know that the system (3) has non uniform decay of energy even with both linear force and moment controls: $f_1(t) = y_{xt}(L, t)$, $f_2(t) = y_t(L, t)$ (*cf.* Rao [4]). Hence an estimate of type:

$$\int_0^T \|\psi(t)\|_H^2 dt \leq C\|\psi_0\|_H^2$$

is not true in the present case, since such an estimate would imply the uniform decay of energy.

4. Stabilization with an a priori Bounded Force Control

In this case the *a priori* bounded controls are chosen:

$$f_1(t) = 0, \qquad f_2(t) = \begin{cases} -r \text{ sgn } y_t(L,t), & \text{if} \quad |y_t(L,t)| \geq r; \\ -y_t(L,t), & \text{if} \quad |y_t(L,t)| < r. \end{cases} \tag{15}$$

Thus we obtain the following closed-loop control system:

$$\frac{d}{dt} u(t) + Au(t) = B_2 f_2(t), \qquad u(0) = u_0 \in H. \tag{16}$$

Theorem 3. Let $\mu > 0$ be the largest constant such that

$$\mu \varphi_{xx}^2(1) \leq \int_0^1 \varphi_{xxxxx}^2 dx \tag{17}$$

for all $\varphi \in H^5(0,L)$ such that $\varphi(0) = \varphi_x(0) = \varphi_{xxxx}(0) = \varphi(1) = \varphi_{xxx}(1) = \varphi_{xxxx}(1) = 0$. Assume that $L^6 \leq 3\mu$. Then for all $u_0 \in H$, the system (16) admits a unique weak solution satisfying:

$$\|u(t)\|_H \to 0 \quad \text{as} \quad t \to +\infty.$$

Proof. As proceeded in the proof of theorem 2, it is sufficient to prove the uniqueness result for the equation $B_2^* e^{-tA} \psi_0 = 0$. We first establish for smooth initial data $\psi_0 \in D(A^2)$ the energy estimate:

$$\int_0^T \|\psi(t)\|_H^2 dt \leq C \|\psi_0\|_{D(A^2)}^2, \tag{18}$$

where $\psi(t) = e^{-tA} \psi_0$ satisfying $B_2^* \psi(t) = 0$.

Let $\psi(t) = (w(t), v(t), a(t), b(t))$, then w satisfies the equation:

$$\begin{cases} w_{tt} + w_{xxxx} = 0, & 0 < x < L, \\ w(0,t) = w_x(0,t) = 0, \\ w_{xtt}(L,t) + w_{xx}(L,t) = 0, \\ w_t(L,t) = w_{xxx}(L,t) = 0. \end{cases} \tag{19}$$

First multiplying the equation (19) by $x^2(x - 3L)$ and integrating by parts, we obtain:

$$\left[\int_0^L (x^3 - 3x^2 L) w_t dx \right]_0^T = \int_0^T \left(-3L^2 w_{xx}(L,t) + 6w(L,t) \right) dt$$

$$= \int_0^T (3L^2 w_{xtt}(L,t) + 6w(L,t))dt = 3L^2[w_{xt}(L,t)]_0^T + 6Tw(L,t).$$

We deduce that

$$|w(L,t)| \leq \frac{C}{T}\|\psi_0\|_H,$$

which implies that

$$w(L,t) = 0, \qquad \forall t \geq 0.$$

Next multiplying the equation (19) by xw_{xxx} and integrating by parts, we obtain

$$\frac{1}{2}\int_0^T \int_0^L (3w_{xt}^2 + w_{xxx}^2)dxdt = \frac{L}{2}\int_0^T w_{xt}^2(L,t)dt + \left[\int_0^L w_t x w_{xxx} dx\right]_0^T. \tag{20}$$

Since $w_{xxx}(L,t) = 0$, it follows that

$$\int_0^L w_t x w_{xxx} dx \leq C \int_0^L (w_t^2 + w_{xxxx}^2)dx \tag{21}$$

$$= C \int_0^L (w_t^2 + w_{tt}^2)dx \leq C(\|\psi(t)\|_H^2 + \|\psi_t(t)\|_H^2) = C\|\psi_0\|_{D(A)}^2.$$

Inserting (21) into (20) gives:

$$\int_0^T \int_0^L (3w_{xt}^2 + w_{xxx}^2)dxdt \leq L \int_0^T w_{xt}^2(L,t)dt + C\|\psi_0\|_{D(A)}^2. \tag{22}$$

Now replacing w by w_t in (22) and using the equation (19), we obtain:

$$\int_0^T \int_0^L (3w_{xxxxx}^2 + w_{xxxt}^2)dxdt \leq L \int_0^T w_{xx}^2(L,t)dt + C\|\psi_0\|_{D(A^2)}^2. \tag{23}$$

Since $w(0,t) = w_x(0,t) = w_{xxxx}(0,t) = w(L,t) = w_{xxx}(L,t) = w_{xxxx}(L,t) = 0$, we deduce from (17) that

$$\mu w_{xx}^2(L,t) \leq L^5 \int_0^L w_{xxxxx}^2 dx. \tag{24}$$

Hence for all $L^6 \leq 3\mu$, it follows from (23) and (24) that

$$\int_0^T \int_0^L w_{xxxt}^2 dxdt \leq C\|\psi_0\|_{D(A^2)}^2,$$

which, together with the conditions: $w(0,t) = w_x(0,t) = w(L,t) = 0$, implies that

$$w_{xt}^2(L,t) \leq C \int_0^T \int_0^L w_{xxxt}^2 dx dt \leq C\|\psi_0\|_{D(A^2)}^2. \tag{25}$$

Finally inserting (25) into (22) gives the estimate (18)

$$\int_0^T \|\psi(t)\|_H^2 dt = \int_0^T \Big(\int_0^L (w_t^2 + w_{xx}^2)dx + w_{xt}^2(L,t) \Big)dt \leq C\|\psi_0\|_{D(A^2)}^2. \tag{18}$$

The remaind part of the proof is the same as that one of theorem 2. The proof is thus complete.

Remark 3. A straightforward computation shows that $\mu > 25$. We obtain thus a sufficient condition: $L \leq 2$ for the validity of theorem 3.

Remark 4. The hypothesis $L < 3$ in theorem 2, (resp. $L^6 \leq 3\mu$ in theorem 3) is by no means optimal. But as shown in Slemrod [5], we know that there exist large values L for which the system (6) or (16) can't be strongly stabilized.

References

[1] P.-O. GUTMAN ; P. HAGANDER, *A new design of constrained controllers for linear systems*, IEEE Trans. Automate. Control, 30 (1985), 22-23.

[2] W. LITTMAN ; L. MARKUS, Exact boundary controllability of a hybrid system of elasticity, Arch. Rational Mech. Anal. 103 (1988) 193-236.

[3] A. PAZY, *Semigroups of linear operators and applications to partial differential equations*, Springer–Verlag, New York, (1983).

[4] B.P. RAO, *Stabilisation uniforme d'un système hybride en élasticité*, C. R. Acad. Sci. Paris, 316, Sér. I, (1993) 261-266.

[5] M. SLEMROD, *Feedback stabilization of a linear system in Hilbert space with an a priori bounded control*, Math. Control Signals Systems, 2 (1989), 265-285.

NONLINEAR BOUNDARY STABILIZATION OF A VON KÁRMÁN PLATE VIA BENDING MOMENTS ONLY

Mary Ann Horn

Institute for Mathematics and its Applications, University of Minnesota
Minneapolis, Minnesota 55455

Continued study in the areas of exact controllability and uniform stabilization of partial differential equations has been spurred by current research problems such as the deployment of large scale flexible structures in space. How can vibrations be damped out or suppressed in a newly designed space station? Plate equations such as the von Kármán plate can be used as a model for this problem as well as others. At the same time, problems such as this, which are motivated by real world applications, raise new questions in the area of mathematics and, in particular, stability theory for partial differential equations.

Recently, much attention has been directed toward the problem of uniform stabilization of plate equations. A qualitative definition of uniform stability of a plate is as follows: *If a damping term is introduced into the system, preferably acting through all or part of the boundary, the energy of the system, defined in an appropriate function space, decays uniformly with respect to the energy of the initial state of the plate as time increases.* The preference for boundary control in the above definition, as opposed to interior control, is motivated by the fact that boundary controls, though mathematically more challenging, are more easily implemented as they need to act only on the boundary of the spatial domain.

1 Statement of the Problem

We consider a fully nonlinear von Kármán system with, in addition to the nonlinearity which appears in the equation, a nonlinear feedback control acting through the boundary as a moment. Let Ω be an open bounded domain in R^2 with a sufficiently smooth boundary, Γ. In Ω, we consider the following von Kármán system in the variables $w(t,x)$ and $\chi(w(t,x))$ with a nonlinear feedback control, g:

$$w_{tt} - \gamma^2 \Delta w_{tt} + \Delta^2 w + b(x)w_t = [w, \chi(w)] \quad \text{in } Q_\infty = (0,\infty) \times \Omega \qquad (1.a)$$

$$\left.\begin{array}{l} w(0,\cdot) = w_0 \\ w_t(0,\cdot) = w_1 \end{array}\right\} \quad \text{in } \Omega \qquad (1.b)$$

$$w = 0 \quad \text{on } \Sigma_\infty = (0,\infty) \times \Gamma \qquad (1.c)$$

$$\Delta w + (1-\mu)Bw = -g(\frac{\partial}{\partial\nu}w_t) \quad \text{on } \Sigma_\infty = (0,\infty) \times \Gamma, \qquad (1.d)$$

where $b(x) \in L^\infty(\Omega)$ satisfies $b(x) > 0$ a.e. in Ω, $0 < \mu < \frac{1}{2}$ is Poisson's ratio, and the parameter, γ, is proportional to the thickness of the plate and is therefore assumed to be small. The operator B is given by

$$Bw \equiv -\frac{\partial^2}{\partial\tau^2}w - k\frac{\partial}{\partial\nu}w = -k\frac{\partial}{\partial\nu}w, \qquad (1.2)$$

where k is the geodesic curvature of the boundary and the second equality follows from (1.1.c). Additionally, we assume the control, g, is a continuous, monotone increasing function and is subject to the following constraints:

$$\left.\begin{array}{ll} g(s)s > 0 & \text{for } s \neq 0 \\ m|s| \leq |g(s)| \leq M|s| & \text{for } |s| > 1. \end{array}\right\} \quad (H)$$

Notice that no growth assumptions are made on the behavior of g at the origin. This is in contrast to most of the literature related to the subject (see [6], [7], etc.).

In (1.1), $\chi(w)$ satisfies the system of equations

$$\left.\begin{array}{l} \Delta^2\chi = -[w,w] \quad \text{in } \Omega \\ \chi = \frac{\partial}{\partial\nu}\chi = 0 \quad \text{on } \Gamma, \end{array}\right\} \qquad (1.3)$$

where

$$[\phi,\psi] = \frac{\partial^2\phi}{\partial x^2}\frac{\partial^2\psi}{\partial y^2} + \frac{\partial^2\phi}{\partial y^2}\frac{\partial^2\psi}{\partial x^2} - 2\frac{\partial^2\phi}{\partial x\partial y}\frac{\partial^2\psi}{\partial x\partial y}. \qquad (1.4)$$

Physically, the boundary conditions represent a "simply supported" plate, i.e., the position of the boundary remains fixed while the plate is allowed to rotate about the tangent to the boundary. The term corresponding to γ^2 in (1.1.a) represents the rotational inertia of the plate. We note that without this term, i.e., when $\gamma \equiv 0$, only recently has the solution to system (1.1) been proven to be unique (see [3]).

Our goal is to show that the boundary control, g, causes the energy of our system, $E_w(t)$, defined in some appropriate topology, to decay uniformly with respect to the initial energy as time increases. Although our model contains some light internal damping in the interior, represented by the term $b(x)w_t$, alone it is not enough to uniformly stabilize the model. However, this mild damping term plays a critical role in the proof of the compactness/uniqueness argument. Without it, there is no apparent way to show that the boundary feedback alone is sufficient to uniformly stabilize the model.

We define the energy functional by

$$E_w(t) \equiv \frac{1}{2} \int_\Omega \{|w_t|^2 + \gamma^2 |\nabla w_t|^2 + |\Delta w|^2 + |\Delta \chi(w)|^2\} d\Omega \equiv E_{w,1}(t) + E_{w,2}(t), \quad (1.5)$$

where $E_{w,2}(t)$ is defined by

$$E_{w,2}(t) \equiv \frac{1}{2} \int_\Omega |\Delta \chi(w)|^2 d\Omega. \quad (1.6)$$

In view of this, the associated space of finite energy is $\mathcal{H} \equiv H^2(\Omega) \times H_0^1(\Omega)$ with the norm

$$\|(w, w_t)\|_{\mathcal{H}}^2 \equiv \|w\|_{H^2(\Omega)}^2 + \|w_t\|_{L_2(\Omega)}^2 + \gamma^2 \|\nabla w_t\|_{L_2(\Omega)}^2. \quad (1.7)$$

The following well-posedness theorem for problem (1.1)-(1.3) is a very special case of the result in [8].

Theorem 1.1 *(See [8].) For any $w_0 \in H^2(\Omega)$, $w_1 \in H^1(\Omega)$, and $T > 0$, there exists a unique solution to (1.1), $w \in C(0, T; H^2(\Omega)) \cap C^1(0, T; H^1(\Omega))$, such that*

$$\frac{\partial}{\partial \nu} w_t|_\Gamma \in L_2(0, T; L_2(\Gamma)). \quad (1.8)$$

Remark: Notice that the regularity property in (1.8) does not follow from a priori interior regularity of w (i.e., $w_t \in H^1(\Omega)$).

1.1 Literature

Boundary stabilization of thin plates has attracted considerable attention in recent years (e.g., see [6], [10], [11], [12]). We shall briefly concentrate on results which apply to model (1.1).

In the context of control theory and, in particular, stabilization theory, the von Kármán model was introduced for the first time in [6]. In fact, in [6], the uniform decay rates for the solutions to the von Kármán plate equation with $\gamma = 0$ and with *linear feedbacks* which act through higher boundary conditions than those in (1.1.c) and (1.1.d) (i.e., through moments and torques) were established. This result of [6] was derived under geometric conditions on Γ which required that Ω be "star-shaped." Subsequently, in [2], the results of [6] were extended to the case when:

    i.) $\gamma \neq 0$, *i.e., the rotational forces are taken into account;*

    ii.) no geometric conditions are imposed on Γ.

These generalizations required techniques which were based on microlocal estimates combined with a nonlinear compactness/uniqueness argument, rather than the Lyapunov techniques which were used in [6]. In [5], further generalizations of the results of [2] are established by additionally allowing the feedback controls to be nonlinear, provided they satisfy appropriate growth conditions *away from the origin.*

Initial results for the von Kármán plate which has boundary conditions (1.1.c) and (1.1.d) can be found in [1], where uniform stabilization has been established assuming *no geometric conditions on the domain* when control is acting through the entire boundary.

We extend the results of [1] by additionally assuming that the feedback control can be nonlinear, provided it satisfies appropriate growth conditions *away from the origin.* In this fully nonlinear case, we do not have, in general, smooth solutions even if the initial data are assumed to be very regular. However, rigorous derivation of the estimates needed to solve the stabilization problem requires a certain amount of regularity of the solutions which is not guaranteed. To deal with this problem, we introduce a regularization/approximation procedure which leads to an "approximating" problem for which partial differential equation calculus can be rigorously justified. Passage to the limit on the approximation reconstructs the needed estimates for the original nonlinear problem.

1.2 Statement of Main Results

To state our stability result, we will need the following notation. Let the function $h(x)$ be a concave, strictly increasing function with $h(0) = 0$ such that

$$h(sg(s)) \geq s^2 + (g(s))^2 \qquad \forall |s| < 1. \tag{1.9}$$

Such a function can be easily constructed (see [9]). Define

$$\tilde{h}(x) \equiv h(\frac{x}{mes\Sigma_T}). \tag{1.10}$$

Since $\tilde{h}$ is monotone increasing, for every $c \geq 0$, $cI + \tilde{h}$ is invertible. Setting

$$p(x) \equiv (cI + \tilde{h})^{-1}(Kx), \tag{1.11}$$

where K is a positive constant, we see that p is a positive, continuous, strictly increasing function with $p(0) = 0$.

We are now in a position to state our result.

Theorem 1.2 *Assume hypothesis (H) holds. Let w be the solution to system (1.1). Then for some $T_0 > 0$,*

$$E_w(t) \leq S(\frac{t}{T_0} - 1) \text{ for } t > T_0, \tag{1.12}$$

where $S(t) \to 0$ as $t \to \infty$ and is the solution (contraction semigroup) of the differential equation

$$\begin{cases} \frac{d}{dt}S(t) + q(S(t)) = 0 \\ S(0) = E_w(0), \end{cases} \tag{1.13}$$

and $q(x)$ is given by

$$q(x) \equiv x - (I + p)^{-1}(x) \text{ for } x > 0. \tag{1.14}$$

In this case, the constant K will generally depend on $E_w(0)$ and the constant $c = \frac{1}{mes\Sigma_T}(m^{-1} + M)$, but will not depend on the parameter γ.

2 Preliminary Energy Estimates

Our goal is to prove energy decay rates for problem (1.1). In order to do this, we wish to use multiplier methods which require regularity of the solutions higher than is available from Theorem 1.1. Since our nonlinear problem may not have a sufficiently regular solution (even if the initial data are smooth), we resort to an approximation argument (this argument was used in the context of wave equations in [9]). In fact, the idea here is to approximate solutions to the nonlinear problem (1.1) by solutions to different (linear) problems. Since this linear problem admits regular solutions for smooth initial data, the partial differential equation calculations can be performed on this problem. Final passage to the limit on the approximation problem allows us to obtain needed energy identities for the original nonlinear problem.

To follow our program, we start by defining the following approximations. Hypothesis (H) with (1.8) of Theorem 1.1 implies

Corollary 2.1 *Let w be the solution to (1.1). Then*

$$g(\tfrac{\partial}{\partial \nu} w_t) \in L_2(0, T; L_2(\Gamma)). \tag{2.1}$$

Let w be the solution of the original problem, (1.1). By using the regularity properties in (1.8) and (2.1), along with density of approximate (see below) Sobolev spaces, we are in a position to define

$$f_n \in H^{1,1}(Q_T); \quad \|f_n - [w, \chi(w)]\|_{L_2(0,T;H^{-1}(\Omega))} \longrightarrow 0 \tag{2.2}$$

$$g_n \in H^{1,1}(\Sigma_T); \quad \|g_n - g(\tfrac{\partial}{\partial \nu} w_t)\|_{L_2(\Sigma_T)} \longrightarrow 0 \tag{2.3}$$

$$\alpha_n \in H^{1,1}(\Sigma_T); \quad \|\alpha_n - \tfrac{\partial}{\partial \nu} w_t\|_{L_2(\Sigma_T)} \longrightarrow 0, \tag{2.4}$$

where $Q_T \equiv \Omega \times (0, T)$ and $\Sigma_T \equiv \Gamma \times (0, T)$. We consider the following approximating problem:

$$\begin{cases} w_{n,tt} - \gamma^2 \Delta w_{n,tt} + \Delta^2 w_n + b w_{n,t} = f_n \\ w_n(0) = w_{n,0}; \quad w_{n,t}(0) = w_{n,1} \\ w|_\Gamma = 0 \\ \Delta w_n + (1-\mu)B_1 w_n + \tfrac{\partial}{\partial \nu} w_{n,t}|_\Gamma = -g_n + \alpha_n \end{cases} \tag{2.5}$$

where

$$\|w_{n,0} - w_0\|_{H^2(\Omega)} \to 0; \quad \|w_{n,1} - w_1\|_{H^1(\Omega)} \to 0, \tag{2.6}$$

and $(w_{n,0}, w_{n,1}) \in \mathcal{D}$, where $\mathcal{D}$, as dense set of $\mathcal{H}$, consists of $w_{n,0} \in H^4(\Omega)$, $w_{n,1} \in H^3(\Omega)$, where $w_{n,0}, w_{n,1}$ satisfy the appropriate compatibility conditions on the boundary. By standard linear semigroup methods, one easily shows that the linear problem, (2.5), admits a classical solution,

$$w_n \in C(0, T; H^4(\Omega)) \cap C^1(0, T; H^3(\Omega)). \tag{2.7}$$

The following proposition, which is analogous to Proposition 2.1 of [5] and is proven in the same manner, plays a critical role in our development.

Proposition 2.1 *Let w_n (respectively, w) be a solution of (2.5) (respectively, (1.1)). Then as $n \to \infty$, the following convergence holds.*

$$w_n \to w \text{ in } C(0,T; H^2(\Omega)) \cap C^1(0,T; H^1(\Omega)) \tag{2.8}$$

$$\tfrac{\partial}{\partial \nu} w_{n,t}|_\Gamma \to \tfrac{\partial}{\partial \nu} w_t \text{ in } L_2(\Sigma_T). \tag{2.9}$$

Multiplier methods and the convergence properties in Proposition 2.1 allow us to prove the following fundamental energy relation for problem (1.1).

Lemma 2.1 *(Energy Identity) Let w be the solution to (1.1). Then the following energy identity holds:*

$$E_w(T) - E_w(0) + \int_{\Sigma_T} g(\tfrac{\partial}{\partial \nu} w_t)\tfrac{\partial}{\partial \nu} w_t \, d\Gamma dt + \int_{Q_T} b(x) w_t^2 \, d\Omega dt = 0. \tag{2.10}$$

3 A Priori Estimates

To prove Theorem 1.2, the following inequality is needed.

Lemma 3.1 *Let w be the solution to (1.1), $0 < \alpha < T/2$ and $\epsilon > 0$ be arbitrary. Then there exist constants, C and $C(E_w(0))$ such that*

$$\int_\alpha^{T-\alpha} E_w(t)dt - CE_w(0) \le C(E_w(0))\{\|\tfrac{\partial}{\partial \nu} w_t\|^2_{L_2(\Sigma_T)} + \|g(\tfrac{\partial}{\partial \nu} w_t)\|^2_{L_2(\Sigma_T)} + \int_{Q_T} b(x) w_t^2 d\Omega dt\}, \tag{3.1}$$

where $C(E_w(0))$ is an increasing function of $E_w(0)$.

Proof: For the sake of brevity, we include only a brief sketch of the proof, particularly as the proof is very similar to that of Lemma 3.1 in [5] with appropriate adjustments for the change in boundary conditions.

Step 1: Using the multipliers w_n and $(x - x_0) \cdot \nabla w_n$, where $x_0 \in R^2$ on the approximation problem, (2.5), we estimate the energy of that system in terms of the norms of f_n, g_n, α_n, w_n and $w_{n,t}$.

Step 2: Second order traces of the w_n in the estimate derived in Step 1 may be bounded as follows (see [4], Proposition 3.1): Assume $0 < \alpha < T/2$. Then

$$\|\tfrac{\partial^2 w_n}{\partial \tau \partial \nu}\|^2_{L_2(\alpha, T-\alpha, \Gamma)} \le C\{\|\tfrac{\partial}{\partial \nu} w_{n,t}\|^2_{L_2(\Sigma_T)} + \|g_n\|^2_{L_2(\Sigma_T)} + \|\alpha_n - \tfrac{\partial}{\partial \nu} w_{n,t}\|^2_{L_2(\Sigma_T)}$$
$$+ \|f_n\|^2_{H^{-3/2+\epsilon}(Q_T)} + l.o.(w_n)\},$$

where $l.o.(w_n) \equiv \|w_n\|_{L_2(0,T;H^{2-\epsilon}(\Omega))} + \|w_{n,t}\|_{L_2(0,T;H^{-\epsilon}(\Omega))}$ and $f \equiv -bw_{n,t} + f_n$.

Step 3: The convergence properties of Proposition 2.1 allow us to obtain the following estimate for the original problem, (1.1), by passing with the limit as $n \to \infty$ on the estimate obtained for w_n and noting (2.10):

$$\int_0^T E_w(t)dt - C_1 E_w(0) \leq C_2 \int_{\Sigma_T} |\tfrac{\partial}{\partial\nu}w_t|^2 d\Gamma dt + C_3 E_w^2(0) \int_{\Sigma_T} |\Delta\chi(w)| d\Gamma dt$$
$$+ C_4 \int_{Q_T} b(x)w_t^2 d\Omega dt + C_5 l.o.(w).$$

Step 4: Remaining terms, i.e., those involving $\chi(w)$ and $l.o.(w)$, are absorbed by using a nonlinear compactness/uniqueness argument as in [5]. $\square$

4 Final Estimates

Rewriting the L_2-norm of the control in the following manner,

$$\int_{\Sigma_T} |g(\tfrac{\partial}{\partial\nu}w_t)|^2 d\Gamma dt = \int_{\Sigma_{A_1}} |g(\tfrac{\partial}{\partial\nu}w_t)|^2 d\Gamma dt + \int_{\Sigma_{B_1}} |g(\tfrac{\partial}{\partial\nu}w_t)|^2 d\Gamma dt, \qquad (4.1)$$

where $\Sigma_{A_1} \equiv \{(t,x) \in \Sigma_T : |\tfrac{\partial}{\partial\nu}w_t| \leq 1\}$ and $\Sigma_{B_1} \equiv \Sigma_T \backslash \Sigma_{A_1}$, we use hypothesis (H) on Σ_{B_1} and the properties of the function $h(x)$ to find

$$\int_{\Sigma_T} |\tfrac{\partial}{\partial\nu}w_t|^2 d\Gamma dt + \int_{\Sigma_T} |g(\tfrac{\partial}{\partial\nu}w_t)|^2 d\Gamma dt$$
$$\leq \int_{\Sigma_{A_1}} [|\tfrac{\partial}{\partial\nu}w_t|^2 + |g(\tfrac{\partial}{\partial\nu}w_t)|^2] d\Gamma dt + (M + \tfrac{1}{m}) \int_{\Sigma_{B_1}} g(\tfrac{\partial}{\partial\nu}w_t)\tfrac{\partial}{\partial\nu}w_t d\Gamma dt$$
$$\leq \int_{\Sigma_{A_1}} h(\tfrac{\partial}{\partial\nu}w_t g(\tfrac{\partial}{\partial\nu}w_t)) d\Gamma dt + (M + \tfrac{1}{m}) \int_{\Sigma_T} g(\tfrac{\partial}{\partial\nu}w_t)\tfrac{\partial}{\partial\nu}w_t d\Gamma dt.$$
$$(4.2)$$

Denoting $\mathcal{F} \equiv \int_{\Sigma_T} g(\tfrac{\partial}{\partial\nu}w_t)\tfrac{\partial}{\partial\nu}w_t d\Gamma dt + \int_{Q_T} b(x)w_t^2 d\Omega dt$, we obtain from Lemma 3.1 and (4.2), using the monotonicity of $\tilde{h}$ to include $\int_{Q_T} b(x)w_t^2 d\Omega dt$,

$$\int_\alpha^{T-\alpha} E_w(t)dt - C_1 E_w(0) \leq C_{T,\alpha,\epsilon}(E_w(0))[\mathcal{F} + \tilde{h}(\mathcal{F})]. \qquad (4.3)$$

Since

$$\int_0^\alpha E_w(t)dt + \int_{T-\alpha}^T E_w(t)dt \leq 2\alpha E_w(0), \qquad (4.4)$$

we find

$$\int_0^T E_w(t)dt - C_{1,\alpha}E_w(0) \leq C_{T,\alpha,\epsilon}(E_w(0))[\mathcal{F} + \tilde{h}\mathcal{F}], \qquad (4.5)$$

and by Lemma 2.1,

$$\int_0^T E_w(t)dt \leq C_{T,\alpha,\epsilon}(E_w(0))[\mathcal{F} + \tilde{h}\mathcal{F}] + C_{1,\alpha}E_w(0)$$
$$\implies (T - C_{1,\alpha})E_w(T) \leq C_{T,\alpha,\epsilon}(E_w(0))[\mathcal{F} + \tilde{h}(\mathcal{F})] \qquad (4.6)$$
$$\implies E_w(T) \leq C_T(E_w(0))[\mathcal{F} + \tilde{h}(\mathcal{F})].$$

Hence, recalling (2.1),

$$(I + \tilde{h})^{-1}(\frac{E_w(T)}{C_T(E_w(0))}) \leq \mathcal{F} = E_w(0) - E_w(T). \tag{4.7}$$

Setting

$$p(s) \equiv (I + \tilde{h})^{-1}(\frac{s}{C_T(E_w(0))}), \tag{4.8}$$

we have proven the following proposition.

Proposition 4.1 *Let w be the solution to (1.1) and $E_w(t)$ be the corresponding energy at time t. If T is sufficiently large, then there exists a monotone increasing function, p, such that*

$$p(E_w(T)) + E_w(T) \leq E_w(0). \tag{4.9}$$

To arrive at the conclusion of Theorem 1.2, we need to apply the result of Lemma 3.3 in [9].

Lemma 4.1 ([9], **Lemma 3.3**) *Let p be a positive, increasing function such that $p(0) = 0$. Since p is increasing, we can define a function q such that $q(x) = x - (I + p)^{-1}(x)$. Notice that q is also an increasing function. Consider a sequence s_n of positive numbers which satisfy:*

$$s_{m+1} + p(s_{m+1}) \leq s_m. \tag{4.10}$$

Then $s_m \leq S(m)$, where $S(t)$ is a solution of a differential equation

$$\begin{cases} \frac{d}{dt}S(t) + q(S(t)) = 0 \\ S(0) = s_0. \end{cases} \tag{4.11}$$

Moreover, if $p(x) > 0$ for $x > 0$, then $\lim_{t \to \infty} S(t) = 0$.

Applying the result of Proposition 4.1, we obtain

$$E_w(m(T+1)) + p(E_w(m(T+1))) \leq E_w(mT), \tag{4.12}$$

for $m = 0, 1, \ldots$ Thus, applying Lemma 4.1 with

$$s_m \equiv E_w(mT), \tag{4.13}$$

yields

$$E_w(mT) \leq S(m), \quad m = 0, 1, \ldots \tag{4.14}$$

Setting $t = mT + \tau$, $0 \leq \tau < T$, and recalling the evolution property gives

$$E_w(t) \leq E_w(mT) \leq \mathcal{S}(m) \leq \mathcal{S}(\frac{t-\tau}{T}) \leq \mathcal{S}(\frac{t}{T} - 1) \ for \ t > T, \qquad (4.15)$$

which completes the proof of Theorem 1.2. $\square$

References

[1] M. E. Bradley and M. A. Horn. Global stabilization of the von Kármán plate with boundary feedback acting via bending moments only. Technical Report 1085, Institute for Mathematics and its Applications, Minneapolis, Minnesota, December 1992.

[2] M. E. Bradley and I. Lasiecka. Global decay rates for the solutions to a von Kármán plate without geometric conditions. *Journal of Mathematical Analysis and Applications*. To appear.

[3] A. Favini, M. A. Horn, I. Lasiecka, and D. Tataru. Global existence, uniqueness and regularity of solutions to the dynamic von Kármán system with nonlinear boundary dissipation. Manuscript.

[4] M. A. Horn and I. Lasiecka. Asymptotic behavior with respect to thickness of boundary stabilizing feedback for the Kirchoff plate. *Journal of Differential Equations*. To appear.

[5] M. A. Horn and I. Lasiecka. Global stabilization of a dynamic von Kármán plate with nonlinear boundary feedback. Technical Report 1086, Institute for Mathematics and its Applications, Minneapolis, Minnesota, December 1992.

[6] J. E. Lagnese. *Boundary Stabilization of Thin Plates*. Society for Industrial and Applied Mathematics, Philadelphia, 1989.

[7] J. E. Lagnese and G. Leugering. Uniform stabilization of a nonlinear beam by nonlinear boundary feedback. *Journal of Differential Equations*, 91(2):355–388, 1991.

[8] I. Lasiecka. Existence and uniqueness of the solutions to second order abstract equations with nonlinear and nonmonotone boundary conditions. *Journal of Nonlinear Analysis, Methods and Applications*. To appear.

[9] I. Lasiecka and D. Tataru. Uniform boundary stabilization of semilinear wave equations with nonlinear boundary damping. *Differential and Integral Equations*. To appear.

[10] I. Lasiecka and R. Triggiani. *Differential and Algebraic Riccati Equations with Application to Boundary/Point Control Problems: Continuous Theory and Approximation Theory*. Springer-Verlag, New York, 1991.

[11] J. L. Lions. *Contrôlabilité exacte, perturbations et stabilization de systèmes distribués*, volume 1. Masson, Paris, 1989.

[12] W. Littman. Boundary control theory for beams and plates. *Proceedings of 24th Conference on Decision and Control*, pages 2007–2009, 1985.

Design of a feedback controller for wave generators in a canal using H_∞ methods.

Stéphane Mottelet, Ghislaine Joly Blanchard, Jean-Pierre Yvon
Division Mathématiques Appliquées - Université de Technologie de Compiègne BP 649
60206 COMPIEGNE Cedex, FRANCE

e-mail : mottelet@sloane.univ-compiegne.fr
tel : (033) 44234644   fax : (033) 44234477

Abstract

We study the problem of designing a feedback controller for wave generators in a canal. The goal is to obtain a progressive wave at the surface by acting on plane generators at each side of a rectangular canal. The model of such a system is given by a classical linear model of shallow waters which leads to a damped "hyperbolic" equation. We formulate the problem within the additive uncertainty framework for Callier-Desoer systems and we then use the standard algorithm of Doyle and al. to derive a finite dimensional controller that ensures robust stability of the closed loop system and a prespecified performance level in a narrow frequency band. Numerical results show that we are able to create and maintain a progressive wave on the surface of the canal with this controller. We also show that the closed loop system remains stable with variations of the damping parameter.

Key words : Control applications, control system synthesis, distributed parameter systems, robustness.

1   Introduction

Canals with wave generators are commonly used to study the behaviour of floating bodies in the sea. Classical methods to obtain a progressive wave on the surface of the canal work by acting on a generator at one end and by absorbing waves at the other end with a *passive* absorber (usually a shore). Unfortunately such a method needs a very long canal. This length can be significantly reduced by using a generator at both ends, the second generator playing the role of an *active* absorber. Open loop optimal control techniques have been successfully applied to this problem (see [13] and [10]), and real experiments have shown the validity of this approach. In this paper we want to show how feedback techniques can be applied to the same problem. The actual difficulty is due to the fact that the system to be controlled is infinite dimensional, and that we obviously need a finite-dimensional controller. Fortunately, in the last five years, major advances have been made in infinite-dimensional robust control theory, and powerfull tools are now available to design controllers for this kind of systems (see [5] for an overview). The main idea that can be retained is that it is possible to know whether a controller based upon a finite-dimensional approximation of a infinite dimensional system will stabilize this system or not. Some useful theoritical results can be used to check the stability of a feedback system *a priori* , or can be used as constraints when designing a controller.

This paper is organized into six sections. In section 2 we will recall some elements about the model of the canal (see [13] for more details). We introduced an artificial damping, which is not a usual thing in hydrodynamics because the viscosity of water is usually neglected. Taking viscosity into account naturally introduces damping on the poles of the canal transfer matrix. This was necessary because the theoritical framework we use is unable to deal with transfer matrices having an infinite number of poles on the imaginary axis. In section 3 we will formulate our controller design problem within the additive uncertainty framework for Callier-Desoer systems after recalling some elements of theory (see [6] and [4] for more details). We will then define the finite dimensional approximation of the system that we will use in the sequel. In section 4, we state a finite dimensional H_∞ mixed sensitivity problem. We solve this problem with the standard algorithm (see [7]) and we obtain a controller that achieves tracking of a monochromatic reference input. In section 5 we present some numerical simulations. We show that our controller is able to create a monochromatic progressive wave on the surface of the canal. We also show that variations of the damping parameter in a given interval do not destroy stability of the feedback system. Section 6 contains concluding remarks.

2 Model of the canal

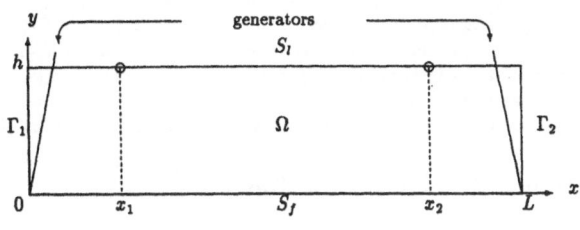

Figure 1

The canal is supposed wide enough to assume that waves are plane. This allows to use a bidimensional model, where the domain Ω is the rectangle $[0, L] \times [0, h]$ (L lenght, h depth). Generators are plane and can rotate around an axis perpendicular to the (x, y) plane at the bottom of the canal. Under the hypothesis of irrotationality and incompressibility, the velocity of the fluid $\vec{V}(x, y, t)$ is represented by its potential $\varphi(x, y, t) : \vec{V}(., ., t) = \vec{grad}\ \varphi(., ., t)$. The equations governing such a system, under the classical assumptions of linearity, are

$$\Delta \varphi = 0 \quad \text{in } \Omega, \tag{1}$$

$$\frac{\partial \varphi}{\partial y} = 0 \quad \text{on } S_f, \tag{2}$$

$$\frac{\partial^2 \varphi}{\partial t^2} + g_0 \frac{\partial \varphi}{\partial y} + \varepsilon \frac{\partial^2 \varphi}{\partial y \partial t} = 0 \quad \text{on } S_l, \tag{3}$$

where ε is a small positive number. Boundary conditions are the following

$$-\frac{\partial \varphi}{\partial x}(0, y, t) = y\ v_1(t) \quad \text{on } \Gamma_1, \tag{4}$$

$$\frac{\partial \varphi}{\partial x}(L, y, t) = y\ v_2(t) \quad \text{on } \Gamma_2, \tag{5}$$

where v_1 et v_2 (control inputs) are the angular velocities of generators. The fluid elevations y_1, y_2 (measured outputs) on the surface at $x = x_1, x_2$ is given by

$$y_i = -\frac{1}{g_0}\left(\frac{\partial \varphi}{\partial t}(x_i, h, t) + \varepsilon \frac{\partial \varphi}{\partial y}(x_i, h, t)\right), \quad i = 1, 2. \tag{6}$$

Equations (1)-(6) have been studied by several methods. For the case $\varepsilon = 0$ see [11], [13] and [14], [1] for unbounded domains. For the case $\varepsilon > 0$, which corresponds to a Kelvin-Voight type damping, see [12]. The essential point consists in converting the original problem into a new problem, posed on the free surface S_l. The solution φ is then decomposed on the eigenfunctions of an operator $\mathcal{A}$ whose eigenvalues are $\lambda_k = \alpha k \tanh(\alpha k h)$, $k \geq 0$. After some calculations (see [12]) the transfer matrix (from generators velocities v to surface elevation measurements y) takes the form

$$G(s) = -\frac{2}{\pi(\varepsilon s + g_0)} \frac{\delta_0 + \theta_0 s}{s} G_0 - \frac{4s}{\pi(\varepsilon s + g_0)} \sum_{k=1}^{\infty} \frac{\delta_k + \theta_k s}{s^2 + \varepsilon \lambda_k s + g_0 \lambda_k} G_k$$

where

$$G_0 = \begin{bmatrix} 1 & 1 \\ 1 & 1 \end{bmatrix}, \quad G_k = \begin{bmatrix} \cos(\alpha k x_1) & (-1)^k \cos(\alpha k x_1) \\ \cos(\alpha k x_2) & (-1)^k \cos(\alpha k x_2) \end{bmatrix},$$

and

$$\delta_0 = g_0 \alpha h^2 / 4, \quad \delta_k = \frac{g_0}{2k}\left[h \, \tanh(\alpha k h) + \frac{1}{\alpha k}\left(\frac{1}{\cosh(\alpha k h)} - 1\right)\right], \quad \theta_k = \frac{2\varepsilon(1 - \lambda_k)}{\alpha(1 - 4k^2)}$$

If we set $\varepsilon = 0$, then G has an infinite number of poles on the imaginary axis at $p_k = \pm j\sqrt{g_0 \lambda_k}$, $k = 0, 1, \cdots$, and does not have a limit as $|s| \to \infty$. One can show that for any $\varepsilon > 0$, then $G(s) \to 0$ as $|s| \to \infty$ for $\mathrm{Re}(s) \geq 0$ (G is strictly proper). Moreover $G(s)$ has a finite number of complex poles at $p_k = -\varepsilon\lambda_k/2 \pm j\sqrt{g_0 \lambda_k - \varepsilon^2 \lambda_k^2 / 4}$ which alternate along the circle $|s + g_0/\varepsilon| = g_0/\varepsilon$. One branch of real poles tends to $-\infty$ and the other branch tends to $-g_0/\varepsilon$. Since $G(s)$ has an infinite sum representation, it seems hard to establish an explicit relationship between x_1, x_2 and the location of the zeros of $G(s)$ in the complex plane. It seems that $G(s)$ has right half plane zeros, even if the sensors and actuators are collocated in some sense ($x_1 = 0, x_2 = L$). Figure 4a shows a singular value bode plot of $G(s)$.

3 Formulation of the control problem

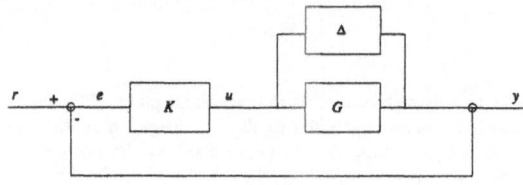

Figure 2

3.1 Basic definitions

The name H_∞ ('H-infinity') refers, for control engineering purposes, to the set of exponentially stable (maybe irrationnal) transfer functions. A transfer matrix (a matrix of transfer functions) is said to be stable when every one of its elements is a stable transfer function. H_∞ methods of control design, broadly speaking, work by minimizing the norms of certain closed-loop transfer matrices. The H_∞ norm of a stable transfer matrix G is the largest value of its spectral norm, that is,

$$\|G\|_\infty = \sup_\omega \bar{\sigma}\left[G(j\omega)\right]$$

where the symbol $\bar{\sigma}[.]$ represents the maximum singular value. In the scalar case the H_∞ norm is simply the maximum frequency response magnitude

$$\|G\|_\infty = \sup_\omega |G(j\omega)|$$

We must also define what is called stability of a feedback system. The feedback system of figure 2 with $\Delta = 0$ is said to be stable if the four transfer matrices S, KS, SG, $(I - KSG)$ are stable, where $S = (I + GK)^{-1}$. The feedback system is said to be *robustly* stable if it remains stable for all perturbation Δ in a predefined set.

3.2 The additive uncertainty framework for Callier-Desoer systems

We summarize the main results from the theory of robust controllers given in Curtain and Glover [6]. Suppose, as in Figure 2, we have a possibly unstable infinite-dimensionnal plant with transfer matrix G, and we wish to design a finite dimensional controller with transfer matrix K so that it stabilizes the class of perturbed plants $G + \Delta$ such that Δ belongs to a predefined set. A convenient class of infinite dimensional transfer matrices is the class of $p \times m$ transfer matrices $\hat{B}_0^{p \times m}$ introduced by Callier and Desoer. The simplest and most intuitive way of defining the algebra $\hat{B}_0^{p \times m}$ is to use Theorem 3.3 and

Corollary 2.2D of Callier and Desoer [3]. Then $G \in \hat{B}_0^{p \times m}$ iff G has the decomposition $G = G_s + G_u$, where G_s is a stable transfer matrix and G_u is a rational transfer matrix whose poles are in $\text{Re}(s) \geq 0$. In other words, $\hat{B}_0^{p \times m}$ is the class of $p \times m$ transfer matrices with a stable infinite dimensional part and a finite dimensional unstable part. We need a few more conditions before recalling the theoritical results that we will use in the sequel.

- $G \in \hat{B}_0^{p \times m}$ is proper with no poles on the imaginary axis,

- $\Delta \in \hat{B}_0^{p \times m}$ is proper, and G and $G + \Delta$ have an equal number of poles in $\text{Re}(s) \geq 0$,

- There exists a rational, stable, minimum phase transfer function w such that $\|w^{-1}\Delta\|_\infty < 1$ (this means we have a frequency dependent bound on Δ).

Under the assumptions made above the following theorem holds [4] :

Theorem 2. If a controller K stabilizes G, and if

$$\|wK(I + GK)^{-1}\|_\infty \leq 1, \tag{7}$$

then K will also stabilize the perturbed system $G + \Delta$.

Remark 1 Condition (7) is known to be conservative. It means that a particular controller K can violate (7) but nevertheless stabilize $G + \Delta$. So if (7) is used as a constraint when designing the controller, the feedback system will surely have poor performances.

Theorem 2 can be used in the following context [2]. Let G^N be a finite dimensionnal approximation of G. We can write $G = G^N + (G - G^N)$ and consider $\Delta = (G - G^N)$ as a perturbation of G^N. If we have a rational, stable, minimum phase transfer function $w(s)$ such that $\|w^{-1}(G - G^N)\|_\infty < 1$, and a (finite dimensionnal) controller K stabilizing G^N, then K will stabilize G, if

$$\|wK(I + G^N K)^{-1}\|_\infty \leq 1. \tag{8}$$

In fact any finite dimensional method can be used to do the synthesis of K. The advantage of the standard algorithm of Doyle et al., is the possibility of directly using (8) as a constraint in the design of the controller while shaping the frequency response of the closed loop system.

3.3 Finite dimensional approximation of G

Since we have a inifinite sum representation for $G(s)$, we shall choose the most natural finite dimensional approximation $G^N(s)$ by truncating this sum at a finite rank N :

$$G^N(s) = -\frac{2}{\pi(\varepsilon s + g_0)} \frac{\delta_0 + \theta_0 s}{s} G_0 - \frac{4}{\pi(\varepsilon s + g_0)} \sum_{k=1}^{N} \frac{\delta_k s + \theta_k s^2}{s^2 + \varepsilon \lambda_k s + g_0 \lambda_k} G_k, \tag{9}$$

where the number N must be chosen with respect to the desired frequency bandwidth. The rational transfer matrix G^N admits a finite dimensional state space representation, say (A_N, B_N, C_N, D_N) (where $D_N = 0$ because G^N is strictly proper). When looking at the zeros of G^N, it is straightforward to see that there are right-half plane zeros, which can be an undesirable property (see [9] and [8] Chapter 9). We can write $G = G^N + \Delta^N$ with

$$\Delta^N(s) = -\frac{4}{\pi(\varepsilon s + g_0)} \sum_{k=N+1}^{\infty} \frac{\delta_k s + \theta_k s^2}{s^2 + \varepsilon \lambda_k s + g_0 \lambda_k} G_k. \tag{10}$$

The transfer matrix $G^N(s)$ has one simple pole at $s = 0$. Since we want G^N to belong to $\hat{B}_0^{p \times m}$, it is necessary to move this pole to the left at $s = -\zeta$, where $\zeta > 0$ is small, by replacing $(\delta_0 + \theta_0 s)/s$ by $(\delta_0 + \theta_0 s)/(s + \zeta)$ in (9). We now have to bound $\bar{\sigma}[\Delta^N(j\omega)]$ by a reasonably low-order scalar rational transfer function, i. e. find $w_2(s)$ such that $\|w_2^{-1}\Delta^N\|_\infty < 1$. As we will see it in the next section, the order of w_2 directly influences the order of the controller.

4 A mixed sensitivity problem

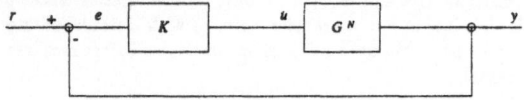

Figure 5

Consider the feedback system of Figure 5. Let us define the three transfer matrices $T = G^N K(I + G^N K)^{-1}$, $S = (I + G^N K)^{-1}$ and $R = K(I + G^N K)^{-1}$. The three transfer matrices T, S and R represent the transfer from the reference input r to y, e and u (note that $S + T = I$). The tracking problem formulates as follows : we want T to be close to indentity in a given frequency band (we want y to be close to r). So $\bar{\sigma}(S(j\omega))$ has to be as small as possible in that frequency band. This performance specification takes the form

$$\bar{\sigma}(S(j\omega)) < \frac{1}{|w_1(j\omega)|}, \tag{11}$$

where $w_1(s)$ is a scalar rational transfer function chosen such that $|w_1(j\omega)|$ is large in the desired frequency band. The inequality (11) can be rewritten in the following way

$$\|w_1 S\|_\infty < 1. \tag{12}$$

A controller K will stabilize G^N and $G = G^N + \Delta$, with $\|w_2^{-1}\Delta\|_\infty < 1$, if

$$\|w_2 R\|_\infty < 1. \tag{13}$$

We can summarize the problem as follows : find a controller K which verifies (12) and (13). A closely related problem can be solved by the standard algorithm (see [7]) : find K such that

$$\left\| \begin{matrix} w_1 S \\ w_2 R \end{matrix} \right\|_\infty < 1. \tag{14}$$

This form is almost equivalent in the sense that (14) implies (12) and (13). We can get extra freedom in the design by solving the problem of optimal nominal performance (in the sense of w_1), under robust stability constraint, say

$$\min_{K \in \mathcal{K}_\gamma} \gamma > 0,$$

where $\mathcal{K}_\gamma$ is the set

$$\mathcal{K}_\gamma = \left\{ K \ stabilizing \ G^N; \ \left\| \begin{matrix} w_1 S \\ \gamma w_2 R \end{matrix} \right\|_\infty < \gamma \right\}.$$

Practically one uses an iterative procedure (gamma-iterations) to find the minimum value of γ, say γ_{min}, and the controller is obtained in state-space form. If the order of G^N is n_G, the order of the controller that one obtains is equal to $n_G + n_{w1} n_o + n_{w_2} n_c$, where n_{w1} and n_{w_2} are the order of w_1 and w_2, n_c the number of control inputs and n_o the number of measured outputs.

Remark 2 The difficulty of solving a mixed sensitivity problem (γ_{min} is very large) can reflect an incompatibility between w_1 and w_2. In particular, $|w_2(j\omega)|$ has to be small (the system is known precisely) in the frequency band where $|w_1(j\omega)|$ is large (good performance expected). Having a tight bound on Δ needs a high order w_2, which leads to a high order controller. It is often simpler to take a higher order approximation of G, and a low order w_2.

Remark 3 Another difficulty raises when the system has right half-plane zeros, because it creates a constraint on the shape of the sensibility function. This can be simply described as follows : the smaller $\bar{\sigma}[S(j\omega)]$ will be in an interval of the form $[0,\omega_1]$, the higher it will be outside this interval, which can lead to an unstable closed loop system. In other words, if w_1 has the classical shape of a low-pass filter, the mixed sensitivity problem may have no solution.

5 Numerical results

The data come from a reduced scale canal at ENSTA-GHN (Palaiseau). Dimensions are L=6.93 m, h=0.35 m. The sensors (resistive probes) are located at $x_1 = 1$ m and $x_2 = 2$ m. The damping parameter has been arbitrarily chosen to be ε=0.001. Our goal is to obtain a monochromatic progressive wave of the form $a\ \sin(\omega_0 t - \mu x)$ (ω_o and μ are chosen with respect to the relation $\omega_0^2 = \mu g_0 \tanh(\mu h)$). Since we only have two sensors measurements y_1, y_2, our goal will be to track this progressive wave at the two positions x_1, x_2, and we will define the reference input r as

$$r = \left[\begin{array}{c} a\ \sin(\omega_0 t - \mu x_1) \\ a\ \sin(\omega_0 t - \mu x_2) \end{array} \right].$$

The number of modes N in G^N has to be chosen with respect to ω_0 because N directly determines its bandwidth. With $\omega_0 \leq 2\pi$, N has to be greater than 16 if we want frequencies up to ω_0 to be represented in G^N. We took $N = 19$, which yields a state-space representation (A_N, B_N, C_N, D_N) of order 41. Figures 4a and 4b represents singular values bode plot of G and G^N and Figure 4c the corresponding Δ^N. The shaping transfer functions have been chosen as follows :

$$w_1(s) = \frac{s}{s^2 - 2a + \omega_0^2 + a^2} + \frac{10}{s + 10} , \quad w_2(s) = \left(\frac{g_\infty^{1/4} s + g_0^{1/4} \omega_c}{s + \omega_c} \right)^4 .$$

In all designs we took $g_\infty = 5$, $g_0 = 0.001$, $a = -0.01$. The transfer function $w_1(s)$ has a peak value at $\omega = \omega_0$. The frequency ω_c will take a different value in our two designs.

5.1 A first design

Remark 2 emphasises on the fact that w_2 directly influences the achievable performance. To illustrate that, a first design can be done as follows. Figure 4c shows the singular values of Δ^N and the magnitude of w_2 (solid line). The frequency $\omega_c = 2\pi$ has been chosen with respect to the relation $\|w_2^{-1}\Delta^N\|_\infty < 1$. The shape of $w_1(s)$ is shown in figure 4d. It is interesting to see the influence of ω_0 on γ_{min}. For example, if we do not allow γ_{min} to be greater than 10 (this corresponds to a correct level of performance), then numerical computations show that ω_0 cannot be greater than 1.7 (solid line), which is considerably lower than the bandwidth of G^N. The obtained controller is of order 55. This means that the robustness condition (8) leads to controllers with poor performance with respect to their order. Figure 4e shows the resulting sensitivity and complementary sensitivity functions, respectively $S = (I + GK)^{-1}$ and $T = GK(I + GK)^{-1}$. Figure 5a shows the first 128 poles of S. We can see that the controller does not affect any mode past the 8th.

5.2 A second design

Following common practice [2], a way of obtaining more performance is to deliberately violate condition (8). The drawback of such a method is that robustness is no longer guaranteed *a priori*. One has to compute "by hand" the poles of the closed loop system, and check if all of them lie in Re(s) < 0. We now take $\omega_c = 10\pi$, which yields a w_2 (dash-dot line on Figure 4c) that frankly violates (8). With that w_2 we can shift the peak frequency of w_1 (dash-dot line on Figure 4d) at $\omega_0 = 2\pi$ and still have $\gamma_{min} = 10$. Figure 4f shows the resulting sensitivity and complementary sensitivity functions. Figure 5b shows the first 128 poles of S. It is easy to check that the resulting closed loop system is stable. We can see that the controller does not affect any mode past the 36th. We will use this controller for the rest of the simulations.

5.3 Simulations

Our goal is to show the effiency of the controller, by making two numerical experiences. In the first one we try to obtain a monochromatic progressive wave. In the second one we try to drive the canal at rest. For the simulations, we took 128 modes for the transfer function of the canal, which yields a state-space representation of order 259 $(A_{218}, B_{128}, C_{128})$. The controller is the one obtained in section 5.2, and has a state space representation (A_K, B_K, C_K) of order 55. We form the following closed-loop system

$$\begin{bmatrix} \dot{x} \\ \dot{\xi} \end{bmatrix} = \begin{bmatrix} A_{128} & B_{128}C_K \\ -B_K C_{128} & A_K \end{bmatrix} \begin{bmatrix} x \\ \xi \end{bmatrix} + \begin{bmatrix} 0 \\ B_K \end{bmatrix} r. \tag{15}$$

5.3.1 Tracking

We solve (15) for $t \in [0, T]$ with $T = 60$ s, $x(0) = \xi(0) = 0$, and $r = 0.01 \begin{bmatrix} \sin(2\omega_0 t - \mu x_1) \\ \sin(2\omega_0 t - \mu x_2) \end{bmatrix}$, with $\omega_0 = 2\pi$, $\mu = 4.40937$. Figure 6a shows the elevation of the surface at x_1 and x_2. Figure 7a shows the angular velocities of generators. Figure 8a shows the elevation of the surface at various instants. Figure 9a shows the elevation on the whole surface as a function of x and t. As a comparison, Figure 9b shows what happens when one tries to make a progressive wave with only one generator at the left, without using the right generator.

5.3.2 Stabilization

We now try to drive the canal at rest. We solve (15) for $t \in [60, 120]$, with $r = 0$. Figure 6b shows the elevation of the surface at x_1 and x_2. Figure 7b shows the angular velocities of the generators. Figure 8b shows the elevation of the surface at various instants. Figure 9c shows the elevation on the whole surface as a function of x and t. As a comparison, figure 9d shows what happens after the same duration when one lets the canal go back at rest by itself, without using the controller.

Remark 4 Since we are supposed to have a certain amount of robustness, it is interesting to see in what interval can vary the damping parameter ε without destroying the stability of the closed loop system when we use the same controller : numerical tests show that we can take $\varepsilon \in [0.0005, 0.02]$ and still have a stable closed loop system. This interval is quite large, and this is a very interesting robustness property, since the viscosity parameter corresponds to an arbitrary way of modelling the observed damping of the canal.

6 Conclusions

This work is an interesting application of infinite dimensional system theory to a non-academic example. The obtained finite dimensional controller allows to create a progressive wave on the whole surface of the canal, by mean of only two sensors. The relationship between the positions of sensors on the surface and the ability of the controller to create a progressive wave has not been clearly established, but it seems that the chosen positions are near optimal. Some other positions are not optimal at all, in the sense that the controller creates a mixed progressive/stationnary wave. The original system exhibits non-linearities under the form of parasitic waves whose frequency is a multiple of input frequency. It will be interesting to see if one can design a controller to reject these parasitic waves by considering them as disturbances on the input of the system. The main problem that we have met is an undesirable property of systems whose transfer matrix has right half plane zeros. The presence of these zeros, when solving a mixed sensitivity problem, creates a bound on the achievable performance, at the frequency of those zeros. In particular, the weight w_1 on the sensitivity function cannot have the classical shape of a low-pass filter. If the mixed sensitivity is expected to have a solution, then w_1 must have the shape that we have used, say, tracking can be achieved only in a narrow frequency band.

References

[1] J.T. Beatle. Eigenfunction expansion for object floating in an open sea. *Comm. Pure Appl. Math.*, 30:283–313, 1977.

[2] J. Bontsema and R. F. Curtain. A note on spillover and robustness for flexible systems. *IEEE Transactions on Automatic Control*, 33:567–569, 1988.

[3] F. M. Callier and C. A. Desoer. An algebra of transfer functions for distributed linear time-invariant systems. *IEEE Transactions on Circuits and Systems*, 27:320–323, 1980.

[4] M. J. Chen and C. A. Desoer. Necessary and sufficient conditions for robust stability of linear distributed feedback systems. *International Journal of Control*, 35:255–267, 1982.

[5] R. F. Curtain. A synthesis of time and frequency domain methods for the control o infinite-dimensional systems: A system theoritic approach. In H. T. Banks, editor, *Control and Estimation in Distributed Parameter Systems*, volume 11. Frontiers in applied mathematics, SIAM, 1992.

[6] R. F. Curtain and K. Glover. Robust stabilization of infinite dimensional systems by finite dimensional controllers. *Systems and Control Letters*, 7:41–47, 1986.

[7] J.C Doyle, K. Glover, P. P. Khargonekar, and B. A. Francis. State-space solutions to standard h_2 and h_∞ control problems. *IEEE Transactions on Automatic Control*, 34:831–847, 1989.

[8] B. A. Francis. *A Course in H_∞ Control Theory*. Lecture notes in control and information sciences, Springer-Verlag Berlin, 88.

[9] J. S. Freudenberg and P. D. Looze. Right half plane poles and zeros and design tradeoffs in feedback systems. *IEEE Transactions on Automatic Control*, 30:555–565, 1985.

[10] G. Joly-Blanchard, F. Quentin, and J.P. Yvon. Optimal control of waves generators in a canal. In *Proc. IFIP Conf. on System Modelling and Opt. 1991*, Zurich, 1991. Springer-Verlag.

[11] J.L. Lions. *Quelques methodes de résolution des problèmes aux limites non linéaires*. Dunod, Paris, 1969.

[12] S. Mottelet. *Méthodes de Contrôle Robuste pour les systemes distribués*. PhD thesis, Université de Technologie de Compiègne (to appear).

[13] F. Quentin. *Contrôle Optimal de batteurs à houle*. PhD thesis, Université de Technologie de Compiègne, 1992.

[14] A.J. Roberts. Transient free surface flows generated by moving vertical plate. *Q.J. Mech. Appl. Math.*, 40, 1987.

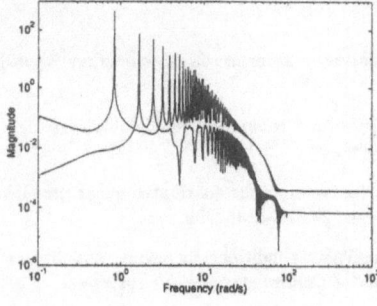

Figure 4a : singular values of G

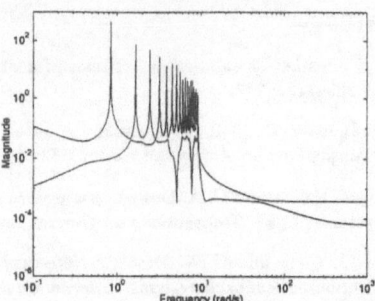

Figure 4b : singular values of G^N

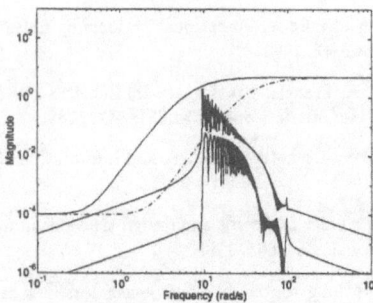

Figure 4c : singular values of Δ^N and magnitude of w_2 (solid : first design, dash-dot : second design)

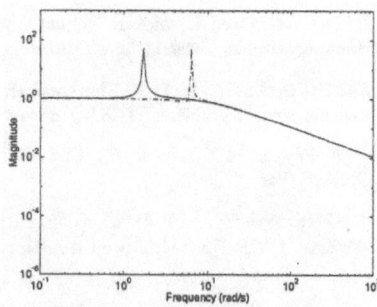

Figure 4d : magnitude of w_1 (solid : first design, dash-dot : second design)

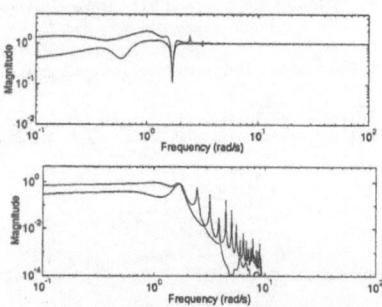

Figure 4e : sensitivity (top) and complementar sensitivity (bottom) obtained with the first controller ($w_0 = 1.7$)

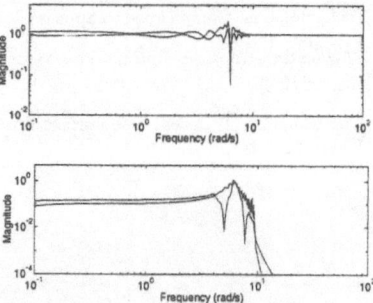

Figure 4f : sensitivity (top) and complementar sensitivity (bottom) obtained with the second controller ($w_0 = 2\pi$)

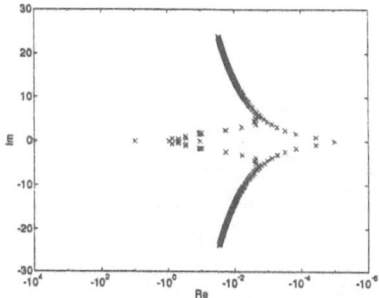

Figure 5a : poles of $(I + GK)^{-1}$ with the first controller
($\omega_0 = 1.7$)

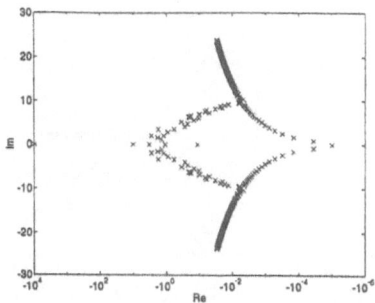

Figure 5b : poles of $(I + GK)^{-1}$ with the second controller
($\omega_0 = 2\pi$)

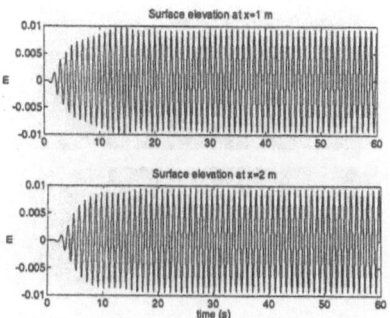

Figure 6a : (tracking) surface elevations.

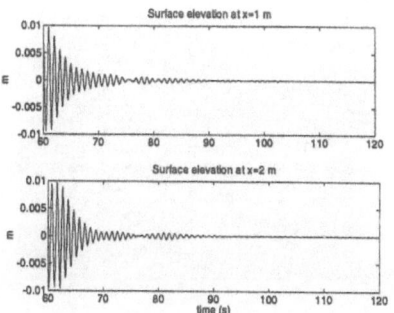

Figure 6b : (stabilization) surface elevations.

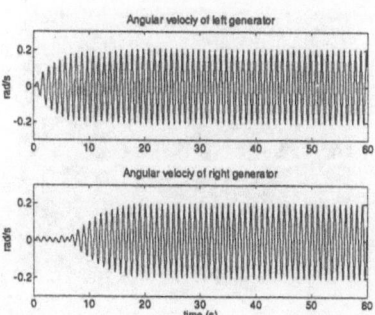

Figure 7a : (tracking) angular velocity of generators.

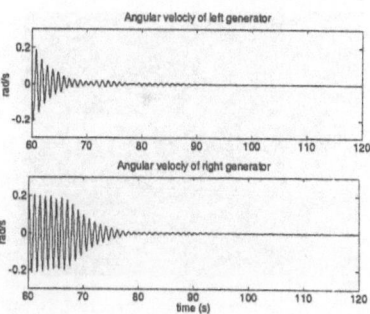

Figure 7b : (stabilization) angular velocity of generators.

726

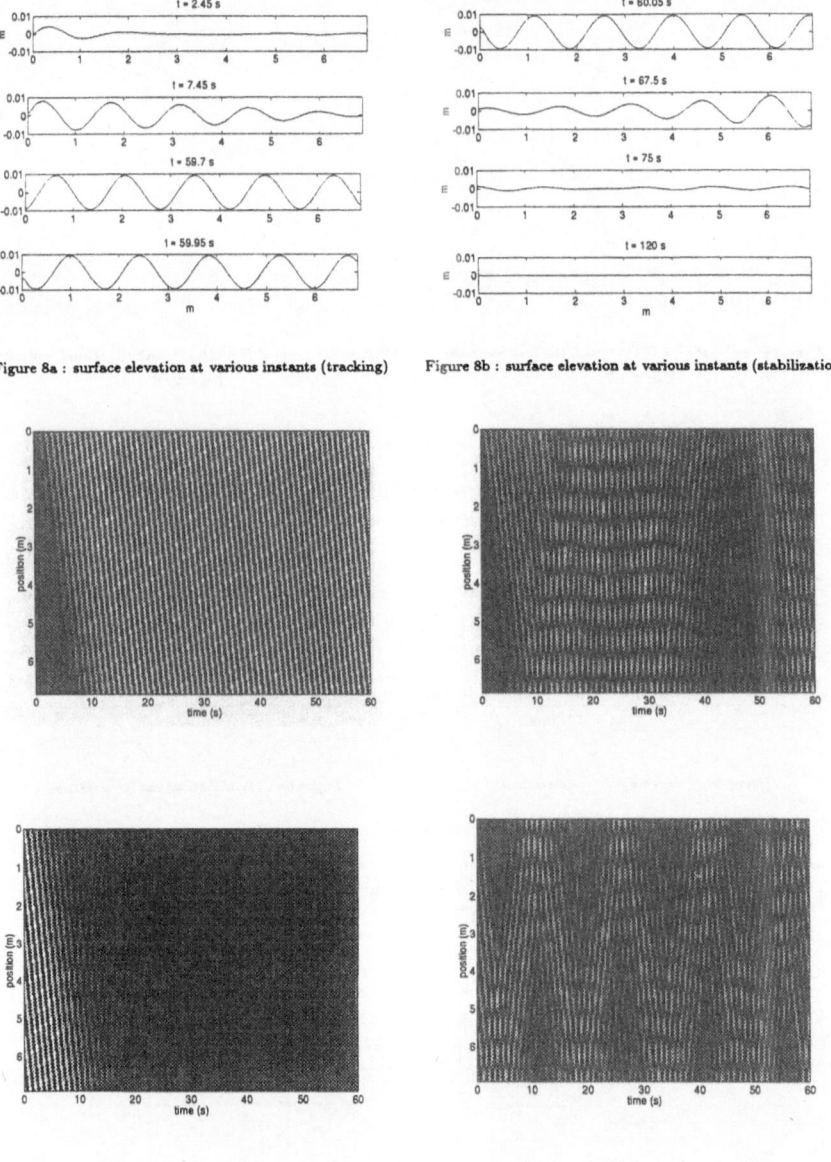

Figure 8a : surface elevation at various instants (tracking)   Figure 8b : surface elevation at various instants (stabilization)

Figure 9c : stabilization   Figure 9d : the canal returns at rest itself

Figure 9 : surface elevation as a function of x and t

Shape optimization

SHAPE OPTIMIZATION of NONLINEAR
CONTACT PROBLEMS with PRESCRIBED FRICTION

Andrzej Myslinski
System Research Institute
01-447 Warsaw, ul. Newelska 6, Poland

1 Introduction

The paper deals with a shape optimization problem of a hyperelastic body in unilateral contact with a body having prescribed deformation. It is assumed a contact with prescribed friction [5] occurs at a portion of the boundary of the body. The equlibrium state of nonlinear contact problem is described by a nonlinear elliptic variational inequality of the second order where the first Piola–Kirchhoff stress tensor is a coefficient [2, 3, 6, 9]. This problem has at least one global solution , in general, nonunique [2, 3, 9]. Conditions of the existence of the local unique solutions to this problem and its regularity are studied in [2, 15, 16]. Numerical algorithms for solving nonlinear contact problems [2, 6, 9, 1, 15] are based on the existence of local unique solutions.

Shape optimization problem for nonlinear contact problem consists in finding , in a contact region, such shape of the boundary occupied by the body in its reference configuration that the normal contact stress is minimized.

Shape optimization of contact problems was considered in [7, 12, 13, 17] for linear elliptic variational inequalities only. Necessary optimality conditions for these problems were formulated in [12, 17], convergence of the finite–dimensional approximation was studied in [7, 13] and numerical examples are provided in [7, 12].

In this paper we shall formulate a necessary optimality condition for the shape optimization problem of nonlinear contact problem with prescribed friction. We formulate the contact problem employing mixed variational approach [8], i.e., where the deformation and the tangential contact traction are independent variables. To define tangential traction components we introduce, following [10], a convected coordinate chart. Using this mixed variational formulation and material derivative method [17] as well as the results of differentiability of solutions to the variational inequality [17] we calculate the directional derivative of the cost functional and we formulate a necessary optimality condition for this problem. Note, that the calculated directional derivative may be used in numerical algorithm in descent direction finding procedure [6]. For the sake of simplicity we shall deal in this paper with static model only, however our approach can be extended to quasistatic contact models [6, 9, 10, 14] as well.

2   Formulation of the contact problem

Consider a bounded connected and open subset $\Omega \in R^3$. The closure $\overline{\Omega}$ of Ω represents the reference configuration of a deformable body. The body is subjected to body forces of intensity f per unite volume in the reference configuration and to surface tractions g measured per unite area in the reference configuration. The boundary Γ of Ω is assumed to be Lipschitz continuous and split into three disjoint parts Γ_0, Γ_1, Γ_2. The displacement takes on prescribed value φ_0 on Γ_0 and the traction g is prescribed on Γ_1. The body is assumed to be in contact with the obstacle body C along the boundary Γ_2. The body C is assumed to be rigid.

The deformation of the hyperelastic body is a regular map [1, 2, 6, 9, 10] :

$$\varphi \: : \: X \ni \overline{\Omega} \to x \in R^3 \tag{1}$$

The displacement u of the body is determined by [2, 6] :

$$u(X) = \varphi(X) - X \tag{2}$$

The deformation gradient F is given by [2, 6, 9] :

$$F = \nabla\varphi(X) \tag{3}$$

where $\nabla\varphi(X)$ is a gradient of the function φ with respect to a variable X. We shall assume that the deformation φ is one–to–one and is orientation preserving, i.e., the determinant $detF$ of F satisfies [2, 6] :

$$detF > 0 \quad \text{a.e. in} \quad \Omega \tag{4}$$

Moreover, during the deformation the mass conservation law is satisfied [2, 6]. The mechanical response of a hyperelastic body is characterized by a stored energy function $W = W(X, F)$ [2, 6]. Assuming that W is enough regular, the first Piola–Kirchhoff stress tensor $P = P(\varphi) = P(X, \varphi(X))$ is defined by [1, 2, 6, 9] :

$$P(\varphi) = \frac{\partial W(X, F)}{\partial F} \tag{5}$$

The equlibrium state of the hyperelastic body is determined by [1, 2, 6] : *find φ satisfying*:

$$divP(\varphi) + f = 0 \text{ in } \Omega \tag{6}$$

Moreover, the following boundary conditions are given :

$$\varphi \: = \varphi_0 \text{ on } \Gamma_0 \tag{7}$$
$$Pn \: = g \text{ on } \Gamma_1 \tag{8}$$

where n is outward normal versor to the boundary Γ. We shall consider deformation of a hyperelastic body in contact with a rigid obstacle C along the boundary Γ_2. In order to formulate the contact conditions we introduce a convected coordinate chart [10, 11]. Let us denote by $S \subset R^3$ large enough neighbourhood of the boundary ∂C of the obstacle body C. Let $(\psi, \Lambda_1, \Lambda_2) = \Lambda : S \to R^3$ be a mapping such that :

$$\nabla \psi \nabla \Lambda_i = 0, \quad i = 1, 2 \tag{9}$$

$$\psi(\varphi(X)) = \begin{cases} > 0 & X \in C \cap S \\ = 0 & X \in \partial C \\ < 0 & X \in S \setminus C \end{cases} \tag{10}$$

Let us denote :

$$w_N = w_N(X) = \psi(\varphi(X)), \quad w_T^i = \Lambda_i(\varphi(X)) \ i = 1, 2 \tag{11}$$

Let $t(X)$, $X \in \Gamma_2$ be a traction due to contact interaction with the surface ∂C. There are uniquely defined scalar fields $p_N = p_N(X)$ and $p_T^i = p_T^i(X), i = 1, 2$ on Γ_2 such that :

$$t = -p_N \nabla \psi - \sum_{i=1}^{2} p_T^i \nabla \Lambda_i \overset{\text{def}}{=} t_N + t_T \tag{12}$$

The contact with prescribed friction is given by [5, 8, 10] :

$$w_N \leq 0, \quad p_N \geq 0, \quad w_N p_N = 0 \text{ on } \Gamma_2 \tag{13}$$

$$w_T p_T + \mid w_T \mid = 0, \quad \mid p_T \mid \leq 1 \text{ on } \Gamma_2 \tag{14}$$

We shall consider problem (6) with boundary conditions (7), (8), (13), (14) in variational form. Let us introduce the notation :

$$V = \{\varphi \in W^{1,p}(\Omega) : \varphi = 0 \text{ on } \Gamma_0\} \tag{15}$$

$$K = \{z \in V : z = \varphi_0 \text{ on } \Gamma_0, \ z_N \leq 0 \text{ on } \Gamma_2 \} \tag{16}$$

$$E(\varphi) = \int_\Omega W(\nabla \varphi) d\Omega - \int_\Omega f\varphi d\Omega - \int_{\Gamma_1} g\varphi d\Gamma + \int_{\Gamma_2} \mid w_T \mid d\Gamma \tag{17}$$

where $f \in L^{p^*}(\Omega)$ and $g \in L^{p^*}(\Gamma_1)$, $p^* = \frac{p}{p-1}$, $\varphi_0 \in W^{1,p}(\Omega)$ are given. For detailed description of Sobolev spaces $W^{m,p}(\Omega)$ and the spaces $L^p(\Omega)$ see [2].

The system (6), (7), (8), (13), (14) is equivalent to the following optimization problem [1, 2, 3, 6, 9, 10, 16] :

$$\textit{Find } \varphi \in K \textit{ such that: } E(\varphi) \leq E(z) \quad \forall z \in K \tag{18}$$

We shall consider problem (18) as a saddle point problem [7]. Let us introduce a set L :

$$L^{\sim} = \{\mu \in L^{p^*}(\Gamma_2) \ : \ \int_{\Gamma_2} \mu(x)\varphi(x) d\Gamma \geq 0 \ \forall \varphi(x) \geq 0 \text{ for a.e. } x \in \Gamma_2\}$$

$$L = \{\mu \in L^{\sim} \ : \ -1 \leq \mu \leq 1 \text{ on } \Gamma_2\} \tag{19}$$

and a Lagrangian $\mathcal{L} \ : \ K \times L \to R$ defined as follows :

$$\mathcal{L}(\varphi, \tau) = \int_\Omega W(\nabla \varphi) d\Omega - \int_\Omega f\varphi d\Omega - \int_{\Gamma_1} g\varphi d\Gamma - \int_{\Gamma_2} \tau w_T d\Gamma \tag{20}$$

Note that the Lagrange multiplier τ corresponds to t_T defined by (12) [5, 8, 17]. The problem (18) can be written in an equivalent form :

$$Find \ (\varphi, p_T) \in K \times L \ satisfying$$

$$\int_\Omega P(\varphi) \bigtriangledown \eta d\Omega - \int_\Omega f\eta d\Omega - \int_{\Gamma_1} g\eta d\Gamma - \int_{\Gamma_2} p_T \bigtriangledown \Lambda\eta d\Gamma \geq 0 \ \forall\eta \in V \qquad (21)$$

$$\int_{\Gamma_2} p_T \Lambda(\varphi)d\Gamma \leq \int_{\Gamma_2} \tau\Lambda(\varphi)d\Gamma \ \ \forall\tau \in L \qquad (22)$$

From [2, 3] it follows, that if : f and g do not depend on φ, the stored energy function W(F) is *(i)* polyconvex in the sense of Ball [1] *(ii)* measurable on Ω *(iii)* coercive in the sense of Ball [1] and satisfies *(iv)* $W(F) \rightarrow \infty$ if $det F \rightarrow 0^+$ then the problem (21), (22) has at least one global solution $\varphi \in W^{1,p}(\Omega)$, $\tau \in L$, $p \geq 2$. This global solution is in general nonunique and it is not clear whether it is smooth enough to satisfy operator system (6)–(8) with the boundary conditions (13)–(14).

Since we shall deal with differentiability of solutions to system (21), (22) we need a local existence result. From [2, 6, 9, 15, 16] it follows that if : *(i)* all data in the problem (21),(22) are smooth enough, i.e. $f \in W^{2,p}(\Omega)$, $\varphi_0 \in W^{3,p}(\Omega)$, $g \in W^{3,p}(\Omega)$ *(ii)* the Hessian of the function W is locally strictly convex then there exists a unique solution $(\varphi, \tau) \in W^{2,p}(\Omega) \times L$ to the problem (21), (22) in the neighbourhood of a linearization point of the problem (21), (22). It is enough to chose $p > 3$ [2, 4] or under aditional assumptions on the fourth order derivatives of the stored energy function W $p \geq 2$ [16]. Note [15] that the existence of the local solution to the problem (21), (22) is the foundation to apply the Newton type algorithm for numerical solving the problem (21), (22) [2, 6, 9, 10, 15].

From now on $(\varphi, p_T) \in (W^{2,p}(\Omega)\cap K) \times L$ satisfying the system (21), (22) will denote a unique local solution to this system.

3 Shape optimization problem formulation

The contact problem (21), (22) is formulated in reference configuration Ω. Let Ω be variable subject to optimization. Consider a family $\{\Omega_\varepsilon\}$ of the domains Ω_ε depending on parameter ε. For each Ω_ε we formulate a variational problem corresponding to (21), (22). We formulate a shape optimization problem for this family of variational problems. The domain Ω_ε we shall consider as an image of a reference domain Ω under a smooth mapping T_ε. To describe the transformation T_ε we shall use the speed method [17].

Let us denote by $\mathcal{F}(\varepsilon, x)$ enough regular vector field depending on a parameter $\varepsilon \in [0, \sigma)$, $\sigma > 0$:

$$\mathcal{F}(.,.) \ : \ [0, \sigma) \times R^3 \rightarrow R^3 \qquad (23)$$

$$\mathcal{F}(\varepsilon,.) \in C^3(R^3, R^3) \ \forall\varepsilon \in [0, \sigma), \ \ \mathcal{F}(., x) \in C([0, \sigma), R^3) \ \forall x \in R^3$$

Let $T_\varepsilon(\mathcal{F})$ denotes the family of mappings : $T_\varepsilon(\mathcal{F}) : R^3 \ni X \rightarrow x(\varepsilon, X) \in R^3$ where the vector function $x(., X) = x(.)$ satisfies the system of ordinary differential equations :

$$\frac{d}{ds}x(s, X) = \mathcal{F}(s, x(s, X)), \ s \in [0, \sigma), \ x(0, X) = X \in R^3 \qquad (24)$$

The family of domains $\{\Omega_\varepsilon\}$ depending on parameter $\varepsilon \in [0, \sigma)$, $\sigma > 0$ is defined as follows : $\Omega_0 = \Omega$ and

$$\Omega_\varepsilon = T_\varepsilon(\Omega)(\mathcal{F}) = \{x \in R^3 \ : \ \exists X \in R^3 \text{ such that, } x = x(\varepsilon, X)$$
$$\text{where the function } x(., X) \text{ satisfies equation (24) for } 0 \le s \le \varepsilon\} \tag{25}$$

Consider the system (21), (22) in the domain Ω_ε. Let V_ε, K_ε, L_ε be defined, respectively, by (15), (16), (19) with Ω_ε instead of Ω. We shall write $\varphi_\varepsilon = \varphi(\Omega_\varepsilon)$, $p_{T\varepsilon} = p_T(\Gamma_{2\varepsilon})$. The problem (21), (22) in the domain Ω_ε takes on the form :

Find $(\varphi_\varepsilon, p_{T\varepsilon}) \in K_\varepsilon \times L_\varepsilon$ *satisfying* :

$$\int_{\Omega_\varepsilon} P(\varphi_\varepsilon) \nabla \eta_\varepsilon d\Omega - \int_{\Omega_\varepsilon} f\eta_\varepsilon d\Omega - \int_{\Gamma_{1\varepsilon}} g\eta_\varepsilon d\Gamma -$$
$$\int_{\Gamma_{2\varepsilon}} p_{T\varepsilon} \nabla \Lambda(\varphi_\varepsilon)\eta_\varepsilon d\Gamma \ge 0 \ \forall \eta_\varepsilon \in V_\varepsilon \tag{26}$$

$$\int_{\Gamma_{2\varepsilon}} p_{T\varepsilon}\Lambda(\varphi_\varepsilon)d\Gamma \le \int_{\Gamma_{2\varepsilon}} \tau_\varepsilon\Lambda(\varphi_\varepsilon)d\Gamma \ \forall \tau_\varepsilon \in L_\varepsilon \tag{27}$$

For every $\varepsilon \in [0, \sigma)$ the problem (26), (27) has a unique local solution $(\varphi_\varepsilon, p_{T\varepsilon}) \in K_\varepsilon \times L_\varepsilon$ [2, 4, 8].

Let us formulate the shape optimization problem. By $\Omega^\sim \subset R^3$ we denote a reference configuration domain such that for all $\varepsilon \in [0, \sigma), \sigma > 0$ we have $\Omega_\varepsilon \subset \Omega^\sim$. Let $\phi \in M$ be a given function. The set M is determined by :

$$M = \{\phi \in W^{1,p}(\Omega^\sim), \ \phi \le 0 \text{ on } \Omega^\sim, \ \| \phi \|_{W^{1,p}(\Omega^\sim)} \le 1\} \tag{28}$$

We shall consider the following shape optimization problmem :

For given $\phi \in M$, find the boundary $\Gamma_{2\varepsilon}$ of the domain Ω_ε
occupied by the hyperelastic body in its reference configuration,
minimizing the cost functional :

$$J_\phi(\varphi_\varepsilon) = \int_{\Gamma_{2\varepsilon}} p_N(\varphi_\varepsilon)\phi d\Gamma \tag{29}$$

over the set of admissible configurations U.

The set U is assumed to be nonempty. Moreover U is the family of admissible domains compact in the sense of the convergence of the characteristic functions of the elements of U [4]. Note [12] the goal of the shape optimization problem (29) is to find such boundary Γ_2 of Ω that the normal contact stress is minimized.

Lemma 1 *There exists an optimal domain* $\Omega_o \subset U$ *to the problem (29).*

Proof : follows from [4] and the compactness property of U.

$\square$

4 Necessary optimality condition

Our goal is to calculate the directional derivative of the cost functional (29) with respect to parameter ε. This derivative will be used to formulate necessary optimality condition as well as to find a descent direction in the numerical optimization method. First, let us recall from [17] the notion of the Euler derivative of the cost functional depending on Ω :

Definition 1 *Euler derivative $dJ(\Omega; \mathcal{F})$ of the cost functional J at a point Ω in the direction of the vector field $\mathcal{F}$ is given by :*

$$dJ(\Omega; \mathcal{F}) = \lim_{\varepsilon \to 0} \frac{J(\Omega_\varepsilon) - J(\Omega)}{\varepsilon} \tag{30}$$

Calculating the Euler derivative of the cost functional (29) we shall use the notion of a shape derivative of a function depending on a parameter. Let us recall from [17] :

Definition 2 *The shape derivative $\varphi' \in V$ of a function $\varphi_\varepsilon \in V_\varepsilon$ is determined by :*

$$(\varphi_\varepsilon^\sim)_{|\Omega} = \varphi + \varepsilon\varphi' + o(\varphi) \tag{31}$$

where $\| o(\varphi) \|_V / \varepsilon \to 0$ for $\varepsilon \to 0$, $\varphi = \varphi_0 \in V$, $\varphi_\varepsilon^\sim \in V(R^3)$ is an extension of the function $\varphi_\varepsilon \in V_\varepsilon(\Omega)$ into the space $V(R^3)$. $V(R^3)$ is defined by (15) with R^3 instead of Ω.

Let us calculate the Euler derivative of the cost functional (29) :

Lemma 2 *The directional derivative $dJ_\phi(\varphi; \mathcal{F})$ of the cost functional (29), for $\phi \in M$ given, at a point $\varphi \in V$ in the direction of the vector field $\mathcal{F}$ is determined by :*

$$dJ_\phi(\varphi; \mathcal{F}) = \int_\Omega \frac{\partial P(\varphi)}{\partial \varphi} \varphi' \bigtriangledown \phi d\Omega - \int_{\Gamma_2} \{p_T' \bigtriangledown \Lambda(\varphi)\phi +$$

$$p_T \bigtriangledown \Lambda'(\varphi)\phi + p_T \frac{\partial}{\partial \varphi}[\bigtriangledown \Lambda(\varphi)]\varphi'\phi\} d\Gamma + I_1(\varphi, \phi) - I_2(p_T, \varphi, \phi) \tag{32}$$

$$I_1(\varphi, \phi) = \int_\Gamma \{P(\varphi) \bigtriangledown \phi - f\phi - [(\bigtriangledown gn)\phi + g(\bigtriangledown \phi n) + g\phi H]\}\mathcal{F}(0)n d\Gamma \tag{33}$$

$$I_2(p_T, \varphi, \phi) = \int_{\Gamma_2} \{(\bigtriangledown p_T n) \bigtriangledown \Lambda\phi + p_T[\bigtriangledown(\bigtriangledown \Lambda)n]\phi +$$

$$p_T \bigtriangledown \Lambda(\varphi)\phi H + p_T \frac{\partial}{\partial \varphi}(\bigtriangledown \Lambda(\varphi))(\bigtriangledown \varphi n)\phi +$$

$$p_T \bigtriangledown \Lambda(\bigtriangledown \phi n)\}\mathcal{F}(0)n d\Gamma \tag{34}$$

where φ' is the shape derivative of φ_ε with respect to ε defined by (31), $\mathcal{F}(0) = \mathcal{F}(0, X)$, H is a mean curvature of the boundary Γ [2, 17], n is outward normal versor to Γ.

Proof : Note [5, 8], that :

$$J_\phi(\varphi_\varepsilon) = \int_{\Omega_\varepsilon} P(\varphi_\varepsilon) \bigtriangledown \phi d\Omega - \int_\Omega f\phi d\Omega - \int_{\Gamma_{1\varepsilon}} g\phi d\Gamma - \int_{\Gamma_{2\varepsilon}} p_{T\varepsilon} \bigtriangledown \Lambda(\varphi_\varepsilon)\phi d\Gamma \qquad (35)$$

Proof follows from the application of the mappings (23), (25) and the definition (30) to the cost functional (35). Note, that by standard arguments [2, 5] and from [2, 9, 16] it follows that the pair $(\varphi_\varepsilon, p_{T\varepsilon}) \in K_\varepsilon \times L_\varepsilon$, $\varepsilon \in [0, \sigma), \sigma > 0$, satisfying the system (26), (27) is Lipschitz continuous with respect to ε. For details see [14].

$\square$

In order to use the derivative (32) of the cost functional (29) we have to determine the shape derivative $(\varphi', p_T') \in V \times L$ of a solution $(\varphi_\varepsilon, p_{T\varepsilon}) \in K_\varepsilon \times L_\varepsilon$ to the system (26), (27). Let us recall the notion of the material derivative [17] :

Definition 3 *The material derivative* $\varphi^\bullet \in V$ *of the function* $\varphi_\varepsilon \in V_\varepsilon$ *is defined by :*

$$\lim_{\varepsilon \to 0} \| [(\varphi_\varepsilon \circ T_\varepsilon - \varphi]/\varepsilon - \varphi^\bullet \|_V = 0 \qquad (36)$$

where $\varphi \in V$, $\varphi_\varepsilon \circ T_\varepsilon \in V$ *is an image of the function* $\varphi_\varepsilon \in V_\varepsilon$ *in the space* V *under the mapping* T_ε.

Using (36) let us calculate the material derivative $(\varphi^\bullet, p_T^\bullet) \in V \times L$ of the solution to the system (26), (27).

Lemma 3 *The material derivative* $(\varphi^\bullet, p_T^\bullet) \in V \times L$ *of a solution* $(\varphi_\varepsilon, p_{T\varepsilon}) \in K_\varepsilon \times L_\varepsilon$ *to the system (26), (27) is a unique solution to the following system :*

$$\int_\Omega [\frac{\partial P(\varphi)}{\partial \varphi}\varphi^\bullet \bigtriangledown \eta + P(\varphi) \bigtriangledown \eta^\bullet - P(\varphi)^\bullet D\mathcal{F}(0) \bigtriangledown \eta - f^\bullet \eta - f\eta^\bullet +$$

$$(P(\varphi) \bigtriangledown \eta - f\eta) div\mathcal{F}(0)]d\Omega - \int_{\Gamma_1} [g^\bullet \eta + g\eta^\bullet + g\eta DZ]d\Gamma -$$

$$\qquad\qquad\qquad\qquad\qquad\qquad\qquad\qquad\qquad\qquad (37)$$

$$\int_{\Gamma_2} \{p_T^\bullet \bigtriangledown \Lambda(\varphi)\eta + p_T \bigtriangledown \Lambda^\bullet(\varphi)\eta + p_T \bigtriangledown \Lambda(\varphi)\eta^\bullet +$$

$$p_T \frac{\partial}{\partial \varphi}(\bigtriangledown\Lambda(\varphi))\varphi^\bullet \eta - p_T^\bullet D\mathcal{F}(0) \bigtriangledown \Lambda(\varphi)\eta +$$

$$p_T \bigtriangledown \Lambda(\varphi)\eta DZ\}d\Gamma \geq 0 \ \forall \eta \in O_1$$

$$\int_{\Gamma_2} \{(p_T^\bullet - \tau)\Lambda(\varphi) + (p_T - \tau^\bullet)\Lambda(\varphi) + (p_T - \tau)\frac{\partial \Lambda}{\partial \varphi}(\varphi)\varphi^\bullet +$$

$$p_T\Lambda^\bullet + (p_T - \tau)\Lambda(\varphi)DZ\}d\Gamma \leq 0 \ \forall \tau \in O_2 \qquad (38)$$

$$DZ = div\mathcal{F}(0) - (D\mathcal{F}(0)n, n) \qquad (39)$$

$$A_0 = \{x \in \Gamma_2 \ : \ w_N = 0\} \qquad (40)$$

$$A_1 = \{x \in A_0 \ : \ p_N > 0\}, \ A_2 = \{x \in A_0 \ : \ p_N = 0\} \qquad (41)$$

$$B_0 = \{x \in \Gamma_2 \ : \ p_T = \pm 1, w_T \neq 0\} \qquad (42)$$

$$B_1 = \{x \in \Gamma_2 \ : \ p_T = -1, \ w_T = 0\} \tag{43}$$

$$B_2 = \{x \in \Gamma_2 \ : \ p_T = +1, w_T = 0\} \tag{44}$$

$$O_1 = \{\eta \in V \ : \ \eta = \varphi_0 - D\mathcal{F}(0)\varphi_0 \text{ on } \Gamma_0 \ ;$$
$$\eta n \geq n D\mathcal{F}(0)\varphi \text{ on } A_1 \ ; \ \eta n = n D\mathcal{F}(0)\varphi \text{ on } A_2\} \tag{45}$$

$$O_2 = \{\tau \in L^\sim \ : \ \tau \geq 0 \text{ on } B_2 \ ; \ \tau \leq 0 \text{ on } B_1 \ ; \ \tau = 0 \text{ on } B_0\} \tag{46}$$

Proof : Let $\varphi^\varepsilon = \varphi_\varepsilon \circ T_\varepsilon \in V$ and $p_T^\varepsilon = p_{T\varepsilon} \circ T_\varepsilon \in L$. Note [15], that φ^ε may not belong to K. Let us introduce new variable $z^\varepsilon = DT_\varepsilon^{-1}\varphi^\varepsilon \in K$ and replace φ^ε by z^ε in (26), (27). Using the variable z^ε as well as formulae for transformation the function and its gradient into a reference domain [17] we write the system (26), (27) in the reference domain Ω. Applying to this system the result concerning differentiability of solutions to the variational inequality [17] we obtain (37), (38). Moreover using the same arguments as in [2, 9, 16] we can show that $(\varphi^\bullet, p_T^\bullet) \in O_1 \times O_2$ is a unique solution to the system (37), (38) in the neighbourhood of the linearization point.

□

Recall [17], if the shape derivative $\varphi' \in V$ of $\varphi_\varepsilon \in V_\varepsilon$ exists, then it satisfies,

$$\varphi' = \varphi^\bullet - \nabla\varphi\mathcal{F}(0) \tag{47}$$

From [2, 9, 16] as well as (23), (25) it follows that the solution $(\varphi_\varepsilon, p_{T\varepsilon}) \in K_\varepsilon \times L_\varepsilon$ to the system (26), (27) is enough regular that (47) holds.

Integrating by parts system (37), (38), using (47) we obtain the system characterizing the shape derivative $(\varphi', p_T') \in$ of the solution $(\varphi_\varepsilon, p_{T\varepsilon}) \in K_\varepsilon \times L_\varepsilon$ to the system (26), (27) :

$$\int_\Omega \frac{\partial P}{\partial \varphi}(\varphi)\varphi' \nabla \eta d\Omega - \int_{\Gamma_2} (p_T' \nabla \Lambda + p_T \nabla \Lambda' +$$

$$p_T \frac{\partial}{\partial \varphi}(\nabla\Lambda)\varphi'\eta)d\Gamma + I_1(\varphi, \eta) - I_2(p_T, \varphi, \eta) \geq 0 \quad \forall \eta \in N_1 \tag{48}$$

$$\int_{\Gamma_2} (\Lambda'(\tau - p_T) + \frac{\partial\Lambda}{\partial\varphi}(\varphi)\varphi'(\tau - p_T) - \Lambda p_T')d\Gamma + I_3(\varphi, \tau - p_T) \geq 0 \ \forall \tau \in O_2 \tag{49}$$

where

$$I_3(\varphi, \tau - p_T) = \int_{\Gamma_2} [(\nabla\Lambda n)(\tau - p_T) + \frac{\partial\Lambda}{\partial\varphi}(\nabla\varphi n)(\tau - p_T) + \Lambda(\nabla\tau n) -$$
$$\Lambda(\nabla p_T n) + \Lambda(\tau - p_T)H]\mathcal{F}(0)n d\Gamma \tag{50}$$

and

$$N_1 = \{\eta \in V \ : \ \eta = \lambda - \nabla\varphi\mathcal{F}(0), \ \lambda \in O_1\} \tag{51}$$

For the sake of simplicity using (47) in (48), (49) we assumed that the shape derivatives of given and test functions are equal to zero, i.e. these functions do not depend on ε.

In order to eliminate the shape derivatives φ' and p'_T from (32) let us introduce an adjoint state $(r, q) \in K_1 \times L_1$ defined as a solution to the following system :

$$\int_\Omega \frac{\partial P}{\partial \varphi}(\varphi)(\nabla \phi + \nabla r)\zeta d\Omega + \int_{\Gamma_2} \frac{\partial \Lambda}{\partial \varphi}(\varphi)(q - p_t)\zeta d\Gamma - $$
$$\int_{\Gamma_2} p_T \frac{\partial}{\partial \varphi}(\nabla \Lambda(\varphi))(\phi + r)\zeta d\Gamma = 0 \ \forall \zeta \in K_1 \tag{52}$$

$$\int_{\Gamma_2} [\nabla \Lambda(\varphi)(\phi + r) + \Lambda(\varphi)]\delta d\Gamma = 0 \ \forall \delta \in L_1 \tag{53}$$

where

$$K_1 = \{\zeta \in K \ : \ \zeta n = 0 \text{ on } A_0\} \quad L_1 = \{\delta \in L \ : \ \delta = 0 \text{ on } A_0 \cap B_0\}$$

Since $\phi \in M$ is given then by [8] it follows the existence of a unique solution $(r, q) \in K_1 \times L_1$ to the system (52), (53).

From (32), (48), (49), (52), (53) it follows :

$$dJ_\phi(\varphi; \mathcal{F}) = I_1(\varphi, \phi + r) - I_2(p_T, \varphi, \phi + r) + I_3(\varphi, q - p_T) +$$
$$\int_{\Gamma_2} \Lambda'(\varphi)(q - p_T)d\Gamma - \int_{\Gamma_2} p_T \nabla \Lambda'(\varphi)r d\Gamma \tag{54}$$

Assuming $\Lambda \neq \Lambda(\varepsilon)$, i.e, $\Lambda' = 0$, and $\nabla \Lambda' = 0$ we obtain :

$$dJ_\phi(\varphi; \mathcal{F}) = I_1(\varphi, \phi + r) - I_2(p_T, \varphi, \phi + r) + I_3(\varphi, q - p_T) \tag{55}$$

Using (55) we can formulate a necessary optimality condition :

Lemma 4 *If* $\varphi_o = \varphi(\Omega_o)$ *is an optimal solution to the problem (29) then for all vector fields* $\mathcal{F}$ *satisfying (23), (25) the following condition holds :*

$$dJ_\phi(\varphi_o; \mathcal{F}) \geq 0 \tag{56}$$

Proof : see [2, 9].

$\square$

The numerical realization of the problem (29) is being carried out. The mixed finite element method is used as the discretization method. For details concerning the discretization see [6, 7, 8, 9, 10, 11, 13]. The Newton type algorithm [6, 10, 11] is employed to solve the discretized state system (21), (22). The conjugate gradient algorithm with projection [6] is used as an optimization method for problem (29). The calculated directional derivative (55) of the cost functional (29) is being used to find a descent direction in the optimization algorithm.

References

[1] Ball, J.M., Convexity Conditions and Existence Theorems in Nonlinear Elasticity , *Archive for Rational Mechanics and Analysis*, vol 63, 1977, pp.337–299.

[2] Ciarlet, Ph., *Mathematical Elasticity. Vol 1 : Three Dimensional Elasticity*, North-Holland, Amsterdam, 1988.

[3] Ciarlet, Ph., Necas, J., Injectivity and Self Contact in Nonlinear Elasticity , *Archive for Rational Mechanics and Analysis*, vol 97, 1987, pp.171–188.

[4] Chenais, D., On the Existence of a Solution in a Domain Indentification Problem , *Journal of Mathematical Analysis and Applications*, vol 52, 1975, pp.189–289.

[5] Duvaut, G., Lions, J.L., *Inequalities in Mechanics and Physics*, Springer, Berlin, 1976.

[6] Glowinski, R., LeTallec, P., *Augmented Lagrangian and Operator–Splitting Methods in Nonlinear Mechanics*, SIAM Studies in Applied Mathematics, SIAM, Philadelphia, 1989.

[7] Haslinger, J., Neittaanmaki, P., *Finite Element Approximation for Optimal Shape Design. Theory and Application.*, J.Wiley and Sons, 1988.

[8] Hlavacek, I., Haslinger, J., Necas, J., Lovisek, J., *Solution of Variational Inequalities in Mechanics*, Springer, Berlin, 1988.

[9] Kikuchi, N., Oden, J.T., *Contact Problems in Elasticity: A Study of Variational Inequalities and Finite Element Methods*, SIAM Studies in Applied Mathematics, SIAM, Philadelphia, 1988.

[10] Klabring, A., Bjorkman, G., Solution of Large Displacement Contact Problems with Friction using Newton's Method for Generalized Equations , *International Journal for Numerical Methods in Engineering*, vol 34, 1992, pp.249–269.

[11] LeTallec, P., Numerical Analysis of Equilibrium Problems in Finite Elasticity , *Preprint*, No 9021, CEREMADE, Paris, 1990.

[12] Myslinski, A., Shape Optimization of Contact Problems using Mixed Variational Formulation , in : *System Modelling and Optimization*, P. Kall ed., Lecture Notes in Control and Information Sciences, Vol.180, Springer, Berlin, 1992, pp. 414-423.

[13] Myslinski, A., Mixed Finite Element Approximation of Shape Optimization Problem for Contact Systems with Prescribed Friction , *Preprint*, 1992.

[14] Myslinski, A., *Shape optimization of large displacement contact problems with prescribed friction*, in: Contact Mechanics, A. Curnier ed., Presses Polytechniques et Universitaires Romandes, Lousanne, Switzerland, 1992, pp. 305-318.

[15] Necas, J., *Introduction to the Theory of Nonlinear Elliptic Equations*, Wiley, 1986.

[16] Necas, J., On the Regularity of Weak Solutions to Variational Equations and Inequalities for Nonlinear Second Order Elliptic Systems , *Lecture Notes in Mathematics*, vol 703, Springer, Berlin, 1979, pp.286–299.

[17] Sokolowski, J., Zolesio, J.P., *Introduction to Shape Optimization.Shape Sensitivity Analysis*, Springer, Berlin, 1992.

Discrete optimization

A Modified Benders' Decomposition Technique for Solving Large Scale Unit Commitment Problems

Risto Lahdelma, Sampo Ruuth
Systems Analysis Laboratory
Helsinki University of Technology
Otakaari 1M, 02150 Espoo, Finland

Keywords: Decomposition, Linear Programming, Mixed Integer Programming, Unit commitment

1.   Introduction

Multiperiod unit commitment problems form a subclass of linear mixed integer models, where 0/1-variables are used for encoding the off/on-states of each unit at each period. The model may contain additional integer variables for encoding non-linearities elsewhere in the model. Due to the combinatorial complexity of the problem, traditional techniques cannot always be used for solving large unit commitment problems.

We have developed a modified Benders' decomposition algorithm /Tah75/, exploiting the special structure of the unit commitment problem. Benders' algorithm decomposes the mixed integer model into a continuous and a pure integer (combinatorial) model. In our algorithm it is the continuous primal model with bounded variables that is solved instead of the dual problem. Moreover, the logical constraints between the 0/1-variables are appended to the integer model.

The software implementation is based on the MME - Mathematical Modelling Environment /Lah91,Ruu91/, which is an interactive software environment supporting mathematical modelling and algorithm design. MME provides an interpreted, high level programming language with built-in vector- and matrix-arithmetics and efficient large scale LP/MIP optimization (the LP2 optimization package).

2.   A Multi-Period Unit Commitment Problem

Our target is to model and optimize an industrial power plant at a typical pulp and paper factory. We also include the Thermo Mechanical pulping process (TMP) in the model, because the process is a major electricity consumer. A multi-period unit commitment model is based on the static models for successive time periods. The decision variables are the fuel flows of the boilers, steam flows, pulp mass flow (TMP), the electric power generated by the turbine and the zero-one integer variables. The multi-period model is formed by combining the static models of different periods together with recursive equations for cumulating the level of the pulp container.

The cost of thermal and power generation is determined from the fuel costs and the efficiency of the boilers and turbines. Several kinds of fuel such as oil, natural gas and coal are used. Piecewise linear electricity tariffs together with variable power demand make it important to optimize when to use the TMP-process and when to use local power generation facilities.

The static model for each time period can be written as a mixed integer model (MIP) of the form

$$\min cx + dy$$

subject to

$$Ax + Ey = b$$

$$\underline{x} \le x \le \overline{x}$$

$$y = \text{zero-one vector}$$

(1)

where

c $=(c_1,..., c_n)$ cost coefficients for purchased fuel and power.
d   constant operating costs, shut-down and set-up costs.
x $=(x_1,..., x_n)$ continuous decision variables (fuel and steam flows, power) including the slacks.
y $=(y_1,..., y_k)$ zero-one variables used for representing on-off states of devices and for encoding other non-linearities.
k   number of zero-one variables.
m   number of constraints.
n   number of continuous variables.
A   m*n parameter matrix of model constraints due to mass and energy balances, boiler, turbine and valve characteristics, physical limits of devices and needs for thermal generation and electrical power.
E   m*k parameter matrix of model constraints due to the physical limits of devices.
b   right hand side of the equation system including thermal and power requirements for the period.
$\underline{x}$ lower bounds for x.
$\overline{x}$ upper bounds for x.

The multi-period unit commitment problem is formed by combining the static models together with pulp stock equations and interperiod logical constraints. The structure of the resulting MIP-model matrix is shown in Figure 1.

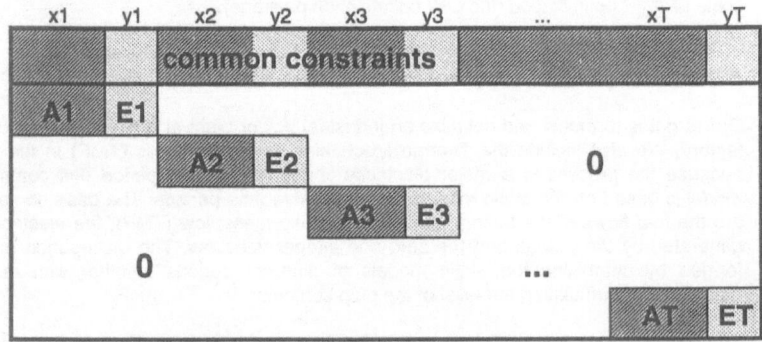

Figure 1. The structure of the multi-period model

3. Modified Benders' Algorithm

Benders' decomposition algorithm /Tah75/ splits a nonlinear optimization problem into a dual LP-problem and a nonlinear problem, which are solved in turn. A unit commitment problem of type (1) is split into a pure LP-model and a MIP-model containing the y-variables and one additional continuous variable. At each iteration the objective function of the dual LP model is modified based on the solution of the previous MIP model. A new constraint (the Benders' Cut) is then added to the MIP-model based on the dual LP-solution.

Our modified algorithm uses the special structure of the unit commitment problem. Most of the constraints containing integer variables $y_{j,t}$ are of the following two types. The variable bound constraints

$$l_j y_{j,t} \le x_{j,t} \le u_j y_{j,t} \tag{2}$$

where the physical lower and upper bounds l_j and u_j for device j are valid when the device is on ($y_{j,t}=1$) but $x_{j,t}$ is forced to zero when the device is shut down ($y_{j,t}=0$).

The logical constraints

$$0 \le y_{j,t-2} - y_{j,t-1} + y_{j,t} \le 1 \tag{3}$$

ensure that a device cannot be shut down or set up for a single period. Similar constraints can be written for restricting the minimun shutdown or up-time of a device to 3 or more periods. We write the logical constraints in the general form

$$C(y) = g \tag{4}$$

We can now write the problem in the form

$$\max z(x,y) = cx + dy$$

subject to

$$Ax + Ey = b$$
$$C(y) = g$$

$$(5)$$

$$ly \le x \le uy$$
$$\underline{x} \le x \le \overline{x}$$
$$y = \text{zero-one vector}$$

where the physical limits (2) $ly \le x \le uy$ and the logical constraints (4) $C(y) = g$ have been extracted from the main constraint equations $Ax + Ey = b$. The structure is illustrated in Figure 2. The storage equations are purely linear and they connect the subsequent period models together. The logical constraints contain only integer variables and they combine three (or more) subsequent period models together. Each period model may contain pure LP-constraints. The variable bounds are always MIP-constraints.

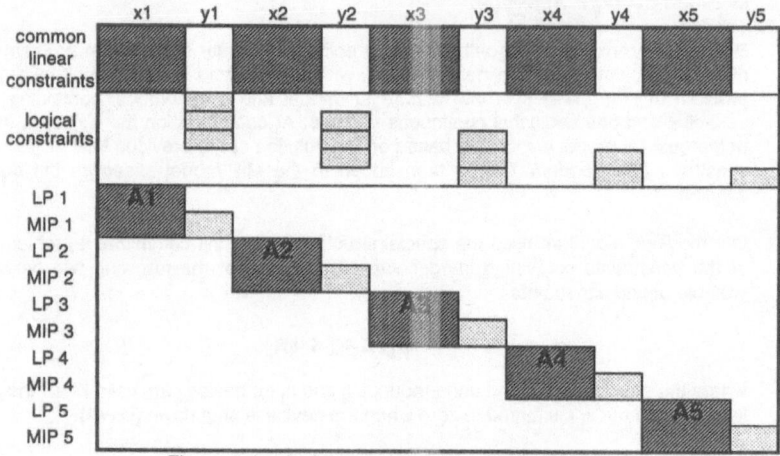

Figure 2. Logical constraints and variable bounds

If y is fixed at given values y' such that the logical constraints (4) are satisfied, then the problem reduces to the form

$$\max_{x} z(x,y') = cx + dy'$$

subject to

$$Ax = b - Ey' \tag{6}$$

$$\underline{x}' \leq x \leq \overline{x}'$$

where $\underline{x}'$ and $\overline{x}'$ are the joint lower and upper bounds of $ly' \leq x \leq uy'$ and $\underline{x} \leq x \leq \overline{x}$. Let $z^*(y')$ be the optimal objective value for the given y'. The original problem can now be written as

$$\max_{y} z^*(y)$$

subject to $$\tag{7}$$

$$C(y) = g$$
$$y = \text{zero-one vector}$$

The dual of the LP-problem (6) is

$$\min f(u,v,w,y') = dy' + u(b - Ey') + v\overline{x}' - w\underline{x}'$$

subject to

$$uA + v - w = c \tag{8}$$

$$u = \text{unrestricted}$$
$$v \geq 0$$
$$w \geq 0$$

where $[u,v,w]$ is the dual vector corresponding restrictions $Ax = b - Ey$, $x \leq \overline{x}'$ and $-x \leq -\underline{x}'$, respectively. The feasible region of the dual problem is independent from the choise of y'. Let $[u^k, v^k, w^k]$ $k=1,...K$, be the extreme points of the feasible region (need not be known explicitly). Then the dual problem can be replaced with

$$f^*(y') = dy' + \min_k [u^k(b - Ey') + v^k\overline{x}' - w^k\underline{x}'] \tag{9}$$

The optimal solution of the dual problem equal to the primal solution

$$f^*(y') = z^*(y') \tag{10}$$

is substituted in (7) giving

$$\max_{y'} z^*(y')$$

subject to

$$C(y') = g \tag{11}$$

$$z^*(y') = dy' + \min_k [u^k(b - Ey') + v^k\overline{x}' - w^k\underline{x}']$$

$$y' = \text{zero-one vector.}$$

This can equivalently be written as

$$\max z$$

subject to

$$C(y') = g$$

$$z \leq dy' + u^k(b - Ey') + v^k\overline{x}' - w^k\underline{x}', \quad k = 1,...,K \tag{12}$$

$$y = \text{zero-one vector.}$$

As in Benders' decomposition algorithm, two problems are solved in turn. Initially the MIP-model (12) is solved with an empty set of cut-constraints, as if K=0. Then the solution $[u,v,w]$ to the dual problem (8) corresponding to this y is calculated. A new cut-constraint (Benders' cut) is formed and added to (12). This process is repeated until the new cut-constraint does not worsen the z-value.

This decomposition differs from Benders' decomposition in that we have moved the logical constraints (4) into the MIP-problem. Thus the MIP-problem will at every stage give a solution which satisfies these constraints. This makes the combinatorial problem faster to solve and also decreases the number of main iterations. The second difference is that we solve the primal problem (6) instead of the dual (8), which gives the advantage that the variable bounds. (2) become fixed when y is fixed. This makes the LP-model smaller and faster to solve using efficient bounded variable algorithms. The structure of the resulting LP- and MIP-model is illustrated in Figure 3. Note that all MIP-constraints are variable bound constraints, which in this decomposition become fixed bounds for the linear problem.

744

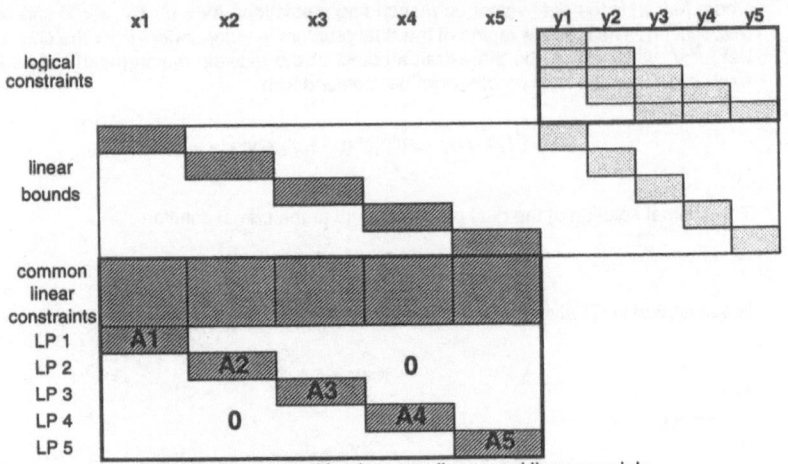

Figure 3. Decomposition into non-linear and linear models

4. Modified Decomposition Algorithm

The steps of the algorithm can now be summarized as follows:

Step 0 Determine $[u^1,v^1,w^1]$, any feasible solution to the dual problem (8). Instead of solving the dual problem, we solve the primal problem (6), and obtain the dual solution from the optimal primal table. If no solution is found, then the original model is infeasible and the algorithm stops. Otherwise, set r:=1 and proceed from step 1.

Step 1 Solve the MIP-problem (12) in z and y:

$$\max z$$

subject to

$$z \le dy + u^k(b - Ey) + v^k \overline{x}^{k-1} - w^k \underline{x}^{k-1}, \quad k = 1,...,r \tag{14}$$

$$C(y) = g$$
$$y = \text{zero-one vector.}$$

Let (z^r, y^r) be the optimum solution. The optimal objective values z^r form a non-increasing sequence $\overline{z}$ of upper bounds on the true optimum value z^*.

Step 2 Solve the LP-problem (6) with $y' = y^r$:

$$\max cx + dy^r$$

subject to

$$Ax = b - Ey^r \tag{15}$$

$$\underline{x}^r \le x \le \overline{x}^r$$

Only the bounds or rhs-coefficients change between iterations, making it efficient to reuse the basic solution from the previous iteration. Let x^r be the optimum primal solution and

$[u^{r+1}, v^{r+1}, w^{r+1}]$ the optimum dual solution (obtained from the optimal primal tableau).
Then

$$f^r = dy^r + u^{r+1}(b - Ey^r) + v^{r+1}\overline{x}^r - w^{r+1}\underline{x}^r$$

is a sequence of lower bounds on the true optimum value z^*.

Step 3 If $f^r = \overline{z}$ then $x^* = x^r$ and $y^* = y^r$ gives the optimum solution to (5); otherwise, set $r:=r+1$ and repeat from step 1.

5.   Conclusions

We have presented a modified Benders' decomposition technique suitable for solving multi-period unit commitment problems used for optimizing industrial power plants. The problem is formulated as a large MIP-model, where 0/1-integer variables are used for encoding on/off-states of devices and other non-linearities.

The problem is decomposed into an LP-problem and a MIP-problem, which are solved in turn. We solve the primal problem instead of the dual to obtain the dual variables needed for forming the Benders' cuts. The advantage is that a large number of variable bounds become fixed bounds in the LP-problem. This reduces the size of the LP-problem and makes it possible to use efficient upper/lower-bound algorithms and to reuse the basis from the previous iteration.

Logical constraints between integer variables are included to the MIP-model where the Benders' cuts are added. Thus the mixed integer programming problem will at every stage produce a solution which satisfies these constraints.

Another interesting decomposition algorithm using a combination of MIP and dynamic programming is described in /Ant91/. This algorithm is suitable for national long-term energy optimization and planning when several thermal and hydro power plants must be considered.

References

/Ant91/   Antila H., Lautala P., Ruuth S., Lahdelma R.: Decomposition Technique and Coordination of Optimal Energy Production; EURO XI, July 16-19, Aachen 1991

/Lah91/   Lahdelma R.: MME - a Mathematical Modelling Environment; Helsinki Univ. of Technology; Systems Analysis Laboratory; Licentiate's dissertation, 1991

/Ruu91/   Ruuth S., Lahdelma R., Holm R.: MME - a Rapid Modelling and Algorithm Prototyping Environment; Advances in Methodology and Software of Decision Support Systems, ed: M. Makowski, Y. Sawaragi, IIASA, 1991

/Tah75/   Taha H.A.: Integer Programming; Academic Press, New York, 1975

Development of a Mixed Integer Programming Model for the Optimal Recycling of Demolition Waste

Th. Spengler, M. Nicolai, O. Rentz, M. Ruch

German-French Institute for Environmental Research (DFIU)
University of Karlsruhe (TH)
Hertzstr. 16, D-76187 Karlsruhe, Federal Republic of Germany

Abstract

Demolition waste ranges with an amount of about 23 million tons per year in Germany and 25 million tons in France in the same order as municipal waste. Recycled demolition waste can be reused not only in landfilling but also in road construction, in the production of paving blocks, in the manufacturing of wall blocks as well as a concrete aggregate. These secondary raw materials mainly should be used in order to replace the natural raw materials in civil engineering and underground construction. The total dismantling costs of buildings and recycling costs of the resulting materials are a function of the considered dismantling techniques, the sorting techniques, the recycling techniques and the transportation distances. Furthermore, the recycling capacities of the different reuse options may be limited in the considered geographical region. In this paper the integrated dismantling and recycling planning problem will be described and a sophisticated mixed integer linear programming model will be formulated. Different solution procedures will be analysed and due to the high complexity of the planning problem, a heuristic decomposition algorithm will be developed. In order to obtain integrated dismantling and recycling strategies for the Upper Rhine Valley (Alsace and Baden), the model will be implemented on a Personal Computer and evaluated with specific regional database.

1 Introduction

Although recycling and reuse is generally possible, in Germany only 16% of the building materials were re-cycled in 1989. Construction waste was almost completely disposed. The objectives of the German legis-lation are to achieve a recycling rate of 60% for building rubble and 40% for construction waste for the year 1995. Initiatives in France aim at the direct reuse of materials from the dismantling of buildings.

In reality, reuse options are often limited to road construction or soundproof barriers which represent a downcycling and are therefore insufficient from an economical point of view. Only recycling-products of a high standardized quality level can compete with traditional materials. Therefore, environmental legislation aims at selective dismantling and recycling of buildings. Major instruments are higher disposal prices for demolition waste, the integration of dismantling schemes into pulling-down permits and fees on natural construction materials.

The target of the presented research project is the development and application of an integrated dismantling and recycling planning system for the region of the Upper Rhine Valley. With this system, the costs and interactions of dismantling and recycling are evaluated and cost-efficient combinations of dismantling procedures, recycling techniques and reuse options are determined. The effects of external factors (e.g. legislation, disposal costs, waste duties) will be examined.

Due to the high complexity of the planning problem, a mixed integer linear programming model has to be formulated and implemented on a Personal Computer (cf. Figure 1).

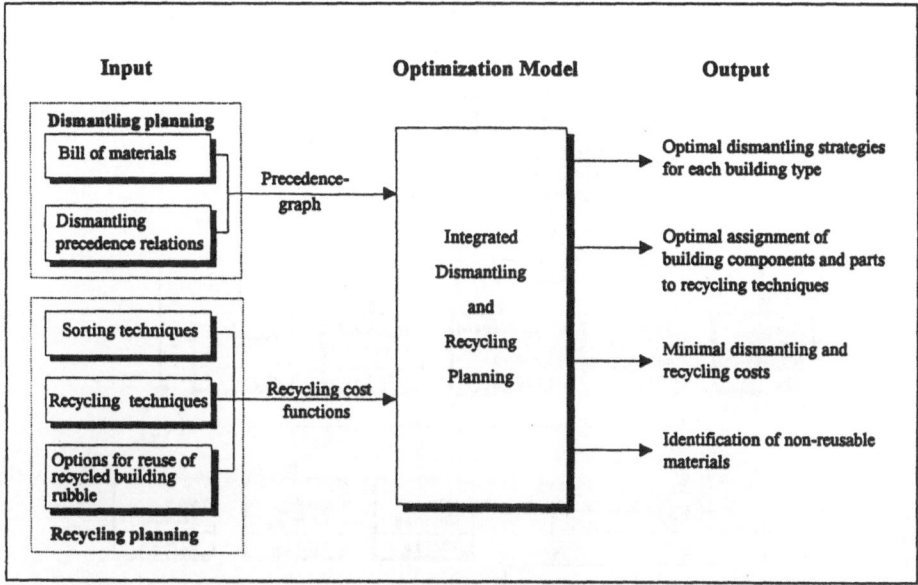

Figure 1: Structure of the integrated dismantling and recycling planning system

2 Dismantling planning

As the amount and composition of the materials from demolition shows important variation with the kind of building, the buildings have to be classified into different types. An analysis of the domestic buildings in the region of the Upper Rhine Valley showed, that eleven types of buildings predominate and can be classified as shown in the following Table 1.

Table 1: Types of domestic buildings in the region of the Upper Rhine Valley

	Skeleton construction				Massive construction			
	wood		steel		concrete		masonry	
	single	multiple	single	multiple	single	multiple	single	multiple
before 1930	X						X	X
1930 - 1960						X	X	X
after 1960	X				X	X	X	X

For each of these types a representative building will be chosen and a detailed bill of materials will be prepared. On the basis of these bills of materials the different dismantling techniques will be analysed and aggregated into so called dismantling activities. For these dismantling activities the associated dismantling costs and the precedence-relations will be determined, so that for each identified class of buildings the following dismantling-precedence-graphs can be developed (cf. Figure 2).

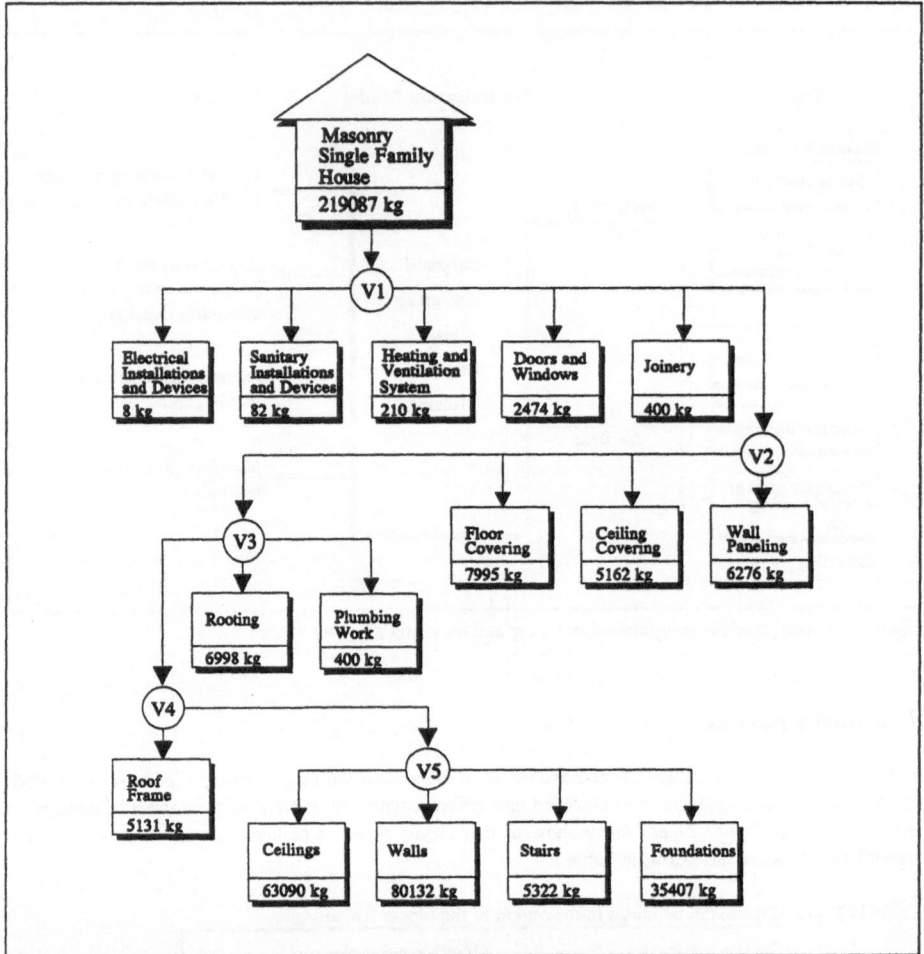

Figure 2: Dismantling-precedence-graph for a single family masonry house

Following the engineering industry, where the concept of the precedence-graphs is used in the Materials-Requirement-Planning, the precedence-graph is the starting point of the dismantling planning process. By the application of a set of dismantling activities $v_1,...,v_5$ the corresponding building is disassembled into building components and parts.

The decision whether a certain dismantling activity will be applied or not depends on the comparison of the alternative recycling costs of the corresponding building component with the dismantling costs and the minimal recycling costs of the resulting building components and parts.

3 Recycling planning

The objective of recycling planning is the design of optimal recycling techniques for processing dismantled materials and building components into reusable materials. Depending on the stage of dismantling, the feed can be either a single material or a mix of all the building materials. For some materials such as glass and metals, classic recycling techniques already exist and recycling planning is a simple coordination or pretreatment. For other materials, techniques are not yet developed or have to be modified according to the composition of the specific feed, e.g. for plastic and composite materials. Toxics like asbestos require specific treatments for disposal.

The recycling of used materials can be split into two major parts:

- recycling techniques to process used materials and
- reuse options for the processed materials.

Both are interdependant which means that the development of sophisticated recycling techniques requires reuse option for high quality products whereas standardized high quality levels for recycling materials can only be set if techniques to realise them already exist.

An overview of possible reuse options for the mineral fraction is given in Figure 3:

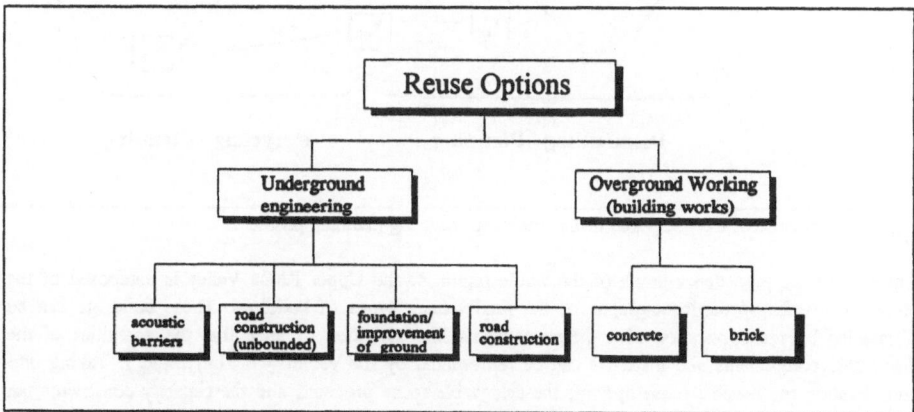

Figure 3: Reuse options for the mineral fraction of demolition waste

4 Integrated dismantling and recycling planning

The dismantling planning process only can be carried out if the recycling costs for all building components and building materials are already determined. This means the recycling planning process has to precede the dismantling planning process. On the other hand the recycling planning process requires information about the amount and composition of the dismantled building components and materials. Due to the fact that the recycling capacities of the different reuse options are limited in the considered geographical region, an optimal allocation of the dismantled components and materials to the available reuse options is necessary in order to determine the minimal recycling costs. Following this idea the dismantling planning process has to precede the recycling planning process. Because of these interdependances the dismantling and recycling planning process, can't be carried out independently; furthermore an integrated approach is desirable (c.f. Figure 4).

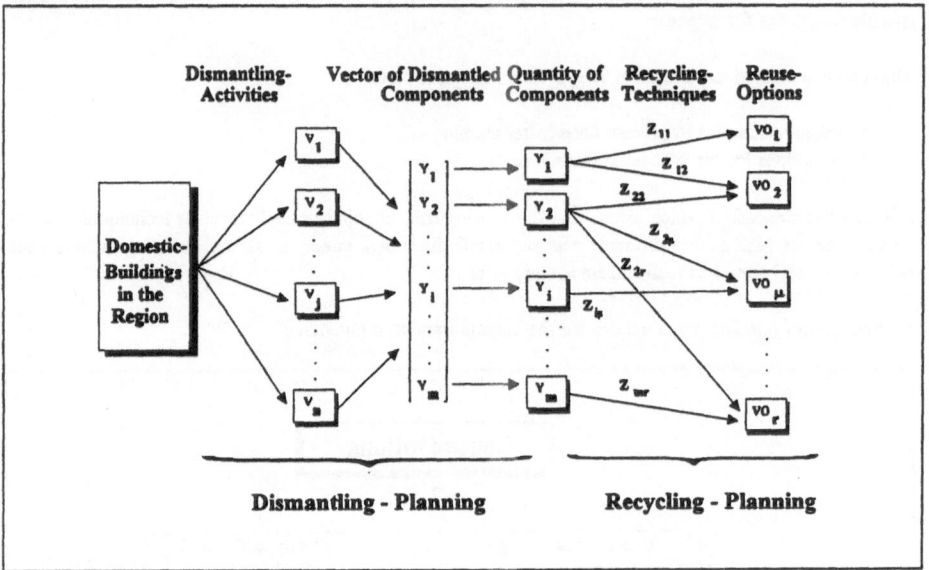

Figure 4: Structure of the integrated dismantling and recycling planning problem

The dismantling-precedence-graph of the whole region of the Upper Rhine Valley is composed of the eleven dismantling-precedence-graphs of the predominant types of buildings. These buildings can be dismantled by the application of n different dismantling activities v_j such that the quantities of the dismantled components and materials can be represented by the vector $y^T = (y_1,...,y_m)$. Taking into consideration the feasible reuse options, the achievable reuse proceeds and the capacity constraints, an optimal allocation has to be determined.

The integrated dismantling and recycling planning problem can be formulated as a mixed-integer linear programming model (MIP):

$$\text{MAX} \quad \sum_{i=1}^{m} \sum_{\mu \in T_i} c_{i\mu}^{v} \cdot z_{i\mu} - \sum_{j=1}^{n} c_{j}^{z} \cdot x_j \tag{4.1}$$

Subject to

$$y_i = y_i^a + \sum_{j=1}^{n} x_j \cdot v_{ij} \qquad\qquad i = 1,...,m \tag{4.2}$$

$$\sum_{\mu \in T_i} z_{i\mu} - \gamma_i \cdot y_i = 0 \qquad\qquad i = 1,...,m \tag{4.3}$$

$$\sum_{i \in I_\mu} z_{i\mu} - Q_\mu \leq 0 \qquad\qquad \mu = 1,...,r \tag{4.4}$$

$$x_j \in IN_0 \qquad\qquad j = 1,...,n \tag{4.5}$$

$$z_{i\mu} \geq 0 \qquad\qquad i = 1,...,m \ ; \ \mu \in T_i \tag{4.6}$$

$$y_i \geq 0 \qquad\qquad i = 1,...,m \tag{4.7}$$

where

i : index for the different components of the product ($i = 1,...,m$),

j : index for the different dismantling activities ($j = 1,...,n$),

μ : index for the different reuse options ($\mu = 1,...,r$),

$c_{i\mu}^{v}$: proceeds per ton of component i which is recycled by reuse option μ,

c_{j}^{z} : dismantling cost of dismantling activity j,

y_i : number of dismantled components i,

y_i^a : initial number of dismantled components i,

v_{ij} : number of dismantled components i resulting from one application of dismantling activity v_j,

γ_i : tons per unit of component i,

Q_μ : capacity limit of reuse option m,

T_i : $\{\mu$ / component i can be recycled by reuse option $\mu\}$,

I_μ : $\{i$ / component i can be recycled by reuse option $\mu\}$,

$z_{i\mu}$: decision variable for the amount [t] of component i to be recycled by reuse option μ,

x_j : decision variable for the number of applications of dismantling activity j.

The subject of the optimization problem is the maximization of the total achievable marginal income (reuse proceeds - dismantling costs) (5.1) subject to

- dismantling constraints (4.2),
- complete recycling of the dismantled components and materials (4.3),

- capacity constraints of the reuse options (4.4),
- integer and non-negativity constraints (4.5), (4.6), (4.7).

5. Solution procedures of the planning problem

The problem can be solved by the application of various kinds of solution procedures for combinatorial optimization problems, such as

- Cutting-Plane-Algorithms
- Branch-and-Bound-Algorithms
- Dynamic-Programming-Algorithms
- Decomposition Algorithms.

General mixed - integer problems are NP-complete so that the required solution time will grow exponentially with the complexity of the considered planning problem. Standard optimization software packages (e.g. LINDO, [Schrage 91]) generally use sophisticated Branch-and-Bound-Techniques based on the Algorithm of Dakin [Dakin 65]. Upper bounds are generated by an **LP-Relaxation** of the mixed-integer problem. The upper bounds computed in this way are not often strong enough to reduce the required solution time drastically. In this case it can be advantageous to introduce so called upper-bound-functions such as the **Lagrangean function** [Geoffrion 74]. The restrictions (4.3) in (MIP) contain integer variables $x_j \in IN_0$ as well as continuous variables $z_{i\mu} \geq 0$. The **Lagrangean Relaxation** of the restrictions leads to the following modified optimization problem:

$$MAX \qquad L(x,z,\lambda) = \qquad\qquad\qquad (5.1)$$

$$\sum_{i=1}^{m}\sum_{\mu \in T_i} c_{i\mu}^v \cdot z_{i\mu} - \sum_{j=1}^{n} c_j^z \cdot x_j - \sum_{i=1}^{m}\lambda_i \cdot \left[\sum_{\mu \in T_i} z_{i\mu} - \gamma_i \cdot \left(y_i^a + \sum_{j=1}^{n} x_j \cdot v_{ij}\right)\right]$$

Subject to

$$y_i^a + \sum_{j=1}^{n} x_j \cdot v_{ij} \geq 0 \qquad\qquad i = 1,...,m \qquad (5.2)$$

$$\sum_{i \in I_\mu} z_{i\mu} - Q_\mu \leq 0 \qquad\qquad \mu = 1,...,r \qquad (5.3)$$

$$x_j \in IN_0 \qquad\qquad j = 1,...,n \qquad (5.4)$$

$$z_{i\mu} \geq 0 \qquad\qquad i = 1,...,m \ ; \ \mu \in T_i \qquad (5.5)$$

$$\lambda_i \in IR \qquad\qquad i = 1,...,m \qquad (5.6)$$

where

λ_i: Lagrangean multiplier $(i = 1,...,m)$

$L(x,z,\lambda)$: Lagrangean function.

Both kinds of Branch-and-Bound-Algorithms don't take any advantage of the special structure of the considered dismantling and recycling planning problem, which can be partitioned into an all-integer linear dismantling problem and a linear recycling problem. The decomposition algorithm of Benders [Benders 62] uses this special feature in order to find an optimal solution by the iterative solution of the all-integer and the linear problem [Salkin 89].

For this the mixed-integer problem (MIP) will be transformed as follows:

$$\underset{x_j \in IN_0}{MAX} \left(-\sum_{j=1}^{n} c_j^z \cdot x_j + MAX \left\{ \begin{array}{l} \sum_{i=1}^{m} \sum_{\mu \in T_i} c_{i\mu}^v \cdot z_{i\mu} \Big/ \sum_{\mu \in T_i} z_{i\mu} = \gamma_i \cdot \left(y_i^a + \sum_{j=1}^{n} x_j \cdot v_{ij} \right); \\ \sum_{i \in I_\mu} z_{i\mu} \le Q_\mu ; z_{i\mu} \ge 0 \ \forall i = 1,...,m; \mu \in T_i \end{array} \right\} \right) \quad (5.7)$$

The interior linear optimization problem can be dualised for given integer variables $\hat{x}_j$ $(j = 1,...,n)$ (DL):

$$MIN \quad \sum_{i=1}^{m} \gamma_i \cdot \left(y_i^a + \sum_{j=1}^{n} \hat{x}_j \cdot v_{ij} \right) \cdot u_i + \sum_{\mu=1}^{r} Q_\mu \cdot \pi_\mu \quad (5.8)$$

Subject to

$u_i + \pi_\mu \ge c_{i\mu}^v$ $i = 1,...,m ; \mu \in T_i$ (5.9)

$\pi_\mu \ge 0$ $\mu = 1,...,r$ (5.10)

$u_i \in IR$ $i = 1,...,m$ (5.11)

where

π_μ: dual variable of the capacity constraints $(\mu = 1,...,r)$
u_i: dual variable of the dismantling constraints $(i = 1,...,m)$

With the help of the extreme points[1] u_i^p, π_μ^p $(p = 1,...,P)$ of the convex polyhedron (5.9),...,(5.11), the following all-integer problem (IP), which is equivalent to the mixed-integer problem (MIP), can be formulated:

---

[1] It is assumed that a feasable solution of the dual linear problem (DL) exists.

$$MAX \quad DB \tag{5.12}$$

Subject to

$$DB \leq -\sum_{j=1}^{n} c_j^z \cdot x_j + \sum_{i=1}^{m} \gamma_i \cdot \left(y_i^a + \sum_{j=1}^{n} x_j \cdot v_{ij} \right) \cdot u_i^p + \sum_{\mu=1}^{r} Q_\mu \cdot \pi_\mu^p$$

$$p = 1,...,P \tag{5.13}$$

$$x_j \in IN_0 \qquad\qquad j = 1,...,n \tag{5.14}$$

where

$$DB := \left(-\sum_{j=1}^{n} c_j^z \cdot x_j + \underset{p=1,...,P}{MIN} \left\{ \sum_{i=1}^{m} \gamma_i \cdot \left(y_i^a + \sum_{j=1}^{n} x_j \cdot v_{ij} \right) \cdot u_i^p + \sum_{\mu=1}^{r} Q_\mu \cdot \pi_\mu^p \right\} \right) \tag{5.15}$$

In order to find the optimal solution of (MIP), it is often not necessary to generate all restrictions (5.13) of (IP). The idea of Benders'Decomposition is an iterative solution of (DL) and (IP) whereby each optimal solution of (DL) leads to a new restriction (5.13) of (IP). A new lower bound is generated by each solution of (DL) and a new upper bound by each solution of (IP). The iteration will stop with the optimal solution if the lower bound equals to the upper bound. In each iteration an optimal solution of the all - integer problem (IP) has to be computed. The required solution time can be reduced drastically, if only feasible solutions of (IP) are computed [Hummeltenberg 85], [Geoffrion 74]. This heuristic modification of Benders'Decomposition Algorithm will compute "good" solutions of the considered dismantling and recycling problem in a reasonable time.

6 Implementation of the planning model and further research

A prototype of the above presented dismantling planning model has been implemented on a Personal Computer (cf. Figure 5) [Spengler 93], [Nicolai 93].

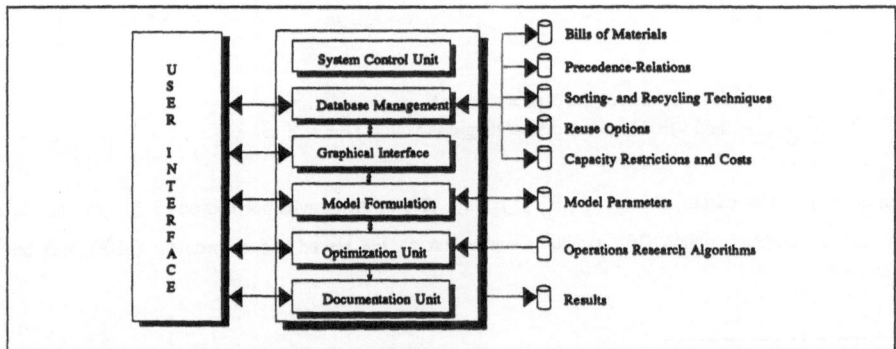

Figure 5: Structure of the implemented prototype

In order to obtain integrated dismantling and recycling strategies for the region of the Upper Rhine Valley, the model has to be evaluated with the specific regional database. Different environmental control strategies such as waste duties, taxes on natural raw materials and recycling objectives will be considered and their influence on the optimal solution will be analysed.

Literature

[Benders 62]   Benders, J.F.:
Partitioning Procedures for Solving Mixed-Variables Programming Problems, in: Numerische Mathematik 4, 1962, S. 238 - 252

[Dakin 65]   Dakin, R.:
A Tree Search Algorithm for Mixed Integer Programming Problems, in: Computer Journal 8(3) , 1965, S. 250 -255

[Geoffrion 74]   Geoffrion, A.:
Lagrangean Relaxation for Integer Programming, in: Mathematical Programming Study 2, S. 82 - 114

[Hummeltenberg 85]   Hummeltenberg, W.:
Capital Budgeting with Benders Decomposition, in: EJOR, 1985, S. 318 - 329

[Nicolai 93]   Nicolai, M.; Spengler, Th.; Ruch, M.; Valdivia, S.; Hamidovic, J.; Rentz, O.:
Minimization, Recycling and Reuse of Demolition Waste - A Dismantling and Recycling Planning System for the Region of the Upper Rhine Valley, in: Proc. of International Recycling Congreß Re'93, Genf, 1993

[Salkin 89]   Salkin, H.; Mathur, K.:
Foundations of Integer Programming, New York, Amsterdam, London, 1989

[Schrage 91]   Schrage, L.:
LINDO: An Optimization Modeling System, Chicago, San Francisco, 1991

[Spengler 93]   Spengler, Th.; Hamidovic, J.; Nicolai, M.; Ruch, M.; Valdivia, S.; Rentz, O.:
Konzeption eines EDV gestützten Planungssystems zur optimalen Demontage- und Verwertungsplanung von Wohngebäuden im Oberrheingraben, in: Proc. Informatik im Umweltschutz, 7. Symposium, Ulm 1993, Springer Verlag, 1993

ON THE L-STRUCTURE OF INTEGER LINEAR
PROGRAMMING PROBLEMS

Alexander A. Kolokolov

Institute of Information Technologies and Applied
Mathematics, Russian Academy of Sciences, pr. Mira,
19-a, Omsk, 644050, Russia

The approach to investigation and solving of integer programming problems (IPP) was developed in [1-8] and in other papers. This approach are connected with the application of the special regular set partitions in space R^n. In particular, some algorithms and classes of cuts, bounds of iterations number for dual fractional cutting plane algorithms, for branch and bound method have been obtained on this base.

Some new results for L-partition are given here. We have found tight connection between compleces of L-classes and vertex set of the relaxation polyhedron for boolean programming problems. It is proved, that L-covering cardinality of the problem depends on the number of the verteces in the L-covering. This connection gives new upper bounds of iteration number for the fractional cutting plane algorithms and for other algorithms.

First we give some necessary definitions and facts. Let Z^n be the set of all integer n-vectors, $M=\{x: Ax \leqslant b, x \geqslant 0\}$, where $x=(x_1,\ldots,x_n)^T$, A - real-valued (m×n)-matrix, $b \in R^m$ and exists x^* - lexmax M. Consider the following lexicographical IPP:

$$\text{find} \quad z_* - \text{lexmax} \ (M \cap Z^n) \qquad (1)$$

For any $X \subseteq R^n$ denote $X_0 - X \cap Z^n$. The set

$$M_* - \{x \in M: x \succ z \ \ \forall z \in M_0\}$$

is called a *fractional covering* of problem (1). Here and further $\succ, \not\succ$ are symbols of the lexicographical order. Evidently, if $M_0 - \varnothing$, then $M_* - M$.

The L-partition is determined as follows. The points $x,y \in R^n$ $(x \succ y)$ belong to the same class of partition, if $z \in Z^n$ does not exist so that $x \succ z \succ y$. We denote X/L via factor-set, generated by L-partition for arbitrary set $X \subseteq R^n$. Elements of X/L are called *L-classes*. This partition has some important properties:

1) any $z \in Z^n$ is a class of partition; other classes have only noninteger points and are called *fractional*;

2) if X is bounded, then X/L is finite;

3) for non-empty $X,X' \subset R^n$ we write $X \succ X'$, if $x \succ x'$ for any $x \in X$, $x' \in X'$; this relation for X/L is a linear order; for example, if X is bounded, then

$$X/L = \{V_1, \ldots, V_p\}, \quad V_i \succ V_{i+1}, \quad i=1,\ldots,p-1.$$

The factor-set M_*/L is called *L-covering* of problem (1). Let $x^* \notin Z^n$ and $V_{x^*}(M_*)$ be an element from M_*/L, contained x^*. A linear inequality $(g,x) \leqslant g_0$ is called *regular cut* (for L-partition), if

a) $(g,x') > g_0$ for any $x' \in V_{x^*}(M_*)$,

b) $(g,z) \leqslant g_0$ for any $z \in M_0$.

There are methods of such cuts construction. Particularly some Gomory cuts are regular. *The depth* of the cut is determined as a number of elements of M_*/L, which are exluded by this cut.

Discribe the dual fractional cutting plane process D.

Step 0. Put $M^{(1)} = M$, $t=1$.

Iteration t $(t \geqslant 1)$.

Step 1. Find $x^{(t)} = \text{lexmax } M^{(t)}$. If $x^{(t)} \in Z^n$ or $M^{(t)} = \emptyset$, then process D is termined. In the first case an optimal solution of the problem (1) has been obtained, in the second one the problem has no solution.

Step 2. Exlude some additional inequalities (cuts) from the current constraints system. Denote new polyhedron $M'^{(t)}$. The following condition must be valid:

$$x^{(t)} = \text{lexmax } M'^{(t)}.$$

Step 3. Construct a regular cut $(g^{(t)}, z) \leqslant g_0^{(t)}$. Add it to the constraints of the current problem and define

$$M^{(t+1)} = M'^{(t)} \cap \{x : (g^{(t)}, z) \leqslant g_0^{(t)}\}.$$

Go to the iteration t+1 (to step1).

In particular, this process has following properties: $M_0 \subseteq M^{(t)}$ and $x^{(t)} \succ x^{(t+1)}$ for any t. Let $I_D(M)$ be a number of its iterations for the solving of (1), $H_D(M)$ be an upper bound of the cuts depth in D. Earlier we have obtained the following fundamental relations:

$$(1/H_D(M))|M_*/L| \leqslant I_D(M) \leqslant |M_*/L|, \qquad (2)$$

where $|M_*/L|$ is the cardinality of the M_*/L.

In connection with (2) we study the structure of L-partition (*L-structure*) of any non-empty set $X \subseteq R^n$. The subset Q of the fractional classes from X/L is called a *complex*, if for any $V, V' \in Q$ $(V \succ V')$ $z \in X_0$ does not exists so that $V \succ z \succ V'$. Denote $C(X)$ set of all the compleces, generated by set X . A complex $Q \in C(X)$ is called *full*, if it is not subset of another complex of $C(X)$. Let

$$\Phi(X) = \max\{|Q|: Q \in C(X)\}.$$

Evidently, that $|M_*/L| = \Phi(M_*) \leqslant \Phi(M)$.

There are some bounds of $\Phi(M)$ for various polyhedra. For example, if $A \geqslant 0$, $b \geqslant 0$, then $\Phi(M) \leqslant n$.

Some classes of polyhedra have $\Phi(M) \leqslant 1$ and integer lexicographical maximum and minimum in M/L (if these elements exist). Such L-structure is called *alternating*. If X has alternating L-structure and $M_* \subset X$, then $|M_*/L| \leqslant |X_0|-1$. This inequality was used for estimation of iterations number. For example, the alternating L-structure have parallelepiped with integer vertices, polyhedra of set packing and set covering problems, polyhedron M with absolute unimodular matrix A and integer b.

Consider L-structure of boolean programming problem (BPP). Let
$$B^n = \{x: 0 \leqslant x_i \leqslant 1, i=1,\ldots,n\}$$
and polyhedron $M \subset B^n$. An element $V \in M/L$ is called *essential*, if lexicographical maximum or minimum point $x \in V$ exists. It is not difficult to show, that any such point is a vertex of M.

Theorem 1. *If $M \subset B^n$, then any full complex from $C(M)$ with q elements has at least $\lceil q/2 \rceil$ essential L-classes.*

Proof. Let $Q = \{V_t\}t=1,\ldots,q$ be a full complex from $C(M)$, where $q>1$ and $V_1 \succ \ldots \succ V_q$. Choose any L-classes V_k, V_{k+1} and prove, that at least one of them is essential. For $x \notin Z^n$ define $\phi(x)=\min\{i: x_i \neq [x_i], i=1,\ldots,n\}$ and $r(t)=\phi(x)$, $x \in V_t$, $t=1,\ldots,q$. The convexity M and

fullness of Q follows $r(k) \neq r(k+1)$. Consider consecutively other cases:
        a) $r(k) < r(k+1)$,   b) $r(k) > r(k+1)$.
Formulate the linear programming problem:

$$x_{r(k+1)} \dashrightarrow \text{lexmax} \qquad (3)$$

s.t.

$$x \in M, \ x_i = x'_i, \ i = 1, \ldots, r(k+1)-1. \qquad (4)$$

Here x' is an element of V_{k+1}. Let x'' be an optimal vertex of this problem. The condition "a" and the neighbourhood of V_k, V_{k+1} follows $x''_{r(k)} = 0$. Therefore $V_k \succ x''$. These two L-classes are contained in the same complex, consequently x'' is not integer point. Then from the fullness of Q we obtain that $x'' \in V_{k+1}$.

The case "b" is analysed analogously with the help of the problem:
        $$x_{r(k)} \dashrightarrow \text{lexmin}$$
    s.t.
        $$x \in M, \ x_i = x'_i, \ i = 1, \ldots, r(k)-1,$$

where x' is an element of V_k.

From this theorem it follows, that $\text{conv}(M \cap Z^n)$ has the alternating L-structure. This property is not valid for the general integer programming problem (1).

Evidently, that M_* / L is a full complex. Also the lexicographical maximal element of the L-covering is essential. Let $v(X)$ be a number of essential classes in X/L for any $X \subseteq B^n$. By using of theorem 1 one can easy obtain the following statment.

Theorem 2. *For the problem (1) with $M \subset B^n$ the inequality $|M_*/L| \leq 2v(M_*)$ takes place.*

This estimate is exact. Consider a corresponding example. Let $M' = \{x \in B^n : 2x_n = 1\}$, $z^1 = (0, \ldots, 0)^T$, $z^2 = (0, \ldots 0, 1)^T$ and $M = \text{conv}(\{z^1, z^2\} \cup M')$. Denote $v^1, \ldots, v^s$ the verteces of M'. One can suggest that $v^1 \succ \ldots \succ v^s$. It is not difficult to show that any vertex of M' is of the type $(z_1, \ldots, z_{n-1}, 1/2)^T$ and $v^1, \ldots, v^{s-1}$ are all fractional verteces of M. Evidently, $s = 2^{n-1}$ and $v^{s-1} \succ z_* = z^2 \succ v^s$. All verteces $v^1, \ldots, v^{s-1}$ are contained in the different elements of M_*/L. The convex combination $w^k = (1/2)v^k + (1/2)v^{k+1}$ is a representative of a class, which is not essential. The combination $w = (1/2)v^{s-1} + (1/2)z^*$ gives else one such L-class. Therefore $|M_*/L| = 2(s-1) = 2v(M_*)$.

From (2) and from theorem 2 the upper bound of iterations number for BPP follows: $I_D(M) \leqslant 2v(M_*)$. This means, that the BBP with a little number of verteces in M_* are solved efficiently.

Consider an example of BPP. Let A be a boolean $(m \times n)$-matrix, $m \leqslant n$, rankA=m, all columns of A are nonzero, $e=(1,\ldots,1)^T$, $M'=\{x: Ax=e, x \geqslant 0\}$ and $M' \neq \emptyset$. The number of verteces of M' is not greater than C^m_n, therefore $v(M'_*) \leqslant v(M') \leqslant C^m_n$. Consequently, for problem (1) with M' and process D the upper bound $I_D(M') \leqslant 2C^m_n$ takes place.

Recently we have obtained analogous results for branch and bound algorithms by the application of some other partitions [6].

References

1. Kolokolov A.A. Regular cuts by solving of the integer optimization problems. In: Controllable systems. Novosibirsk, 1981, N21, p.18-25.

2. Kolokolov A.A. A study of integer programming problems and methods on the base of L-partition. 33 Inter. Wiss. Koll. TH Ilmenau, 1988, 53-55.

3. Kolokolov A.A., Tcepkova E.V. A study of the knapsack problem on the base of L-partition. Cybernetics, 1991, N2, p. 38-43.

4. Kolokolov A.A. Regular partitions in integer programming. In: Methods of solving and analysis for the discrete optimization problems. Omsk, OmGU, 1992, p. 67-93.

5. Kolokolov A.A. Upper bound of L-covering power in the boolean programming problems. III international seminar on global optimization. Irkutsk, 1992, p.108.

6. Kolokolov A.A. Regular partitions and branch and bound method. Conference "Mathematical programming and applications", Ekaterinburg, 1993, p.61.

7. Zablotcskaja O.A., Kolokolov A.A. Fully regular cuts in boolean programming. In: Controllable systems. Novosibirsk, 1983, N23, p.55-63.

8. Zabudsky G.G. On an integer programming formulation of the facilities layout problem on the line. In: Controllable systems, Novosibirsk, 1990, N30, p. 35-45.

This work was supported by the Russian Fund of Fundamental Research.

AN ARTIFICIAL NEURAL NETWORK APPROACH FOR NONLINEAR OPTIMIZATION WITH DISCRETE DESIGN VARIABLES

Jian-Bo Yang and Pratyush Sen
Engineering Design Centre, University of Newcastle upon Tyne
Newcastle upon Tyne NE1 7RU, United Kingdom

Abstract — An artificial neural network based approach is presented for dealing with discrete nonlinear optimization problems. The basic idea involved in the new approach is to use quadratic approximations of a nonlinear function to be minimized with discrete-valued design variables so that a sequence of artificial neural networks can be constructed which share the basic features of the Hopfield and Tank neural network. With the discrete values of a design variable being represented by the states of the neurons, these networks are then used to generate a sequence of designs which quadratically converge to the discrete minima of the original function. Constrained problems are converted into unconstrained ones by using a quadratic extended interior penalty technique which preserves the continuity of the second derivatives throughout the design space. The simple implementation process and the potential of the proposed approach are demonstrated by two examples. The nature of this approach provides scope for parallelization and consequent gains in computational efficiency.

1. Introduction

In the development of optimization techniques for engineering design, a significant amount of attention has been paid in recent years to unconstrained or constrained nonlinear optimization with discrete-valued design variables. The importance of dealing with this type of combinatorial optimization problems cannot be overemphasized because of the fact that in many real-world engineering optimization problems the designer must choose values of design variables from a list of permissible values. For instance, design variables such as the number of stiffeners, cross-sectional areas of sections and available thicknesses of plates may fall into this category.

Some techniques exist for dealing with such problems, for example, the branch and bound method[5], the sequential linear discrete programming method[8], the round-off method[9], the dual method[10], and the penalty method[11]. However, the combinatorial nature of discrete nonlinear optimization problems often presents computational problems. To date, no completely satisfactory methods have been suggested for discrete nonlinear optimization problems[8]. This paper proposes a new approach which is based on artificial neural networks and appears to have certain advantages.

It has been shown that some combinatorial optimization problems could be solved using the Hopfield and Tank artificial neural network system by minimizing a quadratic energy function[1, 4]. It has also been indicated that the Hopfield and Tank network is effective for quadratic discrete optimization[2, 6]. The neural network circuits for linear programming and nonlinear programming have been explored as well[12]. The aim of the work to be reported in this paper is to utilize these basic but important features of the Hopfield and Tank network so that a new artificial neural network based algorithm could be developed for dealing with general unconstrained and constrained discrete nonlinear optimization problems. In addition to being simple in implementation, the new approach can always provide a sequence of designs which quadratically converge to (at least local) discrete optima of a nonlinear problem. Furthermore, the nature of the algorithm provides scope for parallelization and consequent gains in computational efficiency.

In this paper, the Hopfield and Tank network is presented along with a discrete computation model of the network. Number representation techniques for representing a design variable in terms of neurons' states are also discussed. The artificial neural network approach for unconstrained discrete nonlinear optimization is then explored. Constrained problems are connected to unconstrained ones by using the exterior penalty method[7] and a quadratic extended interior penalty technique[3], which preserve the continuity of the second derivatives throughout the design space. Two examples are then presented to demonstrate the implementation process and the potential of the proposed approach.

2. Hopfield and Tank Neural Network

2.1 Hopfield and Tank network

A neural network consists of mutually interconnected neurons, which collectively change their states until stable states are reached at which the outputs of all neurons remain constant.

Hopfield and Tank showed that some combinatorial optimization problems could be solved effectively on their continuous neural network model[4, 12]. In their model, the system behavior of nonlinear analog neurons is formulated by the following input-output equations and the continuous state equations (1) describing the time evolution of the neural network

$$v_i(t) = \Phi[u_i(t)], \quad \frac{du_i(t)}{dt} = \sum_{j=1}^{N} T_{ij} v_j(t) + I_i \quad i=1, \cdots, N \tag{1}$$

where $v_i(t)$ is the state/output of neuron i at instant t, $u_i(t)$ the input of neuron i at instant t, N the number of neurons, T_{ij} the synaptic interconnection strength from neuron j to neuron i, and I_i the bias input of neuron i.

The input-output function Φ is a sigmoid monotonically increasing relation which takes values between zero and one. $\Phi(u_i)$ can be taken to be the following function[1, 4]

$$\Phi(u_i(t)) = \frac{1}{2}\left[1 + tanh(u_i(t)/u_{i0})\right] \tag{2}$$

where u_{i0} is called gain width for v_i. $\Phi(u_i)$ approaches a unit function as u_{i0} tends to zero.

In the following sections of this paper, the Hopfield and Tank neural network described above will be denoted by $NN(V,U,T,I,N)$, where $V = [v_1 \cdots v_N]^T$, $U = [u_1 \cdots u_N]^T$, $T = (T_{ij})_{N \times N}$, and $I = [I_1 \cdots I_N]^T$. The $NN(V,U,T,I,N)$ is obviously a continuous dynamic model. To facilitate the implementation of the network on a digital computer, the continuous model may be approximated by the following discrete model under the assumption that the states of all neurons remain constant in a small time interval Δt, where $t_k = t_{k-1} + \Delta t$

$$u_i(t_k) = u_i(t_{k-1}) + \left[\sum_{j=1}^{N} T_{ij} v_j(t_{k-1}) + I_i\right]\Delta t, \quad v_i(t_k) = \Phi(u_i(t_k)) \quad i=1, \cdots, N \tag{3}$$

In $\Phi(u_i(t))$ the gain width u_{i0} may be set to be a constant. In the new method, however, we will adopt a so-called "annealing" schedule given by[4]

$$u_{i0}(t) = u_0 / \ln(1 + t) \tag{4}$$

where u_0 is a constant gain width.

In the $NN(V,U,T,I,N)$, the neurons may continuously change their states which take values between zero and one, i.e. $0 \leq v_i \leq 1$. Thus the state space of the neurons is a N-dimensional hypercube. At the corners of the hypercube, $v_i = 0$ or 1; and in the interior, $0 \leq v_i \leq 1$.

2.2 Basic features of the $NN(V,U,T,I,N)$

The $NN(V,U,T,I,N)$ has several important features. Summarized below are only those which have been commonly accepted and will be used for development of the new approach. Firstly, the so-called energy function of the $NN(V,U,T,I,N)$ is defined by[1,4]

$$E(V) = -\frac{1}{2}\sum_{i=1}^{N}\sum_{j=1}^{N} T_{ij} v_i v_j - \sum_{i=1}^{N} I_i v_i \tag{5}$$

The basic features of the $NN(V,U,T,I,N)$ may then be summarized as follows.

Feature 1: If the synaptic interconnection strengths of the $NN(V,U,T,I,N)$ are symmetric, that is, $T_{ij}=T_{ji}$, then the network always converges to stable states, which are minima of the energy function defined by (5).

Feature 2: In the low-gain limit where the gain widths u_{i0} $(i=1,\cdots,N)$ are large, the stable states of the $NN(V,U,T,I,N)$ or the minima of $E(V)$ may occur in the interior or at the corners of the state space of V. In the high-gain limit where u_{i0} $(i=1,\cdots,N)$ are all sufficiently small, the minima only occur at the corners of this space.

Feature 3: Every time the states of the neurons in the $NN(V,U,T,I,N)$ change, the energy defined by (5) stays the same or becomes lower. Besides, the $NN(V,U,T,I,N)$ alters states in a relatively irreversible manner, i.e., it has the tendency not to return to a previously experienced state.

Feature 4: The interconnection strength T_{ij} is a quantitative measure of the desirability that the connection between neuron i and neuron j denoted by $\{v_i, v_j\}$ be activated. If $T_{ij}=0$, it means that neuron i is disconnected from neuron j; if $T_{ij}>0$, it is desirable that $\{v_i, v_j\}$ be activated; if $T_{ij}<0$, it is undesirable.

2.3 Number representations

To apply the $NN(V,U,T,I,N)$ to deal with an optimization problem with discrete-valued design variables, it is essential to represent the design variables in terms of the states of the neurons. Let $X=[x_1\cdots x_i\cdots x_n]^T$ be a design where x_i is a design variable which takes one of a range of discrete values. A design variable may generally take either regularly or irregularly spaced discrete values. The so-called distributed representation and intermediate representation[6] may then be used to represent these two types of design variables.

Distributed representation: For a variable x_i which takes one of (N_i+1) regularly spaced discrete values, e.g., $d_{i0}, d_{i0}+d_i,\cdots, d_{i0}+N_id_i$ with $d_i>0$ and $d_{i0}=0$, the distributed representation is given by

$$x_i = \psi_i(V_i) = d_i\sum_{p=1}^{N_i}v_{ip} \tag{6}$$

where $V_i=[v_{i1}\cdots v_{ip}\cdots v_{iN_i}]^T$ is the vector of the states of the neurons for the variable x_i.

If $d_{i0}\neq 0$, it is always possible to take the linear displacement transformation, $x_i' = x_i - d_{i0}$. In (6) the neurons are independent of one another.

Intermediate representation: For a variable x_i which takes one of (N_i+1) irregularly spaced discrete values, e.g., $d_{i0}, d_{i1},\cdots,d_{i,p-1}, d_{ip},\cdots,d_{iN_i}$ with $d_{ip}>d_{i,p-1}$ and $d_{i0}=0$, the intermediate representation is given by

$$x_i = \psi_i(V_i) = \sum_{p=1}^{N_i}(d_{ip} - d_{i,p-1})v_{ip} \qquad v_{ip} \leq v_{i,p-1} \tag{7}$$

If $d_{i0}\neq 0$, the transformation, $x_i' = x_i - d_{i0}$, may be used. In (7) the neurons are dependent on each other, that is $v_{ip}\leq v_{i,p-1}$. To make it convenient to describe a design in terms of neurons' states, let $X=\Psi(V)$ where $\Psi(V)=[\psi_1(V_1)\cdots\psi_n(V_n)]^T$ and $V=[V_1^T\cdots V_n^T]^T$.

3. A Neural Network Approach for Unconstrained Optimization

3.1 Nonlinear optimization and quadratic approximation

An unconstrained discrete nonlinear optimization problem may be formulated as follows

$$\begin{cases} min. \ f(X) \\ s.t. \ X \in \mathbf{X}_1 \times \cdots \times \mathbf{X}_i \times \cdots \times \mathbf{X}_n \end{cases} \tag{8}$$

where $X=[x_1 \cdots x_n]^T$, $x_i \in \mathbf{X}_i = \{x_i \mid x_i = \{d_{i0}\, d_{i1} \cdots d_{ip} \cdots d_{iN_i}\}$ with $d_{ip} > d_{i,\,p-1}\}$ and $f(X)$ is a twice continuously differentiable nonlinear function for all $x_i \in [d_{i0}\ d_{iN_i}]$. Without loss of generality, $d_{i0}=0$ is assumed.

To apply the $NN(V,U,T,I,N)$ to deal with problem (8), the function $f(X)$ needs to be represented by an energy function similar to that defined by (5). Therefore, the basic idea of applying the $NN(V,U,T,I,N)$ to search for the minima of $f(X)$ consists of replacing, in the neighbourhood of a seed point X^k, the function $f(X)$ by its quadratic approximation $q(X, X^k)$

$$q(X, X^k) = q_E(X, X^k) + \bar{q}(X^k) \tag{9}$$

$$q_E(X, X^k) = \frac{1}{2} X^T \nabla^2 f(X^k) X + \left[\nabla f^T(X^k) - (X^k)^T \nabla^2 f(X^k)\right] X \tag{10}$$

$$\bar{q}(X^k) = f(X^k) - \nabla f^T(X^k) X^k + \frac{1}{2}(X^k)^T \nabla^2 f(X^k) X^k \tag{11}$$

$\bar{q}(X^k)$ is a constant for a given X^k. Thus, $q(X, X^k)$ decreases (or increases) with $q_E(X, X^k)$. In other words, the descent direction of $f(X)$ is determined by that of $q_E(X, X^k)$ in the neighbourhood of X^k. $q_E(X, X^k)$ is then used as a basis to construct an artificial $NN(V,U,T,I,N)$.

3.2 Construction of artificial neural networks

To construct an artificial $NN(V,U,T,I,N)$, let's first expand $q_E(X, X^k)$ defined by (10) and denote it as follows

$$q_E(X, X^k) = -\frac{1}{2} \sum_{i=1}^{n} \sum_{j=1}^{n} -\frac{\partial^2 f(X^k)}{\partial x_i \partial x_j} x_i x_j - \sum_{i=1}^{n} \left[\sum_{j=1}^{n} \frac{\partial^2 f(X^k)}{\partial x_i \partial x_j} x_j^k - \frac{\partial f(X^k)}{\partial x_i}\right] x_i \tag{12}$$

Suppose the variable x_i is represented using the number representation technique (6) or (7), i.e., $X=\Psi(V)$. Then, an energy function $E(V, V^k)$ can be formulated by

$$E(V, V^k) = q_E(\Psi(V), X^k) = -\frac{1}{2} \sum_{i=1}^{n} \sum_{p=1}^{N_i} \sum_{j=1}^{n} \sum_{q=1}^{N_j} T^k_{ip,jq} v_{ip} v_{jq} - \sum_{i=1}^{n} \sum_{p=1}^{N_i} I^k_{ip} v_{ip} \tag{13}$$

where V^k represents the starting states of the neurons V. In (13), for regularly discrete-valued variables x_i and x_j represented by (6)

$$T^k_{ip,jq} = -d_i d_j \frac{\partial^2 f(X^k)}{\partial x_i \partial x_j}, \quad I^k_{ip} = d_i \left[\sum_{j=1}^{n} \frac{\partial^2 f(X^k)}{\partial x_i \partial x_j} x_j^k - \frac{\partial f(X^k)}{\partial x_i}\right]; \tag{14}$$

for irregularly discrete-valued variables x_i and x_j represented by (7)

$$T^k_{ip,jq} = -(d_{ip}-d_{i,p-1})(d_{jq}-d_{j,q-1}) \frac{\partial^2 f(X^k)}{\partial x_i \partial x_j}, \quad I^k_{ip} = (d_{ip}-d_{i,p-1}) \left[\sum_{j=1}^{n} \frac{\partial^2 f(X^k)}{\partial x_i \partial x_j} x_j^k - \frac{\partial f(X^k)}{\partial x_i}\right]; \tag{15}$$

and for a regularly discrete-valued variable x_i and for an irregularly discrete-valued variable x_j

$$T^k_{ip,jq} = -d_i(d_{jq} - d_{j,q-1}) \frac{\partial^2 f(X^k)}{\partial x_i \partial x_j} \tag{16}$$

The energy function $E(V, V^k)$ can be further transformed into the following form

$$E(V, V^k) = -\frac{1}{2} \sum_{r=1, s=1}^{N} \sum^{N} T^k_{rs} v_r v_s - \sum_{r=1}^{N} I^k_r v_r \tag{17}$$

$$I^k_r = \frac{1}{C_n} I^k_{ip}, \quad T^k_{rs} = \frac{1}{C_n} T^k_{ip,jq}, \quad v_r = v_{ip}, \quad r = \sum_{l=1}^{i-1} N_l + p, \quad s = \sum_{l=1}^{j-1} N_l + q, \quad N = \sum_{i=1}^{n} N_i$$

where C_n is a positive constant and may be calculated as follows

$$C_n = \max_{i,j,p,q} \{|T^k_{ip,jq}|, |I^k_{ip}|\} \tag{18}$$

We are then in a position to construct an artificial neural network defined as follows

$$v_r(t) = \frac{1}{2}\left[1 + tanh\left(\frac{u_r(t)}{u_{r0}}\right)\right] \quad \frac{du_r(t)}{dt} = \sum_{s=1}^{N} T_{rs}^k v_s(t) + I_r^k, \quad r=1, \cdots, N \tag{19}$$

and denoted by $ANN(V,U,T^k,I^k,N)$ where $T^k=(T_{rs}^k)_{N \times N}$ and $I^k=(I_r^k)_{N \times 1}$.

If C_n is determined using (18), $|T_{rs}^k| \le 1$, r, $s=1, \cdots, N$. So, the interconnection strength T_{rs}^k in the $ANN(V,U,T^k,I^k,N)$ may also be explained as the desirability of the connection between neuron r and neuron s being activated. It is easy to show that in the $ANN(V,U,T^k,I^k,N)$ T^k is symmetric, that is, $T_{rs}^k=T_{sr}^k$ for all r, $s=1, \cdots, N$. Thus, we can conclude that the $ANN(V,U,T^k,I^k,N)$ given by (19) shares the four basic features of the Hopfield and Tank network.

3.3 Algorithm and implementation

If the $ANN(V,U,T^k,I^k,N)$ begins to operate from a starting state V^k, it will reach a stable state, denoted by $\hat{V}^k$. Let $\hat{X}^k=\Psi(\hat{V}^k)$. If the following starting conditions are satisfied,

$$v_{ip}^k = \Phi(u_{ip}^k), \quad x_i^k = \psi_i(V_i^k), \quad p=1, \cdots, N_i; \quad i=1, \cdots, n \tag{20}$$

$q_E(\hat{X}^k, X^k) \le q_E(X^k, X^k)$ is then held to be true. $\hat{X}^k$ may thus be used as a new seed point, that is, $X^{k+1}=\hat{X}^k$, at which $f(X)$ is re-approximated using its quadratic approximation. Consequentially the $ANN(V,U,T^{k+1},I^{k+1},N)$ can be constructed. As this process continues, a sequence $\{X^k\}$ will be generated. The whole process may be summarized by the following artificial neural network algorithm, which is composed of the following steps.

Step 1: Fist, define a discrete optimization problem as shown in (8). If $d_{i0} \ne 0$, take a linear displacement transformation so that $d_{i0}=0$.

Step 2: Represent each design variable using the number representation (6) or (7).

Step 3: Select a point X^0 as an initial seed point at which $f(X)$ is approximated. Set initial states and inputs of the neurons, denoted by V^0 and U^0. Note that X^0, V^0 and U^0 may be assigned at random and they may not satisfy the starting conditions (20). Let $k=0$.

Step 4: Approximate the nonlinear function $f(X)$ at X^k using its quadratic approximation $q(X, X^k)$ as defined by (9).

Step 5: On the basis of the function $q_E(X, X^k)$ defined by (10), construct an artificial neural network $ANN(V,U,T^k,I^k,N)$ defined by (19).

Step 6: When $k = 0$, operate the $ANN(V,U,T^k,I^k,N)$ from the starting state, denoted by V^k and U^k, until its stable state, denoted by $\hat{V}^k$ and $\hat{U}^k$, is found. When $k > 0$, operate the $ANN(V,U,T^k,I^k,N)$ from its starting state until a new state, denoted by $\tilde{V}^k$ and $\tilde{U}^k$, is reached at which $f(X)$ is reduced by the largest amount available by operating the $ANN(V,U,T^k,I^k,N)$.

Step 7: When $k = 0$, let $V^{k+1} = \hat{V}^k$, $U^{k+1} = \hat{U}^k$, and $X^{k+1} = \hat{X}^k = \Psi(\hat{V}^k)$. When $k > 0$, let $V^{k+1} = \tilde{V}^k$, $U^{k+1} = \tilde{U}^k$, and $X^{k+1} = \tilde{X}^k = \Psi(\tilde{V}^k)$. Then, check if $V^{k+1} = \hat{V}^k$ and if $X^{k+1} = X^k$ (or $f(X^{k+1}) = f(X^k)$). If yes, an equilibrium point of $f(X)$ is found and go on. Otherwise, let $k = k+1$ and go to Step 4.

Step 8: Check if the obtained equilibrium point is at the corners of the hypercube spanned by the state space of the neurons. If yes, a discrete equilibrium point $\bar{X}^*$ is found. Let $\bar{X}^*=\hat{X}^k$ and stop. Otherwise, accelerate the reduction of the gain width u_{r0} if the state of the neuron v_r does not equal to 0 or 1. Let $k=k+1$ and then go to Step 6.

To implement the $ANN(V,U,T^k,I^k,N)$ on a digital computer, the values of certain parameters, such as Δt and u_0, have to be set beforehand. The time interval Δt is concerned with the approximation of the discrete model defined by (3) to the continuous system model defined by (19). Theoretically, the discrete model tends to be identical to the continuous model when $\Delta t \to 0$. If Δt is too small, however, the $ANN(V,U,T^k,I^k,N)$ will need much more calculation effort to reach a stable state while round-off errors are accumulated. u_0 should be so selected that the $ANN(V,U,T^k,I^k,N)$ may initially operate in the low-gain limit in order to increase the possibility of searching for the global minimum of $f(X)$

3.4 Convergence analysis

The artificial neural network algorithm involves three iteration loops. The inner loop, composed of Step 5 and Step 6, concerns the operation of the $ANN(V,U,T^k,I^k,N)$. The outer loop, consisting of Step 4 to Step 7, is used for nonlinear optimization. The third loop, composed of Step 6 to Step 8, may be necessary for the discrete optimization if the outer loop only reaches a continuous equilibrium point.

The following criterion may then be used to test the convergence of the inner loop

$$\varepsilon_E = \left[(E(V(t_k), V^k) - \hat{E}(V(t_k))) / E(V(t_k), V^k)\right]^2 \leq \delta_E \tag{21}$$

where δ_E is a small positive number and $\hat{E}(V(t_k))$ is the mean energy defined as

$$\hat{E}(V(t_k)) = \frac{1}{L_k+1}\sum_{i=0}^{L_k} E(V(t_k-(L_k-i)\Delta t), V^k) \tag{22}$$

where L_k is a positive integer, for example, $L_k = 10$.

To test the convergence of the outer loop, let's first define the equilibrium point.

Definition 1 (equilibrium point): Suppose $f(X)$ is approximated at X^k using its quadratic approximation $q(X, X^k)$ defined by (9). Suppose the $ANN(V,U,T^k,I^k,N)$ is constructed on the basis of the $q_E(X, X^k)$, as defined by (19), and V^k and U^k represent its starting state corresponding to X^k, that is, $X^k = \Psi(V^k)$ and $V^k = \Phi(U^k)$. Let $\hat{V}^k$ be a stable state of the $ANN(V,U,T^k,I^k,N)$. Then, $\hat{X}^k = \Psi(\hat{V}^k)$ is an equilibrium point of $f(X)$ if $\hat{X}^k = X^k$

As it is easier to test $f(\hat{X}^k) = f(X^k)$ than to test $\hat{X}^k = X^k$, the convergence of the outer loop may then be examined using the following criterion

$$\varepsilon_f = \left[(f(\hat{X}^k) - f(X^k)) / f(\hat{X}^k)\right]^2 \leq \delta_f, \qquad \delta_f > 0 \tag{23}$$

To find a discrete equilibrium point, the states of all the neurons need to be stablized at either of the two states, 0 and 1. Suppose the $ANN(V,U,T^k,I^k,N)$ is a finally constructed network at a point X^k and $\hat{V}^k$ is its stable state at which both the inner and the outer loops have converged. Then, the following criteria may be used to examine if $\hat{V}^k$ is at a corner of the neurons' state space, that is, for any $r=1,\cdots,N$,

$$either \quad \varepsilon_r^1 = (\hat{v}_r^k - 1.0)^2 \leq \delta_r^1 \quad or \quad \varepsilon_r^0 = (\hat{v}_r^k - 0.0)^2 \leq \delta_r^0 \quad for \quad \delta_r^1, \delta_r^0 > 0 \tag{24}$$

If all the neurons can pass the test, $\hat{V}^k$ is at the corner and $\hat{X}^k = \Psi(\hat{V}^k)$ is a discrete equilibrium point of $f(X)$. If any neuron, say v_r, fails to pass the test, we may further reduce the gain width u_{r0} using the following strategy in addition to the "annealing" schedule

$$u_{r0}^n = \tau u_{r0}^c \qquad 0 < \tau < 1 \tag{25}$$

where u_{r0}^c is the current gain width for v_r and u_{r0}^n is the new one. τ may be referred to as a control parameter for regulating the speed of the move of the $ANN(V,U,T^k,I^k,N)$'s stable state from the interior to a corner of the neurons' state space.

Obviously, the inner loop can always converge as $T_{ij} = T_{ji}$ and so can the third loop when all u_{r0} $(r=1,\cdots,N)$ are reduced to sufficiently small values. The quadratic convergence of the outer loop and the optimality of the equilibrium point can be proved as follows, though the proofs of the following Lemma 1, Lemma 2 and Theorem 1 are not listed because of space limitation.

Lemma 1: Suppose the sequence $\{X^k\}$ is generated by the artificial neural network algorithm. Then $\{X^k\}$ is characteristic of the following property, i.g., $\nabla f(X^k) + \nabla^2 f(X^k)(X^{k+1} - X^k) = 0$

Lemma 2: If $f(X)$ is twice continuously differentiable, then the equilibrium point $\hat{X}^k$ obtained by using the artificial neural network algorithm and defined by Definition 1 is a minimum of $f(X)$.

Theorem 1: If $\nabla^2 f(X)$ satisfies, in the neighbourhood Λ of the minimum X^*, the following Lipschitz condition, for any $X, Y \in \Lambda$, $\|\nabla^2 f(X) - \nabla^2 f(Y)\| \leq c\|X - Y\|$ where c is a constant, then the sequence $\{X^k\}$ generated by using the artificial neural network algorithm converges to X^* quadratically.

4. Constrained Optimization Based on Neural Network Algorithm

4.1 Constrained optimization and penalty method

A constrained discrete nonlinear optimization problem may be defined by

$$
\begin{aligned}
&\textit{min. } f(X) \\
&\textit{s.t. } X \in \Omega \cap X_1 \times \cdots \times X_i \times \cdots \times X_n \\
&\Omega = \{ X \mid h_j(X) = 0, j=1, \cdots, m_1; \ g_i(X) \geq 0, i=1, \cdots, m_2; \ X = [x_1 \cdots x_n]^T \} \\
&X_i = \{ x_i \mid x_i = \{ d_{i0}\, d_{i1} \cdots d_{ip} \cdots d_{iN_i} \}, d_{ip} > d_{i,p-1} \} \quad i=1, \cdots, n
\end{aligned}
\tag{26}
$$

where $f(X)$, $g_i(X)$ and $h_j(X)$ are twice continuously differentiable nonlinear functions.

Problem (26) can be transformed into a series of unconstrained optimization problems by the sequential unconstrained minimization technique[7]. A modified problem may then be written as

$$
P(\gamma_1, \gamma_2) \quad
\begin{cases}
\textit{min. } \Gamma(X, \gamma_1, \gamma_2) = f(X) + \gamma_1 H_1(X) + \gamma_2 H_2(X) \\
\textit{s.t. } X \in X_1 \times \cdots \times X_i \times \cdots \times X_n
\end{cases}
\tag{27}
$$

where Γ is the pseudo-objective function, H_1 the penalty function for the equality constraints $h_j(X) = 0$, $j=1, \cdots, m_1$, H_2 the penalty function for the inequality constraints $g_i(X) \geq 0$, $i=1, \cdots, m_2$, and γ_1 and γ_2 are the penalty coefficients.

To apply the artificial neural network algorithm to deal with the modified problem (27), $\Gamma(X, \gamma_1, \gamma_2)$ needs to preserve the continuity of the second derivatives. Thus, the exterior penalty method may be used to define the penalty function H_1, i.e.

$$
H_1(X) = \sum_{j=1}^{m_1} [h_j(X)]^2
\tag{28}
$$

To construct $H_2(X)$ for the inequality constraints, the quadratic extended interior penalty technique is adopted[3], which is defined by

$$
H_2(X) = \sum_{i=1}^{m_2} p_i(X), \quad p_i(X) =
\begin{cases}
1/g_i(X) & \text{if } g_i(X) \geq g_0 \\
1/g_0 \left[\left[g_i(X)/g_0 \right]^2 - 3g_i(X)/g_0 + 3 \right] & \text{if } g_i(X) \leq g_0 \\
g_0 = c_2 \sqrt{\gamma_2}
\end{cases}
\tag{29}
$$

g_0 is a positive transition parameter and c_2 is a positive constant[3].

The penalty function $H_2(X)$ is continuous up to its second derivatives throughout the design space[3]. It can then be concluded that the pseudo-objective function $\Gamma(X, \gamma_1, \gamma_2)$ preserves the continuity of the first and the second derivatives of the functions in the original problem (26).

4.2 Algorithm and convergence

It has been proven that the minima of the modified problem (27) converge to the minima of the original problem when $\gamma_1 \to \infty$ and $\gamma_2 \to 0$[7]. It has also been proven that $H_1(X) \to 0$ when $\gamma_1 \to \infty$ and $\gamma_2 H_2(X) \to 0$ when $\gamma_2 \to 0$[7]. In numerical calculation, the convergence is normally tested by comparing the penalty functions with the original objective function. The following criteria are thus adopted to test the convergence of the constrained optimization

$$
\varepsilon_{\gamma_1} = \left[H_1(X)/f(X) \right]^2 \leq \delta_{\gamma_1}, \quad
\varepsilon_{\gamma_2} = \left[\gamma_2 H_2(X)/f(X) \right]^2 \leq \delta_{\gamma_2}
\tag{30}
$$

where δ_{γ_1} and δ_{γ_2} are small positive number.

The algorithm for constrained discrete nonlinear optimization based on the artificial neural network algorithm can then be summarized as follows.

Step 1: Define a constrained discrete optimization problem as shown in (26). If $d_{i0} \neq 0$, take a linear displacement transformation so that $d_{i0} = 0$ for $i=1, \cdots, n$.

Step 2: Set initial values for the penalty coefficients γ_1^0 and γ_2^0, say, $\gamma_1^0 = \gamma_2^0 = 1$. Then, construct the modified problem $P(\gamma_1^0, \gamma_2^0)$ as shown in (27) where $H_1(X)$ and $H_2(X)$ are defined by (28) and (29). Let $\rho = 0$.

Step 3: Implement the artificial neural network algorithm to solve $P(\gamma_1^\rho, \gamma_2^\rho)$. The discrete minimum of $P(\gamma_1^\rho, \gamma_2^\rho)$ is denoted by $\bar{X}^\rho = \bar{X}(\gamma_1^\rho, \gamma_2^\rho)$. Note that in the algorithm X^0, V^0 and U^0 may be selected arbitrarily.

Step 4: If the convergence criteria (30) are both satisfied, $\bar{X}^\rho$ is regarded as a good approximation to the discrete minimum $\bar{X}^*$ of (26). So, let $\bar{X}^* = \bar{X}^\rho$ and stop.

Step 5: If the first part of (30) is not satisfied, let $\gamma_1^{\rho+1} = S_f \gamma_1^\rho$; if the second part of (30) is not satisfied, let $\gamma_2^{\rho+1} = \gamma_2^\rho / S_f$, where S_f is a scale factor larger than one, say $S_f = 10$. Let $\rho = \rho+1$ and $X^0 = \bar{X}^\rho$ and then go to Step 3.

In Step 5, $X^0 = \bar{X}^\rho$ can be chosen as a starting point for solving the newly constructed modified problem $P(\gamma_1^{\rho+1}, \gamma_2^{\rho+1})$. The computation experience has shown that such a choice of X^0 for $P(\gamma_1^{\rho+1}, \gamma_2^{\rho+1})$ can save considerable computation effort compared with assigning X^0 arbitrarily.

5. Examples

5.1 An unconstrained optimization problem

The following discrete optimization problem is artificially constructed

$$min. \ f(X) = \prod_{i=1}^{4}(x_i - x_i^*)^2 + \sum_{i=1}^{4}(x_i - x_i^*)^4 - 1/(1 + \sum_{i=1}^{4}(x_i - x_i^*)^2) + e^{\sum_{i=1}^{4}(x_i - x_i^*)^2}$$

$$s.t. \ X \in X_1 \times X_2 \times X_3 \times X_4, \quad X = [x_1 \ x_2 \ x_3 \ x_4]^T$$

$X_1 = \{ x_1 \mid x_1 = \{ 0, 0.025, 0.05, 0.075, \cdots, 2.5 \} \}$, $X_2 = \{ x_2 \mid x_2 = \{ 0, 0.05, 0.1, 0.15, \cdots, 5.0 \} \}$,

$X_3 = \{ x_3 \mid x_3 = \{ 0, 0.025, 0.05, 0.075, \cdots, 2.5 \} \}$, $X_4 = \{ x_4 \mid x_4 = \{ 0, 0.05, 0.1, 0.15, \cdots, 5.0 \} \}$

Obviously, the global minimum of this problem is $X^* = [x_1^*, x_2^*, x_3^*, x_4^*]^T$ and $f(X^*) = 0$.

Suppose $X^* = [1.73 \ 2.03 \ 1.23 \ 2.53]^T$. The problem is then solved using the artificial neural network algorithm. The distributed representation (6) is used to represent the four design variables. The values of the parameters in the algorithm are set as follows, $\delta_E = \delta_f = 1.0 \times 10^{-5}$, $\delta_r^1 = \delta_r^0 = 1.0 \times 10^{-6}$, $u_0 = 0.01$, $\Delta t = 0.005$, $\tau = 0.75$, $L_k = 10$.

Table 1   Convergence Process of $f(X)$

k	\multicolumn design variables X^k				objective function $f(X^k)$	k	design variables X^k				objective function $f(X^k)$
	x_1^k	x_2^k	x_3^k	x_4^k			x_1^k	x_2^k	x_3^k	x_4^k	
1	0.146	0.091	0.123	0.071	763043.65	13	0.566	1.354	0.238	1.819	30.522880
2	0.176	0.170	0.131	0.170	312845.73	14	0.699	1.493	0.296	1.972	14.500503
3	0.196	0.248	0.133	0.286	128856.94	15	0.875	1.616	0.392	2.103	6.718945
4	0.214	0.328	0.133	0.413	53269.14	16	1.080	1.726	0.545	2.218	2.888361
5	0.231	0.412	0.134	0.546	22107.20	17	1.299	1.843	0.753	2.340	1.037300
6	0.250	0.502	0.135	0.686	9220.45	18	1.552	1.970	1.021	2.470	0.165335
7	0.272	0.596	0.140	0.831	3879.37	19	1.693	2.030	1.188	2.532	0.006309
8	0.296	0.698	0.147	0.984	1651.34	20	1.725	2.029	1.222	2.530	0.000192
9	0.325	0.809	0.155	1.143	713.14	21	1.732	2.029	1.228	2.530	0.000014
10	0.361	0.931	0.166	1.308	313.48	22	1.732	2.029	1.228	2.530	0.000014
11	0.408	1.064	0.181	1.478	140.78						
12	0.473	1.207	0.204	1.650	64.80	$\bar{X}^*$	1.725	2.050	1.225	2.550	0.001710

The optimization process is shown in Table 1. It is clear from the table that 22 approximations of $f(X)$ are made so that 22 artificial neural networks are sequentially constructed. The generated sequence $\{X^k\}$ $(k=1,\cdots,21)$ converges to the global continuous minimum of $f(X)$, that is, $X^* \approx X^{22} = X^{21}$, and $\{f(X^k)\}$ $(k=1,\cdots,21)$ is a strictly monotonously decreasing sequence. The global discrete minimum $\bar{X}^* = [1.725\ \ 2.05\ \ 1.225\ \ 2.55]^T$ is obtained from X^* by operating the finally constructed $ANN(V,U,T^{22},I^{22},400)$ with $u_{r,0}$ being reduced by (25).

5.2 Three-bar truss problem

A three bar truss (TBT) design problem can be written as follows as shown in reference [11]

$$TBT^0 \begin{cases} min.\ f(X) = 2x_1 + x_2 + \sqrt{2}x_3 \\ s.t.\ \ X \in \Omega \cap X_1 \times X_2 \times X_3, \quad X = [x_1\ \ x_2\ \ x_3]^T \end{cases}$$

$$\Omega = \left\{ X \left| \begin{array}{l} g_1(X) = 1 - (\sqrt{3}x_2 + 1.932x_3)/(1.5x_1x_2 + \sqrt{2}x_2x_3 + 1.319x_1x_3) \geq 0 \\ g_2(X) = 1 - (0.634x_1 + 2.828x_3)/(1.5x_1x_2 + \sqrt{2}x_2x_3 + 1.319x_1x_3) \geq 0 \\ g_3(X) = 1 - (0.5x_1 - 2x_2)/(1.5x_1x_2 + \sqrt{2}x_2x_3 + 1.319x_1x_3) \geq 0 \\ g_4(X) = 1 + (0.5x_1 - 2x_2)/(1.5x_1x_2 + \sqrt{2}x_2x_3 + 1.319x_1x_3) \geq 0 \end{array} \right. \right\}$$

$$X_i = \{\ x_i\ |\ x_i = \{\ 0.1\ \ 0.2\ \ 0.3\ \ 0.5\ \ 0.8\ \ 1.0\ \ 1.2\ \}\ \} \quad i=1,2,3$$

As the design variables x_i $(i=1, 2, 3)$ take irregularly discrete values with $d_{i0} = 0.1 \neq 0$, we first augment the discrete values as follows

$$X_i = \{\ x_i\ |\ x_i = \{\ 0.1\ \ 0.2\ \ 0.3\ \ \underline{0.4}\ \ 0.5\ \ \underline{0.6}\ \ 0.8\ \ 1.0\ \ 1.2\ \}\ \} \quad i=1,2,3$$

Then, let's take the linear displacement transformation, $x_i' = x_i - 0.1$ for $i=1, 2, 3$, or $X = X' + e(0.1)$ where $X' = [x_1'\ \ x_2'\ \ x_3']^T$ and $e(0.1) = [0.1\ \ 0.1\ \ 0.1]^T$. The original TBT^0 problem is thus transformed into the following identical problem

$$TBT^1 \begin{cases} min.\ f(X' + e(0.1)) \\ s.t.\ \ X' \in \Omega' \cap X_1' \times X_2' \times X_3' \end{cases}$$

$$\Omega' = \{\ X'\ |\ g_i(X' + e(0.1)) \geq 0, \quad j=1,\cdots,4 \}$$

$$X_i' = \{\ x_i'\ |\ x_i' = \{\ 0.0\ \ 0.1\ \ 0.2\ \ 0.3\ \ 0.4\ \ 0.5\ \ 0.7\ \ 0.9\ \ 1.1\ \}\ \} \quad i=1,2,3$$

The TBT^1 is solved using the algorithm for constrained discrete nonlinear optimization with the augmented and transformed design variables being represented by the intermediate representation (7). The values of the parameters required in the algorithm are set as in the first example. In addition, let $\gamma_2^1 = 1$, $\delta_{\gamma_2} = 0.05$ and $S_f = 10$.

Table 2   Convergence Process of $\Gamma(X', \gamma_2^3)$ for γ_2^3=0.01

k	design variables X'^k			$\Gamma(X'^k, \gamma_2^3)$	$\varepsilon_{\gamma_2}^k$	k	design variables X'^k			$\Gamma(X'^k, \gamma_2^3)$	$\varepsilon_{\gamma_2}^k$
	$x_1'^k$	$x_2'^k$	$x_3'^k$				$x_1'^k$	$x_2'^k$	$x_3'^k$		
1	0.674	0.529	0.581	819.533	67575.589	16	1.005	0.615	0.226	3.628	0.005
2	0.796	0.577	0.595	99.552	775.400	17	1.012	0.605	0.204	3.602	0.005
3	0.846	0.603	0.601	11.726	5.153	18	1.019	0.593	0.185	3.579	0.005
4	0.879	0.614	0.607	4.563	0.059	19	1.025	0.582	0.167	3.556	0.006
5	0.905	0.621	0.614	4.233	0.017	20	1.031	0.570	0.152	3.536	0.006
6	0.930	0.633	0.615	4.142	0.008	21	1.036	0.558	0.137	3.516	0.006
7	0.948	0.647	0.603	4.104	0.005	22	1.041	0.546	0.124	3.499	0.006
8	0.957	0.659	0.560	4.042	0.004	23	1.046	0.535	0.112	3.482	0.006
9	0.958	0.661	0.493	3.949	0.004	24	1.051	0.524	0.102	3.467	0.006
10	0.962	0.658	0.432	3.870	0.004	25	1.055	0.514	0.092	3.453	0.007
11	0.968	0.654	0.384	3.811	0.004	26	1.059	0.504	0.083	3.440	0.007
12	0.975	0.649	0.343	3.764	0.005	27	1.062	0.496	0.074	3.429	0.007
13	0.983	0.642	0.308	3.723	0.005	28	1.065	0.488	0.066	3.418	0.007
14	0.991	0.634	0.277	3.688	0.005						
15	0.998	0.625	0.250	3.657	0.005	$\bar{X}'^3$	1.100	0.400	0.000	3.358	0.011

For $\gamma_2^1 = 1$, the modified problem $P(\gamma_2^1)$ attains its discrete optimum after 17 iterations (k=17) and $\bar{X}(\gamma_2^1) = [1.1 \ 1.1 \ 0.7]^T$. For $\gamma_2^2 = 0.1$, k=4 and $\bar{X}(\gamma_2^2) = [1.1 \ 1.1 \ 0.0]^T$. For $\gamma_2^3 = 0.01$, the optimization process is demonstrated in Table 2. The starting point X^0 for the modified problem $P(\gamma_2^{p+1})$ is set by $X^0 = \bar{X}^p$ where $\bar{X}^p$ is the discrete minimum of $P(\gamma_2^p)$. For each of the three given penalty coefficients, i.e., $\gamma_2^p = 1, 0.1, 0.01$, the pseudo-objective function values $\{\Gamma(X'^k, \gamma_2^p)\}$ form a strictly monotonously decreasing sequence before the continuous minimum is obtained. The algorithm reaches the global discrete minimum of $f(X)$ in three response surfaces (ρ=3), that is, $\bar{X}^* = \bar{X}^3 + e(0.1) = [1.2 \ 0.5 \ 0.1]^T$.

6. Concluding Remarks

The artificial neural network approach proposed in this paper provides an alternative way for dealing with a discrete nonlinear optimization problem. This approach is easy to implement and can always generate a sequence of designs which quadratically converge to the discrete minima of the problem. In addition, the nature of the algorithm provides scope for parallel computing and consequent gains in computational efficiency. The two examples presented in this paper have demonstrated these features of the new approach. As the number of the neurons required only increases linearly with the number of design variables, this new approach may be applied to deal with large discrete nonlinear optimization problems.

Acknowledgment

This work is part of a project to develop a general multiple criteria decision support environment supported by the UK Science and Engineering Research Council under Grant no.GR/F 95306

References

[1] Aleksander, I. and Morton, H., An Introduction to Neural Computing, Chapman and Hall, London, 1990.

[2] Chiu, C. Maa, C.-Y. and Shanblatt, M.A., "An Artificial Neural Network Algorithm for Dynamic Programming," International Journal of Neural Systems, Vol.1, No.3, 1990, pp.211-220.

[3] Haftka, R.T. and Starness Jr, J.H., "Applications of a Quadratic Extended Interior Penalty Function for Structural Optimization," AIAA Journal, Vol.14, No.6, 1976, pp.718-724.

[4] Hopfield, J.J. and Tank, D.W., ""Neural" Computation of Decisions in Optimization Problems," Biological Cybernetics, Vol.52, 1985, pp.141-152.

[5] John, K.V., Ramakrishnan, C.V. and Sharma, K.G., "Optimum Design of Trusses from Available Sections — Use of Sequential Linear Programming with Branch and Bound Algorithm," Engineering Optimization, Vol.13, 1988, pp.119-145.

[6] Kishi, M. Suzuki, T. and Hosoda, R., "Structural Design by Neuro-optimizer," Proceedings of the PRADS'92 Conference, Newcastle upon Tyne, U.K., Elsevier Publications, 1992, pp. 2.940-2.952.

[7] Minoux, M., Mathematical Programming: Theory and Algorithms, John Wiley and Sons, 1986.

[8] Olsen, G.R. and Vanderplaats, G.N., "Method for Nonlinear Optimization with Discrete Design Variables," AIAA Journal, Vol.27, No.11, 1989, pp.1584-1589.

[9] Ringertz, U.T., "On Methods for Discrete Structural Optimization," Engineering Optimization, VOl.13, 1988, pp.47-64.

[10] Schmit, L. and Fleury, C., "Discrete-Continuous Variable Structural Synthesis Using Dual Methods," AIAA Journal, Vol.18, 1980, pp.1515-1524.

[11] Shin, D.K. Gurdal, Z. and Griffin, Jr., O.H., "A penalty Approach for Nonlinear Optimization with Discrete Design Variables," Engineering Optimization, Vol.16, 1990, pp.29-42.

[12] Tank, D.W. and Hopfield, J.J., "Simple "Neural" Optimization Networks: An A/D/ Converter, Signal Decision Circuit, and a Linear Programming Circuit," IEEE Transactions on Circuit and Systems, Vol.CAS-33, No.5, 1986, pp.533-541.

OPTIMIZING THE STRUCTURE OF A PARTITIONED POPULATION

Andrzej M.J. Skulimowski

Institute of Automatic Control

University of Mining & Metallurgy, Cracow, Poland

and

Institut fuer Wirtschaftsinformatik, Hochschule St. Gallen,
St. Gallen, Switzerland

Abstract. In this paper we propose a mathematical model for the controlled evolution of partitioned population. We assume that the values of classification criteria may change in a direct response to external actions. The change of attributes may be controlled by the decision-maker whereby an improvement of the criteria values bears certain cost. Thus we get a bilevel multicriteria optimization problem : an optimal allocation of resources at the lower level, and finding the related nondominated outputs surpassing a reference point q at the higher level. A concrete problem of this type, motivated by ecological and economical applications, will be discussed in more detail, namely optimizing the structure of a finite population Ω by assuring that after a fixed time T a maximal number of its elements is characterised by nondominated values of criteria. Assuming that Ω consists of N elements, the solution to this problem is equivalent to solving parallelly N discrete dynamic programming problems sharing the same resources.

AMS Subject Classification. Primary : 90C29, **Secondary :** 90C35, 90A06, 90C39

1. Introduction

A new family of mathematical models for the dynamical decision problems with alternatives which change their properties in a direct response to external actions, has been described in [1]. As an application of the theory there presented, and motivated by real-life ecological and economical applications, in this paper we will formulate and investigate the problem of optimizing the structure of a given finite population Ω by assuring that after a fixed time T a maximal number of elements of Ω is characterized by nondominated values of criteria. Assuming that Ω consists of N elements, the solution to this problem is equivalent to solving·parallelly N discrete dynamic programming problems sharing the same resources. We give also examples of other structural optimality conditions to be fulfilled by Ω.

We assume that the elements of a finite population Ω are classified according to N classification criteria $F:=(F_1,\ldots F_N)$ which admit their values from a

*/

Research supported in part by the Swiss National Fund for Scientific Research, Grant No. 12-30240.90

given partially ordered finite set V. Let $V_i := \{v_{i,1}, v_{i,2}, \cdots v_{i,c(i)}\}$ denote the set of values of F_i ordered from the least to the most preferred one. Then V is the Cartesian product $V_1 \times \ldots \times V_N$ with the coordinatewise partial order "$\{$". The evolution of values of F for a fixed $\omega \in \Omega$ can be represented as a sequence of transitions (v->w) which are results of the control actions $u(t) \in U$ undertaken by the system's supervisor on a discrete time interval $[t_0, T]$:

$$F(t+1)(\omega) = \phi(F(t)(\omega), u(t)), \quad \text{for each } \omega \in \Omega, \ t \in [t_0, T-1]. \tag{1}$$

Each element ω_α of Ω may pass to another class on the time interval $[t, t+1]$ if a control $u_\alpha(t) \in U(t)$ has been applied to ω_α according to the control law (1).

Besides of controlled transitions we will distinguish the deterministic uncontrolled transitions which may not be influenced, as e.g. passing to the following age classes, random transitions occurring spontaneously, and non-admissible transitions. The classes of controllable and random transitions need not be disjoint. Thus, the evolution of attributes may be modelled in a manner similar to the discrete-event systems described in [3], whereas the values of F play the role of states. Since the set of alternatives Ω, the values of criteria V, and time are all discrete, such systems are called *D-D-D-systems*.

2. Evolution of attributes : the transition and cost patterns

The task of supplying all necessary information concerning the transitions between the values of F might be considerably simplified if it were possible to find a convenient description of the transfer function ϕ from (1), and to identify non-admissible transitions before starting the quantitative solution process. Further, we would like to reduce the computational complexity of the general problem by decomposing it into several subproblems, each one of them referring to the single criterion F_i, i=1,..N. This would be possible if the characterization of transitions between the values of $(F_1, \ldots F_N)$ could be derived from the properties of the single criteria, $F_1(t)$ through $F_N(t)$, considered separately. Below we will show that this goal can be achieved under some additional assumptions concerning the set Ω and the criterion F by introducing so-called *transition* and *cost patterns*.

Let us fix the moment of time $t \in [t_0, T]$ and let $V_i := \{v_{i,1}, v_{i,2}, \cdots v_{i,c(i)}\}$ denote the set of values of the criterion F_i ordered from the least to the most preferred one. If we know which transitions between the values of the criterion F_i for an $\omega \in \Omega$ are at all possible on the time interval $(t, t+1]$, we can define for F_i and x the *transition pattern* as a quadratic 0-1 matrix $P(F_i)(\omega) = [p^i_{jk}(\omega)]$ with the following coefficients :

$$p^i_{jk}(\omega) = \begin{cases} 1 & \text{iff } \omega \in \Omega \text{ may change its classification in one} \\ & \text{time step from j-th to k-th attribute of } F_i \\ 0 & \text{otherwize.} \end{cases} \tag{2}$$

for $j, k = 1, \ldots c(i)$, $i = 1, \ldots N$.

Observe that the dimension of $P(F_i)(\omega)$ equals to the number of elements of V_i, $c(i)$, and its columns indicate the admissible transitions from an appropriate fixed starting value of F. To each transition pattern $P(F_i)$ one can associate the digraph $G(F_i)$ such that $P(F_i)$ is its structural matrix. In general, the transition patterns may vary on the interval $[t_0, T]$, being thus functions of both, ω and t.

Transitions from v to w on the time interval $(s, t]$ may be regarded as pairs (v, w) and will be denoted by $v\rightarrow w$. By a *superposition* of the transitions $\xi_1 := v_1\rightarrow v_2$ on the interval $(t_1, t_2]$, and $\xi_2 := v_2\rightarrow v_3$ on the interval $(t_2, t_3]$ we will mean the transition $\xi : = v_1\rightarrow v_3$ on $(t_1, t_2]$, and denote it by $\xi = \xi_1 \circ \xi_2$. Suppose now that $v_1 = (v_{11}, v_{12})$. By the *composition* of the transitions $\xi_1 := v_{11}\rightarrow v_2$ on $(t_1, t_2]$ and $\xi_2 := v_{12}\rightarrow v_3$ on the same interval $(t_1, t_2]$, we will mean the transition $v_1\rightarrow (v_2, v_3)$ on $(t_1, t_2]$. To denote compositions we will use the notation $\xi_1 c \xi_2$. Let $\tilde{\xi}_1 := (v_{11}, v_{12})\rightarrow (v_2, _{12})$ and $\tilde{\xi}_2 := (v_{11}, v_{12})\rightarrow (v_{11}, v_3)$. Then, formally, $\xi_1 c \xi_2 = \tilde{\xi}_1 \circ \tilde{\xi}_2$, the diversity between composition and superposition being expressed by the associated time intervals. Observe that the superpositions and compositions describe sequential and parallel processing of transitions, respectively. By definition, the compositions are always admissible in one time step, the superpositions may, but need not necessarily have this property. To assure a minimal number of non-zero coefficients in $P(F)(\omega, t)$, it is convenient to include in the transition patterns only the transitions which may not be represented as compositions of other admissible transitions.

Transitions lasting several time steps may often be represented as superposition of one-step transitions. If it is not so, they can still be considered within the same framework by introducing the *intermediate* or *wait values* of F. This question is considered in more detail in [4].

Now we will introduce several properties of the decision process (Ω, F, ϕ) which will be used in the further analysis of the problem.

Definition 2.1. We will say that the set of alternatives Ω is *homogeneous* with respect to F at the moment $t\in[t_0, T]$, iff

$$\forall 1\leq i\leq N \ \forall x, y\in\Omega : P(F_i)(x) = P(F_i)(y). \tag{3}$$

If (3) is satisfied for all $t\in[t_0, T]$, we will call Ω simply homogeneous. ∎

If Ω is not homogeneous but handling a separate transition pattern for all alternatives would be computationally inefficient then one may consider instead the boolean product of transition patterns for all $x\in\Omega$. Generally, in models of real-life dynamical systems the transition patterns depend on the discretization of time. Moreover, as we already noted, they may depend on time itself. In the sequel we will usually admit the assumption that the decision process (Ω, F, ϕ) is *stationary*, according to the following definition :

Definition 2.2. If for each $\omega\in\Omega$ and $1\leq i\leq N$, $P(F_i)(\omega)$ remains constant on the interval $[t_0, T]$ then the decision process (Ω, F, ϕ) will be called *stationary*. ∎

Observe that the stationarity assumption is equivalent to the fact that the function ϕ from (1) does not depend on time t. Another important set of properties concerns the independence of the criteria $F_1 \ldots F_N$.

Definition 2.3. The criteria $F_1, \ldots F_N$ are *evolution independent* at $x \in \Omega$ and $t \in [t_0, T]$, by definition it means that any transition

$$F(t)(x) := (v_{1,i(1)}, \ldots v_{N,i(N)}) \rightarrow (v_{1,j(1)}, \ldots v_{N,j(N)}) = F(t+1)(x)$$

is admissible iff for each $1 \leq k \leq N$ the transitions $v_{k,i(k)} \rightarrow v_{k,j(k)}$ are admissible, i.e. iff $p_{i(k),j(k)}^k(x)(t) = 1$. ∎

The criteria $F_1, \ldots F_N$ will be called simply *evolution independent* iff the above holds for all $x \in \Omega$ and $t \in [t_0, T]$. Roughly speaking, the criteria are evolution independent iff the admissibility of transitions between the values of F_k, $1 \leq k \leq N$, is not affected by the present values of all remaining criteria. It is easy to observe that the following fact is true :

Proposition 2.1. If the decision process (Ω, F, ϕ) is homogeneous and stationary, and for certain $x_0 \in \Omega$, $t \in [t_0, T]$, $F_1, \ldots F_N$ are evolution independent at (x_0, t) then $F_1, \ldots F_N$ are evolution independent. ∎

In the sequel we will always assume that the criteria concerned are evolution independent.

Example 2.1. Suppose that Ω is the population of citizens of a city and one of the objectives s is the age scale with the attributes $v_1 = [0, 20]$, $v_2 = (20, 40]$, $v_3 = (40, 60]$, and $v_4 = (60, \infty)$, denoting the age in years of a single individual. If all the time discretization steps in the process (1) are less than 20 years, which is usually the case, then the transition pattern $P(s) = [p_{jk}]$ is the matrix

$$P(s) = \begin{bmatrix} 1 & 1 & 0 & 0 \\ 0 & 1 & 1 & 0 \\ 0 & 0 & 1 & 1 \\ 0 & 0 & 0 & 1 \end{bmatrix}.$$

No transitions are controllable, unless we dispose a relativistic vehicle to force remaining within the same age class (elements on the main diagonal would then correspond to controllable transitions). ∎

An important feature of the transition patterns for evolution independent criteria consists in the fact that it is sufficient to determine the patterns for the single criteria only, while the transition patterns for the vector criterion may be calculated basing on the following

Proposition 2.2. Assume that the evolution independent criteria F_i and F_j are defined on a homogeneous population Ω with the transition patterns

$$P_i = [p_{k,1}^{(i)}] := P(F_i)(t) \in M_{c(i),c(i)} \text{ and } P_j = [p_{m,n}^{(j)}] := P(F_j)(t) \in M_{c(j),c(j)},$$

for certain fixed $t \in [t_0, T]$, respectively. Then the transition pattern $P_{ij} := P(F_i, F_j)(t)$ of the vector criterion $F_{ij} := (F_i, F_j)$, is the block matrix

$$P_{ij} = [p_{k,1}^{(i)} P_j]_{k,1=1}^{c(i)} \in M_{c(i)c(j), c(i)c(j)} \qquad (4)$$

(the block product of P_i and P_j), where $c(i)$ and $c(j)$ denote the number of admissible values of the criteria F_i and F_j, respectively. The values of F_{ij}, $v_{kl} = (v_k^{(i)}, v_l^{(j)})$, labelling the rows and columns in P_{ij}, are ordered lexicographically with the first coordinate more relevant than the second. ∎

Proof : Suppose first that the transitions $(v_{i,k(i)} \rightarrow v_{i,1(i)})$ and $(v_{j,k(j)} \rightarrow v_{j,1(j)})$ are both admissible, i.e. $p_{k(i),1(i)}^{(i)} = 1$ and $p_{k(j),1(j)}^{(j)} = 1$. Then from the evolution independency assumption (Def. 2.3) it follows that the transition $\xi := (v_{i,k(i)}, v_{j,k(j)}) \rightarrow (v_{i,1(i)}, v_{j,1(j)})$ is admissible. According to the construction of P_{ij} (cf. (4)), the element corresponding to ξ in P_{ij}, $p_{m,n}^{(ij)}$ with $m = k(i)c(j) + k(j)$ and $n = 1(i)c(j) + 1(j)$, is the $(k(j), 1(j))$-th coefficient of the block $p_{k(i),1(i)}^{(i)} P_j$, which is equal to 1 since $p_{k(i),1(i)}^{(i)} = 1$, and $p_{k(i),1(i)}^{(i)} P_j = P_j$.

If at least one from the above simple transitions is non-admissible then $p_{k(i),1(i)}^{(i)} = 0$ or $p_{k(j),1(j)}^{(j)} = 0$. In the first case the block $p_{k(i),1(i)}^{(i)} P_j$ of P_{ij} contains only zero elements, in the second, its $(k(j),1(j))$-th coefficient, $p_{m,n}^{(ij)}$, is equal to zero. However, from the definition of the evolution independent criteria it follows that any transition between values of (F_i, F_j) must be represented as a superposition of simple admissible transitions cannot be admissible, therefore the transition ξ corresponding to the zero coefficient $p_{m,n}^{(ij)}$ is not admissible.

Now, let us fix a coefficient $p_{m,n}^{(ij)}$ of P_{ij}. Then there exist integers $k1(m,n), 11(m,n) \in [1, c(i)]$ and $k2(m,n), 12(m,n) \in [1, c(j)]$ such that $p_{mn}^{(ij)}$ is the $(k2(m,n), 12(m,n))$-th coefficient of the block $p_{k1(m,n),11(m,n)}^{(i)} P_j$ of P_{ij}, i.e. to $p_{m,n}^{(ij)}$ there can be associated the transition $(v_{i,k1(m,n)}, v_{j,k2(m,n)}) \rightarrow (v_{i,11(m,n)}, v_{j,12(m,n)})$, and $p_{m,n}^{(ij)}$ determines its admissibility, as shown in the first part of the proof. Hence we conclude that the above characterization of the of P_{ij} as transition pattern (2) for (F_i, F_j) is complete. ∎

Corollary 2.1. If the transition patterns of two evolution independent criteria s and q, $P(s)$ and $P(q)$, have $p(s)$ and $p(q)$ non-zero elements, respectively, then the transition pattern for (s,q), $P(s,q)$ contains at most $p(s)p(q)$ non-zero elements. ∎

Consequently, the transition patterns for any finite number of evolution independent criteria, $F_1 \ldots F_n$, are sparse block matrices which can be constructed making recursively use of Proposition 2.2.

2.1. Assignment of controls and costs to admissible transitions.

Let us fix $\omega\in\Omega$, and $t\in[t_0,T]$, and assume that the indices of the coefficients of $P(F)(\omega,t)$ are ordered lexicographically. Then to each admissible transition $v\to w$ we can associate a control $u_m(t)$ and its cost $J(v,w,\omega,t) := J(u_m,\omega,t)$, where the cardinal $m:=m(v,w)$ is the ordinal number of that control from the list $U:=\{u_1,\ldots u_M\}$ which is responsible for the change from v to w.

If $J(v,w,\omega,t)$ does not depend on the past transitions then, analogously to the transition patterns, for each $\omega\in\Omega$, $t\in[t_0,T]$, and each criterion F_i, one can define the *cost pattern* $J(F_i)(\omega,t)$, as a function associating to each transition $v_{i,k}\to v_{i,l}$ realizable between t and $t+1$, the cost of applying the control $u_m:=u(i,k,l,\omega,t)$ causing the change from $v_{i,k}$ to $v_{i,l}$,

$$J(F_i,v_{i,k},v_{i,l},\omega,t):=\tilde{J}_i(u_m,\omega,t).$$

Hence, the cost pattern for F_i can be represented as the $c(i)\times c(i)$ real matrix defined as follows :

$$J_{kl}^i(\omega,t) := \begin{cases} \tilde{J}_i(u(i,k,l,\omega,t),\omega,t) & \text{iff } v_{i,k} \; v_{i,l} \text{ is admissible} \\ \infty & \text{otherwise} \end{cases} \tag{5}$$

Consequently, the cost pattern for $F_1,\ldots F_N$, $J(F_1,\ldots F_N)(\omega,t)$, is the $c(1)\ldots c(N) \times c(1)\ldots c(N)$ real matrix storing the costs of transitions between the values of the vector criterion $F:=(F_1,\ldots F_N)$, i.e. $J_{i(1),\ldots,i(N),j(1),\ldots,j(N)}(t,\omega)$ is the cost of changing the value $(v_{1,i(1)},\ldots,v_{N,i(N)})$ of F to $(v_{1,j(1)},\ldots v_{N,j(N)})$, or it is undefined iff such transition is non-admissible. Hence it follows that the structure of the cost patterns is closely related to the transitions patterns whereby only those coefficients of J which correspond to a "1" in P are finite. Thus, in a machine implementation of the above decision process the transition patterns may serve as addresses of those elements of $J(F_1,\ldots F_N)$ which has to be stored in the memory. Moreover, observe that the zero coefficients of J correspond usually to non-controllable transitions.

For the evolution independent criteria $F_1,\ldots F_N$ an important role is played by the following condition :

Definition 2.4. The cost function $J(F_1,\ldots F_N)$ satisfies the *cost additivity condition* iff for any $\omega\in\Omega$, $t\in[t_0,T]$, the cost of any admissible transition $v\to w$, where $v:=(v_{1,i(1)},\ldots v_{N,i(N)})$ and $w:=(v_{1,j(1)},\ldots v_{N,j(N)})$, is the sum of changing the single criteria values, i.e.

$$J(v,w,\omega,t) = \sum_{k=1}^{N} J(v_{k,i(k)},v_{k,j(k)},\omega,t). \quad \blacksquare \tag{7}$$

The above condition lets us consider each transition as a composition of simple transitions while computing the optimal evolution strategy for the elements of Ω. This, in turn, allows to omit the operations on $J(F)$, using instead $J(F_i)$, for $i=1,..N$.

3. The evolution problem for a single population element

In this section we will discuss in more detail the following basic problem corresponding to the trivial population Ω ($\#\Omega=1$) :

Problem 3.1. For an $\omega\in\Omega$ find the optimal allocation of resources $u(1),...u(\tau)$, to achieve or surpass in the minimal time τ and at minimal cost one of the reference points $q\in Q$ by the value of $F(\tau)(\omega)$.

In a more rigorous setting, let Ω be the finite set of admissible alternatives at time $t_0:=1$, and $F_1,..F_N$ the criteria functions defined on Ω with values in the discrete sets V_i with the partial order "$\{_i$" for $i=1,..N$. Similarly as in the previous sections denote by F the vector criterion $F:=(F_1,..F_N)$, $F:\Omega \to V$, valued in the cartesian product $V:=V_1\times...\times V_N$ with the coordinatewise partial order "$\{$". For a fixed $\omega\in\Omega$, the values of criteria on ω may vary according to (1), i.e.

$$F(t+1)(\omega)=\phi(F(t)(\omega),u(t),t), \quad \text{for } t\in[t_0,T-1].$$

Our task consists in finding an $\omega\in\Omega$, a $\tau\in[t_0,T]$, and a sequence of controls $u(t_0),...u(\tau)$, so that :

(i) $((F_1(t_0),..F_N(t_0))(\omega)$ is nondominated in V and $((F_1(\tau),..F_N(\tau))(\omega) \{ q$ (8)

for certain reference point $q\in Q$;

(ii) τ with the property (8) is minimal in $[t_0,T]$ for a fixed ω; such value will be denoted $\tau(\omega)$;

(iii) $$\sum_{t_0\le t\le\tau(\omega)} J(u(t),\omega,t)$$ (9)

is minimal on the set Λ defined as follows :

$\Lambda:= \bigcup_{w\in\Omega} \{w\}\times\{y\in U^s : \tau(w) \text{ and } F(\tau(w))(w) \text{ satisfy (i) and (ii)}\}$,

where $s:= \tau(w)-t_0$, and $y:=(u(t_0),...u(\tau(w)))$.

Observe that, according to (1), $F(\tau(w))(w)$ and $\tau(w)$ are indirect functions of y. The minimal value of (9) will be denoted by $J_{min}(\omega)$.

(iv) $((F_1(\tau(\omega)),..F_N(\tau(\omega)))(\omega), \tau(\omega), J_{min}(\omega))$ is nondominated in $V\times R^2$ with the coordinatewise partial order.

The general Problem 3.1 consists of two tasks : finding an optimal alternative ω, and a sequence of controls y assuring the achievement of q at a minimal cost. Each alternative $\omega\in\Omega$ is characterized by the minimal time $\tau(\omega)$ and the minimal cost $J_{min}(\omega)$ of achieving or surpassing q. Consequently, if for each $\omega\in\Omega$ one knows these minimal parameters, then the final choice of ω is a bicriteria trade-off between the cost and time, which can be made using one of well-known interactive bicriteria decision-making methods.

We will present a solution to the above problem for the evolution processes satisfying the following assumptions :

(i)   the decision process (Ω, F, ϕ) is stationary and homogeneous;

(ii)   the criteria F are evolution independent;

(iii) the costs of transitions satisfy the cost additivity condition;

(iv) the reference set $Q \subset V$ can be represented in the form

      $Q := \{p \in V : p \nmid q\}$ for certain $q \in V$.

As the first step of the solution, below we will show, how can one determine $\tau(\omega)$ and $J_{min}(\omega)$ for a fixed ω.

3.1. Solving the single evolution problem.

Let us admit the above assumptions (i)-(iv), fix an $\omega \in \Omega$, and let $f_0 :=$ $F(t_0)(\omega)$. Further, let us consider a directed network $G = (V, E)$, where the nodes V can be identified with the set $V = V_1 \times \ldots V_N$ of potential values of F, while the edges $e \in E \subset V^2$ are determined by the transition patterns $P_1, \ldots P_N$, $P_i := P(F_i)(\omega)$, for $i = 1, \ldots N$, in the following way :

$$e = (f, g) \in E \Leftrightarrow [f \neq g \text{ and } \exists! \ j \in \{1, \ldots N\} \ f_i = g_i \text{ for } i \in \{1, \ldots N\} \setminus \{j\} \text{ and } P_j(f_j, g_j) = 1]$$
$$\text{or} \quad [f = g \text{ and } \forall \ i \in \{1, \ldots N\} \ P_i(f_i, g_i) = 1.] \tag{10}$$

Thus, the edges of G correspond to the simple transitions between the values of F or may be loops. Additionally, the edges of G are equipped with quantitative labels describing the time θ_i and the cost of transition J_i, and qualitative labels c_i indicating whether the corresponding transition is forced or controllable.

Hence, the following observation is straightforward :

Proposition 3.1. The transition between two values of F, f and g, is possible iff the nodes corresponding to f and g in G can be connected by a path. ∎

As a corollary from Prop. 2.2 we get

Proposition 3.2. The graph G is the cartesian product of graphs $G_1, \ldots G_N$, corresponding to the criteria $F_1, \ldots F_N$ and their transition patterns $P_1, \ldots P_N$, respectively. Its structural matrix is given as the block product of $P_1, \ldots P_N$, $P_{1, \ldots N}$. ∎

Hence it follows

Theorem 3.1. The solution to the Problem 3.1 for a single alternative $x \in \Omega$ can be found as a bicriteria shortest path in G between f_0 and the reference set $Q := \{v \in V : q \nmid v\}$. ∎

The solution algorithm which can be derived from the above Prop. 3.1. and 3.2 and Thm. 3.1 may be presented as follows :

Algorithm 3.1.

The input data :

The transition and cost patterns for $F_1, \ldots F_N$, $P_1, \ldots P_N$, and $J_1, \ldots J_N$, respectively. The time horizon T, the starting value f_0, the reference point $q \in V$, the reference set Q.

Step 1. Augment the transition patterns by the time-distributed transitions, applying the procedure presented in Sec. 2.3.

Step 2. Check whether the criteria are evolution independent.
If yes : construct the network G using the Prop. 2.2 and 3.2;
otherwise : set manually all edges of G.

Step 3. Check whether it exists a path joining f_0 and q, or any other $p \in Q$
(i.e. check whether Q is attainable from f_0).

If not, return to the communication shell to let the decision-maker define new reference point or to undertake another action.

Step 4. Determine the set D containing all bicriteria shortest paths between f_0 and all $p \in Q$, using the bicriteria shortest path algorithm.

Find the set of nondominated points of D, P(D).

Step 5. Select a compromise strategy from P(D) using any bicriteria trade-off procedure. ∎

3.2. Choice from among multiple evolving alternatives.

In the present setting we assume that at the moment t_0 the decision-maker should choose that alternative $\omega_0 \in \Omega$ which gives the best chances to be improved till the time T so as it were not worse than q. After simulating the evolution of F(t)(ω) over time $t \in [t_0, T]$, one chooses ω_0, which will be called *prospectful alternative*, and starts investing in its development, by undertaking the actions $u(t_0), \ldots u(T-1)$, without taking care what happens with all remaining alternatives. This solution procedure implies the following :

Theorem 3.2. To select the prospectful alternative and the best strategy in Problem 3.1 for stationary homogeneous processes with evolution independent criteria, it is necessary to solve the simultaneous bicriteria shortest path problem for the set of starting points $V_0 := \{f \in V : f = F(t_0)(\omega)$ for all $\omega \in \Omega\}$ and Q as the set of terminal points. ∎

As the simultaneous shortest path algorithm one can apply a combination of the well-known Dijkstra algorithm and the bicriteria shortest path method [2]. The above presented procedure will be illustrated by the following example.

The solution methods for decision processes which do not satisfy the stationarity, homogenity, or evolution independence assumptions are discussed in [4].

4. Optimizing the population structure

The above presented framework may be applied to solve a variety of decision problems. Here, we will focus our attention on the problem of optimizing the structure of a population Ω by assuring that after a fixed time T a maximal number of elements of Ω is characterized by nondominated values of criteria.

We assume that the elements of a finite population Ω are classified according to N ordered classification criteria $F_1, \ldots F_N$. Each element ω_α of Ω may pass to another class on the time interval $[t, t+1]$ if according to (1) a control $u_\alpha(t) \in U(t)$ has been applied to ω_α. All transitions can be described by $M := \#\Omega$ equations of type (1) sharing the same resources :

$$\sum_{1 \leq \alpha \leq M} b_\alpha(t) J(u_\alpha(t), \omega_\alpha, t) \leq u_t, \tag{11}$$

where $b_\alpha(t) \geq 0$ for all α. Let us note that the case $b_\alpha(t) := 1/x_{ti}$ with $x_{ti} := \#\{w \in \Omega: F(t)(w) = F(t)(\omega)\}$ corresponds to the situation where the same control $u(t)$ acts simultaneously on all elements of Ω characterized by the same values of criteria.

At the macroscopic level the evolution of Ω may be described by the following discrete controlled dynamical system

$$\begin{aligned} x_{t+1} &= A_t x_t + B_t w_t + \eta_t, \\ z_{t+1} &= C_t x_t + D_t w_t + \zeta_t, \end{aligned} \tag{12}$$

for $t = t_0, \ldots T-1$, subject to the constraints

$$\sum_{t_0 \leq t \leq T} (u_t + w_t) \leq \xi(t_0, T), \tag{13}$$

where x_t, w_t, and z_t are the state, macroscopic control, and observation vectors, respectively, A_t, B_t, C_t, and D_t are real matrices, and η_t and ζ_t are random factors perturbing the growth/migration and observation processes for $t = t_0, \ldots T$. The matrices A_t and B_t may be derived by aggregating the equations (1) for all $\omega \in \Omega$ and $t \in [t_0, T]$. The macroscopic controls w_t allow an "external" migration by attaching to (or removing from) Ω elements independently from the "internal" transitions controlled at the lower level (1), and may bear certain additional costs.

The state vectors $x_t = (x_{1t}, \ldots x_{nt})^T$ contain the numbers of elements of Ω characterized by the same values of $F_1, \ldots F_N$, for $t = t_0, \ldots T$ (cf. [5]). Thus, there is a one-to-one correspondence I between the indices of the state variables and the elements of V, so for each $t \in [t_0, T]$ one may order the state variables $x_{t1}, \ldots x_{tn}$ by the partial order generated from V. The values of F corresponding to the state variables are called the *interpretation vector* (cf. the above quoted paper) and denoted by $I(F, x_t)$. Assuming that a population Ω is characterized by certain distribution of attributes at an initial moment t_0 represented by the state vector $x^0 := x(t_0)$, the aim of control is to achieve an optimal distribution of elements of Ω at time T, using for that a minimal quantity of resources represented by u_t and w_t. Below we propose two of a variety of possible optimization problem statements. According to an initial remark, the first one of them is related to the nondominated values of F.

Let $K:=\{v \in V : \exists \omega \in \Omega$ such that $v=F(T)(\omega)\}$, and let $P(K)$ be the set of nondominated values of K.   Denote by $\Pi(K)$ the set of nondominated indices of corres-
ponding state variables, i.e. $\Pi(K):=I^{-1}(P(K))$.   By definition, the relative
population structure will be optimal if

$$(\sum_{j \in \Pi(T)} x_{Tj}) / (\sum_{1 \leq k \leq n} x_{Tk}) \text{ is maximal,} \tag{14}$$

$$\sum_{t_0 \leq t \leq T} (q_t u_t + r_t w_t) \text{ is minimal,} \tag{15}$$

and

$$m_0 \leq \sum_{1 \leq k \leq n} x_{Tk} \leq m_1, \tag{16}$$

where q_t and r_t are positive real coefficients.   The above problem has ecological or social motivation, namely, assuming that a population Ω remains stable if under a classification F a maximal number of its members cannot get in
touch with another individuals which are better (in the partial order in V)
than themselves in all relevant aspects (represented as the criteria
$F_1, \ldots F_N$).   In this setting, it is less important what is the shape of K and
where it is situated at time T.

Introducing a *loss function* $\psi : V \times V \to \mathbb{R}_+$, which is right strictly order
increasing, i.e. $v_1\langle v_2$, $v_2\langle v_3$, and $v_1 \neq v_3 \Rightarrow \psi(v_1,v_2) \leq \psi(v_1,v_3)$, (as ψ one can
choose e.g. a strictly convex distance function), we can evaluate the
deviations from the ideal value $v^* := (v_{1,c(1)}, v_{2,c(2)}, \ldots v_{N,c(N)})$ for each $\omega \in \Omega$
at time T.   Con-sequently, the deviation of the whole set Ω can be
characterized by the following criterion σ :

$$\sigma(u(t_0)(\Omega), \ldots u(T-1)(\Omega), w_{t_0}, \ldots w_{T-1}) := \sum_{1 \leq \alpha \leq M} \psi(v^*, F(T)(\omega_\alpha)) \to \min, \tag{17}$$

which may be more suitable for economical applications such as e.g. balancing
the portfolio structure than (14)-(16).   While optimizing σ, we strive to
approach the most preferred element of V for a possibly maximal number of
elements of Ω.   As the result, the set of alternatives actually characterized
by nondominated values of F need not be numerous, but in average, their values
are better approximating the ideal value v than in case of optimizing the
criterion (14).   Let us note that always

$$\sigma = \sum_{1 \leq k \leq n} x_{Tk} \psi(v^*, I(F, x_{Ti})). \tag{18}$$

From a computational point of view a solution to the above problems consists
in solving parallelly N discrete optimal control problems coupled by the common resource or expense limitation (11).   Thus, this problem requires nonstandard solution algorithms based on dynamical programming which have been
proposed in Sec. 3.2.   Roughly speaking, if the decision process is homogeneous and the criteria are evolution independent, one can construct the network
G presented in Secs. 3.1 and 3.2, assigning additionally the varying labels
x_{ti} to the nodes $v \in V$ determined by the interpretation vector $I(F, x_t)$.   The
further procedure consists in finding shortest paths (in terms of the cost
function J) to the nondominated values of $F(T)(\Omega)$, calculating the values of
the macroscopic criterion σ, and choosing a subset $\Omega_1 \subset \Omega$ consisting of those
elements of Ω which values are to be improved.

5. Concluding remarks

The above description and assumptions reflects a complicated nature of certain real-life systems, where the growth coefficients may be derived a posteriori from empirical experience. As examples of such systems may serve e.g. the populations of inhabitants of a town, portfolio of a company, or a wildlife reservation. Further application of the above theory include the decision problems, related to the migration models, or personnel choice, where the classical modelling and decision support methods do not allow to include the structure evolution aspects into the problem analysis. The solution method applied to the Problem 3.1 shows a remarkable coincidence with the optimal control of discrete-event systems presented in [3]. On the other hand, however, the approach to simultaneously control the evolution of a population Ω outlined in Sec. 4, resulting in a discrete-time control system model (12) could be applied to control large-scale discrete-event systems which allow an appropriate decomposition of the state-space.

In the present paper we concentrated our attention on deterministic processes, although in real-life situations some of the transitions may be stochastic. The analysis of such systems which involves the optimal control of discrete Markov processes (cf. e.g. [1]) may be considered as generalization of the methods here presented and needs further investigation.

R e f e r e n c e s :

[1] D.P. Bertsekas (1987). Dynamic Programing : Deterministic and Stochastic Models. Prentice-Hall, Inc., Englewood Cliffs, p.376.

[2] M.I. Henig (1985). The shortest path problem with two objective functions. *European J. Oper. Res.*, **25**, 281-291.

[3] A.M.J. Skulimowski (1991). Optimal Control of a Class of Asynchronous Discrete-Event Systems. *Proceedings of the 11th IFAC World Congress*, Tallinn (Estonia), August 1990, Vol.3, pp. 489-495; Pergamon Press, London.

[4] Making optimal decisions in multicriteria problems with varying attributes. Technical Report IWI HSG, No. 17/1993, March 1993.

[4] A.M.J. Skulimowski, B.F. Schmid (1992). Redundance-free description of partitioned complex systems. *Mathl. Comput. Modelling*, **16**, No.10, 71-92.

Combinatorial optimization and manufacturing

The Minimization of Resource Costs in Scheduling Independent Tasks with Fixed Completion Time

L. Bianco

I.A.S.I. - C.N.R. - Viale Manzoni, 30 - I-00185 Rome

Dept. of Electronic Eng. - Univ. of Rome "Tor Vergata"

P. Dell'Olmo

I.A.S.I. - C.N.R. - Viale Manzoni, 30 - I-00185 Rome

Abstract

We consider the problem of scheduling a set of independent tasks within a given time limit in order to minimize the cost of required resources. Tasks have different processing times and are non-preemptable (i.e. cannot be interrupted and restarted later). Each task requires a set of dedicated processors and some additional resources of different costs simultaneously available for its processing. Given a fixed set of processors and unlimited resources availability, the problem is to find the minimum resource cost schedule respecting the time constraint. We study this problem by means of a graph model which represents compatibilities between tasks, as that of finding the optimal reduction to an interval graph. Results show that dominant schedules correspond to a family of graphs which is contained in the constraints structure of the problem. Structural properties of the problem are discussed and solution algorithms are proposed.

1 Introduction

We study a class of non preemptive scheduling problems which arise in the project scheduling and multiprocessors task scheduling areas. The main problem can be stated as follows: given a set of independent tasks with different processing times, each one requiring one of more dedicated processors and a given number of additional resources units for its processing, find the minimum resources cost non pre-emptive schedule to meet a given time limit. The requirement of a set of machines and additional resources at the same time for the processing of each task, makes this model different from those studied in classical scheduling theory as in Baker (1974) and Coffman (1976) where a task is processed by one machine at a time. With the development of new technologies, this assumption is becoming not so obvious. A meaningful example is the task scheduling in computers with parallel dedicated processors and additional resources (input devices, communication lines, etc.), in order to meet fixed performances requirements in processing a given task set. Similarly, one can consider concurrent resources machine scheduling in flexible manufacturing systems context (see for instance Luggen (1991)). An other example is the minimum cost assignment of resources to a group of activities, which can occur both in services and production systems. In these cases the deadline for the completion of the activities is mandatory, as usually it happens in projects and production plans, while it is asked to reduce costs the most it is possible.

At the best of our knowledge, the problem, in this formulation, has never been studied before. In previous works, we investigated the problem of scheduling independent tasks in order to minimize the maximum completion time (see Bianco et al (1991) for the general case, and Blazewicz et al (1992) for the particular case of three processors). Mohring (1983) has studied the problem of minimizing resources cost in project networks, when tasks are related by precedence constraints, (i.e. partially ordered). As in the case here considered, tasks are not related by precedence constraints, we cannot directly apply that model, and search for the optimal extension of a given partially ordered set.

For this reason, adopting an approach similar to Mohring (1983), we study the relationships between the intersection representation of processors requirements, and the interval orders of minimum resource cost respecting the time constraint.

The proposed formulation is based on undirected graphs in spite of partial orders. This allows to represent the cost and the maximum completion time a solution without considering any partial order among the tasks (i.e. schedules). Set of schedules with the same behaviour are then replaced by an interval graph, and the whole scheduling problem is formulated as that of searching for an optimal interval graph reduction of a given graph. As a consequence, we reduce considerably the dimension of the search space and make it possible to apply the data structures and the algorithms related to interval graphs for solving a complex scheduling problem.

The paper is organized as follows. In Section 2 the problem is defined. In Section 3 the problem is investigated from the theoretical point of view by means of a graph model. In particular, some relatively simple graph structures corresponding to the problem parameters have been identified. Finally, a general scheme of a Branch and Bound algorithm together with a numerical example are presented in section 4.

2 Notation and Problem Definition

Let be given a set of n tasks $T = \{T_1, \ldots, T_n\}$, a set of m dedicated machines (processors) $\mathcal{M} = \{M_1, \ldots, M_m\}$ and different kinds $k = 1, \ldots, r$ of additional resources. Resources are assumed to be available at any time and in any quantity at cost c_k per unit of resource of kind k. To each task T_i is associated a vector of resources requests $R(T_i)$ and a non empty set of machines $M(T_i) \subseteq \mathcal{M}$. Each element $R_k(T_i)$ $k = 1, \ldots, r$, denotes the units of resources of type k required by

T_i for its processing. The processing of each task T_i requires the simultaneously availability of both the set of processors $M(T_i)$ and of the sets of resources $R(T_i)$ for all its duration, denoted by d_i. Processors are dedicated and available in one unit each, thus tasks sharing at least one processor must be sequenced. Moreover, scheduling is non-preemptive, that is, tasks cannot be interrupted and restarted later. Let B denote the time limit for the execution of the task set. A feasible schedule is a vector of starting times defined by a function $S : \mathcal{T} \to \{t_1, \ldots, t_n\}$ satisfying the following conditions:

$$S(T_i) + d_i \le B \qquad i = 1, \ldots, n \tag{1}$$

$$M(T_i) \cap M(T_j) = \emptyset \qquad \forall i, j \in E[S, S(T_j)] \qquad j = 1, \ldots, n \qquad i \ne j \tag{2}$$

The set $E[S, S(T_i)] \subseteq \mathcal{T}$ denotes the set of tasks being executed at time $S(T_i)$ under the schedule S. To each schedule can be associated the following cost functions for each resource type $k = 1, \ldots, r$

$$z_k = \max_j \sum_{T_i \in E[S, S(T_j)]} c_k |R_k(T_i)| \qquad j = 1, \ldots, n \qquad i \ne j \tag{3}$$

and the total cost function

$$z = \sum_{k=1}^{r} z_k \tag{4}$$

The problem is then that of finding a schedule respecting constraints in (1) and (2), and minimizing the cost function defined in (4). The problem is NP-hard in the strong sense as it contains a multiprocessor scheduling problem on dedicated processors, which was proved to be NP-hard in the strong sense in Blazewicz et al (1992).

In Figure 1.a it is shown a problem instance with 6 tasks, 4 processors and (for sake of simplicity) only one additional type of resource of cost $c_1 = 1$, for each unit. Each column of the matrix denotes, for each task, the processors required, the units of additional resource and the processing time. The time limit B is equal to 7 time units. Figure 1.b presents a feasible schedule while, in Figure 1.c, it is shown an optimal one. Note that both schedules have the same maximum completion time, but different costs.

3 Graph Model and Related Problems

As shown in the example, schedules can be used for representing solutions to this problem. In general, if the problem admits a feasible schedule, then there will be many others, often only slightly different one to the other, and having the same cost and the same duration. In order to avoid examining huge numbers of equivalent performance schedules, it would be convenient to adopt a more suitable representation for problem solutions.

It is known that an *Interval Graph* is an undirect simple graph whose vertices correspond to intervals on the real line and with edges between vertices if and only if

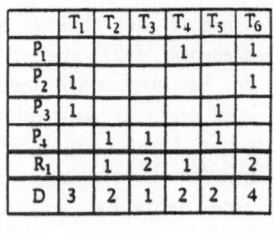

	T_1	T_2	T_3	T_4	T_5	T_6
P_1				1		1
P_2	1					1
P_3	1			1		
P_4		1	1	1		
R_1	1	2	1			2
D	3	2	1	2	2	4

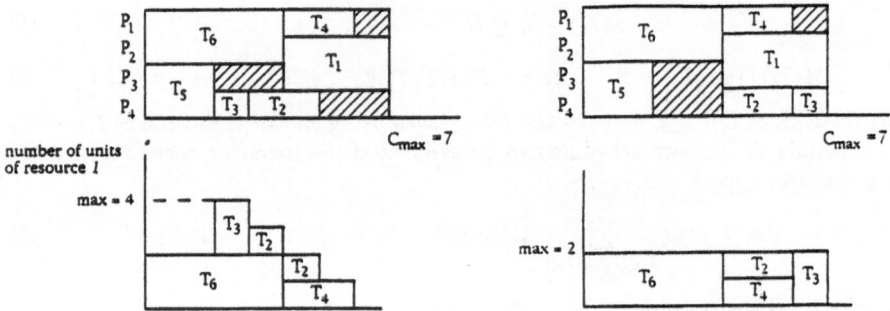

Figure 1: A problem instance (a), a feasible schedule (b), and an optimal one (c)

the corresponding two intervals have non empty intersection (Golumbic (1980)). It is also known that given an interval graph it is possible to derives different *Interval Orders*, with the same intersection behaviour, (Fishburn (1985), Mohring (1985)). We want to take advantage from the relationship between the two structures. Our basic idea is to replace schedules (in which the order of the tasks is important) by interval graphs. This can be accomplished by taking an interval order $P(V, \prec)$ associated to the interval graph and, successively, build the Earliest Start Time (EST) schedule corresponding to it.

Firstly we want to recall some properties related to the interval graph associated to a schedule S. Let S be an earliest start time schedule for the set of tasks T with durations $\{d_1, \ldots, d_n\}$, and $\mathcal{H} = (V(\mathcal{H}), E(\mathcal{H}))$ be an interval graph with $V(\mathcal{H}) = T$ and $(i, j) \in E(\mathcal{H})$ if and only if tasks T_i, T_j are executed in parallel (at least for one time instant), under the schedule S. Let $D : V \to \Re^+, K : V \to \Re^+$, with $\Re^+$ the set of positive real numbers, be the weight functions (duration and cost, respectively) such that $D(V_i) = d_i$ and $K(V_i) = c_k |R_k(T_i)|$. Moreover, let $MWIS_D(\mathcal{H})$ and $MWC_K(\mathcal{H})$ denote the maximum weighted independent set, with respect to function D, and the maximum weighted clique, with respect to function K, of $\mathcal{H}$ respectively.

Proposition 1 *For the schedule S and the corresponding interval graph $\mathcal{H}$ it holds:*

$$(a) \qquad \max_i \{S(T_i) + d_i\} = MWIS_D(\mathcal{H})$$

$$(b) \qquad \max_k z_k = MWC_K(\mathcal{H})$$

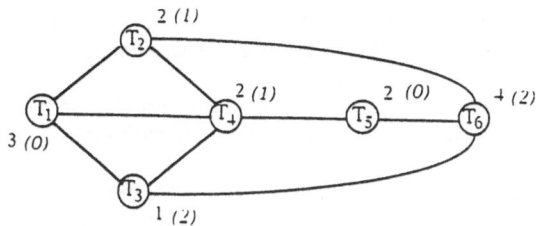

Figure 2: Compatibility Graph

Proof:(a) Consider the interval representation of the EST schedule . The maximum completion time is given by the longest sequence of tasks which are executed consecutively (if not so, the schedule is not an EST one), that is non overlapping. By definition of interval graph, these sets induce the maximum weighted independet set of $\mathcal{H}$. (b) Maximum usage of resource arises when maximal set of tasks are executed in parallel. In $\mathcal{H}$, set of pairwise overlapping tasks induces cliques. For any of this tasks set the Helly's property holds (Golumbic (1980)). Hence there exists at least one time instant in which all tasks of a clique are executed in parallel. Then the maximum weighted clique corresponds to the maximum resource usage of a set of parallel tasks. □

Viceversa, given an interval graph, we ask under what conditions there exists a feasible schedule for the problem which can be obtained considering the interval order $P(V, \prec)$ associated with the graph. This results is more important for our purposes, but less obvious. In order to define such conditions, let us observe that to each problem instance we may associate an undirect simple graph as follows:

Definition 1 *A Compatibility Graph $\mathcal{G} = (V(\mathcal{G}), E(\mathcal{G}))$ with set of vertices $V(\mathcal{G}) = \{V_1, \ldots, V_n\} = \mathcal{T}$ and edges $(i,j) \in E(\mathcal{G})$ if and only if $M(T_i) \cap M(T_j) = \emptyset$, that is if and only if tasks T_i, T_j can be processed in parallel, because they do not share any processor.*

Definition 2 *Given a graph $\mathcal{G} = (V(\mathcal{G}), E(\mathcal{G}))$ we denote as Spanning Interval Graph (SIG) of $\mathcal{G}$ any interval graph $\mathcal{H} = (V(\mathcal{H}), E(\mathcal{H}))$, which is contained in $\mathcal{G}$, in the sense that $V(\mathcal{G}) = V(\mathcal{H})$ and $E(\mathcal{H}) \subseteq E(\mathcal{G})$.*

In Figure 2 the Compatibility Graph corresponding to the problem instance of Figure 1.a, is shown. The nodes are labelled with duration and cost (in brackets).

Lemma 1 *Let $\mathcal{G} = (V(\mathcal{G}), E(\mathcal{G}))$ be the Compatibility Graph of the problem. Let $\mathcal{H} = (V(\mathcal{H}), E(\mathcal{H}))$ be a Spanning Interval Graph of $\mathcal{G}$ and let $P = (V(\mathcal{H}), \prec)$ be an interval order associated with $\mathcal{H}$. Then:*

(a) P induces a schedule S on the task set $\mathcal{T} = \{T_1, \ldots, T_n\}$, which respects the processors requirements (see constraint (1) in Section 2).

(b) The maximum weigthed independent set of $\mathcal{H}$, with respect to function D, is equal to the value of the maximum completion time of the schedule S.

(c) The maximum cost of resource of type k for the schedule S is bounded from above by the maximum weighted clique of $\mathcal{H}$ with respect to weight function K.

Proof:(a) The schedule S, corresponding to the partial order relation $\prec_P$ induced by P on T, can be computed recursively in $O(|V|^2)$ time by the following algorithm. Assign $S(T_i) = 0$ if V_i has no predecessors in P. Otherwise, assign $S(T_i) = \max_{T_j \prec_P T_i}\{S(T_j) + d_j\}$. Now, consider that tasks which are joined by and edge in $\mathcal{H}$ are also joined in $\mathcal{G}$, as $\mathcal{H}$ is a spanning subgraph of $\mathcal{G}$, and hence are compatible. In general, any clique of $\mathcal{H}$ identifies a set of tasks compatible with regard to processors requirements. We recall that cliques of $\mathcal{H}$ are antichains in P (Mohring 1985), such that for any set of tasks $T' \subseteq T$ processed in parallel under S, there exists an antichain $\mathcal{A}$ in P with $T' \subseteq \mathcal{A}$.

(b) It can be easily proved considering that, all interval orders $P(V, \prec)$ obtained from $\mathcal{H}$, have the same comparability graph, $G(P)$, which is the complement of $\mathcal{H}$, namely $C(\mathcal{H})$. As the length of maximal chains in P is a comparability invariant of $C(\mathcal{H})$, all maximal chains have the same length. In particular the size of the chain of maximum weight of $P(V, \prec)$, corresponds to the $MWC_D(C(\mathcal{H}))$, that is the $MWIS_D(\mathcal{H})$.

(c) Following the same reasoning of (a), maximal set of tasks processed in parallel under S are contained in the corresponding antichains of P, that is cliques of $\mathcal{H}$. In, particular the result holds for the set of parallel tasks of maximum cost, which is contained in the clique of $\mathcal{H}$ of maximum weight with respect to function K. $\square$

Note that the bound given by the $MWC_K(\mathcal{H})$ specified in point (c) the may be not tight. Infact this depends on tasks duration, i.e. the interval lengths (Fishburn (1985)). However, we will prove that for a spannig interval graph of minimum value of $MWC_K(\mathcal{H})$ the bound is tight.

Feasible schedules can then be substituted by interval graphs $\mathcal{H}$ with $MWIS_D \leq B$. (see Figures 3.a and 3.b, where the interval graphs corresponding to the schedules of Figures 1.b and 1.c are shown). Now, we can consider the scheduling problem as that of finding an optimal interval graph $\mathcal{H}$ in a given graph family.

Theorem 1 *The minimum cost of requirements of resource k for a feasible schedule is equal to the $\min(MWC_K(\mathcal{H}))$ of a Spanning Interval Graph $\mathcal{H}$ of the Compatibility Graph $\mathcal{G}$ with $MWIS_D \leq B$.*

$$z_k^* = \min\{MWC_K(\mathcal{H})|\mathcal{H} \text{ is an SIG of } \mathcal{G} \text{ and } MWIS_D(\mathcal{H}) \leq B\}$$

Proof: Denoting by z_k^* the optimal value of the objective function and by $MWC_K(\mathcal{H}^*)$ the value of the $MWC_K(\mathcal{H}^*)$ for the optimal graph $\mathcal{H}^*$, we show that $z_k^* \leq MWC_K(\mathcal{H}^*) \leq z_k^*$.

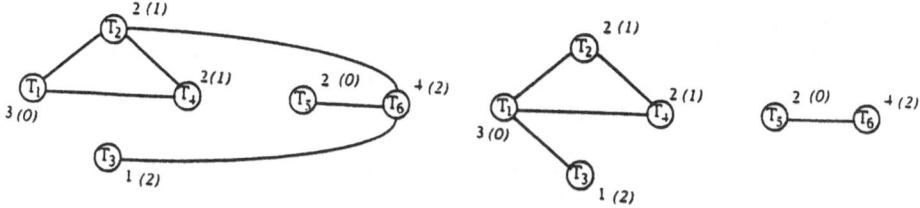

Figure 3: The interval graph reduction corresponding to the feasible schedule (a), and to the optimal one (b)

We first show that $MWC_K(\mathcal{H}^*) \leq z_k^*$. Suppose $z_k^* > MWC_K(\mathcal{H}^*)$ and consider the interval graph $\mathcal{H}$ associated with S and the durations $\{d_1, \ldots, d_n\}$. For Proposition 1 such a graph has $MIS_D(\mathcal{H}) \leq B$, hence is feasible, and $MWC_K(\mathcal{H}) < MWC_K(\mathcal{H}^*)$ (contradiction).

We now show that $z_k^* \leq MWC_K(\mathcal{H}^*)$. Suppose $MWC_K(\mathcal{H}^*) > z_k^*$ and consider the schedule obtained by the interval order associated to $\mathcal{H}^*$. Then we can consider the schedule corresponding to the interval order. For Lemma 1, this is feasible and its cost z_k would be equal to $MWC_K(\mathcal{H}^*) < z_k^*$. Then z_k^* would not be optimal (contradiction). □

Definition 3 *An interval graph reduction $\mathcal{H}$ is called feasible if and only if $MWIS_D(\mathcal{H}) \leq B$. Let $\mathcal{F}$ denotes the family of feasible interval graphs.*

Theorem 2 *$\mathcal{F}$ is finite and convex. Moreover for any $\mathcal{H}', \mathcal{H}' \in \mathcal{F}$, if $\mathcal{H}' \subseteq \mathcal{H}''$, then any schedule obtained from $\mathcal{H}'$ dominates all schedules obtained by $\mathcal{H}''$.*

Proof: As $\mathcal{G}$ is finite, then the space of its spanning subgraphs, which contains $\mathcal{F}$, is also finite. Let $\mathcal{H}$ be an interval graph such that $\mathcal{H}' \subseteq \mathcal{H} \subseteq \mathcal{H}''$. As $\mathcal{H}$ contains $\mathcal{H}'$, it follows that $\mathcal{H}'$ is a spanning interval graph of $\mathcal{H}$, which is also feasible because the maximum weighted independent set is a non-increasing function on the number of edges and $|E(\mathcal{H}')| \leq |E(\mathcal{H})|$, thus $MWIS_D(\mathcal{H}) \leq MWIS_D(\mathcal{H}') \leq B$. Thus $\mathcal{F}$ is convex. Moreover, as the the maximum weighted clique is a non decreasing function in the number of the edges, it follows also that $MWC_K(\mathcal{H}') \leq MWC_K(\mathcal{H}'')$. □

We briefly recall here some of complexity results related to interval graph recognition an interval order representation. An interval order $P(V, \prec)$ of an interval graph $\mathcal{H}$ can be constructed in $O(|V(\mathcal{H})| + |E(\mathcal{H})|)$ time and $O(|V(\mathcal{H})|)$ space by using algorithm based on PQ-trees (Boot and Lueker (1976)). Finding the maximum weighted clique and the maximum weighted independent sets are NP-hard problems for general graphs, but can be solved efficently in the case of interval graphs. In particular the $MWC(\mathcal{H})$ can be found in $O(|V(\mathcal{H})| + |E(\mathcal{H})|)$. The $MWIS(\mathcal{H})$ can be computed considering the $MWC(C(\mathcal{H}))$ of the comparability graph $C(\mathcal{H})$ in

$O(|V|^4)$ (Golumbic (1980)) or in $O(|V(\mathcal{H})| + |E(\mathcal{H})|)$ as in Mohring (1985). These observations show that the hardness in solving the scheduling problem is that of finding the optimal interval graph reduction of $\mathcal{G}$. Considering the above points, we introduce the following problem formulation:

Definition 4 *Interval Graph Reduction Problem: Given a graph $\mathcal{G}$ and weight functions D and K find a spanning interval graph $\mathcal{H}$ of $\mathcal{G}$ such that $MWIS_D(\mathcal{H}) \leq B$ and $MWC_k(H)$ is as small as possible.*

Theorem 3 *The scheduling problem and the interval graph reduction problem are polynomially equivalent.*

Proof: It follows directly for the complexity results mentioned above. $\square$

The above formulation can be easily extended for the case of multiple resources considering $z = \sum_{k=1}^{r} MWC_K(\mathcal{H})$ as measurement criterion for the interval graph $\mathcal{H}$.

4 Solution Algorithm

In this section we present the scheme of a Branch and Bound algorithm which permits to find optimal solution to the problem. This is a very general procedure and additional criteria, both for pruning and for variable selection, can be considered according to the characteristics of $\mathcal{G}$. However, our purpose is to show that, relaying on the graph model presented in the previous section, it is possible to design a relative simple exact algorithm, applyable to any configuration of machines, resources, processing times, machines and resources requirements. This is in contrast with a standard machine-resources scheduling approach which has the natural tendency of proposing different solution algorithms, applicable only for fixed configurations of problem parameters.

Next we sketch the Branch and Bound algorithm for finding an optimal interval graph. The branching variables are the edges $(i, j) \in E(\mathcal{G})$. The algorithm, starting from the Compatibility Graph $\mathcal{G}$, generates the enumeration tree in Bread-First Search, by successively deleting edges of $E(\mathcal{G})$ in order to find a feasible interval graph $\mathcal{H}$ with $MWC_D(\mathcal{H})$ minimum. Nodes are fathomed when $MWIS_D(\mathcal{H}) > B$.

Algorithm 1

Step 1: Start with the compatibility graph $\mathcal{G}$.

Step 2: At each stage of the algorithm we obtain a reduced graph $\mathcal{G}^-$ by deleting one edge. We test this graph with the algorithm for interval graph recognition. This can be performed using PQ trees and algorithm presented in Boot and Lueker (1976) and Mohring (1985) in $O(|V(\mathcal{G}^-)| + |E(\mathcal{G}^-)|)$.

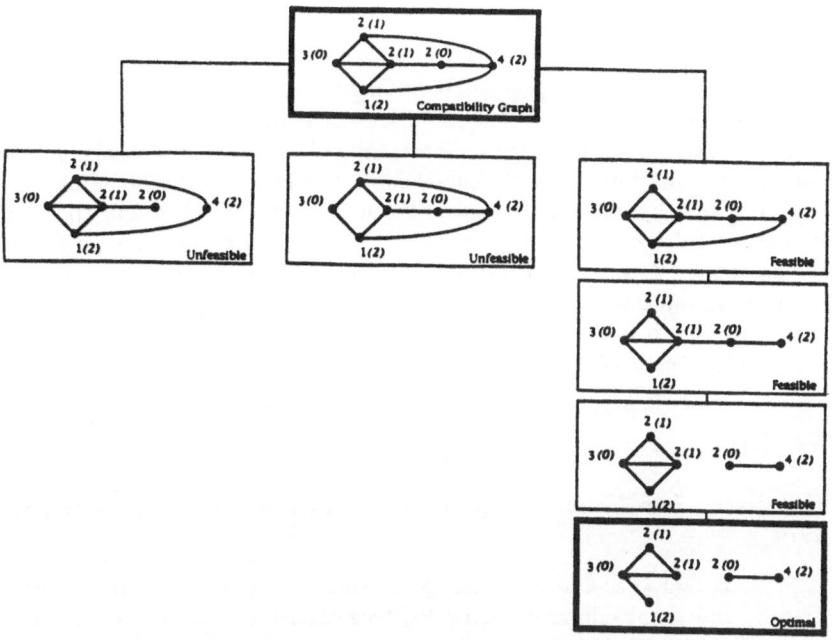

Figure 4: The Branch and Bound tree for the example

Step 3: If the current graph $\mathcal{G}^-$ is an interval graph, we then test if it is feasible by determining $MWIS_D(\mathcal{G}^-)$. This can be done in $O(|V(\mathcal{G}^-)| + |E(\mathcal{G}^-)|)$. If feasible, we compute the $MWC_K(\mathcal{G}^-)$ and compare this value with the current solution.

Step 4: If the graph is not an interval graph, we select an edge to be deleted. Different criteria can be applied.

Step 5: The deletion of an edge can increase the size of the maximum independent set and cause infeasibility for any successive graph reduction. We can compute the $MWID_D(\mathcal{G}^-)$ approximatively, using heuristic algoritms, or exactly if $\mathcal{G}^-$ is is triangulated in $O(|V(\mathcal{G}^-)| + |E(\mathcal{G}^-)|)$. If $MWIS_D(\mathcal{G}^-) > B$ we fathomed the node.

Further considerations can be added to the above algorithm. The upper bound of the objective function is given by $MWC_K(\mathcal{G})$. An easy to compute lower bound of the objective function is the maximum resource usage for a task, that is $max_j\{c_k|R_k(T_j)|\}$. In Figure 4 it is shown an enumeration tree generated by the above procedure for the example presented in Figure 1. Note that in this simple case the optimal solution is found, and recognized, for which $z_k = max_j\{c_k|R_k(T_j)|\}$.

5 Conclusions

In this paper we examined the problem of finding a minimum resource cost schedule for a set of non preemptable task. The simultaneous requirement of a set of machines and a set of additional resources for the processing of each task, makes this problem different from the machine scheduling models studied in classical scheduling, where a task is processed by one processor at a time. We presented a graph theoretical formulation for this complex problem which allow to design an exact algorithm, that can be applied to any configuration of machines and resource requirements, so avoiding to examine a taxonomy of different configuration of problem parameters. Further researchs will be devoted in testing the performances of this approach on practical problem instances.

6 References

K.R. Baker, *Introduction to sequencing and scheduling*, John Wiley and Sons, 1976.

K.S. Boot and G.S. Lueker, Testing for consecutive one's property interval graphs, and graph planarity using PQ-tree algorithms. *J. Comput. System Sci.* 13 (1976) 93–109.

L. Bianco, P. Dell'Olmo and M.G. Speranza, Nonpreemptive Scheduling of Independent Tasks with Dedicated Resources, Technical Report 320, I.A.S.I. - C.N.R. - Rome, 1991.

J. Blazewicz, P. Dell'Olmo, M. Drozdowski and M.G. Speranza, Scheduling Multiprocessor Tasks on Three Dedicated Processors, *Information Processing Letters*, 41 (1992), 275–280.

E.G. Coffman Jr (Ed.), *Computer and job-shop scheduling theory*, John Willey an sons. 1976.

P.C. Fishburn, Interval Graphs and Interval Orders *Discrete Mathematics*, 55, (1985), 135–149.

M.C. Golumbic, *Algorithmic Graph Theory and Perfect Graphs*, Academic Press, New York, 1980.

W. Luggen, *Flexible Manufacturing Systems*, Prentice All (1991).

R.H. Mohring, Minimizing Costs of Resource Requirements in Project Networks Subject to Fixed Completion Time, *Operations Research* (1983), 4, 88–119.

R.H. Mohring, Algoritmic Aspects for Comparability Graphs and Interval Graph, in *Graphs and Orders*, ed. I.Rival (Reidel, Dordrecht), (1985), 41–101.

A STOCHASTIC SCHEDULING PROBLEM ISSUED FROM IRON AND STEEL INDUSTRY : SOME RESULTS

Brigitte Finel
INRIA-Lorraine, Project SAGEP, CESCOM, Technopôle 2000,
4 rue Marconi, 57 070 Metz (France)

Marie-Claude Portmann
Ecole des Mines de Nancy, I. N. P. L., Parc de Saurupt
54 042 NANCY CEDEX (France)
& INRIA-Lorraine, Project SAGEP

Abstract

A stochastic scheduling problem with resource constraints arising in iron and steel industry is discussed. It concerns the utilization of highly flexible automated and integrated bricklaying systems inside converters located on different sites. The stochastic aspect concerns the time windows to which one task of bricklaying has to be assigned. The resource constraint comes from the limited number of bricklaying systems. For this problem simulation is used first in order to compare some strategies which can be used for the on-line decision problem. We then try to analyse them : was the chosen number of bricklaying systems adequate ? To answer this question two ways are proposed : the search of a priori lower and upper bounds under particular assumptions and the study of the deterministic problem a posteriori after a simulated execution. The first section describes the considered problem. The second one presents the lower and upper bounds. The third section presents very briefly the simulation strategies for the on-line decisions. The fourth one gives the general approach for an "a posteriori" analysis and the fifth section presents a numerical sample.

1. The problem considered

In iron and steel companies, the inside surface of converters is covered by rows of refractory bricks. The lifetime of these refractors is limited, thus they must be regularly changed. So far, this operation is made by human workers. It is long, painful and expensive. So, a group of partners has designed and built a highly flexible automated and integrated bricklaying system (at least the first prototype [WUR 89]). This system, called a "machine" throughout this paper, is very expensive. It is also voluminous, but nevertheless, in order to maximize its utilization, it may be displaced from one site (a group of several converters in a given plant) to another one with a very large and very long truck. The successive states of a converter are described by the figure 1.

796

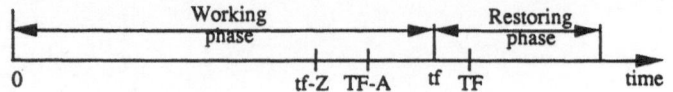

figure 1 : successive states of a converter

We denote by
TF   the maximal lifetime of a converter (TF = 480 hours in the experiments, value obtained from industrial data analysis),

TF - A the minimal lifetime of a converter : after this duration, a breakdown may happen which stops the life of the converter before the duration TF (TF-A = 240 hours in the experiments),

tf   the moment where the converter is effectively stopped (i.e. the exact lifetime of a converter between two bricklaying changes, in the simulation experiments, we use either a constant TF-A plus a random variable UNIFORM[0,A] or GAUSS[TF-A/2,50] truncated to [TF-A,TF], the results are quite similar),

tf - Z the moment when the moment tf the converter will be stopped is known (Z = 72 hours in the experiments).

Converters are grouped on s production sites (s varies from 4 to 6 in the different experiments). As the steel production must be regular, at each site i ($1 \leq i \leq s$) which contains n_i converters : at each instant, $n_i - 1$ converters are making steel and the last one is in restoring phase. The normal sequence of converter states for a given site is illustrated by the figure 2. A restoring phase is divided into sub phases :
- a cooling phase (which immediately follows a working phase),
- a demolishing phase (human task),
- a bricking phase (with the new bricklaying system),
- a heating phase (which must immediately precede the working phase).

Between two sub phases there may exist a waiting phase, especially before the heating phase in order to begin the heating phase just in time. The duration of the various phases (with the numerical values used in our experiments) are given by :
K   the constant duration of the cooling phase (K=12),
D   the constant duration of the demolishing phase (D=24),
B_{min} the minimal duration of the bricking phase using a maximal number of human resources in order to diminish its duration (B_{min} = 30),
B   the chosen duration of the bricking phase,
B_{max} the maximal duration of the bricking phase using a minimal number of human resources (B_{max} = 72),
C   the constant duration of heating phase (C=20).

When one of the converters stops, the converter which was under repair must be ready to be used and start a working phase immediately. Normally, inside a site, we have a cycle of repairing tasks : converter n° 1, n° 2, ... , n° n_i, n° 1, n° 2, ... and so on, as illustrated in the figure 2. Nevertheless, if accidentally a converter has a very short lifetime, then the order of repairing tasks can vary. If the lifetime of the brick rows were deterministic and identical for all the converters (normally equal to TF-A/2), then we would have a cyclic scheduling problem and we have analysed the properties of this very specific deterministic problem (see [FIN 92]). In fact, the converters and their refractory brick rows can be considered as identical on a given site, but their lifetime is given by a random variable and so the time windows into which the bricklaying tasks must be scheduled are randomly placed on the time axis.

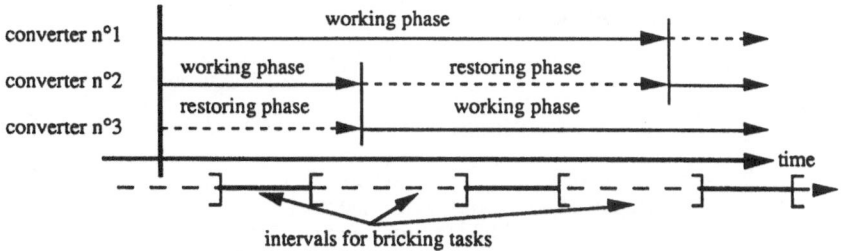

<u>figure 2</u> : normal sequence of converter states for a given site

If the number of machines m is strictly less than the number of sites s, then we have to move the machines from site to site in order to do the repairing tasks inside stochastic time windows. This is a stochastic scheduling problem with resource constraints. The costs coming from the moves of the machines will be called "transportation costs". Due to the stochastic nature of the problem, we are not sure that the converter being repaired will be able to begin work when another one is stopped (and this situation may happen even if we have an infinite number of machines). When two converters are being repaired simultaneously on the same site, it induces an extremely large cost for the steel company's drop of steel production ; we call it "cost of malfunction time". In order to try to prevent moves of machines or malfunction times, it is possible to use extra work hours or to accelerate the process which will decrease the duration of some bricklaying tasks. This induces a "cost of overtime hours". The problem has three criteria, but we aggregate them using weights which represent the cost for each unity of time or each move depending on the criterion unity : Total cost = Malfunction Cost + Overtime Hour Cost + Transport Cost

Moreover, there are some tight specific constraints or conditions, such as a bricklaying task which must begin at the beginning of a morning shift and the total number of extra hours authorized for a bricklaying task which decreases on-line during the execution of the task with very precise rules. The considered problem has some similarities with some scheduling problems with time windows or with vehicle scheduling problems with time windows (see for example : [BAK 83], [BAR 88], [BER 91], [DES 88], [FER 89], [HAO 90], ...), but unlike most of these studied problems it is stochastic, with particular conditions or constraints which need to be taken into account. For the deterministic a posteriori analysis, we used a structural decomposition which was first presented by B. ROY ([ROY 70]), and already used several times for this type of scheduling problem (see for example [GUI 84]), we adapted this approach to our specific problem.

2. Lower and upper bounds

In order to obtain a lower bound for the maximal number of required machines independently from the random values of the lifetime converters, we assume that we know the maximal and minimal duration of the repairing time windows and we relax some constraints of the problem. In order to obtain an upper bound independent of the random values, we consider the worse situation for the reassignment of the machines and construct a feasible solution which provides us with an upper bound.

2.1. Lower bounds

a) notations
Let us denote by :

TR_{max} the longest duration of a repairing time window,
TR_{min} the shortest duration of a repairing time window,
TB_{min} the shortest length for a bricklaying time window (TR_{min} - K - D - C),
s the number of sites,
H the length of the considered horizon,
τ_{min} the shortest transport duration for moving a machine from one site to another one,
δ the minimal length of time interval between two utilizations of a machine, which may either remain on the same site waiting to repair another converter or be moved to another site ($\delta = \min(K+D+C, \tau_{min})$),

b) remark
Malfunction time costs can be caused by :
- a lack (or a bad utilization) of bricklaying machines which creates a waiting time which is too long before the bricklaying phase,
- a random very short lifetime for one of the converters, which reduces the repairing window length of the previous stopped converter below the duration of the repairing tasks. In order to obtain correct bounds for the number of machines, we do not take into account the second possibility which is not linked with the machine utilization and then suppose that we have always the condition (1) :

$$B_{min} \quad \leq \quad TR_{min} \tag{1}$$

c) first lower bound
We first compute a lower bound denoted by NTB_{min} for the total number of bricking tasks which must be made on the horizon H and then an upper bound denoted by NBM_{max} for the total number of bricklaying tasks assumed by a machine during the same horizon H :

$$NTB_{min} = \frac{H}{TR_{max}} \times s \qquad\qquad NBM_{max} = \frac{H+\delta}{B_{min}+\delta}$$

We then obtain a lower bound (denoted by UPM_H) for the minimal number of machines required to assume the whole set of bricklaying tasks during the horizon H :

$$UPM_H = \frac{NTB_{min}}{NBM_{max}} = \frac{H}{TR_{max}} \times s \quad \times \frac{H+\delta}{B_{min}+\delta}$$

As we suppose that the horizon H is infinite (or at least very large compared with the length of any task), we obtain :

$$UPM_\infty = \frac{s\,(B_{min} + \delta)}{TR_{max}}$$

(2)

d) improved lower bound

One of the constraints given by the iron and steel enterprise was that a bricklaying task always begins at the beginning of a morning shift. In this case, the lower bound can be increased using the following formulas :

$$NBM'_{max} = \frac{H + \delta}{24\left\lceil \dfrac{B_{min} + \delta}{24} \right\rceil} \qquad\qquad UPM'_\infty = \frac{s}{TR_{max}} \cdot 24\left\lceil \dfrac{B_{min} + \delta}{24} \right\rceil \qquad (2')$$

2.2. Upper bounds

a) general scheme

In order to obtain upper bounds, we will use the following scheme. If we suppose that the set of sites S is partitioned into K subsets of sites : $S_1, S_2, \ldots , S_K$ and if we prove that one machine is sufficient to assume all the bricklaying tasks of each subsets of sites, then K is an upper bound for the number of machines. We can obtain different upper bounds depending on the chosen partition of the set of sites, we keep the best one obtained.

b) local analysis

In order to obtain an upper bound, we must solve the following decision problem .

Considering a subset of sites S and denoting by $TR_{max}(S)$, $TR_{min}(S)$, $card(S)$, $\tau_{min}(S)$, $\tau_{max}(S)$ respectively the longest and shortest duration of a repairing interval, the number of sites and the minimal and maximal transport times between sites when we consider only the sites of S,

is a machine sufficient to assume all the bricklaying tasks of the subset S ?

Lemma : If the following condition (3) is verified, then a machine is sufficient to assume all the bricking tasks :

$$card(S) \cdot B_{min} + [card(S) - 1] \cdot \tau_{max}(S) + max\,[K+D+C, \tau_{max}(S)] \leq TR_{min}(S) \quad (3)$$

The lemma can be proved by considering successively particular cases that are generalized until obtaining all the possible cases.

So, if we combine a partition scheme and the condition (3), we can obtain a first upper bound for the number of machines.

c) computation of an upper bound

We consider first the whole set of sites and the condition (3). If the condition (3) is verified for the whole set of sites, then one machine is sufficient and we stop the computation of the upper bound. Otherwise, we examine the partitions of the set of sites into two subsets : $\Pi_2 = \{ \pi_2 = (S_1, S_2), S_1 \cup S_2 = S, S_1 \cap S_2 = \varnothing \}$. We stop the search as soon as we find a partition π_2 such that the condition (3) holds for S_1 and S_2 and then concludes that two machines are sufficient. Otherwise, we examine the partitions of the set of sites into three subsets, etc.

The total number of different partitions into m subsets is equal to :

$$\left(\sum_{(k_1,k_2,...k_m)\in E_m} \frac{1}{k_1! \, k_2! \, ... k_m!} \right) * \frac{s!}{m!}$$

with
$$E_m = \left\{ (k_1,k_2,...k_m) \; / \; \sum_{k=1}^{m} k_i = s \text{ and } k_i > 0 \text{ for } i=1, ..., m \right\}$$

But the total number of sites does not exceed five or six for this industrial concrete problem and so all the permutations can be tested in order to find a good upper bound.

d) some computations of lower and upper bounds

Table 1 : computation of lower and upper bounds

nb sites	TR_{max}	TB_{min}	B_{min}	τ_{min}	δ	$B_{min} + \theta$	LW_1	UP_1	LW_2	UP_2
3	240	70	30	10	10	48	*1*	*1*	1	2
4	240	70	30	10	10	48	1	2	1	3
5	160	70	30	10	10	48	*2*	*2*	2	3
6	120	70	30	10	10	48	2	3	3	6
3	120	70	30	10	10	48	1	2	2	3
3	240	110	50	10	10	72	1	2	1	3
4	240	110	50	10	10	72	1	2	2	3
5	160	110	50	10	10	72	2	3	3	5
6	120	110	50	10	10	72	*3*	*3*	4	6
3	120	110	50	10	10	72	*2*	*2*	2	3
3	240	110	72	10	10	96	2	3	2	3
4	240	110	72	10	10	96	2	3	2	4
5	160	110	72	10	10	96	3	5	3	5
6	120	110	72	10	10	96	5	6	5	6
3	120	110	72	10	10	96	*3*	*3*	*3*	*3*
3	240	110	30	60	50	96	1	3	2	3
4	240	110	30	60	50	96	2	3	2	4
5	160	110	30	60	50	96	3	5	3	5
6	120	110	30	60	50	96	4	6	5	6
3	120	110	30	60	50	96	2	3	*3*	*3*
3	240	70	50	60	50	120	2	3	impossible	

Table 1 gives a numerical sample of lower and upper bounds when the maximal and minimal duration of repairing intervals are assumed to be known with :

LW1 and UP1 : lower and upper bounds without the constraint of morning shift

LW2 and UP2 : lower and upper bounds with the constraint of morning shift

K+D+C = 50 and $\tau_{min} = \tau_{max}$

The bounds are in bold when lower and upper bounds are equal for the same hypotheses for the constraints.

3. On-line strategies

For the on-line strategies, the number of machines is fixed (the previous lower and upper bounds can help in order to choose this number). A discrete event simulation was developed to test different dispatching rules concerning :
- the moving of the machines (immediately after an assignment or just in time for their next task),
- the reassignment of the machines (allowed or not),
- the use of extra work hours (which can be forbidden and with on-line rules when the bricklaying task is begun),
- the possibility of stopping a converter sooner in order to repair it completely before stopping one of the following ones whose date is already known
...

We so obtained 10 heuristics methods (see [CHU 92] & [FIN 92]). Some of them are better for the overtime hour cost, others for the malfunction cost and others for the transport cost (see the numerical sample). We do not detail here these methods, their more original part consists in deciding when a converter must be stopped and repaired in advance in order to replace another one which will stop earlier than expected and in trying to adjust the extra hours to the minimal value while avoiding malfunction cost either on the considered site by using earlier the repaired converter or on another one by urgently sending the machine to another site.

4. Approaches for the deterministic problem

The simulation enabled us to compare different strategies with a random generation of data. For each simulation, we fixed the number of machines and we used as many machines as were available. So, at the end of a simulation : we did not know if this number was or not the right one : if this number is too small, then the whole cost is very high and particularly the malfunction cost, but if the number of machines is too big, we have no means of measuring that except for the cost of the machines which is very expensive. So, after a given simulation, we try to analyse **a posteriori**if the used total number of machines was correct. We consider a given simulation with the dates of the beginning and end of the time windows where the bricklaying tasks are placed and study the corresponding deterministic scheduling problem : if the minimal number of machines found for the deterministic scheduling problem is equal or just a little smaller than the number of machines used in the simulation, then the number of machines used in the simulation was adequate considering it stochastic nature. The deterministic problem is NP-hard ([GAR 79]). We use a structural decomposition approximation scheme, i.e. an iterative algorithm in which we fix alternatively the values of the unknown times and of the unknown assignments. The method is briefly described below.

Step 1 :
We build a first schedule of the machines using a constructive method in which we consider successively machine 1, 2, ... and try to assign to machine i as much of the

unscheduled tasks as possible (using a percentage ξ of extra hours for each task and scheduling the tasks as soon as possible).

Step 2 :
Leave the machine assignment and keep only the starting and ending dates of every bricklaying task which are considered here as fixed. Minimize the total number of machines by using a polynomial simple assignment method ([KUH 55]])

Step 3 :
Analyse the dual problem in order to know where new immediate re-employment of the machines between two bricklaying tasks is the most interesting and try to obtain at least one new re-employment by modifying the starting and ending dates of some tasks (introducing eventually new extra hours). If we have modified at least one re-employment of the machines, and we have not yet met the lower bound for the number of machines and we have not yet decided to stop the search then we go to step 2.

Remark
The constraints of beginning every bricklaying task at the beginning of a morning shift was introduced without any major difficulties in this approximation scheme.

We apply this approximation scheme to several simulation results and can see that for some of them, the used number of machines seems adequate and for other ones this number was surely too great.

5. Numerical examples

In this section we will present some numerical results concerning the bounds, the simulations and the deterministic problem. Most of the numerical values have been given in the paper, we have still to describe the sites. We consider here an example with 4 sites, one site has 4 converters, two sites have 3 converters and one site has two converters. The simulations were made using 4 machines. The lower bound UP2 is computed by knowing the minimal length of transport and the maximal length of repairing intervals found inside the simulations. The length of the considered horizon is 20 000 hours. We repeated the simulation experiments 25 times with the same values of the parameters and computed the corresponding average for all the measures of performance. The total computation time to obtain this whole set of results was 385 minutes. A set of experimentation is provided in the table 2.

Meaning of the column of table 2 :
1. number of the simulation strategy.
2. average value of the converter lifetime.
3. average value of bricklaying task duration depending on the extra hours used.
4. malfunction time.
5. total number of bricklaying tasks.
6. average number of extra hours per bricklaying task in the simulation solution.
7. average number of extra hours per bricklaying task in the deterministic solution.

8. average minimal number of machines after the first call of assignment method in the deterministic solution.
9. average minimal number of machines at the end of the improvement loop on the assignment method in the deterministic solution.
10. ratio which is used in the deterministic method in order to define the initial quantity of extra hours in the constructive initial method.
11. lower bound when we know the maximal length of the repairing intervals.

We can note for this series of experiments that the number of machines obtained by the deterministic post-optimization is the same as the number of machines used by the simulation as soon as the ratio of extra hours is less than 0,75 (less than 31 extra hours per bricklaying task). When we use in the simulation a maximal quantity of extra hours, then for a great number of strategies, the corresponding deterministic problems can be solved by using only three machines.

table 2 : simulation and deterministic results

	SIMULATION RESULTS						DETERMINISTIC RESULTS			
	duration	of	malfunc	nb	extra		number of			
str.	conv. life	brickl.	time	of	hours / task		machines		ξ	UP2
	Average	/ cycle		brick	simul	deter	1	2		
1	363,37	47,98	79,79	357	23,98	29,05	3,04	2,84	1	2
2	363,34	39,86	82,33	355	32,06	28,73	3	2,84	1	2
3	360,67	47,22	89,83	351	24,68	27,52	3	2,68	1	2
4	360,15	40,03	75,6	364	31,92	29,48	3	2,88	1	2
5	365,56	57,48	13,35	418	15,13	32,46	3,08	3	1	2
6	365,99	43,11	13,34	418	29,50	32,20	3,2	3,04	1	2
7	360,39	57,8	8,97	430	14,68	32,40	3,16	3,04	1	2
8	359,76	42,8	8,55	431	29,68	32,22	3,16	3,04	1	2
1	363,37	47,98	79,79	357	23,98	21,00	3,32	3,24	0,5	2
2	363,34	39,86	82,33	355	32,06	21,00	3,32	3,16	0,5	2
3	360,67	47,22	89,83	351	24,68	21,00	3,16	3,04	0,5	2
4	360,15	40,03	75,6	364	31,92	21,00	3,36	3,28	0,5	2
5	365,56	57,48	13,35	418	15,13	21,00	4	3,8	0,5	2
6	365,99	43,11	13,34	418	29,5	21,00	4	3,88	0,5	2
7	360,39	57,8	8,97	430	14,68	21,00	4	3,84	0,5	2
8	359,76	42,8	8,55	431	29,68	21,00	4	3,76	0,5	2
1	363,37	47,98	79,79	357	23,98	10,50	3,36	3,28	0,25	2
2	363,34	39,86	82,33	355	32,06	10,50	3,6	3,28	0,25	2
3	360,67	47,22	89,83	351	24,68	10,50	3,32	3,2	0,25	2
4	360,15	40,03	75,6	364	31,92	10,50	3,56	3,44	0,25	2
5	365,56	57,48	13,35	418	15,13	10,50	4	4	0,25	2
6	365,99	43,11	13,34	418	29,5	10,50	4	4	0,25	2
7	360,39	57,8	8,97	430	14,68	10,50	4	4	0,25	2
8	359,76	42,8	8,55	431	29,68	10,50	4	4	0,25	2
9	365,14	72,5	26,51	402	0,00	0	4	4	0	2
10	361,72	72,48	22,81	410	0,00	0	4	4	0	2

804

6. Conclusion

In this paper, we have considered a particular stochastic problem with resource allocation. We have successively considered three cases. Making some assumptions about the size of the time windows where the bricklaying tasks will be put : we can obtain **a priori** lower and upper bounds for the number of resources needed. Considering the concrete industrial problem with all its constraints, we used discrete simulation in order to test different strategies to control on-line the utilization of the resources and the production system (stopping the converters before their expected stop). Analysing each simulation and considering the deterministic problem **a posteriori** in order to examine if the number of used machines was adequately chosen. Even if the number of machines is infinite, malfunction costs persist due to too small successive repairing intervals. In this case, the expected value of the malfunction cost can be minimized by choosing a good strategy for the anticipated stop of converters : we are now studying this problem with stochastic models.

References

[BAK 83]   E. BAKER, An Exact Algorithm for the Time Constrained Traveling Salesman Problem, Oper. Res. , 31, 938-945, 1983

[BAR 88]   M.BARTUSCH, R.H. MÔHRING, F.J. RADERMACHER, Scheduling project networks with resource constraints and time windows, Annals of Operations research, Vol 16, 201-240, 1988

[BER 91]   D. J. BERTSIMAS, G. VAN RYSIN, A stochastic and dynamic vehicule routing problem in the euclidean plane, Oper. Res. , 39 n° 4, 601-615, 1991

[CHU 92]   C. CHU, B. FINEL, M.C. PORTMANN, J. M. PROTH, N. SAUER, Simulation de politiques d'ordonnancement de machines, INRIA-LORRAINE, Rapport n°2, Projet EUREKA, mars 1992

[DES 88]   M. DESROCHERS, F. SOUMIS, A reoptimization algorithm for the shortest path problem with time windows, European Journal of Operational Research, Vol 35, 242-254, 1988

[FER 89]   J.A. FERLAND, L. FORTIN, Vehicles scheduling with sliding time windows, European Journal of Operational Research, Vol 38, 213-226, 1989

[FIN 92]   B. FINEL, Ordonnancement à contrainte cumulative en univers déterministe et stochastique : allocation dynamique de machines à briqueter à des convertisseurs, Mémoire CNAM, CAMOS, Metz, 16 octobre 1992

[GAR 79]   M.R. GAREY, R.L. GRAHAM, D.S. JOHNSON, Computers and intractability : a guide to the theory of NP-completness, Freeman, San Francisco, 1979

[GUI 84]   A. GUINET, Transports industriels routiers, un problème d'affectation avec réemplois sous contraintes, R.A.I.R.O. Recherche opérationnelle, Vol 18, 353-379, n° 4, novembre 1984

[HAO 90]   M. HAOURI, P. DEJAX, M. DESROCHERS, Routing problems with time windows constraints, state of art, Rairo Recherche opérationnelle, Vol 24, n° 3, 217-244, 1990

[KUH 55]   H.W. KUHN, The hungarian method for the assignment problem, Naval Research Logist. Quart., n° 2, 83-97, 1955

[ROY 70]   B. ROY, Algèbre moderne et théorie des graphes, Tome 2, ch 8, DUNOD, 1970

[WUR 89]   P. WURTH S.A., Highly flexible automated and integreted bricklaying system, FAMOS project, Definition Study, septembre 1989

A decision aid tool for deliveries planning in a cement plant

Marc J. A. LESCRENIER

ARBED s.a. Service de l'Informatique
19, avenue de la Liberté
L-2930 Luxembourg

Abstract

Daily planning of cement deliveries in a cement plant aims at just-in-time deliveries of customers orders using a minimum number of lorries. Planning involving about 100 customers orders can require more than one hour work for an experimented employee. Customers from the building domain frequently ask for late modifications of their orders such as ordered quantity or delivery time. Such modifications imply almost permanent adaptation of the initial planning which is time consuming and spoils the benefits of the initial optimization attempt.

In contrast, a computerized decision aid tool would allow for fast and permanent optimization in the course of the day. This paper presents first results of a study for determining an adequate optimization method. Simulated annealing techniques outperform the classical integer linear programming method: both approaches are described and discussed.

1 The industrial background

The ARBED group, primarily engaged in iron and steel production, preserves a diversification of its activities inside and beyond the frontiers of Luxembourg. For instance, the group is also active in Luxembourg in cement and bricks production and supplies customers involved in building or road construction.

In this production area, each customer order specifies the quality of cement, quantity in tons and delivery delay in the form of "not before 7.00 am and not after 7.30 am". Delivery of each order consists in the loading of a lorry at the cement plant, transportation to the customer, unloading of the lorry and return journey. The lorries are ordered at a nearby company and are available on the cement plant in very short delay.

2 Daily planning

The customers orders are registered in the course of the day. At the end of the day, planning of deliveries is established so that lorry drivers are informed of their next day schedule. The number of drivers and the starting time of their working day can vary from day to day.

Here is an example of the result of such a planning. Each line stands for a delivery and mentions successively: the customer identification - delivery period imposed by the customer - loading time - unloading time - time for a one-way journey between the cement plant and the customer - delivery time planned by the cement plant.

LORRY No 1

Smith -[5.30 am, 6.30 am]- 15' - 45' - 30' - 6.00 am

Duff -[7.45 am, 8.00 am]- 15' - 45' - 30' - 8.00 am

Reid -[10.30 am, 11.00 am]- 15' - 45' - 15' -10.30 am

Johnson -[1.00 pm, 2.30 pm]- 15' - 45' - 45' - 1.00 pm

Gould -[3.30 pm,4.30 pm]- 15' - 45' - 30' - 4.00 pm

LORRY No 2

Kain -[4.30 am, 5.30 am]- 15' - 45' - 150' - 5.00 am

Bensy -[10.30 am, 11.00 am]- 15' - 45' - 60' -10.30am

Ford -[12.30 am,1.00 pm]- 15' - 45' - 15' - 1.00 pm

LORRY No 3

Royce - [5.30 am, 7.00 am] - 15' - 45' - 60' - 6.00 am

Mill - [9.30 am, 10.30 am] - 15' - 45' - 30' - 10.00 am

Suey - [11.30 am, 12.30 am] - 15' - 45' - 30' - 12.00 am

Light- [2.00 pm,4.00 pm] - 15' - 45' - 50' - 2.30 pm

3 Problem formulation

At the light of this example, we can now formulate the problem faced by the cement plant every day. Given a set of deliveries to satisfy, determine the schedule of a minimum number of lorries so that:

1. a lorry is loaded for only one customer at a time and returns to the cement plant after each delivery.

2. the delivery period specified by the customer is satisfied.

3. each lorry (or driver) does not work more than the legal 8 hours per day.

4. limited overtime is allowed for each lorry but total overtime is minimized.

5. idle time for each driver between deliveries is minimized.

In the course of the day, initial planning is regularly overthrown due to late order modifications asked by customers concerning the ordered quantity or delivery period for instance, not to mention cancellation of orders or registration of new orders for the same day.

In that case, new planning is required taking into account the remaining deliveries, the number of worked hours for each driver and possible delay for obtaining new lorries on the cement plant if necessary. Significant modifications of the schedule for each driver are permitted in order to optimize the criteria mentioned above.

We will now discuss and compare two approaches proposed for use in a computerized decision aid tool for optimized planning determination.

For this purpose, we will restrict our attention from now on to a simplified version of the initial problem: given a set of deliveries to satisfy, determine the schedule of a minimum number of lorries so that:

1. a lorry is loaded for only one customer at a time and returns to the cement plant after each delivery.

2. the delivery time (under the form of a precise hour in contrast to a period) specified by the customer is satisfied.

3. each lorry (or driver) does not work more than the legal 8 hours per day.

4 A linear programming model

A model using mixed integer linear programming has first been established.

A decision variable is introduced for each {delivery, lorry} pair modelizing the decision of assigning the delivery to the lorry.

Additional real variables are introduced to compute departure and return time of each lorry so as to take into account the legal 8 hours work per day.

A last set of decision variables is used in order to count the number of lorries that are used in the planning.

The corresponding mixed integer linear programming model has been optimized using a commercialized modelizer and solver implementing a branch and bound technique when dealing with integrity constraints and allowing definition of SOS sets of type 1 and 2.

Tests on a set of problems of real size (typically of the order of 60 deliveries for 20 lorries) show that computation time of about 30 minutes on a DIGITAL VAXstation 3100 are necessary to find a good solution (not necessarily the optimal solution).

Such computation times are acceptable if no more than one planning is carried out once a day. In the context of permanent reoptimization due to orders modifications by customers, these computation times are prohibitive which is why an alternative method is to be found.

5 A model using graph theory

5.1 Model description

The graph coloring problem is one of the most popular problems in graph theory. Given a graph $G = (V, E)$, the problem consists in finding a minimal coloring, i.e. a mapping f from V into $\{1, 2, \ldots k\}$ such that k is minimal and

$$f(u) \neq f(v), \forall u, v \in V \mid \{u, v\} \in E$$

This problem is encountered in a large number of conflicts resolution applications since it can be viewed as the problem of finding a partition of the sets of vertices into a minimal number of independent sets. The simplified version of the problem we face can be modelized in terms of graph coloring as follows.

The deliveries form the set E of vertices and lorries correspond to colors in {1, 2, ... k}. An edge {u, v} is built between two vertices if and only if at least one of the following conditions is satisfied:

1. the deliveries u and v overlap in time (considering departure and return times for u and v) and therefore cannot be taken in charge by the same lorry.

2. the deliveries u and v, if taken in charge by the same lorry would lead to overtime.

For instance, here is the colored graph corresponding to this set of deliveries:

Smith - [6.00 am, 6.00 am] - 15' - 45' - 30' - 6.00 am

McKain - [5.00 am, 5.00 am] - 15' - 45' - 150' - 5.00 am

Duff - [8.00 am, 8.00 am] - 15' - 45' - 30' - 8.00 am

Light - [4.00 pm, 4.00 pm] - 15' - 45' - 50' - 4.00 pm

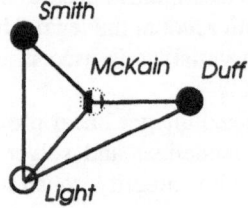

The minimal number of colors is 3, deliveries for Smith and Duff being taken in charge by the same lorry.

5.2 Solving techniques

Most graph coloring problems are NP-complete so that optimal algorithms are likely to be exponential. And indeed, practice shows that classical exact methods developed for solving graph coloring problems require unacceptable computation times when dealing with graphs of more than 100 vertices (see [2] for instance). Our experience with a mixed integer linear programming model is an illustration.

As the problem under consideration may involve about 100 deliveries it appears wiser to turn to heuristics for solving it. Computational experiments show that tabu search techniques [3] and simulated annealing techniques are often adequate to tackle large graph coloring problems.

A simulated annealing algorithm for solving general graph coloring problems has then been implemented. We refer to [1] for an introduction to simulated annealing techniques. Performance of such techniques is strongly related to the choice of parameters such as the cooling schedule and the cost function. These parameters have been chosen as suggested by Aarts and Korst in [1]. For instance, the cost function to maximize has been defined as

$$f=\sum_{i=1}^{kmax} w_i(|V_i| - p|E_i|)$$

where *kmax* is the maximal number of colors, w_i is the penalty cost for using color i, V_i is the set of vertices colored with color i, E_i denotes the set of edges {u, v} in E with u and v in V_i and p is the penalty factor for unfeasible solutions.

5.3 Implementation and computational results

The algorithm has been implemented in standard FORTRAN 77 so as to be available on PC DOS or DIGITAL VAX VMS machines. We report in this section results obtained for this implementation when applied to the graph coloring model corresponding to the simplified version of the deliveries planning problem.

Computation times of about 10 seconds are observed for problems of real size (typically of the order of 60 deliveries for 20 lorries) on a DIGITAL VAXstation 3100 or a PC DOS 486 50 Mhz. The solutions obtained are often optimal. In all cases, they are nearly optimal in the sense that they do not differ significantly in number of lorries that are needed with respect to the optimal solution.

Besides, such computation times now permit successive planning during the day so that permanent reoptimization is made possible in case of orders modifications by customers.

6 Model and algorithm extensions for solving the initial problem

In order to prevent cases where the simulated annealing algorithm does not find any feasible coloring, it is wise to compute a good initial feasible solution using sequential coloring algorithms [4] for instance.

If we now turn back to the initial problem, we need to extend our model and/or algorithm so that the complete set of specifications is taken into account. We will not enter into fine details of such adaptations, most of them are minor and too specific to the problem to present any real interest in the context of this general presentation. We will pay however particular attention to the way definition of delivery times intervals for each order can be handled.

For each delivery, a discrete set of possible delivery times is considered in accordance to the delivery times interval imposed by the customer. The graph is then extended in the sense that a vertex is now considered for each possible delivery time. Vertices corresponding to the different possible delivery times for a given order form what we will call a supervertex.

Edges are build between vertices as is described for the simplified model but no edge is created between vertices of the same supervertex.

An extended definition of the graph coloring problem for this graph is then introduced: find a minimal coloring, i.e. a mapping f from V into {0, 1, 2, ... k} such that

1. k is minimal,

2. all vertex in a supervertex are colored with color 0, except one which is colored with a color between 1 and k,

3. and for all u and v in V linked by an edge $\{u,v\}$, either

$$f(u) \neq f(v)$$

or

$$f(u) = f(v) = 0$$

During the simulated annealing process, a solution (not necessarily feasible, see 5.2) is generated in the neighborhood of the current solution in the following way:

1. choose arbitrarily a vertex of the extended graph

2. if this vertex is colored with a color $c>0$, choose arbitrarily another color d in $[0,k]$ and color the vertex with this color. If $d=0$, choose arbitrarily another vertex colored with color 0 in the same supervertex and color it with an arbitrarily nonzero color.

3. if this vertex is colored with the color 0, choose arbitrarily another color in $[1,k]$ and assign this color to the vertex. Change also to 0 the color of the only vertex of the supervertex initially colored with a nonzero color.

7 Conclusions

A prototype implemented for solving a simplified version of the deliveries planning problem shows that a computerized decision aid tool allows for fast and permanent optimization.

Experiments for this simplified problem show that simulated annealing techniques outperform the classical integer linear programming method.

If the simplified problem has been modelized as a classical graph coloring problem, the actual problem to solve presents more specific constraints and objectives. Simulated annealing techniques present a high degree of flexibility that proved to be of great help for handling these particular specifications.

Experiments with the extended model and algorithm are not available yet.

8 Bibliography

[1] Aarts E. and Korst J., "Simulated Annealing and Boltzmann Machines", John Wiley & Sons, 1990.

[2] Campers G., Henkes O. and Leclercq J.-P., "Sur les méthodes exactes de coloration de graphes", Cahiers du Centre d'Etudes et de Recherche Opérationnelle, Vol. 19, No. 1-2, 1987, Ceuterik, Bruxelles.

[3] Hertz A. and de Werra D., "Using Tabu Search Techniques for Graph Coloring", Computing 39, 345-351 (1987).

[4] Leighton F.T., "A Graph Coloring Algorithm for Large Scheduling Problems", Journal of Research of the National Bureau of Standards, Vol. 84, No. 6, November-December 1979.

An Estimation of Energy Saving Potential by the Allocation of Co-generation Systems

Shigeo Sagai[*1], Yoshifumi Fujii[*2], and Takayuki Shiina[*1]

*1: Communication and Information Research Laboratory,
Central Research Institute of Electric Power Industry,
2-11-1 Iwado-kita, Komae-shi, Tokyo 201, JAPAN
*2: Bunkyo University, 1100 Namegaya, Chigasaki-shi, Kanagawa 253, JAPAN

1 Introduction

Estimation of energy saving potential is highlighted again by the Global Warming problem. Research to estimate energy saving potentials treating individual technical field or each industry as one unit have been carried out. Especially, many efforts are performed to improve energy efficiency of energy supply equipments such as generators, turbines, and so on.

On the other hand, little researches has been done to evaluate total integrated energy saving potential by introducing systems such as co-generation systems into districts. Two points are included in this problem(See Fig. 1). The first problem is to establish a method of evaluation for total energy supply system. We need to take into account the 'integrated effect' of connected equipment, so the problem is more difficult than evaluating each component equipment and then summing the results. The second problem is how to evaluate "regional" energy saving potential. A region contains many different types of load factors. We have to evaluate the mixture of complicated energy load in the region.

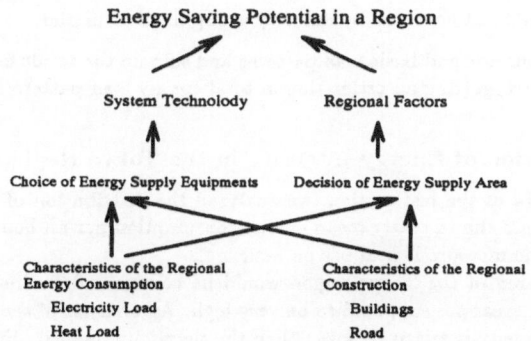

Figure 1: An Overview of Research

We chose the Tokyo region as an object for this case study. And we also have developed a method to evaluate district energy saving potential by the setting up of suitable co-generation system into the district. The result of preliminary analysis obtained by applying this method will be shown.

2 An Overview of Energy Load and Geographic Characteristics in Tokyo

2.1 About Energy Load Data and Their Estimation

We selected Tokyo as a target area for a case study. In this study, we treat an about 1km square area called "standard district mesh" in Japan as a unit and call it as a "district". Six hundred districts are included in the target area and the region spreads about 30 km from north to south and 20 km from east to west. We estimated commercial/business and residential electricity and heat load patterns of a typical day of each month for each district in the region. The estimated data includes:

- the electricity load pattern of each district,

- the heat load pattern of each district, including the cooling, heating, and hot-water supplying.

So, each districts has 48 load patterns — 4types of load pattern for each month of a year. The outline of estimations is as follows:

1. estimate energy intensity per unit floor size for each industry group, The details of each step are omitted, but we checked the consistency of the estimation, comparing with the macro energy statistics data of Tokyo published by the government.

2. estimate basic energy load pattern by the real data,

3. estimate the total floor size by industry group in each district,

4. multiply floor size and basic load patterns and sum up the result for all the industry group in the district, getting an estimation of total energy load pattern in the district.

2.2 Distribution of Energy Intensity in the Tokyo Region

Using the data of the last section, we analyzed the distribution of energy intensity in the Tokyo region. Fig.2 shows yearly mean energy consumption per an hour in each district in the Tokyo region. The measure is kcal per an hour.

The central area of the Tokyo region would be commercial/business districts and energy intensity of those areas are supposed to be very high. An analysis of energy consumption shows that the heat intensity is relatively lower than the electricity intensity in the central area. So, if co-generation is planed to supply all the electricity in the district, the heat surplus would occur and the surplus would be abandoned outside, causing the total energy efficiency keep low. It is expected that co-generation systems with efficient electricity generation are suitable for the energy supply of such districts.

On the other hands, resident area spreads outside of Tokyo surrounding commercial and business districts. The energy intensity of the district is relatively low and the heat intensity is higher than the electricity intensity, opposite to the central district.

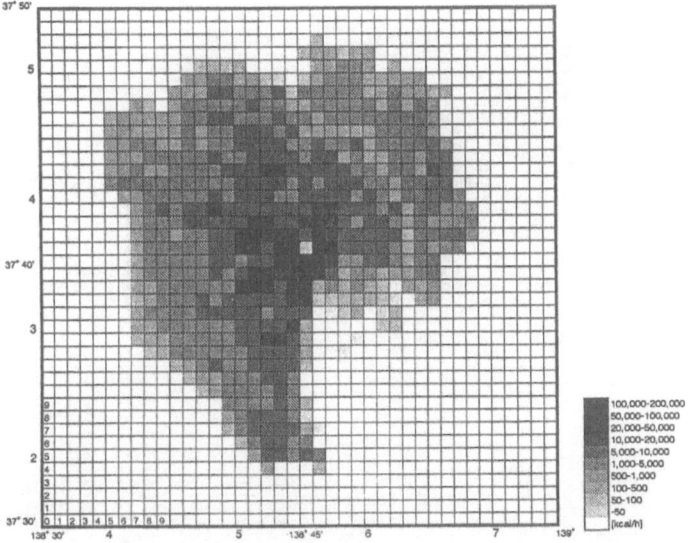

Figure 2: Energy Intensity in the Tokyo Region

2.3 Classification of Districts by Energy Load Patterns

We tried to classify the energy load pattern of each district to get an overview of energy consumption characteristics of the Tokyo region. The distribution of energy consumption characteristics shows how the region is used. The method is as follows:

- Define the basic energy load patterns. The basic energy load patterns are collected by an actual survey. We have five basic load patterns, including

  1. commercial,
  2. business,
  3. mixed 1 (shows tendency to be more similar to business),
  4. mixed 2 (shows tendency to be more similar to residential), and
  5. residential.

- classify all the districts into five groups to the nearest basic load patterns

Fig.3 shows the result of classification of districts in the Tokyo region. Checking the map, classification result is not far different from the real constructions of those districts. From the view point of energy consumption, the constructions in the districts which have similar characteristics is gathering.

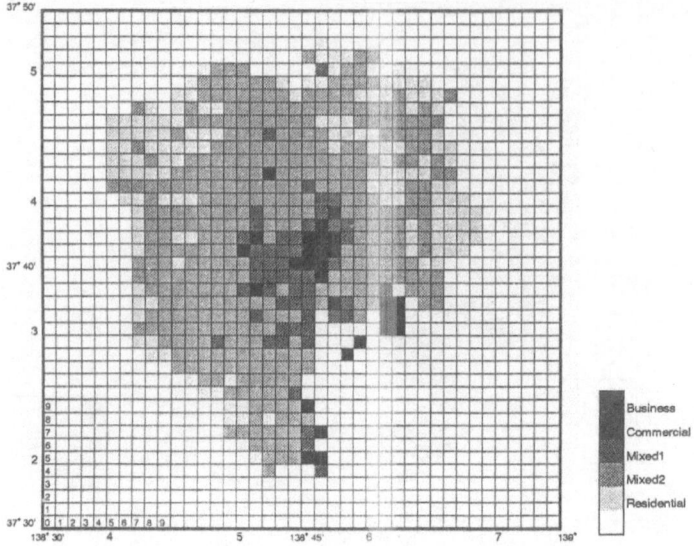

Figure 3: Classification of Districts by Energy Load Patterns

3 Model Formulation

3.1 The Problem

Energy supply by co-generation systems is efficient if it is optimizing well-balanced to both the electricity consumption and heat consumption patterns, because it generates both heat and electricity at the same time. Evaluation of the improvement in energy efficiency is obtained by :

1. evaluating the energy saving rate attained when an co-generation system *suitable for the energy load characteristics in each area* is built,

2. deciding *the appropriate supply area* for the co-generation systems.

3.2 Energy Saving Potential by Introducing Co-generation System into Districts

To describe the characteristics by the co-generation types, we set parameters of energy conversion rates for 4 types of co-generation systems.

And to describe the operation patterns of co-generation systems, we prepare 3 operation patterns. We evaluate the energy saving rate of all the 12 combination of systems and operations

Table 1: Model variables, data, and parameters

Parameter and Data	
name	meaning
i	suffix for time
j	suffix for area
k	suffix for co-gen. type(1=GT,2=GE,3=DE,4=FC)
η^B	heat transfer efficiency of exclusive boiler
η_k^C	cold heat transfer efficiency of k-type co-gen.
η_k^H	hot heat generation efficiency of k-type co-gen.
η^P	electricity generation efficiency of commercial line
η_k^e	electricity generation efficiency of k-type co-gen.
η_k^h	heat generation efficiency of k-type co-gen.
η_k^a	heat effective use rate of k-type co-gen.
α_k	$= \eta_k^h/\eta_k^e$
η^f	heat effective use rate of exclusive boiler
E_{ij}	electricity demand of j-area in i-time
H_{ij}^h	hot heat demand of j-area in i-time
H_{ij}^c	cold heat demand of j-area in i-time
variable	
x_{ij}	electricity generation by co-gen. of j-area in i-time
y_{ij}	heat generation by co-gen. of j-area in i-time $y_{ij} = \alpha_k x_{ij}$

to select the best combination for the energy load pattern of a district. Co-generation system types are followings:

1. Gas Turbine(GT), 2. Gas Engine(GE), 3. Diesel Engine(DE), 4. Fuel Cell(FC).

Followings are the operation patterns:

1. Square Shaped Approximation Operation

2. Square Shaped Approximation + Base Operation

3. Electricity Demand Following Operation

We assume that only one co-generation system is built in a district and the system supplies as much energy as it can within the range that it generates no excessive electricity. This assumption is set because the reverse current of electricity would not be occur.

$$i.e., x_{ij} \leq E_{ij} (\text{See Table 1}) \tag{1}$$

Typical operating pattern of electricity generation are shown in Fig.4. The Square Shaped Approximation Operation is a simple model of operation in which the co-generation is started and stopped once in a day. The output level of a model generator is optimized in the above constraint. The Base Operation is the operation in which the co-generation keeps working all day and the output level is a constant. The constant is the minimum demand in a day. The mixed patterns of above two is decided to supply as much as electricity to the district. So first,

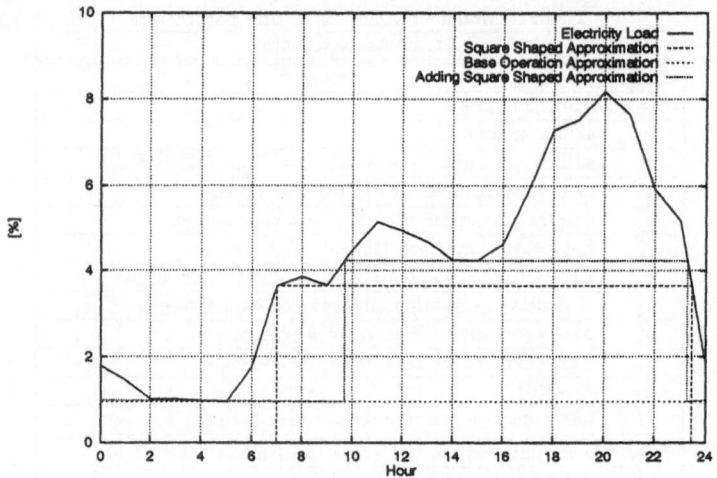

Figure 4: Co-generation System Operation Pattern Models

the minimum generation level is covered by the base operation, and the optimum square shaped pattern of the rest load shape can be decided by the same algorithm to decide the optimal square shaped approximation. In any model operation, if heat surplus is generated, the surplus is assumed to be abandoned. And on the electricity demand following operation, the generator is assumed to be operated to meet the electricity demand, following the load curve.

Formerly said, a district has electricity load pattern and heat load patterns of a typical day of each month of a year. To evaluate the energy saving rate, this model calculates the primary energy consumption of both by a conventional way of supply and by co-generation systems. By the conventional way, heat is supplied by the exclusive boiler and electricity is supplied by the commercial line. On the other hand, by the co-generation system, only shortage of generated heat and electricity is supplied by the conventional way.

When the co-generation type introduced in a region is defined, the primary energy consumption[1] by a conventional system ($= Q_{1j}$) is calculated as follows:

$$Q_{1j} = \sum_{i=1}^{24} (\frac{E_{ij}}{\eta^P} + \frac{H_{ij}^h}{\eta^B \eta^f} + \frac{H_{ij}^c}{\eta^B \eta^f}). \tag{2}$$

There are two cases of heat balance according to whether the heat generated by the co-generation is enough for the demand or not. The primary energy consumption to supply both heat and electricity consumption in a district by a co-generation system ($= Q_{2j}$) is defined as

---

[1] total energy consumption to supply heat or electricity concerning energy transforming efficiency from fuel such as oil or gas

follows:

$$q_{ij2} = \begin{cases} \dfrac{(E_{ij} - x_{ij})}{\eta^P} + \dfrac{x_{ij}}{\eta^e_k} & : \text{if heat generation is enough} \\[3mm] \dfrac{(E_{ij} - x_{ij})}{\eta^P} + \dfrac{x_{ij}}{\eta^e_k} + \dfrac{(\frac{H^h_{ij}}{\eta^H_k} + \frac{H^c_{ij}}{\eta^C_k} - \eta^a_k y_{ij})}{\eta^B \eta^f} & : \text{Otherwise.} \end{cases}$$

(3)

$$Q_{2j} = \sum_{i=1}^{24} (q_{ij2})$$

(4)

The energy saving rate of j-district($= R_j$) is defined comparing above Q_{1j}, Q_{2j} as follows:

$$R_j = \frac{(Q_{1j} - Q_{2j})}{Q_{1j}}.$$

(5)

We need attention in that R_j is calculated from just an estimation of the primary energy to supply all the energy demand of a district. Other factors such as energy loss during distributing heat or energy to pressure heat conductor are not taken into account.

3.3 Making Combination of Districts to Improve Heat Efficiency

We also tried to evaluate the energy saving by "connecting districts" to decide the energy efficient supply area. We can expect improvement in energy efficiency by connecting districts into one large area and supply energy from one bigger equipment. Because the component districts have compliment energy load patterns each other making a good pattern for energy supply by a co-generation system. This model estimates the effect of the reformation of whole load shape of such combined districts.

We regard a district as a node, and the border of two adjacent districts as an edge of a tree.

The most strict and intuitive method to get the optimal connection set of districts is to allocate 0-1 variables for all the border between two districts and calculate energy saving rate for all the combinations of 0-1s. Suppose n is the number of districts, then total number of estimation of all the subsets of combination of districts becomes 2^n.

We developed a simple greedy algorithm and applied it. So we can get feasible solution in the problem in practical time.

The algorithm first decides the candidates for connection from all the borders of the areas. These candidates compose a set of trees (= forest). Then the algorithm selects a tree in the forest and decide which connection to leave in the tree by a greedy type algorithm.

When this algorithm stops, a forest which indicates the energy saving is constructed. From the viewpoint of energy consumption, one large connected components of this forest by the above algorithm means that load pattern of each areas compensates others to good patterns in their neighborhood.

4 Model Simulation Studies

4.1 Assumptions

Many complications must be considered in real operation of a co-generation system. But to simplify the situation, we set next assumptions:

1. electricity or heat generation is in proportion to operation time of each equipment (i.e. energy transformation/generation efficiency is a constant),

2. energy consumption in a district is supplied by the co-generation system set up in the district, except for the districts included in a combination,

3. equipment set up cost is not considered,

4. any energy storage is not considered,

5. environmental regulation is not considered.

4.2 Energy Saving Potential in Each District in the Tokyo Region

We applied the model for the above data of the Tokyo region and estimate energy saving potentials when a co-generation system is introduced (or both heat and electricity is supplied imaginary from one equipment at the same time) in each district.

We estimate all the 12 combinations of the equipments and the operations and choiced one which attain the maximum energy saving rate for each district. The selected combination is thought to be the most adapted to the energy load characteristics of the district. For most districts, the most efficient combination is introducing fuel cell system in the electricity demand follow operation. The result is because setting of the electricity generation efficiency for the fuel cell system is higher than other systems. Though the result show us the high efficiency of fuel cell system, we need to further check for the estimation of those efficiency parameters.

Fig.5 shows the energy saving potential by introducing co-generation system in the Tokyo Region.

Next points become clear by above analysis:

1. The area which has high energy saving ability lies around the central area of Tokyo in a doughnut shape.

2. The average energy saving potential by introducing co-generation systems in Tokyo district is about 20% by the above rough evaluation.

This result of average energy saving potential is lower than we expected before the estimation. As we have not done cost analysis, we cannot judge whether this energy effect counterbalance the investment to the co-generation yet.

4.3 Energy Savings by Connecting Districts

We tried to make combination of district to save energy for the above 12 combinations of the equipment and the operations.

Fig.6 shows an example of the combination of districts in the Tokyo region. This figure shows the case of setting a gas engine co-generation systems in each area which is operated following the electricity demand. The edges between areas indicate connection and the isolated points indicate the nodes left alone. From the Fig.6, we see that the connected area is not so large, and the maximum width is about 3 km. And the connected areas spread near the boundary between the business/commercial and the resident district (Compare with Fig.2.) .

The total amount of energy saving caused by connecting districts is relative very small compared with the improvement in efficiency of energy supply equipment. The result is , we suppose, because a districts of 1 km square shape is large and the energy load shape improvement is already attained within the district.

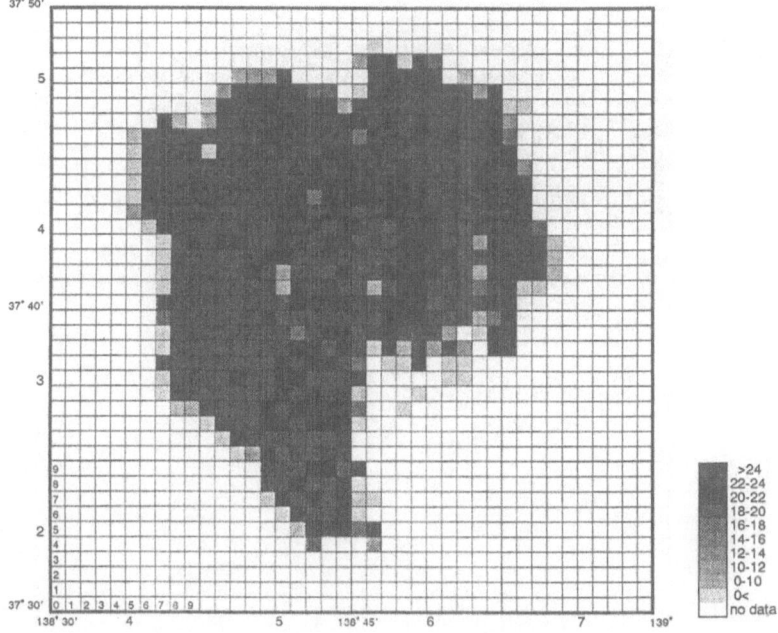

Figure 5: Energy Saving Potential by introducing co-generation system in the Tokyo Region

5 Concluding Remarks and Future Subjects

We developed a model to evaluate the energy saving potential by introducing co-generation systems into districts. The above analysis is an estimation of one upper limit of energy saving potential without cost terms.

We made another analysis including the estimation of the ability to set heat distribution pipes in the Tokyo Region. The result shows the structure of each district, such as the average size of buildings and the length of roads, quite affect the possibility to set up a co-generation system in the district. Greater sophistication for this method based on the cost-benefit analysis will be needed to obtain more practical estimation.

ACKNOWLEDGMENTS

The authors thank Dr. Yutaka Tonooka of the Institute of Behavioral Sciences for his insights on the results and helping to collect the energy data.

820

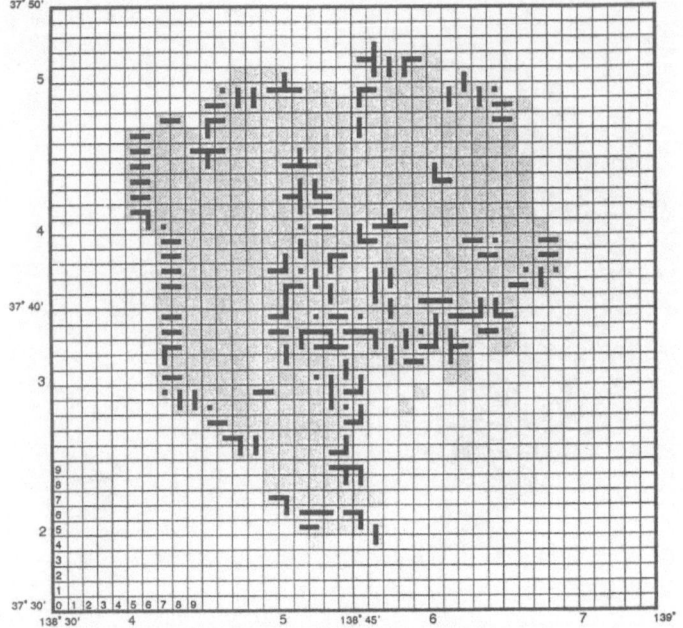

Figure 6: An Example of area connection to increase the effect of energy saving

References

[1] S. Sagai and T. Shiina: An Estimation of Energy Saving Potential by introduction of Co-generation Systems into Tokyo District, CRIEPI ERC Rep., No. Y92004, 1992 (in Japanese).

Combinatorial geometry

AUTOMATIC LAY PLANNING FOR IRREGULAR SHAPES ON PLAIN FABRIC.
SEARCH IN DIRECT GRAPH AND A_ε ε-ADMISSIBLE RESOLUTION

G. ROUSSEL, S. MAOUCHE
Centre d'automatique de Lille
Université des Sciences et Technologies de Lille
59655 Villeneuve d'Ascq

Abstract: *The paper deals in optimal allocation of irregular shapes on a single sheet. The proposed formulation is particularly adapted to fit in an integrated continuous cutting system in garment industry. After a description of the shape modelling, the mathematical operators to handle the shapes and the space state representation, one resolution is given using the search of a minimal cost way through out the graph with A_ε , an ε -admissible algorithm.*

Keywords: ε-admissible algorithm, irregular shapes, optimal allocation, waste optimization, marker making, clothing industry, flexible cutting cell.

I. Introduction

The automatic allocation problem of irregular shapes gives rise to interest from cutting industry. As a matter of fact, whatever material one have to cut textile, leather, paper, wood or steel, the step of fitting irregular shapes on a sheet is complex. The quality of the lay plan and the necessary time to reach it are some antagonistic criteria which must adjusted according to the branch of industry.

The goal of the paper is to present an algorithm for template layout problem particularly fitted to the cutting of garment pieces by integrated continuous cutting system. This is the marker making. The proposed formulation is adapted in order to provide some partial layouts in phase with the cutting module rhythm. A single sheet of material is enrolled and put forward on the cutting table. This work takes place in the flexible cutting cell development already described in [1], [2]. This frame is composed of three successive modules: inspection, marker making and cutting. Using information from a collection of shapes whose references are given in the order book, the marker making module have to be able to process the absolute location of all the shapes to lay. When a subset of the order book allows to build a layout having satisfying length and quality, the layout is then communicated to the cutting module, and the marker making module starts again with the updated order book. The inspection module fits to characterise the stuff, i-e the periodicity of patterns, the boundaries and the location of flaws. But, this study is bounded to the allocation problem on plain fabric without flaws. It remains to define the idea of layout quality. There, the quality of a layout is measured by a material consummation cost, i.e. the length of the strip of cloth used to fit all the shapes. But, if the order book "$\mathcal{F}$" hold many references, it cannot find a minimal cost solution in an acceptable time. The layout of all the pieces is then realised in yielding some successive partial layouts P_q until the order book is empty. Each unitary layout P_q contains a subset $\mathcal{F}_q$ of shapes. The following algorithm allows to yield the unitary layouts

822

P_q with a near optimal efficiency in combining a subset of shapes included in the current order book

$$F_k \in \left(F - \bigcup_{r=0}^{q-1} F_r \right)$$

Several authors have already worked on the problem of allocation of irregular shapes. Among the automatic lay planning, we discern:

1) The "two stages", method based on a first step to build rectangular modules followed by a rectangular layout step [3], [4], [5], [6], [7].

2) The heuristic algorithm based on the using of sequential and deterministic rules [8], [9], [10], [11], [12], [13], [14], [15], [16].

3) The non deterministic resolution whose the main feature is to allow the backtracking when the expanded way is not valuable. In order that, they use some operational research theories [17], [18], [19], stochastic combinatorial optimization methods [20], [21], [22] or the search in graph [23]. At end, we have find some publications dealing in on some attempts with expert system and simplified shapes [24], [25]. Our algorithm is also based on search in graph method as [23]. However, it is different by the shapes description technique, the space of possible configurations, the control operators, the data structure, the oriented object choice, the goal to reach, the strategy of research and the guiding heuristics.

II Constraints and objective expressions

Let $C = \{\ C_k,\ k \in [0,N]\}$ be the set of $(N+1)$ varieties of closed boundaries, each of them different of the others, such, being given the current order book $OB= (a_0,a_1,...,a_k...a_N)$, we have to yield a strength of a_k pieces of boundary C_k. We define by $\mathcal{F} = \{F_{k,i}\ /\ k \in [o,a_k]\}$ the set of forms to yield having a boundary C_k and a mark i. Each form $F_{k,i}$, $i \in [o,a_k]$, can be characterised by the following assumptions:

C_k = boundary of $F_{k,i}$

R_k = rectangular enclosure of C_k

L_k = length of R_k

A_k = area of $F_{k,i}$

$O_{k,i}, X_{k,i},\ Y_{k,i}$= orthogonal basis linked to $F_{k,i}$

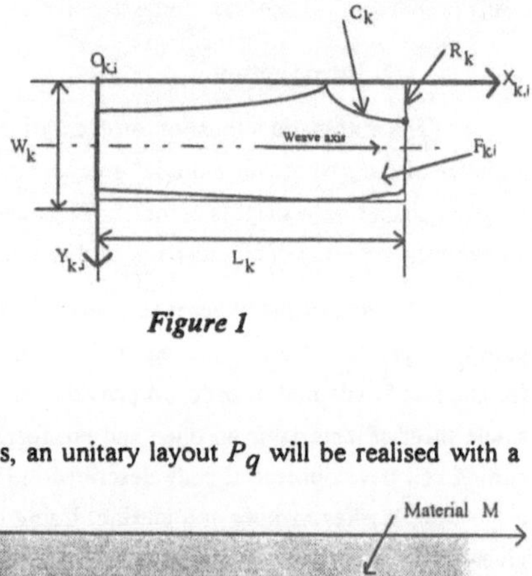

Figure 1

Let $\cup$ be the internal operator of clustering of shapes, an unitary layout P_q will be realised with a clustering of subset $\mathcal{F}_q$ of shapes included in $\mathcal{F}$ We will write $P_q = \bigcup_{F_{k,i} \in \mathcal{F}_q \subset \mathcal{F}} F_{k,i}$

Finally, the global layout will hold all the unitary layouts, i.e. $P = \bigcup_q \mathcal{F}_q$.

The layout P fits on a sheet of plain material M (see figure 2), characterised by:

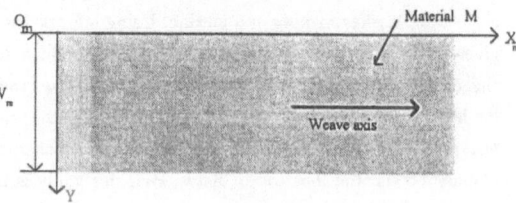

Figure 2

-W_M width of the sheet

-O_M, X_M, Y_M the orthogonal basis linked to the material. The material has a fibre axis considered parallel with the $O_M X_M$ axis

The sheet has a non bounded length relative to $O_M X_M$ direction. The clustering operator will be preceded by a local nesting search, which must provide the relative co-ordinates between two shapes. This clustering operator have to satisfy the following constraints:

1) *Non-overlapping constraint*

The inside of the two shapes are disjointed. We can write:

$$\forall F_{k,i}, F_{l,j} \in \mathscr{S}(k,i) \neq (l,j), INS (F_{j,i} \cap F_{l,j}) = \varnothing \qquad (1)$$

where $INS(.)$ denotes the inside of its argument

2) *Width constraint*

It is defined by the inequality:

$$W_q \leq W_M \qquad (2)$$

2) *orientation constraint*

$$\forall F_{k,i} \in \mathscr{S}, \; Ori(F_{k,i}) = \lambda \pi/2, \lambda \in Ori_{ad}(k) \subset \{0, 1, 2, 3\}. \qquad (3)$$

with $Ori(.)$ denotes the orientation of its argument and $Ori_{ad}(k)$ the set of admissible λ defined for the shapes $F_{k,i}$. The orientation of the shapes is defined by the angle between $O_{k,i} X_{k,i}$ axis of $F_{k,i}$ and $O_M X_M$ axis which indicates the material weave direction.

The aim of marker making is to provide a successive placements P_q, such:

$$Waste(q) = \underset{F_q \subset F - \bigcup_{r=0}^{q-1} F_r}{MIN} (W_M L_q - \underset{F_{k,i} \in F_q}{\sum S_k}) \qquad (4)$$

where $Waste(q)$ denotes the waste resulting of the placement P_q on the fabric of width W_M. We notice that W_M can take different values according to the irregularity of the edges. The formula (4) give the expression of the waste, i.e. the difference between the used area of fabric and the area of the forms. In order to remain compatible with continuous cutting process, the marker making will be yielded in successive set of strip whose length L_q must respect the following constraint:

$$\forall q \; \underset{F_k \in (F - \bigcup_{r=0}^{q-1} F_r)}{MAX \; L_k} \leq L_q \leq (1+\alpha) \underset{F_k \in (F - \bigcup_{r=0}^{q-1} F_r)}{MAX \; L_k} \qquad (5)$$

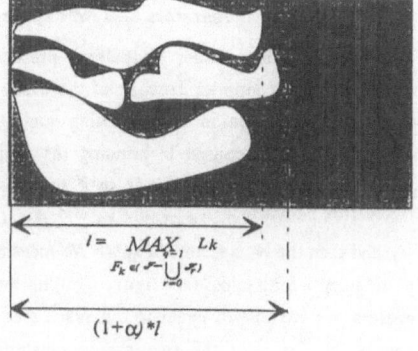

$$l = \underset{F_k \in (F - \bigcup_{r=0}^{q-1} F_r)}{MAX \; L_k}$$

$(1+\alpha)^* l$

Figure 3

(5) express the idea that the length of the unitary layout must range between 1 and $(1+\alpha)$ multiplied by the longest shape among the yet unallocated shapes.

This condition gives the lower and upper bounds of layout length allowing to allocate the longest shape of the current order book. This is experimentally favourable regarding the efficiency and the complexity of the algorithm. The parameter α is chosen in order to tolerate a small unbalancing of some shapes if this decreases the final waste (see figure 3). If the shapes are not too irregularly, $\alpha = 0.05$ gives good results.

III Shapes description

Here, we remind the principle of irregular shapes description developed in [22]. This technique has been worked out to process the local nesting of two shapes having a parallel direction. This technique is based on the coding of the visible boundary sub-curves relative to the four sides of the rectangular enclosure R_k. Let $C_{k,c}$, $c \in [0,3]$ be the four boundary sub-curves. We determine $\tilde{A}(C_{k,c})$, $c \in [0,3]$ be the four boundary sub-curves when $C_{k,c}$ are watched perpendicularly to the side c of R_k. Next, these sub-curves are sampled by a Dirac distribution of period T (see figure 4.a). Each boundary sub-curves is then simplified in keeping only the vertex of the visible segment (see figure 4.b). The list $S_{k,c}$ holds the X,Y co-ordinates of the previous vertex relative to the orthogonal basis (O_c, X_c, Y_c) of the considered side. $S_{k,c}$ are called reduced_Shape_Comb (see figure 4.c) . The vertex are selected by a polygonal rough estimate of $\tilde{A}(C_{k,c})$.

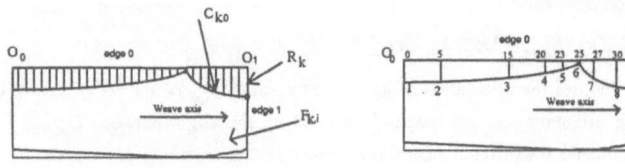

| Figure 4a | Figure 4b |

$$S_{k,0}=((0,5),(5,5),(15,3),(20,2),(23,1),(25,0),(27,3),(30,6),(35,7))$$

Figure 4c

IV Control operators and state space

The irregular shapes modelling presented previously has been adopted because of the easily to provide a precise evaluation of the nesting quality of two shapes. The idea consist in bringing the shapes $F_{k,i}$ and $F_{l,j}$ nearer (in the mind) until there is a contact vertex between $\tilde{A}(C_{k,c})$ of $F_{k,i}$ and $\tilde{A}(C_{l,j})$ of $F_{l,j}$, and then the X-axes are Nd apart. Nd measures the nesting depth for a gap S (see figure 5). With these parameters, we are able to evaluate the space area of rectangular enclosure of the two shapes clustering.

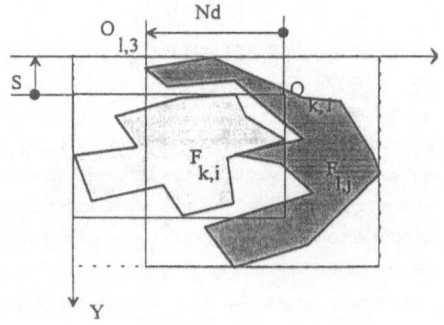

Figure 5

Several successive tries guided by a heuristic allows to bring the relatives co-ordinates $(Nd_{opt},\ S_{opt})$ giving the optimal nesting location . This couple of parameters will be useful to process the clustering hull of the two shapes. In nesting search, each admissible-orientations will be considered.

Therefore, we have built the following operators.

$\underline{F_{k,i}\ orientate:\ \lambda.}$ $\lambda \in \text{Ori}_{ad}(k)$: This operator orientates the receiver $F_{k,i}$ according the argument λ such that the angle between $X_{k,i}$ and X_M is $\lambda\pi/2$.

$\underline{F_{k,i}\ nestingSearch:\ F_{l,j}\ side:\ c}$: This operator returns the couple $(Nd_{opt},\ S_{opt})$ indicating the optimal relative location so as the rectangular enclosure area of the clustering of $F_{k,i}$ and $F_{l,j}$ be minimum. The parameter c indicates the side of contact about $F_{k,i}$. Notice that this operator is built such that all the constraints are intrinsically satisfied.

$\underline{F_{k,i}\ cluster:\ F_{l,j}\ configuration:\ (Nd_{opt},S_{opt},c)}$: This internal operator returns a new shape built in processing the clustering of the two shapes $F_{k,i}$ and $F_{l,j}$ for the considered configuration given in argument.

With all these operators, the problem can be represented with the help of a state graph $G = (\mathcal{S},O)$, where $\mathcal{S}$ is the set of possible states and O the set of operators transforming one state into another. Let u,v be two nodes corresponding to two states, hence the (u,v) arc will exist if and only if there is an operator $o_{i(u)}=v$. We will say that v is a successor of u, and $O(u)$ the set of successors of u. The problem can be explicitly formulated by:

1) an initial node $u_0 \in \mathcal{S}$ u_0 represents the empty lay-plan where any shapes has been allocated.

2) Each node u correspond to an intermediary layout $P_{q,u}$ brought after the clustering of $F_{k,i}$ and the antecedent node. Then, the layout $P_{q,v}$ has the following characteristics: $L_{q,u}$, $W_{q,u}$, $a_{k,u}$ $(k \in\ [0,N])$ respectively the length, the width and the order at the node u.

3) a unique operator $O_{k,l,c}(u)$ depending of three parameters (k,l,c). This operator carries out the following sequential steps.

if $a_k \neq 0$,

<u>step one</u> : $F_{k,i}$ $orientate:\ \lambda$;

<u>step two</u> : $(Nd_{opt},\ S_{opt}):=F_{k,i}$ $nestingSearch:\ F_{l,j}\ side:\ c$;

<u>step three</u> : $P_{q,v} := P_{q,u}$ $cluster:\ F_{l,j}\ configuration:\ (Nd_{opt},S_{opt},c)$

These successive operations build implicitly a new node v associated to the intermediary layout $P_{q,v}$. But, in order to limit the size of the successors set, a new shape can be allocate only at the right or at left of the intermediary layout relative to $O_q X_{q,v}$ and c takes its value in $\{0,2\}$

4) Each node u is associated to the cost $g(u) = waste(u)$. The cost has been presented in formula (4). Let us notice that $waste(u_0) = W_M \times MAX\left(\underset{n}{\underbrace{\bigcup_{n} k}} L_k\right)$ represents the maximal waste at the beginning of the

search of the layout P_q. On the other hand, while the length L_q of P_q is a constant, then $waste\ (u)$ is a monotonous decreasing function along a descending way of the tree.; that is not necessary true in case of L_q

increasing. This property is interesting to detect the sudden degradation of the efficiency for the visited node of the graph.

5) The sub-set $T \subset \mathcal{S}$ of terminal states. T is here implicitly known by the termination rules expressed by:

$$u \in T \Leftrightarrow \left\{ \begin{array}{l} \forall\, v \in O(u) \\ v \text{ does not satisfy the width constraint } \left(W_{q,v} > W_M\right) \\ \text{or } v \text{ does not respect the length bound } \left(L_{q,v} > (1+a) \times MAX(L_k)\right)_{F_k \in \left(\mathcal{F} - \bigcup\limits_{r=0}^{q-1}\mathcal{F}_r\right)} \\ \text{or } waste(u) \text{ become non monotonous } (g(v) > g(u)) \\ \text{or the order book is empty } (\forall\, k, \; a_k = 0) \end{array} \right\}$$

(6)

a terminal node t correspond to a candidate layout $P_{q,t}$ for P_q elections.

Let us precise that the previous graph is connected without cycle and looks like a tree, where u_O is the root and the terminal nodes are the leaves of the tree (see **figure 6**). The number of nodes is important, but however limited because only the states verifying the constraints are represented.

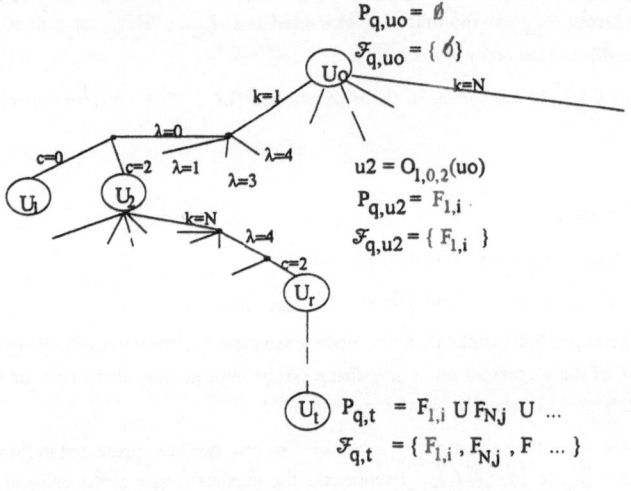

Figure 6

V Combinatory optimization and object structure

In this part, we are going to describe the data structure. Each shape or clustering of shapes can be considered as an entity linked to the data used to describe the processing and the visual model, the relative location model (Nd_{opt}, S_{opt}, c) relative to the connected shape, the absolute location model relative to the material orthogonal basis. Hence, a class called Shape_Model has been built with all previous data as instance variables (**see figure 7**). A lay-plan is an instance of Shape_Model class which is brought after a clustering of two instances of Shape_Model: a shape and an intermediary layout. All the instances of Shape_Model can be the receiver of the instance methods such as (*receiver rotate: an Integer,   receiver nestingSearch: aShape_Model_Instance side: an_Integer,   receiver cluster: aShape_Model_Instance configuration: a_3_uple*). These methods respect Smalltalk syntactic. The tree representing the multiple arrangement of the problem can be also modelled as an instance of the class *Graph* composed of instances of the class *Node*.

Each node contains as instance variables: the current layout (instance of *Shape_Model*), the costs of it (presented later) and a collection of variables storing the current order book and the location models. The main instance method of the class *Node* is: *aNode expand* which develops all the successors in ranging the parameters k, λ, c.

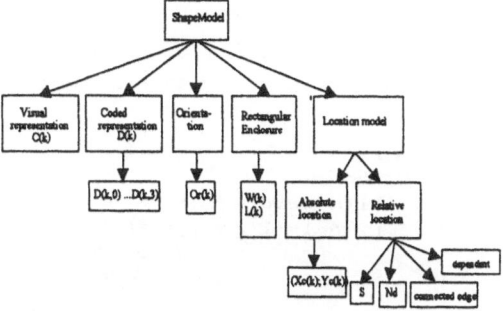

Figure 7

VI solving algorithm

6.1 Search method

The time imperative peculiar to the continuous production structure and to the requested debit induce to accept a good solution without certainty to be admissible but however with an acceptable waste. Among the resolution technique replying to this goal, we can prefer the non-deterministic methods, allowing to manage the alternatives with the possibility to backtrack on a previous choice. However, without heuristic to guide the search, the algorithm has to visit a lot of nodes. To avoid this, one have to inform the choice procedure of the successor to develop in estimating at best the terminal node cost. When the complexity must be low, it is useful to estimate the number of step until the terminal node by a second heuristic. This function, weighted by the cost function, guides the search to have an acceptable solution in reasonable time. At end, the parameter ε allows to adjust the compromise admissibility-complexity. These considerations led us to use the ε -admissible algorithm A_ε

6.2 Principle of A_ε

This algorithm is presented in [26]. As to the algorithm A^*, A_ε is based on the estimated cost at terminal nodes. Let $g(u)$ be the cost of the node u. $\hat{h}(u)$ be the estimated cost of the way between u and the best terminal node t accessible from u. then $f(u) = g(u) + \hat{h}(u)$ represents the estimated cost between u_0 and t going through out u. We say that an algorithm is ε -admissible if and only if it provides an acceptable solution t of cost $f(t) \leq$ Threshold where *Threshold* $= (1+\varepsilon) f^*(u_0)$. $f^*(u_0)$ is the minimum cost of ways going from u_0 to the best terminal node t^*. $f^*(u)$ being not a priori known, a lower bound of $f^*(u_0)$ is used [27]. ε fits in the minimum relative gap tolerated by the user in the expected solution. Then, when a node u is selected to be developed ($u \in$ *Generated*) we will look priority for those among the acceptable successors ($u \in$

Acceptable_Successors $\subset$ *Generated* i.e. $f(u) \leq (1+\varepsilon) f(u_O)$). The acceptability threshold is updated each time that a higher lower bound is known, i.e. $f(u_O)$ nearest of $f^*(u_O)$. This highest lower bound is giving by the node from *Generated* list with f minimum. The choice of the node to develop will be carry out with the help of a selection procedure (*select_in_Generated* or *select_in_Acceptable_Successors*) returning a node with f_c minimum. The used function is as $f_c(u) = f(u) + \beta \hat{h}_c(u)$. The term $\hat{h}_c(u)$ correspond to the number of arcs to arrive at the best reachable terminal node. The parameter β already allow to adjust the compromise admissibility-complexity. Higher is β, bigger are the probabilities to reach rapidly a terminal node. On the other hand, it is possible to include some improvements such as a limited deep search which request investing in nearer acceptable way in order to increase the acceptable threshold and perhaps to decrease the complexity.

In the following algorithm, the list *Acceptable_Successors* and *Solutions* are typically some LIFO stacks. *Generated* and *Successors* are some list ordered in f increasing.

A_ε
entry: ε
1° Initialisation
    1.1 *Generated* := $\{u_O\}$;
    1.2 $g(u_O) := W_M \times \underset{F_k \in \left(\mathcal{F} - \overset{q-1}{\underset{r=0}{\bigcup}} \mathcal{J}_r \right)}{MAX(L_k)}$;

    1.3 $f(u_O) := g(u_O) + \hat{h}(u_o)$;
    1.4 *Threshold* := $(1+\varepsilon)f(u_O)$;
    1.5 *Solutions* := $\varnothing$;
    1.6 *develop* (u_O);
    1.7 *Acceptable_Successors* := $\{v \in Successors / f(v) \leq Threshold\}$;
2° While *Generated* $\neq \varnothing$ and $\forall$ $t \in$ *Solutions*: $f(t) > Threshold$
    2.1 if *Acceptable_Successors* $\neq \varnothing$ then $u := select\_in\_Acceptable\_Successors$
        else $u := select\_in\_Generated$;
    2.2 *develop* (u);
    2.3 While *Persevere* and *Successors* $\neq \varnothing$ and $(\forall v \in Successors : f(v) > Threshold)$
        2.3.1 *develop* $(\hat{v})$ $\{\hat{v} =$ first element of *Successors*$\}$;
    EndWhile;
    2.4 *Acceptable_Successors* := $\{v \in Successors / f(v) \leq Threshold\}$;
EndWhile;
3° Exit : $t \in$ *Solutions* $/ g(t)$ minimum
    $\varepsilon' := [(1+\varepsilon)/Threshold]^*f(t) - 1$
    Exit $P_{q,t}$ and ε' ; Rem: ε' being the real gap relative to the acceptability threshold.

The development of a node is carried out by:
develop(u)
    1° suppress u of *Generated* ; *Successors* := $\varnothing$;
    2° For $\{v / v = O_{k,l,c}(u)$ $k \in [0,N]$, $a_k > 0$, $\lambda \in Ori_{Ad}(k)$, $c \in [0,2]\}$
        calculate $g(v)$;
        $f(v) := g(v) + \hat{h}(v)$;
        $father(v) := u$;
        if v respects all the constraints (6) put v in *Successors* and in *Generated*
    EndFor
    3° if *Successors* = $\varnothing$ then u is terminal, put u in *Solutions*
    4° *Threshold* := $MAX\{Threshold; (1+\varepsilon)f(\hat{v})\}$, $\hat{v}$ is the first element of *Generated*

6.3 estimating cost heuristic $\hat{h}(u)$

Let us remind that g(u) is the current waste of the node u relative to the future layout P_q. So as to decrease the waste we have to fill P_q with another shapes. It comes to the same thing as to add a negative waste $h(u)$ to reach the final allocation. As $h(u)$ is unknown, we estimate it by $\hat{h}(u)$. Intuitively, this function depends on the available area around the current layout $P_{q,u}$ and on the distribution of the current order book. Let $A_a(u)$ be the available area (see figure 8) enclosed by the material rectangle of size $(W_M, L_{q,u})$. So $A_a(u) = (W_M - W_{q,u})L_{q,u} + A_0(u) + A_2(u)$, where $A_0(u)$ and $A_2(u)$ are the areas of the reduced_Shape_Comb of the sides 0 and 2 of the composed shape $P_{q,u}$

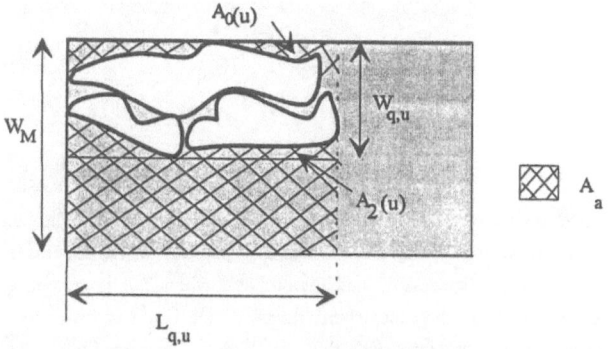

Figure 8

Let us assume that the algorithm is admissible, then $f(u)$ have to be a lower bound on the cost of the minimal cost path from node u_0 to a goal node. $\hat{h}(u)$ being negative, $|\hat{h}(u)|$ have to be a higher bound on the waste to remove, for an optimistic filling of *90%* of the available area. However, this estimating will be weighted by a *Correcting(u)* function taking into account the total area of the "small" shapes. If $A_a(u)$ is more than *1/5* of the future consumed area $Wm*L_{q,u}$, the filling will be good and *Correcting(u)* equal to 1 else, *Correcting(u)* will be smaller but always more than *0.9*. These values has been defined experimentally. Of course, we consider that there are enough shapes to fill $A_a(u)$. We give one example of the typical correcting function. Notice that the shapes are considered small, when their area are less than *1/3* of the largest shape of the current order book. Finally , $\hat{h}(u) = -A_a(u)*0.90*correcting(u)$.

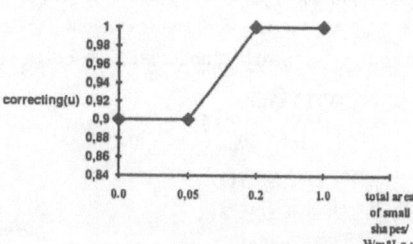

Figure 9: used correcting(u) function

6.4 Path length heuristic $\hat{h}_c(u)$

The heuristic estimates the minimal number of arcs between the current node u and a terminal node. In knowing the estimated filling $|\hat{h}(u)|$, we can process the number of shapes which is possible to allocate.

Of course, the minimal path length is obtained in laying the largest shapes at first. Initially, let *Filling* be equal with $|\hat{h}(u)|$. The following heuristic consist in subtracting the successive shapes area from *Filling* while it is positive, in beginning with the shapes $F_{k,i}$ of S_k maximum.

$\hat{h}_c(u)$

1) Initialisation

    1.1 $\hat{h}_c := 0$; *Filling* $:= |\hat{h}(u)|$; *aBoolean* $:= true$;

2) While *aBoolean*

    2.1 $B := \underset{F_k \in \left(\mathcal{F} - \overset{q-1}{\underset{r=0}{\cup}} \mathcal{S}_r \right)}{MIN} \left(Filling - S_k \right)$

    2.2 if $B>0$ then $B := Filling$; $\hat{h}_c := \hat{h}_c + 1$;

        else *aBoolean* $:= false$;

EndWhile

3) Exit $\hat{h}_c(u) := \hat{h}_c$

6.5 Search with persevering strategy

If all the successors v of the last developed node u are not acceptable but close to acceptability, it could be profitable in persevering on this way. In this case, the best of the successors is expanded and we expect finding a new acceptable successor, which is probable whether $\hat{h}$ is monotonous. In case of check, this investing has used to increase the acceptability threshold and hence to lower the complexity. The loop (2.3) of the main algorithm manages this strategy. The condition *Persevere* is true if $\left(\dfrac{f(\hat{v}) - Threshold}{Threshold} \right) \leq 10\%$ where $\hat{v}$ is the node from *Successors* with f minimum. However, the descent must not to exceed 3 levels so as to avoid developing a costly path further.

6.6 Generated nodes limitations

Because of the memory size, we had to limit the number of nodes from Generated list. So, after each development, the nodes whose cost $f(u)$ are equal with another one are removed; it comes to the same thing as to suppress the identical layouts holding some symmetrical shapes. In addition, if the nodes are still numerous, only the 40 best nodes are kept, i.e. those having $f(u)$ minimum.

VII Results

In this table, we present the efficiency in % in function of the number of expanded nodes, for the four different values of ε. For each value

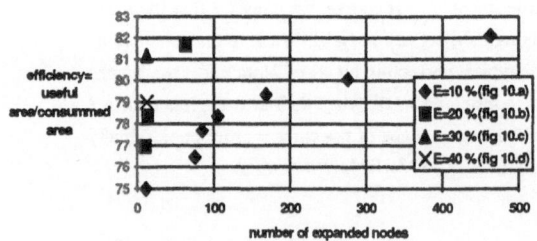

of ε (10, 20, 30, 40 %) ,we give the layouts resulting of the search respectively in figures 10.a, b, c, d . The table shows that lower is ε, higher is the complexity.

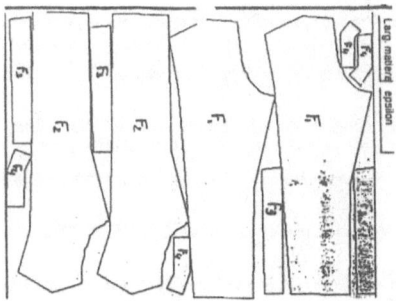

figure 10.a : ε = 10% ε' = 9.92%

efficiency = 82% expanded nodes: 475

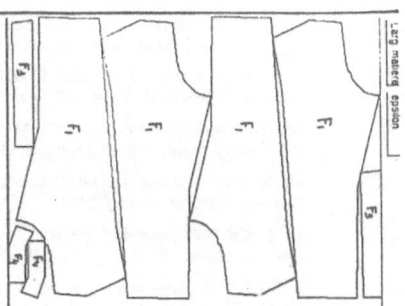

figure 10.b: ε= 20% ε' = 19,31%

efficiency : 81% expanded nodes 64

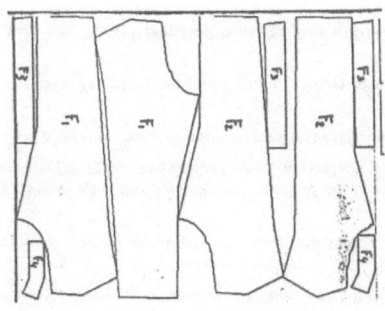

figure 10.c : ε = 30% ε' = 27,75%

efficiency = 81,17% expanded nodes: 13

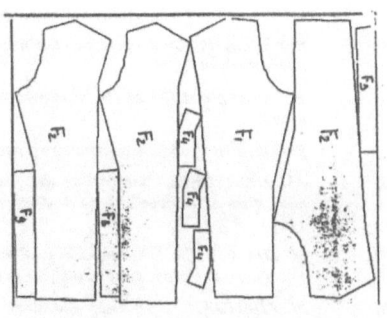

figure 10.d: ε= 40% ε' = 38,05%

efficiency : 78,92% expanded nodes 13

VIII Conclusion

The algorithm has been implemented on SUN 4 station computer and programmed in Smaltalk 80 release 4 language. The execution time is near the goal and some improvements are being studied to decrease the complexity again. Among the possibility, one of them would consist in estimating the cost $g(u)$ before developing u. It would allow to avoid investing in bad expanding.

References

[1]:   S MAOUCHE "Flexible cutting frame in real time", , 1er C.I.N.T.T., Mulhouse 1985.

[2]:   S MAOUCHE "A continuous cutting system in the suit make up industry. XI International Conference on Systems Science, sept 1992

832

[3]: M.J. HAIMS, "On the optimum two-dimensionnal allocation problem", Ph D dissrtation, dept of elec engrg, New York University, Bronx, techn rept 400-136, june 1966

[4]: M.J. HAIMS and H. FREEMAN, "A multistage solution of the template-layout problem", IEEE Trans. on Syst. Scien. and Cybern., vol ssc-6, n°2, April 1970

[5]: M. ADAMOWICZ and A. ALBANO, "A two-stage solution of the cutting-stock problem", Information processing 71, North Holland publishing company (Proc IFIP congress 71), Amsterdam (1972) pp 1086-1091

[6]: M. ADAMOWICZ and A. ALBANO, "Nesting two dimensional shapes in rectangular modules", Computer Aided Design, vol 8, N°1, january 1976

[7]: M. ADAMOWICZ and A. ALBANO, "A solution of the rectangular cutting stock problem", IEEE Trans. on Syst. Scien. and Cyber, SMC-6, pp 302-310, 1976

[8]: R.C. ART, "An approach to the two-dimensional irregular cutting stock problem", IBM Campbridge Scientific Center, Campbridge, MA, Rep N° 320-2006, 1966

[9]: O. GÜREL, "Circular graph of marker layout", IBM techn, report N° 320-2965, feb 1969

[10]: G.R. MOREAU and P.J. De SAINT HARDAVIN, "Marker layout problem: an experimental attempt", IBM New York, N.Y., Rep N° 320-2978, 1969

[11]: J.C. DIVERREZ, B. Du Bois de MEYRIGNAC, R. SOENEN, "Recherche d'un classement hierarchique de surfaces planes et application à un probleme de placement selon une méthode heuristique, revue RAIRO, novembre 1973, J-3, pp 93- 106

[12]: G. MOREAU, "Méthodes pour la résolution des problèmes d'optimisation de découpe", Thèse de Docteur Ingénieur, Université Claude Bernard, Lyon, 1973

[13]: H. FREEMAN, "On the packing of arbitrary-shaped templates", division of applied science, New York University, internal report, 1974

[14]: M. TANAKA and T. WACHI, "Computerized marker making", J. Textil machinery society of japan, vol 19, pp 74-81, 1973

[15]: R. SOENEN, "Conduite automatisée d'une unité de coupe en confection", Automatisme, mai 1977

[16]: J.L. DELAPORTE, "Intégration des fonctions de conception et de préparation de la fabrication pour les entreprises de découpe", Thèse de doctorat de 3ème cycle, Université de Valenciennes et du Hainaut Cambraisis, 1989

[17]: M. ADAMOWICZ, "The optimal two-dimensional allocation of irregular, multiply-connected shapes with linear, logical and geometric constraints", Ph D thesis, Dept of E-E, N.Y.U. (Dec 1969)

[18]: M. ADAMOWICZ, "Optimal allocation of two-dimensional shapes", in Proc Int'l Design, Automation Conf, Toronto, Canada, 1971

[19]: G. KUNTZ, J.P. UHRY, "Optimisation des pertes de matière en industrie textile", Actes du congrès de l'AFCET 1981, Edition Hommes et Techniques

[20]: F. PREMPTI, "Méthodes stochastiques dans les problèmes de placement", Thèse de 3ème cycle, Université scientifique et médicale de Grenoble", 1983

[21]: C.H. DAGLI and A. HAJAKBARI, "A parallel architecture for stock cutting problem", IEEE International conference on PARBASE-90 : database, parallel architecture and their applications,

[22]: G. ROUSSEL, S. MAOUCHE, P. VIDAL, "A new description of irregular shapes boundary. Application to the nesting degree processing and shapes clustering for the optimal two dimensional marking problem in clothing industry", XI conference on Systems Sciences, Wroclaw, Poland, sept 1992.

[23]: A. ALBANO and G. SAPUPPO, "Optimal allocation of two-dimensional irregular shapes using heuristic search method", IEEE Trans. on Syst., Man and Cybern., volSMC-10, N° 5, may 1980

[24]: L. M. PEREIRA, "Artificial intelligence techniques in automatic layout design, Artificial Intelligence and Pattern recognition on computer aided design, North Holland, 1978, pp 159-173

[25]: T. CLERC et M. SAGDAL, "Prototype d'un système de placement intelligent dans le plan", MICAD 87, pp 91-103

[26]: H. FARRENNY, M. GHALLAB, Eléments d'Inteligence Artificielle, Hermes, 1987

[27]: NJ. NILSON, Problem Solving Methods in Artificial Intelligence, Mc Graw Hill.

Computational Geometry

Computational Geometry

AN ALGORITHM FOR FINDING THE CHEBYSHEV CENTER OF A CONVEX POLYHEDRON

N.D.Botkin and V.L.Turova-Botkina [1]
Institut für Angewandte Mathematik und Statistik der Universität Würzburg,
Am Hubland, D-8700 Würzburg, Deutschland

Many problems of control theory and computational geometry require an effective procedure for finding the Chebyshev center (the center of the smallest enclosing ball) of a convex polyhedron. Particularly, it is natural to consider the Chebyshev center as the center of an information set in control problems with uncertain disturbances and state information errors. Chebyshev centers are also useful as an auxiliary tool for some problems of computational geometry. These are the reasons to propose a new algorithm for finding the Chebyshev center (see also [1,4]). The algorithm terminates after a finite number of iterations. In each iteration of the algorithm the current point is projected orthogonally onto the convex hull of a subset of the polyhedron's vertices.

1. Formulation of the Problem. Optimality Conditions

Let $Z = \{z_1, z_2, ..., z_m\}$ be a finite set of points in the Euclidean space R^n. A point $x_* \in R^n$ is called Chebyshev center of the set Z, if

$$\max_{z \in Z} \|x_* - z\| = \min_{x \in R^n} \max_{z \in Z} \|x - z\| \tag{1}$$

It is easily seen that the point x_* satisfying relation (1) is unique and $x_* \in coZ$, where symbol "co" means the convex hull of a set.

For any point $x \in R^n$ we denote

$$d_{max}(x) = \max_{z \in Z} \|x - z\|.$$

Let symbol $E(x)$ denote the subset of points of Z which have the largest distance from x, i.e.

$$E(x) = \{z \in Z : \|x - z\| = d_{max}(x)\}.$$

The optimality conditions are given by the following theorem [2].

Theorem 1. *A point $x_* \in R^n$ is the Chebyshev center of Z iff $x_* \in coE(x_*)$.*

This fact will be used as a criterion of the algorithm termination.

2. The Idea of the Algorithm

Choose an initial point $x_0 \in coZ$ and find the set $E(x_0)$. Assume that $x_0 \notin coE(x_0)$. Otherwise $x_0 = x_*$.

Let

$$J = \{j \in \overline{1, m} : z_j \in E(x_0)\}, \quad I = \{i \in \overline{1, m} : z_i \in Z \setminus E(x_0)\}.$$

[1] On leave from the Institute of Mathematics and Mechanics, Ural Department of Russian Academy of Sciences, 620219 Ekaterinburg, S.Kovalevskaya str.16, Russia.

Find a point $y_0 \in coE(x_0)$ nearest to the point x_0. For any $\alpha \in [0, 1]$ consider the point $x_\alpha = x_0 + \alpha(y_0 - x_0)$. Obviously, for $\alpha = 0$ the following inequality holds

$$\max_{j \in J} \|x_\alpha - z_j\| - \max_{i \in I} \|x_\alpha - z_i\| > 0 \tag{2}$$

When we increase α, the point x_α moves from the point x_0 towards y_0. We will increase α until the left-hand side of the inequality (2) becomes zero. If we reach the point y_0 before the left-hand side of (2) is zero, then the point y_0 is the desired solution x_*. If the left-hand side of (2) is equal zero for some $\alpha = \alpha_0 < 1$, then we take the point x_{α_0} as new initial point and repeat the above process. Note that finding the value α_0 does not require solving of any equation. The algorithm provides explicit formula for α_0.

Thus, the algorithm resembles the simplex method in linear programming. The set $E(x_0)$ is an analog of the support basis in the simplex method. After one iteration we obtain a new set $E(x_{\alpha_0})$ by adding to the set $E(x_0)$ some new points and removing some old points. Besides the "objective" function $d_{max}(\cdot)$ decreases, i.e. $d_{max}(x_{\alpha_0}) < d_{max}(x_0)$.

3. The Algorithm

Algorithm 1

Step 0: *Choose an initial point $x_0 \in coZ$ and set $k := 0$.*

Step 1: *Find $E(x_k)$.*
If $E(x_k) = Z$, then $x_* := x_k$ and stop.

Step 2: *Find $y_k \in coE(x_k)$ as the nearest point to x_k.*
If $y_k = x_k$, then $x_* := x_k$ and stop.

Step 3: *Calculate α_k by the formula:*

$$\alpha_k = \min_{i \in I_k^-} \frac{\|z_i - x_k\|^2 - d_{max}^2(x_k)}{2\langle y_k - x_k, z_i - y_k \rangle}, \tag{3}$$

$$I_k^- = \{i : z_i \in Z \setminus E(x_k), \langle y_k - x_k, z_i - y_k \rangle < 0\}$$

(we set formally $\alpha_k = +\infty$ whenever $I_k^- = \varnothing$).
If $\alpha_k \geq 1$, then $x_* := y_k$ and stop.

Step 4: *Let*

$$x_{k+1} := x_k + \alpha_k(y_k - x_k), \quad k := k + 1$$

and go to Step 1.

Remark. Note that Algorithm 1 comprises the non-trivial operation of finding the distance to convex hull of the point set (Step 2). In our latest program realizing Algorithm 1 we used the recursive algorithm of [3] for the implementation of this operation. However, we will see below that the point y_k of Step 2 is the Chebyshev center of $E(x_k)$. So we come to the following recursive algorithm.

Algorithm 2

Step 0: *Choose an initial point $x_0 \in coZ$ and $k := 0$.*

Step 1: *Find $E(x_k)$.*
If $E(x_k) = Z$, then $x_* := x_k$ and stop.

Step 2: *Call Algorithm 2 with $Z = E(x_k)$ and let y_k be an output of this call (the Chebyshev center of $E(x_k)$).*
If $y_k = x_k$, then $x_* := x_k$ and stop.

Step 3: *Calculate α_k by the formula:*

$$\alpha_k = \min_{i \in I_k^-} \frac{\|z_i - x_k\|^2 - d_{max}^2(x_k)}{2\langle y_k - x_k, z_i - y_k \rangle},$$

$$I_k^- = \{i : z_i \in Z \setminus E(x_k), \langle y_k - x_k, z_i - y_k \rangle < 0\}$$

(we set formally $\alpha_k = +\infty$ whenever $I_k^- = \varnothing$).
If $\alpha_k \geq 1$, then $x_* := y_k$ and stop.

Step 4: *Let*

$$x_{k+1} := x_k + \alpha_k(y_k - x_k), \quad k := k + 1$$

and go to Step 1.

First we give the proof of Algorithm 1. The proof of Algorithm 2 will simply follow from the fact that y_k obtained at Step 2 of Algorithm 2 is the nearest point to x_k.

4. Auxiliary Propositions

Consider k-th iteration of Algorithm 1. We have the current approximation x_k and the point $y_k \in coE(x_k)$ nearest to x_k. Let us assume further that $x_k \neq y_k$. Otherwise $x_* = y_k$ by Theorem 1.

Introduce some notations. For any subset $S \subset Z$ we put

$$\text{num}(S) = \{l \in \overline{1, m} : z_l \in S\}.$$

Define the following sets of indices

$$J_k = \text{num}\Big(E(x_k)\Big),$$
$$J_k^0 = \{l \in J_k : \langle y_k - x_k, z_l \rangle = \min_{j \in J_k}\langle y_k - x_k, z_j \rangle\},$$
$$I_k = \text{num}(Z) \setminus J_k,$$
$$I_k^- = \{i \in I_k : \langle y_k - x_k, z_i - y_k \rangle < 0\},$$
$$I_k^+ = I_k \setminus I_k^-.$$

Note that the set I_k^- is the same as in formulas (3).
We now prove two propositions characterizing the point y_k.

Proposition 1. *The following equality holds*

$$\langle y_k - x_k, y_k \rangle = \min_{j \in J_k}\langle y_k - x_k, z_j \rangle.$$

Proof. Choose an arbitrary point $z \in \mathrm{co}E(x_k)$ and consider the function $\delta(\lambda) = \|(1 - \lambda)y_k + \lambda z - x_k\|^2, \lambda \in [0, 1]$. Because of the definition of y_k the function $\delta(\cdot)$ takes its minimum at $\lambda = 0$. Hence $\delta'(0) \geq 0$, which implies

$$\langle y_k - x_k, z - y_k \rangle \geq 0.$$

Therefore,

$$\langle y_k - x_k, y_k \rangle = \min_{z \in \mathrm{co}\, E(x_k)} \langle y_k - x_k, z \rangle.$$

Taking into account the obvious equality

$$\min_{z \in \mathrm{co}\, E(x_k)} \langle y_k - x_k, z \rangle = \min_{j \in J_k} \langle y_k - x_k, z_j \rangle$$

we obtain the desired result.

Proposition 2. *The following inclusion holds*

$$y_k \in \mathrm{co}\{z_j : j \in J_k^0\}.$$

Proof. Since $y_k \in \mathrm{co}E(x_k)$, we have

$$y_k = \sum_{j \in J_k} \lambda_j z_j, \quad \lambda_j \geq 0, \quad \sum_{j \in J_k} \lambda_j = 1.$$

Proposition 1 gives

$$\langle y_k - x_k, \sum_{j \in J_k} \lambda_j z_j \rangle = \min_{j \in J_k} \langle y_k - x_k, z_j \rangle$$

or

$$\sum_{j \in J_k} \lambda_j [\langle y_k - x_k, z_j \rangle - \min_{l \in J_k} \langle y_k - x_k, z_l \rangle] = 0.$$

This implies

$$\lambda_j [\langle y_k - x_k, z_j \rangle - \min_{l \in J_k} \langle y_k - x_k, z_l \rangle] = 0$$

for any $j \in J_k$. Hence $\lambda_j = 0$ for any $j \notin J_k^0$ (the term in the square brackets is positive for $j \notin J_k^0$). Thus, Proposition 2 is proved.

Suppose that $I_k^- \neq \emptyset$ and consider for $\alpha \geq 0$ the following function

$$g(\alpha) = \max_{j \in J_k} \|x_k + \alpha(y_k - x_k) - z_j\|^2 - \max_{i \in I_k^-} \|x_k + \alpha(y_k - x_k) - z_i\|^2. \tag{4}$$

Lemma 1. *The function $g(\cdot)$ is monotone decreasing and has an unique zero $\alpha_k > 0$ given by formula (3).*

Proof. First we note that $g(0) > 0$ by definition of the sets J_k, I_k^-. Rewrite $g(\cdot)$ as follows

$$g(\alpha) = \min_{i \in I_k^-} \max_{j \in J_k} [\|x_k + \alpha(y_k - x_k) - z_j\|^2 - \|x_k + \alpha(y_k - x_k) - z_i\|^2] =$$

$$\min_{i \in I_k^-} \max_{j \in J_k} [\langle 2x_k + 2\alpha(y_k - x_k) - (z_j + z_i), z_i - z_j \rangle] =$$

$$\min_{i\in I_k^-}\max_{j\in J_k}[2\alpha\langle y_k - x_k,\ z_i - z_j\rangle + \langle 2x_k - (z_j + z_i),\ z_i - z_j\rangle] =$$

$$\min_{i\in I_k^-}\max_{j\in J_k}[2\alpha\langle y_k - x_k,\ (z_i - y_k) + (y_k - z_j)\rangle + \langle (x_k - z_j) + (x_k - z_i),\ (z_i - x_k) - (z_j - x_k)\rangle] =$$

$$\min_{i\in I_k^-}\max_{j\in J_k}[2\alpha\langle y_k - x_k,\ z_i - y_k\rangle + 2\alpha\langle y_k - x_k,\ y_k - z_j\rangle + \|x_k - z_j\|^2 - \|x_k - z_i\|^2] =$$

$$\min_{i\in I_k^-}[2\alpha\langle y_k - x_k,\ z_i - y_k\rangle - \|x_k - z_i\|^2] + \max_{j\in J_k}[2\alpha\langle y_k - x_k,\ y_k - z_j\rangle + \|x_k - z_j\|^2].$$

Since $\|x_k - z_j\| = d_{max}(x_k)$ for any $j \in J_k$, and

$$\max_{j\in J_k}\langle y_k - x_k, y_k - z_j\rangle = \langle y_k - x_k, y_k\rangle - \min_{j\in J_k}\langle y_k - x_k, z_j\rangle = 0$$

due to Proposition 1, we obtain the following representation of $g(\cdot)$

$$g(\alpha) = \min_{i\in I_k^-}[2\alpha\langle y_k - x_k, z_i - y_k\rangle - \|x_k - z_i\|^2] + d_{max}^2(x_k). \tag{5}$$

Since $\langle y_k - x_k, z_i - y_k\rangle < 0$ for any $i \in I_k^-$, the function $g(\cdot)$ is the minimum of a finite set of strictly decreasing affine functions of α. Hence $g(\cdot)$ is strictly decreasing. Using inequality $g(0) > 0$ we obtain that there exists an unique zero $\alpha' > 0$ of $g(\cdot)$. Due to the monotonicity of the function $g(\cdot)$ the value α' can be found by

$$\alpha' = \max\{\alpha \geq 0 : g(\alpha) \geq 0\}.$$

Let us prove now that α' coincides with α_k defined by formula (3). Choose an arbitrary $\bar{\alpha} > \alpha_k$. Then from definition of α_k there exists $i_* \in I_k^-$ such that

$$\frac{\|z_{i_*} - x_k\|^2 - d_{max}^2(x_k)}{2\langle y_k - x_k, z_{i_*} - y_k\rangle} < \bar{\alpha}.$$

Since $\langle y_k - x_k, z_{i_*} - y_k\rangle < 0$ we obtain

$$2\bar{\alpha}\langle y_k - x_k, z_{i_*} - y_k\rangle - \|z_{i_*} - x_k\|^2 + d_{max}^2(x_k) < 0.$$

A comparison with (5) gives $g(\bar{\alpha}) < 0$. Using the monotonicity of $g(\cdot)$ we get $\bar{\alpha} > \alpha'$. Thus, for any $\bar{\alpha} > \alpha_k$ we have $\bar{\alpha} > \alpha'$. Hence $\alpha_k \geq \alpha'$.

Let us prove the opposite inequality. For any $i \in I_k^-$ we have

$$\frac{\|z_i - x_k\|^2 - d_{max}^2(x_k)}{2\langle y_k - x_k, z_i - y_k\rangle} \geq \alpha_k.$$

This implies

$$2\alpha_k\langle y_k - x_k, z_i - y_k\rangle - \|z_i - x_k\|^2 + d_{max}^2(x_k) \geq 0 \quad \text{for any } i \in I_k^-.$$

A comparison with (5) gives $g(\alpha_k) \geq 0$. Hence $\alpha_k \leq \alpha'$ and Lemma 1 is proved.

Lemma 2. *For any $\alpha > 0$ the maximum in the expression*

$$\max_{j\in J_k}\|x_k + \alpha(y_k - x_k) - z_j\|^2$$

is attained on the subset J_k^o, and all $j \in J_k^o$ are maximizing.

Proof. Let $\alpha > 0$, $s, q \in J_k$. Consider the function

$$\varphi(\alpha) = \|x_k + \alpha(y_k - x_k) - z_s\|^2 - \|x_k + \alpha(y_k - x_k) - z_q\|^2,$$

with $\varphi(0) = 0$. Since $\varphi'(\alpha) = \langle y_k - x_k, z_q - z_s \rangle > 0$ for $s \in J_k^o$, $q \notin J_k^o$, $\varphi(\alpha)$ increases strictly with $\alpha > 0$. On the other hand $\varphi'(\alpha) \equiv 0$ for $s, q \in J_k^o$, which proves the lemma.

Lemma 3. *For any $i \in I_k^+$ and any $\alpha \geq 0$ the following inequality holds*

$$\max_{j \in J_k} \|x_k + \alpha(y_k - x_k) - z_j\|^2 > \|x_k + \alpha(y_k - x_k) - z_i\|^2.$$

Proof. Let $i \in I_k^+$ and consider a fixed arbitrary index $s \in J_k^o$. One can easily see from the definition of I_k^+ and Proposition 1 that the function

$$\varphi(\alpha) = \|x_k + \alpha(y_k - x_k) - z_s\|^2 - \|x_k + \alpha(y_k - x_k) - z_i\|^2$$

satisfies $\varphi(0) > 0$ and $\varphi'(\alpha) = \langle y_k - x_k, z_i - z_s \rangle = \langle y_k - x_k, z_i - y_k \rangle + \langle y_k - x_k, y_k - z_s \rangle \geq 0$, hence $\varphi(\alpha)$ is increasing with $\alpha \geq 0$. This gives the proof.

Now define x_{k+1} by

$$x_{k+1} = \begin{cases} x_k + \alpha_k(y_k - x_k) & ; \quad \alpha_k \leq 1 \\ y_k & ; \quad \alpha_k > 1 \end{cases}. \tag{6}$$

Lemma 4. *The following equality holds*

$$\operatorname{num}(E(x_{k+1})) = \begin{cases} J_k^o \cup I_k^{-o} & ; \quad \alpha_k \leq 1 \\ J_k^o & ; \quad \alpha_k > 1, \end{cases} \tag{7}$$

where

$$I_k^{-o} = \{l \in I_k^- : \|x_{k+1} - z_l\|^2 = \max_{i \in I_k^-} \|x_{k+1} - z_i\|^2\}.$$

Proof. Consider first the case $\alpha_k \leq 1$. Since α_k is a zero of the function $g(\cdot)$ defined by (4), we have

$$\max_{j \in J_k} \|x_{k+1} - z_j\|^2 = \max_{i \in I_k^-} \|x_{k+1} - z_i\|^2. \tag{8}$$

Taking into account the definition of the set I_k^{-o}, we obtain

$$\max_{j \in J_k} \|x_{k+1} - z_j\|^2 > \max_{i \in I_k^- \setminus I_k^{-o}} \|x_{k+1} - z_i\|^2.$$

Using Lemma 3, we get

$$\max_{j \in J_k} \|x_{k+1} - z_j\|^2 > \max_{i \in (I_k^- \setminus I_k^{-o}) \cup I_k^+} \|x_{k+1} - z_i\|^2.$$

With Lemma 2 we get for any $s \in J_k^o$

$$\|x_{k+1} - z_s\|^2 = \max_{j \in J_k} \|x_{k+1} - z_j\|^2 > \max_{i \in (I_k^- \setminus I_k^{-o}) \cup I_k^+ \cup (J_k \setminus J_k^o)} \|x_{k+1} - z_i\|^2. \tag{9}$$

Taking into account the definition of the set I_k^{-0} and (8), we obtain that relation (9) is valid for any $s \in J_k^0 \cup I_k^{-0}$.

Since

$$J_k^0 \cup I_k^{-0} \cup (I_k^- \setminus I_k^{-0}) \cup I_k^+ \cup (J_k \setminus J_k^0) = \operatorname{num}(Z),$$

we conclude that

$$\operatorname{num}(E(x_{k+1})) = J_k^0 \cup I_k^{-0}.$$

Let now $\alpha_k > 1$ but $I_k^- \neq \emptyset$. Then $g(1) > 0$ which means that

$$\max_{j \in J_k} \|x_{k+1} - z_j\|^2 > \max_{i \in I_k^-} \|x_{k+1} - z_i\|^2.$$

With Lemma 3 this implies

$$\max_{j \in J_k} \|x_{k+1} - z_j\|^2 > \max_{i \in I_k^- \cup I_k^+} \|x_{k+1} - z_i\|^2.$$

Using Lemma 2, we obtain for any $s \in J_k^0$

$$\|x_{k+1} - z_s\|^2 = \max_{j \in J_k} \|x_{k+1} - z_j\|^2 > \max_{i \in I_k^- \cup I_k^+ \cup (J_k \setminus J_k^0)} \|x_{k+1} - z_i\|^2.$$

Since

$$J_k^0 \cup I_k^- \cup I_k^+ \cup (J_k \setminus J_k^0) = \operatorname{num}(Z),$$

we have

$$\operatorname{num}(E(x_{k+1})) = J_k^0.$$

If $I_k^- = \emptyset$ then $I_k = I_k^+$, and from Lemma 3 we get

$$\max_{j \in J_k} \|x_{k+1} - z_j\|^2 > \max_{i \in I_k} \|x_{k+1} - z_i\|^2.$$

With the help of Lemma 2 this gives for any $s \in J_k^0$

$$\|x_{k+1} - z_s\|^2 = \max_{j \in J_k} \|x_{k+1} - z_j\|^2 > \max_{i \in I_k \cup (J_k \setminus J_k^0)} \|x_{k+1} - z_i\|^2.$$

Since

$$J_k^0 \cup I_k \cup (J_k \setminus J_k^0) = \operatorname{num}(Z),$$

we conclude that

$$\operatorname{num}(E(x_{k+1})) = J_k^0,$$

which completes the proof.

Lemma 2 shows that for any $j \in J_k^0$ the expression $\|y_k - z_j\|$ assumes the same value. By Proposition 2 this value is the Chebyshev radius of the set $\{z_j : j \in J_k^0\}$, which we denote by r_k.

Lemma 5. *The following formulas for r_k are valid*

$$r_k = \sqrt{d_{max}^2(x_k) - \|y_k - x_k\|^2}, \qquad (10)$$

$$r_k = \sqrt{d_{max}^2(x_{k+1}) - \|y_k - x_{k+1}\|^2}. \tag{11}$$

Proof. Choose an arbitrary $s \in J_k^o$. Then we have

$$z_s - x_k = y_k - x_k + z_s - y_k.$$

By squaring both sides of this expression we obtain

$$\|z_s - x_k\|^2 = \|y_k - x_k\|^2 + r_k^2 + 2\langle y_k - x_k, z_s - y_k \rangle.$$

From Proposition 1 and the definition of the set J_k^o we have

$$\langle y_k - x_k, z_s - y_k \rangle = 0. \tag{12}$$

Using that $\|z_s - x_k\| = d_{max}(x_k)$ for any $s \in J_k^o \subset J_k$, we obtain (10).
As for (11), we have

$$z_s - x_{k+1} = y_k - x_{k+1} + z_s - y_k.$$

Squaring gives again

$$\|z_s - x_{k+1}\|^2 = \|y_k - x_{k+1}\|^2 + r_k^2 + 2\langle y_k - x_{k+1}, z_s - y_k \rangle.$$

Since $y_k - x_{k+1} = (1 - \alpha_k)(y_k - x_k)$, we get analogously to (12) $\langle y_k - x_{k+1}, z_s - y_k \rangle = 0$. Since $s \in J_k^o \subset \mathrm{num}(E(x_{k+1}))$ (see (7)), we conclude that $\|z_s - x_{k+1}\|^2 = d_{max}^2(x_{k+1})$. Thus we obtain (11) and lemma is proved.

Lemma 6. *If $\alpha_k < 1$ then $r_{k+1} > r_k$.*
Proof. In fact, from (10) we have

$$r_{k+1} = \sqrt{d_{max}^2(x_{k+1}) - \|y_{k+1} - x_{k+1}\|^2}, \tag{13}$$

where $y_{k+1} \in \mathrm{co}E(x_{k+1})$ is the nearest point to x_{k+1}. Comparing (11) and (13), we see that we have to prove inequality

$$\|y_{k+1} - x_{k+1}\|^2 < \|y_k - x_{k+1}\|^2. \tag{14}$$

To do this we note that from Proposition 2 and (7) the point $y_k \in \mathrm{co}E(x_{k+1})$. Choose an arbitrary point z_q, $q \in I_k^{-o}$. It is seen from (7) that $(1-\lambda)y_k + \lambda z_q \in \mathrm{co}E(x_{k+1})$ for any $\lambda \in [0, 1]$. Consider the expression

$$\|x_{k+1} - (1 - \lambda)y_k - \lambda z_q\|^2 = \|x_{k+1} - y_k + \lambda(y_k - z_q)\|^2 =$$

$$\|x_{k+1} - y_k\|^2 + 2\lambda(1 - \alpha_k)\langle y_k - x_k, z_q - y_k \rangle + \lambda^2 \|y_k - z_q\|^2.$$

Since $q \in I_k^{-o} \subset I_k^-$ the following inequality holds

$$\langle y_k - x_k, z_q - y_k \rangle < 0.$$

Hence, taking small enough $\lambda' > 0$, we obtain

$$\|x_{k+1} - (1 - \lambda')y_k - \lambda' z_q\|^2 < \|x_{k+1} - y_k\|^2.$$

Since $y_{k+1} \in \mathrm{co}E(x_{k+1})$ is the nearest point to x_{k+1} and $(1 - \lambda')y_k + \lambda'z_q \in \mathrm{co}E(x_{k+1})$, we get

$$\|x_{k+1} - y_{k+1}\|^2 \leq \|x_{k+1} - (1 - \lambda')y_k - \lambda'z_q\|^2 < \|x_{k+1} - y_k\|^2.$$

Thus (14) is true and inequality $r_{k+1} > r_k$ also holds.

5. The Properties of Finiteness and Monotonicity

Theorem 2. *The Algorithm 1 terminates within a finite number of iterations.*

Proof. Notice that the inequality $r_{k+1} > r_k$ of Lemma 6 was obtained under assumptions $x_k \notin \mathrm{co}E(x_k)$ and $\alpha_k < 1$. It is easily seen that after a finite number of iterations one of these conditions will be violated. Indeed, there is only a finite number of distinct sets $\{z_j : j \in J_k^o\}$, $k = 1, 2, \ldots$. This number does not exceed the number of all subsets of Z. By the properties of the algorithm each set $\{z_j : j \in J_k^o\}$ can arise only once, because the corresponding Chebyshev radius r_k increases strictly with k. Hence, after a finite number of iterations we obtain either $x_k \in \mathrm{co}E(x_k)$, or $\alpha_k \geq 1$. The first case means termination of the algorithm.

Consider the case $\alpha_k \geq 1$. Let x_{k+1} be defined by formula (6). Then $x_{k+1} = y_k$ and from Proposition 2 and (7) we have

$$x_{k+1} = y_k \in \mathrm{co}\{z_j : j \in J_k^o\} \subset \mathrm{co}\, E(x_{k+1}).$$

Thus, by Theorem 1 y_k is the Chebyshev center and the algorithm stops. Theorem 2 is proved.

Theorem 3. *The Algorithm 2 terminates within a finite number of iterations.*

Proof. If $y_k \in \mathrm{co}E(x_k)$ is the nearest point to x_k, then Lemma 2 (with $\alpha = 1$) implies that $\{z_j : j \in J_k^o\}$ is the subset of such points of $E(x_k)$ which have the largest distance from y_k. Then from Proposition 2 and Theorem 1 we obtain that y_k is the Chebyshev center of $E(x_k)$. So, with the uniqueness of the Chebyshev center we deduce that y_k obtained at Step 2 of Algorithm 2 is the nearest point to x_k.

It remains to prove that there is an exit from the recursion. In fact, if the point x_k is not the Chebyshev center of Z, then the set $E(x_k)$ has fewer points than Z has. So, the depth of recursion less or equal m. This proves Theorem 3.

Theorem 4. *The Algorithms 1,2 have the property of monotonicity, that is*

$$d_{max}(x_{k+1}) < d_{max}(x_k).$$

Proof. From (10) and (11) we have

$$d_{max}(x_k) = \sqrt{r_k^2 + \|x_k - y_k\|^2},$$

$$d_{max}(x_{k+1}) = \sqrt{r_k^2 + \|x_{k+1} - y_k\|^2} = \sqrt{r_k^2 + (1 - \alpha_k)^2\|x_k - y_k\|^2}.$$

This gives the desired result.

6. Numerical Examples

Let us consider some numerical examples characterizing the efficiency of the algorithm.

844

These examples were implemented with a QuickBASIC program on a PC (80486, 33 MHz).

Example 1. $Z = \{z : z = (\delta_1, \delta_2, ..., \delta_n)^T, \ \delta_j = \pm 1, j = 1, ..., n\}$ (vertices of the unit cube in R^n). The number of vertices is $m = 2^n$. The Chebyshev center is $x_* = 0$. The program realizing Algorithm 1 was run with the initial point $x_0 = (\frac{1}{1}, \frac{1}{2}, ..., \frac{1}{n})^T$ for dimensions $n = 2, 3, ..., 10$. For each n the solution was obtained just in n iterations. The CPU-time for $n = 10$ was 8.62 s.

Example 2. $Z = \{z_i : z_i = (\xi_1, ..., \xi_n)^T, i = 1, ...m\}$, where ξ_j are random values uniformly distributed on the interval $[-1, 1]$. The results obtained for Algorithm 1 are presented in the following table. The initial point was the same as in Example 1. The average values were computed on the basis of 20 testruns.

n=10	Number of iterations			Time (s)		
m	min.	av.	max.	min.	av.	max.
100	8	12.55	18	1.1	3.4	14.66
200	8	13.45	18	1.49	5.7	12.97
500	9	15.65	24	2.85	10.26	25.04
1000	11	17.4	26	8.56	16.77	33.61
5000	13	21.6	29	71.46	115.06	151.21

The average speed of Algorithm 2 is approximately 1.5 times lower then the speed of Algorithm 1, but the corresponding computer program is shorter and looks more elegant. Computer tests show that the efficiency of the algorithms proposed is comparable with the rapidity of the algorithm of [4]. Our test results compare favourable with the test results reported in [1] for various methods to compute the Chebyshev center proposed by other authors.

Acknowledgement

This work was partly supported by the Alexander von Humboldt- Foundation, Germany. The authors thank Prof. Dr. J.Stoer for his warm hospitality at the University Wuerzburg, Germany and for helpful discussions. Authors also thank Dr. M.A.Zarkh for his help with computer program and valuable remarks.

References

1. Goldbach R. (1992) Inkugel und Umkugel konvexer Polyeder. Diplomarbeit, Institut für Angewandte Mathematik und Statistik Julius-Maximilians- Universität Würzburg.

2. Pschenichny B.N. (1980) Convex analysis and extremal problems, Nauka, Moscow.

3. Sekitani K., Yamamoto Y. (1991) A recursive algorithm for finding the minimum norm point in a polytope and a pair of closest points in two polytopes, Sci. report No.462, Institute of Socio-Economic Planning, University of Tsukuba, Japan.

4. Welzl E. (1991) Smallest enclosing disks (balls and ellipsoids). Lecture Notes in Computer Science, Springer-Verlag, New York 555:359-370.

ZONOHEDRA, ZONOIDS AND MIXTURES MANAGEMENT[1].

B. Lacolle,
Université Joseph Fourier,
Laboratoire LMC-IMAG, BP 53 X, F-38041 Grenoble Cedex,
e-mail : lacolle@imag.fr,
N. Szafran,
Université Joseph Fourier, CNRS,
Laboratoire LMC-IMAG, BP 53 X, F-38041 Grenoble Cedex,
et
P. Valentin
ELF-FRANCE,Centre de Recherche ELF SOLAIZE, BP 22,
69360 St Symphorien D'Ozon, France.

Key words and phrases : Computational Geometry, Zonohedron, Zonoid, Projective diagram, Separation theory, Mixtures management.

I.- Introduction.

In this paper, we are concerned with an actual application in the field of chemical engineering, more precisely in some features of the theory of separation initiated by P. Valentin [10,11]. We are interested in producing mixtures by mixing parts of basic products. A mixture is defined by quantities of n physico-chemical species subjected to conservation laws. So, we use a geometrical model in which any product can be viewed as a vector in $\mathbb{R}^n_+$, one component for each species. Another fundamental hypothesis is that the mixing procedure corresponds to vector addition in the vector space $\mathbb{R}^n$. For instance, for a typical example about gasoils in the oil industry, the species we use could be mass, volume, mass of sulphur. However, using particular transformations, it is possible to take into account more sophisticated species (cloud point index of a gasoil, for example) [8].

[1]This work was supported by Université Joseph Fourier de Grenoble (France), Centre National de la Recherche Scientifique (CNRS) and Société Nationale ELF-Aquitaine.

In first, we deal with the following discrete model : using a set of given basic products $\{n_1, ..., n_p\}$, the mixture obtained by taking the proportion λ^j of product n_j $(j=1, ..., n)$ is represented by :

$$\sum_{j=1, ..., p} \lambda^j n_j.$$

A product m is feasible if and only if m lies in the set :

$$Z\{n_1, ..., n_p\} = \{ \sum_{j=1, ..., p} \lambda^j n_j, \ 0 \leq \lambda^j \leq 1 \ (1 \leq j \leq p) \}.$$

This set is a centrally symmetric convex set, so called a zonotope [1,3,7]. This paper is concerned both with the feasibility of one mixture and with the joint problem of computing a decomposition of this mixture (the coefficients λ^j). In the general case, there is not a unique mixing procedure leading to a given mixture and some mixing procedures may be better than others. Optimization of mixing procedure was studied in [5]. The previous problems can be solved by using classical linear or convex programming methods, but these approaches give no real understanding about the solutions.

We present here a geometrical approach which is quite different. When the value of the number n of species equals two, the mixtures are called binary mixtures and we have complete solutions by using classical methods in the field of computational geometry [6]. This paper deals with particular problems in the case of three species ($n = 3$). Then, a zonotope is called a zonohedron and it can be constructed in time $O(p^2)$, using the well known equivalence with arrangements of lines in the plane [4].

In the first part of the paper, we deal with an important particular case in which the set of basic mixtures verifies a property we name "regular selectivity". Taking into account this fact, we set up a particular algorithm for feasibility and decomposition problems, running in $O(p)$ time only.

In the second part of this paper, we study a generalization to continuous families of basic mixtures. The set of basic products is represented by a "differential family" $\{f(t)dt\}_{t \in I}$ where I is a real interval and f a function in $\mathcal{C}^0[I, \mathbb{R}^3]$. Then, the set of feasible mixtures, is defined as :

$$Z(f) = \{ m = \int_I \lambda(t) f(t) dt : \lambda \in L^1[I, [0,1]] \},$$

and it is called a zonoid [1,9,12]. As we have done in the discrete model, we consider particular process which verify the property of "regular selectivity" and we study feasibility and decomposition in the case of "differential families".

Note 1. We do not make any distinction between the Euclidean affine space $\mathbb{R}^n$ and the associated vector space. We denote o the origin and a point m corresponds to the vector $\boldsymbol{m}$. The segment from o to m is denoted $[o,m]$.

Note 2. Due to commutativity of the Minkowski sum, zonotopes and zonoids do not depend on the order of the generating families.

II. Zonohedron and regular selectivity.

Making up a zonohedron, from the set of vectors $\{\boldsymbol{n}_1, ..., \boldsymbol{n}_p\}$, requires $O(p^2)$ time, because a natural connection with an arrangement of p lines in the plane [4]. In fact, this arrangement can be seen as the dual of the second projective diagram of a zonohedron, introduced by Coxeter [3]. In the following, we use some geometric representation of the first projective diagram (FPD) in the Coxeter description, as the set of points $\{\boldsymbol{v}_1, ..., \boldsymbol{v}_p\}$, intersection of the lines $\mathbb{R}\boldsymbol{n}_j$ with a plane which does not contain the origin. Although not popular in computational geometry, the first projective diagram gives more direct information about the relative disposition of vectors $\{\boldsymbol{n}_1, ..., \boldsymbol{n}_p\}$. So, it is of interest for our application.

The "regular selectivity".

Definition 1.- We say that the family $\{\boldsymbol{n}_1, ..., \boldsymbol{n}_p\}$ verifies the property of "regular selectivity" (more precisely, "3-regular selectivity" in separation theory), if up to a permutation :
- there exist $\boldsymbol{m} \in \mathbb{R}^3$ so that for every i, $1 \leq i \leq p$, $\langle \boldsymbol{n}_i, \boldsymbol{m} \rangle$ is strictly positive,
- and for any (i,j,k) such that $1 \leq i < j < k \leq p$ the determinant $(\boldsymbol{n}_i, \boldsymbol{n}_j, \boldsymbol{n}_k)$ is strictly positive.

This property arises in a natural way in modelling some actual applications as distillation process [10,11]. If we use a geometric representation of the FPD in a plane of oriented normal $\boldsymbol{m}$ which does not contain the origin, the property of "regular selectivity" implies that the points $\{v_1, ..., v_p\}$ in the FPD, are the vertices of a convex polygon. The point v_j characterizes the composition of the basic product $\boldsymbol{n}_j$, and in some way, each vector $\boldsymbol{n}_j$ is "extreme" in the family.

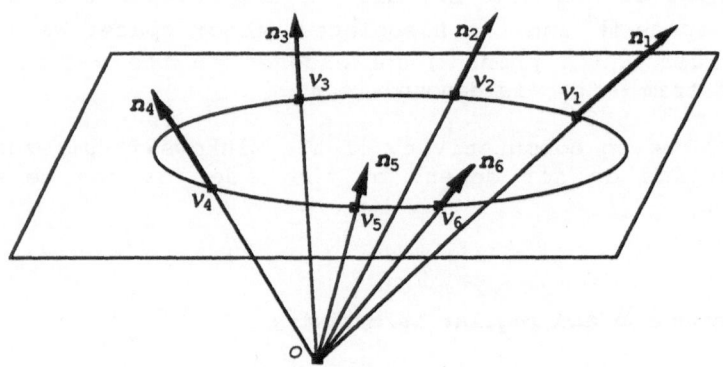

Figure 1 : An example of first projective diagram
in the case of regular selectivity.

Property 2.- If the family $\{\boldsymbol{n}_1, ..., \boldsymbol{n}_p\}$ verifies the "regular selectivity" property, whether the mixture $\boldsymbol{m}$ is feasible or not can be determined in $O(p)$ time, and if $\boldsymbol{m}$ is feasible a particular decomposition can be found in $O(p)$ time.

Demonstration : We note $\boldsymbol{s} = \sum_{j=1, ..., p} \boldsymbol{n}_j$. For a given vector $\boldsymbol{m}$, we consider the plane Π defined by o, s and $\boldsymbol{m}$. We can show that the section of the zonohedron $Z\{\boldsymbol{n}_1, ..., \boldsymbol{n}_p\}$ by this plane is a convex centrally symmetric polygon in $\mathbb{R}^2$, so it is a zonogon [1]. The following method, worked out by N. Szafran [9], consists in computing this zonogon directly from the set $\{\boldsymbol{n}_1, ..., \boldsymbol{n}_p\}$, without building the corresponding zonohedron. The hypothesis implies that there exists no set of three colinear vectors. So, any face of zonohedron is a parallelogram.

At any pair $(\boldsymbol{n}_i, \boldsymbol{n}_j)$, $i<j$, we associate the two faces :

$$F_{i,j} = \sum_{i<k<j} \boldsymbol{n}_k + [o, n_i] + [o, n_j],$$

and :
$$\overline{F}_{i,j} = s - F_{i,j} = \sum_{k<i} \boldsymbol{n}_k + \sum_{k>j} \boldsymbol{n}_k + [o, n_i] + [o, n_j],$$

which are symmetric. Using the first projective diagram, any face $F_{i,j}$ can be represented as a directed line through two points v_i and v_j. The face is the parallelogram built on $\boldsymbol{n}_i$ and $\boldsymbol{n}_j$ translated by the sum of vectors $\boldsymbol{n}_k$ corresponding to the points m_k lying at the right of the directed line. It follows from the "regular selectivity" property, that the points $\{v_1, ..., v_p\}$ are the vertices of a convex polygon and can be numbered in counterclockwise order. With the natural conventions :

$$\boldsymbol{n}_0 = \boldsymbol{n}_p, \ \boldsymbol{n}_{p+1} = \boldsymbol{n}_1,$$

it is easy to see that the face $F_{i,j}$ has four adjacent faces which are in the general case $i < j-1$:

$$F_{i+1,j}, \ F_{i-1,j}, \ F_{i,j+1}, \ F_{i,j-1},$$

and in the particular case $i = j-1$:

$$F_{i+1,i+2}, \ F_{i-1,i}, \ F_{i,i+2}, \ F_{i-1,i+1},$$

We note :

$$F_{i,j} = \overline{F}_{j,i}, \text{ if } i > j.$$

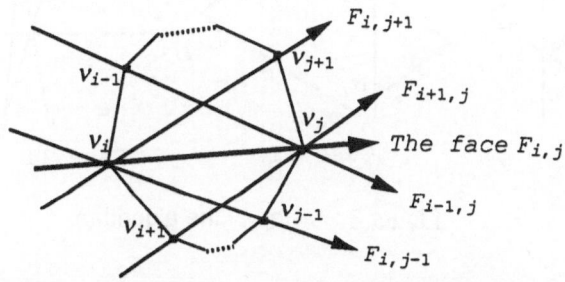

Figure 2 : The general case of the four adjacent faces to $F_{i,j}$, represented in the first projective diagram.

At the step q of the algorithm, we will consider a face described on the form :

$$\omega_q + [o, n_{u_q}] + [o, n_{1_q}],$$

and we note the intersection of this face with the plane Π as the segment starting at the point σ_q and ending at the point σ'_q.

Initialization : We suppose the intersection non degenerated, and the first step consists in finding an initial face of the zonohedron cut by the plane Π. If we note μ a normal vector of Π, starting with $i = 1$, we choose this first face $F_{i,i+1}$ so that :

$$<n_i, \mu> \; <0 \text{ and } <n_{i+1}, \mu> \; \geq 0,$$

and we note : $\omega_0 = o$, $\sigma_0 = o$, $l_0 = i$ and $u_0 = i+1$.

Step q : In order to determine, the next face, we look for a point σ_q', verifying :

$$<\sigma_q', \mu> \; = 0,$$

among one of the following expression, with λ in $[0,1]$:

$$\sigma_q' = \omega_q + n_{u_q} + \lambda \, n_{l_q} \text{ (case I)},$$

or :

$$\sigma_q' = \omega_q + \lambda \, n_{u_q} + n_{l_q} \text{ (case II)}.$$

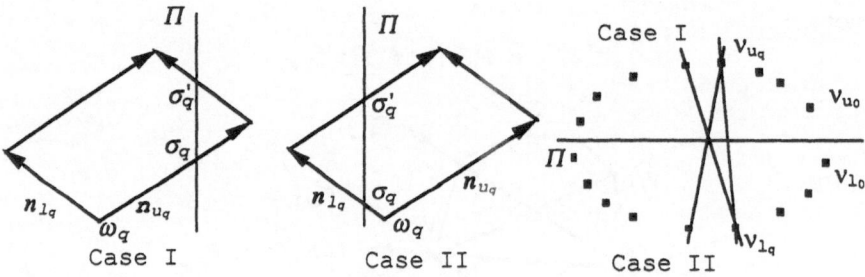

Case I Case II Case II

Figure 3 : A step of the algorithm.

In the case I, ω_{q+1}, u_{q+1} and l_{q+1} take respectively the values $\omega_q + n_{u_q}$, $u_q + 1$ and l_q, and in the case II, they take respectively the values $\omega_q + n_{l_q}$, u_q and $l_q - 1$. More, σ_q becomes σ_q'. We note $\bar{n}_q = \sigma_q' - \sigma_q$ and we keep in memory the expression :

$$\bar{n}_q = \lambda_q^u n_{u_q} + \lambda_q^l n_{l_q}, \quad 0 \leq \lambda_q^u \leq 1, \; 0 \leq \lambda_q^l \leq 1.$$

The algorithm ends when σ_q' reaches s. Then, taking into account the symmetry, the zonogon $\Pi \cap Z\{n_1, ..., n_p\}$ is known in a standard form. Since, it is easy to determine whether or not m belongs to this zonogon. If it is true, a particular decomposition over the set of vectors $\{\overline{n}_1, ..., \overline{n}_q\}$ is easy to compute, for example by determining the intersection on the semi-line through m and the boundary of the zonogon.

Complexity analysis : Finding the first face requires time $O(p)$ and the number q of steps is less than p. Whether m belongs or not to this zonogon can be found in $O(p)$ time. We have the same complexity for a decomposition of m on the vectors $\{\overline{n}_1, ..., \overline{n}_q\}$ which leads directly to a decomposition over the set $\{n_1, ..., n_p\}$.   □

III. The zonoid associated to a differential family.

Definition 3.- Let I a compact interval in $\mathbb{R}$, and f be a function in $\mathcal{C}^0[I, \mathbb{R}^3]$. The zonoid associated to the "differential family" $\{f(t)dt\}_{t \in I}$, is defined as the set :

$$Z(f) = \{m = \int_I \lambda(t) f(t) \, dt \; : \; \lambda \in L^1[I, [0,1]]\}$$

This is a natural extension of the previous discrete model. Moreover, this generalization is important for actual applications [9,10,11], like some distillation process in chemical engineering. For simplicity of the presentation, we only consider the case of a compact interval, ($I = [0,1]$, for example), but generalizations are possible.

III.1. Zonoids and sequences of zonohedra.

Property 4.- With $I = [0,1]$, $Z(f)$ can be expressed as the limit in the Haussdorf metric of the following sequence of zonohedra :

$$Z_p(f) = Z \{ \int_{t_{i-1}^p}^{t_i^p} f(t) \, dt, \; i = 1, ..., p\},$$

where $(t_0^p, t_1^p, ..., t_{p-1}^p, t_p^p)$ verify :

$$0 = t_0^p < t_1^p < ... < t_{p-1}^p < t_p^p = 1,$$

and : $$\lim_{p \to \infty} \underset{1 \le i \le p}{\text{Max}} \, (t_i^p - t_{i-1}^p) = 0.$$

This property is a result which directly issues from more general properties [2], and allows to solve approximately the main problems of feasibility and decomposition.

III.2. Zonoids and regular selectivity.

Definition 5.- As a direct extension of the definition 1, we say that the zonoid $Z(f)$ verifies the property of "regular selectivity" if :
- there exists $m \in \mathbb{R}^3$ such that for every $t \in I$, $<f(t),m>$ is strictly positive,
- for every (t,u,v), $0 \leq t < u < v \leq 1$, the determinant $(f(t),f(u),f(v))$ is strictly positive.

Example : In the case of a batch distillation process (Fenske equation), we can show that the function f defined by :

$$f(t) = (exp(-\alpha t), exp(-\beta t), exp(-\gamma t)), \quad t \in I \ (= \mathbb{R}^+),$$

where α, β et γ are real numbers such that $0<\alpha<\beta<\gamma$, implies the property of "regular selectivity".

Property 6.- If $Z(f)$ verifies the property of regular selectivity then it is the same for the zonohedra $Z_p(f)$.

The proof is obvious. $\qquad\qquad\qquad\qquad\qquad\qquad\qquad\qquad\square$

Numerical experiments : The algorithm in the property 2, provides an efficient tool for solving in an approximate way the feasibility and decomposition problems. Numerical tests have been carried out for some values of p up to 2000.

III.3.- Parametrization of the boundary of a zonoid with regular selectivity.

Property 7.- If $Z(f)$ verifies the property of regular selectivity, the boundary of $Z(f)$ is decomposable into two symmetrical patches with the following parametric equations :

$$s_1(u,v) = \int_u^v f(t)\,dt, \quad 0 \leq u \leq v \leq 1,$$

and :
$$s_2(u,v) = \int_0^u f(t)\,dt + \int_v^1 f(t)\,dt, \quad 0 \leq u \leq v \leq 1.$$

Demonstration : Using a natural extension of the discrete case, the FPD becomes a convex parametric arc $\{v(t), \ t \in I\}$. Then, we obtain the boundary of the zonoid by considering any directed line and by summing the function $f(t)$ over the set of parameters t such that $v(t)$ lies on the right of this line. In order to obtain a patch, we have to consider the lines which cut the continuous projective diagram in two points, as shown in figure 4. □

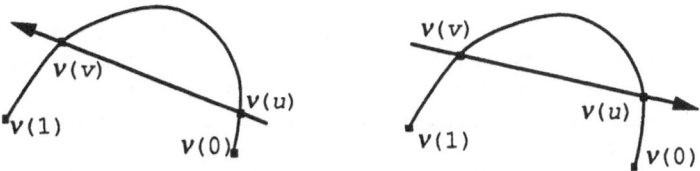

<u>Figure 4</u> : Parametrization of the boundary of a zonoid
with the property of regular selectivity.

III.4.- Towards continuous methods for zonoids.

Now, we consider the problem of determining the section of the previous zonoid by the plane of equation :

$$\alpha x + \beta y + \gamma z = 0.$$

This case includes the extension to the continuous case of the method described in the property 2. If we consider a patch of the first form, and if the function $f(t)$ is given by :

$$f(t) = (f_x(t), f_y(t), f_z(t)),$$

then we have to determine a function $v = g(u)$ such that :

$$\int_u^{g(u)} \phi(t) \, dt = 0,$$

with : $\qquad\qquad \phi(t) = \alpha f_x(t) + \beta f_y(t) + \gamma f_z(t).$

This amounts to solve the equivalent differential equation :

$$g'(u) \ \phi(g(u)) - \phi(u) = 0,$$

with some suitable initial conditions. These interpretations can be used to set up numerical algorithms for the problems of feasibility and decomposition of a mixture m.

Conclusion.

We presented different problems in separation theory and showed the simplicity of some geometrical solutions in the particular case we have named "regular selectivity". Using the same geometrical framework, extensions of these solutions to some more complicated cases, will be the aim of future works.

Acknowledgements

This work was supported by Université Joseph Fourier de Grenoble (France), Centre National de la Recherche Scientifique (CNRS) and Société Nationale ELF-Aquitaine. The authors gratefully acknowledge Société Nationale ELF-Aquitaine for kind permission to publish it.

References.

[1] E.D. Bolker, "A class of convex bodies", *Trans. Amer. Math Soc.*, Vol. 145, Nov. 1969.
[2] J. Bourgain, J. Lindenstrauss, V. Milman, "Approximation o: zonoids by zonotopes", *Acta Math.*, 162, (1989).
[3] H.S. Coxeter, "The classification of zonohedra by means of projective diagrams", *Journ. de Math*, tome XLI., Fasc. 2, 1962
[4] H. Edelsbrunner, "Algorithms in Combinatorial Geometry", *EATCS Monographs on Theoretical Computer Science*, Vol. 10, Springer Verlag, 1987.
[5] D. Girard et P. Valentin, "Zonotopes and Mixtures management", *New Methods in optimization and their industrial uses*, *International Series of Numerical mathematics*, Birkhäuser Verlag, 1988.
[6] B. Lacolle et P. Valentin, "Les mélanges binaires : modélisation géométrique et algorithmes", *Rapport de Recherche IMAG, RR. 841-M.*, Février 1991.
[7] P. Mc Mullen, "On zonotopes", *Trans. Amer. Math. Soc.*, 159 (1971), pp. 91-109.
[8] N. Odeh, "Modélisation mathématique des propriétés de mélanges : B-splines et optimisation avec condition de forme", *Thèse de l'Université Joseph Fourier*, 19 Mars 1990.
[9] N. Szafran, "Zonoèdres : de la géométrie algorithmique à la théorie de la séparation", *Thèse de l'Université Joseph Fourier*, Grenoble, Octobre 1991.
[10] P. Valentin, "Theory of Zonoids : I A Mathematical Summary", *NATO Institute of Advanced studies*, Advances in Theory of Chromatography, Ferrare, Aug. 1991, F. Dondi, G. Guiochon Ed., Kluwer, 1992.
[11] P. Valentin, "Theory of Zonoids : II Application to Chromatography", *NATO Institute of Advanced studies*, Advances in Theory of Chromatography, Ferrare, Aug. 1991, F. Dondi, G. Guiochon Ed., Kluwer, 1992.
[12] W. Weyl, "Centrally symmetric bodies and distributions", Israel Jour. Math., 24 (1976), pp. 352--367.

k-Violation Linear Programming

(Extended Abstract)

Thomas Roos*   Peter Widmayer*

Abstract

We introduce the notion of *k-violation linear programming*. Given a set of n halfplanes, we want to compute an optimal solution with respect to a given linear functional. However, in opposite to classical linear programming [MaShWe 92], we allow to violate at most k of the n constraints, for some fixed $k \in \{0, \ldots, n-1\}$. We solve this problem in $O(\beta_k(n))$ time and $O(n)$ space, where $\beta_k(n) := n \log n + k \log^2 k$. This is optimal if $k \in O(n^\alpha)$ for any fixed positive $\alpha < 1$.

The general idea behind our approach is a new technique for computing a minimum of the k-level of an arrangement. Based on recent slope selecting techniques by Cole et al. [Co 89] and Matoušek [Ma 91], we develop an algorithm for computing a minimum k-level point in $O(\beta_k(n))$ time and linear space. Our result is by a $O(\sqrt{k} \log n)$ factor better than the application of an algorithm for computing the entire k-level by Edelsbrunner and Welzl [EdWe 86]. The presented technique is of independent interest and can be applied to several other problems, as well.

1   Introduction

In this paper, we study a theoretically interesting problem that has a number of practical applications. The applications range from the computation of minimum size geometric objects that contain exactly k out of n given points, to the solution of a linear programming problem where a certain amount of the linear constraints may be violated. That amount may be expressed by associating a penalty with each constraint; whenever a constraint is violated, the associated penalty has to be paid. We show how to solve the linear programming problem for a given, fixed budget to pay penalties.

The instances of the problems we study can be modeled by arrangements of lines. For an arrangement of lines in the plane, we ask for the point that has minimum y-coordinate among all points lying above k lines (or lying above lines with at least a certain sum of penalties, in the weighted case), the so-called y-minimum of the k-level of the arrangement. It turns out that the point we are looking for lies on the intersection of two lines, under the assumption that we bound the space we consider, or on the space boundary. Edelsbrunner and Welzl [EdWe 86] have shown how to compute each line segment of the k-level of n lines in time $O(\log^2 n)$; since the k-level consists of at most $O(\sqrt{k}\, n)$ line segments, the entire k-level can be computed in time $O(\sqrt{k}\, n \log^2 n)$.

Interestingly, it turns out that we need not even compute the entire k-level to find its minimum. Instead, we can make use of recent slope selecting techniques [Co 89, Ma 91] to probe the arrangement with horizontal lines at positions that follow the pattern of a binary search. At each probing position, a scan through the arrangement along the probing line determines where the binary search continues.

---

*Departement Informatik, ETH Zentrum, CH–8092 Zürich, Switzerland. Email: {roos,widmayer}@inf.ethz.ch

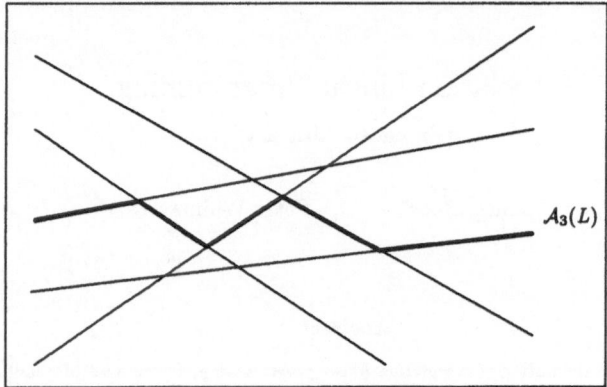

Figure 1: The 3-level of an arrangement in the plane.

2 Preliminaries

Given a set $L := \{l_1, \ldots, l_n\}$ of n lines in the plane $\mathbb{R}^2$ and a parameter $k \in \{1, \ldots, n\}$. In order to simplify the formulation, let us assume a certain kind of *general position*: we demand that no line is vertical and no two lines are parallel. In addition, the y-coordinates of all points of intersection are assumed to be disjoint. Notice that there are well-known techniques for removing these slight restrictions (see, e.g., [EdMü 90]). Now, let $(x_{ij}, y_{ij}) \in \mathbb{R}^2$ denote the point of intersection of the two lines l_i and l_j, and $P := \{(x_{ij}, y_{ij}) \mid i \neq j\}$ the set of intersection points for any two lines $l_i, l_j \in L$.

The dissection of the plane induced by the lines of L is called the *arrangement* $\mathcal{A}(L)$ (cf. [Ed 87]). (In our context, $\mathcal{A}(L)$ will usually denote the embedding of the union of all lines in the plane.) Now, the *k-level* $\mathcal{A}_k(L)$ of the arrangement $\mathcal{A}(L)$ is defined to be the set of points

$$\mathcal{A}_k(L) := \{x \in \mathcal{A}(L) \mid below(x) \leq k - 1 \text{ and } above(x) \leq n - k\}$$

where $below(x)$ and $above(x)$ denote the number of lines lying below or above the point x, respectively. Figure 1 shows the 3-level of an arrangement in the plane indicated by solid lines.

Now, we face the problem to determine a point of the k-level $\mathcal{A}_k(L)$ which has minimum y-coordinate. Since we are not interested in a solution at infinity, let us restrict the problem in order to obtain a constrained solution. This is done by considering the k-level inside a vertical strip which is determined by the smallest and largest x-coordinate of the points in P. Thus, we consider the following *restricted k-level minimum problem*:

$$\min_{\substack{(x,y) \in \mathcal{A}_k(L) \\ x_{min} \leq x \leq x_{max}}} y \quad \equiv \quad \min_{(x_{ij}, y_{ij}) \in \mathcal{A}_k(L) \cap P} y_{ij}$$

Thereby, it is obvious that there is always a point of P on the k-level $\mathcal{A}_k(L)$ which achieves this minimum.

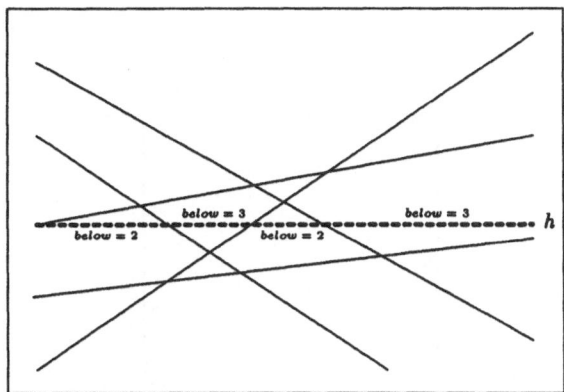

Figure 2: Deciding if the k-level lies completely above h.

3 Computing a Minimum of the k-Level

Now, a first algorithm for computing the (restricted) minimum of the k-level calculates all parts of the k-level and minimizes their y-coordinates. It is well-known that the *complexity* of the k-level, i.e. the number of pieces of the k-level, is bounded by $O(\sqrt{k}\,n)$ (see [Ed 87]). Since $O(\log^2 n)$ time suffices for computing each part of the k-level (cf. [EdWe 86]), we immediately get an $O(\sqrt{k}\,n \log^2 n)$ algorithm. This algorithm, however, is far from optimal as k increases.

Another approach is a *bottom-up sweepline algorithm* which outputs the appearance of a lowest point of the k-level in $O(k\,n \log n)$ time and $O(n)$ space. The claimed runtime can be achieved by maintaining the *below*-function for the horizontal sweepline. Notice that the y-coordinate $\bar{y}_i$ of the y-minimum point of the i-level $\mathcal{A}_i(L)$ is higher than the y-coordinate $\bar{y}_{i-1}$ of the y-minimum point of the $(i-1)$-level $\mathcal{A}_{i-1}(L)$. Thus, the y-coordinates of the y-minima of all levels form an increasing sequence for increasing levels, i.e. $\bar{y}_1 \leq \ldots \leq \bar{y}_n$.

As a consequence, our sweep-line may only meet intersection points of levels less than k before meeting the lowest point of level k. The number of these points is bounded by $O(k\,n)$ due to a random-sampling lemma by Clarkson and Shor [ClSh 89]. In a sweepline algorithm where we maintain the value of the *below*-function for each of the intervals induced by the intersections of all lines in L with the sweepline, an intersection point can be processed in time $O(\log n)$. Finally, we obtain an $O(k\,n \log n)$ algorithm; this is optimal for constant k.

A starting observation to improve this bound in the general case is the following slope selection technique[1] due to Cole et al. [Co 89] and Matoušek [Ma 91]. They proved that given an arrangement of n lines, the point of intersection with kth smallest y-coordinate can be computed in optimal $O(n \log n)$ time and $O(n)$ space. Using this, we will compute a point of the k-level with minimum y-coordinate in $O(n \log^2 n)$ time and $O(n)$ space. For this, we apply *binary search* on the set P in the following way. First, we find a horizontal line h that partitions the $\binom{n}{2}$ points of P into two halves by means of a horizontal line h. With the help

---

[1]In fact, the original papers treat the dual problem, i.e. computing the kth smallest slope of a set of $\binom{n}{2}$ lines through n given points in the plane. Cole et al. [Co 89] presented a deterministic algorithm for this problem which was later simplified by Matoušek [Ma 91] at the cost of randomization.

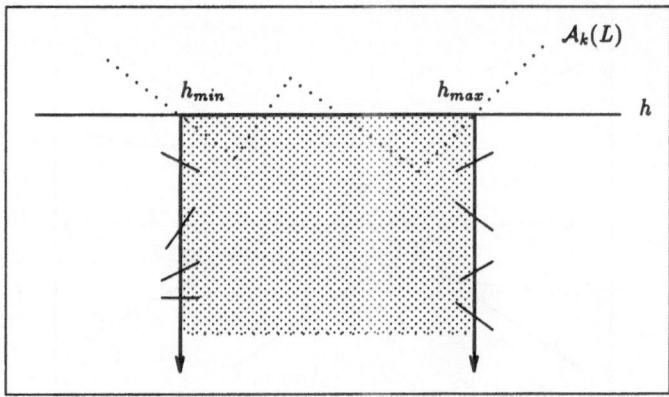

Figure 3: Reducing the problem size.

of the slope selection technique this can be achieved in $O(n \log n)$ time and $O(n)$ space. Next, we test on which side of h the desired y-minimum k-level vertex lies.

This can be done by computing all intersections of h with the n lines in L. Afterwards, these intersection points are sorted by increasing x-coordinate and processed by a sweepline algorithm in order to test if there is *no* point on h with at least k lines of L below (cf. Figure 2). With that, we can decide in $O(n \log n)$ time on which side of h the binary search has to continue. Thus, $O(n \log^2 n)$ time suffices to process the $O(\log n^2) = O(\log n)$ recursions of the binary search to compute a lowest point of P which belongs to the k-level $\mathcal{A}_k(L)$.

We can still improve this binary search by the following observation. Any horizontal probing line h which hits the k-level $\mathcal{A}_k(L)$ determines two points h_{min} and h_{max} on h with minimum and maximum x-coordinate which belong to the k-level. Now, it is obvious that the minimum of the k-level must be contained in the south-grounded rectangle given by h_{min} and h_{max} (cf. Figure 3) or lie at position x_{min} or x_{max}. In order to determine an initial line h intersecting the k-level $\mathcal{A}_k(L)$, we choose an arbitrary line of L and sort the intersection points of this line with all other lines in L. After that, a sweepline algorithm determines a point on L which belongs to the k-level and, with that, the horizontal line h passing through that point in total $O(n \log n)$ time.

Now, if we consider the two bounding vertical halflines of the shaded region (see Figure 3), there are exactly k lines of L intersecting each of them below h. In fact, these are the only lines which can contribute to the minimum of the k-level in the shaded region. Thus we have reduced our problem size from n to at most $2k$. Applying the binary search technique above to the reduced problem, we finally obtain an $O(n \log n + k \log^2 k)$ time algorithm.

Theorem 1 *Given an arrangement of n lines, a point of the (restricted) k-level with minimum y-coordinate can be computed in $O(\beta_k(n))$ time and $O(n)$ space, where $\beta_k(n) := n \log n + k \log^2 k$.*

This result should be contrasted with the $\Omega(n \log n)$ lower bound for selecting the kth lowest point of intersection[2] of an arrangement of n lines in the plane (see [Co 89]). In fact, this lower bound applies to our problem, as well. This can be seen by choosing $k = 1$, in which case the point of the (restricted) 1-level with minimum y-coordinate is exactly the point of P with minimum y-coordinate. Notice that our algorithm is already optimal if $k \in O(n^\alpha)$ for some fixed positive $\alpha < 1$, since

$$k \log^2 k \ \leq \ C \alpha^2 n^\alpha \log^2 n \ \leq \ C \alpha^2 n \log n$$

for some constant C. However, it is an open problem whether the running time of the algorithm can be improved so that it matches the $\Omega(n \log n)$ lower bound for any k.

Our Theorem still holds for the generalization of the problem in which we do not look for the y-minimum under the restriction that it be above or on the k-level, but instead we assume that each line carries with it a non-negative weight, and the restriction is that the y-minimum be above at least a certain total weight of the lines below it. If we define the total weight as the sum of the weights of the lines, then Theorem 1 as stated above takes care of the unit weight case, that is, all weights are 1. It should be clear that the algorithm for solving this *weighted y-minimum problem* is essentially the same as the one given above.

4   k-Violation Linear Programming

An interesting application of the minimum k-level problem is the two-dimensional *k-violation linear programming problem*. Here we are given a set $\mathcal{H} := \{H_1, \ldots, H_n\}$ of n halfplanes in $\mathbb{R}^2$, a linear objective function (which we assume without loss of generality to be y) and a parameter $k \in \{0, \ldots, n-1\}$. In opposite to classical linear programs (see, e.g., [Se 90] and [MaShWe 92]), we are not interested in an optimum solution which meets all constraints, but allow that at most k of the n restrictions may be violated. This leads to the following series of linear programs:

$$(LP_k) \qquad \min_{(x,y) \in P_k} y \qquad \text{where} \qquad P_k := \bigcup_{\substack{\mathcal{H}' \subset \mathcal{H} \\ |\mathcal{H}'| \geq n-k}} \ \bigcap_{H_i \in \mathcal{H}'} H_i$$

Thereby the feasible region P_k of the kth linear program is defined to be the set of points in the plane violating at most k constraints. In addition, the y-coordinates $\bar{y}_k$ of the optimal solutions of (LP_k) form a decreasing sequence, i.e. $\bar{y}_0 \geq \ldots \geq \bar{y}_{n-1}$.

Notice that we obtain classical linear programs by choosing $k = 0$. Figure 4 displays the arrangement $\mathcal{A}(\mathcal{H})$ with the boundaries of the feasible regions P_0 and P_2. To get some more geometric insight, we prove that the feasible regions P_i possess the following properties:

(1) $P_i \subset P_{i+1}$, for all $i \in \{0, \ldots, n-2\}$.

(2) P_i is star-shaped with respect to any point $p \in P_0$, for $i \in \{0, \ldots, n-1\}$,
       i.e. the line $\overline{pq}$ is entirely contained in P_i for any $p \in P_0$ and any $q \in P_i$.

---

[2]This can be shown by transforming an instance of the *element uniqueness problem* to the problem of selecting the smallest slope, i.e. the lowest point of intersection.

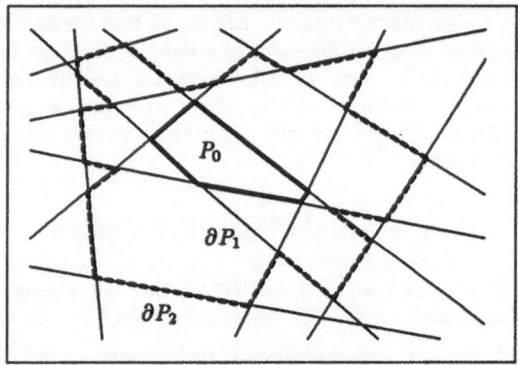

Figure 4: The boundaries ∂P_i forming onion-skins around P_0.

The first part follows directly from the definition of P_i. The second part can be shown by considering a point $q \in P_i$ and a subset $\mathcal{H}' \subset \mathcal{H}$ with $|\mathcal{H}'| \geq n - i$ and $q \in A := \bigcap_{H_j \in \mathcal{H}'} H_j \subset P_i$. Now, $P_0 \subset A$ and A is convex. Thus, for any $p \in P_0$ the line $\overline{pq}$ is entirely contained in A and, with that, also in P_i. This completes the proof.

Using property (1), the sequence of boundaries $\partial P_0, \ldots, \partial P_{n-1}$ of the feasible regions form onion-skins around the feasible region P_0. In addition, the star-shapedness of P_i allows a parametrization of ∂P_i in the following way: Choose an arbitrary point of the interior of P_0 and shoot polar rays from that point towards infinity until they hit the boundary ∂P_i. More formally, we assume that the interior P_0° of the feasible region P_0 is not empty and without loss of generality that $0 \in P_0^\circ$. Then, each restriction $a_i x + b_i y + c_i \geq 0$ of $\mathcal{H}$ is associated to a curve

$$f_i(\phi) := \frac{c_i}{a_i \cos \phi + b_i \sin \phi}$$

which describes the distance of 0 to ∂P_i in the polar direction ϕ when restricted to regions where $f_i(\phi) > 0$. (This selection guarantees the proper choice of ϕ in a region of 180°). These functions $f_i(\phi)$ form a new arrangement $\mathcal{A}(\mathcal{H}_\phi)$ of trigonometric curves, in which the k-level corresponds to the boundary of the kth feasible region. Although the functions f_i are no longer straight lines, the incidence relations of the arrangement $\mathcal{A}(\mathcal{H})$ are preserved.

Now, in order to find a point of minimum y-coordinate on the boundary ∂P_i (again we are only interested in a constrained solution), we do a binary search in the arrangement $\mathcal{A}(\mathcal{H}_\phi)$ with probing curves of type

$$f(\phi) := \frac{c}{\sin \phi} \quad \text{with } c \in \mathbb{R}$$

corresponding to horizontal lines in the arrangement $\mathcal{A}(\mathcal{H})$, again restricted to regions where $f(\phi) > 0$.

In fact, the technique of the previous section applies to this curvilinear arrangement $\mathcal{A}(\mathcal{H}_\phi)$ as well, because the topology is the same as in $\mathcal{A}(\mathcal{H})$.[3] Notice that due to the special type of the functions f_i and f, their unique pairwise intersection point can be computed in constant time. Thus, we obtain the following main result:

Theorem 2 *Given a set of n linear constraints in the plane, an optimal solution of the k-violation linear program can be computed in $O(\beta_k(n))$ time and $O(n)$ space, where $\beta_k(n) := n \log n + k \log^2 k$.*

5 Remarks and Open Problems

We presented in this paper an efficient technique for computing a minimum of the k-level of an arrangement of straight lines which can also be extended to more general arrangements, as well. However up to now, it is an open problem if our technique can be generalized to higher dimensions. The bottleneck here would probably be the counting problem, i.e. given an arrangement of hyperplanes and a horizontal hyperplane, how many points of intersection lie above and below this hyperplane. We applied our technique to k-violation linear programs which can be solved in $O(\beta_k(n))$ time and $O(n)$ space, where $\beta_k(n) := n \log n + k \log^2 k$. This turned out to be optimal if $k \in O(n^\alpha)$ for some fixed positive $\alpha < 1$.

Again, it is surprising that all $\binom{n}{k} \in O(n^k)$ possible selections of violated restrictions can be handled so efficiently. In this context, another interesting problem appears when we drop the restriction that the feasible region of the original problem is non-empty. In this case we can ask for the smallest k which generates a non-empty feasible region P_k. This problem can be solved in $O(n^2)$ time and space by first computing the arrangement $\mathcal{A}(\mathcal{H})$ (cf. [Ed 87]). Afterwards, we assign to each point of intersection of the arrangement $\mathcal{A}(\mathcal{H})$ the number of restrictions which are violated at that point. This can be done, e.g., by sweeping all points of intersection along the lines of the arrangement $\mathcal{A}(\mathcal{H})$. Thus, the minimum k with non-empty feasible region P_k and an optimum solution of (LP_k) can be derived in $O(n^2)$ time.

Acknowledgement

The authors would like to thank Herbert Edelsbrunner, Jiří Matoušek and the anonymous referees for their helpful comments.

References

[ClSh 89]   K.L. Clarkson and P.W. Shor,
Applications of random sampling in computational geometry, II, Discrete & Comput. Geometry, Vol. 4, 1989, pp 387–421

---

[3]This is necessary, because the slope selection techniques above are based on a technique for counting inversions of permutations. In other words, intersecting an arrangement by two simple disjoint curves generates two orders (permutations) of the curves of the arrangement. Now, the critical point is that the topology of the arrangement allows to derive the number of intersection points between the two curves by counting inversions of the corresponding two permutations. To achieve this situation, any two curves of the arrangement may have only one point of intersection and any probing curve must intersect any curve of the arrangement.

862

[Co 89]   R. Cole, J. Salowe, W. Steiger and E. Szemerédi,
            An Optimal-Time Algorithm for Slope Selection, SIAM J. Comp., Vol. 18, No. 4,
            1989, pp 792–810

[Ed 87]   H. Edelsbrunner,
            Algorithms in Combinatorial Geometry, EATCS Monographs in Computer Sci-
            ence, Springer, Berlin, 1987

[EdMü 90]   H. Edelsbrunner and P. Mücke,
            A Technique to Cope with Degenerate Cases in Geometric Algorithms, ACM
            Trans. Computer Graphics, Vol. 9, 1990, pp 66–104

[EdWe 86]   H. Edelsbrunner and E. Welzl,
            Constructing Belts in Two-Dimensional Arrangements with Applications, SIAM
            J. Comput., Vol. 15, No. 1. 1986, pp 271–284

[Ma 91]   J. Matoušek,
            Randomized Optimal Algorithm for Slope Selection, Info. Proc. Letters, Vol. 39,
            1991, pp 183–187

[MaShWe 92] J. Matoušek, M. Sharir and E. Welzl,
            A Subexponential Bound for Linear Programming, Proc. 8th Annual ACM Symp.
            on Computational Geometry, Berlin, 1992, pp 1–8

[Mu 92]   K. Mulmuley,
            On Levels in Arrangements and Voronoi Diagrams, Discrete & Comput. Geom.,
            to appear

[Se 90]   R. Seidel,
            Linear Programming and Convex Hulls Made Easy, Proc. 6th ACM Symposium
            on Computational Geometry, Berkeley, California, 1990, pp 211–215

[ShWe 92]   M. Sharir and E. Welzl,
            A Combinatorial Bound for Linear Programming and Related Problems, Proc.
            Symposium on Theoretical Aspects of Computer Science STACS'92, LNCS 577,
            pp 569–579

Networks

Evaluation of Telecommunication Network Performances [(1)]

Jacques Carlier & Yu Li
Université de Technologie de Compiègne
Heudiasyc CNRS 817
60206 Compiègne cedex France
Fax: 33-44234477
e-mail: yliu@hds.univ-compiegne

Jean-luc Lutton
CNET/PAA/ATR
38-40 rue du Général-Leclerc
92131 Issy les Mx. France
Fax: 33-45296069
e-mail: jean-luc.lutton@issy.cnet.fr

Abstract: In this paper, a methodology for evaluating performances of telecommunication networks based on optical fiber systems and DCS nodes is presented. Availability and lost traffic are chosen as performance measures. The methodology includes two aspects: the probabilistic calculation is done by partial enumeration and the rerouting calculation (restoration from failures) is accomplished by a heuristic method based on K-shortest paths. This rerouting is made very rapidly thanks to a sophisticated data structure. Consequently, the availability of networks up to 26 nodes can be evaluated accurately.

Key words: telecommunication network, availability, lost traffic, rerouting, K-shortest paths.

1. INTRODUCTION

Recently, two technological advances have had a radical impact on the telecommunication networks. One is the extensive deployment of optical fiber transmission systems carrying multiple gigabits per second, which makes telecommunication network structures less and less meshed. Thus the failure of an optical fiber system or of a node will result in serious loss of services. Fortunately, the other advance, which is the deployment of reconfigurable digital cross-connect (DCS) nodes also known as facility switches, makes possible a rapid restoration from failures with the spare capacity of optical fiber links [1,2,3,4]. Therefore, the problems of evaluating network performances and studying the tradeoff between reliability and spare capacity assignment become essential for both operation and design of the networks.

Traditionally, most network performance measures are related to the network connectivity and the associated problems are NP-hard [6,7]. In the case of networks such as electric ones, a network can be considered as a flow network carrying commodities from origin to destination nodes. The performance measures are related to the ability of the network to transmit a level of flow. In fact, this is a single-flow problem [8]. The methods for calculating the two kinds of performance measures mainly include the enumeration of minimal sets, the techniques of reduction, factorisation, and decomposition [9].

The situation is particularly difficult in a telecommunication network based on optical fiber systems and DCS nodes because the network elements have high reliability and consequently the network generally remains connected even after failures of several elements. Such a network is often very large (its size can reach 100 nodes) and not very meshed. Moreover, the new restoration technique as well as the traffic which it transports makes it necessary to take into account the multiflow characteristics. So far, we have not seen many results in this domain. The work done by Sanso consists in evaluating the performance measures of telecommunication networks carrying traffics between all pairs of nodes [10,11]. It takes into account only the case of single failures, but, in practice two or three simultaneous failures are not rare. Moreover, the restoration from failures is not in the context of DCS nodes and optical fibers.

The aim of this article is to decribe an efficient program for evaluating performances of telecommunication networks based on optical fiber systems and DCS nodes. In section 2, we

---

(1) This work has been supported by the Centre National d'Etudes des Télécommunications (contract 7758c).

give a model for telecommunication networks. Then in Section 3, we present a method to evaluate their performances as well as a restoration technique based on DCS nodes. It uses a sophisticated data structure which is described. Computational results on networks of medium size are reported in Section 4. Finally, conclusions and directions for future work are given in Section 5.

2. MODELLING TELECOMMUNICATION NETWORKS

In this section, a model for telecommunication networks having optical fiber systems and DCS nodes is proposed. After describing briefly this model, we present the way of the failure treatment.

2.1 Model Description

This model consists of two layers. The first layer is called the "physical network" and it corresponds to the "civil engineering" structure. An arc of this network corresponds to a transmission cable. The second layer is called the "logical network" and it is routed along the physical network. An arc of the logical network is constituted by a set of optical fiber transmission systems which is routed through the cables (this set is also called a system arc in the following discussion). A system arc has a given capacity which consists of a used capacity and a spare capacity. The nodes correspond to DCS nodes. The figure 1 shows an example which comes from [4,5]. We take their network as "logical network" (the figure 1(b)), then we construct a physical network (the figure 1(a)) [5].

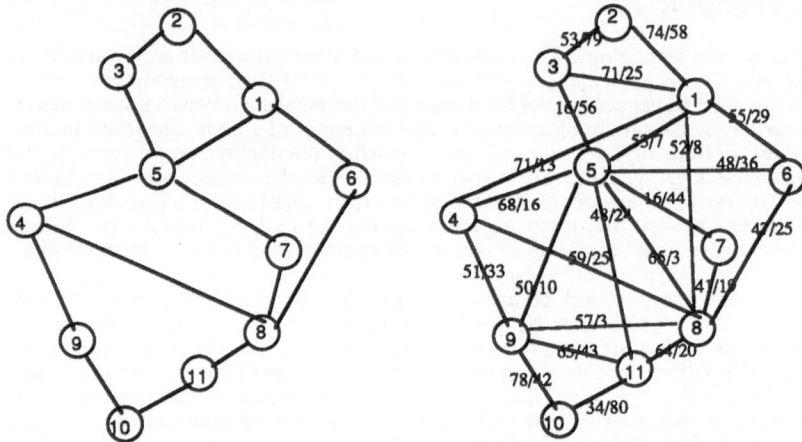

Figure 1 (a) Physical network	Figure 1 (b) Logical network

The traffic demands between the nodes are represented by a matrix. It is then routed on the logical network with some routing strategy.

2.2 Failure Treatment

Three types of failures are taken into account: node failures, cable failures and system arc failures. A system arc failure causes cancellation of its capacity, thus all demands passing through this arc cannot be transmitted. A cable failure makes all the system arcs using this cable out of work. A node failure causes the failure of all the system arcs connected to it. Each failure is repaired by its own repairer. Component failure and repair rates are supposed constant

(Markov hypothese). For simplicity, it is supposed that the system arcs, the cables and the nodes fail and are repaired independently.

After a failure, the perturbed demands are rerouted on alternative paths through the spare capacity on the surviving system arcs. This technique is called rerouting or restoration. Until now, many restoration algorithms, using centralized or distributed control, are proposed [2,3,4]. Here, we use a centralized control algorithm which restores from failures by rerouting the perturbed demands on the K-shortest paths. This algorithm is simple and effective in practice.

The rerouting strategy used is partial: only the set of routes passing through the failed system arcs is concerned. The rerouting is accomplished through the logical network. We distinguish two categories of rerouting: local rerouting also known as *link restoration*, and end-to-end rerouting known as *path restoration* or global rerouting. When only one system arc fails, the local rerouting is used that allocates restoration paths between the two extremities of the failed system arc. When several system arcs fail, the global rerouting is used that allocates restoration paths between two nodes of a failed route passing through failed system arcs.

Local rerouting is a single-flow problem, whereas global rerouting is a multi-flow problem. A heuristic method is proposed for treating uniformly local rerouting and global rerouting: the end-to-end demands affected are ranked by some criterion, e.g. , in decreasing order of these demands, then we sequentially treat them in this order. However, the essential problem for realizing our method is to find the K-shortest paths in a graph.

In this model, we want to know the proportion of time in which the traffic demands can be transmitted, whether or not the failures are arrived, as well as the lost demands per unit of time.

3. METHODOLOGY

3.1 Availability and Lost Traffic: Performance Measures

It appears to us that the *availability* of a network subject to failures and repairs is an important performance measure. By availability we mean the probability that the network is operational at any given time. The measure related to availability is the lost traffic, which represents the demand that cannot be transmitted because of failures.

In order to define the availability and the lost traffic, we need to introduce the notions of binary component states and binary network states.

Binary Component State

We assume that components are in a binary state: operational state or failed state, and that components are statistically independent.

Suppose that λ_i is the failure rate of component i and μ_i the repair rate of component i, the probability that component i operates is calculated by $p_i = \dfrac{\mu_i}{\lambda_i + \mu_i}$. Note that the components have high reliability, that is, p_i is very close to 1.

Binary Network State

A state of the network is determined by the states of its components, its topology, the rerouting strategy and the capacities of the system arcs. We can associate a rerouted traffic rate with each state of the network:

$$\partial(e_i) = \frac{\text{demand\_traffic - lost\_traffic}(e_i)}{\text{demand\_traffic}}$$

where *demand_traffic* is the sum of all demands between all pairs of nodes, and *lost_traffic(e_i)* is the sum of the unsatisfied demands between two nodes for state e_i, which is obtained by the rerouting.

For the network in state e_i, if $\partial(e_i)$ after the rerouting is greater than or equal to a given rate τ (τ is called *availability-coefficient*), the network operates (in other words, it is in operational state); otherwise, the network fails (in failed state).

Availability And Lost Traffic

The availability is the probability that the network operates at the given time t. At steady state, it is equal to the probability that the network operates per unit of time. The lost traffic is the mathematical expectation of the lost traffic over all network states.

3.2 Method for Calculating Availability and Lost Traffic

We propose a mixed approach to calculate the availability and the lost traffic: the probabilistic calculation is done in an analytical way and the rerouting calculation is done by simulating the rerouting of the network.

Probabilistic Calculation

Because of the assumption of independent components and the fact that the components have high reliability (p_i is close to 1), we can use the method referred as *partial enumeration* [12]. This method permits to obtain an interval for the calculated measures.

Let m be the number of components in the network. So, there are 2^m states in the network. These states can be divided into m+1 classes: states without failure, states with one failure, ... , states with m failures. Since the components have high reliability, we can neglect the states with more than f failures (f = 3 or 4). For each state, we calculate its probability $P(e_i)$ and the lost_traffic(e_i) with some rerouting strategy. Then the interval of confidence of the lost traffic is given by:

$$E(\text{lost\_traffic})_{up} = \sum_{e_i \in \Omega} P(e_i)\text{lost\_traffic}(e_i) + (1 - \sum_{e_i \in \Omega} P(e_i))\text{lost\_traffic}(e_{2m});$$

$$E(\text{lost\_trafic})_{low} = \sum_{e_i \in \Omega} P(e_i)\text{lost\_traffic}(e_i) + (1 - \sum_{e_i \in \Omega} P(e_i))\,\text{lost\_traffic}(e_1).$$

Here Ω is the set of states which are enumerated; e_{2m} is the state where all the components fail and the lost traffic is maximal; e_1 is the state without failure where the lost traffic is minimal and equal to zero.

Similarly, summing up the probabilities of the operating states of the network, we obtain an interval of confidence for the availability:

$$\text{Availability}_{up} = 1 - \sum_{\partial(e_i) < \tau} P(e_i);$$

$$\text{Availability}_{low} = \sum_{\partial(e_i) \geq \tau} P(e_i).$$

Rerouting Calculation

The rerouting is the principal module of our method. In addition, it can be used for other problems such as spare capacity assignment. Thus, it is important to develop an efficient rerouting procedure. In fact, the basic difficulty to achieve an effective rerouting is to develop an efficient algorithm for searching K-shortest paths in a graph. For this reason, we propose a simple and efficient solution using an optimal data structure for the K-shortest paths problem, as it is discussed below.

3.2 The K-Shortest Path Problem

We have developed an algorithm that searches K-shortest paths and stores them definitively in an optimal data structure. With this sophisticated data structure, we are able to get access to a path of a given length in linear time. Thus we have a very efficient rerouting algorithm. Moreover, using this structure, a distributed rerouting algorithm can be more easily written.

3.2.1 Principle of the Algorithm

We restrict our discussion to undirected graphs in which there are no loop nor parallel edges. Such a graph can be represented by $G = (V, E)$, where V is the set of vertices and E the set of edges. The sizes of V and E are denoted by N and M respectively. We assume that the vertices are numbered 1, 2,..., N and the edges, 1, 2, ..., M. Generally, an undirected graph can be considered as a directed graph by replacing an edge (i, j) with two arcs (i, j) and (j, i). The basic concepts of graph theory can be found in [13].

We want to find all the elementary paths from i to j of length smaller than or equal to L for every pair of vertices (i, j) and to store them in an increasing length order, where the length of a path is the number of edges in it and L is a given small integer (often, L<5). We solve the problem by running a single-destination algorithm, once from each vertex, i.e. by finding the elementary paths of length smaller than or equal to L to a given vertex i from every vertex j, j>i.

The algorithm is divided into two phases:
(1)   search the elementary paths to a given vertex i from every vertex j of length smaller than or equal to L;
(2)   store all these paths in an increasing order of length.

The first phase is accomplished by developing a breadth-first tree. This tree represents all paths to a given vertex i from every vertex j of maximal length L.

We illustrate the algorithm by an example. The figure 2 shows an undirected graph G, where N=10, M=14.

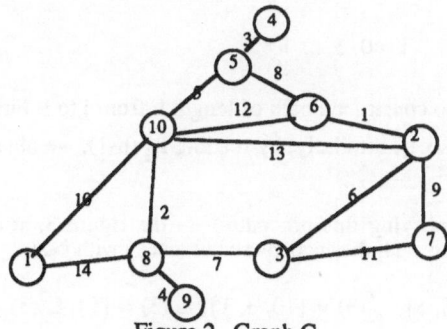

Figure 2   Graph G

870

The inverse breadth-first tree representing the paths to i=2 from every vertex with L=3 is given in figure 3.

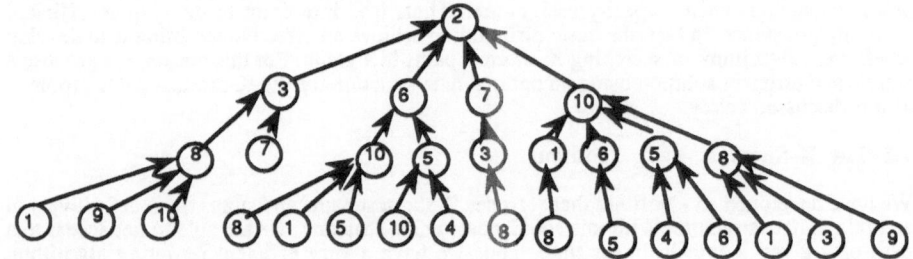

Figure 3 Breadth-first tree

The second phase is done by associating with each vertex k in the above breadth-first tree the sets of vertices defined by:

$L_i^k(m)$ = {successors of k from which i is reachable by a path of length m-1}, where m=1, 2,..., L.

Particularly,

$L_i^k(1)$ = {i}, if i is a successor of k;
$L_i^i(m)$ = ø.

Now, we explain how to calculate $L_i^k(m)$. This is accomplished by a breadth-first searching of the tree created in the phase (2). During this searching, when vertex i_h is met at depth h, its successor vertex i_{h-1} is put in the set $L_i^{i_h}(h)$ and so on. At the end of the searching, we obtain $L_i^k(m)$ for each vertex k (k≠i).

Next, we explain how to construct a path, from j to i of length h, denoted by j=$j_0,j_1,...,$ j_{h-1}, j_h=i with the sets $L_i^k(m)$.

By the definition of $L_i^k(m)$, the following relation holds:

$$j_{t+1} \in L_i^{j_t}(h-t) \qquad t = 0, 1, ..., h.$$

This relation allows us to construct a path of length h from j to i. First, j_0=j, then by reading $L_i^j(h)$, we obtain the vertex j_1. Similarly, by reading $L_i^{j_1}(h-1)$, we obtain the vertex j_2, and so on.

We give an example clarifying this procedure. In the figure 3, at depth 3 of the tree, we associate the following sets with the encountered vertices j with j>2:

$L_2^9(3) = \{8\}, L_2^{10}(3) = \{8, 5\}, L_2^8(3) = \{10, 3, 1\}, L_2^6(3) = \{5\}, L_2^5(3) = \{10, 6\}, L_2^4(3) = \{5\}.$

Similarly, at depth 2, we have:

$L_2^{10}(2) = \{6\}$, $L_2^8(2) = \{3, 10\}$, $L_2^7(2) = \{3\}$, $L_2^6(2) = \{10\}$, $L_2^5(2) = \{6, 10\}$, $L_2^3(2) = \{7\}$, $L_2^1(2) = \{10\}$.

And at depth 1, we have:

$L_2^{10}(1) = \{2\}$, $L_2^7(1) = \{2\}$, $L_2^6(1) = \{2\}$, $L_2^3(1) = \{2\}$.

From the above sets, we can find all the elementary paths from j to 2 (j>2) of maximal length 3. For example, if we look for the paths from 10 to 2, we begin by constructing the paths from 10 to 2 of length 1. Since $L_2^{10}(1) = \{2\}$, there exists a path from 10 to 2.

Similarly in constructing the paths from 10 to 2 of length 2, $L_2^{10}(2) = \{6\}$, $L_2^6(1) = \{2\}$, thus we have a path: 10->6->2.

Finally in constructing the paths from 10 to 2 of length 3, $L_2^{10}(3) = \{8, 5\}$, $L_2^8(2) = \{3, 10\}$, $L_2^3(1) = \{2\}$, $L_2^{10}(1) = \{2\}$, then we have the first path:10->8->3->2, the second path:10->8->10->2. $L_2^5(2) = \{6, 10\}$. $L_2^6(1) = \{2\}$, $L_2^{10}(1) = \{2\}$, we have the third path:10->5->6->2 and the fourth path:10->5->10->2.

The drawback of this algorithm is the risk of constructing a path having a cycle in the case of a path $j=j_0,j_1,..., j_p,...,j_q,...,j_h=i$ where $j_p \in L_i^{j_q}(h-q)$. It could happen when L-2 is greater than the shortest path between j and i. In the example above, when we search the paths from 10 to 2 of length 3, we obtain two paths having a cycle:10->8->10->2 and 10->5->10->2. To avoid this, we test the existence of cycle during the construction of a path.

We have proposed an optimal data structure to store the sets $L_i^k(m)$ associated with paths. The advantage of this data structure is that it uses little memory space and allows direct access to a path. In addition, it may be potentially distributed.

3.2.2 Data Structures

We create a linked data structure to represent $L_i^j(m) = \{i_1,i_2,...,i_h\}$ as shown in the following figure.

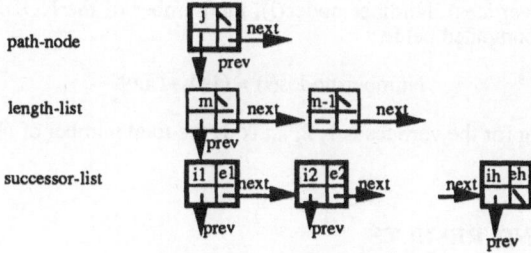

Figure 4   Data structure of $L_i^j(m)$

Each element in this representation is an object *x* called NODE with two data fields: *int1* and *int2*, and two other pointer fields: *prev* and *next*. The pointer field *next* is used to make up a linked list.

In the above figure, the NODE *x* at the first level is called *path-node: int1[x]* is the vertex j. The linked list at the second level is called *length-list: int1[x]* is the length of paths. The linked list at the third level is called *successor-list: int1[x]* is a successor *it* of the vertex j, *int2[x]* is the edge corresponding to (j, it), and *prev[x]* points to the length-list of $L_j^{it}(m-1)$.

With this representation, the paths from j to i (i given and j>i) can be constructed by linking all the path-nodes. The figure 5 represents the structure corresponding with the paths from j to 2 of the graph G.

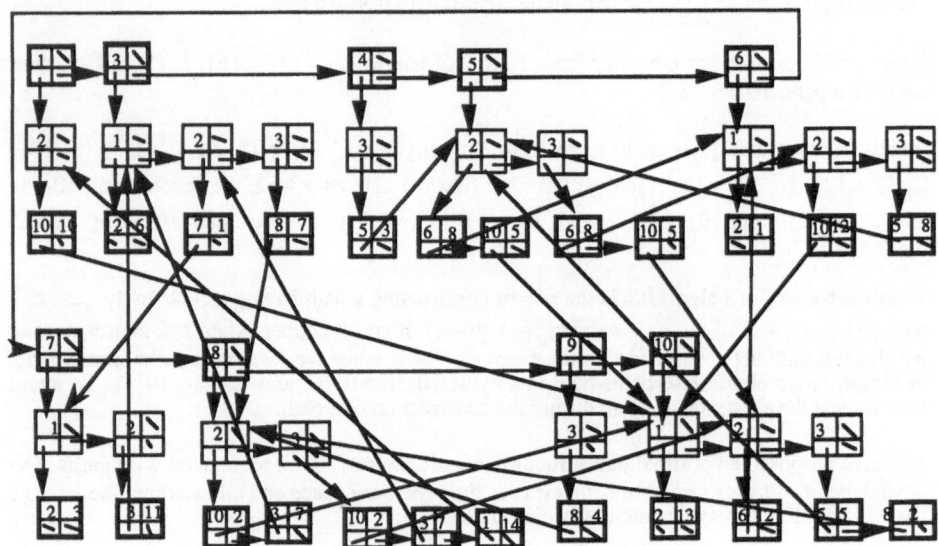

Figure 5 Represntation of paths from j to i

We have analyzed the memory space complexity in the worst case. In this case, the size of *length-list* is L (L is the maximal length of paths) and the size of *successor-list* is D (D is the maximal degree of vertices). Number-nodes(i), the number of the NODEs used to store the paths from j to i are computed below:

$$\text{Number-nodes(i)} = (1+L+LD)N$$

Doing the same thing for the vertices i=1, 2, ..., N-1, the total number of NODEs used for G is then $(1+L+LD)N^2$.

4. PROGRAM AND RESULTS

Using the method presented above, we have implanted a prototype program which permits us to calculate intervals for the lost traffic and the availability. Its results show that the chosen approach is realizable in pratice. We can treat medium networks with 26 nodes, 30 cables and 42 system arcs in 30 minutes CPU on SUN station SPARC 2 when 3 failures are considered and the maximal length of paths is 5. Moreover, the program has a good modularity. We can easily modify some hypothesis in the model. Thus, the rerouting module can be used for other problems, such as the problem of spare capacity assignment to a network.

In the following table, we give one group of results by applying the program to a network with 11 nodes (in the figure 1), and to a network with 26 nodes, 42 system arcs and 30 cables [10], where the operational probabilities of nodes, cables and system arcs are pn, pc, ps respectively. f is the number of treated failures, L is the maximal length of the paths for restoration, p_accum is the probability of the network states without failure and with f failures, the *availability_coefficient* is 0.99. This table gives the intervals of confidence for the availability and the lost_traffic and the running times for calculating these intervals.

Network	f	p_accum	L	Intervals of confidence	Time (second)
net_11 pn=0.9999 pc=0.999 ps=0.99 demand_ traf = 799	3	0.999828	5	avail = [0.969932, 0.970104] lost_traf = [2.263118, 2.400650]	85.8
			10	avail = [0.970343, 0.970515] lost_traf = [2.223253, 2.360785]	342.9
net_26 pn=0.9999 pc=0.999 ps=0.99 demand_ traf = 4142	3	0.998612	5	avail = [0.791330, 0.792719] lost_tra f= [46.137259, 51.887208]	1128.3
			10	avail = [0.907208, 0.908596] lost_traf = [22.628767, 28, 378716]	9081.4

By varying the network parameters, the program can give much information on a network, for example, the tradeoff between the spare capacity and the availability, the relation between the maximal length of restoration paths and the availability, etc ...

5. CONCLUSION

Calculating the availability of telecommunication networks based on optical fiber and DCS nodes is a relatively complex process. Factors such as demand matrix, restoration technique and network sizes, have to be considered.

This paper presents an efficient and flexible methodology to calculate the availability and associated measures of a network. It is made efficient by a sophisticated data structure. The rerouting module can also be used for the problem of assigning spare capacity in the network. We believe that its modularity, its simplicity, and its realizability are the main contributions of our method.

At present, we can treat networks of medium size (26 nodes). But we are currently working on approximate methods such as stratification and such a method is so powerful that it can treat very large network (100 nodes).

6. REFERENCES

[1] R.J.Boehm, Y.C.Ching, C.G.Griffith and F.A.Saal, Standardized fiber optic transmission systems - a synchronous optical network view. *IEEE J. on Select. Areas Communication.* SAC-4 (1986) 1424-1431.

[2] C.Palmer and F.Hummel, Restoration in a partitioned multi-bandwidth cross-connect network. *IEEE Global Telecommunication Conf.* (1990) 301.7.1-301.7.5.

[3] W.D.Grover, The selfhealing network: A fast distributed restoration technique for networks using digital cross-connect machines. *IEEE Global Telecommunication Conf.* (1987) 28.2.1-28.2.6.

[4] W.D.Grover, T.D.Bilodeau and B.D.Venables, Near optimal spare capacity planning in a mesh restoration network. *IEEE GLOBECOM'91* 57.1.1-57.1.6.

[5] J.Carlier, Y.Li, J.-L.Lutton, Etude et calcul de la disponibilité des réseaux de télécommunication. Rapport du CNET, Dec., 1992, NT/PAA/ATR/ORI/3275.

[6] M.O.Ball, Computational complexity of network reliability analysis: an overview. *IEEE Trans. Reliability* **35** (1986) 230-239.

[7] A.Satyanarayana, A unified formula for analysis of some network reliability problems. *IEEE Trans. Reliability* **31** (1982) 23-32.

[8] K.K.Aggarwal, Integration of reliability and capacity in performance measures of a telecommunication network. *IEEE Trans. Reliability* **34** (1985) 184-186.

[9] O.Theologou, J.Carlier, Factoring & Reductions for Networks with Imperfect Vertices. *IEEE Trans. Reliability* **40** (1991) 210-217.

[10] B.Sanso, F.Soumis, Communication & transportation network reliability using routing models. *IEEE Trans. Communication* **40** (1991) 29-38.

[11] B.Sanso, F.Soumis, M.Gendreau, On the evaluation of telecommunication networks reliability using routing models. *IEEE Trans. Communication* **39** (1991) 1494-1501.

[12] V.O.K.Li, J.A.Silvester, Performance analysis of networks with unreliable components. *IEEE Trans. Communication,* **32** (1984) 1105-1110.

[13] S.Even, *Graph algorithns.* Computer Science Press (1979).

NONLINEAR MULTICOMMODITY NETWORK FLOWS THROUGH PRIMAL PARTITIONING AND COMPARISON WITH ALTERNATIVE METHODS

Jordi Castro & Narcís Nabona
Statistics and Operations Research Dept., Universitat Politècnica de Catalunya
c. Pau Gargallo 5, 08071 Barcelona
e-mail: jcastrop@eio.upc.es

Abstract : *This paper presents a specialized code for solving the linear and non-linear multicommodity network flow problem with linear side constraints using primal partitioning techniques. The computational performance of the program developed is reported and compared with that of another specialized code for the same problem —based on price-directive decomposition— and also compared to the performance of a general purpose nonlinear optimization package. Test problems of many sizes corresponding to real cases of nonlinear multicommodity network flows have been employed.*
Keywords: Linear and Nonlinear Network Flows, Multicommodity Network Flows, Network Simplex Methods, Nonlinear Optimization, Side Constraints.

1. Introduction.

Primal partitioning is described and used in [5] to solve the linear multicommodity network flow problem(LMP). An specialization of the simplex method using primal partitioning has been developed for linear problems. Furthermore, this methodology has been extended to minimize a continuous nonlinear objective function (1) and to include linear side constraints (5). The formulation studied considers a mutual capacity constraint at each arc of the network and a certain number of side constraints. The nonlinear multicommodity problem (NMP) can be formulated as:

$$\min_{X_1, X_2, \ldots, X_K} h(X_1, X_2, \ldots, X_K) \tag{1}$$

$$\text{subj. to} \quad AX_k = R_k \qquad k = 1, \ldots, K \tag{2}$$

$$\underline{0} \leq X_k \leq \overline{X}_k \qquad k = 1, \ldots, K \tag{3}$$

$$\sum_{k=1}^{K} X_k \leq T \tag{4}$$

$$L \leq \sum_{k=1}^{K} L_k X_k \leq U \tag{5}$$

where $X_k \in \mathbb{R}^m$, (m: number of arcs) is the vector of flows of commodity k ($k = 1, \ldots, K$), and h being a $\mathbb{R}^{K \times m} \to \mathbb{R}^1$ real valued function. $A \in \mathbb{R}^{n \times m}$ (n: number of nodes) is a network matrix. Constraints (3) are simple bounds with $\overline{X}_k \in \mathbb{R}^m, k = 1, \ldots, K$ being upper limits and constraints (4) are the mutual capacity constraints with

$T \in \mathbb{R}^m$ and (5) are the linear side-constraints defined by matrices L_k: $L_k \in \mathbb{R}^{p \times m}$, $k = 1, \ldots, K$ with elements of any type, and $L, U \in \mathbb{R}^p$ (p: number of side constraints).

Every basis using the primal partitioning method can be written as:

$$
B = \begin{array}{|c|c|c|}
\hline
L_1 & R_1 & 0 \\
\hline
L_2 & R_2 & 0 \\
\hline
L_3 & R_3 & 1 \\
\hline
\end{array}
$$

being L_1, R_2 and $\boldsymbol{1}$ square matrices, and where: L_1 refers to the network constraints and arcs of the K spanning trees. The topology of this matrix is:

$$
L_1 = \begin{pmatrix} B_1 & & & \\ & B_2 & & \\ & & \ddots & \\ & & & B_K \end{pmatrix}
$$

being each B_k a nonsingular matrix associated with the kth spanning tree. L_1 can be represented at every iteration by K spanning trees following the methodology described in [2]. R_1 refers to the network constraints and complementary arcs of the K commodities. Complementary arcs are required to preserve the nonsingularity of the basis. L_2 refers to saturated mutual capacity and side constraints, for the arcs of the spanning trees. R_2 refers to saturated mutual capacity and side constraints, for the complementary arcs. L_3 refers to unsaturated mutual capacity and side constraints, for the arcs of the spanning trees. R_3 refers to unsaturated mutual capacity and side constraints, for the complementary arcs. $\boldsymbol{1}$ refers to the slacks of the unsaturated mutual capacity and side constraints. (It has to be noticed that constraints whose slacks are in matrix $\boldsymbol{1}$ are treated as unsaturated constraints, even though the values of slacks are zero).

2. Influence of the side constraints.

To perform the required operations with the basis B (compute the Lagrange multipliers $\pi^t B = c^t$, determine the descent direction of basic variables $B p_B = b$) it is only necessary to store and use a nonsingular working matrix, that we will denote by Q. The expression of this matrix is $Q = R_2 - L_2 L_1^{-1} R_1$. This result can be obtained directly solving the previous systems —considering the partition of the basis exposed before— and doing some algebraic manipulations (a detailed explanation can be found in [5]). The fact of inverting Q instead of the whole basis B improves substantially the efficiency of the method.

If we denote the set of saturated mutual capacity constraints by S_{mc}, the set of saturated side constraints by S_{sc} and the number of elements of a set S by $|S|$ then the dimension of the matrix Q ($dim(Q)$) can be expressed as $dim(Q) = |S_{mc}| + |S_{sc}|$. Since Q has full rank it follows than the number of complementary arcs in the basis must be equal to $|S_{mc}| + |S_{sc}|$. We can consider this matrix divided into two submatrices

$Q = \begin{bmatrix} Q_{mc} \\ Q_{sc} \end{bmatrix}$ where Q_{mc} is the submatrix whose rows refer to constraints $\in S_{mc}$, and Q_{sc} is the submatrix whose rows are associated with constraints $\in S_{sc}$.

The expression for computing Q involves the calculation of $L_1^{-1}R_1$. Since L_1 is a block diagonal matrix where the kth block is a minimum spanning tree for the kth commodity, and R_1 expresses for each complementary arc of the kth commodity its connection to the kth minimum spanning tree, then solving $L_1^{-1}R_1$ is equivalent to having the paths (denoted by $P_j, j = 1 \ldots dim(Q)$) of complementary arcs in their associated spanning trees. Given an arc $a \in P_j$, we will say than a has normal orientation if it points to the origin node of the complementary arc j; otherwise, it has reverse orientation.

If we denote by:
- a_j the arc associated with the jth column of Q, $j = 1 \ldots dim(Q)$.
- sc_i the side constraint of the ith row of $Q, i = |S_{mc}| + 1 \ldots dim(Q)$.
- $B(a, n)$ a logical function than returns *true* if the arc a appears in the side constraint n, and *false* otherwise.
- $c_{a,n}$ the coefficient of the arc a in the side constraint n.

Then we can compute directly the matrix Q as follows:

Submatrix Q_{mc} can be calculated knowing the mutual capacity constraints associated to the first $|S_mc|$ rows of Q, and the complementary arcs a_j, following [5].

Submatrix Q_{sc}:

$$
\begin{array}{l}
Q_{ij} \\
{\scriptstyle i=|S_{mc}|+1\ldots dim(Q)} \\
{\scriptstyle j=1\ldots dim(Q)}
\end{array}
=
\left\{
\begin{array}{l}
\text{Following next 4 steps:} \\
1) \quad \text{Set } Q_{ij} = 0 \\
2) \quad \text{if } B(a_j, sc_i) \text{ then } Q_{ij} = c_{a_j, sc_i} \\
\qquad \text{for each } a \in P_j, \text{ do next 2 steps} \\
3) \quad \text{if } B(a, sc_i) \text{ and } a \text{ has normal orientation then} \\
\qquad\qquad Q_{ij} = Q_{ij} + c_{a, sc_i} \\
4) \quad \text{if } B(a, sc_i) \text{ and } a \text{ has reverse orientation then} \\
\qquad\qquad Q_{ij} = Q_{ij} - c_{a, sc_i}
\end{array}
\right.
$$

It is clear from this procedure than the information of the mutual capacity constraints is not stored and is implicitly assumed in the construction of the Q_{mc} submatrix. However when computing submatrix Q_{sc} a function such as the logical function $B(a, sc_i)$ employed before, is required. In this implementation, information about the side constraints is stored in sparse form by columns (that is, for each arc we have the side constraints where it appears), and sorted by the number of side constraint. Thus the boolean function $B(a, sc_i)$ is reduced to a binary search.

3. The Algorithm

The algorithm presented has three different phases called phase 0, 1 and 2. In phases 0 and 1 the algorithm finds a feasible starting point, while phase 2 tries to achieve the optimizer without leaving the feasible region. Phases 0 and 1 are common to the linear and nonlinear problem. We will describe each phase in next three subsections. When referring to phase 2, only that of the nonlinear problem will be adressed.

For computational purposes the inequality constraints (4) and (5) in the original NMP problem are substituted by equality constraints by adding slacks variables.

$$\sum_{k=1}^{K} X_k + s = T \ ; \quad \underline{0} \le s \tag{6}$$

$$\sum_{k=1}^{K} L_k X_k + t = U \ ; \quad \underline{0} \le t \le U - L \tag{7}$$

where $s \in \mathbb{R}^m$ and $t \in \mathbb{R}^p$. In this formulation equations (6) and (7) substitute original equations (4) and (5). Then the formulation of the problem considered by the algorithm (that will be referred to as NMP2) is the minimization of (1) subject to (2), (3), (6) and (7)

3.1. Phase 0.
In phase 0 the algorithm considers only the network constraints and bounds of the varibles of the problem. Without any constraint linking the flows of differents commodities, it solves for each commodity $k, k = 1 \dots K$ the linear network problem: $\{\min_{X_k} C_k^t X_k \ ; \ \text{subj.to } A X_k = R_k \ \text{and} \ \underline{0} \le X_k \le \overline{X}_k\}$ where the costs' vector C_k can be introduced by the user. This can be useful to guide the algorithm towards good initial feasible points, if information about the model is known before the execution of the program. Once phase 0 has finished, the algorithm has a minimum spanning tree for each commodity.

3.2. Phase 1.
The K points obtained in phase 0 will not satisfy in general the mutual capacity and side constraints, thus having a pseudo-feasible point. That implies than some slack variables s for the mutual capacity constraints or t for the side constraints will be out of bounds. Let $\hat{X}_k, k = 1 \dots K$ be the pseudo feasible point obtained, then, the following index sets are defined:

- $s^- = \{i : (\sum_{k=1}^{K} \hat{X}_k)_i > T_i \Leftrightarrow s_i < 0\}$.
- $t^- = \{i : (\sum_{k=1}^{K} L_k \hat{X}_k)_i > U_i \Leftrightarrow t_i < 0\}$.
- $t^+ = \{i : (\sum_{k=1}^{K} L_k \hat{X}_k)_i < L_i \Leftrightarrow t_i > (U - L)_i\}$

Introducing new artificial variables e and f, and fixing initial values for s and t such that:

- $(\sum_{k=1}^{K} \hat{X}_k)_i + s_i - e_i = T_i \ ; \quad s_i = 0 \ ; \quad \forall i \in s^-$
- $(\sum_{k=1}^{K} L_k \hat{X}_k)_i + t_i - f_i = U_i \ ; \quad t_i = 0 \ ; \quad \forall i \in t^-$

- $(\sum_{k=1}^{K} L_k \hat{X}_k)_i + t_i + f_i = U_i \;\; ; \;\; t_i = (U - L)_i \;\; ; \;\;\;\; \forall i \in t^+$

The problem solved in phase 1 is:

$$\min_{X_1, X_2, \ldots, X_K, s, t, e, f} \quad \sum_{i \in s^-} e_i + \sum_{i \in t^-} f_i + \sum_{i \in t^+} f_i \tag{8}$$

$$\text{subj. to} \quad (1) \quad \text{and} \quad (2)$$

$$\sum_{k=1}^{K} X_k + s + \mathbf{1}^e e = T \tag{9}$$

$$\sum_{k=1}^{K} L_k X_k + t + \mathbf{1}^f f = U \tag{10}$$

$$\underline{0} \leq t \leq U - L \;\; ; \;\; \underline{0} \leq s \;\; ; \;\; \underline{0} \leq e \;\; ; \;\; \underline{0} \leq f$$

Where both matrices $\mathbf{1}^e \in \mathbb{R}^{m \times m}$ and $\mathbf{1}^f \in \mathbb{R}^{p \times p}$ in (9) and (10) are diagonal and defined as follows:

$$(\mathbf{1}^e)_{ii} = \begin{cases} -1 & \text{if } i \in s^- \\ 0 & \text{otherwise} \end{cases} \qquad (\mathbf{1}^f)_{ii} = \begin{cases} -1 & \text{if } i \in t^- \\ +1 & \text{if } i \in t^+ \\ 0 & \text{otherwise} \end{cases}$$

Problem NMP2 will be feasible if, at phase 1, a point where the value of (8) is 0, is achieved.

3.3. Phase 2

Once a feasible point has been obtained, phase 2 tries to achieve the optimizer of the nonlinear function (1). The primal partitioning method, as presented in [5], was thought for linear objective functions. However, when optimizing nonlinear functions, primal partitioning can be applied together with the Murtagh and Saunders' strategy —described in [6]— of dividing the set of variables in Basic, Superbasic and Nonbasic variables: $\hat{A} = [B|S|N]$, being $\hat{A}$ the matrix of constraints (2), (3), (6) and (7). The efficiency in managing the working matrix Q with respect the whole basis B is preserved in the nonlinear case. Also the structure of network, mutual capacity and side constraints, can be exploited, improving the computation time with respect general methods of optimization where this constraints are treated in a general way.

The current implementation of the program solves system $Z^t H_k Z P_S = g_z$ (Newton's projected direction) by using a truncated-newton algorithm, following the description in [3]. One of the next tasks to do would be testing the behaviour of the program with a quasi-newton update of $Z^t H_k Z$ as described in [6].

When a basic variable hits its bound a column of the basis B is removed and replaced by a column of the superbasic set S. The new basis (denoted by B_n) could be expressed as $B_n = B\eta$ being η a convenient eta-matrix. However the algorithm does not work with the whole basis B. For our purposes it is necessary to reflect how this change in the basis affects the K spannings trees and the working matrix Q. Given that the $dim(Q)$ is equal to the number of saturated mutual capacity and side constraints it is clear than $dim(Q)$ can increase or decrease at each iteration (In this context "saturated constraint" means "constraint whose associated slack is not in the basis B". Of course, in the original formulation of the primal partitioning when a slack was not in the basis it was a nonbasic variable at value 0. In the nonlinear extension with side constraints, when an slack is not basic it can be superbasic or nonbasic at its upper bound, having a non-zero value. Thus it is not correct talking about "saturated constraints", which is a reminiscence of the linear problem. This expression should be understood as "constraints whose slacks are not basic"). Considering that the variables of the problem can be arcs or slacks (and the arcs of the basis B can be subdivided into arcs of the K spanning trees or complementary arcs), then, depending on the case of the variable entering and leaving the basis, the following 6 cases must be observed with the related operations (denoting by "E:-" the case of a variable entering and by "L:-" the case of a variable leaving):

• *E: slack–L: slack.* The row of Q associated with the leaving slack is removed and substituted by a new row for the entering slack. $Dim(Q)$ is not modified.

• *E: slack–L: complementary arc.* The row and column of Q associated with the entering slack and leaving complementary arc respectively are removed. Update $dim(Q) = dim(Q) - 1$.

• *E: slack–L: arc of kth tree.* A complementary arc of the kth commodity, e.g. the jth complementary arc, having the leaving arc in its path P_j, must be found to replace the leaving arc in the kth tree. This complementary arc will always exist (otherwise the basis would become singular). The row an column of Q associated with the entering slack and the jth complementary arc are removed. Update $dim(Q) = dim(Q) - 1$.

• *E: arc–L: slack.* A new row associated with the entering slack is added to Q. To maintain the nonsingularity of Q a new column for the entering arc —which will become complementary arc— is also added to the working matrix. Update $dim(Q) = dim(Q) + 1$.

• *E: arc–L: complementary arc.* The column of Q associated with the leaving complementary arc is removed, and substituted for a column for the entering arc, which will become complementary arc. $Dim(Q)$ is not modified.

• *E: arc–L: arc of kth tree.* A complementary arc of the kth commodity, e.g. the jth complementary arc, having the leaving arc in its path P_j, is searched for. If this arc is found, it will replace the leaving arc in the kth tree, and the entering arc will become complementary arc. If no complementary arc is found, then the entering arc will substitute the leaving arc in the kth tree. One of the two last cases must be always possible to preserve the nonsigularity of the basis. $Dim(Q)$ is not modified.

It has not been explicited, but it must be noticed that, when rows of matrix $Q = \begin{bmatrix} Q_{mc} \\ Q_{sc} \end{bmatrix}$ are removed of added, depending on the kind of the associated slack (if it is an slack of mutual capacity or side constraints) the operations will afect submatrix Q_{mc} or/and Q_{sc}.

4. Updating matrix Q.

The efficiency of the method is directly related to the efficiency of the routines than manage matrix Q. Obviously the first step is having a way of updating matrix Q

instead of recalculate it at each iteration. It is not the purpose of this work to prove all the formulae required to obtain the expressions for updating Q_{k+1} from Q_k (where the subindex k refers to the current iteration). An approach of how to obtain it can be found in [5]. Two important remarks have to be made on this approach:

• It only considers the updating of the Q matrix with mutual capacity constraints. The updating used by the algorithm here presented is extended to include side constraints.

• It considers an updating of Q^{-1} instead of Q. The difficulty of the variable dimension of Q at each iteration makes that updating Q^{-1} is a costly operation if it is stored as a sparse matrix. On the other hand, it seems not appropiate to store Q^{-1} as a dense matrix (tests made with problems of many sizes showed that the number of nonzero elements of Q^{-1} was always less than 10%). This led us to work with an update of Q, and not with its inverse.

Consider that at iteration k we recalculate the working matrix Q_k, which has dimension $dim(Q_k) = n_k$, and that it will be recalculated after i iterations (that is, at iteration $k + i$), where it will have dimension n_{k+i}. Once the working matrix is recalculated, a LU decomposition of the same is performed. The LU routine developed tries to exploit the sparse structure of Q by using either the $P3$ Hellerman-Rarick algorithm [4] or an adhoc variant of the same. Given that at each iteration the dimension of the working matrix can, at most, increase only in one column and row, then it follows than $n_j \leq n_k + i$, $k \leq j \leq k + i$. The method proposed consists in working with an *extended matrix* $\overline{Q}_j$ at each iteration j, $k \leq j \leq k + i$, where $\overline{Q}_j$ is defined as:

$$\overline{Q}_j = \begin{matrix} n_j \\ l_j \end{matrix} \begin{pmatrix} \overset{n_j}{Q_j} & \overset{l_j}{0} \\ 0 & \mathbf{1} \end{pmatrix}$$

Where the dimensions n_j and l_j of matrices Q_j and identity $\mathbf{1}$ satisfy $n_j + l_j = n_k + i$. That is, the extended matrix has the maximum dimension that the working matrix can achieved at iteration $k + i$, in the worst case of increasing one column and row at each iteration j. This form of the extendend matrix must be kept invariant at each iteration when the working matrix is updated (pre and post multiplying by eta and permutation matrices). This enables the code to work with the original LU decomposition obtained when the working matrix was recalculated, even if its dimension is modified. It must be noticed than solving systems $Q_j x = b$ and $x^t Q_j = b^t$ is equivalent to solving $\overline{Q}_j x = b$ and $x^t \overline{Q}_j = b^t$ (only the dimension of the independent term b must be increased adding l_j zeros and, once the system solved, considering only the first n_j components of the solution vector x).

5. Computational results.

Problems of Long-Term Hydro-Thermal Coordination of Electicity Generation, modeled as nonlinear multicommodity network flow problems with side constraints have been used as test (a complete description of the objective function and constraints can be found in [8]). For these problems 4 commodities are normally used although a higher number of commodities could have also been envisaged. It must be noticed that in this problem $\overline{X}_k$ of (3) is $\overline{X}_k = T$, $k = 1, \ldots, K$, which means that, for each arc of the network, each single commodity can saturate the mutual capacity limit and, as a consequence, there are as many potentially active mutual capacity constraints as there are arcs

Four models of different sizes have been used. Each model has been tried with three objective functions: a linear function, a convex nonlinear function and a highly nonlinear (and locally nonconvex) function (this last is the objective function of the Electricity Generation problem). The description of the models (that will be referred to as M1, M2, M3 and M4) is displayed in Table 1, where for each model the number of nodes, arcs, commodities and side constraints is given, together with the total number of variables and constraints.

Table 1. Definition of the models.

	# nodes	# arcs	# comm.	# SC	# var.	# cons.
M1	37	117	4	2	468	267
M2	37	153	4	12	612	313
M3	99	315	4	3	1260	714
M4	685	2141	4	3	8564	4884

Results obtained for each objective function with the algorithm proposed have been compared with those obtained with the Minos.5.3 general purpose optimization package [7]. The executions have been made on a SUN Sparc-10/41 machine. Table 2 shows the CPU time in seconds, number of iterations at phase 1, number of iterations at phase 2, optimal value of the objective function ($h(x^*)$), dimension of matrix Q at the optimizer ($\dim(Q)$), number of superbasics variables (# SB VAR.) and number of objective function evaluations (# O.F.EVAL) for Primal Partitioning algorithm (referred to as PP) and Minos.5.3 package (referred to as MI) for each objective function.

As it can be appreciated, for the linear objective function the gain of CPU time of the PP algorithm with respect to the general optimization package is considerable. For the convex nonlinear function this effect is still preserved. However in the highly nonlinear function of the Electricity Generation problem, Minos is faster than the PP algorithm. It must be noticed than the number of evaluations of the objective function is higher in the PP algorithm than in Minos. This could explain the low performance of the PP algorithm for the highly nonlinear function, given that it is a costly function to compute. The fact of finding descent directions by a truncated-newton algorithm could explain why the PP algorithm requires more function evaluations than Minos (which uses a quasi-newton method). The preparation of a revised version of the PP algorithm employing a quasi-newton update routine is in project.

Another important point is the behaviour of primal partitioning as compared to another specialized multicommodity network algorithms: price directive decomposition. The computational performance of multicommodity network flows through primal partitioning and through price directive decomposition (referred to as PDD) has been compared, for linear test problems, by Ali & al. [1]. This comparison indicates that PDD performs much better than PP. The experience of the authors does not agree with that of Ali & al. It appears that for nonlinear problems PP is better than the other method. In general PDD has a very efficient phase 1 but as the number of active

Table 2. Results using the different objective functions.

	iter. sec.	iter. Phase 1	iter. Phase 2	$h(x^*)$	dim(Q)	# SB VAR.	# O.F. EVAL
Linear function							
M1 (PP)	0.4	69	26	1057375.00	28		
M1 (MI)	2.5	238	43	1057375.07			
M2 (PP)	0.4	59	18	2.27×10^{-12}	20		
M2 (MI)	2.6	190	30	2.36×10^{-11}			
M3 (PP)	2.1	219	276	4477156.58	82		
M3 (MI)	8.8	569	265	4477156.60			
M4 (PP)	83.0	2081	1711	249354122.	616		
M4 (MI)	745.0	12712	3009	249354124.			
Convex nonlinear function							
M1 (PP)	10.5	69	1203	5.42797×10^9	36	249	4511
M1 (MI)	20.0	256	887	5.42796×10^9		249	1754
M2 (PP)	12.9	59	1491	3.97185×10^9	34	238	5323
M2 (MI)	23.0	163	1165	3.97183×10^9		231	2315
M3 (PP)	89.8	219	4540	6.04754×10^9	96	453	16798
M3 (MI)	105.9	617	1991	6.04651×10^9		449	3862
M4 (PP)	4249	2081	35306	1.94319×10^{14}	771	1292	120176
M4 (MI)	14308	17008	16498	1.94319×10^{14}		2077	23766
Highly nonlinear function (locally non convex)							
M1 (PP)	3.5	69	364	-2.5164×10^{11}	52	2	490
M1 (MI)	3.4	261	202	-2.5164×10^{11}		2	211
M2 (PP)	8.1	59	630	-7.9577×10^{10}	47	2	941
M2 (MI)	5.8	103	354	-7.9578×10^{10}		2	365

vertices [5,9] increases in phase 2 the iterations become less effective and, after a given number of iterations —or after an analogous number of function evaluations— PP has reduced more the objective function than PDD. Reference [9] describes the principles and performance of a nonlinear multicommodity network flow code implementing PDD. Although some improvements have been recently introduced by the authors in this code, its convergence pattern remains the same as described in [9]. This code follows the same stages of phase 0, phase 1 and phase 2, as the PP algorithm of this work (the code of phase 0 is the same in both programs). Figure 1 shows the behaviour of each code with model M1 and the highly nonlinear function for phase 1 (infeasibilities) and phase 2 (objective function $h(x)$). Time in Figure 1 corresponds to system time and not CPU time (flat parts in phase 2 plots are due to system tasks performed periodically). PDD is better than PP at phase 1 but it is clearly outperformed at phase 2.

It can thus be said that, problems of linear and nonlinear multicommodity network flows with linear side constraints can be solved efficiently with the variant put forward of the existing PP algorithms. Instrumental in this efficiency is the treatment of phases 0 and 1 described and the special manipulation of matrix Q presented.

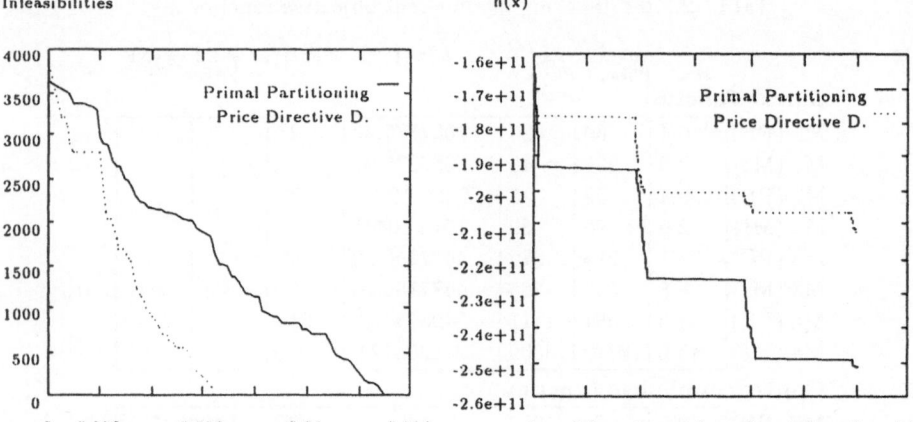

Infeasibilities   h(x)

(seconds)   (seconds)

Figure 1. Infeasibilites and objective function versus time

REFERENCES

[1] Ali, A., R.V. Helgason, J.L. Kennington & H. Lall. 1980. *Computational comparison among three multicommodity network flow algorithms*. Operations Research, v. 28, pp. 995-1000

[2] Bradley, G.H.; G.G. Brown & G.W. Graves. 1977. *Design and implementation of large scale primal transshipment algorithms*. Management Science, Vol.24, N.1, pp. 1-34.

[3] Dembo, R.S. and T. Steihaug. 1983. *Truncated-Newton algorithms for large-scale unconstrained optimization*. Mathematical Programming, v.26, pp. 190-212.

[4] Hellerman, E and D. Rarick. 1971. *Reinversion with the preassigned pivot procedure*. Mathematical Programming, v. 1, pp. 195-216.

[5] Kennington, J.L. and R.V. Helgason. 1980. *Algorithms for network programming*. John Wiley & sons.

[6] Murtagh, B.A. and M.A. Saunders. 1978. *Large-scale linearly constrained optimization*. Mathematical Programming, v. 14, pp. 41-72.

[7] Murtagh, B.A. and M.A. Saunders. 1983. *MINOS 5.0. User's guide*. Dpt. of Operations Research, Standford University, CA 9430, USA.

[8] Nabona, N. 1993. *Multicommodity network flow model for long–term hydro– generation optimization*. IEEE Trans. on Power Systems, v. 8, num. 2, pp. 395-404.

[9] Nabona, N. and J.M. Verdejo. 1992. *Numerical implementation of nonlinear multicommodity network flows with linear side constraints through price–directive decomposition*, in P. Kall (Ed.) System Modelling and Optimization, Proccedings of the 15th IFIP Conference, Zurich. Springer–Verlag. pp. 311-320.

A Flow Network Model based on Information, and Its Stability: An Application to Ecological Systems

Hironori Hirata, Tatsuya Uno and Seiichi Koakutsu

Department of Electrical and Electronics Engineering
Chiba University
1-33 Yayoi-cho, Inage-ku, Chiba-shi 263
JAPAN

Abstract

We propose a model of flow networks based on information coding and channel to discuss the stability. To study the relation between the structure and the stability of networks, we regard a flow network as an information source, and perturbation from the environment as disturbance in channel. We define an information index (H^2-information) and the related indeces to evaluate the structure of networks. H^2-information shows the degree of organization of the network. We apply the proposed information indices to ecological systems and study some properties between the stability and the structure of ecological flow networks.

1. Introduction

We evaluate the structure of flow networks using the information based on information coding (Hirata, 1993a) and discuss the relation between the stability and the structure of them using information channel.

First we define an information index (H^2-information) to evaluate the structure of networks. H^2-information shows the degree of organization of the network. It is defined by evaluating the bias of the distribution of elements and the strength of dependency between them. Practically, thinking of an itinerary of flows as a sequence of letters makes it possible to regard a flow network as an information source and define an information index.

Secondly we use an analogy of disturbance in the channel to express the perturbation on the network. Using Shannon's second theorem, we get the trend of structure which stable networks should have.

Using the model of Volterra-like flow ecological systems, we discuss the relation between the structure of ecological systems and stability: what kind of structure is suitable to be stable?

We also apply the results to ecological systems and give some theoretical support to simulation results and field insights about the relation between the structure of ecological systems and environments.

2. Organization of the Structure of Flow Networks

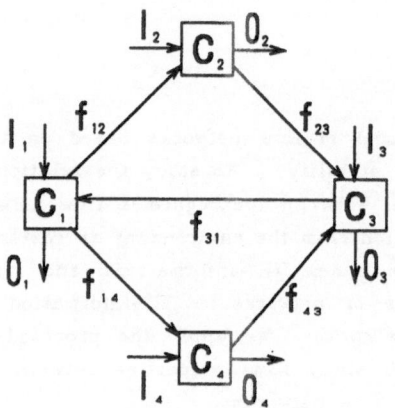

Fig.1 A flow network.

We define a flow network FN like Fig.1 consisting a set of compartments, C, and a set of flows, F, as follows.

$$FN = \{ C, F \} \qquad (1)$$

where $C = \{ c_i \ (i=0,\cdots,n)\}$. c_i is the symbolic name of the i-th element, (and especially c_0 is considered as environment.)
$F = \{ f_{ij} \ (i,j=1,\cdots,n), \ I_i \ (i=1,\cdots,n), \ O_i \ (i=1,\cdots,n)\}$.
f_{ij} shows the flow from the i-th compartment to the j-th compartment;

I_i, the input to the i-th compartment $(I_i = f_{0i})$;
O_i, the output from the i-th compartment $(O_i = f_{i0})$.

When we consider a material flowing in a network, it has inherent itineraries in the network. Regarding the name of any compartment, c_i, as a letter, we may think of an itinerary as a sequence of letters, $c_i c_j c_k c_\ell \cdots$. Since a set of itineraries corresponds to a set of sequences of letters, we may regard a flow network as an information source. We define the information contained in the structure of flow networks using the information content of sequences of letters (Hirata, 1993a). The information contained in the itineraries of a network consists of two different kinds of information: information generated by the divergence from equiprobability and information generated by the divergence from independence. We calculate them using the concept of entropy and define H^2-information and related indices.

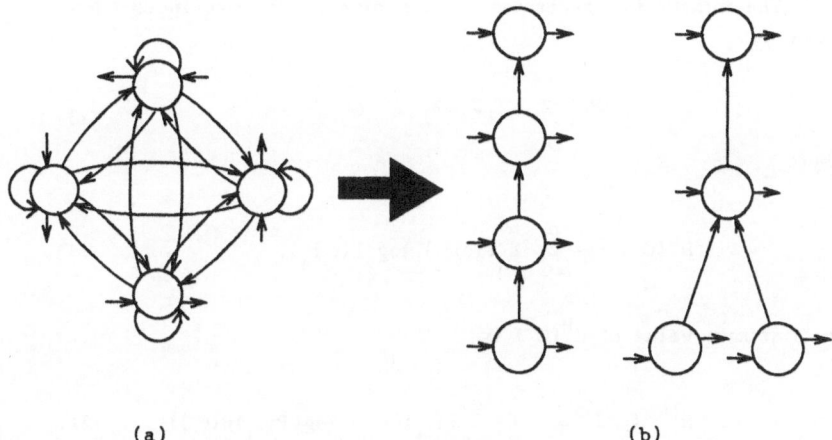

(a)   (b)

Fig.2 Organization of structure.

We calculate the difference of entropy between the random structure and the observed one to evaluate the degree of organization of structure. Random structurte corresponds to a sample itinerary of random transition structure (complete connected structure with equal probability) like Fig.2(a). Fig.2 (b) shows examples of an ordered structure organized from Fig.2(a).

3. Information of Flow Networks

We define two levels of information using the set of compartments, C, and the set of its doublets, $C^2 = \{c_i c_j\}$, as follows.

The information generated by the divergence from equiprobability, D_1, is given by

$$D_1 = H^{max}(C) - H(C) \qquad (2)$$

where

$$H(C) = - \sum_{i=0}^{n} P(c_i) \log P(c_i). \qquad (3)$$

The information generated by the divergence from independence, D_2, is given by

$$D_2 = H^{IND}(C^2) - H^D(C^2) \qquad (4)$$

where

$$H^D(C^2) = - \sum_{i=0}^{n} \sum_{j=0}^{n} P(c_i c_j) \log P(c_i c_j). \qquad (5)$$

The maximum value of $H^D(C^2)$ is

$$H^{IND}(C^2) = - \sum_{i=0}^{n} \sum_{j=0}^{n} P(c_i) P(c_j) \log\{P(c_i) P(c_j)\} \qquad (6)$$

when c_i and c_j are independent. The divergence from independent state is the difference between the entropy of the independent state, $H^{IND}(C^2)$, and that of the dependent state, $H^D(C^2)$.

The total information I_H2 is given by

$$I_H2 = D_1 + D_2 \tag{7}$$

Here D_1 shows the contribution of the compartments themselves to organization. In contract, D_2 shows mainly the contribution of flow structure to organization. Therefore I_H2 shows the total information contained in the structure of a flow network and expresses the degree of organization. We call this total information (I_H2) 'H^2-information' because it evaluates information based on two kinds of entropy. Since D_2 evaluates mainly the connectedness of the flow network, we may think of D_2 as showing information of higher order than that of D_1. At steady state D_2 is the same as mutual information term defined by information channel theory (Hirata and Ulanowicz, 1984 and 1985, Hirata, 1993b). After calculation we can easily see that I_H2 is expressed as

$$I_H2 = D_1 + D_2 = \{H^{max}(S) - H(S)\} + \{H(S) - H_M\} \tag{8}$$

$$= H^{max}(S) - H_M \tag{9}$$

where

$$H_M = - \sum_{i=0}^{n} \sum_{j=0}^{n} P(c_i)P(c_j/c_i) \log P(c_j/c_i) \tag{10}$$

In flow networks we define the probabilities as follows (Hirata, 1993a and b).

$$P(c_j|c_i) = f_{ij} / F_i \tag{11}$$

$$P(c_i) = F_i / \sum_{j=0}^{n} F_j \tag{12}$$

where

$$F_i = \sum_{j=0}^{n} f_{ij} \tag{13}$$

4. Information versus Stability

Let us study the relation of H^2-information and the stability against perturbation. We regard a flow network as an information source and perturbation as noise in the information channel.

Let us describe Shannon's second theorem:

[Shannon's Second Theorem]

If U < C (where U is the emission rate of information source and C is channel capacity), there is a code such that transmission over the channel is possible with an arbitrarily small number of errors.

The essential concept of Shannon's second theorem is this: we cannot eliminate noise in the channel, but we can under certain conditions (U<C) transmit a message without error in spite of this noise if the message has been properly encoded at the source.

In studying flow networks, the emission rate U is expressed as

$$U = k \, H_M \qquad\qquad (14)$$

where k is the average turnover rate in all transition between elements for flow structure. We adopt k as a global time factor of flow networks. Therefore an interpretation of the Shannon's second theorem for flow systems is represented as follows:

Proposition 1

A flow network is stable against perturbation under the condition

$$k \, H_M < C \qquad\qquad (15)$$

if it has been properly organized or structured. Here C is a decreasing function of the strength of perturbation. C shows the capacity of the environment.

Although condition (15) is not a necessary and sufficient condition, we use it as a rule of thumb to study the trend of structure which stable flow networks should have. Condition (15) means that flow networks should keep U, or H_M and k small to combat perturbation, i.e., to keep itself stable.

Proposition 2

To be stable flow networks should keep H_M and/or k small.

Since we can easily notice from equation (9) that H_M and I_H2 have the following relation

$$H_M = H^{max}(S) - I_H2 \qquad (16)$$

i.e. the smaller H_M means the larger I_H2, we obtain the trend of H^2-information about stability.

Proposition 3

To be stable flow networks should keep H^2-information I_H2 large.

Since large H^2-information means that structure is simple and relatively linear or cyclic.

Corollary 1

To be stable the structure of flow networks should be simple and relatively linear or cyclic.

In practice Proposition 3 implies the next proposition.

Proposition 4

(a) Flow networks should keep H^2-information large to defend against severe perturbation.

(b) Flow networks can permit H^2-information to be small to defend against mild perturbation.

5. Application to Ecological Systems
5.1. Through simulation

Using flow networks based on the predator–prey Volterra model, we will verify the theoretical results. The Volterra type flow network is expressed as

$$dN_i/dt = a_i N_i + \sum_{j=1}^{n} \mu_{ij} b_{ij} N_i N_j \qquad (17)$$

where $b_{ij}=-b_{ji}$ $(i \neq j)$ $b_{ii}<0$, $0<\mu_{ij}\leq 1$ $(b_{ij}>0)$ $\mu_{ij}=1$ $(b_{ij}<0)$. N_i is the population of the i-th species. a_i is the growth rate (or death rate). b_{ij} shows the predator-prey relation and if species i preys on species j, b_{ij} is positive and b_{ji} negative, since species i is benefiting at the expense of species j. μ_{ij} shows the efficiency.

Model (17) is regarded as the followimg flow network model.

$$dN_i/dt = \sum_{j=0}^{n} f_{ji} - \sum_{j=0}^{n} f_{ij} \qquad (18)$$

The flow terms f_{ji} and f_{ij} in (18) are expressed by the corresponding terms in (17). We omit the detailed about it here due to the lack of space.

We will study the stability of (17) or (18) at the equilibrium. We will simulate the case of 4 species and 5 species. As concerns 4 species case, there are 8 patterns and 64 cases for 5 species. The main pathes are shown

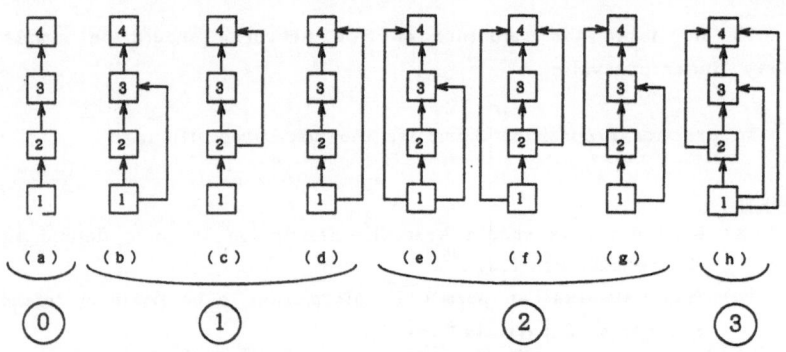

Fig.3   8 patterns of main pathes in 4-species flow network.

in Fig.3. In this figure we have omitted the pathes to or from the environment. We can classify them into 4 groups according to the number of feedforward paths. The feedforward path (except main pathes) means omnivory.   For each pattern we choose the coefficients of (17) randomly from the ranges of $0<|a_i|\leq1$, $0\leq|b_{ij}|\leq1$ and $0<\mu_{ij}\leq1$, and analyze the stability and calculate the information index through (11) and (12).   We continue until feasible cases become 20,000.   The word "feasible" means having a positive equilibrium.   We set here that $b_{11}=-1$ and $b_{jj}=0$ ($j\neq1$).   We define the degree of stability, S, as

$$S = \frac{[\text{the number of stable cases}]}{[\text{the number of feasible cases}]} \times 100 \quad (\%) \tag{19}$$

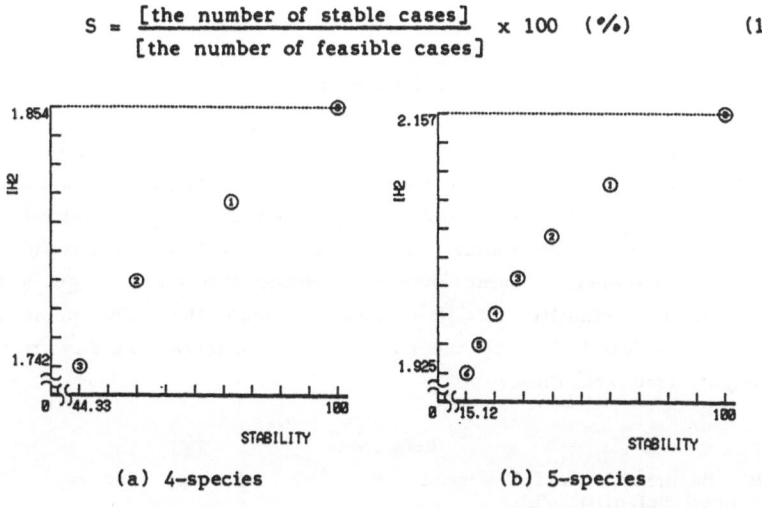

(a) 4-species   (b) 5-species

Fig.4   The relation between H^2-information and stability.

Fig.4(a) shows the relation between the stability and the H^2-Information for 4-species flow model, and Fig.4(b) is for 5-species flow model.   We can easily notice that there is a positive correlation between the stability and I_H2.   The simulation results verify Proposition 3 and Corollary 1.

5.2. Through field insights
Since large H^2-information means that structure is simple and relatively linear or cyclic, Propositions 3 and 4 give some theoretical support to experimental insights such as

(1) The structure of foodwebs under fluctuation is simpler than that under constant environments (Briand, 1983).
(2) Arctic aquatic ecosystems are relatively simpler than temperate ones in their structure (Johnson, 1990).

(3) The complex, fragile communities of relatively constant environments are more susceptible to disturbance than the simpler, more robust communities that are more used to disturbance (Begon et. al, 1990).

(4) Generally, ecosystems tend to become more complex in benign physical environments than when subjected to stochastic input disturbances (Odum, 1983).

6. Conclusion

To study the evaluation of the structure of flow networks and the relation between the structure and the stability, we proposed a model of flow networks based on information coding and channel. We regarded a flow network as an information source, and perturbation from the environment as disturbance in channel. Using Shannon's second theorem, we got a rule of thumb about the stability of flow networks from the view point of the structure. We verified it by the simulation of Volterra like flow model and some insights from real fields.

References

Begon,M., Harper,J.L. and Townsend,C.R. (1990). Ecology (2nd ed.). Oxford: Blackwell Scientific Pub.

Brooks,D.R. and Wiley,E.O. (1986). Evolution as Entropy. Chicago and London: The University of Chicago Press.

Hirata,H. and Ulanowicz R.E. (1984). Information theoretical analysis of ecological networks. Int. J. Systems Sci. 15, 261-170.

Hirata,H. and Ulanowicz R.E. (1985). Information theoretical analysis of the aggregation and hierarchical structure of ecological networks. J. theor. Biol. 116, 321-341.

Hirata,H. (1993a). Information of organization in ecological systems: Nutrient > Energy > Carbon. J. Theor. Biol. (in press).

Hirata,H. (1993b). Information theory and ecological networks. In Patten,B.C. and Jorgensen,S.E. (Editors), Complex Ecology: The Part-Whole Relation in Ecosystems. Englewood Cliffs: Prentice Hall (in press).

Johnson,L. (1990). The thermodynamics of ecosystems. In pp.1-47: Hutzinger, O.(Editor), Handbook of Environmental Chemistry, Vol.1, Part E, The Natural Environment and Biogeochemical Cycles: Heidelburg, Springer-Verlag.

Odum,E.P. (1983). Basic Ecology. Fort Worth: Saunders College Pub.

SUPER LOW FREQUENCY RESPONSE OF WATER DISTRIBUTION NETWORKS WITH APPLICATION

Kotaro ONIZUKA

Department of Agricultural Engineering

Tokyo University of Agriculture and Technology

3-5-8 Fuchu, Tokyo, 183 Japan

Abstract: A frequency response analysis approach to capacity design of terminal reservoirs for branching pipe networks is presented. The state equation is linearized into a pair of differential equations of the 2nd order for the forced oscillation. Inertia terms are ignored to get approximate amplitudes of reservoir water levels in terms of reservoir surface areas. This facilitates the process of getting the reservoir surface areas that yield desired system performance. An example for the network with 5 terminal reservoirs is presented.

1 Introduction

Hydraulic design of water distribution networks having terminal reservoirs, as shown in Fig. 1, usually requires unsteady simulations over a period of 24 hours or more to evaluate system performance including the changes in pipe flow rates and reservoir water levels under varying conditions of loading. An essential point is to see whether the networks have adequate reservoir capacities to meet varying load flows, since we have no definite analytical means of determining the reservoir capacities [1]. A rule of thumb is to provide each terminal with a capacity of about a half of average daily supply, which is often insufficient. Also, the rule does not refer at all to the reservoir surface areas that dominate the changes in the water levels.

As to simulation methods, the writer developed a reliable method based on rigid water column theory to evaluate the performance of large networks with pumps, tanks and reservoirs [2]. The essence of the method is a loop analysis, equivalent to that of electrical networks,

to get the state equation, a system of nonlinear ordinary differential
equations of the first order, which is numerically integrated over a
given period of time under varying conditions of loading. Another mer-
it is that the state equation is given by the Lagrange's equations of
motion in terms of the Mixed Potential [3] of the networks. Actually,
the use of recent symbolic computing systems such as REDUCE has drastically facilitated the derivation of the nonlinear state equation with
the Lagrangian formulation, as well as automatic generation of FORTRAN
programs for the unsteady simulations of the networks [4].

However, when the loading conditions change periodically with a
very long period, for example with a basic period of 24 hours as usually the case, the unsteady simulations over many cycles of the time tend
to be time consuming and impractical. In such cases, what we need to
know is how the amplitudes of reservoir water levels depend on the reservoir surface areas, since these two factors substantially determine
the adequate reservoir capacities. Clearly the best policy is to apply
ordinary frequency response analysis to the distribution networks, if
the nonlinear systems behave like linear systems in the low frequency
range. Fortunately, experience with the simulations has indicated that
the networks as shown in Fig. 1 are almost linear in the low frequency
range. The present approach stands on the experience.

In the following, the nonlinear state equation is linearized into
a pair of differential equations of the second order for the forced
oscillation. A sufficiently accurate approximation to the amplitudes
of the water levels in the low frequency range will be presented in
terms of the reservoir surface areas. This approximation provides the
basis for determining the adequate reservoir capacities. Although the
present approach is still under development, an application to the network in Fig. 1 will be presented.

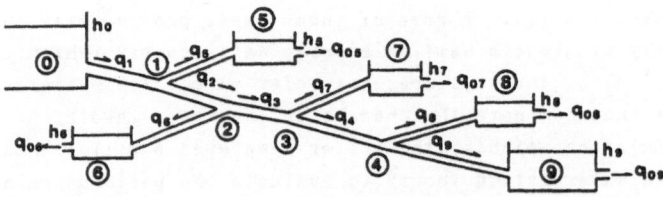

Fig. 1   Water Distribution Network with Terminal Reservoirs

2 Nonlinear State Equation and Linear System Model

The state equation for the network in Fig.1 describes time changes in terminal pipe flow rates, $q5, \ldots, q9$, and reservoir water levels, $h5, \ldots, h9$. The Lagrangian formulation [2] gives the state equation:

$$d(\mathrm{grad}\ L")/dt = \mathrm{grad}\ P \qquad (1)$$

where L" is the Lagrangian of the network explicitly given by

$$L" = T - U = -\frac{1}{2}W\sum_{i=1}^{9}L_i q_i^2 - -\frac{1}{2}W\sum_{j=5}^{9}A_j h_j^2 \qquad (2)$$

in which T is the kinetic energy of flow in the network, U the potential energy stored in the reservoirs measured from a datum plane, W the unit weight of water, q_i the pipe flow rate, L_i the pipe inertia, h_j the water level of reservoir, A_j the surface area of reservoir that is constant with the water level h_j, and P is the Mixed Potential, a generalized dissipation function by Brayton and Moser [3] given by

$$P = \sum_{i=1}^{9}W q_i (h_m - h_n) - -\frac{1}{3}\sum_{i=1}^{9}W K_i |q_i| q_i^2 + \sum_{j=5}^{9}h_j qo_j \qquad (3)$$

where h_m and h_n are the hydraulic heads at both ends of each pipe, K_i the coefficient of pipe flow resistance, and qo_j the load flow rate.

Thus the state equation in the descripter form is given by

$$[A]\dot{H} = Q - Qo \qquad \text{for reservoirs}$$

$$[L]\dot{Q} = Ho - H - [K(Q)]Q \qquad \text{for pipe network} \qquad (4)$$

where [A] is the diagonal matrix with the elements, $A5, \ldots, A9$, H the water level vector for $h5,\ldots,h9$, Ho the constant vector whose each element is ho, Q the terminal flow rate vector for $q5, \ldots, q9$, Qo the load flow vector for $qo5,\ldots,qo9$, [L] the inertia matrix, and [K(Q)] the flow resistance matrix dependent on Q. All the matrices are symmetric. Nonlinearity of Eq.(4) stems from the flow resistance term.

The state equation, Eq.(4), can be linearized with respect to the steady state corresponding to the 24-hour average load flows as

$$[A]\dot{X} = Y - Z \qquad \text{for reservoirs}$$

$$[L]\dot{Y} = -[R]Y - X \qquad \text{for pipe network} \tag{5}$$

where [R] is the constant resistance matrix determined with the steady state flow rates, X the water level vector denoting the variations from the steady state levels, Y the terminal flow rate vector for the variations from the steady state flow rates, and Z the load flow rate vector for the variations from the average load flow rates.

From Eq.(5) we get the following equations of the second order:

$$[L][A]\ddot{X} + [R][A]\dot{X} + X = -[R]Z - [L]\dot{Z}$$

$$[A][L]\ddot{Y} + [A][R]\dot{Y} + Y = Z \tag{6}$$

Adequateness of Eqs.(5),(6) in the low frequency range has been confirmed with perturbation techniques applied to the nonlinear state equation, Eq.(4), under steady sinusoidal excitation Z.

Note that Eq.(6) can be rewritten in the following symmetric form:

$$[AL]\ddot{X}' + [AR]\dot{X}' + X' = - [AR]Z' - [AL]\dot{Z}'$$

$$[AL]\ddot{Y}' + [AR]\dot{Y}' + Y' = Z' \tag{7}$$

where $[AL] = [A^{\frac{1}{2}}][L][A^{\frac{1}{2}}]$ and $[AR] = [A^{\frac{1}{2}}][R][A^{\frac{1}{2}}]$. Both [AL] and [AR] are real symmetric and positive definite matrices obtained by a simple similarity transformation:

$$X = [A^{\frac{1}{2}}]^{-1}X', \quad Y = [A^{\frac{1}{2}}]Y', \quad Z = [A^{\frac{1}{2}}]Z' \tag{8}$$

Under steady excitation with $Z' = Zo \exp(iw_o t)$, where w_o is the angular frequency, we put $X' = Xo \exp(iw_o t)$ and $Y' = Yo \exp(iw_o t)$ to get

$$\{(I - w_o^2 [AL]) + iw_o [AR]\}Xo = -\{[AR] + iw_o [AL]\}Zo$$

$$\{(I - w_o^2 [AL]) + iw_o [AR]\}Yo = Zo \tag{9}$$

When [A], [L], [R] and Zo are known, numerical solution to Eq.(9) for a specific w_o is readily obtained.

In actual design, however, pipe lengths and diameters, which determine the inertia matrix [L] and the resistance matrix [R], are calculated and fixed to meet the average steady state flow conditions specified in the design. Thus [L] and [R] are fixed system parameters, but the elements of [A], the reservoir surface areas, are not determined. This is why we repeat the time consuming simulations with different surface areas, searching for such reservoir surface areas that yield desired system performance. Thus, from a practical point of view, we need an explicit formula giving a good approximate value of the amplitude Xo in terms of w_o , Zo and [A]. Such a formula will facilitate the proccess of determining the adequate reservoir capacities.

3 Amplitudes of Water Levels in Super Low Frequency Range

The system given by Eq.(6) or Eq.(7) is over damping under the sufficient condition:

$$x^T \{[R][A][R] - 4 [L]\}x > 0 \quad \text{for } x \neq 0$$

or
$$(10)$$

$$x^T \{[AR]^2 - 4 [AL]\}x > 0 \quad \text{for } x \neq 0$$

This condition is well satisfied with ordinary systems, since the elements of [A], the surface areas, are large enough. Note that Eq.(10) can be used to get nearly minimum surface areas for the system to be non oscillatory. For large networks, however, the minimum areas are very small and of little practical interest.

From Eq.(10), with the maximum eigenvalues λ_{AR} and λ_{AL} of [AR] and [AL] respectively taken as the norm, the relation between the minimum natural frequency w_n and a standard frequencey w_s of the system is given by

$$w_s = (\alpha/4) w_n^2 , \quad 0 < \alpha = 4\lambda_{AL}/(\lambda_{AR})^2 < 1 ,$$
$$(11)$$

$$w_s = 1/\lambda_{AR}^2 , \quad w_n = 1/\lambda_{AL}$$

and a normalized frequency w for the system can be defined as

$$w = w_o / w_s \qquad (12)$$

When $w \approx 1$ and $\alpha /4 \ll 1$, the effect of [AL] on Xo and Yo in Eq.(9) is negligible, and [AL] can be equated to 0. This condition specifies "super low frequency range". Usually, the value of w_o for the 24 hour period is several times w_s , but α is very small compared to 1. Thus, we can well ignore the inertia term, w_o^2 [AL], in Eq.(9). The relative error of Xo by this approximation is evaluated by

$$\|\Delta Xo\| / \|Xo\| = w_o^2 \lambda_{AL} / (1 + w_o^2 \lambda_{ARmin}^2)^{1/2} \qquad (13)$$

where λ_{ARmin} is the minimum eigenvalue of [AR].

Ignoring the inertia terms in Eq.(6), we get simple equations:

$$(I + iw_o [R][A])Xo = -[R]Zo$$
$$\qquad (14)$$
$$Xo = -[R]Yo$$

where Xo, Yo and Zo are the complex amplitudes of X, Y and Z respectively defined by

$$X = Xo \exp(iw_o t), \quad Y = Yo \exp(iw_o t), \quad Z = Zo \exp(iw_o t) \qquad (15)$$

From Eq.(14) we obtain

$$iw_o [A]Xo = Yo - Zo$$
$$\qquad (16)$$
$$Yo = -[R]^{-1} Xo$$

In the super low frequency range, where [A] is sufficiently large, the amplitude Yo of the terminal pipe flow rates is very small compared to the amplitude Zo of the load flow rates. Thus, from Eq.(16), we get the first approximation to Xo and Yo as

$$Xo' = - (1/iw_o)[A]^{-1} Zo$$
$$\qquad (17)$$
$$Yo' = (1/iw_o)[R]^{-1} [A]^{-1} Zo$$

From Eqs.(16),(17), the second approximation is obtained as

$$Xo" = (1/iw_o)[A]^{-1}(Yo' - Zo)$$

(18)

$$Yo" = - (1/iw_o)[R]^{-1}[A]^{-1}(Yo' - Zo)$$

Similarly, the third approximation to Xo is given by

$$Xo'" = (1/iw_o)[A]^{-1}(Yo" - Zo)$$

(19)

This gives a sufficiently accurate approximation to Xo.

When Zo is real, the real and imaginary part of Xo'", denoted by Xor and Xoi respectively, are given by

$$Xor = - (1/w_o^2)[A]^{-1}[R]^{-1}[A]^{-1}Zo$$

(20)

$$Xoi = (1/w_o)[A]^{-1}\{I - (1/w_o^2)[R]^{-1}[A]^{-1}[R]^{-1}[A]^{-1}\}Zo$$

This formula is robust enough to calculate ill conditioned metworks with considerable difference in the average load flows at the terminals, as shown in the following.

4 Application: Adequate Reservoir Surface Areas and Capacities

Using a symbolic computing system such as REDUCE, we can readily obtain, by Eq.(20), an explicit FORTRAN equation giving the amplitude of each water level in terms of the symbolic variables w_o, Zo and reservoir surface areas, Aj's, with numerical coefficients and constants that reflect the network configuration and other fixed parameters. Similar FORTRAN equations for the derivatives of the amplitude with respect to Aj's are obtained. These FORTRAN equations constitute a key subroutine program for calculating adequate reservoir surface areas.

The following is an application to the network in Fig. 1. Suppose that the cyclic load flow qoj at each terminal varies with time in the same stepwise shape as shown in Fig. 2. The dimensions of the network and the average load flow $\overline{q}oj$ at the terminals are given in Table 1.

The problem is to determine the reservoir surface areas A5, ..., A9 so that the maximum change in the water level may be close to but not exceed 3.0 m at every reservoir.

For each qoj, we calculate a short Fourier series for the variation Zoj from $\overline{q}oj$ (shown in Fig. 2 in dotted line) as

$$Zoj = \overline{q}oj \ (1.654 \sin(w_o t + s1) + 0.827 \sin(2w_o t + s2)) \qquad (21)$$

where w_o the angular frequency for the basic period of 24 hours, and s1 and s2 are the phase angles which will be ignored in the following.

The maximum amplitude Xoj of the water level caused by Zoj will be

$$Xoj \stackrel{<}{=} 1.654 \ Xoj1 + 0.827 \ Xoj2 \qquad (22)$$

where Xoj1 is the amplitude by the single input $\overline{q}oj \sin(w_o t + s1)$, and also Xoj2 is the amplitude by the single input $\overline{q}oj \sin(2w_o t + s2)$.

On the other hand, for sufficiently large [A], we can estimate by Eq.(20) that for each Xoj2

$$Xoj2 \doteqdot 0.5 \ Xoj1 \qquad (23)$$

Thus, from Eq.(25) we get

$$Xoj \stackrel{<}{=} 2.068 \ Xoj1 \qquad (24)$$

Since Xoj should be less than 3/2 m or 1.5 m,

$$Xoj1 \stackrel{<}{=} 1.5/2.068 \ m = 0.725 \ m \qquad (j = 5,6,7,8,9) \qquad (25)$$

Ignoring Yo in Eq.(16), and assuming that Xoj1 = 0.725, we guess

$$Aj = \overline{q}oj/(0.725 \ w_o) = 18970 \ \overline{q}oj \quad (j = 5,6,7,8,9) \qquad (26)$$

Thus we calculate Xoj1's by Eq(20), and get values greater than 0.725.

Assuming that all Xoj1's are the same, and decreasing Xoj1 little by little and repeating the above process, we get to a smaller value, 0.70, at wich we quess

$$Aj = \overline{q}oj/(0.70 \ w_o) = 19600 \ \overline{q}oj \quad (j = 5,6,7,8,9) \qquad (27)$$

and obtain, by Eq.(20), a satisfying result: $Xoj1 \stackrel{<}{=} 0.70$. Rounding the factor 19700 to 20000, we have an approximate estimate:

$$Aj = 20000 \ \overline{q}oj \qquad (j = 5,6,7,8,9) \qquad (28)$$

Now we can simulate the linear system model in Eq.(6) with Z given by Eq.(21). The steady oscillations of the water levels are shown in Fig. 3. in dotted line. Also, results of nonlinear system simulation by Eq,(4) with the load flows in Fig. 2 are shown in thick line. Both results are in good agreement and satisfactory. A summary is given in Table 2, where Xojl by Eq.(20) is compared with the exact value by Eq. (9) in parenthesis, and Xoj is the result by the nonlinear simulation.

Pipe No.	1	2	3	4	5	6	7	8	9
Diameter(m)	1.80	1.80	1.65	1.10	0.60	0.20	0.60	0.35	1.10
Length (m)	3113	1826	2703	1593	360	150	165	618	1111
$\overline{q}oj$ (m^3/s)	3.484	3.050	2.958	2.508	0.434	0.092	0.450	0.121	2.387

Table 1. Dimensions of the Network

Reservoir No.	5	6	7	8	9
Xojl (m)	0.63	0.70	0.61	0.70	0.69
	(0.673)	(0.689)	(0.667)	(0.690)	(0.689)
Xoj (m)	1.45	1.40	1.45	1.50	1.50
Aj (m^2)	8680	1840	9000	2420	27740

Table 2. Amplitudes and Reservoir Surface Areas

The present application is perhaps the simplest but instructive: The adequate capacity of each reservoir can now be given by

$$Vj = 3 \times 20000 \, \overline{q}oj = 60000 \, \overline{q}oj \quad (j = 5,6,7,8,9) \qquad (29)$$

Note that Vj is greater than two third of average daily supply, i.e., 57600 qoj that corresponds to the shaded area in Fig. 2.

In conclusion, determining adequate reservoir capacities based on the super low frequency response is promising in the network design.

904

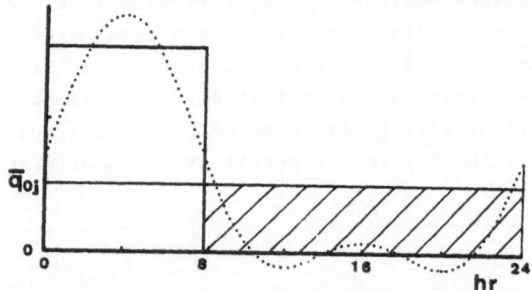

Fig. 2 Load Flow Variation with Time

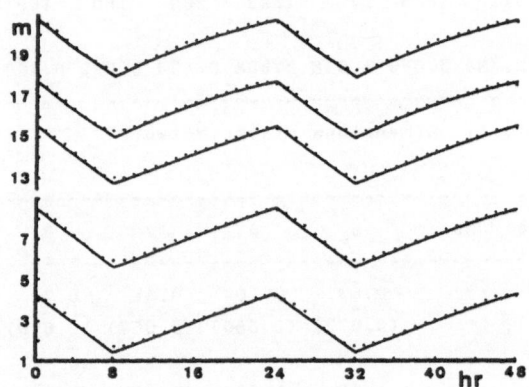

Fig. 3 Changes in Reservoir Water Levels

References

[1] Walski, T.M. et al., Journal of Water Resources Planning and
    Management, ASCE, Vol. 113, No. 2, 1987.
[2] Onizuka, K., Journal of Hydraulic Engineering, ASCE, Vol. 112,
    No. HY 8, 1986.
[3] Brayton, R.K. and Mozer, J.K., Quarterly of Applied Mathematics,
    Vol. 22, No. 1, 1964.
[4] Onizuka, K., IMACS Annals on Computing and Applied Mathematics,
    Vol. 10, 1991.

Transportation systems

A MARGINAL-VALUE APPROACH TO AIRLINE ORIGIN & DESTINATION REVENUE MANAGEMENT

Robert L. Phillips
Decision Focus Incorporated

INTRODUCTION

At the highest level, the airline revenue management problem consists of six components:

1. **Overbooking.** How many total bookings should we accept on each flight leg?

2. **Booking Class Management.** How many bookings should we accept in each booking class on each flight leg?

3. **Origin and Destination (O&D) Management.** How many bookings should we accept for passengers for each Origin and Destination combination served?

4. **Dynamics.** How should our decisions be updated as bookings are received?

5. **Forecasting.** How can we forecast all future variables of interest including bookings, cancellations, and no-shows to support the yield management process?

6. **Implementation.** How can we develop an effective computerized system and business process to make yield management decisions given the capabilities of our reservation system and the availability of data?

All of these aspects of the revenue management problem must be addressed to develop an effective system. However, since the Overbooking and Booking Class Management aspects for a single flight leg have received considerable attention in the literature, we will not address these issues here. In this paper, we focus instead on the Origin-Destination and Dynamic aspects of the overall revenue management problem.

THE DETERMINISTIC INDEPENDENT BOOKING CLASS CASE

Consider the case of an airline with many different departures on some future day. We refer to a flight *leg* as a single departure and arrival. The *schedule* is the collection of all flight legs that the airline will operate on a given date. Each flight leg has an associated maximum seating capacity. A *market* is any origin-destination pair that is served by the airline, either by an individual flight leg (non-stop service) or by two or more legs (connecting service). An *itinerary* is a set of one or more connecting legs serving a market. For example, a passenger seeking to travel from San Francisco to Chicago (i.e., a passenger in the San Francisco to Chicago market) might have the choice to take Flight 020 direct from San Francisco to Chicago or take Flight 100 from San Francisco to Denver connecting with flight 120 from Denver to Chicago. Associated with each itinerary are one or more *booking classes*. We assume that passengers within the same booking class on the same itinerary all represent the same net revenue to the airline. For example, one booking class may include all discount fare passengers on an itinerary while another includes all full-fare passengers. For each booking class on each itinerary, the *unconstrained demand* is the total number of booking requests that the airline will receive for a booking class/itinerary combination.

The *O&D Revenue Management* problem is how to manage the mix of passengers to maximize net revenue given the seating capacities allocated to each flight leg. The mechanism by which the airline manages its mix of passengers is dictated by the capabilities of its reservation system. For the purposes of what follows, we will assume that the airline has full control over its bookings at the level of itinerary and booking class. The airline receives booking requests for future departures with associated itinerary and booking class information. Based on this information, the airline must decide whether or not to accept each booking as it is received. Alternatively, an airline might establish "booking limits" so that bookings in every itinerary/booking class combination would be accepted until the corresponding booking limit is reached. Assuming that booking limits can be freely adjusted over time, the two approaches are equivalent.

We will now make two assumptions that will be relaxed later:

1. Booking classes are independent. This means that the unconstrained demand for a booking class on an itinerary is independent of whether other booking classes on that itinerary are open or closed.

2. Unconstrained demand in each booking class/itinerary combination is fully known.

Under these assumptions, the *Deterministic Independent Revenue Management Problem (DIRM)* is:

$$\text{Maximize} \atop x \qquad \sum_{i,j,k,s} P_{ijks} x_{ijks} \qquad\qquad\qquad \text{(DIRM)}$$

$$\text{Subject to} \qquad \sum_{i,j,k,s} \delta_{ijkm} x_{ijks} \leq c_m$$

$$x_{ijks} \leq d_{ijks}$$

$$x_{ijks} \geq 0$$

where:

x_{ijks} = Booking limit for itinerary k serving the O&D market from i to j in booking class s.

P_{ijks} = Net revenue associated with passengers on itinerary k serving the O&D market from i to j in booking class s.

d_{ijks} = Total demand associated with itinerary k serving the O&D market from i to j in booking class s.

c_m = Seating capacity of the aircraft associated with leg m.

δ_{ijkm} = 1 if leg m is part of itinerary k serving the O&D market from i to j and 0 otherwise.

Even under the simplifying assumption of deterministic demands and no cancellations or no-shows, it would be difficult or impossible for a large airline to directly solve DIRM. For one thing, the dimensionality of DIRM is immense. It requires setting separate allocations for *every* feasible combination of passenger origin and destination, itinerary, and booking class. It also requires explicit forecasts for each of these combinations. The marginal value approach to DIRM is motivated by Theorem 1.

Theorem 1. For any feasible and bounded realization of DIRM, there exists a set of marginal values λ^* associated with each leg such that any x^* satisfying:

1.) $x_{ijks}^* > 0$ only if $p_{ijks} \geq \sum_m \delta_{ijkm} \lambda_m^*$

2.) For all $p_{ijks} \geq \sum_m \delta_{ijkm} \lambda_m^*$, either:

     a.) $x_{ijks}^* = d_{ijks}$ or,

     b.) $\sum_s x_{ijks} \delta_{ijkm}^* = c_m$ for some $\delta_{ijkm} = 1$.

3.) $\sum_s \delta_{ijkm} x_{ijks}^* \leq c_m$ for all m.

4.) $0 \leq x_{ijks}^* \leq d_{ijks}$ for all i,j,k,s.

is an optimal solution to DIRM.

Proof. The Dual of DIRM is:

$$\underset{\underline{u},\ \underline{\lambda}}{\text{Maximize}} \qquad \sum_{i,j,k,s} d_{ijks} u_{ijks} + \sum_m c_m \lambda_m$$

$$\text{Subject to} \qquad u_{ijks} + \sum_m \delta_{ijkm} \lambda_m \geq p_{ijks}$$

$$u_{ijks} \geq 0, \quad \lambda_m \geq 0$$

where λ_m and u_{ijks} are the dual variables coresponding to constraints (1.b) and (1.c) respectively in DIRM. Constraint (2.b) implies:

$$u_{ijks} \geq p_{ijks} - \sum_m \delta_{ijkm} \lambda_m.$$

Combining (3) with the non-negativity constraint on u_{ijks} and noting that $d_{ijks} > 0$, gives:

$$u_{ijks} = \max \left[p_{ijks} - \sum_m \delta_{ijkm} \lambda_m ,\ 0 \right]$$

The Complementary Slackness Theorem guarantees that, for dual optimal values of u^*_{ijks} and λ^*_m and primal optimal values of x^*_{ijks}:

$$u^*_{ijks} > 0 \rightarrow x^*_{ijks} = d^*_{ijks}$$

$$u^*_{ijks} + \sum_m \delta_{ijkm} \lambda^*_m > p_{ijks} \rightarrow x^*_{ijks} = 0$$

Substituting, we get

$$p_{ijks} > \sum_m \delta_{ijkm} \lambda^*_m \rightarrow x_{ijks} = d_{ijks}$$

$$p_{ijks} < \sum_m \delta_{ijkm} \lambda^*_m \rightarrow x_{ijks} = 0$$

Since DIRM is bounded and feasible, there must exist at least one $\underline{x}^*$ and one $\underline{\lambda}^*$ optimal for the dual and primal respectively. And, any feasible $\underline{x}^*$ satisfying the Complimentary Slackness conditions must be optimal for the primal.

Based on the results of Theorem 1, we can specify the following algorithm for O&D revenue management:

1. Forecast demand for every itinerary/booking class combination. (For the DIRM, this forecast is exact.)

2. Find $\underline{\lambda}^*$.

3. Initialize $b_m = 0$ for all m. (b_m is a count of all bookings already received on a leg.)

4. When a booking request is received for itinerary I_{ijk}, calculate its net revenue (NR) and either accept or reject it according to the following rule:

    a.) If $NR > \sum_m \delta_{ijkm} \lambda_m$ and $b_m < c_m$ for all m such that $\delta_{ijkm} = 1$, then accept the booking request and set $b_m = b_m + 1$ for all m such that $\delta_{ijkm} = 1$.

    b.) If $NR < \sum_m \delta_{ijkm} \lambda_m$ or $b_m = c_m$ for some m such that $\delta_{ijkm} = 1$, then reject the booking request.

5. Go to 4 and continue until departure time.

Independent of booking order or other considerations, this algorithm will result in a set of bookings that maximizes system revenue.

THE GENERAL DETERMINISTIC CASE

We have demonstrated the main result in the case when booking classes constitute discrete independent packets of demand. In other words, we have assumed that, for a given itinerary, the unconstrained demand in a booking class is independent of the states of other booking classes in the same itinerary (i.e., whether they are opened or closed.) This is equivalent to assuming that the partition into booking classes represents "perfect" discrimination that is, customers who happen to fall into a particular booking class will not book in another class on the same itinerary, even if their booking class is not available.

While the assumption of independent booking class demands is often made in designing yield management systems, it is quite restrictive. It is easy to conceive of situations in which potential customers who find their desired booking class unavailable, may elect to "buy-up" to a more expensive but available booking class. Similarly, customers in a higher-price booking class may "dilute" to a lower-priced booking class on the same itinerary if it is available. Ideally, a yield management system should be able to accommodate these types of quite reasonable demand behaviors.

To develop a more general model, we assume that, for each itinerary, an airline can choose to be in one of N possible discrete demand states. Conceptually, a demand state represents a combination of open and closed booking classes. If the airline has S booking classes that can each be either on or off, then $N = 2^S$. If the airline is in demand state n on itinerary k for the i to j O&D market, then the airline will realize unconstrained demand on that itinerary of $\hat{d}_{ijkn}$ and unconstrained revenue of r_{ijkn}. The corresponding average price is:

$$q_{ijkn} = r_{ijkn} / \hat{d}_{ijkn}$$

We will assume that, for each itinerary, the demand states can be ordered such that:

$$\hat{d}_{ijkN} > \hat{d}_{ijkN-1} > \hat{d}_{ijkN-2} > ... > \hat{d}_{ijk1} > 0$$

$$r_{ijkN} > r_{ijkN-1} > r_{ijkN-2} > ... > r_{ijk1} > 0$$

$$q_{ijkN} < q_{ijkN-1} < q_{ijkN-2} < ... < q_{ijk1}$$

(For convenience, we will define $\hat{d}_{ijk0} = 0$ and $r_{ijk0} = 0$ for all itineraries—i.e., demand state 0 corresponds to all booking classes closed. q_{ijk0} is undefined.)

Note that we have defined a demand relationship in which states with more demand produce more revenue but at a lower average price. Although these conditions may seem to be restrictive, in fact they characterize <u>all</u> demand states of interest for a revenue-maximizing airline. For example, for two states m and n such that:

$$\hat{d}_{ijkm} > \hat{d}_{ijkn} \text{ and } q_{ijkm} > q_{ijkn}$$

then, for a revenue-maximizing airline, demand state m would dominate demand state n and demand state n would never be chosen as part of an optimal solution. Similarly, for two demand states m and n such that:

$$\hat{d}_{ijkm} > \hat{d}_{ijkn} \text{ and } r_{ijkm} < r_{ijkn}$$

state n would dominate state m and would demand state m never be part of a revenue-maximizing solution. Thus, the ordering conditions above are non-restrictive.

With this general characterization of demand, the problem facing the airline is: What set of demand states should I choose and how much demand from each state should I allow to book in order to maximize revenue? For each itinerary, the airline must choose exactly one among H+1 possible demand states. Of course the demand state 0 — accepting no demand from that itinerary — can be chosen. For each itinerary, the airline must also determine the total number of bookings to accept from the chosen demand state. The total bookings accepted for state n can range from 0 to $\hat{d}_{ijkn}$. We can write the *General Deterministic Revenue Management Problem* (GDRM) as:

$$\text{Maximize} \atop \hat{\underline{x}}, \underline{y}} \qquad \sum_{i,j,k,n} \hat{x}_{ijkn} y_{ijkn} q_{ijkn} \qquad \text{(GDRM)}$$

$$\text{Subject to} \qquad \sum_{i,j,k} \delta_{ijkm} \sum_{n} \hat{x}_{ijkn} y_{ijkn} \leq c_m$$

$$y_{ijkn} \leq \hat{x}_{ijkn} \hat{d}_{ijkn}$$

$$y_{ijkn} \geq 0, \ \hat{x}_{ijkn} \in \{0,1\}$$

where $\hat{x}_{ijkn}$ is a binary variable such that $\hat{x}_{ijkn} = 1$ means that demand state n is chosen for itinerary ijk, $\hat{x}_{ijkn} = 0$ means that demand state n is not chosen for that itinerary and y_{ijkn} is the amount of demand in state n that is allowed to book on that itinerary.

It is not clear how to derive an efficient approach to solving GDRM. GDRM is not a linear program and its objective function is not guaranteed to be concave. However, we can define a simple condition under which GDRM can be solved. To do so, define the marginal revenue p_{ijkn} and the marginal demand e_{ijkn} associated with state ijkn by:

$$p_{ijkn} = (r_{ijkn} - r_{ijkn-1})/(\hat{d}_{ijkn} - \hat{d}_{ijkn-1})$$

$$e_{ijkn} = \hat{d}_{ijkn} - \hat{d}_{ijkn-1}$$

It is straightforward to verify that:

$$\hat{d}_{ijkn} = \sum_{e=1}^{n} e_{ijkl}$$

$$r_{ijkn} = \sum_{e=1}^{n} p_{ijkl} e_{ijkl}$$

Our main result concerns the case in which the demand states for each itinerary demonstrate non-increasing marginal revenues, that is:

$$p_{ijkn} \leq p_{ijkn-1} \leq \ldots \leq p_{ijk1}$$

In this case, we can find the optimal booking policy by solving problem DIMRRM:

$$
\begin{array}{lll}
\text{Maximize} & \sum_{i,j,k,n} p_{ijkn} z_{ijkn} & \text{(DIMRRM)} \\
z & & \\
\end{array}
$$

$$
\begin{array}{ll}
\text{Subject to} & \sum_{i,j,k,n} \delta_{ijkm} z_{ijkn} \leq c_m \\
\end{array}
$$

$$z_{ijkn} \leq e_{ijkn}$$

$$z_{ijkn} \geq 0$$

With the simple change of notation, DIMRRM is clearly equivalent to DIRM. Thus, we immediately obtain Theorem 2.

Theorem 2. For any feasible and bounded realization of DIMRM, there exists a set of marginal values such that:

1.) $z_{ijkn}^{*} > 0$ only if $p_{ijkn} \geq \sum_{m} \delta_{ijkm} \lambda_{m}^{*}$

2.) For all $p_{ijkn} \geq \sum_m \delta_{ijkm} \lambda_m^*$, either:

    a.) $\quad z_{ijkn} = e_{ijkm}^*$ or,

    b.) $\quad \sum_s x_{ijkn}^* \delta_{ijkm} = c_m$ for some $\delta_{ijkm} = 1$

3.) $\sum_s \delta_{ijkm} z_{ijkn}^* \leq c_m$ for all m

4.) $0 \leq z_{ijkn}^* \leq e_{ijkn}$ for all i,j,k,n

THE CASE OF DEMAND UNCERTAINTY

The previous sections treated the case in which demand was deterministic and known beforehand. We showed that, in these cases, a set of fixed marginal values by leg could be used as criteria when to accept bookings in a way that maximized system profit. Although this result is very powerful, the assumption of perfect foreknowledge of demand is extremely unrealistic. All experience has shown that a crucially important determinant of the effectiveness of any approach to revenue management is its ability to cope with uncertainty.

The *Stochastic O&D Revenue Management* Problem (SORM) is to determine which bookings to accept in order to maximize <u>expected</u> system profit. Assume that we wish to find the set of allocations to passenger O&D market/itinerary/booking-class combinations that maxmizes expected profit given our current uncertainty about future demands. We can write SORM as a stochastic equivalent of DIMRRM:

$$\text{Maximize} \atop \underline{b} \qquad \sum_{i,j,k,n} p_{ijkn} E(z_{ijkn} \mid b_{ijkn}) \qquad\qquad \text{(SORM)}$$

$$\text{Subject to} \qquad \sum_{i,j,k,n} \delta_{ijkm} b_{ijkn} \leq c_m$$

$$b_{ijkn} \geq 0$$

Where:

z_{ijkn} = A random variable representing demand on the k'th itinerary serving the i to j passenger market in demand state n.

b_{ijkn} = Booking limit (allocation) set on the k'th itinerary serving the i to j passenger market in demand state n.

$E(z_{ijkn} \mid b_{ijkn})$ = Expected value of z_{ijkn} given b_{ijkn}.

By standard duality theory, the result of Theorem 1 holds for SORM, if the objective function is concave. Theorem 3 shows this to be the case.

Theorem 3. The objective function of SORM is concave.

Proof. Since p_{ijkn} is positive, we merely need to show that $d^2E[z_{ijkn}|b_{ijkn}]/db^2_{ijkn} \leq 0$ for all i,j,k,n. For what follows, we drop all the subscripts for simplicity.

$$E[z|b] = \int_0^b y \, f(y)dy + (1 - F(b))b$$

where $f(\cdot)$ = probability density function on demand and $F(\cdot)$ is the cumulative distribution function corresponding to $f(\cdot)$. Then:

$$dE[z|b]/db = (1 - F(b))$$

$$d^2E[z|b]/db^2 = -f(b) \leq 0.$$

This shows that, at any time, there exists a marginal value associated with each leg such that it is profitable to accept a booking if and only if the net revenue of the booking is greater than the marginal value associated with each leg (and there is at least one seat remaining unbooked each leg in the itinerary. The leg marginal values need to be updated as bookings are received. This suggests the following algorithm:

1. For each leg departure, derive a forecast of demand by revenue range. (In the deterministic case, we assume that this forecast is exact.)

2. Find an initial set of leg marginal values $\underline{\lambda}$.

3. As each booking request is received, with associated net revenue of p_{ijkn}, accept it if:

$$p_{ijkn} \geq \sum_m \delta_{ijkm} \lambda_m$$

    Otherwise, reject it.

4. When a booking is accepted, update the marginal values on the corresponding legs and go to 3.

Assuming that the appropriate updating procedures are used, this procedure will maximize expected profit at each time. This procedure is, in effect, the Dynamic Marginal Value Approach.

CALCULATING THE MARGINAL VALUES

We have shown that a set of leg marginal values exists and can be used to effectively manage the O&D bookings mix. To develop an effective system, we must be able to rapidly compute and update the leg marginal values as new information is received. Although we will not give any details here, we will sketch out an approach that DFI has developed to perform these computations and updates very rapidly.

Without much effort, we can define a continuous mapping H, such that:

$$\underline{\lambda}^* = H(\underline{\lambda}^*)$$

Calculating $\underline{\lambda}^*$ is then equivalent to determining the fixed point of a continuous mapping on a closed convex set. Under suitable, non-restrictive conditions, we can show that H is a contraction mapping. This means that:

1.   Values of $\underline{\lambda}^*$ can be easily and rapidly computed from distant starting points using Gauss-Seidel-like methods.

2.   Updates of $\underline{\lambda}^*$ can also be rapidly performed using a Gauss-Seidel approach.

In fact, in all practical cases, we have s(H') < 1, where H' is the Jacobian of the mapping H and s(H') is the spectral radius (absolute magnitude of the real component of the largest eigenvalue) of H'. This means that Gauss-Seidel-like schemes are guaranteed to converge very quickly to the appropriate fixed point of H.

Sensitivity Analysis for Degradable Transportation Systems

Zhen-Ping Du and **Alan J Nicholson**

Department of Civil Engineering, University of Canterbury
Private Bag 4800, Christchurch, New Zealand

Abstract This paper describes sensitivity analysis for evaluating the socioeconomic impacts of degradation in transportation systems, based on an integrated equilibrium model with elastic traffic demands.

Key Words   Degradable Transportation Systems; Integrated Equilibrium Model; Nonlinear Programming; Sensitivity Analysis.

1 Introduction

So far, a substantial number of papers have addressed the analysis of transportation systems (TS's). Most of them assume that components (nodes and links) in a TS never fail (i.e., they always operate at their initial capacities). Actually, the assumption is not always appropriate because operation of the TS components can be affected for a period of time varying from a few hours to a few years, by events such as earthquakes, floods, traffic accidents and industrial action, or inadequate maintenance to redress the effect of deterioration caused by TS use. The direct impact of such events is that the component capacities are reduced, the generalised travel costs increase, and some travel is consequently deterred. The socioeconomic impacts of TS degradation can be very substantial [1]. Therefore, a TS should be regarded as a degradable transportation system (DTS), and analysed and designed accordingly.

It is necessary to develop an appropriate method to assess the socioeconomic impacts of TS degradation quantitatively, so that system performance can be optimised in the presence of degradation. Du and Nicholson have developed an integrated steady-state equilibrium model [1] to predict macroscopic traffic behaviour, and to evaluate system performance in a large-scale, complex, multi-modal TS (including road, rail, water and air transport). The model allows for components of a DTS to exhibit different degrees of capacity degradation (a completely failed component is described by a capacity of zero), resulting in different degrees of system performance degradation. We are interested in obtaining the values of the system performance indices corresponding to various combinations of component capacity degradation. Rather than having to re-solve the integrated equilibrium model, which generally requires a substantial computational effort, the above "what if" analysis can be undertaken using the methods of sensitivity analysis.

Although the integrated equilibrium model [1] is equivalent to a concave programming problem, where the objective function is the system performance index (the system surplus) and the decision variables are path flows subject to non-negative constraints, the existing results of sensitivity analysis for general nonlinear programming [2] can not be used directly, because the basic solution (the path flow vector), which is generally contained in a convex set, is not unique. Fortunately, for the analysis of the socioeconomic impacts of TS degradation, only the origin/destination (OD) flows and the link flows are essential, and they are unique under some weak assumptions about the monotonicity and differentiability of traffic demand functions and travel time functions. Tobin and Friesz [3] studied sensitivity analysis for equilibrium network flow, assuming fixed demands. In this paper, we extend their work to analyse the sensitivity of the system state vector (the OD flow vector and link flow vector) and the system performance index (the system surplus) of a DTS, with respect to the component state vector (the component capacity vector), based on the integrated equilibrium model [1] with elastic demands.

The remainder of this paper is comprised of five sections. In Section 2, the integrated equilibrium model [1] is described briefly. In Section 3, a unique path flow vector is selected from the convex set of equilibrium path flow vectors by solving a simple quadratic programming problem. Then the stability of the selected path flow vector with respect to the component state vector is discussed. It is shown that the zero elements in the selected path flow vector under infinitesimal perturbation are unchangeable. In Section 4, a Jacobi matrix, consisting of the second partial derivatives of the system performance index with respect to the positive elements in the selected path flow vector, is shown to be nonsingular. Therefore, the well-defined partial derivatives of the selected path flow vector with respect to the component state vector can be calculated. Consequently, in Section 5 the partial derivatives of the system state vector and system performance index with respect to the component state vector are obtained analytically, to give first-order approximations of the system state vector and system performance index after component state vector perturbation. Some discussion and conclusions are given in Section 6.

2 The Integrated Equilibrium Model

The integrated equilibrium model developed assumes that all movement is between OD centroids, and considers the TS as a directed graph, in which the nodes correspond to OD centroids (eg. cities and towns), major highway intersections, railway stations and junctions, harbours and airports, and the links correspond to major highways, railways, shipping and airline routes. It is assumed that only link capacities in the TS network can be degraded (the capacity degradation of a node can be converted to capacity degradations of the links connecting the node). Generally, several paths are available for each movement, and a path is essentially a combination of modes and routes. It is assumed that each individual user chooses a path to minimize their total generalised cost of travel (or carriage) from the origin centroid to the destination centroid. According to Wardrop's user-optimum principle, a "traffic equilibrium" exists when the generalised travel costs on all used paths between any given OD pair are equal, and not greater than those on unused paths. From a more macroscopic viewpoint, the level of traffic flow between each OD pair is a direct outcome of the interaction between traffic demand and supply. As the traffic flow increases the generalised travel cost increases, and as the generalised travel cost increases the number of users willing to pay the cost decreases. The steady state traffic flow between each OD pair is that at which demand equals supply, and we call such a balance "market equilibrium".

The integrated equilibrium model combines the four traditional, sequential steps in TS modelling (i.e. trip generation, trip distribution, mode split and traffic assignment), by taking account of both market equilibrium and traffic equilibrium, and choosing the system surplus (the sum of the user and producer surpluses) as a suitable performance measure [4]. Based on some weak assumptions on the monotonicity and differentiability of traffic demand functions and travel time functions, the integrated equilibrium model established is shown [1] to be equivalent to the following concave programming problem:

$P(x)$: *Given* $x=[x_a|a\in A]$, *find* $q=[q_h|h\in H]$ *to maximize*

$$SS[q,x] = \sum_{k\in K} \int_0^{f_k[q]} D_k^{-1}(f_k)df_k - \sum_{a\in A} \int_0^{v_a[q]} [\varphi t_a(v_a,x_a)+e_a^0]\rho_a\mu_a dv_a \qquad (1)$$

$$s.t. \quad q_h \geq 0 \qquad h\in H \qquad (2)$$

$$where \quad f_k[q] = \sum_{h\in H} \zeta_{kh}q_h, \quad k\in K \quad and \quad v_a[q] = (\rho_a\mu_a)^{-1} \sum_{h\in H} \xi_{ah}q_h, \quad a\in A \qquad (3)$$

with the symbols defined as follows:
K is an OD pair set with K elements and indicator k;
A is a directed link set with A elements and indicator a;
H is a directed path set with H elements and indicator h;
$Z=[\zeta_{kh}|k\in K,h\in H]$ is an OD pair-path incidence matrix, $\zeta_{kh}=1$ if path h connects the k-th OD pair; otherwise 0;
$B=[\xi_{ah}|a\in A,h\in H]$ is a link-path incidence matrix, $\xi_{ah}=1$ if link a is on path h; otherwise 0;
$x=[x_a|a\in A]$ is a link capacity vector, x_a is an upper bound to the traffic flow (vehicles/unit time) that may pass over link a;
$q=[q_h|h\in H]$ is a path flow vector, q_h is the traffic flow (persons or tonnes/unit time) on path h;
$f=[f_k|k\in K]$ is an OD flow vector, f_k is the traffic flow (persons or tonnes/unit time) between the k-th OD pair;
$v=[v_a|a\in A]$ is a link flow vector, v_a is the traffic flow (vehicles/unit time) on link a;
μ_a is average capacity (persons or tonnes/vehicle) of vehicles moving on link a;
ρ_a is average loading ratio of vehicles on link a;
$f_k=D_k(c_k)$ is the demand function for the k-th OD pair, c_k is a generalised travel cost (\$/person or tonne) between the k-th OD pair, f_k is decreasing and differentiable for c_k, $c_k=D_k^{-1}(f_k)$ is the inverse of $f_k=D_k(c_k)$, where c_k is decreasing and differentiable for f_k;
$t_a=t_a(v_a,x_a)$ is the travel time function for link a, t_a is the travel time on link a (unit time/

vehicle, or equivalently, unit time/person or tonne), is increasing and differentiable for v_a, and is non-increasing and differentiable for x_a, $|\partial t_a(v_a, x_a)/\partial x_a|$, $a \in A$ are positive and finite;

φ is an average value of time (\$/unit time), and $\varphi t_a(v_a, x_a)$ is the variable cost which depends on the link flow and link capacity;

e_a^0 is the fixed cost (\$/person or tonne) on link a, and does not depend on the link flow and link capacity;

$SS[q,x]$ is the system surplus function (\$/unit time).

For the given link capacity vector x, let $q(x)=[q_h(x)|h \in H]$, $f(x)=[f_k(x)|k \in K]$ and $v(x)=[v_a(x)|a \in A]$ be the solutions of Problem $P(x)$ (i.e, the equilibrium path flow vector, OD flow vector and link flow vector respectively), and $SS(x)=SS[q(x),x]$ be the optimal value of the objective function. Then we have the following result [1]:

Theorem 1

(1)   For Problem $P(x)$, there exists a solution $q(x)$; the equilibrium OD flow vector $f(x)$, equilibrium link flow vector $v(x)$ and system surplus $SS(x)$ are unique; however the equilibrium path flow vector $q(x)$ is generally not unique and is contained in a convex set

$$\Gamma(x) = [q(x)| \; Zq(x)=f(x), \; \Xi q(x) = \pi v(x), \; q(x) \geq 0] \tag{4}$$

where $\pi = diag[\rho_1 \mu_1, ..., \rho_A \mu_A]$

(2)   The equilibrium OD flow vector $f(x)$, the equilibrium link flow vector $v(x)$ and the system surplus $SS(x)$ are continuous functions of the link capacity vector x, i.e., there exist constants $C_1, C_2 > 0$ such that

$$|f(x+\Delta x) - f(x)|^2 + |v(x+\Delta x) - v(x)|^2 \leq C_1 |\Delta x|^2 \tag{5}$$

$$|SS(x+\Delta x) - SS(x)|^2 \leq C_2 |\Delta x|^2 \tag{6}$$

where $|\cdot|$ is an Euclidean norm.

The link capacity vector x is called the component state vector, and all the possible component state vectors constitute a component state vector space $X \subset R^A$; a pair of the equilibrium OD flow vector and link flow vector $[f(x),v(x)]$ is called a system state vector, and all the possible system state vectors constitute a system state vector space $\Pi(X) \subset R^K \times R^A$; all the possible values of the objective function in Problem $P(x)$ constitute a system performance index space $\Lambda(X) \subset R^I$; where R^n means an n-dimensional Euclidean space.

Hence, the integrated equilibrium model can be regarded as a mapping from the component state vector space X to the system state vector space $\Pi(X)$, and a mapping from the system state vector space $\Pi(X)$ to the system performance index space $\Lambda(X)$ (see Fig.1). Conclusion 2 of Theorem 1 means small changes in the component state vector x induce small changes in the system state vector $[f(x),v(x)]$, and small changes in the system performance index $SS(x)$.

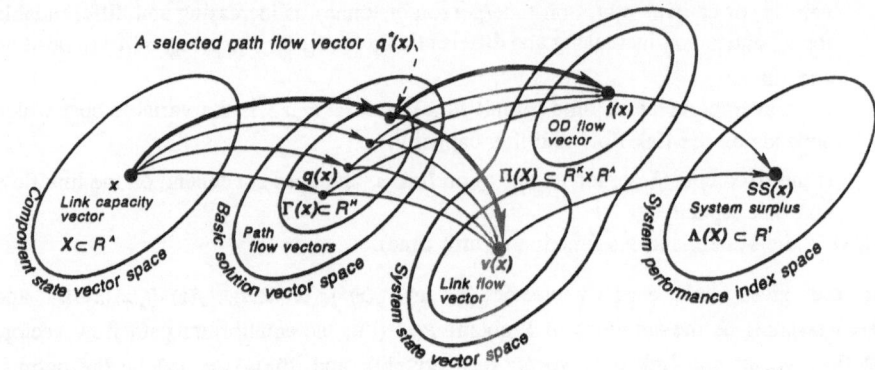

Fig.1

3 Selection of a Unique Path Flow Vector

3.1 A Quadratic Programming Problem

Since the basic solution $q(x)$ of the integrated equilibrium model is generally non-unique and forms a convex set $\Gamma(x)$ in (4), we now try to select a unique path flow vector, denoted as $q^*(x)$, from the set $\Gamma(x)$ (see Fig.1), by solving the following quadratic programming problem:

$P^*(x)$: *Find* $q(x)=[q_h(x)]$ *to minimize*

$$SEN[q(x)] = \sum_{h \in H} [q_h(x)]^2 \tag{7}$$
$$s.t. \quad q(x) \in \Gamma(x) \tag{8}$$

<u>*Theorem 2*</u> *There exists a unique solution* $q^*(x)=[q_h^*(x)]$ *for Problem* $P^*(x)$.

<u>*Proof*</u> Since $SEN[q(x)]$ is continuous on the closed set $\Gamma(x)$, there exists a $q^*(x)=[q_h^*(x)]$ in $\Gamma(x)$ such that $SEN[q^*(x)]$ is minimum over $\Gamma(x)$. Moreover, since $SEN[q(x)]$ is a strictly convex function defined on the convex set $\Gamma(x)$, the solution $q^*(x)$ of $P^*(x)$ is unique. ∎

Obviously, the unique selected path flow vector $q^*(x)$ satisfies

$$\sum_{h \in H} \zeta_{kh} q_h^*(x) = f_k(x), \quad k \in K \text{ and } \sum_{h \in H} \xi_{ah} q_h^*(x) = \rho_a \mu_a v_a(x), \quad a \in A \tag{9}$$

where $f_k(x)$, $k \in K$ and $v_a(x)$, $a \in A$ can be obtained uniquely by solving $P(x)$.

3.2 Stability of the Selected Path Flow Vector to the Component State Vector

Theorem 3 The selected path flow vector $q^*(x)$ is a continuous function of link capacity vector x, i.e., there exists a constant $\hat{C}>0$ such that

$$\| q^*(x+\Delta x) - q^*(x) \|^2 \leq \hat{C} \| \Delta x \|^2 \tag{10}$$

if there exist $q(x)\in\Gamma(x)$ and $q(x+\Delta x)\in\Gamma(x+\Delta x)$ such that $q(x)>0$, $q(x+\Delta x)>0$, where $|\cdot|$ is an Euclidean norm.

Proof Since the path flow sets $\Gamma(x)$ and $\Gamma(x+\Delta x)$ are not empty and there exist $q(x)\in\Gamma(x)$, $q(x+\Delta x)\in\Gamma(x+\Delta x)$ such that $q(x)>0$, $q(x+\Delta x)>0$, from Theorem 4.3 of J.W. Daniel [5], we know that there exists a constant $C_0>0$ such that

$$\sum_{h\in H} [q_h^*(x+\Delta x) - q_h^*(x)]^2 \leq C_0\Big\{ \sum_{k\in K} [f_k(x+\Delta x) - f_k(x)]^2 + \sum_{a\in A} (\rho_a\mu_a)^2[v_a(x+\Delta x) - v_a(x)]^2\Big\} \tag{11}$$

Let $\sigma = max_{a\in A}[\rho_a\mu_a]$ and $\overline{C} = max[C_0, C_0\sigma^2]$. Then from (11)

$$\sum_{h\in H} [q_h^*(x+\Delta x)-q_h^*(x)]^2 \leq \overline{C} \Big\{ \sum_{k\in K} [f_k(x+\Delta x)-f_k(x)]^2 + \sum_{a\in A} [v_a(x+\Delta x)-v_a(x)]^2 \Big\} \tag{12}$$

From (12), (5) and letting $\hat{C}=\overline{C}C_l>0$, we obtain (10). ∎

3.3 Unchangeability of the Zero Elements in the Selected Path Flow Vector

Let $Y_+(x) = [h \mid q_h^*(x) > 0]$ and $Y_0(x) = [h \mid q_h^*(x) = 0]$ $\qquad$ (13)

Then $Y_+(x) \cup Y_0(x) = H$ since $q_h^*(x) \geq 0$, $h\in H$ $\qquad$ (14)

Theorem 4 For $x+\Delta x$ in a neighbourhood of x, if the solutions $q(x)$, $q(x+\Delta x)$ of Problems $P(x)$, $P(x+\Delta x)$ satisfy the following strict complementarity conditions respectively, i.e.

$$\begin{cases} \textit{if } q_h(x) = 0, \textit{ then } \ \partial SS[q,x]/\partial q_h \big|_{q=q(x)} < 0 \\ \textit{if } q_h(x+\Delta x) = 0, \textit{ then } \ \partial SS[q,x]/\partial q_h \big|_{q=q(x+\Delta x)} < 0 \end{cases} \quad h\in H \tag{15}$$

then the solutions $q^*(x)$, $q^*(x+\Delta x)$ of Problems $P^*(x)$, $P^*(x+\Delta x)$ have the following properties respectively

$$Y_+(x+\Delta x) = Y_+(x) \ \textit{ and } \ Y_0(x+\Delta x) = Y_0(x) \tag{16}$$

The above conditions (15) mean that the generalised travel costs on all used paths between any given OD pair are equal, and less than those on unused paths.

Proof From Theorem 3, $q^*(x)$ is a continuous function of x. Then for $x+\Delta x$ in a neighbourhood of x, we have

$$q_h^*(x+\Delta x) > 0 \quad \textit{if} \quad q_h^*(x) > 0 \qquad h\in H \tag{17}$$

Besides, from the necessary and sufficient conditions that $q(x)$ and $q(x+\Delta x)$ are solutions of Problems $P(x)$ and $P(x+\Delta x)$ respectively, as well as the strict complementarity condition (15), we have

$$\begin{cases} \partial SS[q,x]/\partial q_h \big|_{q=q(x)} = 0 \textit{ iff } q_h(x) > 0 \textit{ i.e. } h\in Y_+(x) \\ \partial SS[q,x]/\partial q_h \big|_{q=q(x)} < 0 \textit{ iff } q_h(x) = 0 \textit{ i.e. } h\in Y_0(x) \end{cases} \tag{18}$$

$$\begin{cases} \partial SS[q,x]/\partial q_h\big|_{q=q(x+\Delta x)} = 0 \quad \text{iff} \quad q_h(x+\Delta x) > 0 \quad \text{i.e.} \quad h\in Y_+(x+\Delta x) \\ \\ \partial SS[q,x]/\partial q_h\big|_{q=q(x+\Delta x)} < 0 \quad \text{iff} \quad q_h(x+\Delta x) = 0 \quad \text{i.e.} \quad h\in Y_0(x+\Delta x) \end{cases} \tag{19}$$

where $\partial SS[q,x]/\partial q_h\big|_{q=q(x)} = \sum_{k\in K} \zeta_{kh}D_k^{-1}[f_k(x)] - \sum_{a\in A} \xi_{ah}\{\varphi t_a[v_a(x), x_a] + e_a^0\}, \quad h\in H$ (20)

Since $q^*(x)$ and $q^*(x+\Delta x)$ are also solutions of Problems $P(x)$ and $P(x+\Delta x)$ respectively, they satisfy (18) and (19), i.e.

$$\begin{cases} \partial SS[q,x]/\partial q_h\big|_{q=q^*(x)} < 0 \quad \text{if} \quad q_h^*(x) = 0 \\ \\ q_h^*(x+\Delta x) = 0 \quad \text{if} \quad \partial SS[q,x+\Delta x]/\partial q_h\big|_{q=q^*(x+\Delta x)} < 0 \end{cases} \tag{21}$$

From (20) and (9), it follows that $\partial SS[q,x]/\partial q_h\big|_{q=q^*(x)} = \partial SS[q,x]/\partial q_h\big|_{q=q(x)}$. Therefore $\partial SS[q,x]/\partial q_h\big|_{q=q^*(x)}$ is a continuous function of $[f(x),v(x)]$, and from (5), a continuous function of the component state vector x. Thus, for $x+\Delta x$ in a neighbourhood of x, we have

$$\partial SS[q,x+\Delta x]/\partial q_h\big|_{q=q^*(x+\Delta x)} < 0 \quad \text{if} \quad \partial SS[q,x]/\partial q_h\big|_{q=q^*(x)} < 0 \tag{22}$$

It follows from (21) and (22) that for $x+\Delta x$ in a neighbourhood of x,

$$q_h^*(x+\Delta x) = 0 \quad \text{if} \quad q_h^*(x) = 0 \qquad h\in H \tag{23}$$

and (16) follows from (17) and (23). ∎

4 Partial Derivatives of the Selected Path Flow Vector to the Component State Vector

From Theorem 4, it simply follows that

$$\partial q_h^*(x)/\partial x_a = 0 \qquad h\in Y_0(x), \; a\in A \tag{24}$$

Now we try to obtain $\partial q_h^*(x)/\partial x_a$, $h\in Y_+(x)$, $a\in A$. By differentiating the first equation in (18) with respect to x_a, $a\in A$, and from (20) and (9), we obtain the following equation

$$\sum_{r\in Y_+(x)} \partial^2 SS[q,x]/\partial q_h\partial q_r\big|_{q=q^*(x)} \partial q_r^*(x)/\partial x_a = -\sum_{b\in A} \partial^2 SS[q,x]/\partial q_h\partial x_b\big|_{q=q^*(x)} \partial x_b/\partial x_a, \quad h\in Y_+(x), \; a\in A \tag{25}$$

which can be written as

$$\{\nabla_q^2 SS[q^*(x),x]\}_{Q\times Q}\{\nabla_x q^*(x)\}_{Q\times A} = -\{\nabla_{qx}^2 SS[q^*(x),x]\}_{Q\times A}\{\nabla_x x\}_{A\times A} \tag{26}$$

where Q is the number of elements in the indicator set $Y_+(x)$, and

$$\{\nabla_x q^*(x)\}_{Q\times A} = \left\{ \partial q_r^*(x)/\partial x_a \mid r\in Y_+(x), \; a\in A \right\} \tag{27}$$

is an unknown matrix. Now

$$\{\nabla_q^2 SS[q^*(x),x]\}_{Q\times Q} = \left\{ \partial^2 SS(q,x)/\partial q_h\partial q_r\big|_{q=q^*(x)} \mid h, \; r\in Y_+(x) \right\}$$

$$= \left\{ \sum_{k\in K} \zeta_{kh}\zeta_{kr}dD_k^{-1}(f_k)/df_k\big|_{f_k=f_k(x)} - \sum_{a\in A} \xi_{ah}\xi_{ar}\varphi(\rho_a\mu_a)^{-1}\partial t_a(v_a,x_a)/\partial v_a\big|_{v_a=v_a(x)} \mid h, \; r\in Y_+(x)\right\} \tag{28}$$

$$\{\nabla_{qx}^2 SS[q^*(x),x]\}_{Q\times A} = \left\{ \partial^2 SS[q,x]/\partial q_h\partial x_a\big|_{q=q^*(x)} \mid h\in Y_+(x), \; a\in A \right\}$$

$$= \left\{ \; -\xi_{ah}\varphi \partial t_a(v_a,x_a)/\partial x_a \big|_{v_a=v_a(x)} \; \middle| \; h \in Y_+(x), \; a \in A \right\} \tag{29}$$

and $\{ \nabla_x x \}_{A \times A} = \left\{ \; \partial x_b/\partial x_a \; \middle| \; b, \; a \in A \right\}$ (30)

which can be used to describe the interactions between the link capacity degradations. The following theorem shows the non-singularity of the Jacobi matrix.

Theorem 5 $\{ \nabla_q^2 SS[q^*(x),x] \}_{Q \times Q}$ is negative-definite, i.e.

$$\hat{q}_{Q \times 1}^T \{ \nabla_q^2 SS[q^*(x),x] \}_{Q \times Q} \hat{q}_{Q \times 1} < 0 \; \text{for any column vector } \hat{q}_{Q \times 1} = [\hat{q}_h \mid h \in Y_+(x)] \neq 0 \tag{31}$$

Proof Let $\hat{f}_k = \sum_{h \in Y_+(x)} \zeta_{kh} \hat{q}_h$, $k \in K$ and $\hat{v}_a = (\rho_a \mu_a)^{-1} \sum_{h \in Y_+(x)} \xi_{ah} \hat{q}_h$, $a \in A$ (32)

From (28) it can be shown [1] that

$$\hat{q}_{Q \times 1}^T \{ \nabla_q^2 SS[q^*(x),x] \}_{Q \times Q} \hat{q}_{Q \times 1} = \sum_{k \in K} dD_k^{-1}(f_k)/df_k \big|_{f_k = f_k(x)} (\hat{f}_k)^2 - \sum_{a \in A} \varphi \rho_a \mu_a \partial t_a(v_a,x_a)/\partial v_a \big|_{v_a = v_a(x)} (\hat{v}_a)^2 \tag{33}$$

$\because \; dD_k^{-1}(f_k)/df_k \big|_{f_k = f_k(x)} < 0$, $k \in K$ and $\partial t_a(v_a,x_a)/\partial v_a \big|_{v_a = v_a(x)} > 0$, $a \in A$ (34)

Thus $\hat{q}_{Q \times 1}^T \{ \nabla_q^2 SS[q^*(x),x] \}_{Q \times Q} \hat{q}_{Q \times 1} < 0$ if we can show that not all $\hat{f}_k$, $k \in K$ and $\hat{v}_a$, $a \in A$ are equal to zero for all $\hat{q}_{Q \times 1} \neq 0$. We prove this by contradiction. Suppose $\hat{f}_k = 0$, $k \in K$ and $\hat{v}_a = 0$, $a \in A$ for all $\hat{q}_{Q \times 1} \neq 0$. Let $\hat{q}_h = 0$ for $h \in Y_0(x)$, $q^1(x) = [q_h^1(x) \mid h \in H]$ and $q^2(x) = [q_h^2(x) \mid h \in H]$, where

$$q_h^1(x) = q_h^*(x) + \alpha \hat{q}_h \quad \text{and} \quad q_h^2(x) = q_h^*(x) - \beta \hat{q}_h \qquad h \in H \tag{35}$$

For $\alpha, \beta > 0$ chosen small enough, it can be easily shown that $q^1(x), q^2(x) \in \Gamma(x)$, and from (35) we know $q^1(x) \neq q^2(x)$. That is, $q^1(x)$, $q^2(x)$ are two individual feasible solutions of Problem $P^*(x)$. From (35), however

$$q_h^*(x) = \beta/(\alpha+\beta) q_h^1(x) + [1 - \beta/(\alpha+\beta)] q_h^2(x) \qquad h \in H \tag{36}$$

which means an arbitrary point on the segment between $q^1(x)$ and $q^2(x)$ is equal to the solution $q^*(x)$ of Problem $P^*(x)$. This contradicts Theorem 2 that $q^*(x)$ is a unique solution of Problem $P^*(x)$, and completes the proof. ∎

Based on Theorem 5, it follows that $\{ \nabla_q^2 SS[q^*(x),x] \}_{Q \times Q}$ is nonsingular. Thus its inverse exists, and from (26) we have

$$\{ \nabla_x q^*(x) \}_{Q \times A} = -\{ \nabla_q^2 SS[q^*(x),x] \}_{Q \times Q}^{-1} \{ \nabla_{qx}^2 SS[q^*(x),x] \}_{Q \times A} \{ \nabla_x x \}_{A \times A} \tag{37}$$

5 The First-Order Approximations of the System State Vector and System Performance Index

Based on the results (24) and (37), the partial derivatives of the system state vector with respect to the component state vector are obtained from (9)

$$\begin{cases} \partial f_k(x)/\partial x_a = \sum_{h \in Y_+(x)} \zeta_{kh} \partial q_h^*(x)/\partial x_a & k \in K \\ \\ \partial v_b(x)/\partial x_a = (\rho_b \mu_b)^{-1} \sum_{h \in Y_+(x)} \xi_{bh} \partial q_h^*(x)/\partial x_a & b \in A \end{cases} \qquad a \in A \tag{38}$$

As for the partial derivatives of the system performance index to the component state vector, we have following result:

Theorem 6

$$\partial SS(x)/\partial x_a = -\varphi\sum_{b\in A} \rho_b\mu_b\partial x_b/\partial x_a \int_0^{v_b(x)} \partial t_b(v_b, x_b)/\partial x_b dv_b \qquad a\in A \tag{39}$$

Proof Obviously from (1) and (9) we have $SS(x)=SS[q^*(x), x]$. Differentiating $SS(x)$ with respect to x_a, $a\in A$ gives

$$\partial SS(x)/\partial x_a = \sum_{h\in Y_r(x)} \partial SS[q,x]/\partial q_h\Big|_{q=q^*(x)} \partial q_h^*(x)/\partial x_a + \sum_{h\in Y_d(x)} \partial SS[q,x]/\partial q_h\Big|_{q=q^*(x)} \partial q_h^*(x)/\partial x_a$$

$$+ \sum_{b\in A} \partial SS[q,x]/\partial x_b\Big|_{q=q^*(x)} \partial x_b/\partial x_a \qquad a\in A \tag{40}$$

Since $q^*(x)$ satisfies the first condition in (18), and (24), it follows that the first two terms on the right of (40) vanish. From (1) and (9)

$$\partial SS[q,x]/\partial x_b\Big|_{q=q^*(x)} = -\varphi\rho_b\mu_b\int_0^{v_b(x)} \partial t_b(v_b, x_b)/\partial x_b dv_b \qquad b\in A \tag{41}$$

and (39) can be obtained from (40) and (41). ∎

Based on these partial derivatives, the first-order approximations of $[f(x+\Delta x), v(x+\Delta x)]$ and $SS(x+\Delta x)$ for $x+\Delta x$ in a neighbourhood of x are given by

$$\begin{cases} f_k(x+\Delta x) \doteq f_k(x) + \sum_{b\in A} \partial f_k(x)/\partial x_b \Delta x_b & k\in K \\ v_a(x+\Delta x) \doteq v_a(x) + \sum_{b\in A} \partial v_a(x)/\partial x_b \Delta x_b & a\in A \end{cases} \tag{42}$$

$$SS(x+\Delta x) \doteq SS(x) + \sum_{b\in A} \partial SS(x)/\partial x_b \Delta x_b \tag{43}$$

6 Discussion and Conclusion

In this paper, the difficulty that the basic solution (the path flow vector) is not unique, has been overcome by selecting a unique path flow vector, to obtain the partial derivatives of the system state vector and system performance index with respect to perturbation of the component state vector. The partial derivatives can be used to predict the direction of change in the system state vector and the system performance index, resulting from changes in the component state vector. Once they are calculated, first order numerical approximations of the new system state vector and the corresponding system performance index can be easily constructed for any combination of element perturbations in the component state vector. Any perturbation of the component state vector will generally result in changes in the system state vector and in the system performance index, because of the inherent nonlinearity of the integrated equilibrium model. In a DTS, users are affected in two distinctly different ways. Firstly, some users may cancel or postpone their travel, because of the extra cost due to travel delays or the need to divert to another path, and there is a consequent loss of user surplus for those users. Secondly, those users who do not cancel or postpone their travel will experience increased generalised costs, arising from travel delays or re-routing. This will reduce the user surplus for those users.

Since the integrated equilibrium model has a spatial network structure describing the road, rail, water and sea sub-networks simultaneously, the impacts of degradations in the individual sub-networks on the whole system performance, and the interaction among them, can be evaluated using sensitivity analysis. Generally, the elements in a component state vector are inter-dependent. For example, two bridges over a river may be washed away or damaged together if the river is in flood, since the debris of the failed upstream bridge moves downstream, increasing the probability of the downstream bridge failing. The preparation of input data for sensitivity analysis should take account of any relationship between the elements in the component state vector.

In the analysis of a DTS, a component is called an important component if the impact of its degradation on the performance of the whole system is high (i.e., the system performance index is sensitive to degradation of the component). A component is called a weak component if its probability of degradation is high (i.e., a weak component is more likely to be damaged than others). Furthermore, a component is called a critical component if it is both important and weak. The critical components in a DTS can be identified by the results of sensitivity analysis, which also provides considerable information for the optimization of the DTS to minimise the socioeconomic impacts of degradation.

References

[1]   Zhen-Ping Du and Alan J. Nicholson, Degradable Transportation Systems—Performance, Sensitivity and Reliability Analysis, Civil Engineering Research Report, University of Canterbury, New Zealand, 1993.

[2]   A.V. Fiacco, Introduction to Sensitivity and Stability Analysis in Nonlinear Programming, Academic Press, Inc. New York, 1983.

[3]   R.L. Tobin and T.L. Friesz, Sensitivity Analysis for Equilibrium Network Flow, Transportation Science, Vol.22, No.4, 242-250, 1988.

[4]   N.H. Gartner, Optimal Traffic Assignment with Elastic Demands: A Review Part I. Analysis Framework, Transportation Science, Vol.14, No.2, 174-191, 1980.

[5]   J.W. Daniel, Stability of the Solution of Definite Quadratic Programs, Mathematical Programming, Vol.5, 41-53, 1973.

Convergence and optimality of two-step algorithms for public transportation system optimization.

Wojciech Grega

Institute of Automatics
Technical University of Mining and Metallurgy
30-059 Krakow, Al.Mickiewicza 30, Poland.

1. Introduction

This study is concerned with the urban public transportation system. The increasing difficulties in many countries in balancing their public budgets, coupled with an increasing demand for better public transportation, is very likely to increase the demand for improved planning methods. The need for efficient transportation planning is not new: long ago it was noted that even small relative improvements of an urban transportation system may imply large absolute savings. Unfortunately, before the advent of high-speed computers it was not possible to collect, store and analyze the vast quantities of data needed to carry out urban transportation planning within a reasonable time period and a cost less then experimenting with the real transportation system.

Among many problems related to the planning of the urban public transportation system [1] this paper examines the particular aspect of the short range transit planing: allocation of vehicles. Once the planner has decided which routes to operate, which stops to serve and the hours of service, the next broad decision is how the operating resources are to be distributed over the routes. This decision determines the headway for each route, which is subsequently used as a basis for the bus driver scheduling. The purpose of this paper is to suggest models and present solution algorithms developed for determination of optimal allocation of buses among routes in order to minimize travel time for passengers. The algorithms described in the paper exploit the specific problem structure for the reduction of the optimization problem to a sequence of related but simpler and more easily solvable problems.

The transportation subsystem of interest can be diagramed as in Fig.1

in the most general way. Two kinds of relationships should be characterized by appropriate models:

i) The pattern of flows in the transportation network is determined by both the transportation system and travel demands.

ii) The current flow pattern will cause changes in the transportation system through the control system and through the resources consumed in providing better service.

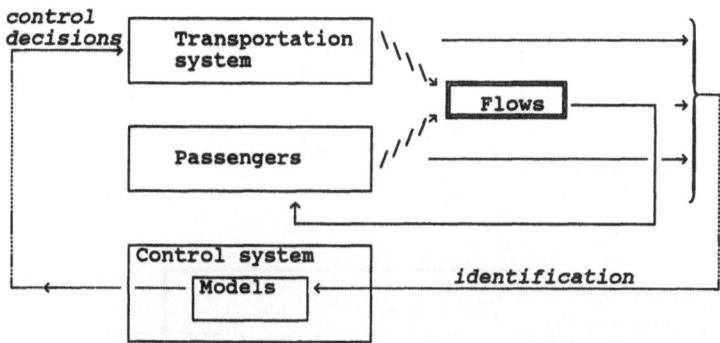

Fig. 1. Basic relationships in a public transportation system.

Such a description of the public transportation process, adapted for the allocation problem, can be given in the following general terms. Passengers flows on routes are a function of control decisions due to variable travel times between some node pairs. Hence, the allocation of buses affects the assignment of passengers to routes and passenger flows affect the allocation of vehicles. Thus two cooperative models can be created. These are: the allocation model and assignment model (Fig.2).

Models for the assignment-allocation problem are often large and structured. Although existing optimization algorithms are efficient, in many cases it is necessary to decompose the models. This can be done mainly in two ways. A natural strategy for the problem is to apply a

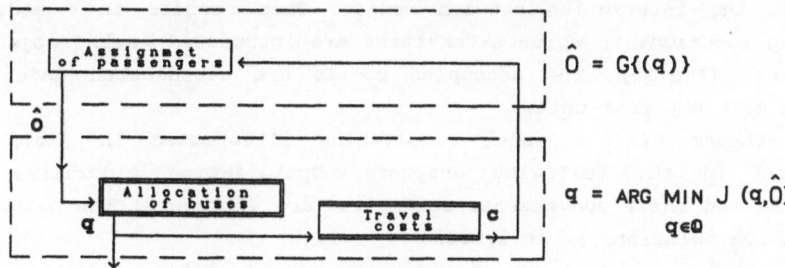

$$\hat{O} = G\{(q)\}$$

$$q = \underset{q \in Q}{\text{ARG MIN}} \ J(q, \hat{O})$$

Fig 2. Assignment-allocation problem.

930

repeated two-step technique in which the output of the upper model gives
the parameters for the lower model (Fig.3a). The two models interact and
several iterations between models may be required until an equilibrium
point is found. The risk of dividing the model into separate parts can
be seen from the fact that the successive solutions of the allocation
problem do not necessarily converge to a solution of the
assignment-allocation problem.

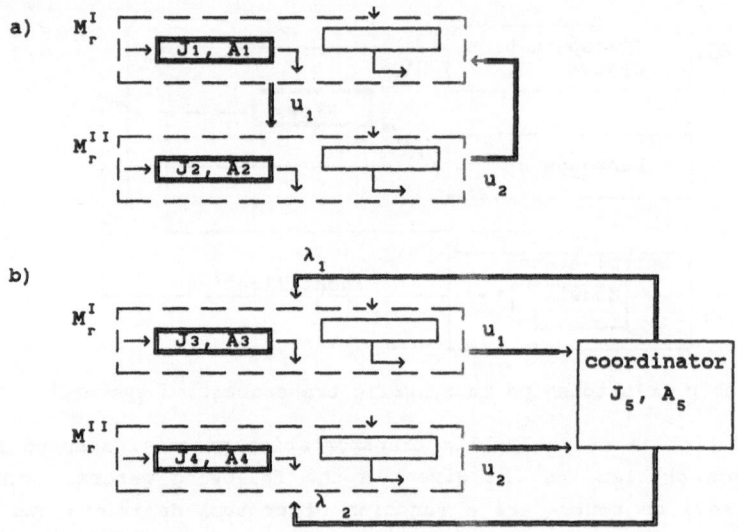

Fig.3. Cooperation of transportation models: J_i - objective
function, A_i - algorithm.

An iterative computational algorithms from Figure 3a may be viewed as
a sequence of action-reaction of the two players of the game. In this
respect, convergence of the algorithm corresponds to some stability
notions of the game theory [4].

Secondly, the hierarchical system theory [5] can be applied to
decouple the interaction between models. This results in a modified
two-step technique in which extra terms are introduced to the objective
functions. (Fig.3b). The decoupled models are coordinated under the
command of a supremal unit.

Two schemes of the model cooperation illustrated in Fig.3 are
developed in the following chapters. Optimization algorithms are
suggested and their convergence conditions are discussed. The nature of
equilibrium solutions is analyzed.

2. The model

Let L denote a set of routes in a transportation network, each defined as a sequence of nodes at which passengers may board and alight. The following attributes are associated with the bus route l:

q_l - service frequency on the bus route l, $l \in L$,

q_l^o - minimum allowed allocation on the route l,

τ_l - round trip running time on the route l.

Data on passenger travel demands are given in a form of a origin-destination matrix $O = \{O_{kj}\}$, $(k,j) \in I_o$, where I_o is a set of node pairs in the network. At each stop where he waits, the passenger considers a subset of attractive routes L_{kj} and boards the first vehicle of these routes to be served. Thus, we assume that the passenger behaves in a way that minimizes his travel time, using a *fixed travel strategy* [6].

It is assumed that at each node served by a route, the distribution of interarrival times of vehicles of that route is known. Hence, we can calculate the average waiting time for a passenger (AWT_{kj}) for each subset of routes L_{kj}, as well as the probability of each route arriving first. Based on this probability, the number of passengers using each route can be calculated.

In a urban transportation network, a node pair (k,j) may be connected by the set of routes offering a *direct* (non-transfer) service between node pairs or *transfer trips* should also be used to characterize paths between a given origin and destination. The set of attractive routes L_{kj} can thus be classified into two mutually exclusive sets:

L_{kj}' - set of routes offering direct connections between nodes (k,j),

L_{kj}'' - set of transfer routes, $L_{kj} = L_{kj}' \cup L_{kj}''$.

Therefore, passengers flows must be split between competing sets of routes. In the model it will be assumed that the network is strongly connected and each node can be reached using one transfer at most. The number of transfer trips will be a function of frequency on competing transfer routes and will create the *variable flow* in the network. The variable flow in the network can be modeled by an *assignment procedure*, creating a *modified origin-destination* matrix in the form: $\hat{O} = O + O''(q)$, where $O''(q) = \{O_{kj}''(q)\}$ is a transfer flow matrix.

The given above notation and assumptions helps to formulate the optimal allocation problem as:

$$\text{MIN} \sum_{\substack{i,j \in I_o}} \hat{O}_{ij}(q) \cdot AWT_{ij}(q) \qquad (1)$$
$$q \in Q$$

$$Q = \{q \in R_+^L: q_1 \geq q_1^0 > 0, \; 1 \in L, \sum_{l \in L} \tau_1 q_1 \leq N\}, \qquad (2)$$

$$L = |L|, \; \tau > 0,$$

where the assignment model is in the form:

$$\hat{O} = G(q), \qquad G: Q \rightarrow \hat{O}, \qquad (3)$$

$\hat{O}$ -set of modified origin-destination matrices, $\hat{O} \subset R_+^{IxI}$, $I=|I_0|$.

It should be noted that a non-integer number of buses may be allocated on a route because of the possibility of interlining.

3. Solution technique

The following assumptions on the problem (1), (2), (3) will be used in most of the results of this section:

Assumption 1. A unique solution of (1), (2) exists for each fixed $\hat{O} \in \hat{O}$.
Assumption 2. The function G is continuous on Q.
Assumption 2a. The function G is continuously differentiable on Q.

An application of the procedure illustrated in Fig.3a can be briefly outlined by the following:

| Algorithm 1 |

STEP 1: Guess an initial allocation on routes $q^{(0)} \in Q$. Compute an initial assignment of passengers in the form of the modified origin-destination matrix $O(q^{(0)})$. Set the iteration index n=0.

STEP 2: Solve the problem (1), (2) for fixed $\hat{O}(q^{(n)})$, obtaining:

$$q^{(n+1)} = \arg \min_{q \in Q} \sum_{i,j \in I} \hat{O}_{ij}(q^{(n)}) \cdot AWT_{ij}(q) \qquad (4)$$

Step 4: From the equation (3) compute elements of the modified origin-destination matrix $\hat{O}_{ij}(q^{(n+1)})$.

STEP 4: Check the convergence criterion:
if $\|\hat{O}(q^{(n+1)}) - \hat{O}(q^{(n)})\| \leq \varepsilon$, stop
otherwise set n=n+1 and go to **STEP 1**.

Proposition 1. Suppose that Assumptions 1 and 2 hold. Then, the mapping F, defined by Algorithm 1 in the form:

$F : \mathbf{Q} \rightarrow \mathbf{Q}$, $F(q) = \arg \min\limits_{q\in\mathbf{Q}} f(y,G(q))$ has, at least, one fixed point in $\mathbf{Q}$. ∎

The proof of the proposition is given in [7]. An interpretation of the solution is given in Fig.4.

The model given by (1)-(3) can be formulated in an equivalent decomposed form by introducing interconnection variables x, y_j:

$$\min_{q} f(q,\hat{O}_1 \ldots \hat{O}_I),$$
$$q\in\mathbf{Q} \tag{6}$$

subject to: $y_j = G_j(x)$, $y_j = \hat{O}_j$, $x = q$,

where the matrix $\hat{O}$ and matrix assignment function $G(x)$ were rewritten as column matrices in the form:

$$\hat{O} = [\hat{O}_1, \hat{O}_2, \ldots \hat{O}_j, \ldots \hat{O}_I],$$

$$G = [G_1, G_2, \ldots G_j, \ldots G_I],$$

$$\hat{O}_j = \begin{bmatrix} \hat{O}_{1j} \\ \hat{O}_{2j} \\ \vdots \\ \hat{O}_{Ij} \end{bmatrix}, \quad G_j = \begin{bmatrix} G_{1j} \\ G_{2j} \\ \vdots \\ G_{Ij} \end{bmatrix}, \quad j\in\mathbf{I}, \quad I = |\mathbf{I}_o|, \quad L = |\mathbf{L}|$$

$x_j \in \mathbf{X} \subset \mathbf{R}^L_+$, $\qquad y_j \in \mathbf{Y} \subset \mathbf{R}^I$, $\mathbf{X}$, $\mathbf{Y}$ - open sets.

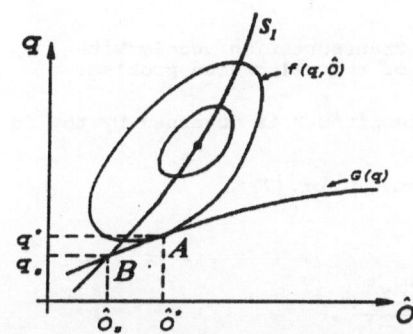

Fig.4. Interpretation of the solutions of the problem (1)-(3). The fixed-point solution is attained at point B. Point A is an optimal solution of the problem (1)-(3), $f(q^*)\geq f(q^\bullet)$.

The following two-step coordinated procedure can be suggested (Fig.5):

Algorithm 2

STEP 0: Start with an initial guess for the frequency $x^{(0)}\in \mathbf{Q}$. Set the iteration index $n=0$.

STEP 1: Compute an assignment of passengers as the elements of the modified origin-destination matrix: $y_j^{(n)} = G_j(x^{(n)})$ and appropriate Lagrange multipliers in the form:

$$\lambda^{(n)} = -\sum_j \frac{dG_j^T(x^{(n)})}{dx}\frac{\partial f(x,^{(n)}y_j^{(n)})}{\partial y_j}, \quad j \in I_o, \tag{7}$$

STEP 2: Specify the solution of the modified allocation problem:

$$q^{n+1} = \arg\left(\min_{q \in Q}\left\{\sum_{i,j \in I} y_{ij}^{(n)} AWT_{ij}(q) - \lambda^{(n)T}q\right\}\right) \tag{8}$$

STEP 3: Check the convergence criterion:

if $\|q^{(n+1)} - x^{(n)}\| \le \varepsilon$ stop, otherwise calculate:

$x^{(n+1)} = x^{(n)} + K(q^{(n+1)} - x^{(n)})$,

set n=n+1 and go to **STEP 1**, $K \in \mathbb{R}^1$, $0<K<1$.

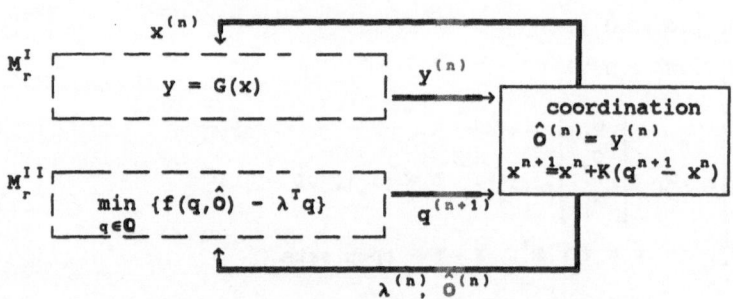

Fig.5. Cooperation of transportation models with a modification of the allocation problem.

The convergence of the Algorithm 2 is governed by the following result [7].

Let us denote by $r:Q \to \mathbb{R}^L$ r.h.s. of (7):

$$\lambda_a^{(n)} = -r(x^{(n)}), \tag{9}$$

where:

$$r(x^{(n)}) = \sum_{j \in I} \frac{dG_j^T(x^{(n)})}{dx}\frac{\partial f(x,^{(n)}y_j^{(n)})}{\hat{\partial} y_j}. \tag{10}$$

Proposition 2.

Suppose that:

- assumption 2a holds,
- exists an unique solution of (8) together with Lagrange multipliers in some neighborhood of the solution point,
- the function r(x) satisfies:

$$(r(x+\Delta) - r(x))^T\Delta \ge \beta\|\Delta\|^2, \quad \forall\, x,\, x+\Delta \in X, \ \beta>0. \tag{11}$$

$$\|r(x + \Delta) - r(x)\| \leq \gamma \|\Delta\|, \quad \forall x, \ x+\Delta x \in X, \ \gamma > 0, \tag{12}$$

then Algorithm 2 fulfills existence and convergence conditions of the dual price method.■

Some remarks about assumptions are in order.

(i) If the Assumption 1 holds, then the solution of (7) is unique.

(ii) Assumption (11) on $r(x)$ is an uniform monotonity condition. Assumption (12) is a standard convergence condition of descent algorithms [8].

4. Example application

The algorithms described above were developed as a part of the study of the Kielce public transit system. The trial network (Fig.6) consists of four routes and nine nodes with a demand matrix given in Table 1. It was also assumed that all link travel times are 10 minutes.

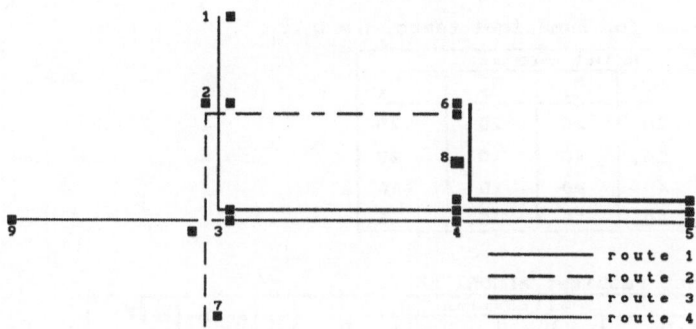

Fig.6. Example network.

The algorithms described in the previous section have been used to allocate N=80 vehicles between four routes. Remaining constraints in (2) were assumed as: $q_1^0=10$, $q_2^0=12$, $q_3^0=4.5$, $q_4^0=9$. The results of this example are shown in Tables 3 and 4.

First it should be noted that, hopefully, ALGORITHM 1 finds a fixed point solution (4-6 iterations were necessary in this example). The solutions produced by ALGORITHM 2 appear to differ significantly from the outputs of ALGORITHM 1. However, the differences of the objective function value at the solution point are rather slight, a reallocation

Table 1. Demand matrix

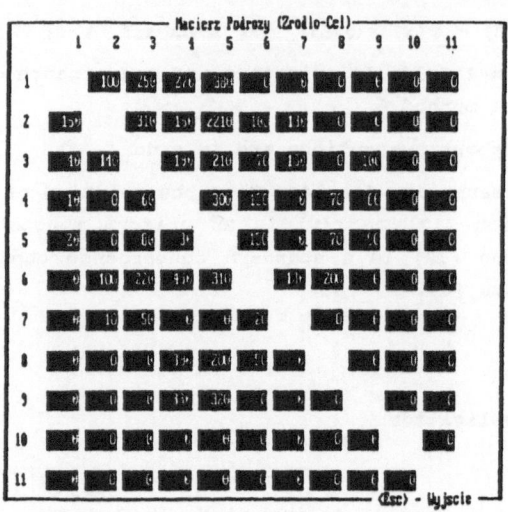

of more than one bus can be observed. This suggests that the positions of the fixed-point type solution (ALGORITHM 1) and optimal solution (ALGORITHM 2) differ significantly in the solution space.

Table 2. Data for numerical tests, $n_1 = q_1 \cdot \tau_1$

N = 80	Initial solution			
test no.	n_1	n_2	n_3	n_4
1	20	20	20	20
2	10	40	10	20
3	40	20	10	10
4	40	30	5	5

Table 3. Test results: ALGORITHM 1.

N = 80	allocation				objective	stop test
test no.	n_1	n_2	n_3	n_4	function	(STEP4)
1	26.44	17.91	19.93	15.73	396.71	1.0
2	26.53	17.86	19.92	15.68	396.57	1.41
3	26.29	17.91	20.00	15.79	396.66	1.41
4	26.29	17.91	20.01	15.79	396.65	1.0

Table 4. Test results: ALGORITHM 2.

N = 80	allocation				objective	stop test
test no.	n_1	n_2	n_3	n_4	function	(STEP3)
1	26.57	16.05	21.00	16.37	396.23	0.277
2	26.20	16.15	21.10	16.54	396.45	0.250
3	26.92	16.04	20.81	16.23	395.99	0.570
4	26.57	16.05	21.00	16.37	396.23	0.277

5. Conclusions

Two algorithms that can be used to optimize public transportation network taking account of passenger interests were presented in this paper. The first formulation of the assignment - allocation problem is natural and simple, but carries serious difficulties as far as convergence properties are concerned. In this case the equilibration procedure is based on the computation of fixed points.

The second formulation requires several assumptions on the assignment function, but offers a convergence proof based on a dual formulation of the generally better value of the objective function in a solution point.

References.

[1] Grega W. (1989). *Decomposition method for generation of optimal timatables for urban bus network* In: Control, Computers, communication in Transportation, CCCT'89 (ed.J.P.Perrin), 433-437.

[2] Florian, M., M. Gaudry (1979). *A conceptual framework for the "supply side" in transportation systems.* Montreal University, Department of Computer Studies, Publication 324, Montreal.

[3] Ceder, A., N.H.M. Wilson (1986). *Bus network design.* Transportation Research, 20B, 331 - 343.

[4] Basar, T.,J.G. Olsder (1982). *Dynamic noncooperative game theory.* Academic Press, London.

[5] Sandrin, P. (1991). *Duality-based decomposition methods for optimal operation of the EDF generation systems.* Engineering Optimization, 18, 93-106.

[6] Spies H.,M.Florian (1989). *Optimal strategies: a new assignment model for transit networks.* Transportation Research, 22B, 83-102.

[7] Grega W. (1993). *Decomposition methods in optimization of urban public transportation.* Automatics 62, Scientific Bulletins of Stanislaw Staszic Academy of Mining and Metallurgy, Kraków.

[8] Bertsekas D.P. (1982). *Constrained optimization and Lagrange multipler methods,* Academic Press, London.

Reliability

A MATHEMATICAL FORMULATION OF RELIABILITY OPTIMIZED DESIGN

by

Martin L. Shooman
Department of Computer Science
Clifford Marshall
Department of Mathematics
Polytechnic University
Six Metrotech Center, Brooklyn, NY 11201

Abstract

The problem of optimization of system reliability through an optimum choice of parallel redundancy has been extensively researched in the literature. Many mathematical techniques have been adapted or developed to solve this integer programming problem, however, because of mathematical complexity few of these are widely known or used by reliability engineers.

It is shown that most systems are designed using hierarchical decomposition, which greatly reduces the complexity of the problem and allows simple bounding and enumeration methods. If the entire problem is called Level 0, the next level down, Level 1, is designed by apportioning the reliability goal among perhaps 3-10 subsystems. At Level 2, we would attempt to meet the apportioned reliability goals by an optimum assignment of redundant components. A set of upper and lower bounds on the number of components are developed which establishes a search space. As feasible candidate combinations are evaluated, the bounds are recomputed, further limiting the search space.

The paper also develops solution techniques for standby redundancy, discusses the component improvement approach, and develops criteria for comparing the efficiency of component improvement with parallel redundancy.

A few examples are analyzed to compare the methods, their complexity, and speed of solution.

1.0 INTRODUCTION

This paper develops a set of simple techniques which yield an optimum or approximately optimum design which should be understandable, and provide insight which can be conveniently applied by practicing engineers for optimizing the reliability of a system through the use of: component improvement, redundancy, standby systems, repair, mixed strategies, etc.

The goal is to maximize the reliability under several constraints, cost, weight, volume, etc., but a single constraint, cost will be used in this paper. Bounds on the optimal solution are computed and an approximate solution is devised using allocation techniques. A set of strategies is computed within the bounds and the best strategy is selected. Component redundancy, standby redundancy, and component improvement techniques for improving reliability are considered. In some cases an optimum is achieved and in other situations a good suboptimal solution is obtained. Because of practical considerations, design proceeds via a hierarchical decomposition resulting in a modest number of cases to investigate. Furthermore, the bounds constrain the number of feasible solutions such that a personal computer program takes a few seconds to a few minutes to compute all allowable solutions. Thus, elaborate theoretical models are not needed in many practical cases.

1.1 Mathematical Statement of the Problem

The mathematical statement of the problem is to maximize the reliability under one or more constraints of a system which can be defined by its associated reliability graph or block diagram which includes all the k major components or assemblies in the system. The classical formulation of such problems is to consider the optimum arrangement of additional parallel components where n_1 is the number of parallel elements for component one. ($n_1 = 1$ means a single component, $n_1 = 2$ means two in parallel, etc.) Exact optimization requires the solution of an integer programming optimization problem, which is very difficult and approximations are generally used to effect a solution. Theoretical approaches are given in [Ashrafi 92, Dinghua 87, El-Neweihi 86, Hwang 84, Kuo 87, Pram 91, Prasad 91, Tillman 80.

1.2 Engineering Statement of the Problem

In an engineering design the numbers of redundant components n_1, n_2, ... n_i are limited to small integers (1,2,3,4) by various practical considerations. Thus if $n_i \leq 4$ the maximum number of combinations to consider $\leq 4^k$ and more likely closer to 2^k or 3^k on average. Furthermore, the next section will explain why in many practical problems k is in the range of 3-10. Thus, the gross number of combinations to be considered is not as large as one might initially assume, and bounding significantly reduces the number of feasible combinations. The approach in this paper is to bound, enumerate, and compute candidate solutions in a systematic fashion, providing optimum or near optimum solutions.

2.0 HIERARCHICAL DECOMPOSITION

A designer approaches the problem of reliability optimization in a much different manner than an optimization theorist would. The theorist begins with the system structure, the resource constraints, and *synthesizes* the system which achieves the highest reliability through the addition of redundant components or component improvement. The practical designer begins by decomposing the system into 3 to 10 subsystems and apportions the reliability among the various sub-systems. The number of decomposed subsystems is suggested by Miller [1956]. We can relate the number of subsystems or subsubsystems to the number of levels, ℓ, down in the decomposition process. The entire system is represented by level $\ell = 0$, the apportionment level is $\ell = 1$, and the subsubsystems are level $\ell = 2$, etc. This decomposition continues until the lowest level, ℓ_{max} is reached, where all the elements are components or as they are called in practice, *line replaceable units*, LRU's. We assume that there are the same number of decompositions, D, at each subsystem or subsubsystem. The number of subsubsystems, N, at each level ℓ and the number of LRU's, N_{LRU}, are given by:

$$N = D^\ell \quad (1) \qquad N_{LRU} = D^{\ell_{max}} \quad (2)$$

It is assumed that all of the subsystems must function for the system to function, thus, at this level all the subsystems are in series in the reliability graph. In turn, each subsystem designer takes the apportioned reliability goal and repeats the process in a recursive manner. This process is used for almost all large, complex problems and is called hierarchical decomposition or sometimes "divide-and-conquer". One can consider meeting the apportioned reliability goals by using redundant sub-systems, and assuming that D=5, and $\ell=1$, from Eq. (1) we have that N = 5. If we consider parallel subsystem redundancy and a maximum of 4 parallel elements for each subsystem, then using the notation of Sec. 1.1, $n_1 = n_2 = n_3 = n_4 = n_5 = 4$ and there are $4^5 = 1024$ different combinations to consider. In practice not all the n_i will be as large as 4 and the bounds which will be introduced further limit the number of combinations to be considered. However, it is easy to show that better reliability is obtained if the redundancy is applied at a lower level. [Shooman 1990, p. 281-286]. Thus, if we go to one level lower, $\ell = 2$, there are 25 subsubsystems. If we optimize over all these subsubsystems, there are $4^{25} = 1.13 \times 10^{15}$ different combinations to consider. There are 5 separate optimizations to be performed of the type given above, and there are 5 x 1024 = 5120 different combinations. Again we state that the number of combinations is significantly lower in practice because the value of $n_i < 4$ for most of the subsubsystems and because many combinations can be excluded by the bounds which we will develop.

3.0 PARALLEL REDUNDANCY

3.1 Introduction

We assume that the system is to be apportioned at level 1 and that our optimization is to be performed at the subsystem level, level 2. (Actually, this same technique will be repeated about 5 times by each of the subsystem designers.) If the system is really very huge, one can apportion both levels 1 and 2 and perform optimization at level 3. An alternate is to apportion at level 1, and perform recursive optimization at both levels 2 and 3.

3.2 Apportionment Techniques

We define the reliability goal of the complete system at level 0 to be R_g^0. At level 1 we have n subsystems each with a reliability of R_i. Since all the subsystems are needed for operation and they are assumed independent, the system reliability is given by the product of all the subsystem reliabilities:

$$R_g^0 = \prod_i R_i \tag{3}$$

Thus, the apportionment problem can be stated as the choice of a set of numbers $R_1, R_2, ..., R_n$ such that their product is $\geq R_g^0$. The apportionment problem is generally solved by one of three procedures:

a. An equal apportionment technique where all the R_i are set equal to p, leading to the solution

$$\log p = R_g^0/n \tag{4}$$

b. An apportionment technique due to Albert. [Albert 58, Lloyd 77, pp. 267-271]. This technique assumes that all the components of interest require the same amount of effort to increase their reliability as well as other assumptions.

c. A practical apportionment technique based on experience, negotiation among the subsystem contractors or managers, and trial and error.

3.3 Lower Bounds

We assume that apportionment has taken place at level zero using one of the above methods. We now have a set of apportioned reliability numbers at level one for the subsystems: R_1^1, R_2^1, R_3^1, ... R_k^1 we now wish to optimize R_i^1 by assigning parallel reliability at the subsubsystem level. Specifically we compute the maximum value of R_1^1 which can be obtained by choosing a set of parallel redundant subsystems, $n_1, n_2, ... n_k$. The reliability expression becomes

$$R_1^1 = \prod_{i=1}^{k} R_i^2 \tag{5}$$

where R_i^2 is the reliability of the parallel subsubsystems at level two.
Since Eq. (5) is a product of probabilities, each must be equal to or larger than R_1^1, thus we have a set of lower bounds

$$R_i^2 \geq R_1^1 \quad (i - 1,...,k) \tag{6}$$

Eq. (6) of course holds for the case of no redundancy, i.e. $n_1, n_2, ... n_k = 1$. In the case of n_1 redundant subsystems R_1^2 fails when all n_1, items fail, thus the probability of failure is $(1 - R_1^2)^{n_1}$ and the probability of success is:

$$1 - (1 - R_i^2)^{n_1} \quad (i - 1,...k) \tag{7}$$

Substitution of Eq. (7) into Eq. (6) yields:

$$1 - (1 - R_i^2)^{n_i} \geq R_i^1 \quad (i = 1,...k) \tag{8}$$

Solution of Eq. (8) yields a set of lower bounds on n_1, n_2, ... n_k.

$$n_{il} \geq \frac{\log(1 - R_1^1)}{\log(1 - R_i^2)} \quad (i = 1,...k) \tag{9}$$

3.4 Upper Bounds

One can obtain a set of upper bounds on the redundancy numbers n_1, n_2, ... n_k by combining the lower bounds of Eq. (9) with the constraint. Assume that the constraint is a maximum cost of c_0 and that each subsubsystem costs c_i units, thus,

$$n_1 c_1 + n_2 c_2 + ... n_k c_k \leq c_0 \tag{10}$$

We know that the value of n_1 must be $\geq n_{1\ell}$, thus we can write in general

$$n_i = n_{i\ell} + \Delta n_i \tag{11}$$

Substitution of Eq. (11) into Eq. (10) for each value of i yields

$$\Delta n_1 c_1 + \Delta n_2 c_2 + ...\Delta n_k c_k \leq c_0 - (c_1 n_{1\ell} + c_2 n_{2\ell} + ... c_k n_{k\ell}) \tag{12}$$

The largest value of Δn_1 occurs when $\Delta n_2 = \Delta n_3 = ... \Delta n_k = 0$ and an upper bound is

$$\Delta n_i \leq \left[c_0 - (c_1 n_{1\ell} + c_2 c_{2\ell} + ... + c_k n_{k\ell}) \right] / c_i \quad (i = 1,...k) \tag{13}$$

Thus,

$$n_{iu} = n_{i\ell} + \Delta n_i \tag{14}$$

The bounds given in Eqs. (8) and (13) considerably limit the search space for an optimal solution. Furthermore, as computations progress, various values of the Δn_i are selected and the values of the upper bounds n_{iu}, decrease, further narrowing the search space. If during the search along an allocation sequence $n_{ju} = n_{j\ell}$ then the sequence terminates and the values of Δn_i for $j+1 < i < k$ are zero. This will be illustrated in the problems in the following section.

3.5 Examples - Parallel Redundancy

The procedures described in the preceding sections are illustrated with two examples, which indicate the reduction in computational complexity due to bounds and approximations and the closeness of the suboptimum solutions which are obtained to the true optimum solution.

Example 1: Consider a system which has been apportioned at level 0 so that each level 1 subsystem has a reliability goal of 0.90. Subsystem 1 is composed of 3 components (subsubsystems) with reliabilities of 0.85, 0.5, 0.3, and normalized costs of one unit each, and a total cost budget of 16 units for the final design. The minimum value of redundancy, n_1, n_2, n_3 for the example is given by solving Eq. (8) and yields values of 1.21, 3.32, and 6.46 which become the integers: $n_1 = 2$, $n_2 = 4$, $n_3 = 7$. Thus, the minimum system has 13 components with a total cost of 13 units leaving 3 additional units of cost available for additional redundancy. Thus using Eqs. (12) and (13), the maximum values of n_1, n2, and n_3 are 5, 7, and 10. The total number of combinations which must be evaluated to find the optimum is $4 \times 4 \times 4 = 64$. However, as was discussed, the upper bounds are dynamic and are recomputed after each assignment. The results are illustrated by the search tree given in Fig. 1 and are listed in Table 1. Note that the 64 search cases have been reduced to 10.

A simple program has been written on a personal computer which computes all combinations between the upper and lower bounds and displays those solutions where total cost $\leq c_o$ and resulting reliability $\geq R_j$. For Example (1), the computations (on a 386/15 mhz. computer with a coprocessor) takes less than 2 seconds to compute the total number of trials, 1024, and about 13 seconds to compute and display the second through fifth cases in Table 1. (Cases 6-11 are not displayed by the program because the resulting reliability < 0.9.) Clearly the second case in Table 1 is the optimum, however, in a practical case one might wish to explore other possibilities. For example, if the program is rerun with resulting reliability ≥ 0.86 and cost ≤ 15 one obtains 8 other solutions including $[n_1 = 2, n_2 = 5, n_3 = 7, c = 14, R = 0.86895]$ and $[n_1 = 2, n_2 = 5, n_3 = 8, c = 15, R = 0.89235]$. It is possible that the $c = 15$ solution yielding $R = 0.89235$ may be superior to the $c = 16$, $R = 0.9098$ solution. Practical considerations might dictate such a choice. Furthermore, the solutions which require $n_3 \geq 7$ would probably be rejected on a practical basis. It is unreasonable to use 7 parallel components to raise the reliability of a poor component, and a better choice would be product improvement. Thus, based on this criterion, none of the solutions in Table 1 would be acceptable.

The computer program also computes the appointment solution which is $n_1 = 3$, $n_2 = 6$, and $n_3 = 11$ at a cost of 22, which exceeds the maximum cost and yields a reliability of 0.96163.

Example 2: As a slightly larger example consider a case of 5 components with costs of 2, 2, 2, 3, 3 and reliabilities of .8, .8, .8, .9, and .9. The maximum cost budget is 36 units and the reliability goal is 0.95. In about one second, the program computes that there are 8,575 cases to enumerate and that the apportionment solution is $n_1 = n_2 = n_3 = 3$ and $n_4 = n_5 = 2$, with a total cost of 30 and a reliability of 0.95677. The complete enumeration and display results in 31 cases and takes about 75 seconds. The 31 cases are given in Table 2. In a practical case, a designer might choose Case 31 rather than Case 1. Since it costs 5 units less and still meets the reliability goal of $R_g{}^\circ \geq 0.95$. The advantage of Case 1 over Case 31 can be stated as a decrease in $(1-R_g{}^\circ)$ from .04323 to .0246 (43%) which is certainly significant, but may not be cost effective in a highly competitive environment.

4.0 STANDBY REDUNDANCY

4.1 Introduction

In many cases systems are designed with standby redundancy (often called cold standby). In such a case the on-line component has power applied and has a certain failure rate during operation. A sensing scheme is used to automatically detect failure of on the on-line component and to switch signal inputs and power to the standby component. The governing failure law (assuming no standby failures and a perfect switch) is the Poisson Distribution rather than the binomial distribution which governed in the case of parallel redundancy. Thus, the probability of x failures is given by

$$P(x;\mu) = \frac{\mu^x e^{-\mu}}{x!} \qquad (15)$$

where μ is the expected number of failures.

4.2 Lower Bounds

The lower bound solution for the standby case is similar to that of Sec. 3.0, however, the probabilities of success, Eq. (7), become

$$\sum_{i=0}^{n_j-1} P(i,\mu) \qquad (j = 1,...,k) \qquad (16)$$

and the lower bound equations are

$$\sum_{i=0}^{n_j-1} P(i,\mu) \geq R_j^1 \quad (j = 1,...k) \tag{17}$$

The above equations are solved numerically for n_1, n_2, ... n_k.

4.3 Upper Bound

The upper bound equations given in Eqs. (10)-(14) are the same, except Eqs. (17) rather than Eqs. (9) are used to compute the lower bounds.

4.4 Examples - Standby Redundancy

We now consider applying standby redundancy to Example 1, where the cost of each unit is raised to 1.5 units to account for the increased cost of standby equipment. We begin the example by computing the corresponding values of μ from $\mu = - \ln(p)$ (equating probabilities of no failure for the binomial and Poisson distributions) yielding: $\mu_1 = 0.1625189$, $\mu_2 = 0.6931471$, $\mu_3 = 1.2039728$. Substitution in Eq. (17) yields:

$$e^{-\mu_1} [1 + \mu_1 + \mu_1^2/2! + \cdots] \geq 0.85 [1 + 0.1625189] = 0.9881 \geq 0.9 \tag{18a}$$

$$e^{-\mu_2} [1 + \mu_2 + \mu_2^2/2! + \cdots] \geq 0.5 [1 + 0.6931471 + 0.2402264] = 0.9667 \geq 0.9 \tag{18b}$$

$$e^{-\mu_3} [1 + \mu_3 + \mu_3^2/2! + \cdots] \geq 0.3 [1 + 0.12039728 + 0.7247752 + 0.2908735] = 0.9659 \geq 0.9 \tag{18c}$$

Thus, the minimum values of standby redundancy are $n_1 = 2$, $n_2 = 3$, $n_3 = 4$. The minimum cost is $9 \times 1.5 = 13.5$ and the reliability of the lower bound is

$$R_\ell = 0.9881 \times 0.9667 \times 0.9659 = 0.9226 > 0.9 \tag{19}$$

Thus, the lower bound is a solution where in addition to the three on-line elements there is one standby for component one, two for component 2 , and 3 for component 3. This solution would generally be preferred in practice to any of the solutions given in Table 1.

The standby case can represent or approximate in the limit many other schemes of reliability improvement. Suppose we have n replaceable components and the on-line system provides ample visual and audible warnings when the on-line system fails and replacement of components is rapid. If the small amount of downtime during component replacement is insignificant, then such a component replacement scheme can be modeled by a standby system. If the replacement down time is unacceptable then there can be one on-line unit, one standby unit, and (n-1) spares. The system automatically switches from the on-line to the standby unit, and the failed on-line is replaced. In such a case, the standby switch must be designed to toggle back and forth.

Another similar case is where the success of the system depends on r out of n components working. In such a case, we can treat the problem as was done in Eq. (17), however, instead of summing Poisson probabilities binomial probabilities are used, B(i, n; p), and the sum goes from i = r to n. Similarly, in a digital system where we are using n-modular redundancy (see Shooman, 1990, Appendix H), n is generally an odd number, binomial probabilities are used, and the summation goes from i = (n-1)/2 to n.

5.0 COMPONENT IMPROVEMENT

Another major technique for improving system reliability is to improve the reliability of a single component (subsystem). In general there are extra costs involved (development and production). Typically such an improved design begins by listing all the ways in which the subsystem can fail in order of frequency. Design engineers then propose schemes to eliminate, mitigate, or reduce the frequency of occurrence. The design changes are made (sometimes one-at-a-time or a few-at-a-time) and the prototype is tested to confirm the success of the redesign. Sometimes over-stress (accelerated) testing is used in such a process to demonstrate unknown failure modes which must be fixed.

In general, such a component improvement process follows a growth curve where the reliability improves as a function of time, which is generally proportional to development cost and often production cost. (See Fig. 2a.)

Models of this type have been applied to hardware development (see O'Connor 1985, Duane Method), software development (see Shooman 1983, Chap. 5), and to growth processes in general (Nathan 84).

One can develop models which correspond to Fig. 2(a,b) by starting with the general equation for reliability as a function of operating time t, in terms of hazard function (failure rate), z,

$$R(t) = e^{\int_o^t z(x)dx} \tag{20}$$

Suppose that the failure rate is proportional to the number of failure modes in the design,

$$z(\tau) = k_1 M(\tau) \tag{21}$$

m and that as development time (τ) progress this decays exponentially,

$$M(\tau) = M_o e^{-k_2 \tau} \tag{22}$$

(Identical models can be constructed for software where the number of residual error, $E(\tau)$ is substituted for $M(\tau)$. If we assume that the cost of design changes and increased production costs are proportional to τ

$$c = k_3 \tau \tag{23}$$

we can now combine Eqs. (20)-(23) and since $z(\tau)$ is independent of t we obtain

$$R(t) = e^{-\left[K_1 M_o e^{-K_2 c/K_3}\right]t} \tag{24}$$

Since the probability of failure $F(t) = 1 - R(t)$, and for short time periods the first exponential can be approximated by two terms in the series

$$F(t) = 1 - R(t) = (K_1 M_o t)e^{-K_2 c/K_3} \tag{25}$$

Assuming that the time interval t_o is fixed we can absorb the constants into α and β so that

$$F(t_o) = \alpha e^{-\beta c} \tag{26a}$$

$$\ln F(t_o) = \ln \alpha - \beta c \tag{26b}$$

We can write a similar cost model for parallel redundancy where we have n items in parallel and the system only fails if all the n components fail

$$F(t_o) = 1 - R(t_o) = q^n = q \times q^{n-1} \tag{27a}$$

where n-1 are the number of added components

$$\ln F(t_o) = \ln q + (n-1)\ln q \tag{27b}$$

and since the total cost of the added components is $c = (n-1)c_i$

$$\ln F(t_o) = \ln q + \left(\frac{\ln q}{c_i}\right)c \tag{28}$$

If we assume that α is the failure probability of the component before we begin development, i.e. q, then Eq. (26b) becomes

$$\ln F(t_o) = \ln q - \beta c \tag{29}$$

Thus, component improvement is more cost effective than parallel redundancy if

$$\frac{K_2}{K_3} t_o = \beta > \frac{|\ln q|}{q} \tag{30}$$

Empirical data which allows us to evaluate K_2 and K_3 is needed to evaluate the inequality of Eq. (30) to determine which scheme is best.

6.0 ADDITIONAL CONSIDERATIONS

Several additional considerations can be addressed to extend the results of this paper:
1. Include the effects of common mode failures.

2. Include the effects of imperfect standby switches. (See Shooman 1990.)

3. Treat the case of repairable systems, where the repair downtimes are significant.

4. Consider effects of multiple constraints such as cost, c, weight, w, and volume, v.

REFERENCES

Albert, A., "A Measure of the Effort Required to Increase Reliability," Technical Report No. 43, Nov. 5, 1958, Applied Mathematics and Statistics Lab., Stanford University, Contract No. N6onr-25140 (NR 342-022).

Ashrafi, N. and O. Berman, "Optimization Models for Selection of Programs Considering Cost & Reliability", IEEE Transactions on Reliability, Vol. 41, No. 2, June 1992.

Dinghua, S., A New Heuristic Algorithm for Constrained Redundancy-Optimization in Complex Systems," IEEE Trans. on Reliability, Vol. R-36, No. 5, Dec. 1987, pp. 621-

El-Neweihi, E., F. Proschan, J. Sethuraman, "Optimal Allocation of Components in Parallel-Series and Series-Parallel Systems," J. Applied Probability, Vol. 23, 1986, pp. 770-777.

Hwang, C. L., H. B. Lee, F. A. Tillman and C. H. Lie, "Nonlinear Integer Goal Programming Applied to Optimal System Reliability," IEEE Trans. on Reliability, Vol. R-33, No. 5, Dec. 1984, pp. 431-

Kuo, W., H-H. Lin, Z. Xu and W. Zhang, "Reliability Optimization with the Lagrange-Multiplier and Branch-and-Bound Technique," IEEE Trans. on Reliability, Vol. R-36, No. 5, Dec. 1987, pp. 624-

Lloyd, D. K. and M. Lipow, Reliability: Management, Methods, and Mathematics, Second Ed., ASQC, 1977, Appendix 9A.

Manber, U., Introduction to Algorithms, Addison-Wesley, 1989, pp. 210-212.

Messinger, M. and M. L. Shooman, "Techniques for Optimum Spares Allocation: A Tutorial Review", IEEE Transactions on Reliability, Vol. R-19, No. 4, November 1970.

Miller, G. A., "The Magical Number Seven, Plus or Minus Two: Some Limits on Our Capacity for Processing Information," The Psychological Review, Vol. 63, No. 2, March 1956, p. 81.

Nathan, I., "Study of Markov Defusion Growth Models," Ph.D. Dissertation, Operations Research, Polytechnic Institute of New York, May 1984.

O'Connor, P.D.T., Practical Reliability Engineering, Second Ed., John Wiley & Sons, 1985.

Pham, H. and M. Pham, "Optimal Designs of {k,n-k+1}-out -of-n:F Systems (Subject to 2 Failure Modes), IEEE Trans. on Reliability, Vol. 40, No. 5, Dec. 1991, pp. 559.

Prasad, V. R., Y. P. Aneja and K. P. K. Nair, "A Heuristic Approach to Optimal Assignment of Components to a Parallel-Series Network," IEEE Trans. on Reliability, Vol. 40, No. 5, Dec. 1991. pp. 555.

Shooman, M.L., Software Engineering: Design, Reliability, and Management, McGraw-Hill, 1983.

Shooman, M. L., Probabilistic Reliability: An Engineering Approach, Second Ed., Kreiger, 1990.

Shooman, M. L. and C. Marshall, "Reliability Optimized Design", Polytechnic University Research Report, Oct, 1992.

Tillman, F. A., C. L. Hwang, W.Kuo, Optimization of Systems Reliability, Marcel Dekker, 1980.

TABLE 1
Parallel Redundancy Optimum and Suboptimum Solutions for Problem 1

n_1	n_2	n_3	Cost	Reliability	Comments
2	4	7	13	.8409	Lower Bound
3	5	8	16	.9098	Optimum
2	5	9 ·	16	.9087	Sub Optimum
2	6	8	16	.9067	"
3	6	7	16	.9002	"
3	4	9	16	.8966	"
2	4	10	16	.8905	"
2	7	7	16	.8900	"
4	5	7	16	.8885	"
4	4	8	16	.8830	"
5	4	7	16	.8602	"

TABLE 2
Parallel Redundancy Optimum and Suboptimum Solutions for Problem 2

Rank	n_1	n_2	n_3	n_4	n_5	Cost	Reliability
1	4	4	4	2	2	36	.97540
2	3	3	3	3	3	36	.97424
3	3	3	4	2	3	35	.97169
4	3	3	4	3	2	35	.97169
5	3	4	3	2	3	35	.97169
6	3	4	3	3	2	35	.97169
7	4	3	3	2	3	35	.97169
8	4	3	3	3	2	35	.97169
9	3	4	5	2	2	36	.97039
10	3	5	4	2	2	36	.97039
11	4	3	5	2	2	36	.97039
12	5	3	4	2	2	36	.97039
13	5	4	3	2	2	36	.97039
14	4	5	3	2	2	36	.97039
15	4	3	4	2	2	34	.96915
16	3	4	4	2	2	34	.96915
17	4	4	3	2	2	34	.96915
18	3	3	3	4	2	36	.96633
19	3	3	3	2	4	36	.96633
20	3	3	3	3	2	33	.96546
21	3	3	3	2	3	33	.96546
22	3	3	6	2	2	36	.96442
23	6	3	3	2	2	36	.96442
24	3	6	3	2	2	36	.96442
25	5	3	3	2	2	34	.96417
26	3	3	5	2	2	34	.96417
27	3	5	3	2	2	34	.96417
28	4	3	3	2	2	32	.96294
29	3	3	4	2	2	32	.96294
30	3	4	3	2	2	32	.96294
31	3	3	3	2	2	30	.95677

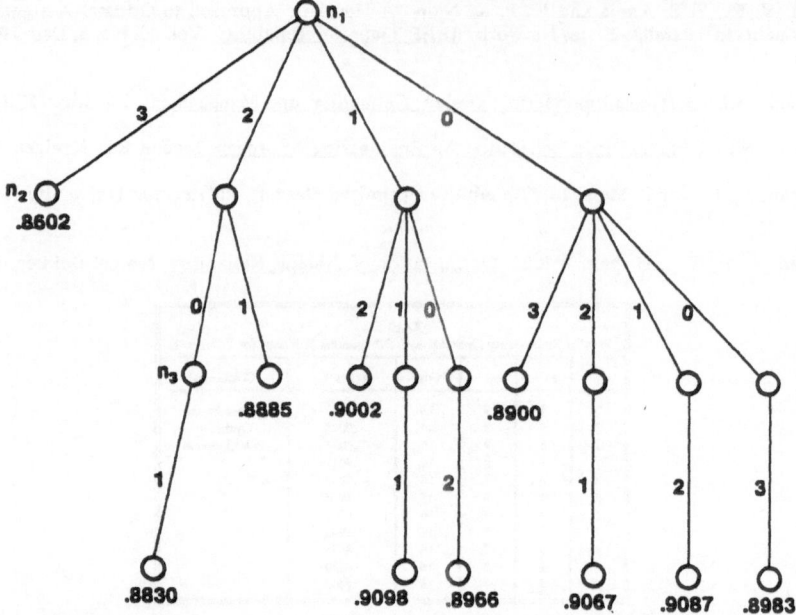

Fig. 1. A search tree for Example 1. Branches radiating from the start node represent the number of additional components assigned to subsubsystem one, $\Delta n_1 = 3,2,1,0.0$, and similarly for the second and third levels. Thus, the optimum solution is $\Delta n_1 = 1$, $\Delta n_2 = 1$, $\Delta n_3 = 1$ and R = 0.9098.

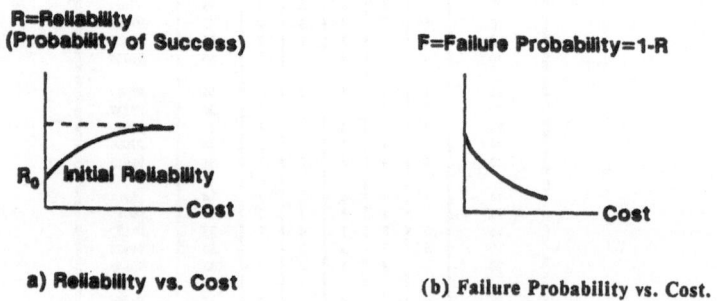

a) Reliability vs. Cost

(b) Failure Probability vs. Cost.

Fig. 2. Reliability and failure probability vs. cost.

AN EFFICIENT METHOD FOR PROBABILITY EVALUATION OF A FAULT-TREE

K.Odeh & N.Limnios
Division de Mathématiques Appliquées
Université de Technologie de Compiègne
B.P. 649, 60206 Compiègne cedex, France

Key Words-- Fault-tree, Factorization, Exact evaluation, Complexity

Abstract

This paper proposes a new method for fault-tree probabilities evaluation. The method is based on Shanon's factorization of boolean binary functions using the repeated basic events as pivot. Examples taken from the literature are used to show the efficiency of our method in comparison with other methods.

1. Introduction

It is well known that fault-tree probabilities evaluation is a NP-difficult problem. This evaluation follows two axes: a direct evaluation which gives us the exact value or an approximate evaluation by fixing a maximum error.
Our work concerns the exact evaluation , while, reducing the complexity of the problem. So we propose a new method which consists of factorizing the tree with respect to repeated and complementary basic events.
By introducing a strategy in the choice of these events, we compare this method with the other methods.

A fault-tree is an 1-graph strong connected without loop and without circuit. It consists of three kinds of basic components: AND gates, OR gates and basic events. The basic events are thought of as primitive inputs. The fault-tree are usually viewed graphically as trees whith the top-event (root of the tree) representing the event of primary interest, and the leaf nodes corresponding to the basic events. Thus AND and OR gates correspond to nonleaf nodes in the tree.
A cut set is a set of basic events which together cause the Top-event to occur.
Fault-tree probabilities evaluation consists of calculating the top-event probability occurrence given its primary event probabilies.

There is a big variety of approaches of fault-tree probabilities evaluation. The most of methods use minimal cut sets, usually followed by some kind of approximation scheme to bypass the often computationally horrendous problem posed by exact computation.
Tow types of truncation are used, one based on minimal cut sets size and the other on minimal cut sets probability[3].
If we are not interested by a qualitative analysis, it is profitable to work directly with elementary properties of probabilities and avoid the idea of cut sets altogether. We cite some of these

methods:
- The complete factorization of the fault-tree using Shanons's expression which generates the structure function in sums of disjoint product [5]
- The method of inclusion-exclusion based on the recursive application, starting from the top-event, of:

If A and B are s-independent

For 'OR' operator, we have:

$$P(A \cup B) = P(A) + P(B) - P(A) \times P(B) \qquad (1.1)$$

For 'AND' operator, we have:

$$P(A \cap B) = P(A) \times P(B). \qquad (1.2)$$
$$P((A \cup B) \cap C) = P(A \cap C) + P(B \cap C) - P(A \cap B \cap C) \qquad (1.3)$$

We cite TDPP[4] as algorithm based on this method.
- The method of recursive disjoint products which consists to generates recursively the structure function in sums of product. [2].

Complexity: If n is the number of gates in the fault-tree, the complexity of top-event probabilities evaluation, with respect to n, is linear if we have a simple tree and it is exponential if not.

Notation and Definitions

Repeated event : an event is repeated if there is more than one path from the root to this event. The number of these paths defines the order of the event $d(i)$.
Complementary event : two basic events are complementary if they are logical complement, ie, exactly one of the two must occur.
Node : a node refers to an AND gate, an OR gate, or a basic event in a fault-tree.
    The basic events corresponds to leaf nodes in the fault-tree.
R : repeated and complementary events set.
    $R = \{e_i \mid d(e_i) > 1 \text{ or } \exists e_j \text{ complementary to } e_i\}$.
Simple fault-tree : is a fault-tree for which R is empty.
No-simple fault-tree : is a fault-tree for which R is not empty.
$L(i)$: node level is the path length (number of vertices) between the root and the node i.
$NL(i)$: basic event level:

$$NL(i) = \begin{cases} L(i) & \text{if } d(i) = 1 \\ Min\{L_{i_j} : 1 \leqslant j \leqslant d(i)\} & \text{if } d(i) > 1 \end{cases}$$

x_i : indicator variable of the event e_i.
$$x_i = \begin{cases} 1 \text{ if occurrence of event } i \\ 0 \text{ if non-occurrence of event } i \end{cases}$$
T_i : The subtree of node G_i.
$\underline{x} = (x_1 , \ldots, x_n)$ $(._i\underline{x})$ if we fixe the ith component of vector $\underline{x}$, $(._{i_{..j}}\underline{x})$ if we fixe the ith and jth component of vector $\underline{x}$, etc . . .
$\varphi : \{0,1\}^n \to \{0,1\}$ Structure function of a fault-tree.
$$\varphi(\underline{x}) = \begin{cases} 1 \text{ if occurrence of Top event} \\ 0 \text{ other} \end{cases}$$

2. The proposed method

We try to simplify a no-simple fault-tree by applying Shanon factorization with respect to R events. This procedure leads us to have simple fault trees.
For $e_i \in R$, we can write:

$$\varphi(\underline{x}) = x_i \varphi(1_i, \underline{x}) + (1 - x_i) \varphi(0_i, \underline{x}) \qquad (2.1)$$

This factorization generates simple fault-trees which can be treated by simple methods[1].
Let $R = \{e_{i_1}, \ldots, e_{i_r}\}$ the set of repeated events, then we can demonstrate that the probability of top-event $E[\varphi(\underline{X})]$ (where E means expectation) is given by the following relation:

$$E[\varphi(\underline{X})] = \sum_{(\alpha_{i_1}, \ldots, \alpha_{i_r}) \in \{0,1\}^r} E\varphi(\alpha_{i_1}, \ldots, \alpha_{i_r}, \underline{X}) \prod_{k=1}^{r} EX_{i_k}^{\alpha_{i_k}} \qquad (2.2)$$

$$X_{i_k}^{\alpha_{i_k}} = \begin{cases} X_{i_k} & \text{if } \alpha_{i_k} = 1 \\ 1-X_{i_k} & \text{if } \alpha_{i_k} = 0 \end{cases}$$

In (2.2) $\varphi(\alpha_{i_1}, \ldots, \alpha_{i_r}, \underline{X})$ are strucure functions of simple fault-trees and expectations can be calculated by simple algorithms. Thus the product part of (2.2) can be calculated in linear time but the sigma part contrains 2^r terms.
The order in choosing the pivot has an important role in the factorization because a good choice can reduce the number of fault-trees generated.

What criteria should we use in the choice of the pivots to reduce the number of generated simple fault trees?
We introduce the following criteria:

1- We built the set A : $A = \{ e_i : e_i \in R \text{ and } NL(i) = Min\{NL(j) : e_j \in R\}\}$.
2- If $|A| = 1$ we choose this element of A.
   If not   we chooes i such that $d(i) = Max\{d(j) : e_j \in A\}$.
The algorithm is given in the Appendix.

3. Exemples

Let us take the no-coherent tree of the figure 1. Then, in the first step of factorization:

$R = \{e_2, e_3, e_6, e_7\}$
$A = \{e_2, e_3\}$
$NL(2) = 2 \qquad d(2) = 2$
$NL(3) = 2 \qquad d(3) = 3$

So, we select e_3 as a pivot.
In the figure 2, we have the factorization tree in which the leafs represent simple fault-trees.

4. The complexity of the algorithm

Let T be a tree with n gates and $|R| = r$. We can have a maximum of 2^r generated trees by the algorithm *prob_factoris* (see Appendix). So, the complexity is $o(2^r n)$.
Let us consider the case $r = \frac{n}{2}$. In this case, the complexity rate between our methods and standard one is:

$$t = \frac{2^{n/2} n}{2^n} = \frac{n}{2^{n/2}}$$

For example, if n = 100 then $t = 1.5 \times 10^{-6}$.
If all the basic events are repeated, then t = 1. It's obvious that for a small number of repeated events in a fault-tree, the proposed algorithm is efficient.

954

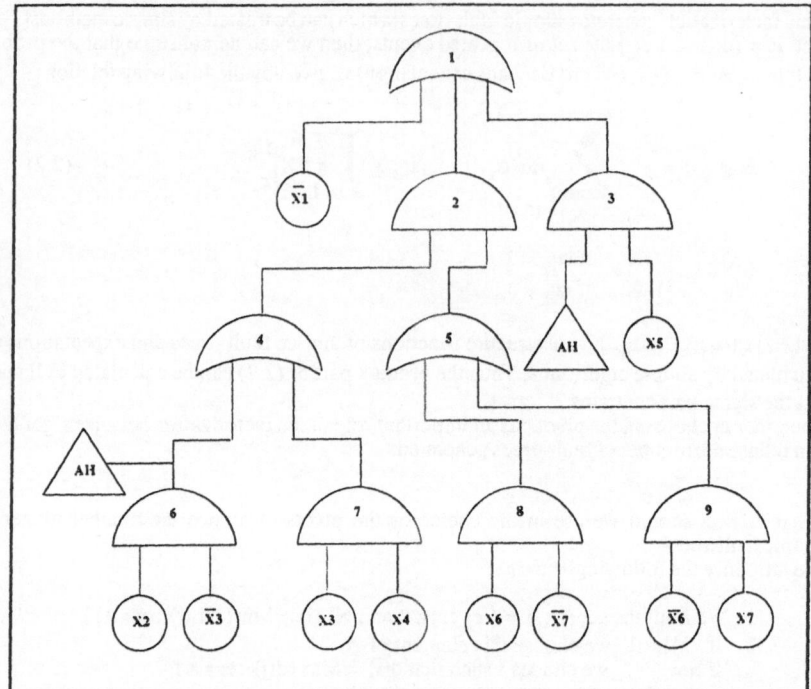

Figure 1.

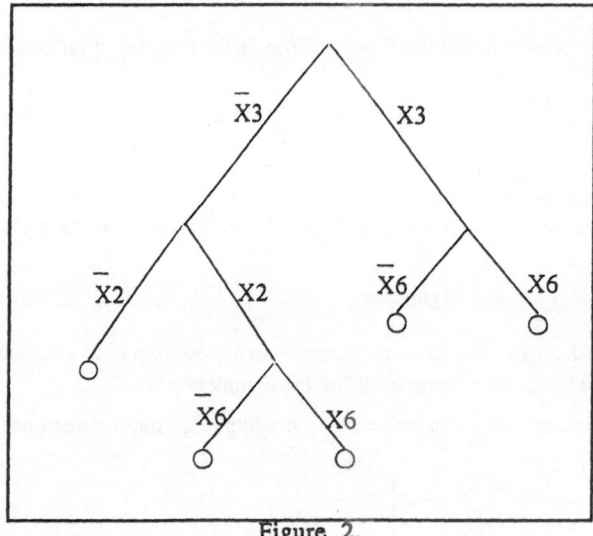

Figure 2.

To test the efficiency of our algorithm, we have implemented it with the algorithms of other methods on a Sun station in C language.
Table 1. contains the measures obtained for 6 test tree after a partial modularization.

	FACT		PDR		TDPP		PNCR	
	Time	Calls	Time	Calls	Time	Calls	Time	Calls
11(5)	6	10	16	25	9	46	19	-
19(6)	14	10	16	24	52	244	6	-
24(3)	12	3	237	39	167	648	147	-
13(2)	13	4	36	26	17	123	23	-
22(6)	14	10	16	24	49	244	7	-
18(9)	94	36	310	48	69	289	100	-

Table 1.

Where:
Time : Execution time in milliseconds.
Calls : The number of recursive calls of the function.
The trees are described by $n(m)$ where n represents the number of gates, m represents the number of repeated and complementary events.
FACT : Our method.
PDR : Recursive disjoint product method.
TDPP : Page and Perry method.
PNCR : Poincarré developpement (we consider only the first two terms in the development. The minimal cut set obtention time is not taken into account).

In conclusion FACT is faster. We can observe that in the case of fault-tree 3 and 4 the repeated events number is small, FACT gives the best result. We note the efficiency of the pivot criteria choice because, the calls number is à 2^m in 4 trees among 6.

REFERENCES

[1]   K.K.Lee, "A compilation technique for exact system reliability".
IEEE Trans. Reliab., Vol R-30, 1981 Aug, pp. 284-288.

[2]   N. Limnios "Arbres de défaillance",
Editions Hermes, 1991.

[3]   M. Modarres, H. Dezfuli, " A truncation methodology for evaluating large fault trees", IEEE Trans. Reliab., Vol. R-33, 1984 Oct., pp 325-328.

[4]   L.B.Page, J.E. Perry, "An algorithm for exact fault-tree probabilities without cut sets", IEEE Trans. Reliab. Vol. R-35, N°5, Dec. 1986, pp. 544-558.

[5]   W.G.Shneeweiss, "Disjooint boolean product via Shanon's expansion", IEEE Trans. Reliab. Vol. R-33, N°4, Oct. 1984, pp. 329-332.

Appendix

The   algorithm

1. Factorization   method
   To introduce the factorization algorithm, we use the followed functions:'

- *Tikrar(porte, liste_pivot)*
  1. Porte: the code of the gate.
  2. Liste_pivot: list of the basic events and their associated values.
  Tikrar: returns the domain of the gate *porte* by replacing the values of the basic events represented by
        liste_pivot.
  The returned domain contains:
                  1- the basic event code (i).
                  2- NL(i).
                  3- d(i).

- *Choix_pivot(liste)*
  Liste: basic event set.
  Choix_pivot: returns a basic event shoosed by the criteria of our proposed method.
- *Prob_factoris(porte, liste_pivot)*
  Porte: the code of the gate.
  Liste_pivot : it has the same significance as in Tikrar procedure.
  Prob_factoris: returns the probabilitie of occurrence of the gate *porte* corresponding to liste_pivot.
      Prob_factoris(porte, liste_pivot)
            Domaine = Tikrar(porte, liste_pivot)
            Pivot = Choix_pivot(Domaine)
            if (pivot = 0) then { The case in which the tree becomes simple}
                  Prob_factoris < - - calculte the probability by a simple method
        else
                Liste_pivot_1 = liste_pivot + < pivot,1 >
                Liste_pivot_2 = liste_pivot + < pivot,0 >
                prob_factoris =
                Pb(pivot) * Prob_factoris(porte, liste_pivot_1)
                                  +
                (1-Pb(pivot)) * Prob_factoris(porte, liste_pivot_2)

OPTIMAL STRATEGIES FOR THE PREVENTIVE MAINTENANCE

OF REAL-TIME REPAIRABLE SYSTEMS

Eugene Levner and Dror Zuckerman
The Hebrew University of Jerusalem
School of Business Administration
Mount Scopus, 91905 Jerusalem, Israel

George Gens
Moscow Institute of Automation
Laboratory of Expert Systems
4b, Sivashskaya St., 113149 Moscow, Russia

Alexander Ptuskin
Central Economic Mathematical Institute
Russian Academy of Sciences
32, Krasikova St., 117418 Moscow Russia

INTRODUCTION

Consider a complex technological system (for example, an aircraft, a submarine, a nuclear reactor) which is composed of independent stochastically deteriorating components, the breakdown of which, albeit rare, can lead to devastating consequences.

We are interested in finding optimal on-line strategies for the periodic maintenance of such systems to ensure the repair/replacement of risky components in order to prevent the catastrophic breakdown. The problem posed by this task can be decomposed in two subproblems:

Subproblem 1: For the system in good working order, to localize and recondition the most risky components in order to maximize the reliability of the system.

Subproblem 2: For the system known to be faulty, to schedule the search and repair of a defective component in order to minimize the cost of search/repair.

For Subproblem 1 we develop a fast on-line hybrid algorithm providing an almost-optimal solution running in "almost-linear" time. For Subproblem 2 we find a fast on-line local algorithm which is proved to be globally-optimal.

Results of computational experiments are presented and discussed.

This research was supported by the Ministry of Science and Technology of Israel (Research Project "Maintenance Strategies with General Degree of Repair for Stochastically Failing Equipment", Grant # 3864-1-91).

PROBLEM DESCRIPTION

Extending the maintenance models proposed by Kadane [5] and Zuckerman [10] we consider a repairable technological system composed of n independent stochastically deteriorating components. In order to prevent the system's breakdown, preventive maintenance actions are performed periodically with period T, the value of T being determined exogenously by standard methods of reliability theory (see, e.g., Barlow and Proschan [1]). The sequence of preventive maintenance actions is interrupted when a controller discovers that the system is faulty, and the defective component is identified and repaired.

Specifically, the following maintenance procedure is employed: The first preventive maintenence action is performed at t=0. It starts with malfunction diagnostics which predicts, for all i=1,...,n:

p_i, the probability that component i will fail within interval [0,T];
D_i, the decrement of probability p_i which is expected as a result of inspecting and repairing component i; and
t_i and c_i, the duration and cost of inspecting/repairing component i.

Next, using these data, a real-time schedule (i.e., a schedule subject to given time and cost limits) is found determining which of the components are to be inspected and repaired so as to ensure the maximal reliability of the system within interval [0,T]. The system's components are then inspected according to the schedule and reconditioned if a malfunction is discovered.

If, up to time t=T, the system is found to be in good working order, the next cycle starts at t=T, that is, the preventive maintenance action is performed again in order to predict new, age-dependent parameters p_i and D_i, for interval [T,2T], and to arrive at the next schedule of inspecting/correcting components that will maximize the system's reliability on [T,2T].

However, if at some random $t=T_1<T$ the system is found to be out of order (but the identity of the defective component is unknown), a series of inspections of risky components is initiated at time T_1 in order to identify and repair the faulty component with the minimal cost. Thereafter, the system begins its life anew, and the preventive maintenance cycle is renewed.

Thus, for each cycle, the problem is decomposed into two subproblems: (1) for the system in good working order, to localize and recondition the most risky components; (2) for the system known to be out of order, to schedule the search/repair of a faulty component.

ALMOST-OPTIMAL HYBRID ALGORITHM FOR SUBPROBLEM 1

In Subproblem 1, a schedule for the preventive maintenance of the most risky components may be presented as a finite sequence of components, s, to be inspected one after another, and, if necessary, repaired within the given time limit, $T_{p.m.}$, and the given budget, B.

We assume that failures are extremely rare, and failure probabilities p_i are very small, $p_i << 1/n$. Then the reliability of the system in any period $[kT, (k+1)T]$, $k=0,1....$, is approximated by

$$1- \sum_{i \in E} p_i \text{ (where } E=\{1,...,n\}\text{), before preventive maintenance action,}$$

and

$$1- \sum_{i \in s} (p_i-D_i)- \sum_{i \in E \backslash s} p_i, \text{ after preventive maintenance action.} \qquad (1)$$

(Remind that in the model, p_i values are dependent on the age, i.e., on k, but for simplicity of presentation, we do not supply p_i by index k).

Taking (1) into account, the problem of maximizing the system's reliability within a cycle is reduced to the following knapsack-type integer programming problem:

$$\text{Maximize } D(x) = \sum_{i=1}^{n} D_i x_i$$

$$\text{subject to } t(x) = \sum_{i=1}^{n} t_i x_i \leq T_{p.m.},$$

$$\sum_{i=1}^{n} c_i x_i \leq B, \quad x_i = 0 \text{ or } 1, \quad i=1,2,...n.$$

(We interpret $x_i=1$ iff component i is to be included into s). In what follows, we shall be interested in studying a special class of this optimization problem in which D_i, c_i and t_i are dependent: $D_i = kt_i$, $c_i = mt_i$, $i=1,...n$, where k and m are known (this problem, known as "the subset-sum problem", is denoted below as problem P):

$$\text{Maximize } D(x)= \sum_{i=1}^{n} kt_i x_i \text{ s.t. } t(x) \leq T_o=\min(T_{p.m.},B/m); \quad x_i = 0 \text{ or } 1.$$

The design of almost-optimal algorithms for the subset-sum problem is generally based on either partitioning items t_i into "large" and "small" ([4], [6]), or eliminating close solutions ([2], [8], [9]). We present here an improved hybrid algorithm, HA(e) which is based on the combined use of the both approaches.

We shall use the following notations:

x^*: the optimal solution of problem P; $D^*=D(x^*)$;

$\bar{x}$: the e-approximate solution of problem P: $D^*-D(\bar{x}) \leq eD^*$; $\bar{D}=D(\bar{x})$;

b: the bound such that $b \leq D^* \leq 2b$, which is found as follows:

$$b = kQ, \quad Q = \max(\sum_{i=1}^{z} t_i; \ \max_i\{t_i\}), \text{ where } z = \max\{j : \sum_{i=1}^{j} t_i \leq T_0\}$$

GL(e): the fast e-approximation algorithm exploited in A(e) as a subprocedure (for its description see ([2], [8]).

Auxiliary Algorithm A(e)

Input: n; (t_i), $1 \leq i \leq n$; T_0, e; Q.

Output: $\bar{x}$, an e-approximate solution with value $\bar{D}=D(\bar{x})$.

Step A1. [Forming lists L and S of "large" and "small" items].

Set $t:=eQ$; $K:=et/2$. Set $L:=\{i|t_i>t\}$; $S:=E \backslash L=\{s_1, s_2, \ldots, s_{|S|}\}$;
$m=\sum_{i \in L} t_i$. If $m \leq T_0$ go to Step A2, otherwise, go to Step A3.

Step A2. [Forming a solution $\bar{x}$]. Determine $d=\max\{j: m + \sum_{k=1}^{j} t_{s_k} \leq T_0\}$.

Form $x^0=\{x_i^0=1, \text{ if } i \in LU\{s_1,s_2,\ldots,s_d\}; \ x_i^0=0, \text{ otherwise}\}$.
Set $\bar{x}:=x^0$, $t:=t(\bar{x})$. Stop.

Step A3. [Using subprocedure GL(e) to "large" items].

If $m > T_0$, for t_i ($[2^k t, 2^{k+1}t]$, $k=0,1,\ldots, [\log_2(T_0/t)]$,
set $q_j:=\lfloor t_j/2^k K \rfloor 2^k$. Among the items with the same q_i values
choose $\lceil T_0/Kq_i \rceil$ items with smallest t_i values, and discard all
others as superfluous. Denote remaining "large" items by RL.

Use GL(e) for solving the following problem RLP:

$$\text{Problem RLP: Maximize } t_{RL}(x)=\sum_{i \in RL} t_i x,$$
$$\text{subject to } \sum_{i \in RL} t_i x_i \leq T_0,$$
$$x_i=0 \text{ or } 1, \quad i \in RL.$$

Let $\bar{x}_{RL}=(\bar{x}_i)$ be a solution of problem RLP generated by GL(e).

Step A4. [Forming $\bar{x}$]. Determine $r= \max\{j: t_{RL}(\bar{x})+\sum_{k=1}^{j} t_{s_k} \leq T_0\}$.

Form $x^1=\{x_i^1=\bar{x}_i \text{ if } i \in RL; \ x_i^1=1 \text{ if } i \in \{s_1,\ldots,s_r\}; \ x_i^1=0, \text{ otherwise}\}$.
Set $\bar{x}:= x^1$, $\bar{D}:=kt(\bar{x})$. Stop.

Algorithm HA(e)

Step 1. [Initialization].

Step 1.1. [Solving the continuous problem]. Solve the following continuous problem associated with problem P:

Maximize $t(x)$, subject to $t(x) \leq T_o$, $0 \leq x_i \leq 1$, $i=1,\ldots,n$.

Step 1.2. [Integrality test].
If the continuous solution is integer, stop.

Step 1.3. [Finding an upper bound of the optimum].
Otherwise, find $\lfloor T_o \rfloor$, an upper bound of the optimum.

Step 1.4. [Performing the e-test].
Find a current "champion" solution, x_o, by exploiting A(e), with e_o fixed, say, $e_o=0.1$. Let e be an allowed error.
If $\lfloor T_o \rfloor - t(x_o) \leq et(x_o)$, stop.

Step 2. [Reduction of variables].

Step 2.1. [Finding the pivot element].
Find the "pivot" element, x_p, $p \in \{1,\ldots,n\}$, such that

$$\sum_{i=1}^{p-1} t_i \leq T_o \text{ and } \sum_{i=1}^{p} t_i > T_o.$$

Step 2.2. [Performing new e-tests].
For $i=1$ to p (resp., $i=p+1$ to n) set $x_i=0$ (resp., $x_i=1$);
find $C(i)$, the optimal solution of the continuous problem associated with problem P, with x_i fixed;
find $x_o[i]$, a "current champion" solution, with x_i fixed, (by exploiting A(e) again with error value e_o fixed).

If $\lfloor T_o \rfloor - t(x_o[i]) \leq et(x_o[i])$, stop.

Step 2.3. [Reduction of variables].
For $i=1$ to p:
    If $\lfloor C(i) \rfloor - t(x_o[i]) \leq et(x_o[i]$ then $x_i^*:=1$;
    $E:=E\setminus\{i\}$; $n:=n-1$; $T_o:= T_o -t_i$.

For $i=p+1$ to n:
    If $\lfloor C(i) \rfloor - t(x_o[i]) \leq et(x_o[i]$ then $x_i^*:=0$; $E:=E\setminus\{i\}$; $n:=n-1$.

Step 3. [Solving the problem in residual variables].
For residual variables remained after the reduction at Step 2, solve e-approximately the corresponding knapsack problem, using Algorithm A(e). Stop.

Computational complexity of the algorithm: The time and space of algorithm HA(e) at Steps 1 and 2 is $O(n)$. The time and space of algorithm A(e) at Steps A1, A2 and A4 is $O(n)$. Cardinality $|RL|$ of the list RL at Step A3 does not exceed $O(1/e^2)$, due to the rounding-off of the values t_i [6]. Procedure GL(e) is known to have $O(\min(n/e, |L|/e))$ time and space ([2], see also [8]). Therefore, A(e) at Step A3 has $O(1/e^3)$ time. Then the total time of A(e) at Steps A1-A4 is, at most, $O(n + |L|/e) = O(n + 1/e^3)$. Using the backtracing scheme described by Lawler in [6], at Step 3 we obtain the space bound of $O(n+1/e^2)$. Thus, in the worst case, A(e) (and hence HA(e)) has the "almost-linear" time bound $O((\min(n/e, n+1/e^3)))$, and space bound $O(\min(n/e, n+1/e^2))$.

Computational experiments: For n between 30 and 1000, t_i values were selected randomly, independently and in accordance with a uniform law between 1 and 100; e values were selected between 0.01 and 0.10. For each n, 50 samples were solved using IBM PC AT (12 MHz, 80286 CPU).

As Table 1 shows, for e>0.01, as n becomes larger, our hybrid algorithm works better (the average and maximal number of residual variables decreases). The same tendency manifests itself as e becomes larger, and for e>0.05 there were virtually no residual variables remained after Steps 1 and 2.

Table 2 gives numerical results showing the accuracy with which algorithm A(e) solves the subset-sum problem with 10 to 40 residual variables at Step 3 of HA(e). Algorithm A(e) has far fewer actual errors than the theoretical (worst-case) guaranteed error e. These results agree with the computational experiments presented by Martello and Toth for algorithm GL(e) in [8].

Table 1. Behavior of Hybrid Algorithm HA(e)

n	e=0 ANRV	e=0 MNRV	e=0.01 ANRV	e=0.01 MNRV	e=0.02 ANRV	e=0.02 MNRV	e=0.03 ANRV	e=0.03 MNRV	e=0.05 ANRV	e=0.05 MNRV
30	4.4	13	1.0	9	0.44	5	0.24	4	0.08	3
50	7.1	16	0.46	8	0.14	5	0.08	3	0.06	2
100	7.1	17	0.20	6	0.06	3	0.0	0	0.0	0
200	9.9	21	0.06	3	0.04	2	0.0	0	0.0	0
300	14.7	28	0.0	0	0.0	0	0.0	0	0.0	0
400	13.8	37	0.0	0	0.0	0	0.0	0	0.0	0
500	14.5	31	0.0	0	0.0	0	0.0	0	0.0	0
1000	4.5	40	0.0	0	0.0	0	0.0	0	0.0	0

ANRV=average number of residual variables;
MNRV=maximal number of residual variables.

Table 2. Accuracy of Algorithm A(e)

	e=0.01	e=0.02	e=0.03	e=0.05	e=0.10
n	a.e.	a.e.	a.e.	a.e.	a.e.
10	exact	exact	exact	0.01	0.02
20	exact	exact	exact	0.01	0.02
30	exact	exact	0.01	0.02	0.03
40	exact	0.01	0.01	0.03	0.03

e=theoretical (worst-case) error;
a.e.=actual error;
n=number of residual variables.

OPTIMAL ON-LINE ALGORITHM FOR SUBPROBLEM 2

At some random $t=T_1<T$, the system is found to be out of order (but the component responsible for the failure is not known). A series of inspections of risky components is performed in order to identify and repair the faulty component. Our objective is to find optimal inspection strategy that will minimize inspection/repair costs.

For each component i, the following parameters are known:

prior failure probability b_i;
probability a_i^j of overlooking the failure in the faulty component i while inspecting it for the j-th time;
time t_i^j and cost c_i^j of inspecting component i for the j-th time;
time u_i and cost r_i needed to repair component i; and
"discounting" factor f.

The components are inspected sequentially until the faulty component is identified and repaired. The positive probability a_i^j of overlooking the failure while inspecting a faulty component indicates that the inspections may be imperfect and, therefore, each component may have to be inspected more then once.

Each sequential inspection strategy specifies an infinite sequence s=(s[1], s[2],..., s[n],...) denoting that the component inspected at the n-th step of s, is s[n]. Specifically, the sequence s defines the following procedure: start by inspecting component s[1], and if it is discovered to be in good working order, select s[2], and so forth. The search is terminated when the faulty component is discovered.

Assume, w.l.g., that at the k-th step of s component i is searched for the j-th time. For any given s, we shall use the following notation:

$a_{s[k]}$ and $c_{s[k]}$: The overlooking probability and the search cost for this component i (being looked for the j-th time);

$P_{[k,s]}$: The probability of the failure being detected at the k-th step of s (assuming that s defines the strategy employed):

$$P_{[k,s]} = 1 - b_{s[k]}(1 - a_{s[k]});$$

$t_{[k,s]}$: Time spent for inspecting component s[k] in s.

$R_{[k,s]}$: Expected "discounted" cost of inspecting/repairing s[k] in s:

$$R_{[k,s]} = (1 - a_i^j) b_i \exp[f(t_i^j + u_i)] r_i + c_i^j \exp(f t_i^j). \qquad (2)$$

By using a simple probabilistic argument (see, e.g., [3], [7]), it can be shown that the total expected cost of inspecting/repairing under strategy s is given by

$$V(s) = \sum_{k=1}^{\infty} R_{[k,s]} \exp\left(f \sum_{m=1}^{k-1} t_{[m,s]}\right) \prod_{m=1}^{k-1} (1 - P_{[m,s]}), \qquad (3)$$

where, by definition, $\sum_{m=1}^{0} t_{[m,s]}) \prod_{m=1}^{0} (1 - P_{[m,s]}) = 1.$

The optimal policy s^* minimizes V(s) over all possible (infinite) sequences of elements in E any of which may be repeated infinitely many times. This model is an infinite-horizon generalization of the Granot-Zuckerman search model studied in [3].

Let us define, for all values i and j, the following ratio

$$r_{ij} = R_i^j / [1 - (1 - Q_{ij}) \exp(f t_i^j)],$$

where parameters $R_i^j = R_{[k,s]}$ (see (2)) and Q_{ij} are:

$$Q_{ij} = 1 - b_i(1 - a_i^j); \quad i = 1, 2, \ldots, n; \quad j = 1, 2, \ldots \qquad (4)$$

Notice that parameters Q_{ij} do not depend of any sequence s.

The optimal search strategy is determined by the following theorem:

THEOREM 1. Let the problem of minimizing objective (3) be regular (it means, by definition, that ratio r_{ij} does not decrease as j increases for any i). Let the values of ratio r_{ij}, for all i and j, be arranged in non-decreasing order, and the sequence s^* be such that:

      if a ratio r_{ij} is the k-th smallest value in the ordering,
      then the k-th step in s^* is the j-th search of component i,
      k=1,2,..., i=1,...,n.

Then the sequence s^* is optimal.

Proof. First, we have to show that the problem of minimizing costs V(s) is well-defined for any given input data. In fact, the minimum in this problem can not be greater than the expected cost of the procedure under which components 1 through n are searched cyclically until the fault is located. Using standard probability arguments, we obtain that the latter cost cannot be more than

$$n \max_{1 \le i \le n} R_i \exp[f(\sum_{i=1}^{n} t_i)]/(1 - \prod_{i=1}^{n} (1-b_i(1-a_i))) < \infty .$$

Clearly, the sequence s^* is feasible. Indeed, since the problem is regular, the procedure s^*, due to its structure, for any component i, first assigns its j-th inspection, and only thereafter does it assign the next, (j+1)-th, inspection, j=1,2,3,....

Assume now that for a given optimal sequence r^*, $r^*=s^*$, we have that, at a certain step k,

$$R_{[k,r^*]})]/[1-(1-Q_{[k,r^*]}\exp(ft_{[k,r^*]})] <$$

$$R_{[k+1,r^*]})]/[1-(1-Q_{[k+1,r^*]}\exp(ft_{[k+1,r^*]})]. \quad (5)$$

Assume, w.l.g., that component i is inspected at the k-th step of sequence s^*, and component j is inspected at the (k+1)-th step (notice that i=j, due to relation (3)). "Interchange" the i and j, that is, consider a new sequence, q, in which component j is inspected at the k-th step and component i is inspected at the the (k+1)-th step, all other steps remaining as in sequence r^*.

From (3) and (5), we immediately obtain $V(q)-V(r^*)<0$, which contradicts the optimality of r^*. This completes the proof.

The regularity condition, though restrictive, is realized in various search scenarios, for example, in the independent search [5].

Notice that if we set all $a_i=0$, Theorem 1 degenerates into the O(n log n)-time algorithm by Granot and Zuckerman [3].

CONCLUDING REMARKS

This paper introduced a new class of preventive maintenance problems and demonstrated that they can be solved optimally, or almost optimally, by fast on-line algorithms.

We believe that the above combinatorial approach can be applied to effectively solve a wide class of preventive maintenance problems involving a variety of inspection scenarios: multiple simultaneously occuring failures; resource-constrained and precedence-constrained inspections; inspections of changeable and intelligent systems, etc.

Along with developing efficient numerical methods of optimal preventive maintenance, perhaps, no less interesting is the theoretical study related to the classification of these problems from the viewpoint of their computational complexity.

REFERENCES

[1] Barlow, R.E. and Proschan, F. (1975) Statistical Theory of Reliability and Life Testing, Holt, Rinehart and Winston, N.Y.

[2] Gens, G.V. and Levner, E. V. (1978) Approximation algorithms for certain universal problems in scheduling theory. Engineering Cybernetics. Soviet Journal of Computer and Systems Science, 6, 31-36.

[3] Granot, D. and Zuckerman, D. (1991). Optimal sequencing and resource allocation in research and development projects. Management Science, 37, 2, 140-156.

[4] Ibarra, O.H. and Kim C.E. (1975) Fast approximation algorithms for the knapsack and sum of subset problems. Journal of ACM, 22, 463-468.

[5] Kadane, J.B. (1969) Discrete search and the Neyman-Pearson lemma. Journal of Mathematical Analysis and Applications, 22, 156-171.

[6] Lawler, E.L. (1979) Fast approximation algorithms for knapsack problems. Mathematics of Operations Research, 4, 339-356

[7] Levner, E., Vainberg Y., and Zuckerman D. (1993) Optimal sequencing of R&D projects with learning, Presented at the National Meeting - 1993 of Operations Research Society of Israel, Naharia, May 9-10.

[8] Martello, S. and Toth, P. (1990) Knapsack Problems. Algorithms and Computer Implementations. Wiley, New York.

[9] Sahni, S. (1975) Approximate algorithms for the 0-1 knapsack problem. Journal of ACM, 22, 115-125.

[10] Zuckerman, D. (1989) Optimal inspection policy for a multi-unit machine. Journal of Applied Probability, 26, 543-551.

SIMULATION OF STRUCTURAL MEMBERS TAKING INTO ACCOUNT THE MATERIAL DISTRIBUTION AND THE CORRELATION.

H. M. Kessel*, P. Haller**, F. Bertolino***

* Labor für Holztechnik, FH Hildesheim, Germany
** Department of Civil Engineering, IBOIS, EPF Lausanne, Switzerland
***Department of Civil Engineering, IMAC, EPF Lausanne, Switzerland

Abstract

Within the scope of a Swiss national research project, the load carrying behaviour of one hundred Swiss spruce columns was examined. The columns were chosen randomly from different regions of Switzerland and tested under combined axial and lateral loading.

The columns were modelled individually in a geometrically and physically nonlinear finite element analysis using the statistical data. In order to reduce the computational effort required by the complex finite element model, a probabilistic approach by means of the Point-Estimate-Method was chosen. By only a few computations this method permitted the determination of the normal distribution of the load capacity. The method considered the effect of scattering and correlation of the input data.

1. Introduction

A geometrically and physically nonlinear two dimensional finite element was developped [1] on the basis of the B2000 program system in order to take into consideration the biaxial stress situation. This plane element permits a realistic modelling of the geometrical and statical boundary conditions.

Moreover, the present study follows a reliability based approach in the structural analysis in civil engineering that take into account the scattering of the data as well as the correlations. According to the complex nonlinear model and the important scattering of material properties of construction lumber, preference was given to the Point-Estimate-Method [2][3][4] over other approaches that are more time consuming.

Photograph 1 Use of timber in large structures

2. Experimental study

The experimental set-up and the load deflection diagram are shown in Figure 1 and 2 respectively. The axial force was kept constant during the test. Three test series were done for axial forces of 20, 40 and 60 kN. The determination of the Young's modulus was done following the recommendation of ISO 3133 and CIB-RILEM 3 tt-3.

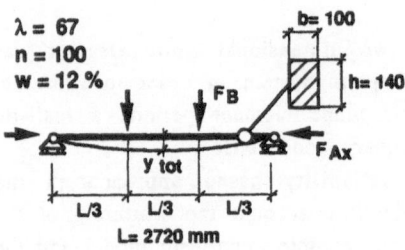

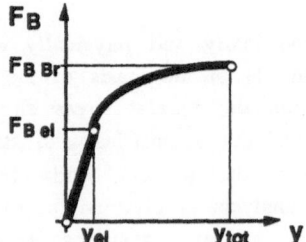

Figure 1 Experimental set-up

Figure 2 Load deflection curve

2.1. Determination of the Material properties

The determination of the material properties was done by means of a separate program individually for each column. This program assumed a linear material behaviour in the traction zone and a nonlinear behaviour based on equation (1) in the compression zone.

$$\sigma = E\varepsilon - A\varepsilon^n \qquad (1)$$

The deformation of the beam section was increased incrementally until the internal forces were in equilibrium with the external moment and the axial force. Hence, the stresses at the boundary were taken as compression and tensile strength in the finite element model discussed later on. The yield stress in the compression zone was determined following equation (2).

$$\sigma_{Drel} = \frac{F_{Bel}}{F_{BBr}} \sigma_{DrBr} \qquad (2)$$

Photograph 2 Rupture of structural timber

2.2. Statistical data and correlation matrix

The statistical as well as the computed data are shown in Figure 4. In addition, the correlations between the different parameters and their coefficients are given in the so called correlation matrix. This matrix will play a central role in the following.

3. Numerical study

There is a tendancy in structural analysis in civil engineering for a reliability based approach that consider the scattering of any structural data such as load, geometry and material properties.

Different methods have been developped to achieve these requirements. First, the Monte-Carlo-Method is proposed to consider the scattering of the input data by means of randomly chosen values which are then treated deterministically. The disadvantage of the Monte Carlo method is its slow convergence which is detrimental for complex models. Furthermore, an additional programming effort is to be done in order to take into consideration the correlation of the data.

Another probabilistic approach is the First-Order-Second-Moment method (FOSM) that is based on a multivariate Taylor expansion around the mean values of the variables. Unfortunately, the correlations between these variables need an important computational effort. In many cases the partial derivations are difficult to provide.

According to the complexity of the finite element model studied here, an alternative approach to the above mentionned procedures was found in the Point-Estimate Method.

3.1. The Point Estimate Method

This method is based on the assumption of the normal distribution of the variables and provides a normal distribution of the function, in our case the ultimate load, after 2^n simulations for n variables. These are already correlated which represent an enormous advantage in the case of timber.

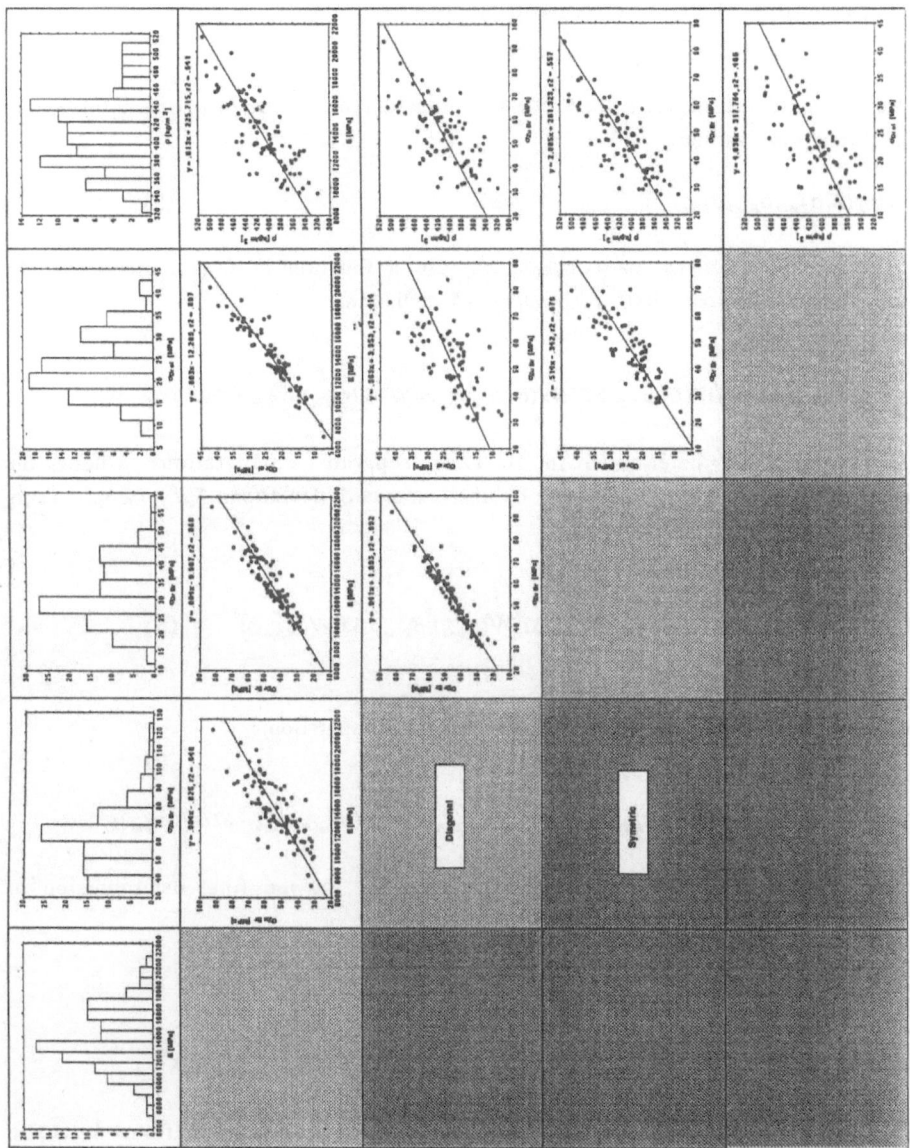

3.2. Application of the Point Estimate Method

In our case the ultimate load F_{Br} is a function of 4 variables (the influence of the density is ignored).

$$F_{Br} = f(E, \sigma_{ZuBr}, \sigma_{Drel}, \sigma_{DrBr}) \qquad (3)$$

The partial ultimate loads $F_{Br}\ {\pm\pm\pm\pm}$ are a function of a permutation of the mean values plus or minus their standard deviation.

$$\qquad (4)$$

$$F_{Br}\ {\pm\pm\pm\pm} = g(E \pm STDV(E), \sigma_{ZuBr} \pm STDV(\sigma_{ZuBr}), \sigma_{Drel} \pm STDV(\sigma_{Drel}), \sigma_{DrBr} \pm STDV(\sigma_{DrBr}))$$

$F_{Br}\ {\pm\pm\pm\pm}$ are determined in 16 (2^4) separate computations where the variables are reduced or increased by their standard deviation. For example using as input data the tupel:

$$\{E \pm STDV(E), \sigma_{ZuBr} \pm STDV(\sigma_{ZuBr}), \sigma_{Drel} \pm STDV(\sigma_{Drel}), \sigma_{DrBr} \pm STDV(\sigma_{DrBr})\}^T \qquad (5)$$

F+-+- is obtained by means of a finite element computation.

$$\qquad (6)$$

$$F_{Br}\ {+-+-} = g(E + STDV(E), \sigma_{ZuBr} - STDV(\sigma_{ZuBr}), \sigma_{Drel} + STDV(\sigma_{Drel}), \sigma_{DrBr} - STDV(\sigma_{DrBr}))$$

The partial ultimate loads have to be weighted for the final determination of F_{Br} by means of the correlation coefficients.

The evaluation of the weighting factors is done by equation (7).

$$p{\pm\pm\pm\pm} = \frac{1}{16}(1 \pm \rho_{12} \pm \rho_{13} \pm \rho_{14} \pm \rho_{23} \pm \rho_{24} \pm \rho_{34}) \qquad (7)$$

The sign of the correlation coefficients r_{ij} (i,j = {1...4}) result from the basic arithmetic sign rule. For instance, p++-- leads to $r_{12} = + \cdot + = +$, $r_{23} = + \cdot - = -$, $r_{34} = - \cdot - = +$. Due to the symmetry of the weighting factors $p{\pm\pm\pm\pm}$ only (2^{n-1}) factors have to be determined. For instance, the weighting factor for F+-+- is determined by:

$$p+-+-=\frac{1}{16}(1-\rho_{12}+\rho_{13}-\rho_{14}-\rho_{23}+\rho_{24}-\rho_{34}) \qquad (8)$$

If the result of a weighting factor is negative, it is supposed to be zero and the remaining factors are normalized.

For the case of a normal distribution of the ultimate load, its mean value $E[F_{Br}]$ can be determined using equation (9). The standard deviation (STDV) of the distribution results from equation (10).

$$E[F_{Br}] = p++++F_{Br}, +++++p+++-F_{Br}, +++-...+p----F_{Br}, ---- \qquad (9)$$

$$STDV[F_{Br}] = \sqrt{E[F_{Br}^2] - E[F_{Br}]^2} \qquad (10)$$

3.3. The finite element model

The finite element model is based on the assumption of a monotonous material law shown in Figure 3.The following relationship has been adopted for this law:

$$
\begin{aligned}
\varepsilon_1 &= \varepsilon_{DrBr} = 3.5\varepsilon_{Drel} \\
\varepsilon_2 &= \varepsilon_{Drel} + 0.75(\varepsilon_{DrBr} - \varepsilon_{Drel}) \\
\varepsilon_3 &= \varepsilon_{Drel} + 0.50(\varepsilon_{DrBr} - \varepsilon_{Drel}) \\
\varepsilon_4 &= \varepsilon_{Drel}
\end{aligned}
\qquad (11)
$$

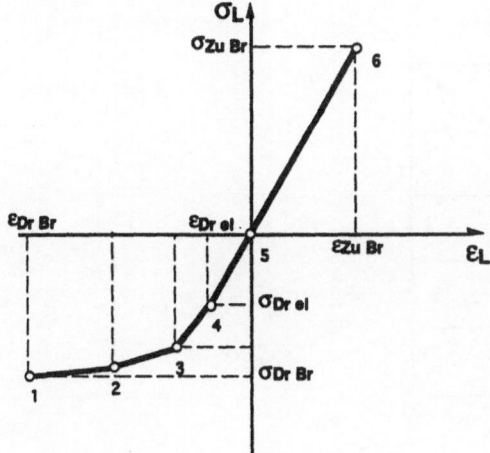

Figure 3 Behaviour law

The numerical study was done by means of the element mesh shown in Figure 5. According to the calibration of the model, the 100 columns were computed individually. The failure was predicted by means of a Tsai-Wu criteria.

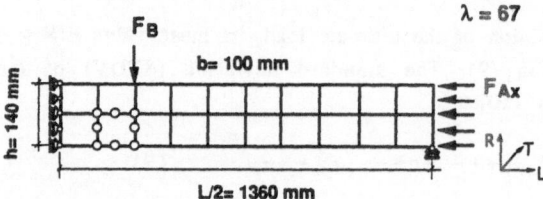

Figure 5 Finite element mesh

4. Resultats

4.1. Individual computations

The numerical results of the 100 columns is given in Figure 6. The simulated ultimate load showed good correlation with experimental results. The prediction of the deflection especially for columns presenting a noticeable deflection prior to failure is inaccurate which is due to an insufficient plastic behaviour of the material law used in equation (1).

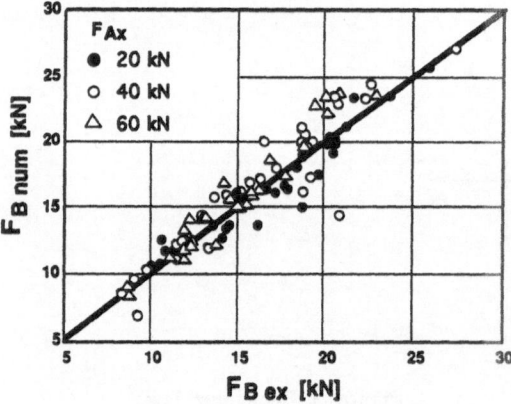

Figure 6 Comparison between numerical and experimental results

4.2. Results of Point Estimate Method

The results of the Point Estimate method are given in Figure 7 for an axial load of 20 kN. The comparison shows smaller mean values and a more important standard deviation. The reason for this phenomena is due to the fact that the exponential behaviour law in equation (1) provides not enough plasticity for the compression zone which leads numerically to a premature failure of the column. Moreover, the parameters of the material law had to be taken from literature that add another uncertainty.

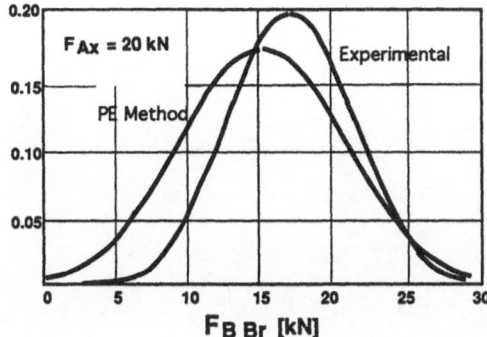

Figure 7 Experimental and numerical normal distribution of the ultimate load

5. Conclusion

The load carrying behaviour of 100 spruce columns under combined loading was studied experimentally and numerically.

A physically and geometrically nonlinear finite element model was developped in order to simulate the load carrying behaviour of the columns.

The scattering of the input data as well as their correlations regarding a reliability based approach was taken into account by means of the Point Estimate Method. The advantages of this method were discussed. The results showed a satisfactory correlation with experimental data, considering the very few computations that are necessary for this method.

6. References

[1]   F. Bertolino, internal report 1990, EPF Lausanne DGC, IMAC

[2]   E. Rosenblueth, Proc. Nat. Acad. Sci, USA, 72 (1975) 0

[3]   E. Rosenblueth, Appl. Math. Modelling, 5 (1981)

[4]   M. E. Harr, Symposium on Reliability Based Design, EPFL 1988, Switzerland

[5]   J. Natterer, H. M. Kessel, P. Haller, Final Report Swiss National Science
      Foundation, Nr. 20-5'308.87, 1990

Lecture Notes in Control and Information Sciences

Edited by M. Thoma

Vol. 181: Drane, C.R.
Positioning Systems - A Unified Approach.
168 pp. 1992 [3-540-55850-0]

Vol. 182: Hagenauer, J. (Ed.)
Advanced Methods for Satellite and Deep
Space Communications. Proceedings of an
International Seminar Organized by Deutsche
Forschungsanstalt für Luft-und Raumfahrt
(DLR), Bonn, Germany, September 1992.
196 pp. 1992 [3-540-55851-9]

Vol. 183: Hosoe, S. (Ed.)
Robust Control. Proceesings of a Workshop
held in Tokyo, Japan, June 23-24, 1991.
225 pp. 1992 [3-540-55961-2]

Vol. 184: Duncan, T.E.; Pasik-Duncan, B.
(Eds.)
Stochastic Theory and Adaptive Control.
Proceedings of a Workshop held in Lawrence,
Kansas, September 26-28, 1991.
500 pages. 1992 [3-540-55962-0]

Vol. 185: Curtain, R.F. (Ed.); Bensoussan, A.;
Lions, J.L.(Honorary Eds.)
Analysis and Optimization of Systems: State
and Frequency Domain Approaches for Infinite-
Dimensional Systems. Proceedings of the 10th
International Conference, Sophia-Antipolis,
France, June 9-12, 1992.
648 pp. 1993 [3-540-56155-2]

Vol. 186: Sreenath, N.
Systems Representation of Global Climate
Change Models. Foundation for a Systems
Science Approach.
288 pp. 1993 [3-540-19824-5]

Vol. 187: Morecki, A.; Bianchi, G.;
Jaworeck, K. (Eds.)
RoManSy 9: Proceedings of the Ninth
CISM-IFToMM Symposium on Theory and
Practice of Robots and Manipulators.
476 pp. 1993 [3-540-19834-2]

Vol. 188: Naidu, D. Subbaram
Aeroassisted Orbital Transfer: Guidance and
Control Strategies.
192 pp. 1993 [3-540-19819-9]

Vol. 189: Ilchmann, A.
Non-Identifier-Based High-Gain Adaptive
Control.
220 pp. 1993 [3-540-19845-8]

Vol. 190: Chatila, R.; Hirzinger, G. (Eds.)
Experimental Robotics II: The 2nd International
Symposium, Toulouse, France, June 25-27
1991.
580 pp. 1993 [3-540-19851-2]

Vol. 191: Blondel, V.
Simultaneous Stabilization of Linear Systems.
212 pp. 1993 [3-540-19862-8]

Vol. 192: Smith, R.S.; Dahleh, M. (Eds.)
The Modeling of Uncertainty in Control
Systems.
412 pp. 1993 [3-540-19870-9]

Vol. 193: Zinober, A.S.I. (Ed.)
Variable Structure and Lyapunov Control
428 pp. 1993 [3-540-19869-5]

Vol. 194: Cao, Xi-Ren
Realization Probabilities: The Dynamics of
Queuing Systems
336 pp. 1993 [3-540-19872-5]

Vol. 195: Liu, D.; Michel, A.N.
Dynamical Systems with Saturation
Nonlinearities: Analysis and Design
212 pp. 1994 [3-540-19888-1]

Vol. 196: Battilotti, S.
Noninteracting Control with Stability for
Nonlinear Systems
196 pp. 1994 [3-540-19891-1]